Introduction to Materials for Advanced Energy Systems

Colin Tong

Introduction to Materials for Advanced Energy Systems

Springer

Colin Tong
Chicago, IL, USA

ISBN 978-3-319-98001-0 ISBN 978-3-319-98002-7 (eBook)
https://doi.org/10.1007/978-3-319-98002-7

Library of Congress Control Number: 2018954936

This Springer imprint is published by the registered company Springer Nature Switzerland AG
The registered company address is: Gewerbestrasse 11, 6330 Cham, Switzerland

This book is dedicated to my wife Dali. She has given me a beautiful family and a happy life. Her love has always filled my heart and encouraged me during the long journey of my work.

Preface

Energy is the single most valuable resource for human activity and the basis for all human progress. Materials play a key role in enabling technologies that can offer promising solutions to achieve accessible, renewable, and sustainable energy pathways for the future. However, energy conversion and storage processes have limited efficiencies, cost factors, and environmental effects. Research on fundamental science and engineering of micro- and nano-materials, including the exchange of photons and electrons among quantized energy levels of semiconductors, molecules, and metals at nanoscale spatial scales and at fast or ultrafast time scales, is key to improving these factors for sustainable energy production. To make sustainable energy technologies competitive with fossil fuel technologies requires probing and understanding these quantum phenomena, and this understanding must then be translated into effective control of the functionality and performance of the materials through optimized manufacturing processes and proper integration methods.

For enabling today's students to understand current and future energy challenges, to acquire skills for selecting and using materials and manufacturing processes in the design of energy systems, and to develop a cross-functional approach to materials, mechanics, electronics, and processes of energy production, this textbook provides a comprehensive introduction to the range of materials used for advanced energy systems, including fossil, nuclear, solar, bio, wind, geothermal, ocean and hydropower, hydrogen, and nuclear, as well as thermal energy storage and electrochemical storage in fuel cells. A separate chapter is devoted to emerging energy harvesting systems. The final chapter presents future perspectives and developing trends of advanced energy materials for accessibility, affordability, and sustainability of future energy systems.

Integrated coverage is achieved through the consistent application of scientific and engineering principles regardless of the material or enable system under consideration. Properties, performance, modeling, fabrication, characterization and application of structural, functional, and hybrid materials are described for each energy system. The complex relationships among materials selection, optimizing design, and component operating conditions are also addressed, as well as research and development trends of novel emerging materials for future hybrid energy harvesting and storage systems. Throughout the text, images, illustrations, and terminology boxes clarify core concepts, techniques, and processes. At the end of each chapter, references serve as a gateway to the primary literature in the field, and exercises are provided to aid understanding. Each chapter is basically a self-contained unit, easily enabling instructors to adapt this book for coursework.

It is my great pleasure to acknowledge the help and support I have received from my colleagues and friends. I would like to express my sincere gratitude to Dr. David Packer and all other editing staff who have done a fantastic job on the publication of this book.

Chicago, IL, USA Colin Tong

Contents

Abbreviations

AC	Asbestos cement
ACC	Active combustion control
ACN	Acrylonitrile
AD	Anaerobic digestion, or air drilling
ADF	Annular dark field
AFA	Alumina-forming austenitic
AFPM	Axial flux permanent magnet
AGR	Advanced gas-cooled reactor
ALD	Atomic layer deposition
AM	Additive manufacturing
AnMBR	Anaerobic membrane bioreactor
APR	Aqueous phase reforming
ASCPMM	Advanced sensors, control, platforms and modeling for manufacturing
ATC	Air traffic control
ATES	Aquifer thermal energy storage
ATL	Automated tape laying
ATP	Automated tow placement
ATR	Autothermal reforming
ATRP	Atom transfer radical polymerization
AUSC	Advanced ultra-supercritical combustion system
AWES	Airborne wind energy system
BCC	Body-centered-cubic
BET	Brunauere emmette teller
BHE	Borehole heat exchanger
BHJ	Bulk hetero-junction
BIG	Biomass integrated gasifier
BIPV	Building-integrated photovoltaic
BMC	Bulk molding compound
BMI	Bismaleimide
BTL	Biomass to liquid
BWR	Boiling water reactor
CAC	Calcium aluminate cement
CAES	Compressed air energy storage

CANDU	Canada Deuterium Uranium (reactor)
CAPEX	Capital expenditure
CBTLE	Coal and biomass to produce liquid fuels and electricity
CCD	Charge-coupled device
CCS	Carbon capture and storage
CDP	Composite drill pipe
CFB	Circulating fluidized bed
CFRD	Concrete-face rock-fill dam
CFRP	Carbon fiber reinforced polymer
CHP	Combined heat and power
CI	Compression-ignition (engine)
CIGS	Copper indium gallium selenide
CIS	Containment isolation system, or copper indium diselenide
CM	Conventional subtractive manufacturing
CMP	Coupled magnetic polariton
CMS	Carbon molecular sieves
CNT	Carbon nanotubes
COF	Covalent organic framework
COD	Chemical oxygen demand
COP	Coefficient of performance
CPOX	Catalytic partial oxidation
CPVC	Chlorinated polyvinyl chloride
CQD	Colloidal quantum dot
CR	Chloroprene rubber
CRA	Corrosion-resistant alloy
CSH	Calcium silicate hydrate
CSLB	Coincident site lattice boundary
CSP	Concentrating solar power system
CSV	Comma-separated values
CTE	Coefficient of thermal expansion
CTL	Coal to liquid (fuel)
CVD	Chemical vapor deposition
DBTT	Ductile-to-brittle transition temperature
DCFC	Direct carbon fuel cell
DCL	Direct coal liquefaction
DFIG	Doubly fed induction generator
DLC	Diamond-like carbon
DMF	Dimethylformamide
DMS	Dynamical mechanical spectroscopy
DSSC	Dye-sensitized solar cell
EBPVD	Electron beam-physical vapor deposition
ECCS	Emergency core cooling system
EDC	Endocrine disrupting chemical
EELS	Electron energy loss spectroscopy
EGS	Engineered geothermal system

EHPT	Evacuated heat pipe tube
EPDM	Ethylene-propylene-diene terpolymer
EWT	Emitter wrap through
FAME	Fatty acid methyl ester
FBC	Fluidized bed combustor
FBES	Flow battery energy storage
FBR	Fast breeder reactor
FC	Fuel cell
FCC	Face-centered-cubic
FCEV	Fuel cell electric vehicle
FED	Field emission display
FEP	Fluorinated ethylene propylene
FET	Field effect transistor
FO	Forward osmosis
FOG	Fats, oils, and grease
FRP	Fiber-reinforced plastic
GBE	Grain boundary engineering
GCR	Gas cooled reactor
GFR	Gas fast reactor
GHG	Greenhouse gas
GHT	Gorlov helical turbine
GSHP	Ground source heat pump
GTCC	Gas turbine combined cycle
GRP	Fiberglass reinforced plastic
GS	Galvanized steel
HAWT	Horizontal axis wind turbine
HCPV	High concentration photovoltaic
HC	Hot carrier (solar cell)
HCP	Hexagonal close packed
HDI	Hexamethylene diisocyanate
HDPE	High-density polyethylene
HEMT	High-electron-mobility transistor
HES	Hydrogen energy storage
HFP	Hexafluoropropylene
HHV	Higher heating value
HIP	Hot isostatic pressing
HPC	High-performance computing
HPHT	High-pressure/high temperature
HSE	Health, safety, and environmental
HTG	High-temperature gasification
HTGR	High temperature gas-cooled reactor
HTN	High temperature nylon
HTPEM	High temperature proton exchange membrane
HTSE	High temperature steam electrolysis
IASCC	Irradiation assisted stress corrosion cracking

IBPCS	Integrated biomass-production conversion system
ICE	Internal combustion engine
ICL	Indirect coal liquefaction
ICT	Information and communication technology
ICME	Integrated Computational Materials Engineering
ICP	Internal concentration polarization
IGCC	Integrated gasification combined cycle
IHX	Intermediate heat exchanger
IR	Infrared
ISS	Instrumentation safety (monitoring) system
ITER	International Thermonuclear Experimental Reactor
ITO	Indium tin oxide
KTL	Electrophoretic cathode metal coating, or E-coat
LAES	Lead acid battery energy storage, or liquid air energy storage
LAOS	Large amplitude oscillatory shear
LCA	Life-cycle assessment
LCD	Liquid-crystal display
LCOE	Levelized cost of electricity
LED	Light emitting diode
LENR	Low energy nuclear reaction
LFR	Lead fast reactor
LHV	Lower heating value
LIES	Lithium ion battery energy storage
LMFBR	Liquid-metal cooled fast breeder reactor
LPG	Liquefied petroleum gas
LSC	Luminescent solar concentrator
LTG	Low-temperature gasification
LWGR	Light water graphite reactor
LWR	Light water reactor
MA	Mechanical alloying
MBR	Membrane bioreactors
MCFC	Molten carbonate fuel cell
MD	Membrane distillation
MDA	Methylene-dianiline
MDI	Methylene diisocyanate
MEA	Membrane-electrode assembly
MEMS	Micro-electro-mechanical systems
MF	Microfiltration
MFC	Microbial fuel cell
MM	Molecular mechanics
MMM	Mixed matrix membrane
MO	Molecular orbital
MOF	Metal-organic framework
MP	Magnetic polariton
MPD	M-phenylenediamine or managed pressure drilling

MSR	Molten salt reactor
MSW	Municipal solid waste
MTBF	Mean time between failure
MWCNT	Multi-walled carbon nanotube
NAE	Nuclear active environment
NaSES	Sodium sulfur battery energy storage
NBR	Nitrile butadiene rubber
NF	Nanofiltration
NIPS	Non-solvent induced phase separation
ODS	Oxide dispersion strengthened
OLP	Organic liquid product
OPC	Operating point control
OPEX	Operational expenditure
OPV	Organic photovoltaic
ORC	Organic Rankine cycle
OTEC	Ocean thermal energy conversion
OWC	Oscillating water columns
OWSC	Oscillating wave surge converter
PAFC	Phosphoric acid fuel cell
PAH	Polycyclic aromatic hydrocarbon
PAN	Polyacrylonitrile
PBR	Pebble bed reactor
PCD	Polycrystalline diamond drill
PCI	Pellet-cladding mechanical interaction
PCM	Phase change material
PDA	Polydopamine
PE	Polyethylene
PEG	Polyethylene glycol
PEI	Polyetherimide
PEM	Proton exchange membrane, or polymer electrolyte membrane
PEMFC	Polymer electrolyte membrane fuel cell
PEO	Polyethylene oxide
PES	Poly (ether sulfone)
PEV	Plug-in electric vehicle
PEX	Crosslinked polyethylene
PF	Pulverized fuel
PhAC	Pharmaceutically active compound
PHES	Pumped hydro energy storage
PIP	Piperazine
PMC	Polymer matrix composite
PMG	Permanent magnet generator
PMMA	Polymethyl acrylate
POP	Persistent organic compounds
POX	Partial oxidation
PP	Polypropylene

PPO	Pure plant oil
PRO	Pressure-retarded osmosis
PSf	Polysulfone
PSS	Pressure suppression system
PTFE	Polytetrafluorethylene
PTO	Power take off
PUF	Polyurethane foam
PV	Photovoltaic
PVC	Polyvinyl chloride
PVD	Physical vapor deposition
PVDF	Poly(vinylidene fluoride)
PWHT	Post weld heat treatment
PWR	Pressurized water reactor
QD	Quantum dot
QM	Quantum mechanics
R2R	Roll-to-roll
RCS	Radar cross section
RDD&D	Research, development, demonstration, and deployment
RDF	Refuse derived fuel
RED	Pressure-retarded osmosis
REE	Rare earth element
RF	Radio frequency
RIS	Radiation-induced segregation
RLC	Resistor (R), inductor (L), and capacitor (resonant circuit)
RO	Reverse osmosis
ROP	Rate of penetration
RTFDDA	Real-time four-dimensional data assimilation
RTM	Resin transfer molding
RTRP	Reinforced thermosetting resin pipe
RVE	Representative volume element
SAE	Styrene acrylic emulsion
SANS	Small-angle neutron scattering
SAOS	Small amplitude oscillatory shear
SCC	Stress corrosion cracking
SCH	Solar heating and cooling
SCR	Split-coin resonator
SCWR	Supercritical water cooled reactor
SEM	Scanning electron microscopy
SEER	Seasonal energy efficiency rating
SES	Supercapacitor energy storage
SFR	Sodium fast reactor
SLS	Selective laser sintering
SMC	Sheet molding compound
SMES	Superconducting and magnetic energy storage
SNIPS	Self-assembly and non-solvent-induced phase separation

SOFC	Solid oxide fuel cell
SPF	Seasonal performance factor
SPP	Solid phase polymerization, or surface plasmon polariton
SPR	Surface plasmon resonance
SR	Synthetic rubber, or silicone rubber, or steam reforming
SSAS	Sodium silicate-activated slag
STC	Standard testing conditions
STEM	Scanning transmission electron microscope
STIG	Steam injected gas (turbine)
SUMR	Segmented ultralight morphing rotor
SVO	Straight vegetable oil
SWCNT	Single-walled carbon nanotube
TBC	Thermal barrier coating
TCS	Thermo-chemical storage
TDI	Toluene diisocyanate
TE	Transverse electric
TEM	Transmission electron microscopy
TES	Thermal energy storage
TF	Thin film
TFC	Thin-film composite
TFE	Tetrafluoroethylene
TFN	Thin-film nanocomposite
TFPM	Transverse flux permanent magnet
THAWT	Transverse horizontal axis water turbine
TIM	Transparent insulation material
TIPS	Thermal-induced phase separation
TM	Transverse magnetic
TMC	Trimesoyl chloride
TPB	Triple phase boundary
TRISO	Tri-structural isotropic
TSR	Tip speed ratio
UBD	Underbalanced drilling
UF	Ultrafiltration
UPHES	Underground pumped hydroelectric energy storage
UPS	Uninterruptible power supply
USC	Ultra-supercritical
UTS	Ultimate tensile strength
UV	Ultra-violet
VARI	Vacuum-assisted resin injection
VARTM	Vacuum-assisted resin transfer molding
VAWT	Vertical-axis wind turbine
VB	Valence bond
VHTR	Very high temperature reactor
VLS	Vapor-liquid-solid
VRE	Variable renewable energy

WBG	Wide bandgap
WEC	Wave energy converter
WEST	Wood–epoxy saturation technique
WGS	Water gas shift
WRF	Weather research and forecasting
WT	Wind turbine
YSZ	Yttria-stabilized zirconia

Materials-Based Solutions to Advanced Energy Systems

1

Abstract

Energy is one of the critical issues that directly impact the economy, the environment, and the security of human beings. All energy technologies require materials; therefore, the types and amounts of materials consumed vary widely. While materials science and engineering are only one aspect of the response to the energy challenges, it primarily has a crucial part to play in creating the advanced energy systems. In the past, it has contributed significantly to advances in the safe, reliable, and efficient use of energy and available natural resources. Now materials research is being performed from structural materials, functional materials to high photon energies, which can offer promising solutions to achieve accessible, renewable, and sustainable energy pathways for the future. Particularly, the growing importance of environmental issues is such that energy generation, conservation, storage, and security of supply will continue to be major drivers for materials technology. Sustainable energy production and use are needed while at the same time meeting socioeconomic and environmental targets: The high priority of energy makes it important to sustain research, development, and modeling of materials for energy applications; the knowledge-base of high-integrity structural and functional materials should be recovered, captured, and developed for future power generation; transferable material solutions and methods across the complete energy portfolio should be examined to attain maximum efficiency and competitive advantages. With the advent of nanomaterials and innovative multifunctional materials, materials science and engineering is expected to play an increasing role in sustainable technologies for energy generation, storage, and distribution, as well as efficient utilization of future energy. Principal areas of advanced materials development include but not limited to sustainable structural and functional materials for fossil power, solar energy, wind energy, geothermal energy, biofuels, ocean energy and hydropower, nuclear power, as well as advanced energy-harvesting technologies. This chapter will introduce fundamentals and basic design guidelines of advanced energy systems with accompany of materials solutions and environmental compliance of energy materials.

C. Tong, *Introduction to Materials for Advanced Energy Systems*,
https://doi.org/10.1007/978-3-319-98002-7_1

1.1 Energy Resource and Energy System Development

As one of the critical issues, energy directly impacts the economy, the environment, and the security of human beings. Since ancient times, advances in the development of materials and energy have defined and limited human social, technological, and political aspirations. The modern era, with instant global communication and the rising expectations of developing nations, poses energy challenges greater than ever seen before. Access to energy is critical to the wealth, lifestyle, and self-image of every country (TMS 2010).

Current consumption of fossil fuels by low-efficiency engines in power generation and transportation significantly contributes to economic problems and environmental issues including possible climate change. Increasing rates of energy consumption around the world have led to a corresponding rise in concerns about where the sustainable energy comes from and what kind of future energy systems should be developed. A great many materials-based solutions have been explored for enabling various energy technologies to achieve accessible, renewable, and sustainable energy pathways for the future. New materials and technologies are needed for advanced energy devices and innovative approaches to harvesting, distributing, and storing energy.

IN DEPTH: Energy Technology

Energy technology is an interdisciplinary engineering science having to do with the efficient, safe, environmentally friendly and economical generation, conversion, distribution, storage and use of energy, targeted towards yielding high efficiency while skirting side effects on humans, nature and the environment. Advanced energy technologies have mainly dealt with the following aspects (Baldwin et al. 2015):

- Systems integration—Appropriate application of systems integration requires understanding, control, and optimization across multiple energy systems, time frames, and spatial scales. An integrated systems approach can address complexity and enable more efficient deployment of advanced energy technologies.
- Cybersecurity—Cybersecurity is essential to the increased use of Information and Communication Technology (ICT) in modernizing energy systems. The proliferation of computer controlled, wireless, and/or Internet-connected energy devices creates a growing attack surface for cyber intrusion.
- Energy-water nexus—Science and technology advancements at the intersection of energy and water can reduce energy use and increase water availability for human consumption, other non-energy uses, and natural systems.

(continued)

- Subsurface—Understanding and controlling fractures, fluid flow, and complex chemistry in subsurface rock formations on timescales of microseconds to millennia are important for oil and gas production, geothermal energy, carbon capture and storage (CCS), and nuclear waste disposal.
- Materials—Across all energy systems, advancements in materials could dramatically accelerate and reduce the cost of developing new energy technologies. Examples include development of materials for extreme working conditions, advanced processing of them, and their rapid qualification.
- Fuel-engine co-optimization—With bio-derived and/or other synthetic fuels there is an opportunity to optimize the end-to-end fuel-vehicle system for improved efficiency and reduced environmental impacts.
- Energy storage—Efficient, durable storage could enable transformational change across multiple energy systems, including transportation, and the electricity system.
- Grid modernization—Advanced grid technologies are developed to improve the agility and flexibility of the system to better integrate the changing characteristics of devices and technology systems on both the supply and demand sides.
- Enabling tools:

  (a) Computational modeling and simulation—Advances in high-performance computation have enabled simulation of increasingly complex physical phenomena. High-fidelity simulations, in turn, inform models that improve and accelerate the research, development, demonstration, and deployment (RDD&D) phases of the energy innovation cycle.
  (b) Data and analysis—Opportunities to apply advanced analytics transect the entire clean energy economy. The emerging science of extracting actionable information from large data sets is both an opportunity to accelerate RDD&D and a research need.
  (c) Analysis of complex systems—Increasing complexity resulting from the convergence of the energy systems introduces a need for foundational, conceptual research on integrated, networked, and complex systems.
  (d) Characterization and control of matter at multi-scales—Advances in characterization and modeling of materials and chemistry have paved the way for manipulating and synthesizing materials at the atomic-, nano-, and mesoscale to create new tailored functionalities. The research spans a range of dimensions from the atom, to biological cells, to macroscopic structures, with applications across many scientific and engineering disciplines.

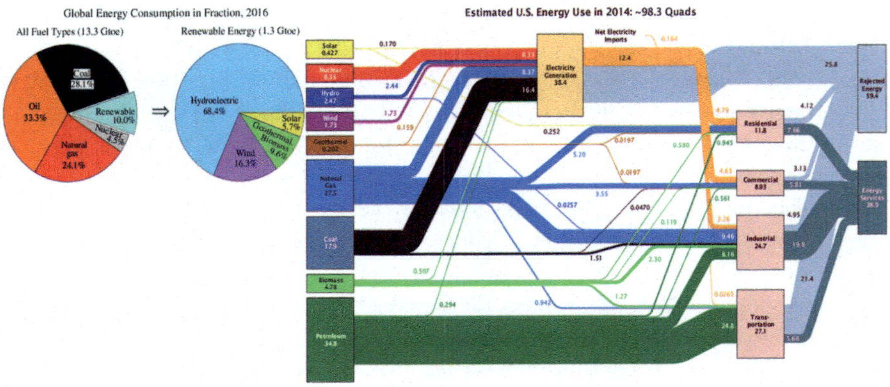

Fig. 1.1 Global energy consumption in fraction (2016) and energy flow charts showing the relative size of primary energy resources and end uses in the United States in 2014, with fuels compared on a common energy unit basis (Adapted from Lawrence Livermore National Laboratory 2015. Credit: Lawrence Livermore National Laboratory)

1.1.1 Energy Resources

Energy resources may be classified as primary resources, suitable for end use without conversion to another form, or secondary resources, where the usable form of energy required substantial conversion from a primary source. As shown in Fig. 1.1, examples of primary energy resources are solar power, wind power, wood fuel, fossil fuels such as coal, oil, and natural gas, and uranium. Secondary resources are those such as electricity, hydrogen, or other synthetic fuels. Among those energy resources, renewable resources are those that recover their capacity in a time significant by human needs. Examples are hydroelectric power or wind power, when the natural phenomena that are the primary resource of energy are ongoing and not depleted by human demands. Nonrenewable resources are those that are significantly depleted by human usage and that will not recover their potential significantly during human lifetimes. An example of a nonrenewable energy resource is coal, which does not form naturally at a rate that would support human use. In addition, sustainable energy generally indicates a form of energy which is considered sustainable, meaning that the usage of such energy can potentially be kept up well into the future without causing harmful repercussions for future generations. Sustainable energy resources are most often regarded as including all renewable resources, such as hydroelectricity, solar energy, wind energy, wave power, geothermal energy, artificial photosynthesis, biofuels, and tidal power. It usually also includes technologies that improve energy efficiency, such as energy generation, carbon management, energy storage and distribution, and energy consumption and utilization with enhanced energy conservation, efficiency, and environmental stewardship.

A vast and complex array of systems and associated technologies extract energy resources, convert them into usable forms of energy, and deliver them to end users to provide desired services such as manufactured goods, thermal comfort, lighting, and mobility. For instance, the overall flow of energy through the U.S. energy system is illustrated in Fig. 1.1 (right side). It illustrates the initial energy resources, their conversions into fuels and electricity, and their use in the buildings, industry, and transportation systems to provide the energy services that support national economy and human being's way of life. It also illustrates energy losses (rejected energy) that result from the fact that energy conversion processes are never 100% efficient (QTR 2015). Fossil fuels make up the bulk of the world's current primary energy sources. Fossil fuel sources burn coal or hydrocarbon fuels, which are the remains of the decomposition of plants and animals. There are three main types of fossil fuels: coal, petroleum, and natural gas. Another fossil fuel, liquefied petroleum gas (LPG), is principally derived from the production of natural gas. Heat from burning fossil fuel is used either directly for space heating and process heating, or converted to mechanical energy for vehicles, industrial processes, or electrical power generation. These fossil fuels are part of the carbon cycle and thus allow stored solar energy to be used today. Nuclear power is the use of nuclear fission to generate useful heat and electricity. Fission of uranium produces nearly all economically significant nuclear power. Radioisotope thermoelectric generators form a very small component of energy generation, mostly in specialized applications such as deep space vehicles. Renewable energy comes from resources which are naturally replenished on a human timescale such as sunlight, wind, rain, tides, waves, and geothermal heat. Renewable energy replaces conventional fuels in four distinct areas: electricity generation, hot water/space heating, motor fuels, and rural (off-grid) energy services (Outhred et al. 2007).

There are many pathways to produce electricity, with the generation mix currently dominated by coal, natural gas, and nuclear resources. The global use of electricity captures the triumph and the challenge of energy. Since Edison, Tesla, and Westinghouse installed the first primitive electricity grids 130 years ago, electrical technology has undergone many revolutions. From its initial use exclusively for lighting, electricity now symbolizes modern life, powering lights, communication, entertainment, trains, refrigeration, and industry. In the past century, 75% of the world has gained access to this most versatile energy carrier. Such changes in human lives do not come from incremental improvements, but from groundbreaking research and development on materials that open new horizons. Tremendous opportunities currently exist for transitioning from carbon-based energy sources such as gasoline for engines to electric motors for transportation, as well as from coal-fired electric power generation to renewable, clean solar, nuclear, and wind energy sources for electricity, and thereby dramatically increasing the capacity and reliability of urban grids in high density. These advances will require a new generation of advanced materials, such as battery/supercapacitor materials for massive electrical energy storage, high-efficiency and low-cost solar cells, corrosion-resistant alloys for high-temperature power conversion, strong and lightweight composites for turbine blades, superconducting power distribution cables, as well as functional materials for advanced electronics. Moreover, revolutionary advancements in materials, including lightweight aerospace alloys, high-temperature engine materials, and advanced composites, have been a critical part of improving the capability, safety, and energy efficiency of transportation vehicles. Therefore, the connection is clear between materials research and the energy technologies that people rely on today and those need for future. Materials research covers a broad set of science and engineering disciplines, and seeks to understand fundamental physical and chemical properties, and then use that understanding to improve the technology base that people count on to meet the needs for energy, national security and defense, information technology and telecommunications, consumer products, healthcare, and more (TMS 2010).

1.1.2 The Role of Advanced Materials in Energy Systems

All energy technologies require materials, but the types and amounts of materials consumed vary widely. While materials science and engineering is only one aspect of the response to the energy challenges, it primarily has a crucial part to play in creating the advanced energy systems. In the past, it has contributed significantly to advances in the safe, reliable, and efficient use of energy and available natural resources. With the advent of nano-materials and innovative multifunctional materials, materials research is expected to play an increasing role in sustainable technologies for energy harvesting, conversion, storage, savings, and usages. Principal areas of advanced materials development include but not limited to: solar cells, batteries and supercapacitors, fuel cells, thermoelectrics, superconductors, more efficient lighting, hydrogen, and advanced energy-harvesting technologies. In most of these areas, incremental improvements of current technologies are not sufficient to address the critical issues of durability, efficiency, and costs. New materials research avenues are therefore needed to design, elaborate, and integrate materials for energy applications. Nanotechnologies and modeling activities have been instrumental in this respect; for example, in the development of new electrocatalysts for fuel cells membrane electrode assembly (MEA) and in the development of new materials for solid-state storage of hydrogen. New energy-efficient devices also call for improved processing technologies for materials elaboration and integration. This is the case for superconductor tapes formed with complex thin-film architectures where there is a need to control each step of the fabrication—the final product being tape of several hundred meters in length with a nanostructured superconducting active layer of 1–3 μm thickness (Vouldis et al. 2008).

IN DEPTH: Five Grand Challenges for Basic Energy Sciences (BESAC 2007)

- *How do we control material processes at the level of electrons?*
  Electrons are the negatively charged subatomic particles whose dynamics determine materials properties and direct chemical, electrical, magnetic, and physical processes. If we can learn to direct and control material processes at the level of electrons, where the strange laws of quantum mechanics rule, it should pave the way for artificial photosynthesis and

(continued)

other highly efficient energy technologies, and could revolutionize computer technologies.

- *How do we design and perfect atom- and energy-efficient synthesis of revolutionary new forms of matter with tailored properties?*
  Humans, through trial and error experiments or through lucky accidents, have been able to make only a tiny fraction of all the materials that are theoretically possible. If we can learn to design and create new materials with tailored properties, it could lead to low-cost photovoltaics, self-repairing and self-regulating devices, integrated photonic (light-based) technologies, and nano-sized electronic and mechanical devices.
- *How do remarkable properties of matter emerge from complex correlations of the atomic or electronic constituents and how can we control these properties?*
  Emergent phenomena, in which a complex outcome emerges from the correlated interactions of many simple constituents, can be widely seen in nature, as in the interactions of neurons in the human brain that result in the mind, the freezing of water, or the giant magneto-resistance behavior that powers disk drives. If we can learn the fundamental rules of correlations and emergence and then learn how to control them, we could produce, among many possibilities, an entirely new generation of materials that supersede present-day semiconductors and superconductors.
- *How can we master energy and information on the nanoscale to create innovative technologies with capabilities rivaling those of living things?*
  Biology is nature's version of nanotechnology, though the capabilities of biological systems can exceed those of human technologies by a vast margin. If we can understand biological functions and harness nanotechnologies with capabilities as effective as those of biological systems, it should clear the way towards profound advances in a great many scientific fields, including energy and information technologies.
- *How do we characterize and control matter away-especially very far away-from equilibrium?*
  All natural and most human-induced phenomena occur in systems that are away from the equilibrium in which the system would not change with time. If we can understand system effects that take place away—especially very far away—from equilibrium and learn to control them, it could yield dramatic new energy-capture and energy-storage technologies, greatly improve our predictions for molecular-level electronics, and enable new mitigation strategies for environmental damage.

The modern world runs on energy that, in the main, was captured millions of years ago from the Sun by photosynthesis in trees and other prehistoric flora. Today, by burning coal, oil, and natural gas, this energy is used to fuel transport systems, electricity grids, industry, and agriculture, and to heat and cool homes, work, and leisure places. It is clear by now that this supply of fossil energy is limited. Furthermore, returning massive quantities of carbon dioxide to the atmosphere increases the amount of energy that it absorbs from the Sun with consequences for the balance of the planet's climate systems. There is a need to develop alternative sources of energy and to ensure that energy, irrespective of its source, is used as efficiently as possible, which then results in that energy having to be stored, delivered, and converted into heat, light, or motion with minimal losses. Materials research is at the heart of developments for modern energy systems. Photovoltaic panels convert sunlight into electricity that can be used locally or fed into the grid. The most common material currently used is bulk silicon, although there are worldwide efforts to produce more highly performing systems based on other semiconductors, thin films or dyes, etc. Fuel cells require a fuel, such as hydrogen, alcohols or hydrocarbons that are oxidized electrochemically to produce electricity, which can then be used to power a vehicle with an overall improved efficiency as compared to burning the fuel in an internal combustion engine. The fuel is oxidized at the anode and is separated from the oxidant at the cathode by an electrolyte that allows ions to flow internally between the electrodes. The performance of the cell depends critically on these components and many new materials are being designed for optimum cost and efficiency. Similarly, new materials for rechargeable and/or printed batteries, especially those based on lithium and solid-phase electrolyte, are under development. If hydrogen is used as a fuel, a means to store it on a vehicle is needed. Hydrogen-storage materials, based on light elements that bind and release hydrogen under mild conditions are actively sought. Similarly, materials that can adsorb carbon dioxide or other pollutants can be used to help clean up the products of combustion from a fossil-fuel power plant, perhaps to be buried so that it does not enter the atmosphere (Fitch and Reichert 2010).

Materials research on energy-related materials is being performed from structural materials, functional materials to high photon energies. Studies by a series of characterization methods such as diffraction and spectroscopy can be used to reveal the composition, atomic structure, and crystalline state of materials, and imaging allows defects and aggregates to be observed. In situ studies can be particularly effective to follow materials or even complete devices under operating conditions to watch the evolution of the components, e.g., the electrolyte in a working fuel cell or lithium-ion battery during the electrochemical cycle. From such observations, a better understanding of the performance can be obtained and thereby improvements and modern designs envisaged. A significant increase in capacities with particularly an extension into the nanoworld, has become ever more important for the development of high-performance materials (Fitch and Reichert 2010). For example, the base materials can be polymers, elastomers, composites, high-temperature alloys, powder metals, thin films, and others, while the related devices can be sensors and controllers, optics, etc. used in the fabrication of the various types of sustainable energy systems.

Energy is an area where materials technology will play a particularly significant role in meeting the needs of the future. The growing importance of environmental issues is such that energy generation, conservation, storage, and security of supply will continue to be major drivers for materials technology. Sustainable energy production and use are needed while at the same time meeting socioeconomic and environmental targets: The high priority of energy makes it important to sustain research, development, and modeling of materials for energy applications; it should be set out to recover, capture, and develop the knowledge-base of high-integrity structural materials for future power generation; transferable material solutions and methods across the complete energy portfolio should be examined to attain maximum efficiency and competitive advantages; the global focus on energy is an opportunity to develop world-class knowledge and capability in materials for energy generation, low energy processing and energy conservation, and to extend commercial markets (Wadsworth 2012).

1.1.3 Solutions for Future Energy Systems

Within and between the electric grid, power, buildings, manufacturing, fuels, and transportation systems, increasing interconnectedness and complexity are creating opportunities. Systematical approaches can help to identify critical technology needs and can also be used to develop solutions to complex energy challenges. The most promising way towards a sustainable energy system is to integrate energy efficiency and renewable energies as much as possible and economically benign at project, regional, national, and international level. Such integration will improve security of supply, minimize risk, keep the energy bills lower, and thus enhance the ability to finance the necessary investments. As the importance of sustainable energy and the immediate need for new and more efficient technologies have been clearly recognized, a broad range of technologies are employed for developing advanced energy systems (Farley 2008).

1.1.3.1 Enabling Modernization of the Electric Power System

The traditional electricity infrastructure has provided reliable electricity for more than a century, but today's energy requirements are rapidly changing. Evolution of the electric power grid is shown in Fig. 1.2 (OECD/IEA 2011). The broad

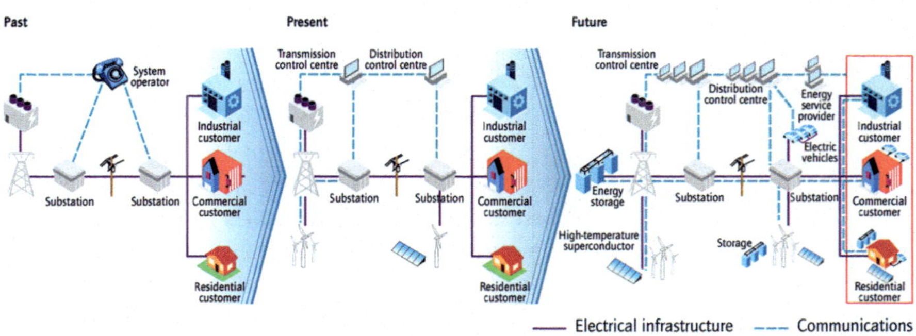

Fig. 1.2 Evolution of the electric power grid (Adapted from OECD/IEA 2011 with license: https://www.iea.org/media/copyright/TermsandconditionsFinal11Jan2017.pdf)

deployment of variable generation resources (most notably, solar and wind), distributed energy resources, energy storage, electric vehicles, and actively managed loads could substantially alter how the system will need to be designed, operated, and protected. Changes in the supply and generation mix, evolving demand loads, and the transition of consumers to active "prosumers" are all creating technical challenges for an aging electricity infrastructure. The proliferation of new digital control and communication devices brings new opportunities for managing distributed generation and storage, but creates new security and integration challenges. Simultaneously, growing dependence on highly reliable electricity for national and economic security makes electricity resilience a top priority. A modern electric grid must be more flexible, agile, and dynamic—able to integrate and optimize a wide mix of generators, loads, and storage capabilities. These trends create new technical requirements for the power grid and redefine its fundamental design and operational structures. Profoundly different generation and load characteristics will affect power system behavior and overall operational performance. Therefore, advanced technologies to plan, manage, monitor, and control electricity delivery are needed to enable safe and reliable two-way flow of electricity and information, support growing numbers of distributed energy resources, and support customers participating in electricity markets as both power suppliers and demand managers. The addressment of these issues requires to accomplish the following research, development, demonstration, and deployment (Orr 2015; Baldwin et al. 2015):

(a) Develop and refine interoperable grid architectures and new system designs.
(b) Develop software and visualization tools that use new data from transmission and distribution system devices for enhanced, real-time operations and control.
(c) Research material innovations and develop transmission and distribution component designs for higher performance, reliability, and resilience.
(d) Embed intelligence, communication, and control capabilities into distributed energy resources and systems such as micro-grids to support grid operations.
(e) Improve energy storage capabilities and systems designs that lower costs while increasing capacity and performance, and facilitating integration.
(f) Develop high-fidelity planning models, tools, and simulators and a common framework for modeling, including databases.
(g) Design innovative technologies and resilient and adaptive control systems to improve physical- and cyber-security of the grid.

1.1.3.2 Advancing Clean Electric Power Technologies

Clean electric power is paramount to meet today's interdependent security, economic, and environmental requirements. While promoting aggressive emission reductions, the reliability, safety, and affordability of the electric power must be maintained and enhanced. The current portfolio of electric production includes a combination of reliable, but aging, baseload generation, evolving renewable resources, and new natural gas resources. Complementing this evolving generation mix are technologies to enable higher efficiencies and pollution control. In addition, electric power generation technologies are maturing to a new level of integration and interdependence that requires an expanded system approach and a global view to optimize integration, minimize risks, and maintain reasonable costs (Outhred et al. 2007, Baldwin et al. 2015).

There is potential in each of the clean technologies: more efficient coal and natural gas generation with carbon capture; advanced nuclear reactors; rapidly advancing renewable technologies, such as wind and solar; and developing technologies, such as fuel cell and marine hydrokinetic power. Meanwhile, common component developments offer opportunities for breakthroughs, such as advances in high temperature and pressure steam turbines, new supercritical carbon dioxide power cycles, hybrid systems matching renewables with nuclear or fossil, and energy storage. Moreover, advanced capabilities in materials, computing, and manufacturing can significantly improve electric power technologies cost and performance. A systematical approach for each power systems also enables innovation at the technology level, such as by identifying key characteristics needed in supply technologies to meet the changing requirements of the grid, including such factors as cost, efficiency, emissions, ramping rates, turn-down ratios, water use, and others. These can be approached through multivariable portfolio analysis. Furthermore, international cooperation greatly expands the collective research, development, demonstration, and deployment (RDD&D) investment in clean power technologies by governments and industry, accelerating the successful completion of demonstrations and full commercial deployment (Orr 2015; Baldwin et al. 2015).

1.1.3.3 Advancing Systems and Technologies to Produce Cleaner Fuels

A "fuel" is defined as a carrier of chemical energy that can be released via reaction to produce work, heat, or other energy services. Fuel resources include oil, coal, natural gas, biomass, and hydrogen. Each fuel has strengths and shortcomings, and the fuel system must meet several challenging needs (Orr 2015; Baldwin et al. 2015):

- Economic prosperity requires low-cost fuels; energy security requires stable, abundant domestic resources; and meeting environmental goals requires reduction of greenhouse gas emissions and other externalities.
- In the long term, to reduce greenhouse gas (GHG) emissions, significant deployment of carbon capture, utilization, and storage (CCS), coal/biomass to liquids (CBTL) and/or bioenergy with carbon capture and storage (BECCS) will be needed to enable fossil fuels to continue to be robust contributors to energy needs.
- Renewable fuels show promise, but biofuels face land constraints, and hydrogen production from renewables is currently expensive; significant research, development, demonstration, and deployment (RDD&D) remains to solve the challenges associated with scale and cost for these fuels.
- In the near to midterm, multiple technological pathways need to be explored to serve as bridges to a low-carbon future. Particular focus should be given to interim technologies that help alleviate GHG challenges while minimizing

embedded infrastructure changes that would inhibit the transition to sustainable solutions.

Fuel sources such as natural gas and first-generation biofuels, if utilized properly, could help enable this transition. Each type of fuel has an associated system to produce the resource, upgrade, and transport it to a facility for cleanup and/or conversion into its final form for distribution to the end user. Although many of these steps are unique for each fuel, some do interconnect, particularly as they enter distribution systems.

While fossil fuels have advantages from an economic and security perspective, their emissions of greenhouse gases, chiefly CO_2, and methane (CH_4), are the primary contributor to global warming. Potential impacts on water systems are also a growing concern. This has led to increased investment, development, and commercialization of fuels that would reduce climate, water, and/or other impacts.

Some fuels, such as hydrogen and alcohols, can be derived from both renewable and fossil resources. Hydrogen is an energy carrier that can be produced from a variety of energy resources. It is produced in massive quantities today from natural gas. Technology options such as electrolysis from low-carbon electricity, direct reforming of fossil fuels with CCS, or production from biomass (possibly with CCS to achieve negative carbon emissions) can produce hydrogen for fuel with a very low-carbon footprint from domestically available energy resources. Challenges include technology costs of these low-carbon resources, as well as distribution and fueling infrastructure. Hydrocarbon fuels that are compatible with the existing fossil fuel infrastructure can also be synthesized from renewable resources. These fuels have immense potential as environmentally sound, sustainable, and domestic resources.

Bioenergy from a variety of feedstocks can be converted to a wide variety of products and liquid fuels and offer the potential to significantly reduce the GHG emissions associated with liquid fuel use. While ethanol from corn is an established industry, advanced pathways to use cellulosic, lignin, and waste inputs are just now beginning to enter the market but could scale up domestic low-carbon fuel production if key technology cost, scalability, and land use challenges can be met.

In addition to security concerns for imported oil and economic concerns over fuel prices and price volatility, environmental concerns are important for the entire global fuel enterprise. For fossil fuels used in buildings and some industries, CCS systems near the point of use may often not be possible. This provides motivation for converting fossil resources to low-carbon energy carriers, such as electricity or hydrogen, at a central location where CCS can be deployed, and then using these energy carriers at the distributed locations. Concurrently, development of carbon-neutral fuels utilizing biomass or renewable energy sources is needed (Baldwin et al. 2015).

1.1.3.4 Advancing Clean Transportation Systems

Transportation provides essential services for the economy, but also produces significant negative impacts, including economic costs and risks of dependence on oil, environmental impacts on air quality and health, and greenhouse gas (GHG) emissions. The need for a safe, fuel-efficient, operationally efficient, low-emission, and flexible transportation system can be addressed by advanced technology throughout the transportation system, such as efficient vehicle drivetrains, including efficient combustion vehicles, plug-in electric vehicles, and fuel cell vehicles. Future personal vehicle markets—in which these efficient vehicle technologies compete—may be transformed, and perhaps reduced in size, by information technologies as well as by social and demographic trends. Under the right circumstances, information technology can offer a less costly and less energy-intensive alternative to vehicular transportation. Low or zero tailpipe emissions technologies, including plug-in electric vehicles and fuel cell vehicles, address the need for cleaner transportation to improve air quality, especially in busy metropolitan areas and ports. Technologies, including plug-in electric vehicles and fuel cell vehicles, offer the potential for greater integration between energy systems for transportation, electricity, and building, which could be pursued to improve overall efficiency and reduce emissions.

Therefore, research on reducing the petroleum use and GHG emissions of the transportation systems have been focused on the following (Orr 2015; Baldwin et al. 2015):

(a) Combustion efficiency—Improving internal combustion engines requires research in simulation, sensors, controls, materials, and engine waste heat recovery, as well as new combustion strategies.
(b) Co-optimization of fuels and engines—Current fuels constrain engine design due to knock performance. New high-performance, low-carbon fuels that are co-optimized (designed in tandem) with engines could improve both performance and efficiency.
(c) Light weighting—Reduced vehicle weight improves vehicle efficiency and range. Research focuses on new materials such as advanced, high-strength steel, aluminum, magnesium alloys, and carbon fiber polymer matrix composites.
(d) Plug-in electric vehicles (PEVs)—PEVs have efficient drivetrains and allow for petroleum-free and lower-carbon fueling options, but need further improvements that will require research in new battery designs and chemistries to reduce cost and recharge time while improving energy density, power electronics and motors, and system design.
(e) Fuel cell electric vehicles (FCEVs)—FCEVs can be refueled in minutes, meet a wide range of performance requirements, achieve a better than 300-mile driving range, and have zero emissions from the tailpipe, while offering large potential petroleum and GHG reductions. Key issues are fuel cell cost and durability, and on-board hydrogen storage.
(f) Other modes—Projected activity growth in off-road transportation (e.g., air, rail, and marine) will make efficiency of these modes increasingly important.
(g) Connected and automated vehicles: Vehicle connectivity and automation present a variety of potential energy benefits and risks. Research opportunities

include supporting technologies (sensors, computation, communications, and control) as well as system research to improve energy outcomes.

(h) Transportation systems—A systems perspective on transportation, incorporating the interactions between (for example) vehicles, infrastructure, information technology, and human behavior, will enable future investment to optimize energy use through smarter transportation systems and technologies.

1.1.3.5 Innovating Clean Energy Technologies in Advanced Manufacturing

Clean energy manufacturing involves the minimization of the energy and environmental impacts of the production, use, and disposal of manufactured goods, which range from fundamental commodities such as metals and chemicals to sophisticated final-use products such as automobiles and wind turbine blades. Manufacturing affects the way products are designed, fabricated, used, and disposed; hence, manufacturing technologies have energy impacts extending beyond the industrial system. Furthermore, life-cycle analysis is essential to assess the total energy impact of a manufactured product. As shown in Fig. 1.3, state-of-the-art technologies available today could provide energy savings, but many have not yet penetrated the market due to barriers such as high capital intensity and lack of knowledge.

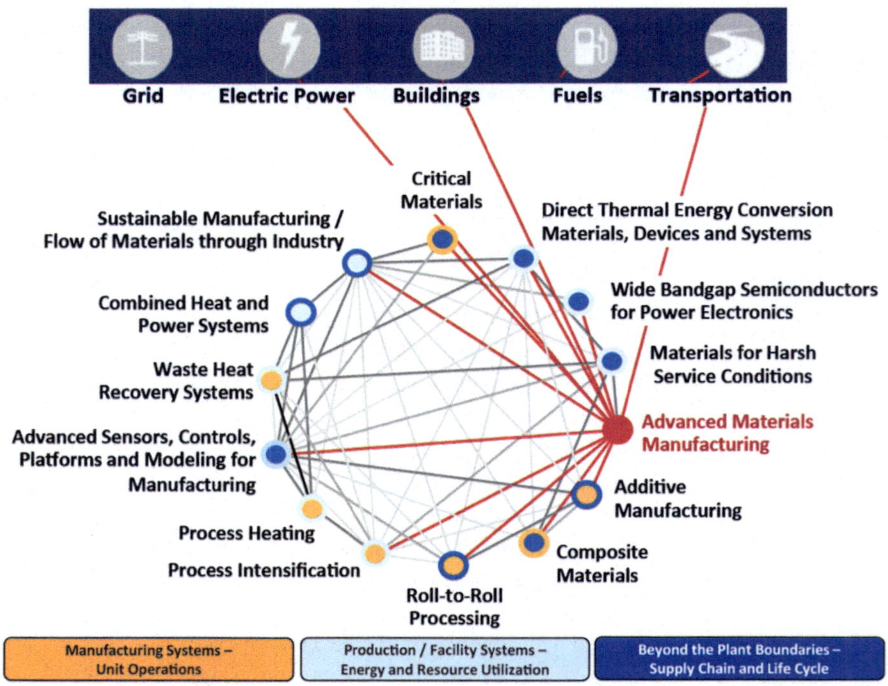

Fig. 1.3 Manufacturing technologies and their potential energy impacts (Adapted from Baldwin et al. 2015. Credit: US Department of Energy)

These barriers can be overcome, increasing technology uptake. Moreover, transformative manufacturing processes, materials, and technologies can provide advantages over the practices widely in use, and in many cases, enable the fabrication of innovative new clean energy products. Industrial-scale energy systems integration technologies, such as waste heat recovery and distributed energy generation, can reduce the manufacturing system's reliance on the electric grid and increase industrial efficiency. In addition, data, sensors, and models can improve design cycles and enable real-time management of energy, productivity, and costs, increasing manufacturing efficiency while improving product quality and throughput (Baldwin et al. 2015).

Therefore, the manufacture of clean energy products impacts the entire energy economy, with cross-sectoral and life-cycle energy benefits. Opportunities beyond the plant boundaries include improvements to the networks of facilities, business processes, and operations involved in moving materials through industry, from extraction of raw materials to the production of finished goods. The manufacturing system also supports nation's economic growth, as a strong manufacturing base can lead to competitive advantages gained through manufacturing innovations (Orr 2015).

1.1.3.6 Increasing Efficiency of Building and Hybrid Systems

Building efficiency must be considered as improving the performance of a complex system designed to provide occupants with a comfortable, safe, and attractive living and work environment. This requires superior architecture and engineering designs, quality construction practices, and intelligent operation of the structures. Increasingly, operations will include integration with sophisticated electric utility grids. The major areas of energy consumption in buildings are heating, ventilation, and air conditioning-35% of total building energy; lighting-11%; major appliances (water heating, refrigerators and freezers, dryers)-18% with the remaining 36% in miscellaneous areas including electronics. In each case there are opportunities both for improving the performance of system components (e.g., improving the efficiency of lighting devices) and improving the way they are controlled as a part of integrated building systems (e.g., sensors that adjust light levels to occupancy and daylight). Therefore, key research areas to increase efficiency of building and hybrid systems have been focus on (U.S. Department of Energy 2014): (a) High-efficiency heat pumps that reduce or eliminate the use of refrigerants that can lead to GHG emissions; (b) Thin insulating materials; (c) Windows and building surfaces with tunable optical properties; (d) High-efficiency lighting devices including improved green light-emitting diodes, phosphors, and quantum dots; (e) Improved software for optimizing building design and operation; (f) Low-cost, easy to install, energy-harvesting sensors and controls; (g) Interoperable building communication systems and optimized control strategies; (h) Decision science issues affecting purchasing and operating choices.

Energy use in buildings depends on a combination of good architecture and energy systems design and on effective operations and maintenance once the building is occupied. Buildings should be treated as sophisticated, integrated, interrelated systems. Different climates probably require assorted designs and equipment, and the performance and value of any component technology depends on the system in which it is embedded. Attractive lighting depends on the performance of the devices that convert electricity to visible light, as well as on window design, window and window covering controls, occupancy detectors, and other lighting controls. As the light fixture efficiency is greatly increased, lighting controls will have a reduced net impact on energy use. In addition, the thermal energy released into the room by lighting would decrease, which then affects building heating and cooling loads.

Since buildings consume a large fraction of the output of electric utilities, they can greatly impact utility operations. Specifically, buildings' ability to shift energy demand away from peak periods, such as on sweltering summer afternoons, can greatly reduce both cost and GHG emissions by allowing utilities to reduce the need for their least efficient and most polluting power plants. Coordinating building energy systems, on-site generation, and energy storage with other buildings and the utility can lower overall costs, decrease GHG emissions, and increase system-wide reliability (Sawyer 2014).

In addition, hybrid energy systems can be utilized to improve overall energy efficiency by combining multiple inputs and/or outputs, or using waste from one system as input to another. Examples include combined heat and power in buildings and manufacturing; polygeneration of electricity and fuels, chemicals or fresh water; and hybrid systems that combine nuclear with renewables, or co-fire coal and biomass (potentially with CCS). With respect to the consumer, examples include hybrid space and/or water heating, solar PV and/or thermal heating, daylight/task lighting systems, and combination appliances or electronics (Cates 1996).

One example of a dual-fuel, polygeneration hybrid system that has received significant attention for its potential to mitigate GHG emissions is combining coal and biomass to produce liquid fuels and electricity with CCS (CBTLE-CCS). There are several ways in which such a system could be configured, but the underlying principle is to integrate the heat and mass flows of (a) a gasification and Fischer-Tropsch fuel synthesis plant with (b) a combined-cycle cogenerating electricity plant. Depending on the ratios of biomass-to-coal in the plant inputs and fuels-to-electricity in the plant outputs, the net carbon emissions of the polygeneration system (including the CO_2 from the fuel it produces) could be positive, negative or zero. Assuming the system displaces electricity that would otherwise have been generated by a carbon-emitting power plant, the net environmental benefits could be significant. Detailed cost and energy analysis down to the equipment level shows the cost and performance benefits of an integrated system over separate fuels- and power-production systems competing in the same markets. Scenario analysis has shown that such carbon negative systems are important options for achieving global emissions goals (Liu et al. 2011; IPCC 2014).

1.1.3.7 Enabling Capabilities for Science and Energy

Basic science, including the tools needed to facilitate discovery, expands human being's understanding of the natural world and forms the foundation for future energy technology. The current imperative-energy systems that meet nation's energy security, economic, and environmental challenges-requires advances in energy generation, storage, efficiency, and security that demand a new generation of materials (including biological and bio-inspired materials) that may not be naturally available. However, creating these new materials requires a level of understanding of the relationships between structure and function, and across many spatial scales, which is not yet supported by present understanding of the physical world. Basic scientific research is necessary to fill these knowledge gaps and enable creation of new materials with the specific characteristics needed for next-generation energy technology. For instance, in nanoscience research for energy needs, all elementary steps of energy conversion take place at the atomic and nanoscale. The ability to rationally tailor matter at such scale would enable production of new materials for energy applications, including photovoltaics, electrodes and electrolytes, smart membranes, separators, superconductors, catalysts, fuels, sensors, and piezoelectrics. By extension, tailoring biological materials—from microbes to plants—at the genomic and subcellular levels would enable more efficient means of conversion, including those required to produce renewable and sustainable biofuels and bioproducts (Alivisatos et al. 2004).

The current challenge in materials science is to understand how nanoscale phenomena translate to properties at the mesoscale and beyond. Quantum mechanics describes atomic, molecular, and nanoscale phenomena, while classical mechanics describes macroscale behavior. The organizing principles governing emergent phenomena at the mesoscale, where classical properties first begin to emerge out of the quantum world, is only now being revealed. As systems grow in size from the nanoscale to the mesoscale, defects, interfaces, and fluctuations emerge that could be manipulated to program the various desired functionalities of materials, including specific thermal, electronic, and mechanical properties at the bulk level. In this way, nanoscale design can result, at the mesoscale and beyond, in the creation of radically new materials, with properties and functionalities that expand upon, or fundamentally differ from, those found in nature. Analogous to inorganic materials, living systems demonstrate properties and functionalities that go beyond the additive functions of their constituent parts. The challenge for systems biology is to understand how particular changes to metabolic pathways—often stemming from small changes at the genome scale—play out at the level of the whole organism or an entire microbial community. For example, this latter understanding is critical for achieving effective conversion of biomass into biofuels. Finally, this new energy research agenda is being shaped by dramatic advances in computation. Today's high-performance computers allow complex real-world phenomena to be studied virtually, including phenomena at the nano- and mesoscale, at very high spatial and temporal fidelity, and at a much-accelerated pace. Critically, these tools are giving access to the properties of systems too dangerous to study experimentally, or too costly to develop by trial-and-error (BESAC 2012).

Taken together, these developments have put science and technology on the threshold of a transformation from observation to control and design of new systems. This paradigm shift is transforming the processes by which new materials and bio-systems are predicted, designed, and created. This revolution represents a convergence of theory, modeling, synthesis, and characterization, and will enable predictive modeling of materials, control of chemistry, and synthetic biology. The paradigm of "control" and "design" requires a diverse suite of experimental tools for spatial and temporal characterization and computational tools for theory, modeling, and simulation of complex phenomena. Furthermore, the new energy systems that will usher in a low-carbon, high-efficiency, environmentally sustainable future require a strong disciplinary base and sustained support for new scientific discoveries (Matlock & Speer 2007, DOE 2010).

Consequently, the prospections of future energy systems can be summarized as (Baldwin et al. 2015):

(a) The convergence of energy systems
Virtually all aspects of the energy system are becoming more interdependent. Information and communications technologies, advanced sensors and controls, and market phenomena are enabling the proliferation of advanced technologies that overlap the power generation, electricity transmission and distribution, buildings, manufacturing, fuels, and transportation systems. Furthermore, energy systems are increasingly coupled to water systems, material flows, waste products, and financial markets. Properly tuned and integrated systems have the potential to improve their overall operations, increase their efficiencies, and enable fundamentally new concepts in the structure of the economy and urban environments.

(b) The potential of increased diversification of energy resources, carriers, and uses
Many energy systems in the have the opportunity for multiple technology pathways and the potential of increased diversification. This diversification creates challenges to energy infrastructures. In transportation, electric vehicle offerings and new developments in fuel cells complement existing alternatives to petroleum such as (natural) gas and biofuels, but complicates refueling infrastructure. In the power generation, retiring units are being replaced with a mixture of natural gas, wind, and solar generation, among others, increasing the complexity of electric grid management. Diversification can also be advantageous by giving the energy system resource flexibility and consumer choice. These multiple resource options can potentially have stabilizing effects on the marketplace and enhance energy security.

(c) Improvement of energy efficiency
Energy efficiency has a long and well-established record of success in reducing energy use, as well as associated factors, such as water use and waste generation. Efficiency improvements can significantly benefit national security, the economy, and the environment, for example, by reducing oil use, business and consumer costs, and environmental emissions, respectively. Advance cost-effective efficiency technologies abound throughout all energy systems. The delivery of energy services typically goes through a sequence of energy conversion steps from the initial energy resources to the final delivered energy services, each with associated energy losses. Improving efficiency at any step in the

energy services chain can proportionately reduce the energy use and associated losses at each of the upstream steps. Energy efficiency can thus provide high leverage in reducing energy use and cost.

(d) The growing confluence of computational and empirical capabilities to enable a new era of "systems by design"
This confluence includes scientific theory, modeling, simulation, high-performance computing, data management and analysis, algorithms, software, and high-throughput experimental techniques to enable the prediction, design, engineering, and experimental characterization of materials and systems from the atomic through the nano-, meso-, and macroscale to manufacturing. These capabilities offer the potential to develop new materials, technologies, and systems more rapidly and at lower cost than traditional approaches.

1.2 Fundamentals of Energy Systems

Energy is the capability of matter to do work. All available energy forms may be classified as (a) accumulated or stored energy, such as chemical energy of fossil fuels, nuclear energy for binding nuclei, internal energy of a substance, and potential energy associated with position or a mass in a force field; and (b) transitional energy, like thermal energy, heat and work, electromagnetic radiation, and electrical energy. Any one form of energy can be converted to any other form. Extent of energy conversion can be complete or partial. For example, mechanical, chemical, and electrical energy can be completely converted to thermal energy. The conversion of heat to mechanical energy, however, is only partial and occurs in a conversion system such as a turbine or internal combustion engine using a working fluid, gas or steam, with a cyclic change of its state. The efficiency of this energy conversion depends on the temperature difference between the working fluid in the system and the surroundings.

When considering energy generation, production, conversion, and utilization, one uses such terms as conventional and alternative energy sources, primary, secondary, end use energy, useful energy, and sustainable energy. Conventional energy sources are fossil fuels (coal, lignite, pit, fuel oil, natural gas, and wood), as well as artificially produced fuel types such as coal gas, liquefied gas, coke, char, and combustible waste. Alternative or renewable energy sources are solar, wind, hydro, geothermal, wave and tidal, and biomass energy. Primary energy is the energy of an energy source without conversion, such as solar energy and chemical energy of fossil fuels. Secondary energy forms are the energy forms produced by conversion of primary energy. These are electrical energy, work, and thermal energy for heating, cooking, and cooling. To calculate the amount of secondary energy produced from a given amount of primary energy, a conversion efficiency of a converter facility such as a steam or gas turbine in a power station, and internal combustion engine is used. The end use energy is the energy that is available for the energy user. Its amount for a particular application is equal to the secondary energy losses in the energy transport and distribution systems. The useful energy is the energy that is required to be supplied for a particular purpose, such as vehicles, motors, lighting, heating, cooling, cooking, or industrial processes (Kharchenko 1998).

There are many different definitions of sustainability. Most of these the definitions contain elements for allowing life to survive and thrive for longer and longer periods of time, living within limits, understanding the inter-dependencies and inter-connectedness of environment, economy and society and, the appropriate and equitable distribution of resources and opportunities within the parts of any given whole. Sustainable energy is the provision of energy that meets the needs of the present without compromising the ability of future generations to meet their needs. Sustainable energy sources are most often regarded as including all renewable sources, such as plant matter, solar power, wind power, wave power, geothermal power and tidal power. It usually also includes technologies that improve energy efficiency, such as low-carbon energy system. Conventional fission power is sometimes referred to as sustainable, but this is controversial politically due to concerns about peak uranium, radioactive waste disposal and the risks of disaster due to accident, terrorism, or natural disaster. Solar Energy is the primary energy and all other energy sources are secondary energy or after effects and resultant of solar energy.

1.2.1 Forms of Energy

Energy is a quantitative property of a system which may be kinetic, potential, or other in form. Energy comes in many forms and can in principle be transformed from one form to another. The laws of thermodynamics govern how and why energy is transferred. Some of the common forms of energy include (Ristinen and Kraushaar 2005):

(a) **Chemical Energy**
Chemical energy is the energy stored in certain chemicals or materials that can be released by chemical reactions, often combustion. The burning of wood, paper, coal, natural gas, or oil releases chemically stored energy in the form of heat energy. Heating houses, powering automobiles, and turning the generators that provide electricity are primarily with chemical energy. Other examples of chemical energy sources are hydrogen, charged electric batteries, and food in the stomach. Chemical reactions release this energy for use.

(b) **Thermal Energy**
Thermal energy interchangeable with heat energy, which is the energy associated with random molecular motions within any medium. Thermal energy is related to the concept of temperature. Increases of thermal energy contained in any substance result in a temperature increase and, conversely, a decrease of thermal energy produces a decrease of temperature.

(c) **Mass Energy**

Base on Einstein's theory, there is equivalence between mass and energy. Energy can be converted to mass, and mass can be converted to energy. The formula (Ristinen and Kraushaar 2005)

$$E = mc^2 \tag{1.1}$$

gives the amount of energy, E, represented by a mass, m. This energy is often referred to as the mass energy. The symbol c stands for the speed of light. The most dramatic examples of this equivalence are in nuclear weapons and nuclear reactors, but entire existence of human beings depends on nuclear reactions in the sun. Here atomic nuclei come together in a reaction with the resulting products having less mass than what went into the reaction. The mass that is lost in the reaction appears as energy according to the Einstein equation (Ristinen and Kraushaar 2005)

$$\Delta E = \Delta mc^2 \tag{1.2}$$

where Δm is the missing mass, and c is the speed of light. The energy that appears, ΔE, is in joules if Δm is in kilograms and c is in meters per second. Because c is such a very large number, 3×10^8 m/s, a small loss of mass results in a huge release of energy. At a detailed level, any reaction, of any type, chemical or nuclear, which releases energy does so in association with a loss of mass between the inputs and outputs, according to the Einstein equation. Einstein put forth the mass energy equation in the early 1900s. It was not until the 1920s and 1930s that the nuclear fusion processes in stars were first understood, and in the 1940s that energy release from man-made nuclear fission reactions was first demonstrated.

(d) **Kinetic Energy**

Kinetic energy is a form of mechanical energy. It has to do with mass in motion. An object of mass m, moving in a straight line with velocity v, has kinetic energy given by Ristinen and Kraushaar (2005)

$$K_E = \frac{1}{2}mv^2 \tag{1.3}$$

If the object in question is an automobile, work must be done to bring the auto up to speed, and, conversely, a speeding car does work in being brought to rest. The work done on the accelerating car is derived from the fuel, the work done by the stopping car will appear mainly as heat energy in the brakes, if the brakes are used to stop the car. In an equivalent manner, an object rotating around an axis has kinetic energy associated with the rotation. It is just a matter of all the mass elements which make up the object each having velocity and kinetic energy. These combined kinetic energies make up the kinetic energy of the rotating object. Someday rapidly rotating flywheels may provide the stored energy needed to power a car.

(e) **Potential Energy**

Potential energy is associated with position in a force field. An obvious example is an object positioned in the gravitational field of the earth. If an object has weight w at a height h above the earth's surface, it will have potential energy (Ristinen and Kraushaar 2005)

$$P_E = wh \tag{1.4}$$

relative to the earth's surface. If the object is released and falls to the earth, it will lose its potential energy but gain kinetic energy in the same amount. Another example would be at a hydroelectric dam where water is effectively, but usually not literally, dropped onto a turbine below. In this example, the water hitting the blades of the turbine has kinetic energy equal to the potential energy it would have had at the top of the reservoir surface. This potential energy is measured relative to the turbine's location. The kinetic energy of the water becomes electric energy as the turbine spins a generator.

(f) **Electric Energy**

Electrical energy is the presence and flow of an electric charge. The energy portion of electricity is found in a variety of phenomena such as static electricity, electromagnetic fields and lightning. Humans have found the ability to harness these phenomena and store the electrical charge for later use. The concept of electrical energy is defined using a variety of different terminologies such as charge, current, and potential. If an electric charge q is taken to a higher electric potential (higher voltage) V, then it can release its potential energy, given by $P_E = qV$, in some other form such as heat or mechanical energy. A battery is a common device for storing electric energy. The chemicals in a battery have an inherent difference of electric potential. When the battery is charged, electric charges are brought to the higher potential so that energy is stored as chemical energy for later use as electric energy. Thus, a battery works both ways; it can convert electric energy to chemical energy, or chemical energy to electric energy.

Mechanical energy is converted to electric energy in a generator, where conductors are forced to move through a magnetic field to induce a voltage between the ends of the conductor. And, if a voltage is applied to the terminals of a common type of generator, it can function as a motor, thereby converting electrical energy to mechanical energy.

(g) **Electromagnetic Radiation**

The energy radiated by the sun travels to the earth and elsewhere by electromagnetic radiation. That part of the spectrum of electromagnetic energy to which human eyes are sensitive is known as visible light, and a large fraction of the solar energy that is received is in the form of visible light. As shown in Fig. 1.4, the electromagnetic spectrum covers a very wide range of frequency, and visible light is only a small part of the entire spectrum (Faezams 2013). Electromagnetic radiation is characterized by a wavelength, λ, and a frequency, f. In a free

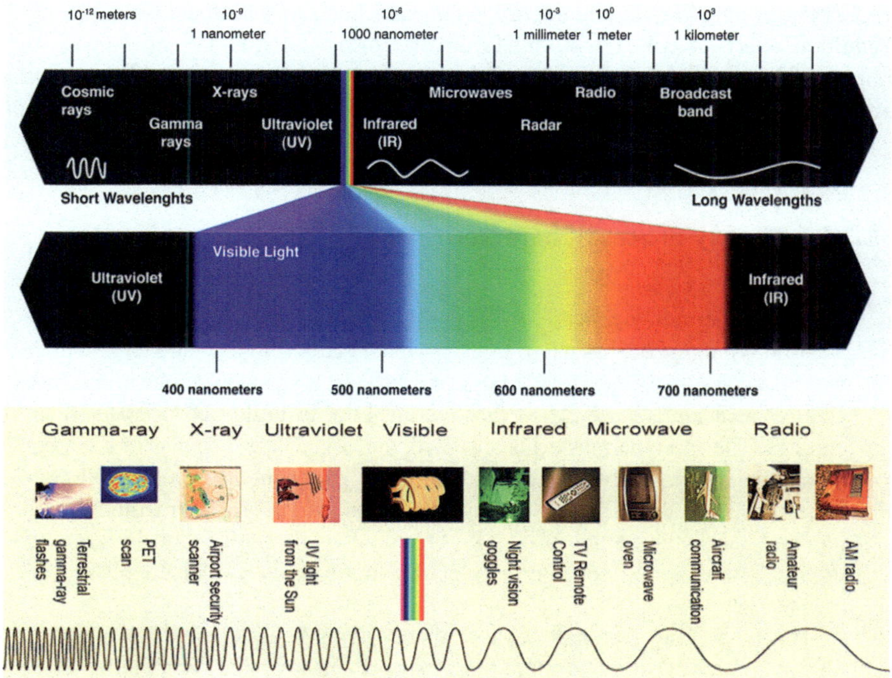

Fig. 1.4 Illustration of electromagnetic spectrum (Modified from https://imagine.gsfc.nasa.gov/science/index.html. Credit: National Aeronautics and Space Administration Goddard Space Flight Center)

space, the velocity of light, c, is related to these quantities by the equation $c = f\lambda$. The numerical value of c is 3×10^8 m/s. The electromagnetic spectrum ranges from radio waves ($\lambda = 200$ m) to microwaves ($\lambda = 0.1$ m), to light ($\lambda = 5 \times 10^{-7}$ m), to X-rays ($\lambda = 1 \times 10^{-8}$ m) and beyond. Various portions of the electromagnetic spectrum are important to the transformation and use of energy on earth. The portion that includes radio waves and microwaves is generated by electronic devices. Light and X-rays have their origin in atomic excitations and radiating electrons. Gamma rays are produced by the decay of excited states of atomic nuclei.

Fundamentally, energy can be summarized as two basic types: potential energy and kinetic energy. Potential energy is the energy of state or position. Potential energy is thought as stored energy. For example, consider water backed up by a large dam. While the water isn't really doing anything, it has the potential to move very quickly and with great force, if the dam were removed. Kinetic energy, on the other hand, is the energy of motion. A freely flowing river, for example, has kinetic energy in the movement of its water. It is important to consider the different forms potential and kinetic energy may take, in order to fully understand their importance. Potential energy, for example, may include the energy stored in chemical bonds. Kinetic energy may include electric energy, light, and heat, in addition to movement. In addition, energy may be transformed from one form to another. Potential energy may be converted into kinetic energy, or kinetic energy may be converted into potential energy.

1.2.2 Energy Process Characterization

INDEPTH: Thermodynamics

Thermodynamics is a branch of physics which deals with heat and temperature and their relation to energy and work of a system. It was born in the nineteenth century as scientists were first discovering how to build and operate steam engines. Thermodynamics deals only with the large-scale response of a system which people can observe and measure in experiments. Small-scale gas interactions are described by the kinetic theory of gases. The methods complement each other; some principles are more easily understood in terms of thermodynamics and some principles are more easily explained by kinetic theory.

There are four principal laws of thermodynamics, defining fundamental physical quantities (temperature, energy, and entropy) that characterize thermodynamic systems. Each law leads to the definition of thermodynamic properties which help people to understand and predict the operation of a physical system. The zeroth law of thermodynamics involves some simple definitions of thermodynamic equilibrium. Thermodynamic equilibrium leads to the large-scale definition of temperature, as opposed to the small-scale definition related to the kinetic energy of the molecules. The first law of thermodynamics relates the various forms of kinetic and potential energy in a system to the work which a system can perform and to the transfer of heat. This law is sometimes taken as the definition of internal energy, and introduces an additional state variable, enthalpy. The first law of thermodynamics allows for many possible states of a system to exist. But experience indicates that only certain states occur. This leads to the second law of thermodynamics and the definition of another state variable called entropy. The second law stipulates that the total entropy of a system plus its environment cannot decrease; it can remain constant for a reversible process but must always increase for an irreversible process. The Third Law states that the entropy of a pure crystal at absolute zero is zero. This allows an absolute scale for entropy to be established that, from a statistical point of view, determines the degree of randomness or disorder in a system.

The science of thermodynamics has been developed over centuries, and its principles apply to nearly every device ever invented. Its importance in modern technology cannot be overstated.

1.2.2.1 The Laws of Thermodynamics

Thermodynamics can be defined as the study of energy, energy transformations and its relation to matter. Regardless of any types of energy, all energy is governed by the laws of thermodynamics. The four laws of thermodynamics define fundamental physical quantities (temperature, energy, and entropy) that characterize thermodynamic systems. The laws describe how these quantities behave under various circumstances, and forbid certain phenomena, such as perpetual motion (Kondepudi 2008).

(a) Zeroth law of thermodynamics:
If two systems are both in thermal equilibrium with a third system, then they are in thermal equilibrium with each other. Thermal equilibrium means that when two bodies are brought into contact with each other and separated by a barrier that is permeable to heat, there will be no transfer of heat from one to the other. This is, the three bodies are all the same temperature. The Zeroth law establishes that temperature is a fundamental and measurable property of matter. The temperature of two systems is the only thing people need to know in order to determine which direction heat. In other words, the temperature defines the direction of heat flow, and it does not depend directly on the amount of energy that's involved.

(b) First law of thermodynamics:
Like the Zeroth Law, which defined a useful property, "temperature," the First Law defines a useful property called "energy." Any thermodynamic system in an equilibrium state possesses a state variable called the internal energy. The system energy can be considered as a sum of internal energy, kinetic energy, potential energy, chemical energy, and other forms of energy. The system energy is invariably conserved; however, the internal energy of a closed system may change as heat is transferred into or out of the system or work is done on or by the system. The first law of thermodynamics is a version of the law of conservation of energy, adapted for thermodynamic systems, and states that energy can neither be created nor destroyed, rather it can only be transformed from one form to another. In other words, during an interaction, energy can change from one form to another, but the total amount of energy remains constant. The first law is often formulated by stating that the change in the internal energy of a closed system is equal to the amount of heat supplied to the system, minus the amount of work done by the system on its surroundings.

The internal energy is just a form of energy like the potential energy of an object at some height above the earth, or the kinetic energy of an object in motion. In the same way that potential energy can be converted to kinetic energy while conserving the total energy of the system, the internal energy of a thermodynamic system can be converted to either kinetic or potential energy. A carbohydrate, for instance, holds potential energy in its chemical bonds. As those bonds are broken, the potential energy could be transferred into the chemical bonds of another molecule, or could be released as the kinetic energy of heat.

Like potential energy, the internal energy can be stored in the system. It is noticed, however, that heat and work cannot be stored or conserved independently since they depend on the process. The first law of thermodynamics allows for many possible states of a system to exist, but only certain states are found to exist in nature. The second law of thermodynamics helps to explain this observation.

(c) Second law of thermodynamics:
There is a useful thermodynamic variable called entropy. A natural process that starts in one equilibrium state and ends in another will go in the direction that causes the entropy of the system plus the environment to increase for an irreversible process and to remain constant for a reversible process. The second law of thermodynamics states that the total entropy of an isolated system can only increase over time. It can remain constant in ideal cases where the system is in a steady state (equilibrium) or undergoing a reversible process. The increase in entropy accounts for the irreversibility of natural processes, and the asymmetry between future and past. An isolated system, if not already in its state of thermodynamic equilibrium, spontaneously evolves towards it. Thermodynamic equilibrium has the greatest entropy among the states accessible to the system. The second law of thermodynamics states that when energy is transferred, either from one form to another or between carriers, some of the usable energy will be lost. In other words, no transfer of energy is 100% efficient. At the molecular level, for example, this means that if the potential energy stored in the chemical bonds of one molecule is transferred into the chemical bonds of another molecule, the new bonds formed will contain less energy than the original bonds.

While the first law of thermodynamics provides the basic definition of internal energy, associated with all thermodynamic systems, and states the rule of conservation of energy; the second law is concerned with the direction of natural processes. It asserts that a natural process runs only in one sense, and is not reversible. For example, heat always flows spontaneously from hotter to colder bodies, and never the reverse, unless external work is performed on the system. The principles of the laws of thermodynamics can be used to describe biological systems, and to predict the outcomes of chemical reactions. In any system, the total energy is equal to the amount of usable energy plus the amount of unusable energy present. Chemical reactions are reversible, and in many cases the reactants and products will reach equilibrium, rather than all the reactants being used up to form products.

(d) Third law of thermodynamics:
The most accepted version of the third law of thermodynamics, the unattainability principle, states that any process cannot reach absolute zero temperature in a finite number of steps and within a finite time. It can be restated as: the entropy of a system approaches a constant value as the temperature approaches zero. The entropy of a system at absolute zero is typically zero, and in all cases, is determined only by the number of different ground states it has. Specifically, the entropy of a pure crystalline substance at absolute zero temperature is zero.

In actuality, no object or system can have a temperature of zero Kelvin, because of the Second Law of Thermodynamics. The Second Law, in part, implies that heat can never spontaneously move from a colder body to a hotter body. As a system approaches absolute zero, it will eventually have to draw energy from whatever systems are nearby. If it draws energy, it can never obtain absolute zero. Therefore, this state is not physically possible, but is a mathematical limit of the universe. At absolute zero temperature, theoretically all modes of motion stops (no vibration, no rotation, and no translation). Molecules near these temperatures have been called the fifth state of matter: Bose-Einstein Condensates. Awesome things like super-fluidity and superconductivity happen at these temperatures.

Basically, thermodynamics deals with matter as a continuous system rather than as discrete or granular. A system is defined as a quantity of matter or a region in space chosen for study. The mass or region outside the system is called the surroundings. Boundary is a real or imaginary surface that separates the system from its surroundings. The boundaries of a system can be fixed or movable. Mathematically, the boundary has zero thickness, no mass, and no volume. A closed system or control mass consists of a fixed amount of mass, and no mass can cross its boundary. But, energy in the form of heat or work, can cross the boundary, and the volume of a closed system does not have to be fixed. An open system or control volume is a properly selected region in space. It usually encloses a device that involves mass flow such as a compressor. Both mass and energy can cross the boundary of a control volume. As some thermodynamics relations that are applicable to closed and open systems are different. Thus, it is extremely important to recognize the type of system before start analyzing it. In more details, an isolated system is a closed system that does not communicate with the surroundings by any means. A rigid system is a closed system that communicates with the surroundings by heat only. An adiabatic system is a closed or open system that does not exchange energy with the surroundings by heat (Bahrami 2009).

1.2.2.2 Macroscopic and Microscopic Energy Systems

In thermodynamics, total energy of a system can be categorized as one of two basic types: macroscopic and microscopic. Macroscopic forms of energy indicate the forms of energy that a system possesses as a whole with respect to some outside reference frame, such as kinetic and potential energy. The macroscopic energy of a system is related to motion and the influence of some external effects such as gravity, magnetism, electricity, and surface tension. Microscopic forms of energy are those related to molecular or atomic structure of a system. The sum of microscopic energy forms the internal energy, U. The total energy of a system consists of the kinetic, potential, and internal energies (Bahrami 2009):

$$E = U + K_E + P_E = U + \tfrac{1}{2}mv^2 + wh\ (\mathrm{kJ}) \tag{1.5}$$

where the contributions of magnetic, electric, nuclear energy are neglected. Internal energy is related to the molecular structure and the degree of molecular activity and it may be viewed as the sum of the kinetic and potential energies of molecules. The sum of translational, vibrational, and rotational energies of molecules is the kinetic energy of molecules, and it is also called the sensible energy. At higher temperatures, system will have higher sensible energy. Internal energy associated with the phase of a system is called latent heat. The intermolecular forces are strongest in solids and weakest in gases. The internal energy associated with the atomic bonds in a molecule is called chemical or bond energy. The tremendous amount of energy associated with the bonds within the nucleolus of atom itself is called atomic energy.

Energy interactions with a closed system can occur via heat transfer and work. A system is said to be in thermodynamic equilibrium if, over a lengthy period of time, no change in the character or state of the system is observed. Any characteristic of a system is called a property. In classical thermodynamics, the substance is assumed to be a continuum, homogenous matter with no microscopic holes. This assumption holds as long as the volumes, and length scales are large with respect to the intermolecular spacing. Intensive properties are those that are independent of the size (mass) of a system, such as temperature, pressure, and density. They are not additives. Extensive properties are values that are dependent on size of the system such as mass, volume, and total energy U. They are additives. Extensive properties per unit mass are called specific properties. At a given *state*, all the properties of a system have fixed values. Thus, if the value of even one property changes, the state will change to different one. In an equilibrium state, there are no unbalanced potentials (or driving forces) within the system. A system in equilibrium experiences no changes when it is isolated from its surroundings (Bahrami 2009).

Any change a system undergoes from one equilibrium state to another is called a *process,* and the series of states through which a system passes during a process is called a *path*. Quasi-equilibrium process can be viewed as a sufficiently slow process that allows the system to adjust itself internally and remains infinitesimally close to an equilibrium state at all times. Quasi-equilibrium process is an idealized process and is not a true representation of the actual process. Actual processes are modeled with quasi-equilibrium ones. Moreover, they serve as standards to which actual processes can be compared. Process diagrams are used to visualize processes. The process path indicates a series of equilibrium states, which cannot be specified for a non-quasi-equilibrium process. Isothermal is a process during which the temperature remains constant; Isobaric is a process during which the pressure remains constant; Isometric is process during which the specific volume remains constant. A system is said to have undergone a cycle if it returns to its initial state at the end of the process. The state of a system is described by its properties. The state of a simple compressible system is completely specified by two independent, intensive properties (Bahrami 2009).

A system is called simple compressible system in the absence of electrical, magnetic, gravitational, motion, and surface tension effects (external force fields). Independent properties are independent if one property can be varied while the other one is held constant. For example, pressure is the force exerted by a fluid per unit area. For a fluid at rest, the pressure at a given point is the same in all directions. The actual pressure at a given position is called the *absolute pressure*, and it is measured relative to absolute vacuum. Gauge pressure is the difference between absolute pressure and atmospheric pressure. In thermodynamics calculations, always use absolute pressure. Most pressure measuring devices are calibrated to read zero in the atmosphere (they measure P_{gauge} or P_{vac}). In accordance with the Zeroth law, any system that possesses an equation of state that relates temperature T to other accurately measurable properties can be used as a thermometer, e.g., an ideal gas obeys the equation of state (Bahrami 2009):

$$mRT = PV \tag{1.6}$$

Experimentally obtained Temperature Scales, the *Celsius* and *Fahrenheit* scales, are based on the melting and boiling points of water.

From basic mechanics, *work*, *W*, is defined as the energy provided by an entity that exerts a force, *F*, in moving one or more particles through a distance, *x*. Thus, work must be done by an external agent to decrease the volume, *V*, of a system of molecules. In a piston-cylinder arrangement, an infinitesimal volume change of the system due to the motion of the piston is related to the differential work through the force-distance product (Bahrami 2009):

$$dew = Fdx = pAdx = pdV \; [\mathrm{N-m}] \tag{1.7}$$

or

$$dw = pdv \; [\mathrm{kJ/kg}] \tag{1.8}$$

where *p* is the system pressure, and *A* is the piston cross-sectional area. Note that in Eq. (1.8), the lower-case letters *w* and *v* denote work and volume on a unit mass basis. All extensive properties, i.e., those properties of state that are proportional to mass, are denoted by lowercase characters when on a unit mass basis. These are so-called *specific properties*. Thus, if *V* represents volume, then *v* denotes *specific volume*. Although work is not a property of state, it is dealt with in the same way. When work decreases the volume of a system, the molecules of the system move closer together. The moving molecules then collide more frequently with each other and with the walls of their container. As a result, the average forces (and hence pressures) on the system boundaries increase. Thus, the state of the system may be changed by work done on the system.

Given a system immersed in a container of hot fluid, by virtue of a difference in temperature between the system and the surrounding fluid, energy passes from the fluid to the system. That is, *heat*, *Q* [kJ], is transferred to the system. The system is observed to increase in temperature or to change phase or both. Thus, heat transfer to or from the system, like work, can also change the state of the matter within the system. When the system and the surrounding fluid are at the same temperature, no heat is transferred. In this case the system and surroundings are said to be in *thermal equilibrium*. The term *adiabatic* is used to designate a system in which no heat crosses the system boundaries. A system is often approximated as an adiabatic system if it is well insulated. If a system undergoes a process in which temperature and pressure gradients are always small, the process may be thought of as a sequence of near-equilibrium states. If each of the states can be restored in reverse sequence, the process is said to be internally reversible. If the environmental changes accompanying the process can also be reversed in sequence, the process is called externally reversible. Thus, a reversible process is one that is both internally and externally reversible. The reversible process becomes both a standard by which the success of real processes is measured in avoiding losses and a tool is used to derive thermodynamic relations that approximate reality. All real processes fail to satisfy the requirements for reversibility and are therefore irreversible. Irreversibility occurs due to temperature, pressure, composition, and velocity gradients caused by heat transfer, solid and fluid friction, chemical reaction, and high rates of work applied to the system (Cengel and Boles 2010).

Mechanically, work can change the kinetic energy of mass and can change the elevation or potential energy of mass in a gravitational field. Thus, work performed by an outside agent on the system boundary can change the energy associated with the particles that make up the system. Likewise, heat is energy crossing the boundary of a system, increasing or decreasing the energy of the molecules within. Thus, heat and work are not properties of state but forms of energy that are transported across system boundaries to or from the environment. They are sometimes referred to as *energy in transit*. Energy conversion engineering is vitally concerned with devices that use and create energy in transit (Cengel and Boles 2010).

1.2.2.3 Entropy and Enthalpy

Entropy and *enthalpy* are thermodynamic properties that, like internal energy, usually appear in the form of differences between initial and final values. The entropy change of a system, Δs [kJ/kg K], is defined as the integral of the ratio of the system differential heat transfer to the absolute temperature for a reversible thermodynamic path, that is, a path consisting of a sequence of well-defined thermodynamic states. In differential form this is equivalent to (Weston 2000):

$$ds = dq_{\mathrm{rev}}/T \ [\mathrm{kJ/kg - K}] \tag{1.9}$$

where the subscript *rev* denotes that the heat transfer must be evaluated along a reversible path made up of a sequence of neighboring thermodynamic states. It is implied that, for such a path, the system may be returned to its condition before the process took place by traversing the states in the reverse order. An important example of the use of Eq. (1.9) considers a thermodynamic cycle composed of reversible processes. The network of the cycle can be expressed as (Weston 2000):

$$w_n = \oint dq = \oint Tds \ [\mathrm{kJ/kg}] \tag{1.10}$$

This shows that the area enclosed by a plot of a reversible cyclic process on a temperature-entropy diagram is the network of the cycle.

The *enthalpy*, *H*, is a property of state defined in terms of other properties (Weston 2000):

$$H = u + pv \ [\mathrm{kJ/kg}] \tag{1.11}$$

where *h*, *u*, and *v* are, respectively, the system specific enthalpy, specific internal energy, and specific volume, and *p* is the pressure. These properties can be used to describe the First Law.

For a reversible process (Weston 2000)

$$Tds = du + pdv = dH - vdp \ [\mathrm{kJ/kg}] \tag{1.12}$$

Equation (1.12) may be regarded as relating changes in entropy for reversible processes to changes in internal energy and volume in the former and to changes in enthalpy and pressure in the latter. The fact that all quantities in these equations are properties of state implies that entropy must also be a thermodynamic property. Because entropy is a state property, the entropy change between two equilibrium states of a system is the same for all processes connecting them, reversible or irreversible. To use Eq. (1.11) directly or as in Eq. (1.12), a reversible path must be employed. Because of the path independence of state property changes, any reversible path will do.

While Eq. (1.11) may be used to determine the entropy change of a system, the *Second Law of Thermodynamics*, is concerned with the entropy change of the universe, i.e., of both the system and the surroundings. Because entropy is an extensive property, the entropy of a system is the sum of the entropy of its parts. Applying this to the universe, the entropy of the universe is the sum of the entropy of the system and its surroundings. The Second Law may be stated as "The entropy change of the universe is nonnegative" (Weston 2000):

$$\Delta S_{\mathrm{univ}} \geqq 0 \ [\mathrm{KJ/K}] \tag{1.13}$$

However, the entropy change of a system may be negative (entropy decrease) if the entropy change of its environment is positive (entropy increase) and sufficiently large that inequality is satisfied. The Second Law implies that, for the combined process to be possible, the environmental entropy change must exceed the magnitude of the system entropy change. The First Law of Thermodynamics deals with how the transfer of heat influences the system internal energy but says nothing about the nature of the heat transfer, i.e., whether the heat is transferred from hotter or colder surroundings. In fact, the environment must be hotter to transfer heat to a cooler object, but the First Law is indifferent to the condition of the heat source. However, calculation of the entropy change for heat transfer from a cold body to a hot body yields a negative universe entropy change, violates the Second Law, and is therefore impossible. Thus, the Second Law provides a way to distinguish between real and impossible processes. Moreover, the entropy change of the universe produced by a process is a measure of the irreversibility of the process. For an isolated system, there is no change in the entropy of the surroundings. Hence the system entropy change is the entropy change of the universe and therefore must be nonnegative. In other words, the entropy of an isolated system can only increase or at best stay constant.

1.2.2.4 Chemical Kinetics

Chemical kinetics is the study of factors that affect how fast a reaction occurs and the step-by-step processes involved in chemical reactions. The factors that affect reaction rate include concentration of reactants, catalyst, temperature, and surface area of reactants. Reaction rates range from the almost instantaneous, as in an explosion, to the almost unnoticeably slow, as in corrosion. The aim of chemical kinetics is to make predictions about the composition of reaction mixtures as a function of time, to understand the processes that occur during a reaction, and to identify what controls its rate. The rate of a chemical reaction is the rate of change of the concentration of one of its components, either a reactant or a product. The experimental investigation of reaction rates therefore depends on being able to monitor the change of concentration with time. Classical procedures for reactions that take place in hours or minutes make use of a variety of techniques for determining concentration, such as spectroscopy and electrochemistry. Very fast reactions are studied spectroscopically. Spectroscopic procedures are available for monitoring reactions that are initiated by a rapid pulse of electromagnetic radiation and are over in a few femtoseconds (Atkins and Jones 2010).

The analysis of kinetic data commonly proceeds by establishing a *rate law,* a mathematical expression for the rate in terms of the concentrations of the reactants (and sometimes products) at each stage of the reaction. For instance, a reaction with a rate law of the form

$$\text{Rate} = k[\text{Reactant } A]^a[\text{Reactant } B]^b \tag{1.14}$$

is said to be of order a in A, of order b in B, and to have an overall order of $a + b + \cdots$. k is called rate constant. Some rate laws are far more complex than this example and many involve the concentrations of the products.

The advantage of identifying the reaction order is that all reactions with the same rate law (but different characteristic rate constants) behave similarly. For example, the concentration of a reactant in a first-order reaction decays exponentially with time at a rate determined by the rate constant (Atkins and Jones 2010)

$$[A] = [A]_0 e^{-kt} \tag{1.15}$$

where $[A]_0$ is the initial concentration of the reactant. Figure 1.5 illustrates the first-order reaction profile (Zumdahl and Zumdahl 2014). On the other hand, all second-order reactions lead to the time-dependence of the concentration. It is common to report the time-dependence of first-order reactions in terms of the *half-*

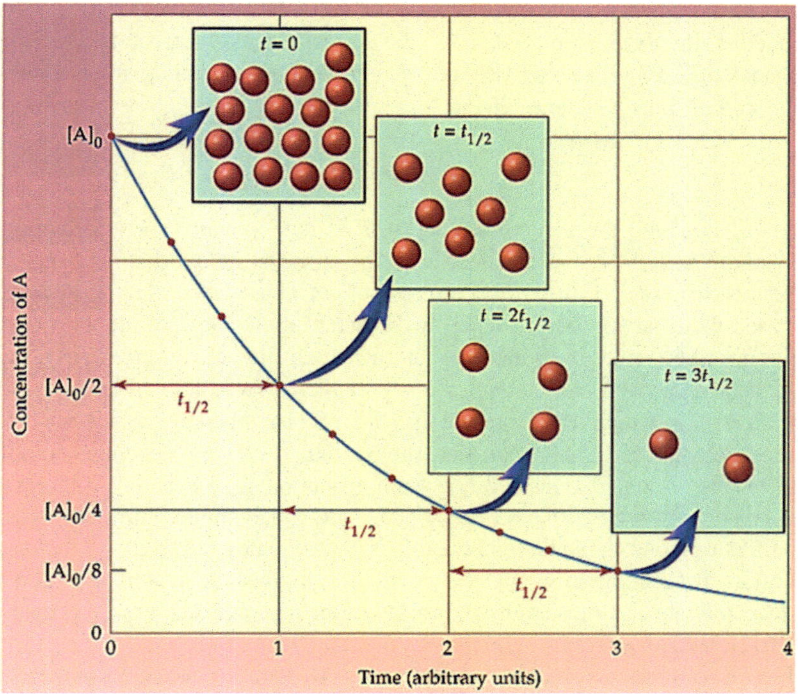

Fig. 1.5 Illustration of first-order reaction profile (Adapted with permission from Zumdahl and Zumdahl 2014 (Brooks/Cole, A Division of Cengage Learning))

life, $t_{½}$, of the reactant, the time needed for its concentration to fall to half its initial value. Reactions with large rate constants have short half-lives. The identification of a rate law provides valuable insight into the reaction mechanism, the sequence of elementary steps by which a reaction takes place. The aim is to identify the reaction mechanism by constructing the rate law that it implies. This procedure may be simplified by identifying the *rate-determining step* of a reaction, the slowest step in a sequence that determines the overall rate. Thus, if the proposed mechanism is $A \rightarrow B$ followed by $B \rightarrow C$, and the former is much faster than the latter, then the overall rate of the reaction will be equal to the rate of $A \rightarrow B$, for once B is formed, it immediately converts into C.

In general, for a mechanism of many steps (including their reverse), the construction of the overall rate law is quite difficult, requiring an approximation or a computer for a numerical analysis. One common approximation is the *steady-state assumption,* in which the net rate of formation of any intermediate is set equal to zero. A hazard of using kinetic information to identify a reaction mechanism, however, is that more than one mechanism might result in the same rate law, especially when approximate solutions are derived. For this reason, proposed reaction mechanism must be supported by additional evidence. Once a reaction mechanism has been identified, attention turns to the molecular properties that govern the values of the rate constants that occur in the individual elementary steps. A clue to the factors involved is provided by the experimental observation that the rate constants of many reactions depend on temperature according to the Arrhenius expression (Atkins and Jones 2010)

$$k = Ae^{-\mathrm{Ea/RT}} \tag{1.16}$$

where E_a is called the activation energy. R is gas content (8.314 J/K-mol), T is temperature in K, A is frequency factor. The simplest model that accounts for the Arrhenius expression is the *collision theory* of gas-phase reaction rates, in which it is supposed that reaction occurs when two reactant molecules collide with at least a minimum kinetic energy (which is identified with the activation energy). A more sophisticated theory is the *activated complex theory* (also known as the *transition state theory*), in which it is supposed that the reactants encounter each other, form a loosened cluster of atoms, then decompose into products. Reactions in solution require more detailed consideration than reactions in gases. It is necessary to distinguish between "diffusion-controlled" and "activation-controlled" reactions. In a diffusion-controlled reaction, the rate is controlled by the ability of the reactants to migrate through the solvent and encounter each other. In an activation-controlled reaction, the rate is controlled by the ability of the reactants that have met each other to acquire enough energy to react.

The rate of a reaction may also be increased by finding a catalyst, a substance that takes part in a reaction by providing an alternative pathway with lower activation energy but is regenerated in the process and is therefore not consumed. Catalysis is the foundation of the chemical and energy industry and a great effort is made to discover or fabricate efficient, economical catalysts. It is also the foundation of life, because the biological catalysts known as enzymes (elaborate protein molecules) control almost every aspect of an organism's function (Atkins and Jones 2010).

1.2.2.5 Energy Evolution and Availability

Energy has always been among the most fundamental elements for the survival, reproduction, and evolution of human society. The sun is the ultimate source of energy. Nonrenewable fossil fuels are formed by solar energy that has been captured over extremely long geological periods. What is more, renewable energy sources being directly (photovoltaic systems) and indirectly (wind, water, etc.) interrelated with the sun. Inevitably, almost all organisms rely, either directly or indirectly, on solar energy for their survival and maintenance. Life on earth would be impossible without the photosynthetic conversion of solar energy into plant biomass. The sun provides approximately 1366 watts per square meter per second ($W/m^2/s$), hence, about 170,000 terawatts ($TW/m^2/s$) on the Earth's surface. In the food chain, solar energy flows are captured and converted through the complex process of photosynthesis. Part of this energy is used by organisms, while a great proportion is lost as heat and a small portion is passed down the food chain as one organism digests another. Apart from the food chain, intelligent human systems utilize the solar energy embodied in fossil fuels and the renewable energy sources as the essential power, the "engine" of modern civilization (Bithas and Kalimeris 2015).

The first era of mankind's quest for new energy resources, from the early discovery of fire to the agricultural (and farming) revolution, could be briefly described as the Organic Energy Economy. This solar-based energy economy was intimately based on intensive land use and biomass consumption. The milestone that determined the transition from the organic economy to the fossil fuel economy, the invention that characterized the era called "The Industrial Revolution," was the steam engine. The unique process that the steam engine initiated was the conversion of chemical energy (heat) into mechanical energy (motion). The biomass energy stocks accumulated in the earth's crust for hundreds of millions of years were now available to serve human needs for the first time in mankind's history, to such an extent that the dawn of the fossil fuel era was about to begin. While the early steam engine was mainly used for pumping water out of coal mines, it soon became (thanks to the efficiency improvements made by James Watt, a Scottish inventor and mechanical engineer) a valuable tool which increased human muscle and animal power for extracting more coal, drove the manufacturing industry, moved ships and trains, and laid the foundation for today's complex and energy-intensive human (economic) systems (Fouquet 2011).

Population growth has always been and will remain one of the key drivers of energy demand, along with economic and social development. Meanwhile, the supply and use of energy have powerful economic, social, and environmental impacts. Figure 1.6 illustrates energy resource evolution and availability in the history before 2015. The energy resources and availability include fossil, renewable, and fissile energy sources. The fossil energy sources are oil, gas, and coal; the renewables mainly cover wind, solar, and biomass; and the fissile energy sources are uranium and thorium. These energy sources are bounded by geographical constraints, e.g., wind and solar energy, as (long-distance) transport of electricity based on wind or solar energy is very costly. Fossil energy sources—oil, gas, coal—and sources of fissile energy—uranium and thorium (a source of fissile uranium)—have in common that they are exhaustible (Lake and Kets 2005).

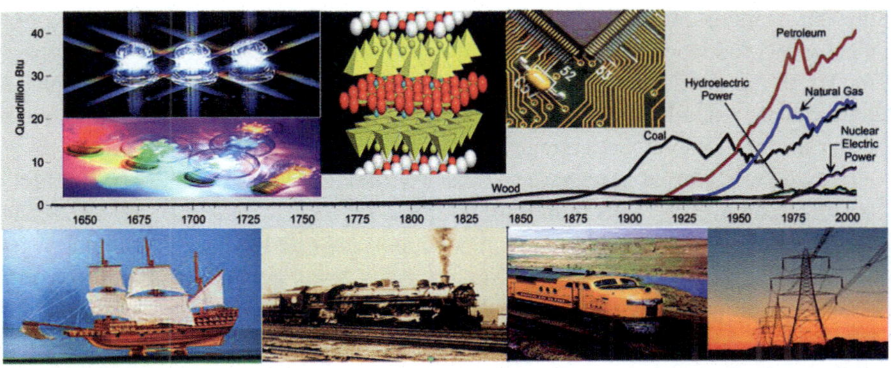

Fig. 1.6 Energy resource evolution and availability

If the unconventional oil resources, including oil shale, oil sands, extra heavy oil and natural bitumen are considered, the global oil reserves will be four times larger than the current conventional reserves. Oil still remains the premier energy resource with a wide range of possible applications. Its main use however, will be shifting towards transport and the petrochemical industries. In future oil's position at the top of the energy ladder will face a strong challenge from other fuels such as natural gas. The oil resource assessments have increased steadily between 2000 and 2009, and about a half of this increase is due to the reclassification of the Canadian oil sands and the revisions undertaken in major OPEC countries: Iran, Venezuela, and Qatar (WEC 2013).

Natural gas is yet another fossil fuel resource that will continue making significant contribution to the world energy economy. The cleanest of all fossil-based fuels, natural gas is plentiful and flexible. It is increasingly used in the most efficient power generation technologies, such as, Combined Cycle Gas Turbine (CCGT) with conversion efficiencies of about 60%. The reserves of conventional natural gas have grown by 36% over the past two decades and its production by 61%. Natural gas is expected to continue its growth spurred by falling or stable prices, and thanks to the growing contribution of unconventional gas, such as shale gas. In addition to power generation, natural gas is expected to play an increasing role as a transport fuel (WEC 2013).

Coal is playing a significant role in delivering energy access, because it is widely available, safe, reliable, and relatively low cost. Despite its poor environmental credentials, coal remains a crucial contributor to energy supply in many countries. Coal is the most wide-spread fossil fuel around the world, and more than 75 countries have coal deposits. Coal reserves are rather evenly spread around the globe: 25% are in the USA, 16% in Russia, and 11.5% in China (Lake and Kets 2005; WEC 2013).

Uranium is the main source of fuel for nuclear reactors. Worldwide output of uranium has recently been on the rise after a lengthy period of declining production caused by oversupply following nuclear disarmament. The present survey shows that total identified uranium resources have grown by 12.5% since 2008 and they are sufficient for over 100 years of supply based on current requirements (WEC 2013).

Hydropower provides a significant amount of energy throughout the world and is present in more than 100 countries, contributing approximately 15% of the global electricity production. In many cases, the growth in hydro power was facilitated by the lavish renewable energy support policies and CO_2 penalties. The development of renewables, excluding large hydro, has been considerably slower than expected 20 years ago. Despite the exponential growth of renewable resources in percentage terms, in particular wind power and solar PV, renewable energy still accounts for a small percentage of TPES in most countries. The continuing growth of renewables strongly depends on subsidies and other support provided by governments. Integration of intermittent renewables in the electricity grids also remains an issue, as it results in additional balancing costs for the system and thus higher electricity bills. Through all energy resources, however, energy efficiency provides an immediate opportunity to decrease energy intensity. This will achieve energy savings and reduce the environmental impacts of energy production and use (WEC 2013).

1.2.3 Energy Calculations and Accounting

Energy accounting is a system to record, analyze, and report energy consumption and cost on a regular basis. Just as financial accounting is used for the effective management of an organization, energy accounting is critical to energy management. It can be one of the most cost-effective tools that school districts, cities, counties, colleges, and other organizations can use to cut energy costs.

1.2.3.1 Energy Efficiency

One of the major criteria guiding the design of energy conversion systems is efficiency. Efficiency is a measure of the quality of an operation or of a characteristic of a device. It indicates how much of a given energy is converted into a desired form of energy. In real life, the efficiency of an energy conversion system is always less than 100%. It does not mean that the part of the energy that is not converted into the desired form is destroyed. It means that the device or plant failed to transform 100% of the input energy into the desired form of energy. The remainder either remains the original form or gets converted into some other form.

Several types of energy conversion efficiencies are widely used, as shown in Fig. 1.7 (Weston 2000). The efficiency of a machine that transmits mechanical power is measured by its mechanical efficiency, the fraction of the power supplied to the transmission device that is delivered to another machine attached to its output (Fig. 1.7a). Thus a gearbox for converting rotational motion from a power source to a device driven at another speed dissipates some mechanical energy by fluid and/or dry friction, with a consequent loss in power transmitted to the second machine. The efficiency of the gearbox is the ratio of its power output to the power input, a value less than one. For example, a turboprop engine with a gearbox efficiency of 0.95 will transmit only 95% of its power output to its propeller. Another type of efficiency that measures internal losses of power is used to indicate the quality of performance of turbomachines such as pumps, compressors, and turbines. These devices convert flow energy to work (power), or vice versa. Here the efficiency compares the output with a theoretical ideal in a ratio (Fig. 1.7b). The resulting efficiency ranges from 0 to 1 as a measure of how closely the process approaches a relevant isentropic process. A turbine with an efficiency of 0.9, will, for example, deliver 90% of the power of a perfect (isentropic) turbine operating under the same conditions. Another form of isentropic efficiency, sometimes called *compressor efficiency* (or pump efficiency), is defined for compressors (or pumps). It is the ratio of the isentropic work to drive the compressor (or pump) to the actual work required. Because the actual work required exceeds the isentropic work, this efficiency is also less than to 1. A third type of efficiency compares the magnitude of a useful effect to the cost of producing the effect, measured in comparable units. An example of this type of efficiency compares the network output, w_n, of a heat engine to the heat supplied, q_a, to operate the engine. This is called η_{th} ($\eta_{th} = w_n/q_a$) (Fig.1.7c). For example, the flow of natural gas to an electrical power plant provides a chemical energy flow rate or heat flow rate to the plant that leads to useful electric power output. It is known from basic thermodynamics that this efficiency is limited by the Carnot efficiency. Another

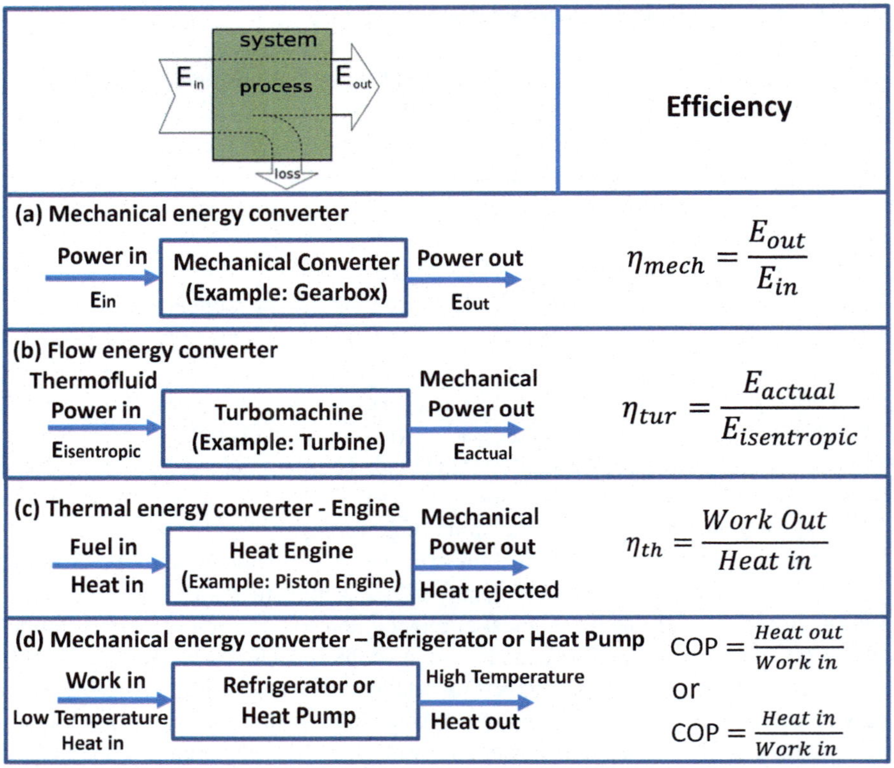

Fig. 1.7 Types of energy conversion efficiencies

example of this type of efficiency as applied to refrigerators and heat pumps (Fig.1.7d) is called the *coefficient of performance,* COP. In this case the useful effect is the rate of cooling or heating, and the cost to produce the effect is the power supplied to the device. The term "coefficient of performance" is used instead of efficiency for this measure of quality because the useful effect usually exceeds the cost in comparable units of measure. Hence, unlike other efficiencies, the COP can exceed unity. As seen in Fig.1.7d, there are two definitions for COP, one for a refrigerator and the other for a heat pump. It may be shown using the First Law of Thermodynamics, a simple relationship exists between the two definitions (Weston 2000): $COP_{hp} = COP_{refr} + 1$.

The Second Law of thermodynamics relevant to heat engines, which, in modern form is that it is impossible for a device which operates in a cycle to receive heat from a sole source and convert the heat *completely* to work. Carnot put forward the theoretical limit for efficiency of heat engine cycles that operate between two given temperature levels: the *Carnot cycle*. The *Carnot cycle* consists of two reversible, isothermal processes separated by two reversible adiabatic or isentropic processes. All of the heat transferred to the working fluid is supplied isothermally at the high temperature, and all heat rejected is transferred from the working medium at the low temperature. No heat transfer takes place, of course, in the isentropic processes. The thermal efficiency of the Carnot cycle, like that of other cycles, is given by w_n/q_a. A heat engine should operate between the widest possible temperature limits. Thus the efficiency of a heat engine will be limited by the maximum attainable energy-source temperature and the lowest available heat-sink temperature. It is impossible for any engine operating in a cycle between two reservoirs at different temperatures to have an efficiency that exceeds the Carnot efficiency corresponding to those temperatures. It can also be shown that all reversible engines operating between two given reservoirs have the same efficiency and that all irreversible engines must have lower efficiencies. Thus, the Carnot efficiency sets an upper limit on the performance of heat engines and therefore serves as a criterion by which other engines may be judged (Weston 2000).

Instead of comparing the work output of the power plant with the energy supplied from fuel to run the plant, it is instructive to compare it with the maximum work achievable by a reversible heat engine operating between the appropriate temperature limits, the *reversible work*. It has been established that any reversible engine would have the same efficiency as a Carnot engine. The Carnot engine provides a device for determining the reversible work associated with a given source temperature, and a lower sink temperature. The *irreversibility*, *I*, of a process is defined as the difference between the reversible work and the actual work of a process (Weston 2000):

$$I = W_{\text{rev}} - WL_{\text{act}}[\text{kJ}] \tag{1.17}$$

It is seen that the irreversibility of a process vanishes when the actual work is the same as that produced by an appropriate Carnot engine. Moreover, the irreversibility of a nonwork-producing engine is equal to the reversible work. It is clear that $I \geq 0$, because no real engine can produce more work than a Carnot engine operating between the same limiting temperatures. A second-law efficiency, η_{II}, as the ratio of the actual work of a process to the reversible work, can be expressed as (Weston 2000):

$$\eta_{II} = W_{\text{act}}/W_{\text{rev}} \tag{1.18}$$

This is an efficiency that is limited to 100%, as opposed to the thermal efficiency of a heat engine, sometimes referred to as first-law efficiency, which may not exceed that of the appropriate Carnot engine. An engine which has no irreversibility is a reversible engine and has a second-law efficiency of 100%.

Therefore, the thermal efficiency, or first-law efficiency, of an engine is a measure of how well the engine converts the energy in its fuel to useful work. It says nothing about energy loss, because energy is conserved and cannot be lost: it can only be transformed. The second-law efficiency, on the other hand, recognizes that some of the energy of a fuel is not available for conversion to work in a heat engine and therefore assesses the ability of the engine to convert only the available work into useful work. This is a reason why some regard the second-law efficiency as more significant than the more commonly used first-law efficiency. Thermodynamics guarantees that no closed system can be over unity or have efficiency above 100%, at least when all inputs are fully accounted for. Energy sinks can have negative efficiencies, for instance. Some current efficiency figures of interest (Lancaster 2002; Golay et al. 2010): The latest of utility power plants are nearing 60% by combined gas turbine plus steam turbine cycling (from chemical and thermal energy to electrical energy). Auto internal combustion engines have a thermal efficiency of about 25–50% (chemical to mechanical). Hydro water turbines can reach an efficiency of 60–90% (gravitational to electrical). Wind turbines have an efficiency of 30 up to 60% (Kinetic to mechanical to electrical). The estimated efficiency of geothermal is 6–13% (thermal to mechanical to electrical), and about 1–3% for ocean thermal. Photovoltaic cells can generally reach 10–15% with around 85–90 theoretical limit (radiative to electrical). Before amortization, heat pumps can output more heat energy than they are input as electrical or mechanical energy, but both the energy forms have to be fully included when doing any true thermodynamic accounting. The COP or coefficient of performance of a heat pump is a ratio of the heat energy that moves to the input electrical energy. A COP of six is easily reached if the temperature difference is reasonably low. The seasonal energy efficiency rating (SEER) is an alternative to a COP. Modern air conditioners have a SEER value ranging from 13 or 14 SEER as a minimum to a maximum of 21 or 25 SEER (based on current technology limitations).

1.2.3.2 Heating Values

The heating value is the amount of heat produced by combustion of a unit quantity of a fuel, which can be differentiated with higher and lower heating values (Barabas and Todorut 2011). The higher heating value (HHV; also known as the gross calorific value or gross energy) takes into account the latent heat of vaporization of water in the combustion products, and is useful in calculating heating values for fuels where condensation of the reaction products is practical (e.g., in a gas-fired boiler used for space heat). In other words, HHV assumes all the water component is in liquid state at the end of combustion (in product of combustion). The higher heating value of a fuel is defined as the amount of heat released by a specified quantity (initially at 25 °C) once it is combusted and the products have returned to a temperature of 25 °C. The higher heating value is experimentally determined in a bomb calorimeter by concealing a stoichiometric mixture of fuel and oxidizer (e.g., two moles of hydrogen and one mole of oxygen) in a steel container at 25° is initiated by an ignition device and the combustion reactions completed. When hydrogen and oxygen react during combustion, water vapor emerges. Subsequently, the vessel and its content are cooled down to the original 25 °C and the higher heating value is determined as the heat released between identical initial and final temperatures (Sundar 2013).

When the lower heating value (LHV) is determined, cooling is stopped at 150 °C and the reaction heat is only partially recovered. The limit of 150 °C is an arbitrary choice. The difference between the two heating values depends on the chemical composition of the fuel. In the case of pure carbon or carbon dioxide, both heating

Table 1.1 Heating values for selected fuels

Name	HHV (MJ/kg)	LHV (MJ/kg)	HHV/LHV	LHV/HHV
Coal	34.1	33.3	1.024	0.977
CO	10.9	10.9	1.000	1.000
Methane	55.5	50.1	1.108	0.903
Natural gas	42.5	38.1	1.115	0.896
Propane	50.3	46.3	1.086	0.920
Gasoline	46.7	42.5	1.099	0.910
Diesel	45.9	43.0	1.067	0.937
Hydrogen	141.9	120.1	1.182	0.846
Methanol	22.9	20.1	1.139	0.878
Ethanol	29.9	26.9	1.112	0.900
Butane	49.2	45.2	1.088	0.919
Isobutane	49.1	44.9	1.094	0.914
Isobutylene	48.2	44.8	1.076	0.929
Propane	50.2	46.3	1.084	0.922
Farmed trees	20.6	19.6	1.051	0.951
Herbaceous biomass	18.1	17.2	1.052	0.950
Com stover	17.4	16.4	1.061	0.943
Forest residue	16.5	15.4	1.071	0.933

values are almost identical, the difference being the sensible heat content of carbon dioxide between 150 °C and 25 °C (sensible heat exchange causes a change of temperature. In contrast, latent heat is added or subtracted for phase changes at constant temperature. Examples: heat of vaporization or heat of fusion). For hydrogen the difference is much more significant as it includes the sensible heat of water vapor between 150 °C and 100 °C, the latent heat of condensation at 100 °C and the sensible heat of the condensed water between 100 °C and 25 °C. Table 1.1 shows heating values for selected fuels. All in all, the higher heating value of hydrogen is 18.2% above its lower heating value (142 MJ/kg vs. 120 MJ/kg). For hydrocarbons the difference depends on the hydrogen content of the fuel. For gasoline and diesel the higher heating value exceeds the lower heating value by about 10% and 7%, respectively, for natural gas about 11% (Boundy et al. 2011).

1.2.4 General Energy Devices and Process Approaches

Industrial restructuring, international competition, and environmental restrictions that make the construction of new power plants and transmission lines prohibitively expensive, as well as the need to import foreign oil, are significant parameters that continuously shape people's attitudes towards electric energy. More than ever before, there is a need to conserve electric energy, and focus on new visions, to invent, to improve on older energy conversion technologies, to optimize designs, and to develop better active materials and devices that lead to more compact and more efficient equipment. On the other hand, the same opportunities sometimes will prompt the development of novel energy systems, the energy savings capabilities of which are questionable and cannot be substantiated (Emanuel 1997).

1.2.4.1 Electricity and Its Utilization

Electricity is a secondary source of energy that is a little different from the other primary sources of energy. The primary energy source that is used to make electricity may be renewable or nonrenewable, but the electricity is neither. In fact, electricity is a general term encompassing a variety of phenomena resulting from the presence and flow of electric charge, such as lightning and static electricity, electromagnetic field and electromagnetic induction. Electricity's extraordinary versatility as a source of energy means it can be put to an almost limitless set of applications which include transport, heating, lighting, communications, and computation. Electrical power is the backbone of modern industrial society and is expected to remain so for the foreseeable future.

Electricity is an energy carrier that can be transformed into useful work through electric circuits. An electric circuit is an interconnection of electric components such that electric charge is made to flow along a closed path (a circuit), usually to perform some useful task. The components in an electric circuit can take many forms, which can include elements such as resistors, capacitors, switches, transformers, and electronics. Electronic circuits contain active components, usually semiconductors, and typically exhibit nonlinear behavior; and passive and linear components: while they may temporarily store energy, they contain no sources of it, and exhibit linear responses to stimuli. The resistor is perhaps the simplest of passive circuit elements: it resists the current through it, dissipating its energy as heat. The resistance is a consequence of the motion of charge through a conductor: in metals, for example, resistance is primarily due to collisions between electrons and ions. Ohm's law is a basic law of circuit theory, stating that the current passing through a resistance is directly proportional to the potential difference across it. The resistance of most materials is relatively constant over a range of temperatures and currents; materials under these conditions are known as "ohmic." The capacitor is a device capable of storing charge, and thereby storing electrical energy in the resulting field. Conceptually, it consists of two conducting plates separated by a thin insulating layer; in practice, thin metal foils are coiled together, increasing the surface area per unit volume and therefore the capacitance. A capacitor connected to a voltage supply initially causes a current as it accumulates charge; this current will however decay in time as the capacitor fills, eventually falling to zero. A capacitor will therefore not permit a steady state current, but instead blocks it. The inductor is a conductor, usually a coil of wire that stores energy in a magnetic field in response to the current through it. When the current changes, the magnetic field does too, inducing a voltage between the ends of the conductor. The induced voltage is proportional to the time rate of change of the current. The constant of proportionality is termed the inductance. The inductor's behavior is in some regards converse to that of the capacitor: it will freely allow an unchanging current, but opposes a rapidly changing one.

Electronic devices make use of the transistor, perhaps one of the most important inventions of the twentieth century, and a fundamental building block of all modern circuitry. A modern integrated circuit may contain several billion miniaturized transistors in a region only a few centimeters square (Joseph 1965).

Electrical power is usually generated by electromechanical generators driven by steam produced from fossil fuel combustion, or the heat released from nuclear reactions; or from other sources such as kinetic energy extracted from wind or flowing water. The modern steam turbine invented by Sir Charles Parsons in 1884 today generates about 80% of the electric power in the world using a variety of heat sources. Such generators bear no resemblance to Faraday's homopolar disk generator of 1831, but they still rely on his electromagnetic principle that a conductor linking a changing magnetic field induces a potential difference across its ends. Since electrical energy cannot easily be stored in quantities large enough to meet demands on a national scale, at all times exactly as much must be produced as is required. This requires electricity utilities to make careful predictions of their electrical loads, and maintain constant coordination with their power stations. A certain amount of generation must always be held in reserve to cushion an electrical grid against inevitable disturbances and losses.

Electricity is an extremely flexible form of energy, and has been adapted to a huge, and growing, number of uses. The invention of a practical incandescent light bulb in the 1870s led to lighting becoming one of the first publicly available applications of electrical power. Although electrification brought with it its own dangers, replacing the naked flames of gas lighting greatly reduced fire hazards within homes and factories. Public utilities were set up in many cities targeting the burgeoning market for electrical lighting. The Joule heating effect employed in the light bulb also sees more direct use in electric heating. While this is versatile and controllable, it can be seen as wasteful, since most electrical generation has already required the production of heat at a power station. A number of countries, such as Denmark, have issued legislation restricting or banning the use of electric heating in new buildings. Electricity is however a highly practical energy source for refrigeration, with air conditioning representing a growing sector for electricity demand, the effects of which electricity utilities are increasingly obliged to accommodate (Baigrie 2006).

Electricity is used within telecommunications, and indeed the electrical telegraph, demonstrated commercially in 1837 by Cooke and Wheatstone, was one of its earliest applications. With the construction of first intercontinental, and then transatlantic, telegraph systems in the 1860s, electricity had enabled communications in minutes across the globe. Optical fiber and satellite communication technology have taken a share of the market for communications systems, but electricity can be expected to remain an essential part of the process (Baigrie 2006).

The effects of electromagnetism are most visibly employed in the electric motor, which provides a clean and efficient means of motive power. A stationary motor such as a winch is easily provided with a supply of power, but a motor that moves with its application, such as an electric vehicle, is obliged to either carry along a power source such as a battery, or to collect current from a sliding contact such as a pantograph, placing restrictions on its range or performance (Baigrie 2006).

1.2.4.2 Energy Conversion Devices and Technical Barriers

In all the stages of energy transfer or conversion from electrical to mechanical, to light, to thermal, to chemical, or vice versa, the law of conservation will always be supreme. Mathematically, this law is expressed by the equation (Emanuel 1997)

$$W_i = W_o + \Delta W \tag{1.19}$$

where W_i is input energy; W_o is output energy; ΔW is energy lost, i.e., converted in useless forms of energy. The efficiency of an energy converter is (Emanuel 1997)

$$\eta = W_o/W_i = 1/(1+\delta), \quad \delta > 0 \tag{1.20}$$

As technologies are progressing, the values of δ for different applications will decrease, but will never be nil. Today, by every 1.0 J of mechanical energy delivered by a medium-size motor, 3–4 J of fuel energy must be used at the power plant. Conversely, for every 1.0 J saved at the electromechanical energy converter's output, 3–4 J of fuel energy are spared. Energy savings and conservation can be achieved by many legitimate venues. Here are some important technologies and methods known today that will help reduce the end user's energy bill (Emanuel 1997).

1. Methods that directly help to reduce the input kilowatt hours.
   (a) Renewable energy sources—Wind power generation, photovoltaics, geothermal, and biomass are the most developed technologies that allow cogeneration at competitive cost in favorable locations.
   (b) Energy conversion optimizers—These are power electronics devices that create conditions for optimum transfer of energy to the loads. Three major categories of devices are identified.
      (i) Adjustable-speed drives—These are ideally suited to reduce energy consumption for pumps, compressors, blowers, and fans by eliminating the throttling energy loss by means of adjusting the velocity of the motor.
      (ii) Optimum voltage controllers—Such devices automatically adjust the input voltage to the best value that yields maximum motor efficiency for a given mechanical power.
      (iii) Electronically ballasted lamps—The efficacy (lm/W) of electric discharge lamps is greatly increased when supplied at a higher frequency.
   (c) Use of higher efficiency equipment—New ferromagnetic materials, such as amorphous steel for transformer cores, lower resistivity conductors, reduced air gap designs, permanent-magnet DC and synchronous motors, better bearings, and better insulators (higher dielectric strength and temperature) allow for designs of equipment with better efficiency.

2. Methods that help the electric utilities to reduce energy.
   (a) Demand-side management programs—The utility shifts load at certain times of the day, reducing the need for energy at particular times.
   (b) Energy storage devices—Conventional batteries, fuel cells, flywheels, and superconductive coils can be used to store energy during the low demand and release it during the peak hours. Such devices help the utilities to reduce the generation reserve margin, to operate the power plants with improved efficiency, and to reduce the power loss in the network.
   (c) Active filters and dynamic compensators—These provide current harmonic cancellation, help restore voltage balance, improve the power factor, and eliminate voltage dips and surges or momentary interruptions.
   (d) Passive filters and power factor correction capacitors—These devices help to improve the power factor, hence, reduce the power system's energy losses. These devices are time-honored and proven methods. They are inexpensive and reliable. Their main drawbacks are lack of continuous adjustment and the possibility of harmonic resonances. Passive filters also may become overloaded by sinking harmonics generated elsewhere.

Energy conversion is heart part of an energy system. As the automation of human society progresses, there is a need for energy to power more and more modern electrical devices of numerous sizes. Typical energy conversion devices include reciprocating, Stirling engines, micro-turbines, full cells, and nuclear batteries (Scoditti and Barker 2003).

The wide power range and operating flexibility make reciprocating engines suitable for substations and small municipalities plus commercial, industrial, institutional, and even residential applications. Reciprocating engines are the lowest-cost distributed energy generation technology in the world today. So far gas engines have been developed, which can utilize an unmatched range of different gases such as biogas, pyrolysis gas, and almost any other combustible gas. The heating value of gases that can be turned into electricity at an efficiency of up to 40% lies between 0.5 and 34 kWh/m^3. However, the feeding of producer gas to the reciprocating engine needs a deep purification with the elimination of the following pollutants (Scoditti and Barker 2003): (a) Dust (< 10–20 mg/m^3); (b) Tar (<100 mg/m^3 in suction engines); and (c) Acids (<20 mg/m^3 in turbo-compressed engines). Dust can obstruct the device that controls the fuel rate in function of the requested power. In this way it is kept steady the rotating speed independently by the load conditions. Furthermore, when the particles reach the combustion chamber can provoke some damage to the pistons sealing, to the inner surface of the cylinders and to the valves. In order to prevent these problems, the gas produced by the gasifier must be opportunely treated. To this purpose before the engine a cooling system is installed which lowers the gas temperature to about 0 °C, once more the gas is condensed; following an electric resistance heats the gas up to 40 °C required for a good combustion into the engine. In this way the gas which is obtained presents a humidity degree almost null.

A Stirling engine is a heat engine operating by cyclic compression and expansion of air or other gas, the working fluid, at different temperature levels such that there is a net conversion of heat energy to mechanical work. The Stirling engine has advantages of high efficiency compared to steam engines, quiet operation, and the ease with which it can use almost any heat source. This compatibility with alternative and renewable energy sources has become increasingly significant as the price of conventional fuels rises, and also in light of concerns such as peak oil and climate change. This engine is currently exciting interest as the core component of micro-combined heat and power (CHP) units, in which it is more efficient and safer than a comparable steam engine (Organ 2007). The primary disadvantage of Stirling Engines is undoubtedly manufacturing cost. It appears that will be difficult to produce a Stirling engine that costs less than twice the price of diesel engine of equivalent power. The high manufacturing cost is mostly due to the expense to manufacture the heat exchanger. The engine thermal efficiency is highly dependent on the maximum cycle temperatures so to sustain a reasonable level (30%) of efficiency it is necessary to use relatively expensive materials for the hot parts, such as stainless steel or high-temperature alloys. Although the sterling engine is inherently more efficient than the internal combustion engine, in practice even the best Stirling engines can only achieve thermal efficiencies similar to those routinely achieved in modern diesel engines (up to 50%). Another contribution to increased cost is that Stirling engines require a cooling system having a thermal capacity approximately twice that of the thermal combustion engine of comparable power output. Some modifications have been made mostly with the costs reduction, efficiency improvement, and the environment saving (Scoditti and Barker 2003).

Operation and maintenance costs of microturbine have a key role to determine the development of these devices in the turbine market. Both the primary operational and maintenance costs are relatively low, in fact the former deals only with the fuel costs which vary from country to country, the latter deals with the replacement of consumables. What affects the total operation and maintenance costs is the recuperator considered the critical component in terms of durability as well. The future developments shall focus on increase of capacity, increase inlet temperatures, options for the gas turbine cycle, performance improvement of industrial gas turbines through use of advanced materials, low emissions technology (target <5 ppm NO_x), consideration for transition to back-up fuels, durable for at least 8000 h, no more than 10% cost add-on, and no negative impacts on gas turbine performance (Scoditti and Barker 2003).

The introduction of fuel cells requires, beyond the development of a favorable context to the distributed energy generation or cogeneration that users take confidence with the technology, overcoming the preoccupations connected with the novelty in terms of safety, modalities and management costs, reliability, maintenance, etc., and feel themselves guaranteed by the continuity and quality of the plant operations (Scoditti and Barker 2003).

Currently available nuclear conversion technologies suffer from inherent inefficiencies and problems. For example, the two-step nature of the conversion process and the limitations of the thermodynamic cycle consume as much as 90% of the initial nuclear energy in its conversion to electricity. One area of ongoing research is in the area of nuclear batteries. Based on the discovery in 1954 that p–n junctions can generate electric current from beta particles and alpha particles emitted from radioactive materials (Rappaport 1954), research has continued for many years investigating a wide variety materials and techniques to construct nuclear batteries with higher efficiencies (Bower et al. 2002). The physics of direct conversion from nuclear energy to electric current is based on a potential difference maintained by a voltage source between a positive and a negative electrode while a charged particle is emitted by an unstable nucleus of a radioactive material. The emitted charged particle creates electron/hole pairs that migrate towards the positive and negative electrodes. A resistive load completes the circuit so that the positive and negative charges which have migrated recombine and power is generated by this induced current flow in the completed circuit (Gadeken et al. 2007).

1.2.4.3 Energy Storage Process

Energy storage is accomplished by devices or physical media that store some form of energy to perform useful operation at later time. Energy storage materials comprise those materials that can provide or store energy in a useful form of oil, coal or uranium etc. All forms of energy are either potential energy or kinetic energy. A wind-up clock stores potential energy (in this case mechanical, in the spring tension), a battery stores readily convertible chemical energy to operate a mobile phone, and a hydroelectric dam stores energy in a reservoir as gravitational potential energy. Ice storage tanks store ice (thermal energy) at night to meet peak demand for cooling. Fossil fuels such as coal and gasoline store ancient energy derived from sunlight by organisms that later died, became buried and over time were then converted into these fuels. Even food (which is made by the same process as fossil fuels) is a form of energy stored in chemical form (Roebuck 2012).

Energy storage as a natural process is as old as the universe itself—the energy present at the initial formation of the universe has been stored in stars such as the Sun, and is now being used by humans directly through solar heating, or indirectly by growing crops or conversion into electricity in solar cells. As a purposeful activity, energy storage has existed since pre-history, though it was often not explicitly recognized as such. An example of deliberate mechanical energy storage is the use of logs or boulders as defensive measures in ancient forts-the logs or boulders were collected at the top of a hill or wall, and the energy thus stored used to attack invaders who came within range. A noticeable application is the control of waterways to drive water mills for processing grain or powering machinery. Complex systems of reservoirs and dams were constructed to store and release water and the potential energy it contained when required. Storing energy allows humans to balance the supply and demand of energy. Energy storage systems in commercial use today can be broadly categorized as mechanical, electrical, chemical, biological, and thermal. Specifically, energy storage methods can be classified as (Huggins 2010): (a) Chemical methods including Hydrogen, Biofuels, Liquid nitrogen, Oxy-hydrogen, and Hydrogen peroxide; (b) Biological methods such as Starch and Glycogen; (c) Electrochemical methods including Batteries, Flow batteries, and Fuel cells; (d) Electrical methods including Capacitor, Supercapacitor, Superconducting and magnetic energy storage (SMES); (e) Mechanical methods such as Compressed air energy storage (CAES), Flywheel energy storage, Hydraulic accumulator, Hydroelectric energy storage, and Spring and Gravitational potential energy (device); and (f) Thermal methods including Phase Change Materials, Ice Storage, Molten salt, Cryogenic liquid air or nitrogen, Seasonal thermal store, Solar pond, Hot bricks, Steam accumulator, Fireless locomotive, Eutectic system; and Fuel Conservation storage.

Energy storage became a dominant factor in economic development with the widespread introduction of electricity and refined chemical fuels, such as gasoline, kerosene, and natural gas in the late 1800s. Unlike other common energy storage in prior use such as wood or coal, electricity must be used as it is being generated, or converted immediately into another form of energy such as potential, kinetic or chemical. Until recently electrical energy has not been converted and stored on a major scale, however new efforts to that effect began in the twenty first century with enhanced research on energy storage and its integration with smart electrical grids. Electricity is transmitted in a closed circuit, and for essentially any practical purposes cannot be stored as electrical energy. This means that changes in demand cannot be accommodated without either cutting supplies as by brownouts or blackouts or by storing the electric energy in another medium. Even renewable energy must be stored in order to make it reliable. Wind blows intermittently and so some form of storage is required to compensate for calm periods. Solar energy is equally not available on cloudy days and during the nighttime, so stored energy must be available to compensate for the loss of sunlight (Huggins 2010).

An early solution to the problem of storing energy for electrical purposes was the development of the battery as an electrochemical storage device. Batteries have previously been of limited use in electric power systems due to their relatively small capacity and excessive cost, however since about the middle of the first decade of the twenty first century newer battery technologies have been developed that can now provide significant utility-scale load-leveling capabilities. A similar possible solution to deal with the intermittency issue of solar and wind energy is found in the capacitor (Huggins 2010).

Since the 1980s, thermal energy storage (TES) has been developed to meet the growing demand for air conditioning during peak hours. The most popular form of thermal energy storage for cooling is ice storage, since it can store more energy in less space than water storage and it is also less costly than energy recovered via fuel cells or flywheels. Thermal storage has cost-effectively shifted Gigawatts of power away from daytime peak usage periods, and in 2009 was used in over 3300 buildings in over 35 countries. It works by creating ice at night when electricity is usually less costly, and then using the ice to cool the air in buildings during the hotter daytime periods (Huggins 2010).

Chemical fuels have become the dominant form of energy storage, both in electrical generation and energy transportation. Chemical fuels in common use are processed coal, gasoline, diesel fuel, natural gas, liquefied petroleum gas (LPG), propane, butane, ethanol, and biodiesel. All of these materials are readily converted to mechanical energy and then to electrical energy using heat engines via turbines or other internal combustion engines, or boilers or other external combustion engines used for electrical power generation. Heat-engine-powered generators are nearly universal, ranging from small engines producing only a few kilowatts to utility-scale generators with ratings up to 800 MW. Liquid hydrocarbon fuels are the most commonly used forms of energy storage for use in transportation, but because the byproducts of the reaction that utilizes these liquid fuels' energy (combustion) produce greenhouse gases other energy carriers like hydrogen can be used to avoid production of greenhouse gases. Some areas of the world have used geographic features to store large quantities of water in elevated reservoirs, using excess electricity at times of low demand to pump water up to the reservoirs, then letting the water pass through turbine generators to retrieve the energy when electrical demands peak (Huggins 2010).

Electrochemical devices fuel cells were invented about the same time as the battery in the 19th Century. However, for many reasons, fuel cells were not well-developed until the advent of manned spaceflight when lightweight, nonthermal (and therefore efficient) sources of electricity were required in spacecraft. Fuel cell development has also attempted to increase conversion efficiency of chemical energy stored in hydrocarbon or hydrogen fuels into electricity. Several other technologies have also been investigated, such as flywheels, which can store kinetic energy, and compressed air storage that can be pumped into underground caverns and abandoned mines. Another method uses molten salt to store solar power and then dispatch that power as needed. The system pumps molten salt through a tower heated by the sun's rays. Insulated containers store the hot salt solution, and when needed water is then used to create steam that is fed to turbines to generate electricity.

Grid energy storage (or large-scale energy storage) lets energy producers send excess electricity over the electricity transmission grid to temporary electricity storage sites that become energy producers when electricity demand is greater. Grid energy storage is particularly important in matching supply and demand over a 24-h period of time. A proposed variant of grid energy storage is called Vehicle-to-Grid energy storage system, where modern electric vehicles that are plugged into the energy grid can release the stored electrical energy in their batteries back into the grid when needed.

Hydrogen is also being developed as an electrical power storage medium. Hydrogen is not a primary energy source, but a portable energy storage method, because it must first be manufactured by other energy sources in order to be used. However, as a storage medium, it may be a significant factor in using renewable energies. On a weight-for-weight basis, hydrogen has the highest energy content of any element and it is the most abundant element in the universe, which makes it attractive as an energy source. Unfortunately, it is also the lightest element and, therefore, the least densely concentrated element in the universe, so it is difficult to concentrate sufficient hydrogen in one place to make it economically efficient to use as an energy source for most human activities. Existing technologies to generate hydrogen are economically and environmentally expensive, devices to concentrate and store it are typically too big, too heavy, inefficient or otherwise economically unacceptable, and the infrastructure to transport it and to dispense it are typically inefficient and unfriendly to end consumers. Many metals, with their relatively larger molecular voids, and many chemicals, have an affinity to absorb hydrogen into, or onto, themselves (to "take up" hydrogen). Therefore, metal hydride and chemical hydride materials are the focus of much of the attention in hydrogen storage materials. None of these materials have reached the density of storage that is necessary to enable commercial acceptance. Therefore, early portable power and transportation power fuel cell designers are using liquid fuel alternatives, such as direct methanol fuel cells, where the hydrogen is "stored" in a fuel like methanol and taken directly into the fuel cell. Other fuel cell systems use a liquid or gaseous hydrocarbon fuel and "reform" it into hydrogen, which can then be taken up by the fuel cell. Liquid fuel hydrogen storage leaves a waste product that must be disposed or a spent fuel that must be "regenerated" or recycled to refill it with hydrogen. In a couple of cases, liquid hydrogen is stored cryogenically on vehicles for direct combustion, rather than in a fuel cell. Liquid fuels, cryogenic storage, and reforming of fuels add complexity to the overall fuel cell system (Huggins 2010).

Underground hydrogen storage is the practice of hydrogen storage in underground caverns, salt domes and depleted oil and gas fields. Massive quantities of gaseous hydrogen are stored in underground caverns for many years without any difficulties. The storage of large quantities of hydrogen underground can function as grid energy storage which is essential for the hydrogen economy. By using a turboexpander, the electricity needs for compressed storage at 200 bars amounts to 2.1% of the energy content. With intermittent renewables such as solar and wind, the output may be fed directly into an electricity grid. At penetrations below 20% of the grid demand, this does not severely change the economics; but beyond about 20% of the total demand, external storage will become important. If these sources are used for electricity to make hydrogen, then they can be utilized fully whenever they are available, opportunistically. Broadly speaking, it does not matter when they cut in or out; the hydrogen is simply stored and used as required. Energy losses are involved in the hydrogen storage cycle of hydrogen production for vehicle applications with electrolysis of water, liquification or compression, and conversion back to electricity (Huggins 2010).

Various biofuels such as biodiesel, straight vegetable oil, alcohol fuels, or biomass can be used to replace hydrocarbon fuels. Various chemical processes can convert the carbon and hydrogen in coal, natural gas, plant and animal biomass, and organic wastes into short hydrocarbons suitable as replacements for existing hydrocarbon fuels. Examples are Fischer-Tropsch diesel, methanol, dimethyl ether, or syngas. Some of the energy in the original source is lost in the conversion process. Historically, coal itself has been used directly for transportation purposes in vehicles and boats using steam engines. And compressed natural gas is being used in exceptional circumstances fuel, for instance in busses for some mass transit agencies.

Carbon dioxide in the atmosphere has been, experimentally, converted into hydrocarbon fuel with the help of energy from another source. To be useful industrially, the energy will probably have to come from sunlight using, perhaps, future artificial photosynthesis technology. Another alternative for the energy is electricity or heat from solar energy or nuclear power. Compared to hydrogen, many hydrocarbon fuels have the advantage of being immediately usable in existing engine technology and existing fuel distribution infrastructures. Manufacturing synthetic hydrocarbon fuel reduces the amount of carbon dioxide in the atmosphere until the fuel is burned, when the same amount of carbon dioxide returns to the atmosphere. Methane is the simplest hydrocarbon with the molecular formula CH_4. Methane could be produced from electricity of renewable energies. Methane can be stored more easily than hydrogen and the transportation, storage, and combustion infrastructure are mature (pipelines, gasometers, power plants) (Huggins 2010).

1.2.4.4 Systems Engineering

Systems engineering is an interdisciplinary field of engineering that focuses on how complex engineering projects including energy system should be designed and managed. Issues such as logistics, the coordination of different teams, and automatic control of machinery become more difficult when dealing with large, complex projects. Systems engineering deals with work-processes and tools to handle such projects, and it overlaps with both technical and human-centered disciplines such as control engineering, industrial engineering, organizational studies, and project management.

Systems engineering focuses on analyzing and eliciting customer needs, and required functionality early in the development cycle and documenting requirements, then proceeding with design synthesis and system validation while considering the complete problem, the system lifecycle. The systems engineering process can be decomposed into (a) Systems Engineering Technical Process; and (b) Systems Engineering Management Process. The goal of the Management Process is to organize the technical effort in the lifecycle, while the Technical Process includes assessing available information, defining effectiveness measures, to create a behavior model, create a structure model, perform trade-off analysis, and create sequential build and test plan. Depending on their application, although there are several models that are used in the industry, all of them aim to identify the relation between the various stages mentioned above and incorporate feedback (Ryen 2008).

Energy system development often requires contribution from diverse technical disciplines. By providing a system (holistic) view of the development effort, systems engineering helps meld all the technical contributors into a unified team effort, forming a structured development process that proceeds from concept to production to operation and, in some cases, to termination and disposal.

1.2.5 Sustainable Energy Development

The relation between sustainable development and the use of resources, particularly energy resources, is of great significance to societies (Dincer and Rosen 2007). A supply of energy resources is generally agreed to be a necessary, but not sufficient, requirement for development within a society. Societies, such as countries or regions that undergo significant industrial and economic development almost always have access to a supply of energy resources. Sustainable development requires not just that sustain able energy resources be used, but that the resources be used efficiently. Even if one or more energy resources eventually become inexpensive and widely available, increase in efficiency will likely remain sought to reduce the associated environmental impacts, and the resource requirements including energy and materials to create and maintain systems to harvest the energy.

The environmental impacts of energy use merit further consideration, as a substantial portion of the environmental impact in a society is associated with energy resource utilization. Ideally, a society seeking sustainable development utilizes only energy resources which cause no environmental impact. Such a condition can be attained or nearly attained by using energy resources in ways that cause little or no wastes to be emitted into the environment, and/or that produce only waste emissions that have no or minimal negative impact on the environment. This latter condition is usually met when relatively inert emissions that do not react in the environment are released, or when the waste emissions are in or nearly in equilibrium thermally, mechanically and chemically with the environment, i.e., when the waste exergy emissions are minimal (Dincer and Rosen 2007).

In reality, however, all resource use leads to some degree of environmental impacts. A direct relation exists between exergy efficiency and environmental impact, in that through increased efficiency, a fixed level of services can be satisfied with less energy resources and, in most instances, reduced levels of related waste emissions. Therefore, it follows that the limitations imposed on sustainable development by environmental emissions and their negative impacts can be in part overcome through increased efficiency, i.e., increased efficiency can make development more sustainable (Dincer and Rosen 2007).

1.3 Development of Key Materials for Advanced Energy Systems

The effectiveness and practicality of many critical energy solutions will depend on advancements in materials and their manufacturing processes. Next-generation materials can be thought of as those that hold promise for step-change impacts in the economic, engineering, and environmental performance of materials across their entire life cycles (i.e., extraction, manufacturing, use and reuse, and end-of-life) as compared to historical performance improvement rates within specific materials classes. Furthermore, materials innovations can occur at many different scales, including improved structural properties at the nanometer scale, novel surface geometries at the micrometer scale, and creation of new materials markets and applications at the global scale. Key examples include new catalysts for more profitable and sustainable fuels and chemicals, advanced surface coatings and geometries for improving materials durability and reducing friction, lightweight metal alloys and composites for more fuel-efficient vehicles, and net-shape and near-net-shape techniques for less wasteful and more profitable materials processing. Next-generation innovation materials technologies can be briefly categorized as

(DOE 2015): (a) Materials and process development acceleration tools; (b) High-performance structural materials; (c) Functional materials and interfaces for energy conversion; (d) Functional materials for separations or isolation; and (e) New paradigm materials manufacturing processes.

1.3.1 Materials and Process Development Acceleration Tools

Conventional materials development typically progresses through an iterative cycle of experimentally producing and testing variations of a material until the desired properties for the specific use are achieved. Then the material must go through steps to ensure its performance and lifetime in the technology system under expected conditions, certify it for use, scale-up production (while maintaining its desirable properties), manufacture the technology system, and deploy it. It can take 10–20 years or more for a new material to advance from initial discovery of the material, through its development, to initial commercialization. Advanced computational design of materials and systems, high-throughput experimental testing of material properties, and Big Data analytical tools, have the potential to significantly reduce the time and cost to develop a new material and integrate it in an energy technology. These capabilities offer the opportunity to accelerate the development of advanced energy technologies that can help meet the onrushing energy-linked economic, security, and environmental challenges (Wadsworth 2012).

However, the science and engineering challenges of developing, verifying, and validating advanced computational, experimental, and data tools to aid in materials development are substantial. Materials must be modeled across a broad range of spatial and temporal scales. Spatial scales vary from clusters of atoms (10^{-9} m) to complete engineered systems (10 m or more)—a range of ten orders-of-magnitude. The temporal scale ranges even more, from roughly 10^{-12} s at the atomic level to 10^9 s for the 30-year life of a power plant component (a difference of more than 20 orders-of-magnitude) (LeSar 2013).

Individual existing models cannot predict behavior and properties across this broad range of spatial and temporal scales. For example, modeling the inter-atomic binding strength of iron atoms does not capture the complex grain structure necessary to maximize the performance of high-strength steels. Understanding the meso-scale transition from atomic-level modeling to macro-scale performance is critical. Similarly, modeling the picosecond dynamics of atoms does not address how the properties of the bulk material will change over a 30-year component lifetime under extreme stress and corrosion conditions. A variety of models are used for these different scales, including density functional theory; molecular dynamics; kinetic Monte Carlo; thermochemistry and mean field [rate] theory; macroscopic finite element models; and more, as sketched in Fig.1.8 (Stan 2009).

Advances in scientific theory, modeling, simulation, high-performance computing, algorithms, software, data analysis, and experimental techniques are merging to create the ability to design and engineer materials and systems more rapidly and at lower cost than traditional approaches. Integrated Computational Materials

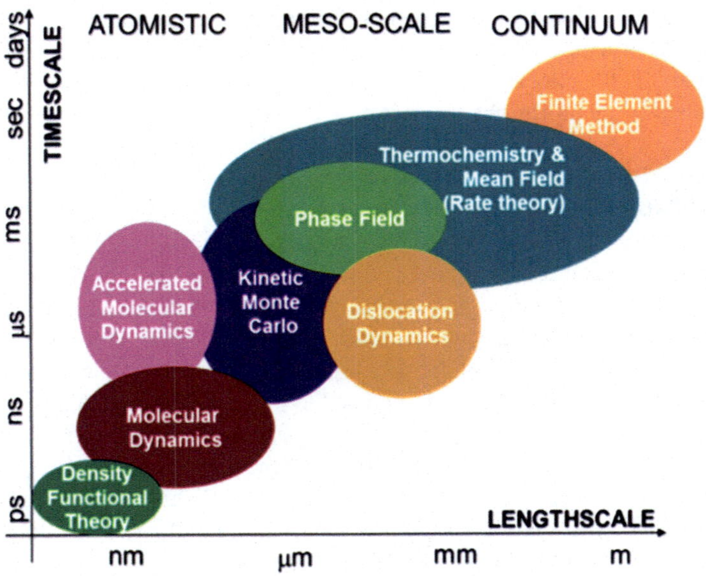

Fig. 1.8 Models bridging the spatial and temporal scales from atomic to large structures (Adapted with permission from Stan 2009 (Elsevier))

Engineering (ICME) is an emerging discipline that integrates. Computational predictive-modeling tools with system-level computational engineering design optimization and manufacturing design techniques to provide a unified design and manufacturing process to accelerate development of materials into engineering systems. The top-down computational system engineering focuses on determining what material capabilities are required to meet system-level performance needs; the bottom-up computational materials discovery, design, and development focuses on identifying the material and determining how to produce it for the system. The full spectrum from discovery science to deployment of innovative technologies include (Panchal et al. 2013):

(a) Systems Modeling and Simulation
From a top-down perspective, computational modeling and simulation of an energy technology or system is a crucial step in determining the requirements for the materials and components within the system, and for evaluating operations, safety, and performance. Advanced computing capabilities for modeling, simulation, and other purposes are important across every energy systems. The results of such modeling, conducted iteratively with computational materials development, can help identify requirements for materials design.
(b) Predictive Theory and Modeling and Born Qualified
At the other end of the scale from systems modeling is bottom-up predictive materials modeling. Advanced computing is increasingly able to characterize new materials and materials processes, as well as complex chemical systems and

processes. High-performance computing can now enable simulations of matter at scales from a few atoms to millions of atoms with increasing fidelity. An important next step now underway is extending these computational and other capabilities from the atomic level into the meso-scale, where many material performance characteristics are determined, and key functionalities of the materials are designed in. This upscaling from an atomic quantum mechanical level to the meso-and macro-level is essential to support engineering development and characterization of large, complex engineered systems. Models identified above-molecular dynamics, kinetic Monte Carlo, etc.—extend capabilities across portions of the mesoscale. These capabilities need further development to advance their performance and support accelerated development of new materials and capabilities. Evaluating the characteristics of a material from first principles computational analysis enables searching a large space for materials with the desired characteristics. High-throughput computational approaches range from systematically testing all of the potential materials in an area to using computational approaches such as genetic algorithms (Leaf et al. 2011).

Significantly extending predictive theory and modeling would include developing the capability to computationally design functional materials with sufficient fidelity that the material is qualified to meet performance standards for various applications—"Born Qualified"—before it is ever produced, and that the production process is scalable and well-characterized. It would also characterize the material's (and component's) performance under expected operating conditions over its lifetime. This will require substantial further development of computational modeling capabilities and validation of their performance across performance parameters such as material strength, hardness, ductility, corrosion resistance, catalytic activity, or others depending on the specific applications of the material. Achieving this would significantly accelerate the design, development, manufacturing, and introduction of new materials, avoiding the development cycle of repeated testing and then looping back into further development that can take years at high costs (Larzelere et al. 2012).

(c) Digital Manufacturing and Design

With top-down modeling and simulation and bottom-up predictive theory and modeling of materials, developing capabilities for computational manufacturing is also important. Since materials design and manufacturing design are intrinsically integrated, digital capabilities can accelerate modern design, manufacturing, and product support to reduce cycle time and improve productivity and quality. Factories of the future will be networked, data-driven systems that use automation, advanced metrology and sensing, and control systems to improve productivity and competitiveness (CoC 2014).

(d) Advanced Experimentation and Model Verification and Validation

Capabilities are needed for the synthesis, characterization, manufacturing scale-up, and performance validation of new materials. Further, capabilities such as in situ metrology, real-time process characterization, and process control are needed. Experimentation (high throughput and otherwise) will always continue to be a critical part of the material RDD&D and validation process; this includes development of improved synthesis and characterization tools and capabilities, as well as the associated expertise. In many cases (such as for functional materials interfaces), existing computation and simulation tools have known weaknesses in predicting characteristics and performance. Traditional experimentation can also be very slow or expensive in these cases. Synergistic development of computational and experimental capabilities is needed to accelerate the discovery and development of new materials systems and products. An important aspect of computational materials development is validation that the model accurately represents the intended design and the physical world. Experimental analysis of the material modeled is thus essential. Fundamental experimental capabilities to support experimental validation and other needs include world class light sources, neutron sources, and nanoscale science facilities. These can provide high-resolution (spatial and temporal) analysis of atomic-scale and larger structures and dynamics of materials, chemicals, and chemical processes, and can be tightly integrated with theoretical analysis and with high-performance computational modeling to validate and advance the modeling simulations.

(e) Uncertainties in Models and Physical Systems
There are many different uncertainties in systems and materials models, as well as the actual physical systems. It is important to characterize, quantify, and track these uncertainties at different spatial and temporal scales in order to assess overall component and system performance and cost. For models, uncertainties are inherent in numerical solution methods, and numerical error can potentially be amplified as calculations are carried through a sequence of models. For materials, uncertainties include the materials' measured parameters, the processing of the material, and the operating conditions of the material. For instance, the failure of a material will typically begin from a localized point due to a combination of microstructural failure and an extreme event. Further, many material failures occur at the largest flaw; variations and uncertainty in processing and input materials dictate the distribution of largest flaws and hence the distribution of performance characteristics among groups of manufactured components. Uncertainty analyses need to consider such stochastic effects to adequately characterize material performance over time (Panchal et al. 2013).

(f) Data and Data Analytics
Identifying materials with the right properties for a particular application has often been handled by searches across large databases of material properties, including materials informatics approaches with data mining and visualization and combinatorial search. Going forward, the predictive modeling, Born Qualified, systems simulation, design, and demonstration tools described above will generate large volumes of data, amplified by the verification and validation of these computational results using high-throughput experimentation and measurement. Altogether, this will require extensive digital data management and curating. Software can be used to aggregate huge quantities of materials data and

then use these data to develop machine-learning models of materials behavior. These models could then guide research in academia and industry towards advanced materials with the desired capabilities. Deep learning tools can have significant benefits for discovery science. Building broad capabilities for such data tools could have substantial benefits for extracting diverse data sets from across energy system engineering and related materials science and other issues. This will also be important for managing uncertainty in the data and extracting key information to provide useful feedback to help improve high-fidelity simulations. Data mining technologies are also needed, for example, in wind resource assessment to improve integration of wind power into the grid (Kamath and Fan 2012; Janssen et al. 2014).

(g) Demonstrations

Demonstrations of energy technologies can pose particular challenges when the necessary scale is large. For example, full-scale demonstrations of fossil power with carbon capture and storage (CCS), nuclear, concentrated solar power, and others can cost billions of dollars. Some of the difficulty and cost of achieving full-scale performance might be addressed through computational techniques such as predictive materials modeling, systems simulation, and integrated computational materials engineering, which can potentially provide extensive information on the performance of materials and the systems themselves. Vehicle simulations, for example, are used to characterize crashworthiness of vehicles, allowing a much broader range of designs to be tested and saving substantial time and cost by reducing the amount of prototyping and physical testing needed. For energy supply technologies, simulation may also reduce the design-test time, enable a wider exploration of the design space, and may allow some of the scale-up demonstration stages to be reduced or bypassed. Even with these potentially substantial savings in time, large-scale demonstrations will still be necessary, and these will require significant investments of time and capital. One approach that can substantially reduce the development time to produce a demonstration system at scale is hybrid additive manufacturing integrated with necessary procession subtractive manufacturing. Not only can this technology be used to manufacture parts with designs not attainable by conventional manufacturing methods, additive manufacturing can also reduce the time to develop and review initial prototypes, accelerate the process of getting a design ready for production, and enable rapid iterative prototyping. However, additive manufacturing itself requires extensive computational support to model systems and to understand the dynamics of the process, for example, the nonequilibrium dynamics of deposition.

There are many energy technologies that could benefit from the above tools of predictive modeling and development of materials. The typical approaches, summarized in Fig. 1.9, offer pathways to significantly accelerate the development and manufacturing of innovation materials for advanced energy technologies (NSTC 2014). This will drive a fundamental paradigm shift in which computational, experimental, and data tools will accelerate the discovery, development, and optimization of new materials and systems to meet national and global energy challenges.

Fig. 1.9 Tools and techniques for accelerated development and manufacturing of advanced materials and energy technologies. From discovery to deployment, there are numerous feedback loops, reversals, additions of new technologies and processes from other sources, and many other nonlinear aspects (Adapted from NSTC 2014. Credit: National Science and Technology Council, Committee on Technology, Subcommittee on the Materials Genome Initiative)

1.3.2 High-Performance Structural Materials

Higher performance structural materials cover metals, polymers, ceramics and glasses, and composites with improved strength and engineering performance in various applications. For some applications, materials that meet performance requirements are unavailable and new materials must be developed and qualified. In other cases, materials that meet the application's stringent operation requirements are available, but costs are too high to justify use of the material. In those cases, research is needed to improve the efficiency of processing techniques and equipment in order to bring the manufacturing costs of these materials down.

For many energy systems, the path to realizing greater energy efficiency requires operation in harsher environments characterized by varying thermal, chemical, mechanical, and radiation stresses. Materials are frequently the limiting factor for pushing energy systems to these extremes. In these environments, structural materials must maintain the chemical and physical properties necessary to enable the energy systems to operate efficiently without reducing component and system life. Although surface aspects of materials play a significant role in determining material performance in extreme environments, improvements in bulk material properties are also needed to maximize the efficiency of energy systems operating with extreme environments. To develop technologies that can improve industrial and vehicle energy efficiency; advance nuclear fission, fusion, and solar technologies; and enable the affordable use of hydrogen fuel cells, it is necessary to develop breakthrough materials that can maintain their structural integrity under these extreme conditions. The breakthroughs needed may include (Robinson 2012): (a) Increasing the efficiency of industrial combustion and conversion systems requires corrosion- and chemical-resistant materials capable of withstanding higher temperatures and aggressive chemicals such as sulfur, hydrogen, chlorine, and water. (b) Nuclear fission and fusion requires radiation tolerant materials and new fuel systems to improve energy efficiency. These materials must be cost-effective and may require new materials manufacturing techniques. (c) The cost-effective development of new materials for tomorrow's energy systems requires improved methods of predicting material properties and performance with limited operating data and design experience. (d) New models are needed for researchers to better understand and predict environmental degradation modes on material lifetimes. And (c) Methods to rapidly detect damage to energy system materials in service will enable systems to maintain their maximum operation efficiencies. As a result, examples of the breakthrough materials technologies have been addressed as below.

1.3.2.1 Lightweight and High-Strength Ductile Materials

Lightweight, damage-tolerant, and corrosion-resistant materials serve roles in transportation body and structural applications. These high-strength materials include composites, aluminum, magnesium, titanium, and high-strength steel alloys, hybrid materials/gradient metallic systems, and polymer-based materials. Advancements in high-strength, lightweight materials for automotive applications can lead to energy savings in the transportation system through lighter weight bodies, chassis, and drivetrains and more efficient engines. In addition, structural materials resistant to high-pressure hydrogen enable advancements in hydrogen storage and handling applications, such as hydrogen containers and compressors that could enable the more rapid commercial deployment of hydrogen fuel cell vehicles. However, the use of such materials is limited by a variety of factors, including their resistance to corrosion, durability in high-friction environments, forming and assembly challenges, joining and repair challenges, and end-of-life materials management challenges (Thomsen 2009).

1.3.2.2 Phase-Stable Metallic Materials

The route to greater energy efficiency in many energy systems must travel through brutal operating conditions, including extreme heat, intense radiation, punishing wear, and highly corrosive environments. Increasing the efficiency of industrial combustion and conversion systems, for example, requires higher temperatures and the use of aggressive chemicals that can degrade materials and cause them to fail. Advancements in other energy-related processes and technologies, such as nuclear fission and fusion, solar technologies, and fuel cells, similarly push materials to their limits. Next-generation energy systems will only be possible when they are constructed from advanced metallic materials that can retain their strength and stability under the most challenging of conditions. More specifically, increasing the efficiency of industrial combustion and conversion systems requires corrosion- and chemical-resistant materials capable of withstanding temperatures greater than 650 °C and aggressive chemicals such as sulfur, hydrogen, chlorine, and water. High-strength, phase-stable alloys are needed for extended service at temperatures greater than 650 °C. These alloys will be manufactured by advanced techniques, such as combining strong, corrosion-resistant steel with advanced surface engineering processes (Robinson 2012).

Basically, the groundwork for this has already been laid through steady advancement in high-temperature materials, including improvements in certain chromium alloys and nickel cobalt alloys. Other phase-stable materials include high-strength, low-alloy steels; high-pressure steels; irradiation-resistant steels; stress, corrosion, and cracking-resistant stainless steels; corrosion-resistant zirconium alloys with reduced hydrogen pickup; oxidation and corrosion-resistant refractory alloys; and alternative fuel cladding materials. Advanced, phase-stable materials that can withstand the harsh operating environment of an advanced ultra-supercritical steam turbine could increase the energy efficiency of electricity generation.

1.3.2.3 Irradiation-Resistant Structural Alloys for Nuclear Applications

Nuclear applications require structural alloys that retain their strength, ductility, and dimensional stability when exposed to radiation levels up to 100 displacements per atom (dpa). These alloys must not only resist irradiation, but also offer superior mechanical strength and resistance to swelling, similar to oxide dispersion strengthened steels. For instance, conventional nuclear fuel cladding materials are unstable at very high temperatures (in excess of normal core operating conditions) and limit operating temperatures and thermal efficiency. Phase transitions and reactivity of zirconium alloys may contribute to nuclear core damage in loss-of-coolant accidents. Improved irradiation-resistant and phase-stable nuclear fuel cladding materials could mitigate the consequences of accidents at nuclear facilities.

1.3.2.4 Composite Materials

Lightweight, high-strength, and high-stiffness composite materials have been identified as an important crosscutting technology in clean energy manufacturing. These materials have the potential to substantially improve the efficiency of the transportation systems, enable efficient power generation, improve the storage and transport of reduced-carbon fuels, and increase renewable power production. In order to reach this potential, advanced manufacturing techniques are required that will enable an expansion of cost-competitive production of advanced composite materials at commercial volumes (TMS 2010).

A composite can be defined as a combination of two or more materials that retain their macro-structure, resulting in a material that can be designed to have improved properties compared to the constituents alone. Structural composite materials are often composed of a reinforcement material and a matrix material. The reinforcement material provides mechanical strength and transfers most of the loads in the composite, while the matrix material maintains alignment or spacing and protects the reinforcement from abrasion and the environment. The combination of a reinforcement material with an appropriate matrix material can enable products that are lighter-weight or have other unique properties relative to monolithic materials (like metals), while providing similar or better performance properties. There are many methods to manufacture composites owing to the diversity of composite materials and combinations. While composites encompass a wide range of matrix/reinforcement combinations, fiber-reinforced polymer (FRP) have been widely applied to key clean energy applications. FRP composites are made by combining a polymer resin with strong, reinforcing fibers. These lightweight composites can enable energy savings in applications where large amounts of energy use and carbon emissions occur in the use phase, such as fuel savings in lighter-weight vehicles. Other energy benefits of FRP composites include more-efficient wind turbine operation at a lower installed cost, and compressed gas storage tanks for natural gas (and ultimately, hydrogen) that enable increased use of fuels with a lower life cycle environmental impact. Other types of composites, such as metal-matrix composites, offer advantages specific to the application (such as conductivity, durability, hardness, radiation resistance, high strength at higher temperatures) and have different manufacturing challenges than FRP composites. Lower cost, high strength and stiffness, corrosion-resistant, and lightweight composite materials could also provide benefits in diverse applications including industrial equipment and components, pipelines, structural materials for buildings, fly-wheels for energy storage, support structures for solar energy systems, shipping containers, and the use in aerospace applications (Thomsen 2009, Wheatley et al. 2014).

1.3.2.5 Low-Carbon Cements

Cement is the "glue" in concrete. Concrete is an essential building material used across the world in buildings and infrastructure. Concrete combines simplicity, durability, strength, affordability, and infinite ability to be molded. It provides the solid foundations and essential built environment for society. Concrete is the third most used substance in the world after air and water, a staple of modern life and society.

Cement manufacturing is an energy and carbon-intensive industry. The cement industry contributes approximately 5% of the global man-made carbon dioxide (CO_2) emissions and is thus becoming the second largest CO_2 contributor in industry after power plants. The major sources of CO_2 emissions at all stages of cement manufacturing include (a) raw material preparation (grinding and transportation), (b) clinker production and the combustion of fuels in the kiln, and (c) the production of cement final product (milling, blending, mixing, packaging, and transportation) (Ishak and Hashim 2015).

A wide range of CO_2 mitigation strategies have been developed to considerably reduce CO_2 emissions, such as (a) energy efficiency improvements; (b) waste heat recovery; (c) the substitution of fossil fuel with renewable energy; (d) the production of low-carbon cement by replacing ordinary Portland cement with alternative materials, i.e., geo-polymers, blast furnace slag, coal fly ash, and natural pozzolanic materials; and (e) carbon capture and storage (Ishak and Hashim 2015).

1.3.2.6 Multi-Material Joining

Joining of dissimilar materials enables the use of lighter-weight or greater strength materials with conventional metals. New joining processes stand to improve the corrosion-resistance properties, strength, and performance of the joined connections. A challenge to integrating lightweight materials into complex vehicular structures is joining them effectively to dissimilar materials that also comprise the system. Advancements in joining processes—using lasers, electron beams, adhesives, heat treatments, or chemical reactions—are key to the mass production and increased use of multi-material structures in the transportation systems. This includes improving the robustness, life, and strength of joining processes that preserve core materials properties and eliminate defects. Advanced joining processes would enable more seamless construction of vehicle structures, while also permitting the use of lighter weight materials in more demanding operating environments, such as the higher temperatures near the engine (Robinson 2012).

Key joining processes that have been developed with proper heat compatibility may include: joining of steel to dissimilar materials (e.g., polymers), solid-state joining processes, rapid thermal cycle processing, next-generation adhesive technology, and additive manufacturing.

1.3.3 Functional Materials for Energy Conversion

Materials that convert energy from one form to another can provide energy benefits by enabling unused or underutilized energy sources to be harnessed for useful purposes. Examples of high-performance functional materials for energy conversion can be categorized as below.

1.3.3.1 Thermoelectric Materials

Thermoelectric materials hold significant promise in converting waste heat into useful electricity. Development of low-cost, stable thermoelectric materials with low thermal conductivities and simultaneous high electric conductivities (measured as ZT [figures of merit] values) can offer an efficient alternative to processes such as mechanical generation and refrigeration, while also improving the harvesting of waste heat. To fully enable effective implementation of thermoelectric materials in industrial processes, sealants need to be developed that can protect thermoelectric elements from degrading when exposed to air, moisture, and extreme temperatures. Work is also underway to identify potential substitutes for common thermoelectric materials that are less susceptible to oxidation and better suited for high-temperature operating environments (Robinson 2012).

For example, approximately 40% of a vehicle's energy input is lost as waste heat in the exhaust gas. Thermoelectric materials can address this issue by converting the waste heat into useful electricity without releasing CO_2 emissions, and improve vehicle fuel economy by reducing vehicle electrical power requirements placed on the engine for such functions as lights, pumps, and electronic braking. Essential to large-scale market penetration of thermoelectric materials is improvement of their manufacturing processes, which are currently complex, labor-intensive, and expensive. Development of thermoelectric devices with low thermal conductivities and simultaneous high electric conductivities can improve the efficiency of waste heat harvesting even further. In fact, advances in thermoelectric technologies and processing techniques have the potential to nearly double the current ZT of commercial thermoelectric applications due to new methods of raw material purification and advances in nano-manufacturing techniques (Hendricks and Choate 2006).

Therefore, thermoelectric materials with greatly enhanced conversion efficiency would represent a breakthrough in product performance in the efficient conversion of waste heat into useful electricity. Developing thermoelectric materials with a figure of merit, or ZT, greater than 3.0 will allow these materials to become cost-effective energy recyclers. In addition to improved conversion efficiency, these materials also require improved mechanical performance, lower toxicity, and low-cost processing methods (Robinson 2012).

1.3.3.2 Wide Bandgap Semiconductors

Wide bandgap (WBG) semiconductor devices have higher switching frequencies than their silicon-based counterparts, enabling the use of smaller inductors and capacitors in power circuits, and resulting in weight, volume, and cost reductions. Wide bandgap devices also provide higher power density and can reduce energy requirements as a result of their high efficiency. The use of WBG semiconductors in variable frequency drives controlling motors and generators can result in significant levels of energy reduction as well as enabling substantial decreases in the weight and volume of the drive electronics. Other applications where WBG power electronics could achieve appreciable energy savings include hybrid and electric vehicles, lighting, data servers, AC adapters, solar inverters, power supplies, charging circuits, and grid control (Extance 2013).

Two major WBG materials with the potential to enable significant advances in power electronics are silicon carbide (SiC) and gallium nitride (GaN). To date, the predominant use of SiC in electronics is as a substrate for GaN light emitting diodes (LEDs). While SiC currently has limited use in power electronics, its role is expected to grow as it becomes the prevailing WBG replacement for Si in applications requiring device ratings in excess of 600 V. A major challenge to widespread adoption of SiC power electronics devices is the high cost of substrates and epitaxial materials. These high costs are tied to small production volumes and high manufacturing costs. Significant markets are expected for SiC devices in hybrid and electric vehicles as well as solar inverters and power supplies. SiC diodes are already used with companion Si transistors in photovoltaic (PV) inverters and hybrid vehicle chargers. Their greatest revenue-generating applications are expected to be in industrial motor drives and hybrid and electric vehicles (Eden 2013).

GaN is currently widely used in LEDs and radio frequency (RF) amplifiers, and also emergences in power electronics. Challenges for GaN-on-Si semiconductors, the current most cost-effective method for fabricating GaN power devices, are mostly related to their lack of maturity. Issues include overcoming material challenges, such as the high lattice strain at the GaN and Si interface owing to mismatches in the coefficient of thermal expansion. GaN high-electron-mobility transistors (HEMTs) are expected to be the dominant WBG semiconductor replacement for Si in applications requiring device ratings less than 600 V, whereas impacts have already been realized in RF and power-supply applications. For example, the high-frequency capabilities of WBG semiconductors might benefit imaging and sensing devices, and have smaller WBG LEDs improve minimally invasive surgical probes. GaN-based devices could enable MRI systems to operate at higher frequencies, thereby reducing noise; also save space by integrating the control equipment, which is currently placed in a separate room, into the MRI system. The size reductions made possible by WBG devices could also enable a wider array of implantable medical devices. In addition, the Ga-N-based devices can power themselves through energy harvesting, such as self-powering pacemakers. The improved efficiencies that WBG power electronics allow could mean that devices with even higher power requirements could eventually be made to last longer and be more compact by eliminating the need for batteries (Millan et al. 2014).

1.3.3.3 Photovoltaic Materials that Utilize a Broader Spectrum of Light

In order to achieve high conversion efficiency in producing electricity from solar energy, solar cells must absorb photons from a broad portion of the solar spectrum. Each photovoltaic material responds and absorbs a narrow range of solar energies (low-energy infrared to high-energy ultraviolet), corresponding to its characteristic band gap. Photons with energy lower than the band gap escape unabsorbed; photons with higher energy are absorbed, but most of their energy is wasted as heat. New photovoltaic materials could allow for tuning the band gaps to enable solar cells to absorb and utilize a broader and more controllable spectrum of light in order to increase the amount of energy collected (Robinson 2012).

Solar photovoltaic technologies can be used to harness the direct and diffuse components of solar radiation to generate electricity, displacing energy generated through fossil fuel combustion and reducing greenhouse gas emissions and air pollutants. Although most widely available and robust materials have a wide band gap, the narrower the band gap, the more solar energy is captured. Most light escapes because low-energy photons do not have enough energy to excite electron–hole pairs across the energy gap, and high-energy photons excite pairs with energy above the gap resulting in heat energy loss rather than useable electrical energy. New materials development should focus on (Robinson 2012): (a) Identifying new, low-cost materials with tunable band gap; better interface and control for energy conversion;

(b) Developing new processing techniques for such materials; (c) Address the issue of power loss due to loss of low-energy and excessively high-energy photons; and (d) Establishing enhanced photonics modeling capabilities with improved computational modeling tools.

Progress has already been made in the area of photovoltaic (PV) solar cells—which convert solar radiation into electricity using semiconductors—with a number of promising breakthroughs on the near horizon. Thanks to processing and manufacturing changes, as well as the development of new and alternative materials, the efficiency of certain types of first- and second-generation PVs has increased by about 5% over the past 20 years. With further advancements, they are expected to increase efficiency continuously. Currently, the efficiency rate—the ratio of electricity generated to sunlight captured—of commercial PVs stands at about 17–19%. In the near future, advances in manufacturing methods and materials properties are expected to yield flat-plate modules with efficiencies of 25% by 2030 and 40% by 2050; solar panel operational lifetime of 35 years by 2030 (IEA 2010; NREL 2011).

1.3.3.4 Catalysts

Catalysts are substances that alter the rate of a chemical reaction, but are chemically unchanged at the end of the reaction, so they can be used repeatedly. They are important enabling technologies for many energy systems and an integral part of the production of more than 90% of all industrial chemicals, including ammonia, ethylene, and methanol (Armor 2011). Catalytic processes can yield products at a relatively constant rate over the life of a catalyst. However, as the catalyst ages, its reaction temperature increases, resulting in a decrease in selectivity—the measure of the percentage of reactants that are converted to useful products—and conversion efficiency. Advanced catalysts with higher initial selectivity and conversion efficiency can improve industrial processes and manufacturing by effectively boosting the yield of chemical production over the catalyst's life (Robinson 2012).

The chemicals industry consumes more than 3000 trillion British thermal units (TBtu) of onsite energy per year, of which 104 TBtu of energy is estimated to be lost from catalyst non-selectivity in 42 high-volume production petrochemical processes (Neelis et al. 2007). Advanced high-volume catalysts with increased selectivity can reduce these losses by requiring less process heating fuel for catalysis, resulting in increased energy efficiency and a drop in carbon dioxide (CO_2) emissions and fuel costs.

Catalysts to drive chemical-to-chemical and chemical-to-power reactions represent an important technical area with cross-cutting applicability. Product performance breakthroughs in catalytic materials with higher selectivity and conversion efficiency can improve efficiency in industrial, hydrogen fuel cells, solar, and carbon management applications. For industrial applications, higher selectivity with higher conversion efficiencies would offer significant performance improvement opportunities. Reducing operating temperatures in chemical production processes would save significant amounts of energy and associated carbon emissions. In addition, replacement or extension of noble metals used in catalysts with non-noble metals will improve the cost-effectiveness of resulting products. Efficient photocatalytic materials such as titanium dioxide and better alternatives, more efficient catalysts for reduction reactions, and nanostructured catalysts can enable more cost-effective hydrogen generation. Additionally, there is a need to identify more efficient catalysts for reduction reactions and nanostructured catalysts for hydrogen generation and carbon dioxide reduction. A robust catalyst for fuel conversion is a high priority to support carbon management (Robinson 2012).

1.3.4 Functional Materials for Separation and Isolation

Functional materials for separations or isolation influence every energy systems, from coatings that protect structural steel in buildings and bridges, to surface treatments and/or modifications to improve the performance of drives, pistons, and bearings in vehicles and machinery, to membranes that assist chemical separations. Enhanced functional surface technologies enable surfaces to interact with process conditions to improve efficiency, speed reaction times, produce and store energy more efficiently, and withstand demanding operating conditions longer. Functional surface innovations can improve renewable energy technologies, storage and distribution, transportation and industrial efficiency, and carbon management applications throughout the energy sector. Specific examples of envisioned advances include (Robinson 2012): (a) Improving conversion efficiencies in industrial processes and hydrogen fuel cells combined with highly reactive catalysts; (b) Enabling vehicle light-weighting through the use of cost-effective, high-strength lightweight materials that can withstand demanding operating conditions; (c) Radically extending component life through innovative surface restoration; (d) Enabling large-scale and prolific carbon management through selective separation advances enabled by functional surfaces (e.g., membranes); and (e) Improving the efficiency of solar photovoltaics by creating surfaces that absorb light more effectively and increase conversion efficiency.

1.3.4.1 Gas Separating Membranes

Carbon management requires materials for CO_2 separation, capture, or both. Improved selectivity of separation decreases the energy cost associated with carbon capture and therefore improves the economics. High-flux membranes address the intrinsic challenges of efficiently separating CO_2 from N_2, such as their similar molecular sizes (0.33 nm vs. 0.36 nm, respectively) and CO_2 molecules' tendency to move through membranes relatively slowly. Conventional technologies to remove CO_2 and other impurities from air waste streams rely on expensive, energy-intensive processes that change the gas to a liquid state. Gas-separating membranes eliminate this step. A high-pressure gas mixture to be purified passes through the membrane, which has been designed to sift out the molecules of the substances that need to be captured or eliminated. Gas-separating membranes offer significant potential to make carbon capture much more efficient and affordable. In terms of environmental impact, if advanced membrane-enabled carbon capture technology reduced CO_2

emissions from coal-fired power plants by 10%, it could decrease CO_2 emissions by more than 180 MMT (Robinson 2012).

In addition to their use for carbon capture, advanced membranes can also enable more efficient separation of oxygen and hydrogen, reducing energy use, CO_2 emissions, and fuel costs in new power plants and coal plant retrofits. The gas separating membranes are potentially valuable tools in industrial processes requiring separation of hydrogen and oxygen from gas mixtures, as well as the production of pure hydrogen. Used to some extent in such areas as the production of ammonia and natural gas, the benefits of gas-separating membrane technology could be extended to other processes by optimizing selectivity involving similar-sized molecules, reducing maintenance, and designing membranes that can be more easily retrofitted into existing systems. Polymer membranes are currently in the widest use, but membranes comprised of ceramics, metals, and composites have demonstrated great promise to address a wider variety of filtration needs, operating temperatures, and service conditions (Robinson 2012).

1.3.4.2 Coatings and Surface Treatment

Corrosion and wear affect the metallic surfaces of industrial equipment and lead to progressive deterioration that can reduce plant efficiency and cause equipment failures and/or plant shutdowns. Advanced protective coatings have been used with great success to protect surfaces from wear and corrosion in harsh environments. Wear-resistant coatings, for instance, are ideal for use with system components that operate in high-friction situations. These coatings can help extend component life, reduce the amount of material required for an application, and decrease the use of in-service materials, such as lubricants in machining operations. In other applications, coatings can protect components that need to operate at high temperatures, reducing the occurrence of thermally induced failure, as well as oxidation in metals that typically deteriorate at higher temperatures. Chemical, structural, and processing innovations in coatings are necessary, however, to reduce corrosion in biomass systems and improve oxidation resistance in many industrial processes (Robinson 2012).

For instance, lightweight alloyed materials in vehicles are subject to degradation by friction and wear. Smart coating/lubricant systems offer thermal stability and wear resistance in order to broaden the opportunities for use of lightweight alloys that are typically inadequate for such high-wear environments as vehicle applications. These coatings offer good coating-substrate adhesion, do not degrade the substrate during deposition, and are strain-tolerant allowing for the retention of critical dimensions of moving sealing surfaces. In addition, lightweight automotive materials can experience severe galvanic corrosion, posing a major challenge to the use of alloys for light-weighting. Coatings and coating deposition processes that inhibit galvanic corrosion of materials such as aluminum (Al) and magnesium (Mg) allow the more widespread use of these materials in automotive and other applications by improving the surface properties of lightweight materials to increase their adequacy for high-wear environments over the long term (Liu & Heinimann 2010, Robinson 2012).

Surface treatments on steam and gas turbine blades can offer great potential in enhancing the efficiency of electricity generation. Protective coatings applied by surface treatment processes significantly improve the tolerance of these components to wear, corrosion, and fatigue. This extends turbine life and decreases downtime due to repair or replacement. The performance and durability of these surface treatments have improved five times since the 1950s, when thermal spraying coating process that involves spraying melted materials onto a surface was first utilized (Khanna et al. 2009). Continued innovations in surface treatment processes and coatings—including the development of materials that are highly resistant to hot corrosion—will give turbine blades the strength and durability they need to operate effectively under the demanding conditions of high-efficiency electricity generation processes.

A significant advancement in the battery of surface treatment technologies was the introduction of laser cladding in the early 2000s. This process involves depositing a layer of powder on a surface and then fusing the two materials metallurgically with a laser beam. Future development on the surface treatment will focus on identification of materials for high temperature, conductivity, and oxidation resistance; low-cost laser processing; high-accuracy nonplanar surface treatments; and ultra-high-temperature thermal barrier coatings for oxy-combustion turbines.

In addition, surface treatment processes can extend the service life of products by addressing surface fatigue, environmental protection, and damage tolerance issues. Process innovation in this area offers the potential for performance breakthroughs in products while providing substantial energy and carbon savings. New repair and remanufacturing processes are needed for advanced surface coating materials and alloys. Promising techniques include new surface treatment processes that utilize a diffusion process as well as nanoparticles to repair damage and self-healing materials. Smart materials with the ability to detect damage are also important for improving industrial energy efficiency.

1.3.5 Critical Materials

Advanced energy technologies—both new energy sources and novel ways to store, transmit, transform, and conserve energy—are enabled by the unique chemical and physical properties of a multitude of specific materials. A material's criticality is usually determined by considering its importance to clean energy applications, as well as any supply challenges, such as a small global market, lack of supply diversity, market complexities caused by co-production, or geopolitical risks. A materials shortage—exhibited through physical unavailability of a material, or high or volatile prices—may inhibit the widespread deployment of advanced energy technologies, potentially causing adverse consequences to the economy, environment, security, and competitiveness (Zepf et al. 2014).

A variety of critical materials enable clean energy technologies such as photovoltaics, wind turbines, electric vehicles, and energy-efficient lighting, as shown in Fig. 1.10. These clean energy technologies in turn reduce carbon pollution

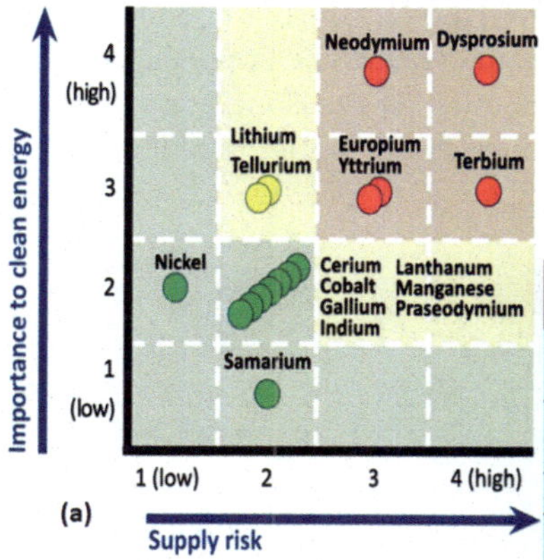

Critical
Near-Critical
Not Critical

Technology	Key Elements (b)
Catalytic converters	Platinum, palladium, rhodium, cerium, lanthanum
Fuel cells	Platinum, palladium, rhodium, lanthanum, cobalt, cerium, yttrium, gadolinium
Gas turbines	Yttrium, rhenium, hafnium
Batteries for electric vehicles and storage	Lithium, vanadium, graphite, cerium, cobalt, manganese, nickel, terbium
Hydrogen electrolysis	Platinum, palladium, rhodium
Nuclear power	Indium, cobalt, gadolinium
Thermoelectrics	Tellurium, bismuth, cerium, cobalt, lanthanum, lead, tellurium, ytterbium
Vehicle light-weighting	Magnesium, titanium, gadolinium

Fig. 1.10 Critical elements important to wind turbines, electric vehicles, photovoltaic cells, and fluorescent lighting (**a**) and key materials for other energy technologies (**b**) (Adapted from U.S. DOE 2011. Credit: US Department of Energy)

that contributes to climate change. The functionality of many clean energy applications depend upon the unique properties of rare earth elements (REEs). Rare earth oxides for clean energy applications are mainly utilized by permanent magnets (such as for wind turbines and electric vehicles), fluid catalytic cracking catalysts (namely, lanthanum and cerium), and phosphors (for energy-efficient fluorescent lighting, lasers, cathode ray tubes, etc.). For metal alloys, the increasing deployment of lithium ion batteries in electric vehicles may impact the demand of lanthanum (used in nickel metal-hydride batteries). The other category includes agriculture and water treatment applications, and demand for this category may increase with, for example, the development of cerium-based water treatment chemicals. Diversifying the source of supply reduces the criticality of a material, moving it towards the left in the criticality matrix (Fig.1.10a). Metal and alloy production should also be made more efficient to save the usage of the critical elements. Another way to reduce a material's criticality is to develop substitutes: although this does not directly reduce the supply risk of a material, substitutes can reduce the dependence of a clean energy technology on a particular material, moving it downward in the criticality matrix. Moreover, recycling of the critical materials that currently feed into clean energy products and their manufacture may ensure that the deployment of the relative products is not limited by materials supply (Navarro and Zhao 2014).

Material criticality is dynamic—while REEs are a challenge today, other materials may become critical in the future. Vigilant scrutiny of potential material criticality is required to avoid future materials supply disruptions. When considering other energy technologies, some key materials have emerged as candidates for criticality (Fig.1.0b). Further, materials essential to the manufacture of clean energy technologies, but are not present in the final products, may also require oversight. Examples of such manufacturing materials include tungsten, bismuth, helium, and catalytic materials for chemical production (Graedel et al. 2013).

INDEPTH: Materials Integration in Sustainable Energy Systems
Every energy system requires different materials working together to deliver desired functionality. Integrating new materials, including those with intrinsic heterogeneity like composites and smart materials, creates design and manufacturing challenges, especially as systems become more complex and service environments become more demanding. However, this approach also holds promise to yield significant energy and carbon emissions reductions in many energy systems. Specific examples may include (Robinson 2012):

- Broad use of low-cost carbon fibers and composites in vehicle manufacturing can reduce body weight by up to 50% and correspondingly reduce vehicle fuel consumption by 20–30%.
- Advances in low-cost, abuse tolerant, and long life cycle cathode materials for batteries can lead to transformational improvements in cost and efficiency of battery technologies.
- Low-cost fuel cells utilizing cheaper alternatives to platinum can promote a wider adoption of fuel cells for vehicular use. Together with improved vehicular efficiency and increased use of biofuels or synthetic fuels, this technology can lead to over 90% reduction in petroleum use by light duty vehicles.
- Improved computational modeling capabilities in multi-materials science can allow researchers to simultaneously model material capabilities, engineering product performance and associated manufacturing processes, compressing overall development time for new materials and processes by up to 50%. In the best case, a 20-year development cycle can be reduced to less than 5 years by using new modeling capabilities.

Designing materials that can work reliably and deliver the desired functionality when integrated into a single system requires many innovations, including (Matlock & Speer 2007, Robinson 2012):

- New joining processes for disparate materials (e.g., by using lasers, plasma or chemical reaction to join materials).
- Creating low-cost composite materials with wide applicability.
- Effective thermal management of materials and processes to improve the efficiency and reliability of energy systems.
- Design and manufacturing of smart materials such as shape memory alloys and materials with embedded sensors for monitoring and verification.

(continued)

- Sensors to enable self-healing materials that can repair damage incurred during normal use.
- Improving the ability to accurately predict the performance and reliability of disparate material systems to allow designers of energy systems to take full advantage of all material capabilities in order to maximize the efficiency and performance of the entire system.

1.3.6 New Paradigm Materials Manufacturing Processes

Energy-efficient, advanced materials manufacturing processes provide an opportunity to advance the state-of-the-art and competitiveness of industry for several reasons (DOE 2015). First, novel materials innovations may require entirely new production processes, such as a shift from sputtering-based coating applications to colloid-based applications for new energy-efficient electrochromic window coatings. Moreover, improvements to high-volume processes for producing metals, polymers, and other bulk materials in a more energy- and materials-efficient fashion can reduce both the operating costs and the energy, resource, and waste footprints of modern industry. For example, next-generation metals industry processes, including intelligent casting, improved controls, blank geometry optimization, and inert atmosphere melting can substantially reduce yield losses in the production of metal products. In addition, next-generation manufacturing processes such as nanofabrication, roll-to-roll processing, and additive manufacturing may lead to dramatic improvements in manufacturing flexibility, lead time, and productivity while also opening up opportunities for new materials innovations to support the processes (e.g., advanced powders for additive manufacturing).

1.3.6.1 Net-Shape Processing

Net-shape processing refers to any manufacturing method that can produce a component very close to its final shape, reducing material waste and often eliminating the need for costly secondary processing and finish machining. In addition to saving time and money, net-shape processing offers an avenue for tremendous energy savings, as well as reduction in associated CO_2 emissions, by eliminating or combining energy-intensive processing steps. For example, near net-shape strip casting is a net-shape processing technique that integrates casting and hot rolling into one step, reducing the need to reheat metal before rolling it (Worrell et al. 2004). Net-shape processing approaches have been found to be particularly effective in the manufacture of hard-to-form materials, such as high-performance, lightweight metals and composites, producing components with improved materials properties, and offering downstream savings opportunities for lightweight transportation manufacturing (Robinson 2012).

Net-shape processing of metals such as Al, Ti, and Mg include thixocasting, rheocasting, and power metallurgy—all at lower cost and with better material properties. Other processes include net-shape forging, castings, and laser processing.

1.3.6.2 Magnetic Field Processing

The extreme high magnetic field (B) environment, generally coupled with elevated temperature (T), provides an enabling disruptive technology for developing the next generation of novel structural and functional materials. All materials are impacted by high magnetic fields and so all material systems from metallic through polymeric and protein will respond to a BT environment.

Major improvements in performance (from 15% to 300%) can be manifested in mechanical and/or physical properties as well as the development of nanocrystalline or textured microstructures or reaction paths made easier/faster through the synthesis/catalytic chemical effect of the extreme BT environment. In addition, the deformation behavior (magnetoplasticity) of materials appears to be impacted by high fields potentially enabling high and low cycle fatigue damage mitigation superplastic behavior at ambient temperature, residual stress relief, and other visionary applications. Perhaps classically brittle materials can be made to be formable under high magnetic fields. The BT environment therefore impacts phase equilibria and kinetics, is a new synthesis/catalysis paradigm, and a deformation/life enhancement processing technology.

1.3.6.3 Additive Manufacturing

Additive manufacturing (AM) is the process of producing objects from computer-aided design (CAD) model data, usually adding layer upon layer, in contrast to conventional subtractive manufacturing (CM) methods that involve the removal of material from a starting work piece. AM is also called 3-D printing, additive fabrication, or free-form fabrication. These new techniques, while still evolving, are projected to exert a profound impact on manufacturing. They can give industry new design flexibility, reduce energy use, and shorten time to market. Interest in additive techniques has grown swiftly as applications have progressed from rapid prototyping to the production of end-use products. Additive equipment can now use metals, polymers, composites, or other materials to "print" a range of functional components, layer upon layer, including complex structures that may be difficult or impossible to manufacture by other means. However, AM processes may not be applicable to all manufactured products produced using CM processes today. There are size, material property, and cost limitations that restrict its use (at least currently) to certain applications (Wohlers 2014).

AM and CM face different trade-offs, with each process likely to play a role in future manufacturing capabilities. AM has the potential to minimize materials and energy usage, reduce waste, accelerate innovation, and compress supply chains. Some benefits of AM technology include (Petrovic et al. 2011; Huang et al. 2016):

(a) Innovation. AM enables designs with novel geometries that would be difficult or impossible to achieve using CM processes. The novel designs enabled by AM can improve a component's engineering and cost performance and can also lead to performance and environmental benefits in a component's product application. AM systems can also be programmed to tailor material properties within a component as desired to increase performance beyond that achievable through conventional processing.

(b) Part consolidation. The ability to design products with fewer, more complex parts rather than a large number of simpler parts may be the most important benefit of AM. This enables designs that are optimized for performance at a system level without making compromises for the sake of manufacturability at the subsystem level. Reducing the number of parts in an assembly may cut the overhead associated with documentation, production planning, and control. Also, fewer parts can mean that less time and labor is required to assemble the product, again contributing to a reduction in overall manufacturing costs. The "footprint" of the assembly line may also become smaller, further cutting costs.

(c) Lower energy consumption. AM can save energy by eliminating production steps, using substantially less material, enabling reuse of by-products, and producing lighter products.

(d) Less waste. Building objects up layer upon layer, instead of traditional machining processes that cut away material, can reduce material needs and costs by up to 90%. AM can also reduce the "cradle-to-gate" environmental footprints of component manufacturing through avoidance of the tools, dies, and material scrap associated with CM processes. Additionally, AM could reduce waste by lowering human error in production.

(e) Reduced time to market. Items can be fabricated as soon as the 3-D digital description (3-D scanning or 3-D imaging to construct a Standard Tessellation Language [STL] file) of the part has been created, eliminating the need for expensive and time-consuming part tooling and prototype fabrication.

(f) Light weighting. With the ability to create complex shapes, AM enables the design of parts that can often be made to the same functional specifications as conventional parts, but with less material.

(g) Agility of manufacturing operations. AM enables rapid response to markets and creates new production options outside of factories, such as mobile units that can be placed near the source of local materials. Parts manufactured directly by AM do not need the (often) expensive tooling used in conventional operations, which must be amortized over long production runs, but the relatively slow production speed of AM is not currently effective for high volume production. In some cases, AM is used indirectly in the manufacturing operation to fabricate the molds, dies, and tooling used in the production process. Tooling can be designed, printed, and delivered to the shop floor faster and more cost-effectively than by traditional methods, which are often outsourced. Spare tooling parts can be produced on demand, reducing or eliminating the need for stockpiles and complex supply chains. However, the AM technologies required to manufacture metal tooling have relatively high capital costs and steep learning curves that need to be overcome to increase the rate of uptake by manufacturers.

As a widely applicable process technology, AM will impact energy in many different systems. AM may also contribute to energy efficiency, whether through the rapid development and fabrication of prototypes to reduce the cost and lead time of new products or more directly through the production of energy-saving products, such as compact, high-surface area heat exchangers, that are more efficient than heat exchangers made by conventional methods.

Within the energy system, AM is being explored for a range of applications spanning energy production, storage, and delivery. For example, selective laser sintering (SLS)-based processes have been used to fabricate graphite composite bipolar plates, which is one of the most important components in polymer electrolyte membrane (PEM) fuel cells. By using SLS, the cost and lead-time to develop new bipolar plates can be reduced dramatically compared with conventional methods, such as injection molding and compression molding, in which expensive metal molds must be manufactured. AM applications in electricity production include turbine blades for windmills, custom electrical components used in substations, high-temperature alloys for combustion and steam turbines, tooling for large castings, radiation-tolerant materials for nuclear applications, high-toughness materials for heat engines, and other complex parts used in power plants (Bourell et al. 2011; Guo and Leu 2012).

1.3.6.4 Roll to Roll Processing

Roll-to-roll (R2R) is a family of manufacturing techniques involving continuous processing of a flexible substrate as it is transferred between two moving rolls of material. R2R is an important class of substrate-based manufacturing processes in which additive and subtractive processes can be used to build structures in a continuous manner. Other methods include sheet to sheet, sheets on shuttle, and roll to sheet R2R is a hybrid process comprising many technologies that, when combined, can produce rolls of finished material in an efficient and cost-effective manner with the benefits of high production rates and in mass quantities. High throughput and low cost are the factors that differentiate R2R manufacturing from conventional manufacturing which is slower and higher cost due to the multiple steps involved, for instance, in batch processing. Initial capital costs can be high to set up such a system; however, these costs can often be recovered through economy of scale (Willmann et al. 2014).

R2R processing have been applied in numerous manufacturing fields such as flexible and large-area electronics devices, flexible photovoltaics, printed/flexible thin-film batteries, fuel cells, fibers and textiles, metal foil and sheet manufacturing, medical products, energy products in buildings, and membranes etc. In the field of electronic devices, R2R processing is a method of producing flexible and large-area electronic devices on a roll of plastic or metal foil. Substrate materials used in R2R printing are typically paper, plastic films or metal foils. Stainless steel is sometimes used because it is durable and has a high-temperature tolerance.

Silicon wafers, cadmium-telluride solar cells, battery electrodes, fuel cell membranes, and high-performance window films are just a few examples of materials that have clean energy applications and are characterized by a two-dimensional functional surface, often with one or more coated or deposited layers. Not surprisingly, these materials are often made using similar processes—namely continuous roll-to-roll, belt-fed, or conveyor-based processes that enable successive steps to build a final construction at high throughput. A variety of processes that can be used for R2R manufacturing, such as (Sondergaard et al. 2012; Anderson et al. 2014):

(a) Deposition—Evaporation, sputtering, chemical vapor deposition (CVD), and atomic layer deposition (ALD) can all be easily implemented in R2R processing. This technology has been used for superconductor tape production and nanomaterial synthesis and is growing in popularity for thin-film solar deposition.
(b) Gravure—A type of printing process which involves engraving the image onto an image carrier.
(c) Flexographic Printing—A form of printing process which utilizes a flexible relief plate. It is essentially a modern version of letterpress which can be used for printing on almost any type of substrate, including plastic, metallic films, cellophane, and paper.
(d) Flatbed and Rotary Screen Printing—In flatbed printing, a squeegee, moves relative to a mesh, then forces the ink through the open area and onto the substrate. The wet layer thickness is defined by the thickness as well as the open area of the mesh and generally relative thick wet layers can be achieved (10–500 μm). In rotary screen printing, the substrate moves through rollers past the squeegee forcing the ink onto the substrate.
(e) Imprint or Soft Lithography—In soft lithography (e.g., self-aligned imprint lithography (SAIL)), multiple mask levels are imprinted as a single three dimensional (3-D) structure.

    The photopolymer layer is heated above its glass transition temperature to allow it to flow into the crevices of the stamp. The stamp/ polymer sandwich is cured with ultraviolet (UV) light as the polymer cools and hardens, allowing the stamp to pull off cleanly. The process is completed with standard wet and dry etch processes, leaving an accurately reproduced 3-D, high-resolution pattern on the substrate. The technology is called self-aligning because the mask would deform with the substrate during the embossing heat treatment step.
(f) Laser Ablation—A technique that would eliminate both the photoresist coating and wet etching steps is called laser photo-ablation. This technique is used to write directly into a polymer layer using a high-powered laser. The photo-ablation works by breaking molecular bonds in polymer layer, fracturing the polymer into shorter units that are "kinetically ejected" upon removal.
(g) Offset Printing—A commonly used technique in which the inked image is transferred (or "offset") from a blanket cylinder that bridges the plate cylinder and the substrate. The pattern is transferred to the blanket (usually made of rubber), and then transferred to the substrate.
(h) Inkjet and aerosol-jet Printing—While laser ablation may be called a subtractive technique, inkjet aerosol-jet printing can be considered an additive technique,

using an array of piezoelectric print heads to deposit conducting, semiconductor or dielectric organic solutions at precise locations.

Further development of R2R production capabilities that are energy efficient, low environmental impact and lower cost and that are employed to manufacture technologies and products for clean energy applications will have a "global impact" in the manufacturing industry. There are huge savings in energy just from higher throughputs since the tools and equipment used in R2R manufacturing (per unit area of manufactured roll) are using less energy for a much shorter period of time relative to conventional manufacturing processes. Additionally, efficiencies are obtained from more efficient deposition processes, for example, that would provide additional savings in energy. Breakthroughs that will have high impact, and therefore high value, are in the nano-manufacturing community (Bae et al. 2010).

1.3.6.5 Next-Generation Metals Processing

Continued reductions in energy use and carbon emissions associated with the metals production industries require efficiency improvements across the spectrum of metals processing. These include streamlining processing steps and equipment needs, reducing reheating frequencies, and increasing production yields. Specific process improvements that could make a significant impact include the development of new materials for anodes and cathodes used in aluminum production; insulating materials for furnaces and reactors; recycling processes better capable of converting dirty, impure materials to high-grade product; and technologies that can produce titanium in a continuous process, rather than in batches. Material advances leading to energy and cost reductions in the primary production of lightweight metals can have even more far-reaching impact by making the use of these materials more affordable for the vehicle manufacturing sector (Liu & Heinimann 2010, Robinson 2012).

Development of new, low-cost, energy-efficient, primary reduction and processing technology for titanium, aluminum, and magnesium may focus on (a) Develop improved electrodes for processing technologies; (b) Develop novel, low-temperature electrolytes; and (c) Advance carbothermic reductions (high-temperature chemical reactions that use carbon as the reducing agent) of aluminum.

Selective separation of materials allows for increased recycling rates and reduced solid waste streams. Future development may include (a) Develop better technologies for screening and sorting of materials; (b) Develop high-throughput, multi-sensor approaches to identify mixed materials (e.g., metal alloys, plastics); and (c) Develop low-temperature recycling processes (Liu & Heinimann 2010).

Advanced characterization and sensors are needed for damage detection in the processing and service of high-performance materials. Multi-sensor, integrated spectrum devices and software would serve to enable solid and liquid process control and material identification (e.g., metal ion sensors for molten metal processing). Real-time sensor technologies facilitate improved control in the various steps involved in the processing of gases and metals, such as melting, heat treating, and combustion, across the manufacturing sector.

1.3.6.6 Energy Feedstock Conversion Technologies for Industry

Conversion of organic materials to energy can proceed along three main pathways--thermochemical, biochemical, and physicochemical. Useful forms of energy that can be produced include heat, steam, electricity, renewable natural gas, and fuels. Currently, all three pathways are utilized or have been demonstrated using mixed or separated municipal solid wastes (sometimes in combinations with industrial or petroleum refining residues).

(a) Biochemical
Biochemical conversion processes include anaerobic digestion (which occurs in landfills and controlled reactors or digesters) and anaerobic fermentation (for example, the conversion of sugars from cellulose to ethanol). Biochemical conversion proceeds at relatively low temperatures and low reaction rates. Higher moisture feedstocks are generally good candidates for biochemical processes. The lignin fraction of biomass cannot be economically converted by anaerobic biochemical means. As a consequence, a significant fraction of woody and some other fibrous feedstocks exits the process as a residue that may or may not have market value. The residue, called digestate, can be composted.

(b) Thermochemical
Thermochemical conversion processes include gasification and pyrolysis. Thermochemical conversion is characterized by higher temperatures and faster conversion rates. It is best suited for lower moisture feedstocks. Thermochemical routes can convert all of the organic portion of suitable feedstocks and create marketable co-products. The inorganic fraction (ash) of a feedstock does not contribute to the energy products and may contribute to fouling of high-temperature equipment, increased nutrient loading in wastewater treatment and disposal facilities, and increased disposal costs. Inorganic constituents may also accelerate some of the conversion reactions.

(c) Physiochemical
Physiochemical conversion involves the physical and chemical synthesis of products from feedstocks (for example, biodiesel from waste fats, oils, and grease--known as FOG) and is primarily associated with the transformation of fresh or used vegetable oils, animal fats, greases, tallow, and other suitable feedstocks into liquid fuels or biodiesel.

1.4 Summary

Energy is one of the critical issues that directly impact the economy, the environment, and the security of human beings. Development and application of advanced sustainable energy systems have been paid great attention for generation, storage, transportation and utilization of future energy. Sustainable energy is considered sustainable, accessible, and affordable. The usage of such energy can potentially be kept up well into the future without causing harmful repercussions for future generations. Sustainable energy sources are most often regarded as including all renewable sources, such as hydroelectricity, solar energy, wind energy, wave power, geothermal energy, artificial photosynthesis, biofuels, and tidal power. It usually also includes technologies that improve current and future energy efficiency. Advanced sustainable energy systems generally include energy generation, carbon management, energy storage and distribution, and energy consumption and utilization with enhanced energy conservation, efficiency, and environmental stewardship. Materials play a key role in enabling technologies that can offer promising solutions to achieve accessible, renewable, and sustainable energy pathways for the future. New materials and technologies are needed for advanced energy devices and new approaches to obtaining, distributing, and storing energy.

Growing concerns about energy security and climate change have heightened interest in harnessing sustainable and renewable energy resources as a response to these critical issues. Electricity generated using these resources will in the most part be delivered to the point of use via large-scale transmission and distribution systems. Consequently, the successful integration of renewable energy generation into large power systems has become fundamental to successfully addressing climate change and energy security concerns. More broadly, these issues translate into sustainability challenges for the stationary energy sector and specifically the electricity industry to contribute to: Societal sustainability, through good industry governance processes that deliver reliable, affordable and sustainable electrical energy and in the process foster social cohesion and consensus and provide a benchmark for other sectors of the economy; Economic sustainability by delivering economic efficiency, particularly dynamic efficiency in achieving rapid, effective and efficient innovation in the electricity industry; Environmental sustainability, achieved via effective, market-compatible environmental regulation for local, regional, and global impacts, particularly climate change; Technological sustainability, through rapid and effective innovation to a more sustainable set of technologies, while not compromising energy security and affordability. Key technology options to be considered for a more sustainable resource portfolio include enhanced end-use efficiency and substitution for electricity by other energy vectors (fuel switching), responsive electricity demand, and low emission generation, including renewable energy, carbon capture and sequestration, and nuclear energy. The target portfolio must be technically effective as well as economically efficient and environmentally sound.

Energy is the capability of matter to do work. Al available energy forms may be classified as (a) accumulated or stored energy, such as chemical energy of fossil fuels, nuclear energy for binding nuclei, internal energy of a substance, and potential energy associated with position or a mass in a force field; and (b) transitional energy, like thermal energy, heat and work, electromagnetic radiation, and electrical energy. Any one form of energy can be converted to any other form. Extent of energy conversion can be complete or partial.

The effectiveness and practicality of many critical energy solutions will depend on advancements in materials and their manufacturing processes. In other words, the key to making critical energy and carbon reduction solutions more effective, affordable, and widely implemented are materials and processing breakthroughs focused on removing barriers to progress and optimizing efficiencies. These innovations are not limited to making significant enhancements in energy products—more efficient solar cells and longer range car batteries, for instance. Also of great value is the power of materials technologies to vastly improve the productivity and profitability of manufacturing industries by enabling them to capture lost sources of energy, reduce wear on equipment and processing infrastructure, and turn out products more quickly and with less impact on the environment. The demand for the new materials responsible for this shift in manufacturing capabilities will, in turn, create businesses and industries focused on their production and distribution. Many materials technologies offering the best opportunities to change the energy and manufacturing landscape are on the cusp of realizing their fullest potential, such as functional materials and surface technologies, materials integration in sustainable energy systems, higher-performance structural materials, new paradigm materials manufacturing processes, and materials and process development acceleration tools.

Exercises

Part 1: General Questions

1.1 How do you define the energy? What's the general classification of energy forms?
1.2 Describe the possible solution to increase energy efficiency.
1.3 What's the relationship between four laws of thermodynamics?
1.4 What are the advantages and disadvantages of using coal?
1.5 Does an electric car reduce the use of fossil fuels?
1.6 Is a fuel oil heater or an electric resistance heater the best for the environment?
1.7 Is a natural gas heater or a geothermal heating system the best for the house?
1.8 Why is electrical energy so useful? What is the best energy source to convert to electricity?
1.9 How can the energy in the wind be used? How can wind power help conserve oil supplies? How might using wind energy help reduce the air pollution?
1.10 Do the white-colored roof tiles keep houses cool?
1.11 How can energy from the sun be used to heat water?
1.12 How can using solar energy help reduce pollution in the atmosphere and help conserve oil supplies?
1.13 Why is the process of photosynthesis so valuable?
1.14 Why are battery-powered vehicles considered to be the transport of the future? Do you have different opinions?
1.15 Why is chemical energy useful? What other forms of energy can be produced from chemical energy? Name three examples of other fuels that contain chemical energy.
1.16 A car's daily traveling distance is about 80 km/day. A car has a city mileage of 20 km/kg. If the car is replaced with a new car with a city mileage of 30 km/kg and the average cost of gasoline is \$4.50/kg, estimate (a) the amount of fuel,

energy, and money conserved with the new car per year, (b) reduction in CO_2 emission.

1.17 A car consumes about 6 gallons a day, and the capacity of a full tank is about 15 gallons. The density of gasoline ranges from 0.72 to 0.78 kg/l. The lower heating value of gasoline is about 44,000 kJ/kg. Assume that the average density of gasoline is 0.75 kg/l. If the car was able to use 0.2 kg of nuclear fuel of uranium-235, estimate the time in years for refueling.

1.18 When a hydrocarbon fuel is burned, almost all of the carbon in the fuel burns completely to form CO_2 (carbon dioxide), which is the principle gas causing the greenhouse effect and thus global climate change. On average, 0.59 kg of CO_2 is produced for each kWh of electricity generated from a power plant that burns natural gas. A typical new household uses about 7000 kWh of electricity per year. Determine the amount of CO_2 production that is due to the refrigerators in a city with 100,000 households.

1.19 A large public computer lab operates Monday through Saturday. There the computers are either being used constantly or remain on until the next user comes. Each computer needs around 240 W. If the computer lab contains 53 computers and each is on for 12 h a day, during the course of the year how much CO_2 will the local coal power plant have to release to the atmosphere in gram moles to keep these computers running?

1.20 How can you control your carbon footprint?

1.21 A 150-W electric light bulb is used on average 10 h per day. A new bulb costs \$2.0 and lasts about 5000 h. If electricity cost is \$0.15/kWh, estimate the yearly cost of the bulb.

1.22 A 20-hp electric motor is used to pump ground water into a storage tank 4 h every day. Estimate the work done by the pump in kW every year and the cost of electricity every year. Assume that the electricity unit cost is \$0.1/kWh.

1.23 Describe the process of how natural gas goes from its natural state to the market?

1.24 Some people like to have background noise when they are falling asleep. Many choose to listen to their television. The television will usually run on about 340 W and will run during the 8 h that you are asleep. With electricity costing \$0.20/kWh, calculate how much this will cost you if you do this for five days a week for an entire year.

1.25 What are the advantages and disadvantages of electrical energy in an alternating current? What are the advantages and disadvantages of electrical energy flowing in direct current?

1.26 In the search for new sources of energy that are renewable and emit less greenhouse gases, carbon-based biofuels are of major interest. These fuels are still carbon-based and must undergo combustion to release the chemical energy. Why is this process being looked at as a reasonable energy source?

1.27 The mass contained between an insulated piston and an insulated cylinder decreases in internal energy by 50 Btu. How much work is involved, and what is the sign of the work term? What does the sign indicate?

1.28 Derive an equation for the pressure drop for a loss-free incompressible flow in a varying-area duct as a function of area ratio.
1.29 Two units of work are required to transfer 10 units of heat from a refrigerator to the environment. What is the COP of the refrigerator? Suppose that the same amount of heat transfer instead is by a heat pump into a house. What is the heat pump COP?
1.30 A steam turbine has an efficiency of 90% and a theoretical isentropic power of 100 kW. What is the actual power output?
1.31 Discuss the role of materials and process development acceleration tools for enabling advanced energy systems. What's the major difference between conventional materials development approaches and advanced computational design methodology?
1.32 List the energy systems that need breakthrough structural materials?
1.33 What's the current major focus on development of functional materials for energy conversion?
1.34 List the current major critical materials in USA to enable clean energy technologies. How is a material's criticality determined?
1.35 Describe your understanding about new paradigm materials manufacturing processes. What are their major differences with traditional manufacturing processes?

Part 2: Thought-Provoking Questions

1.36 Describe the major aspects dealt with advanced energy technologies. Which area would you like to specialize in?
1.37 Discuss the effect of energy supply and end-use technologies on the global climate change. There are some arguments regarding the reasons that cause the global climate changes, such as, does CO_2 causes global warming? What's your opinion?
1.38 How many materials-based solutions can you find for enabling various energy technologies to achieve accessible, renewable, and sustainable energy pathways for the future?
1.39 Discuss the major challenges to successfully integrate the renewable energy generation into current large power systems. What's the current progress have you seen?
1.40 Describe your understanding about the five grand challenges for basic energy sciences. Do you have different opinions or thoughts?
1.41 What's the clean electric power technology? Briefly describe the current status and future trends of advancing systems and technologies to produce cleaner fuels.
1.42 With the clear advantages of nuclear power, why is it not more commonly used? Is it possible to develop nuclear power batteries in the future?
1.43 What are your prospections of future energy systems?

References

Alivisatos, P., Cummings, P., De Yoreo, J., Fichthorn, K., Gates, B., Hwang, R., et al.: Nanoscience Research for Energy Needs. Report of the National Nanotechnology Initiative Grand Challenge Workshop, March 16–18, 2004. DOESC (USDOE Office of Science (SC)) (2004)

Anderson, T.R., et al.: Scalable, ambient atmosphere roll-to-roll manufacture of encapsulated large area, flexible organic tandem solar cell modules. Energy Environ. Sci. **7**, 2925–2933 (2014)

Armor, J.: A history of industrial catalysts. Catal. Today. **163**(1), 3–9 (2011)

Atkins, P., and Jones, L.: Chemical Principles (5th Ed). New York: W.H. Freeman, New York (2010)

Bae, S., Kim, H., Lee, Y., Xu, X., Park, J.S., Zheng, Y., Balakrishnan, J., Lei, T., Kim, H.R., Song, Y.I., Kim, Y.J., Kim, K.S., Ozyilmaz, B., Ahn, J.H., Hong, B.H., Iijima, S.: Roll-to-roll production of 30-inch graphene films for transparent electrodes. Nat. Nanotechnol. **5**(8), 574–578 (2010). https://doi.org/10.1038/nnano.2010.132

Bahrami, M.: Basic concepts of thermodynamics. http://www.sfu.ca/~mbahrami/ENSC388/Notes/IntroandBasicConcepts.pdf. Accessed 17 June 2013 (2009)

Baigrie, B.: Electricity and Magnetism: A Historical Perspective. Greenwood Press, Westerport (2006)

Baldwin, S., et al.: Quadrennial technology review: an assessment of energy technologies and research opportunities. https://energy.gov/sites/prod/files/2015/09/f26/Quadrennial-Technology-Review-2015_0.pdf (2015). Accessed 16 April 2017

Barabas, I., Todorut, I.-A.: Biodiesel quality, standards and properties. http://cdn.intechopen.com/pdfs/23666/InTech-Biodiesel_quality_standards_and_properties.pdf (2011). Accessed 23 June 2013

BESAC Directing Matter and Energy: Five Challenges for Science and the Imagination. A Report from the Basic Energy Sciences Advisory Committee (BESAC). US Department of Energy. https://science.energy.gov/~/media/bes/pdf/reports/files/Directing_Matter_and_Energy_rpt.pdf (2007). Accessed 24 April 2017

BESAC: From quanta to the continuum: opportunities for mesoscale science. Refort for the Basic Energy Sciences Advisory Committee Mesoscale Science Subcommittee, September 2012. http://science.energy.gov/~/media/bes/pdf/reports/files/OFMS_rpt.pdf (2012). Accessed 1 May 2017

Bithas, K., Kalimeris, P.: A brief history of energy use in human societies. Revisiting the Energy-Development Link. Part of the series Springer Briefs in Economics, pp. 5–10 (2015)

Boundy, B., Diegel, S.W., Wright, L., Davis, S.C.: Biomass energy data book: Edition 4. http://large.stanford.edu/courses/2013/ph241/kallman1/docs/BEDB4_Full_Doc.pdf (2011). Accessed 26 June 2014

Bourell, D.L., Leu, M.C., Chakravarthy, K., Guo, N., Alayavalli, K.: Graphite-based indirect laser sintered fuel cell bipolar plates containing carbon fiber additions. CIRP Ann. Manuf. Technol. **60**(1), 275–278 (2011)

Bower, K.E., et al. (eds.): Polymers, phosphors, and voltaic for radioisotope microbatteries. CRC Press, Boca Raton (2002)

Cates, M.R.: Hybrid Lighting: Illuminating Our Future. Oak Ridge National Laboratory Review (29:3), March 1996. Oak Ridge, TN. http://web.ornl.gov/info/ornlreview/rev29_3/text/contents.htm (1996). Accessed 30 April 2017

Cengel, Y., Boles, M.: Thermodynamics: an engineering approach. 7th ed. McGraw-Hill Science/Engineering/Math (2010)

CoC: Council on Competitiveness, "Solve. The Exascale Effect: The Benefits of Supercomputing Investment for U.S. Industry." http://www.compete.org/publications/all/2695 (2014). Accessed 12 May 2017

Dincer, Rosen: Energy, Environment and Sustainable Development. Elsevier, Boston (2007)

DOE: Computational materials science and chemistry: accelerating discovery and innovation through simulation-based engineering and science. Report of the Department of Energy Workshop on Computational Materials Science and Chemistry for Innovation, July 26–27, 2010. http://science.energy.gov/~/media/bes/pdf/reports/files/cmsc_rpt.pdf (2010). Accessed 2 May 2017

DOE: Quadrennial Technology Review 2015—Advanced Materials Manufacturing. https://energy.gov/sites/prod/files/2016/04/f30/QTR2015-6B-Advanced-Materials-Manufacturing.pdf (2015). Accessed 10 May 2017

Eden, R.: The World Market for Silicon Carbide & Gallium Nitride Power Semiconductors—2013 Edition, vol. 9790. IHS, Wellingborough (2013)

Emanuel, A.E.: True and false energy-saving devices. IEEE Trans. Ind. Appl. **33**(6), 1439–1443 (1997)

Extance, A. SiC and GaN power devices jostle to grow their role. Power Dev, pp. 1–4. http://www.yole.fr/iso_upload/mag/powerdev_april2013_ir.pdf (2013). Accessed 16 May 2017

Faezams Electromagnetism spectrum. http://faezams.edublogs.org/6scg/ (2013). Accessed 17 June 2013

Farley, J.M.: Energy in the future—new challenges for NDT. In: 17th World Conference on Nondestructive Testing, 25–28 Oct 2008, Shanghai, China (2008)

Fitch, A., Reichert, H.: Energy-related studies a high priority. ESRFnews. (53), 5 (2010)

Fouquet, R.: A brief history of energy. In: Evans, J., Hunt, L.C. (eds.) International handbook on the economics of energy, pp. 1–19. Edward Elgar Publishing, Cheltenham (2011)

Gadeken, L.L. et al.: Methods of making energy conversion devices with a substantially contiguous depletion regions. US Patent 7250323 B2 (2007)

Golay, M., Green, Jr. W., Wright, J.C.: Introduction to Sustainable Energy: Energy Transfer and Conversion Methods. https://ocw.mit.edu/courses/nuclear-engineering/22-081j-introduction-to-sustainable-energy-fall-2010/lectures-and-readings/MIT22_081JF10_lec03a.pdf (2010). Accessed 9 May 2017

Graedel, T.E., Harper, E.M., Nassar, N.T., Reck, B.K.: On the materials basis of modern society. Proc. Natl. Acad. Sci. **112**, 6295–6300 (2013)

Guo, N., Leu, M.C.: Experimental study of polymer electrolyte membrane fuel cells using a graphite composite bipolar plate fabricated by selective laser sintering. In: Proceedings of the Solid Freeform Fabrication Symposium. Austin, TX, 2012 (2012)

Hendricks, T., Choate, W.T. Engineering scoping study of thermoelectric generator systems for industrial waste heat recovery. http://www1.eere.energy.gov/industry/imf/pdfs/teg_final_report_13.pdf (2006). Accessed 4 July 2013

Huang, R., Riddle, M., Graziano, D., Warren, J., Das, S., Nimbalkar, S., Cresko, J., Masanet, E.: Energy and emissions saving potential for additive manufacturing: the case of lightweight aircraft components. J. Clean. Prod. **135**, 1559–1570 (2016)

Huggins, R.A.: Energy Storage. Springer, New York (2010)

International Energy Agency (IEA): Technology roadmap: solar photovoltaic energy. 2010. Available from https://www.iea.org/publications/freepublications/publication/pv_roadmap.pdf (2010). Accessed 18 May 2017

IPCC Climate change 2014: synthesis report, summary for policymakers. In: Fifth Assessment Report of the Intergovernmental Panel on Climate Change. IPCC, November 2014. Available http://www.ipcc.ch/pdf/assessment-report/ar5/wg1/WG1AR5_SPM_FINAL.pdf (2014). Accessed 30 April 2017

Ishak, S.A., Hashim, H.: Low carbon measures for cement plant—a review. J. Clean. Prod. **103**, 260–274 (2015)

Janssen, M., Estevez, E., Janowski, T.: Interoperability in big, open, and linked data—organizational maturity, capabilities, and data portfolios. IEEE Comput. **47**(10), 44–49 (2014)

Joseph, E.: Electric Circuits. McGraw-Hill, New York (1965)

Kamath, C., Fan, Y.J.: Using data mining to enable integration of wind resources on the power grid. Statist. Anal. Data Mining. **5**, 410–427 (2012)

Khanna, A.S., et al.: Hard coatings based on thermal spray and laser cladding. Int. J. Refract. Met. Hard Mater. **27**(2), 485–491 (2009)

Khartchenko, N.V.: Advanced Energy Systems. Taylor & Francis, Washington, DC (1998)

Kondepudi, D.: Introduction to Modern Thermodynamics. Wiley, Chichester (2008). ISBN 978-0-470-01598-8

Lako, P., Kets, A.: Resources and future availability of energy sources—a quick scan. ftp://energie.nl/pub/www/library/report/2005/c05020.pdf (2005)

Lancaster, D.: Some energy fundamentals. Blatant Opportunist 71.1–71.7. http://www.tinaja.com/glib/energfun.pdf (2002). Accessed 21 June 2013

Larzelere, A. et al.: Grand Challenges of Advanced Computing for Energy Innovation: Report from the Workshop Held July 31–August 2, 2012. USDOE. http://www.pnnl.gov/main/publications/external/technical_reports/PNNL-22265.pdf. Accessed 12 May 2017 (2012)

Lawrence Livermore National Laboratory: Estimated U.S. Energy Use in 2014. https://flowcharts.llnl.gov/ (2015). Accessed 17 January 2018

Leaf, C.E.W.M., Lee, C.Y., Farha, O.K., Hauser, B.G., Hupp, J.T., Snurr, R.Q.: Large-scale screening of hypothetical metal-organic frameworks. Nat. Chem. **6**, 83–89 (2011)

LeSar, R.: Introduction to Computational Materials Science: Fundamentals to Applications. Cambridge University Press, New York (2013)

Liu, J., Heinimann, M.: ALCOA aerospace technology portfolio: solutions to meet the challenges of current and future applications. In: AeroMat 2010 Conference and Exposition, Bellevue, WA, June 22, 2010 (2010)

Liu, G., Larson, E.D., Williams, R.H., Kreutz, T.G., Guo, X.: Making Fischer-tropsch fuels and electricity from coal and biomass: performance and cost analysis. Energy Fuel. **25**, 415–437 (2011)

Matlock, D., Speer, J.: Third generation of AHSS: microstructure design concepts. In: Haldar, A., Suwas, S., Bhattacharjee, D. (eds.) Microstructure and Texture in Steels, pp. 185–208. Springer, London (2007)

Millan, J., Godignon, P., Perpina, X., Perez-Tomas, A., Rebollo, J.: A survey of wide bandgap power semiconductor devices. IEEE Trans. Power Electron. **29**(5), 2155–2163 (2014)

Navarro, J., Zhao, F.: Life-cycle assessment of the production of rare-earth elements for energy applications: a review. Front Energy Res. **2**, Article 45 (2014). http://journal.frontiersin.org/article/10.3389/fenrg.2014.00045/full. Accessed 17 May 2017

Neelis, M., et al.: Approximation of theoretical energy-saving potentials for the petrochemical industry using energy balances for 68 key processes. Energy. **32**(7), 1104–1123 (2007)

NREL: 2010 Solar technologies market report. http://www.nrel.gov/docs/fy12osti/51847.pdf (2011). Accessed 5 July 2013

NSTC: Materials genome initiative strategic plan. National Science and Technology council, Committee on Technology, Subcommittee on the Materials Genome Initiative, December 2014, https://www.whitehouse.gov/sites/default/files/microsites/ostp/NSTC/mgi_strategic_plan_-_dec_2014.pdf (2014). Accessed May 15, 2017

OECD/IEA: Technology Roadmap: Smart Grids, IEA Publishing. License: http://www.iea.org/t&c/termsandconditions/ (2011). Accepted on 27 April 2017

Organ, A.J.: The Air Engine: Stirling Cycle Power for a Sustainable Future. Woodhead Publishing, Cambridge (2007)

Orr, F.M.: 2015 Quadrennial technology review. In: Proceeding SC '15 Proceedings of the International Conference for High Performance Computing, Networking, Storage and Analysis Austin, Texas (November 15–20, 2015). ACM, New York, NY (2015)

Outhred H et al.: Meeting the challenges of integrating renewable energy into competitive electricity industries. From debate to design: a report on the work of the reil network 2007–08. Yale School of Forestry & Environmental Studies (2007)

Panchal, J.H., Kalidindi, S.R., McDowell, D.L.: Key computational modeling issues in integrated computational materials engineering. Comput.-Aid. Des. **45**, 4–25 (2013)

Petrovic, V., Gonzales, J., Ferrando, O., Gordillo, J., Puchades, J., Grinan, L.: Additive layered manufacturing: sectors of industrial applications shown through case studies. Int. J. Prod. Res. **49**, 1061–1079 (2011)

Rappaport, P.: The electron voltaic effect in p-n junctions induced by beta particle bombardment. Phys. Rev. **93**, 246 (1954)

Ristinen, R.A., Kraushaar, J.P.: Energy and the Environment, 2nd ed. Wiley (2005)

Robinson, L. Materials: foundation for the clean energy age. http://energy.tms.org/docs/pdfs/Materials_Foundation_for_Clean_Energy_Age_Press_Final.pdf (2012). Accessed 2 July 2013

Roebuck, K.: Energy Storage: High-impact Strategies—What You Need to Know: Definitions, Adoptions, Impact, Benefits, Maturity, Vendors. Emereo Publishing, Australia (2012)

Ryen, E.: Overview of the system engineering process. http://www.dot.nd.gov/divisions/maintenance/docs/overviewofsea.pdf (2008). Accessed 27 June 2013

Sawyer, K.: Windows and building envelope research and development: roadmap for emerging technologies. DOE Building Technologies Office, 2014. http://energy.gov/sites/prod/files/2014/02/f8/BTO_windows_and_envelope_report_3.pdf (2014). Accessed 30 April 2017

Scoditti, Barker Review of energy conversion devices. http://www.ghp-books.com/snb98hgt7q6a1/IEA_ReviewofEnergyConversionDevices.pdf (2003). Accessed 24 June 2013

Sondergaard, R., et al.: Roll-to-roll fabrication of polymer solar cells. Mater. Today. **15**(1–2), 36–49 (2012)

Stan, M.: Discovery and design of nuclear fuels. Mater. Today. **12**(11), 20–28 (2009)

Sundar, A.S.: Potassium alum effect on performance and emissions of diesel in an I.C engine. Int. J. Adv. Res. Technol. **2**(5), 126–168 (2013)

Thomsen, O.T.: Sandwich materials for wind turbine blades- present and future. J. Sandw. Struct. Mater. **11**(1), 7–26 (2009)

TMS: The advanced materials for our energy future. https://www.tms.org/portal/downloads/publications/studies/MaterialsandEnergyEfficiency/1005AdvMatlsforEnergy.pdf (2010). Accessed 16 January 2018

U.S. Department of Energy: Windows and buildings envelope research and development. Office of Energy Efficiency and Renewable Energy, Washington, DC (2014). http://energy.gov/eere/buildings/downloads/research-and-development-roadmap-windows-andbuilding-envelope. Accessed April 30, 2017

U.S. DOE: The Department of Energy's Critical Materials Strategy. http://energy.gov/epsa/initiatives/department-energy-scritical-materials-strategy (2011). Accessed 17 May 2017

Vouldis, I., Millet, P., Valles, J.L.: Novel Materials for Energy Applications—A Decade of EU-Funded Research. Office for Official Publications of the European Communities, Belgium (2008). http://ec.europa.eu/research/industrial_technologies/pdf/novel-materials-for-energy-applications_en.pdf. Accessed 12 June 2013

Wadsworth, J.: Forging the solution to the energy challenge: the role of materials science and materials scientists. Metall. Mater. Trans. A. **41**(5), 1047–1062 (2012)

WEC: World Energy Resources: A Summary World Energy Council 2013. ISBN: 978 0 946121 29 8. https://www.worldenergy.org/wp-content/uploads/2013/09/Complete_WER_2013_Survey.pdf (2013). Accessed 8 May 2017

Weston, K.C.: Energy conversion. http://onlinebooks.library.upenn.edu/webbin/book/lookupid?key=olbp33597 (2000). Accessed 18 June 2013

Wheatley, A., Warren, D., Das, S.: Low-cost carbon fibre: applications, performance and cost models. In: Elmarakbi, A. (ed.) Advanced Composite Materials for Automotive Applications: Structural Integrity and Crashworthiness. Wiley (2014)

Willmann, J., Stocker, D., Dorsam, E.: Characteristics and evaluation criteria of substrate-based manufacturing. Is roll-to-roll the best solution for printed electronics? Org. Electron. **15**, 1631–1640 (2014)

Wohlers: 3D Printing and additive manufacturing state of the industry. Annual Worldwide Progress Report, Wohlers Associates. http://www.wohlersassociates.com/state-of-the-industry-reports.html (2014). Accessed 18 May 2017

Worrell, E., et al.: Emerging Energy-Efficient Technologies in Industry: Case Studies of Selected Technologies. Lawrence Berkeley National Laboratory, Berkeley, CA (2004). http://ies.lbl.gov/iespubs/54828.pdf. Accessed 3 July 2013

Zepf, V., Simmons, J., Reller, A., Ashfield, M., Rennie, C.: Materials Critical to the Energy Industry: an Introduction, 2nd edn. BP, London (2014). http://www.bp.com/content/dam/bp/pdf/sustainability/group reports/ESC_Materials_handbook_BP_Apr2014.pdf. Accessed 17 May 207

Zumdahl, S.S., Zumdahl, S.A.: Chemistry, 9th edn. Indepence, Cengage Learning (2014)

Fundamentals of Materials and Their Characterization Methodologies for Energy Systems

2

Abstract

Fundamentals of materials can explain how structure leads to function—from the atomic- and nanoscale to the mesoscale and beyond. Materials structure can be manipulated to construct materials and inspired devices with desired properties and behaviors to meet the requirements of the next generation of energy technologies. Moreover, materials characterization methodology plays an essential role in any materials design and processing developments. This chapter will provide a brief review about the fundamentals of energy materials and their characterization methodologies, mainly focusing on materials structures and property behaviors, advanced characterization methods, computational modeling, and integrated process control and sensing in energy systems. Advanced characterization methods can play an important role in helping us to understand fundamental mechanisms and develop new materials with unique properties. Computational modeling may be may be used to quantitatively predict the chemistry, mechanical, and physical performance properties under various conditions of temperature, pressure, and strain as a function of time. As these tools improve, new materials can be developed faster and with greater control over properties, processing, and, ultimately, material performance in end-use applications. In addition, integrated process control and sensors can be used to achieve maximum efficiency from both materials manufacturing processes and energy systems that demand careful process monitoring and control with sensor technology.

2.1 Introduction

The advanced energy systems require advances in energy generation, storage, efficiency, and security that demand a new generation of materials that may not be naturally available. However, creating these new materials requires a level of understanding of the relationships between structure and function, and across

C. Tong, *Introduction to Materials for Advanced Energy Systems*,
https://doi.org/10.1007/978-3-319-98002-7_2

many spatial scales. Basic materials research is necessary to fill these knowledge gaps and enable creation of new materials with the specific characteristics needed for next-generation energy technology. In fact, tools for the synthesis, processing, characterization, and simulation and modeling of materials and chemical systems at the nanoscale have been developed and deployed. These tools, which include advanced X-ray and neutron sources, nanoscale science facilities, and high-performance computers, provide an unprecedented view of the atomic-scale structure and dynamics of materials and the molecular-scale basis of chemical processes. This has paved the way for manipulating materials at the nanoscale and the mesoscale to create new tailored functionalities.

In particularly, computational power has increased by a factor of a million due to advances in hardware and software, and enabled the development of computer simulations and models of unprecedented fidelity. Emerging capabilities in predictive modeling and simulation have the potential to revolutionize the development of new materials and chemical processes. Coupled with advanced materials characterization and nanoscale science facilities, this predictive capability provides the foundation for an innovation ecosystem that can accelerate the discovery, development, and deployment of new technologies, including advanced energy systems.

Furthermore, the applications of advanced networked data and information technologies for manufacturing, i.e., Advanced Sensors, Control, Platforms and Modeling for Manufacturing (ASCPMM), have the potential to transform the entire manufacturing supply chain—from extraction of materials at mines, through commodities, to finished products. ASCPMM technologies enable the extensive application of data for the optimization of enterprises and multicompany supply chain ecosystems, where data from advanced sensors and sensor systems form the basis for process control applications, decision work flows, and enterprise and supply chain optimization. A networked, open architecture, open access, and open application data platform combined with "plug-and-play" capabilities facilitates the application of ASCPMM technologies. ASCPMM technologies include infrastructure, software, and networked solutions for sensing, instrumentation, control, modeling, and platforms for manufacturing applications. These technologies interact in machine-to plant-to-enterprise-to-supply-chain ecosystem applications of real-time data and models that are networked for enterprise and ecosystem optimization, as well as monitoring, diagnostics, enterprise/ecosystem analytics, and integrated performance metrics. These technologies come together to perform tasks and realize benefits often associated with smart manufacturing (Rogers 2014).

2.2 Structure of Materials

From the periodic table, as shown in Fig. 2.1, it can be seen that there are about 100 different kinds of chemical elements in the entire Universe (Helmenstine 2015). These chemical elements form thousands of different substances ranging from the air, biomass, metals, polymers to ceramics. Metals behave differently than ceramics, and ceramics behave differently than polymers. The property of each substance

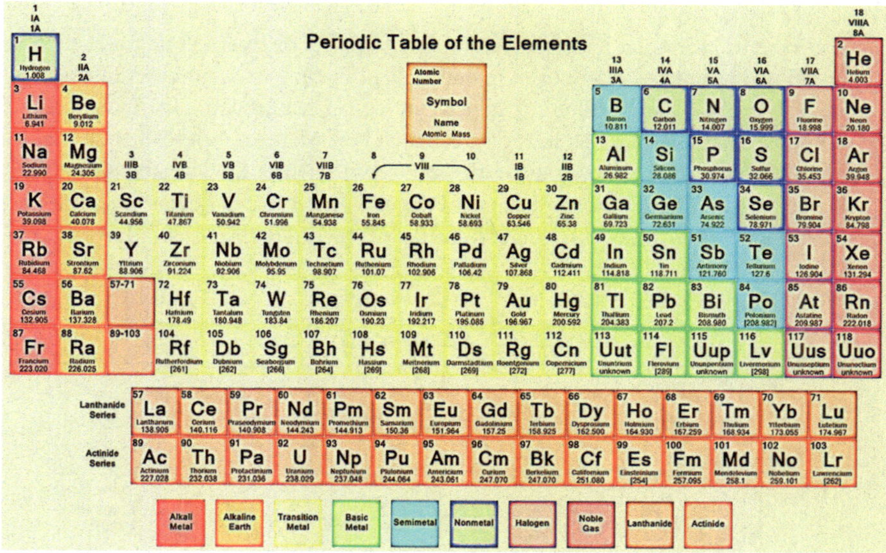

Fig. 2.1 Periodic table of the elements

depends on how it is structured with what elements how the elements are bonded together. The structure of materials can be classified by the general magnitude of various features being considered. The most common major classification of structural, listed generally in increasing size, are: nuclear structure, atomic structure, nanostructure, microstructure, mesostructure, and macrostructure.

The atomic structure primarily affects the chemical, physical, thermal, electrical, magnetic, and optical properties. The microstructure and macrostructure can also affect these properties but they generally have a larger effect on mechanical properties and on the rate of chemical reaction. The effects of nanostructure and mesostructure on materials properties and functions have also been widely explored, but not as matured as that of atomic and microstructures. In general, the properties of a material offer clues as to the structure of the material. For instance, the strength of metals suggests that these atoms are held together by strong bonds. However, these bonds must also allow atoms to move since metals are also usually formable. To understand the structure of a material, the type of atoms present, and how the atoms are arranged and bonded must be firstly known.

IN DEPTH: Materials Classifications

There are thousands of materials available for use in energy system applications. Most materials fall into one of three classes that are based on the atomic bonding forces of a particular material. These three classifications

(continued)

are metallic, ceramic, and polymeric. Additionally, different materials can be combined to create a composite material. Within each of these classifications, materials are often further organized into groups based on their chemical composition or certain physical or mechanical properties. Composite materials are often grouped by the types of materials combined or the way the materials are arranged together. Below is a list of some of the commonly classification of materials within these four general groups of materials (NDT 2014):

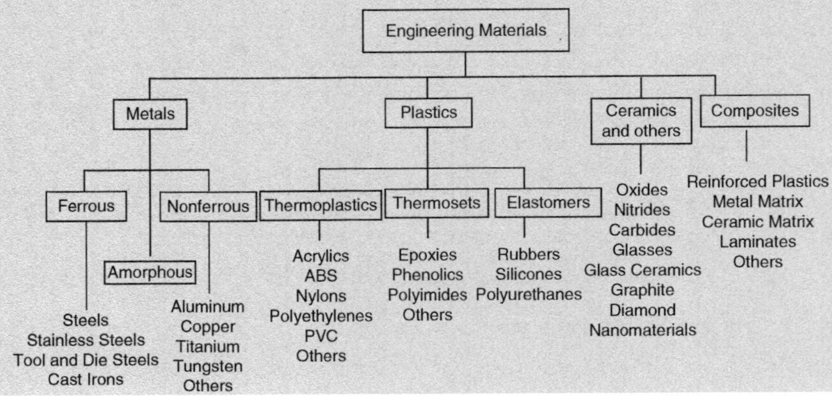

2.2.1 Nuclear Structure

The atom consists of a small but massive nucleus surrounded by a cloud of rapidly moving electrons. The nucleus is composed of protons and neutrons. Total number of protons in the nucleus is called the atomic number of the atom and is given the symbol Z. The total electrical charge of the nucleus is therefore +Ze, where e (elementary charge) equals to 1.602×10^{-19} coulombs. From a mass point of view, the mass of a proton is roughly equal to the mass of a neutron and each of these is about 2000 times the mass of an electron. Therefore, most of the mass of an atom is concentrated in the region at its core. From an electrical point of view, the proton is positively charged and the neutron has no charge. In a neutral atom, there are as many electrons as protons moving about nucleus. It is the electrons that are responsible for the chemical behavior of atoms, and which identify the various chemical elements. The charge of an ion is the difference between the number of protons and electrons in an atom. Ions with more protons than electrons are positively charged and ions with more electrons than protons are negatively charged.

It is possible for nuclei of a given element to have the same number of protons but differing numbers of neutrons, i.e., to have the same Atomic Number but different Mass Numbers. Such nuclei are referred to as Isotopes. All elements have isotopes and the number ranges from three for hydrogen to over 30 for elements such as cesium and barium.

There are about 2450 known isotopes of the approximately one hundred elements in the Periodic Table. The unstable isotopes attempt to reach the stable energy level by splitting into fragments, in a process called Fission, or by emitting particles and/or energy in the form of radiation. This latter process is called Radioactivity. Radioactivity, interaction of radioactivity and material, and nuclear reactions are nuclear properties of materials. The most general types of radioactivity are alpha particles, beta particles, and gamma photons. When bombarded by subatomic particles, elements undergo nuclear reactions, changing their elemental identities. Interactions of radioactivity with material cause chemical and physical changes. Nuclear reactions also cause changes that affect properties of materials.

2.2.2 Atomic Structure

The atom is considered to be the basic building block of all matter. It consists of two components: a nucleus surrounded by an electron cloud. From a size point of view, the radius of an atom is about 10^{-10} m while the radius of a nucleus is about 10^{-14} m, i.e., about ten thousand times smaller. From an electrical point of view, the nucleus is said to be positively charged and the electrons negatively charged. Atoms intend to have a balanced electrical charge. Therefore, they usually have negatively charged electrons surrounding the nucleus in numbers equal to the number of protons. Electrons are present with different energies and it is convenient to consider these electrons surrounding the nucleus in different energy "shells." For example, magnesium, with an atomic number of 12, has two electrons in the inner shell, eight in the second shell and two in the outer shell.

Chemical phenomena can be thought of as interactions between the electrons of individual atoms. Radioactivity on the other hand can be thought of as changes which occur within the nuclei of atoms. All chemical bonds involve electrons. Atoms will stay close together if they have a shared interest in one or more electrons. Atoms are at their most stable when they have no partially filled electron shells. If an atom has only a few electrons in a shell, it will tend to lose them to empty the shell. These elements are metals. When metal atoms bond, a metallic bond occurs. When an atom has a nearly full electron shell, it will try to find electrons from another atom so that it can fill its outer shell. These elements are usually described as nonmetals. The bond between two nonmetal atoms is usually a covalent bond. Where metal and nonmetal atom come together an ionic bond occurs. In general, the chemical bonding can be classified into three basic mechanisms, corresponding to the three different ways in which the electron clouds on adjacent atoms can interact (Morris, Jr. 2007).

(a) Shared valence electrons: covalent and metallic bonding
    In this mechanism the valence electron clouds interpenetrate so that, effectively, the atoms share electrons. This is the most potent mechanism of bonding. When an electron in an orbital about one nucleus penetrates the electron cloud of the other it feels the electrostatic attraction of both nuclei, and its energy is lowered substantially. However, the Pauli exclusion principle severely restricts the extent

to which the electron clouds can penetrate one another. When an electron on one atom penetrates the electron cloud of another its charge density overlaps into the occupied orbitals of that atom. This is forbidden unless the orbitals are only partly filled; atoms can only share electrons if there are empty states in the outer, valence shells of each. The number of empty states in the valence shell determines the number of electrons that can be accommodated, that is, the number of shared electron bonds that can be formed. The shared electron bond is the most important type of bond in engineering solids. Shared electron bonds are conventionally separated into two categories, covalent and metallic bonding, which lead to qualitatively different properties in the solid state.

(i) Covalent bonding

Covalent bonding describes a situation in which all available valence states on the bonding atoms are used by shared electrons; the bonds are said to be saturated. The covalent bond is the usual bond type in molecular solids, semiconducting solids, and elemental insulators. Materials with saturated covalent bonds are invariably either semiconductors or insulators. Covalently bonded solids also tend to be difficult to deform mechanically. Without invoking the detailed mechanism of deformation, it can easily be seen that the shape of a solid can only be changed if the atoms move with respect to one another. Since covalent bonds are directional, it is relatively difficult to produce the local atomic reconfigurations that are required for deformation. The diamond modification of carbon, a covalently bonded material, is the hardest substance known.

(ii) Metallic bonding

Metallic bonds are unsaturated; only some of the available valence states are needed when all valence electrons are shared. Since the number of valence electrons in a metallic solid is much less than the number of available bonding states, it is somewhat meaningless to associate a particular electron with a particular bond. It is common to visualize metallic bonding by imagining the ion cores of the individual atoms suspended in a "sea" of valence electrons that bind them to one another and hold them in place. Because of the relative freedom enjoyed by the bonding electrons, metallic bonding leads to high electrical conductivity. Moreover, the metallic bond is relatively nondirectional. The large number of empty bonding states makes it possible for the electron cloud to relax in response to local perturbations in the configuration of atoms. Hence metals are often ductile. Metallic gold can be beaten into sheet or drawn into wire; covalent diamond cannot.

(b) Ionic bonding

In the ionic bonding one or more electrons are transferred from one atom to another so that the neutral atoms are converted into ions with alternate positive and negative charges. The ions then bond electrostatically to one another. It predominates in molecules and solids that are made up of two or more kinds of atoms with very different ionization potentials, so that the electron transfer is energetically preferred. In pure ionic bonding the transferred charge is just sufficient to strip the valence electrons from one species, called the cation, and fill the valence shell of the other, called the anion, so that both have filled outer shells. Most metal oxides and halides have predominantly ionic bonding. Because the valence electrons are localized and tightly bound to individual ions, an ionically bonded solid is usually an electrical insulator.

Saturated bonds between dissimilar atoms often have a mixed character that is partly covalent and partly ionic. Electrons are shared between neighboring atoms, but the sharing is asymmetric. The shared electron density is biased towards one of the two atoms, creating a net charge separation that is equal to a fraction of an electron charge. The degree of ionicity can be related to the difference in the electronegativity of the atoms, which is a semi-empirical parameter that measures the tendency of an atom to acquire an extra electron. The electronegativity decreases monotonically as one moves from right to left across a row of the periodic table, for example, from Cl to Na. As a consequence the ionic character of a heteronuclear bond increases as the atoms are drawn from successively separated columns of the periodic table (for example, NaCl is more ionic than ZnS). The electronegativity decreases as one moves down a column of the periodic table, essentially because the large number of electrons in a relatively heavy atom creates a diffuse charge distribution that partly shields the valence electrons from the nuclear charge. The ionic character of a heteronuclear bond decreases as the atoms become heavier (for example, ZnO is more ionic than CdS).

(c) Dipole bonding

Dipole bonding is the dominant source of bonding between atoms of the noble gases and between neutral molecules. The dipole bond involves no charge transfer. It has its source in the distribution of charge in the neutral atom or molecule. Dipole-bonded solids are inevitably electrical insulators since the electrons are localized on atoms or molecules.

Many molecules contain permanent dipoles that are due to the inhomogeneity of the internal charge distribution. Organic molecules often contain ions that create permanent internal dipoles, and bond into solids by aligning these dipoles with one another. Local dipoles in many organic molecules are associated with dangling hydrogen atoms. These form local dipoles because the electron distribution about the hydrogen atom is strongly polarized towards the atom to which it is bonded, leaving a nearly bare nucleus on the external side. The local dipole interacts significantly with ions of O, N, and F, which are relatively small and negatively charged, so that they sense the intense local field of the H nucleus and interact with it. This type of bonding is particularly important in biological systems.

A second kind of dipole bond, called the Van der Waals bond, is always formed between atoms, and is the primary source of bonding in electrically homogeneous molecules and noble gases that have no other bonding mechanism. Van der Waals bonding is due to transient dipole moments. The electrons that orbit the nucleus are in constant motion. When neutral atoms or electrically homogeneous molecules are close to one another, the interaction between the electrons causes their orbital motion to become correlated; they avoid one another to the extent possible. This correlated motion produces aligned, transient dipole moments that have a weak attractive interaction. This interaction is responsible, for example, for the solidification of inert gases at sufficiently low temperature.

The different kinds of bonding occur simultaneously in most molecules and solids. However, one type usually predominates, and it is, therefore, convenient to classify materials as covalent, metallic, ionic or polar according to the dominant bonding mechanism.

During materials fabrication, atoms can be gathered together as an aggregate through a number of different processes, including condensation, pressurization, chemical reaction, electrodeposition, and melting. The process usually determines, at least initially, whether the collection of atoms will take to form of a gas, liquid or solid. The state usually changes as its temperature or pressure is changed. Melting is the process most often used to form an aggregate of atoms. When the temperature of a melt is lowered to a certain point, the liquid will form either a crystalline solid or/and amorphous solid.

INDEPTH: Ceramic Structures

A ceramic has traditionally been defined as "an inorganic, nonmetallic solid" that is usually prepared from powdered materials and is generally fabricated into products through the application of heat. Most ceramics are made up of two or more elements, which is called a compound. For example, alumina (Al_2O_3) is a compound made up of aluminum atoms and oxygen atoms.

The two most common chemical bonds for ceramic materials are covalent and ionic. The bonding of atoms together is much stronger in covalent and ionic bonding than in metallic. This is why ceramics generally have high hardness, high compressive strength, and chemical inertness. This strong bonding also accounts for the less attractive properties of ceramics, such as low ductility and low tensile strength. The absence of free electrons is responsible for making most ceramics poor conductors of electricity and heat.

However, the crystal structures of ceramics are many and varied and this results in a very wide range of properties. For example, while ceramics are perceived as electrical and thermal insulators, ceramic oxide (initially based on Y-Ba-Cu-O) is the basis for high-temperature superconductivity. Diamond and silicon carbide have a higher thermal conductivity than aluminum or copper. Control of the microstructure can overcome inherent stiffness to allow the production of ceramic springs, and ceramic composites which have been produced with a fracture toughness about half that of steel. Also, the atomic structures are often of low symmetry that gives some ceramics interesting electromechanical properties like piezoelectricity, which is used in sensors and transducers.

(continued)

The structure of most ceramics varies from relatively simple to very complex. The microstructure can be entirely glassy (glasses only); entirely crystalline; or a combination of crystalline and glassy. In the latter case, the glassy phase usually surrounds small crystals, bonding them together. The main compositional classes of engineering ceramics are the oxides, nitrides, and carbides (NDT 2014).

2.2.3 Nano- and Microstructure

Microstructure is defined as the structure of a prepared surface of material as revealed by a microscope above 25× magnification. The microstructure of a material (such as metals, polymers, ceramics or composites) can strongly influence physical properties such as strength, toughness, ductility, hardness, corrosion resistance, high/low temperature behavior or wear resistance. These properties in turn govern the application of these materials in advanced energy systems. Microstructure at scales smaller than can be viewed with optical microscopes is often called nanostructure, while the structure in which individual atoms are arranged is known as crystal structure, quasicrystal structure and/or amorphous structure. The mesostructure of a material, if it has one, also has a strong impact on material performance. Many mesostructural parameters are analogous to microstructural parameters, though they typically two orders of magnitude larger in size.

2.2.3.1 Nanostructure

Nanomaterials describe, in principle, materials of which a single unit is sized (in at least one dimension) between 1 and 1000 nm (10^{-9} m) but usually is 1–100 nm (the usual definition of nanoscale). Most are synthetic and can be engineered to a wide range of physical properties. Nanosurfaces, cylindrical nanotubes, and nanospheres are common nanostructures with different dimensions. Nanosurfaces indicates nanostructured surfaces, the thickness of the surface on an object is generally between 0.1 and 100 nm. Nanotubes, nanometer-scale tube-like structures, usually refer to Carbon nanotube, inorganic nanotube, DNA nanotube, and Membrane nanotube with a diameter of 0.1 and 100 nm, and a much greater length. Nanospheres is defined as a small particulate object (between 0.1 and 100 nm in each spatial dimension) that behaves as a whole unit in terms of its transport and properties.

Actually, nanomaterials can be divided into diverse forms, such as (a) Nanoparticle, a particle is defined as a nano-sized object; (b) Nano-fiber, fibers with diameters less than 1000 nm, which can be produced by interfacial polymerization and electro-spinning; (c) Nano-flower, refers to a compound of certain elements in chemistry that results in formations which in microscopic view resemble flowers or, in some cases, trees that are called nano-bouquets or nano-trees; (d)

Nano-fabrics, an nanotechnology that deals with building specialized fabrics; (e) Nanocomposite, defined as a multiphase solid material where one of the phases has one, two, or three dimensions of less than 100 nm; (f) Nano-cages usually indicate hollow porous metal (like gold) nanoparticles ranging in size from 10 to over 150 nm, which are created by reacting silver nanoparticles with chloroauric acid ($HAuCl_4$) in boiling water, for example; (g) Nano-foam, a class of nanostructured porous materials/foams, containing a significant population of pores with diameters less than 100 nm (Aerogels are one example of nano-foam); (h) Nano-flake, a perfect crystalline structure that can absorb whole spectrum of the light, which have the potential to convert up to 30% of the solar energy into electricity; (i) Nano-platelet, nanoparticles consisted of small stacks of graphene that are 1–15 nm thick, with diameters ranging from sub-micrometer to 100 μm; (j) Nano-pillar, a type of metamaterial with pillar-shaped nanostructures approximately 10 nm in diameter that can be grouped together in lattice like arrays; (k) Nano-mesh, inorganic nanostructured two-dimensional material (similar to graphene), which consists of a single layer of boron and nitrogen atoms to form into a highly regular mesh by self-assembly; (l) Nano-pin film, covered with nanoscale topped off pins or cones perpendicular to the surface with unusual superhydrophobic properties (for example, a droplet of water makes contact with the surface of this film and forms an almost perfect sphere with a contact angle of 178°) (Hosono et al. 2005); (m) Nano-shell, or rather a nanoshell plasmon, is a type of spherical nanoparticle consisting of a dielectric core which is covered by a thin metallic shell (usually gold). These nanoshells involve a quasiparticle called a plasmon which is a collective excitation or quantum plasma oscillation where the electrons simultaneously oscillate with respect to all the ions (Loo et al. 2004); (n) Nano-ring, a small ring formed crystal, the first one is a zinc oxide nanoring made by a spontaneous self-coiling process of nanobelts (Kong et al. 2004); (o) Nano-rod, chemically synthesized from metals or semiconducting materials with dimensions range from 1 to 100 nm, and standard aspect ratios (length divided by width) of 3–5; (p) Nano-tip, a nanoscale tip, especially used for scanning tunneling microscope; (q) Quantum dot, a semiconductor nanostructure that confines the motion of conduction band electrons, valence band holes, or excitons (bound pairs of conduction band electrons and valence band holes) in all three spatial directions; (r) Sculptured thin film, nanostructured materials with unidirectionally varying properties that can be designed and realized in a controllable manner using variants of physical vapor deposition (Lakhtakia and Messier 2005); (s) Gradient multilayer or GML nano-film, an assembly of quantum dot layers with a built-in gradient of nanoparticle size, composition or density, with properties for applications in design of solar cells and energy storage devices; and (t) Quantum heterostructure, a heterostructure in a substrate (usually a semiconductor material), where size restricts the movements of the charge carriers forcing them into a quantum confinement. This leads to the formation of a set of discrete energy levels at which the carriers can exist. Quantum heterostructures have sharper density of states than structures of more conventional sizes (Bastard et al. 1991).

The properties of nanomaterials change significantly from those at larger scales, which are size-dependent in the nanoscale range. In the size scale, quantum effects rule the behavior and properties of the materials. Thus, when particle size is made to be nanoscale, properties such as melting point, fluorescence, electrical conductivity, magnetic permeability, and chemical reactivity change as a function of the size of the particle. Accordingly, the electron states of nanostructures are quantized, leading to new and usually striking electrical, thermal, magnetic, optical, and mechanical properties at the nanoscale. Therefore, nanostructures are of both basic and practical interest since their physicochemical properties can be tailored by controlling their size and shape at the nanoscale, leading to improved and/or novel applications.

2.2.3.2 Crystal Structure

A crystalline structure is any structure of ions, molecules, or atoms that are held together in an ordered, three-dimensional arrangement. Crystalline structure is one of three common types of structural ordering of atoms, the other being the amorphous structure and quasicrystal structure. Additional possible structures include modulated structures, incommensurately modulated structures, and liquid crystals.

The key difference in the crystalline and amorphous structure is the ordering of the structure. Crystalline structure can be thought of as the highest level of order that can exist in a material, while an amorphous structure is irregular and lacks the repeating pattern of a crystal lattice. A quasicrystal, or quasiperiodic crystal, is a structure that is ordered but not periodic. A quasicrystalline pattern can continuously fill all available space, but it lacks translational symmetry. While crystals can possess only two-, three-, four-, and sixfold rotational symmetries, the Bragg diffraction pattern of quasicrystals shows sharp peaks with other symmetry orders, for instance fivefold. The materials whose atoms or molecules fill space by periodic repetition are called crystalline solids. Most metallic and inorganic materials fall into this class. The materials that have approximately random distributions of atoms or molecules are called amorphous solids, or glasses. They are mostly materials that are cooled from a liquid or gaseous state so quickly that a nearly random atom distribution is preserved. Polymeric and other organic solids that are made of large molecules are often amorphous. Inorganic compounds, such as the silica glasses can often be made amorphous by rapid cooling from the melt. Many semiconductors and some metal alloys can be made amorphous by solidifying them very rapidly or depositing them from the vapor onto a cold substrate. Quasicrystals exist universally in many metallic alloys and some polymers. They generally have ceramic-like properties including high thermal and electrical resistance, hardness and brittleness, resistance to corrosion, nonstick, hard-wearing, low-friction, and unique optical properties.

Crystal structures are usually specified by the unit cell lattice structures describing the arrangement within the solid of a small representative group of atoms or molecules. By multiplying identical unit cells in three directions, the location of all the particles in the crystal is determined. As shown in Fig. 2.2, 14 different types of unit cell lattice structures are found in nature. The simplest crystalline unit cell to picture is the cubic, where the atoms are lined up in a square, 3D grid. The unit cell is simply a box with an atom at each corner. Simple cubic crystals are relatively rare, mostly because they tend to easily distort. However, many crystals form body-centered-cubic (bcc) or face-centered-cubic (fcc) structures, which are cubic with

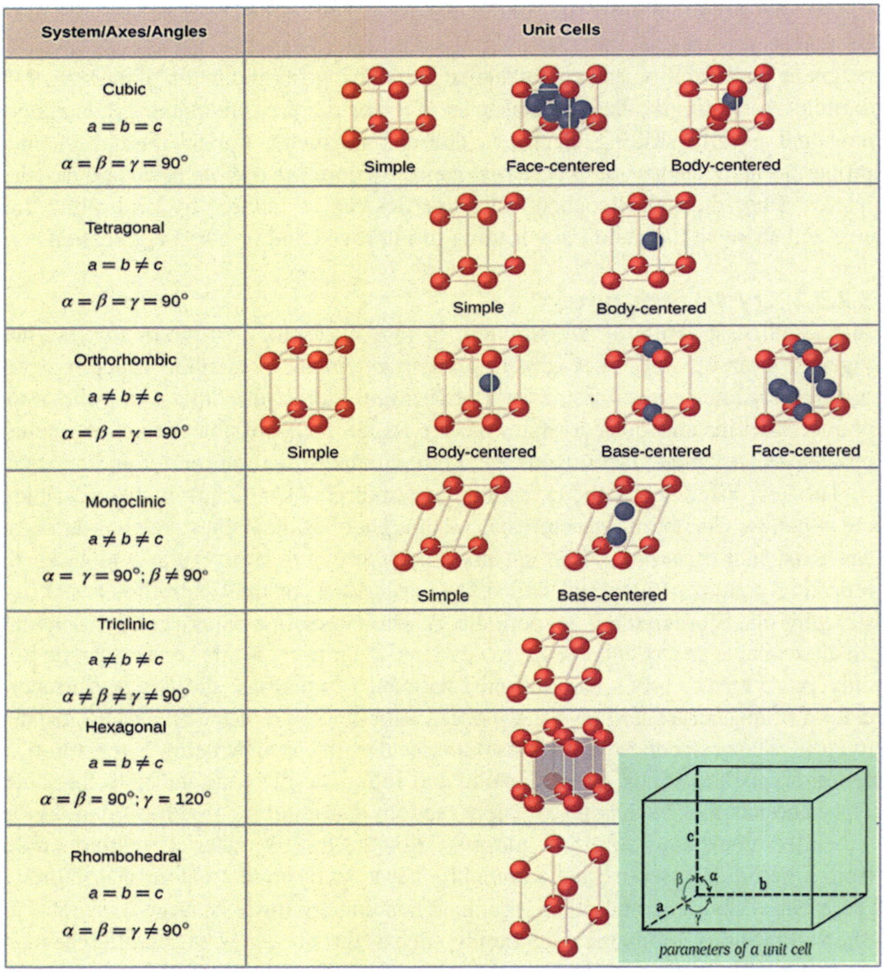

Fig. 2.2 Fourteen different types of crystal unit cell structures

either an extra atom centered in the cube or centered in each face of the cube, as shown in Fig. 2.3. Most metals form bcc, fcc or Hexagonal Close Packed (hcp) structures; however, the structure can change depending on temperature.

The body-centered cubic unit cell has atoms at each of the eight corners of a cube (like the cubic unit cell) plus one atom in the center of the cube (top image of Fig. 2.3). Each of the corner atoms is the corner of another cube so the corner atoms are shared among eight unit cells (a coordination number of 8). The bcc unit cell consists of a net total of two atoms; one in the center and eight eighths from corners atoms. The bcc arrangement does not allow the atoms to pack together as closely as the fcc or hcp arrangements. The bcc structure is often the high temperature form of metals that are close-packed at lower temperatures. The volume of atoms in a cell per

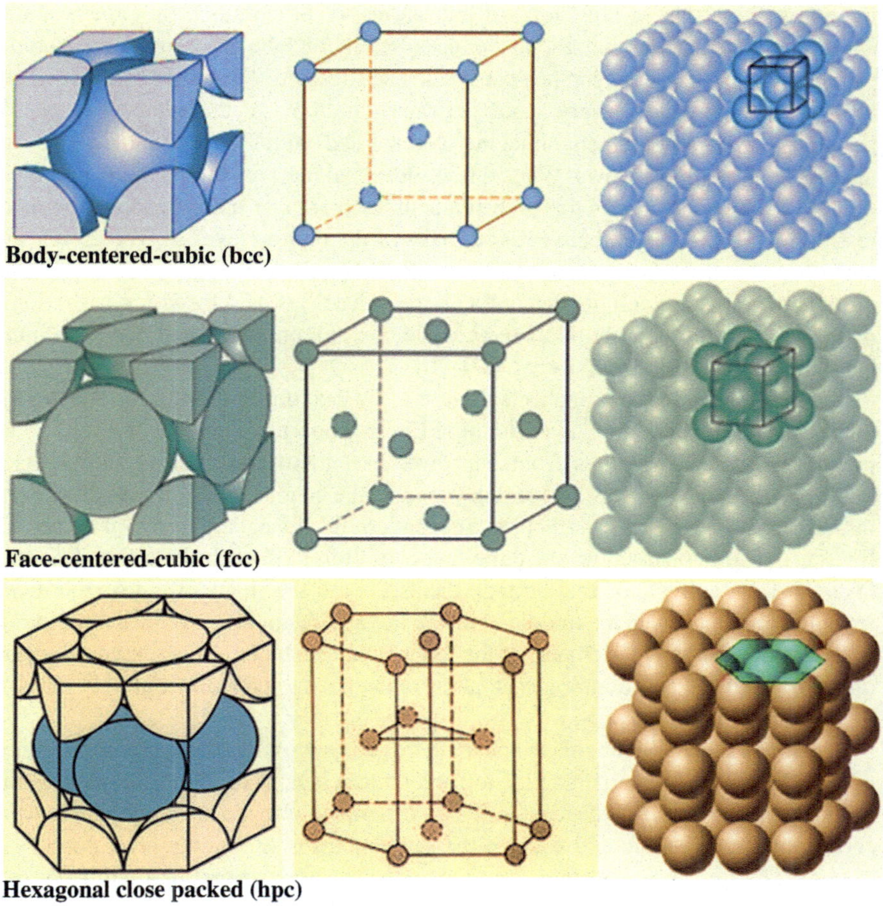

Fig. 2.3 Primary metallic crystalline structures (BCC, FCC, HCP)

the total volume of a cell is called the **packing factor**. The bcc unit cell has a packing factor of 0.68. Some of the materials that have a bcc structure include lithium, sodium, potassium, chromium, barium, vanadium, alpha-iron and tungsten. Metals which have a bcc structure are usually harder and less malleable than close-packed metals such as gold. When the metal is deformed, the planes of atoms must slip over each other, and this is more difficult in the bcc structure. In addition, there are other important mechanisms for hardening materials, such as introducing impurities or defects which make slipping more difficult (NDT 2014).

The face-centered-cubic structure has atoms located at each of the corners and the centers of all the cubic faces (Middle image of Fig. 2.3). Each of the corner atoms is the corner of another cube so the corner atoms are shared among eight unit cells. Additionally, each of its six face-centered atoms is shared with an adjacent atom. Since 12 of its atoms are shared, it is said to have a coordination number of 12. The fcc unit cell consists of a net total of four atoms; eight eighths from corners atoms and six halves of the face atoms. In the fcc structure (and the hcp structure) the atoms can pack closer together than they can in the bcc structure. The atoms from one layer nest themselves in the empty space between the atoms of the adjacent layer. To picture packing arrangement, imagine a box filled with a layer of balls that are aligned in columns and rows. When a few additional balls are tossed in the box, they will not balance directly on top of the balls in the first layer but instead will come to rest in the pocket created between four balls of the bottom layer. As more balls are added they will pack together to fill up all the pockets. The packing factor (the volume of atoms in a cell per the total volume of a cell) is 0.74 for fcc crystals. Some of the metals that have the fcc structure include aluminum, copper, gold, iridium, lead, nickel, platinum, and silver (NDT 2014).

Another common close packed structure is the hexagonal close pack. The hexagonal structure of alternating layers is shifted so its atoms are aligned to the gaps of the preceding layer. The atoms from one layer nest themselves in the empty space between the atoms of the adjacent layer just like in the fcc structure. However, instead of being a cubic structure, the pattern is hexagonal (bottom image of Fig. 2.3). The hcp structure has three layers of atoms. In each the top and bottom layer, there are six atoms that arrange themselves in the shape of a hexagon and a seventh atom that sits in the middle of the hexagon. The middle layer has three atoms nestle in the triangular "grooves" of the top and bottom plane. Note that there are six of these "grooves" surrounding each atom in the hexagonal plane, but only three of them can be filled by atoms.

As shown in the middle image above, there are six atoms in the hcp unit cell. Each of the 12 atoms in the corners of the top and bottom layers contribute 1/6 atom to the unit cell, the two atoms in the center of the hexagon of both the top and bottom layers each contribute ½ atom and each of the three atom in the middle layer contribute 1 atom. The image on the right above attempts to show several hcp unit cells in a larger lattice. The coordination number of the atoms in this structure is 12. There are six nearest neighbors in the same close packed layer, three in the layer above and three in the layer below. The packing factor is 0.74, which is the same as the fcc unit cell. The hcp structure is very common for elemental metals and some examples include beryllium, cadmium, magnesium, titanium, zinc, and zirconium (NDT 2014).

Crystalline structure plays an important role in influence on the properties of a material, especially on its mechanical behavior. For example, it is easier for planes of atoms to slide by each other if those planes are closely packed. Therefore, lattice structures with closely packed planes allow more plastic deformation than those that are not closely packed. Additionally, cubic lattice structures allow slippage to occur more easily than non-cubic lattices. This is because their symmetry provides closely packed planes in several directions. A face-centered cubic crystal structure will exhibit more ductility (deform more readily under load before breaking) than a body-centered cubic structure. The bcc lattice, although cubic, is not closely packed and forms strong metals. Alpha-iron and tungsten have the bcc form. The fcc lattice is both cubic and closely packed and forms more ductile materials. Gamma-iron,

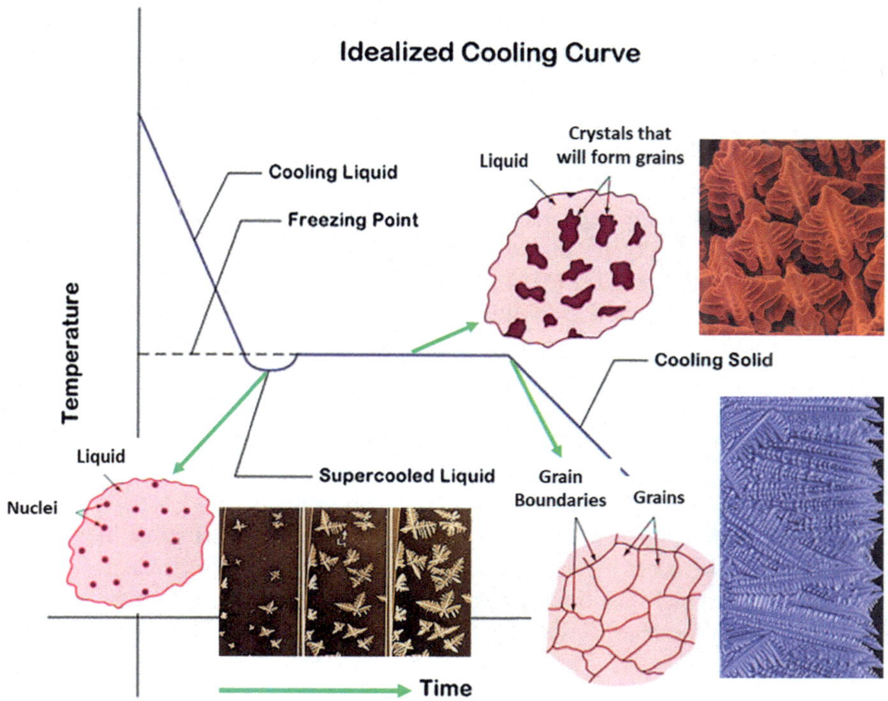

Fig. 2.4 Schematic microstructure variations during crystallization or solidification process

silver, gold, and lead have fcc structures. Finally, HCP lattices are closely packed, but not cubic. HCP metals like cobalt and zinc are not as ductile as the fcc metals (NDT 2014).

2.2.3.3 Crystallization and Solidification

The crystallization of a large amount of material from a single point of nucleation results in a single crystal. In engineering materials, single crystals are produced only under carefully controlled conditions. The expense of producing single crystal materials is only justified for special applications, such as turbine engine blades, solar cells, and piezoelectric materials. Normally when a material begins to solidify, multiple crystals begin to grow in the liquid and a polycrystalline (more than one crystal) solid forms, as shown in Fig. 2.4.

The moment a crystal begins to grow is known as nucleation and the point where it occurs is the nucleation point. At the solidification temperature, atoms of a liquid, such as melted metal, begin to bond together at the nucleation points and start to form crystals. The final sizes of the individual crystals depend on the number of nucleation points. The crystals increase in size by the progressive addition of atoms and grow until they impinge upon adjacent growing crystal.

In engineering materials, a crystal is usually referred to as a grain. A grain is merely a crystal without smooth faces because its growth was impeded by contact with another grain or a boundary surface. The interface formed between grains is called a grain boundary. The atoms between the grains (at the grain boundaries) are said to be disordered or formed into defected structure. Grains are sometimes large enough to be visible under an ordinary light microscope or even to the unaided eye. The spangles that are seen on newly galvanized metals are grains. Rapid cooling generally results in more nucleation points and smaller grains (a fine grain structure). Slow cooling generally results in larger grains which will have lower strength, hardness, and ductility (NDT 2014).

In metals, the crystals that form in the liquid during freezing generally follow a pattern consisting of a main branch with many appendages. A crystal with this morphology slightly resembles a pine tree and is called a dendrite, which means branching, as shown in Fig. 2.4. The formation of dendrites occurs because crystals grow in defined planes due to the crystal lattice they create. Secondary dendrite arms branch off the primary arm, and tertiary arms off the secondary arms and etcetera. During freezing of a polycrystalline material, many dendritic crystals form and grow until they eventually become large enough to impinge upon each other. Eventually, the interdendritic spaces between the dendrite arms crystallize to yield a more regular crystal. The original dendritic pattern may not be apparent when examining the microstructure of a material. However, dendrites can often be seen in solidification voids that sometimes occur in castings or welds (NDT 2014).

In addition, most materials contract or shrink during solidification and cooling, due mainly to (a) Contraction of the liquid as it cools prior to its solidification; (b) contraction during phase change from a liquid to solid; (c) contraction of the solid as it continues to cool to ambient temperature. Shrinkage can sometimes cause cracking to occur in component as it solidifies. Since the coolest area of a volume of liquid is where it contacts a mold or die, solidification usually begins first at this surface. As the crystals grow inward, the material continues to shrink. If the solid surface is too rigid and will not deform to accommodate the internal shrinkage, the stresses can become high enough to exceed the tensile strength of the material and cause a crack to form. Shrinkage cavitation sometimes occurs because as a material solidifies inward, shrinkage occurred to such an extent that there are not enough atoms present to fill the available space and a void is left (NDT 2014).

2.2.3.4 Crystal Defects

A perfect crystal, with every atom of the same type in the correct position, does not exist. All crystals have some defects. Defects contribute to the mechanical properties of metals. In fact, using the term "defect" is sort of a misnomer since these features are commonly intentionally used to manipulate the mechanical properties of a material. Adding alloying elements to a metal is one way of introducing a crystal defect. Crystal defects are generally classified as (NDT 2014):

(a) Point defects, which are places where an atom is missing or irregularly placed in the lattice structure. Point defects include lattice vacancies, self-interstitial atoms, substitution impurity atoms, and interstitial impurity atoms.
(b) Linear defects, which are groups of atoms in irregular positions. Linear defects are commonly called dislocations.
(c) Planar defects, which are interfaces between homogeneous regions of the material. Planar defects include grain boundaries, stacking faults, and external surfaces.

In general, the movement of dislocations results in plastic deformation in a material. Millions of dislocations result from plastic forming operations such as rolling and extruding. Any defect in the regular lattice structure disrupts the motion of dislocation, which makes slip or plastic deformation more difficult. These defects not only include the point and planer defects, but also other dislocations. Dislocation movement produces additional dislocations, and when dislocations run into each other it often impedes movement of the dislocations. This drives up the force needed to move the dislocation or, in other words, strengthens the material (NDT 2014).

2.2.3.5 Diffusion and Phase Equilibria

Diffusion or mass transfer is the migration of atoms or molecules from a region of high concentration to a region of low concentration. In a homogeneous material, atoms or molecules are routinely moving around but the movement is random (i.e., there is always an equal number of atoms or molecules moving in all directions). In an inhomogeneous material, all the atoms or molecules are moving near randomly, but there is a migration of atoms or molecules to areas where their concentrations are lower. In other words, there is a net diffusion or mass transfer.

Atom diffusion in a solid can occur by the motion of host or substitutional atoms to vacancies (vacancy diffusion), or interstitial impurities atoms to different interstitial positions (interstitial diffusion). In order to move, an atom must overcome the bond energy due to nearby atoms. This is more easily achieved at high temperatures when the atoms are vibrating strongly (NDT 2014).

A collection of atoms or molecules is in equilibrium with its surroundings if there is no net diffusion or transfer of energy or matter across the boundaries to the environment. Individual atoms or molecules may constantly be zinging across the boundaries, or bumping into neighbors and transferring heat, due to thermal motion. But if over millions of atoms or molecules are averaged, the number of atoms or molecules leaving the control volume gets canceled out by an equal number returning to the control volume, so the net motion across the boundary is zero. This situation is analogous to chemical reaction equilibrium—when equilibrium is reached, the rate of the forward reaction is equal to the rate of the reverse reaction, and there is no net measurable change in the concentration of reactants or products with time. In each case, the equilibrium is dynamic, meaning that forward and reverse reactions are constantly occurring. However, the overall measurable properties of the system do not change with time, and the system is considered at equilibrium.

Diffusion or mass transfer is inherently a nonequilibrium process, characterized by the net motion of atoms or molecules down a concentration gradient. It is usually assumed that diffusion or mass transfer at the continuum scale (averaged over zillions of atoms or molecules), and the concentration profile of the diffusing solute is continuous within a single phase. Often, mass transfer occurs across the interface between two phases—solid/solid (e.g., intermetallic compound/alloy solution), liquid/liquid (e.g., oil/water), liquid/gas (e.g., air/water), liquid/solid (e.g., water/polymer film), or gas/solid (e.g., air/polymer film). In this class, only diffusion of dilute solutes and "slow" mass transfer rates are generally considered. The nature of the interface is not disrupted due to convective flow of material across the interface under these conditions.

At the interface between two phases, the chemical potential of the diffusing solute in phase I must equal that in phase II at an equilibrium state. The interface between the two phases can be thought as an infinitely thin plane and that the phases are in equilibrium in this plane. There is typically a discontinuity in concentration at the interface. Here at the interface, the concentration of the diffusing solute A in Phase I is related to that in Phase II via a partition coefficient, Henry's law, coefficient, vapor pressure, etc. The concentration of A at the interface is often well below its solubility limit, but that partition coefficients, etc. are simply the relative concentrations in each phase at equilibrium.

INDEPTH: Polymer Structure

Engineering polymers include natural materials such as rubber and synthetic materials such as plastics and elastomers. Polymers are very useful materials because their structures can be altered and tailored to produce materials (a) with a range of mechanical properties (b) in a wide spectrum of colors and (c) with different transparent properties.

A polymer is composed of many simple molecules that are repeating structural units called monomers. A single polymer molecule may consist of hundreds to a million monomers and may have a linear, branched, or network structure. Covalent bonds hold the atoms in the polymer molecules together and secondary bonds then hold groups of polymer chains together to form the polymeric material. Copolymers are polymers composed of two or more different types of monomers.

A polymer is an organic material and the backbone of every organic material is a chain of carbon atoms. The carbon atom has four electrons in the outer shell. Each of these valence electrons can form a covalent bond to another carbon atom or to a foreign atom. The key to the polymer structure is that two carbon atoms can have up to three common bonds and still bond with other atoms. The elements found most frequently in polymers: H, F, Cl, Bf, O, S, C, and Si. The ability for molecules to form long chains is a vital to producing polymers. The polymer chain is often shown in two dimensions,

(continued)

but they have a three dimensional structure. Each bond is at 109° to the next and, therefore, the carbon backbone extends through space like a twisted chain of TinkerToys. When stress is applied, these chains stretch and the elongation of polymers can be thousands of times greater than it is in crystalline structures. The length of the polymer chain is very important. As the number of carbon atoms in the chain is increased to beyond several hundred, the material will pass through the liquid state and become a waxy solid. When the number of carbon atoms in the chain is over 1000, the solid material polyethylene, with its characteristics of strength, flexibility, and toughness, is obtained. The change in state occurs because as the length of the molecules increases, the total binding forces between molecules also increases.

Thermoplastic materials, such as polyethylene, can be pictured as a mass of intertwined worms randomly thrown into a pail. The binding forces are the result of van der Waals forces between molecules and mechanical entanglement between the chains. When thermoplastics are heated, there is more molecular movement and the bonds between molecules can be easily broken. This is why thermoplastic materials can be re-melted. There is another group of polymers in which a single large network, instead of many molecules is formed during polymerization. Since polymerization is initially accomplished by heating the raw materials and bringing them together, this group is called thermosetting polymers or plastics. For this type of network structure to form, the mers must have more than two places for boning to occur; otherwise, only a linear structure is possible. These chains form jointed structures and rings, and may fold back and forth to take on a partially crystalline structure. Since these materials are essentially comprised of one giant molecule, there is no movement between molecules once the mass has set. Thermosetting polymers are more rigid and generally have higher strength than thermoplastic polymers. Also, since there is no opportunity for motion between molecules in a thermosetting polymer, they will not become plastic when heated (NDT 2014).

2.2.4 Electronic Structure

Electronic structure is the state of motion of electrons in an electrostatic field created by stationary nuclei. The term encompass both the wave functions of the electrons and the energies associated with them. Electronic structure is obtained by solving quantum mechanical equations for the clamped-nuclei problem. Electronic structure problems arise from the Born–Oppenheimer approximation. Except for a small number of simple problems such as hydrogen-like atoms, the solution of electronic structure problems require modern computers (Simons 2003).

2.2.4.1 Valence Bond Theory

In chemistry, valence bond (VB) theory is one of two basic theories—along with molecular orbital (MO) theory—that use quantum mechanics to explain chemical bonding. It focuses on how the atomic orbitals of the dissociated atoms combine to give individual chemical bonds when a molecule is formed. In contrast, molecular orbital (MO) theory, has orbitals that cover the whole molecule (Magnasco 2009).

The VB theory describes the formation of covalent bonds from the overlap of atomic orbitals on two different atoms. Because of the overlap, it is highly probable that a pair of electrons are found in the physical region or space where the orbitals overlap. A covalent bond results when two conditions are met (OpenStax 2016): (a) an orbital on one atom overlaps an orbital on a second atom and (b) the single electrons in each orbital combine to form an electron pair. The mutual attraction between this negatively charged electron pair and the two atoms' positively charged nuclei serves to physically link the two atoms through a force defined as a covalent bond (as an example, Fig. 2.5 illustrates how the sum of the energies of two hydrogen atoms changes as they approach each other). The strength of a covalent bond depends on the extent of overlap of the orbitals involved. Orbitals that overlap extensively form bonds that are stronger than those that have less overlap.

There are two types of overlapping orbitals: sigma (σ) and pi (π), as shown in Fig. 2.5. Both bonds are formed from the overlap of two orbitals, one on each atom. σ bonds occur when orbitals overlap between the nuclei of two atoms, also known as the internuclear axis. π bonds occur when two (unhybridized) p-orbitals overlap. The

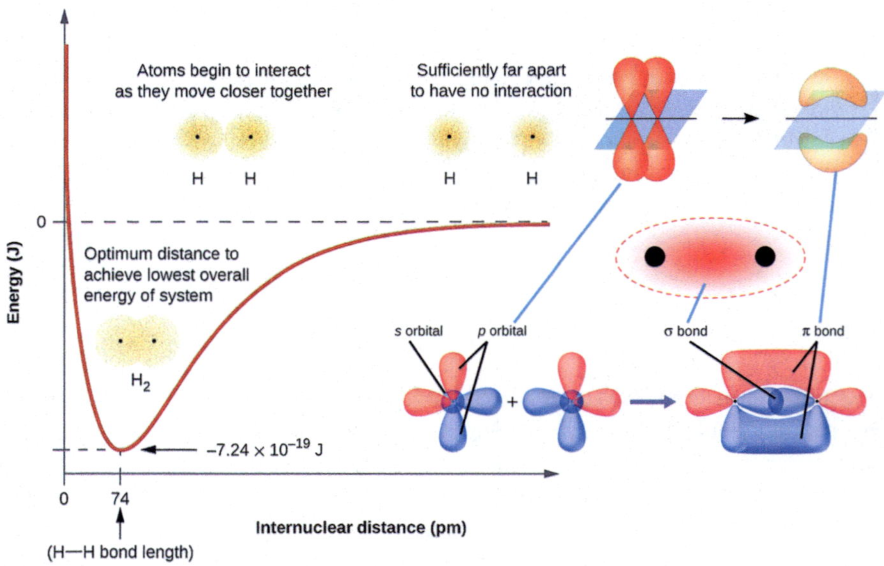

Fig. 2.5 Illustration of the σ bond and π bond formation, and how the sum of the energies of two hydrogen atoms changes as they approach each other (Modified from OpenStax 2016 ©1999–2018 Rice University, licensed under a Creative Commons Attribution 4.0 License): The energy of the system changes as the atoms interact—the lowest (most stable) energy occurs at a distance of 74 pm, which is the bond length observed for the H_2 molecule

p-orbitals, in one π bond, are located above and below the nuclei of the atoms. By occupying the region of space that is above, below, and on the sides of an atom's nuclei, two π bonds can form. Both types of overlapping orbitals can be related to bond order. Single bonds have one sigma bond. Double bonds consist of one σ and one π bond, while triple bonds contain one σ and two π bonds.

Modern valence bond theory now complements molecular orbital theory, which does not adhere to the valence bond idea that electron pairs are localized between two specific atoms in a molecule but that they are distributed in sets of molecular orbitals which can extend over the entire molecule. Molecular orbital theory can predict magnetic and ionization properties in a straightforward manner, while valence bond theory gives similar results but is more complicated. Modern valence bond theory views aromatic properties of molecules as due to spin coupling of the π orbitals. Valence bond treatments are restricted to relatively small molecules, largely due to the lack of orthogonality between valence bond orbitals and between valence bond structures, while molecular orbitals are orthogonal. On the other hand, valence bond theory provides a much more accurate picture of the reorganization of electronic charge that takes place when bonds are broken and formed during the course of a chemical reaction. In particular, valence bond theory correctly predicts the dissociation of homonuclear diatomic molecules into separate atoms, while simple molecular orbital theory predicts dissociation into a mixture of atoms and ions. In addition, modern valence bond theory replaces the overlapping atomic orbitals by overlapping valence bond orbitals that are expanded over a large number of basic functions, either centered each on one atom to give a classical valence bond picture, or centered on all atoms in the molecule (Shaik and Hiberty 2008).

2.2.4.2 Molecular Orbital Theory

Molecular orbital (MO) theory is a method for determining molecular structure in which electrons are not assigned to individual bonds between atoms, but are treated as moving under the influence of the nuclei in the whole molecule. The spatial and energetic properties of electrons within atoms are fixed by quantum mechanics to form orbitals that contain these electrons. While atomic orbitals contain electrons ascribed to a single atom, molecular orbitals, which surround a number of atoms in a molecule, contain valence electrons between atoms.

MO theory provides a global, delocalized perspective on chemical bonding. In MO theory, any electron in a molecule may be found anywhere in the molecule, since quantum conditions allow electrons to travel under the influence of an arbitrarily large number of nuclei, as long as they are in eigenstates permitted by certain quantum rules. Thus, when excited with the requisite amount of energy through high-frequency light or other means, electrons can transition to higher-energy molecular orbitals. For instance, in the simple case of a hydrogen diatomic molecule, promotion of a single electron from a bonding orbital to an antibonding orbital can occur under UV radiation. This promotion weakens the bond between the two hydrogen atoms and can lead to photodissociation—the breaking of a chemical bond due to the absorption of light. Although in MO theory some molecular orbitals may hold electrons that are more localized between specific pairs of molecular atoms, other orbitals may hold electrons that are spread more uniformly over the molecule. Thus, overall, bonding is far more delocalized in MO theory, which makes it more applicable to resonant molecules that have equivalent non-integer bond orders than valence bond (VB) theory. This makes MO theory more useful for the description of extended systems.

As in benzene, in substances such as beta carotene, chlorophyll, or heme, some electrons in the π orbitals are spread out in molecular orbitals over long distances in a molecule, resulting in light absorption in lower energies (the visible spectrum), which accounts for the characteristic colors of these substances. This and other spectroscopic data for molecules are well explained in MO theory, with an emphasis on electronic states associated with multicenter orbitals, including mixing of orbitals premised on principles of orbital symmetry matching. The same MO principles also naturally explain some electrical phenomena, such as high electrical conductivity in the planar direction of the hexagonal atomic sheets that exist in graphite. This results from continuous band overlap of half-filled *p* orbitals and explains electrical conduction. MO theory recognizes that some electrons in the graphite atomic sheets are completely delocalized over arbitrary distances, and reside in very large molecular orbitals that cover an entire graphite sheet, and some electrons are thus as free to move and therefore conduct electricity in the sheet plane, as if they resided in a metal (Miessler et al. 2014).

2.2.4.3 Electronic Band Structure

Electronic band structure of a solid describes the range of energies that an electron within the solid may have (called energy bands, allowed bands, or simply bands) and ranges of energy that it may not have (called bandgaps or forbidden bands). Band theory derives these bands and bandgaps by examining the allowed quantum mechanical wave functions for an electron in a large, the electrons of a single, isolated atom occupy atomic orbitals each of which has a discrete energy level. Band theory has been successfully used to explain many physical properties of solids, such as electrical resistivity and optical absorption, and forms the foundation of the understanding of all solid-state devices (transistors, solar cells, etc.).

The electrons of a single, isolated atom occupy atomic orbitals each of which has a discrete energy level. When two atoms join together to form into a molecule, their atomic orbitals overlap. The Pauli exclusion principle dictates that no two electrons can have the same quantum numbers in a molecule. So if two identical atoms combine to form a diatomic molecule, each atomic orbital splits into two molecular orbitals of different energy, allowing the electrons in the former atomic orbitals to occupy the new orbital structure without any having the same energy.

Similarly if a large number N of identical atoms come together to form a solid, such as a crystal lattice, the atoms' atomic orbitals overlap. Since the Pauli exclusion principle dictates that no two electrons in the solid have the same quantum numbers, each atomic orbital splits into N discrete molecular orbitals, each with a different energy. Since the number of atoms in a macroscopic piece of solid is a very large number ($N \sim 10^{22}$) the number of orbitals is very large and thus they are very closely spaced in energy (of the order of 10^{-22} eV). The energy of adjacent levels is so close together that they can be considered as a continuum, an energy band (Holgate 2009).

This formation of bands is mostly a feature of the outermost electrons (valence electrons) in the atom, which are the ones responsible for chemical bonding and electrical conductivity. The inner electron orbitals do not overlap to a significant degree, so their bands are very narrow.

Bandgaps are essentially leftover ranges of energy not covered by any band, a result of the finite widths of the energy bands. The bands have different widths, with the widths depending upon the degree of overlap in the atomic orbitals from which they arise. Two adjacent bands may simply not be wide enough to fully cover the range of energy. For example, the bands associated with core orbitals (such as *1 s* electrons) are extremely narrow due to the small overlap between adjacent atoms. As a result, there tend to be large bandgaps between the core bands. Higher bands involve comparatively larger orbitals with more overlap, becoming progressively wider at higher energies so that there are no bandgaps at higher energies (Holgate 2009).

Although electronic band structures are usually associated with crystalline materials, quasi-crystalline and amorphous solids may also exhibit band structures. These are somewhat more difficult to study theoretically since they lack the simple symmetry of a crystal, and it is not usually possible to determine a precise dispersion relation. As a result, virtually all of the existing theoretical work on the electronic band structure of solids has focused on crystalline materials.

A solid has an infinite number of allowed bands, just as an atom has infinitely many energy levels. However, most of the bands simply have too high energy, and are usually disregarded under ordinary circumstances. Conversely, there are very-low-energy bands associated with the core orbitals (such as *1 s* electrons). These low-energy core bands are also usually disregarded since they remain filled with electrons at all times, and are therefore inert. Likewise, materials have several bandgaps throughout their band structure. The most important bands and bandgaps—those relevant for electronics and optoelectronics—are those with energies near the Fermi level. The bands and bandgaps near the Fermi level are given special names, depending on the material (Yu and Cardona 2010):

(a) In a semiconductor or band insulator, the Fermi level is surrounded by a bandgap, referred to as *the* bandgap (to distinguish it from the other bandgaps in the band structure). The closest band above the bandgap is called the conduction band, and the closest band beneath the bandgap is called the valence band. The name "valence band" was coined by analogy to chemistry, since in many semiconductors the valence band is built out of the valence orbitals.

(a) In a metal or semimetal, the Fermi level is inside of one or more allowed bands. In semimetals the bands are usually referred to as "conduction band" or "valence band" depending on whether the charge transport is more electron-like or hole-like, by analogy to semiconductors. In many metals, however, the bands are neither electron-like nor hole-like, and often just called "valence band" as they are made of valence orbitals. The bandgaps in a metal's band structure are not important for low-energy physics, since they are too far from the Fermi level.

Band structures have been modeled with different attempts. Each model describes some types of solids very well, and others poorly. For instance, the nearly free electron model works well for metals, but poorly for nonmetals. The tight binding model is extremely accurate for ionic insulators, such as metal halide salts (e.g., NaCl).

INDEPTH: Composite Structures

A composite material is basically a combination of two or more materials, each of which retains its own distinctive properties. Multiphase metals are composite materials on a micro scale, but generally the term composite is applied to materials that are created by mechanically bonding two or more different materials together. Composite materials mostly have a bulk phase, which is continuous, called the matrix; and a dispersed, noncontinuous, phase called the reinforcement.

Metal parts can be replaced with lighter weight parts manufactured from advanced composites. For example, carbon-epoxy composites are two thirds the weight of aluminum, and two and a half times as stiff. Composites are resistant to fatigue damage and harsh environments, and are repairable.

Composites meeting the criteria of having mechanical bonding can also be produced on a micro scale. For example, when tungsten carbide powder is mixed with cobalt powder, and then pressed and sintered together, the tungsten carbide retains its identity. The resulting material has a soft cobalt matrix with tough tungsten carbide particles inside. This material is used to produce carbide drill bits and is called a metal-matrix composite. A metal matrix composite is a type of metal that is reinforced with another material to improve strength, wear or some other characteristics.

2.3 Analytical Techniques of Nano- and Micro-Scale Material Structures and Properties

Experimental tools to characterize nano- and micro-scale material structures and properties are fundamental to building materials science and engineering understanding and developing new materials with unique properties. As these tools improve, new materials can be developed faster and with greater control over properties, processing, and, ultimately, material performance in energy systems.

Achieving maximum efficiency from both materials manufacturing processes and energy systems demands careful process monitoring and information they reveal complement the progress in modeling to result in synergistic acceleration of materials development. The ability to characterize and manipulate materials, particularly at the nanoscale, offers promise of breakthrough discoveries and large-scale use of new materials. Advances in characterization methods such as light, electron, ion, and atomic force microscopy; X-ray, neutron, and other source-based methods; and electron- and ion-beam instruments for microchemistry, structure, and texture identification and surface structure analysis are providing increased understanding of materials at the atomic level. Nanomechanical evaluation methods provide data on the performance of materials at the nanoscale (DOE 2010).

2.3.1 Dynamic Mechanical Spectroscopy

Dynamical Mechanical Spectroscopy (DMS) is one fundamental rheological technique used to characterize the viscoelastic properties of complex fluids and to investigate their microstructure. Complex fluids can be polymeric liquids and melts, suspensions, emulsions, gel, micellar solutions, foams, and composites. All these systems present constitutive elements organized into a precise microstructure, which determines the macroscopic properties of the material. During DMS experiments, the material is subjected to a sinusoidal shear deformation and the corresponding stress response is measured. There are two different type of DMS tests, depending on the amplitude of the applied deformation, namely SAOS (Small Amplitude Oscillatory Shear) and LAOS (Large Amplitude Oscillatory Shear). SAOS is a robust, largely adopted rheology technique in which the strain is small enough to produce a linear response of the material. The stress is proportional to the oscillatory deformation and is represented by a sinusoidal wave with the same frequency. LAOS is a much more recent technique in which the deformation is large enough to determine a nonlinear response. The stress is still periodic, but not sinusoidal and can be deconvoluted in terms of a fundamental harmonic and its odd multiples. Stress is often analyzed in the Fourier space leading to the so-called Fourier Transform Rheology. Nonlinear measurements are, obviously, more complex than linear ones. However, they carry much richer information about the morphology of micro-structured systems (Carotenuto 2007).

2.3.1.1 Instrumentation

The instrumentation of a DMA consists of a displacement sensor such as a linear variable differential transformer, which measures a change in voltage as a result of the instrument probe moving through a magnetic core, a temperature control system or furnace, a drive motor (a linear motor for probe loading which provides load for the applied force), a drive shaft support and guidance system to act as a guide for the force from the motor to the sample, and sample clamps in order to hold the sample being tested. Depending on what is being measured, samples will be prepared and handled differently. A general schematic of the primary components of a DMA instrument is shown in Fig. 2.6 (Menard 1999).

There are different types of dynamic mechanical analyzers, namely the stress-controlled, which more realistically mimic the real-life conditions, the strain-controlled, the axial and the torsional type. In the latter two types, twisting motion can be applied to the sample and additional shear measurements can be performed.

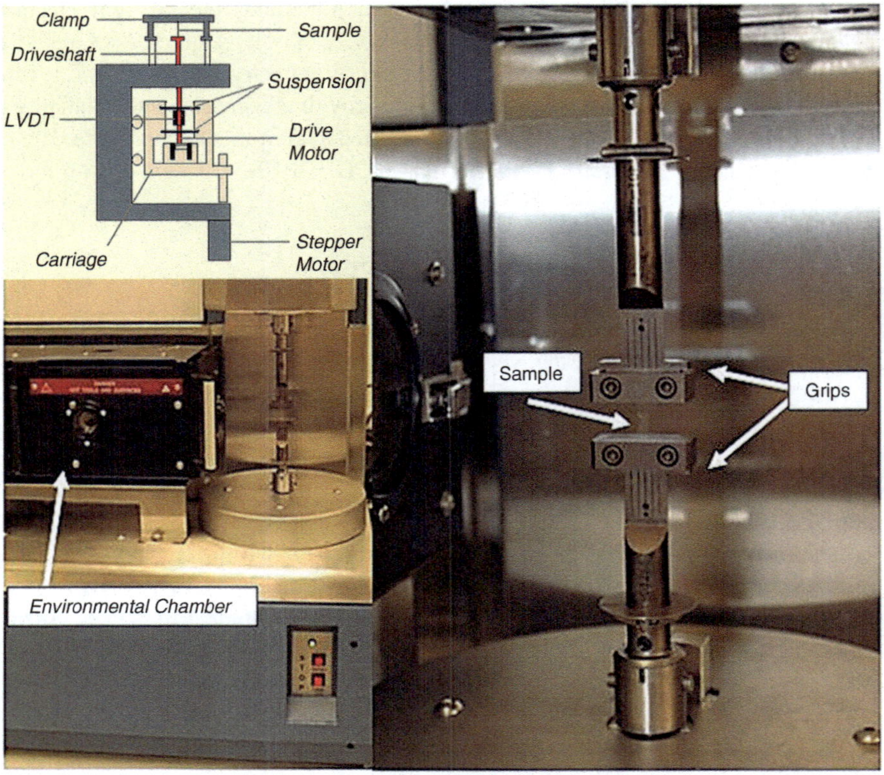

Fig. 2.6 A typical DMA tester with grips to hold sample and environmental chamber to provide different temperature conditions. A sample is mounted on the grips and the environmental chamber can slide over to enclose the sample (Modified with permission from Menard Menard 1999 (Taylor & Francis))

Moreover, there are different fixtures or testing geometries under which the DMS measurements are made. For example, there are three-point and four-point bending modes, the dual and single cantilever modes, parallel plates, the extension or tensile mode, the shear plates, and the cone-and-plate geometry. With the large variety of testing geometries and the different type of dynamic mechanical analyzers available, different forms of sample, be it liquid, rigid rod or thin film, can be tested accordingly. Furthermore, DMS usually does not require a large amount of sample, making it an increasingly popular characterization technique for mechanical testing, especially for nanomaterials (Wee 2004).

2.3.1.2 Linear Dynamic Mechanical Spectroscopy

SAOS measurements have been adopted to characterize linear properties of viscoelastic fluids: from polymer melts to liquids crystalline, from gels to polymer blends. The typical shear stress response of a SAOS test is (Wee 2004):

$$\sigma(t) = \sigma_0 \sin(\omega t + \varphi) \tag{2.1}$$

The stress oscillates with the same frequency ω of the imposed deformation, but it is shifted by an angle $\varphi \in [0, \pi/2]$. The value of φ is equal to 0 for elastic solids, so the stress is in phase with the strain. The value of φ is equal to π/2 for viscous liquids, in this case, the stress is in phase with the shear rate. In the linear regime, the phase angle φ depends upon strain frequency only, and not on the amplitude of the applied strain γ_0. Eq. (2.1) is usually written in the following way (Wee 2004):

$$\sigma(t) = \gamma_0 (G' \sin(\omega t) + G'' \cos(\omega t)) \tag{2.2}$$

where

$$G' = \frac{\sigma_0}{\gamma_0} \cos(\varphi) \tag{2.3}$$

$$G'' = \frac{\sigma_0}{\gamma_0} \sin(\varphi) \tag{2.4}$$

The term G' is in phase with the strain and is called elastic or storage modulus, while the term G'' is in phase with the shear rate and is called viscous or loss modulus. In linear regime the amplitude of the stress is proportional to the amplitude of the imposed deformation, consequently G' and G'' do not depend on γ_0 but on frequency only. For the "liquid-like" fluid, the elastic modulus is much lower than the viscous one and it scales with frequency as $G' \propto \omega^2$, the viscous modulus is linear in frequency, $G'' \propto \omega$. For "solid-like" fluid, $G \gg G''$, and G' is nearly frequency-independent. Real complex fluids often show an intermediate viscoelastic behavior. At low frequencies, the system is "liquid-like," with $G' << G''$, while more nearly "solid-like" behavior, with $G' > G''$, is found at high frequencies.

2.3.1.3 Nonlinear Dynamic Mechanical Spectroscopy

Nonlinear dynamic mechanical spectroscopy is intrinsically more complex than linear one, from both the experimental and the theoretical point of view. Many polymeric systems are processed in the nonlinear regime, therefore, LAOS experiments can give useful information on the material properties just in that range. Furthermore, nonlinear tests can provide a very detailed characterization of complex fluids microstructure. The shear stress response of a LAOS test is still periodic, with the same frequency ω of the imposed oscillation, but not sinusoidal. It can be described by a Fourier series of odd harmonics (Wee 2004):

$$\sigma(t) = \sum_{\substack{k=1 \\ k,\text{odd}}}^{\infty} a_k \cos(k\omega t) + \sum_{\substack{k=1 \\ k,\text{odd}}}^{\infty} b_k \sin(k\omega t) = \sum_{\substack{k=-\infty \\ k,\text{odd}}}^{\infty} I_{k\omega} e^{-jk\omega t} \tag{2.5}$$

where a_k and b_k represent the amplitude of the cosine and sine terms of the kth harmonic and depend on both frequency and strain amplitude. $I_{k\omega}$ is the complex coefficient of the kth harmonic in the Fourier domain, $I_{k\omega} = a_k - jb_k$. It should be remarked that shear stress is an odd function of the strain. They can be calculated as follow (Wee 2004):

$$a_k = \frac{2}{T}\int_0^T \sigma(t)\cos(k\omega t)dt \tag{2.6}$$

$$b_k = \frac{2}{T}\int_0^T \sigma(t)\sin(k\omega t)dt \tag{2.7}$$

$$I_{k\omega} = \frac{2}{T}\int_0^T \sigma(t)e^{-jk\omega t}dt \tag{2.8}$$

where T is the period of the imposed strain oscillation. LAOS technique requires a sophisticated experimental apparatus, an elaborated data manipulation, and a theoretical model able to interpret nonlinear results.

2.3.2 Nanoindentation

Designing energy materials based on the knowledge of mechanics of their building blocks and microstructure manipulations at nanometer scale have become a reality with the advance of micro-level experimental techniques that revolutionize the material science. Nanoindentation, as a leading micro-level mechanical testing technique, has attracted wide attention in numerous research fields and applications. Nowadays, an extensive variety of testing areas ranging from classical thin coatings in machinery engineering, electronics and composites to far fields of civil engineering, biomechanics, implantology or even agriculture can be covered with this universal testing tool. The most common use of nanoindentation is for the measurement of hardness and elastic modulus, and there has been considerable progress in the measurement of other mechanical parameters as well, including hardening exponents, creep parameters, residual stresses, and property extraction. However, nanoindentation also lends itself to more fundamental inquiries in materials science. New capabilities in in situ and ex situ imaging, acoustic emission detection, and high-temperature testing are now being used to probe nanoscale phenomena such as defect nucleation and dynamics, mechanical instabilities or strain localization, and phase transformations (Schuh 2006).

2.3.2.1 Principle of the Nanoindentation Instrument

The principal components in a nanoindentation experiment are the test material, the sensors and actuators used to apply and measure the mechanical load and indenter displacement, and the indenter tip. The latter component is conventionally made of diamond, formed into a sharp, symmetric shape such as the three-sided Berkovich pyramid as shown in Fig. 2.7 (Pharr 2013). The pyramidal shape is chosen at least in part for its nominal geometric self-similarity, which makes for relatively simpler analysis using the methods of continuum mechanics. However, because of the very

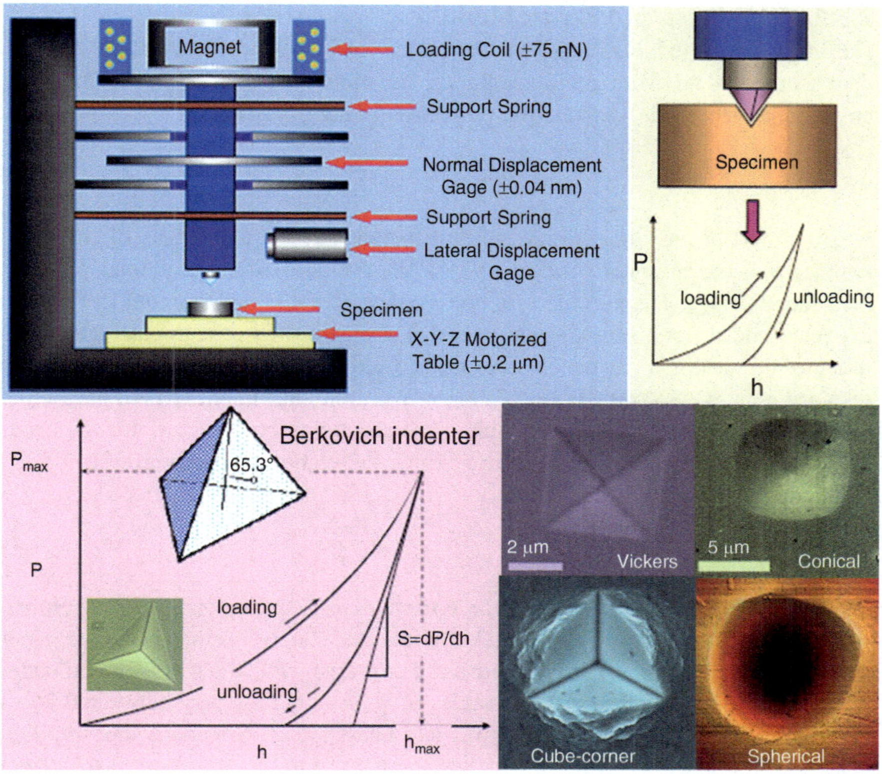

Fig. 2.7 Illustration of nanoindentation with various indenter geometries (Adapted from Pharr Pharr 2013. Credit: Oak Ridge National Lab)

fine scale of nanoindentation testing, imperfections in the pyramidal tip shape are of paramount importance in such analysis, and much effort has been focused upon methods of characterizing and cataloging tip shapes for more exact quantitative measurements. Of particular relevance in this regard is the nature of the tip apex, which is never atomically sharp and exhibits significant blunting. During a typical nanoindentation test, force and displacement are recorded as the indenter tip is pressed into the test material's surface with a prescribed loading and unloading profile. The response of interest is the load-displacement curve, often called the P-h curve. The global shape of the P-h curve differs from one material to the next, and these variations usually reflect different mechanical properties. The local details in the P-h curve, which may signal the operation of discrete physical events beneath the indenter tip. Because of the very small volume of material sampled in a nanoindentation, these events are detected in a discrete fashion in real time and can be isolated and studied in detail (Schuh 2006).

2.3.2.2 Hardness and Elastic Modulus

Hardness and modulus can be determined by proper interpretation of load-displacement data from nanoindentation as shown in Fig. 2.7. The unloading curve follows a power law relation (Oliver and Pharr 1992)

$$P = \rho(h - h_f)^m \tag{2.9}$$

where ρ and m are dimensionless fitting parameters. The stiffness of the elastic contact, S, is the derivative of Eq. (2.9), dP/dh, evaluated at the peak load, P_{max}, and peak depth, h_{max}. Displacement of the surface surrounding the contact region is expected and differs from sink-in, in that it does not affect the contact area. If the indenter tip used is a flat punch, S could be used directly to extrapolate the surface displacement, h_s, as shown in Fig. 2.7. However, for conical indenters or paraboloids of revolution, S must be adjusted with an intercept correction factor, fint, to account for the effects of indenter geometry, yielding the relation (Wheeler 2009)

$$h_{max} - h_s = f_{int}\frac{P_{max}}{S} \tag{2.10}$$

Typical values of f_{int} is 0.72 for cone indenter shape; and 0.75 for sphere, Berkovich, cube corner Vickers and knop indenter shapes. Values of f_{int} are dependent on the indenter shape and the material constants which describe the unloading behavior of the material. However, values of f_{int} usually only vary between 0.74 and 0.79 over the range of materials (Oliver and Pharr 1992). Assuming sink-in/pile-up is negligible, subtracting the surface displacement from the maximum displacement yields the depth of the contact impression $h_c = h_{max} - h_s$. The contact area is dictated by the indenter geometry (Wheeler 2009)

$$A = F(h_c) \tag{2.11}$$

where F is a function relating penetration depth to projected contact area for the indenter. This projected contact area is the cross-sectional area of the indenter at the depth of interest. This is used, instead of the surface area in contact with the material, because it has a physical relationship to the stress applied to the sample. The various common indenter shapes have well-defined functions relating their ideal depth-contact area relationship. However, inevitable variation occurs in the manufacture of indenters, so the experimental determination of the shape function for each indenter is recommended.

Hardness is defined as the load supported by the area of material in contact with the indenter (Wheeler 2009)

$$H = P/A \tag{2.12}$$

This is essentially an empirical ranking exercise, although the hardness value will depend on constitutive properties such as the yield strength and work hardening parameters of the material.

Determining the elastic modulus of the sample requires another extra consideration. The indenter tip, while preferably made of diamond, is not perfectly rigid. It elastically deforms simultaneously with the sample, so account must be taken of this. This reduced modulus can be determined from the sample stiffness and the contact area by Wheeler (2009)

$$E_r = \frac{\sqrt{\pi}}{2B} \frac{S}{\sqrt{A}} \tag{2.13}$$

Here another correction factor, B, is introduced. This factor addresses the axial variation in stress introduced by non-axisymmetric, polygonal, indenter shapes. Values for B (Wheeler 2009): 1 for sphere and cone indenter shapes; 1.05 for Berkovich; 1.034 for cube corner; 1.012 for Vickers and Knoops. For computational simplicity, pyramidal indenters are frequently modeled as cones with an angle which gives them equivalent depth-area functions as the pyramid. Spherical indenter tips manufactured for nano-scale indentation are generally conospheroids, a cone with a rounded tip which has a uniform radius of curvature until it becomes tangent with the cone. This tangent depth is the maximum depth at which the indenter can be accurately analyzed as a sphere, and, for a blunted pyramidal indenter, it is the absolute minimum depth at which the indenter can be analyzed as a pyramid.

2.3.2.3 Incipient Plasticity

Plastic yield on the nanoscale has been investigated with incipient plasticity through the very earliest stages of the mechanical contact, where the transition from elastic to plastic deformation can be observed. When the Berkovich tip is used in the earliest stages of contact at which plastic yield first occurs, the geometry of the nanoindenter can often be approximated as spherical. With this fortuitous geometry, it becomes possible to predict the expected elastic response using the Hertzian law for mechanical contacts, based on isotropic continuum elasticity. This law predicts a simple power-law form for the elastic portion of the load-displacement curve, $P \propto h^{3/2}$, with a proportionality constant that is fully specified by the radius of the blunted indenter tip and the elastic properties of the two contacting materials (Schuh 2006).

With theoretical expectations for the elastic response given by the Hertzian theory, the onset of plastic deformation during nanoindentation can nominally be identified by the first point at which the experimental data deviate from the elastic curve. In a great variety of materials, including metals, alloys, intermetallics, and ceramics, this yield point has been found to occur at a discontinuity in the P-h curve, sometimes referred to as a "pop-in" event. During this event the indenter travels without a measured increase in applied load (for a load-controlled experiment), or the load is rapidly released at a constant displacement (for a displacement-controlled experiment). Indentations performed to lower, subcritical loads, usually exhibit ideal reversibility and leave no trace of a residual impression on the specimen surface, confirming that the first pop-in marks the onset of irreversible flow. Furthermore, the burst-like character of the first pop-in suggests that strain is accommodated by an abrupt avalanche of atomic activity beneath the indenter, such as might be expected for activation of a dislocation source. This has been verified by simulation work and experiments—significant undetectable dislocation activity may precede the first pop-in, which would suggest that the pop-in event is not, in fact, a homogeneous dislocation nucleation event, but rather a heterogeneous process of, for example, dislocation source activation or multiplication. In this case, the nominally "elastic" portion of the P-h curve, which fits the predictions of Hertzian contact theory, apparently would contain some amount of superimposed plastic displacement too subtle to detect. Further quantitative details on the nature of incipient plasticity have also been revealed through the use of high-temperature nanoindentation. Higher temperatures generally lower the load required to initiate yield, which conforms with expectations for a thermally activated deformation mechanism. In the case of a single-crystal Pt surface prepared by electrochemical polishing, such analysis has revealed very low values for both the characteristic energy and volume of the rate-limiting process for yield. This suggests that the first pop-in may be associated with a heterogeneous event, such as the activation of a dislocation source from preexisting surface features or subcritical dislocations (Mason et al. 2006).

Beyond the initial yield point in nanoindentation, additional pop-in events are usually observed at higher loads, which are associated with further dislocation motion, multiplication, and the evolution of a complex defect structure. Such structural evolution is generally quite complicated and varies significantly between materials of different structure, chemistry, and crystallographic orientation. Because of these complexities, the P-h curve can be augmented by additional capabilities, such as a high-temperature testing stage, in situ electron microscopy, or ex situ measurements by scanning probe microscopy or Raman spectroscopy. For example, scanning tunneling microscopy on indented surfaces of Au(100) was used to reveal the mechanisms by which dislocation activity relocates matter around the impression site. The material ejected during indentation "piles up" in a series of geometric terraces reflective of the underlying crystal symmetry, and the termination points of some individual screw dislocations can be discerned. The geometry of such surface traces (and the geometry of the impression itself) can be linked to the dislocation slip systems of the crystal (Carrasco et al. 2004).

2.3.2.4 Mechanical Instabilities

In crystalline metals, the motion and mutual interaction of dislocations gives rise to work hardening mechanisms that encourage stable plastic flow. In contrast, in amorphous metals (or metallic glasses) there are no dislocations per se, and plastic deformation is inherently unstable, occurring in bursts of highly localized strain referred to as "shear banding" events. Nanoindentation testing has become an important tool for fundamental studies of shear banding in metallic glasses, owing to its ability to resolve individual shear events under well-controlled conditions (Schuh 2006).

Quantitative analysis of the nanoindentation P-h data has led to significant insight into the process of shear localization in metallic glasses. For example, a collection of high-temperature experiments was used to identify a "critical" submicron-scale size for shear banding, below which strain is localizing and the shear band is growing, and above which strain has completely localized into the now rapidly propagating shear band. In complementary work combining indentation with ex situ TEM observations, deformation-induced structural evolution and nanocrystallization events have been studied in metallic glasses. These studies represent initial efforts in the exploration of a very complex topic, and there is not yet broad agreement as to the mechanisms of shear localization in metallic glasses, or how the glass structure and structural evolution impacts the deformation. Nonetheless, these kinds of studies clearly represent an avenue for significant scientific advances in the future (Schuh 2006).

Shear banding in amorphous metals is just one example of a mechanical instability that can be studied using nanoindentation, and similar instabilities in other materials are amenable to the same kinds of experimental approaches. A good example of this is the so-called "jerky flow" of solid solution alloys. Depth-sensing indentation P-h curves have been obtained with various degrees of flow serration beneath the indenter. The instability that gives rise to discontinuities in this case results from the interaction of lattice dislocations with solute atoms, which leads to instantaneous negative strain rate sensitivity and localization of flow. These kinds of data have been used to study the statistics of flow serration, as well as the influence of solute concentration on the instability (Chinh et al. 2004).

2.3.2.5 Phase Transformations

Many materials undergo phase transformations when subjected to large hydrostatic stresses, and the pressure beneath a nanoindenter is generally quite high (on the order of several gigapascals). In the case of some diamond-cubic semiconductors—including Si and Ge—the mean contact pressure of hardness indentations closely matches the critical pressure to trigger a structural transformation (Schuh 2006).

For Si for example, when sharp indentation impressions are examined ex situ, thin sheets of material extruded around the periphery of the indenter can be observed. This extensive plastic flow has been interpreted as evidence for the transformation from diamond cubic Si to the metallic (and malleable) β-Sn phase under the pressure of the indentation. More direct evidence for a semiconducting/metallic transition under the indenter has also been provided by in situ electrical measurements. The pop-out phenomenon observed during unloading is attributed to the reversion of the high-pressure β-Sn phase to metastable rhombohedral or body-centered cubic phases. The transformation carries an attendant increase in volume that can do mechanical work against the indenter tip, producing the pop-out effect. Deviatoric stresses or plastic deformation may impact the evolution of metastable phases under the indenter. Thus, nanoindentation may offer the possibility of studying nonequilibrium phase transformations (so-called "driven" transformations) with new levels of consistency and control (Tachi et al. 2002; Schuh 2006).

2.3.2.6 MEMS Nanoidentification

Microdevices such as micro-electro-mechanical systems (MEMS) has been developed and used for various applications that impact energy systems and many industrial sectors including automotive, consumer electronics, telecommunication, aerospace, and medical. MEMS are usually composed of micrometer-sized mechanical structures, such as suspended bridges, cantilevers, membranes, and fluid channels, and often integrated with analog and digital circuitry. MEMS can act as sensors, receiving information from their environment, or as actuators, responding to a decision from the control system to change the environment. The majority of MEMS products include accelerometers, pressure and chemical flow sensors, optical MEMS, micromirrors, gyroscopes, fluid pumps, magnetometers, and inkjet print heads. MEMS can undergo early failure during the operation due to mechanical stresses, which can be induced by a single high stress or cyclic low stresses compared with the strength of the component. An accurate knowledge of the mechanical properties of micro- and nanomaterials, especially thin films, which form mechanical structures of MEMS systems, is necessary to minimize or virtually eliminate failures. Nanoindentation has emerged as a leading technique for the investigation of mechanical properties on small volumes of material. Mechanical testing of MEMS using nanoindentation is related to the bending tests of samples such as cantilever and bridge beams. The purpose is to extract the elastic and plastic characteristics as well as the failure mechanism of the mobile parts of the MEMS (Diop 2012).

Common nano-indentation instruments/systems feature, in general, high resolution for generating the indentation force and for measuring the indentation depth, especially when the specimens under test are of hard materials. On the other hand, they also share common shortcomings, such as high cost, difficulties for on-site applications or for applications to test large work pieces, and so on. All these disadvantages strongly restrict the potential widespread application of the current commercially available nano-indentation systems. In the meantime, noting that almost all commonly used nano-indentation instruments are "soft" force controlled machines, i.e., the indentation force is applied in such a way that small changes in displacement should not change the force significantly, therefore, nanoindentation instruments often fail to measure the mechanical properties of materials which demonstrate heavy creep, plasticity, etc., since the indentation depth in these cases (under the force control cycles) grows quickly and thereby the applied indentation force cannot be maintained any more. Under consideration that more and more soft materials, such as various biological specimens and organic materials, are now being required to measure their mechanical properties with the nano-indentation method, this disadvantage of the current nano-indentation instruments becomes more and more unacceptable. In order to overcome the disadvantages of nano-indentation instruments, it has been more and more demanded to further develop the nano-indenters. Well-developed MEMS technique can be employed to develop the next generation of nano-indentation systems. Micro-electro-mechanical systems (MEMS) enable suspended microstructures to be driven precisely for linear or angular motion. Various actuation mechanisms, such as thermal/bimetallic bimorph, electromagnetic, piezoelectric and electrostatic actuation, have been established and applied fundamentally in MEMS-based devices where mechanical actuation is required. Among them electrostatic actuators form the most ubiquitous MEMS family, having a force output lying mid-range at 10^{-6} N–10^{-3} N and offering relatively large displacements with adequate resolution. The comb-drive-type electrostatic microactuator might be one of the most important electrostatic MEMS actuators, which has in general a large number of fine interdigitated fingers to generate the actuated force. MEMS-based micro miniature nanoindentation instrument has been developed, which features high force resolution (down to 1 nN) and a force range up to 1 mN. This device is prepared to incorporate an external indenter by means of post-assembly. Similar to other MEMS products/systems, this microminiature indentation system has the following advantages, including small size, low power consumption, high precision in manufacture, the potential for low cost through batch fabrication, and the ability for on-site applications (Li et al. 2007).

2.3.3 Optical Microscopy

Optical microscopy, or light microscopy, refers to the sample inspection with a type of microscope which uses visible light and a system of lenses to magnify images of small samples. A basic optical microscope has the following parts (Tong 2011): (a) a lamp to illuminate the specimen; (b) a nose piece to hold 4–5 objectives used in changing the viewing magnification; (c) an aperture diaphragm to adjust the resolution and contrast; (d) a field diaphragm to adjust the field of view; (e) an eye piece to magnify the objective image (usually by 10×); and (f) a stage for manipulating the specimen. During optical microscope inspection, the specimen is positioned perpendicularly to the axis of the objective lens. Light is then shown on the sample, which reflects some light back to the lens. The image seen in the microscope depends not only on how the specimen is illuminated and positioned, but on the characteristics of the specimen as well. Optical microscopes are commonly classified as either low-power or high-power microscopes. Low-power microscopes are those which typically magnify the specimen at 5× to 60×, although some can magnify up to 100×. High-power microscopes, on the other hand, typically magnify the specimen at 100× to 1000×.

There are three modes by which optical microscopy is commonly conducted, namely, bright field illumination, dark field illumination, and interference contrast. Bright field illumination is the normal mode of viewing with an optical microscope. This mode provides the most uniform illumination of the sample. Under this mode, a full cone of light is focused by the objective on the sample. The image observed results from the various levels of reflectivities exhibited by the compositional and topographical differences on the surface of the sample. Under darkfield illumination, the inner circle area of the light cone is blocked, such that the sample is only illuminated by light that impinges on its surface at a glancing angle. This scattered reflected light usually comes from feature edges, particulates, and other irregularities on the sample surface. Dark field illumination is therefore effective in detecting surface scratches and contamination. Interference contrast makes use of polarized light that is divided by a Wollaston prism into two orthogonal light packets. These slightly displaced light packets hit the specimen at two different points and return to the prism through different paths. The differences in the routes of the reflected packets will produce interference contrasts in the image when the packets are recombined by the prism upon their return. Surface defects or features such as etch pits and cracks that are difficult to see under bright field illumination can stand out clearly under Noma ski mode (Tong 2011).

The stereo microscope is designed differently from the compound optical microscopes, and serves a different purpose. It uses two separate optical paths with two objectives and two eyepieces to provide slightly different viewing angles to the left and right eyes, which produces a three-dimensional visualization of the sample being examined. Unlike compound microscopes, illumination in a stereo microscope most often uses reflected (episcopic) illumination rather than transmitted (diascopic) illumination, that is, light reflected from the surface of an object rather than light transmitted through an object. Use of reflected light from the object allows examination of specimens that would be too thick or otherwise opaque for compound microscopy. However, stereo microscopes are also capable of transmitted light illumination as well, typically by having a bulb or mirror beneath a transparent stage underneath the object, though unlike a compound microscope, transmitted illumination is not focused through a condenser in most systems. Stereoscopes with specially equipped illuminators can be used for dark field microscopy, using either reflected or transmitted light (Tong 2011).

Various video dual CCD camera pickups have been fitted to stereo microscopes, allowing the images to be displayed on a high-resolution LCD monitor. Software converts the two images to an integrated Anachrome 3D image, for viewing with plastic red/cyan glasses, or to the cross converged process for clear glasses and somewhat better color accuracy. The results are viewable by a group wearing the glasses. These files may be recorded as well.

In addition, digital microscope with built-in digital camera, functions as a standard microscope, but when connected to the computer, the included software allows the operator to show the microscope image on a screen, capture and save images and video, as well as make measurements.

2.3.4 Electron Microscopy

Spurred on by the enormous versatility of electron microscopy, steady improvement in spatial resolution and a variety of spectroscopy tools have been achieved. Widely available capabilities include energy-filtered electron energy loss spectroscopy (EELS) and sub-nanometer diameter probe energy dispersive spectroscopy that can provide detailed information on the chemical segregation of solute to interfaces and the composition of precipitates, along with a broad range of electron diffraction conditions that provide crystallographic space group and lattice parameter information. These capabilities are routinely being used to provide detailed characterization on advanced materials of interest for energy systems, such as nanocomposited oxide dispersion strengthened ferritic alloys and vanadium alloys (Zinkle et al. 2009).

Whereas optical microscopes have been operating near the resolution limit set by the wavelength of light for over 100 years, the achievable resolution of electron microscopes was limited to more than 50 times the electron wavelength through the end of the twentieth century. This resolution breakthrough has been enabled by the successful correction of the dominant aberrations present in electron lenses. It has also been aided by placing the microscopes in specialized rooms isolated from humans, vibration, electromagnetic fields and thermal gradients. Resolution has now crossed well into the sub-Ångstrom regime. The benefits are much greater, however, than just the ability to resolve smaller atomic distances. It is now possible to image individual heavy atoms on surfaces and inside a bulk material, even to perform a spectroscopic identification of a single atom. These instrumental advances allow new insights into long-standing issues in materials research for solving energy problems (Pennycook et al. 2009).

The electron microscope uses electrostatic and electromagnetic lenses to control the electron beam and focus it to form an image. These electron optical lenses are analogous to the glass lenses of a light optical microscope. There are several major types of electron microscope, such as: Transmission Electron Microscope (TEM), Reflection Electron Microscope (REM), Scanning Electron Microscope (SEM), Scanning Transmission Electron Microscope (STEM), and others. The main objective lens in an electron microscope is a round lens, which has an intrinsically high spherical aberration, of the order of 50 wavelengths compared to the roughly one wavelength typical of a light optical lens. There are presently two designs of aberration corrector available for electron microscopes, a quadrupole/octopole design exclusively used in scanning transmission electron microscopy (STEM) and a hexapole design used in both STEM and transmission electron microscopy (TEM). These correctors are able to shape the electron wave front to a degree of perfection better than a quarter wavelength (~0.5 pm) over 70 μm, a level of performance that exceeds that of the Hubble Space Telescope. However, the correctors only compensate for the geometric aberrations of electron lenses, not for any chromatic aberration, which is the focusing of electrons of different energies at different points. Chromatic aberration correction has been successfully demonstrated so far only for scanning electron microscopy (SEM) which operates at significantly lower accelerating voltages. Efforts are currently underway to extend this correction to the higher accelerating voltages normally needed for TEM. The normal STEM imaging mode uses an annular dark field (ADF) detector that collects a large fraction of the scattered electrons. If the inner detector angle is sufficiently high, the scattered intensity varies approximately as Z^2, where Z is the atomic number. This mode of operation provides a High Angle ADF image, commonly referred to as a Z-contrast image. One of the key advantages of STEM has always been the ability to have simultaneous detection of a variety of signals, for example, simultaneous Z-contrast and bright field images that allow pixel to pixel correlation. Replacing the bright field detector with an electron spectrometer, simultaneous Z-contrast imaging and electron energy loss spectroscopy becomes possible. Aberration correction has enabled dramatic gains in sensitivity due to the availability of smaller probes containing the same current. More current can channel down an atom column of interest with less wasted illumination of neighboring columns. This has allowed the spectroscopic identification of a single atom inside a bulk material, true two-dimensional spectroscopic maps with atomic resolution. In addition, just as with X-ray absorption spectroscopy, the fine structure at an absorption edge gives an indication of the local electronic structure (Kimoto et al. 2007).

One of the advantages of aberration correction has been the greatly reduced depth of field, which opens the possibility for three-dimensional (3D) imaging. Similar to a camera aperture, the depth of field decreases as the inverse square of the aperture angle. Although the aperture angles are still quite small by light optical standards, the typical depth of field has reduced dramatically as result of aberration correction, becoming less than the thickness of a typical TEM specimen. Therefore, electron microscopy no longer provides the simple two-dimensional projection that it did in the past. Now, a through focal series becomes a through depth series of images, providing three-dimensional information on the specimen with depth resolution at the scale of a few nanometers and atomic resolution laterally (Hashimoto et al. 2009).

The next generation of aberration correctors is beginning to appear, correcting all geometric aberrations up to fifth order. Another jump in resolution is to the level of 0.5 Å, with a concomitant increase in single atom sensitivity and depth resolution. It should become possible to probe single impurity atoms at grain boundaries and dislocation cores, to probe their electronic environment and link it to macroscopic mechanical and electronic properties in a rigorous manner. With these new eyes, it should finally become possible to see the ultimate atomic origins of materials properties. Another important development is the area of in situ observation stages. It is now routinely possible to obtain high-resolution images of nanoscale defect cluster motion during straining in an electron microscope at a wide range of temperatures. There is considerable activity underway to develop improved specimen stages that will enable high-resolution imaging to be obtained during exposure of the sample to gaseous or liquid environments at a range of temperatures and pressures (Zinkle et al. 2009).

2.3.5 Atom Probe Tomography

Atom probes are unlike conventional optical or electron microscopes, in that the magnification effect comes from the magnification provided by a highly curved electric field, rather than by the manipulation of radiation paths. Technically, the method is destructive in nature removing ions from a sample surface in order to image and identify them, generating magnifications sufficient to observe individual atoms as they are removed from the sample surface. Through coupling of this magnification method with time of flight mass spectrometry, ions evaporated by application of electric pulses can have their mass-to-charge ratio computed. Through successive evaporation of material, layers of atoms are removed from a specimen, allowing for probing not only of the surface, but also through the material itself. Computer methods are utilized to rebuild a three-dimensional view of the sample, prior to it being evaporated, providing atomic scale information on the structure of a sample, as well as providing the type atomic species information. The instrument allows the three-dimensional reconstruction of up to hundreds of millions of atoms from a sharp tip (corresponding to specimen volumes of 10,000–1000,000 nm^3) (Miller 2000).

In the world of tomographic imaging, atom probe tomography (APT) occupies the high-spatial-resolution end of the spectrum. It is highly complementary to electron tomography and is applicable to a wide range of materials. The use of a replaceable 20–50 μm-diameter local electrode in close proximity to the apex of the needle-shaped specimen enables the use of lower voltages to field evaporation atoms from the specimen. Lower voltage permits the pulse repetition rate of the field evaporation pulse to be increased from 1 to 2 kHz to over 200 kHz to dramatically reduce the data acquisition time. The local electrode atom probe is also equipped with a wide area crossed delay line type of single atom, position-sensitive detector to increase the field of view. Typical datasets contain between 5 and 100 million atoms. Focused-ion-beam-based specimen preparation methods enable atom probe specimens to be fabricated from site specific locations, a wider range of materials, and may also reduce the volume and hence the activity of radioactive samples. Although primarily driven by the need to quantify dopant profiles in semiconductors, the reintroduction of laser-assisted field evaporation with more reliable pulsed picosecond and femtosecond lasers enables low electrical conductivity materials, including SiC and zirconium alloys, to be characterized. In addition, laser-assisted field evaporation has benefits for the analysis of brittle materials due to the lower stress that is applied to the specimen. Moreover, the incorporation of a wide acceptance angle reflection in the mass spectrometer improves the mass resolution so that the base of the mass peaks of all the isotopes can be separated in most materials. This higher resolution reduces the complexity of peak deconvolution and background noise subtraction for more reliable concentration measurements. In addition, significant advances have been made in the quantification of three-dimensional atom probe data. In particular, friends-of-friends methods have been developed to quantify the size, number density and composition of nanoscale clusters, precipitates and solute segregation to dislocations, and the isoconcentration-based proximity histogram method has been developed to estimate solute profiles across interfaces and grain boundaries (Miller and Russell 2007).

2.3.6 Advanced X-ray Characterization

X-ray characterization of materials is revolutionized by intense synchrotron sources with orders-of-magnitude greater flux and brilliance (photons/s/μm^2/mrad^2) than conventional sources. In the past decades, achievable X-ray source brilliance has increased by approximately 14 orders of magnitude. As brilliance is the figure-of-merit for most characterization methods, this has enabled powerful new techniques that were previously impractical. For example, the energy tenability of X-ray beams from early (first-generation) synchrotron sources fostered the development of absorption-fine-structure methods that can determine the local chemistry, near-neighbor atomic coordination and bond distances for dilute elements in crystalline or amorphous materials. Similarly, the availability of intense and highly collimated radiation from second-generation synchrotron sources enabled new surface diffraction and microfluorescence probes. The stability of second-generation sources also allowed for widespread use of X-ray scattering contrast altering methods and/or near-resonance and absorption contrast techniques that highlight specific atoms in absorption or scattering (Chang 2004).

The success of high-brilliance X-ray sources has also stimulated a renaissance in X-ray optics and detectors. These developing X-ray tools leveraged with vastly more powerful sources provide even more characterization opportunities. In the last decade, the achievable focal-spot area has decreased by more than two orders of magnitude; ~2.5×10^5 nm^2 to ~600 nm^2. Small diameter X-ray beams allow for spatially resolved measurements with resolution sufficient to resolve inhomogeneities in most real materials. Coupled with intense sources that provide sufficient flux into small phase-space volumes, it is possible to perform experiments on submicron samples that once required mm or cm sized samples. Similarly, fast and efficient area detectors have emerged that parallelize data collection for fast data acquisition and reduced sample damage. These detectors can be used to make millisecond to femtosecond measurements, and integrate spectroscopy into area detection for improved experimental signal-to-noise and additional new characterization opportunities (Zinkle et al. 2009).

Although powerful characterization methods based on synchrotron sources are already deployed at facilities around the world, there have been three directions of intense activity that extend opportunities for materials characterization. These areas can be summarized as (Zinkle et al. 2009): (a) spatially resolved methods for 3D mapping of materials properties; (b) phase contrast and coherent imaging methods for the study of structures in thick samples and for nm resolution of 3D structures, and (c) high-energy experiments that enable nondestructive measurements of local strain, texture, and phase distributions in cm thick samples and for precision analysis of point and other defects. All three of these methods are interrelated and ultimately based on a need to understand how real materials are organized in three dimensions and at different length scales. Polychromatic microdiffraction, which is one important example of spatially resolved methods, allows for nondestructive characterization of local phase, crystal orientation, elastic strain, and dislocation tensor distributions in 3D with submicron resolution. Together with microfluorescence, it allows people to nondestructively map chemistry and local crystal structures to provide new insights into materials behavior. In the development of advanced nuclear materials, there are a number of situations with an obvious need for advanced X-ray characterization methods. For example, at the interface of dissimilar materials, spatially resolved measurements of phase and elemental distributions can provide important new information about interface phase evolution. Similarly, measurements of damage and defect evolution in irradiated materials can be carried out on polycrystalline samples with μm-sized grains. This will enable studies of real materials and the role of grain boundaries, crystalline orientation and other inhomogeneities in defect organization and will simplify measurements by using small-volume samples with orders of magnitude lower activity than traditional specimens. In addition, studies of mesoscale structure evolution during processing will guide the development of new grain boundary engineered and other high-performance materials. Ultimately, emerging real-time, high-resolution images of 3D materials properties will enable a new kind of materials exploration where people can move through a virtual sample with critical materials parameters encoded in the visualization. As advanced characterization techniques provide detailed tensor information in three dimensions, fundamentally new ways of visualizing data will be required to take full advantage of this emerging technology.

2.3.7 Neutron Scattering

One of the unique features of the neutron scattering technique is the deep penetration compared to electrons and X-rays. For laboratory- sized samples, the measurements are therefore representative of the bulk rather than from surface areas. Moreover, sample environments can be readily implemented to allow in situ examination of structure evolution under extreme conditions. Because neutrons can penetrate deep inside most materials, a volumetric spatial mapping technique has been developed as a nondestructive means to determine residual stress or damage inside industrial-sized components (Zinkle et al. 2009).

Neutrons have wavelengths on the order of 1–10 Å, which is ideally suited for characterizing the structure of materials. Local atomic structures (e.g., atomic positions within a crystal lattice) are probed by wide-angle diffraction, whereas nano-scale microstructures are determined by small-angle neutron scattering (SANS) in the forward direction. The latter has proven to be particularly useful in the study of precipitation or phase transformation in alloys, such as in situ observation of the formation of nanocrystalline phases in multicomponent metallic glass and investigation of nanoscale solute clusters in unirradiated and irradiated steels. Three-dimensional residual stress mapping is now routine and has made a significant impact on industrial applications. While much of the work was strongly oriented towards mechanical engineering, involving the determination of the residual stress distribution and the use of residual stress data in design and life prediction, opportunities for fundamental research began to emerge when it became evident that some of the experimental data could not be understood within the framework of the continuum theory and simple thermal-mechanical simulations. For example, in situ measurements of the grain-orientation-dependent intergranular strains during uniaxial loading have been extensively used to identify the deformation mechanisms in polycrystalline materials. This type of measurement has been extended to materials subject to cyclic loading to study the damage mechanism by fatigue. In addition, high flux will make it possible to carry out in situ time-resolved measurements to study transient behavior. Simultaneous wide-angle diffraction and small-angle scattering will be particularly useful to examine structure evolution at multiple length scales during exposure to mechanical stress or chemical environments. In situ wide-angle and small-angle scattering measurements of phase transformation kinetics at local and nano-length scales are expected to shed light on classic problems in materials science, such as the mechanism of precipitate nucleation and growth, particularly in the early stage of phase transformation. On the application side, time-resolved measurements will allow real-time study of synthesis or processing (Wang et al. 2006).

2.4 Characterization of Physical and Chemical Properties

Physical properties are those that can be observed without changing the composition/identity of the substance. The general properties of matter such as color, density, hardness, conductivity, temperature are examples of physical properties. Properties that describe how a substance changes into a completely different substance are called chemical properties. Flammability and corrosion/oxidation resistance are examples of chemical properties.

2.4.1 Phase Transformation Temperatures

When temperature rises and pressure is held constant, a typical substance changes from solid to liquid and then to vapor or even plasma. Transitions from solid to liquid, from liquid to vapor, from vapor to solid and vice versa, are called phase transformations or transitions. Since some substances have several crystal forms, technically there can also be solid to another solid form phase transformation.

Phase transitions from solid to liquid, and from liquid to vapor absorb heat. The phase transition temperature where a solid changes to a liquid is called the **melting point**. The temperature at which the vapor pressure of a liquid equals 1 atm (101.3 kPa) is called the **boiling point**. Some materials, such as many polymers, do not go simply from a solid to a liquid with increasing temperature. Instead, at some temperature below the melting point, they start to lose their crystalline structure but the molecules remain linked in chains, which results in a soft and pliable material. The temperature at which a solid, glassy material begins to soften and flow is called the **glass transition temperature** (NDT 2014).

2.4.2 Density and Specific Gravity

Mass (m) is a fundamental measure of the amount of matter. Weight (w) is a measure of the force exerted by a mass and this force is force is produced by the acceleration of gravity. Therefore, on the surface of the earth, the mass of an object is determined by dividing the weight of an object by 9.8 m/s^2 (the acceleration of gravity on the surface of the earth). The density (d) of a material depends on the phase it is in and the temperature. (The density of liquids and gases is very temperature dependent.) Water in the liquid state has a density of 1 g/cm^3 = 1000 kg/m^3 at 4 °C. Ice has a density of 0.917 g/cm^3 at 0 °C, and this decrease in density for the solid phase is unusual. For almost all other substances, the density of the solid phase is greater than that of the liquid phase. Water vapor (vapor saturated air) has a density of 0.051 g/cm^3 (NDT 2014).

Specific gravity is the ratio of density of a substance compared to the density of fresh water at 4 °C (39 °F). At this temperature the density of water is at its greatest value and equal 1 g/cm^3. Since specific gravity is a ratio, so it has no units.

2.4.3 Thermal Conductivity and Thermal Expansion

Thermal conductivity (λ) is the intrinsic property of a material which relates its ability to conduct heat. Conductive heat flow occurs in the direction of decreasing temperature because higher temperature equates to higher molecular energy or more molecular movement. Energy is transferred from the more energetic to the less energetic molecules when neighboring molecules collide. Thermal conductivity is defined as the quantity of heat (Q) transmitted through a unit thickness (L) in a direction normal to a surface of unit area (A) due to a unit temperature gradient (ΔT) under steady state conditions and when the heat transfer is dependent only on the temperature gradient: $\lambda = Q \times L/(A \times \Delta T)$.

When heat is added to most materials, the average amplitude of the atoms' vibrating within the material increases. This, in turn, increases the separation between the atoms causing the material to expand. If the material does not go through a phase change, the expansion can be easily related to the temperature change. The linear coefficient of thermal expansion (α) describes the relative change in length of a material per degree temperature change: $\alpha = \Delta L/(L \times \Delta T)$, where ΔL is the ratio of change in length, L is the total starting length, and ΔT is the change in temperature.

2.4.4 Electrical Conductivity and Resistivity

Electrical conductivity is a measure of how well a material accommodates the movement of an electric charge. It is the ratio of the current density to the electric field strength. Its SI derived unit is the Siemens per meter, but conductivity values are often reported as percent IACS. IACS is an acronym for International Annealed Copper Standard, which was established by the 1913 International Electrochemical Commission. The conductivity of the annealed copper (5.8001×10^7 S/m) is defined to be 100% IACS at 20 °C. All other conductivity values are related back to this conductivity of annealed copper. Therefore, iron with a conductivity value of 1.04×10^7 S/m, has a conductivity of approximately 18% of that of annealed copper and this is reported as 18% IACS.

Electrical resistivity is the reciprocal of conductivity, which is the opposition of a body or substance to the flow of electrical current through it, resulting in a change of electrical energy into heat, light, or other forms of energy. The amount of resistance depends on the type of material. Materials with low resistivity are good conductors of electricity and materials with high resistivity are good insulators.

The conductivity and resistivity of material is temperature dependent. The conductivity of most materials decreases as temperature increases. Alternately, the resistivity of most material increases with increasing temperature.

2.4.5 Permittivity and Permeability

Permittivity and permeability are two different measures used in electromagnetism. Permittivity measures the ability of a material to store energy within the material. Permeability, on the other hand, is a measure of the ability of a material to support the formation of a magnetic field within the material. The permittivity of a material is related to the polarization of the material whereas the permeability of a material related to the magnetization of the material. Therefore, permittivity and permeability have very different and particular meanings in electromagnetism.

Specifically, the permittivity of material is a measure of the capability of the material to support the formation of an electric field within the material in response to an external electric field. It is commonly denoted by the symbol ε.

The permittivity of free space, also known as vacuum permittivity or electric constant, is usually denoted by the symbol ε_0. Its value is 8.85×10^{-12} Fm^{-1}. The permittivity of a homogeneous isotropic material is equal to the ratio of the electric displacement field to the electric field. It can be expressed as $\varepsilon = D/E$, where D is the electric displacement field. The permittivity of a material depends on several factors such as frequency of the applied electric field, temperature, humidity, and strength of the applied electric field. It has a complex relationship with the frequency of the applied electric field. The static permittivity of a material is a special case, which is the permittivity of a material under the influence of a static electric field. Usually, the permittivity of a material is expressed as a relative permittivity, which is a dimensionless quantity. Relative permittivity, also known as the dielectric constant, is the ratio of the absolute permittivity of a material to the vacuum permittivity: $\varepsilon_r = \varepsilon/\varepsilon_0$. As a result, the relative permittivity of free space is equal to 1. Usually, the materials having higher values of permittivity are highly polarizable. The higher the permittivity of a medium, the more energy is stored in the medium. Therefore, high-permittivity materials are used as dielectric materials in capacitors.

In electromagnetism, the magnetic permeability of a material is a measure of the capability of the material to support the formation of a magnetic field within the material in response to an external magnetic field. In general, the permeability of a material depends on several factors such as the temperature, magnetic field strength, humidity, and the frequency of the magnetic field. The permeability of a material is usually denoted by the symbol μ and is equal to the ratio of the magnetic flux density to the magnetic field strength: $\mu = B/H$. The permeability of free space, also known as the permeability constant, vacuum permeability or magnetic constant of free space, is usually denoted by the symbol μ_0. Its value is $4\pi \times 10^{-7}$ Hm^{-1}. The ratio of the permeability of a given medium to the permeability of free space is known as relative permeability, which can be expressed as $\mu_r = \mu/\mu_0$. The relative permeability of free space is 1. Usually, the permeability of a material is expressed as a relative permeability. The relative permeability of a paramagnetic material is slightly higher than 1. The relative permeability of a diamagnetic material, on the other hand, is slightly less than 1. There is another type of magnetic materials called ferromagnetic materials. The relative permeability of a ferromagnetic material is notably higher than 1. For example, a material with high magnetic permeability is usually used for designing transformer cores and inductors.

2.4.6 Corrosion and Oxidation

Corrosion and oxidation are both similar processes that can happen under natural or forced conditions, but there is a difference between corrosion and oxidation processes. Both processes can be accelerated using external factors; corrosion rate can be increased by wet atmospheric conditions and oxidation rate can be increased using selective catalysts. Corrosion can be considered as a part of oxidation process; it is actually one of the most disastrous disadvantages of oxidation. A key difference between corrosion and oxidation is, corrosion mostly happens in metals and metallic materials, but oxidation occurs in many materials including living and nonliving substances. Clearly oxygen leads to oxidation, while corrosion is definitely the term employed in a comparable electrochemical method brought on by a number of other atoms and also molecules.

Corrosion is a natural process which degrades the useful properties of a material such as strength, structure, appearance, and permeability. This mainly occurs in metals, but it can also take place in ceramics and certain polymers. Corrosion starts when metals or metallic materials are exposed to the atmosphere and aqueous environments. Some corrosion processes are controlled by itself forming a protective layer on the surface; however, in some cases it completely destroys the original material. But, several precautions available can be taken to avoid or control this problem. In aluminum metal, the coverage associated with aluminum oxide is incredibly tough and also heavy which usually shields the particular metal surface area from the, later on, attack of air and surrounding. Many architectural metals corrode basically through contact with wetness, humidity, and moisture in air and atmosphere; however, the method could be highly impacted by the experience of particular substances. Corrosion could be focused locally to make a hole or maybe split, or maybe it could possibly prolong throughout a broad region pretty much evenly corroding the outer lining and surface of the metal. Due to the fact deterioration can be described as a diffusion-controlled method, it happens to uncovered materials. Because of this, techniques to lessen the exercise for the uncovered surface area, for example, passivation and chromate transformation, can easily improve a material's deterioration level of resistance. Nonetheless, a number of rust mechanisms are usually much less noticeable and much less foreseeable.

Oxidation is an electrochemical reaction between Oxygen molecules and some other substances it may contact, including metals and living tissues. It is the method in which electrons tend to be attracted aside simply by free oxygen molecules that happen to be comparatively volatile and searching for accessible electrons.

Theoretically, nevertheless, together with the breakthrough of electrons, oxidation came into existence a lot more specifically thought as the losing of a minimum of one electron when several materials interact. These materials might or might not consist of oxygen. At times, oxidation is beneficial, like for example the development involving super-durable anodized aluminum. Sometimes, oxidation may be damaging, like the rusting of the automobile or even the spoiling involving clean fresh fruit.

2.5 Characterization of Mechanical Properties

2.5.1 Loading

The application of a force to an object is known as loading. Materials can be subjected to many different loading scenarios and a material's performance is dependent on the loading conditions. There are five fundamental loading conditions; tension, compression, bending, shear, and torsion. Tension is the type of loading in which the two sections of material on either side of a plane tend to be pulled apart or elongated. Compression is the reverse of tensile loading and involves pressing the material together. Loading by bending involves applying a load in a manner that causes a material to curve and results in compressing the material on one side and stretching it on the other. Shear involves applying a load parallel to a plane which caused the material on one side of the plane to want to slide across the material on the other side of the plane. Torsion is the application of a force that causes twisting in a material. If a material is subjected to a constant force, it is called static loading. If the loading of the material is not constant but instead fluctuates, it is called dynamic or cyclic loading. The way a material is loaded greatly affects its mechanical properties and largely determines how, or if, a component will fail; and whether it will show warning signs before failure actually occurs (NDT 2014).

2.5.2 Tensile Properties

Tensile properties indicate how the material will react to forces being applied in tension. A tensile test is a fundamental mechanical test where a carefully prepared specimen is loaded in a very controlled manner while measuring the applied load and the elongation of the specimen over some distance. Tensile tests are used to determine the modulus of elasticity, elastic limit, elongation, proportional limit, reduction in area, tensile strength, yield point, yield strength and other tensile properties.

The main product of a tensile test is a load versus elongation curve which is then converted into a stress versus strain curve, as shown in Fig. 2.8. The stress and strain initially increase with a linear relationship, which indicates that no plastic deformation has occurred. In this region of the curve, when the stress is reduced, the material will return to its original shape; the line obeys the relationship defined as Hooke's Law where the ratio of stress to strain is a constant. The slope of the line in this

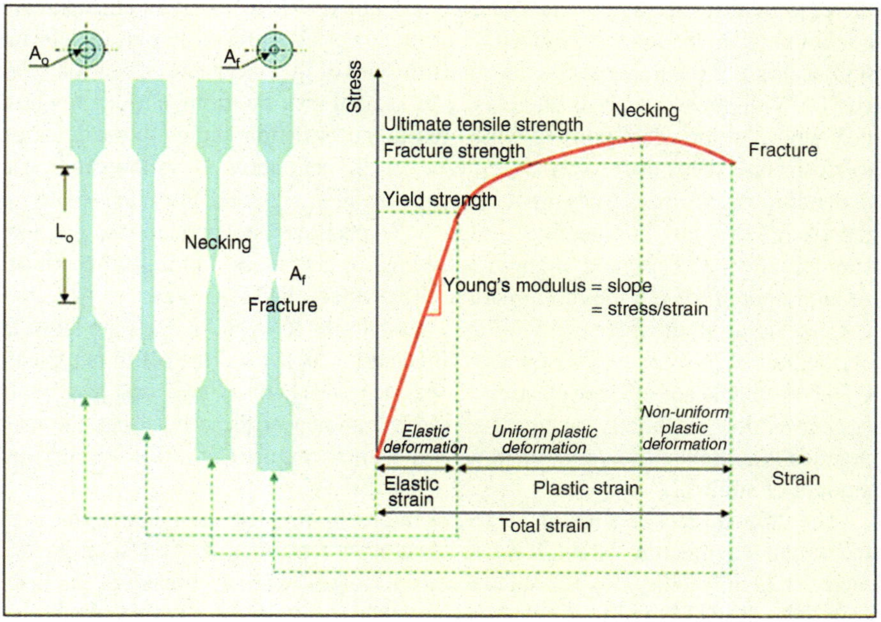

Fig. 2.8 Stress-strain relationship under uniaxial tensile loading (Adapted with permission from Faridmehr et al. 2014 (Science and Education Publishing))

region where stress is proportional to strain and is called the modulus of elasticity or Young's modulus. The modulus of elasticity (E) defines the properties of a material as it undergoes stress, deforms, and then returns to its original shape after the stress is removed. It is a measure of the stiffness of a given material. There are several different kinds of moduli depending on the way the material is being stretched, bent, or otherwise distorted. When a component is subjected to pure shear, for instance, a cylindrical bar under torsion, the shear modulus describes the linear-elastic stress-strain relationship. During tensile test, axial strain is always accompanied by lateral strains of opposite sign in the two directions mutually perpendicular to the axial strain. Strains that result from an increase in length are designated as positive (+) and those that result in a decrease in length are designated as negative (−). Poisson's ratio is defined as the negative of the ratio of the lateral strain to the axial strain for a uniaxial stress state. For a perfectly isotropic elastic material, Poisson's Ratio is 0.25, but for most materials the value lies in the range of 0.28–0.33. Only two of the elastic constants are independent so if two constants are known, the third can be calculated using the following formula: $E = 2\ (1 + n)\ G$, where E is modulus of elasticity (Young's modulus); n is Poisson's ratio; and G is modulus of rigidity (shear modulus).

In ductile materials, at some point, the stress-strain curve deviates from the straight-line relationship and Law no longer applies as the strain increases faster than the stress. From this point on in the tensile test, some permanent deformation occurs in the specimen and the material is said to react plastically to any further increase in load or stress. The material will not return to its original, unstressed condition when the load is removed. In brittle materials, little or no plastic deformation occurs and the material fractures near the end of the linear-elastic portion of the curve. With most materials there is a gradual transition from elastic to plastic behavior, and the exact point at which plastic deformation begins to occur is hard to determine. Therefore, various criteria for the initiation of yielding are used depending on the sensitivity of the strain measurements and the intended use of the data. For most engineering design and specification applications, the yield strength is used. The yield strength is defined as the stress required to produce a small, amount of plastic deformation. The offset yield strength is the stress corresponding to the intersection of the stress-strain curve and a line parallel to the elastic part of the curve offset by a specified strain (in the US the offset is typically 0.2% for metals and 2% for plastics). Some materials such as gray cast iron or soft copper exhibit essentially no linear-elastic behavior. For these materials the usual practice is to define the yield strength as the stress required to produce some total amount of strain.

The ultimate tensile strength (UTS) or, more simply, the tensile strength, is the maximum engineering stress level reached in a tension test. The strength of a material is its ability to withstand external forces without breaking. In brittle materials, the UTS will at the end of the linear-elastic portion of the stress-strain curve or close to the elastic limit. In ductile materials, the UTS will be well outside of the elastic portion into the plastic portion of the stress-strain curve. On the stress-strain curve above, the UTS is the highest point where the line is momentarily flat. Since the UTS is based on the engineering stress, it is often not the same as the breaking strength. In ductile materials strain hardening occurs and the stress will continue to increase until fracture occurs, but the engineering stress–strain curve may show a decline in the stress level before fracture occurs. This is the result of engineering stress being based on the original cross-section area and not accounting for the necking that commonly occurs in the test specimen. The UTS may not be completely representative of the highest level of stress that a material can support, but the value is not typically used in the design of components anyway. For ductile metals the current design practice is to use the yield strength for sizing static components. However, since the UTS is easy to determine and quite reproducible, it is useful for the purposes of specifying a material and for quality control purposes. On the other hand, for brittle materials the design of a component may be based on the tensile strength of the material.

The ductility of a material is a measure of the extent to which a material will deform before fracture. The amount of ductility is an important factor when considering forming operations such as rolling and extrusion. It also provides an indication of how visible overload damage to a component might become before the component fractures. Ductility is also used a quality control measure to assess the level of impurities and proper processing of a material. The conventional measures of ductility are the engineering strain at fracture (usually called the elongation) and the reduction of area at fracture. Both of these properties are obtained by fitting the specimen back together after fracture and measuring the change in length and cross-sectional area. Elongation is the change in axial length divided by the original length of the specimen or portion of the specimen. It is expressed as a percentage. Because an appreciable fraction of the plastic deformation will be concentrated in the necked region of the tensile specimen, the value of elongation will depend on the gauge length over which the measurement is taken. The smaller the gauge length the greater the large localized strain in the necked region will factor into the calculation. Therefore, when reporting values of elongation, the gauge length should be given. Reduction of area is the change in cross-sectional area divided by the original cross-sectional area. This change is measured in the necked down region of the specimen. Like elongation, it is usually expressed as a percentage.

Tension is just one of the way that a material can be loaded. Other ways of loading a material include compression, bending, shear, and torsion, and there are a number of standard tests that have been established to characterize how a material performs under these other loading conditions (NDT 2014).

2.5.3 Compressive and Shear Properties

In theory, the compression test is simply the opposite of the tension test with respect to the direction of loading. In compression testing the sample is squeezed while the load and the displacement are recorded. Compression tests result in mechanical properties that include the compressive yield stress, compressive ultimate stress, and compressive modulus of elasticity. For some materials, such as concrete, the compressive strength is the most important material property that engineers use when designing and building a structure. Compressive strength is also commonly used to determine whether a concrete mixture meets the requirements of the job specifications (NDT 2014).

Bearing properties and shear properties are applications related to compression loading. Bearing properties are used when designing mechanically fastened joints. The purpose of a bearing test is to determine the deformation of a hole as a function of the applied bearing stress. A shearing stress acts parallel to the stress plane, whereas a tensile or compressive stress acts normal to the stress plane. Shear properties are primarily used in the design of mechanically fastened components, webs, and torsion members, and other components subject to parallel, opposing loads. Shear properties are dependent on the type of shear test and there is a variety of different standard shear tests that can be performed including the single-shear test, double-shear test, blanking-shear test, torsion-shear test, and others. The shear modulus of elasticity is considered a basic shear property. Other properties, such as the proportional limit stress and shear ultimate stress, cannot be treated as basic shear properties because of "form factor" effects.

2.5.4 Creep and Stress Rupture Properties

Creep is a time-dependent deformation of a material while under an applied load that is below its yield strength. It is most often occurs at elevated temperature, but some materials creep at room temperature. Creep terminates in rupture if steps are not taken to bring to a halt. Creep data for general design use are usually obtained under conditions of constant uniaxial loading and constant temperature. Results of tests are usually plotted as strain versus time up to rupture. Creep often takes place in three stages (NDT 2014): In the initial stage, strain occurs at a relatively rapid rate but the rate gradually decreases until it becomes approximately constant during the second stage. This constant creep rate is called the minimum creep rate or steady-state creep rate since it is the slowest creep rate during the test. In the third stage, the strain rate increases until failure occurs. Creep in service is usually affected by changing conditions of loading and temperature and the number of possible stress-temperature-time combinations is infinite. While most materials are subject to creep, the creep mechanisms is often different between metals, plastics, rubber, and concrete. Stress rupture testing is similar to creep testing except that the stresses are higher than those used in a creep testing. Stress rupture tests are used to determine the time necessary to produce failure so stress rupture testing is always done until failure.

2.5.5 Toughness

The ability of a metal to deform plastically and to absorb energy in the process before fracture is termed toughness. Comparably, ductility is a measure of how much something deforms plastically before fracture, but just because a material is ductile does not make it tough. The key to toughness is a good combination of strength and ductility. A material with high strength and high ductility will have more toughness than a material with low strength and high ductility. Therefore, one way to measure toughness is by calculating the area under the stress strain curve from a tensile test. This value is simply called “material toughness” and it has units of energy per volume. Material toughness equates to a slow absorption of energy by the material, which is mainly influenced by strain rate (rate of loading), temperature, and notch effect.

A metal may possess satisfactory toughness under static loads but may fail under dynamic loads or impact. As a rule ductility and, therefore, toughness decrease as the rate of loading increases. Temperature is the second variable to have a major influence on its toughness. As temperature is lowered, the ductility and toughness also decrease. The third variable is termed notch effect, due with the distribution of stress. A material might display good toughness when the applied stress is uniaxial; but when a multiaxial stress state is produced due to the presence of a notch, the material might not withstand the simultaneous elastic and plastic deformation in the various directions.

2.5.6 Fatigue Properties

Fatigue cracking is one of the primary damage mechanisms of structural components. Fatigue cracking results from cyclic stresses that are below the ultimate tensile stress, or even the yield stress of the material. The fatigue life of a component can be expressed as the number of loading cycles required to initiate a fatigue crack and to propagate the crack to critical size. Therefore, fatigue failure occurs in three stages—crack initiation; slow, stable crack growth; and rapid fracture.

Dislocations play a major role in the fatigue crack initiation phase. In the first stage, dislocations accumulate near surface stress concentrations and form structures called persistent slip bands (PSB) after a large number of loading cycles. PSBs are areas that rise above (extrusion) or fall below (intrusion) the surface of the component due to movement of material along slip planes. This leaves tiny steps in the surface that serve as stress risers where tiny cracks can initiate. These tiny crack (called microcracks) nucleate along planes of high shear stress which is often 45° to the loading direction. In the second stage of fatigue, some of the tiny microcracks join together and begin to propagate through the material in a direction that is perpendicular to the maximum tensile stress. Eventually, the growth of one or a few crack of the larger cracks will dominate over the rest of the cracks. With continued cyclic loading, the growth of the dominate crack or cracks will continue until the remaining uncracked section of the component can no longer support the load. At this point, the fracture toughness is exceeded and the remaining cross-section of the material experiences rapid fracture. This rapid overload fracture is the third stage of fatigue failure (NDT 2014).

For some components the crack propagation life is neglected in design because stress levels are high, and/or the critical flaw size small. For other components the crack growth life might be a substantial portion of the total life of the assembly. Moreover, preexisting flaws or sharp design features may significantly reduce or nearly eliminate the crack initiation portion of the fatigue life of a component. The useful life of these components may be governed by the rate of subcritical crack propagation.

2.6 Materials Design with Computational Modeling

Simulation-based engineering and science offers a significant opportunity to increase industrial competitiveness by reducing design times, accelerating the development and incorporation of new materials and processes, and minimizing testing requirements. The development and application of a predictive capability to facilitate the development of industrial products, processes, and technologies are transformational. Predictive capability for the performance of materials and chemical systems is essential to computational modeling. Predictive capability also drives technological innovation. Advanced technologies typically require increasingly complex materials systems and processes. This complexity is a challenge for traditional development strategies due to the extensive parameter space that must be explored. Predictive capability based on simulation-based engineering and science offers the opportunity to significantly expand this parameter space while lowering both costs and development times. This will accelerate the replacement of rare or nondomestic source materials (such as rare earths for magnets or lithium for batteries) with abundant materials, the discovery of new materials with tailored properties, the deployment of "green" technologies and processes with lower environmental impact and improved performance, and the development of advanced manufacturing technologies with improved efficiency and flexibility. By accelerating the development and deployment of complex materials systems and processes, predictive capability will drive innovation and economic competitiveness (DOE 2010).

Integrated Computational Materials Engineering (ICME) is a major crosscutting set of tools underlying and in some cases enabling successful development in varies energy systems. Key elements are not only fundamental materials and processing models informed by critical experiments and fueled by robust databases, but also predictive performance models again tuned by experimental data. Through successful integration of existing tools as well as yet-to-be-developed capabilities, ICME can accelerate and enhance the probability of successful development and commercial implementation of product and process innovations (Robinson 2012).

As one of key elements of ICME (Fig. 2.9), modern materials modeling techniques allow both interpretation of experimental data and prediction of new materials properties. Improvements in nano-scale synthesis and characterization are bringing experimental techniques towards quantitative analysis at the nano-scale. At the same time, improvements in computational modeling methods allow to directly correlate atomic structural models with experiment (Mokrani and Ewels 2013; Paddison 2013).

2.6.1 The Challenge of Complexity

Advanced materials are complex. Achieving the required performance gains depends on exploiting the many degrees of freedom of materials development including multiple chemical components, nanoscale architectures, and tailored electronic structures. This introduces enormous complexity in the discovery process, complexity that must be understood and managed. Early steels consisted of three to four essential chemical components and a relatively simple microstructure. Today's advanced high-strength, high-temperature steels average six to eight chemical components and require complex, multiphase nanostructures. The parameter space for exploration has increased enormously, making continued development by trial and error impractical (DOE 2010).

For example, new catalysts are needed to improve the efficiency of industrial processes, make effective use of bioenergy, and drive energy conversion and environmental mitigation processes. There are billions of options: chemical combinations, local morphologies, atomic-scale structure. Sifting through the options using predictive modeling is the only intelligent and efficient path forward.

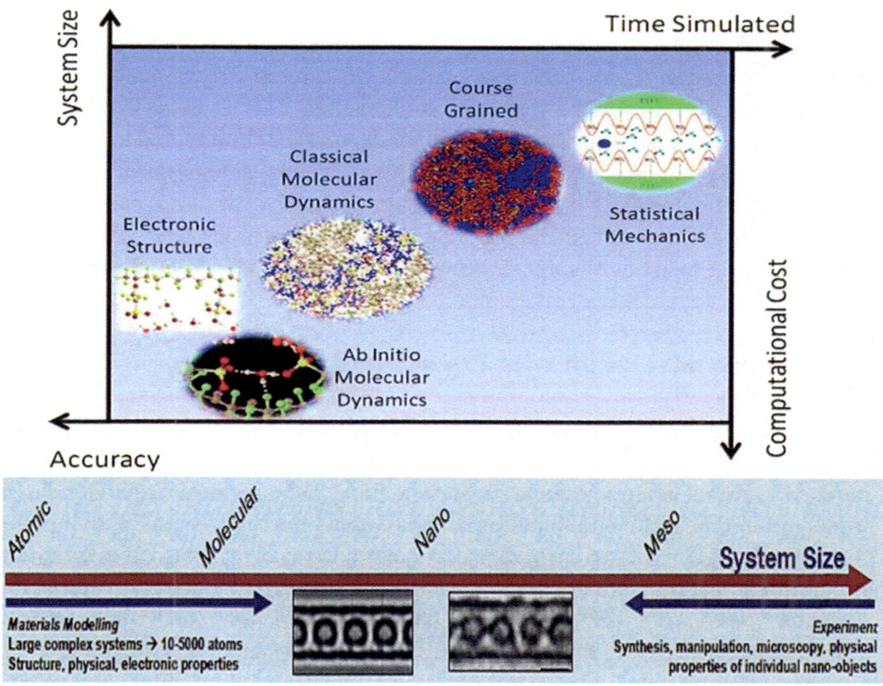

Fig. 2.9 Overview of computational materials modeling (Modified from Mokrani and Ewels 2013. Credit: Institut des Matériaux Jean Rouxel)

The superconductors of the 1980s were typically two-component systems with a simple crystal structure. Today's high-temperature superconductors boast four or more chemical components, layered architectures, and sensitive electronic doping. This trend is also apparent in new high-field magnetic materials that derive their properties from the interaction of multiple chemical elements in complex microstructures. Therefore, high performance comes with a significant increase in complexity, and with a corresponding need to narrow discovery possibilities to a manageable number of the most promising options (DOE 2010).

Across the spectrum of new materials and chemical processes, discovery is increasingly confronted with complexity. It is not practical any more to explore all the options experimentally. The only solution is materials and chemistry by design, using new synthesis and characterization tools, theory, and simulation and modeling to understand complex materials and chemical systems and predict the most promising research directions (DOE 2010).

Materials science and engineering are at the threshold of a new era where the integrated synthesis, characterization, and modeling of complex materials and chemical processes will transform one's ability to understand and design new materials and chemistries with predictive power. This has profound implications for the pace of discovery and the creation of new technologies.

Moreover, simulation-based engineering and science has accelerated progress in scientific understanding and technology development by enabling complex systems such as astrophysical and climate phenomena, aircraft wings, and integrated manufacturing to be explored rapidly and efficiently. This leads to new scientific understanding of systems that are too large for experimental study, reduced time and cost of prototyping, and accelerated deployment of new technologies. Harnessing the potential of computational science and engineering for the discovery and development of materials and chemical processes is essential to proceeding acceleratively in these foundational fields and their downstream energy and industrial applications.

2.6.2 Materials Design with Predictive Capability

2.6.2.1 Controlling Microstructures in Materials for Extreme Conditions

The availability of structural materials that can operate at extreme values of temperature, stress and strain, pressure, radiation flux, and chemical reactivity is the principal limiting factor in the performance of many energy systems. Fossil power plants, nuclear plants, and transportation systems all operate at lower efficiencies due to the limitations of existing structural materials. Impressive gains in efficiency of 30% and more can be achieved by the development of new materials capable of withstanding these demanding conditions. The failure of materials, often at one-tenth or less of their intrinsic limits, is not understood. Understanding failure and achieving intrinsic properties require bridging length and time scales from molecular structures and their interactions to continuum models of bulk components. Central to this challenge is predicting and controlling the microstructure—the complicated arrangement of crystalline grains, defects, interfaces, and impurities that make up the micro-scale structure. Microstructure is the key to understanding damage processes, preventing failure, and enhancing performance (Brady 2010).

The design space of modern structural materials is huge—much too complex to explore by trial and error. Predictive modeling is needed to guide experiments in the most productive directions, to accelerate design and testing, and to understand performance. State-of-the-art computational tools allow scientists to calculate from first principles the interactions that dominate microstructural behavior, while experimental tools can now provide time resolved measurements on real materials to validate these models. This integration of theory, simulation, and experiment will accelerate materials discovery and innovation. Key to achieving these advances is verification, validation, and uncertainty quantification of the computer models. Physical measurements must be made at relevant length and time scales and compared directly with theory and simulation. The time is ripe for development of a sustained effort in integrated computational materials engineering. The lack of new materials is a critical factor in design and manufacturing, and a barrier to sustained competitiveness. The outcome of this effort will be the rapid development and deployment of new materials that can be incorporated in energy systems and manufacturing. This can increase the efficiency of power plants and transportation systems, significantly reduce requirements for physical testing, and increase competitiveness by achieving improved functionality and reduced time-to-market of a wide variety of products and technologies (DOE 2010).

The microstructure of a material controls a wide range of important properties, including strength, fatigue, high-temperature performance, corrosion, and radiation resistance. While there is substantial qualitative understanding of microstructural evolution, there are no predictive models that link materials processing to resultant microstructures. Further, there is a lack of understanding of the connections between microstructure and materials performance. The new generation of synchrotrons, neutron sources, and synthesis and characterization equipment, together with computational and algorithm advances, provides an opportunity to envision designing microstructures for specific purposes and bringing them to fruition in real materials. The potential impact of optimizing performance through engineered microstructures by design is huge. Using integrated computational materials engineering to develop quantitative prediction of materials properties based on processing history, a substantial return on investment can be provided by reducing design times, lowering development costs, and accelerating the product cycle. The capability to tailor materials for specific applications, such as radiation environments, will become increasingly important as materials properties are pushed to their theoretical limits. Realizing this vision will require integrating and linking models that capture the multitude of individual physical phenomena that dictate material performance (micro-porosity, grain size, multiple phases, precipitates, etc.) and their effect on properties (fatigue, creep strength, corrosion, radiation resistance, etc.). This will also require close collaboration across the synthesis, characterization, theory, and computational communities, as well as sustained efforts in the related computer science, mathematics, and information science fields. Finally, integration with industry will be essential to develop and transfer the new computational tools and ensure their applicability to industry needs (Cummings and Glotzer 2010).

2.6.2.2 Controlling Self-Assembly in Designing and Engineering Materials at the Nanoscale

Designing materials made of preprogrammed building blocks that spontaneously organize into structures with unique and complex properties currently exhibited by biological systems have long been a dream of technologists. Structures whose constituents can assemble, disassemble, and reassemble autonomously or on command enable materials capable of self-repair, multitasking, and even shape-shifting-properties known throughout the biological world. Imagine coatings that can change color or toggle between translucent and opaque on cue; sensors that can detect, trap, and dispose of pathogens; materials that can self-regulate porosity, strength, water or air resistance, elasticity, or conductivity—all these and much more are possible through self-assembly (DOE 2010).

Importantly, self-assembly also permits material structures far more complex than traditional metals, ceramics, and polymers, with many levels of hierarchical organization and compartmentalization typical of biological structures such as cells and organelles. Such structural complexity is demanded by the sophisticated properties and behavior of next-generation materials capable of meeting future energy demands—especially active materials, which must perform functionally in ways not possible today for traditional, nonbiological matter. Advances in nanoscience have made possible the creation, imaging, characterization, and manipulation of highly complex building blocks ranging from single molecules to supramolecular objects nanometers to microns in size-precisely the size range needed for the "bricks and mortar" of next-generation, self-assembled materials. Nanoparticles and colloids of nearly any shape, made of metals, semiconductors, and/or polymers, and functionalized with organic molecules and biomolecular ligands—including proteins, viruses, and DNA—as well as other chemical "hooks" are now possible. As a result, vast palettes of designer building blocks, in many cases coupled with solvents that play an active role in mediating interactions, are at hand, with the propensity for self-assembly into structures of unprecedented complexity and function (Damasceno et al. 2012).

The design space for self-assembled materials is now so vast that computational tools are required for the rapid screening and prototyping of building blocks that will predictably self-assemble into desired structures. High-performance computing (HPC) have produced computing platforms that are now fast enough to permit predictive simulations of self-assembly for complex building blocks, and new experimental probes promise the needed resolution of nanoscale structure to monitor assembly processes in situ, parameterize models, and validate simulations. These advances in synthesis, characterization, and modeling capabilities set the stage for a tipping point in our ability to discover the underlying principles controlling self-assembly, and to develop robust, predictive simulation-based tools to achieve materials by design (Glotzer et al. 2004).

To fully master the science and engineering of self-assembly requires rapid and integrated progress on several related fronts of discovery and innovation. It requires harnessing the often competing theoretical principles of thermodynamics and kinetics in materials comprised of complex molecular and supramolecular building blocks whose shapes and interactions can result in kinetic traps that prevent assembly into equilibrium structures. It also requires discovering and applying the principles of statistical thermodynamics in active systems and systems driven far from equilibrium, as in biological systems, in which the constant input of energy creates and stabilizes structures. Finally, it requires developing simulation-based design tools that enable both the prediction of structures and their properties from building blocks and the rapid prototyping and reverse engineering of building blocks designed and preprogrammed to assemble into target structures. The advanced computation and simulation technology will provides revolutionized capability to predict, model, and design self-assembled materials with desired functional properties for a broad host of energy and other applications (Nguyen and Glotzer 2010; DOE 2010).

2.6.2.3 Photons to Energy in Light Harvesting

Developing future molecular, polymeric, and hybrid materials for harvesting and converting energy from sunlight requires sophisticated computational search strategies to find the optimal combination of organic and inorganic materials that can harvest light from the entire solar spectrum. Designing materials for such light collection requires capture of the infrared and ultraviolet parts of the solar spectrum and is expected to involve strategies based upon intermediate bandgap materials, plasmonic excitations in clusters and molecules, optical metamaterials, and optimized thermoelectrical conversion complexes. Requisite to finding the optimal solar-harvesting "needle" in the materials genome "haystack" is the ability to accurately determine how any material specimen absorbs energy at every color of light (frequencies) delivered by the sun. Additional challenges are to determine the mechanisms by which the absorbed energy (exciton) migrates through the system prior to splitting into charges (electrons and holes) that are converted to electricity or chemical bonds. In addition to the computational design of materials for solar cells, photosynthetic mimics, and photochemical pathways to fuels, the need to computationally predict and optimize a material's or chemical's propensity towards absorption and transfer of optical energy is ubiquitous and relevant to several energy technologies including (DOE 2010): (a) Radiative energy transfer in combustion, (b) Photocatalysis, and (c) Interfacial charge transfer.

Computer simulation and prediction of electronic excited states and their interactions with light and vibrations in molecules and materials can be pursued using several different techniques, such as the solution of the time-dependent hybrid density-functional theory coupled cluster equations, the Bethe-Salpeter equations, and the so-called GW equations. For light-harvesting systems, interest is in understanding phenomena associated with timescales for charge/energy transport that are comparable to the reorganization time of the surrounding solvent. Accounting for such dynamics, especially in a realistic environment containing defects, is difficult but can be obtained from first-principles calculations such as quantum mechanics/molecular mechanics (QM/MM). However there is no unified or universal code that embodies these powerful methods, and the development of such a tool would enable the routine simulation of light harvesting in realistic systems. To directly simulate and understand the transfer of electrons from an adsorbed photoexcited dye molecule at the solid-liquid interface, semiclassical dynamical approaches relying upon publicly available force fields and quantum-mechanical methods are required. Additionally, one must account for dissipation and transport through the electrode (Panitchayangkoon et al. 2011).

As hundreds of thousands of possible combinations of elements can be scanned across the entire periodic table, many new materials solutions may far exceed the traditional intuition of experts in these fields. Even incomplete and low-level theories have suggested novel combinations of materials for new energy technologies. In principle, finding the best solution to solar harvesting and other issues related to composition-dependent process optimization can now be accomplished using this approach. The challenge of solar harvesting cross-cuts the traditional domains of chemistry and physics as it requires coupling together capabilities for localized excitations, band-to-band transitions, and electron transport (DOE 2010).

2.6.2.4 Controlling Chemical Reactions with Combustion and Catalysis

The chemical bond represents the most compact nonnuclear energy storage medium known. Gasoline, for example, has more than 30 times the energy density of lithium-ion batteries. For this reason, chemical reactions hold center stage in the storage and release of energy in both current devices and those envisioned for a secure and sustainable future. The robust nature of chemical energy storage is embodied in the millennia between the photosynthesis of prehistoric plants, which stored the energy of the sun, and the recovery and combustion of the resulting fossil fuels today, which releases that energy. The challenge for the future is to capture, store, and release energy on an immediate timescale and in a sustainable way. The predictive simulation of chemical transformations will accelerate the transition from utilizing energy stored in prehistory to establishing a sustainable cycle of energy storage and utilization. There are three principal areas of opportunity where design using predictive simulation will impact energy technologies (DOE 2010): (a) Artificial Photosynthesis and Solar Fuels; (b) Efficient Combustion of Low-Carbon Fuels; and (c) Converting Biomass to Transportation Fuels.

For the first of these, the sun offers an abundant and sustainable source of energy that can be used to convert water or water and carbon dioxide to fuels or intermediates used in the production of fuels. The essential barrier to developing this technology to industrial scales is finding appropriate photocatalysts that work together to perform the complete photosynthetic chemical cycle. In the case of low-carbon fuels, efficient combustion technologies make use of turbulent reacting flows. Development of these technologies requires a combination of complex fluid flow simulations and high-accuracy chemical kinetic reaction mechanisms that capture the complex chemistry of current and future fuels: hundreds of reactants and thousands of reactions. Finally, the challenge in the production of biofuels is to convert the principal components of biomass, a combination of cellulose, hemicelluloses, and lignin, into products that can be used as a substitute for gasoline and diesel fuel for various forms of transportation. The difficulty in meeting this challenge is the intrinsic complexity of biomass. Further development of these three essential energy technologies requires predictive simulation that crosses many length and time scales. Combustion science is on the threshold of a new era of predictive modeling and simulation based on the convergence of new computer resources and the high-fidelity simulation codes ready to exploit them. Artificial photosynthesis, splitting water with sunlight, and biomass conversion into fuels all require new catalysts. The process of screening potential candidates, or synthesizing new ones, in the near future will be accelerated by predictive simulation of their properties (Ejzak and Sullivan 2013).

Simulation and screening of photocatalysts for solar fuel production require the computational treatment of a large number of elementary processes covering a wide spectrum of timescales. Simulation of such processes requires methods to calculate the dynamics of charge pairs formation, separation, and mechanisms used to drive catalytic chemical reactions. Complete and accurate modeling of surface catalysis at the molecular level is required to predict reaction rates accurately. The tools of modern quantum chemistry and molecular dynamics form the basis for these needed new capabilities, but they must be extended, verified, and deployed on the most powerful available computing resources. These tools bridge the gap between high-fidelity, first principles-based direct numerical simulation tools and Reynolds-averaged Navier-Stokes tools used in current industrial design. In addition, new methods are required to automatically generate high-accuracy reaction rate sets for arbitrary fuels. For biomass conversion, efficient simulation methods are required to describe the multi-scale dynamics and thermodynamics of solvated biopolymers. Quantum chemical simulation of biomass-catalyst interactions including the effects of solvent are also needed to help guide the selection of catalysts that will facilitate the conversion of sugars to suitable substitutes for gasoline and diesel (DOE 2010).

2.6.2.5 Design Fluids for Separations and Carbon Capture

Gases and liquids are collectively called fluids. Many of most important global climate and energy challenges relate to fluids and separating them from other fluids or from solids. Examples include removing CO_2 from fossil power plant emissions; purifying water, since sufficient water supplies is one of the biggest challenges to ensure future global health and prosperity; and extracting high-energy-content liquid fuels for transportation applications from crude oil, or from the liquid mixtures obtained from biomass conversion (DOE 2010).

Today, the number of possible fluids and fluid mixtures that have the potential to revolutionize the world's energy and environmental future is virtually limitless. The discovery of new, large classes containing over a billion completely tailorable (i.e., designer) liquids, such as room-temperature ionic liquids, and new, equally large classes of tailorable solids that can be used in adsorption-based separation processes, such as metal-organic frameworks, has made the design space of possible fluids and separations processes essentially infinite. Thus, computational screening of these processes for energy and environmental applications is no longer simply desirable-it is imperative. However, the properties of fluids, which in turn determine the nature and efficiency of the processes used to separate them, depend, often very subtly, on the atomic- and molecular-level forces within and between molecules (known as force fields). Given accurate force fields, methods exist (molecular simulation, statistical mechanics) to computationally derive the properties needed to evaluate the potential applicability and separability of fluids. However, current predictive capabilities are limited because of no robust methods for predicting force fields in an arbitrary fluid mixture. Specifically, force fields need to be predicted in the absence of experimental data (for systems not yet synthesized or characterized). Data-free force field prediction (DF3P) must be achieved by using first-principles calculations (quantum chemistry, density functional theory, etc.) as input, but given the high computational costs of such methods, this is an enormous challenge. Success in DF3P will enable true fluid properties prediction to optimize and invent new energy and environmentally relevant systems. The flow of information can be envisioned beginning with the chemistry of a system (its chemical composition) and ending up with optimized processes. As DF3P is realized, the flow of information can be inverted by the use of stochastic optimization (e.g., simulated annealing) at each step. For example, given a desired process, the properties needed to achieve that process can be determined; once the required properties are established, the optimal fluids for the process can be determined. The result will be unlimited innovation in energy-relevant processes enabled by designer fluids (Li et al. 2011).

Data-free force field prediction (DF3P) will enable the development of computational screening of solvents and adsorbents for carbon capture, leading to new, cost-efficient processes. Likewise, optimizing the extraction of liquid fuels from the mixtures resulting from biomass conversion and the purification and/or desalination of water are just two of many other profound energy/environmental challenges that can be addressed by computational modeling made quantitative by a robust DF3P capability. DF3P is also a key enabling technology across all of chemistry, geochemistry, biochemistry, and engineering: any field in which molecular interactions in fluids in bulk or at interfaces are important is currently limited by the availability of robust, accurate force fields. For DF3P to become a reality, force field development must become an automated workflow utilizing petascale and exascale computational resources. The outputs of a force field-based molecular simulation or statistical mechanical calculation of most interest to a process designer are thermophysical and themochemical properties; however, for experimental validation purposes, such methods also predict structures that are much more sensitive measures of force field accuracy (DOE 2010).

2.6.2.6 Design Interfaces from Interfacial Materials to Advanced Batteries

Fuel cells offer a way to extract twice the energy from hydrocarbons as internal combustion engines. Photovoltaics provide direct access to the most abundant energy source available on the Earth. Fiber-reinforced polymer matrix composites achieve strength and stiffness characteristics paralleling those of steel at a fraction of the weight, thus providing for lighter, fuel-efficient vehicles. These are but a few examples of devices and structures in which functionality is achieved by deliberately juxtaposing disparate types of materials, and where the processes that are fundamental to the device performance occur at the interfaces between constituents. For example, fine tuning the bonding interactions between fibers and matrix provides for simultaneous high strength and toughness of the composite. In solar cells and solid-state lighting, the separation and recombination of charge carriers at interfaces provide the basis for converting light into electricity, or vice versa, and the efficiency of these processes depends on the quality and perfection of the interfacial region. The long-term stability of the interfaces between electrodes and separator membranes is essential for sustaining the electrochemical processes responsible for the generation or storage of electric energy in fuel cells or batteries (Xia and Jiang 2008).

Because interfaces are imperfect and degrade with time, the highest energy conversion efficiencies cannot be achieved or the strongest composites created now. The key to producing materials systems with the desired performance characteristics is the ability to fabricate them consistently and with nanoscale precision to design specifications. This entails the realization of atomically sharp definition and long-term stability of interfaces. Hence, detailed knowledge about the structure of interfaces and the mechanisms of interfacial phenomena, both as they govern the functional response of a device and contribute to the deterioration of the interface, is essential for advancing many critical technologies. Given their small extent in one dimension and because they are typically buried within bulk materials, interfaces are difficult to resolve or access by experimental means. Simulation and modeling is therefore ideally suited to complement experiments and supply the missing information. Computational science and its infrastructure will impact research and development with respect to interface science in two transformative ways (DOE 2010): (1) Modeling and simulation can be used not simply to recreate experimental observations, and thereby provide for detailed interpretation, but to generate new information that is inaccessible by experiments; (2) through combinatorial and high throughput, rational exploration of interfacial phenomena, and modeling and simulation will yield the necessary fundamental insights to develop a predictive design toolset. With the advancement of HPC, modeling and simulation can guide and accelerate materials development. Materials building blocks can be conceived, encoded to self-assemble into desired configurations, numerically test and optimize design criteria for specific interfacial functionality, and achieve high figures of merit for multiple performance criteria simultaneously, for example, electrolytes with high ionic mobility and stiffness, electrodes with high intercalation capacity and phase stability, and low-density composites with high strength and thermal conductivity.

Unlike with new experimental probes, where breakthroughs occur somewhat sporadically, computational power increases at a predictable rate due to relatively steady progress in chip circuit density and computer architectures. Based on the projected performance growth, resources need to be mobilized to harness this power and engage in the creative process that leads to establishing new theoretical frameworks, modeling approaches, algorithms, and codes for increasingly realistic and demanding simulations that allow people to predict interfacial phenomena. Without these advances in modeling and simulation, the crucial knowledge gap that prevents people from developing technological materials with precise control of those properties governed by interfaces will persist (DOE 2010).

To account for the important phenomena occurring at interfaces, features such as the electron distributions within the irregular gaps between adjacent materials must be resolved, while interfacial reconstructions, defects, and roughness are described at scales encompassing millions to billions of atoms. To accurately predict the behavior and properties of interfaces, realistic structural models must first be generated. This requires specialized procedures, such as acceleration algorithms, heuristic schemes for advancing structural evolution, and statistical sampling techniques. Such an inherently multi-scale computational challenge must include electronic structure calculations, atomistic simulations, and continuum methods, and key improvements in the available computational infrastructure still need to be achieved. Simulation algorithms and statistical analysis formalisms must be adapted to account for lack of periodicity and symmetry of interfacial regions. Quantum mechanical, particle-based deterministic, statistical, and geometric simulation techniques must be effectively integrated or coupled through adaptable information conduits. Workflow integrators that autonomously balance multi-scale simulation tasks, better order-N methods for first-principles calculations, and more sophisticated reactive force fields will allow people to eliminate the trade-off between computational speed and accuracy.

Validation of these simulation approaches will be possible with in situ experimental characterization of buried interfaces using powerful new neutron and synchrotron sources, Z-contrast high-resolution electron microscopy, and nano-probe transport measurements that have only recently become available. Modeling and simulation will provide predictive capabilities for the design of interfaces in materials. As a result, people will be able to improve the efficiency of solar panels, solid-state lighting, thermoelectric generators, and fuel cells; create high-capacity batteries; fabricate lightweight composites with increased strength, stiffness, and toughness; and extend the lifetime of materials systems, devices, and components (DOE 2010).

2.6.2.7 Controlling Electronic Structure Through Modeling Strongly Correlated Electrons

Strongly correlated materials, whose behavior is dominated by the Coulomb repulsion among electrons, are exceptionally rich in dramatic behavior and useful functionality. Magnets that enable digital memory, superconductors that carry electricity without loss above liquid nitrogen temperature, actinides that power nuclear reactors, and quantum dots that enable nanotechnology are all strongly correlated materials. These materials typically display a host of competing phases closely spaced in energy leading to "colossal" response to small external stimuli, a feature with endless potential for technological exploitation (Ferrero et al. 2009).

Although some of these materials on an empirical basis have been used, people cannot model, predict, or control their properties as we do for semiconductors like silicon. This lack of predictive capability is a severe bottleneck—strongly correlated materials are so common and their compositions and structures so complex that they simply overwhelm serendipity as a discovery and development tool. Reliable predictions of the behavior of strongly correlated materials are critical for designing the next generation of globally competitive information, communication, sensing, and clean energy technologies. In the past decade, however, strongly correlated materials have begun to yield their secrets to a host of modeling approaches that go beyond existing density functional theory. The challenge set by strongly correlated electrons is to develop methods that can simultaneously describe the strong correlations that can lead to localization of electrons on atomic sites and the weak correlations that allow itinerant electrons to move freely throughout a material (DOE 2010).

The density functional approach excels where dynamical correlations are modest in size, which is in the weakly correlated materials. Static extensions of density functional theory have proven useful, but the real need for the future is to treat strong dynamic correlations directly. Impressive advances have recently been made in this area, with dynamical mean field implementations providing the solution to several long-standing, classic problems. Quantum Monte Carlo techniques promise essentially exact solutions that, due to technical challenges and issues of scaling with system size, will require a longer time frame for broad application. Low-dimensional correlated systems present additional challenges, and density matrix methods are most promising for this class. Crucial extensions of density functional theory are required, with the dynamical treatment of intra-atomic interactions and of interactions of localized and itinerant electrons comprising the minimum requirements. The energy resolution of correlated electron treatments needs to be increased in order to treat the low-energy states, where electrons fluctuate on short timescales between local and itinerant character. This rapid fluctuation between dual existences lies at the very heart of strongly correlated behavior. In addition, experimental validation is essential to developing predictive capabilities for strongly correlated electron materials. Many strongly correlated electron phenomena take place at ultra-small spatial scales and ultrafast timescales. Continuing advances in experimental tools such as X-ray free electron lasers, aberration-corrected electron microcopy, high-resolution photoemission, diffuse neutron scattering, and scanning probe microscopy are critical to probe, verify, and refine the theory and modeling predictions (Yoo et al. 2005; DOE 2010).

The rewards of predicting the behavior and functionality of strongly correlated materials are enormous: next-generation magnetic memories, spintronic and metal-insulator ("Mott-tronic") digital logic, high-resolution sensors for electromagnetic radiation and environmental chemicals, and better performing high-temperature superconductors, among others (DOE 2010).

2.6.3 Materials Modeling Approaches

Modeling and simulation is today a critical part of every discipline of science and engineering. However, while models and methods are well understood in areas such as structural engineering and modeling tools are embedded within several related industry sectors, simulation-based engineering and science capabilities for materials research are much less mature. Although scientific models and algorithms are in hand for many materials applications, critical information is missing for others, and there is only a handful of simulation-based materials design tools sufficiently predictive and robust for industrial use. As a result, despite many successful examples of the use of simulation-based engineering and science, it has only just begun to exploit its full capabilities and promise for discovery and innovation when it comes to the critical pacing technologies of new materials. Emerging capabilities in predictive modeling and simulation have the potential to revolutionize the development of new materials and chemical processes. Coupled with materials characterization and experimental validation, this predictive capability provides the foundation for an innovation ecosystem that can accelerate discovery and the development of new technologies, including advanced energy systems (DOE 2010; Vaz Jr. et al. 2011).

Draw upon advances in measurement sciences, biological sciences, and information technology, as shown in Fig. 2.10, multi-scale simulation methods have been developed, which are validated by critical experiments across a wide range of time and length scales. For example, key structure property relationships are being addressed by atomistic and continuum methods that include molecular dynamics, Monte-Carlo simulation, micromechanics and finite element methods. Advances to date include constitutive relationships and effective-continuum representations

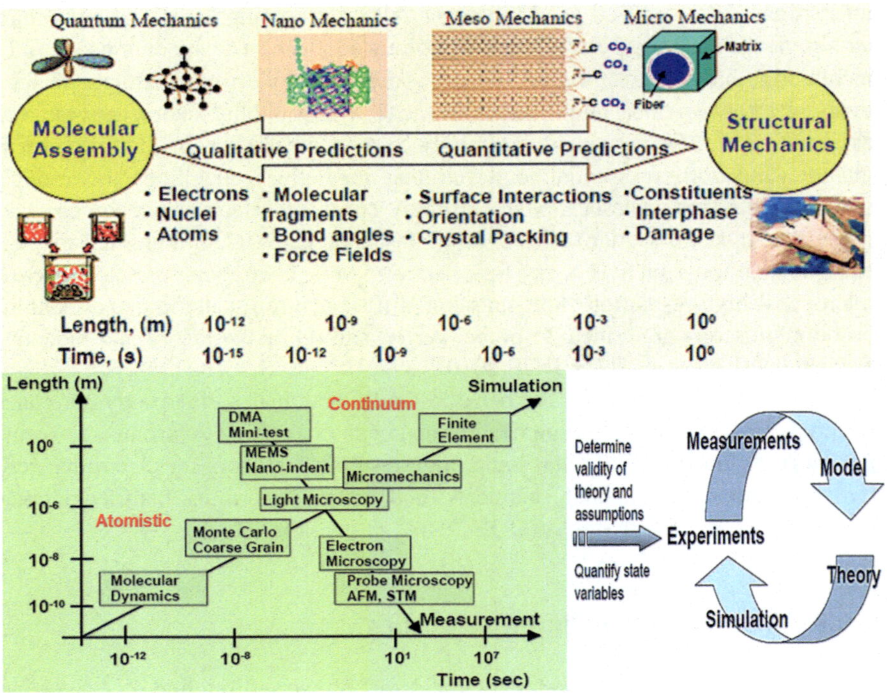

Fig. 2.10 Illustration of relationships between time and length scales for the multi-scale simulation methodology (Modified from Gates and Hinkley 2003. Credit: NASA Langley Research Center)

of polymers and polymer/nanotube composite materials. Critical issues to be resolved include seamless transfer of data between the nano-to-meso-scale models and experimentally validating simulations of atomistic behavior (Gates and Hinkley 2003).

2.6.3.1 Atomistic and Molecular Methods

The approach is to formulate a set of integrated predictive models that bridge the time and length scales associated with material behavior from the nano through the meso scale. At the atomistic or molecular level, the reliance is on molecular mechanics, molecular dynamics, and coarse-grained, Monte-Carlo simulation. Molecular models encompassing thousands and perhaps millions of atoms can be solved by these methods and used to predict fundamental, molecular level material behavior. The methods are both static and dynamic. For example, molecular mechanics can establish the minimum-energy structure statically and molecular dynamics can resolve the nanosecond-scale evolution of a molecule or molecular assembly. These approaches can model both bonded and nonbonded forces (e.g., Van der Waals and electrostatic) but cannot explicitly account for bond cleavage. Molecular dynamics simulations of organic and inorganic material systems address a variety of issues including the thermodynamics of biological process, polymer chemistry and crystal structure (Young et al. 2000). The number of simulation techniques has greatly expanded; there exist now many specialized techniques for particular problems, including mixed quantum mechanical—classical simulations. Molecular dynamics simulation techniques are widely used to help interpret experimental results from X-ray crystallography and nuclear magnetic resonance spectroscopy, and for atomistic simulations of carbon nanotube behavior at the nano-scale, for instance. Molecular dynamics simulations generate information at the nano-level, including atomic positions and velocities. The conversion of this information to macroscopic observables such as pressure, energy, and heat capacities requires statistical mechanics. An experiment is usually made on a macroscopic sample that contains an extremely large number of atoms or molecules, representing an enormous number of conformations. In statistical mechanics, averages corresponding to experimental measurements are defined in terms of ensemble averages. For example, the average potential energy of the system is defined as (Gates and Hinkley 2003):

$$V = \frac{1}{M}\sum\nolimits_{i=1}^{M} V_i \tag{2.14}$$

where M is the number of configurations in the molecular dynamics trajectory and V_i is the potential energy of each configuration. Similarly, the average kinetic energy is given by Gates and Hinkley (2003):

$$K = \frac{1}{M}\sum\nolimits_{j=1}^{M}\left\{\sum\nolimits_{i=1}^{N}\frac{m_i}{2}v_i^2\right\}_j \tag{2.15}$$

where M is the number of configurations in the simulation, N is the number of atoms in the system, m_i is the mass of the particle i and v_i is the velocity of particle i. To ensure a proper average, a molecular dynamics simulation must account for a large number of representative conformations.

By using Newton's second law to calculate a trajectory, one only needs the initial positions of the atoms, an initial distribution of velocities and the acceleration, which is determined by the gradient of the potential energy function. The equations of motion are deterministic; i.e., the positions and the velocities at time zero determine the positions and velocities at all other times, t. In some systems, the initial positions can be obtained from experimentally determined structures. In a molecular dynamics simulation, the time dependent behavior of the molecular system is obtained by integrating Newton's equations of motion. The result of the simulation is a time series of conformations or the path followed by each atom. Most molecular dynamics simulations are performed under conditions of constant number of atoms, volume, and energy (N,V,E) or constant number of atoms, temperature, and pressure (N, T, P) to better simulate experimental conditions. The basic steps in the molecular dynamics simulation include Gates and Hinkley 2003): (a) Establish initial coordinates. (b) Minimize the structure. (c) Assign initial velocities. (d) Establish heating dynamics. (e) Perform equilibration dynamics. (f) Rescale the velocities and check if the temperature is correct. (g) Perform dynamic analysis of trajectories.

Current generation force fields (or potential energy functions) provide a reasonably good compromise between accuracy and computational efficiency. They are often found empirically and calibrated to experimental results (e.g., X-ray crystallography) and quantum mechanical calculations of small model compounds. The development of parameter sets that define these force fields may require extensive optimization. One of the most important limitations imposed on a force field is that no drastic changes in electronic structure are allowed, i.e., no events like bond making or breaking can be modeled. The most time consuming part of a molecular dynamics simulation is the calculation of the nonbonded terms in the potential energy function, e.g., the electrostatic and van der Waals forces. In principle, the nonbonded energy terms between every pair of atoms should be evaluated. This requirement would imply that the number of computations increases as the square of the number of atoms for a pair-wise model. To speed up the computation, the interactions between two atoms separated by a distance greater than a predefined distance, the cutoff distance, are ignored (Gates and Hinkley 2003).

2.6.3.2 Monte Carlo Simulation

Although molecular dynamics methods provide the kind of detail necessary to resolve molecular structure and localized interactions, however, both the size and time scales of the model are limited by numerical and computational boundaries. To help overcome these limitations, coarse-grained methods are available that represent molecular chains as simpler, bead-spring models. A comparison to molecular dynamics has shown up to four orders of magnitude decrease in CPU time through the use of the simpler models (López-Lemus and Alejandre 2002). Although the coarse-grain models lack the atomistic detail of molecular dynamics, they do preserve many of the important aspects of the chemical structure and allow for simulation of material behavior above the nano-scale (Kremer and Müller-Plathe 2001). The connection to the more detailed atomistic model can be made directly through an atomistic-to-coarse-grain mapping procedure that when reversed allows one to model well equilibrated atomistic structures by performing this equilibration by using the coarse-grain model. This mapping and reverse mapping helps to overcome the time-scale upper limits of molecular dynamics simulations. Several approaches to coarse graining have been proposed and include both continuous and lattice models. The continuous models seem to be preferable for dynamic problems such as might occur when considering dynamic changes in volume. The systematic development of the coarse-grain model requires three principal steps (Kremer and Müller-Plathe 2001): (a) Determine the degree of coarse-graining and the geometry of the model. (b) Choose the form of the intra- and inter-chain potentials. (c) Optimize the free parameters, especially for the nonbonded interactions.

Coarse-grain models are often linked to Monte Carlo (MC) simulations to provide a timely solution. The MC method is used to simulate stochastic events and provide statistical approaches to numerical integration. There are three characteristic steps in the MC simulation (Gates and Hinkley 2003): (a) Translate the physical problem into an analogous probabilistic or statistical model. (b) Solve the probabilistic model by a numerical sampling experiment. (c) Analyze the resultant data by using statistical methods. Monte Carlo simulation methods are roughly grouped into four categories: Weighted and non-weighted sampling methods, lattice type, spin model, and energy operator.

2.6.3.3 Continuum Methods Combined with Micromechanics

Despite the importance of understanding the molecular structure and nature of materials, at some level in the multi-scale analysis the behavior of collections of molecules and atoms can be homogenized. At this level, the continuum level, the observed macroscopic behavior is explained by disregarding the discrete atomistic and molecular structure and assuming that the material is continuously distributed throughout its volume. The continuum material is assumed to have an average density and can be subjected to body forces such as gravity and surface forces such as the contact between two bodies. The continuum can be assumed to obey several fundamental laws. The first, continuity, is derived from the conservation of mass. The second, equilibrium, is derived from momentum considerations and Newton's second law. The third, the moment of momentum principle, is based on the model that the time rate of change of angular momentum with respect to an arbitrary point is equal to the resultant moment. The next two laws, conservation of energy and entropy are based on the first and second laws of thermodynamics, respectively. These laws provide the basis for the continuum model and must be coupled with the appropriate constitutive equations and equations of state to provide all the equations necessary for solving a continuum problem. The state of the continuum system is described by several thermodynamic and kinematic state variables. The equations of state provide the relationships between the nonindependent state variables. The continuum method relates the deformation of a continuous medium to the external forces acting on the medium and the resulting internal stress and strain. Computational approaches range from simple closed-form analytical expressions to micromechanics to complex structural mechanics calculations based on beam and shell theory. The continuum-mechanics methods rely on describing the geometry (i.e., a physical model), and must have a constitutive relationship to achieve a solution (Mase 1970). For a displacement-based form of continuum solution, the principle of virtual work is assumed valid. In general, this is given as Gates and Hinkley (2003):

$$\delta W = \oiint \int_V \sigma_{ij} \delta \varepsilon_{ij} dV = \oiint \int_V P_j \delta u_j dV \oiint \int_S T_j \delta u_j dS + F_j \delta u_j \tag{2.16}$$

where W is the virtual work which is the work done by imaginary or virtual displacements, ε is the strain, σ is the stress, P is the body force, u is the virtual displacement, T is the tractions and F is the point forces. The symbol δ is the variational operator designating the virtual quantity. For a continuum system, a necessary and sufficient condition for equilibrium is that the virtual work done by sum of the external forces and internal forces vanish for any virtual displacement (Yang 1986).

Homogenization of a multi-constituent material requires the combination of the continuum method and a micromechanics model to provide a transition from the microscale to the macroscale. Micromechanics assumes small-deformation continuum mechanics as outlined in the preceding section. Continuum mechanics, in general, assumes uniform material properties within the boundaries of the problem. At the microscale, this assumption of uniformity may not hold and hence the micromechanics method is used to express the continuum quantities associated with an infinitesimal material element in terms of the parameters that characterize the structure and properties of the micro-constituents of the element. A central theme of micromechanics models is the development of a representative volume element (RVE) that is a statistical representation of the local continuum properties. In this sense, the RVE may include material boundaries, voids, and defects apparent at the microscale. The RVE is constructed to ensure that the length scale is consistent with the smallest constituent that has a first-order effect on the macroscopic behavior. The RVE is then used in a repeating or periodic nature in the full-scale model. The approach to the micromechanics solution therefore requires a RVE and a suitable averaging technique. The volume average of a typical, spatially variable, integrable quantity $T(x)$ is (Gates and Hinkley 2003)

$$\langle T \rangle = \frac{1}{V}\int_V T(x)dV \tag{2.17}$$

where V is the volume of the RVE. Then, the unweighted volume average stress and strain are given by Gates and Hinkley (2003)

$$\overline{\sigma} \equiv \langle \sigma \rangle \quad \text{and} \quad \overline{\varepsilon} \equiv \langle \varepsilon \rangle \tag{2.18}$$

respectively. The principle of virtual work is assumed to be valid. The micromechanics method can account for interfaces between constituents, discontinuities, and coupled mechanical and nonmechanical properties.

2.6.3.4 Finite Element Methods

Finite element methods (FEM) have a long history of development for a wide variety of applications including problems in mechanical, biological, and geological systems. The FEM goal is to provide a numerical, approximate solution to initial-value and boundary-value problems including time-dependent processes. The method uses a variational technique for solving the differential equations wherein the continuous problem described by the differential equation is cast into the equivalent variation form and the solution is found to be a linear combination of approximation functions. In the FEM, the physical shape of the domain of interest is broken into simple subdomains (elements) that are interconnected and fill the entire domain without overlaps. A displacement-based form of the FEM starts with the principle of virtual work for a continuum with the following steps (Gates and

Hinkley 2003): (a) Replace the continuum domain with an assemblage of subdomains. (b) Select the appropriate constitutive laws. (c) Select the interpolation functions necessary to map the element topology. (d) Describe the problem by using the variational principle and divide the system level integral into subintegrals over the elements. (e) Replace continuum state variables by interpolation functions. (f) Assemble element equations. (g) Assemble global system equations. (h) Solve global system of equations, taking into account the prescribed boundary conditions. (i) Calculate the state equation values from state variables.

The coupled atomistic-continuum modeling approach illustrates one of the primary challenges associated with hierarchical modeling of materials; namely, the accurate prediction of physical/chemical properties and behavior from nanoscale to macroscale without loss of intrinsic structural information. The time and length scales associated with the simulation methods have been illustrated in Fig. 2.10, with each method placed according to the upper range of its resolution. As one moves across a scale, overlaps on both time and length resolution occur, but the overall trend is consistent. It is recognized that at each level of homogenization or scale-up, the risk of losing the key structural information increases. The way to provide an accurate check and balance against these losses is to establish verification of analysis methods and validation of simulations at both the atomic and bulk scales (Gates and Hinkley 2003).

2.7 Integrated Materials Process Control and Sensing in Energy Systems

Monitoring and controlling not only materials production processes (Fig. 2.11), but also the processes occurring in the energy systems (Fig. 2.12) is an important foundational area. From a materials science and engineering perspective, the development and utilization of robust sensors are a key component. Sensors that can tolerate harsh environments are especially important for materials production processes as well as a number of energy generation systems. Sensor materials development, design, and implementation, including in a wireless mode, apply across the energy source and use spectrum. Smart materials are materials that have one or more properties that can be significantly changed in a controlled fashion by external stimuli, such as stress, temperature, moisture, pH, electric fields, or magnetic fields. Coupling sensing and actuation functions into materials with structural or other functional purposes can facilitate materials and system performance. Achieving maximum efficiency from both materials manufacturing processes and energy systems demands careful process monitoring and control, often in extreme environments not conducive to using existing sensor technology. Materials-enabled innovations in integrated process control and sensors will provide improvements in energy and resource efficiency across the energy sector. This section will take combustion process as an example to demonstrate the integrated process control with sensing technologies.

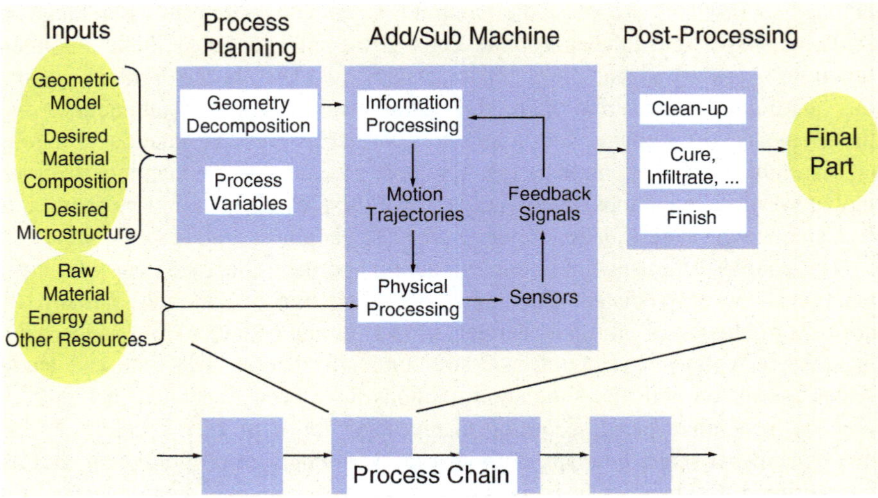

Fig. 2.11 Illustration of additive/subtractive control system for materials processing (Adapted from Rosen 2004. Credit: World Technology Evaluation Center (WTEC))

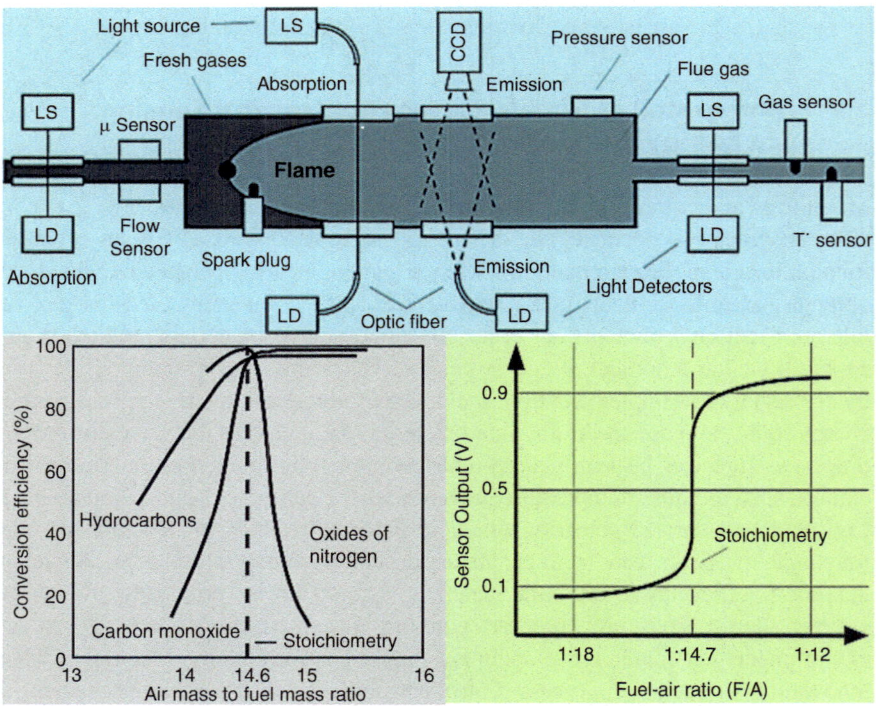

Fig. 2.12 Typical combustion process, sensors and diagnostics for combustion control (*LD* light detector, *LS* light source) (Modified with permission from Docquier and Candel 2002 (Elsevier))

2.7.1 Process Control and Its Constituents

The term control means methods to force parameters in the environment to have specific values. This can be as simple as making the temperature in a room stay at 25 °C or as complex as manufacturing an integrated circuit or guiding a spacecraft to Jupiter. In general, all of the elements necessary to accomplish the control objective are described by the term control system.

The basic strategy by which a control system operates is quite logical and natural. In fact, the same strategy is employed in living organisms to maintain temperature, fluid flow rate, and a host of other biological functions. This is natural process control. The technology of artificial control was first developed with a human as an integral part of the control action. When it is learnt how to use machines, electronics, and computers to replace the human function, the term "automatic control" came into use. The process may be controlled by measuring a variable representing the desired state of the product and automatically adjusting one of the other variables of the process. In process control, the basic objective is to regulate the value of some quantity. To regulate means to maintain that quantity at some desired value (reference value or set point) regardless of external influences. During the first industrial revolution the work done by human muscles was gradually replaced by the power of machines. Process control opened the door to the second industrial revolution, where the routine functions of the human mind and the need for the continuous presence of human observers were also taken care of by machines. In true sense process control had made optimization, and thereby, the beginning of the third industrial revolution possible. Here the traditional goal of maximizing the quantity production is gradually replaced by the goal of maximizing the quality and durability of the produced goods, while minimizing the consumption of energy and raw materials and maximizing recycling and reuse (Krishnaswamy 2007).

For instance, control of a combustion process relies on several elements which are essentially similar for the automotive and the gas turbine applications. In these cases, combustion takes place between fresh reactants supplied (at least for the fuel) by flow metering injectors. Exhaust gases are produced and pass downstream of the combustor through a manifold (Docquier and Candel 2002): (a) In active combustion control (ACC), the controller output is used to modulate the flow properties such as fuel flow rate modulation to avoid or limit pressure oscillations, or to improve the combustion characteristics. (b) In operating point control (OPC), the injection of fuel is regulated in order to maintain certain flame parameters like the fuel to air equivalence ratio in a prescribed range of values.

2.7.1.1 Sensing Techniques

Sensors and analyzers are a control system's window to the world. A sensor is a device that converts a physical stimulus into a readable output, which would preferably be electronic or optoelectronic, but can also be communicated via other means, such as visual and acoustic. Sensors usually designed with one or more of surprisingly many performance features and attributes, such as signal-to-noise ratio, reliability, safety and intrinsic safety, accuracy, response time, dynamic range, cost, power consumption, size, and electromagnetic interference immunity. Developments in sensor technology have been driven by the increasing use of signal processing for compensation, typically used for reducing cross-sensitivity to secondary variables; multivariable inferential sensing, which allows sensing solutions to be developed for parameters that are infeasible to directly measure online; and self-checking and self-compensating sensors that enhance reliability and reduce maintenance costs. There are many kinds of sensors, which are usually classified based on the form of energy being transduced and on whether the transduction mechanism is self-generating (like a thermocouple or a piezoelectric material) or a modulating mechanism (like a thermistor or a piezoresistor). For instance, examples of sensors based on human-made versus biological include photodiode (light intensity)/Retina (light intensity), psychrometer (humidity)/Cochlea (sound), barometer (pressure)/Ear canal (level, rotation), phenolphthalein (pH)/Skin (temperature), timer (time)/Skin and hair (air flow), and odometer (distance)/Olfactory cells (gas composition). A classification focuses on the physical effects that sensors can respond to. For examples, in view of their principle of operation being based on mixed effects, the sensors can be classified as (Zook et al. 2000): (a) Mechanical motion including mechanical resonance, such as pendulum-clock, quartz clock, spring balance, odometer, piezoresistive pressure sensor, accelerometer, and gyro. (b) Thermal including temperature differences, such as thermometer, thermocouple, thermistor, thermal conductivity detector, transistor built-in voltage, and air flow sensors. (c) Optical energy (photons), such as photodiode, CCD camera, Geiger-Mueller tube, color sensor, and turbidity sensor. (d) Magnetic field, such as compass, Hall-effect, magnetoresistance, and inductive proximity sensor. (e) Electric field, such as electrostatic voltmeter, and field-effect transistor.

Figure 2.12 shows a generic combustion process with available or promising sensing techniques. Sensors may be placed at various locations. In some cases, it might be useful to sense the flow properties in the upstream manifold, in fresh reactants, e.g., to evaluate the mixing quality and improve it if necessary, determine fuel or air composition, or estimate the flow velocity. Most of the sensors however are located further downstream to observe the flame or to analyze the flue gases. The number of flame parameters, which can be measured is relatively small and the sensing techniques differ in terms of time and spatial resolution as well as in accuracy. Table 2.1 classifies the diagnostic techniques sketched in Fig. 2.12 according to their time response, principle of operation and sensed parameter. The selection of a diagnostic technique not only depends on the type of controller one would like to design, but also and mainly on practical aspects. As an example, optical techniques generally feature a good spatial and temporal resolution and are nonintrusive, but require at least one optical access, which is difficult to incorporate in a practical device. Indeed, windows and window mounts have to sustain the pressure and temperature of the flow. For in situ measurements, film cooling with purge gas might also be necessary to cool the windows and keep soot and particles from impacting and depositing on the surface. Therefore, combustors might be sophisticatedly designed to provide parts for optical access. However, development of optical fibers opens possibilities to the application of optical fiber diagnostics to

Table 2.1 Optical and solid-state sensors for combustion control (Docquier and Candel 2002)

Sensor	Frequency	Technique	Detector	Parameter
Optical sensors	Low	UV-Vis emission	CCD camera	OH, CH, C_2, CO_2, temperature (soot)
			Wide-band spectrometer	
		IR emission	CCD camera	Temperature (T)
	High	UV-Vis emission	Photo multiplier tube and filter	OH, CH, C_2, CO_2
			Photo diode and filter	
		Vis-IR emission	Photo diode and filter	Temperature (T)
			Narrow-band spectrometer	
		IR absorption	Laser diode and photo diode	H_2O, CH, CO_2, T, pressure (p)
Solid-state sensors	Low	Electrochemical	–	O_2, CO, HC, NO_x
		Catalytic	–	CO, Nox
		Thermocouple, resistive temperature detector	–	Temperature (T)
		Viscosimeter	–	Wobbe index (W_O)
	High	Spark plug	–	Ion current
		Electric, resistive	–	Pressure (p)

industrial energy systems. Fibers are available for UV or IR light channeling and feature a reduced cross section. Their mechanical resistance may be augmented by a protective envelope. The sensor and its equipment may be installed at a distance from the combustor thus simplifying the setup for in-cylinder and for gas turbine measurements. Whereas spatial and time resolution might be essential for controller design, the performance will be measured in terms of global emissions of the combustion process. In this respect, the line of sight methods might lead to errors in the evaluation of the system output if the flow is highly nonhomogeneous. In this case, the global performance could be evaluated with, for example, solid-state gas sensors exposed to combustion products in the exhaust stream. This will induce however a time lag due to gas transport and it will slow down the controller response to changes. In addition, due to the hostile conditions prevailing in combustors, the sensors would withstand exposure to this environment. The sensor not only has to accurate enough for control but also has to operate safely for long periods of time. Automobile sensors have to reliable for more than 160,000 km and the mean time between failure for gas turbine sensors should be about 25,000 h. Sensor and process aging are also likely to modify the controlled system behavior (Docquier and Candel 2002).

While techniques presented in Table 2.1 can be implemented simultaneously, the controller may also rely on a complex modeling effort of the combustion process.

This allows predictions of the system response up to a certain limit and makes use of sensor data to correct the model and take decisions for process optimization. Whereas a complete controller design also requires efficient actuators and control algorithms, a central issue is to identify parameters which are best suited for precise and robust control with proper selection of the diagnostic techniques.

2.7.1.2 Input Parameters for Combustion Control

The selected parameters which could be used as input to the control system have to be detectable, but also have to be closely related to the property to be controlled.

1. Operating point control
   In most cases, the control of the fuel to air equivalence ratio Φ of the mixture is crucial for maintaining emissions as a low level. In the case of lean premixed combustion, the flame temperature T_f is one of the physical parameters responsible for the rapid change of NO_x and CO with Φ. Therefore, sensing of Φ or T_f would be directly useful for control. However, for some cases it is not always possible to detect the appropriate flame characteristic and one has to combine several parameters. For example, the equivalence ratio Φ, flame temperature T_f and oxygen mole fraction in the flue gas X_{O_2} can easily be liked together (Docquier and Candel 2002):

$$\Phi = \alpha \frac{1 - (1 + \beta) X_{O_2}}{\alpha + X_{O_2}} \tag{2.19}$$

   where $\alpha = 2$ and $\beta = 3.76$ for a fuel lean mixture of methane and air:

$$\Phi CH_4 + \alpha(O_2 + \beta N_2) \overset{\text{yields}}{\rightarrow} \Phi CO_2 + \alpha(\Phi H_2O + \beta N_2) + \alpha(1 - \Phi)O_2 \tag{2.20}$$

2. Performance optimization
   The optimization of combustion performance is often associated to pollutant emissions reduction or combustion efficiency improvement. In each case, the observation of combustion parameters directly related to parameter one would like to improve should be performed. For example, major pollutants such as NO_x, CO, and soot should be sensed to achieve pollutant emissions reduction. This has been performed using classical gas analyzers, calibrated optical sensors, or optical sensors delivering uncalibrated signals. Combustion efficiency optimization has also been performed using optical measurements of H_2O and temperature or gas analyzers (Docquier and Candel 2002).
3. Active combustion control
   Active control generally requires an observation of the dynamic state of the system and a performance index. In many applications, this index is deduced from the signals describing the state. In some cases, the performance index is obtained from separate measurements. In closed loop active control the state

information is fed back by the controller to an actuator or a set of actuators, and the performance index is used to adjust the controller parameters in order to optimize the combustor operation. Combustion control strategies are needed to fulfill future objectives in pollutant emissions reduction and instability alleviation. This will require monitoring of a broad range of parameters. Detection should be carried out with fast, reliable, and preferably low-cost sensors (Docquier and Candel 2002).

2.7.2 Diagnostic Techniques

Apart from classical devices such as thermocouples, pressure sensors, and optical detectors (photodiodes, photomultiplier tubes, and spectrometers), advanced diagnostic techniques mainly include optical diagnostic techniques and solid-state sensors.

2.7.2.1 Optical Diagnostics

Optical diagnostic operate in a broad spectral range, from UV (~200 nm) to IR (~10,000 nm), most sensors operating in the UV to near IR (~2000 nm) range. In each spectral domain, the optical sensors can either monitor emission, absorption, scattered light or fluorescence. Qualitative or quantitative data may be collected. Sensors detecting the light emission essentially provide qualitative information. A calibration procedure may be used to extract indirect information on the flame parameters (equivalence ratio or heat release). Absorption sensors on the other hand may be used to determine the mole fraction of combustion intermediates products as well as the temperature and sometimes even the pressure or the velocity of the probed medium (Docquier and Candel 2002).

1. Optical absorption sensors
   Optical absorption sensors are usually made based on near IR and visible absorption. Commercially available devices with sensing range from 0.63 to 2.0 μm (AlGaAs, InGaAsP, InGaAs/InP) allow exploitation of absorption bands of species like H_2O, NO_x, CO or CO_2. Processing of the absorption data collected at several wavelengths may be used to calculate the temperature. The light is provided by diode lasers and the detection system includes optical isolators, fibers, and photodetectors which are most of the time photodiode elements (InGaAs, Si). Taking advantage of fiber components for wavelength division multiplexing, multiplexing lasers may be combined into common signal and reference fibers. Each laser may be swept at different phases of a common ramp function. This provides a time-domain multiplexing where a single photodetector observes the sequential absorption features as each laser is swept across its corresponding line shape. Wavelength-domain multiplexing has also been used in diode-laser combustion sensors. In this configuration, the lasers can be ramped across their respective absorption lines simultaneously. Multiple wavelengths appearing in the transmitted beam are separated onto multiple

detectors using a frequency-dispersive unit such as a diffraction grating (Allen 1998).

2. Optical emission sensors
   The light emitted by the flame may be used to monitor and control combustion. Information may be gathered from UV and IR wavelengths. Several phenomena contribute to the light emission process. Discrete emission spectra correspond to the chemical reactions producing electronically excited radicals in the flame front in the UV-visible range (chemiluminescence). Combustion products in thermal equilibrium such as H_2O, CO_2 as well as CO are also responsible for rotation-vibration emission bands in the near-IR and IR range (IR emission). Continuum emission may also be considered whenever soot is produced and radiates in the combustion process (black body emission). Laser diodes, fibers, fiber splitters, and photo-detectors have been used to develop optical sensors for practical combustion control applications. Because they are simple, naturally robust, and relatively easy to set up, light emission sensors can play an important role in the estimation of combustion parameters like the equivalence ratio, the temperature or the heat release fluctuations (Docquier and Candel 2002).

2.7.2.2 Solid-State Sensors

Various types of solid-state sensors have been developed mainly for combustion control applications. The sensors can withstand severe operating conditions, and offer long term reliability at a low cost.

1. Gas sensors
   The initial solid-state gas sensors were designed to monitor automotive combustion stoichiometry and were operating on a binary basis. To improve control performance, wide-band oxygen sensors have been utilized to monitor variable amounts of oxygen mixed with combustion products. In order to cope with more stringent pollutant emissions regulations, gas sensors are devised to measure regulated species such as CO, unburnt hydrocarbons, and NO_x (Docquier and Candel 2002). Sensors capable of providing three-dimensional maps of emission profiles will allow for feedback control systems of combustion processes, resulting in lower emissions and efficient use of fuels. Emissions-monitoring sensors for these applications include those for CO, NO_x, O_2, CO_2, hydrocarbons, and volatile organic compounds (Akbar et al. 2006).

   Gas species monitoring is mainly achieved with semiconductor sensors. These probes use a semiconducting material to detect a particular species. Materials employed are (a) oxide semiconductors, such as zirconium (ZrO_2), titanium (TiO_2), and tin (SnO_2) oxides doped with other oxides; or (b) ion-conducting ceramics with bulk electrolytic properties. In TiO_2 and SnO_2 sensors, the principle of operation relies on the change of the semiconductor resistance with oxygen partial pressure whereas in ZrO_2 sensors, the presence of the probed species is determined through voltage or current changes across the semiconductor. TiO_2 and SnO_2 sensors have also been designed for NO_x and CO monitoring but are less attractive than zirconium-oxide-based sensors which have greater selectivity and feature better stability under elevated operating temperature conditions. Other types of gas sensors use catalytic effects. These devices generally comprise two resistive temperature detectors, one of which is coated with a catalytic material, which activates reactions with the gas to detect. The heat released at one probe induces a difference in resistance between the elements of two resistive temperature detectors. Sensors of this type have been developed for CO and NO_x (via ammonia addition) monitoring but feature relatively poor selectivity and are subject to poisoning. Therefore, most ceramic gas sensors are rely on the electrochemical properties of a solid solution of zirconium oxide doped with other oxides such as CaO, MgO, or Y_2O_3 (Docquier and Candel 2002).

2. Ion current probes

   While solid-state gas sensors provide useful information concerning the combustion process, they suffer from major disadvantages with respect to optical diagnostics (Docquier and Candel 2002): (a) Gas sensors are operated downstream of the process. This induces a time lag between combustion and corresponding probe measurements. This delay affects the control algorithm stability and design. (b) Gas sensors have a slow time response and only provide global information about the combustion process. As a result, they cannot be used to monitor transients like combustion instabilities, knock or misfire in spark ignition engines. In this case also, gas sensors cannot provide cylinder-to-cylinder information. To address these problems, optical techniques may be used advantageously. However, optical access is required together with specific optical elements (adapters, filters, detectors). Ion current probes or ionization sensors provide alternative possibilities.

   Ion current has long been investigated as a combustion diagnostic tool. It works as follows (Gazis et al. 2006): The prevailing conditions during the combustion causes ionization of the gases inside the cylinder. This ionization occurs mainly in two phases. The first phase occurs during combustion as fuel reacts with oxygen. It is defined as the chemical phase. The second phase occurs as the already burnt gases are compressed by rising pressure created by the fuel farther away from the spark plug combusting. It can therefore be defined as the thermal phase. By applying a voltage through these gases, a current will be observed. It is up to the engine controller then to decide what kind of information can be deduced from this ion current signal. Although this technique seems particularly well adapted for spark ignition engines (localized combustion process, spark plug sensing, fast cylinder-to-cylinder measurements), it could be adapted to other situations where the best ion sensor location can be determined. Ion current signals are generally difficult to interpret, but they clearly correlate with parameters such as pressure and equivalence ratio.

3. Other sensors

   Other sensors including resistive temperature detectors, micromachined sensors can also be used for combustion control. Resistive temperature detectors have been developed to monitor the temperature of automotive engine exhaust gases. Their sensing element is generally made up of a Thin Pt-based film whose resistance increases with temperature (thermally decoupled from the sensor body to minimize conduction), and is exposed to the exhaust gas flowing in an open housing, which also acts like a radiation shield. Micromachined sensors have been used for flow, pressure, and viscosity sensing (Docquier and Candel 2002).

2.8 Summary

The advanced energy systems require advances in energy generation, storage, efficiency, and security that demand a new generation of materials that may not be naturally available. However, creating these new materials requires a level of understanding of the relationships between structure and function, and across many spatial scales. The property and function of each material depend on how it is structured with what elements how the elements are bonded together. The structure of materials can be classified by the general magnitude of various features being considered. The most common major classification of structural, listed generally in increasing size, are nuclear structure, atomic structure, nanostructure, microstructure, mesostructure, and macrostructure. Basic materials research is necessary to enable creation of new materials with the specific characteristics needed for next-generation energy technology.

New materials and processes are critical pacing elements for progress in advanced energy systems and virtually all industrial technologies. Tools for the synthesis, processing, characterization, and simulation and modeling of materials and chemical systems have been developed and deployed. These tools, such as X-ray and neutron sources, nanoscale science facilities, and high-performance computers, provide an unprecedented view of the atomic-scale structure and dynamics of materials and the molecular-scale basis of chemical processes. In particularly, computational power has increased by a factor of a million due to advances in hardware and software, and enabled the development of computer simulations and models of unprecedented fidelity. Emerging capabilities in predictive modeling and simulation have the potential to revolutionize the development of new materials and chemical processes. Coupled with advanced materials characterization and nanoscale science facilities, this predictive capability provides the foundation for an innovation ecosystem that can accelerate the discovery, development, and deployment of new technologies, including advanced energy systems.

Experimental tools to characterize material structures and properties are fundamental to building materials science and engineering understanding and developing new materials with unique properties. As these tools improve, new materials can be developed faster and with greater control over properties, processing, and, ultimately, material performance in end-use applications.

Achieving maximum efficiency from both materials manufacturing processes and energy systems demands careful process monitoring and information they reveal complement the progress in modeling described above to result in synergistic acceleration of progress. The ability to characterize and manipulate materials, particularly at the nanoscale, offers promise of breakthrough discoveries and large-scale use of new materials. Advances in characterization methods such as light, electron, ion, and atomic force microscopy; X-ray, neutron, and other source-based methods; and electron- and ion-beam instruments for microchemistry, structure, and texture identification and surface structure analysis are providing increased understanding of materials at the atomic level. Nanomechanical evaluation methods provide data on the performance of materials at the nanoscale. Increasing capabilities to characterize material in three dimensions are also of value. Extending beyond the 2-D limitations of past methods, these 3-D descriptions add immeasurably to the understanding of materials as well as the ability to more accurately model them.

Monitoring and controlling not only materials production processes, but also the processes occurring in the energy systems is an important foundational area. From a materials science and engineering perspective, the development and utilization of robust sensors are a key component. Sensors that can tolerate harsh environments are especially important for materials production processes as well as a number of energy generation systems. Sensor materials development, design, and implementation, including in a wireless mode, apply across the energy source and use spectrum. Coupling sensing and actuation functions into materials with structural or other functional purposes can facilitate materials and system performance. Achieving maximum efficiency from both materials manufacturing processes and energy systems demands careful process monitoring and control, often in extreme environments not conducive to using existing sensor technology. Materials-enabled innovations in integrated process control and sensors will provide improvements in energy and resource efficiency across the energy sector.

Exercises

Part I: General Questions

2.1 Iron (Fe) crystallizes in a BCC unit cell at room temperature. Calculate the radius of an iron atom in this crystal. At temperatures above 910 °C iron prefers to be FCC. If the temperature dependence of the radius of the iron atom is neglected on the grounds that it is negligible, the density of FCC iron can be calculated. Use this to determine whether iron expands or contracts when it undergoes transformation from the BCC to the FCC structure.

2.2 A metal is found to have BCC structure, a lattice constant of 3.31 Å, and a density of 16.6 g/cm^3. Determine the atomic weight of this element.

2.3 At 100 °C copper (Cu) has a lattice constant of 3.655 Å. What is its density at this temperature?

2.4 Determine the second-nearest neighbor distance for nickel (Ni) (in pm) at 100 °C if its density at that temperature is 8.83 g/cm^3.

2.5 Determine the highest linear density of atoms (atoms/m) encountered in vanadium (V).

2.6 Explain how σ and π bonds are similar and how they are different.

2.7 Explain why bonds occur at specific average bond distances instead of the atoms approaching each other infinitely close.
2.8 Use valence bond theory to explain the bonding in F_2, HF, and ClBr. Sketch the overlap of the atomic orbitals involved in the bonds.
2.9 Use valence bond theory to explain the bonding in O_2. Sketch the overlap of the atomic orbitals involved in the bonds in O_2.
2.10 How many σ and π bonds are present in the molecule HCN?
2.11 How does the arrangement of atoms differ in a crystalline and a noncrystalline substance?
2.12 What evidence do we have for the microscopic crystal nature of some solids? What evidence do we have for the visible crystal structure of some solids?
2.13 Uranium is the heaviest atom found in nature. Why isn't uranium metal the most dense material?
2.14 What is the evidence for the claim that steel is elastic? That putty is inelastic?
2.15 What is Hooke's Law? What is the elastic limit? If the linear dimensions of an object are doubled, how much does the total area increase? How much does the volume increase?
2.16 Briefly explain the band theory of electrical conduction. What is Fermi energy? Why are metals highly conductive? Briefly explain the conduction mechanism in metals?
2.17 How is the conductivity of metals affected by impurity level? What is the role of dislocations on conductivity of metals? Why does the metallic conductivity decrease with increasing temperature?
2.18 What is the typical bandgap in semiconductors? What is intrinsic semi-conductivity? Which factors control the conductivity in these semiconductors? Explain the atomic and band theory models of extrinsic semi-conductivity. What is the effect of temperature on extrinsic semi-conductivity?
2.19 What is dielectric constant? What is polarization? How many types are there?
2.20 What is ferro-electricity? Give some examples of ferroelectric materials.
2.21 What is piezoelectricity?
2.22 Explain the difference between corrosion and oxidation processes.
2.23 What properties can be determined by tensile tests? What's the difference between ductility and toughness? How dislocations affect fatigue cracking?
2.24 Explain the role of computational modeling in new materials design. List major modeling approaches.
2.25 Compare functions and limitations of major experimental techniques for materials characterization.
2.26 Explain process control and its constituents. What's the role of sensing techniques in process control?

Part II: Thought-Provoking Questions

2.27 What are the fundamental limits, in terms of propagation of dislocations and their interaction with matrix obstacles, in strength and toughness for structural materials?

2.28 What are the fundamental phenomena that control the plastic flow and fracture properties, both at low and elevated temperatures, of structural materials under extreme conditions of radiation, stress, and environment?

2.29 In what ways are the principles of macroscopic equilibrium materials science altered at the nanoscale and under nonequilibrium thermodynamic conditions, and how do we simulate such processes?

2.30 What are the steps in developing a fundamental understanding of microstructural evolution and phase stability under nonequilibrium conditions in extreme environments that are needed to enable advances in materials design beyond the current experimental observation-based approach?

2.31 How does the predictive quality of modeling and simulation mean? Take a material system as an example, address how the predictive quality is evaluated through the calculation of reliable thermomechanical properties, equilibrium constants, linear energy transfers, etc., that can interpret or replace experimental measures of these same quantities? For multi-scale simulations, how to control multiple sources of uncertainty that propagate to the final multi-scale results to assigning a predictive quality to the simulation? What are your ideas for developing robust scale-linking strategies with mathematical or physical underpinnings to provide the framework for uncertainty quantification and allow experiment/theory iteration to refine models and input data?

References

Akbar, S., et al.: High-temperature ceramic gas sensors: a review. Int. Appl. Ceram. Technol. **3**(4), 302–311 (2006)

Allen, M.G.: Diode laser absorption sensors for gas-dynamic and combustion flows. Meas. Sci. Technol. **9**(4), 545–562 (1998)

Bastard, G., Brum, J.A., Ferreira, R.: Electronic states in semiconductor heterostructures. In: Ehrenreich, H., Turnbull, D. (eds.) Solid State Physics: Semiconductor Heterostructures and Nanostructures, p. 259. Academic Press, Elsevier, Cambridge (1991)

Brady, M.P.: Overview of alumina-forming austenitic (AFA) stainless steel development. http://web.ornl.gov/sci/physical_sciences_directorate/mst/cst/pdf/Brady/OverviewAFAOct10.pdf (2010). Accessed 17 July 2013

Carotenuto, C.: Advanced mechanical spectroscopy to investigate the microstructure of complex fluids. PhD dissertation, University of PISA, Italy (2007)

Carrasco, E., et al.: Analysis at atomic level of dislocation emission and motion around nanoindentations in gold. Surf. Sci. **572**, 467–475 (2004)

Chang, S.L.: X-Ray Multiple-Wave Diffraction-Theory and Application. Springer, New York (2004)

Chinh, N.Q., et al.: Depth sensing indentation tests in studying plastic instabilities. J. Mater. Res. **19**, 31–45 (2004)

Cummings PT, Glotzer SC Inventing a new America through discovery and innovation in science, engineering, and medicine: a vision for research and development in simulation-based engineering and science in the next decade. http://www.nsf.gov/mps/ResearchDirectionsWorkshop2010/RWD-color-FINAL-usletter_2010-07-16.pdf (2010). Accessed 17 July 2013

Damasceno, P.F., Engel, M., Glotzer, S.C.: Predictive self-assembly of polyhedra into complex structures. Science. **337**, 453–457 (2012)

Diop, M.D.: Characterization of microdevices by nanoindentation. In: Nanoindentation in Materials Science, edited by Jiri Nemecek, IntechOpen (2012). DOI: 10.5772/50997. Available from: https://www.intechopen.com/books/nanoindentation-in-materials-science/characterization-of-microdevices-by-nanoindentation

Docquier, N., Candel, S.: Combustion control and sensors: a review. Prog. Energy Combust. Sci. **28**, 107–150 (2002)

DOE: Computational materials science and chemistry for innovation: accelerating discovery and innovation through simulation-based engineering and science. http://science.energy.gov/~/media/bes/pdf/reports/files/cmsc_rpt.pdf (2010). Accessed 11 July 2013

Ejzak J, Sullivan E Harvesting hydrogen via artificial photosynthesis for the use in hydrogen fuel cells. http://136.142.82.187/eng12/Chair/data/papers/3104.pdf (2013). Accessed 17 July 2013

Faridmehr, I., Osman, M.H., Adnan, A.B., Nejad, A.F., Hodjati, R., Azimi, M.: Correlation between engineering stress-strain and true stress-strain curve. Am. J. Civil Eng. Arch. **2**(1), 53–59 (2014)

Ferrero, M., et al.: Pseudogap opening and formation of Fermi arcs as an orbital-selective Mott transition in momentum space. Phys. Rev. B. **80**, 064501 (2009)

Gates, T.S., Hinkley, J.A.: Computational materials: modeling and simulation of nanostructured materials and systems. NASA/TM-2003-212163. Langley Research Center, Hampton, VA (2003)

Gazis, A., et al.: Computationally inexpensive methods of ion current signal manipulation for predicting the characteristics of engine in-cylinder pressure. Int. J. Engine Res. **7**, 271–282 (2006)

Glotzer, S.C., Solomon, M.J., Kotov, N.A.: Self-assembly: from nanoscale to microscale colloids. AIChE J. **50**(12), 2978–2985 (2004)

Hashimoto, A., et al.: Three-dimensional imaging of carbon nanostructures by scanning confocal electron microscopy. J. Appl. Phys. **106**, 086101 (2009)

Helmenstine AM Printable color periodic table of the elements-2015. https://www.thoughtco.com/printable-periodic-tables-p2-608850 (2015). Accepted 26 May 2017

Holgate, S.A.: Understanding Solid State Physics. CRC Press. 177–178 (2009). ISBN 1420012320

Hosono, E., Fujihara, S., Honma, I., Zhou, H.: Superhydrophobic perpendicular nanopin film by the bottom-up process. J. Am. Chem. Soc. **127**(39), 13458–13459 (2005)

Kimoto, K., et al.: Element-selective imaging of atomic columns in a crystal using STEM and EELS. Nature. **450**(7170), 702–704 (2007)

Kremer, K., Müller-Plathe, F.: Multiscale problems in polymer science: Simulation approaches. MRSBulletin **26**(3), 205–210 (2001)

Kong, X.Y., Ding, Y., Yang, R., Wang, Z.L.: Single-crystal nanorings formed by epitaxial self-coiling of polar nanobelts. Science. **303**(5662), 1348–1351 (2004)

Krishnaswamy, K.: Process control. New Delhi, New Age International (2007)

Lakhtakia, A., Messier, R.: Sculptured Thin Films: Nanoengineered Morphology and Optics. SPIE Press, Bellingham, WA (2005)

Li, Z., et al.: Development of a micro-miniature nanoindentation instrument. www.imeko.org/publications/tc5-2007/IMEKO-TC5-2007-001.pdf. Accessed 1 Aug 2013 (2007)

Li, J.R., et al.: Carbon dioxide capture-related gas adsorption and separation in metal-organic frameworks. Coord. Chem. Rev. **255**(2011), 1791–1823 (2011)

Loo, C., Lin, A., Hirsch, L., Lee, M., Barton, J., Halas, N., West, J., Drezek, R.: Nanoshell-enabled photonics-based imaging and therapy of cancer. Technol. Cancer Res. Treat. **3**(1), 33–40 (2004)

López-Lemus, J., Alejandre, J.: Thermodynamic and transport properties of simple fluids using lattice sums: bulk phases and liquid-vapour interface. Mol. Phys. **100**(18), 2983–2992 (2002)
Magnasco, V.: Methods of molecular quantum mechanics: An introduction to electronic molecular structure. John Wiley & Sons, New York (2009)
Mase, G.E.: Theory and Problems of Continuum Mechanics. McGraw-Hill, New York (1970)
Mason, J.K., et al.: Determining the activation energy and volume for the onset of plasticity during nanoindentation. Phys. Rev. B. **73**(5), 054102 (2006)
Menard, K.P.: Dynamic Mechanical Analysis: A Practical Introduction. CRC Press, Boca Raton (1999)
Miessler, G.L., Fischer, P.J., Tarr, D.A.: Inorganic Chemistry, 5th edn. London, Pearson (2014)
Miller, M.K.: Atom Probe Tomography: Analysis at the Atomic Level. Kluwer Academic/Plenum Publishers (2000). ISBN 0-306-46415-2
Miller, M.K., Russell, K.F.: Embrittlement of RPV steels: an atom probe tomography perspective. J. Nucl. Mater. **371**, 145–160 (2007)
Mokrani, A., Ewels, C.: Computational materials modeling. http://www.ewels.info/science/posters/Ewels_Mokrani.pdf (2013). Accessed 16 July 2013
Morris, Jr. J.W.: A survey of materials science I. Structure http://www.mse.berkeley.edu/groups/morris/MSE200/I-structure.pdf (2007). Accessed 29 May 2017
NDT: Material/processes—structure of materials. NDT Resource Center. https://www.nde-ed.org/EducationResources/CommunityCollege/Materials/Structure/solidstate.htm (2014). Accessed 1 June 2017
Nguyen, T.D., Glotzer, S.C.: Reconfigurable assemblies of shape-changing nanorods. ACS Nano. **4** (5), 2585–2594 (2010)
Oliver, W.C., Pharr, G.M.: An improved technique for determining hardness and elastic modulus using load and displacement sensing indentation experiments. J. Mater. Res. **7**(6), 1564–1583 (1992)
OpenStax: Chemistry. OpenStax CNX. http://cnx.org/contents/85abf193-2bd2-4908-8563-90b8a7ac8df6@9.311 (2016). Accessed 7 June 2017
Paddison SJ Computational materials science: multi-scale modeling. http://www.tnepscor.org/sites/www/Uploads/Files/PaddisonTNSCORE_2013.pdf (2013). Accessed 16 July 2013
Panitchayangkoon, G., et al.: Direct evidence of quantum transport in photosynthetic light-harvesting complexes. Proc. Natl. Acad. Sci. U. S. A. **108**(52), 20908–20912 (2011)
Pennycook, S.J., et al.: Aberration-corrected scanning transmission electron microscopy: from atomic imaging and analysis to solving energy problems. Phil. Trans. R. Soc. A. **367**, 3709–3733 (2009)
Pharr, G.M.: Probing the mechanical properties of materials at small scales with nanoindentation. http://cdp.ornl.gov/EFRCschool/pdf/talks/Pharr.pdf (2013). Accessed 29 July 2013
Robinson, L.: Materials: Foundation for the Clean Energy Age. http://energy.tms.org/docs/pdfs/Materials_Foundation_for_Clean_Energy_Age_Press_Final.pdf (2012). Accessed 11 July 2013
Rogers, E.: The Energy Savings Potential of Smart Manufacturing. Report IE 1403. ACEEE, Washington, DC (2014)
Rosen, D.W.: Process control and metrics. In: WTEC Panel Report on additive/subtractive manufacturing research and development in Europe. http://www.wtec.org/additive/report/additive-report.pdf (2004). Accessed 12 August 2013
Schuh, C.A.: Nanoindentation studies of materials. Mater. Today. **9**(5), 32–40 (2006)
Shaik, S.S., Hiberty, P.C.: A Chemist's Guide to Valence Bond Theory. Wiley-Interscience, New Jersey (2008). ISBN 978-0-470-03735-5
Simons, J.: An introduction to theoretical chemistry. Cambridge University Press, Cambridge (2003). ISBN 0521823609
Tachi, M., et al.: On the dislocation mechanism of amorphization of Si by indentation. Phil. Mag. Lett. **82**(3), 133–139 (2002)
Tong, C.X.: Advanced Materials for Thermal Management of Electronic Packaging (Springer Series in Advanced Microelectronics). Springer, New York (2011)

Vaz Jr., M., et al.: Materials modeling-challenges and perspectives. In: Vaz Jr., M., de Souza Neto, E.A., Munoz-Rojas, P.A. (eds.) Advanced Computational Materials Modeling: From Classical to Multi-Scale Techniques, pp. 1–21. WILEY-VCH Verlag GmbH & Co. KGaA, Weinheim (2011)

Wang, X.L., et al.: VULCAN—the engineering diffractometer at the SNS. Physica B. **385–386**, 673–675 (2006)

Wee G Miscibility, dynamic mechanical and optical limiting behavior of C60-based and multiwalled carbon nanotube-containing polymeric materials. PhD dissertation. University of Singapore (2004)

Wheeler JM Nanoindentation under dynamic conditions, PhD dissertation. University of Cambridge (2009)

Xia, F., Jiang, L.: Bio-inspired, smart, multiscale interfacial materials. Avd. Mater. **20**, 2842–2858 (2008)

Yang, T.Y.: Finite Element Structural Analysis. Prentice-Hall, Englewood Cliffs (1986)

Yoo, C.S., et al.: First-order Isostructural Mott transition in highly compressed MnO. Phys. Rev. Lett. **94**(11), 115502 (2005)

Young, J.A., et al.: Molecular simulations of the imidization of adsorbed polyamic acid. Macromolecules. **33**, 4936 (2000)

Yu, P., Cardona, M.: Fundamentals of Semiconductors Physics and Materials Properties. Springer, New York (2010)

Zinkle, S.J., et al.: Advances in microstructural characterization. J. Nucl. Mater. **386–388**, 8–14 (2009)

Zook, J.D., et al.: Sensors in control systems. In: Samad, T., Weyrauch, J. (eds.) Automation, Control and Complexity: An Integrated Approach, pp. 131–150. Wiley, New York (2000)

3 Advanced Materials Enable Energy Production from Fossil Fuels

Abstract

Development of advanced materials and processes are vital to higher production efficiency and more economical and environmentally sustainable fossil energy systems. There is a need for the discovery of new materials that can withstand higher temperature and severe corrosive environments of the advanced fossil-fueled power generation technologies such as coal gasifiers, turbines, combustors, and coal-based fuel cells. Whether the fuel is converted to electricity in a combined cycle gas turbine or a boiler/steam turbine system, both systems produce higher efficiencies when operated at higher temperatures. Significant efforts have resulted in advanced materials for use in both gas turbines and boiler/steam turbine equipment. Coatings applied to hot components in gas turbines have allowed higher operating temperatures, thereby resulting in efficiency improvements. Advances in materials have also resulted in significant improvements in the overall efficiency of converting the energy in the fuel to electricity. Through the development of advanced coatings and new methods of applying the coatings as well as new materials, significant reductions in fuel consumption with an associated reduction in greenhouse gas and other criteria pollutant emissions have been realized. This chapter will focus on how to overcome materials challenges that are being actively pursued to achieve sustainable fossil energy systems, including ultra-supercritical materials, coatings and protection materials, high-strength and corrosion-resistant alloys, functional materials, as well as sensing materials used in harsh environments.

C. Tong, *Introduction to Materials for Advanced Energy Systems*,
https://doi.org/10.1007/978-3-319-98002-7_3

3.1 Materials Challenges for Fossil Energy Systems

Petroleum, natural gas, and coal are the main sources of energy for modern use. These fuels are classified as fossil fuels. Building high efficiency fossil-fueled power stations with a large-scale carbon capture and storage to reduce CO_2 emissions brings with it new material challenges, as these technologies involve components operating in severe and aggressive environments with significant implications on the materials being used. The key areas for materials technology development are therefore related to (Allen 2007): (a) Increasing plant efficiency; (b) co-firing with renewable fuels; and (c) CO_2 sequestration. The combination of heat and power production can be utilized to reduce overall CO_2 emissions. Coal gasification and the integrated gasification combined cycle can potentially be combined with CO_2 capture and storage. As a result, the generic underpinning technologies and key areas of fossil fuel power generation have been concentrated on boilers, steam turbines, gasifies, and CO_2 capture, as shown in Fig. 3.1. The underpin materials including surface protection coatings, nondestructive technology, lifting, repair, and joining must be developed in parallel with basic component materials to ensure that supporting technology is in place to allow material to be used in products. Each of these technologies is thus an integral part of the individual materials solution for each component.

3.1.1 Boilers

Fossil fuel power stations have rotating machinery to convert the heat energy of combustion into mechanical energy, which then operates an electrical generator. The prime mover may be a steam turbine, a gas turbine, or, in small plants, a

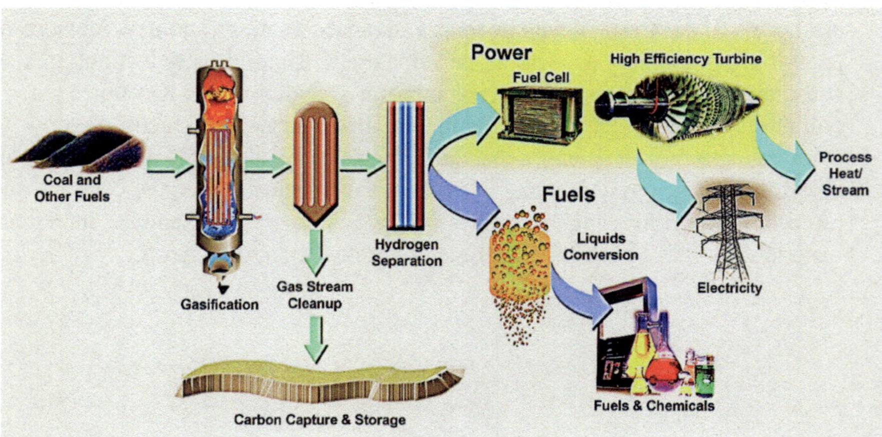

Fig. 3.1 Illustration of advanced power generation, including advanced ultra-supercritical combustion systems (AUSC), Integrated Gasification Combined Cycle (IGCC), fuel cells, gas turbines, and carbon capture and storage (CCS) technology (Adapted from Romanosky and Rawls 2013. Credit: National Energy Technology Laboratory)

reciprocating internal combustion engine. All plants use the energy extracted from expanding gas—steam or combustion gases. Supercritical steam generators are frequently used to produce electric power. They operate at supercritical pressure. In contrast to a subcritical boiler, a supercritical steam generator operates at such a high pressure (over 3200 psi or 22 MPa) that the physical turbulence that characterizes boiling ceases to occur; the fluid is neither liquid nor gas but a super-critical fluid. There is no generation of steam bubbles within the water, because the pressure is above the critical pressure point at which steam bubbles can form. As the fluid expands through the turbine stages, its thermodynamic state drops below the critical point as it does work turning the turbine which turns electrical generator from which power is ultimately extracted. The fluid at that point may be a mix of steam and liquid droplets as it passes into the condenser. This results in slightly less fuel use and therefore less greenhouse gas production. The term "boiler" should not be used for a supercritical pressure steam generator, as no "boiling" actually occurs in this device.

INDEPTH Relationship between operating conditions, plant efficiency, and heat rate for coal-fired power plants (Nicol 2013)

Operating regime	Superheater temperature and pressure	Material in high-temperature components	Efficiency, LHV(net), hard coal, %	Coal consumption, gCOAL/kWh
Subcritical	≤540 °C and < 22.1 MPa	Low-alloy CMn and Mo ferretic steels	<35	≥380
Supercritical	540–580 °C and 22.1–25 MPa	Low-alloy CrMo steels and 9–12% Cr martensitic steel	35–40	380–340
Ultra-supercritical	580–620 °C and 22.1–25 MPa	Improved 9–12% Cr martensitic steels and austenitic steels	40–45	340–320
Advanced ultra-supercritical	700–725 °C and 25–35 MPa	Advanced 10–12% Cr steels and nickel alloys	45–52	320–290

In a thermal power station fuel (coal, oil or gas) is burned in a furnace to produce heat (chemical to heat energy). This heat is then used to change water into steam in the boiler. The steam then drives the turbine (heat to kinetic energy) which in turn drives the generator to produce electricity (kinetic to electrical energy). Ultra-supercritical boiler technology can significantly increase the efficiency of a Rankine cycle power plant, and reduce fuel consumption for a given output, thus proportionally reduce all pollutant and waste streams including CO_2 emissions. However, the benefits of elevated steam conditions must be balanced with plant cost, reliability, and operational flexibility. The limitations for ultra-supercritical boiler technology have been defined in the high gas temperature, high heat flux environment of suspension-fired boilers. Circulating fluidized bed (CFB) boilers have the potential to extend the limits of the technology and further reduce emission levels. These systems include advanced ultra-supercritical combustion systems (AUSC), Integrated Gasification Combined Cycle (IGCC), fuel cells, gas turbines, and carbon capture and storage (CCS) technology to achieve near-zero emissions for power generation. The foundation of this technology, centered in high-temperature materials research, includes the development of new materials that have the potential to improve the performance and/or reduce the cost of existing fossil fuel technologies; development of materials for new systems and capabilities; development of a technology base in the synthesis processing life cycle analysis; and performance characterization of advanced materials. This area has been focused on: (a) Development of new alloys that offer improved corrosion and erosion resistance, and have unique mechanical properties. (b) Development of materials for use under Advanced Ultra-Supercritical (AUSC) conditions of 760 °C and 350 bar pressures (500 pounds per square inch). (c) Development of high-performance materials, particularly alloys, that can perform reliably at temperatures well over 1000 °C. (d) Development of advanced metallic and ceramic coatings to provide thermal and environmental protection. (e) Protection of materials to counter degradation resulting from harsh fossil energy environments. (f) Development of functional materials that serve a unique purpose, such as energy storage. And (g) Pursuit of breakthrough concepts, based on mechanistic understanding from any discipline, for routes to the development of materials with capabilities beyond those currently available. Computational design of materials has the potential to provide major breakthroughs.

3.1.1.1 Furnace Walls

Two design approaches for furnace walls are in common use, spiral and vertical. In both designs a welded membrane joins parallel neighboring tubes. Consequently, weldability and acceptability of welding without post weld heat treatment (PWHT) become factors in the choice of materials together with strength, corrosion resistance, and cost (Allen 2007).

In supercritical steam plant operating in the common regime of 22.1–25 MPa, 540–580 °C, the maximum fluid furnace exit temperature is about 420 °C. This corresponds to a tube mid-wall temperature some 30 °C higher at 450 °C at the start of life and a further 10 °C higher later in the life cycle due to the insulating effect caused by the growth and deposition of magnetic inside the tube. For a design life of 100,000 h both 13CrMo44 and 10CrMo910 have adequate creep strength for these conditions but fireside corrosion must also be considered. Boiler manufacturers offer some of staged combustion to limit NO_x emissions. This requires substoichiometric combustion in the lower portion of the furnace chamber, which produces conditions that can result in rapid thinning and premature failure of the membrane wall tubes. Significant improvements in fireside corrosion resistance are only achieved with much higher chromium levels, >20%. Co-extruded tubing with an austenitic outer layer has been considered but this poses difficulties due to differential thermal expansion and consequential distortion. Boiler manufacturers ensure more oxidizing conditions close to the furnace walls through the arrangement of the burners and thus controlled distribution of the flame and its combustion products. Under conditions where this gives inadequate protection consideration can be given to weld overlay or spray coatings to protect the surface (Allen 2007; Nicol 2013).

For the more advanced steam conditions in ultra-supercritical (22.1–25 and 580–620 °C) steam plant, 7CrWVNb9–6 and 7CrMoVTiB10–10 alloys have been developed to avoid the need for PWHT as this can cause difficulties during erection and for in-service repairs. As steam conditions approach 700 °C nickel alloys, e.g., Inconel 740 or Inconel 617, provide the only suitable material solution. These alloys are significantly more expensive than their ferritic counterparts and a significant improvement in boiler efficiency is required to warrant their use. There is thus a clear opportunity for the development of cost-effective new materials instead of Ni-based alloys for these applications (Nicol 2013).

3.1.1.2 Superheat Tubes

Superheat tubes are designed to operate at temperatures some 35–50 °C above the live steam temperature. Ferritic alloys T91 and T92 are suitable for metal temperatures up to about 615 °C. Higher chromium ferritic alloys such as T122 and VM12 have been shown to suffer from precipitation of a deleterious phase (Z phase) with time which reduces their long-term creep strength resulting in allowable stress not much better than T22 (Allen 2007).

Fireside corrosion of ferritic alloys can also be a problem. Austenitic steels of type 316 and type 347 have been modified by addition of copper, boron, and nitrogen to improve stress rupture capabilities. Modified thermomechanical treatments have also been used to limit grain size and enhance corrosion resistance. Shot peening of tube bore has also limited corrosion. Notwithstanding these improvements, the relatively low (18%) chromium content in these steels limits application to about 620 °C. For higher temperatures, austenitic steels NF709 and 310HNbN (HR3C) are available. Sanicro 25 exhibits superior stress rupture properties matching the target for austenitic alloys of 100 MPa average stress rupture at 700 °C for 100,000 h. The next stage towards higher temperatures and efficiencies is to move to nickel alloys. Fireside corrosion from low-carbon technologies such as oxy-fuel firing requires other materials with properties intermediate between austenitic steels and nickel alloys, such as X7NiCrCeNb32–27 (AC66) (Rautio and Bruce 2008).

3.1.1.3 Steam Separating Vessels

Steam separator vessels are often the heaviest walled components in the boiler circuit and are subject to severe thermal fatigue stresses. They separate the saturated steam, which flows on to the superheaters, from water that is returned to the boiler water feed train. Materials such as P11 have proved satisfactory in the past. However, to minimize wall thickness and hence maximize flexibility, materials with higher yield strength are preferred such as 15NiCuMoNb5–6-4 (WB36) or P91 (Allen 2007).

3.1.1.4 Headers and Steam Pipes

Headers and pipework are situated outside the furnace, so fireside corrosion in not a factor in material selection. However, steam oxidation of the bore must still be considered. Early supercritical plant operating around 540–560 °C used X20CrMoV121, a nominally 12% Cr martensitic steel, for the highest temperature headers and pipework. P91, a modified 9% chromium steel, enabled higher pressures and temperatures to be accommodated; however, the operating temperature was limited to 600 °C due to steam oxidation considerations. Further materials such as P92 and E911 mad possible an increase in metal temperature to around 610 °C. Steam oxidation may be improved through the development of higher chromium (11–12%Cr) variants (P122, NF12, and VM12) led to the formation of Z-phase. Therefore, higher steam temperatures with monolithic piping will require more expensive nickel alloys (Danielsen et al. 2007).

3.1.2 Steam Turbines

A steam turbine is a device that extracts thermal energy from pressurized steam and uses it to do mechanical work on a rotating output shaft. In other words, the steam turbine is a form of heat engine that derives much of its improvement in thermodynamic efficiency from the use of multiple stages in the expansion of the steam, which results in a closer approach to the ideal reversible expansion process.

3.1.2.1 High-Pressure and Intermediate-Pressure Cylinders

Material requirements in the high-pressure (HP) and intermediate-pressure (IP) cylinders depend critically on the steam inlet and reheat temperatures, respectively. For the current generation of steam plant these temperatures are up to 620 °C. Inlet temperatures will increase steadily in increments of about 10 °C as the drive for increased efficiency continues. Materials research is required for plant that will operate at steam inlet temperatures up to 760 °C and above (Allen 2007).

1. Rotor forgings

   Current rotor forgings are based on 9–10% CrMoVNbN steels with either a Mo addition of 1.5% or an addition of 1.0%W in partial substitution of the Mo content. V and N contents have been optimized to provide precipitation strengthening through a dispersion of VN particles and a low level of Nb is incorporated to control grain size during high-temperature heat treatments. For the very highest temperature applications, additions of boron are being made (Kern 2008).

   A step-change to the utilization of Ni-based alloys for rotors is undesirable as it will result in significant cost increases. Consequently, the most advanced steels would be pushed as far as possible to control costs. Engineering advances such as cooling technology will be developed in conjunction with alloy development and this may result in a requirement for thermal barrier coatings for the rotor (Allen 2007).

A major challenge to the introduction of new rotor materials is their inspectability. As the strength of the material increases, the critical defect size is reduced. A suite of inspection techniques for both surface and submerged techniques is required and the development of this technology must evolve alongside material improvements.

2. Castings and valve chest

   The properties required from castings for valve chests and cylinder castings are similar those for rotor forgings. Due to the complex shape, however, castings or multiple forgings are favored. Therefore, the alloys used have additional requirements of good castability or weldability. Similar alloys to those used for rotor forgings are employed but with minor compositional variations, e.g., lower C content to provide improved weldability.

   A further requirement for utilization of cast alloys is availability of appropriate welding processes and consumables. Failure of welded structures at elevated structures usually occurs in the welded region. Measures to improve the properties of weldments either through control of welding process or through material improvements is a key issue for the development of plant to operate at higher temperatures (Allen 2007).

3. Blading

   Blading alloys for operation up to about 600 °C are similar to the rotor forging alloys. At higher temperatures, however, oxidation becomes an issue. Martensitic steels produce a dilemma in that adding chromium to the ally to improve oxidation resistance invariably reduces creep strength. Austenitic alloys are currently used successfully, but as steam inlet temperatures rise to 650 °C and above, these alloys have insufficient oxidation resistance. The oxidation behavior of both martensitic and austenitic alloys can be improved using coatings. However, for plant operating at >700 °C Ni-based alloy will be required to achieve both the strength and oxidation requirements (Allen 2007).

   Solid particle erosion can be an issue in some turbines and erosion-resistant coatings are required to alleviate this problem. These coatings must also be able to meet the oxidation requirements.

4. Sealing

   Current ring and brush seals of steam turbine power generation plant are limited to operation at temperatures below 550–600 °C. Above this temperature range excessive distortion and wear results in efficiency losses and deficient performance that impact upon component design, declared lifetimes and costs of manufacture and operation. The materials requirements for the next-generation sealing systems capable of operating at 650 °C and beyond are (Allen 2007): (a) higher temperature creep strength to prevent loss of sealing due to distortion and enable longer lifetimes for components operating under extreme temperatures and pressures; (b) high-temperature resistance to steam oxidation and wear (use of hard facing treatments) providing lubricant-free abrasion resistance and high load-bearing capability; and (c) effective use of materials in demanding environments providing reduced costs due to improved design, manufacturing and longer periods between overhaul and applicable to retrofit/upgrade of power generation plant.

5. Valve internals

   The key issues for valves are sliding wear and solid particle erosion. Abrasion-resistant coatings or welded inserts are used on a regular basis in current as a palliative measure. As the steam temperature increases, these components will be subjected to more aggressive conditions and it is envisaged that the erosion / corrosion resistance of valve internals will be a key issue in future turbine development (Allen 2007).

6. Bolting

   Meeting the requirements of bolts operating at the very highest temperatures has frequently required exploitation of Ni-based alloys such as Neonic 80A or Refractalloy 26. Even higher strength alloys will be required for >700 °C. Several Ni-based alloys have an inherent problem in that they undergo a lattice transformation during prolonged exposure to a specific temperature range. This results in contraction of the material and overstressing of the bolt. Alloys needed to be identified that have sufficient resistance to stress relaxation and do not undergo this transformation (Lothongkum et al. 2006; Allen 2007).

3.1.2.2 Low-Pressure Turbine Cylinder and Components

Materials requirements for the low-pressure (LP) cylinder are primarily dictated by the need to avoid stress corrosion cracking (SCC) and fatigue. Steam entry and exit temperatures are typically 250 °C and 40 °C, respectively.

LP rotors are usually manufactured from low-alloy NiCrMoV steel. Designs may be either monobloc or welded construction. A key requirement is to avoid SCC in blade attachment areas. However, as a material's strength increases so does its susceptibility to SCC. Design approaches that have been implemented include the use of welded rotors to use material of an appropriate strength in areas where an environment conductive to SCC is present. The use of welded rotors introduces a requirement for inspectability which must be addressed in a similar manner as for HP rotors. An alternative approach to avoiding SCC in rotors is the use of local surface treatment to introduce compressive residual stresses in critical areas. Work is required to optimize these processes to develop a suitable residual stress profile without damaging the alloy surface (Tsuji et al. 1992; Allen 2007).

Castings are not generally highly stressed. Carbon steel or cast iron is usually used subject to flow accelerated corrosion resistance being acceptable.

Blading requirements vary along the steam flow of the LP turbine. Blades near the inlet are relatively small and operate in "dry" steam. Near the outlet the blades are much longer and operate in steam that contains significant moisture. Precipitation hardened stainless steels are usually used for the longest blades although titanium alloys have acceptable properties (albeit much higher cost). Efficacy improvements can be achieved using longer last stage blades. This requires alloys with higher specific strength than those currently used but also having adequate SCC resistance. Alloy development, both of steels or titanium alloys, is required to achieve this balance of properties. Last stage blades are also subject to water droplet erosion. For current alloys the leading edge of the blades may be modified for increased erosion resistance through cladding, welded inserts or local hardening. Any new alloy introduced for last stage blades must either have intrinsic resistance to water droplet erosion or be amendable to modification to achieve the desired level (Brandt et al. 2009).

3.1.3 Gas Turbines

A gas turbine is a type of internal combustion engine. As shown in Fig.3.2, it typically has an upstream rotating compressor coupled to a downstream turbine, and a combustion chamber in-between (Clarke et al. 2012; PCC 2013). The basic operation of the gas turbine is similar to that of the steam power plant except that air

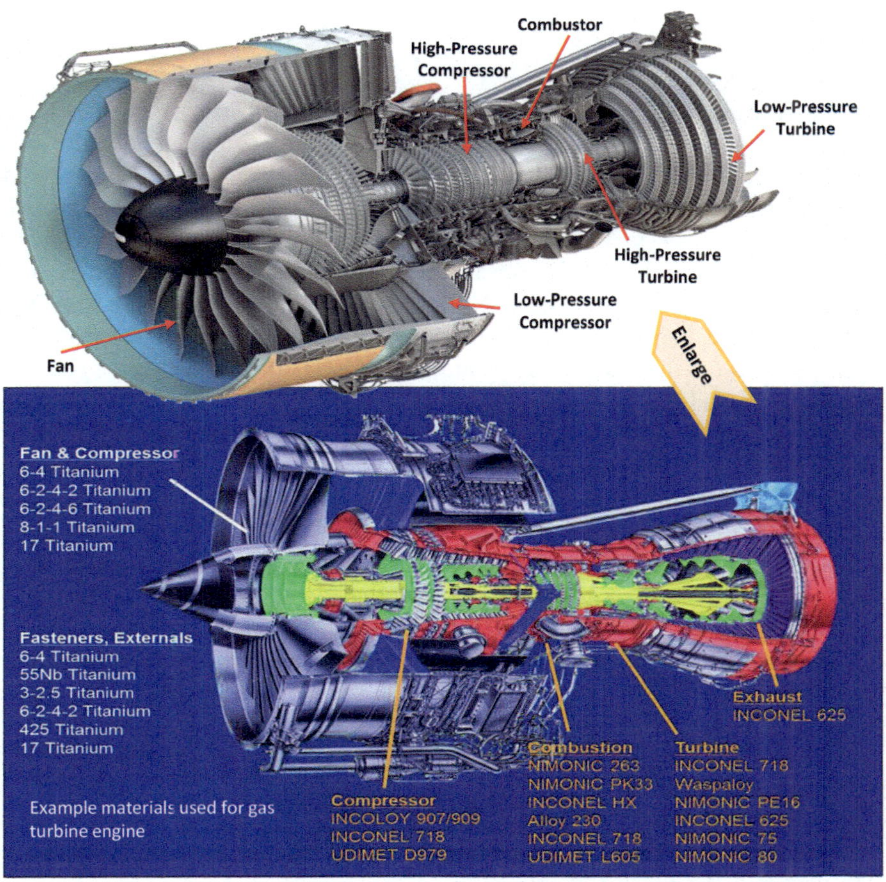

Fig. 3.2 Cutaway view of gas turbine aircraft engine and example materials used (Modified from PCC 2013. Credit: Precision Castparts Corp.)

is used instead of water. Fresh atmospheric air flows through a compressor that brings it to higher pressure. Energy is then added by spraying fuel into the air and igniting it, so the combustion generates a high-temperature flow. This high-temperature high-pressure gas enters a turbine, where it expands down to the exhaust pressure, producing a shaft work output in the process. The turbine shaft work is used to drive the compressor and other devices such as an electric generator that may be coupled to the shaft. The energy that is not used for shaft work comes out in the exhaust gases, so these have either a high temperature or a high velocity. The purpose of the gas turbine determines the design so that the most desirable energy form is maximized. Gas turbines are used to power aircraft, trains, ships, electrical generators, or even tanks (Sonntag and Borgnakke 2006).

The current materials are being pushed to their mechanical and physical limits in terms of both strength and temperature capabilities. Higher temperature materials allow for reduced cooling air requirements and hence higher efficiency can be achieved. Designing with novel materials must take into account key issues such as manufacturability, quality, inspection, and reuse.

3.1.3.1 Compressors

Future compressors need to provide improved cycle efficiency, operability, and reduced costs by optimizing the work done by each stage. This can be achieved by better control of the pressure ratio (from current 15:1 level to 40:1 and beyond) and mass flow through the compressor, improved component reliability and reduced parts count. The need to maintain compressor performance and integrity through life, while reducing parts cost and the use of more effective manufacturing processes is paramount, as is the need to achieve operational lifetimes in excess of 100,000 h. Many of these targets are dependent upon improved design and aero-thermal analysis methods in conjunction with test and validation procedures; however, without suitable high-temperature materials these cannot be achieved (Allen 2007).

For small to intermediate gas turbine compressors, temperature loadings currently range from −50 to about 500 °C. In the short to medium term the continued use of improved low-alloy and ferritic stainless steels will be adequate. However, aero-derivative titanium alloys, nickel alloys and composites will eventually be employed. This would present a significant increase in cost and manufacturing complexity (forgings, machining, joining, component lifting) as well as operational difficulties (component handling, overhaul, repair, cleaning) and may introduce additional problems associated with thermal mismatch and fretting fatigue from adjoining ferritic alloys. Issues associated with rotor corrosion are largely operator dependent, being influenced by the specific nature of the fuel, compressor washing and cleaning practices. These are currently addressed by use of protective coatings. Likewise, abradable tip seal coatings are currently used to provide and maintain efficiency and currently present little technical risk (Allen 2007).

For large utility power generation engines, performance is currently limited by the temperature and strength capability of the rotor steels used. Development of high-nitrogen, nano-precipitate-strengthened steels for high-pressure compressor disk applications could offer equivalent strength and temperature capabilities to some nickel-based alloys with much reduced cost. Application of these high-strength creep-resistant steels necessitates the development of improved large-scale melting (up to 100 tons) and forging capabilities (up to 18 tones) and the development of suitable welding technologies, nondestructive testing methods for large-scale rotors and validated life assessment and risk analysis methods. Successful development of these technologies would negate the need to introduce more expensive (by a factor of 5) nickel alloy technologies. Materials and process developments will be heavily dependent upon successful integrated process modeling that links process developments to materials performance and enables optimized, affordable, manufacturing routes to be developed (Conklin and Szybist 2010).

An added complication for the compressor is the introduction of water/fogging at the intake to improve performance. The presence of water droplets leads to erosion issues on compressor blading. In the short term this will be mitigated through the development of erosion-resistant coatings for existing materials but in the longer term an erosion-resistant materials system will be required (Allen 2007).

3.1.3.2 Combustors

The combustor experiences the highest gas temperatures in a gas turbine and is subject to a combination of creep, pressure loading, high cycle, and thermal fatigue. The materials selected presently are generally wrought, sheet-formed nickel-based superalloys. These provide good thermo-mechanical fatigue; creep and oxidation resistance for static parts and are formable to fairly complex shapes such as combustor barrels and transition ducts. Equally of importance is their weldability, enabling design flexibility and the potential successive repair and overhaul operations, which is crucial to reducing life cycle costs. The high thermal loadings imposed often mean that substantial portions of the combustor hardware need to be protected using thermal barrier coatings (Reed 2006; Allen 2007).

Current temperature loadings experienced by combustors range from 1250 to 1375 °C, depending on engine size and duty cycle. Future developments aim to reduce CO_2, NO_x, and SO_2 emissions to meet environmental legislation. It is demanded for cleaner running by optimizing the distribution of fuel during firing and by using catalytic combustion systems. This will place a limit on the future turbine entry temperature levels and will require control of peak flame temperatures within the combustor. This should also provide more air for cooling the combustor liner; however, other components may be required to run hotter as the demand for combustion flame temperature control increases (Allen 2007).

For future combustor designs, conventional wrought nickel-based products probably would be replaced by higher performance Ni-based alloys, oxide dispersion stringed metallic systems, and ceramic matrix composites. However, there are limitations to these technologies that need to be overcome before they can be deployed on products, such as joining methods, environmental barrier coating systems, robust design, inspection, lifting, and repair capabilities. These materials have also been identified as candidates for efficient, high-temperature heat exchangers for a range of externally fired combined cycle systems that separate the turbine working fluid from the aggressive combustion gases generated by inferior quality fuels. This limits the damage incurred by hot section hardware during engine running and enables the use of a range of low calorific value and biomass fuels with combined heat and power recovery systems (Allen 2007).

The current thermal barrier coatings technology for metallic combustor applications is based exclusively on multilayered systems comprising of a MCrAlY bond coat and a ceramic topcoat applied using plasma spray deposition techniques. Application of this technology generally aims to limit peak metal temperatures between 900 and 950 °C. Future developments are aimed at (a) applying thicker coatings to enable higher flame temperatures and/or reduce metal temperatures further, and (b) increasing the phase stability and resistance to sintering of the ceramic topcoat at temperatures above 1250 °C and to the inclusion of diagnostic sensor layers within the coating that enable the plant and component condition to be actively monitored. New materials will however require new coatings systems that can only be specified and developed in parallel with the substrate development (Clarke et al. 2012).

3.1.3.3 Turbine Blades

Turbine blades are subjected to significant rotational and gas bending stresses at extremely high temperatures. In addition to normal start-up and shutdown operation unexpected interruptions and shutdowns introduce severe thermo-mechanical loading cycles. The turbine entry temperature is typically in excess of 1375 °C, with base metal temperatures ranging from 700 to >1050 °C. The target lifetime under these conditions is dependent on engine type and duty cycle, but can be in excess of 50,000 operating hours. The blades pass through the wake of the combustor and nozzles and are subject to high-frequency excitations, which can lead to high cycle fatigue failure. The high-pressure stages are cooled to withstand the hot gas temperatures and coated to restrict corrosion and erosion of the blade structure (Carter 2005).

The primary consideration in the design of blades has been to avoid the possibility of creep failure. To meet this requirement, an increase in the efficiency by running higher turbine temperatures, more advance materials have been introduced. For vanes and blades there has been a gradual move away from conventionally cast nickel-based superalloys towards directional solidification and single crystals. The increased cost of manufacture is mitigated using recycled materials and increased casting yields, and offset against improved component lifetimes and more efficient running through the higher turbine entry temperatures they make possible. Alloys with greater defect tolerance to low and medium angle boundaries need to be developed and validated with advanced modeling of behavior of defects under load. Equally there is a need to develop alloys which are less susceptible to microstructural and mechanical property degradation. To achieve increased creep strength, successively higher levels of alloying additions have been used to increase the level of precipitate and substitutional strengthening. However, as the level of alloying has increased, chromium levels have had to be significantly reduced to offset the increased tendency to form deleterious phases, which limit ductility and reduce strength. Reduced chromium levels also significantly reduce the corrosion resistance of the alloys. This has necessitated the development of a series of protective coatings that intended to increase component lifetimes, but they often demonstrate low strain-to-failure properties that can impact upon the thermo-mechanical fatigue endurance (Amagasa et al. 1994).

A number of alloys having improved castability, higher corrosion resistance and reduced heat treatment times have been developed, but with limited success. Future emphasis should be on a total system that addresses alloy development, coating, lifting and repair as an entity rather than as a series of unrelated steps. This holistic approach will be critical to the success of future for all gas and steam turbine materials and process development (Allen 2007).

INDEPTH: 3D Printed Gas Turbine Blades (Siebert 2017)

Until now, blades for gas turbines were either cast or forged. The casting of turbine blades requires complex mold construction before each blade can be individually cast—a complex, time-consuming, and costly procedure. Additive manufacturing changes all this. With AM, a laser beam is directed at fine layers of metal powder, which are heated and melted. When the laser is removed, the metal cools. The process is repeated layer by layer until the blade model from the 3D printer is finished.

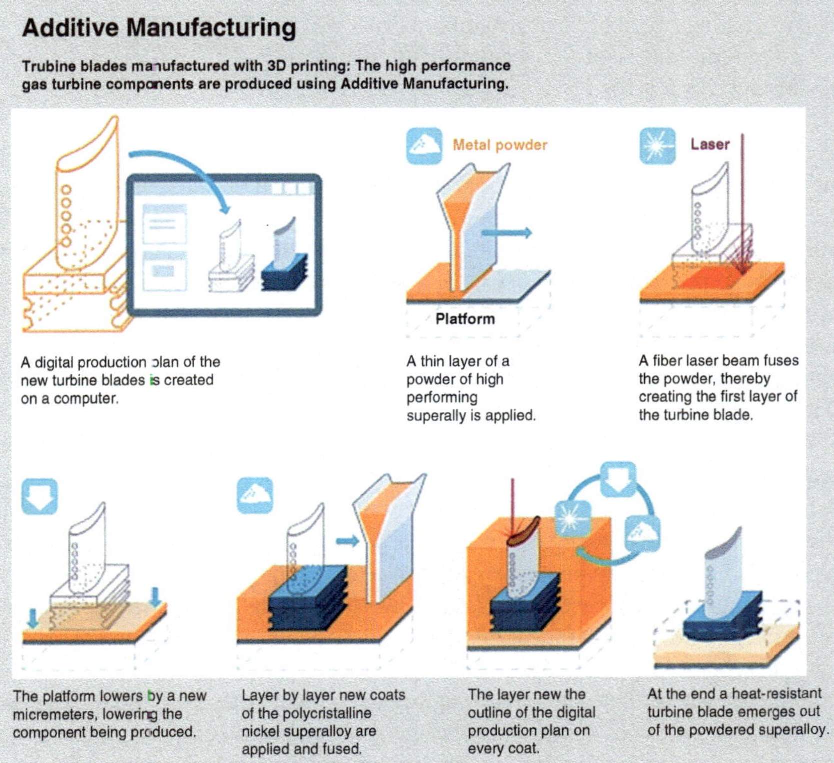

3.1.3.4 Turbine Disks

The main functions of a turbine disk are to locate the rotor blades within the hot gas path and to transmit the power generated to the drive shaft. To avoid excessive wear, vibration, and poor efficiency this must be achieved with great accuracy, while withstanding the thermal, vibrational, and centrifugal stresses imposed during operation, as well as axial loadings arising from the blade set. Under steady-state conditions, current turbine disk temperatures can vary from approximately 450 °C in the hub to in excess of 650 °C close to the rim with a requirement for >50,000 h operating life. These temperature loadings are set to increase further across the disk as the demand for improved efficiencies continues (Cruchley et al. 2016).

Creep and low cycle fatigue resistance are the principal properties controlling turbine disk life and to meet the operational parameters requires high integrity advanced materials. To meet the highest operating temperatures and the component stress levels demanded, it has been necessary to develop a series of progressively higher strength steel and Ni-based super alloys. These are generally manufactured using cast and wrought processing. However, the complex chemistry makes production of segregation-free ingots very difficult. Manufacture of larger components, or more complex alloys, would necessitate a change to atomized powder processing to limit segregation, while dual alloy processing offers the potential for overcoming the variability in strain distribution across the section of large forged turbine wheels. The issue of coatings development for turbine disks is currently treated in the same way as coating for blades but is a much less mature technology. The same holistic systems approach is therefore required (Allen 2007; Cruchley et al. 2016).

3.1.3.5 Sealing

Turbine gas path seals include rotor tip seals and disk rim seals. On unshrouded rotor tips and rim seals, coatings such as MCrAlY and Nickel-Graphite can be employed as an abradable coating to avoid damage in the event of a rub. Erosion resistance and the ability to provide abradability after long-term exposure at high temperature are required for such coatings. On shrouded rotor blading, superalloy honeycomb foil materials are employed as abradable seals and high-temperature oxidation resistance is needed for foil materials to achieve long life at high temperature (Hendricks et al. 2005; Cruchley et al. 2016).

The labyrinth seal is the most common form of air system seal, usually with an abradable stator material. Positive contact carbon seals are often used for bearing chamber seals. The development of brush seals has provided improved sealing efficiency and is displacing the use of labyrinth seals in critical locations. Both metallic and nonmetallic bristles are employed depending on temperature levels and the key requirements for wear and fatigue resistance. Film riding face seals have the potential for even higher sealing efficiencies but place extreme demands on manufacturing technology. Static seals are used to seal gaps and joints between components where there are small relative movements due to thermal expansions and they can be employed on both turbine rotor and turbine and combustor stator components. On cylindrical joints, piston rings and E seals are used and for sealing gaps between adjacent blades and vanes strip seals are employed. On rotor blades it is common for sealing features and blade locking features to be combined. Seal strips and E seals used in high-temperature locations employ wrought sheet materials as used for combustor components. Wear and fretting can be experienced on such seals and the development of hard coating systems can alleviate this. Brush seals, used in a static application, can allow larger thermal movements with low leakage (Hendricks et al. 2005; Allen 2007).

3.1.4 Gasifiers

Materials issues in the major gasification process components are mostly restricted to the gasifier vessel, downstream coolers and gas cleaning vessels, and the gas turbine where the fuel is burned.

3.1.4.1 Refractory Lined Gasifiers

During the gasification process the chamber typically operates at temperatures between 1250 and 1550 °C, and at pressures of 3 MPa or higher, and so it is usually lined with refractory materials to contain the severe environment and to protect the outer steel shell from erosion, corrosion, and the high temperatures. In addition to the lining, the gasifier is usually cooled by air or by water. Water-cooled gasifiers typically have a working force lining of Al_2O_3-SiC refractory, and have a satisfactory service life because slag freezes on the refractory surface, restricting slag penetration and corrosion. Air-cooled gasifiers are also lined with refractory materials—a typical liner would be a high chromium oxide material that contains alumina and may contain other additives like zirconia. The liners can last from between 3 months to 2 years (Carniglia and Barna 1992).

The major challenge for gasifier refractory materials is for them to provide an adequate period of protection to the gasifier structure. Generally, industrial applications run two gasifier units, so that one can continue operation while the other is being refurbished. Therefore, the lifetime of the lining should be predictable so that outages can be planned. A suitable refractory requires the following characteristics (Carniglia and Barna 1992; Allen 2007): (a) Endurance at elevated temperatures; (b) Ability to accommodate thermal shock from temperature transients; (c) Resistance to erosion by particulates; (d) Molten slag resistance; (e) Resistance to corrosive attack from hot gases of varying composition; and (f) ability to withstand variable oxidizing and/or reducing conditions.

Historically a number of refractory compositions have been evaluated for use in these harsh thermal environments, such as sintered and/or fused cast alumina-silicate; high alumina, chromia-alumina and chromia-magnesia spinels; alumina/magnesia, alumina and chromia; and silicon carbide (SiC) refractory compositions. Each composition has its particular benefits and disadvantages (Allen 2007): (a) Fuse-cast refractory materials with little or no porosity have good chemical wear resistance but poor resistance to thermal shock; (b) Materials containing high levels of Al_2O_3 or MgO/Al_2O_3 spinel have very poor wear resistance, as do refractories SiC and Si_3N_4; (c) As a general rule, additions of chromium oxide to a refractory composition tend to improve resistance to chemical attack from slag; (d) Refractories with high chromium oxide content (>85 wt% Cr_2O_3) are used in severe wear areas of a gasifier, while lower chrome oxide materials are used in less severe wear areas. A minimum Cr_2O_3 level of 75 wt% is thought necessary for sustained material performance in slagging gasifiers; and (e) Refractory compositions suitable for economic operation in aggressive gasifier environments are Cr_2O_3/Al_2O_3, $Cr_2O_3/Al_2O_3/CrO_2$, and Cr_2O_3/MgO.

High chromium oxide refractory materials have evolved as the material of choice to line the hot face of gasifiers, but the performance of these materials does not fully meet the desired service requirements of industry. Future developments may include development of refractory material coatings, monolithic linings, and refractory materials that do not contain, or are low in, chromium oxide, such as alkali-aluminate.

3.1.4.2 Gas Coolers

Once generated, the gases need to be cooled and cleaned before being burnt in the gas turbine. The raw hot gas is mainly carbon monoxide (CO) and hydrogen (H_2) with hydrogen sulfide (H_2S) as the major corrosive impurity. The gas mixture is cooled, and water quenched to remove particulates and water-soluble impurities such as ammonia (NH_3) and chlorides. The heat exchanger that cools the raw fuel gas is the first major component downstream of the gasifier vessel. Gas coolers can be either water-tube or shell-boiler in type. Cost is a key issue in gasification; hence most current systems use the cheaper shell-boiler approach, which is suitable for modest stream conditions but increases the risk of fouling and blockages of gas stream (Allen 2007).

The presence of ash/char particles can cause erosion, abrasion or deposition in the gas cooler, leading to blockages, while gaseous species cause corrosion and further deposition through condensation of vapor-phase species such as alkali and trace metal chlorides/sulfides. High-temperature corrosion in syngas coolers is unique to gasifiers, especially when coal is used as the fuel. At present heat exchanger syngas operating temperatures are generally less than 450 °C. This is partly due to concerns over high-temperature corrosion and partly because the use of higher heat exchanger temperatures is not deemed cost effective. One of the critical high-temperature corrosion mechanisms in gas coolers is sulfidation. In carbon and low-alloy steels the rate of attack is too high to allow the use of these low-cost materials. Alloy 800 has adequate properties at an economic price and is generally used. However, the sulfidation potential is not the only parameter to affect the corrosion rate of candidate materials; corrosion rates also depend on the chloride content of the gas. Alloys with relatively low chromium content, such as Alloy 800 and its weld material Inco 82, are particularly susceptible to corrosion while high chromium materials with additional alloying additions such as aluminum and silicon are less affected by the presence of HCl. Stainless steel 310 has an adequate corrosion resistance but is not currently approved due to waterside corrosion issues. Sanicro 28 is currently the favored material (Allen 2007; Zhu 2015).

Aqueous corrosion during downtime will further increase corrosion rates of materials; this is particularly problematic in the presence of chloride-containing deposits. During downtime, chlorides penetrate through cracks in the oxide scale and attack the metal at the oxide/metal interface. Iron chlorides formed by this attack then cause spallation of the oxide scale during start-up, exposing fresh metal for further corrosive attack. In the absence of this spallation, stainless steels, with more than 20 wt% Cr are adequate for gas coolers working between 300 and 400 °C. Where spallation occurs, the corrosion rate of stainless steels is too great and molybdenum containing materials, such as Sanicro 28, offer better protection. If the gas contains high levels of chlorides, stainless steels are unacceptable, and so Inconel Alloy 625, containing molybdenum and a high level of nickel, is the preferred material (Allen 2007).

One of the biggest challenges in gasification systems lies in the development of a reliable and economically viable cooling/cleaning path. This requires development for syngas coolers with novel alloys and manufacturing processes to improve corrosion resistance and lower the cost of these components. The development of improved hot gas cleanup systems could lower the cost of IGCC by providing a cheaper alternative to the conventional low temperature processes currently used (Zhu 2015).

3.1.4.3 Gas Turbines

Although gasification plants use multiple stage cleanup processes for the syngas, a limited amount of impurities do enter the gas turbine hot section flow path after the combustion process. This leads to a risk of damage to the vanes and blades of the turbine. Given the inherent variability of feedstock into the gasification process the composition of impurities varies greatly making it more problematic for materials than conventional gas turbines (Gibbons and Wright 2007).

In general, the issues found in conventional gas turbines can be applied to syngas-fired turbines with the added complexity of the fuel and impurities. At present syngas-fired turbines are operating at similar firing temperatures as natural gas turbines. This is due to the increase in mass flow through the turbine of around 14% compared to natural gas. While this does produce higher turbine outputs it also generates great heat transfer to the hot section vanes and blades. Hence there is likely to be a need for higher temperature materials. Control of NO_x may also cause problems, as water (either used as a dilutant or produced as a by-product) will increase degradation of hot section components. It is likely that the gas inlet temperature of the turbine will be increased to meet efficiency goals. While this will affect the use of high-temperature alloys it will also exaggerate the effect of molten salts and deposits. It has been shown that very low levels of gas stream ash and impurities can produce substantial degradation through corrosion, deposition and erosion (Allen 2007; Gibbons and Wright 2007).

3.1.5 CO_2 Capture and Storage

A range of CO_2 capture technologies have been investigated and can usually classified as (Allen 2007; Leunga et al. 2014): (a) Post combustion—where CO_2 is separated from the process stream downstream of the main process components and after any fuel content has been burned. This approach can also be adopted in a similar way for other industrial process with high CO_2 emissions, such as cement production and steel production. (b) Precombustion, where the carbon-bearing compounds are removed from the gas stream prior to the combustion of the fuel constituents. (c) Oxy-fuel, where the fuel is burned in an enriched oxygen content gas resulting in and exit gas which has high levels of carbon dioxide and steam which are readily separated by a condensation step. (d) CO_2 transport and storage.

3.1.5.1 Post Combustion Capture

A number of technology options exist for capturing CO_2 downstream of the combustion of a fossil or biomass/waste fuel. The basic approach is illustrated in Fig. 3.3. Low-temperature liquid scrubbing, using amine-based solvents, is the most likely technology has been widely used in the oil/gas sector for separation of CO_2 from natural gas, but there have been few demonstrations of its application in large-scale power plants. Oxygen-bearing flue gases with a range of contaminants also pose performance and durability problems not experienced in natural systems. An amine

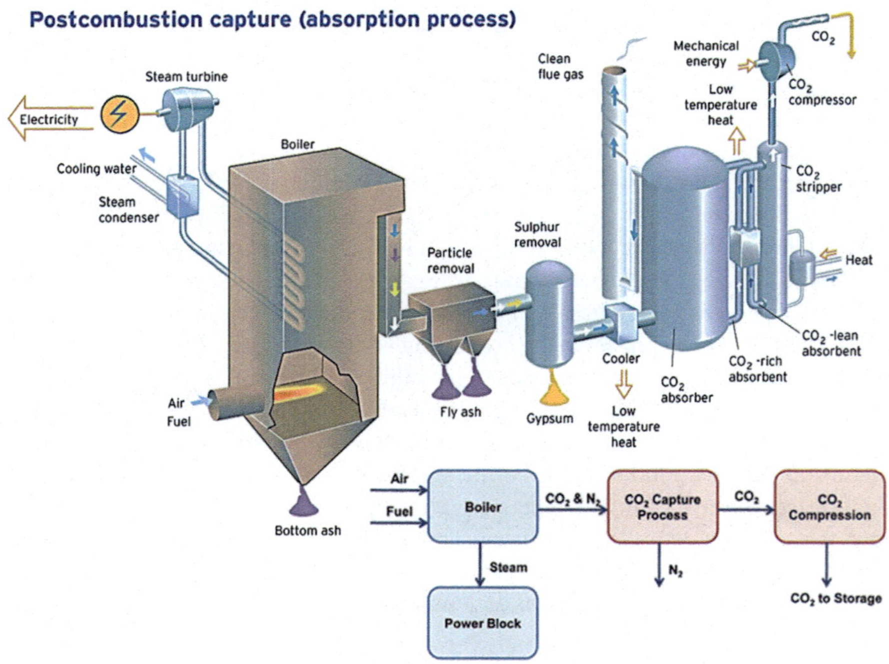

Fig. 3.3 Illustration of post-combustion CO_2 capture (Modified from Vattenfall 2012. Illustration. Courtesy: http://Captureready.com/userfiles/image/Carbon%20Capture/Post-combustion%20Capture%20Process_Vattenfall.jpg)

scrubber comprises two separate units (Allen 2007): (a) An absorber where CO_2-lean solvent reacts with flue gas CO_2 at temperatures typically between 40 and 60 °C; and (b) A stripper (or regenerator) where CO_2-rich solvent is heated to 100–140 °C with steam to strip the gas at close to atmosphere pressure.

In any power plant system, the performance, reliability, and availability of all system components are critically important to economic viability. Critical materials issues in amine scrubbing are (Allen 2007): (a) Corrosion resistance in scrubber environments, including performance of carbon-steel, various grades of stainless steel, ceramics, and plastics. (b) The performance of corrosion inhibitors. (c) Potential surface treatments and coatings which could offer protection and/or repair options. (d) Assessment of whether there are critical fluid velocities above which erosion-corrosion becomes significant. (e) The lack of knowledge of corrosion mechanisms that occur in service, such as pitting, localized corrosion at welds, stress corrosion cracking. (f) The effects of different amine solvents, amine concentrations and degradation products on corrosion. (g) The performance of materials for valve stem packing, pump seals, plate heat exchanger gaskets, etc.

Alternative post-combustion CO_2 separation approaches include the use of ammonia, solid sorbents (e.g., lime or alkali compounds), adsorption on molecular sieves or active carbons (using pressure, temperature, or electrical swing systems) or cryogenics (Leunga et al. 2014).

3.1.5.2 Precombustion Capture

The most common route for precombustion carbon capture is probably the use of Integrated Gasification Combined Cycle (IGCC) technology. Precombustion CO_2 capture from an IGCC plant can be achieved by removing the CO_2 from the syngas within the gasifier (Watson et al. 2007).

Advanced gasification systems are being developed for the generation of power using chemical feedstock, liquid fuels and hydrogen, the latter option requiring the use of a CO_2 capture process. High-pressure, oxygen-blown gasifiers are the most suitable for use with CO_2 capture. To enable CO_2 to be captured in precombustion technology, the fuel gas must be feed to a catalytic shift reactor where most of the CO is reacted with steam to give H_2 and CO_2. Steam must be taken from the steam cycle and added to the fuel gas feed to the shift converter, which will affect the efficiency penalty for CO_2 capture (Allen 2007; Leunga et al. 2014).

The fact that the CO_2 is relatively concentrated in the shifted fuel gas (around 50 vol%) and at high pressure, presents an opportunity for lower CO_2 capture compared to pulverized fuel plants where flue gas is at atmospheric pressure with CO_2 concentration of 10–15 vol%. A number of approaches can be used to separate CO_2 from the shifted fuel gas. Variants include (Allen 2007; Leunga et al. 2014): (a) Use of physical solvents, e.g., the Rectisol process using cold methanol, (b) Selexol process using dimethyl ether of polyethylene glycol, etc., (c) Solid sorbents or adsorbents, (d) Cryogenics, and (e) Advanced separation membranes. The physical solvent approach is well established for ammonia production plants but has a relatively high efficiency penalty, whereas membranes can be tuned to plant conditions and so offer the lowest efficiency penalty.

3.1.5.3 Oxy-combustion

Oxyfuel combustion is more speculative and experimental approach to post-combustion carbon capture. It involves burning the coal in an oxygen and CO_2 rich mixture rather than air. This produces a waste gas stream rich in CO_2 from which the CO_2 much easier to capture. Its main drawback is the need for an expensive, energy intensive air separation unit (Watson et al. 2007).

Combustion of fossil fuels in an oxygen-enriched/low nitrogen environment leads to the gaseous combustion products being mostly a mix of CO_2 and steam, which can readily be separated using a condenser, leaving a high concentration CO_2 steam for further cleanup, compression, transportation and storage. This approach can be used with coal in pulverized fuel boilers. The nitrogen in the air which would previously have been the oxidant is replaced by recycled flue gas, i.e., mostly CO_2 and steam. The materials challenges are centered on the changes to the boiler environment as a result of flue gas recycle, although other parts of the systems, such as the steam condenser/CO_2 separator, could also provide unexpected materials problems in service (Allen 2007; Leunga et al. 2014).

In the pulverized fuel boilers, the combustion chamber environment will contain significantly higher levels of CO_2 and steam than in conventional air-firing, along with much higher levels of contaminants such as SOx; up to 5 times conventional levels depending on where the recycled flue gas is taken from. The cleanest, though most expensive option is to recycle after the flue gas desulfurization system, thus reducing the levels of SOx in the boiler but increasing the size of the FED plan to handle the full flow of cycling flue gas. By comparison, the cheapest option is to recycle to flue gas with little pre-cleaning (may be after some particle removal) giving the dirtiest/highest SOx flue gas. Other contaminants and particulates will also be recycled as a result can lead to increased fouling and corrosion. Water-wall and superheater corrosion are a major research area for such systems, particularly when combined with advanced supercritical plant conditions leading to higher levels of contaminants and temperatures (Leunga et al. 2014).

Oxy-combustion can also be used in fluidized bed systems where the ability to control combustion conditions with a minimum of flue gas recycle means that plant sizes can be significantly reduced thus reducing capital costs. Advanced cycles often use elements of the oxy-combustion approach. Examples include line-based post-combustion capture where oxy-firing is used for the calcination step, or chemical looping where the oxygen is provided from solid oxide (Allen 2007; Leunga et al. 2014).

3.1.5.4 CO_2 Transport and Storage

CO_2 captured will be impure and will contain various other gases depending on the type of power plant and the capture technology used. The CO_2 will contain contaminants such as SOx, HCl, NOx, as well as trace metal compounds due to inefficiencies of the current gas cleaning technologies. Contaminated CO_2 provides a potentially aggressive environment for the pipeline or other transport system to the storage location. While pipeline is used currently to transport CO_2 it is unlikely that capture technologies will be unable to deliver a CO_2 stream which would consistently meet the current specification for CO_2 transport (>95% CO_2, <4%N_2, no free water, <1450 ppm total S, <10 ppm O_2, etc.). Special cleanup measures would be needed, which may deliver additional species, which are currently not specified (Allen 2007; Leunga et al. 2014).

There are a number of established options for transporting carbon dioxide from a fossil fuel power station to a storage site. Pipelines to transport CO_2 are already in use in the USA—in oilfields and at natural gas processing plants. The transport of CO_2 in this way is cheaper than electricity transmission. Tankers could be used as an alternative to pipelines, particularly for long distance transport. The tankers would be similar to those that transport liquefied natural gas. However, the liquefaction of the carbon dioxide would require a significant amount of additional energy (Watson et al. 2007).

The storage of carbon dioxide in depleted oil and gas fields is possible because these fields have remained sealed for long periods of time prior to extraction. In the short to medium term, CO_2 storage in depleted oil fields is thought to be a particularly attractive option for those countries with active oil and gas industries. This is due to the scope for enhanced oil recovery (EOR). An EOR process injects CO_2 into partially depleted fields to extract additional quantities of oil. Deep saline aquifers are estimated to have the largest potential for geological carbon dioxide storage. They consist of large rock formations that contain salt water that is not suitable for drinking. CO_2 could be injected into these formations using similar techniques to those for EOR. The most appropriate aquifers are located beneath rock with low permeability to minimize CO_2 leakage. One further geological storage option merits some attention—the use of difficult or un-mineable coal seams. These could be used to "fix" CO_2 to unmined coal by an adsorption process. One possible advantage of this method is that this CO_2 would displace methane which could then be recovered and used. However, this could then generate more CO_2 which would then have to be dealt with (Watson et al. 2007).

3.1.6 Perspectives

The construction of economically viable and durable fossil fuel power plant components is dependent on the selection of the most appropriate materials and fabrication methods. The implementation of advanced materials in power generation applications has resulted in significant advances in plant performance and hence emissions reductions. Advanced materials are now being deployed in new and existing power plant to improve operating performance and reliability, availability, maintainability, and operability. These materials include an extensive scope, such as ferritic alloys for boilers, ceramics for filter elements, or coatings for the protection of gas turbine blades. These materials can be applied to all the key power technologies are covered from supercritical pulverized coal boilers, through steam turbines and gasifiers to gas turbines and advanced cycles involving fuel cells and

CO_2 management. The major areas may include (Oakey et al. 2003a, b): (a) High-temperature materials for boilers, steam turbines, gas turbines, gasifiers, high-temperature heat exchangers, as well as functional materials such as sorbents, catalysts, and membranes; (b) Protective systems/coatings for the same technology areas as high-temperature materials; (c) Materials for sensing in harsh environments; and (d) Modeling of materials processing, component manufacture and life assessment.

Materials technologies are critical to achieving significant improvements in power plant efficiency. An integrated core material system is needed for power generation covering such key elements as high-temperature materials, protective systems, sensing technology and modeling. The following plant technologies as significant for the timescales considered and can use coal as a fuel (Oakey et al. 2003a, b): Pulverized Fuel (PF) Combustion Gasification-air and oxygen-blown, Fluidized Bed Combustion, Pressurized PF or Fluidized Bed Combustion, Fuel Cells, and Gas Turbines. Regarding these plant technologies, the following key technologies can be identified: Hybrid or novel cycles, Co-fueling with biomass, Hot gas cleanup, Fuel flexible gasification, Combustion, High-temperature heat exchangers, Reformers and fuel cell fueling, Fuel cell electrochemical processes, Membranes, Plant manufacture, Component integrity, CO_2 Management, Plant Modeling, Control Systems, and Power Electronics (Hydrogen Storage and Production if Hydrogen Fueled is included). All of the above have materials implications to greater or lesser extents; CO_2 management can have important impacts on the selection, operation, and performance of the main plant technologies in the medium to long term and may change significantly the operating environments of the materials used.

INDEPTH: Allam Cycle for Zero Emission Fossil Fuel Plant (Service 2017)

The future environment demands a balanced portfolio of renewables and carbon-based methods of power generation, particularly those employing carbon capture. Free emissions capture is the holy grail of carbon-based power generation. Today, power cycles require the addition of expensive, efficiency-reducing equipment to decrease and capture emissions. The Allam Cycle is a new type of power cycle that takes a novel approach to emissions reduction. It uses the oxy-combustion of carbon fuels and a high-pressure supercritical CO_2 working fluid in a highly recuperated cycle that captures all emissions by design. The only by-products are liquid water and a stream of high-purity, pipeline-ready CO_2. The cycle can utilize a variety of fuels, including natural gas, unprocessed raw and sour gas, and gasified solid fuels such as coal or biomass. The Allam Cycle embodies major advantages over conventional systems: attaining high efficiencies at low costs with low to no water consumption. All this with full, free emissions capture.

(continued)

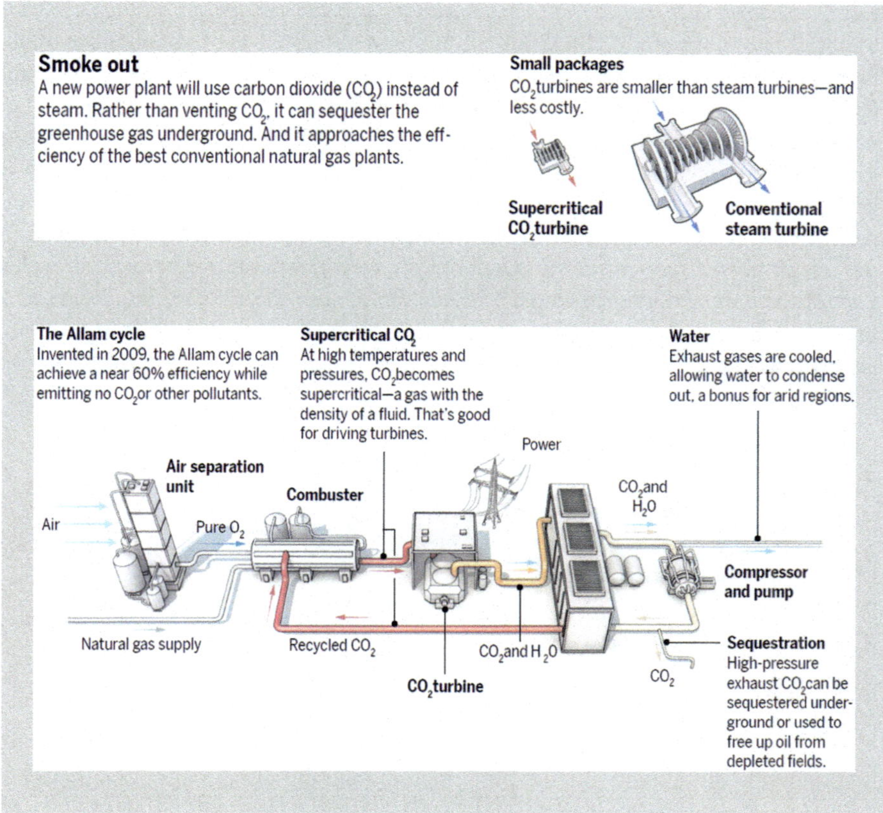

3.2 Materials for Ultra-Supercritical Applications

For fossil fuels, efficiency increases correspond to reductions in all emissions, including the greenhouse gas carbon dioxide, for example, an increase in thermal efficiency from 37 to 52% yields a 29% reduction in CO_2 release. Development of ultra-supercritical (USC) steam generator technology started in the USA in the 1950s. However, a number of problems, including unexpectedly severe superheater corrosion, were encountered during the early stages of USC efforts. Most of the problems were due to the use of austenitic steels for heavy section components operating at high temperatures. These steels have low thermal conductivity and high thermal expansion resulting in high thermal stresses and fatigue cracking. These problems and the general low availability of many supercritical plants due to "teething" problems temporarily dampened utility interest in building super or ultra-supercritical plants and consequently most utilities reverted back to plants with subcritical conditions of about 525 °C (1000 °F) and 17 MPa (2600 psi). The energy crisis in mid 70s re-kindled interests in the development of more efficient pulverized coal power stations, and focused on developing further the existing high-temperature-resistant ferritic-martensitic 9%Cr and 12%CrMoV steels to produce rotors, casings and chests, pipes and headers capable of operating at inlet steam temperatures of up to 650 °C (1200 °F). An improvement in thermal efficiency of the plant not only reduces the fuel costs but also reduces the release of SO_2, NO_x and CO_2 emissions (Wu 2006). Current research and development efforts are focused on enabling the construction of ultra-supercritical plants operating with steam temperatures as high as 760 °C (1400 °F) and steam pressures in excess of 35 MPa (5000 psi), as such conditions can further increase cycle efficiencies and reduce CO_2 emissions. Enhanced reductions in CO_2 emissions can be achieved by the integration of oxy-combustion technology with steam ultra-supercritical cycles. One of the core activities essential to the attainment of these future requirements is that of materials technology. Materials technology is generic in nature and cuts across many energy systems operating at high temperatures. Figure 3.4 illustrates

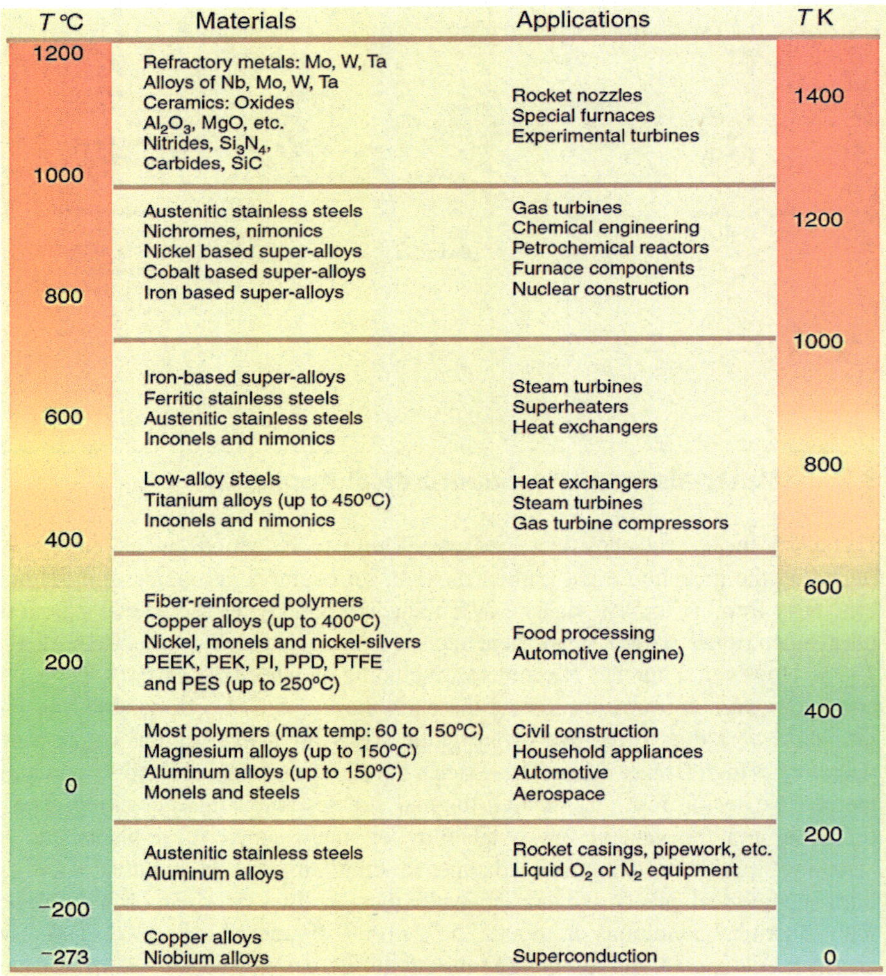

Fig. 3.4 Illustration of materials selection for different temperature regimes (Adapted with permission from Ashby et al. 2007 (Elsevier))

materials selection for different temperature regimes (Ashby et al. 2007). Materials technology developments for high-temperature applications are expected to be of enduring value. In the near term, materials developed for ultra-supercritical plants can be confidently used for retrofit applications in currently operating plants to increase their reliability.

3.2.1 High-Temperature Alloys

In general, ultra-supercritical (USC) steam conditions are defined as being above 24 MPa and 593 °C. The materials, which could be used at such temperatures, are limited mainly due inadequate creep and oxidation/ corrosion resistance. The research goal is that the new high-temperature-resistant alloys have good mechanical and physical properties at high temperatures (650–800 °C). Successful candidate materials must have the following properties (Wu 2006; Lukaszewicz 2012): high creep strength, thermal fatigue strength, good temperature capability in the range between 650 and 750 °C, good thermal conductivity, resistance to fire-side corrosion/erosion and resistance to steam-side oxidation and spallation, as well as good level of weldability, fabricability, and potential to be coated.

INDEPTH: Advantages and disadvantages of advanced ultra-supercritical alloys (Viswanathan et al. 2010; Mukherji et al. 2011)

Material	Example alloys	Maximum operating temperature (at 5000 psi)	Advantages	Disadvantages	Possible applications in an ultra-supercritical turbine
Ferritic steels	SAVE12, NF12, VM12, MARB2	<650 °F (<343 °C)	High strength at low-end temperatures; low cost; can be welded readily	Low temperature resistance and sensitive to oxidation, but could be used in some applications with protective coatings	Low-temperature components such as furnace tubing/ piping
Austenitic steels	Super 304H, HR3C, T92, T22	1000–1270 °F (538–688 °C)	High strength at intermediate temperatures; low cost; can be welded readily	Sensitive to oxidation; low conductivity; high thermal expansion; not suitable for thick-section applications	Mid-temperature applications, including superheater and reheater tubes
Nickel-based alloys	Haynes 230, Inconel	1370–1460 °F (743–793 °C)	High temperature compatibility;	Very high cost; not all alloys are	Highest temperature, highest stress

(continued)

	617, Haynes 740, HR6W		high oxidation resistance	code approved, so extensive testing required	components, such as heavy-wall piping
Advanced aerospace alloys and composites	Cobalt- and rhenium–based alloys	1830–2370 °F (999–1299 °C)	Very high temperature compatibility; high oxidation resistance	Extremely high costs; use of critical materials; early technology readiness level	Highest temperature components, including thermal barrier coatings for turbine blades

The development of the new more resistant alloys has driven design of various ferritic, austenitic and nickel-based alloys, as shown in Table 3.1 and Fig. 3.5. Such alloys are characterized by good thermal conductivity, high-temperature steam oxidation, and creep resistances, as well as significant cost differences—with their price growing in the following order: ferritic, austenitic, and nickel-based alloys. Materials from these three groups are used to manufacture boiler components such as high-pressure steam piping and headers, superheaters (SH) and reheaters (RH), as well as water wall tubing. Some elements operate at even higher temperatures that the steam conditions within them, and therefore the materials used for such components must be able to withstand increased temperatures. SH and RH tubing are examples with the steam flowing within them being around 25–45 °C cooler than the actual temperature of the tube surface (Lukaszewicz 2012).

Table 3.1 Typical high-temperature alloys for ultra-supercritical applications (Viswanathan et al. 2003)

Classification	Designation	Nominal composition	Preferred application	Applied temperature
Ferritic steels	HCM2S	2.25Cr-1.5 W-V	Water wall	Up to 565 °C (1050 °F)
	Tempaloy F-2 W	2Cr-1 W-Mo-V-Nb	Water wall	
	HCM12	12Cr-1Mo-1 W-V-Nb	Water wall	
	NF616 (P-92)	9Cr-2 W-Mo-V-Nb-N	Header	Up to 620 °C (1150 °F)
	HCM12A (P-122)	12Cr-1.5 W-Mo-V-Nb-N-Cu	Header	
	E911	9Cr-1Mo-1 W-V-Nb-N	Header	
	NF12	11Cr-2.6 W-2.5Co-V-Nb-N	Header	Up to 650 °C (1200 °F)
	SAVE12	12Cr-W-Co-V-Nb-N	Header	

(continued)

Table 3.1 (continued)

Classification	Designation	Nominal composition	Preferred application	Applied temperature
Austenitic steels	SAVE25	23Cr-18Ni-Nb-Cu-N	Superheater/reheater tubes	620–675 °C (1150–1250 °F)
	NF709	20Cr-25Ni-Nb-Ti-N	Superheater/reheater tubes	
	HR3C	25Cr-20Ni-Nb-N	Superheater/reheater tubes	
	Super304A	18Cr-8Ni-W-Nb-N	Superheater/reheater tubes	
	347HFG	18Cr-10Ni-Nb	Superheater/reheater tubes	
	800HT	21Cr-32Ni-Al-Ti	Superheater/Reheater tubes	
	HR120	Ni-33Fe-25Cr-N	Superheater/reheater tubes	
Nickel-based alloys	INCO740	25Cr-20Co-2Ti-2Nb-V-Al	Superheater/reheater tubes, pipes and headers	675–788 °C (1250–1450 °F)
	230	22Cr-14 W-2Mo-La	Superheater/reheater tubes, pipes and headers	
	625	21.5Cr-9Mo-5Fe-3.6Nb-Al-Ti	Superheater/reheater tubes, pipes and headers	
	617	22Cr-12.5Co-9Mo-1.2Al	Superheater/reheater tubes, pipes and headers	
	HR6W	23Cr-6 W-Nb-Ti	Superheater/reheater tubes, pipes and headers	
	45TM	27Cr-23Fe-2.75Si	Superheater/reheater tubes, pipes and headers	

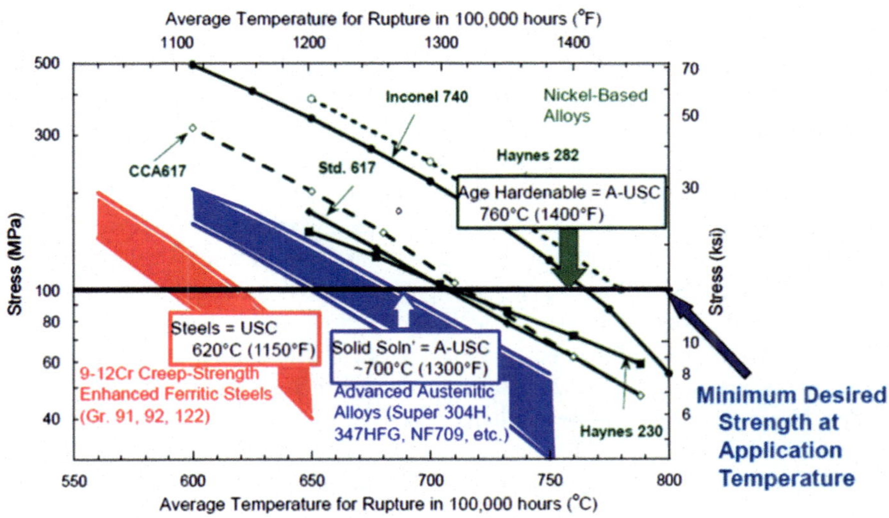

Fig. 3.5 Creep strength comparison among some high-temperature alloys (Adapted from Xie et al. 2015, Licensed by IntechOpen)

3.2.1.1 Ferritic Steels

Ferritic steels are currently widely used for the boilers-application, especially for thick section pipes and headers, as a result of their good creep resistance at current operating temperatures, as well as low coefficients of thermal expansion, high conductivity, fracture toughness and good weldability, as well as relatively low costs which are important from a business perspective. However, application of these steels to higher temperatures is limited by their lack of oxidation resistance at higher temperature and by creep resistance. The strongest of these steels that can be used up to a temperature of 620 °C purely from a creep strength point of view are still limited by fireside corrosion to a metal temperature of 593 °C. This corresponds to a steam temperature of about 565 °C since SH/RH metal temperature can exceed the steam temperature by about 30 °C. Excessive corrosion of ferritic steels caused by liquid iron-alkali sulfates (also known as hot corrosion) in the tube deposits is an acute concern, since high sulfur and corrosive coals are used frequently. Therefore, current research aims to design more oxidation- and creep-resistant ferritic alloys with chromium levels of between 9 and 12% (Wu 2006; Lukaszewicz 2012).

Ferritic steel developments are mostly aimed at their use for thick section pipes (Water walls) and headers. Table 3.1 and Fig. 3.5 show typical ferritic steels for power boilers. HCM12A (P122), NF616 (P92), and E911 emerge as the highest strength alloys suitable for ultra-supercritical plants up to 620 °C, followed by T91, HCM12, EM12, and HT91 suitable for intermediate temperatures up to 593 °C (1100 °F), followed by T22 for use up to 565 °C (1050 °F). Alloy HCM2S has much higher strength than P22, which is weldable and therefore suitable for application as a replacement for P22 (Viswanathan et al. 2003).

Among the 9%Cr steels fully commercialized, the Grade 91(P91 for piping and T91 for tubing) steel has the highest allowable stress and has been extensively used as a material for headers, steam pipes, and tubes in ultra-supercritical plants operating at steam temperatures up to 593 °C (1100 °F). However, this type of material does not have good enough creep and oxidation resistance at higher temperatures and so could not be used for the final stages of the SH/RH in USC plants. To improve the creep and oxidation resistance of the 9%Cr ferritic alloys part of the molybdenum has been substituted with tungsten to give a new alloy (T/P92); such substitution allows this material to be used at 620 °C. Alloy NF616 (P-92), developed by substituting part of the Mo in P91 by W, has an even higher allowable stress and can be operated up to steam temperatures of 620 °C (1150 °F). Beyond 620 °C (1150 °F), the 9%Cr steels become limited by oxidation resistance and 12% Cr steel and austenitic steels must be used. Among the 12%Cr steels, HT91 has been widely used for tubing, headers and piping. Further increases in creep strength by substituting more of the Mo with W and addition of Cu has resulted in alloy HCM12A (P-122), which can be used for header and piping up to 620 °C (1150 °F). Two alloys NF12 and SAVE12 having an even higher creep strength than HCM12A. These steels have very stable mechanical and physical properties at temperatures over 620 °C. The SAVE12 alloy indicates acceptable oxidation rate even at 700 °C, which could increase its application range for SH/RH tubing. NF12 contains 2.5% Co, 2.6%W and slightly higher B compared with HCM12A. SAVE12 contains 3%

Co, 3% W, and minor amounts of Ta and Nb. These latter elements contribute to strengthening by producing fine and stable nitride precipitates (Wu 2006; Lukaszewicz 2012).

The role of alloying elements in development of the ferritic steels has been extensively investigated. W and Mo and Co are primarily solid solution strengtheners. V and Nb contribute to precipitation strengthening by forming fine and coherent precipitation of M(C, N)X carbonitrides in the ferrite matrix. Vanadium also precipitates as VN during tempering or during high-temperature creep. Chromium contributes to solid solution strength as well as to oxidation and corrosion resistance. Nickel improves the toughness but at the expense of creep strength. Partial replacement of Ni by Cu helps stabilize the creep strength. Carbon is required to form fine carbide precipitates, but the amount needs to be optimized for good weldability. Boron enters the structure of $M_{23}C_6$ and segregates to the $M_{23}C_6$–matrix interface. Cobalt is known to delay recovery on tempering of martensitic steels. Cobalt also promotes nucleation of finer secondary carbides on tempering. This is attributed both to its effect on recovery and its effect on the activity of carbon. Cobalt also slows coarsening of alloy carbides in secondary hardening steels. This was suggested to be the result of cobalt increasing the activity of carbon and not being soluble in alloy carbides. The extensive database is available on these ferritic steels (Wu 2006).

3.2.1.2 Austenitic Steels

Austenitic steels are candidates primarily in the finishing stages of superheater/ reheater tubing, where, oxidation resistance and fireside corrosion become important in addition to creep strength. For convenience, austenitic steels can be classified as those containing less than 20% Cr and those containing more than 20% Cr. Alloy modifications based on the 18Cr-8Ni steels, such as TP304H, 316H, 347H, and Tempaloy A-1, and alloys with lower chromium and higher nickel contents, such as 17–14 CuMo steel, Esshete 1250, and Tempaloy A2, fall into the classification of steels with less than 20% Cr. The allowable tensile stresses for steels in this class are intermediate between ferritic steels and high Cr austenitic stainless steels. Several high-creep-strength alloys containing more than 20% Cr, such as NF707, NF709, HR3C, and SAVE 25, offer low-cost alternatives to Incoloy 800 for use in the temperature range from 620 to 675 °C (1150–1250 °F). Clearly, SAVE 25, NF709, HR3C, and Super304H are leading candidates for use in the highest-temperature applications. At temperatures exceeding 675 °C (1250 °F) nickel-based alloys are candidates, as shown in Table 3.1 and Fig. 3.5. The highest temperature capability at a given stress is exhibited by Inco740, Haynes 250, HR6W, and Inco617 in decreasing order. Alloy IN617 possesses higher strength than IN617. From a creep strength point of view, T91 is limited to 565 °C steam (metal 593 °C) and NF616, HCM12A, and E911 are limited to 593 °C steam (metal 620 °C). Even the strongest ferritic steel today is limited to 593 °C (1150 °F) (metal temperature) from an oxidation point of view. At temperatures above these, austenitic steels are required. Hence there has been considerable development with respect to austenitic stainless steels (Viswanathan et al. 2003).

Austenitic steels can also be classified as 15Cr, 18Cr, 20–25Cr, and higher Cr stainless steels.

The various stages in the evolution of these steels have consisted of initially adding Ti and Nb to stabilize the steels from a corrosion point of view, then reducing the Ti and Nb content to promote creep strength rather than corrosion, followed by Cu additions for increased precipitation strengthening by fine precipitation of a Cu rich phase. Further trends have included austenite stabilization using 0.2% nitrogen and W addition for solid solution strengthening. For example, HR6W (0.07C-23Cr-43Ni-6 W-0.1Ti-0.2Nb) and Super 304H (0.1C-18Cr-9Ni-3Cu- Nb-N) have been developed as candidate materials primarily for boiler tubing (Wu 2006).

3.2.1.3 Nickel-Based Alloys

Nickel-based alloys have the highest temperature resistance among the three groups identified for the USC applications, as shown in Table 3.1 and Fig. 3.5. As a result of very good mechanical and physical properties, nickel-based alloys have the highest potential maximum operational temperatures. The nickel-based alloys Inco740, Haynes 230, IN625, IN617, HR6W, and HR120 have much higher temperature capability, in decreasing order as listed compared to austenitic steels, followed by the ferritic steels. Purely from the creep strength point of view, at a pressure of 5500 psi for a 2″ × 0.5″ tube (stress 8.6 ksi), ferritic steels are useful up to about 620 °C (1150 °F) (metal temperature), austenitic steels up to about 675 °C (1250 °F). At metal temperatures higher than about 675 °C (1250 °F), nickel-based alloys are needed. The alloy Inco740 appears capable of reaching 788 °C (1450 °F). Since the thick-walled components are used over a range of conditions, all of the above categories of materials area likely to be used at different locations in an ultra-supercritical plant (Viswanathan et al. 2003).

Applications of nickel-based alloys are, however, significantly limited as a result of their high cost. Despite these high costs nickel-based alloys are believed to be crucial for development of the industrial size USC power plants. Inconel 617 and Inconel 740 are two of the most promising candidates for the new steam boilers. These two Inconels could be used for the SH/RH tubing in the temperature between 675 and 700 °C (Lukaszewicz 2012). As nickel-based alloys have an exceptional combination of high-temperature strength, toughness, and resistance to degradation in corrosive or oxidizing environments, they are widely used in aircraft and power-generation turbines, rocket engines, nuclear power and chemical processing plants and other challenging environments. The availability of nickel-based alloys has led to a steady increase in the turbine entry temperatures, and this trend is expected to continue. New generations of nickel-based alloys can tolerate average temperatures of 1050 °C with occasional excursions to temperatures as high as 1200 °C, which is approximately 90% of the melting point of the material.

3.2.2 Advanced Refractory Materials for Slagging Gasifiers

A gasifier acts as a containment vessel to react liquid fuels, a carbon source (coal and petroleum coke are the most commonly used), water, and oxygen at elevated temperatures and under reducing conditions; producing CO and H_2, as the primary gases, along with CO_2, CH_4, HS and other trace gases. Gasifiers are clarified as fluidized-bed, moving-bed, or entrained-flow systems; with only the entrained-flow gasifier producing a fluid ash (slag) from impurities in the carbon source, the other two gasifiers producing a dry ash by-product. In an entrained-flow gasifier, the bulk of this molten slag flows down the gasifier sidewalls, with only minor amounts associated with the excess carbon and existing as fused particles. The materials issues in the major gasification process components are mostly restricted to the hot parts—the gasifier vessel, downstream coolers, and gas cleaning vessels—and the gas turbine where the syngas or hydrogen will be burned. As shown in Fig. 3.6, during the gasification process the gasification chamber typically operates at temperatures between 1250 and 1550 °C, and at pressures of 3 MPa or higher, and so it is usually lined with refractory materials to contain the severe environment and to protect the outer steel shell from erosion, corrosion, and the high temperatures. Therefore, any materials used in advanced coal gasification systems need to have adequate, reliable, and predictable lives (Bennett and Kwong 2004; Oakey and Fry 2007).

From a process perspective, the duties are often similar to those of a pulverized fuel (PF) boiler and so similar materials could be applied if the fuel gas environment were not so aggressive. However, since the environment is so aggressive it is necessary to line the gasification chamber with refractory materials. In addition to

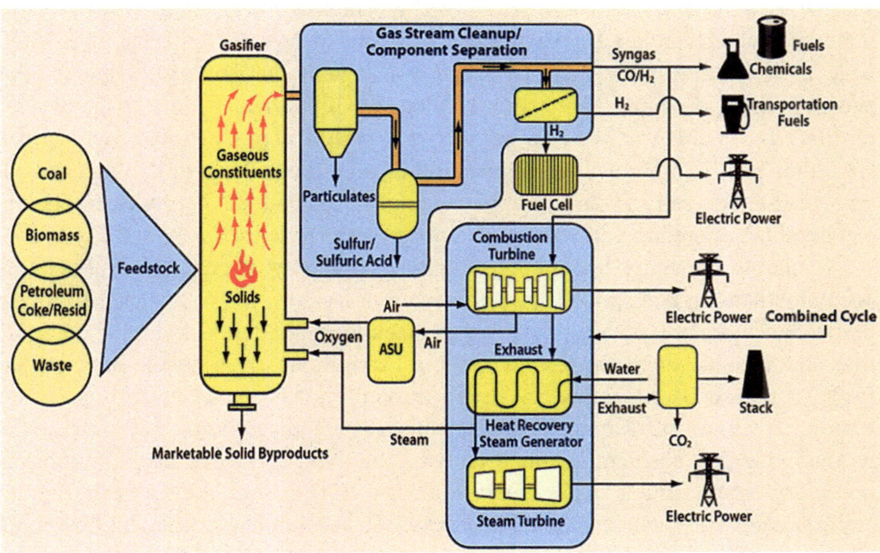

Fig. 3.6 Schematic of gasification-based system (Adapted from DOE 2013. Credit: US Department of Energy)

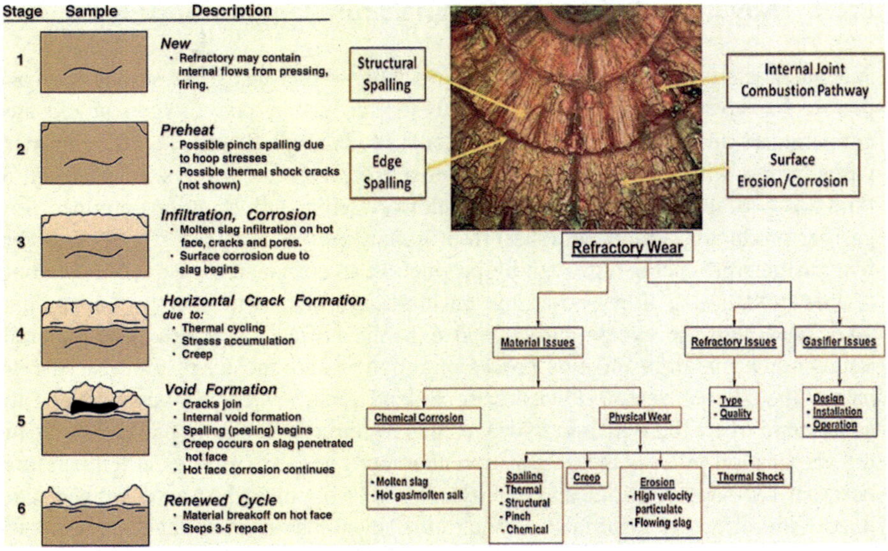

Fig. 3.7 Schematic of refractory wear in high chrome oxide refractories used in slagging gasifiers (Modified from Bennett and Kwong 2011 (Springer). Credit: National Energy Technology Laboratory; open accessed at https://link.springer.com/article/10.1007/s11661-011-0635-x)

the lining, the gasifier is usually cooled by air or by water. Water-cooled gasifiers will typically have a working face lining of Al_2O_3-SiC refractory, and have a satisfactory service life because slag freezes on the refractory surface, restricting slag penetration and corrosion. Air-cooled gasifiers are also lined with refractory materials, which can last anywhere from 3 months to 2 years. A typical liner would be a high chrome oxide material that contains alumina and may contain other additives like zirconia. The major challenge with refractory materials used in gasifiers is for them to provide an adequate period of protection to the gasifier structure. The lifetime of the lining should also be predictable so that outages can be planned. Generally, industrial applications will run two gasifier units, so that one can continue operation while the other is being refurbished (Oakey and Fry 2007).

A suitable refractory is required to endure conditions at elevated temperatures, accommodate thermal shock from temperature transients, and withstand erosion by particulates, molten slag attack, corrosive attack from hot gases of varying composition and variable oxidizing and/or reducing conditions. This hostile environment leads to failure of the refractory lining through corrosion, wear, and spallation. Figure 3.7 shows refractory wear mechanisms in high chrome oxide refractories used in slagging gasifiers (Bennett and Kwong 2011). Historically, a number of refractory compositions have been evaluated for use in these harsh thermal environments and include sintered and/or fused cast alumina-silicate, high alumina, chromia-alumina, chrome-magnesia spinels, alumina and magnesia, alumina and chrome, and SiC refractory compositions. Each particular composition has its benefits and disadvantages. For instance, fuse cast refractory materials with little or no porosity are often found to have good chemical wear resistance but have poor resistance to thermal shock. Materials containing high levels of Al_2O_3 or MgO/Al_2O_3 spinel have been found to produce materials with very poor wear resistance, as have refractories containing SiC and Si_3N_4. These were found to react with components in the slag, causing severe material wear. As a general rule additions of chromium oxide to a refractory composition are found to improve the materials resistance to chemical attack from slag. Indeed, in practice it is found that refractories with high chromium oxide content (greater than 85 wt% Cr_2O_3) are used in the severe wear areas of a gasifier, while the lower chrome oxide containing materials are used in the less severe wear areas of the gasifier. For instance, a minimum Cr_2O_3 level of 75 wt% is necessary in a refractory material for sustained material performance in slagging gasifiers. It is commonly believed that the only compositions of current refractories that could withstand the aggressive environment side a gasifier for a long enough duration to be economically viable are Cr_2O_3/Al_2O_3, $Cr_2O_3/Al_2O_3/CrO_2$, and Cr_2O_3/MgO. While high chromium oxide refractory materials have evolved as the material of choice to line the hot face of gasifiers, the performance of these materials does not meet the desired service requirements of industry (Oakey and Fry 2007).

Future developments in this area would include the development of refractory material coatings, monolithic linings, or refractory materials that do not contain chrome oxide or are low in chromium oxide. The rational of this is cost, health, environmental and processing based. In addition, the high Cr_2O_3 refractories have not met the performance requirements of gasifiers. An alternative refractory material based on alkalialuminate has been developed. More damage tolerant refractories are also developed, which would benefit the industry (Oakey and Fry 2007).

3.2.3 Breakthrough Materials

Some breakthrough materials have been explored and developed with temperature/strength capabilities beyond those currently available such as oxide-dispersion-strengthened (ODS) ferritic steels including nano-size oxide-dispersion-strengthened alloys, as well as alumina-forming austenitic stainless steels.

3.2.3.1 Oxide-Dispersion-Strengthened Steels

Oxide-dispersion-strengthened (ODS) alloys have excellent potential for use in next-generation high-temperature applications where superior creep strength and oxidation resistance compared to current alloys is required. Possible applications include tubing of high-temperature heat exchangers for indirect heating of steam or hydrogen for gasification, for fuel nozzles and sheet for combustor cans for syngas fired gas turbines, SH tubes for steam boilers, steam turbine blades and vanes. ODS alloys are mostly produced by the mechanical alloying of powders. The powder constituents can be in the elemental, intermetallic or pre-alloyed state. The metal and oxide (such as Alumina and Yttria) powders are blended and mechanically alloyed using a ball mill. The repeated impacts of this process cause smearing together of the powders followed by shearing which generates particles that each

Table 3.2 Nominal compositions of typical ODS ferritic alloys (wt%) (Wright et al. 2005)

Alloy	Fe	Cr	Al	Mo	Ti	Si	RE[a]
Kanthal APM	Bal	20.0	5.5	–	0.03	0.23	ZrO_2-Al_2O_3
INCO MA 956	Bal	20.0	4.5	–	0.5	–	Y_2O_3-Al_2O_3
INCO MA 956HT	Bal	21.6	5.9	–	0.4	0.07	Y_2O_3-Al_2O_3
PM2000	Bal	20.0	5.5	–	0.5	–	Y_2O_3-Al_2O_3
Dour Alloy ODM 751	Bal	16.5	4.5	1.5	0.6	–	Y_2O_3-Al_2O_3
ODS-Fe3Al	Bal	2.2	15.9	–	0.07	0.1	Y_2O_3-Al_2O_3

[a]Reactive element addition, in the form of an oxide dispersion

contain a fine mixture of the constituents. The powder is then packaged into sealed containers that are hot worked into simple primary shapes, either by extrusion or hot isostatic pressing (HIP). The resultant product is dense and fine grained (less than 1 μm) but has highly developed, directional residual strain. The alloys are then re-crystallized at about 1280–1350 °C for instance when large and highly elongated grains form. Secondary process can be done to form the tube, sheet, or needed shapes.

The potential advantages in employing ferritic Fe-Cr-Al alloys in high-temperature applications include (Pimentel et al. 2012): lower raw material cost and generally superior oxidation resistance; the alloys have a higher melting point, lower density and lower coefficient of thermal expansion than the current nickel- or cobalt-based alloys. However, the mechanical strengths of the alloys in the cast and wrought condition at temperatures in excess of about 600 °C were too low for them to be considered for critical structural applications. Dispersion strengthening with stable oxide particles is an ideal method for improving high-temperature strength without sacrificing the excellent surface stability of the matrix alloy. Although many oxide dispersoids have been investigated in ferritic steels, Currently, several ODS ferritic steels are commercially available including alloys such as MA 956 and MA 957 (developed by INCO alloys), PM 2000 (developed by Hochtemperatur-Metall GmbH), and ODM 751 (developed by Dour Metal). Table 3.2 gives a summary of the compositions of some commercially available ODS ferritics (Wright et al. 2005).

The oxidation resistance and good creep performance of mechanically alloyed FeCrAl ODS alloys such as PM 2000 and MA956, makes them prime candidates for the heat exchangers. The microstructure following mechanical alloying and consolidation of the resulting powder (for example, by extrusion), consists of fine grains which have a width which is much less than a micrometer and which are cold deformed during the consolidation process. The material in this state is hard and contains an enormous amount of stored energy. For example, the microstructure in the as-extruded and hot-rolled condition of PM 2000 consists of fine (~0.5 mm) grains of ferrite, and ODS particles with sizes ranging from 3 to 40 nm that are homogeneously distributed in the alloy. The ODS particles consist of $Y_3Al_5O_{12}$ garnet (YAG) and $YAlO_3$ perovskite (YAP) phases. Apart from the ODS particles, numerous inclusions or other particles such as γ-Al_2O_3, γ-Al2O3/Y-Al-O, Ti(C,N), or TiN can also be found in the alloy. Only coarse recrystallized grains have adequate high-temperature creep strength. Therefore, the as-extruded and hot-rolled material has to be recrystallized into a coarse-grained microstructure before using. However, the recrystallization behavior of iron-based ODS alloys is peculiar. They recrystallize into a grain structure which resembles that obtained by directional solidification, with coarse, columnar grains which have their longest axes along the extrusion direction. This results in tubes with anisotropic, coarse grained, axially aligned microstructures which exhibit excellent axial creep properties. Furthermore, recrystallization usually does not occur until temperatures close to melting are reached (Pimentel et al. 2012).

A particular feature of the ferritic ODS-FeCrAl alloys is that they exhibit relatively low oxidation rates up to 1200 °C (2192 °F), due to the formation of a protective Al_2O_3 film. In service, the mechanical load applied to these ODS alloys will be below the stress level at which significant creep occurs, so that it is likely that the effective service lifetime will be determined by the rate of environmental degradation. Operation at the very high temperatures possible with these alloys means that, should the ability to form a protective oxide scale be lost, the ensuing oxidation damage could be very rapid. However, since during protective oxidation there is very little loss in alloy section due to the thin alumina scales formed, it is doubtful that mechanical means for monitoring the rate consumption of the alloy would provide sufficient accuracy for assessment of remaining lifetime. Hence, there is a need for some form of lifetime model to provide reliable prediction of remaining lifetime in typical service environments (Wright et al. 2005).

Mechanical creep tests have shown that these alloys offer superior creep properties. For example, a Fe-12Cr-3W-0.4Ti (wt.%) mechanically alloyed with 0.25 wt.% Y_2O_3 (called 12YWT) was shown to fail after 14,500 h at a very high temperature of 800 °C with a 2.3% elongation at a loading stress of 138 MPa. These properties were excellent when compared to a V4Cr4Ti alloy that failed after 4029 h and 52% elongation also tested at 800 °C but at a significantly lower stress of 77 MPa. While the vanadium alloy exhibited tertiary creep, the 12YWT alloy did not. This may make imminent creep failure identification more difficult for the 12YWT alloy. Furthermore, this is not an isolated case. Y or Y-Ti nano-precipitates can reduce creep rates by six orders of magnitude at temperatures from 650 to 900 °C. In addition to superior creep properties, ODS alloys have been shown to have high elevated temperature tensile properties. In these high-performing alloys, ductility remained high (reduction of area > 40%) with increases in strength to greater than 2 GPa (West 2006).

Processing plays a large role in performance. During the alloying process, it is typical to accumulate oxygen and nitrogen interstitials. This can result in a high ductile-to-brittle transition temperature as well as poor mechanical impact properties. Thus, it is important to limit the milling time to reduce interstitial pick-up.

The superior mechanical properties of these materials are well established. However, the structure of the highly stabilized oxide nanoclusters is still not completely understood. These clusters are typically less than 5 nm in size, and have highly defective NaCl structures with a high lattice coherency to the BCC-structured steel matrix. It is this high degree of point defects and the structural affinity of the nanoclusters to the steel matrix that account for the stability. Moreover, this stability is extraordinary when combined with the material's mechanical properties. Even at temperatures of 1400 °C (91% of the melting temperature) and in the presence of intense neutron irradiation fields, stability is maintained (Hirata et al. 2011).

Despite these promising results, challenges remain. First, steel-processing techniques for these highly specialized alloys have not been perfected. Especially when compared to reduced-activation ferritic and martensitic steels, processes like fabrication and welding need to be perfected. Other problems persist. For example, producing these materials in large sizes is difficult. Additionally, a lot of the metal forming processes involved such as rolling and extrusion are directional and result in elongated grain structures. This causes anisotropic behavior that can have detrimental effects such as reduced mechanical properties along certain directions in the material. Finally, the structures of these materials are still not completely understood. As nano-characterization techniques improve, it is possible that understanding can lead to further structural manipulation and even better overall properties.

3.2.3.2 Alumina-Forming Austenitic Stainless Steels

A family of creep-resistant, alumina-forming austenitic (AFA) stainless steel alloys has been developed for structural use in fossil energy conversion and combustion system applications.

The AFA alloys exhibit comparable creep-rupture lives to state-of-the-art advanced austenitic alloys, and superior oxidation resistance in the ~923 K to 1173 K (650–900 °C) temperature range due to the formation of a protective Al_2O_3 scale rather than the Cr_2O_3 scales that form on conventional stainless steel alloys (Yamamoto et al. 2011).

Of particular interest to fossil-fired steam plant applications is the potential for superior oxidation resistance to water vapor/steam containing environments. High-temperature creep strength is achieved in AFA alloys primarily via MC carbide precipitates. The alloys typically contain only 2.5–4 wt pct Al and less than 15 wt pct Cr in order to permit stabilization of an austenitic matrix for high-temperature strength at relatively low levels of Ni additions (20–25 wt pct). A protective alumina scale can be formed at 923–1073 K (650–800 °C) with as little as 2.5 wt pct Al, if Ti, V, and N additions were minimized and relatively high levels of Nb additions were employed (>0.6–1 wt pct Nb). A unique behavior of the AFA alloys is that all compositions exhibit a transition from protective alumina scale formation to internal oxidation of Al with increasing temperature from 1073 to 1273 K (800–1000 °C), depending on composition and oxidizing environment. This behavior is a consequence of composition choices needed to balance alumina scale formation with mechanical properties, particularly creep resistance. In other words, the AFA alloys require a balance of Al and Cr additions to achieve an external, protective alumina scale formation at elevated temperatures, while avoiding δ-Fe and σ-FeCr relative to γ-Fe. High-temperature strength of AFA alloys relies on second-phase precipitates, primarily nanoscale MC-type carbides (M: mainly Nb) in an austenitic single-phase matrix, although NiAl-B_2 and Fe_2(Mo,Nb)-Laves precipitates may also contribute. Maximizing the amount of MC carbides can optimize the creep-rupture life of AFA alloys at around 1023 K (750 °C). The NiAl-B_2 phase precipitates also play an important role for the oxidation resistance, because they act as an Al reservoir for the external protective alumina scale during exposure at elevated temperatures. Higher Nb levels (>0.6–1 wt pct) also correlate with improved oxidation resistance, particularly in water vapor environments. A complete mechanistic understanding of the effects of Nb oxidation has not yet been achieved. Future alloy development directions for the AFA alloy family include both a lower cost, low-Ni/high-Mn grade and a higher performance, Fe-based superalloy AFA grade (Yamamoto et al. 2011).

3.3 Advanced Coatings and Protective Systems

Coatings are often applied to superalloy or other structural substrates operating at high temperatures or high corrosive environments to inhibit the direct interaction of oxidizing or corroding species in the environment with the substrate, thereby minimizing damage to the substrate. Such damage may consist of recession of the substrate surface by oxidation/corrosion (wastage) or of degradation of substrate properties as a result of inward diffusion of undesirable elements (e.g., sulfur and oxygen). The coating is a surface layer of material and may be ceramic, metallic, or combinations of metallic and ceramic layers. These coatings are not inert barriers but provide protection to the substrate by interacting with oxygen in the environment to form dense, tightly adherent oxide films that are resistant to the diffusion of potentially damaging species. Because the protective oxide films can spall as a result of thermal cycling and mechanical damage, there is a requirement that the oxide film be replaceable in situ for the film to remain viable. This replacement is accomplished by making the original coating rich in elements that form the protective oxides (such as aluminum and chromium). The useful life of a coating is governed by its ability to form the desired protective scale and its ability to retain and/or replace the scale during service. In addition, coating life is affected strongly by the composition of the substrate. Also, the possible effects of inter-diffusion between the coating and the substrate must be assessed in selecting a coating (Penfield Jr. and Rittenhouse 2004).

3.3.1 High-Temperature Coatings

High-temperature coatings are used to restrict surface degradation or to thermally insulate the material against the hot environment. Surface degradation takes place through corrosion, oxidation, and solid particle erosion. As the lifetime of structural components is frequently controlled by surface degradation, significant cost and performance improvements are obtained by application of coatings. Coating used for high-temperature applications can be divided into three major types (Eskner 2004): diffusion coatings, overlay coatings, and thermal barrier coatings.

Diffusion coatings are corrosion/oxidation resistant thermally grown oxides by enriching the surface with either Al, Cr, or Si through diffusion. The two most common methods for providing aluminide diffusion coatings are pack cementation and chemical vapor deposition (CVD). In the former method, the component to be coated and the reactants (usually an aluminum-containing powder, a halide that serves as an activator, and inert filler, such as alumina) are heated in a retort in an inert atmosphere. The metal powder and the activator combine to form a vapor that reacts with the surface of the substrate and enriches it with aluminum. A given set of processing conditions will produce different coatings on different alloys, depending on the chemical composition of the alloy. In the CVD process, a vapor of predetermined composition is introduced into the coating chamber where it reacts with the alloy surface. This process has a number of advantages relative to pack cementation, especially in terms of flexibility in selecting the composition of the coating vapor (Penfield Jr. and Rittenhouse 2004). Diffusion coatings are the most widely used types in gas turbine engines and are applied for example on rotating parts like turbine blades. They have a homogeneous microstructure with good thermo-mechanical fatigue properties. A critical limitation is the high ductile-to-brittle transition temperature and the very brittle nature below the transition temperature (Eskner 2004).

Overlay coatings produce a corrosion/oxidation resistant thermally grown oxide by depositing a pre-alloyed material with desired composition on the surface. Typical compositions are based on the MCrAlX alloy system, where M is Ni, Co, Fe, or a combination of these and X is Y, Si, Ta, Hf, etc. Overlay coatings are generally applied by either electron beam-physical vapor deposition (EBPVD) or by plasma spraying processes. These coatings differ from the diffusion coatings in that extensive interdiffusion of the coating with the substrate is not necessary to achieve the appropriate coating composition and structure. Instead, a pre-alloyed material of the composition required to form a protective oxide film is deposited directly onto the substrate surface. Interdiffusion is required only to the extent that it provides an improved bond between the coating and the alloy surface. In the EBPVD process, the coating is provided by condensation of a metal vapor directed onto a preheated surface; in plasma spraying a metal powder is melted in a plasma gas stream and accelerated onto the substrate surface. The main advantage in comparison to diffusion coatings, which have properties that strongly depend on the substrate composition, is that their properties can be better controlled and balanced for a specific application. In general, overlay coatings have better oxidation and corrosion resistance than diffusion coatings and can be used at higher temperatures. They can also be deposited in thicker layers, which may extend the lifetime of the coating. The limitations of overlay coatings are the relatively poor reproducibility of the properties as they are highly dependent on the process. Overlay coatings relative to diffusion coatings, especially those produced by the CVD process, are line-of-sight processes and this invariably causes problems with the uniformity of coatings on complex shapes (Penfield Jr. and Rittenhouse 2004).

Thermal Barrier Coatings (TBCs) are often used in the hot-sections of turbines to reduce the surface temperature of the metallic substrate. The TBC is a multilayer coating system, consisting of a ceramic outer layer (topcoat) and a metallic inner layer (bond coat) between the substrate and the ceramic. The bond coat protects the substrate from oxidation/corrosion and the ceramic top coat provides the thermal protection. Depending on a number of factors, temperature reductions in the range of 100–150 °C are possible using TBCs. Because of its very low thermal conductivity and relatively high coefficient of thermal expansion, the material of choice for the TBC topcoat is Y_2O_3-stabilized (6–8 wt%) ZrO_2. Its thickness will typically be in the range of 125–500 μm. Most sources describe the ZrO_2 topcoat as "transparent" to the transport of oxygen. The metallic bond coat (typically MCrAlY or platinum aluminides) will be 75–100 μm thick. Due to the inherent brittleness of ceramics, thermal barrier coatings mostly are used in nonrotating parts such as combustion chambers, transition ducts, vanes, exhaust nozzles and afterburners. Two major benefits can be obtained with thermal barrier coatings (Eskner 2004): either a prolonged lifetime of parts by reducing the substrate temperature or, alternatively, an increased engine efficiency by reducing the air-cooling flow or increasing the combustion temperature. A limitation to TBC technology is that the coatings are very prone to spallation under the influence of thermal and mechanical stresses. Therefore, there is some danger in relying on the TBC as a "prime reliant" feature for temperature minimization, as it cannot be reformed during service (Penfield Jr. and Rittenhouse 2004; Clarke 2012).

3.3.2 Erosion- and Corrosion-Resistant Coatings

Offshore oil and gas production environments represent aggressive conditions in terms of erosion and corrosion. Consequently, materials selection must be given a detailed attention at every stage of the design, construction and operation of systems and equipment including piping systems and their accessories, such as bends, elbows, tees, and valves. They also include devices that impart movement to fluid such as pumps, impellers, propellers, and blowers, which are exposed to or transport particle-laden fluids such as seawater. Therefore, full attention should be given to general erosion and corrosion resistances, in order to minimize premature failures, which lead to loss of production due to total shut downs and severe economic losses because of the inflating maintenance costs. Even more important is the need to maintain offshore safety (Rayes et al. 2013).

Erosion-corrosion is a process involving both erosion and corrosion in which each of these processes affected by the simultaneous action of the other and thereby accelerated. This synergistic effect is much more serious for material degradation than those caused by pure corrosion or pure erosion individually. Mechanistically, corrosion enhances erosion because of the decrease of strength of materials surface, and erosion enhances corrosion due to the retardation and breakage of the protective film on metal surface (Yang 2013).

Conventional corrosion control techniques, such as cathodic protection and inhibitors, cannot mitigate and prevent erosion-corrosion effectively. The most commonly used technology for erosion-corrosion prevention is high-performance coatings. Coatings could act as barriers between the flowing slurry and the pipe steel.

Table 3.3 Typical coatings for prevention of errosion-corrosion (Yang 2013)

Coating type	Coating material	Composition (wt.%)
Ceramic	Al_2O_3-$2TiO_2$	3.0 wt% SiO_2M, 2.0% FeOM, 2–3.25% TiO_2, 9.0%Al_2O_3, Al_2O_3-TiO_2 balance
	CrN TiN Multilayer	CrN TiN Multilayer
	Cr_3C_2 Ni-Cr sealer	Cr_3C_2 Ni-Cr sealer
	Diamond	Crystalline diamond
Polymeric	Polyurethane hard	Polyurethan
	Epoxy	Epoxy
	Epoxy SiC	Epoxy, amines, SiC, metallic oxides, silicates, ferrosilicon
Metallic	WC Co Cr	86%WC, 10%Co, 4%Cr
	Ni-Cr WC	C, Cr, B, Si, Fe, WC, Ni balance
	Ni-Palloy SiC	10–14%Ni-P alloy, SiC balance
	Ni-WC Co	Nickel-based fusible alloy, WC, Co
	Cr plate (pulsed)	Pulse-plate Cr
	Fe alloy	Fe-based alloy

Table 3.3 shows some typical coatings for prevention of erosion-corrosion, where metal matrix composite or cermet coatings show remarkable advantages (Yang 2013).

There are various types of cermet and also many different grades within the different types. For example, cermet carbide coating, Cr_3C_2-NiCr, is an excellent replacement to hard oxide, Cr_2O_3. Generally, the cermet coatings consist of WC or CrC particles embedded in a metal binder, which can be a pure metal or a mixture consisting of Ni, Cr, and Co. WC-Co and CrC-NiCr systems constitute two main carbide materials used in thermal spraying processes in order to improve the erosion/wear resistance and decrease the friction coefficient between various sliding components. Coatings of the WC-Co system generally have a higher hardness and wear resistance than CrC-NiCr coatings, however, the decarburization of WC into W_2C, W_3C and even metallic W phase leads to the degradation of coating properties and limit the application of these coatings as well as due to the dissolution of Co phase leading to low corrosion resistance. Such coatings are applied by thermal spray technologies. The microstructure of a thermally sprayed coating is usually inhomogeneous and contains discontinuities, such as pores, oxide lamellas, or incompletely molten spray particles all of which may be present in the sprayed coating materials. When the substrate is less noble than the coating, galvanic effects can be found between coating and substrate, resulting in a significant attack of the substrate material. On the other hand, if the substrate is more noble than the coating (i.e., stainless steel) the coating acts as a sacrificial anode accelerating its corrosion. As a result, chemical composition of metallic binder materials and the occurrence of microcracks were the most important factors influencing the corrosion resistance of the HVOF sprayed WC cermet coatings in the strong acidic environment (Rayes et al. 2013).

In addition, protective coatings are seeing increased usage as plant operators seek to reduce the number of forced or planned outages necessary to repair or replace components susceptible to erosion and/or corrosion by extending the component's life. Indeed, a flame spray coating of Al/80Ni20Cr now has a lifetime of at least 3 years when applied to water wall tubes. A number of coatings have been used in both operational and pilot plant, and a number have undergone, or are currently undergoing, laboratory testing, with variable performance in all cases. The choice of erosion-corrosion-resistant coatings should be specific to each individual application. Often a coating will perform well in one FBC (fluidized bed combustion) environment, only to fail in another. This is because the erosion-corrosion behavior of a coating is not only related to the morphology of the coating, but also to the characteristics of the erodent particles. Weld overlay was initially the preferred protective coating, typically forming layers 3 mm thick using 1.85% Cr hard facing material. Durable layers of Alloy 625 have been built up to approximately 7 mm thick on water wall tubes for instance. However, the application of weld overlay is restricted due to the low productivity of the process, erosion of the resulting ledge and thermal shock. Repeated applications at successive outages lead to embrittlement of the old overlay and to cracks that could propagate into the tube proper. This had led to more attention being focused on the application of thermal spray coatings. A range of materials and application techniques (arc spraying, flame spraying, high velocity oxygen fuel spraying [HVOF], sintering) are available. A variety of alloys have been employed in efforts to improve on parameters such as the adhesion strength of the coating to the base material, resistance to spalling, hardness and oxidation in the bonding zone, heat transferability and resistance of the coating to corrodent penetration. It is important to recognize that the coating performance is sensitive to surface preparation and post-processing, with the elimination of surface irregularities, and the critical nature of the contour of the coating from its bottom to top edge. The coating should also be applicable on site at a reasonable cost (Oakey et al. 2003a, b).

3.3.3 Wear-Resistant Coatings

Coatings devised for wear protection share many of the physical and processing characteristics of high-temperature, erosion, and corrosion resistance coatings, but they are usually not oxide based nor does their functioning depend on the formation of oxides in service. They are most often metallic compounds (often incorporating metallic carbides, e.g., CrC-NiCr), carbides (e.g., $Cr_{23}C_6$, Cr_7C_3, and Cr_7C_2), or nitrides (e.g., TiN). Most wear protection coatings have application at low to medium temperatures. However, coatings based on CrC-CrNi might retain abrasion resistance in the range 700–800 °C (Penfield Jr. and Rittenhouse 2004).

3.3.3.1 Classification of Wear-Resistant Coatings

Wear-resistant coatings are generally classified in two categories (Wu 2010): (a) soft low-friction coatings and (b) hard wear-resistant coatings, depending on whether the hardness value is lower or higher than ≈10 GPa. Hard wear-resistant coatings include oxides, carbides, nitrides, borides, and some carbon-based materials (diamond and sp3 dominated diamond-like carbon). Soft low-friction coatings include PTEF, MoS2, and some carbon-based materials (Graphite and sp2 dominated graphite-like carbon).

The combination of basic tribological materials to build various film structures (nanocomposites, superlattice, gradient, etc.) has been explored. One direction is to develop low friction coatings; for instance, sputtered carbon-based coatings have attractive tribological properties, low wear rates and low coefficients of friction. DLC materials are thin films of amorphous metastable carbon-based solids, pure or alloyed with hydrogen having properties similar to that of crystalline diamond. DLC coatings have high hardness (from 20 to 40 GPa), low friction coefficient (0.01–0.02 in vacuum and 0.15 in atmosphere), low wear rate (10–16 $m^3\ mN^{-1}$), and chemical inertness and, therefore, are accepted as outstanding wear-resistant coatings. In order to improve the properties of DLC films, addition of impurities such as silicon, nitrogen, boron, phosphorus, fluorine, and some metals has been deployed. The addition of such impurities reduces the internal stress, electrical resistance and friction coefficient of DLC thin films. Graphit-iCTM coatings are a type of hard carbon coatings with excellent tribological properties, low friction (<0.1), low wear rate (10–17 $m^3\ mN^{-1}$), and high load-bearing capacity (>2.1 GPa). They are normally better than those conventional diamond-like carbon (DLC) or Me:C DLC coating, especially in humid air and in water. Another direction is to improve wear resistance under severe conditions, rather than to decrease friction. From the initial binary hard materials, multicomponent coatings based on TiN and CrN have been developed by alloying these binary structures with metal and/or metalloid components. The nature and amount of alloying elements strongly influence their friction and wear behavior. Such alloying elements include Zr, Hf, V, Nb, Cr, Mo, W, Al, and Si, with various chemical combinations (Wu 2010).

3.3.3.2 Coating Structures

1. Single-component coatings

   Most commercial PVD and CVD coatings consist of a one single layer, often containing one single phase. The commercially successful single layer coatings include TiC, TiN, CrN, CrC, Al_2O_3, DLC, W_2C, WC/C, MoS_2, diamond, soft metals, and some polymers. To fulfill the functional demands, an adhesion interlayer may be necessary. In addition, carbonaceous materials, including DLC and diamond films, have been widely investigated. Among solid lubricants, DLC films probably exhibit the widest range of friction and wear behavior. Diamond films are extremely hard and offer several outstanding properties, such as high mechanical strength, chemical inertness and very attractive friction properties. CVD techniques have been used to produce high-quality diamond films with micro-and nanocrystalline structures on various substrates. The high-quality diamond coatings produced by CVD exhibit most of the desired mechanical and tribological properties of natural diamond. However, the surfaces of diamond coatings are generally rough with sharp facets, which need to be polished using expensive methods (Wu 2010).

2. Multicomponent coatings
   Multicomponent coatings are made up of two or more constituents in the form of grains, particles or fibers. Depending upon the size and distribution of secondary phases in the primary matrix, a coating can either be called multiphase or composite. The term "composite" is used when one phase is dispersed in a matrix which is continuous; the term "multiphase" is used when both are equally present and none of them is continuous. Initially mononitride and monocarbide were produced as coating materials on various substrates to improve tribological and corrosion behavior. They are mainly based on transitional metals and elements. Like metals forming alloys, these mononitrides and monocarbides can form binary nitrides and binary carbides (ternary phase systems) by adding another suitable metallic element to them. One of the ways to improve/optimize properties of a hard material is to substitute the metal lattice of the compound phase with another compatible metal, e.g., the lattice of TiN is partly substituted by Al. Another method is to vary the relative concentrations of the nonmetallic elements. For example, the aluminum in the coating material reacts with the oxygen in the air and forms a passive Al_2O_3 layer, which prevents the film from further intensive oxidation. Moreover, many properties can be achieved in binary and ternary coatings by alloying with Cr which has very high potential for developing multifunctional protective coatings. Cr has a positive effect on the oxidation resistance of titanium carbides, borides, and nitrides and improves their wear resistance, especially at elevated temperatures (Wu 2010).
3. Multilayer coatings
   Multilayered coatings are composed of a periodically repeated structure of lamellae of two or more materials, with thickness up to a few tens of a micrometer (if the thickness of each lamellae is in the nanometer range, these films may become a superlattice). Multilayer coatings can be classified in two categories: isostructural (individual layers have the same structure) and non-isostructural (individual layers have different structure). One example of hard coating multilayers is TiB_2/TiC. Another type of multilayer coating consists of transition metal compounds alternating with carbon-based films, e.g., TiC/DLC (on a Ti/TiC interlayer) or TiN/CNx. In addition, nano-multilayer corresponds to the multilayer concept extrapolated to a thickness of individual layers $\leqq 100$ nm range. The major function of these structures is to significantly enhance hardness, fracture toughness, and adhesion of the coatings, and thus to improve their wear resistance. When the thickness of one layer is several nanometers, a periodic structure of layers of two (or more) materials is a so-called superlattice. Superlattice reflections are those present because the material is ordered such that the actual real space unit cell is layer and thus the reciprocal space cell is smaller. Some superlattices examples include TiAlN/TiAlCrN, TiCN/ZrCN, CrAlYN/CrN, and TiHfN/CrN. The properties of multilayer coatings depend on

the material selection, the deposition processes and kinetics of growth, and on the special nanoscale thin film architecture. The advantage of combining several structures and compositions within one coating include achievement of various individual physical properties (i.e., diffusion barrier + low friction), reduction of the mismatch in mechanical and chemical properties between the substrate and the coating (mainly to enhance adhesion), control of the residual strain and therefore the stress within the coatings, the ability to stop cracks during operation under severe conditions, and enhancement of hardness and/or toughness by allowing layers or phases to slide over each other when they deflect under load (Wu 2010).

4. Gradient and nanostructured coatings
   Nanocrystallized and nanocomposite coatings are aimed to maximize hardness (H) while ensuring an adequately low elastic modulus (E), to provide an appropriate "elastic stain to failure," as determined by the H/E ratio. On the nanometer scale, the mechanism corresponds to higher resistance to dislocation movement by precluding the formation of stable ones. This is achieved by decreasing the grain size, in accordance with Hall-Petch relationship, or by controlling the presence of interfaces between nanocrystallized metal nitride/metal, or between amorphous and nanocrystallized phases as with nc-MnN/α-Si3N4 (where M is Ti, W, V or other transition metal). Control of the structure and composition of coatings can be achieved mainly through three structural configurations. The first configuration is the functionally graded coatings, which are regarded as the logical progression from multilayered coatings. Grading the composition by a drastic control of the deposition process has been achieved to combine a relatively hard TiAlN phase with a softer MoS_2 phase with an increase in the latter phase towards the top surface. One other benefit of grading the composition is that it improves adhesion of DLC-based coatings that contain metal additions. Such improvements have been obtained by Ti- and TiC(H)-graded underlayers to increase the wear resistance of DLC films. Second, nanostructured coatings include nanocrystallized films (with grain sizes in the nanometer range) and nanocomposite films. Nanocomposites include structures that combine amorphous phases with crystallized ones (Wu 2010).

Therefore, an improvement of coating properties and performances can be achieved through (Wu 2010): (1) Multicomponent coatings and graded coatings; (2) Multiple-film coatings and multilayers on the base of hard /hard, hard/soft, and hard lubricating components; and (3) Nano-structured coatings. A rating of the properties of different coating systems is extremely difficult, because they depend not only on the materials and their chemical distribution but most often even more on the deposition processes and parameters. The same holds for the tribological performance as the counter body material, the load and probably lubrication govern to a large part the wear and friction behavior.

INDEPTH: Pipeline Infrastructure (Lahey 2002; Baker and Fessler 2008; Jackson et al. 2014)

Modern pipelines are protected from external corrosion (from the soil or water surrounding the pipeline) through anticorrosion coatings and cathodic protection. However, most pipelines are still unprotected against internal corrosion, the reported cause of 10% of significant pipeline incidents. Corrosion mitigation techniques for legacy pipelines include the introduction of corrosion inhibiters into the pipeline, reduction of moisture in the lines, and the use of robotic devices or "pigs" that detect corrosion failure before it becomes catastrophic. For new pipelines, it is possible to coat the inside of a steel pipeline with a corrosion-resistant coating or paint. While corrosion-resistant coating materials exist, they are costly and may be difficult to apply as a retrofit in the field. Alternatively, a corrosion-resistant material can be selected for the entire pipeline structure. Nonmetallic pipeline materials offer corrosion resistance without the need for coatings and cathodic protection. Fiberglass and polyethylene pipelines have begun entering the market owing to maintenance advantages; however, adoption has been limited by the comparatively high cost of fiberglass and plastic pipelines and by their susceptibility to damage during excavation and digging. Emerging solutions, such as metal/plastic hybrids, are also under active development.

Some of the most important areas for research and development include the following:

(a) Advanced pipeline coating technologies. Coatings must provide uniform corrosion resistance and durability to construction and handling, and the coating should be low-cost and able to be applied in a mill or in the field. Thermal sprayed metallic coatings (aluminum and zinc) are emerging technologies that could provide excellent corrosion resistance at a low cost.

(b) Pipeline corrosion detection. Long-range guided-wave ultrasonic testing is being developed to detect metal loss in pipelines. This technique could be especially valuable in difficult-to-access locations. It is needed to reduce false positives from these devices and enable the calculation of failure pressures.

(c) Computational modeling to support direct assessment of corrosion. Direct corrosion assessment techniques are only effective if the locations that are most susceptible to corrosion are known. It is needed to understand where the likelihood of corrosion is highest and to determine the appropriate intervals for reassessment based on corrosion and crack growth rate modeling.

(d) Prevention of stress corrosion cracking. Stress corrosion cracking (SCC) is known to occur in high pH (pH 9.0–10.5) and near-neutral (pH 6.0–7.0)

(continued)

environments. High-pH stress corrosion cracks are intergranular (propagating along the grain boundaries), while near-neutral stress corrosion cracks are transgranular (propagating through the grains). Internal SCC is emerging as a major concern for the pipeline transport of ethanol because SCC has been observed in ethanol storage tanks. It is needed to prevent internal SCC in pipelines carrying ethanol and ethanol blends and to determine safe conditions for pipeline transport of ethanol.

Corrosion-resistant pipelines could also benefit the development of a hydrogen energy infrastructure. The storage and transportation of hydrogen fuels are complicated by the fact that structural steels are sensitive to hydrogen embrittlement and fatigue fracture, which can lead to hydrogen leakage. The needs for hydrogen-resistant pipelines overlap those for corrosion-resistant natural gas pipelines, including advanced steel and nonferrous pipeline materials, protective coatings, and improved welding techniques.

3.4 Materials for Deep Oil and Gas Well Drilling and Construction

The increasing worldwide demand for oil and gas coupled with the fact that the peak of oil production has been reached or soon will be, has pushed the petroleum industry into drilling ever deeper wells. Well depths of 25,000 ft. (7620 m) and greater are no longer unusual, and even deeper wells are expected. Generally, increasing depth means increasing pressure and temperature. High-pressure/high-temperature (HPHT) wells have generally been considered wells in which temperatures and pressures at the bottom of the well exceed 300–350 °F (149–177 °C) and 10,000 psi (69 MPa), respectively. Drilling these HPHT wells requires specialized methods and considerable planning, but the materials have largely remained steel drill pipe and steel components, although other alloys such as titanium and composite materials are being considered. However, the real materials challenges are in completing and producing the wells after they are drilled (Craig 2008).

For a typical well completion, surface casing and some of the intermediate casing strings are not affected by HPHT conditions, and thus standard steel tubulars function well. The major components that require greater attention and represent materials challenges are the liner at the bottom of the well, the tieback casing string, the tubing (and associated jewelry), and the stack of valves that go on top of the well to control the flow and the pressure of the well. Depending on the temperature, the pressure, the velocity, and the mix of corrodents present in the gas stream (H_2S and CO_2), materials can range from carbon and low-alloy steels to stainless steels, nickel-based alloys, and titanium alloys. The specific materials challenges in HPHT wells can be placed in four broad categories (Craig 2008): mechanical considerations, metallurgical issues, corrosion resistance, and manufacturing capability.

Mechanical considerations for well tubulars generally come down to resisting axial loads, burst, and collapse. These forces define the required strength and wall thickness of the tubulars for a particular application. However, with increasing strength, wall thickness, and rolling practice, anisotropy tends to increase, further complicating casing design. Complications arise because of insufficient data on tensile properties in the transverse and radial directions of not only steel tubulars, but also, even more so, of cold-worked corrosion-resistant alloys (CRAs). In addition, as the temperature of the well rises, strength shows a commensurate loss. This loss is also not well documented, but must be accounted for in the well design. Design guidelines or standards need to be developed for 30,000 psi (207 MPa) surface equipment, especially at expected temperatures of 500–600 °F (260–315 °C). Furthermore, the current choice of Alloy 625 clad steel will be an unacceptable solution in the future, because the simple mass of steel to contain such high pressures would be so large that it would surpass the weight limits on offshore platforms. Moreover, under current standards (NACE and API), no low-alloy steels have sufficient hardenability to achieve the strength needed in steel forgings with wall thickness in excess of 12 inches (30.5 cm), but still maintain good fracture toughness and not exceed the NACE MR0175/ISO 15156 hardness limit of HRC 22. This means that solid forgings of either age hardenable nickel-based alloys or titanium alloys will be required for surface equipment that must contain high pressure with little loss of strength at high temperatures, be corrosion resistant, and display good fracture toughness. Such alloys must also reduce the weight penalty for offshore applications (Craig 2008).

Metallurgy issues are an integral part of solving purely mechanical requirements. The standard steel alloy for casing and tubing that must resist failure from sulfide stress cracking (SSC), as specified by NACE MR0175/ISO 15156, is steel type AISI 4130 and its modifications (such as higher Cr and Mo as well as some Ti, Nb, and B). However, deeper, higher-pressure wells require that wall thickness of the tubulars be increased, and/or the yield strength must be increased. Since NACE MR0175/ISO 15156 limits the yield strength (via the hardness) to approximately 100,000 psi (690 MPa) maximum, the only other available means to increase the resistance to mechanical loads is by increasing wall thickness. In reality, this approach is also limited because of the hardenability limitations of alloys such as modified 4130. After the wall thickness exceeds about one inch (25.4 mm), the inability to fully through-harden steel tubulars lead to poor SSC resistance, which cannot be tolerated in high-pressure wells. To maintain mechanical integrity in HPHT wells and meanwhile meet the strict limitations on yield strength required by NACE MR0175/ISO15156, the currently acceptable alloys are high-strength CRAs such as martensitic and duplex stainless steels, nickel-based alloys, and titanium alloys. However, NACE MR0175/ISO15156 arbitrarily caps the service temperatures of nickel-based alloys and the duplex stainless steels at 450 °F (232 °C) maximum, but has no similar restrictions for carbon and low-alloy steels. Table 3.4 presents some of the many alloys currently available or planned for tubing strings, casing liners, and

Table 3.4 Typical corrosion-resistant alloys for casting liners and tubing (Craig 2008)

Group	Alloy name	Nominal composition, wt%
Martensitic stainless steels	410, 420 (13Cr)	12Cr
Super martensitic stainless steels	Super/Hyper 13Cr	12-13Cr, 4-5Ni, 1-2Mo
Duplex stainless steels	2205	22Cr, 6Ni, 3Mo
Super duplex stainless steels	2507, DP3W	25Cr, 7Ni, 3.5Mo,N,W
Nickel-based alloys	825, 2242	22Cr, 42Ni, 3Mo
	2550	25Cr, 50Ni, 8Mo, 2 W, 1Cu
	G50	20Cr, 52Ni, 9Mo
	C276	15Cr, 65Ni, 16Mo, 4 W
	718	20Cr, 52Ni, 3Mo, 5Nb, 1Ti, 0.6Al
	925	21Cr, 42Ni, 3Mo, 2Ti, 2Cu, 0.4Al
	725	20Cr, 57Ni, 8Mo, 3Nb, 1,5Ti
Titanium alloys	Grade 5 (Ti-6-4)	6Al, 4 V
	Ti 6-2-4-6	6Al, 2Sn, 4Zr, 6Mo
	Grade19 (Beta C)	3Al, 8 V, 6Cr, 4Zr, 4Mo

other accessories. Because of the strength versus-hardness dilemma, current HPHT wells are faced with the possibility of CRA casing tieback strings but at great cost compared to low-alloy steels. Another important metallurgical concern for which little or no data are available is the potential for deleterious phases to form during long term well aging at temperatures in excess of 400 °F (204 °C), primarily for the CRAs. The aging of stainless steels and nickel-based alloys over 20–30 years of exposure may encourage the precipitation of phases such as Sigma, Mu, and Laves, which could lower fracture toughness and strength. Likewise, titanium alloys are expected to be susceptible to further aging under long-term exposure in HPHT well conditions. Coupled with all of the above mechanical considerations is the need for the primary path for conveyance of the oil and gas to the surface (casing liner and tubing) to be corrosion resistant, and most often to meet the requirements of NACE MR0175/ISO 15156. In addition, the ever more severe well conditions may surpass the ability of nickel-based alloys and titanium alloys to function satisfactorily for the long term. Thus, completely new materials that have not been applied in the petroleum industry up to this time may be required (Craig 2008).

As drilling operations are limited by the materials used to manufacture drill pipe, composite drill pipe (CDP) is available or under development that weighs one half or less than the equivalent, highest quality steel drill pipe. Further, with CDP it is possible to incorporate a signal transmission capability within the walls that will allow real-time monitoring of downhole conditions while drilling. Because the CDP has the potential to be lighter (half the weight of steel) and maintain required performance properties, it is considered one of the technologies potentially essential for ultra-deep water resource development. Onshore, it will allow the existing fleet of drill rigs to extensively increase the reach of their drilling capability. And, as a third benefit, it has significant potential to enable high speed communications (for smart drilling) up the drill pipe because of the ease of placing wiring and/or fiber optic leads within the body of the drill pipe. While use of composites has increased dramatically, the cost and performance of these structures has kept them from being used more widely. With the advance of carbon fiber technology, composite structures are beginning to have the performance capabilities to compete with steel. In addition, now that carbon fibers are getting cheaper and pipe designs more sophisticated, the opportunity exists to develop a pipe that can be cost effective when compared with steel (Leslie 2013).

One typical composite drill pipe consists of a composite material tube with steel box and pin connections. The tube is manufactured by winding a composite material consisting of graphite fibers and an epoxy resin around a metal mandrel. This length is then cured before the mandrel is removed and recycled. The cured pipe section is finish machined and coated for abrasion resistance. As a final preparation, normally done in the field, standard elastomeric "centralizers" are added where required. Both the centralizers and the abrasive-resistant coating can be field repaired. More extensive wear, should it occur and not be too severe, can be repaired by recycling the CDP back through the factory. Steel drill pipe protectors (rotating or nonrotating) can also be incorporated into the design. Additional capacity will require the incorporation of automation and continuous operation to the winding, curing and machining functions. A significant feature of CDP is that it can be designed to carry electrical power and/or real-time communication lines embedded in the composite walls (Leslie 2013).

3.5 Materials for Sensing in Harsh Environments

Emerging clean energy technologies offer the potential to operate at high efficiencies while minimizing emissions including approaches that minimize carbon emissions. The conditions under which fossil fuel is converted and thermal energy is generated are increasingly harsh when compared to 1960s boiler/thermal-based systems which are still in operation today. Advanced power generation technologies that are considered part of a clean energy platform include coal gasification, high efficiency combustion turbines, novel oxy-combustion processes, and ultra-supercritical steam cycles. Table 3.5 gives general process conditions for advanced clean energy systems. These conditions challenge the ability to accurately and reliably measure the process, thereby limiting its operational and environmental performance (Maley and Romanosky 2012).

Every aspect of energy conversion from fossil fuels depends on sensors for monitoring and control. Accurate and reliable sensors for high-temperature and/or corrosive environments would have multiple benefits for the thermal plant, such as avoidance of damage to heat-transfer surfaces, combustion controls for individual burners in utility boilers, reduction of noxious emissions, and structural health monitoring of critical components. Standard sensors used for monitoring the operation of the combustion system and the emissions controls systems rely on optical techniques and fiber-optic technology. The materials in most of these optical sensor systems have a temperature dependence that makes them unsuitable for the extreme

Table 3.5 General process conditions for advanced clean energy systems (Maley and Romanosky 2012)

Temperature	Pressure
Coal gasification	
1000–1600 °C	300–1000 psi
Highly reducing, molten coal slag, and corrosive gases	
Combustion turbine	
1000–1500 °C	Pressure Rations up to 30:1
Highly oxidative, high velocities, and rotating components	
Oxy combustion and ultra supercritical steam cycles	
760 °C (Fuel Side)	5000 psi (Steam Side)
Highly Oxidative, Highly Corrosive, High Particulate/Ash in critical combustion zones	

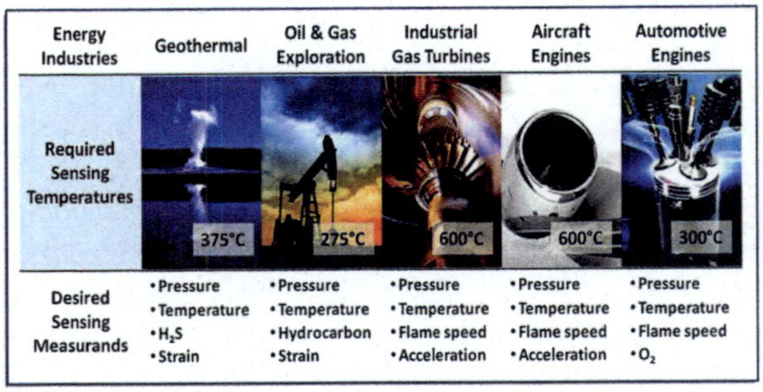

environments of advanced combustion systems. Advanced fiber optic and optical materials for high-temperature (1300–1650 °C) environments are needed, for instance. Moreover, corrosion of steam-plant materials is strongly influenced by the pH of the circulating water. Since the physiochemical properties of water are highly sensitive to temperature, there is a strong incentive to develop online pH sensors that can be used at system temperatures (360–400 °C), rather than relying on analytical extrapolations from grab samples cooled to the ambient temperature (the current practice). However, sensor materials are degraded by hot water, and measured values are sensitive to impurities introduced by even slight dissolution of the sensor in the local cooling water.

Sensing represents one of the vital components of the control schemes with the sensors providing the physical signal to enable the control loop and monitoring of energy system condition. Since the reliability of the sensing element is often the lowest link in the control elements, development of appropriate sensor systems is critical to affordable and reliable implementation of the technologies for more intelligent energy systems. As shown in Fig. 3.8, new sensor materials and designs capable of withstanding these conditions have been developed in harsh environment sensor technology (Romanosky 2008). The integration of the advanced power systems with environmental control technology and carbon capture processes results in a complex plant with new challenges in coordination, operation and optimization

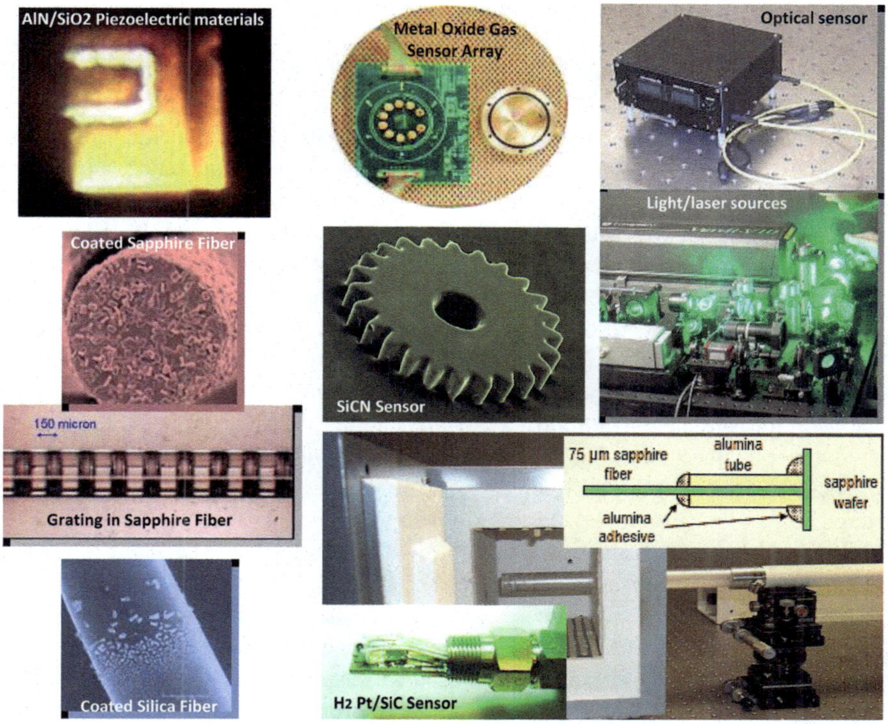

Fig. 3.8 Typical materials for sensing in harsh environments (Modified from Romanosky 2008. Credit: US National Energy Technology Laboratory)

of the plant. The ability sense, assess, and take action in order to achieve an optimized operation results in a challenging information and data management scenarios that traditional process control architectures may not be able to adequately address. The need to manage process complexity, handle high data inputs from sensors and the impetus to achieve optimum environmental performance serve as other motivating factors for new sensor and control technology (Maley and Romanosky 2012).

Two major considerations in the development of harsh environments sensors are the materials used for sensing and the design of the sensing device. Careful consideration needs to be given to (a) the environment in which the on-line sensor will be placed, (b) the measurement of interest, and (c) the accuracy, precision and lifetime needed from the sensor. Sensors materials that have proven most viable in high-temperature combustion type environments include sapphire, yttria stabilized zirconia (YSZ), metal oxides including perovskites, and silica (for certain environments). These materials have been used for the design and fabrication of optically based sensors and micro sensors (capacitance, resistive, mechanical, etc.). One of the biggest challenges for sensor materials in harsh environments is selectivity to a single gas (in a realistic mixed gas environment) that has reasonable and reversible response times. Sensor design also must consider materials of construction. In general, the use of dissimilar materials results in a higher likelihood of failure when exposed to extreme conditions due to thermal mismatch of the materials. The sensor material should be able to be altered or used to fabricate the sensor device (Maley and Romanosky 2012).

If one would consider all possible applications involving sensing, temperature sensing would come without any doubt in the first place. Temperature sensors that are capable to operate at high temperature are available and they have been developed for applications such as metal casting or such. Noncontact colorimetric measurements are also available, but such optical-based measurements require visual access to the hot area of interest. Besides, all such sensors might be incapable to perform within 5 °C resolution at over 1500 °C. Another aspect to consider is the measurement bandwidth, for example, for temperature sensing for burner pattern factor detection and for turbine surface temperature monitoring. Thermal sensors to perform measurements within 0.1 s are challenging, and they are mostly associated with the miniaturization. The technology capable to yield such sensors is foreseen to be based on SiCN. The family of the pressure sensors identified has to cover a very large range of measurement, ranging from 15 to 4000 kPa with resolutions as low as 0.25% and bandwidth ranging from quasi-static to 40 kHz. The temperature that such sensors will face is according to the section of the energy system in which they have to operate. The pressure sensors capable to operate in environment with temperatures from 250 to 750 °C use piezoelectric principle of measurement and therefore cannot measure static pressure, however are mature enough to be used in gas turbine engines. Sensors that operate in environment with temperatures below 250 °C are largely available. The family of sensors that must operate in extremely harsh environment could be achievable from highly refractory materials including SiC and SiCN. Vibration is usually measured by accelerometers. The main type of accelerometers presently used is piezoelectric which could face significantly high temperatures (up to 750 °C). However, the integrated MEMS accelerometers include the electronics; therefore, they can stand only a limited temperature range. Extension of the MEMS principles in conjunction with SiC and SiCN might significantly extend the capability of such systems to operate at high temperatures. Emission species and exhaust gas composition sensors are largely available for the test rig measurements. The measurements are based on gas species detection through the analysis of the energy produced by the reaction of the targeted species when the gas is energized by specific means. Spectrometry provides the most accurate detection of species while metal-oxide-based gas sensors (electronic noses) have been developed. The accuracy of the detection for spectrometry is somewhere in the range of 0.1–1 ppb, depending on the type of gas, while the metal-oxide sensing can reach less accurate levels equivalent to 0.5–1 ppb, also dependent on the type of gas. Improvements in the performance of such sensors are expected to be achieved through the nanotechnologies. Nano-sized particles exhibit much higher surface properties than bulk base properties, so they might be quite appropriate for the gas detection. The high-temperature operating sensors have the same challenges: the existence of an appropriate material that could be used for both sensor substrate and for the conductors. Tip clearance and position/arrival time sensors have been demonstrated on capacitive and microwave principles. The sensors can operate in temperatures as high as 600 °C. Torque sensors are largely available. The existent technology needs to be integrated in specific designs, although improved smaller sized sensor development is still ongoing. There are three types of sensors with an apparent need in the adaptive distributed control propulsion systems: fuel flow sensors for thrust estimation, gas flow in the gas path for the same purpose, and sensors for determination of the fuel properties. Gas flow sensors are available, but they interfere with the continuity of the flow path. The fuel flow sensors are required to operate close to the burner's nozzles, so the high-temperature operation requirements apply to it. General speaking, such sensors might be available to operate under normal temperature range and also for much larger resolution range. Sensors for the measurement of fuel properties are seen as micro-calorimeters that could sample from the used fuel to evaluate the calorimetric properties of the fuel. The challenges to accomplish such type of sensors are significant, which need new sensor materials and integrated device design (Culley et al. 2009).

As sensor materials and designs advance, the opportunities to integrate or embed sensor materials and novel sensor designs into a component or system will become available. One major approach is to utilize sensor materials and designs in a way that enables densely distributed sensor networks. These sensor networks include harsh environment sensors along with low-cost sensors and other types of monitoring technologies for a fully integrated sensor network such that a system or component can be computationally reconstructed and visualized in near real-time using data generated from the sensor networks and other virtual sensing capability. With active smart coatings, spray-on/applied type sensors, and embedded sensors, the use of wireless communication and energy-harvesting technologies can enable the sensor system function within high-temperature environments and communicate sensor data wirelessly to the external environment. Integrating energy-harvesting capability with sensor technology can greatly reduce the cost associated wiring and cabling required for traditional sensing approaches. In addition to these cost savings, power generation systems have many locations where waste heat and vibration are available to power sensors and the associated wireless communication hardware. Integration of these technologies with novel sensor designs will significantly change the way sensor prototypes are prepared and sensor performance evaluation is conducted (Maley and Romanosky 2012).

Networking of sensors with actuation devices and incorporation of advanced process control algorithms are the other key components of harsh environment sensor technology. Appropriate approaches have been explored for linking heterogeneous sensor data to generation actionable generation and the ability to embed artificial intelligence at or near the sensing and actuation devices. Because the actuation hardware directly impacts the process behavior and performance, analysis of how information and decision making can be incorporated with hardware in ways that does not incorporate the complexity of a full-scale integrated advanced energy plant. Approaches to process control including algorithms and system architecture are being considered to enable optimum performance of a complete and highly integrated system. Methods to distribute intelligence to the lowest levels of the process control architecture are utilized along with introducing fundamentally new control formats that mimic the function and control of a biological system. This has led to paradigms in sensing including where to sense, how many sensors are needed, the type of sensors, the sensors included in a fixed or adaptive sensor net and how the number and type of actuation hardware impact the amount of intelligence that can be distributed to the network level. A balance can be achieved between number of sensors and the control objective and this can be an adaptive methodology. It is needed to incorporate how actuation and system nonlinearity can be managed within an adaptive network. These approaches where sensors may not be viable may lead to viable alternatives to sensing in harsh environments through the use of virtual sensing and the incorporation of validated process models to predict operation parameters (Maley and Romanosky 2012).

INDEPTH: 3-D Printed Sensors (Xu et al. 2017)

The 3D-printing process can be started and stopped to incorporate complementary fabrication processes or to embed subcomponents manufactured using traditional methods. Thus, the 3D-printed sensors can be fabricated by either embedding the sensors into printed structures or intrinsically printing the entire sensors. In recent years, a considerable amount of research has focused on 3D-printed sensors of electronics, force, motion, hearing, optics, and so on. Electronic and force sensing modules are particularly well suited for 3D printing, and other sensing categories tend to be manufactured by the integration of commercial components into 3D-printed structures. Advancements in multi-process and hybrid 3D printing technology are leading to the fabrication of sensors that are both geometrically and functionally complex, easily assembled. For example, the harsh environment sensors have been developed to enable smart components with real-time access to operational status, optimized performance, and enhanced reliability and availability.

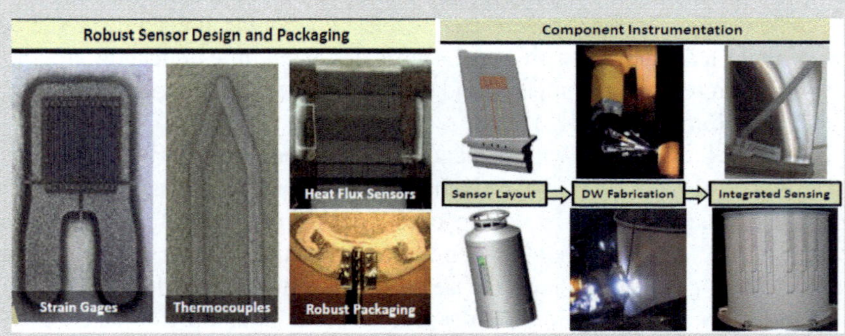

3.6 Summary

Building high efficiency fossil-fueled power stations with a large-scale carbon capture and storage to reduce CO_2 emissions brings with it new material challenges, as all of these technologies involve components operating in severe and aggressive environments with significant implications on the materials being used. The construction of economically viable and durable fossil fuel power plant components is dependent on the selection of the most appropriate materials and fabrication methods. In fact, the implementation of advanced materials in power generation applications has resulted in significant advances in plant performance and hence emissions reductions. Advanced materials are now being deployed in new and existing power plant to improve operating performance and reliability, availability, maintainability, and operability. These materials include an extensive scope, such as ferritic alloys for boilers, ceramics for filter elements or coatings for the protection of gas turbine blades. These materials can be applied to all the key power technologies are covered from supercritical pulverized coal boilers, through steam turbines and gasifiers to gas turbines and advanced cycles involving fuel cells and CO_2 management. The prospected R&D trends include (a) High-temperature materials for boilers, steam turbines, gas turbines, gasifiers, high-temperature heat exchangers, as well as functional materials such as sorbents, catalysts, and membranes; (b) Protective systems/coatings for the same technology areas as high-temperature materials; (c) Materials for sensing in harsh environments; and (d) Modeling of materials processing, component manufacture and life assessment.

In general, ultra-supercritical steam conditions are defined as being above 24 MPa and 593 °C. The materials, which could be used at such temperatures, are limited mainly due inadequate creep and oxidation/ corrosion resistance. The research goal is that the new high-temperature-resistant alloys have good mechanical and physical properties at high temperatures (650–800 °C). Successful candidate materials must have the following properties: high creep strength, thermal fatigue strength, good temperature capability in the range between 650 and 750 °C, good thermal conductivity, resistance to fire-side corrosion/erosion and resistance to steam-side oxidation and spallation, as well as good level of weldability, fabricability, and potential to be coated.

Coatings are often applied to superalloy or other structural substrates operating at high temperatures or high corrosive environments to inhibit the direct interaction of oxidizing or corroding species in the environment with the substrate, thereby minimizing damage to the substrate. Such damage may consist of recession of the substrate surface by oxidation/corrosion (wastage) or of degradation of substrate properties as a result of inward diffusion of undesirable elements (e.g., sulfur and oxygen). The coating is a surface layer of material and may be ceramic, metallic, or combinations of metallic and ceramic layers. These coatings are not inert barriers but provide protection to the substrate by interacting with oxygen in the environment to form dense, tightly adherent oxide films that are resistant to the diffusion of potentially damaging species. Because the protective oxide films can spall as a result of thermal cycling and mechanical damage, there is a requirement that the oxide film be replaceable in situ for the film to remain viable. This replacement is accomplished by making the original coating rich in elements that form the protective oxides (such as aluminum and chromium). The useful life of a coating is governed by its ability to form the desired protective scale and its ability to retain and/or replace the scale during service. In addition, coating life is affected strongly by the composition of the substrate. Also, the possible effects of inter-diffusion between the coating and the substrate must be assessed in selecting a coating.

The increasing worldwide demand for oil and gas has pushed the petroleum industry into drilling ever deeper wells. Well depths of 25,000 ft. (7620 m) and greater are no longer unusual, and even deeper wells are expected. Generally, increasing depth means increasing pressure and temperature. High-pressure/high-temperature (HPHT) wells have generally been considered wells in which temperatures and pressures at the bottom of the well exceed 300 to 350 °F (149–177 °C) and 10,000 psi (69 MPa), respectively. Drilling these HPHT wells requires specialized methods and considerable planning, but the materials have largely remained steel drill pipe and steel components, although other alloys such as titanium and composite materials are being considered. However, the real materials challenges are in completing and producing the wells after they are drilled.

Every aspect of energy conversion from fossil fuels depends on sensors for monitoring and control. Accurate and reliable sensors for high-temperature and/or corrosive environments would have multiple benefits for the thermal plant, such as avoidance of damage to heat-transfer surfaces, combustion controls for individual burners in utility boilers, reduction of noxious emissions, and structural health monitoring of critical components. As sensor materials and designs advance, the opportunities to integrate or embed sensor materials and novel sensor designs into a component or system will become available. One major approach is to utilize sensor materials and designs in a way that enables densely distributed sensor networks. These sensor networks include harsh environment sensors along with low-cost sensors and other types of monitoring technologies for a fully integrated sensor network such that a system or component can be computationally reconstructed and visualized in near real-time using data generated from the sensor networks and other virtual sensing capability. With active smart coatings, spray-on/applied type sensors, and embedded sensors, the use of wireless communication and energy-harvesting technologies can enable the sensor system function within high-temperature environments and communicate sensor data wirelessly to the external environment. Integrating energy-harvesting capability with sensor technology can greatly reduce the cost associated wiring and cabling required for traditional sensing approaches.

Exercises

Part I: General Questions

3.1 What's "fossil fuel?" Name one or more fossil fuels. What are fossil fuels used for?
3.2 Explain the materials challenges that are being actively pursued to achieve sustainable fossil energy systems.
3.3 What are the major materials requirements for a supercritical steam generator? Explain the difference of the materials requirements between a supercritical steam generator and a subcritical boiler.
3.4 Describe main performance requirements of materials used for ultra-supercritical applications.
3.5 Describe key materials used for boilers, furnace walls, superheat tubes, steam separating vessels, and headers and steam pipes.
3.6 What are general requirements for materials used for steam turbines and gas turbines? Compare the commons and differences between these two kinds of turbines.
3.7 What's the Allam Cycle for zero emission fossil fuel plant? Describe its advantages over conventional technology.
3.8 Explain the role of alloying elements in development of the ferritic steels used for boilers.
3.9 What kinds of refractory materials are used for slagging gasifiers? Describe the future development trends in this area.
3.10 Describe the coatings and protective systems available for high-temperature applications, erosion- and corrosion-resistant applications, and wear-resistant applications. What are main effectors on the coating life in these applications?
3.11 Explain main requirements of materials used for deep oil and gas well drilling and construction. What materials have been developed in this area?
3.12 What are major considerations in the development of harsh environments sensors? Describe sensor materials available and future development trends in this area.

Part II: Thought-Provoking Questions

3.13 Explain the advantages and disadvantages of some breakthrough materials that have been explored and developed with temperature/strength capabilities beyond those currently available.
3.14 Explain materials issues in the major gasification process components.
3.15 Discuss the possibility of 3D printing process for printed harsh environments sensors. What are the major challenges?

References

Allen, D.J.: Materials UK Energy Review Report 2-fossil-fueled power generation. http://www.matuk.co.uk/docs/2_Fossil%20Energy%20FINAL.pdf (2007). Accepted 3 Sept 2013

Amagasa, S., Shimomura, K., Kadowaki, M., Takeishi, K., Kawai, H., Aoki, S., Aoyama, K.: Study on the turbine vane and blade for a 1500 °C class industrial gas turbine. J. Eng. Gas Turb. Power. **116**, 597–604 (1994)

Ashby, M., et al.: Materials Engineering, Science, Processing and Design. Butterworth-Heinemann, Oxford (2007)

Baker, M., Fessler, R.R.: Pipeline Corrosion Final Report. U.S. DOT Pipeline and Hazardous Materials Safety Administration. http://primis.phmsa.dot.gov/iim/docstr/FinalReport_PipelineCorrosion.pdf (2008). Accessed 21 June 2017

Bennett, J.P., Kwong, K.S.: Refractory liner materials used in slagging gasifiers. Refract. Appl. News. **9**(5), 20–25 (2004)

Bennett, J.P., Kwong, K.S.: failure mechanisms in high chrome oxide gasifier refractories. Metall. Mater. Trans. A. **42A**, 888–904 (2011)

Brandt, M., Sun, S., Alam, N., Bendeich, P., Bishop, A.: Laser cladding repair of turbine blades in power plants: from research to commercialization. Int. Heat Treat. Surface Eng. **3**(3), 105–114 (2009)

Carniglia, S.C., Barna, G.L.: Handbook of Industrial Refractories Technology—Principles,Types, Properties and Applications. Noyes Publications, Park Ridge, NJ (1992)

Carter, T.J.: Common failures in gas turbine blades. Eng. Failure Anal. **12**(2005), 237–247 (2005)

Clarke, D.R.: Thermal-barrier coatings for more efficient gas-turbine engines. MRS Bullet. **37**, 891–901 (2012)

Clarke, D.R., Oechsner, M., Padture, N.P.: Thermal-barrier coatings for more efficient gas-turbine engines. MRS Bullet. **37**(10), 891–898 (2012)

Conklin, J.C., Szybist, J.P.: A highly efficient six-stroke internal combustion engine cycle with water injection for in-cylinder exhaust heat recovery. Energy. **35**, 1658–1664 (2010)

Craig, B.: Materials for deep oil and gas well construction. Adv. Mater. Processes. **5**, 33–35 (2008)

Cruchley, S., Evans, H., Taylor, M.: An overview of the oxidation of Ni-based superalloys for turbine disc applications: surface condition, applied load and mechanical performance. Materials at High Temperatures (2016)

Culley, D., et al.: More intelligent gas turbine engines. RTO TECHNICAL REPORT TR-AVT-128. https://www.dtic.mil/cgi-bin/GetTRDoc?AD=ADA515458 (2009). Accessed 8 Feb 2014

Danielsen, H.K., Somers, M.A.J., Hald, J.: Z-phase in 9–12% Cr Steels. Kgs. Technical University of Denmark (DTU), Lyngby, Denmark (2007)

DOE: How coal gasification power plants work. http://energy.gov/fe/how-coal-gasification-power-plants-work (2013). Accessed 7 Nov 2013

Eskner, M.: Mechanical Behavior of Gas Turbine Coatings. PhD dissertation. Royal Institute of Technology, Sweden (2004)

Gibbons, T.B., Wright, I.G.: A Review of Materials for Gas Turbines Firing Syngas Fuels. ORNL/TM-2009/137. https://info.ornl.gov/sites/publications/files/pub15496.pdf (2007). Accessed 26 June 2017

Hendricks, R.C., Chupp, R.E., Lattime, S.B., Steinetz, B.M.: Turbomachine interface sealing. In: International Conference on Metallurgical Coatings and Thin Films sponsored by the AVS Science & Technology San Diego, California, May 2–6, 2005. https://ntrs.nasa.gov/archive/nasa/casi.ntrs.nasa.gov/20050175891.pdf (2005). Accessed 16 May 2017

Hirata, A., et al.: Atomic structure of nanoclusters in oxide-dispersion-strengthened steels. Nat. Mat. **10**, 922–926 (2011)

Jackson, R.B., Down, A., Phillips, N.G., Ackley, R.C., Cook, C.W., Plata, D.L., Zhao, K.: Natural gas pipeline leaks across Washington, DC. Environ. Sci. Technol. **48**, 2051–2058 (2014)

Kern, T.-U.: Using creep-resistant steels in turbines. In: Abe, F., Kern, T.-U., Viswanathan, R. (eds.) Creep-resistant steels. CRC Press, Boca Raton (2008)

Lahey, P.: Use of composite materials in the transportation of natural gas. Idaho National Engineering and Environmental Laboratory. http://igs.nigc.ir/STANDS/BOOK/COMPOSIT-PIPE.PDF (2002). Accessed 21 June 2017

Leslie, J.C.: Composite drill pipe for extended-reach and deep water applications. http://www.netl.doe.gov/technologies/oil-gas/publications/EPreports/DCS/40262_OffshoreTechnicalConferencefinaldraft.pdf (2013). Accessed on 22 Dec 2013

Leunga, D.Y.C., Caramanna, G., Maroto-Valerb, M.M.: An overview of current status of carbon dioxide capture and storage technologies. Renew. Sust. Energ. Rev. **39**, 426–443 (2014)

Lothongkum, G., Khuanlieng, W., Homkrajai, W., Wangyao, P.: Effect of aging on stress relaxation of Inconel X-750 bolt at 923 and 1023 K [J]. High Temp. Mater. Processes. **25**(4), 175–185 (2006)

Lukaszewicz, M.: Steam oxidation of advanced high temperature resistant alloys for ultra-supercritical applications. PhD dissertation. Cranfield University, Bedford, UK (2012)

Maley, S.M., Romanosky, R.R.: Sensors for fossil energy applications in harsh environments. In: IMCS 2012—The 14th International Meeting on Chemical Sensors, 2012-05-20–2012-05-23, pp. 60–63. Nürnberg/Nuremberg, Germany (2012)

Mukherji, D., Rosler, J., Strunz, P., Gilles, R., Schumacher, G., Piegert, S.: Beyond Ni-based superalloys: development of CoRe-based alloys for gas turbine applications at very high temperatures. Int. J. Mater. Res. **102**, 1125–1132 (2011)

Nicol, K.: Status of advanced ultra-supercritical pulverized coal technology. ISBN 978–92–9029-549-5. https://www.usea.org/sites/default/files/122013_Status%20of%20advanced%20ultra-supercritical%20pulverised%20coal%20technology_ccc229.pdf (2013). Accessed 12 May 2017

Oakey, J., Fry, T.: MatUK energy materials review-R&D priorities for gasification technologies. http://www.matuk.co.uk/docs/MatUKDraftReview-Gasification.pdf (2007). Accessed 11 June 2013

Oakey, J.E., et al.: Review of status of advanced materials for power generation. Report No. COAL R224, DTI/Pub URN 02/1509. http://webarchive.nationalarchives.gov.uk/+/http://www.berr.gov.uk/files/file18766.pdf (2003a). Accessed 23 Oct 2013

Oakey, J.E., et al.: Review of status of advanced materials for power generation. http://webarchive.nationalarchives.gov.uk/+/http://www.berr.gov.uk/files/file18766.pdf (2003b). Accessed 19 Dec 2013

PCC: Customized product and material solution for aerospace and power generation. http://www.pccforgedproducts.com/products/ (2013). Accessed 12 Oct 2013

Penfield, Jr. S., Rittenhouse, P.: Evaluation of coatings to prevent diffusion of fission products into gas reactor turbomachine blades. EPRI, Palo Alto, CA: 2004. 1009381 (2004)

Pimentel, G., et al.: Advanced FeCrAl ODS steels for high-temperature structural applications in energy generation systems. Resista Metal. **48**(4), 303–316 (2012)

Rautio, R., Bruce, S.: Alloy for ultrasupercritical coal fired boilers. Adv. Mater. Processes. **4**, 35–37 (2008)

Rayes, M.M.E., et al.: Erosion-corrosion of cermet coating. Int. J. Electrochem. Sci. **8**, 1117–1137 (2013)

Reed, R.C.: The Superalloys. Cambridge University Press, Cambridge (2006)

Romanosky, R.R.: Development of harsh environment sensor platform for fossil energy applications. http://citeseerx.ist.psu.edu/viewdoc/download?rep=rep1&type=pdf&doi=10.1.1.207.201 (2008). Accessed 18 Sept 2013

Romanosky, R., Rawls, P.: Advanced research materials program. National Energy Technology Laboratory. http://www.netl.doe.gov/publications/factsheets/program/prog010.pdf (2013). Accessed 20 Sept 2013

Service RF: Goodbye smokestacks: Startup invents zero-emission fossil fuel power. http://www.sciencemag.org/news/2017/05/goodbye-smokestacks-startup-invents-zero-emission-fossil-fuel-power (2017). Accessed 23 June 2017

Siebert M: Additive Manufacturing—Breakthrough with 3D printed Gas Turbine Blades. Pictures of the Future—The Magazine for Research and Innovation (6 February 2017). https://www.siemens.com/innovation/en/home/pictures-of-the-future/industry-and-automation/additive-manufacturing-3d-printed-gas-turbine-blades.html (2017). Accessed 21 June 2017

Sonntag, R.E., Borgnakke, C.: Introduction to Engineering Thermodynamics. Wiley, New York (2006)

Tsuji, et al.: Development of heat resisting steel for high, low pressure steam turbine mono-block rotor. Metals Technol. **62**(11), 64 (1992)

Viswanathan, R., et al.: Materials for ultra-supercritical coal-fired power plant boilers. In: Proceedings of the 2nd Regional Conference on Energy Technology Towards a Clean Environment, 12–14 February 2003, Phuket, Thailand (2003)

Viswanathan, R., Shingledecker, J., Purgert, R.: Evaluating Materials Technology for Advanced Ultrasupercritical Coal-Fired Plants. Power, 1 August 2010. http://www.powermag.com/evaluating-materials-technology-for-advanced-ultrasupercritical-coal-fired-plants/ (2010). Accessed 20 June 2017

Watson, J., et al.: Technology and carbon mitigation in developing countries: Are cleaner coal technologies a viable option? Background Paper for Human Development Report 2007. http://hdr.undp.org/en/reports/global/hdr2007-8/papers/Watson_MacKerron_Ockwell_Wang.pdf (2007). Accessed 22 Oct 2013

West, M.K.: Processing and characterization of oxide dispersion strengthened 14YWT ferritic Alloys. PhD dissertation. University of Tennessee, Knoxville (2006)

Wright, I., et al.: Overview of ODS alloy development. http://cartechcapstone.wikispaces.com/file/view/ODS+overview05.pdf (2005). Accessed 18 Oct 2013

Wu, Q.: Microstructural evolution in advanced boiler materials for ultra-supercritical coal power plants. PhD dissertation. University of Cincinnati (2006)

Wu, W.: Development and characterization of novel low-friction wear-resistant multilayer nanocomposite CrAlTiCN coatings. PhD dissertation. The University of Birmingham, Birmingham, UK (2010)

Xie, X., Wu, Y., Chi, C., Zhang, M: Superalloys for Advanced Ultra-Super-Critical Fossil Power Plant Application, Superalloys, Dr. Mahmood Aliofkhazraei (Ed.), InTech, doi: https://doi.org/10.5772/61139. https://www.intechopen.com/books/superalloys/superalloys-for-advanced-ultra-super-critical-fossil-power-plant-application (2015). Accessed 21 June 2017

Xu, Y., Wu, X., Guo, X., Kong, B., Zhang, M., Qian, X., Mi, S., Sun, W.: The boom in 3D-printed sensor technology. Sensors. **17**, 1166 (2017)

Yamamoto, Y., et al.: Overview of strategies for high-temperature creep and oxidation resistance of alumina-forming austenitic stainless steels. Metall. Mater. Trans. **42A**, 922–931 (2011)

Yang, Y.: Development of a nano-composite coating technology for improvement of carbon steel pipe to erosion-corrosion in oil sands slurry. PhD dissertation University of Calgary, Alberta (2013)

Zhu, Q.: High temperature syngas coolers. IEA Clean Coal Centre. ISBN 978–92–9029-580-8. https://www.usea.org/sites/default/files/media/High%20temperature%20Syngas%20coolers%20-%20ccc257.pdf (2015). Accessed 21 May 2017

Materials-Based Solutions to Solar Energy System

4

Abstract

Solar energy is the cleanest and most abundant renewable energy source available in the world. This energy can be harnessed using a range of ever-evolving technologies, such as photovoltaics (PV), concentrating photovoltaics (CPV), concentrating solar power (CSP), solar thermal, and artificial photosynthesis. Materials-based solutions have been widely explored to improve and renew solar energy systems. Solar energy materials are used to harness the sun's energy with special properties adapted and tuned so that they can absorb, reflect, transmit, or emit light and other electromagnetic radiation in the wavelength ranges for thermal, solar, and visible radiation. PV directly converts sunlight into electrical power. The growth of PV has been driven by lower costs due to increased efficiency, primarily from advances in PV materials mainly including crystalline silicon; thin films such as cadmium telluride (CdTe), copper-indium-gallium-diselenide (CIGS), or amorphous silicon (a-Si); multifunction systems with solar concentrators; and organic flexible molecular, polymeric, or nanoparticle-based cells. The superior optical, electric, and chemical properties of nanomaterials lead to the development of quantum well, quantum dot, dye sensitized, and organic solar cells. CSP uses reflectors to concentrate sunlight to generate high temperatures to heat fluids that drive steam turbines to produce utility-scale electric power. Materials research for CSP has been focused on improving optical materials for reflectors with greater durability and low cost; enhancing absorber materials and coatings with higher solar absorbance and low thermal emittance; develop thermal energy storage materials with improved heat capacity; and improve corrosion resistance of materials in contact with fluids like molten salts. Exceedingly, solar power has a vast resource base and incredible technical potential. Materials science and engineering offers the potential to significantly increase the amount of electricity generated from solar energy. This chapter will provide a brief review on the advanced materials solutions to various solar energy systems.

C. Tong, *Introduction to Materials for Advanced Energy Systems*,
https://doi.org/10.1007/978-3-319-98002-7_4

4.1 The State of Solar Energy Technologies

Solar energy, radiant light and heat from the sun, is harnessed using a range of ever-evolving technologies such as solar heating, solar photovoltaic, solar thermal electricity, solar architecture, and artificial photosynthesis. The amount of solar energy reaching the surface of the Earth is so vast that in one year it is about twice as much as will ever be obtained from all of the Earth's nonrenewable resources of coal, oil, natural gas, and mined uranium combined.

Solar energy technologies are broadly characterized as either passive solar or active solar depending on the way they capture, convert, and distribute solar energy. Active solar techniques include the use of photovoltaic panels and solar thermal collectors to harness the energy and convert sunlight into useful outputs. Passive solar techniques include orienting a building to the Sun, selecting materials with favorable thermal mass or light dispersing properties, and designing spaces that naturally circulate air. Active solar technologies increase the supply of energy and are considered supply side technologies, while passive solar technologies reduce the need for alternate resources and are generally considered demand side technologies.

Solar energy offers a clean, climate-friendly, very abundant, and inexhaustible energy resource to mankind, relatively well-spread over the globe. The development of affordable solar energy technologies will have huge longer-term benefits. It will increase global energy security through reliance on an indigenous, inexhaustible, and mostly import-independent resource, enhance sustainability, reduce pollution, lower the costs of mitigating climate change, and keep fossil fuel prices lower than otherwise.

INDEPTH: Solar Radiation (Le Treut et al. 2007; Voss et al. 2014)

Solar radiation is relevant in three different wavelength regions (Voss et al. 2014): (a) the visual region from 0.4 to 0.7 μm, (b) the solar region from 0.3 to 3.0 μm, and (c) the thermal region with wavelengths larger than 2 μm. As depicted in the following plot, luminous, solar and thermal radiations are confined to these specific wavelength intervals. Planck spectra for two temperatures of practical significance for windows are depicted in the right-hand part. Both spectra are confined to the thermal range. The peak in the spectrum for 50 °C lies at a shorter wavelength than the one for 0 °C, which is a manifestation of Wien's displacement law. At room temperature the peak occurs at about 10 μm. Thermal radiation from a material is obtained by multiplying the Planck spectrum by the emittance, which is generally wavelength dependent and is less than unity. The solid curve in the left-hand part shows a typical solar spectrum for radiation that has already passed the earth's atmosphere. The curve has again a bell shape corresponding with the sun's surface temperature of about 6000 °C. The minima in the spectrum are caused by atmospheric absorption, mainly by water vapor, carbon dioxide, and ozone. The dashed curve shows the relative spectral sensitivity of the human eye in its light-adapted photonic state with the maximum at 0.555 μm. In the darkness-

(continued)

adapted scotopic state, the latter one is displaced about 0.05 μm towards shorter wavelengths (Voss et al. 2014).

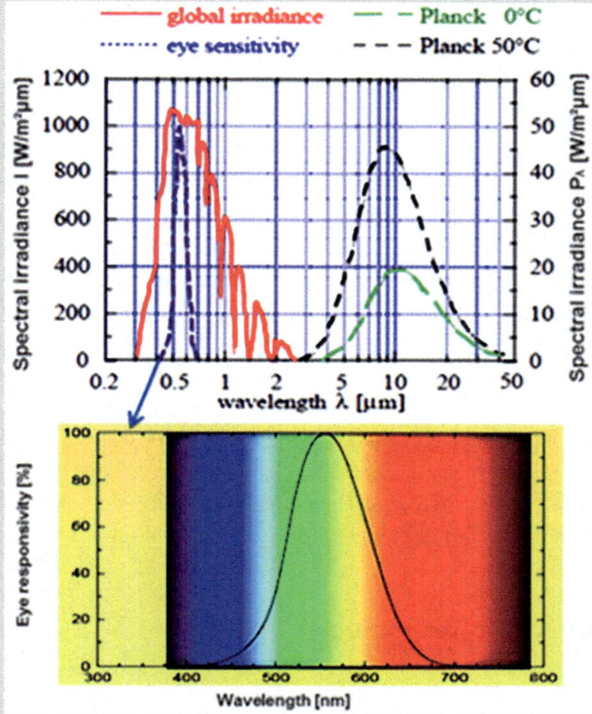

The Earth receives 174 petawatts (PW) of incoming solar radiation/isolation at the upper atmosphere. Approximately 29% is reflected back to space while the rest 71% is absorbed by clouds, oceans, and land masses. Warm air containing evaporated water from the oceans rises, causing atmospheric circulation or convection. When the air reaches a high altitude, where the temperature is low, water vapor condenses into clouds, which rain onto the Earth's surface, completing the water cycle. The latent heat of water condensation amplifies convection, producing atmospheric phenomena such as wind, cyclones, and anti-cyclones. Sunlight absorbed by the oceans and land masses keeps the surface at an average temperature of 14 °C. By photosynthesis, green plants convert solar energy into chemically stored energy, which produces food, wood, and the biomass from which fossil fuels are derived. The total solar energy absorbed by Earth's atmosphere, oceans, and land masses is approximately 3,850,000 exajoules (EJ) per year. Photosynthesis captures approximately 3000 EJ per year in biomass.

The spectrum of solar light at the Earth's surface is mostly spread across the visible and near-infrared ranges with a small part in the near-ultraviolet. Most

(continued)

of the world's population live in areas with insolation levels of 150–300 watts/m^2, or 3.5–7.0 kWh/m^2 per day. The potential solar energy that could be used by humans differs from the amount of solar energy present near the surface of the planet because factors such as geography, time variation, cloud cover, and the land available to humans limit the amount of solar energy that can be acquired. Geography affects solar energy potential because areas that are closer to the equator have a greater amount of solar radiation. However, the use of photovoltaics that can follow the position of the sun can significantly increase the solar energy potential in areas that are farther from the equator. Time variation effects the potential of solar energy because during the nighttime there is little solar radiation on the surface of the Earth for solar panels to absorb. This limits the amount of energy that solar panels can absorb in one day. Cloud cover can affect the potential of solar panels because clouds block incoming light from the sun and reduce the light available for solar cells.

In addition, land availability has a large effect on the available solar energy because solar panels can only be set up on land that is otherwise unused and suitable for solar panels. Roofs have been found to be a suitable place for solar cells, as many people have discovered that they can collect energy directly from their homes this way. Other areas that are suitable for solar cells are lands that are not being used for businesses where solar plants can be established.

4.1.1 Photovoltaic Technologies

Photovoltaic (PV) solar cells directly convert sunlight into electricity, using the photovoltaic effect. The process works even on cloudy or rainy days, though with reduced the production and conversion efficiency. PV cells are assembled into modules to build modular PV systems that are used to generate electricity in both grid-connected and off-grid applications, such as residential and commercial buildings, industrial facilities, remote and rural areas, and power plants (i.e., utility PV systems). Commercial PV technologies include wafer-based crystalline silicon (c-Si) (either mono-crystalline or multi-crystalline silicon) and thin-films (TF) using amorphous Si (a-Si/μc-Si), cadmium-telluride (CdTe), and copper-indium-[gallium]-[di]selenide-[di]sulfide (CI[G]S). The majority of commercial c-Si modules, however, have efficiencies in the range of 13–19% with more than a 25-year lifetime. Commercial TF modules offer lower efficiency between 6 and 12% (with a target of 12–16% by 2020). In addition to the commercial options, a number of new PV technologies is under development (e.g., concentrating PV, organic PV cells, advanced thin films and novel concepts and materials) and hold out the promise of high performance and low costs in the medium-term (Feldman et al. 2014).

Concurrently with the strong growth in PV module production, average PV module prices have dropped sharply over the last decade. Polysilicon, wafer, cell, and module prices dropped especially sharp since around 2010, as shown in Fig. 4.1. Price decreases have long been following a learning curve that has been valid for multiple decades; however, more recently, due to decreased demand and resulting

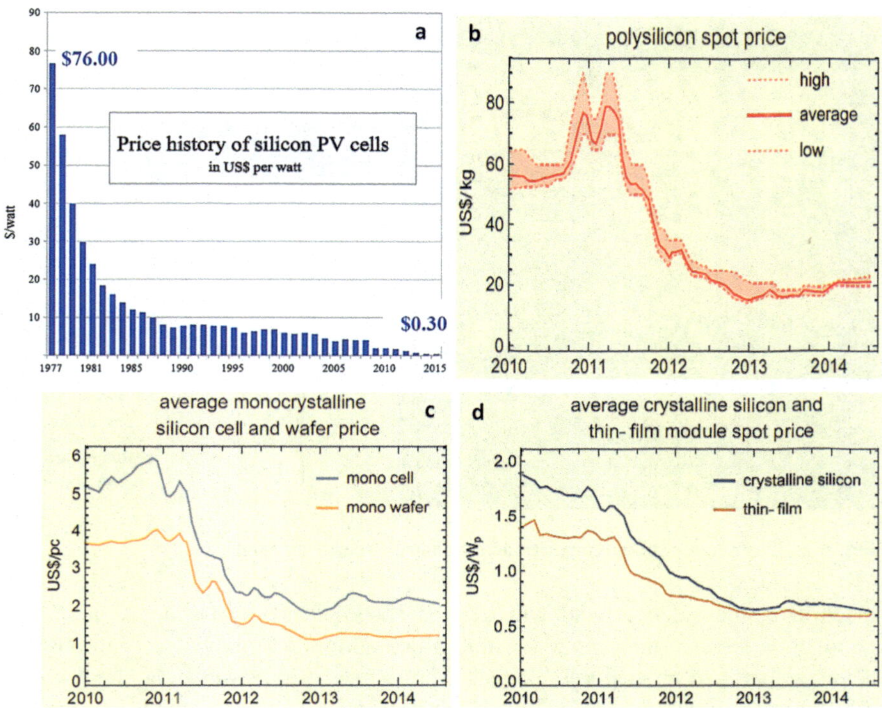

Fig. 4.1 Overview of price history of silicon PV cells (**a**, in USD/Watt); development of prices for polysilicon (**b**, in USD/kg); monocrystalline cells and wafers (**c**, USD/Watt); and crystalline silicon and thin-film modules (**d**, USD/Watt) (Adapted from Louwen et al. 2016 (Elsevier), Licensed under the Creative Commons Attribution International License (CC BY). http://creativecommons.org/licenses/by/4.0/)

oversupply, prices have dropped below what could be extrapolated from the learning curve. As a result, PV producers are scrambling for opportunities to reduce production costs. On the other hand, although residential grid parity has been reached in several countries worldwide, PV electricity is as of yet not competitive with fossil electricity generation. These two factors emphasize the need for further cost reductions in the PV industry, in order to assure PV production that is financially sustainable and competitive with bulk electricity generation. However, Today's PV systems are fully competitive for off-grid electricity generation and with diesel-based on-grid systems in countries with good solar resources. In an increasing number of countries with high cost of electricity and good solar resources, small PV systems are also achieving the so-called grid-parity between the PV electricity cost and the residential retail prices for householders. Furthermore, advantages of PV electricity are that it is usually produced close to the consumption site and can match peak demand profiles (IRENA 2013; Louwen et al. 2016).

The fundamental principle of converting the solar energy into electrical energy is shown in Fig. 4.2. The atoms in a semiconductor placed in the sunlight will absorb photons from the sun's radiation. If these photons are of high enough energy, an

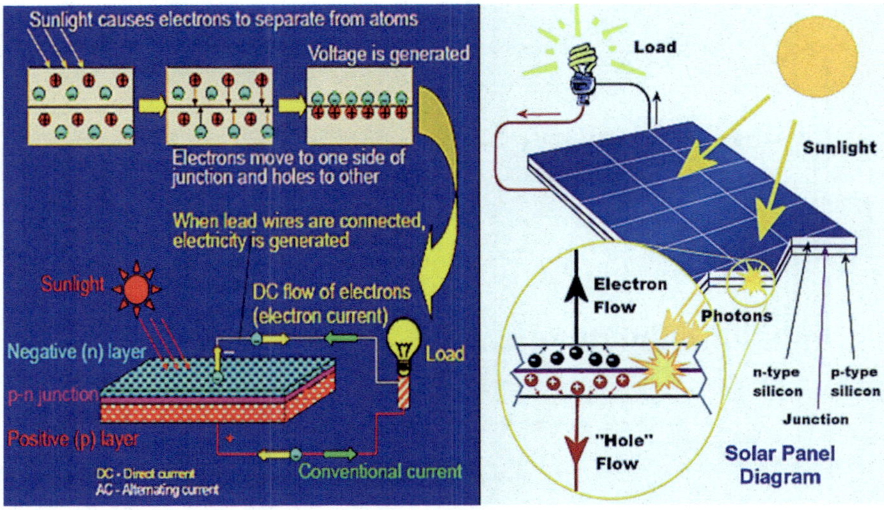

Fig. 4.2 Illustration of photovoltaic principle and solar panel diagram

electron in the valence band will use the absorbed energy to move up into the conduction band of the semiconductor, which allows it to move freely through the semiconductor to an electrical contact from where it is driven by a small voltage through wires as current. The energy the electron needs to move from the valence band to the conduction band is called the band gap of the semiconductor. For this to work in an efficient manner there must be two kinds of semiconductor in the solar cell. One, the n-type semiconductor, is doped with impurities such that it has an excess of electrons. The p-type semiconductor has a deficit of electrons, creating "holes" where it is missing electrons. When the two are placed in contact to create a p–n junction, electrons from the n-type semiconductor diffuse into the p-type semiconductor, leaving holes behind, until enough electrons accumulate on the p-type side of the boundary and enough holes accumulate on the n-type side of the boundary that an opposing electric field is created, and equilibrium is reached. Photons are then absorbed within the p-type semiconductor, freeing electrons which will flow through the p–n junction to the electrical contact, while the hole goes the other way. The electric field at the junction is what drives the electrons around the circuit. The electron will travel into a circuit and do work before recombining with the hole. A solar cell is made of semiconducting material in two layers: P and N. When radiation from the Sun hits the photovoltaic cell in the form of sunlight, the boundary between P and N acts as a diode: electrons can move from N to P, but not the other way around. Photons with sufficient energy hitting the cell cause electrons to move from the P layer into the N layer. An excess of electrons builds up in the N layer while the P layer builds up a deficit. The difference in the amount of electrons is the voltage difference, which can be used as a power source. As long as light continues to hit the panel, the voltage difference is maintained; even on cloudy days, due to diffuse radiation of the light. The amount of electric power that a photovoltaic cell produces depends principally on two factors: (a) the amount of incident sunlight; (b) the efficiency of the photovoltaic in converting this light into electricity. The installed power of a PV system is usually not expressed in watts (W) but in watt-peak (Wp). Since the first solar cell was built in 1883 out of selenium and gold (with an efficiency of 1%), the variety of technologies for converting solar radiation to electricity has greatly expanded and improved.

Solar photovoltaic (PV) is one of the technologies available to use sunlight as an active source. Concentrating solar power systems (CSP) use concentrated solar radiation as a high-temperature energy source to produce electrical power and drive chemical reactions. CSP is typically applied in relatively large-scale plants under very clear skies and bright sun. The availability of thermal storage and fuel backup allows CSP plants to mitigate the effects of sunlight variability. Solar heating and cooling (SHC) uses the thermal energy directly from the sun to heat or cool domestic water or building spaces. These three ways of harnessing the sun are complementary, rather than directly competitive, and their needs and environment should be carefully assessed when choosing which solar technology to use.

Apart from residential photovoltaic, photovoltaic technologies can be divided into two main areas: flat-plate and concentrator. In the flat-plate technologies, semiconductor material is used to cover as much area as possible on a flat plate, while practicing trade-offs between material cost and conversion efficiency of light into electrical power. With concentrators, an additional trade-off is practiced whereby portions of the more expensive semiconductor material in the system are replaced with a system of lenses or reflectors that can be made from less expensive material. This replacement may, however, be at the expense of overall system efficiency, and thus one should consider each system as a whole in evaluating its benefits.

INDEPTH: Semiconductor Junction (Luque and Hegedus 2011)

A p–n junction is a boundary or interface between two types of semiconductor material, p-type and n-type, inside a single crystal of semiconductor. The "p" (positive) side contains an excess of holes, while the "n" (negative) side contains an excess of electrons. The p–n junction is created by doping, for example, by ion implantation, diffusion of dopants, or epitaxy (growing a layer of crystal doped with one type of dopant on top of a layer of crystal doped with another type of dopant). If two separate pieces of material were used, this would introduce a grain boundary between the semiconductors that would severely inhibit its utility by scattering the electrons and holes.

After joining p-type and n-type semiconductors, electrons from the n region near the p–n interface tend to diffuse into the p region leaving behind positively charged ions in the n region and being recombined with holes, forming negatively charged ions in the p region. Likewise, holes from the p-type region near the p–n interface begin to diffuse into the n-type region, leaving behind negatively charged ions in the p region and recombining with electrons, forming positive ions in the n region. The regions near the p–n interface lose their neutrality and most of their mobile carriers, forming the space charge region or depletion layer.

(continued)

The space charge region is a zone with a net charge provided by the fixed ions (donors or acceptors) that have been left uncovered by majority carrier diffusion. When equilibrium is reached, the charge density is approximated by the displayed step function. The following picture shows a p–n junction in thermal equilibrium with zero-bias voltage applied. Under the junction, plots for the charge density, the electric field, and the voltage are illustrated. (The log concentration curves should actually be smoother, like the voltage.)

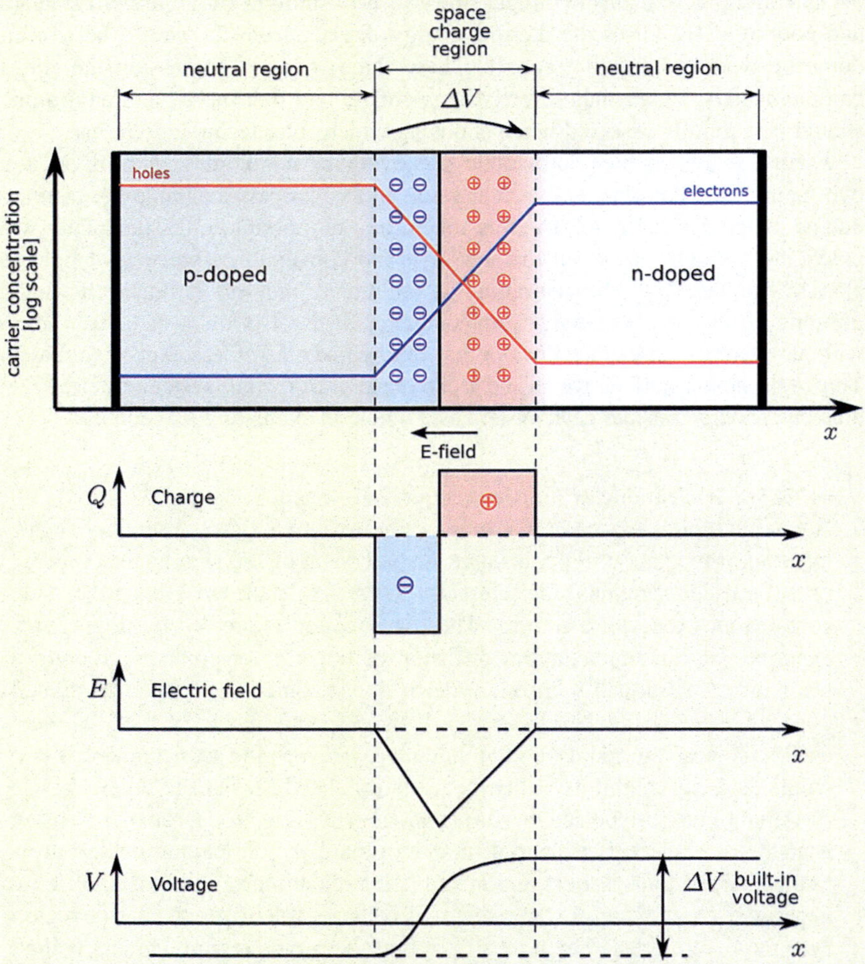

p–n junctions are elementary "building blocks" of most semiconductor electronic devices such as diodes, transistors, solar cells, LEDs, and integrated

(continued)

circuits; they are the active sites where the electronic action of the device takes place. For example, a common type of transistor, the bipolar junction transistor, consists of two p–n junctions in series, in the form n–p–n or p–n–p.

4.1.1.1 Residential Photovoltaic System

A residential PV system enables a homeowner to generate some or all of their daily electrical energy demand on their own roof, exchanging daytime excess power for future energy needs, usually for night time use. The house remains connected to the public electricity grid at all times, thus any power required above what the PV systems can produce is drawn from the grid. PV systems can also include battery backup or uninterruptible power supply (UPS) to operate selected circuits in the residence for hours or days during a grid outage. The emphasis of grid-connected PV is on the built environment, also known as building-integrated PV (BIPV). Most often, PV installations are part of the existing infrastructure, or are integrated into the building structure of residential, office or industrial buildings. Roof-mounted PV systems, for example, are considered a building-integrated application. In most applications, the electrical power generated by solar energy is fed into the internal electrical grid of the building (Hulshorst 2008).

As shown in Fig. 4.3, the three main components of a PV system are the PV cells and panels, the inverter, and the meter that records the amount of power produced.

Fig. 4.3 Schematic residential photovoltaic system

For PV systems without a grid connection—so-called standalone PV—batteries are also a necessary component. Most commonly, photovoltaic cells are produced from monocrystalline or multicrystalline silicon or other materials. The efficiency of monocrystalline silicon cells is significantly greater than that of multicrystalline or polycrystalline silicon. Monocrystalline silicon is produced as single crystal ingots, while multicrystalline manufacturing starts with melting the material, followed by a solidification process with a predetermined crystal orientation structure, resulting in multicrystalline blocks. To produce PV cells, the silicon ingots or blocks are sliced into thin wafers. On top of the cells, a screen of aluminum conductors is usually installed. A PV module is the basic building block of any PV power system. A PV module consists of interconnected cells sealed between a glass cover and weather-proof backing. The modules are typically encased in frames suitable for mounting. Two or more modules can be pre-wired together to be installed as a single unit called a PV or solar panel. Additional PV panels can be added as electricity production needs increase. The entire PV system, consisting of one or more panels, is known as an array.

The PV cells and modules generate direct current (DC). Since most household appliances use alternating current (AC), an inverter is used to convert the DC voltage to AC voltage, matching the frequency and voltage of the local grid. Inverters for PV applications include control functions to optimize the power output, which is referred to as maximum power point tracking (MPPT). The power output is equal to the voltage multiplied by the current ($P = V \times I$), and the MPPT function continuously adjusts the load impedance to guarantee optimal power. Depending on the size (Wp) of the PV installation, smaller units can be connected to the grid by plugging it directly into an electrical socket, whereas larger units can be connected at the meter board where the cables of the public grid enter the house. PV systems with batteries for storage are particularly suitable in areas in which a utility power supply is unavailable or in which utility line extensions would be prohibitively expensive. The ability to store PV-generated electrical energy makes the PV system a reliable source of electric power both day and night, rain or shine. PV systems with batteries can be designed to power equipment that requires DC or AC electricity. PV systems with battery storage are used all around the world to provide electricity for lights, sensors, recording equipment, switches, appliances, telephones, and televisions. A major advantage of PV systems is that they can be easily adopted in existing buildings or homes. PV systems are modular and can be installed anywhere. In addition, these types of systems produce no noise, harmful emissions, or polluting gases, and most importantly the energy produced is free. Manufacturers have designed several different models, which can be placed at a variety of different types of houses or buildings (Hulshorst 2008).

PV panels convert the light that reaches them into electricity. The amount of electricity they produce is roughly proportional to the intensity and the angle of the light that reaches them. The panels, therefore, are positioned to take maximum advantage of available sunlight within the constraints of their placement. Maximum power is obtained when the panels are able to track the sun's movements during the day and throughout the various seasons. These types of panels, referred to as trackers, are usually ground-mounted using a heavy steel pole sunk into a concrete foundation. Roof-mounted tracking units are rare, because they can create structural problems and tend to be noisy during windy weather. The best elevations for PV systems vary by latitude. Along with geographical location, the performance of the system output is also be affected by factors such as the following (Hulshorst 2008): (a) Shading: One of the main factors in the design and placement of a new PV system is that it be free from obstacles that cause shading on part of the PV system. Trees, chimneys, and other roof protrusions, for example, are well known obstacles that can lead to shading losses on roof-mounted PV systems. The problem is that shaded PV cells act like a strong resistor, dissipating the electricity generated by solar units to the remaining, non-shaded, area of the string. This is observable through the high temperature (hot spot) in the shaded modules of a partly shaded system. Frequent high-temperature cycles shorten the lifetime of a cell and module. Currently, most module manufacturers supply their products with bypass diodes to prevent a fully or partly shaded module from sapping the generated energy of the other string modules. (b) Standard test conditions: The output of the solar PV system is rated by manufacturers under standard test conditions. These conditions are easily recreated in a factory and enable consistent comparisons of products, but nevertheless need to be modified to estimate output under common outdoor operating conditions. (c) Temperature: Module output power reduces as module temperature increases (0.5% per degree Celsius). (d) Dirt and dust: Dirt and dust can accumulate on the solar module surface, blocking the sunlight and reducing output. In regions with heavy annual rainfall, the problem is mostly avoided because the dirt and dust is cleaned off by rain showers. (e) Mismatch and wiring losses: The maximum power output of the total PV array is always less than the sum of the maximum output of the individual modules. The difference is the result of slight inconsistencies in performance from one module to the next, which is known as "module mismatch." Power is also lost to resistance in the system wiring. (f) DC to AC conversion losses: The DC power generated by the solar module must be converted into common household AC power by means of an inverter. Some power is lost in the conversion process, and there is additional loss in the wires from the modules at the roof down to the inverter.

Photovoltaic modules can either be integrated into roofing materials or mounted on the ground or pole. Whatever the mounting, the structure should be stable and durable, and be able to support the modules and withstand wind, rain, hail, and other outdoor conditions. PV applications in the built environment, as well as ground-based installations are manifold each requiring a specific type of integration or support structure. A wide range of products has been developed for use in PV module installation. Particularly in the built environment, mounting and support structures are designed in such a way that the PV system is fully integrated into the building and contributes to its aesthetic and architectonic value. PV support structures are available for façades, slanted roofs, flat roofs, and "PV tiles" that can be used to replace conventional roof tiles. Often, the most appropriate and convenient location to place a PV array is on the roof of a building. The PV array can be mounted above and parallel to the surface of the roof with a standoff of several centimeters for cooling purposes. In some cases, such as with flat roofs, a separate structure with a more optimal tilt angle is mounted on the roof. When considering a roof-mounted PV installation, attention must be given to the structure of the roof and weather sealing of the roof (Hulshorst 2008).

Operation and maintenance of a PV system is simple and requires no extensive maintenance or upkeep. PV systems contain no moving parts to wear out, break down, or replace. Operation of the PV system should be checked by measuring the kWh produced by the system. Depending on the amount of dirt and dust build-up, the PV panels should be cleaned annually. Batteries on PV systems require maintenance. The batteries used in PV systems are similar to car batteries, but are built somewhat differently to allow more of their stored energy to be used each day. Batteries must be protected from extremely cold weather and the fluid in unsealed batteries must be checked periodically. Along with investment costs, an economic evaluation of PV systems includes other aspects which should also be taken into account (Hulshorst 2008): (a) Reduction of the annual electricity costs due to the production of electricity by the PV system—future expectations of the electricity price should be taken into account; (b) Possible positive stimulation programs by government for PV systems: for example, subsidies or tax incentives; (c) The costs of saving on other building materials through the use of PV modules; (d) Costs of CO_2 pollution due to the production of electricity—zero for PV systems.

4.1.1.2 Utility-Scale Flat-Plate Photovoltaic System

Flat plate Photovoltaic (PV) modules are the most common solar array designs, and used in fixed systems or integrated into more sophisticated designs that include tracking systems that follow the sun's trajectory across the horizon throughout the day. Flat-plate technologies include thick cells of crystalline silicon (from both ingot and sheet-growth techniques) and thin films (less than 100 micrometers) of various materials, the most common of which is silicon. Silicon can be single (or mono)-crystalline, multicrystalline, or amorphous. Crystallinity is a measure of how perfectly ordered the atoms are in the crystal structure. Most flat plate PV modules use solar cells that are made from either single-crystalline or amorphous silicon. Single-crystalline silicon is composed of a very uniform crystal structure and is ideal for conducting electrons through the material. Solar cells made from this type of silicon are usually more efficient but also tend to be the most expensive because of the purity of silicon material. Solar panels that utilize amorphous silicon solar cells are currently the most common and are usually cheaper; however, they yield lower energy conversion efficiency. To date, crystalline silicon-based flat plate PV technology is able to achieve module conversion efficiencies between 15 and 20%.

Thin-film photovoltaic cells are usually made of a certain type of polycrystalline material. The three most common thin-film materials are amorphous silicon (a-Si), copper indium diselenide (CIS) and its alloys, or cadmium telluride (CdTe). Thin-film solar cells are produced by depositing very thin consecutive layers of atoms on a flexible substrate. Substrate material can be either glass, stainless steel, or various types of polymers. Thin films use much less material during production compared to silicon-based solar cells and can be manufactured in large-area automated continuous-process equipment. One method of production employs roll-to-roll printing technology which further reduces the cost of manufacturing. Thin-film production costs approximately half that of silicon-wafer-based solar cell production. The trade-off is that thin-film PV cells are significantly less efficient compared to single or amorphous-crystalline silicon solar cells. This is mainly because of the red and near-infrared photons, which don't stay trapped inside the thin silicon long enough to get absorbed. CdTe-based thin-film solar cell module efficiency is currently around 13%, the highest of the three material types. CIS produce modules with efficiency around 10% and a-Si around 8 percent. However, research and development in this area is constantly pushing the efficiency of thin-film solar cells closer to that of conventional silicon-based PV.

4.1.1.3 Utility-Scale Photovoltaic Concentrators

Concentrating photovoltaic systems can also be used for utility-scale solar electricity generation, which employ either a large dish of reflective mirrors or concentrating lenses that direct sunlight onto a photovoltaic surface which produces electricity directly from the sun's energy. Either module is installed on a high-precision dual-axis tracking system which ensures optimal operation throughout the day. These systems can be configured to concentrate the sun's energy between 2 and 500 times. High-concentration PV (HCPV) systems favor the use of high efficiency, multi-junction solar cell technology because efficiency of these cells rises faster with concentration than do conventional silicon-based solar cells. Commercially available CPV systems have demonstrated energy conversion efficiencies of approximately 29%.

CPV's main attraction is that it can leverage modest cell production volumes to much larger-scale systems using relatively simple and inexpensive optical concentration. Another attractive feature is that CPV systems can provide higher conversion efficiencies than conventional flat-plate systems—more than 30% (also roughly three-fourths of the theoretical limit) for multi-junction devices incorporating epitaxial layers of Group III–V compounds, such as gallium-aluminum arsenide and gallium-indium antimonide, grown on crystalline substrates.

4.1.2 Solar Thermal Technologies

While photovoltaic technology attempts to achieve efficient conversion of solar energy into electrical energy, solar thermal technology converts it into thermal energy, a much simpler and efficient task. A solar thermal cell utilizes all of the energy of incident photons that it absorbs to heat up a fluid that can then transfer that heat to places where it can be used, such as a swimming pool, a boiler, an absorption chiller, or a power plant turbine. Solar thermal systems can have several advantages over photovoltaic systems (Claremont 2013): (a) Being a simpler technology, they are generally cheaper than photovoltaics, while at the same time being more efficient at converting energy. (b) An advanced solar thermal panel converts solar radiation to thermal energy with around 60% efficiency, although there is a wide variation among the different models available. (c) Solar thermal systems have a wider variety of application than photovoltaic systems, being able to generate steam that can be used to turn a turbine to create electricity, or generate hot water to heat a building, provide domestic hot water, run an absorption chiller to cool a building, or any other application which requires hot water. Some systems can perform multiple functions with the same array of panels. With the use of a heat transfer fluid, the energy that a solar thermal panel generates can be stored in the fluid for extended periods of time, while a photovoltaic panel requires expensive batteries for electricity storage. Thus, solar thermal systems can continue to provide some energy during the night or during cloudy periods.

4.1.2.1 Unglazed Collectors

An unglazed solar collector is a solar thermal collector that consists of an absorber with embedded channels where the fluid circulates as it scrubs heat from the absorber. Unglazed collectors have no insulation and the glass covering like a glazed flat-plate collector; they are usually just metal or plastic pipes placed in the sun with a black coating to enhance absorption and act as an ultraviolet light inhibitor to extend the life of the system, as shown in Fig. 4.4a. Without insulation they cannot hold onto heat at temperatures much above that of the surrounding air. However, when only ambient temperature (or slightly above) needs to be reached, they can be extremely efficient at heating the water within the pipes. Unglazed collectors have been used to heat make-up or ventilation air in commercial, industrial, agriculture, and process applications. The greatest use for unglazed collectors is for heating swimming pools.

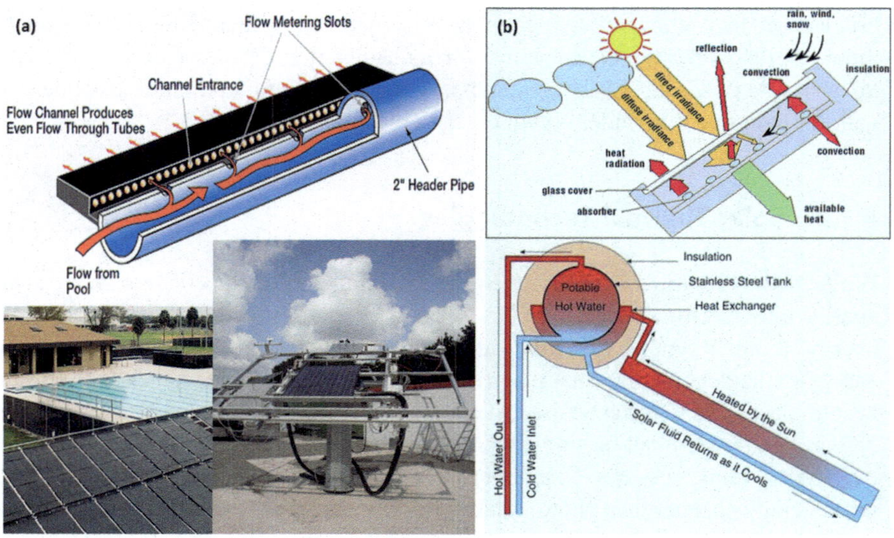

Fig. 4.4 Unglazed collector (**a**) and glazed collector (**b**)

4.1.2.2 Glazed Collectors

Different from the unglazed collectors, glazed collectors usually have a transparent top sheet cover above the absorber, creating a cavity filled with gas (usually air), which generates the greenhouse effect. They are insulated side and back panels to minimize heat loss to ambient air. These systems function well for heating water to moderate temperatures of 100–200 °F (35–100 °C) and are suitable for domestic hot water heating. As shown in Fig. 4.4b, their improvement over unglazed collectors comes from added insulation that allows the water to be heated to higher temperatures without losing much of its energy to the air. The absorber—usually a copper pipe—is coated with a selective surface that collects solar radiation very well, but has low emittance at the operating temperature. Black chrome is a commonly used coating. Glazed collectors are mounted in flat glass-enclosed frames. Thus, solar radiation is absorbed through the cover (usually glass) by the metal absorber, heating up the fluid which circulates within the pipes.

4.1.2.3 Parabolic Trough

A parabolic trough is a type of solar thermal collector that is straight in one dimension and curved as a parabola in the other two, lined with a polished metal mirror. A cylindrical row of mirrors collects the sun's rays and focuses them on the heat collection element that runs along the length of the trough at the focal point of the mirrors, as shown in Fig. 4.5. The heat collection element consists of a stainless steel tube within an insulating evacuated glass tube. The steel tube is coated with a solar-selective absorber surface while the glass tube is coated with an anti-reflective surface. A heat transfer fluid, such as water or molten salts is pumped through the steel tube and is warmed as it passes through the focal point of the trough. Parabolic trough systems are normally designed for utility-scale electricity generation, with the transfer fluid being used to generate steam for a turbine, but there are some roof-top-scale systems available that can run absorption chillers, generate electricity, or produce steam for buildings (Claremont 2013).

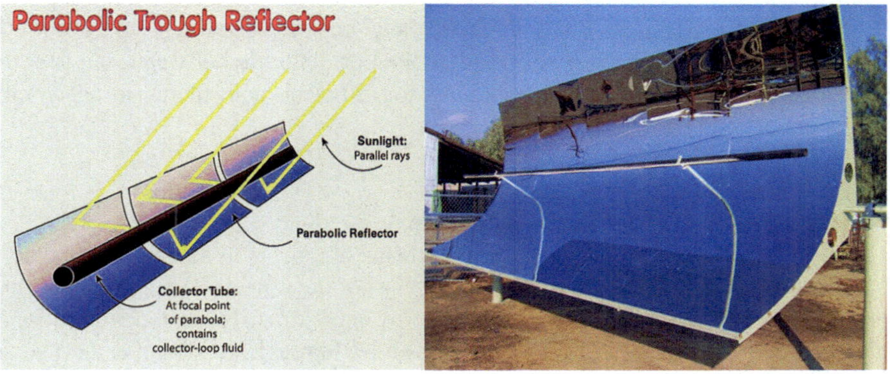

Fig. 4.5 Parabolic trough reflector

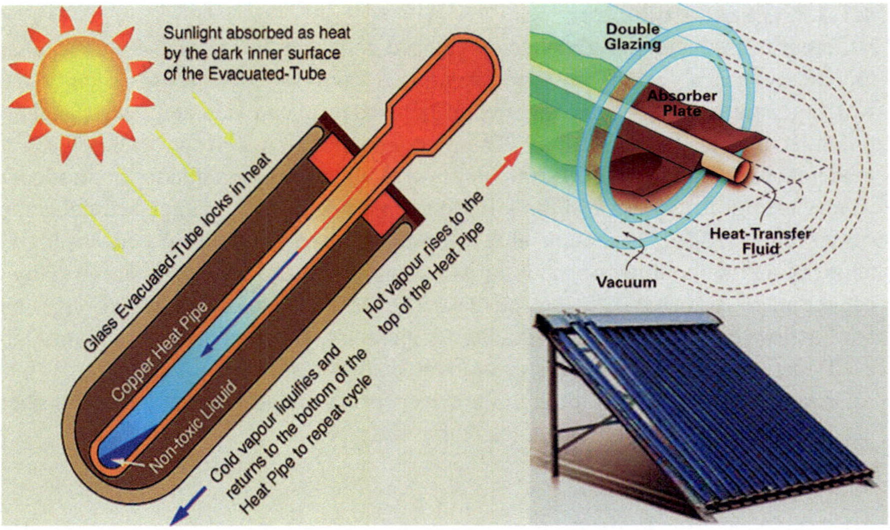

Fig. 4.6 Vacuum tube collectors

4.1.2.4 Vacuum Tube Collectors

A vacuum tuber collector consists of parallel rows of glass tubes connected to a header pipe. Each tube has the air removed from it to eliminate heat loss through convection and radiation. If the mirror were taken out of the parabolic trough system, leaving only the heat collection element, the resulting panels would be a vacuum tube collector. These panels feature a row of metal tubes coated with an especially absorbent surface and insulated by being placed within a glass vacuum tube, as shown in Fig. 4.6. The process of gathering heat is much the same as with parabolic troughs, only the light that is absorbed by the heat collection element is unconcentrated. Because the vacuum tube effectively prevents any heat losses from convection or conduction, the system's efficiency is largely unaffected by changes in the ambient temperature, wind speed, insolation levels or other environmental factors that would otherwise alter the rate of heat loss. Vacuum tube collectors generally heat fluids to moderately high temperatures, up to 400 °F (200 °C), to provide space heating and cooling or domestic hot water, but they do not achieve high enough temperatures for efficient conversion to electricity (Claremont 2013).

4.1.2.5 Linear Fresnel Lens Reflectors

Linear Fresnel reflectors use long, thin segments of mirrors to focus sunlight onto a fixed absorber located at a common focal point of the reflectors. These mirrors are capable of concentrating the sun's energy to approximately 30 times its normal intensity. This concentrated energy is transferred through the absorber into some thermal fluid (this is typically oil capable of maintaining liquid state at very high

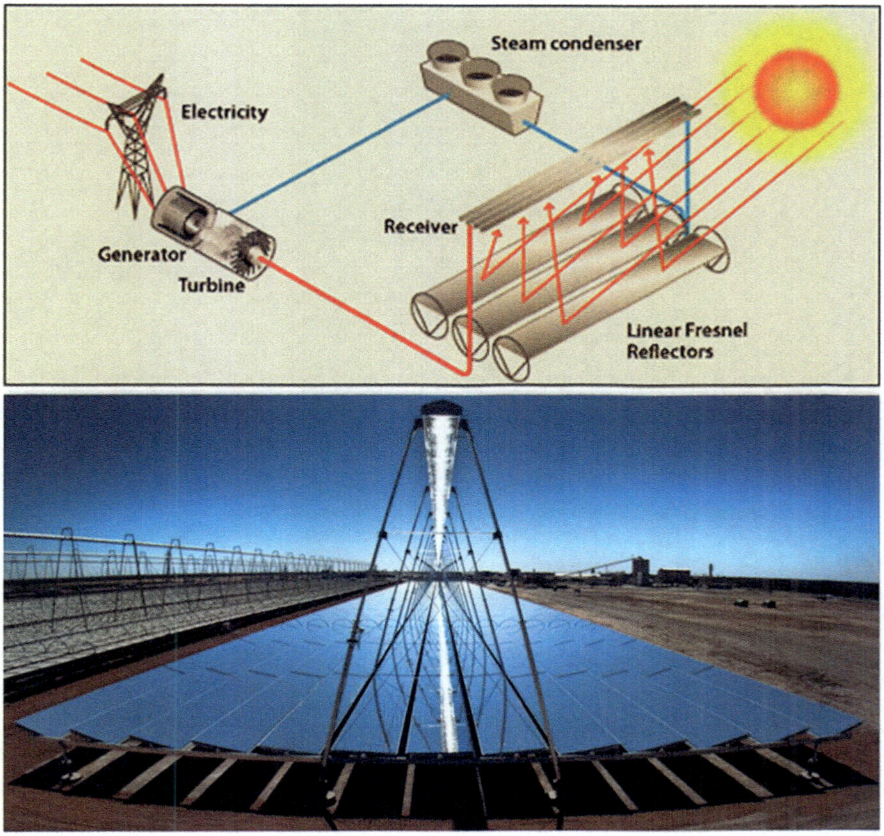

Fig. 4.7 Linear Fresnel lens reflector

temperatures). The fluid then goes through a heat exchanger to power a steam generator.

Instead of having one mirror gathering light and focusing it onto one heat collection element, a linear Fresnel lens reflector system takes an ordered array of flat or slightly curved mirrors and focuses the collected light from the array onto one heat collection element that is perched above the array of mirrors, as shown in Fig. 4.7. The heat collection element can be hundreds of meters long in a large system. By reducing the number of heat collection elements needed and by making the mirror array structurally simpler, the linear Fresnel lens reflector system can achieve lower costs than the parabolic trough system. The linear Fresnel lens reflector design can be improved by alternating the orientation of the mirrors such that one array of mirrors can focus light onto two or more heat collection elements. This method reduces the shading effect that the mirrors have on their neighbors, thus allowing the mirrors to be placed closer together.

Fig. 4.8 Solar Stirling engine

4.1.2.6 Solar Stirling Engine

A solar powered Stirling engine is a heat engine powered by a temperature gradient generated by the sun. Invented in 1816, the Stirling engine features a contained gas which expands and contracts when heat is applied from an external source, pumping a piston which can run an electric generator. The engine is similar to the familiar internal combustion engine, although with an externally applied energy source, rather than igniting an injected fuel. The engine can theoretically be very efficient, operates with almost no maintenance, and can be powered by any source of heat. In most designs, as shown in Fig. 4.8, a large dish mirror focuses sunlight onto the Stirling engine in a setup not unlike a satellite TV dish. The efficiency of solar Stirling engine systems at converting solar energy to grid electricity is the highest of any solar technology (those PV technologies with higher efficiencies are small-scale and not grid-connected). Utility-scale systems such as these are currently the main applications for solar Stirling engines (Claremont 2013).

4.2 Photovoltaic Materials and Devices

The photovoltaic has evolutionarily progressed in using increasingly automated processes to produce ever-thinner cells in greater volumes with higher efficiencies and lower costs. Advances in PV materials and manufacturing will continue to incrementally improve the technology's cost-competitiveness and gradually expand its applications without the need for game changing scientific breakthroughs. If true breakthroughs are made—the type that exploit new understanding of physics and material science to shatter old limits of efficiency and cost—the current solar power

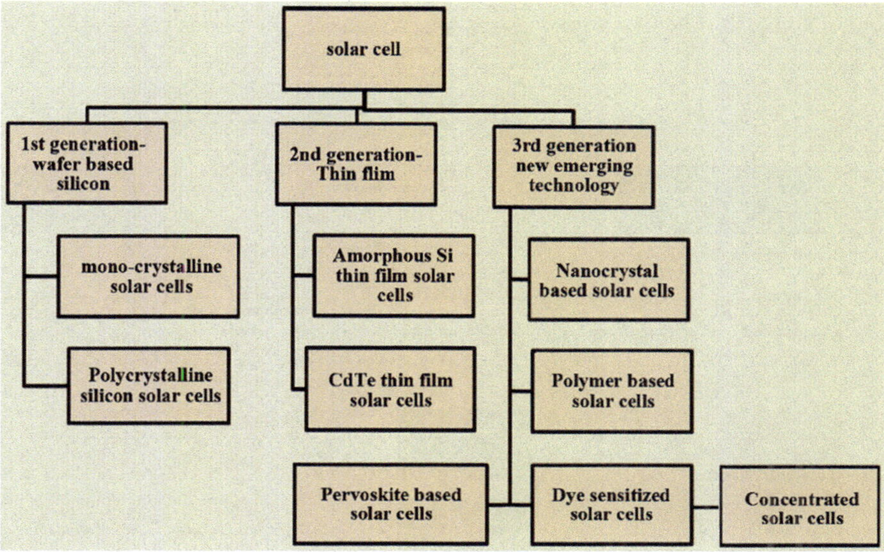

Fig. 4.9 Various types of solar cells classified based on the development generation (Adapted from Sharma et al. 2015 (Scientific Research Publishing), Licensed under the Creative Commons Attribution International License (CC BY). http://creativecommons.org/licenses/by/4.0/)

evolution could become a more rapidly paced revolution. Several broad routes to reduce manufacturing costs have been pursued, including increasing the efficiency of traditional Si-based photovoltaics, utilizing thin-film absorbers (amorphous Si, CdTe, and $CuInxGa_{1-x}$ Se_2; CIGS), and developing novel low-cost platform technologies such as organic photovoltaics (OPVs) and dye-sensitized solar cells (DSSCs). In addition, there has been extensive work on nanocrystal-based PV solar cells. The absorber layers in these devices are fabricated from nanocrystal dispersions—“inks” or “paints”—via scalable and inexpensive methods such as roll-coating or spray-coating (Steinhagen 2013). Various type of solar cells classified based on the development generation and the primary active material are shown in Figs. 4.9 and 4.10, respectively (Sharma et al. 2015; Ibn-Mohammeda et al. 2017).

Owing to parallel research and development with ubiquitous technologies such as integrated circuits, Si-based PVs currently dominate the solar cell market, and have maintained a steady advantage over other device structures and materials for many years. Wafer-based technologies, often called first generation photovoltaic, cut crystalline silicon ingots to get solar cells. The first generation photovoltaic, consists of a large-area, single layer diode, which is capable of generating usable electrical energy from light sources with the wavelengths of sunlight. The second generation, thin-film technology uses amorphous silicon to create a homogeneous thin layer. Consequently, there is no limitation on the size of the cells. One of the greatest advantages is that this process uses much smaller amounts of silicon. Other materials are used as well, like CIS, CIGS, and CdTe, but still silicon is dominating the market. Typically, the efficiencies of thin-film solar cells are lower compared with silicon

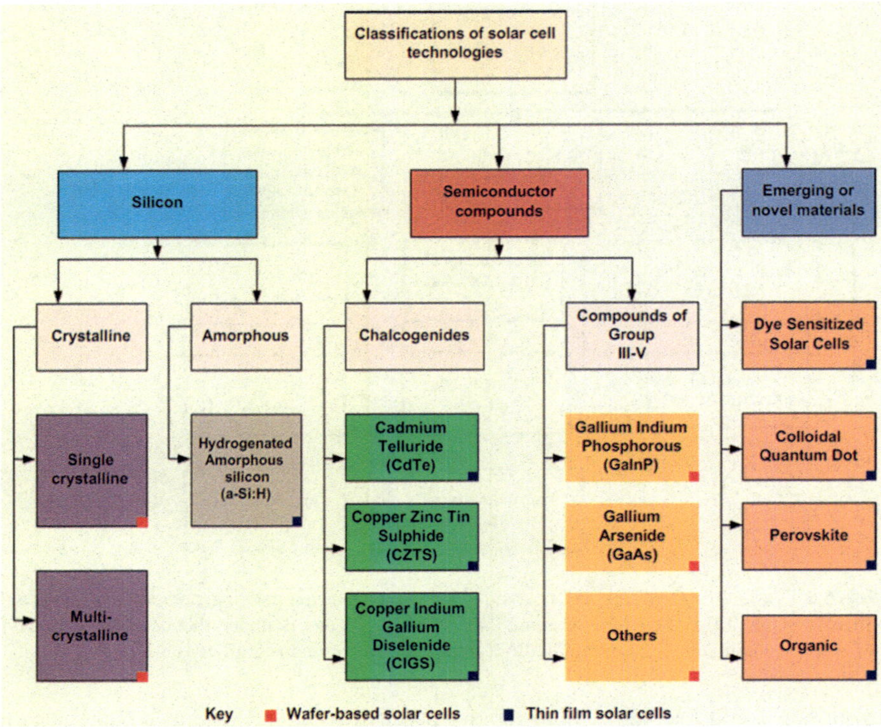

Fig. 4.10 Classification of solar cells based on the primary active material (Adapted with permission from Ibn-Mohammed et al. 2017 (Elsevier))

(wafer-based) solar cells, but manufacturing costs are also lower, so that a lower cost per watt can be achieved. The third generation cells that are used currently in building integration are dye-sensitized solar cells using biomimics through artificial photosynthesis. In this case the semiconductor material is TiO_2. It is a very promising technology that significantly reduces the initial costs of solar cells. The third generation photovoltaic cells are very different from silicon semiconductor devices. These new devices include photo-electrochemical cells, polymer solar cells, and nano-crystal solar cells. A comparison of some typical types of solar cells is summarized in Table 4.1 (Sharma et al. 2015).

Fourth generation mainly uses composite photovoltaic technology. In the fourth generation composite photovoltaic technology with the use of polymers with nanoparticles can be mixed together to make a single multi-spectrum layer. Then the thin multi-spectrum layers can be stacked to make multi-spectrum solar cells more efficient and cheaper based on polymer solar cell and multi-junction technology. The layer that converts different types of light is first, then another layer for the light that passes, and last is an infra-red spectrum layer for the cell, thus converting some of the heat for an overall solar cell composite.

Table 4.1 A comparison of some typical types of solar cells (Sharma et al. 2015)

Cell type	Crystalline		Thin films		Third generation					Perovskites
	Monocrystalline	Polycrystalline	CdTe	CIGS	Amorphous silicon	Nanocrystal	Dye sensitized	Polymer	Concentrated	
Efficiency	14–17%	12–14%	9–11%	10–12%	4–8%	7–8%	≈10%	≈3–10%	≈40%	31%
High-temperature performance	Not good at high temperatures	Not good at high temperatures	Good in cool as well as high temperature conditions	Good in cool as well as high temperature conditions	Good in cool as well as high temperature conditions	Excellent thermal stability	Not good in high temperature conditions	Not good in high temperature conditions	Excellent thermal stability	Excellent thermal stability
Size	Significantly less volume to produce the same amount of power	Significantly less volume to produce the same amount of power	Offering a wide range of product design from flexible, light durable	Offering a wide range of product design from flexible, light durable	Offering a wide range of product design from flexible, light durable	Offering a wide range of product design from flexible, light durable	Offering a wide range of product design from flexible, light durable	Offering a wide range of product design from flexible, light durable	Offering a specialized range of product design	Offering a wide range of product design from flexible, light durable
Cost	Two times more expensive compared to thin-film	Two times more expensive compared to thin-film	50% less expensive than conventional silicon cells	50% less expensive than conventional silicon cells	50% less expensive than conventional silicon cells	50% less expensive than conventional silicon cells	50% less expensive than conventional silicon cells	50% less expensive than conventional silicon cells	50% less expensive than conventional silicon cells	50% less expensive than conventional silicon cells
Additional detail	Oldest TV technology	Economical choice	Toxic due to Cd	Some CIGS have impressive 20% efficiency	Needs long installation time and large space	Needs short installation time and large space	Needs short installation time and large space	Needs short installation time and small space	Needs long installation time and large space	Latest technology. Needs short installation time and minimum space

4.2.1 Crystalline Silicon Photovoltaic Cells

Crystalline silicon PV cells have the longest production history. Silicon is safe for the environment and one of the most abundant resources on Earth, representing 26% of crustal material. The abundance and safety of silicon as a resource grants the silicon solar cell a prominent position among all the various kinds of solar cells in the PV industry (Saga 2010).

Silicon-based PV cells are generally made from layers of silicon a few hundred micrometers in thickness. Silicon for bulk cells is refined and grown into lightly p-type doped crystalline ingots that are then sliced into extremely thin wafers. However, the size that the crystalline wafers can be cut to is still very thick when compared to thin-film solar cells, and when considering the vast areas the wafers must cover it adds up to a highly intensive use of silicon. To make the wafers into solar cells, n-type dopants (often phosphorous) are diffused across the surface, creating the p–n junction. Crystalline silicon PV technology is constantly being developed to improve its capability and efficiency, which covers mono-crystalline, multi-crystalline, and emitter wrap through (EWT) structures.

4.2.1.1 Mono-Crystal Silicon PVs

Mono-crystalline cells are made from a single large crystal wafer of silicon, as shown in Fig. 4.11. These cells have high efficiency, but are expensive due to the demanding production process. These crystals are grown using the Czochralski

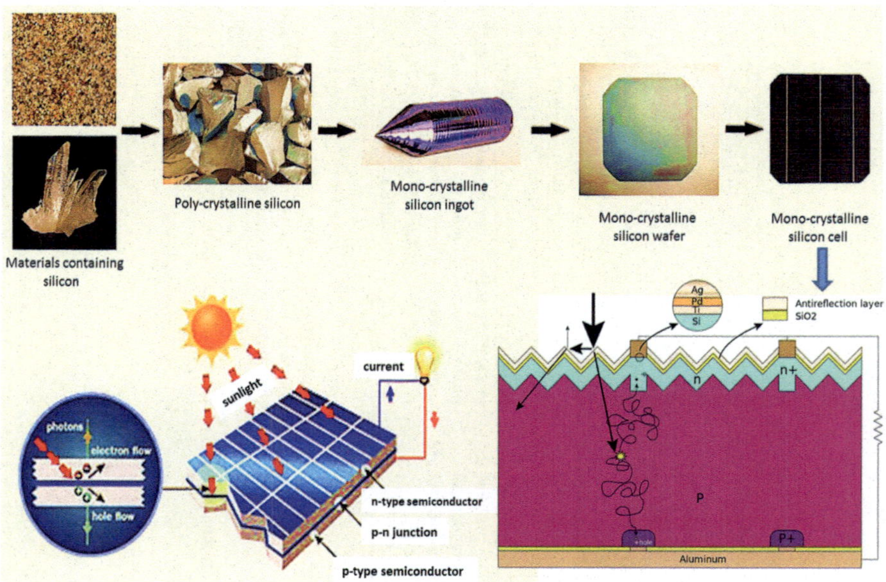

Fig. 4.11 Mono-crystalline silicon PVs

process in cylindrical ingots, and thus do not completely cover all the area of a square solar cell.

This type of cell is the most commonly used, and will continue to be the leader until a more efficient and cost-effective PV technology is developed. It essentially uses crystalline Si p–n junctions. Due to the silicon material, currently attempts to enhance the efficiency are limited by the amount of energy produced by the photons since it decreases at higher wavelengths. Moreover, radiation with longer wavelengths leads to thermal dissipation and essentially causes the cell to heat up hence reducing its efficiency. The maximum efficiency of mono-crystalline silicon solar cell has reached around 23% under STC, but the highest recorded was 24.7% (under STC). Due to combination of solar cell resistance, solar radiation reflection, and metal contacts available on the top side, self-losses are generated. After Si ingot is manufactured to a diameter of 10–15 cm for instance, it is then cut in wafers of 0.3 mm thick to form a solar cell of approximately 35 mA of current per cm^2 area with a voltage of 0.55 V at full illumination. For some other semiconductor materials with different wavelengths, it can reach 30% (under STC). However, module efficiencies always tend to be lower than the actual cell. Solar silicon processing technology has many points in common with the microelectronics industry, and the benefits of the huge improvements in Si wafer processing technologies used in microelectronic applications are to improve the performance of laboratory cells, hence made this technology most favorable (Akarslan 2012).

Current PV production is dominated by mono-junction solar cells based on silicon wafers including mono crystal(c-Si) and multi-crystalline silicon (mc-Si). These types of mono-junction, silicon-wafer devices are now commonly referred to as the first- generation (1G) technology, the majority of which is based on a screen printing-based device.

4.2.1.2 Polycrystalline Silicon PVs

Less expensive than mono-crystalline cells, poly-crystalline cells are also less efficient. They are made of silicon wafers cut from square cast ingots of silicon. The method of casting a poly-crystalline wafer of silicon as opposed to a mono-crystalline wafer requires much less precision and expense.

The advantage of converting the production of crystalline solar cells from mono-silicon to poly-silicon is to decrease the flaws in metal contamination and crystal structure. Poly-crystalline cell manufacturing is initiated by melting silicon and solidifying it to orient crystals in a fixed direction producing rectangular ingot of poly-crystalline silicon to be sliced into blocks and finally into thin wafers, as shown in Fig. 4.12. However, this final step can be abolished by cultivating wafer thin ribbons of poly-crystalline silicon (Akarslan 2012).

4.2.1.3 Emitter Wrap-Through Cells

Emitter wrap-through (EWT) cells (Fig. 4.13) have allowed an increase in efficiency through better cell design rather than material improvements in this technology, small laser drilled holes are used to connect the rear n-type contact with the opposite

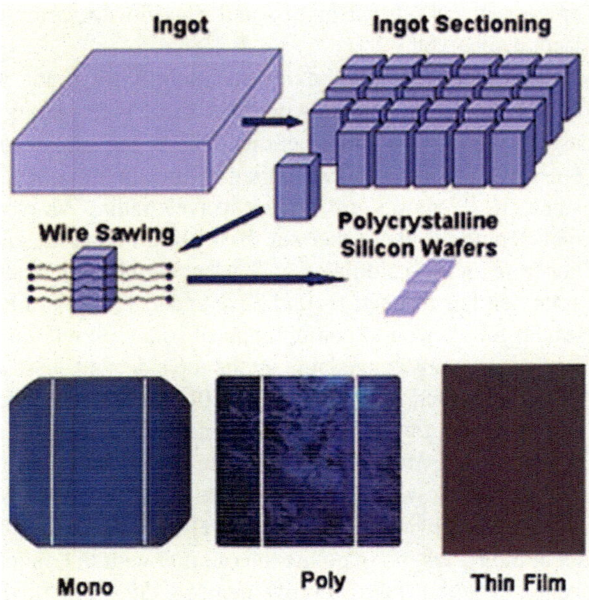

Fig. 4.12 Schematic illustration of production for polycrystalline silicon wafers

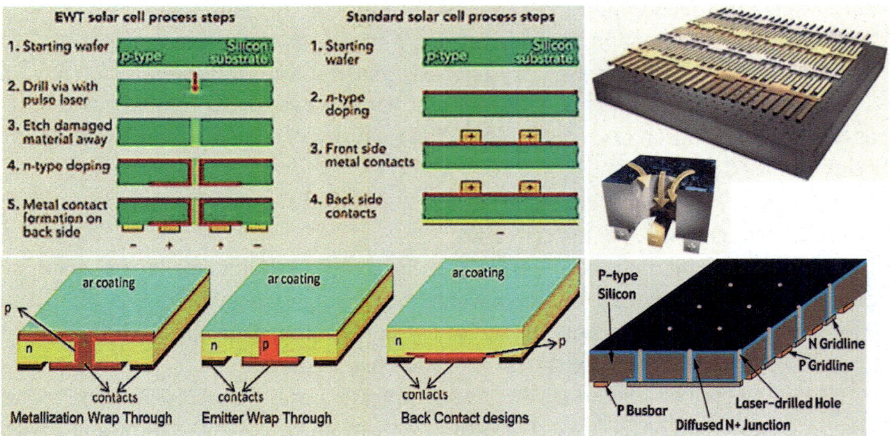

Fig. 4.13 Schematic illustration of emitter wrap-though cells

side emitter. The removal of front contacts allows the full surface area of the cell to absorb solar radiation because masking by the metal lines is no longer present. There are manufacturing gains by putting the contacts on the backs of the cell. One major disadvantage of such a technology is evident on large area EWT cells where this technology suffers from high series resistance which limits the fill factor (Akarslan 2012).

4.2.2 Thin-Film Photovoltaic Cells

In order to counter the processing, materials, and handling costs associated with crystalline silicon cells, much research has gone into perfecting methods of making solar cells with semiconductors only a few micrometers in thickness. These cells are expected to achieve reasonable efficiency while using very little silicon and employing roll-to-roll processing. These cells have lower efficiency than crystalline silicon cells but frequently have costs that are low enough to make them competitive.

Thin-film solar cells are basically thin layers of semiconductor materials applied to a solid backing material. Thin films greatly reduce the amount of semiconductor material required for each cell when compared to silicon wafers and hence lowers the cost of production of photovoltaic cells. Gallium arsenide (GaAs), copper indium gallium selenide ($CuIn_xGa_{1-x}Se_2$ or CIGS), cadmium telluride (CdTe), and titanium dioxide (TiO_2) are materials that have been mostly used for thin-film PV cells.

In comparison with crystalline silicon cells, thin-film technology holds the promise of reducing the cost of PV array by lowering material and manufacturing without jeopardizing the cells' lifetime as well as any hazard to the environment. Unlike crystalline forms of solar cells, where pieces of semiconductors are sandwiched between glass panels to create the modules, thin-film panels are created by depositing thin layers of certain materials on glass or stainless steel (SS) substrates, using sputtering tools. The advantage of this methodology lies in the fact that the thickness of the deposited layers which are barely a few micrometer (smaller than 10 μm) thick compared to crystalline wafers which tend to be several hundred micron thick, in addition to the possible films deposited on SS sheets which allows the creation of flexible PV modules. The resulting advantage is a lowering in manufacturing cost due to the high-throughput deposition process as well as the lower cost of materials. Technically, the fact that the layers are much thinner, results in less photovoltaic material to absorb incoming solar radiation, hence the efficiencies of thin-film solar modules are lower than crystalline, although the ability to deposit many different materials and alloys has allowed tremendous improvement in efficiencies (Chaar et al. 2011; Akarslan 2012).

Four kinds of thin-film cells have emerged as commercially important: the amorphous silicon cell (multiple-junction structure), thin multi-crystalline silicon on a low-cost substrate, the copper indium diselenide/cadmium sulfide hetero-junction cell, and the cadmium telluride/cadmium sulfide hetero-junction cell (Akarslan 2012).

4.2.2.1 Amorphous Silicon Cells

Amorphous silicon ("a-Si")—the noncrystalline form of silicon—can be deposited onto a conductive substrate in a layer a few micrometers thick to create a thin-film solar cell.

The deposition process of applying a-Si allows it to be less than 1% of the thickness of a crystalline cell. Amorphous silicon cells are often built using two or three junctions to increase the amount of the solar spectrum they can utilize. Typically alloys of a-Si and germanium are used to create the additional junctions in the multi-junction cells. These cells are lighter, use much less material, and are less energy intensive to produce than bulk silicon cells. However, the cell efficiency of amorphous silicon is much lower than crystalline silicon due largely to the increased recombination of the electron–hole pairs that results from the lower carrier mobility.

Amorphous (uncrystallized) silicon is the most popular thin-film technology with cell efficiencies of 5–7% and double- and triple-junction designs raising it to 8–10%. Some of the varieties of amorphous silicon are amorphous silicon carbide (a-SiC), amorphous silicon germanium (a-SiGe), microcrystalline silicon (μ c-Si), and amorphous silicon-nitride (a- SiN) (Parida et al. 2011; Akarslan 2012).

Amorphous silicon (a-Si) diverges from crystalline silicon in the fact that silicon atoms are randomly located from each other. This randomness in the atomic structure has a major effect on the electronic properties of the material causing a higher bandgap (1.7 eV) than crystalline silicon (1.1 eV). The larger band gap allows a-Si cells to absorb the visible part of the solar spectrum more strongly than the infrared portion of the spectrum. There are several variations in this technology where substrates can be glass or flexible SS, tandem junction, double and triple junctions, and each one has a different performance (Akarslan 2012).

Amorphous-Si with Double or Triple Junctions

Since a-Si cells have lower efficiency than the mono- and multi-crystalline silicon counterparts. With the maximum efficiency achieved in laboratory currently at approximately 12%, mono junction a-Si modules degrades after being exposed to sunlight and stabilizing at around 4–8%. This reduction is due to the Staebler–Wronski effect which causes the changes in the properties of hydrogenated amorphous Si. To improve the efficiency and solve the degradation problems, approaches such as developing multiple junction a-Si devices have been attempted, as shown in Fig. 4.14. This improvement is linked to the design structure of such cells where different wavelengths from solar irradiation (from short to long wavelength) are captured (Akarslan 2012).

Tandem Amorphous-Si and Multi-Crystalline-Si

Another method to enhance the efficiency of PV cells and modules is the "stacked" or multicrystalline (mc) junctions, also called micro morph thin film. In this approach two or more PV junctions are layered one on top of the other where the top layer is constructed of an ultrathin layer of a-Si which converts the shorter wavelengths of the visible solar spectrum. However, at longer wavelength, microcrystalline silicon is most effective in addition to some of the infrared range. This results in higher efficiencies than amorphous Si cells of about 8–9% depending on the cell structure and layer thicknesses (Akarslan 2012).

4.2.2.2 Cadmium Telluride and Cadmium Sulfide

The crystalline compound cadmium-telluride (CdTe) is an effective solar cell material—it is a very strong absorber of light and has a band gap almost perfectly tuned to match the solar spectrum. To create a p–n junction for solar cells a layer of cadmium

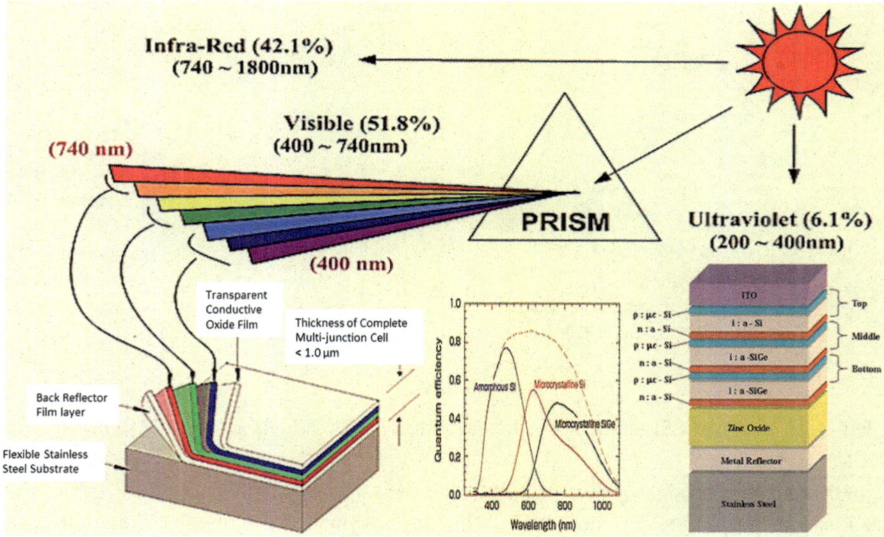

Fig. 4.14 Schematic illustration of solar cell with amorphous-Si, double or triple junctions

sulfide is added to the CdTe. CdTe has a direct bandgap and high optical absorption with the ideal bandgap energy of 1.45 eV. This results in higher current densities and higher open-circuit voltages than with CIS/CIGS. Small-area CdTe cells with efficiencies of greater than 16% and CdTe modules with efficiencies of greater than 9% have been demonstrated. CdTe, unlike the other thin-film technology, is easier to deposit and more apt for large-scale. The potential issue is the availability of Te which might cause some raw material constraints that will then affect the cost of the modules (Chaar et al. 2011).

Another benefit of CdTe solar cells is the significant enhancement of carrier multiplication from CdTe nanocrystals. Carrier multiplication is the process by which inelastic scattering of charge carriers and valence electrons create additional electron–hole pairs (charge carriers). With sufficient enough carrier multiplication, energy lost by photons with energy in excess of the bandgap due to thermalization could be collected. As a binary compound, CdTe is thus easier to produce solar cells than with CIS/CIGS. Typically, a hetero-junction is created on a TCO-coated glass substrate using n-type CdS and p-type CdTe, as shown in Fig. 4.15. Obtaining a uniform deposition of CdS is a critical issue in the production of both CIS/CIGS and CdTe solar cells, particularly since stability and efficiency are dependent on the n-p junction. However, a greater hurtle is the use the Cadmium, particularly in CdTe solar cells. While CdTe and CdS are stable, their production from the hazardous material Cd introduces hazards that require strict regulation (Bertolli 2008).

Because of its effectiveness, a CdTe solar cell uses only about 1% of the semiconductor material that bulk silicon cells use. CdTe solar cells are generally somewhat less efficient than bulk silicon cells, but have lower costs associated with them due to the smaller amount of material used and inexpensive production

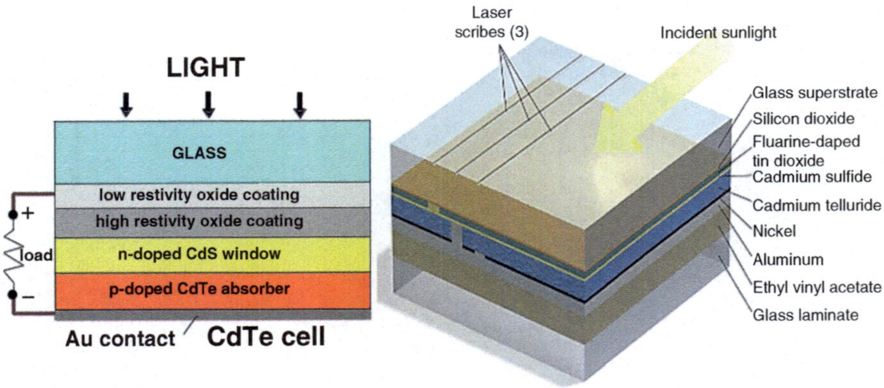

Fig. 4.15 Schematic illustration of cadmium telluride and cadmium sulfide solar cells

methods. While sales of low-priced CdTe cells have sharply increased, the soaring consumption of the very rare metal tellurium has pushed the price of that element up sharply. It remains to be seen how the production of CdTe solar cells will be affected in the future by supply constraints. CdTe is a toxic carcinogen and some concerns have been raised about the danger of solar cells made with CdTe. These concerns have been countered by noting that the Cadmium contained in one square meter of a CdTe cell is less than that within a size-C NiCd flashlight battery and that the CdTe is very well sequestered by the encapsulation of the cell (Sharma et al. 2013).

4.2.2.3 Copper Indium Selenide and Copper Indium Gallium Selenide

A combination of copper-indium-diselenide (CIS) and copper-gallium-diselenide (CIGS) has shown the highest efficiency of any thin-film cell. The mixture of the two materials creates a more complex and effective hetero-junction, where the junction is formed by semiconductors of dissimilar band gaps, rather than the conventional cell junction of p-type silicon and n-type silicon, which have identical band gap energies. The band gap of the semiconductor in CIGS cells can be varied by altering the ratio of indium to gallium, allowing the band gap to range continuously from 1.0 eV to 1.7 eV in the cell, which matches very well with the solar spectrum. CIGS is also a relatively easy material to work with, not requiring the complex vacuum deposition process that some other photovoltaic semiconductors require.

CIS and CIGS are direct-gap polycrystalline p-type semiconductors with high optical absorption. That is, the minimum of the conduction band has the same wave-vector as the maximum of the valence band. CIS and CIGS are used in heterojunctions with n-type layers (commonly CdS or ZnO). The high efficiency of CIS and CIGS is due to the effective bandgap between 1.1 and 1.2 eV. The addition of Gallium in CIGS increases the bandgap and the performance tune by increasing the open-circuit voltage. CIS and CIGS solar cells also show good stability and reliability, except in considerable heat and humidity. While commercial production of these solar cells is projected to be significantly cheaper than for wafer-based crystalline silicon solar cells, it is very likely that full-scale production will experience difficulties with the availability and price of Indium (Bertolli 2008).

CIS and CIGS are photovoltaic devices that contain semiconductor elements from groups I, III, and VI in the periodic table which is beneficial due to their high optical absorption coefficients and electrical characteristics enabling device tuning. Moreover, better uniformity is achieved through the usage of selenide, hence the number of recombination sites in the film is diminished benefiting quantum efficiency and hence the conversion efficiency. CIGS (indium incorporated with gallium—increased band gap) are multilayered thin-film composites. The biggest challenge for CIGS modules has been the limited ability to scale-up the process for high throughput, high yield, and low cost. Several deposition methods are used: sputtering, "ink" printing and electroplating with each having different throughput and efficiencies. Both glass and stainless steel substrates are used, obviously the stainless steel substrates yield flexible solar cells (Akarslan 2012).

4.2.3 Compound Semiconductor Photovoltaic Cells

Compound semiconductor cell is made of a complicated stack of crystalline layers with different band gaps that are tailored to absorb most of the solar radiation. It is more robust when expose to outer space radiation. Since each type of semiconductor has different characteristic band gap energy which then allows the absorption of light most efficiently, at a certain wavelength, hence absorption of electromagnetic radiation over a portion of the spectrum. These hetero-junction devices layer various cells with different bandgaps which are tuned utilizing the full spectrum.

Initially, light strikes a wide bandgap layer producing a high voltage therefore using high-energy photons efficiently enabling lower energy photons transfer to narrow bandgap sub-devices which absorb the transmitted infrared photons. Gallium arsenide (GaAs)/indium gallium phosphide (InGaP) multi-junction devices were originally fabricated on GaAs substrates however, in order to reduce the cost and increase robustness and because it is reasonably lattice-matched to GaAs, germanium (Ge) substrates are being used more often. The first cells had a mono junction much like the Si p–n junction solar cells, however because of the ability to introduce ternary and quaternary materials such as InGaP and aluminum indium gallium phosphide (AlInGaP) dual and triple junction devices were grown in order to capture a larger band of the solar spectrum therefore increasing the efficiency of the cells (Chaar et al. 2011; Akarslan 2012).

4.2.3.1 Space Photovoltaic Cells

Photovoltaic solar generators have been proven to be the optimal option for providing electrical power to satellites, as shown in Fig. 4.16. In 1958, US satellite Vanguard 1 demonstrated the first application. After years of moderate growth of the space PV market, the evolution of large-scale applications has increased in the late nineties, where the main applications are dominated by the telecommunication satellites, military satellites, and scientific space probes. Solar cells which are

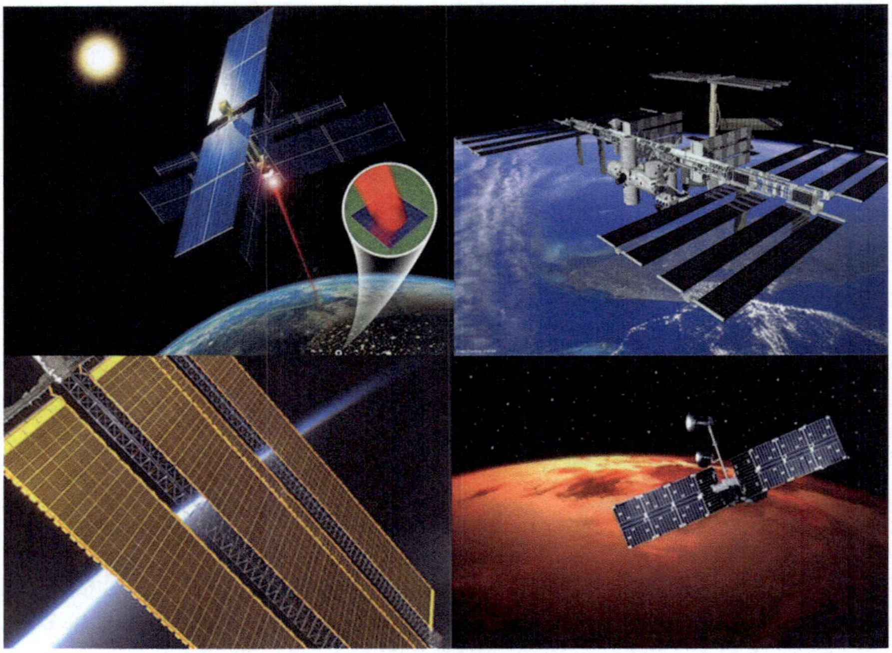

Fig. 4.16 Schematic illustration of space PV cells

designed for space must ensure that their specifications include apriority space environment condition such as spectral illumination and air mass. Issues of concern with terrestrial PV are their high cost while in space, weight, flexibility, efficiency, temperature, and suitable materials. In the 1950s, Si cells were p–n containing base layers of mono crystal N–Si with boron diffused P-emitters with an efficiency around 6%. In the 1960s, efficiency was improved to 12% and CdS was investigated because of its flexibility and lightweight, but, its low efficiency and instability left it unfavorable. In the 1970s, although advances in Si growth by float-zoning (Fz) were promising solutions, space cells made from this material suffered additional degradation after radiation exposure. Despite all competitive approaches, Si remains the leader in PV technology for space. In the 1980s similar technologies to the seventies were used in addition to the deployment in special air force missions with indium phosphate (InP) cells which efficiency reached 18% with high radiation tolerance. In the 1990s, although the high manufacturing cost, GaAs/Ge cells showed significant improvements including reduced area and weight, greater efficiency, and smaller stowage volume per launch. In addition, multi-junction cells have shown great promises with efficiencies reaching almost 30%. Due to the expense of the substrate and the growth process, the cost of these cells is extremely high compared to Si cells. For space applications the expense has been acceptable, however for terrestrial/ commercial application methods had to be developed to make the cost adequate, and the most successful method of reducing the cost has been to use concentration.

Essentially the solar cell wafers are dices into small cells (sometimes as small as 2 mm × 2 mm) and then a large lens is placed above the cell in order to concentrate the solar radiation on the small cell. The cell is placed at the focal length of the lens and the solar radiation incident on the lens will get focused on the PV cell. Effectively the cell is exposed to several times the "normal" radiation which is then quantified by using the terms "100 suns" or "300 suns" which concentrates the sun's radiation 100 and 300 times respectively. The technology is called concentrating PV or CPV. Of course with concentration comes the need for tracking as the lens that is concentrating the sun's radiation needs to track the sun to make sure the radiation is then focused on the cell. The concentrating method has used lenses or mirrors, or a combination of both. The mirrors are curved such that the PV cell is placed at the center of the curvature and the solar radiation is concentrated on the cell (Akarslan 2012).

4.2.3.2 Light Absorbing Dyes

Generally, these types of cells consist of a semiconductor, such as silicon, and an electrolytic liquid, which is a conducting solution commonly formed by dissolving a salt in a solvent liquid, such as water. The semiconductor and electrolyte work in tandem to split the closely bound electron–hole pairs produced when sunlight hits the cell. The source of the photo-induced charge carriers is a photosensitive dye that gives the solar cells their name: "dye sensitized" (most common dye is iodide). In addition, a nanomaterial, most commonly titanium dioxide (TiO_2) is also often used to hold the dye molecules in place like a scaffold (Fig. 4.17). Using dye sensitized cells for photovoltaic application goes back several decades as scientists were trying to emulate chlorophyll action in plants. While the highest efficiency dye-sensitized solar cell ever made is 11%, this technology contains volatile solvents in their electrolytes that can permeate across plastic (i.e., organic compounds) and also present problems for sealing the cells. Cells that contain these solvents are therefore

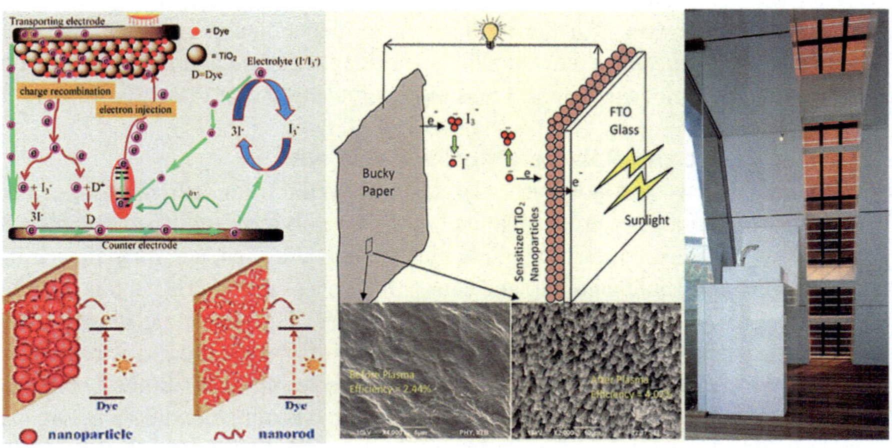

Fig. 4.17 Schematic illustration of dye sensitized PV cells

unattractive for outdoor use due to potential environmental hazards. Researchers have developed solar cells that use solvent-free electrolytes, but the cell efficiencies are too low (Akarslan 2012).

Dye-sensitized solar cells (DSSCs) are quite different in function from the other photovoltaic technologies in existence, and in fact bear a somewhat loose resemblance to photosynthesis. The material components of a DSSC are a thin film of electrolyte sandwiched between two electrodes (the top electrode being transparent to allow light into the cell) with a lattice of dye-coated nano-scale titanium-dioxide (TiO_2) particles coating one of the electrodes. The DSSC works on the principle of splitting the functions performed by the semiconductor in other photovoltaic technologies. In a silicon cell, the silicon both generates the charge and conducts the electrons and holes to the electrodes. In a DSSC, the incident photons excite electrons in the dye molecules. If given sufficient energy, the excited electrons will escape from the dye to the conduction band of the TiO_2 particles and will then diffuse to the electrode, generating a current. The electrons return to the dye through the electrolyte. DSSCs have the ability to provide power under conditions where other solar cells cannot. With most solar cells, if there are not enough electrons flowing through the semiconductor, the few that are will be lost as they recombine with the holes. DSSCs do not have this problem of recombination (often a serious drag on the efficiency of solar cells even during periods of high-intensity light) due to the separation of the electron producer and the electron carrier, and can thus function robustly under limited light; under cloudy skies, and even indoors (Sharma et al. 2013).

Lower processing costs along with flexibility of material and type usage achieved by inkjet/screen printing are the characteristics of dye-sensitized solar cell which depends on a mesoporous layer of nanoparticulate TiO_2 to magnify the surface area (200–300 m^2/g TiO_2, as compared to approximately 10 m^2/g of flat mono crystal). However, heat, ultraviolet (UV) light, and the interaction of solvents within the encapsulation of the cell are negative issues with this technology. Despite all the drawbacks and because of the promise of a low-cost potential for cells and incorporation in paints among other things, this technology's future must be observed. Most of the current work is on the development of more efficient light absorbing dyes and on the improvement of the reliability, as well as the elimination of solvents from the electrolytes while maintaining a reasonable efficiency (Akarslan 2012).

4.2.3.3 Organic and Polymer Photovoltaic Cells

Organic solar cells and polymer solar cells are built from thin films (typically 100 nm) of organic semiconductors such as polymers and small-molecule compounds like pentacene, multiphenylene vinylene, copper phthalocyanine (a blue or green organic pigment) and carbon fullerenes. The interest in this material lies with its mechanical flexibility and disposability. Since they are largely made from plastic opposed to traditional silicon, the manufacturing process is cost effective (lower-cost material, high-throughput manufacturing) with limited technical challenges (not require high-temperature or high vacuum conditions). Organic cells function in a slightly different way than most other cell technologies: instead of semiconductor p–n junctions, organic cells utilize electron donor and acceptor materials. Typical choices are polymers for the electron donors and fullerenes for the electron acceptors. When an electron–hole pair is created by the absorption of a photon in the donor, rather than separating and migrating to opposite sides of the cell, the electron and hole stay together as an exciton. The exciton diffuses through the cell until it reaches the acceptor where the electron is transferred to the acceptor material, creating a current through the acceptor. Because the exciton will only travel for a few nanometers before the electron and hole recombine, it is highly beneficial to have donor and acceptor blended together, forming a bulk heterojunction. Therefore, electron (donor-acceptor) pair forms the basis of organic cell operation where light agitates the donor causing the electron to transfer to the acceptor molecule, hence leaving a hole for the cycle to continue. The photo-generated charges are then transported and collated at the opposite electrodes to be utilized, before they recombine. Typically the cell has a glass front, a transparent indium tin oxide (ITO) contact layer, a conducting polymer, a photoactive polymer and finally the back contact layer (Al, Ag, etc.). Since ITO is expensive, carbon nanotube films have been developed as the transparent contact layer. A typical cross section of an organic solar cell is shown in Fig. 4.18. Nanotechnology seems to support sustainable

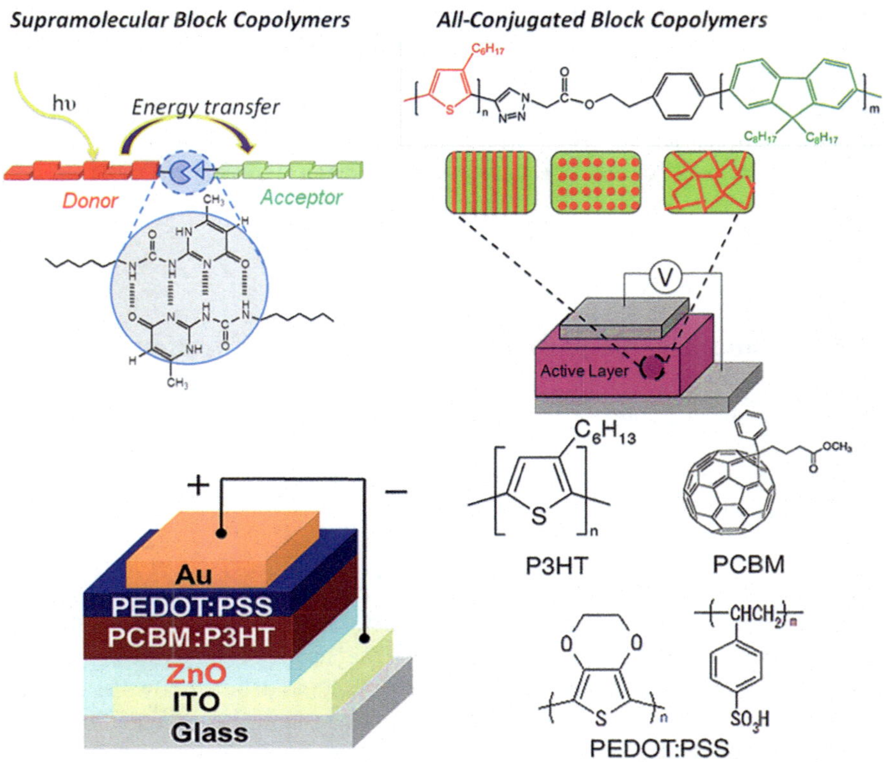

Fig. 4.18 Schematic illustration of organic and polymer solar cells (Modified from Verduzco (2015) All-Conjugated Block Copolymers. Courtesy: Verduzco Laboratory. http://verduzcolab.blogs.rice.edu/research-focus-1/)

economic growth by offering low cost but low efficiency PV which although not ideal offers consumers other alternatives (Akarslan 2012).

Organic photovoltaic (OPV) devices are increasingly pursued in view of their low fabrication costs and fairly easy processing. Their lightweight, mechanical flexibility and large-scale roll-to-roll production capability are additional advantages compared to traditional Si-based photovoltaics. In a typical OPV device, a blend of conjugated polymer (electron donor) and a fullerene derivative (electron acceptor) is normally used as an active layer sandwiched between the cathode and the anode. The interpenetrating network of donor and acceptor components forms the bulk hetero-junction (BHJ) system for the separation of charge carriers upon illumination and subsequently transports the opposite charge carriers towards the electrodes. Among the various active layers, multi(3-hexylthiophene) (P3HT)/–phenyl-C61-butyric acid methyl ester(PCBM) combination remains the promising system researched till date and shows greater than 5% power conversion efficiency. The performance of the device, however, critically depends on the nano-scale morphology and phase separation of the blend components (Akarslan 2012).

The benefit of using of organic materials is that it allows for the simple high volume low-temperature fabrication of flexible solar cells on plastic substrates. If the efficiency of the cells can be improved, organic cell technology will realize extremely low-cost production of very versatile cells. Efficiencies of organic solar cells are currently around 5–6%, although quickly rising. Organic solar cells must also overcome the issue of environmental degradation, a problem that most organic molecules face when exposed to oxygen, water, and solar radiation. There remain many opportunities for breakthroughs in organic solar cell design, given the newness of the technology and great diversity of materials and techniques available (Sharma et al. 2013).

4.2.4 Nanotechnology for Photovoltaic Cell Fabrication

Limitations seen in other PV technologies are lessened by the introduction of nanoscale components due to their ability to control the energy bandgap will provide flexibility and inter-changeability in addition to enhancing the probability of charge recombination (Akarslan 2012).

4.2.4.1 Carbon Nanotubes

Carbon nanotubes (CNT) are constructed of a hexagonal lattice carbon with excellent mechanical and electronic properties. The nanotube structure is a vector consisting of "n" number line and "m" number column defining how the graphene (an individual graphite layer) sheet is rolled up. Nano-tubes can be either metallic or semiconducting and they belong to two categories: single walled or multi-walled. Carbon nanotubes can be used as reasonably efficient photosensitive materials as well as other PV material, as shown in Fig. 4.19. PV nanometer-scale tubes when coated by special p- and n-type semiconductor materials form a p–n junction to

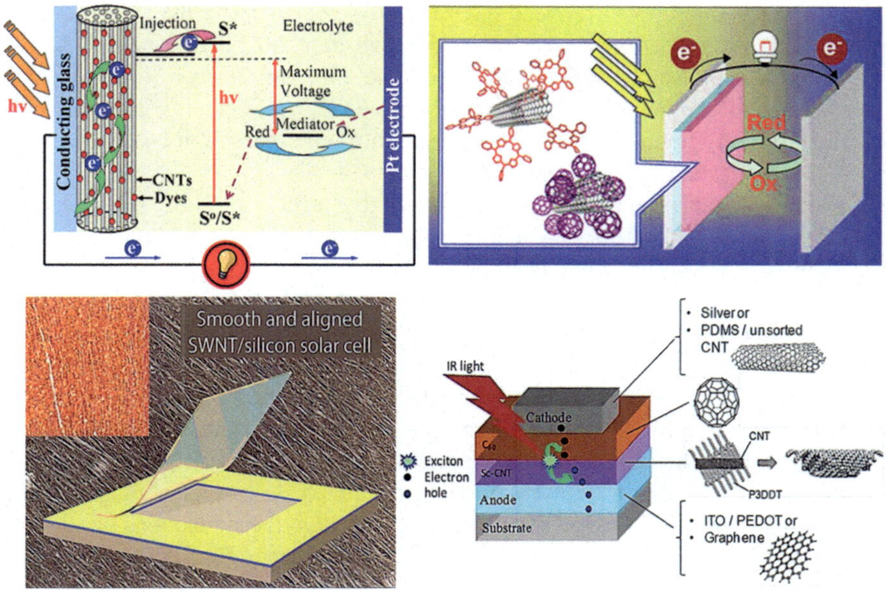

Fig. 4.19 Schematic illustration of carbon nanotubes in solar cells

generate electrical current. Such methodology enhances and increases the surface area available to produce electricity (Akarslan 2012).

1. Single wall carbon nanotubes as light harvesting media
   Single wall carbon nanotubes possess a wide range of direct bandgaps matching the solar spectrum, strong photoabsorption, from infrared to ultraviolet, and high carrier mobility and reduced carrier transport scattering, which make themselves ideal photovoltaic material. Photovoltaic effect can be achieved in ideal single wall carbon nanotube (SWNT) diodes. Individual SWNTs can form ideal p–n junction diodes. Under illumination, SWNT diodes show significant power conversion efficiencies owing to enhanced properties of an ideal diode (Lee 2003).

   SWNTs have been directly configured as energy conversion materials to fabricate thin-film solar cells, with nanotubes serving as both photogeneration sites and a charge carriers collecting/transport layer. The solar cells consist of a semitransparent thin film of nanotubes conformally coated on a n-type crystalline silicon substrate to create high-density p–n heterojunctions between nanotubes and n-Si to favor charge separation and extract electrons (through n-Si) and holes (through nanotubes). Acid infiltration of nanotube networks significantly boosts the cell efficiency by reducing the internal resistance that improves fill factor, and by forming photoelectrochemical units that enhance charge separation and transport. The wet acid-induced problems can be avoided by using aligned CNT film. In aligned CNT film, the transport distance is shortened, and the exciton quenching rate is also reduced. Additionally aligned nanotube film has much smaller void space, and better contact with substrate. So, plus strong acid doping, using aligned single wall carbon nanotube film can further improve power conversion efficiency (Jung et al. 2013). n-SWNT/p-Si photovoltaic device is also made by tuning SWNTs from p-type to n-type through polyethylene imine functionalization (Li and Saini 2010).

2. Carbon nanotube composites in the photoactive layer
Combining the physical and chemical characteristics of conjugated polymers with the high conductivity along the tube axis of carbon nanotubes (CNTs) provides a great deal of incentive to disperse CNTs into the photoactive layer in order to obtain more efficient OPV devices. The interpenetrating bulk donor–acceptor heterojunction in these devices can achieve charge separation and collection because of the existence of a bicontinuous network. Along this network, electrons and holes can travel towards their respective contacts through the electron acceptor and the polymer hole donor. Photovoltaic efficiency enhancement is proposed to be due to the introduction of internal polymer/nanotube junctions within the polymer matrix. The high electric field at these junctions can split up the excitons, while the single-walled carbon nanotube (SWCNT) can act as a pathway for the electrons (Kymakis et al. 2003).

The dispersion of CNTs in a solution of an electron donating conjugated polymer is perhaps the most common strategy to implement CNT materials into OPVs. Generally, poly (3-hexylthiophene) (P3HT) or poly (3-octylthiophene) (P3OT) are used for this purpose. These blends are then spin coated onto a transparent conductive electrode with thicknesses that vary from 60 to 120 nm. These conductive electrodes are usually glass covered with indium tin oxide (ITO) and a 40 nm sublayer of poly (3,4-ethylenedioxythiophene) (PEDOT) and poly(styrenesulfonate) (PSS). PEDOT and PSS help to smooth the ITO surface, decreasing the density of pinholes and stifling current leakage that occurs along shunting paths. Through thermal evaporation or sputter coating, a 20–70 nm thick layer of aluminum and sometimes an intermediate layer of lithium fluoride are then applied onto the photoactive material. Both multi-walled carbon nanotubes (MWCNTs) and single-walled carbon nanotubes (SWCNTs) have been integrated into the photoactive material (Kymakis et al. 2003).

Among the best power conversions achieved to date using CNTs were obtained by depositing a SWCNT layer between the ITO and the PEDOT: PSS or between the PEDOT: PSS and the photoactive blend in a modified ITO/PEDOT: PSS/ P3HT: (6,6)-phenyl-C61-butyric acid methyl ester (PCBM)/Al solar cell. By dip-coating from a hydrophilic suspension, SWCNT were deposited after an initially exposing the surface to an argon plasma to achieve a power conversion efficiency of 4.9%, compared to 4% without CNTs (Chaudhary et al. 2007).

However, even though CNTs have shown potential in the photoactive layer, they have not resulted in a solar cell with a power conversion efficiency greater than the best tandem organic cells (6.5% efficiency). However, there is still room for novel methods to better take advantage of the beneficial properties of CNTs.

3. Carbon nanotubes as a transparent electrode
   ITO is currently the most popular material used for the transparent electrodes in OPV devices; however, it has a number of deficiencies. For one, it is not very compatible with polymeric substrates due to its high deposition temperature of around 600 °C. Traditional ITO also has unfavorable mechanical properties such as being relatively fragile. In addition, the combination of costly layer deposition in vacuum and a limited supply of indium results in high-quality ITO transparent electrodes being very expensive. Therefore, developing and commercializing a replacement for ITO is a major focus of OPV research and development.

   Conductive CNT coatings have recently become a prospective substitute based on wide range of methods including spraying, spin coating, casting, layer-by-layer, and Langmuir–Blodgett deposition. The transfer from a filter membrane to the transparent support using a solvent or in the form of an adhesive film is another method for attaining flexible and optically transparent CNT films. Moreover, films made of arc-discharge CNT can result in a high conductivity and transparency. Furthermore, the work function of SWCNT networks is in the 4.8–4.9 eV range (compared to ITO which has a lower work function of 4.7 eV) leading to the expectation that the SWCNT work function should be high enough to assure efficient hole collection. Another benefit is that SWCNT films exhibit a high optical transparency in a broad spectral range from the UV-visible to the near-infrared range. Only a few materials retain reasonable transparency in the infrared spectrum while maintaining transparency in the visible part of the spectrum as well as acceptable overall electrical conductivity. SWCNT films are highly flexible, do not creep, do not crack after bending, theoretically have high thermal conductivities to tolerate heat dissipation, and have high radiation resistance. However, the electrical sheet resistance of ITO is an order of magnitude less than the sheet resistance measured for SWCNT films. Nonetheless, possibilities exist to develop CNT-based transparent electrodes that exceed the performance of traditional ITO materials (Van de Lagemaat et al. 2006).
4. CNTs in dye-sensitized solar cells
   Improving dye-sensitized solar cells (DSSCs) efficiency has been the subject of a variety of research investigations because it has the potential to be manufactured economically enough to compete with other solar cell technologies. There is a promise to use various CNT-based nanocomposites and nanostructures to direct the flow of photogenerated electrons and assist in charge injection and extraction. To assist the electron transport to the collecting electrode surface in a DSSC, a popular concept is to utilize CNT networks as support to anchor light harvesting semiconductor particles. Including porphyrin and C_{60} fullerene, organization of photoactive donor polymer and acceptor fullerene on electrode surfaces has also been shown to offer considerable improvement in the photoconversion efficiency of solar cells. Therefore, there is an opportunity to facilitate electron transport and increase the photoconversion efficiency of DSSCs utilizing the electron-accepting ability of semiconducting SWCNTs (Hasobe et al. 2006).

DSSCs are also fabricated using the sol-gel method to obtain titanium-dioxide-coated MWCNTs for use as an electrode. Because pristine MWCNTs have a hydrophobic surface and poor dispersion stability, pretreatment was necessary for this application. A relatively low-destruction method for removing impurities, H_2O_2 treatment was used to generate carboxylic acid groups by oxidation of MWCNTs. Another positive aspect was the fact that the reaction gases including CO_2 and H_2O were nontoxic and could be released safely during the oxidation process. As a result of treatment, H_2O_2-exposed MWCNTs have a hydrophilic surface and the carboxylic acid groups on the surface have polar covalent bonding. Also, the negatively charged surface of the MWCNTs improved the stability of dispersion. By then entirely surrounding the MWCNTs with titanium dioxide nanoparticles using the sol-gel method, an increase in the conversion efficiency of about 50% compared to a conventional titanium dioxide cell was achieved. The enhanced interconnectivity between the titanium dioxide particles and the MWCNTs in the porous titanium dioxide film was concluded to be the cause of the improvement in short circuit current density. Here again, the addition of MWCNTs was thought to provide more efficient electron transfer through film in the DSSC (Lee et al. 2007).

4.2.4.2 Graphene-Based Solar Cells

Graphene has a unique combination of high electrical conductivity and optical transparency, which make it a candidate for use in solar cells. A single sheet of graphene is a zero-bandgap semiconductor whose charge carriers are delocalized over large areas, implying that carrier scattering does not occur. Because this material only absorbs 2.3% of visible light, it is a candidate for applications requiring a transparent conductor. Graphene can be assembled into a film electrode with low roughness. However, graphene films produced via solution processing contain lattice defects and grain boundaries that act as recombination centers and decrease the material's electrical conductivity. Thus, these films must be made thicker than one atomic layer to obtain useful sheet resistances. This added resistance can be combatted by incorporating conductive filler materials, such as a silica matrix. Reduced graphene film's electrical conductivity can be improved by attaching large aromatic molecules such as pyrene-1-sulfonic acid sodium salt (PyS) and the disodium salt of 3,4,9,10-perylenetetracarboxylic diimide bisbenzenesulfonic acid (PDI). These molecules, under high temperatures, facilitate better π-conjugation of the graphene basal plane. Graphene films have high transparency in the visible and near-infrared regions and are chemically and thermally stable (Mukhopadhyay 2013).

Graphene's high charge mobilities recommend it for use as a charge collector and transporter in photovoltaics (PV), as shown in Fig. 4.20. Using graphene as a photoactive material requires its bandgap to be 1.4–1.9 eV. Single cell efficiencies of nanostructured graphene-based PVs of over 12% have been achieved. Organic photovoltaics could be devices in which semiconducting graphene is used as the photoactive material and metallic graphene is used as the conductive electrodes (Mukhopadhyay 2013).

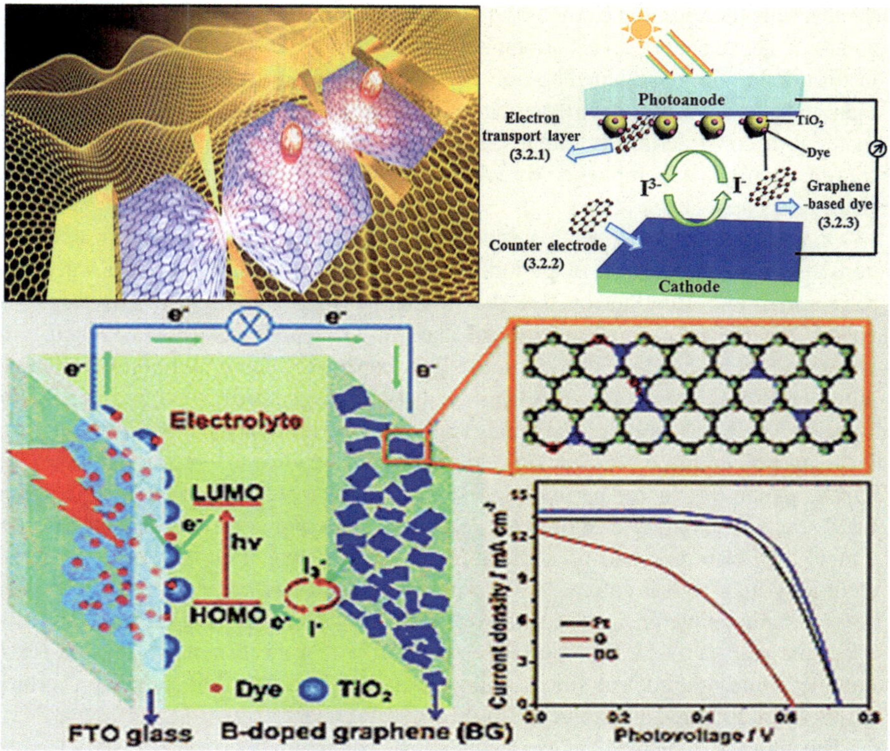

Fig. 4.20 Schematic illustration of graphene-based solar cells

Large-scale production of highly transparent graphene films has been achieved by chemical vapor deposition. In this process, ultra-thin graphene sheets are created by first depositing carbon atoms in the form of graphene films on a nickel plate from methane gas. A protective layer of thermoplastic is laid over the graphene layer and the nickel underneath is dissolved in an acid bath. The final step is to attach the plastic-protected graphene to a flexible polymer sheet, which can then be incorporated into an OPV cell. Graphene/polymer sheets range in size up to 150 square centimeters and can be used to create dense arrays of flexible OPV cells. It may eventually be possible to run printing presses covering extensive areas with inexpensive solar cells, much like newspaper presses print newspapers (roll-to-roll). In addition, silicon generates only one current-driving electron for each photon it absorbs, while graphene can produce multiple electrons. Solar cells made with graphene could offer 60% conversion efficiency—double the widely accepted maximum efficiency of silicon cells.

4.2.4.3 Quantum Dots

Quantum dots are particles of semiconductor material that have been reduced below the size of the Exciton Bohr-radius, and due to quantum mechanics considerations, the electron energies that can exist within them become finite, much alike energies in an atom. These energy levels are tunable by changing the size of quantum dots, and in turn define the bandgap. The dots can be grown over a range of sizes, allowing them to be tuned across a wide variety of bandgaps without changing the underlying material or construction techniques. In typical preparations employing wet chemistry, the tuning is accomplished by varying the duration or temperature of synthesis (Baskoutas and Terzis 2006).

Quantum dot (QD) metamaterials are a special semiconductor system that consists of a combination of periodic groups of materials molded in a variety of different forms. They are on nanometer scale and have an adjustable bandgap of energy levels performing as a special class of semiconductors. The PV cell with larger and wider bandgap absorbs more light hence producing more output voltage, while cells with the smaller bandgap results with larger current but smaller output voltage. The latter includes the bandgap in the red end of solar radiation spectrum. QDs are known to be efficient light emitters with various absorption and emission spectra depending on the particle size. In order to increase the conversion efficiency of PV cells, a 3D array design is needed for strong coupling between QDs in order extend the life of excitons for collecting and transporting "hot carriers" to generate electricity at a higher voltage. The principle of QDs has been implemented using several semiconductor materials and has resulted in the following: when GaAs was used, the cell had a high output advantage but was more expensive than Si semiconductive designs such as silicon–silicon dioxide ($Si–SiO_2$), silicon–silicon carbide (Si–SiC) or silicon–silicon nitrite ($Si–Si_3N_4$) (Akarslan 2012).

The ability to tune the bandgap is what makes quantum dots desirable for solar cell use, as shown in Fig. 4.21. In this respect, they can help replace existing expensive GaAs tandem cells, and in theory can achieve efficiencies on the same order. As for

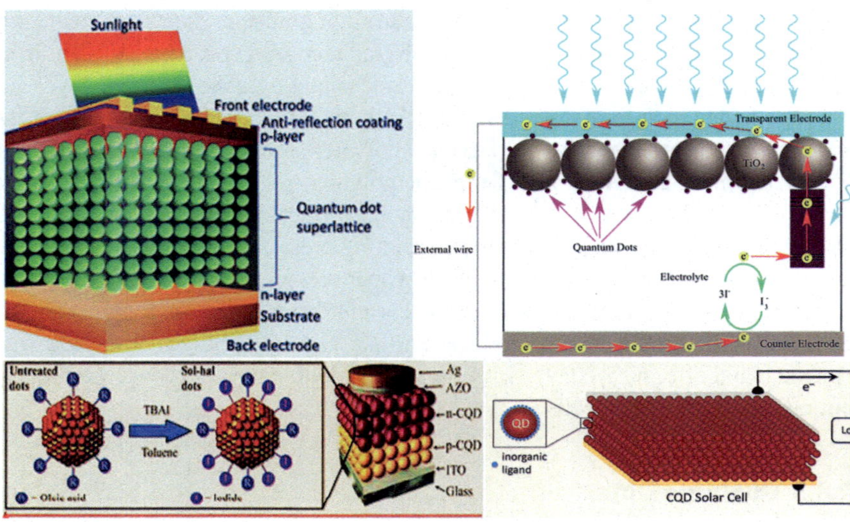

Fig. 4.21 Schematic illustration of quantum dot solar cells (Modified from Okada (2018) Illustration. Courtesy: Okada Laboratory. http://mbe.rcast.u-tokyo.ac.jp/index_eng.html)

single junction implementations, lead sulfide (PbS) colloidal quantum dots (CQDs) have bandgaps that can be tuned into the far infrared, energy levels that are typically difficult to achieve with traditional semiconductor materials. Half of all the solar energy reaching the Earth is in the infrared, most of it in the near infrared region. With a quantum dot solar cell, IR-sensitive materials are just as easy to use as any other, opening the possibility of capturing much more energy cost-effectively (Sargent 2005).

Moreover, CQDs benefit from facile synthesis and preparation. When suspended in a colloidal liquid form they can be easily handled throughout production, with the most complex equipment needed being a fume hood. The CQDs are typically synthesized in small batches, but can be scaled up readily to be mass-produced. The dots can be distributed on a substrate by spin coating, either by hand or in an easily automated process. In large-scale production this technique could be replaced by spray-on or roll-printing systems, which dramatically reduces module construction costs.

4.2.4.4 Hot Carrier Solar Cells

This technique utilizes selective energy contacts to extract light generated by "hot carriers" (HC) (electrons and holes) from semiconductor regions without transforming their extra energies to heat. In other words, "hot carriers" must be collected from the absorber over a very small energy range, with selective energy contacts (Fig. 4.22). This is the most novel approach for PV cell production and it

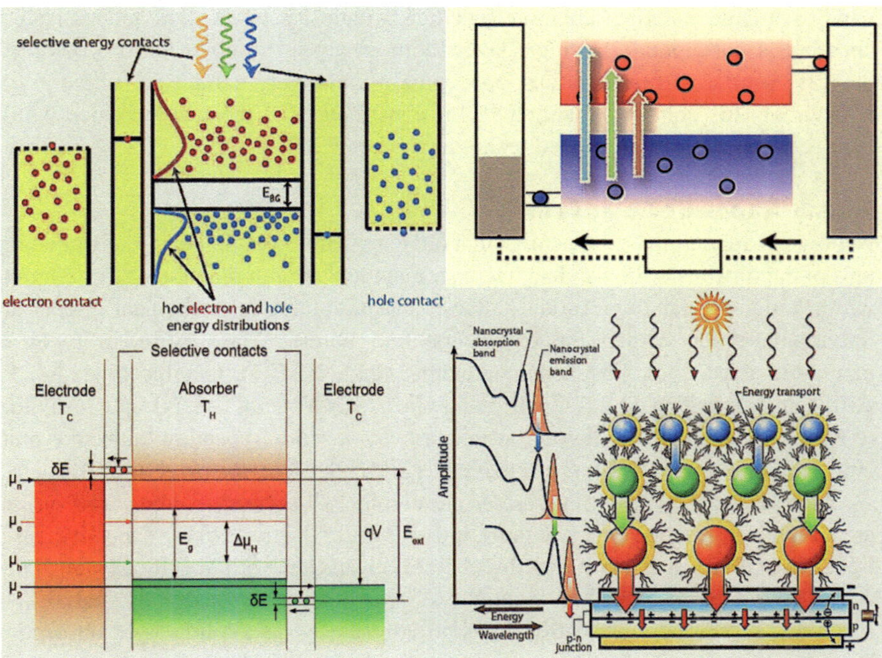

Fig. 4.22 Schematic illustration of hot carrier solar cells (Modified from Conibeer and Green (2008) Hot Carrier Solar Cell. Courtesy: Implementation of the Ultimate Photovoltaic Converter, http://gcep.stanford.edu/pdfs/factsheets/conibeer_etal_solar_08.pdf)

allows the use of one absorber material that yields to high efficiency under concentration.

Hot carrier solar cells offer the possibility of very high efficiencies (limiting efficiency 68% for unconcentrated illumination) but with a structure that could be conceptually simple compared to other very high efficiency PV devices—such as multi-junction monolithic tandem cells. For this reason, the approach lends itself to "thin film" deposition techniques, with their attendant low costs in materials and energy usage and facility to use abundant, nontoxic elements (Green 2003).

The Hot Carrier cell device has the following stringent requirements (Green 2003): (a) Slowing of thermalization of photogenerated electrons (and holes) in the absorber material. (b) Extraction of these "hot carriers" to external contacts over a narrow range of energies, such that excess carrier energy is not lost to the cold contacts. (c) In addition, a working device would require integration of the structures used to tackle (a) and (b) without compromising their performance.

The absorber has a hot carrier distribution at temp T_H. Carriers cool isoentropically in the mono-energetic contacts to T_A: their kinetic energy being converted into useable potential energy. The difference of the Fermi levels of these two contacts manifests as a difference in chemical potential of the carriers at each contact and hence an external voltage. The challenges to produce such devices fall into two categories (Green 2003): (1) keeping carriers hot without heating the lattice and (2) achieving a fast extraction of hot carriers through a narrow allowed energy range. While these challenges are tough, it seems they can be met using newly available materials and more specifically nanostructured semiconductors. The absorber is conceptually a single layer, homogeneous on the scale of a complete device. Similarly, contacts are conceptually relatively simple effectively 2D structures. This conceptual simplicity of construction of the whole device would lend itself to thin-film or related techniques (Green 2003).

4.2.4.5 Nanowire Solar Cells

Nanowires are defined as structures that have a thickness or diameter constrained to tens of nanometers or less and an unconstrained length with the ratio of the length to width being greater than 1000. At these scales, quantum mechanical effects are important, which coined the term "quantum wires." Many different types of nanowires exist, including superconducting (e.g., YBCO), metallic (e.g., Ni, Pt, Au), semiconducting (e.g., silicon nanowires (SiNWs), InP, GaN) and insulating (e.g., SiO_2, TiO_2). Molecular nanowires are composed of repeating molecular units either organic (e.g., DNA) or inorganic (e.g., $Mo_6S_{9-x}I_x$) (Boston et al. 2014).

Nanowires are of great interest in photovoltaics because of their large surface area, high aspect ratio (long and thin), intrinsic antireflection effect (which increases light absorption), and ability to direct light absorption with specifically designed arrays. Moreover, core-shell nanowires, which operate based on radial p–n junctions, represent a revolution in photovoltaics by decoupling light absorption from carrier collection pathways: light is absorbed vertically, whereas carriers are separated radially. This separation eliminates the once-fundamental trade-off between light absorption and carrier collection. By incorporating the superior photovoltaic properties of III–V semiconductors into nanowire structures, nanowire solar cells show great promise for next-generation photovoltaics. However, addressing the many new material and device challenges that arise from these unconventional nanostructures will require more mature simulation, manufacturing, and characterization tools. Because nanowire-based cells use less material than planar devices, such changes can ultimately reduce costs (Wallentin et al. 2013).

4.2.5 Hybrid Solar Cells

Hybrid solar cells combine advantages of both organic and inorganic semiconductors. Hybrid photovoltaics have organic materials that consist of conjugated polymers that absorb light as the donor and transport holes. Inorganic materials in hybrid cells are used as the acceptor and electron transporter in the structure, as shown in Fig. 4.23. The hybrid photovoltaic devices have a potential for not only low-cost by roll-to-roll processing but also for scalable solar power conversion (Milliron et al. 2005).

In hybrid solar cells, an organic material is mixed with a high electron transport material to form the photoactive layer. The two materials are assembled together in a heterojunction-type photoactive layer, which can have greater power conversion

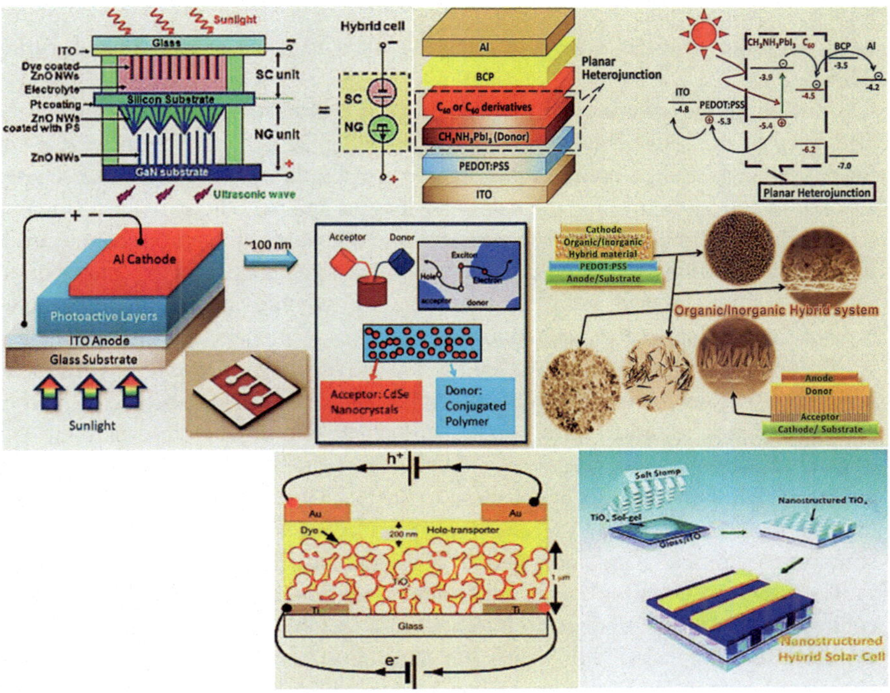

Fig. 4.23 Schematic illustration of hybrid solar cells

efficiency than a single material. One of the materials acts as the photon absorber and exciton donor. The other material facilitates exciton dissociation at the junction. Charge is transferred and then separated after an exciton created in the donor is delocalized on a donor-acceptor complex. The acceptor material needs a suitable energy offset to the binding energy of the exciton to the absorber. Charge transfer is favorable if the following condition is satisfied (Ginger and Greenham 1999):

$$E_A^A - E_A^D > U_D \tag{4.1}$$

where superscripts A and D refer to the acceptor and donor, respectively, E_A is the electron affinity, and U the coulombic binding energy of the exciton on the donor. In commonly used photovoltaic polymers such as MEH-PPV, the exciton binding energy ranges from 0.3 eV to 1.4 eV (Scheblykin et al. 2007).

The energy required to separate the exciton is provided by the energy offset between the LUMOs or conduction bands of the donor and acceptor. After dissociation, the carriers are transported to the respective electrodes through a percolation network. The average distance an exciton can diffuse through a material before annihilation by recombination is the exciton diffusion length. This is short in polymers, on the order of 5^{-10} nm. The timescale for radiative and non-radiative decay is from 1 ps to 1 ns. Excitons generated within this length close to an acceptor would contribute to the photocurrent (Shaw et al. 2008). To deal with the problem of the short exciton diffusion length, a bulk heterojunction structure is used rather than a phase-separated bilayer. Dispersing the particles throughout the polymer matrix creates a larger interfacial area for charge transfer to occur (Saunders and Turner 2008).

Controlling the interface of inorganic–organic hybrid solar cells can increase the efficiency of the cells. This increased efficiency can be achieved by increasing the interfacial surface area between the organic and the inorganic to facilitate charge separation and by controlling the nanoscale lengths and periodicity of each structure so that charges are allowed to separate and move towards the appropriate electrode without recombining. The three main nanoscale structures used are mesoporous inorganic films infused with electron-donating organic, alternating inorganic–organic lamellar structures, and nanowire structures (Saunders and Turner 2008).

Mesoporous films have been used for a relatively high-efficiency hybrid solar cell. The structure of mesoporous thin-film solar cells usually includes a porous inorganic that is saturated with organic surfactant. The organic absorbs light, and transfers electrons to the inorganic semiconductor (usually a transparent conducting oxide), which then transfers the electron to the electrode. Problems with these cells include their random ordering and the difficulty of controlling their nanoscale structure to promote charge conduction (Herman et al. 2011).

The use of alternating layers of organic and inorganic compounds has been controlled through electrodeposition-based self-assembly. This is of particular interest because it has been shown that the lamellar structure and periodicity of the alternating organic-inorganic layers can be controlled through solution chemistry. To produce this type of cell with practical efficiencies, larger organic surfactants that absorb more of the visible spectrum must be deposited between the layers of electron-accepting inorganic (Herman et al. 2011).

In addition, nanostructure-based solar cells have been developed by using ordered nanostructures like nanowires or nanotubes of inorganic surrounding by electron-donating organics utilizing self-organization processes. Ordered nanostructures offer the advantage of directed charge transport and controlled phase separation between donor and acceptor materials. The nanowire-based morphology offers reduced internal reflection, facile strain relaxation, and increased defect tolerance. The ability to make single-crystalline nanowires on low-cost substrates such as aluminum foil and to relax strain in subsequent layers removes two more major cost hurdles associated with high-efficiency cells. There have been rapid increases in efficiencies of nanowire-based solar cells and they seem to be one of the most promising nanoscale solar hybrid technologies (Weickert 2011; Garnett 2011).

Efficiency of hybrid cell efficiency mainly depends on three factors (Shaheen et al. 2005): First, the bandgap should be reduced to absorb red photons, which contain a significant fraction of the energy in the solar spectrum. Current organic photovoltaics have shown 70% of quantum efficiency for blue photons. Second, contact resistance between each layer in the device should be minimized to offer higher fill factor and power conversion efficiency. Third, charge-carrier mobility should be increased to allow the photovoltaics to have thicker active layers while minimizing carrier recombination and keeping the series resistance of the device low.

4.2.5.1 Polymer–Nanoparticle Composite Hybrid Solar Cells

Nanoparticles here indicate a class of semiconductor materials whose size in at least one dimension ranges from 1 to 100 nm, on the order of exciton wavelengths. This size control creates quantum confinement and allows for the tuning of optoelectronic properties, such as band gap and electron affinity. Nanoparticles also have a large surface area to volume ratio, which presents more area for charge transfer to occur (Wu et al. 2005).

The photoactive layer can be created by mixing nanoparticles into a polymer matrix. Solar devices based on polymer-nanoparticle composites most resemble polymer solar cells. In this case, the nanoparticles take the place of the fullerene-based acceptors used in fully organic polymer solar cells. Hybrid solar cells based upon nanoparticles are an area of research interest because nanoparticles have several properties that could make them preferable to fullerenes, such as (Wu et al. 2005; Saunders 2012): (a) Fullerenes are synthesized by a combination of a high-temperature arc method and continuous gas-phase synthesis, which makes their production difficult and energy intensive. The colloidal synthesis of nanoparticles by contrast is a low temperature process. (b) PCBM (a common fullerene acceptor) diffuses during long timespans or when exposed to heat, which can alter the morphology and lower the efficiency of a polymer solar cell. Limited testing of nanoparticle solar cells indicates they may be more stable over time. (c) Nanoparticles are more absorbent than fullerenes, meaning more light can be theoretically absorbed in a thinner device. (d) Nanoparticle size can affect absorption. This combined with the fact that there are many possible semiconducting nanoparticles allows for highly customizable bandgaps that can be easily tuned to certain frequencies, which would be advantageous in tandem solar cells. (e) Nanoparticles with size near their Bohr radius can generate two excitons when struck by a sufficiently energetic photon.

For polymers used in this device, hole mobilities are greater than electron mobilities, so the polymer phase is used to transport holes. The nanoparticles transport electrons to the electrode. The interfacial area between the polymer phase and the nanoparticles needs to be large. This is achieved by dispersing the particles throughout the polymer matrix. However, the nanoparticles need to be interconnected to form percolation networks for electron transport, which occurs by hopping events. Efficiency is affected by aspect ratio, geometry, and volume fraction of the nanoparticles. Nanoparticle structures include nanocrystals, nanorods, and hyperbranched structures. Different structures change the conversion efficiency by effecting nanoparticle dispersion in the polymer and providing pathways for electron transport. The nanoparticle phase is required to provide a pathway for the electrons to reach the electrode. By using nanorods instead of nanocrystals, the hopping event from one crystal to another can be avoided. Fabrication methods include mixing the two materials in a solution and spin-coating it onto a substrate, and solvent evaporation (sol-gel). Most of these methods do not involve high-temperature processing. Annealing increases order in the polymer phase, increasing conductivity. However, annealing for too long causes the polymer domain size to increase, eventually making it larger than the exciton diffusion length, and possibly allowing some of the metal from the contact to diffuse into the photoactive layer, reducing the efficiency of the device (Wu et al. 2005).

Inorganic semiconductor nanoparticles used in hybrid cells include CdSe (size ranges from 6 to 20 nm), ZnO, TiO, and PbS. Common polymers used as photo materials have extensive conjugation and are also hydrophobic. Their efficiency as a photo-material is affected by the HOMO level position and the ionization potential, which directly affects the open circuit voltage and the stability in air. The most common polymers used are P3HT (poly (3-hexylthiophene)), and M3H-PPV (poly [2-methoxy, 5-(2′-ethyl-hexyloxy)-p-phenylenevinylene)]). P3HT has a bandgap of 2.1 eV and M3H-PPV has a bandgap of ~2.4 eV. These values correspond with the bandgap of CdSe, 2.10 eV. The electron affinity of CdSe ranges from 4.4 to 4.7 eV. When the polymer used is MEH-PPV, which has an electron affinity of 3.0 eV, the difference between the electron affinities is large enough to drive electron transfer from the CdSe to the polymer. CdSe also has a high electron mobility (600 cm^2 V^{-1} s^{-1}) (Saunders and Turner 2008). The highest demonstrated efficiency is 3.2%, based upon a PCPDTBT polymer donor and CdSe nanoparticle acceptor. The device exhibited a short circuit current of 10.1 mA cm^{-2}, an open circuit voltage of 0.68 V, and a fill factor of 0.51 (Dayal et al. 2010).

Hybrid solar cells need increased efficiencies and stability over time before commercialization is feasible. In comparison to the 2.4% of the CdSe-PPV system, silicon photodevices have power conversion efficiencies greater than 20%. Problems include controlling the amount of nanoparticle aggregation as the photolayer forms.

The particles need to be dispersed in order to maximize interface area, but need to aggregate to form networks for electron transport. The network formation is sensitive to the fabrication conditions. Dead end pathways can impede flow. A possible solution is implementing ordered heterojunctions, where the structure is well controlled. The structures can undergo morphological changes over time, namely phase separation. Eventually, the polymer domain size will be greater than the carrier diffusion length, which lowers performance. Even though the nanoparticle bandgap can be tuned, it needs to be matched with the corresponding polymer. The 2.0 eV bandgap of CdSe is larger than an ideal bandgap of 1.4 for absorbance of light. The nanoparticles involved are typically colloids, which are stabilized in solution by ligands. The ligands decrease device efficiency because they serve as insulators which impede interaction between the donor and nanoparticle acceptor as well as decreasing the electron mobility. Some, but not complete success has been had by exchanging the initial ligands for pyridine or another short chain ligand. Hybrid solar cells exhibit material properties inferior to those of bulk silicon semiconductors. The carrier mobilities are much smaller than that of silicon. Electron mobility in silicon is 1000 $cm^2\ V^{-1}\ s^{-1}$, compared to 600 $cm^2\ V^{-1}\ s^{-1}$ in CdSe, and less than 10 $cm^2\ V^{-1}$ s^{-1} in other quantum dot materials. Hole mobility in MEH-PPV is 0.1 $cm^2\ V^{-1}\ s^{-1}$, while in silicon it is 450 $cm^2\ V^{-1}\ s^{-1}$ (Wu et al. 2005).

4.2.5.2 Hybrid Solar Cells with Carbon Nanotubes

Carbon nanotubes (CNTs) have high electron conductivity, high thermal conductivity, robustness, and flexibility. Field emission displays (FED), strain sensors, and field effect transistors (FET). Each application shows the potential of CNTs for nanoscale devices and for flexible electronics applications. Photovoltaic applications have also been explored for this material (Charlier et al. 2002).

Mainly, CNTs have been used as either the photo-induced exciton carrier transport medium impurity within a polymer-based photovoltaic layer or as the photoactive (photon-electron conversion) layer. Metallic CNT is preferred for the former application, while semiconducting CNT is preferred for the later.

To increase the photovoltaic efficiency, electron-accepting impurities must be added to the photoactive region. By incorporating CNTs into the polymer, dissociation of the exciton pair can be accomplished by the CNT matrix. The high surface area (~1600 m^2/g) of CNTs offers a good opportunity for exciton dissociation. The separated carriers within the polymer-CNT matrix are transported by the percolation pathways of adjacent CNTs, providing the means for high carrier mobility and efficient charge transfer. The factors of performance of CNT-polymer hybrid photovoltaics are low compared to those of inorganic photovoltaics. SWNT in P_3OT semiconductor polymer demonstrated open circuit voltage (Voc) of below 0.94 V, with short circuit current (Isc) of 0.12 mA/cm^2 (Cinke et al. 2002).

Metal nanoparticles may be applied to the exterior of CNTs to increase the exciton separation efficiency. The metal provides a higher electric field at the CNT-polymer interface, accelerating the exciton carriers to transfer them more effectively to the CNT matrix. In this case, Voc = 0.3396 V and Isc = 5.88 mA/cm^2. The fill factor is 0.3876%, and the white light conversion factor 0.775% (Somani et al. 2008).

CNT may be used as a photovoltaic device not only as an add-in material to increase carrier transport, but also as the photoactive layer itself. The semiconducting single-walled CNT (SWCNT) is a potentially attractive material for photovoltaic applications for the unique structural and electrical properties. SWCNT has high electric conductivity (100 times that of copper) and shows ballistic carrier transport, greatly decreasing carrier recombination. The bandgap of the SWCNT is inversely proportional to the tube diameter, which means that SWCNT may show multiple direct bandgaps matching the solar spectrum. A strong built-in electric field in SWCNT for efficient photogenerated electron–hole pair separation has been demonstrated by using two asymmetrical metal electrodes with high and low work functions. The open circuit voltage (V_{oc}) is 0.28 V, with short circuit current (I_{sc}) 1.12 nA cm^{-2} with an incident light source of 8.8 W cm^{-2}. The resulting white light conversion factor is 0.8% (Chen et al. 2008).

There are several challenges for CNT to be used in photovoltaic applications. CNT degrades overtime in an oxygen-rich environment. The passivation layer required to prevent CNT oxidation may reduce the optical transparency of the electrode region and lower the photovoltaic efficiency. Additional challenges involve the dispersion of CNT within the polymer photoactive layer. The CNT is required to be well dispersed within the polymer matrix to form charge-transfer-efficient pathways between the excitons and the electrode. Challenges of CNT for the photoactive layer include its lack of capability to form a p–n junction, due to the difficulty of doping certain segments of a CNT (A p–n junction creates an internal built-in potential, providing a pathway for efficient carrier separation within the photovoltaic.) To overcome this difficulty, energy band bending has been done by the use of two electrodes of different work functions. A strong built-in electric field covering the whole SWCNT channel is formed for high-efficiency carrier separation. The oxidation issue with CNT is more critical for this application. Oxidized CNTs have a tendency to become more metallic, and so less useful as a photovoltaic material (Collins et al. 2000).

4.2.5.3 Dye-Sensitized Hybrid Solar Cells

Dye-sensitized solar cell consists of a photo-sensitized anode, an electrolyte, and a photo-electrochemical system. Hybrid solar cells based on dye-sensitized solar cells are formed with inorganic materials (TiO_2) and organic materials.

Hybrid solar cells based on dye-sensitized solar cells are fabricated by dye-absorbed inorganic materials and organic materials. TiO_2 is the preferred inorganic material since this material is easy to synthesize and acts as n-type semiconductor due to the donor-like oxygen vacancies. However, titania only absorbs a small fraction of the UV spectrum. Molecular sensitizers (dye molecules) attached to the semiconductor surface are used to collect a greater portion of the spectrum. In the case of titania dye-sensitized solar cells, a photon absorbed by a dye-sensitizer molecule layer induces electron injection into the conduction band of titania, resulting in current flow. However, short diffusion length (diffusivity, Dn $\leq 10^{-4}$ cm^2/s) in titania dye-sensitized solar cells decrease the solar-to-energy conversion efficiency.

To enhance diffusion length (or carrier lifetime), a variety of organic materials are attached to the titania (Lancelle-Beltran et al. 2006).

TiO_2 nanoparticles are synthesized in several tens of nanometer scales (~100 nm). In order to make a photovoltaic cell, molecular sensitizers (dye molecules) are attached to the titania surface. The dye-absorbed titania is finally enclosed by a liquid electrolyte. This type of dye-sensitized solar cell is also known as a Grätzel cell. Dye-sensitized solar cell has a disadvantage of a short diffusion length. Furthermore, supermolecular or multifunctional sensitizers have been investigated so as to enhance carrier diffusion length. For example, a dye chromophore has been modified by the addition of secondary electron donors. Minority carriers (holes in this case) diffuse to the attached electron donors to recombine. Therefore, electron-hole recombination is retarded by the physical separation between the dye-cation moiety and the TiO_2 surface. Finally, this process raises the carrier diffusion length, resulting in the increase of carrier lifetime (Moser 2005).

Mesoporous materials contain pores with diameters between 2 and 50 nm. A dye-sensitized mesoporous film of TiO_2 can be used for making photovoltaic cells and this solar cell is called a "solid-state dye sensitized solar cell." The pores in mesoporous TiO_2 thin film are filled with a solid hole-conducting material such as p-type semiconductors or organic hole conducting material. Replacing the liquid electrolyte in Grätzel's cells with a solid charge-transport material can be beneficial. The process of electron-hole generation and recombination is the same as Grätzel cells. Electrons are injected from photoexcited dye into the conduction band of titania and holes are transported by a solid charge transport electrolyte to an electrode. Many organic materials have been tested to obtain a high solar-to-energy conversion efficiency in dye synthesized solar cells based on mesoporous titania thin film (Lancelle-Beltran et al. 2006).

Liquid organic electrolytes contain highly corrosive iodine, leading to problems of leakage, sealing, handling, dye desorption, and maintenance. Much attention is now focused on the electrolyte to address these problems. For solid-state dye sensitized solar cells, the first challenge originates from disordered titania mesoporous structures. Mesoporous titania structures should be fabricated with well-ordered titania structures of uniform size (~10 nm). The second challenge comes from developing the solid electrolyte, which is required to have these properties (Lancelle-Beltran et al. 2006): (a) The electrolyte should be transparent to the visible spectrum (wide band gap). (b) Fabrication should be possible for depositing the solid electrolyte without degrading the dye molecule layer on titania. (c) The LUMO of the dye molecule should be higher than the conduction band of titania. (d) Several p-type semiconductors tend to crystallize inside the mesoporous titania films, destroying the dye molecule-titania contact. Therefore, the solid electrolyte needs to be stable during operation.

4.2.5.4 Nanostructured Inorganic–Organic Molecules

Nanostructured inorganic–organic molecules have been made to provide an ideal design for bulk heterojunction solar cells. The structure is composed of ZnO and small, conducting organic molecules, which co-assemble into alternating layers of organic and inorganic components. This highly organized structure, which is stabilized by π–π stacking between the organic molecules, allows for conducting pathways in both the organic and inorganic layers. The thicknesses of the layers (about 1–3 nm) are well within the exciton diffusion length, which ideally minimizes recombination among charge carriers. This structure also maximizes the interface between the inorganic ZnO and the organic molecules, which enables a high chromophore loading density within the structure. Due to the choice of materials, this system is nontoxic and environmentally friendly, unlike many other systems which use lead or cadmium (Sofos 2009).

Although this system has not yet been incorporated into a photovoltaic device, preliminary photoconductivity measurements have shown that this system exhibits among the highest values measured for organic, hybrid, and amorphous silicon photoconductors, and so, offers promise in creating efficient hybrid photovoltaic devices.

INDEPTH: Printed Flexible Solar Cells and 3D Solar Collectors (Ruiz-Morales et al. 2017)

Printed electronics is currently evolving towards more complex applications including multilayer flexible solar cells (organic/inorganic) and complex 3D solar collectors. This low-cost and large-area 3D printing process will allow a breakthrough in PV manufacturing, probably enabling the expected revolution of three-dimensional photovoltaic structures that are highly efficient even in the absence of sun tracking. Before this will take place, major advances in 3D printing of flexible (contact) materials and high-quality optics are still required together with the need of developing new printable inks of active materials for growing multilayer devices. Moreover, the development of multi-material 3D printing systems able to integrate focusing optics, coatings, and embedded luminescent–polymer composites for the fabrication of new generations of advanced solar cells based on up-conversion photovoltaics or even Luminescent Solar Concentrators (LSCs) will also be of particular interest. LSCs mainly consist of highly transparent plastic plates, comprising high quantum efficiency luminescent species, for absorbing incident light and re-emitting at a red-shifted wavelength directly allowing concentrated and wave-guided converted radiation. These LSCs could be a cheap alternative to silicon substrates for building integration applications. Finally, it is also interesting to develop environmentally sensitive materials able to actively transform configurations over time in response to external stimuli since the use of shape-memory polymers in so-called 4D printing could be especially interesting for developing self-configurable, active and adaptable harvesting devices such as solar cells.

4.3 Advanced Materials for Solar Thermal Collectors

Solar collectors and thermal energy storage components are the two core subsystems in solar thermal applications. A solar collector which is the special energy exchanger converts solar irradiation energy either to the thermal energy of the working fluid in solar thermal applications. Solar collectors need to have good optical performance in order to absorb heat as much as possible. For solar thermal applications, solar irradiation is absorbed by a solar collector as heat and then is transferred to the working fluid. The heat carried by the working fluid can be used to either provide domestic hot water or to charge a thermal energy storage tank from which the heat can be drawn for use later (Kim and Han 2015).

The most common use of solar collectors is water heating for domestic needs. Solar energy can be trapped more efficiently dependent upon the type of solar collectors used. Each type of solar collector is designed to absorb the shorter wavelengths of light which are received from the sun (0.3–2 mm in length) but prevent heat wavelengths (2–10 mm in length) from escaping by utilizing the greenhouse effect then delivers radiant energy either directly or indirectly to a hot water storage tank. For example, the performance of the flat plate collector depends upon various design parameters, such as the number of covers, type and thickness of glazing, antireflecting coating on cover glass, heat mirror coating on the inner glass, the type of coating on the collector plate, spacing between the collector and the inner glass, an evacuated space between the collector and the inner glass, the arresting of convective movement between the collector plate and the inner glass by using transparent insulation material (TIM), and the type of insulation used, etc. all of which are responsible for the performance of a flat plate collector. There are several other operational parameters, such as the mass flow rate of fluid, solar radiation, inlet temperature, ambient temperature, wind speed, sky conditions, and dust deposition on glass cover, as shown in Fig. 4.24, which also affect the collector performance. The heat loss is indicated by the thermal loss factor or k-value. This is given in watt per square meter of collector surface and the particular temperature difference between the absorber and its surroundings. The higher the temperature difference, then the more heat is lost. Above a specific temperature difference, the amount of heat loss equals the energy yield of the collector, so that no energy at all is delivered to the solar circulation system. A good collector will have a high conversion factor and a low k-value (Alghoul 2005).

The fundamental challenge is to develop a new generation of process heat collectors, raising the temperature, efficiency and performance of current low-temperature collectors. Therefore, the related materials development has focused on (ESTTP 2007): (a) Functional surfaces (such as anti-reflective or low thermal emissivity coatings, self-cleaning surfaces and anti-corrosive coatings); (b) Highly reflective, precise and weather resistant lightweight reflectors; (c) Improved selective absorbers (for example, long-term stability in aggressive climates, such as salt-water droplets from the sea); (c) Cheaper glass with high solar transmittance and exchangeable optical properties; (d) Heat transfer fluids with higher temperature stability (above 160 °C); (e) Temperature-stable and inexpensive

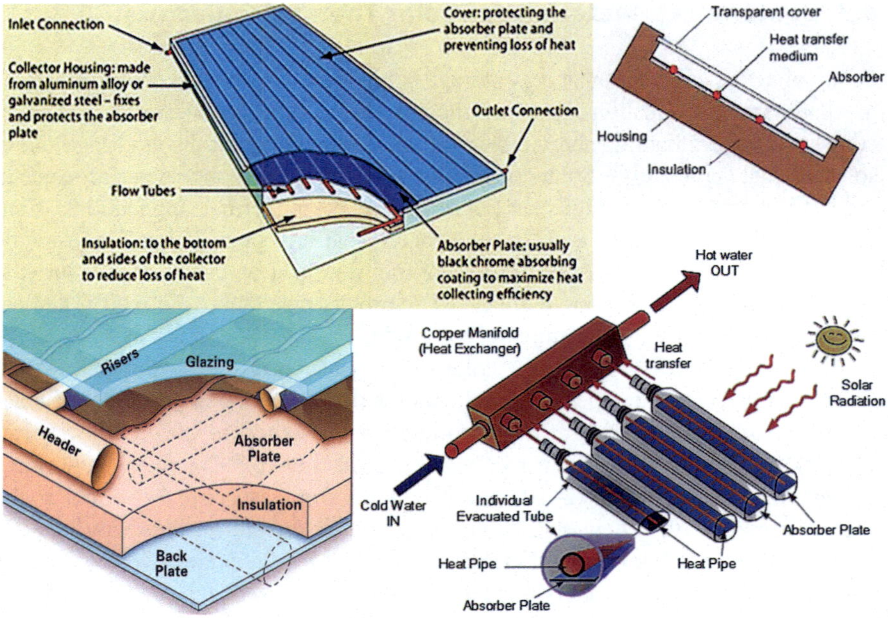

Fig. 4.24 Schematic illustration of the flat solar collector

thermal insulation (for example, vacuum insulation); (f) New design concentrators, favoring roof mounting and low maintenance needs; (g) New heat transfer media, withstanding freezing and high operating temperatures; (h) New concepts for system integration and overheating protection.

Depending on the price evolution of current raw materials, such as copper, aluminum and oil-based plastics, substitution materials may need to be developed. Moreover, new concepts like PV-thermal collectors will require specific basic developments, such as, cheap and reliable collectors based on new materials, which are easy to integrate into various roofs and facades; practical ways to integrate conventional metal collectors onto metal facades and roofs; collectors that can be connected with a heat pump (i.e., condensation inside the collector must not be a problem); improve heat transfer in the collector and heat exchanger; cheap, energy-efficient pumps; connectors that facilitate the quick and easy installation of collectors; evacuated tube collectors with a better price/performance ratio (soda lime glass and anti-reflective coatings); low concentration stationary or quasi-stationary collectors for higher temperature applications in buildings, which can be combined with evacuated tubular collectors; PV-thermal hybrid collectors that save installation costs and increase total conversion efficiency, compared with side-by-side systems; air collectors with improved heat transfer; and cheap sensors and electronics for fault detection and identification.

4.3.1 Desirable Features of Solar Thermal Collectors

4.3.1.1 Transparent Cover

Transparent cover acts as a heat trap for infrared (thermal) radiation. Therefore, it reduces radiation losses and convection to the atmosphere. Together with the frame, the cover protects the absorber from adverse weather conditions (Alghoul 2005).

4.3.1.2 Insulation

Solar collectors are insulated with selective grade of CFC free polyurethane foam (PUF) as insulation material. The use of a PUF-insulated collector and tank ensures superior performance with minimum heat loss. Among expanded polystyrene, extruded polystyrene, polyurethane foam, cellular foam, and mineral wool, rigid PU foam has extremely effective thermal insulation qualities and high strength to weight ratio at low temperatures. These inherent physical characteristics enable it to perform extremely well compared to alternative products. Insulation must be kept dry or it loses all or most of its insulating value. When the collector is assembled, the air trapped inside will contain moisture, which eventually will condense and soak into the insulation. To prevent this, quality collectors contain (Alghoul 2005):

(a) Desiccants. Porous bags of silica gel desiccant to absorb the moisture. If the collector is properly sealed, it is not necessary to have access to the desiccant, as it does not require renewal. Desiccant is also required for the space between the glazings when two covers are used. Typically, the desiccant is contained in the hollow spacers separating the two glazing panes, and small holes on the surface of the spacers facing the space between the panes permit the trapped air to contact the desiccant. If desiccant is not used in either single-glazed or double-glazed collectors, it will become apparent through condensation of drops of water on the inner surface of the glass.
(b) Enclosure. The enclosure is used to contain insulation, provide support for the absorber and glazing, and to protect the collector from heat loss due to wind, plus the important function of keeping moisture out of the insulation from rain and dew. Enclosures are made of an almost endless variety of materials and designs, including wood cases, aluminum extrusions with sheet aluminum back, galvanized steel (GS), welded or formed, and even collectors without back covers.

Whatever the case material and construction, it must be weather resistant, fire-proof, durable, dimensionally stable, strong and completely and permanently sealed against moisture intrusion. As a general rule, the number of joints and seams should be minimized and completely sealed. Steel should be both galvanized and primed before painting and baking and paint should be tough and scratch-resistant. Aluminum should be used with caution in areas exposed to salt air or industrial pollution and smog in the air. Most top quality collectors use enclosures of architectural anodized aluminum similar to those used for exterior windows. Typical frame materials include aluminum and GS, though sometimes fiberglass-reinforced plastic can be used.

4.3.1.3 Evacuated Heat Pipe Tubes

Evacuated heat pipe tubes (EHPTs) are composed of multiple evacuated glass tubes each containing an absorber plate fused to a heat pipe. The heat is transferred to the transfer fluid (water or an antifreeze mix-typically propylene glycol) of a domestic hot water or hydronic space heating system in a heat exchanger called a manifold. The manifold is wrapped in insulation and covered by a protective sheet metal or plastic case. The vacuum that surrounds the outside of the tube greatly reduces convection and conduction heat loss, therefore achieving greater efficiency than flat-plate collectors, especially in colder conditions. This advantage is largely lost in warmer climates, except in those cases where very hot water is desirable, e.g., for commercial processes. The high temperatures that can occur may require special design to prevent overheating.

In an evacuated heat pipe tube collector collector, sunlight enters through the outer glass tube and strikes the absorber, where the energy is converted to heat. The heat is transferred to the liquid flowing through the absorber. The collector consists of rows of parallel transparent glass tubes, each of which contains an absorber covered with a selective coating. The absorber typically is of tin-tube design, although cylindrical absorbers also are used. Evacuated head pipe tube collectors are generally more efficient on an all year round basis as they can still operate under cloudy conditions; however, they are considerably more expensive than flat plate collectors—around 80%—and if the vacuum seal fails then they become inefficient. Glass evacuated tubes are the key component of solar collectors. There are several types of evacuated tube in use in the solar industry. The most widely used is twin-glass tube; this type of tube is chosen for its reliability, performance, and low manufacturing cost. A twin-glass evacuated tube consists of (Alghoul 2005): (a) the outer tube is made of extremely strong transparent borosilicate glass that is able to resist impact from hailstones of up to 25 mm diameter; and (b) the inner tube is also made of borosilicate glass but coated with a special selective coating (Al-N/Al), which features excellent solar heat absorption and minimal heat reflection properties.

4.3.2 Polymer Materials in Solar Thermal Collectors

The cost of energy produced by solar thermal collectors depends on costs of various materials which make up the system, maintenance costs, and obviously the amount of solar energy collected. Serious efforts have been devoted towards making solar collector technology economically more competitive. Polymers are widely available low-cost materials, which lend themselves to a volume production of lightweight low-cost collectors tolerant to corrosion and freezing temperatures. The use of polymer materials reduces the collector weight by 50 percent in comparison with a traditional metal collector; which allows much easier installation.

Polymeric glazing offers significant potential for cost savings, both as (a) direct substitutes for glass cover plates in traditional collector systems; and (b) integral part of all polymeric systems.

Polymer glazing is subject to degradation under the combined effect of elevated temperatures and exposure to UV radiation unless UV absorption additives in the outer cover or special inhibitors are used. Although these additives will certainly appreciably extend the life cycle of collectors, they will also contribute to a proportional increase in cost (Alghoul 2005). For example, thin-walled cellular polycarbonate materials in the form of transparent capillary structures or square honeycombs are being employed extensively as collector glazing materials. Advanced microencapsulated liquid crystal epoxy and polycarbonate polymer films have been developed as potential transmission switching glazing materials, suitable for the design of building construction elements. Among the most suitable polymer materials for absorber design are those of the polyolefin group, like polyethylene and polypropylene and of the ethylene-propylene-diene-monomer (EPDM) group, known as synthetic rubbers. Among them, polyolefins are mainly suitable for the manufacturing of thermally extruded flat rigid absorbers, while EPDM materials are suitable for the production of flexible tubes or tube strips interconnected by flexible webs, mainly suitable as low temperature swimming pool heating collectors. In addition, the thermal conductivity of polymers is substantially (almost three orders of magnitude) lower than that of ordinary metallic absorber materials, something which is very crucial for their use in solar energy applications and makes the redesign of conventional tube and fin metal absorber absolutely necessary.

The primary challenge to design a polymer heat exchanger is selecting a polymer that is (Alghoul 2005): (a) compatible with potable water; and (b) capable of withstanding the high pressure and temperature requirements of domestic hot water systems for up to 10 years without sacrificing thermal performance. Polymer heat exchangers offer the potential advantages of (a) reduced cost of materials and manufacture; (b) resistance to corrosion and mineral build-up, if mineral build-up is reduced, maintenance costs should be lower than those for metal heat exchangers; (c) lower friction coefficients; (d) reduced weight and easy installation; and (e) use of polymers may permit better integration with other components. Polymers certified for tube components and used in heat exchangers and exhibit good high-temperature characteristics are: high-temperature nylon (HTN), polypropylene (PP), and cross-linked polypropylene (PEX).

To reduce the cost of materials to the minimum and improve efficiency, the recommended materials to build solar collector are summarized as (Alghoul 2005): (a) low-iron glass as transparent cover—which is relatively inexpensive; (b) vacuum insulation—a no-cost material; (c) elective coating material—which is relatively cheap, as compared to high output of energy of solar collector; and (d) super conducting working fluid—water (very low cost). Finally, much effort must be invested on improved manufacturing techniques in order to reduce the challenge of high manufacture cost.

4.4 Reflecting Materials for Solar Cookers

A critical task in developing a solar cooker is to identify reflector materials that would be both suitable for this application as well as economical. As shown in Table 4.2, the materials currently being used range from common kitchen aluminum foil to advanced complex films. The perfect material would be one that provides high optical reflectance; is ultra violet ray resistant; is durable in a variety of environmental and abusive conditions; is pliable; can be easily attached to a substrate; is available internationally; and last but not least, is economical (Harrison 2001).

One radiant mirror film has been used as a multilayered polymeric film. The outside layer is polyethylene. It is noncorrosive and nonconducting, and thermally stable to a temperature of 125 °C. For improving ultraviolet ray resistant, a ultraviolet ray protectant can be added on the surface of the film. It can reflect more than 98% of visible light and commonly used as an indoor reflective material for sun tubes and the like. Another one is aluminum vapor coated on 2-mil (0.0508 mm thick) polyester film with an outdoor weathering acrylic coating over the aluminum.

Silver flux material is designed for fluorescent light fixtures, existing luminaires or as interior surfaces for luminaires, 2.5 mil (0.0635 mm) polyester film with a

Table 4.2 Typical reflecting materials for the solar thermal cooker

Material	Reflectivity	Weight	Advantages	Disadvantages
Polished Anodized Aluminum	~95%	Light	Reflective, lightweight, can be salvaged, structurally durable, easily flexed and shaped	Easily scratched, not super cheap
Mylar	>98%	Super light	Super reflective, super light, super cheap	Not good at standing up to the elements, forms "bubbles" if glue starts to give, requires a rigid backing
Aluminum Foil	88% on bright side	Super light	Extremely cheap and widely available	Not so reflective, corrodes when mixed with acidic juices, structurally weak, would only last one or two sessions
Can lids	70–80%	Light	Widely available, effective, salvageable	Not super reflective, non-uniform shape
Acrylic mirror	99%	Medium/ heavy	Very reflective, nearly unbreakable	Comes in plane, likely very difficult to fit to parabola and still maintain reflectivity, super expensive
Glass mirror	99%	Very heavy	Super reflective, widespread—may be salvaged	Very expensive, comes in rigid plane
Astro-foil	76%	Light	Reflectively strong	Not so reflective

specular metallized layer on front surface treated with an acrylic coating to suitable for protecting metallized layer from oxidation/corrosion and polyester film from ultra violet. It maintains a specular reflectance of 85% and an overall reflectance of 90%. It is bonded to 0.7 mm treated aluminum and is ideal for producing rigid precision quality specular reflectors. It does not contain ultra violet ray protection and is unsuitable for outdoor use.

Acrylic mirror material is half the weight of glass. It is shatter resistant and highly reflective. It is an optically perfect acrylic sheet that is vacuum metalized in a vacuum chamber. The sheets themselves are protected on the back by a durable scratch resistant coating and on the front surface be either a clear polyethylene film or a paper masking. The material can also be made with pressure sensitive backing.

Pure (99%+) aluminum develops a protective coating of aluminum oxide immediately on exposure to air (oxygen). This actually protects the aluminum from further oxidation by oxygen in the air. However, pure aluminum has poor structural properties. Therefore, aluminum that is used for anything structural is actually an alloy of aluminum. These alloys can be "polished"; however, without some kind of protection (anodized or coated), they are subject to fairly rapid atmospheric corrosion. Pure aluminum is also subject to severe corrosion from strong acids and bases like cement "lime" products will destroy it in very short order. Thus aluminum products that are expected to last for extended periods of time must be coated with a protective coating.

Mirroflex is a reflective, impact resistant plastic with unusual surface treatments including silver, gold, brass, brushed metallics, wood grains, and granites. In roll form, it has a pressure-sensitive adhesive backing for use in decorative metallic trim applications from 1/4″ (6.35 mm) and up in 1/8″ (3.175 mm) increments.

Metallization is the coating of a material with a fine mist of vaporized metal to create a foil-like effect. NASA used the material as a reflective insulator to protect astronauts from solar radiation, later used as insulated outdoor garments, life rafts, reflective blankets, etc. The material is super reflective, nonporous, waterproof, and rot proof.

4.5 Optical Materials for Solar Absorbers

4.5.1 Metals

Initially the absorber sheet was almost always made entirely of copper, which is one of the best thermal conductors known. However, price increases in copper began to make such collectors too expensive and the attention of designers turned to ways of achieving good collection efficiencies with cheaper materials. Unfortunately, there is nothing like copper for corrosion resistance in the actual waterways and, even now, most solar water heaters still have copper water pipes. The next best readily available conductor for the sheet is aluminum whose conductivity not as good as that of copper, is still quite good. To reduce the cost by replacing copper tubes with galvanized steel (GS) tube and copper plate with aluminum (Al) plate, the aluminum plate is wrapped over the GS tube by a special wire wound technique so that good contact of plate with risers and headers is maintained. To offset the lower conductivity of aluminum, one of the following methods must be applied (Alghoul 2005): (a) A thicker sheet; and (b) The tubes on the collector closer together. Both of these measures tend to offset the initial savings achieved by changing from a copper sheet.

4.5.2 Selective Coatings

A selectively coated absorber exposed to sunlight will get hotter than a simple matt black one, which enables the conversion of a higher proportion of the solar radiation into heat. It reduces radiation losses through heat emission significantly, this in turn means that the lower conductivity of aluminum sheet can be compensated by having it run hotter; so that the copper tubes do not have to be put closer together and the sheet does not have to be made thicker. Another advantage of selective coatings is that they enable the collector to work better in poor conditions (weak sunshine) (Alghoul 2005).

Selective absorber surface coatings can be categorized into six distinct types (Kennedy 2002): (a) intrinsic, (b) semiconductor-metal tandems, (c) multilayer absorbers, (d) multi-dielectric composite coatings, (e) textured surfaces, and (f) selectively solar-transmitting coating on a blackbody-like absorber. Intrinsic absorbers use a material having intrinsic properties that result in the desired spectral selectivity. Semiconductor-metal tandems absorb short wavelength radiation because of the semiconductor bandgap and have low thermal emittance as a result of the metal layer. Multilayer absorbers use multiple reflections between layers to absorb light and can be tailored to be efficient selective absorbers. Metal-dielectric composites-cermets consist of fine metal particles in a dielectric or ceramic host material. Textured surfaces can produce high solar absorptance by multiple reflections among needle-like, dendritic, or porous microstructure. Additionally, selectively solar-transmitting coatings on a blackbody-like absorber are also used but are typically used in low-temperature applications. These constructions are shown schematically in Fig. 4.25a–f, respectively.

4.5.2.1 Intrinsic Absorption Coatings

As shown in Fig. 4.25a, intrinsic or mass absorbers, in which selectivity is an intrinsic property of the materials, are structurally more stable but optically less effective than multilayer stacks examples include, metallic W, MoO_3-doped Mo, Si doped with B, CaF_2, HfC, ZrB_2, SnO_2, In_2O_3, Eu_2O_3, ReO_3, V_2O_5, and LaB_6. No naturally occurring material exhibits intrinsically ideal solar-selective properties, but some roughly approximate selective properties. Intrinsic solar-selective properties are found in transition metals and semiconductors, but both need to be greatly modified to serve as an intrinsic absorber. Hafnium carbide (HfC) could be useful as an absorbing selective surface at elevated temperatures because of its high melting point. However, HfC requires structural and/or compositional changes in the lattice or an antireflective (AR) layer composed of a quarter wavelength of a dielectric

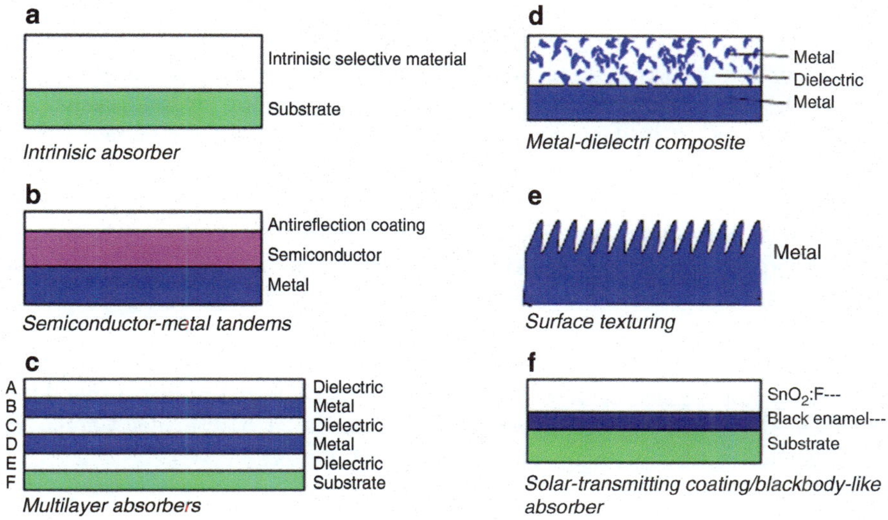

Fig. 4.25 Schematic designs of coatings and surface treatments for selective absorption energy (Adapted from Kennedy 2002. Credit: US National Renewable Energy Laboratory)

material to create the required properties. Single-layer AR coatings that have been used include SiO, SiO_2, Si_3N_4, TiO_2, Ta_2O_5, Al_2O_3, ZrO_2, Nd_2O_3, MgO, MgF_2, and SrF_2. AR coatings can also be made from very thin layers of two materials having properly matched indices of refraction, for example, thallium iodide and lead fluoride. Historically, research in intrinsic absorbers has not been very productive because there are no ideal intrinsic materials; but the intrinsic materials are finding increasing use as a component in high-temperature absorber multilayers and composite coatings (Kennedy 2002).

4.5.2.2 Semiconductor-Metal Tandems

Figure 4.25b illustrates the absorber structured with antireflection coating, semiconductor and metal. Semiconductors with bandgaps from about ~0.5 eV (2.5 μm) to 1.26 eV (1.0 μm) absorb short-wavelength radiation, and the underlying metal provides low emittance to give the desired spectral selectivity to semiconductor-metal tandems. Semiconductors of interest include Si (1.1 eV), Ge (0.7 eV), and PbS (0.4 eV). Thin semiconductor films of high porosity or antireflection coatings are needed because the useful semiconductors have high refractive indices, which result in large detrimental reflectance losses. Si-based designs produced by chemical-vapor deposition (CVD) are well known that are suitable for mid-to high-temperature applications (Kennedy 2002).

4.5.2.3 Multilayer Absorbers

Multilayer absorbers or multilayer interference stacks can be designed so that they become efficient selective absorbers, as shown in Fig. 4.25c. The selective effect is because the multiple reflectance passes through the bottom dielectric layer (E) and is independent of the selectivity of the dielectric. A thin semitransparent reflective layer (D), typically a metal, separates two quarter-wave dielectric layers (C and E). The bottom-reflecting layer (D) has high reflectance in the infrared (IR) region and is slightly less reflective in the visible region. The top dielectric layer (C) reduces the visible reflectance. The thickness of this dielectric determines the shape and position of the reflectance curve. An additional semitransparent (i.e., thin) metal layer (B) further reduces the reflectance in the visible region, and an additional dielectric layer (A) increases the absorption in the visible region and broadens the region of high absorption. The basic physics of the multilayer absorber is well understood, and computer modeling can easily compute the optical properties given by an optimum multilayer design of candidate materials. Multilayer interference stacks have high solar absorption, low thermal emittance, and are stable at elevated temperatures ($\geq$ 400 °C) depending on the materials used. Several multilayer absorbers using different metals (e.g., Mo, Ag, Cu, Ni) and dielectric layers (e.g., Al_2O_3, SiO_2, CeO_2, ZnS) have been used for high-temperature applications (Kennedy 2002).

4.5.2.4 Metal-Dielectric Composite Coatings

Metal-dielectric composite coatings or absorber-reflector tandems have a highly absorbing coating in the solar region (i.e., black) that is transparent in the IR, deposited onto a highly IR-reflective metal substrate. As shown in Fig. 4.25d, the highly absorbing metal-dielectric composite, or cermet, consists of fine metal particles in a dielectric or ceramic matrix, or a porous oxide impregnated with metal. These films are transparent in the thermal IR region, while they are strongly absorbing in the solar region because of interband transitions in the metal and the small particle resonance. When deposited on a highly reflective mirror, the tandem forms a selective surface with high solar absorptance and low thermal emittance. The high absorptance may be intrinsic, geometrically enhanced, or both. The absorbing cermet layer comprised of inherently high-temperature materials can have either a uniform or graded metal content. The metal-dielectric concept offers a high degree of flexibility, and the solar selectivity can be optimized by proper choice of constituents, coating thickness, particle concentration, size, shape, and orientation. The solar absorptance can be boosted with a suitable choice of substrates and AR layers, which can also provide protection (for example, from thermal oxidative degradation). A variety of techniques, such as electroplating, anodization, inorganic pigmentation of anodized aluminum, CVD, and co-deposition of metal and insulator materials by physical vapor deposition (PVD), can produce the composite coatings. A subclass of this category is a powdered semiconductor-reflector combination, where the solar-selective properties of semiconductor, inorganic metal oxides, organic black pigments, and metal-dust-pigmented selective paints can be considered.

Metal-pigmented alumina selective coatings use oxide coatings obtained from the phosphoric anodic anodization of aluminum. The oxide coating consists of a compact barrier layer and a porous alumina layer whose pores are perpendicular to the aluminum. The pores can be impregnated with Ni, V, Cr, Co, Cu, Mo, Ag, and Was rod-like particles 30–50 nm in diameter and 300 nm long (Niklasson and Granqvist 1991).

In a graded cermet, the reflectance from the cermet is reduced by gradually increasing the metal volume fraction, hence the refractive index, as a function of depth from the surface to the base of the film. PVD or CVD techniques can be used for most graded cermets. By controlling the PVD deposition parameters, the microstructure of the oxides can be deposited with a porous to columnar microstructure, and by code position the inclusions or pores can be filled with metal by evaporation or sputtering. For example, in the art of thin-film growth, it is well known that columnar microstructure will grow depending on the material itself and the deposition conditions-substrate temperature, deposition rate, vacuum pressure, and angle of incidence.

In cermets, solar absorptance is mainly determined by the response of the absorbing particles. There is a shift of the absorption and scattering cutoffs to higher wavelengths when the particle radius, *r*, increases this effect is accompanied by a reduction in the maximum of the scattering and absorption efficiencies roughly proportional to r^{-1} (*r* is particle radius). Thicker cermets are needed to reach the same low reflectance in the visible region as seen for larger particles. Thermal emittance strongly increases as the thickness of the cermet increases due to IR absorption. Reducing the thickness and increasing the metallic concentration in the same proportion can reduce emittance. Smaller particles rely on interference effects and are more sensitive to thickness variations. Therefore, the optical properties of the cermets can be improved by using the optimum cermet thickness and particle diameter. Moreover, Alumina is well known as a ceramic stable at high temperature, but SiO_2 and AlN have also been used. In addition, ZrO_2 films could find applications as the dielectric medium in cermets, as multilayer solar-selective absorbers, or as AR coatings because of its high refractive index, high dielectric constant, low thermal conductivity, and corrosion-resistant properties. Using metals with a slower oxidation rate or ceramic binders stable at high temperatures would increase the durability of the cermet (Kennedy 2002).

A double-cermet film structure has been developed to obtain higher photothermal conversion efficiency than surfaces using a homogeneous cermet layer or a graded film structure. Solar radiation is effectively absorbed internally and by phase interference in double-cermet solar coatings. Further, it is easier to deposit the double-cermet selective coating than graded-cermet layer selective surfaces. The typical double-cermet layer film structure from surface to substrate consists of the following: an AR layer that enhances solar absorption; an absorbing layer composed of two homogenous cermet layers, a low-metal-volume fraction (LMVF) cermet layer on a high-metal-volume fraction (HMVF) cermet layer; and a metallic infrared reflector layer to reduce substrate emittance (Zhang and Mills 1992).

4.5.2.5 Surface Texturing

Surface texturing is a common technique to obtain spectral selectivity by the optical trapping of solar energy, as shown in Fig. 4.25e. Properly textured surfaces appear rough and absorb solar energy while appearing highly reflective and mirror-like to thermal energy. The emittance can be adjusted (higher or lower) by modifying the microstructure (microcrystallites) of the coatings with ion-beam treatments. Single-material surfaces can exhibit selective properties if they have the proper roughness, because the selective properties depend on the ratios of mean height deviations and the autocorrelation distance to the wavelength. Properly orienting the textured material can improve the absorption and emissivity of a spectrally selective material. For example, in flat-plate collectors, straight trapezoidal grooves, with the grooves orientated for maximum efficiency (NW to SE), improve the characteristics of a gray absorber plate comparable to that of a flat-plate selective absorber plate. In concentrating applications, to improve the optical properties, the orientation of any grooves in the substrate should be considered (Kennedy 2002).

Needle-like, dendrite, or porous microstructures on the same scale as the wavelength of the incident radiation exhibit both wavelength and directional selectivity. This geometrical selectivity is not very sensitive however to the severe environmental effects (i.e., oxidation, thermal shocks) that has a catastrophic influence on the lifetime of conventional multilayer selective coatings. The surface of the microstructure must be protected from damage caused by surface contact or abrasion. Selection of a material having a high intrinsic absorption coefficient can further optimize the absorptance. Methods to prepare textured microstructures include the following (Seraphin and Meinel 1976): (1) Unidirectional solidification of eutectic alloys—enables formation of a porous, rod-like, or lamellar micromorphology depending on the type of eutectic and solidification parameters (e.g., Al_3Ni fibers in Al-matrix from Al-Ni eutectic, Mg_2Ca and Mg in Mg-Ca eutectic, Ni, Cr-TaC eutectic, Ni-Ta-Cr-Mn alloy). (2) Lithography with X-rays—a mask with a desired microstructure is copied using an X-ray-sensitive resist. (3) Ion-exchange reactions between metals—isothermal transport occurs between two metals where the difference in the work function (ΔEw) is >0.2 eV (e.g., Cu-Ni alloy). (4) Vapor-liquid-solid (VLS) mechanism—the controlled growth of whiskers on substrates from the liquid alloy zone at the interface (e.g., Si, Ge, III–V whiskers). (5) Vapor deposition—the condensation of a metal or compound from the gas phase onto a substrate by CVD or PVD (e.g., Ni-Al_2O_3, Ni). (6) Oxidation of metals at high temperature—the growth of whiskers on metals by the oxidation process in air or O_2 at high temperature (400°–850 °C) (e.g., Fe_2O_3-Fe, steel; CuO-Cu, phosphor bronze; ZnO-Zn, brass; W; Ni; Mo).

Chemically etching a tin-doped, In_2O_3 film to form a transparent microgrid with photolithography gives holes of about 2.5 μm. Reactive-sputter or ion etching with fluorocarbon gases (i.e., CF_4, CH_3), which is primarily a chemical process because the highly reactive species produced on the substrate, has been used with photolithography to produce square-wave gratings with micron and submicron periodicities. Additionally, a vapor-phase transport process using catalyzed epitaxial crystal growth has synthesized high-density arrays of nanowires (e.g., ZnO-Ag) that are hexagonal in cross section and have diameters between 70 and 100 nm. These techniques could be useful in texturing selective surfaces (Kennedy 2002).

In addition, the light trapping surface of micro–nanostructure can significantly improve the performance of optical transmission and can greatly reduce the broadband domain of material surface reflectivity, which results in the enhanced material surface absorptivity on wide-spectrum signal. Several commonly used methods for manufacturing micro–nano surface of light trapping structure with low reflectivity include chemical etching, mechanical grooving, reactive ion etching, common long-pulse laser grooving, and ultra-fast pulse laser processing (Zheng et al. 2015).

4.5.2.6 Selectively Solar-Transmitting Coating on a Blackbody-Like Absorber

As shown in Fig. 4.25f, the selective solar-transmitting coating can be a highly doped semiconductor (e.g., SnO_2:F, SnO_2:Sb, In_2SO_3:Sn, and ZnO:Al) over an absorber with a proven long-term durability. Some low-temperature flat-plate collectors have used black enamel as the absorber material. Highly doped semiconductors may be useful with high-temperature black absorber materials (Kennedy 2002).

4.5.3 Heat Pipes

Another way of compensating for the poor conductivity of the cheaper materials of construction is to use the so-called heat pipe effect. In a heat pipe, high thermal conductivity is achieved by using the evaporation and condensation of a volatile fluid to carry the heat along an evacuated tube. Such a device has a thermal conductivity many hundreds of times that of the same cross-section of pure copper. In its simplest form, a heat pipe is a sealed tube containing a small quantity of a volatile liquid (such as water) with no air or other "permanent" gas present. If such a pipe is placed vertically and the lower end is heated, liquid will evaporate and the vapor so formed will travel to the cooler parts of the pipe where it will condense and give up its latent heat of vaporization. The condensate will then run back to the heated end where it can re-evaporate. This is shown in Fig. 4.26 (Alghoul 2005).

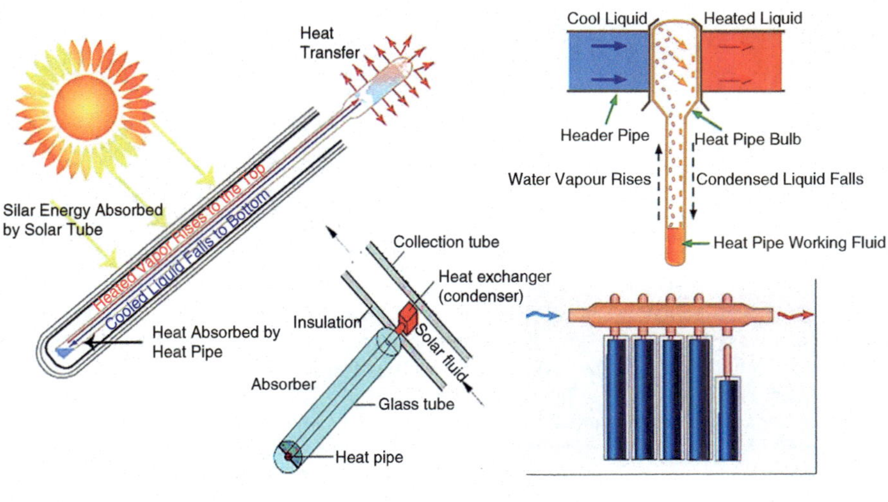

Fig. 4.26 Schematic designs of heat pipes of solar thermal collectors

4.5.4 Metamaterial Solar Absorbers

Metamaterial absorbers can selectively absorb or emit electromagnetic waves by exciting plasmonic resonances at particular wavelengths inside the material structures. The absorbing metamaterials are usually made of micro/nanostructures with subwavelength metallic patterns on a metal film separated by a dielectric spacer. Between the metallic pattern and the metal film, strong electromagnetic coupling can occur at selected wavelengths due to electric and magnetic responses of the metamaterial (Watts et al. 2012; Wang and Wang 2013).

Originally, perfect metamaterial absorbers were made of electric ring resonators coupled to metal wires, exhibiting selective absorption in the terahertz region. By replacing the metal wires with a continuous film, as shown in Fig. 4.27a, the design

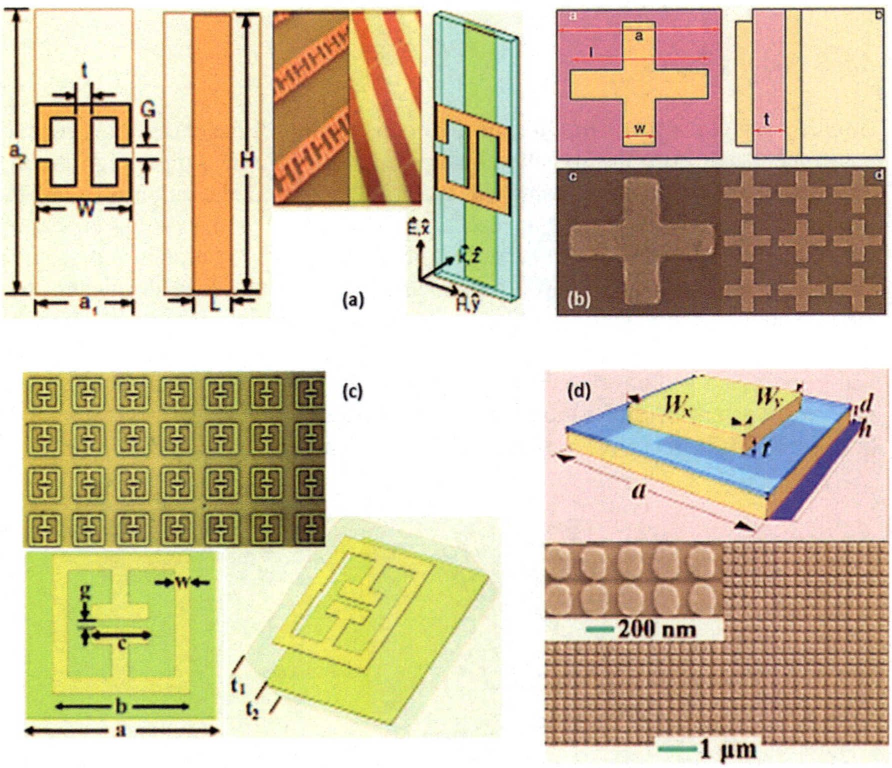

Fig. 4.27 Schematic illustrations of some perfect metamaterial absorbers: (**a**) wide-angle absorption for both transverse electric (TE) and magnetic (TM) polarized waves (Adapted with permission from Landy et al. 2008 (APS Physics)); (**b**) subwavelength perfect absorber made of a film-coupled crossbar structure (Adapted with permission from Tao et al. 2008 (APS Physics)); (**c**) plasmonic absorber made of a layer of gold patch array with a width of less than 200 nm on a thin Al_2O_3 layer over a gold film showed an absorption peak of 88% at the wavelength of 1.58 μm (Adapted with permission from Liu et al. 2010 (APS Physics)); and (**d**) ultra-thin plasmonic absorber in the visible spectrum by depositing a two-dimensional (2D) Ag grating with a period of 300 nm on a 60-nm SiO_2 over an Ag film (Adapted with permission from Hao et al. 2010 (AIP Publishing))

was improved to achieve wide-angle absorption for both transverse electric (TE) and magnetic (TM) polarized waves (Landy et al. 2008). Moreover, different top-layer pattern structure designs such as chiral metamaterial, fishnet structure, and cut-wire array were proposed to achieve omnidirectional and polarization-independent absorption in the terahertz region. Thereafter, by shrinking the size of the metamaterial absorbers, near-perfect selective absorption is achievable in the infrared and visible region. For instance, 97% absorption was demonstrated at the wavelength of 6 μm in a subwavelength perfect absorber made of a film-coupled crossbar structure, as shown in Fig. 4.27b (Tao et al. 2008). Figure 4.27c shows a plasmonic absorber made of a layer of gold patch array with a width of less than 200 nm on a thin Al_2O_3 layer over a gold film, demonstrating an absorption peak of 88% at the wavelength of 1.58 μm (Liu et al. 2010). As shown in Fig. 4.27d, by depositing a two-dimensional (2D) Ag grating with a period of 300 nm on a 60-nm SiO_2 over an Ag film, an ultra-thin plasmonic absorber was made in the visible spectrum (Hao et al. 2010). Strong visible light absorption has also been achieved through film-coupled colloidal nanoantennae, circular plasmonic resonators, and nanoparticles by exciting magnetic resonance inside the metamaterial absorbers.

More specifically, plasmonic metamaterials are typically composed of noble metals in which the features of photonics and electronics are linked by coupling photons to conduction electrons of metal, which is known as surface plasmon. Complete absorption of light is one of the exotic properties of plasmonic metamaterials. This is realized by designing a medium whose impedance matches that of free space while being opaque. If such a medium is filled with some lossy medium, the resulting structure can absorb light totally in a sharp or broad frequency range. Although several types of metamaterials perfect absorber have been demonstrated as shown in Fig. 4.27, perfect absorbers based on nanocomposites have caused great attention, where the total thickness is a few tens of nanometer and the absorption band is broad, tunable and insensitive to the angle of incidence. The nanocomposites consist of metal nanoparticles embedded in a dielectric matrix with a high filling factor close to the percolation threshold. The filling factor can be tailored by the vapor phase co-deposition of the metallic and dielectric components. In addition, novel wet chemical approaches have been used with bio-inspiring or involving synthesis within levitating Leidenfrost drops, for instance (Hedayati et al. 2014).

4.5.4.1 Metal-Dielectric Nanocomposites with Tailored Plasmonic Response

Metal-dielectric nanocomposites can be made of metallic nanoparticles embedded in a dielectric organic or ceramic matrix, with the considerations of (a) a high filling factor of the metallic nanoparticles close to the percolation threshold is required to take advantage of the interaction of plasmon resonances localized at individual nanoparticles, and (b) large area coverage is indispensable in most applications which rules out electron beam lithography and rather calls for self-organized formation of the nanostructures, vapor phase deposition techniques are particularly attractive for tailoring the nanostructure and the resulting properties. Vapor phase deposition, inter alia, allows excellent control of the metallic filling factor and its depth profile as well as the incorporation of alloy nanoparticles with well-defined composition. The metallic nanoparticles typically form via a self-organization during co-deposition of the metallic and matrix components due to the high cohesive energy of the metals and the low metal-matrix interaction energy. Various methods such as sputtering, evaporation, and plasma polymerization have been applied for the deposition of the matrix component, while the metallic component has mostly been sputter-deposited or evaporated. Moreover, gas aggregation cluster sources were utilized to obtain independent control of filling factor and size of the embedded nanoparticles.

Gold Nanocomposite

In the field of plasmonic materials, the gold (in particular in nanoparticles form) is the leading building block not only because of its stability but also due to its plasmon resonance and unique optical properties in visible. Au NPs are among the most stable metal nanoparticles with some unique features and properties such as size-related electronic, optical and magnetic properties as well as application in catalysis and biologic systems. For the mean free path in gold and silver is 50 nm, particles smaller than this size do not experience any bulk scattering and surface effect is dominating. Therefore, the light in resonance with the surface plasmon oscillation causes the free-electrons in the metal (*d* electrons in silver and gold) to oscillate. Since the resonance occurs at the surface, it is called surface plasmon resonance (SPR). Consequently, any effect which changes the surface geometry of the particle (e.g., size or shape) causing a shift in the electric field density on the surface which results in the alteration of oscillation frequency of the electrons (i.e., SPR shift). Changing the surrounding environment of the NPs could also affect the resonance frequency which is the basic principles of plasmonic sensor (Hedayati et al. 2014).

Ultra-thin nanocomposite atop of dielectric-coated metal film could result in complete absorption of light in broad spectrum. For instance, gold was selected as a prime metallic constituent of the structure due to its great stability and unique optical properties; 20 nm nanocomposite (Au-SiO_2) deposited on 25 nm SiO_2-coated gold film (100 nm) is the optimum condition for realization of broad-band perfect absorber in visible frequency. The volume fraction (filling factor) of the gold in composite significantly alters the optical response of the designed metamaterials. ~40% filling factor is the optimum value for high absorption in a wide range of wavelengths making the surface appearance black. The dipole-image interaction is one of the reasons for light trapping in the nanocomposite perfect absorber. For instance, thickening the interlayer results in a drop of the absorption intensity, which can be interpreted as weaker dipole-image interactions due to weaker coupling. Since the resonance of the plasmonic structure is strongly influenced by any changes in the surrounding environment, the peak position of absorption in the plasmonic nanocomposite perfect absorber is also altered by changing the matrix. In the case of the matrix with lower refractive index (PTFE) compared to the SiO_2, the resonance blue shift, while having a higher refractive index material (TiO_2), shifts the peak to a longer wavelength. The retardation effect of the higher refractive-index matrix

(TiO_2) on the resonance of the metallic particles is shifting the absorption band to a longer wavelength while the polymer matrix with less dielectric constant (than that of silicon dioxide matrix) provides the condition for higher resonance frequency.

Copper Nanocomposite

In the field of metamaterials, and in particular, perfect absorbers, copper is a promising candidate due to its significant loss in visible range, though it has been implemented and used for realization of high absorber devices in other frequencies, too. Three-layer absorber has mainly been developed where a dielectric (such as FR4 lossy dielectric) is sandwiched between two metallic structures and/or films. Indeed, the role of dielectric loss in absorption is considerably larger than that of the ohmic losses in metal. Moreover, multilayer of split-ring resonators (SRRs) with different dimension stacked over each other can provide a dispersive refractive index required for anti-reflection (high absorption). High absorption is not because of resonant loss of SRRs but rather due to the anti-reflectivity of the dispersive coating. Integrating resistors into resonators while maintaining an impedance-matched material at normal incidence was another approach in order to have a broader absorption band. Similarly, to broaden the bandwidth of the metamaterial absorber, lumped elements have been incorporated (resistance and capacitance) into a typical three-layer structure (dielectric substrate sandwiched with metal split-coin resonators (SCR)) by welding. In such a system, which is in analogy with an RLC circuit, the incidental EM energy can be converted into electric energy in the circuit, and then electric energy can be subsequently consumed by lumped resistances. In addition, VO_2 is incorporated to the traditional multilayer absorber to enable the resonance tunability by temperature variation. Since the vanadium dioxide phase transition from metallic to insulating occurs around 330 K, its refractive index changes correspondingly. Therefore, the resonance condition inside the structure would change by heating or cooling the sample at this temperature range, giving rise to the realization of a tunable perfect absorber. In spite of the top layers usually made with patterned metallic structures, the triple absorber can also be fabricated with metallic film (nonstructured) as the top and bottom layers surrounding a dielectric film, which could act as the perfect absorber for visible and IR frequency. However, the design is based on Fabry-Perot interferometer principles and therefore the thickness of the interlayer and its refractive index is relatively high. The ease of fabrication of such a class of absorber is beneficial, but its narrow bandwidth and its rather bulky thickness would be its limiting factors for application in nano-optics (Hedayati et al. 2014).

Generally, the metamaterial absorbers made of copper suffer from one or some of the following limitations: fabrication complexity, cost of production, narrow band of resonance, and angular/polarization sensitivity. As dielectric, PTFE was used which has a low refractive index (~1.3) to assess the role of dielectric type on the optical properties of absorber. In spite of Au-SiO_2 absorber where 25 nm SiO_2 interlayer showed the best performance, 20 nm spacer layer of PTFE resulted into the maximum absorption in copper system. Copper particles act as stronger light absorber than gold which can be the cause of the high absorption of structure even without any spacer film. The absorption intensity and broadness of copper perfect absorber is greater than that of its gold counterpart. The difference in the absorption properties of copper and gold base absorber is routed mainly from the fact that copper is more lossy in visible frequency than gold. Nevertheless, the overall behavior of the two systems shows some similar tendencies where the absorption intensity in entire visible spectrum is high. Even though huge absorption was realized in copper-based metamaterials' absorber, the structure and optical properties vary by time mainly due to the probable oxidation of particles via interpenetration of oxygen to the particles surface. Generally, the main difficulty in utilizing copper nanoparticles is their inherent tendency to oxidize in ambient conditions. Applying different barrier layer is known as the major solution to that issue. However, the polymeric (PTFE) matrix which was used as the dielectric could not provide the efficient protective layer and avoid copper NPs oxidation. Copper nanoparticles can be encapsulated in an oxygen impenetrable matrix to guarantee the long-term stability of the final device.

Silver Nanocomposite

Gold have been in use in the perfect absorbers because of ease of fabrication, stability, and high absorption in the middle of visible spectrum. In spite of high damping losses of gold and copper, silver is known as the lowest damping metal in the visible frequency and hence consideration of silver as the constituent of perfect absorber sounds unreasonable. In other words, silver has been implemented in metamaterials for high frequency because of its low loss.

Interband transition from occupied *d* states to unoccupied *p* and *s* states above Fermi level appear at 310 and 350 nm in bulk silver, respectively. However, for silver nanostructures, such electron transitions could occur above 350 nm wavelengths and depend on nanostructure geometry. Therefore, high optical absorption in silver particles is likely to occur at 350 nm wavelengths and above. On the other hand, silver's reflection is close to unity over the whole visible region and hence the absorption of optically thick Ag film is usually below 5% in that region. In comparison with bare silver film, nanocomposite shows higher absorption and its intensity is around 30% in the visible frequency. In spite of the expected low absorption of silver in visible frequency, the high absorption span for the UV up until the green part of spectrum can be realized by silver base plasmonic metamaterial absorber. This is coming from the fact that the resonance of plasmonic materials moves to longer wavelengths when the dielectric constant of the adjacent environment enlarges. This is more prominent if the particle is in proximity to a metallic substrate. Particularly, as a dielectric encloses the metallic nanoparticle, the induced screening charges on the metal–dielectric boundary reduce the plasmon excitation energy resulting in a red-shifting of the resonance. Likewise, for nanoparticles ensembles, the dielectric materials screen and weaken the interaction between the NPs and lessen the shift of the coupled plasmon. Accordingly, the absorption band of the silver absorber moves to the visible range due to the interaction of the composite and the base silver mirror. It is worth mentioning that part of the red-shift originated from the particle dipole which is anti-symmetrically coupled to its image inside the silver mirror (substrate). In analogy with copper perfect absorber system, low reflectivity of the silver absorber is partially routed from the interference. In such layered stacks, the Fabry-Perot cavity is built-up between the top composite and the bottom mirror, and results in a strong interference of the incident and the reflected wave. Specifically, the bounced-back rays from the mirror destructively interfere with the direct reflection from the top (air-composite) interface. If the thickness of the spacer layer is properly selected, the reflected waves cancel each other out which results in negligible reflection. Given that the base silver film is thicker than the skin depth, no light transmits through the layers and thus perfect absorption is achieved. By increasing the thickness of the composite (while keeping the other parameters constant), the absorption band can be widened extensively. Indeed, metamaterial silver absorber which spans the whole visible frequency can be made by increasing the thickness of nanocomposite with 46% filling factor to 30 nm. The high absorption of the silver system in deep UV range shows its potential as UV protection film. The absorption intensity of the metamaterials is higher than the inorganic absorber. Hence, the perfect absorber can be used as a highly efficient UV protective layer (Hedayati et al. 2014).

Therefore, nanocomposite perfect absorber (i.e., nanocomposite-dielectric-metal film stack) shows almost unity absorption in a broad range of frequencies with marginal angular dependency. The absorption peak position and intensity can be tuned by changing the type of nanocomposite, filling factor and the host matrix, and the thickness of the layers. The implementation of perfect absorbing structures on a variety of substrate could be implemented into the new generation of thin-film solar cells for clothing (textile industry). Accordingly, nanocomposite with high filling factor can be deposited even on aluminum kitchen foil, turning it into a black absorber or any other color. This can push the metamaterials absorber for novel energy applications.

4.5.4.2 Nanoporous Composite Absorbers

Lightweight broadband nanocomposite perfect absorbers have been developed base on ensemble of nanoparticles that can strongly confine the light. The light trapping can be intensified when the particles' inner-distance shrinks down to the diameter of the particles. Porous metals are among those structures which provide the small inter particles distance and enable localization of electromagnetic field (via localized surface plasmon resonance). Hence, the tuning of the resonance is directly correlated with the size at dimension of the porosity and the structure itself. For instance, Leidenfrost drop has been used to create a nanoporous gold hybrid structure as well as black plasmonic foam, which absorb the whole visible and NIR electromagnetic waves resulting in a very broadband, perfect plasmonic absorber (Hedayati et al. 2014).

When a water drop touches a plate which is hotter than the boiling point of the drop, the part in contact with the substrate vaporizes and the drop levitates on its own vapor. The remnant solid is left on the surface over which the drop has levitated. Overheating, thermal gradients and charge separation are fundamental to Leidenfrost condition. In addition to the great possibility of nanofabrication under the

Leidenfrost condition such as nanoparticles formation and coating, such an approach can be used for fabrication of 3D metamaterial broadband absorber (400–2500 nm) in a very simple, cost-effective, and an environmentally friendly manner. In this approach, a drop of (1 mL $HAuCl_4$, 20 mM + 700 mL Sod. citrate 1% + 150 mL NaOH 0.5 M, pH ~8.5) was placed on a preheated hot plate (with a constant temperature of 270 °C) which provide a suspension of the desire black porous structure in less than a minute. The black suspension can turn a flexible polymeric substrate to a super absorber by a simple casting method. To realize a macro-scale three-dimensional porous metamaterial, a commercial packaging polymer foam is introduced in a levitated black pool out of the porous gold. Thanks to the dynamic covering potential of Leidenfrost drop, the foam was coated with the metallic spongy structures and a millimeter size black 3D metamaterials is realized. In such a complex polymer-metal structure, the suppression of light reflection is attributed to consolidation of light scattering by the sample roughness, localized and de-localized excitation of plasmons within and on the surface of pores, as well as light trapping inside the gaps. Although broadband absorbers are critical in energy harvesting applications, for more effective use of solar energy it is desirable to develop cost-effective, durable and lightweight systems with improved ability to absorb solar radiation energy particularly at wavelengths below 3 μm. This chemically developed metamaterial absorber can withstand high temperatures, which demonstrates its potential application in energy collecting purposes (Hedayati et al. 2014).

The biologically inspired fabrication approach is another cost-effective method for development of 3D metamaterials perfect absorber. Biological materials are inherently and naturally multifunctional and even smart. For instance, chameleons are well-known for their ability to change color. This stems from the fact that chromatophores (organocell) could either spread the pigment particles all over the cell or concentrate them into a small lump. A quite similar concept is applied for humidity-based color change (light blue to black) in Cryptoglossa verrucosa. The color phases are formed by "waxfilaments" that spread from the tips of miniature tubercles that cover the cuticle surface. A broadband and lightweight, perfect absorber has been made by mimicking beetles using nanocomposites as a standalone matrix. As a result, bio-nanocomposite fabrication method has been introduced wherein the surface color changes from red to black upon wetting thereby enabling realization of an omnidirectional wideband perfect absorber.

In dry state, the sample looks red while the average reflection of the sample is about 35%. On the other hand, by wetting the sample, it turns black while showing low reflectivity and it acts as a swollen, open, porous, nanostructure foam. The resulting porous structure gives rise to the localization of the incident electromagnetic field (in analogy to the nanoporous structures). Therefore, the sample turns black and absorbs the visible energy.

4.5.4.3 Tungsten-Silica-Tungsten Metamaterial Solar Absorbers

Figure 4.28 shows typical structures of the metamaterial solar absorbers made of 2D periodic tungsten gratings on a thin SiO_2 spacer over a tungsten thin film (Wang and Wang 2013). A unit cell of the metamaterial structure with single-sized tungsten

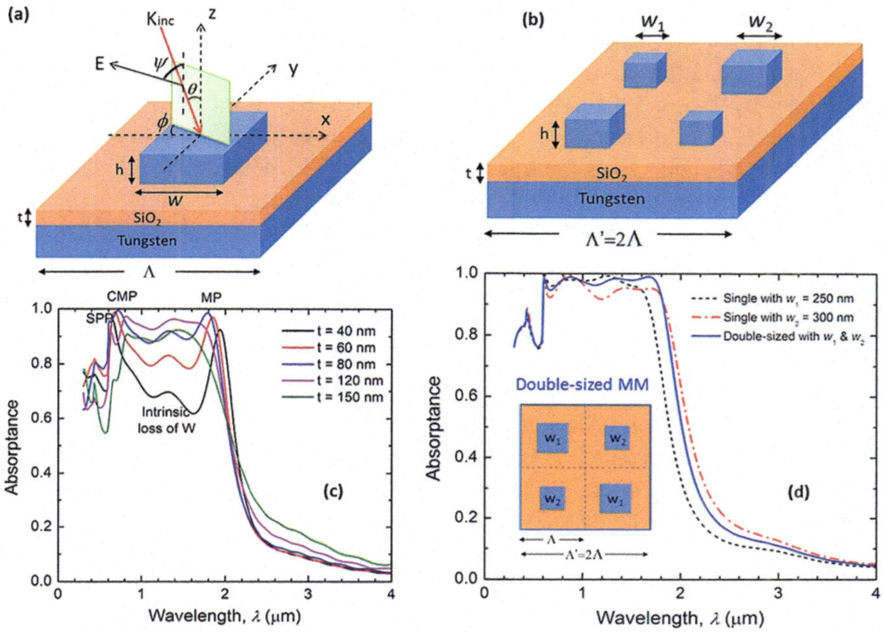

Fig. 4.28 Schematic of metamaterial solar absorbers (Adapted with permission from Wang and Wang 2013 (OSA—The Optical Society)). (**a**) Single-sized, 2D periodic tungsten gratings with period Λ, patch width w and grating height h, on a thin SiO_2 spacer with thickness t over an opaque tungsten thin film. The electromagnetic wave is incident at a polar angle θ, polarization angle ψ, and azimuthal angle φ. The structure is assumed to be geometrically symmetrical in the x and y directions, and φ is taken as 0° for simplicity. (**b**) Double-sized metamaterial solar absorbers with tungsten patches of different widths w_1 and w_2, and period $\Lambda' = 2\Lambda$ Tungsten patches with the same size are arranged diagonally and each patch is centered in its quadrant. (**c**) Spectral absorptance of the single-sized metamaterial solar absorber at normal incidence as a function of spacer thickness t. The base values of the geometric parameters are $\Lambda = 600$ nm, $w = 300$ nm, and $h = t = 60$ nm. The broadband high absorption in the spectral region from 0.3 to 2 μm is due to several physical mechanisms including SPP, CMP, intrinsic loss of tungsten, and MP. (**d**) Spectral normal absorptance in the spectral region from 0.4 to 4 μm for a double-sized metamaterial solar absorber with tungsten patch widths of $w_1 = 250$ nm and $w_2 = 300$ nm, in comparison with that of single-sized metamaterial solar absorbs with w_1 or w_2. Other geometric parameters are the same: $\Lambda = 600$ nm, $h = 150$ nm, and $t = 60$ nm. The inset depicts the arrangement of the tungsten patches for the double-sized absorbers

patches is shown in Fig. 4.25a. The geometric parameters include grating period Λ, tungsten patch width w, grating height (or patch thickness) h, and SiO_2 spacer thickness t. The opaque tungsten thin film may be a couple hundred nanometers thick. The three-layer-structural metamaterial would be deposited on the outside surface of a glass window to provide the anti-frost, anti-fog, and energy-saving capabilities. The 2D periodic gratings are designed with the same geometric parameters (i.e., Λ and w) in the x and y directions because the geometric symmetry is crucial to realizing the polarization independence in the normal direction. Here wavevector K_{inc} represents the electromagnetic wave with a free-space wavelength λ

incident onto the metamaterial structure at a polar angle or incidence angle θ, polarization angle ψ, and azimuthal angle φ. The polar angle θ denotes the angle between K_{inc} and the surface normal of the structure (z direction). The polarization angle ψ is by K_{inc} and the structure surface normal, between electric field vector E and the plane of incidence. $\psi = 0°$ indicates the transverse magnetic (TM) polarized wave while $\psi = 90°$ gives the transverse electric (TE) polarized wave. Azimuthal angle φ is the angle between the x axis and the plane of incidence, and can be taken as $\varphi = 0°$ here for simplicity by ignoring conical diffraction due to the nonzero wavevector components in both x and y directions for the incident wave. Figure 4.28b illustrates a unit cell for the metamaterial absorber with double-sized tungsten patches of different widths w_1 and w_2. The patches with the same width are arranged diagonally such that the structure behaves exactly the same at normal incidence for either TE or TM waves. Each patch is centered in its quadrant, and the period Λ′ of the double-size metamaterials is twice that of single-sized ones, i.e., $\Lambda' = 2\Lambda$ (Wang and Wang 2013).

As shown in Fig. 4.28c, SiO_2 spacer thickness t yields a similar effect as the tungsten grating height h on the normal absorptance of the single-sized metamaterial solar absorber. When the spacer thickness increases from 40 to 150 nm, the magnetic polariton (MP) peak shifts to a shorter wavelength, while the peak amplitude first increases to a maximum value close to 1 with $t = 80$ nm and then drops with thicker spacers. The surface plasmon polariton (SPP) peak locations do not change with spacer thickness, but the amplitudes change with different t values. The coupled magnetic polariton (CMP) peak separates from the SPP peak around $\lambda = 0.6$ μm and shifts gradually towards the longer wavelength as t increases. As a result, the absorbance in the spectral region between 0.6 and 1.8 μm is greatly enhanced with the minimum value of spectral absorbance increasing from 0.6 at $t = 40$ nm to 0.92 at $t = 120$ nm. However, the absorptance starts to decrease with thicker spacers. Therefore, a SiO_2 spacer thickness range of 60–120 nm is preferable when coating the spacer layer on window glass. Figure 4.28d shows the absorptance of both single-sized and double-sized metamaterial absorbers with the same geometric parameters of $\Lambda = 600$ nm, $h = 150$ nm, and $t = 60$ nm but different patch widths $w_1 = 250$ nm and $w_2 = 300$ nm, respectively. The tungsten grating height h can be optimized to make the single-sized metamaterial have close-to-unity absorptance in the partial or full visible and near-infrared region, or in the near-infrared region to let only visible light transmit through window glass for luminosity. The single-sized metamaterial absorber with smaller patch width $w_1 = 250$ nm has a narrower band of absorption but a higher absorptance in the near-infrared than the one with larger patch width of $w_2 = 300$ nm. By comparison, the double-sized metamaterial is more preferable for coating on window glass because of its broader absorption band and higher absorbance than the single-sized one. The minimum absorptance of the double-sized metamaterial is greater than 0.95 in a wide spectral range from 0.6 μm to 1.8 μm while its geometric design together with a couple hundred nanometer thick metamaterial coating would lower luminous transmittance insignificantly.

In addition to the absorptance enhancement of MP, CMP and SPP resonance modes around particular wavelengths, the utilization of the high intrinsic loss of tungsten is another important factor for the broadband high absorption. Compared to the relatively low losses of gold and silver, tungsten has several interband transitions around the wavelengths of 0.4, 0.6, and 1.4 μm. For solar thermal applications, its high intrinsic loss is beneficial to enhance the absorption of solar radiation across a wide spectral range. Therefore, the most important approach to achieving almost perfect absorption in a broad spectral band from visible to the near-infrared region (especially in near-infrared region for transparent window glass application) with the designed metamaterial absorber is the use of the coupling effect between different resonance modes and interband absorption of tungsten. In other words, the absorptance of the metamaterial solar absorbers strongly depends on the geometric parameters (such as the tungsten patch width, grating period, grating thickness, and the spacer thickness), the coupling between MP, CMP, and SPP modes, and the intrinsic loss of tungsten. The peak wavelengths of the MP and CMP modes also strongly depend on the patch width w, grating period Λ, grating height h, and spacer thickness t, which could be potentially employed to further broaden the absorption peak. As a result, the absorption could also be maximized by optimizing the geometric parameters (Wang and Wang 2013).

4.5.4.4 Broadband Solar Absorbing Metasurfaces

Metasurfaces are a kind of 2D metamaterials, which can tailor the in-plane phase front with an extremely thin slab consisting of judiciously designed plasmonic structures. By patterning plasmonic nanostructures and engineering the spatial phase distribution within the metasurfaces, exotic optical phenomena and optical components can be achieved, including negative refraction or reflection, as well as broadband electromagnetic wave including solar absorbers. Metasurfaces can overcome the challenges encountered in bulk metamaterials while their interactions with the incident waves can be still sufficiently strong to obtain very useful functionalities. At microwave and terahertz (THz) frequencies, the design of metasurfaces can take advantage of subwavelength metallic resonators such as split-ring resonators (SRRs) and a variety of elements typically used in frequency selective surfaces. Abrupt and controllable changes of optical properties are achieved by engineering the interaction between light and an array of optical scatters called optical antennas or plasmonic nanostructures, which can take a variety of forms, including metallic or dielectric micro/nanoparticles, apertures formed in metallic films, and their multilayer structures. The most critical feature of metasurfaces is that they provide degrees of freedom in designing spatial inhomogeneity over an optically thin interface. Metasurfaces are able to introduce a spatially varying electromagnetic or optical response (i.e., scattering amplitude, phase, and polarization), and mold wavefronts into shapes that can be designed at will (Chen et al. 2016).

Metasurface perfect absorbers with thickness much smaller than the operational wavelength are attractive in many applications such as sensing, compressive imaging, and thermal management. Broadband absorbers covering the entire solar spectrum are also of great interest in solar energy harvesting. There have been some demonstrations of material structures as high performance solar absorbers, for instance, using dense nanorods and nanotube films, multilayer planar photonic structures, and photonic crystals. Metasurfaces consisting of complex multi-resonator unit-cells have emerged as a powerful and flexible platform to realize multiband and broadband perfect absorption, particularly in microwave, terahertz, and infrared regimes. For instance, broadband omnidirectional metasurface absorber exhibits greater than 90% absorptance in the near infrared and entire visible frequency range. The relatively simple design of the absorber allows scale up to large area fabrication using conventional nano-imprint lithography and advanced printing technologies (Azad et al. 2016).

4.5.4.5 Prospects and Future Trends

The field of metamaterial perfect absorbers is still immature. Much more effort must be made to bridge the gap between the lab-scale fabrication and industrial application. Nevertheless, current achievements both in theory and experiments showed the immense potential of this new type of metamaterials for a variety of applications. As the prime utility of highly absorptive structures, photovoltaic and solar cells are the fields of interest. Other uses of metamaterial perfect absorbers are in fields where huge light confinement is desired. Due to the localized field within the nanostructure and its tunability, the resonance could match the molecular vibrational modes of interest in the analyte which provides the possibility to identify chemical stretches. Microbolometer thermal sensors are another general application of metamaterial absorbers. Gigantic field enhancement achieved by perfect absorber can be applied for Raman spectroscopy of single molecules, too. Cloaking an object by perfect absorber (in reflection mode) is another potential application of this class of material. Any object which can be wrapped by a perfect absorber would be cloaked and turn invisible in reflection due to the suppression of back-scattered light from the wrapped object (Hedayati et al. 2014).

Perfect absorbers designed for low frequency have been shown to be acoustic metamaterials which absorb the airborne sound in the frequency range of 100–1000 Hz. The designed and fabricated structure comprises an elastic membrane decorated with asymmetric rigid platelets. This intelligently designed metamaterial can have a broad range of applications such as reducing the cabin noise in airliners and ships, regulating the acoustic quality of music halls, and environmental noise abatement along highways and railways, among others.

Similar to other plasmonic structures, application of metamaterial perfect absorbers as a sensor is also promising, in particular when the absorber is narrow-band. Tri-layer absorber has been designed and it is shown that the resonance band of such structures can be tuned upon exposure to different liquids or vapor because of the refractive index change of the surrounding environment (Hedayati et al. 2014).

Taking into consideration all the demonstrated applicability of metamaterial perfect absorbers, one could see the progress in this new and fast growing field. However, none of the proposed structures have been used in the currently industrialized solar cells or collectors and therefore their long-term performance and stability needs to be examined. Moreover, up-scaling of the nano-lithographically fabricated system (which is the major fabrication method in metamaterial) in a cost-effective and reproducible way is also in doubt and requires the invention and development of some new robust and cheaper alternatives. Nevertheless, the future of this field is very bright and mass production and implementation of metamaterial perfect absorbers for everyday life is not far away.

4.6 Thermal Storage Materials

After the thermal energy is collected by solar collectors, it needs to be efficiently stored when later needed for a release. Thus, it becomes of great importance to design an efficient energy storage system. Thermal storage is one of the main parts of a solar heating, cooling, and power generating system. If the solar system must operate continuously, the heat storage is necessary. For some applications intermittent operation is acceptable, but most other uses of solar energy require operating at night and when the sun is hidden behind clouds. The energy storage system has an enormous influence on overall system cost, performance, and reliability (Kim and Han 2015).

Thermal energy storage (TES) includes a number of different technologies. Thermal energy can be stored at temperatures from −40 °C to more than 400 °C as sensible heat, latent heat and chemical energy (i.e., thermo-chemical energy storage) using chemical reactions. Thermal energy storage in the form of sensible heat is based on the specific heat of a storage medium, which is usually kept in storage tanks with high thermal insulation. The most popular and commercial heat storage medium is water, which has a number of residential and industrial applications. Underground storage of sensible heat in both liquid and solid media is also used for typically large-scale applications. However, TES systems based on sensible heat storage offer a storage capacity that is limited by the specific heat of the storage medium. Phase change materials (PCMs) can offer a higher storage capacity that is associated with the latent heat of the phase change. PCMs also enable a target-oriented discharging temperature that is set by the constant temperature of the phase change. Thermo-chemical storage (TCS) can offer even higher storage capacities. Thermo-chemical reactions (e.g., adsorption or the adhesion of a substance to the surface of another solid or liquid) can be used to accumulate and discharge heat and cold on demand (also regulating humidity) in a variety of applications using different chemical reactants. At present, TES systems based on sensible heat are commercially available while TCS and PCM-based storage systems are mostly under development and demonstration (Hauer 2013).

4.6.1 Sensible Thermal Energy Storage

The use of hot water tanks is a well-known technology for thermal energy storage. Hot water tanks serve the purpose of energy saving in water heating systems based on solar energy and in co-generation (i.e., heat and power) energy supply systems. Water tank storage is a cost-effective storage option and that its efficiency can be further improved by ensuring optimal water stratification in the tank and highly effective thermal insulation. This technology is also used in solar thermal installations for DHW combined with building heating systems. Large hot water tanks are used for seasonal storage of solar thermal heat in combination with small district heating systems. These systems can have a volume up to several thousand cubic meters (m^3). Charging temperatures are in the range of 80–90 °C. The usable temperature difference can be enhanced by the use of heat pumps for discharging (down to temperatures around 10 °C) (Hauer 2013).

4.6.2 Underground Thermal Energy Storage

Underground thermal energy storage (UTES) is also a widely used storage technology, which makes use of the underground as a storage medium for both heat and cold storage. UTES technologies include borehole storage, aquifer storage, cavern storage, and pit storage. Which of these technologies is selected strongly depends on the local geological conditions. Borehole storage is based on vertical heat exchangers installed underground, which ensure the transfer of thermal energy to and from the ground layers (e.g., clay, sand, rock). Seasonal storage of solar heat in summer has been developed to heat houses or offices in winter. Ground heat exchangers are also frequently used in combination with heat pumps where the ground heat exchanger extracts low-temperature heat from the soil. Aquifer storage uses a natural underground water-permeable layer as a storage medium. The transfer of thermal energy is achieved by mass transfer (i.e., extracting/re-injecting water from/into the underground layer). Most applications deal with the storage of winter cold to be used for the cooling of large office buildings and industrial processes in the summer. A major prerequisite for this technology is the availability of suitable geological formations. Cavern storage and pit storage are based on large underground water reservoirs created in the subsoil to serve as thermal energy storage systems. These storage options are technically feasible, but applications are limited because of the high investment costs.

For high-temperature (i.e., above 100 °C) sensible heat storage, the technology of choice is based on the use of liquids (e.g., oil or molten salts, the latter for temperatures up to 550 °C). For very high temperatures, solid materials (e.g., ceramics, concrete) are also taken into consideration.

However, most of such high-temperature-sensible TES options are still under development or demonstration (Hauer 2013).

4.6.3 Phase Change Materials

Sensible heat storage is relatively inexpensive, but its drawbacks are its low-energy density and its variable discharging temperature. These issues can be overcome by phase change materials.

(PCM)-based TES, which enables higher storage capacities and target oriented discharging temperatures. The change of phase could be either a solid/liquid or a solid/solid process. Melting processes involve energy densities on the order of 100 kWh/m^3 (e.g., ice) compared to a typical 25 kWh/m^3 for sensible heat storage options. Phase change materials can be used for both short-term (daily) and long-term (seasonal) energy storage, using a variety of techniques and materials. For example, the incorporation of micro-encapsulated PCM materials (e.g., paraffin wax) into gypsum walls or plaster can considerably increase the thermal mass and capacity of lightweight building walls. The micro-encapsulated PCMs cool and solidify by night and melt during the day, thus cooling the walls and reducing or avoiding the need for electric chillers (passive cooling). Other applications for active cooling systems involve the use of macro-encapsulated salts that melt at an appropriate temperature. The PCM can be stored in the building's air vent ducts and cold air can be delivered via large-area ceiling and floor ventilation systems. PCM slurries are a promising technology. For example, ice-slurries or water-paraffin dispersions can be used for building or industrial cooling purposes. As slurries can be pumped, they can be used for either storing or distributing thermal energy (Hauer 2013).

PCMs absorb and emit heat while maintaining a nearly constant temperature. Within the human comfort range of 68–868 °F (20–308 °C), latent thermal storage materials are very effective. They store 5–14 times more heat per unit volume than sensible storage materials such as water, masonry, or rock. PCM exhibits the optimum qualities; it provides a minimal amount of volume for its heat of fusion, as well as having a low melting point, and hence PCM can be used as a good heat storage medium. Glauber's salt (sodium sulfate decahydrate), calcium chloride hexahydrate, and paraffin wax are the most commonly used PCMs in solar heating systems. Although these compounds are fairly inexpensive, the packaging and processing necessary to get consistent and reliable performance from them is complicated and costly. Steel and polyethylene are common packaging materials. The use of latent heat storage is especially suited to the storage of solar energy where it can result in high solar collection efficiency, which can mean that solar collector area can be reduced by 30%. Development of solid-liquid PCMs has concentrated on the following materials: linear crystalline alkyl hydrocarbons, fatty acids and esters, polyethylene glycols, long alkyl side chain polymers, the solid-state series of pentaerythritol, pentaglycerine, and neopentyl glycol, low melting metals and alloys, quaternary ammonium clathrates and semiclathrates, and salt hydrides. Additional research has led the development of PCM materials that may be designed for applications in the temperature range from just above 32–2578 °F (0–1258 °C). By blending adjacent alkyl hydrocarbon chains, a mixture having a desired single melting temperature may be produced without significant decrease in thermal energy storage (Alghoul 2005).

4.6.4 Thermal Energy Storage via Chemical Reactions

High-energy-density (i.e., 300 kWh/m^3) TES systems can be achieved using chemical reactions (e.g., thermo-chemical storage, TCS). Thermo-chemical reactions, such as adsorption (i.e., adhesion of a substance to the surface of another solid or liquid), can be used to store heat and cold, as well as to control humidity. Typical applications involve adsorption of water vapor to silica-gel or zeolites (i.e., micro-porous crystalline alumino-silicates). Of special importance for use in hot/humid climates or confined spaces with high humidity are open sorption systems based on lithium-chloride to cool water and on zeolites to control humidity. Interesting fields of application include waste heat utilization. In this context, TCSs are able to store thermal energy with high efficiency and to convert heat into cold (i.e., desiccant cooling) at the same time, which makes these systems very attractive. The high storage capacity of sorption processes also allows thermal energy transportation. The sorption TES (using zeolite/water) is charged at 150 °C, transported over seven kilometers and discharged at 180 °C. Dry and hot air during discharging are directly integrated into the drying process. The higher discharging temperature is made possible because the enthalpy of the humid air from drying is converted into a temperature lift by the adsorption of water vapor. A pilot storage in a standard freight container containing 13 tonnes of zeolite, with a storage capacity of up to three MWh and a charging power of 500 kW, is currently on the road (Hauer 2013).

4.7 Current Challenges and Prospects of Solar Power Technologies

Solar energy is a constant power source that could provide energy security and energy independence to the world. Such a propensity is hugely important not only for individuals but also for the socio-economic prosperity of companies, societies, states, and nations. Nevertheless, solar power is now being adopted as a natural and substantial part of electricity generation in many developed and developing countries to fulfill energy needs. However, there are a number of limitations and challenges as well as benefits associated with its use.

4.7.1 Limitations of Solar Power Technologies

High initial installation cost is one of the most significant flaws of the solar energy system; for example, the average price per watt for solar energy was $3.70 in the USA in early 2016. Based on an average solar energy system of 5 kW per household, the system would cost $13,000 when the Federal solar tax credit is put into consideration (thereby reducing costs by 30%). However, lengthy payback periods and small revenue streams also reduce the value of credits for such systems. Furthermore, the efficiencies of most domestic solar panels are around 10–20% which is another shortcoming of solar technology. However, more efficient (ca. >20%) solar panels are also available at higher prices. The performance limitations of other components such as batteries, inverters, etc. are other areas with considerable room for improvement. Short battery lifetimes and the safe disposal of spent batteries are another concern with regard to solar energy systems.

Moreover, batteries are often large and heavy, thereby requiring large storage space. Additionally, as solar panels are made from rare or precious metals such as silver, tellurium, or indium, in-sufficient facilities exist with which to recycle spent panels. Factors associated with the maintenance of systems such as a shortage of skilled manpower to meet growing demands for installation, maintenance, inspection, repair, and evaluation of solar power systems are another constraint as well. Furthermore, a lack of basic technical knowhow on the user's behalf (especially in rural areas of the developing world) with regard to solar power systems can result in irregular usage, over-charging the battery, polarity reversal, by-passing the charge controller, etc. which can all lead to system damage. Additionally, the plausibility of cracks within the PV module, water intrusion, exposure to dust, and algal growth can greatly lower the performance of the system. Poisonous silicon tetrachloride, a byproduct of the polysilicon production process, is costly (about $84,500 per ton) to process and recycle (Feldman et al. 2014; Kabira et al. 2018).

Another obvious shortcoming is that solar energy can only be harnessed during the day and works most efficiently when it is sunny. Thus, solar energy is likely not the most reliable source of energy in regions with unsustainable weather or climate conditions. Moreover, air pollution levels at the installation area can also influence the effectiveness of the solar cells. Finally, huge plots of land are often required to generate solar power at large scales.

4.7.2 Current Technical Challenges for Solar PV Systems

Solar is being deployed on both utility and distributed scales to provide peak load power, and concentrating solar thermal power (CSP) plants have been coupled with thermal energy storage to provide power into the evening hours. Challenges for solar technologies include reducing "soft costs" (e.g., permitting, financing, interconnection), improving integration into the grid, and increasing reliability, while continuing to lower hardware costs.

Challenges for solar PV system exist across the technology spectrum. Continued module cost reductions and power soft-cost reductions would fuel continued growth of the industry. Increasing cell and module efficiencies and reliability, addressing integration-related challenges associated with high penetration, and streamlining installation through plug-and-play designs will be important. Improving the efficiency of converting sunlight to electricity has the benefit of both reducing the cost per watt (W) of PV modules and many soft costs as well. Building integrated photovoltaics (BIPV) also has the potential to reduce costs by reducing installation labor and building materials costs. Additionally, improving the life-cycle sustainability of PV modules and system components, through either improved recycling techniques or modules made from earth abundant materials, will contribute to reducing the LCOE from PV systems and minimizing the long-term environmental impact of PV as the technology becomes more mature (Friedman et al. 2013).

For CSP, the largest barriers to adoption are the high overall costs of the systems (in particular the collector field and thermal storage systems), and the cost of capital, which increases the overall LCOE of a system. In the long term, technical challenges, including increasing the temperatures at which CSP plants operate, as well as the thermal efficiency of plant materials, such as heat exchangers and receivers, need to be addressed in order to significantly reduce costs. Additionally, increasing the lifetime of plant materials, either through more resilient materials or less corrosive heat transfer fluids, has the potential to significantly decrease O&M costs. Finally, significant challenges exist with respect to integrating solar into the grid and reducing non-hardware "soft costs." A combination of developments in PV and CSP technology and changes to the electric grid will need to be implemented in order to accommodate high penetration levels of solar on both distribution and transmission networks. Additionally, developing innovative and scalable solutions to streamline processes and enable robust and sustainable market solutions, will reduce the soft costs of solar.

Despite the rapid increase in deployment, significant work remains before solar achieves unsubsidized cost competitiveness with conventional energy sources. Novel processes for integrating solar generation into the grid must be developed. Supporting advanced inverter technologies, using next-generation storage, and developing electricity market solutions to ensure that solar energy can be utilized in a safe and reliable manner will become an increasingly important area of focus, as larger amounts of solar energy is deployed. "Soft costs" represent an increasingly large fraction of system cost and must be reduced. Hardware innovations also have the potential to significantly increase solar deployment. For PV, manufacturing improvements could increase efficiencies and reliabilities, and lower costs. CSP has a very large technical potential, but needs significant improvements in performance and cost reductions (Friedman et al. 2013).

4.7.3 Benefits of Solar Power Technologies

As solar power is theoretically abundant enough, it is more than capable of fulfilling the world's electricity demands. Because solar energy is not only sustainable but also renewable, it is not necessary to consider the notion that solar energy may eventually be depleted. Greenhouse gases (GHG) emissions associated with the generation of solar power (including manufacturing, installation, operation, and maintenance) are minimal. On the other hand, electricity generated from solar installations does not require water to operate; additionally, the existence of fuel by-products or the requirement for radioactive waste storage is nonexistent. Hence, solar energy is considered to be a nonpolluting, reliable, and clean source of energy.

From an economical perspective, solar power is beneficial in a variety of ways due to tax incentives, an elimination of electricity bills, increased property values, and high durability. The efficiency of solar power technologies has increased greatly in recent years and has been accompanied by a progressively steady decline in costs, which are projected to drop even further. For instance, the total cost of a PV module has been reduced from USD 1.30 per watt(in 2011) to USD 0.50 per watt (in 2014) (ca. 60% cost reduction).As the solar markets mature and more companies take advantage of the solar economy, the availability and affordability of solar power will grow at an impressive pace. Although solar power systems require an upfront investment for their installation, they otherwise operate at very low costs. Unlike the price of fossil fuels, which are prone to substantial price swings, the financial demand for solar power is relatively stable over long periods. Moreover, there are no (mechanically) moving parts in solar panels, making them free of noise pollution and durable (no wear and tear),with very little in the way of required maintenance. Furthermore, solar power systems are less prone to large-scale failure because they are distributed and composed of numerous individual solar arrays. Therefore, if any section of arrays were found to be faulty, the rest could continue to operate. However, additional solar modules could also be added overtime to improve the energy generation capacity. These notions reveal huge advantages in the ruggedness and flexibility of solar power systems over all other energy sources that have already been established (Feldman et al. 2014; Kabira et al. 2018).

Indeed, solar power has a vast resource base and incredible technical potential. For example, PV panels on 0.6% of the nation's land could supply enough electricity to power the entire United States. PV is flexible in size and deployment and can be integrated into the built environment on building rooftops and facades, parking lots, and abandoned or degraded land close to population centers. Additionally, placing CSP in suitable and available land in seven southwest states could theoretically provide four times the current US annual electricity demand, for example. CSP also provides a stable and cost-effective form of energy storage, and it can cogenerate with on-site fossil energy sources (DOE 2012).

4.7.4 Future Prospects of Solar Power Technologies

Solar power energy is one of the best options to meet future energy demand since it is superior in terms of availability, cost-effectiveness, accessibility, capacity, and efficiency compared to other renewable energy sources. For instance, the flow of solar energy have been successfully measured in and between different parts of a photosynthetic organism, which is a first step in research that could ultimately contribute to the development of technologies that use solar energy far more efficiently than what is currently possible. It has been shown that the lifetime of perovskite solar cells can significantly enhanced by using few-layer MoS_2 flakes as an active buffer interface layer. Furthermore, perovskite-silicon tandem solar cells have been developed with a conversion efficiency of 25.5%. The efficiency of perovskite solar cells was only 3.8% when first appeared in 2009. Hence, semitransparent perovskite solar cells have been created, which demonstrate high-power conversion efficiency and transmit visible light while blocking infrared light, making them great candidates for solar windows. It was demonstrated that the polymer poly (3,4-ethylenedioxythiophene) should have great potential for cost-effective and highly efficient perovskite solar cells as a hole transporting material (Kabira et al. 2018).

In addition, a new solar cell that combines two different layers of sunlight-absorbing material has been developed to harvest a broader range of the sun's energy. Using a heat-resistant device, made of tungsten and alumina layers, the device can absorb the sun's broad spectrum radiation and convert it to electricity. A green polymer derived from bio-waste was applied to the dye-sensitized solar cells. Chitosan obtained from the insects and crustaceans chitin was modified to produce the phthaloylchitosan electrolyte for the dye-sensitized solar cells with efficiency of more than 7%. CdTe and $Cu(In,Ga)Se_2$ thin-film solar cells were also seen to have high efficiencies of around 16.5% and 20%, respectively. The $CH_3NH_3Pb_{0.75}Sn_{0.25}I_3$ perovskite solar cells with inverted structure were consequently found to have a maximum power conversion efficiency of 14.1%. Very high power-conversion efficiencies have also been achieved under ambient light conditions by a dye-sensitized solar cell. More research efforts will be dedicated towards PV technologies to enhance their efficiency, stability, manufacturability, and availability, to reduce balance-of-system (BOS) costs and reduce the costs of modules (Chirumamilla et al. 2016; Kabira et al. 2018).

4.8 Summary

Solar energy, radiant light and heat from the sun, is harnessed using a range of ever-evolving technologies such as solar heating, solar photovoltaic, solar thermal electricity, solar architecture and artificial photosynthesis. Photovoltaic (PV) solar cells directly convert sunlight into electricity, using the photovoltaic effect. The process works even on cloudy or rainy days, though with reduced the production and conversion efficiency. PV cells are assembled into modules to build modular PV systems that are used to generate electricity in both grid-connected and off-grid applications, such as residential and commercial buildings, industrial facilities, remote and rural areas and power plants (i.e., utility PV systems). The photovoltaic has evolutionarily progressed in using increasingly automated processes to produce ever-thinner cells in greater volumes with higher efficiencies and lower costs. Advances in PV materials and manufacturing will continue to incrementally improve the technology's cost-competitiveness and gradually expand its applications without the need for game changing scientific breakthroughs. Several broad routes to reduce manufacturing costs have been pursued, including increasing the efficiency of traditional Si-based photovoltaics, utilizing thin-film absorbers (amorphous Si, CdTe, and $CuInxGa_{1\text{-}x}$ Se_2; CIGS), and developing novel low-cost platform technologies such as organic photovoltaics (OPVs) and dye-sensitized solar cells (DSSCs). In addition, there has been extensive work on nanocrystal-based PV solar cells. The absorber layers in these devices are fabricated from nanocrystal dispersions—"inks" or "paints"—via scalable and inexpensive methods such as roll-coating or spray-coating.

Solar collectors and thermal energy storage components are the two core subsystems in solar thermal applications. A solar collector which is the special energy exchanger converts solar irradiation energy either to the thermal energy of the working fluid in solar thermal applications. Solar collectors need to have good optical performance in order to absorb heat as much as possible. For solar thermal applications, solar irradiation is absorbed by a solar collector as heat and then is transferred to the working fluid. The fundamental challenge is to develop a new generation of process heat collectors, raising the temperature, efficiency and performance of current low-temperature collectors. Therefore, the related materials development has focused on: (a) Functional surfaces (such as anti-reflective or low thermal emissivity coatings, self-cleaning surfaces and anti-corrosive coatings); (b) Highly reflective, precise and weather resistant lightweight reflectors; (c) Improved selective absorbers (for example, long-term stability in aggressive climates, such as salt-water droplets from the sea); (c) Cheaper glass with high solar transmittance and exchangeable optical properties; (d) Heat transfer fluids with higher temperature stability (above 160 °C); (e) Temperature-stable and inexpensive thermal insulation (for example, vacuum insulation); (f) New design concentrators, favoring roof mounting and low maintenance needs; (g) New heat transfer media, withstanding freezing and high operating temperatures; and (h) New concepts for system integration and overheating protection. Depending on the price evolution of current raw materials, such as copper, aluminum, and oil-based plastics, substitution materials may need to be developed. Moreover, new concepts like PV-thermal collectors will require specific basic developments, such as, cheap and reliable collectors based on new materials, which are easy to integrate into various roofs and facades; practical ways to integrate conventional metal collectors onto metal facades and roofs; collectors that can be connected with a heat pump (i.e., condensation inside the collector must not be a problem); improve heat transfer in the collector and heat exchanger; cheap, energy-efficient pumps; connectors that facilitate the quick and easy installation of collectors; evacuated tube collectors with a better price/performance ratio (soda lime glass and anti-reflective coatings); low concentration stationary or quasi-stationary collectors for higher temperature applications in buildings, which can be combined with evacuated tubular collectors; PV-thermal hybrid collectors that save installation costs and increase total conversion efficiency, compared with side-by-side systems; air collectors with improved heat transfer; and cheap sensors and electronics for fault detection and identification.

After the thermal energy is collected by solar collectors, it needs to be efficiently stored when later needed for a release. Thermal storage is one of the main parts of a solar heating, cooling, and power generating system. The energy storage system has an enormous influence on overall system cost, performance, and reliability. Thermal energy storage (TES) includes sensible heat, latent heat and chemical energy (i.e., thermo-chemical energy storage) using chemical reactions. Thermal energy storage in the form of sensible heat is based on the specific heat of a storage medium, which is usually kept in storage tanks with high thermal insulation. The most popular and commercial heat storage medium is water, which has a number of residential and industrial applications. Underground storage of sensible heat in both liquid and solid media is also used for typically large-scale applications. However, TES systems based on sensible heat storage offer a storage capacity that is limited by the specific heat of the storage medium. Phase change materials (PCMs) can offer a higher storage capacity that is associated with the latent heat of the phase change. PCMs also enable a target-oriented discharging temperature that is set by the constant temperature of the phase change. Thermo-chemical storage (TCS) can offer even higher storage capacities. Thermo-chemical reactions (e.g., adsorption or the adhesion of a substance to the surface of another solid or liquid) can be used to accumulate and discharge heat and cold on demand (also regulating humidity) in a variety of applications using different chemical reactants.

Overall, solar is being deployed on both utility and distributed scales to provide peak load power, and concentrating solar thermal power (CSP) plants have been coupled with thermal energy storage to provide power into the evening hours. Challenges for solar technologies include reducing "soft costs" (e.g., permitting, financing, and interconnection), improving integration into the grid, and increasing reliability, while continuing to lower hardware costs. Solar power has a vast resource base and incredible technical potential. PV is flexible in size and deployment and can be integrated into the built environment on building rooftops and facades, parking lots, and abandoned or degraded land close to population centers. Additionally, CSP provides a stable and cost-effective form of energy generation and storage, and it can cogenerate with on-site fossil energy sources. More research efforts will be dedicated towards PV technologies to enhance their efficiency, stability, manufacturability, and availability, to reduce balance-of-system (BOS) costs and reduce the costs of modules.

Exercises

Part I: General Questions

4.1 What are the key pathways to turn energy from the sun's rays into electricity that can be stored and transported?

4.2 Using a solar photovoltaic field of 1 square mile (2.59 km^2) with efficiency of 15%, how many kilowatt-hours will this field generate annually at locations of average daily insolation (on flat ground) of 3 h (Alaska), 4 h (New York), 5 h (Georgia), and 6 h (Arizona)? An average household consumes 1000 kWh per month. How many households can this field support in the four states, respectively?

4.3 A solar oven has a concentration mirror of 1 m^2 with a solar tracking mechanism. If the efficiency is 75%, on a sunny day, how long will it take to melt one kg of ice at 0 °C at the same temperature? How long will it take to heat it to the boiling point? How long will it take to evaporate it at 100 °C?

4.4 A solar heat collector of 2 m^2 using the running water through underground pipes at 15 °C to generate hot water at 55 °C. If the efficiency of the solar collector is 80%, on average, how much hot water can this system generate per day? (a) In a region of 5 h of daily insolation; (b) In a region of 3 h of daily insolation. The standard insolation is 1 kW/m^2.

4.5 The crown of a typical sugar maple tree is roughly a sphere of 5 m radius. If the photosynthesis process of the leaves is devoted to making syrup, on a sunny summer day, how many kilograms of condensed maple syrup (60% sugar in weight) can this tree produce? 1 eV equals 96.5 kJ/mol. Use the experimental value of photosynthesis efficiency (5%) to estimate the solar radiation required to produce 1 kg of syrup.
4.6 A vacuum solar heat collector has an internal diameter of 45 mm and a length of 1800 mm and is filled with water of 20 °C. The axis of the tube is perpendicular to the sunlight. Assuming the efficiency is 90%, with full sunlight, how long does it take to boil the water inside the tube?
4.7 What routes have been pursued to reduce manufacturing costs of photovoltaic solar cells? How different generations of photovoltaic technologies were divided?
4.8 Describe the differences of mono-crystalline, multi-crystalline, and emitter wrap through (EWT) structures. List their advantages and disadvantages.
4.9 What types of thin-film solar cells have been developed? List their advantages and disadvantages.
4.10 Define compound semiconductor photovoltaic cells; what are their major applications?
4.11 Describe the role of nanotechnology for photovoltaic cell fabrication.
4.12 What are hybrid solar cells? List their advantages and disadvantages.
4.13 Describe currents status and future development trends of advanced materials for solar thermal collectors. What are desirable features of solar thermal collector materials?
4.14 What kinds of reflecting materials are used for solar cookers?
4.15 Describe currents status and future development trends of optical materials for solar absorbers.
4.16 Describe applications of different thermal storage materials.

Part II: Thought-Provoking Questions

4.17 Why do new types of electronic materials, based on the unusual properties of quantum physics, exhibit amazing optical properties beyond what is expected?
4.18 How can we design and synthesize materials that exhibit transformative quantum properties for energy applications?
4.19 Discuss current status and future trends of photovoltaic and solar thermal technologies.
4.20 Describe the advantages and disadvantages of 3D printing for solar cell fabrications.
4.21 Discuss the applications of metamaterials for future solar energy systems.

References

Akarslan, F.: Photovoltaic systems and applications. In: Arzu Şencan (Ed.), Modeling and Optimization of Renewable Energy Systems, ISBN: 978–953–51-0600-5, InTech, doi: https://doi.org/10.5772/39244. http://www.intechopen.com/books/modeling-and-optimization-of-renewable-energy-systems/photovoltaic-systems-and-applications (2012). Accessed 24 March 2014

Alghoul, M.A.: Review of materials for solar thermal collectors. Anti-Corrosion Methods Mater. **52** (4), 199–206 (2005)

Azad, A.K., Kort-Kamp, W.J.M., Sykora, M., Weisse-Bernstein, N.R., Luk, T.S., Taylor, A.J., Dalvit, D.A.R., Chen, H.-T.: Metasurface broadband solar absorber. Sci. Rep. **6**, 20347 (2016)

Baskoutas, S., Terzis, A.F.: Size-dependent band gap of colloidal quantum dots. J. Appl. Phys. **99**, 013708 (2006)

Bertolli, M.: Solar cell materials. http://sces.phys.utk.edu/~dagotto/condensed/HW1_2008/BertolliSolarCellMaterials.pdf (2008). Accessed 28 March 2014

Boston, R., Schnepp, Z., Nemoto, Y., Sakka, Y., Hall, S.R.: In situ TEM observation of a microcrucible mechanism of nanowire growth. Science. **344**(6184), 623–626 (2014)

Chaar, L.E., Iamont, L.A., Zein, N.E.: Review of photovoltaic technologies. Renew. Sust. Energ. Rev. **15**, 2165–2175 (2011)

Charlier, J.C., et al.: Enhanced electron field emission in B-doped carbon nanotubes. Nano Lett. **2** (11), 1191–1195 (2002)

Chaudhary, S., et al.: Hierarchical placement and associated optoelectronic impact of carbon nanotubes in polymer-fullerene solar cells. Nano Lett. **7**(7), 1973–1979 (2007)

Chen, C., et al.: Nanowelded carbon-nanotube-based solar microcells. Small. **4**(9), 1313–1318 (2008)

Chen, H.-T., Taylor, A.J., Yu, N.: A review of metasurfaces: physics and applications. Rep. Progress Phys. **79**, 076401 (2016)

Chirumamilla, M., Roberts, A.S., Ding, F., Wang, D., Kristensen, P.K., Bozhevolnyi, S.I., Pedersen, K.: Multilayer tungsten-alumina-based broadband light absorbers for high-temperature applications. Opt. Mater. Express. **6**(8), 2704 (2016)

Cinke, M., et al.: Pore structure of raw and purified HiPco single-walled carbon nanotubes. Chem. Phys. Lett. **365**, 69 (2002)

Claremont: Chapter 2: Basic principles of solar technology. http://sustainability.claremont.edu/solar/chapter2.pdf (2013). Accessed 6 December 2013

Collins, P.G., et al.: Extreme oxygen sensitivity of electronic properties of carbon nanotubes. Science. **287**(5459), 1801–1804 (2000)

Dayal, S., et al.: Photovoltaic devices with a low band gap polymer and CdSe nanostructures exceeding 3% efficiency. Nano Lett. **10**(1), 239–242 (2010)

van de Lagemaat, J., et al.: Organic solar cells with carbon nanotubes replacing In_2O_3:Sn as the transparent electrode. Appl. Phys. Lett. **88**(23), 233503-1–233503-3 (2006)

DOE: SunShot Vision Study. U.S. Department of Energy. http://energy.gov/eere/sunshot/sunshot-vision-study. Accessed 6 July 2017, Washington, DC (2012)

ESTTP: Solar Heating and Cooling for a Sustainable Energy Future in Europe. http://www.estif.org/fileadmin/estif/content/projects/downloads/ESTTP_SRA_RevisedVersion.pdf (2007). Accessed 29 April 2014

Feldman, D., Barbose, G., Margolis, R., Darghouth, N., James, T., Weaver, S., Goodrich, A., Wiser, R.: Photovoltaic System Pricing Trends: Historical, Recent, and Near-Term Projections. 2014 Edition. http://www.nrel.gov/docs/fy14osti/62558.pdf (2014). Accessed 6 July 2017

Friedman, B., Ardani, K., Feldman, D., Citron, R., Margolis, R., Zuboy, J.: Benchmarking Non-hardware Balance-of-System (Soft) Costs for U.S. Photovoltaic Systems, Using a Bottom-Up Approach and Installer Survey, 2nd edn. National Renewable Energy Laboratory, Golden, CO NREL/TP-6A20–60412. Available at: http://www.nrel.gov/docs/fy14osti/60412.pdf (2013). Accessed 6 July 2017

Garnett, E.C.: Nanowire solar cells. Ann. Rev. Mater. Res. **41**, 269 (2011)

Ginger, D.S., Greenham, N.C.: Photoinduced electron transfer from conjugated polymers to CdSe nanocrystals. Phys. Rev. B. **59**(16), 624–629 (1999)

Green, M.A.: Third Generation Photovoltaics. Springer Verlag (2003)

Hao, J., Wang, J., Liu, X., Padilla, W.J., Zhou, L., Qiu, M.: High performance optical absorber based on a plasmonic metamaterial. Appl. Phys. Lett. **96**(25), 251104 (2010)

Harrison, J.: Investigation of reflective materials for the solar cooker. http://www.fsec.ucf.edu/en/research/solarthermal/solar_cooker/documents/reflectivematerialsreport.pdf (2001). Accepted 12 May 2014

Hasobe, T., et al.: Organized assemblies of single wall carbon nanotubes and porphyrin for photochemical solar cells: charge injection from excited porphyrin into single-walled carbon nanotubes. J. Phys. Chem. B. **110**(50), 25477–25484 (2006)

Hauer, A.: Thermal energy storage: Technology brief. IEA-ETSAP and IRENA© Technology Brief E17. http://www.irena.org/DocumentDownloads/Publications/IRENA-ETSAPTechBrief E17ThermalEnergyStorage.pdf (2013). Accessed May 20, 2014

Hedayati, M.K., et al.: Review of plasmonic nanocomposite metamaterial absorber. Materials. **7**, 1221–1248 (2014)

Herman, D.J., et al.: Orienting periodic organic–inorganic nanoscale domains through one-step electrodeposition. ACS Nano. **5**(1), 565–573 (2011)

Hulshorst, W.: Residential photovoltaic systems. http://www.leonardo-energy.org/sites/leonardo-energy/files/root/pdf/2008/howto-photovoltaics.pdf (2008). Accessed 13 February 2014

Ibn-Mohammeda, T., Koha, S.C.L., Reaneyc, I.M., Acquayed, A., Schileoc, G., Mustaphae, K.B., Greenough, R.: Perovskite solar cells: An integrated hybrid lifecycle assessment and review in comparison with other photovoltaic technologies. Renew. Sust. Energ. Rev. **80**, 1321–1344 (2017)

IRENA: Solar photovoltaic—technology brief. IEA-ETSAP and IRENA© Technology Brief E11—January 2013. http://www.irena.org/DocumentDownloads/Publications/IRENAETSAPTechBriefE11SolarPV.pdf (2013). Accessed 12 February 2014

Jung, Y., Li X., Rajan, N.K., Taylor, A.D., Reed, M.A.: Record high efficiency single-walled carbon nanotube/silicon p–n iunction solar cells. Nano Letters **13**(1), 95–99 (2013)

Kabira, E., Kumarb, P., Kumarc, S., Adelodund, A.A., Kim, K.-H.: Solar energy: potential and future prospects. Renew. Sust. Energ. Rev. **82**, 894–900 (2018)

Kennedy, C.E.: Review of mid- to high-temperature solar selective absorber materials. NREL/TP-520-31267. http://www.osti.gov/bridge (2002). Accepted 15 May 2014

Kim, K.H., Han, C.H.: A review on solar collector and solar organic rankine cycle (ORC) systems. J. Autom. Control Eng. **3**(1), 66–73 (2015)

Kymakis, E., et al.: High open-circuit voltage photovoltaic devices from carbon-nanotube-polymer composites. J. Appl. Phys. **93**(3), 1764–1768 (2003)

Lancelle-Beltran, E., Prené, P., Boscher, C., Belleville, P., Buvat, P., Sanchez, C.: All-solid-state dye-sensitized nanoporous TiO_2 hybrid solar cells with high energy-conversion efficiency. Adv. Mater. **18**(19), 2579–2582 (2006)

Landy, N.I., Sajuyigbe, S., Mock, J.J., Smith, D.R., Padilla, W.J.: Perfect metamaterial absorber. Phys. Rev. Lett. **100**(20), 207402 (2008)

Le Treut, H., Somerville, R., Cubasch, U., Ding, Y., Mauritzen, C., Mokssit, A., Peterson, T., Prather, M.: Historical overview of climate change. In: Solomon, S., Qin, D., Manning, M., Chen, Z., Marquis, M., Averyt, K.B., Tignor, M., Miller, H.L. (eds.) Climate Change 2007: The Physical Science Basis. Contribution of Working Group I to the Fourth Assessment Report of the Intergovernmental Panel on Climate Change. Cambridge University Press, Cambridge, United Kingdom and New York, NY (2007)

Lee, J.U.: Photovoltaic effect in ideal carbon nanotube diodes. Appl. Phys. Lett. **87**(3), 073101 (2003)

Lee, T.Y., et al.: Fabrication of dye sensitized solar cell using TiO_2 coated carbon nanotubes. Thin Solid Films. **515**(12), 5131–5135 (2007)

Li, Z., Saini, V.: Polymer functionalized n-type single wall carbon nanotube photovoltaic devices. Appl. Phys. Lett. **96**(3), 033110 (2010)

Liu, X., Starr, T., Starr, A.F., Padilla, W.J.: Infrared spatial and frequency selective metamaterial with nearunity absorbance. Phys. Rev. Lett. **104**(20), 207403 (2010)

Louwen, A., van Sark, W., Schropp, R., Faaij, A.: Solar Energy Mater. Solar Cells. **147**, 295–314 (2016)

Luque, A., Hegedus, S.: Handbook of Photovoltaic Science and Engineering. John Wiley & Sons (2011). ISBN 978-0-470-97612-8

Milliron, D.J., Gur, I., Alivisatos, A.P.: Hybrid organic–nanocrystal solar cells. MRS Bullet. **30**, 41–44 (2005)

Moser, J.: Solar cells: later rather than sooner. Nat. Mater. **4**(10), 723–724 (2005)

Mukhopadhyay, P.: Graphite, Graphene and their polymer nanocomposites, pp. 202–213. Taylor & Francis Group, Boca Raton, FL (2013). ISBN 978-1-4398-2779-6

Niklasson, G.A., Granqvist, C.G.: Selectively solar-absorbing surfaces: optical properties and degradation. In: Granqvist, C.G. (ed.) Materials Science for Solar Energy Conversion Systems, pp. 70–105. Pergamon, Oxford (1991)

Parida, B., Iniyan, S., Goic, R.: A review of solar photovoltaic technologies. Renew. Sust. Energ. Rev. **15**, 1625–1636 (2011)

Ruiz-Morales, J.C., Tarancón, A., Canales-Vázquez, J., Mendez-Ramos, J., Hernandez-Afonso, L., Acosta-Mora, P., Marin Ruedac, J.R., Fernandez-Gonzalez, R.: Three dimensional printing of components and functional devices for energy and environmental applications. Energy Environ. Sci. **10**, 846–859 (2017)

Saga, T.: Advances in crystalline silicon solar cell technology for industrial mass production. NPG Asia Mater. **2**, 96–102 (2010)

Sargent, H.E.: Infrared quantum dots. Adv. Mater. **17**, 515 (2005)

Saunders, B.R.: Hybrid polymer/nanoparticle solar cells: Preparation, principles and challenges. J. Colloid Interface Sci. **369**(1), 1–15 (2012)

Saunders, B.R., Turner, M.L.: Nanoparticle-polymer photovoltaic cells. Adv. Colloid Interf. Sci. **138**(1), 1–23 (2008)

Scheblykin, I.G., et al.: Excited state and charge photogeneration dynamics in conjugated polymers. J. Phys. Chem. B. **111**(23), 6303–6321 (2007)

Seraphin, B.O., Meinel, A.B.: Solar energy conversion and the optical properties of solids. In: Seraphin, B.O. (ed.) Optical Properties of Solids: New Developments, pp. 927–971. North Holland Publishing Co., Amsterdam (1976)

Shaheen, S.E., et al.: Organic-based photovoltaics. MRS Bullet. **30**, 10 (2005)

Sharma, M.K., Tanwar, D., Singh, V.: Study of modification in the efficiency of solar cell. J. Global Res. Comput. Sci. **4**(4), 112–118 (2013)

Sharma, S., Jain, K.K., Sharma, A.: Solar cells: in research and applications—a review. Mater. Sci. Appl. **6**, 1145–1155 (2015)

Shaw, P.E., Ruseckas, A., Samuel, I.D.W.: Exciton diffusion measurements in poly (3-hexylthiophene). Adv. Mater. **20**, 3516–3520 (2008)

Sofos, M.: A synergistic assembly of nanoscale lamellar photoconductor hybrids. Nat. Mater. **8**(1), 68–75 (2009)

Somani, P.R., et al.: Application of metal nanoparticles decorated carbon nanotubes in photovoltaics. Appl. Phys. Lett. **93**(3), 033315 (2008)

Steinhagen, C.R.: $CuInSe_2$ nanowires and earth-abundant nanocrystals for low-cost photovoltaics. The University of Texas at Austin. PhD dissertation (2013)

Tao, H., Bingham, C.M., Strikwerda, A.C., Pilon, D., Shrekenhamer, D., Landy, D.I., Fan, K., Zhang, X., Padilla, W., Averitt, R.: Highly flexible wide angle of incidence terahertz metamaterial absorber: Design, fabrication, and characterization. Phys. Rev. B. **78**(24), 241103R (2008)

Voss, K., Platzer, W., Robinson, P: Education of architects in solar energy and environment—Advanced glazing. http://www.cenerg.ensmp.fr/ease/glazing_overheads.pdf (2014). Accessed 26 June 2017

Wallentin, J., Anttu, N., Asoli, D., Huffman, M., Aberg, I., Magnusson, M.H., Siefer, G., Fuss-Kailuweit, P.: InP nanowire array solar cells achieving 13.8% efficiency by exceeding the ray optics limit. Science. **339**, 1057–1060 (2013)

Wang, H., Wang, L.: Perfect selective metamaterial solar absorbers. Opt. Express. **21**(S6), A1078–A1093 (2013)

Watts, C.M., Liu, X., Padilla, W.J.: Metamaterial electromagnetic wave absorbers. Adv. Mater. **24** (23), OP98–OP120 (2012)

Weickert, J.: Nanostructured organic and hybrid solar cells. Adv. Mater. **23**, 1810 (2011)

Wu, L., Tian, W, Jiang, X.: Silicon-based solar cell system with a hybrid PV module. Solar Energy Materials & Solar Cells **87**, 637–645 (2005)

Zhang, Q.-C., Mills, D.R.: Very low-emittance solar selective surfaces using new film structures. J. Appl. Phys. **72**(7), 3013–3021 (1992)

Zheng, B., Wang, W., Jiang, G., Wang, K., Mei, X.: Research status and application prospects of manufacturing technology for micro-nano surface structures with low reflectivity. Proc. Inst. Mech. Eng. Part B: J. Eng. Manuf. **229**(11), 1877–1892 (2015)

Advanced Materials Enable Renewable Geothermal Energy Capture and Generation

5

Abstract

Geothermal power is a promising renewable energy source able to provide naturally a continuous baseload power. Original exploitation of this technology limited to locations where geothermal heat is easily accessible, such as naturally occurring hot springs, steam vents, or hot fluids at shallow depths. Engineered geothermal systems (EGS) have been developed, which involves deep drilling to exploit the heat from impermeable hot rocks, and could contribute to make geothermal sources more widespread available. To make these incentives also economically viable, innovative materials solutions and an improved understanding of the long-term interaction between the materials and their harsh environment is of key importance. An important contribution comes by researching hard materials for drill bits to extend their lifetime in highly abrasive and corrosive environments at high temperatures and developing site-specific materials for proppants in conjunction with stimulation techniques. Improved monitoring of the downhole requires materials developments to make fiber optic cables and power electronics withstand the hostile environment they should operate in. When assessing the heat reservoir and the subsequent production phase, the accumulated deposition of material inside the pipes (scaling) and the extreme corrosion and temperature problems need to be tackled from a materials' perspective. This involves the development of corrosion resistant materials for the pipes, equipped with protective outer coatings and insulation, and inner liners. Novel polymeric, ceramic, or metallic membranes to separate and reinject gases would make the operation of a zero carbon emission plant possible. During the operation, continuous monitoring of the system should allow for early intervention thus reducing the risk of a fatal breakdown of a well too early in its exploitation life. Also the downtime due to replacement or maintenance of instrumentation such as downhole pumps could be reduced by selecting specific metal alloys. This chapter will provide a brief review on advanced materials that enable renewable geothermal energy capture and generation.

C. Tong, *Introduction to Materials for Advanced Energy Systems*,
https://doi.org/10.1007/978-3-319-98002-7_5

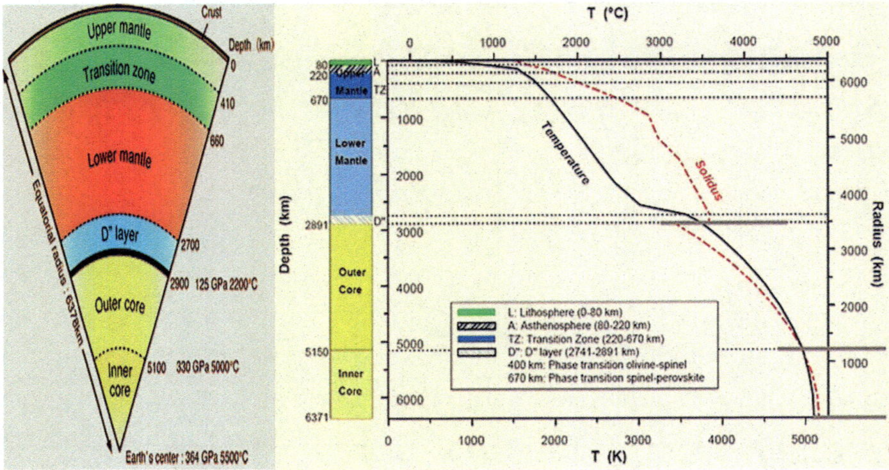

Fig. 5.1 Variation of estimated temperature and melting point of the Earth with depth (Modified with permission from Stacey 1997 (John Wiley & Sons))

5.1 The State of Geothermal Technologies

Geothermal technologies use renewable energy resources to generate electricity and/or heating and cooling while producing very low levels of greenhouse-gas (GHG) emissions. Geothermal energy is stored in rock and in trapped vapor or liquids, such as water or brines, as shown in Fig. 5.1 (Clauser 2006); these geothermal resources can be used for generating electricity and for providing heat (and cooling). Electricity generation usually requires geothermal resources temperatures of over ~150 °C. For heating, geothermal resources spanning a wider range of temperatures can be used in applications such as space and district heating, spa and swimming pool heating, greenhouse and soil heating, aquaculture pond heating, industrial process heating and snow melting. Space cooling can also be supplied through geothermal heat, through the use of heat-driven adsorption chillers as an alternative to electrically driven compression chillers. Even the modest temperatures found at shallower depths can be used to extract or store heat for heating and cooling by means of ground source heat pumps (GSHP). GSHP are a widespread application for geothermal energy, especially in colder climates, but they follow a different concept from deep geothermal heat technologies (Tanaka 2011).

Geothermal typically provides baseload generation, since it is generally immune from weather effects and does not show seasonal variation. The baseload characteristic of geothermal power distinguishes it from several other renewable technologies that produce variable power. Increased deployment of geothermal energy does not impose load-balancing requirements on the electricity system. Geothermal power could be used for meeting peak demand through the use of submersible pumps tuned to reduce fluid extraction when demand falls. However, procedures and methods that allow for a truly load-following system have yet to be developed. Geothermal energy is compatible with both centralized and distributed energy generation and can produce both electricity and heat in combined heat and power (CHP) plants (Tanaka 2011).

INDEPTH: Geothermal Power (Phillips et al. 2013; Jeanloz et al. 2013)
Geothermal power taps into earth's internal heat as an energy source. Vast amounts of heat are contained in the interior of the earth from the slow decay of radioactive elements and the heat remaining from earth's formation. This heat flows to the surface at low rates everywhere on earth (terrestrial average is ~60 mW/m^2). Specific locations, such as settings of crustal extension or near magma emplacement, have anomalously high heat flow. However, the demand for geothermal power generation is tempered by the high up-front costs and early-stage risk associated with geothermal development. Despite the relatively low surface flux of heat, owing to the low conductivity of rocks, high temperatures occur at accessible depths. Several geothermal technologies exist or are in development to capture this heat for power generation (if the fluid temperature is sufficiently high, over ~150 °C, although new power conversion technologies are pushing down the threshold for economic power generation) or direct use for heating, cooling, or industrial uses (typically lower temperature fluids, less ~150 °C). Specific locations have a favorable combination of high heat flow and natural fluid circulation that make them suitable for geothermal power generation. The naturally circulating hot fluid can be tapped to generate power in these naturally occurring hydrothermal systems. Enhanced geothermal systems (EGSs) are engineered reservoirs created to produce electricity from geothermal systems that are not otherwise economical owing to lack of water and/or permeability. In an EGS, fluid is injected into the subsurface, which causes preexisting fractures to reopen. This increases permeability and allows fluid to circulate throughout the rock and transport heat to the surface where electricity can be generated. Though the concept has existed for several decades to engineer fracture networks in the subsurface to mine geothermal heat, an EGS is an immature and high-risk technology.

Geothermal technology development has focused so far on extracting naturally heated steam or hot water from natural hydrothermal reservoirs. However, geothermal energy has the potential to make a more significant contribution on a global scale through the development of the advanced technologies, especially the exploiting of hot rock resources using enhanced geothermal systems (EGS). EGS techniques that would enable energy recovery from a much larger fraction of the accessible thermal energy in the Earth's crust (Tanaka 2011).

5.1.1 Geothermal Resources for Geothermal Energy Development

Although the use of geothermal hot springs has been known since ancient times, active geothermal exploration for industrial purposes started at the beginning of the nineteenth century with the use of geothermal fluids (boric acid) in Larderello (Italy). At the end of the nineteenth century, the first geothermal district heating system began operating in Boise (United States), with Iceland following in the 1920s. At the start of the twentieth century, again in Larderello, the first successful attempt to produce electricity from geothermal heat was achieved. Since then, installed geothermal electricity has steadily increased (Tanaka 2011).

5.1.1.1 Hydrothermal Resources

Originally utilization of geothermal energy was concentrated in areas where geological conditions permit a high-temperature circulating fluid to transfer heat from within the Earth to the surface through wells that discharge without any artificial lift. The fluid in convective hydrothermal resources can be vapor (steam), or water-dominated, with temperatures ranging from 100 °C to over 300 °C. High-temperature geothermal fields are most common near tectonic plate boundaries, and are often associated with volcanoes and seismic activity, as the crust is highly fractured and thus permeable to fluids, resulting in heat sources being readily accessible. Most plate boundaries are below sea level. There are 67,000 km of mid-ocean ridges, of which 13,000 km have been studied, and more than 280 sites with submarine geothermal vents have been discovered. Some submarine vents have been estimated to be able to realize capacities ranging from 60 MWt to 5 GWt. In theory, such geothermal vents could be exploited directly without drilling and produce power by means of an encapsulated submarine binary plant. However, technologies are needed to commercially tap energy from offshore geothermal resources (Tanaka 2011).

Geothermal heat can also be economically extracted from many deep aquifer systems all over the world. Many such locations can be reached within a depth of 3 km, with moderate heat flow in excess of 50–60 MW/m^2 and rock and fluid temperatures of in excess of 60 °C. The actual local performance depends strongly on the natural flow conditions of the geothermal reservoir. Geo-pressured deep aquifer systems contain fluids at pressures higher than hydrostatic (Barnett and Quinlivan 2009).

Water co-produced during oil and gas exploitation is another type of hydrothermal resource. Oil and gas wells can produce warm water that is often seen by operators as a by-product with limited commercial upside. This could be turned into an asset by extracting the energy contained in the produced water by means of binary cycle power plants (Tanaka 2011).

5.1.1.2 Hot Rock Resources

So far, utilization of geothermal energy has been concentrated in areas of naturally occurring water or steam, and sufficient rock permeability. However, the vast majority of geothermal energy within drilling reach—which can be up to 5 km, given current technology and economics—is in relatively dry and low-permeability rock. Heat stored in low-porosity and/or low-permeability rocks is commonly referred to as hot rock resources. These resources are characterized by limited pore space and/or minor fractures and therefore contain insufficient water and permeability for natural exploitation (Tanaka 2011).

Hot rock resources can be found anywhere in the world, although they are found closer to the surface in regions with an increased presence of naturally occurring radioactive isotopes (e.g., South Australia) or where tectonics have resulted in a favorable state of stress (e.g., in the western USA). In stable, old continental tectonic provinces, where temperature gradients are low (7–15 °C/km) and permeability is low but with less favorable state of stress, depths will be significantly greater and developing an EGS resource will be less economic. Technologies that allow energy to be tapped from hot rock resources are still in the demonstration stage and require innovation and experience to become commercially viable. The best-known such technology is enhanced geothermal systems (EGS). Other approaches to engineering hot rock resources, which are still at the conceptual phase, try methods other than fracturing the hot rock. Such technologies aim instead to create connectivity between water inlet and water outlet, for example by drilling a subsurface heat exchanger made of underground tubes or by drilling a 7–10 km vertical well of large diameter that contains water inlet and water outlet at different depths (Barnett and Quinlivan 2009; Tanaka 2011).

5.1.1.3 Resource Assessing and Engineering

Geothermal resources are found deep beneath the surface so exploration is needed to locate and assess them. Exploration consists of estimating underground temperature, permeability, and the presence of fluid, as well as the lateral extent, depth, and thickness of the resource, by using geosciences methods and by drilling exploration wells. The local state of stress must be assessed, too, particularly in the case of EGS. Exploration drilling involves high financial risks as it is expensive and the results are mainly unknown in advance. Wells in sedimentary, hydrothermal reservoirs, where geological formations resemble those exploited for oil and gas, can be drilled using similar methods. In contrast, economic drilling of low-cost exploration-only boreholes and drilling into deep, hard rock formations pose technical challenges requiring new and innovative solutions. Improvement of geophysical data inventories and geoscience exploration methods, as well as innovative geothermal resource assessment tools, will reduce the exploration risk and thus lower a barrier for investment in geothermal energy (Tanaka 2011).

As well as aiding resource assessment, competitive drilling technology will make it easier to access and engineer geothermal resources. Reservoir stimulation technology is also extremely important, both for hydrothermal reservoirs (where the connection of a production well to the reservoir fluids requires improvement) and for creating EGS reservoirs in hot rock resources.

Stimulation techniques to boost the conductivity and connectivity of hot rock resources will make it possible to access larger volumes of rock. Stimulation can be hydraulic, by injecting fluids, or chemical, by injecting acids or other substances that will dissolve the rock or the material filling the fractures. Both hydraulic fracturing and chemical stimulation techniques are similarly deployed in unconventional oil and gas reservoir developments. Hydraulic stimulation creates permeability, releasing seismic energy. In hydraulic fracturing, as in any sort of fluid injection or reinjection that raises underground fluid pressure, there is a risk of inducing microseismic events intense enough to be felt on the surface. Induced seismicity effects also depend on the existing stress field (Tanaka 2011).

5.1.2 Geothermal Electricity

Geothermal power plants harness the Earth's heat by utilizing natural hot springs, drilling a well into a hot aquifer (an underground layer of fluid-bearing rock), or creating artificial aquifers by pumping high-pressure water into the hot rock (see Sect. 5.1.3). The power plants route the steam or hot liquid from the geothermal reservoir through turbine/generator units—either directly or through heat exchangers—to produce electricity. Geothermal power plants today can use water in the vapor phase, a combination of vapor and liquid phases, or liquid phase only. The choice of plant depends on the depth of the reservoir, and the temperature, pressure, and nature of the entire geothermal resource. The three main types of plant are flash steam, dry steam, and binary plants, as shown in Fig. 5.2 (TEEIC 2014).

5.1.2.1 Flash Steam Plants

The most commonly found geothermal resources contain reservoir fluids with a mixture of hot liquid (water) and vapor (mostly steam). Flash steam plants, making up about two-thirds of geothermal installed capacity today, are used where water dominated reservoirs have temperatures above 180 °C. In these high-temperature reservoirs, the liquid water component boils, or "flashes," as pressure drops. Separated steam is piped to a turbine to generate electricity and the remaining hot water may be flashed again twice (double flash plant) or three times (triple flash) at progressively lower pressures and temperatures, to obtain more steam. The cooled brine and the condensate are usually sent back down into the reservoir through injection wells. Combined-cycle flash steam plants use the heat from the separated geothermal brine in binary plants to produce additional power before reinjection (Tanaka 2011).

5.1.2.2 Dry Steam Plants

Dry steam plants, which make up about a quarter of geothermal capacity today, directly utilize dry steam that is piped from production wells to the plant and then to the turbine. Control of steam flow to meet electricity demand fluctuations is easier than in flash steam plants, where continuous up-flow in the wells is required to avoid gravity collapse of the liquid phase. In dry steam plants, the condensate is usually reinjected into the reservoir or used for cooling (Tanaka 2011).

5.1.2.3 Binary Plants

Electrical power generation units using binary cycles constitute the fastest-growing group of geothermal plants, as they are able to use low- to medium-temperature resources, which are more prevalent. Binary plants, using an organic Rankine cycle

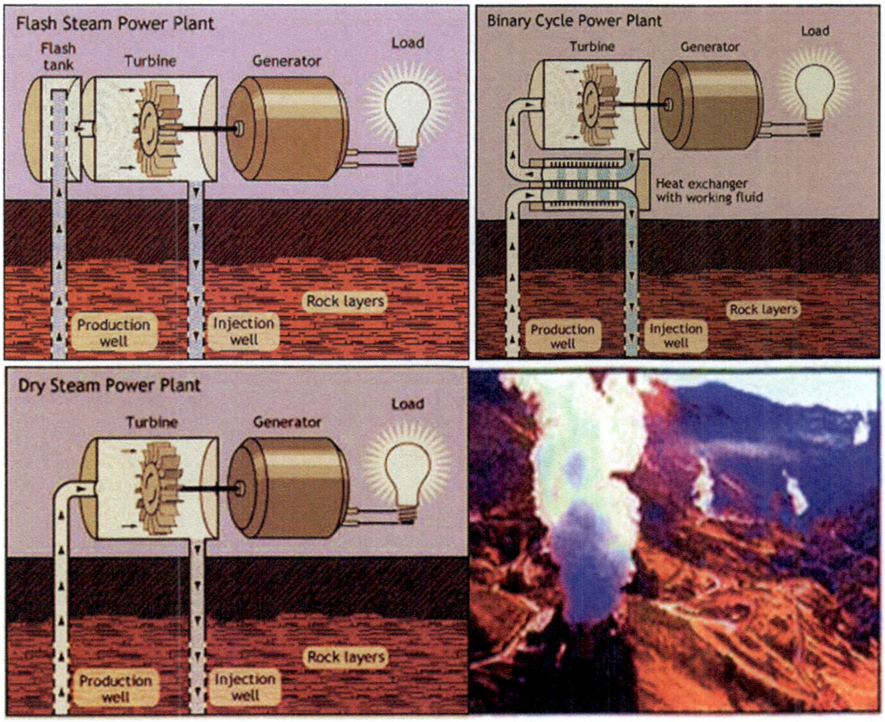

Fig. 5.2 Three main types of geothermal power plants (Modified from TEEIC 2014, available under Creative Commons Zero—OpenEI, developed and maintained by the National Renewable Energy Laboratory with funding and support from the U.S. Department of Energy and a network of International Partners & Sponsors): (**a**) flash steam, (**b**) dry steam, and (**c**) binary plants

(ORC) or a Kalina cycle, typically operate with temperatures varying from as low as 73 °C (at Chena Hot Springs, Alaska) to 180 °C. In these plants, heat is recovered from the geothermal fluid using heat exchangers to vaporize an organic fluid with a low boiling point (e.g., butane or pentane in the ORC cycle and an ammonia-water mixture in the Kalina cycle), and drive a turbine. Although both cycles were developed in the mid-twentieth century, the ORC cycle has been the dominant technology used for low-temperature resources. The Kalina cycle can, under certain design conditions, operate at higher cycle efficiency than conventional ORC plants. The lower-temperature geothermal brine leaving the heat exchanger is reinjected back into the reservoir in a closed loop, thus promoting sustainable resource exploitation (Bertani 2010).

5.1.3 Enhanced Geothermal Systems

Enhanced or engineered geothermal systems (EGS) aim at using the heat of the Earth where no or insufficient steam or hot water exists and where permeability is low, as

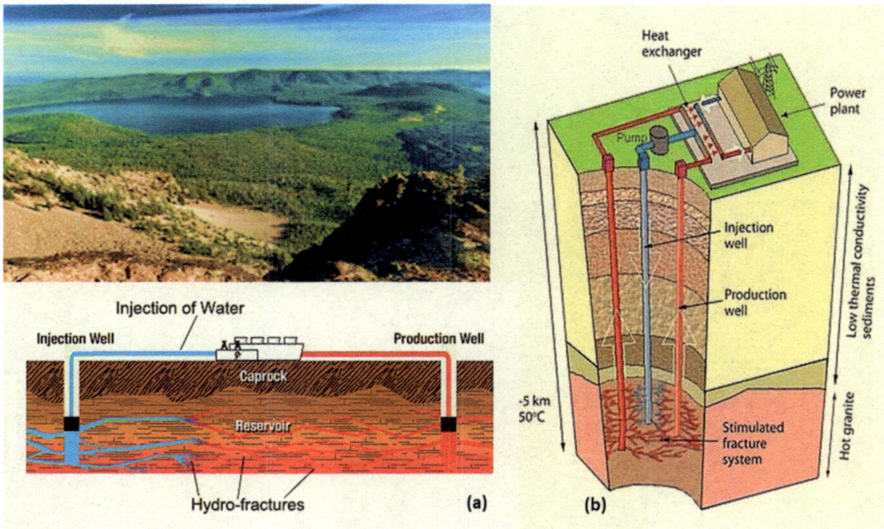

Fig. 5.3 Schematic enhanced geothermal systems (Modified from Right 2012 (Courtesy: www.renewablegreenenergypower.com); and Lakic 2014 (Courtesy: Geothermal Worldwide, Inc.)): (**a**) two-wells; and (**b**) three-wells

shown in Fig. 5.3 (Right 2012; Lakic 2014). EGS technology is centered on engineering and creating large heat exchange areas in hot rock. The process involves enhancing permeability by opening pre-existing fractures and/or creating new fractures. Heat is extracted by pumping a transfer medium, typically water, down a borehole into the hot fractured rock and then pumping the heated fluid up another borehole to a power plant, from where it is pumped back down (recirculated) to repeat the cycle. New and innovative techniques for exploration, stimulation, and exploitation are needed to make EGS technology commercially viable. The key technical and economic challenge for EGS is to achieve efficient and reliable stimulation of multiple reservoirs. This will require well-distributed heat-exchange surfaces that are large enough to attain the volumes needed for long-term production, with low flow impedance, limited short-circuiting fractures, and manageable water loss. One prerequisite for dissemination of EGS technology is proof that the underground heat exchanger can be successfully engineered under all possible site conditions (Tanaka 2011).

The challenges in EGSs are further divided into the following areas (Jeanloz et al. 2013; Ziagos et al. 2013): (a) Subsurface characterization: Efficiently and accurately locate target subsurface geologic environments and quantitatively infer their evolution through time. (b) Accessing: Safely and cost-effectively drill to prospects with high probability for success. (c) Engineering: For an EGS to succeed, methods are needed to create large networks of fractures to create enough surface area per reservoir volume to avoid uneconomically rapid thermal drawdown. An added challenge comes from achieving this goal in challenging high-pressure/high-temperature environments. (d) Sustaining: Maintain these conditions over multi-decadal time frames throughout complex system evolution. (e) Monitoring: Improve observational methods to advance understanding of the multi-scale complexity throughout system lifetimes.

Technologies for subsurface characterization will reduce the high risk associated with finding and developing geothermal reservoirs, which in turn will reduce the cost of geothermal power in new areas. Maturing EGS technologies will help make current or future unviable geothermal resources economical and eventually greatly increase the amount of economically viable geothermal resources. Unlike other renewable energy sources (e.g., solar and wind), a geothermal resource is not confirmed until a well is drilled into a target reservoir and flow is tested, costing millions of dollars. Current exploration success rates for identifying and drilling a viable resource target are such that a developer will typically perform exploration drilling at five sites in order to successfully confirm and develop one commercial project. This low success rate is the source of the very high up-front costs during early development and high associated risk. Therefore, characterization technologies are critical to advancing geothermal energy. Reservoir identification and subsurface characterization rely on four groups of tools based on: geophysics, geochemistry, remote sensing, and geology. These tools are used to assess geothermal potential by identifying target reservoir temperature, permeability, fluid availability, and geologic structures (Mines 2013).

EGS pilot plants are needed in different geological environments. Effort has also expanded to apply EGS techniques to hydrothermal fields. One such concept is to extend existing hydrothermal fields by drilling wells on their boundaries, in appropriate directions with reference to the local stress field, and stimulate them to connect the field to the main hydrothermal reservoir. This would extend the life of a hydrothermal field and deliver more power. Cross-fertilization between EGS and tight oil and gas fields has been explored: the technology is very similar and the hydrocarbon industry is aware of the potential to move EGS much faster and more successfully (Tanaka 2011).

EGS reservoir engineering requires the ability to routinely create EGS reservoirs with sufficient permeability, fracture orientations, and spacing to allow long-term energy extraction. This implies developing improved means of downhole zonal isolation for fracture stimulation and production, including packer, sliding sleeve, and multilateral technologies. New tools are needed to isolate specific zones in a hot borehole for both fracture creation and short-circuit repair, which would enable multiple fracture zones from a single borehole, enhance water circulation rates, and reduce the specific cost of the system creation and activation. Stimulation procedures need to be refined to significantly enhance hydraulic productivity, while reducing the risk associated with induced seismicity. New visualization and measurement methodologies (e.g., imaging of borehole, permeability tomography, tracer technology, coiled tubing technology) should become available for reservoir characterization (Tanaka 2011).

INDEPTH: Subsurface Engineering

EGSs can be developed through directed modifications of the subsurface to engineer desired subsurface conditions. In principle, an EGS encompasses all technologies and methodologies dedicated to improving or creating reservoirs that can provide fluids to the surface at sustained temperatures and flow rates needed for economically viable power generation. The primary pathway to achieve this result is to use thermal and hydraulic stimulation technologies to create an interconnected network of fractures through which water can circulate to extract heat from the reservoir. Significant technical challenges exist in the ability to predict the physical properties associated with the engineered fracture system. EGSs open up much larger potential resources and reduce the risks of drilling by providing options to create a viable energy production system from a much broader set of initial reservoir conditions. An EGS spans the continuum from a reservoir management and enhancement tool for hydrothermal systems to a tool to expand hydrothermal systems at the margins or to develop green field sites that have the appropriate subsurface temperatures but cannot support the required flow rates. EGSs can benefit greatly from advancements in other fields, such as materials science, that can facilitate developing tools that can withstand higher temperature and pressure environments. However, some technology issues to EGSs development include (DOE QTR 2015):

(a) Zonal isolation is critical to EGS development. The ability to stimulate the rock mass, inject fluids, and produce fluids from select regions of the wellbore is critical for the technical and economic viability of EGSs.
(b) Stimulation technologies that can affect the formation and modify permeability needed to supply sustained fluid flow at temperatures are required.
(c) Advances in technologies to measure the stress state in the subsurface are needed to improve risk quantification for induced seismicity and to develop practices to minimize induced seismicity. Appropriate management of induced seismicity is key to public acceptance of EGSs.
(d) Drilling systems that can most favorably orient the wellbore and be actively controlled and interrogated in high-temperature, hard drilling environments are required.
(e) Robust surface-based and subsurface systems to image the development, operation, and evolution of EGS reservoirs are needed.

Developing EGS requires keeping health, safety, and environmental (HSE) risks as low as reasonably practicable. The impact of felt induced seismicity or micro-seismicity at EGS sites in France, Germany, and Switzerland has created public concern. More knowledge about expected micro-seismic activity should result from the first experiences with pilot plant operation. To mitigate risks related to induced seismicity, strategies are needed to set requirements for seismic monitoring and for prolonged field operation. Technologies for imaging fluid pathways induced by hydraulic stimulation treatments would constitute a major improvement of the EGS concept. Geochemical research is needed to better understand the effects and amounts of subsurface precipitation and dissolution of injected fluids during production and injection. Careful monitoring and modeling, as well as energy-efficient utilization, are essential to sustainable geothermal reservoir management. Innovative monitoring techniques (such as interactive calibrated simulations of EGS reservoir behavior) will be needed to reliably forecast future pressure, temperature and flow trends and reservoir life. Modeling of the maximum energy production of EGS plants needs to be developed to ensure longevity of the resource. More knowledge is still needed in effects of cooling down of the host rock and preventing or repairing short-circuiting of flow paths in connected fracture systems, which can reduce the lifetime of reservoirs (Tanaka 2011).

Other conceptual approaches to engineering hot rock reservoirs merit attention. Alternative approaches aim to create connectivity between water inlet and water outlet, for example by drilling a subsurface heat exchanger made of underground tubes or by drilling a long (7–10 km) vertical tunnel of large diameter that contains both water inlet and water outlet.

Theoretical studies are needed to determine the economic and technical feasibility of such approaches, including whether they can achieve an adequate heat exchange potential.

Hot water from gas and oil wells is another source of hydrothermal fluid. Normally regarded as an unwanted waste product, under favorable circumstances this heat could be turned into an asset. In addition, geothermal exploitation has already raised interest from the hydrocarbon industry in a few cases where it has supported remote field operations as well as preheating of unconventional, very viscous, petroleum deposits. Exploration of the feasibility of exploiting geothermal resources with temperatures higher than 375 °C, so-called super-critical fluids, should be continued (Tanaka 2011).

The potential of unconventional hydrothermal resources, such as offshore geothermal vents or even magma, needs to be assessed. The necessary technology for offshore geothermal has not yet been developed, but it may be possible to exploit ocean-floor geothermal vents directly without drilling, using an encapsulated submarine binary plant to produce power. The main challenges for offshore geothermal development include the distance to shore, water depth, grid connection, and disturbance of marine life around hydrothermal vents. Technology developed to exploit offshore wind and ocean energy may help meet these challenges (Tanaka 2011).

5.1.4 Direct Use of Geothermal Energy

Historically, direct use of geothermal resources has been on a small scale and on an individual basis, recently the direct use has focused more on the developments of major district heating systems, greenhouses or aquaculture complexes, or major industrial uses. Heat pumps utilizing very low-temperature geothermal fluids (120°) have extended geothermal developments into traditionally non-geothermal countries such as Denmark, Sweden, Switzerland, and large areas of the mid-western and eastern USA. Internationally the largest uses of geothermal energy are for space heating (37%), 75% of which is in district heating systems and for swimming, bathing and balneology (22%) (Hammons 2011).

The suitable temperature ranges have been determined for various direct uses of geothermal energy. Typically, the agricultural and aquacultural uses require the lowest temperatures, with values from 25 to 90 °C. Space heating requires temperatures in the range of 50–100 °C, with 40 °C useful in some marginal cases and ground-source heat pumps extending the range down to 5 °C. Cooling and industrial processing (such as dehydration) normally require temperatures above 100 °C. Refrigeration based on ammonia absorption is possible at approximately 180 °C. At temperatures over 110–120 °C binary fluid electrical generation also becomes economically viable, and above 140 °C conventional electric generation is viable. There is increased interest in coupling geothermal electrical generation and direct uses. Electrical generation may be either topping or bottoming cycle depending upon the requirements of the direct use application (Hammons 2011).

Geothermal direct heating systems are in operation in at least 12 countries including Iceland, France, Poland, Hungary, Turkey, Japan, China, Romania, Italy, the United States, Sweden, and Denmark. The first known geothermal district heating system was built in Chaudes-Aigues Cantal, France in the fourteenth century and is still in operation today. The first system in the USA was the Artisan Hot and Cold Water Company built in Boise, Idaho, in 1892. The most famous geothermal district heating system in the world is the system supplying nearly 98% of the residents of Reykjavik, Iceland. The installed capacity is 830 MWe and is designed to meet the heating load down to −10 °C. Increased load during colder periods is met by large storage tanks and an oil-fired peaking plant (Hammons 2011).

Agriculture, including greenhouses and soil warming, and aquaculture uses of geothermal energy are increasing rapidly. They are particularly widespread since they require heating at the lower end of the temperature range where there is an abundance of geothermal resources (Hammons 2011).

The potential for substantially increased use seems very promising. Future developments will depend upon (Hammons 2011): (a) Increased resource information; (b) Increased knowledge of potential uses; (c) Prices of competing fuels, e.g., oil and gas; (d) The establishment of clear legal, institutional and regulatory framework conducive to geothermal development on a country by country basis; (e) Availability of capital, especially in the developing countries.

The use of geothermal energy through ground-coupled heat-pump technology has almost no impact on the environment and has a beneficial effect in reducing the demand for electricity. Geothermal heat pumps use the reservoir of constant temperature, shallow groundwater, and moist soil as the heat source during winter heating and as the heat sink during summer cooling. The energy efficiency of geothermal heat pumps is about 30% better than that of air-coupled heat pumps and 50% better than electric-resistance heating. Depending on climate, advanced geothermal heat pump use in the USA reduces energy consumption, and correspondingly, power-plant emissions by 23–44% compared to advanced air-coupled heat pumps, and by 63–72% compared to electric-resistance heating and standard air conditioners (Hammons 2011).

5.2 Geothermal Pump Systems and Used Pipe Materials

Geothermal heat pumps (sometimes referred to as ground source heat pumps, GSHPs) utilize heat retained in the ground for direct heating and cooling applications. GSHPs can be installed almost anywhere in the world, as even in the coldest climates the ground temperature stays at 10–15 °C at relatively shallow depths of below three meters. The GSHP pumps circulate a carrier fluid (usually a water/antifreeze mix) through pipes below the ground. As the fluid circulates it absorbs heat from the ground. The fluid passes through a heat pump and electricity is used to extract heat from it. The re-chilled fluid is sent back through the ground, continuing the cycle. The GSHP system can switch the direction of the heat flow, so in the summer it can act as an air conditioner, transferring the heat from the warm air to the cooler ground. Geothermal heat is often used for industrial processes, but GSHP technology can also be applied at household level, as shown in Fig. 5.4. Current upfront costs for a household system are generally above that of conventional heating systems so GSHPs are usually installed in multifamily developments (Lind 2011).

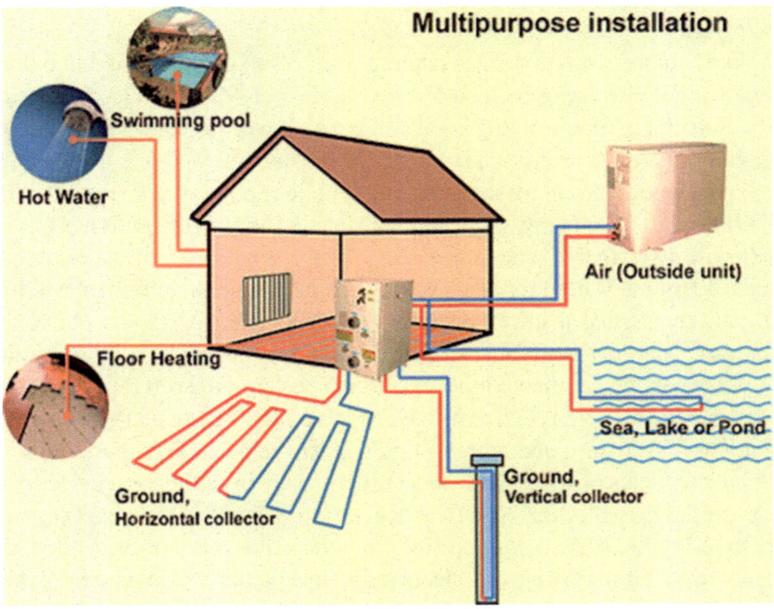

Fig. 5.4 Schematic installation of geothermal heat pumps

Piping materials for geothermal heating systems have been of numerous types with great variation in cost and durability. Some of the materials which can be used in geothermal applications include: asbestos cement (AC), ductile iron (DI), slip-joint steel (STL-S), welded steel (STLW), gasketed polyvinyl chloride (PVC-G), solvent welded PVC (PVC-S), chlorinated polyvinyl chloride (CPVC), polyethylene (PE), cross-linked polyethylene (PEX), mechanical joint fiberglass-reinforced plastic (FRP-M), FRP epoxy adhesive joint-military (FRP-EM), FRP epoxy adhesive joint (FRP-E), FRP gasketed joint (FRP-S), and threaded joint FRP (FRP-T). The temperature and chemical quality of the geothermal fluids, in addition to cost, usually determines the type of pipeline material used (Rafferty 1998).

5.2.1 Pipe Materials

Both metallic and nonmetallic piping can be considered for geothermal applications. Carbon steel is the most widely used metallic pipe and has an acceptable service life if properly applied. Ductile iron has seen limited application. Asbestos cement (AC) material has been the most widely applied product; however, environmental concerns have limited its use and availability. The attractiveness of metallic piping is primarily related to its ability to handle high-temperature fluids. The advantage of nonmetallic materials is that they are virtually impervious to most chemicals found in geothermal fluids (Rafferty 1998).

5.2.1.1 Metallic Materials

Carbon Steel

Steel pipe is manufactured in sizes ranging from 1/4 to over 72 in (from 6.35 mm to over 1.83 m). The joining method for small sizes (<2–1/2 in (<12.7–50.8 mm)) is usually threading, with welding used for sizes above this level. For underground installations, all joints are typically welded when unlined piping is used. For epoxy-lined piping, some form of mechanical joint should be employed so that welding does not interfere with the integrity of the lining material. Commonly used steel pipe ratings are Schedule 40 (standard) and Schedule 80 (extra strong). In most cases, in the U.S., Schedule 40 piping is used for heating applications, although, in Europe and for some newer non-geothermal district systems in the U.S., lighter weights (approximately Schedule 20) are now used. Schedule 80 is employed for high-pressure applications or in cases where higher than normal corrosion rates are expected (Rafferty 1998).

Corrosion is a major concern with steel piping, particularly in geothermal applications. In many geothermal fluids, there are various concentrations of dissolved chemicals or gases that can result primarily in pitting or crevice corrosion. If the potential exists for this type of attack, or if the fluid has been exposed to the air before entering the system, carbon steel should be the material of last resort. Steel piping is used primarily on the clean loop side of the isolation heat exchanger, although in a few cases it has been employed as the geothermal transmission line material (Rafferty 1998).

A distinct disadvantage in using steel pipe is that the buried pipe is also subject to external corrosion unless protected with a suitable wrapping or cathodic protection. External corrosion resulted in a number of failures and all of the steel pipe has been replaced with fiberglass piping located in a utility tunnel. The potential for external corrosion of metallic pipe systems should be considered for all direct buried installations. Various soil types, presence of ground-water, and induced current fields from power lines may accelerate external pipe corrosion and early system failure. If unlined steel piping is employed on the geothermal side of the system, it is most critical to assure a complete internal drying of the material for extended shutdowns. In both buried and aboveground installations, allowances for expansion must be made in the form of expansion joints or loops. These considerations have the effect of increasing both the labor and material costs of the piping system (Rafferty 1998).

Galvanized steel has been employed with mixed success in geothermal applications. Some geothermal fluids have demonstrated the ability to leach zinc from solder and other alloys. Selective removal of the zinc from galvanized pipe could result in severe pitting corrosion. In addition, consideration should be given to the fact that the protective nature of the zinc coating is generally not effective above 135 °F (57 °C) (Rafferty 1998).

Ductile Iron

Ductile iron is similar to cast iron with the exception of the form of the carbon component. In cast iron, the carbon (graphite) is in a flake-like structure. In ductile iron, the structure is more spherical or nodular. This small difference results in the greater strength, flexibility, and machinability from which the product derives its name. Ductile iron has been described as more corrosion resistant than cast iron. However, the slight difference in corrosion resistance would not be of any substantive meaning in most geothermal applications. Cast iron piping was employed for over 80 years in the Warm Springs geothermal system (Boise, ID). As an iron material, ductile iron is susceptible to corrosion from both external and internal sources. External protection generally involves a moisture barrier. For a pre-insulated product, special moisture protection would only be required at the joints and other fittings. Internal corrosion protection is usually provided by a lining. The two most common materials are cement mortar and coal tar epoxy. Coal tar epoxy is limited to a temperature of approximately 120 °F (49 °C). Mortar lining is suitable to a service temperature of 150 °F (66 °C) with a protective seal coat. Without the seal coat, maximum service temperature is 212 °F (100 °C). In some applications with very soft water, a leaching of the mortar lining has been observed when a seal coat is omitted. As a result, a special high-temperature epoxy coating would be required. In applications where water chemistry is such that bare cement lining is accept-able, ductile iron could be an economical piping choice (Rafferty 1998).

Ductile iron is a much-thicker-walled product than standard carbon steel and, for uniform corrosion applications, offers the probability of longer life. In geothermal applications, corrosion occurs by both uniform and pitting modes. Pitting corrosion rates of 70–200 mpy (1.778–5.08 mm/year) in carbon steel have been observed in one low-temperature (<150 °F (66 °C)) system during shutdown periods. Ductile iron piping is cost competitive with asbestos cement material. In addition, its common use in water supply systems results in wider familiarity with its installation practices. The most common method of joining ductile iron piping is through the use of a push-on or Tyton type joint. This is a bell and spigot gasketed joint. In addition, several versions of mechanical joints are available, although these are characterized by higher cost than the push-on joints. It is important to specify gasket materials suitable for the application temperature when using this product. Most suppliers offer EPDM gaskets which are suitable for use to 200 °F (93 °C) (Rafferty 1998).

Copper

Copper piping is generally not acceptable for geothermal applications. Most resources contain very small quantities of hydrogen sulfide (H_2S), the dissolved gas that results in a rotten egg odor. This constituent is very aggressive towards copper and copper alloys. In addition, the solder used to join copper has also been subject to attack in even very low total dissolved solids (TDS) fluids. For these reasons, copper is not recommended for use in systems where it is exposed to the geothermal fluid (Rafferty 1998).

5.2.1.2 Nonmetallic Materials

Fiberglass (RTRP)

Fiberglass piping, commonly referred to as RTRP (reinforced thermosetting resin pipe) or FRP (fiberglass-reinforced plastic), is available in a wide variety of configurations. Epoxy resin and polyester resin are basic materials for the piping. In addition, the piping is available in lined and unlined versions. The epoxy resin piping with an epoxy liner is generally selected for geothermal applications. Both epoxy resin and polyester resin systems can be compounded to be serviceable to temperatures of 300 °F (159 °C). Regardless of the type of fiberglass material used, care must be taken to maintain operating pressure high enough to prevent flashing of hot fluids. At high temperatures (>boiling point), the RTRP systems are susceptible to damage when fluid flashes to vapor. The forces associated with the flashing may spall the fibers at the interior of the pipe surface (Rafferty 1998).

As with all nonmetallic piping, the method of joining is a large consideration with respect to both installation time and expense. With FRP piping, a variety of methods are available, such as spigot and tapered ball joint, threaded joints, keyed mechanical joint, threaded mechanical joint, and flange joint. Among them, the bell and spigot/adhesive has seen the widest application in geothermal systems (Rafferty 1998).

Asbestos Cement (AC)

Asbestos cement pipe has been used for many years in municipal water systems, which had a maximum service temperature of 200 °F (93 °C). For geothermal service, the piping was generally specified with an epoxy lining. In most cases, asbestos cement pressure pipe was the material employed. This piping was available generally in three pressure classifications: 100, 150, and 200, with Class 150 most frequently specified for geothermal service (Rafferty 1998).

Joining of the pipe at couplings and small branch take-offs can be accomplished with O-ring slip together joints made of the same material as the pipe. The generally preferred method is that the pipe be attached to the coupling using the mechanical draw method. Fittings are available in cast iron or steel construction. In sizes applicable to geothermal systems, the cast iron material would be more commonly used. These fittings would be the same as those used in water main construction (AWWA 110). For both the fittings and the couplings, it is important to specify a gasket material that is compatible with the fluid being handled. This compatibility should consider both temperature and fluid chemistry. Most manufacturers can supply EPDM gaskets. Service connections to AC pipe can be accomplished for small diameter service lines with special tapped, AC couplings. In addition, standard mechanical service saddles can be used. Expansion of AC is about the same as that of steel. Because of the construction of the AC couplings, all expansion is compensated for at the joint. As a result, expansion loops or joints are not required. Careful bedding procedures should be employed because of the relatively fragile nature of the pipe material. However, AC pipe can be susceptible to thermal shock with 240 °F geothermal water. At lower system temperatures, the thermal shock is reduced and the AC pipe is not affected (Rafferty 1998).

Polyvinyl Chloride (PVC) and Chlorinated Polyvinyl Chloride (CPVC)

PVC is a low-temperature (maximum service temperature is 140 °F) rigid thermoplastic material. CPVC is a higher temperature rated material with a maximum temperature rating of 210 °F. Pressure handling ability at this temperature is very low (as is PVC at its maximum temperature) and support requirements are almost continuous.

Polyethylene (PE)

Polyethylene is in the same chemical family (polyolefin) as polybutylene and is similar in physical characteristics. It is a flexible material available in a wide variety of sizes from 0.5 to 42 in (12.7 mm to 1.07 m) in diameter. To date, this material has seen little application in direct-use geothermal systems, primarily because of its maximum service temperature of 140–150 °F (60–66 °C). The piping is recommended only for gravity flow applications above this temperature. Very high molecular weight/high-density PE can be employed for low pressure applications up to temperatures as high as 175 °F (79 °C). The SDR (standard dimension ratio—a wall thickness description) requirements under these conditions, however, greatly reduce the cost advantages normally found in polyethylene. This material is serviceable to 194 °F (90 °C) at a pressure of approximately 85 psi (586 kPa). Joining methods for polyethylene pipe is limited to thermal fusion. The pressure ratings of polyethylene piping are a function of SDR and temperature (Rafferty 1998).

Crosslinked Polyethylene (PEX)

Crosslinked polyethylene is a high-density polyethylene material in which the individual molecules are"crosslinked" during the production of the material. This can be accomplished by the use of peroxides, "A20" compounds or exposure to electron bean according to ASTMF 876–70 "Specification for Crosslinked Polyethylene (PEX) Tubing." The effect of the crosslinking imparts physical qualities to the piping which allow it to meet the requirements of much higher temperature/pressure applications than standard polyethylene material. PEX piping carries a nominal rating of 100 psi (690 kPa) at 180 °F (82 °C). Joining the piping is accomplished through the use of specially designed, conversion fittings which are generally of brass construction. Since the piping is designed primarily for use in hydronic radiant floor heating systems, a variety of specialty manifolds and control valves specific to these systems are available. Piping with and without an oxygen diffusion barrier is available. The oxygen barrier prevents the diffusion of oxygen through the piping wall and into the water. This is a necessary corrosion prevention for closed systems in which ferrous materials are included. Larger sizes of the PEX material are available as either bare or pre-insulated. The pre-insulated product is sold in rolls and includes a corrugated polyethylene jacket and a closed-cell polyethylene insulation. Rubber end caps are used to protect the exposed insulation at fittings. The flexible nature of the pre-insulated product offers an attractive option for small-diameter distribution and customer service lines in applications where it is necessary to route the piping around existing utility obstacles (Rafferty 1998).

5.2.2 Geothermal Heat Pump Technology

Geothermal heat pumps are systems with three main components (EGEC 2011): the ground side to get heat out of or into the ground; the heat pump to convert that heat to a suitable temperature level; and the equipment inside the building transferring the heat or cold into the rooms. The heat pump is a device which allows transformation of heat from a lower temperature level to a higher one, by using external energy (e.g., to drive a compressor). The amount of this external energy input, be it electric power or heat, has to be kept as low as possible to make the heat pump ecologically and economically desirable.

The measure for the efficiency of a heat pump system is the Seasonal Performance Factor (SPF), which is the average Coefficient of Performance (COP) in a given plant over a year or a heating/cooling season: SPF = useful heat/(electric) power input. A geothermal heat pump offers the best conditions for achieving high SPF. Shallow geothermal systems are very versatile and can be adapted to almost every subsurface condition. Typically the ground system is linked to a heat pump for achieving sufficiently high temperatures. Ground systems can be classified generally as open or closed systems, with a third category for those not truly belonging to one or the other (EGEC 2011).

To choose the right system for a specific installation, several factors have to be considered (EGEC 2011): Geology and hydrogeology of the underground (sufficient permeability is a must for open systems), area and utilization on the surface (horizontal closed systems require a certain area), existence of potential heat sources like mines, and the heating and cooling characteristics of the building(s). In the design phase, more accurate data for the key parameters for the chosen technology are necessary to size the ground system in such a way that optimum performance is achieved with minimum cost (EGEC 2011).

5.2.2.1 Open Systems

The open system is characterized by the fact that the main heat carrier, ground water, flows freely in the underground, and acts as both a heat source/sink and as a medium to exchange heat with the solid earth. Main technical part of the open system is groundwater wells, to extract or inject water from/to water-bearing layers in the underground ("aquifers"). In most cases, two wells are required ("doublette"), one to extract the groundwater, and one to reinject it into the same aquifer it was produced from (EGEC 2011).

Open systems tend to be used for relatively larger installations. With open systems, a powerful heat source can be exploited at comparably low cost. On the other hand, groundwater wells require some maintenance, and open systems in general are confined to sites with suitable aquifers. The main requirements are (EGEC 2011): (a) Sufficient permeability, to allow production of the desired amount of groundwater with little drawdown. (b) Good groundwater chemistry, e.g., low iron content, to avoid problems with scaling, clogging and corrosion.

5.2.2.2 Closed Systems

Horizontal

The closed system easiest to install is the horizontal ground heat exchanger (synonyms: ground heat collector, horizontal loop). In some areas due to restrictions in the area available, the individual pipes are laid in a relatively dense pattern, connected either in series or in parallel.

For the ground heat collectors with dense pipe pattern, usually the top earth layer is removed completely, the pipes are laid, and the soil is distributed back over the pipes. In the areas where land is cheaper, a wide pattern ("loop") with pipes laid in trenches is preferred. Trenching machines facilitate installation of pipes and backfilling. To save surface area with ground heat collectors, some special ground heat exchangers have been developed. Exploiting a smaller area at the same volume, these collectors are best suited for heat pump systems for heating and cooling, where natural temperature recharge of the ground is not vital. For the trench collector, a number of pipes with small diameter are attached to the steeply inclined walls of a trench some meters deep. Other types include screw type "energy baskets", "Slinky" collectors, etc. (EGEC 2011).

The main thermal recharge for all horizontal systems is provided by the solar radiation to the earth's surface. It is important not to cover the surface above the ground heat collector or to operate it as a heat store, if it has to be located under a building for instance (EGEC 2011).

Vertical

The temperature below a certain depth ("neutral zone") of the earth surface remains constant over the year. This fact and the need to install sufficient heat exchange capacity under a confined surface area, favors vertical ground heat exchangers (borehole heat exchangers). In a standard borehole heat exchanger, plastic pipes (polyethylene or polypropylene) are installed in boreholes, and the remaining room in the hole is filled (grouted) with a pumpable material. However, boreholes in hard, crystalline rock usually are kept open, and the groundwater serves for heat exchange between the pipes and the rock. If more than one borehole heat exchanger is required, the pipes should be connected in such a way that equal distribution of flow in the different channels is secured. Manifolds can be in or at the building, or the pipes can be connected in trenches in the field (EGEC 2011).

Several types of borehole heat exchangers have been used; the two possible basic concepts are (EGEC 2011): (a) U-pipes, consisting of a pair of straight pipes, connected by a 180°-turn at the bottom. One, two or even three of such U-pipes are installed in one hole. The advantage of the U-pipe is low cost of the pipe material, resulting in double-U pipes being the most frequently used borehole heat exchangers in Europe. (b) Coaxial (concentric) pipes, either in a very simple way with two straight pipes of different diameter, or in complex configurations.

Ground source heat pump plants of every size have been realized with borehole heat exchangers, ranging from small houses with just one borehole to large buildings, requiring whole fields of borehole heat exchangers. The heat source for thermal recovery of borehole heat exchangers is solar heat (in the upper part) and the geothermal heat flux (in the lower part), with some influence from flowing ground water or percolating water. However, the influence of groundwater in most cases is not very big, and the thermal conductivity of the ground is the main parameter. The borehole filling and the heat exchanger walls account for a further drop in temperature, which can be summarized as borehole thermal resistance. Values for this parameter usually are on the order of 0.1 K/(W/m); for a heat extraction of 40 W/m, this means a temperature loss of 4 K inside the borehole. Thermally enhanced grouting (filling) materials have been developed to reduce these losses. A special case of vertical closed systems are"energy piles", i.e., foundation piles equipped with heat exchanger pipes. All kind of piles can be used (prefabricated or cast on site), and diameters may vary from 40 cm to over 1 m (EGEC 2011).

Ground Coupling Circuits for Closed Systems

A variation of the horizontal ground source heat pump is direct expansion. In this case, the working medium of the heat pump (refrigerant) is circulating directly through the ground heat collector pipes (in other words, the heat pump evaporator is extended into the ground). The advantage of this technology is the omission of one heat exchange process, and thus a possibility for better system efficiency. Direct expansion requires good knowledge of the refrigeration cycle, and is restricted to smaller units. Also heat pipes as heat source have been tested; they work well for heat extraction, but not at all for heat injection. With the classical brine (liquid) system, the ground can easily be used for cooling, also. Heat is rejected into the ground, either by running the heat pump in reverse, or by directly coupling the building circuit to the ground circuit (EGEC 2011).

5.2.3 Geothermal Heat Pump Applications

Geothermal heat pump (GSHP) systems can be used from small, residential houses to large individual buildings or complexes (offices, hotels, schools, shopping, etc.). In the residential sector, typically heat pumps produced in larger series and in standard heating capacities from about 5–20 W are used. For the commercial sector, all the installation (heat pumps, manifolds) tends to be much larger than for residential houses. Heat pumps with capacities from ca. 50 kW upwards usually are constructed individually or in smaller numbers, adapted to the specific site conditions. The range of applications for GSHP is widely spanned. Besides, geothermal applications can be present everywhere and anytime for heating and cooling. The maximum delivery temperatures typically are in the order of 60–75 °C for refurbishment of older buildings), and in cooling mode ca. 6–7 °C (EGEC 2011).

5.2.3.1 Residential Houses

For small houses, 1–2 borehole heat exchangers or a horizontal collector (brine or direct expansion) are the best suited options. The installation is not visible from outside, the heat pumps (incl. Buffer tanks and/or DHW tank, where appropriate) do not need much space, and a fuel oil tank or connection to the gas grid is not required (EGEC 2011).

5.2.3.2 Offices and Commercial Buildings

For applications in the commercial sector, large borehole heat exchanger (BHE) fields or groundwater wells are the preferred groundside alternative. While BHE are feasible virtually everywhere, and promise maintenance-free operation, their individual capacity is limited, resulting in huge BHE fields for systems with high heating/cooling demand. Groundwater wells, on the other hand, require specific geological site conditions and diligent managing of the wells, but can deliver much higher thermal output per well. So for large installations, ground water use is a favorable option (EGEC 2011).

5.2.3.3 Geothermal Energy Storage

The low temperature in the ground can also be changed artificially by storage of heat or cold, creating geothermal energy storage (GES or Underground Thermal Energy Storage, UTES). The highest storage temperature achieved in GES is about 90 °C, the lowest (for cooling) ca. 5 °C. The heat sources for heat storage can be various, however, waste heat or solar heat are typical. For cold storage, the cold ambient air in wintertime or during night is the cold source (EGEC 2011).

Borehole Thermal Energy Storage: BTES

The system uses borehole heat exchangers (BHE) to store heat and/or cold in solid rock or soil, and in the groundwater that might be found therein. BTES do not need a ground water flow. Best suited is rock and soil with medium thermal conductivity, but high specific heat capacity. For effective storage, relatively large BHE-fields are required (EGEC 2011).

Aquifer Thermal Energy Storage: ATES

In ATES the ground water is used directly as a heat carrier. High porosity and low or none natural groundwater movement are the best conditions. Groundwater chemistry has to be checked closely, in order to secure reliable long-term operation. ATES can provide high thermal output with few wells. The range of applications is large, as for geothermal heat pumps (EGEC 2011): (a) Buildings, through a small district heating network; (b) Shopping malls, airport buildings, fair and convention halls, offices; and (c) Industry like plastics and foundry.

Therefore the main benefits of the geothermal heat pump systems can be summarized as (EGEC 2011): (a) A Renewable Energy Source: the heat from the earth is inexhaustible, delivering heating and cooling 24 h a day throughout the year, and available all over Europe with minor land use. (b) Any geothermal heat pump contributes substantially to the reduction of Green House Gas emissions. (c) A safe and controlled technology: Independent of the season, climatic conditions and time of the day; GSHP are technically proven, with long-term durability of installations; used since >50 years for heating and cooling. (d) An energy adaptable with high performance: an answer to all energy needs: heating, cooling, hot water, energy storage; can be adapted and modulated according to type of resource, to size and nature of equipment, and in order to meet demands. (e) An economically sustainable energy: Indigenous, independent of external supply/demand effects and fluctuations in exchange rates, not sensitive to conventional energy prices; allows "local" fossil resources such as oil, coal, and natural gas to be saved. (f) Furthermore shallow geothermal energy can help to improve the competitiveness of industries, and can have a positive impact on regional development and employment.

5.3 Materials Selection for Heat Exchange Systems

5.3.1 Heat Exchange Fluids

The heat exchange fluid is a fluid, usually liquid which is circulated to transport, extract, and dissipate energy in the form of heat, from an energy source to a thermal load. Each heat exchange fluid is chosen according to its physicochemical properties, such as viscosity, volumetric thermal capacity, potential heat of evaporation or liquefaction in case of a change in phase, and thermal conductivity. The heat exchange fluid is also chosen according to its anti-corrosive properties and its cost.

Although unlikely, the leakage of heat exchange fluids from the geothermal circuit could have significant repercussions on the environment and a building's occupants. A complete analysis of the risks, the life cycle and peripheral accessories necessary to choose a heat transfer fluid are all elements which must be taken into account during its selection (Geo-Energie 2014).

A variety of organic- and water-based heat exchange fluids have been developed to meet the operating needs of diverse applications.

- Organic-based heat exchange fluids include petroleum and synthetic fluids derived from mineral oils, white/paraffinic oils, synthetic aromatics (such as polyphenyls and alkylated benzenes), silicones, and fluorocarbons.
- Water-based heat exchange fluids are formulated primarily with ethylene glycol and propylene glycol.

Temperature range and not just the highest use temperature for a specific fluid type is very important. Viscosity and pumpability are important factors because both impact the economics of the operation. If the viscosity is too high, then a large amount of electricity could be consumed in pumping the fluid through the system. High-temperature fluids operate at temperatures up to 400 °C (752 °F). Above 315 °C, options are limited to synthetics. Below that temperature, mineral oil-based fluids are viable. Low-temperature fluids are water-based and contain either ethylene glycol or propylene glycol. Operating temperatures for water-based fluids on the low end are down to −40 °C. Synthetic fluids can operate at even lower temperatures. One synthetic hydrocarbon-based heat transfer fluid has an operating range down to −112 °C. Specific heat, thermal conductivity, density, and viscosity are also important performance considerations. The key consideration is the efficiency at which the fluid can transfer heat away or towards a specific application. Selection of a fluid having a maximum bulk temperature range marginally above the design operating temperature of the fluid can often significantly extend the fluid life. Different criteria need to be evaluated for high-temperature and low-temperature applications. In the former case, the question needs to be asked about whether the fluid operates in the liquid or vapor phase. Most current applications are in the liquid phase. Heat transfer coefficient is an important parameter but may need to be combined with durability. Silicon-based fluids are very durable but are not nearly as efficient from a heat transfer perspective as other organic-based fluid types. On the low-temperature side, toxicity and environmental friendliness are two important criteria in low-temperature applications because the choice is between fluids based on ethylene glycol and propylene glycol.

Ethylene glycol has better heat transfer properties, including a higher density and lower viscosity, but propylene glycol is essentially nontoxic and more environmentally friendly (Canter 2009).

For geothermal applications, the constant nature of underground conditions can be utilized to heat or cool a house using a heat transfer fluid depending upon whether it is summer or winter. The heat transfer fluid circulates down into the ground and in the summer dissipates heat from the environment to cool the house. In the winter, the process will reverse; the fluid transports heat from the ground into the house. This heat transfer is possible by burying a plastic pipe containing the heat transfer fluid in the ground. The pipe is connected to a heat pump/air conditioning unit and can go forward or in reverse depending upon the season. The heat transfer fluids currently used are water-based due to low-temperature range requirements. Fluids based on methanol, ethanol, ethylene glycol and propylene glycol are commercially available. Heat transfer fluids used in geothermal energy applications must exhibit several key characteristics. Low toxicity, nonflammability, and good corrosion protection are very important because the fluid is being used in a residential setting. The fluid types currently used have negative characteristics. Methanol and ethanol are volatile, flammable, and toxic. Ethylene glycol is also toxic and is viscous at low temperatures. While environmentally friendly, propylene glycol is also viscous at low temperatures and can support microbial growth that could degrade the fluid and generate corrosion. Using the right heat transfer fluid is also important because operating life is expected to be at least 10–12 years. If the heat transfer fluid system is kept closed, properly maintained and the fluid is nonflammable, nontoxic and has a good environmental profile, then there is no limit to the operating life. An alternative product based on potassium formate has been introduced that provides improved thermal conductivity with nonflammability, low toxicity and is environmentally friendly. The potassium formate-based heat transfer fluid can be used down to a temperature of −45 °C. The system can be recharged next to a heat pump without any problem. The potassium formate heat transfer fluid has been commercialized, and a modified formulation is available for industrial food and pharmaceutical applications (Canter 2009).

As energy efficiency becomes more important, heat transfer fluids will play a larger role in providing effective solutions. Particularly important will be applications that can utilize heat in a more productive manner.

INDEPTH: CO_2 as the Geothermal Working Fluid (Brown 2000)

One potentially important technology option is utilizing CO_2 as the geothermal working fluid, or heat transmission fluid. Compared with water, CO_2 is superior as a heat transmission fluid, achieving somewhat larger heat extraction rates when the same injection pressure is applied. At the conditions expected for EGSs—temperatures in excess of 150 °C, pressures of a few hundred bars-CO_2 is a "supercritical fluid," which means that it can exist only as a single fluid phase rather than as liquid and/or gas. CO_2 has a viscosity that is 40% that of liquid water, which means that substantially larger CO_2 volumes would flow for a given injection pressure. Additionally, the pressure difference between injection and production wells makes it more likely that a CO_2 system will establish a thermosiphon effect, where pumping of wells is not required. However, at the temperature and pressure conditions of interest for EGSs, CO_2 has somewhat lower density than water so that there would be less mass

(continued)

transport for a given volumetric flow. Also, CO_2 has a heat capacity two-fifths that of water, meaning that a given mass flow would transport a smaller amount of thermal energy than for water. A number of studies indicate that, overall, CO_2 is superior to water as a heat transmission fluid, achieving somewhat larger heat extraction rates when the same injection pressure is applied. An ancillary benefit to a CO_2 EGS is the potential for CO_2 sequestration as precipitated carbonate minerals and feldspar to clay conversion at the fringes of a CO_2 EGS reservoir. The anticipated challenge associated with the use of CO_2 as a working fluid lies in the likely requirement that the reservoir needs to be completely dried before CO_2 is injected in order to avoid problems associated with the formation of carbonic acid. Technologies coupling CO_2 geologic sequestration and EGSs should be pursued and developed. The vast deep sedimentary basins that are targets for sequestration can also provide geothermal heat.

5.3.2 Heat Exchanger Coatings

Custom heat exchangers are often coated for corrosion protection or for cosmetic purposes. Chemical conversion coating, anodization, E-coating, and painting are four coating options that will help to minimize corrosion and/or result in a more attractive component. One of the most widely used coating options is chemical conversion coating or chromate conversion coating, also known as "Chem-Film" or alodine. Conversion coating helps to minimize surface oxidation and is often specified for military as well as commercial heat exchangers and cold plates. It also sometimes serves as a surface preparation for paint. Another option that can be used to protect aluminum is anodization. Anodizing minimizes corrosion and abrasion by modifying the crystal structure close to the metal surface. It produces a harder part with even greater corrosion protection. However, it is more expensive than chemical conversion coating. A third coating method that provides corrosion protection is called an E-coat, also known as electrodeposition or electrocoating. A DC charge is applied to a metal part immersed in a bath of oppositely charged paint particles. The paint particles are drawn to the metal part and paint is deposited on the part, forming an even, continuous film over the entire surface. Of the four types of coatings described here, it is the most expensive type of corrosion protection. Heat exchangers and cold plates may also be painted for corrosion protection or cosmetic purposes. For example, copper heat exchangers are sometimes painted for aesthetics since uncoated copper may change color over time. Although these are four of the most common coatings for heat exchangers, there are a large number of coating options available to Original Equipment Manufacturers (OEMs) (Lytron 2014).

5.3.3 Polymer Heat Exchangers

Historically, the interest in polymer materials for heat exchanger applications has been driven by their high chemical stability and corrosion resistance. Through the use of a polymer coating or polymer tubing, heat recovery from solvent-laden streams is possible. However, because of the low thermal conductivity of polymers, using polymers in a standard design configuration will likely result in a dominating heat transfer resistance by the polymer walls. But by using very thin polymer structures, both plate and tubular heat exchangers can be successfully designed, constructed, and tested with their performance being comparable to conventional units at lower cost and reduced weight. By using polymer matrix composites, light weight heat sinks have been built that achieve similar performance as the copper equivalent. If one considers the advances made in composite materials, as well as the new emerging technologies such as nanoscale composites, it is clear that, through careful material selection and design modification, the incorporation of polymer materials into geothermal heat exchange applications has a bright future.

5.3.3.1 Liquid-to-Liquid Heat Exchangers

Figure 5.5 shows some liquid-to-liquid heat exchangers designed with polymer materials (T'Joen et al. 2009). The unit (Fig. 5.5a) consisted of 1.3 m long circular tubes with an outer diameter of 34 mm and a wall thickness of 5 mm. Five baffles were placed within the shell. The wall thickness was found to limit the overall heat transfer coefficient to a maximum of 90 W/m^2 K; reducing the thickness would result in a higher maximum value. To enhance the heat transfer rate, double conical PVC turbulators (Fig. 5.5b) were introduced in the tubes. These were held in place using wires. Heat transfer enhancement factors of up to 3.5 were achieved without pressure drop penalty. This is probably due to high dominating pressure drops at the headers.

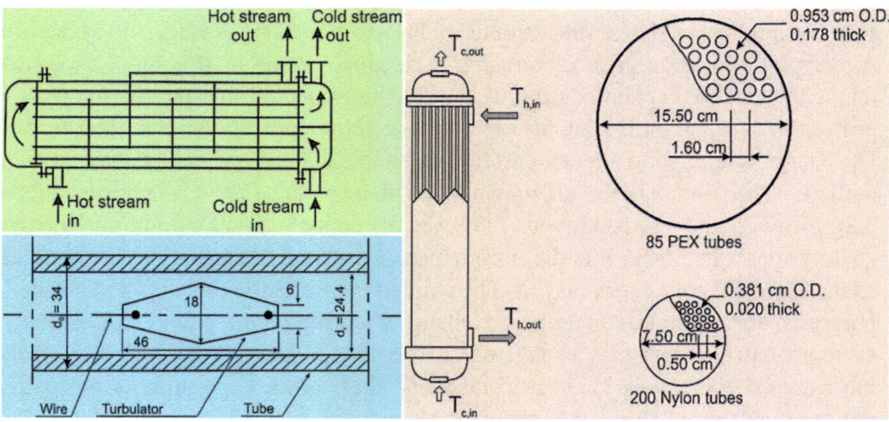

Fig. 5.5 Schematic installation of polymer heat exchangers (Adapted with permission from T'Joen et al. 2009 (Elsevier)): (**a**) PVC shell and tube heat exchanger; (**b**) Turbulator introduced within the heat exchanger; and (**c**) Polymer shell and tube heat exchanger

For the tubes in the heat exchangers, high-temperature nylon (HTN), cross-linked polyethylene (PEX) and polypropylene (PP) may be the best candidates. For the headers, glass-fiber-reinforced polymers are preferred, with HTN, PP, and PPS being the recommended types. For applications with mineral-containing water, Nylon-6.6 has a better combined performance than HTN, PB, PP, and copper.

Figure 5.5c shows the dimensions and layouts of the polymer shell-and-tube heat exchangers designed with two kinds of polymer tubes: PEX and nylon. The thermal conductivities for PEX and nylon were similar, with PEX being slightly more conductive (~ 0.38 W/m K) as compared to nylon (~0.31 W/m K). The PEX heat exchanger used a standard PEX tube with outer diameter 9.53 mm with a wall thickness of 1.78 mm, while the nylon heat exchanger used smaller tubes of outer diameter 3.81 mm and only 0.2 mm thick walls. The thermal resistance of the wall was a dominant limitation of the PEX heat exchanger, while it was not so for the nylon one. At a typical water flow rate of 5.7 L/min, the ratios for tube inside, outside, and wall thermal resistance were 24%, 34%, and 42%, respectively of the total heat transfer resistance for the PEX heat exchanger, and 49%, 26%, and 25%, respectively for the nylon counterpart. Also, the required lengths of the tube for the nylon heat exchanger were ~75% less than the PEX heat exchanger. This is mostly due to decreased thermal resistance due of thinner walls in nylon heat exchanger. Comparably, a copper heat exchanger, with tubes of outer diameter 6.35 mm and wall thickness 0.5 mm, its wall thermal resistance was negligible and its performance was limited only by the flow inside the tubes. The inside, outside, and wall thermal resistance contribution at the same water flow rate of 5.7 L/min were found to be 76%, 24%, and 0.04%, respectively. However, the thin-walled nylon heat exchanger has a very similar thermal performance as the copper heat exchanger (T'Joen et al. 2009).

Therefore, in liquid-to-liquid heat exchangers, thin walls in polymer designs are necessary to achieve thermal performance levels comparable to their metallic counterparts. For instance, a plastic thin-film heat exchanger was designed with a laminated film composed of Mylar/PVDC/adhesive/ polyethylene-EVA. Based on tests of the long-term creep characteristics of a HDPE film, the projected life of this heat exchanger made from HDPE and operating at 25 psi/100 °F (172 kPa/38 °C) would be 20 years (T'Joen et al. 2009).

5.3.3.2 Liquid-to-Gas Heat Exchangers

The use of polymer tubes and polymer-coated metal tubes for a condensing heat exchanger placed in the flue gas channel of a gas fired boiler has been explored. Commercial 1.27 cm diameter tubes of PTFE, PP, and PPS and 1.9 cm diameter tubes of PSU and PEI (polyetherimide) were tested. Aluminum tubes of 1.9 cm diameter were covered with an epoxy or vinyl ester coating of 0.1 mm thickness or a thicker coating of 0.5 mm of FEP (fluorinated ethylene propylene). After 100 days of exposure, none of the full polymer tubes showed any degradation (as indicated by local hardness measurement and infrared scans); however the epoxy and vinyl ester coating had failed. Due to the difference in thermal expansion between the metal and the coating, small cracks had appeared, allowing the corrosive gas to attack the base metal. A coating mixed with aluminum flakes (to increase the thermal conductivity) showed degradation where the flakes made contact with the surface. In addition, thin coatings can also fail due to pinhole effects, indicating the need for thicker extruded coverings (T'Joen et al. 2009).

PTFE plate preheaters and a shell-and-tube evaporator for a single-step mechanical vapor compression unit have been developed. This is a typical desalination application, where the hot brine is very corrosive towards metals. For thin-walled polymer tubes and plates (40–150 mm), spacers are needed to prevent the structure from collapsing and for very fine filtering. Compared to metal heat exchangers made of titanium, high alloy steel, and Cu–Ni alloys, the heat transfer area of the PTFE preheaters and evaporator was 2–4 times larger than that of the metal heat exchangers with varying top brine temperature. However, the polymer heat exchanger had the lowest cost. This would be viable if cheap heat was available, such as a geothermal power source.

A compact plate heat exchanger was made of thin wavy (100 mm) PEEK films with a mean thickness of 53 mm and amplitude of 1 mm. Seven sheets, 13.5 cm wide by 13.5 cm long, were stacked each rotated 90° to one another to provide a cross-corrugated layers for fluid flow, PEEK was selected for its high chemical and fatigue resistance, high working temperature (up to 220 °C), and thermal stability. The hydrophobic and very smooth surface, combined with the thermal expansion properties can help resist fouling. The constructed unit showed remarkable mechanical stability withstanding pressures up to 10 bar at atmospheric conditions. The typical capacity of these units ranged from 10 to 400 W for a heat transfer area of 0.125 m^2. Experimentally measured pressure drops on the liquid side were below 400 Pa for a flow rate up to 500 mL/min and on the gas side below 4.3 kPa for a flow rate up to 10 m^3/h. Overall heat transfer coefficients ranged from 60 to 370 W/m^2 K with drop-wise condensation on the gas side. It was found that a significant volume of liquid condensate remained within the gas layer which decreases as the gas flow rate was increased (T'Joen et al. 2009).

PVDF plate-fin heat exchanger using air-water or air/steam water as fluids is also developed. The overall heat transfer coefficients of air to water ranged from 80 to 130 W/m^2 K; and from 150 to 600 W/m^2 K for air/steam to water. An increase in the inlet steam mass fraction increased the overall heat transfer coefficient. The rise in heat transfer coefficient in the case of heat transfer from air/steam to water was due to drop-wise condensation on the air side, even in the presence of high amounts of non-condensable gas. Very high local gas side heat transfer coefficients were found for air/steam to water heat transfer. By inclining the heat exchanger mildly and changing the locations of the spacers, the heat transfer rate increased by 7%. This was due to enhanced drainage of the condensate. Another interesting approach was to add either HDPE or Nylon inserts between the plates. These random fiber-like structures resulted in a considerable increase in the heat transfer rate. This may be due to a limited fin effect (low thermal conductivity), a large increase of the exterior surface area (159% extra for the HDPE inserts) and an increase of the velocity between the plates. In addition, a micro cross-flow plate heat exchanger made of nickel and PMMA is designed. The overall dimensions of the polymer heat exchanger were 5 mm by 5 mm by 1.8 mm. The performance of the fabricated nickel heat exchanger is significantly better than the fabricated PMMA heat exchanger in terms of heat transfer/frontal area, while the PMMA heat exchanger outperforms the nickel heat exchanger on heat transfer/mass basis, due to the low density of the polymer (T'Joen et al. 2009).

5.3.3.3 Gas-to-Gas Heat Exchangers

PE shell-and-tube heat recovery unit for greenhouses has been designed. Five corrugated PE tubes with a wall thickness of 1 mm were placed within a single shell. The designed unit met the requirements: low cost; ease of assembly, repair, maintenance, and operation; corrosion resistance and satisfactory performance under frosting conditions. In operation, efficiencies up to 84% were measured, with latent heat contributing about 40% of the total heat transfer in some cases. The efficiency was defined as the ratio of the temperature difference between inlet and outlet of the inlet air to the maximum temperature difference. The inlet air passed through the tubes while the shell carried the moist air (T'Joen et al. 2009).

PTFE plate heat exchanger is also developed as a flue gas heat recovery unit. The channels were made of 1.5 mm thick PTFE sheets spaced 1 cm apart. The heat exchanger served as a SO_2 scrubber through condensation. Another heat exchanger consists of a plate heat exchanger using ionomer membranes made of sulfonated or carboxylated polymers. This heat exchanger is intended to be used as "energy recovery units," recovering not only heat but also moisture, as it can pass through the membranes, in contrast to regular "heat recovery units" made of metals (T'Joen et al. 2009).

Because of their low thermal conductivity, polymers are therefore not commonly considered as a material to construct heat exchangers, except for specific applications, e.g., heat recovery from solvent laden streams, where exotic alloys are required to prevent corrosion. However, polymers and polymer matrix composites do hold promise for use in the construction of heat exchangers geothermal applications, but that a considerable amount of research is still required into material properties and life-time behavior. A successful application of polymers or polymer matrix composites is based on careful material selection and modification of the design to fully exploit the material properties.

5.4 Hard Materials for Rock Drilling

Drilling technologies are employed for the exploration of all earth's resources and are the means for extraction for many resources, including geothermal, as shown in Fig. 5.6. Exploration drilling requirements for geothermal development include the construction of temperature gradient wells and resource confirmation wells. For a new resource, temperature gradient wells represent the first foray into the subsurface and are needed to map the temperature subsurface profile of the resource site and to provide reasonable estimates of the resource temperature. Depths of temperature gradient wells vary but can bottom out anywhere between the surface and the target

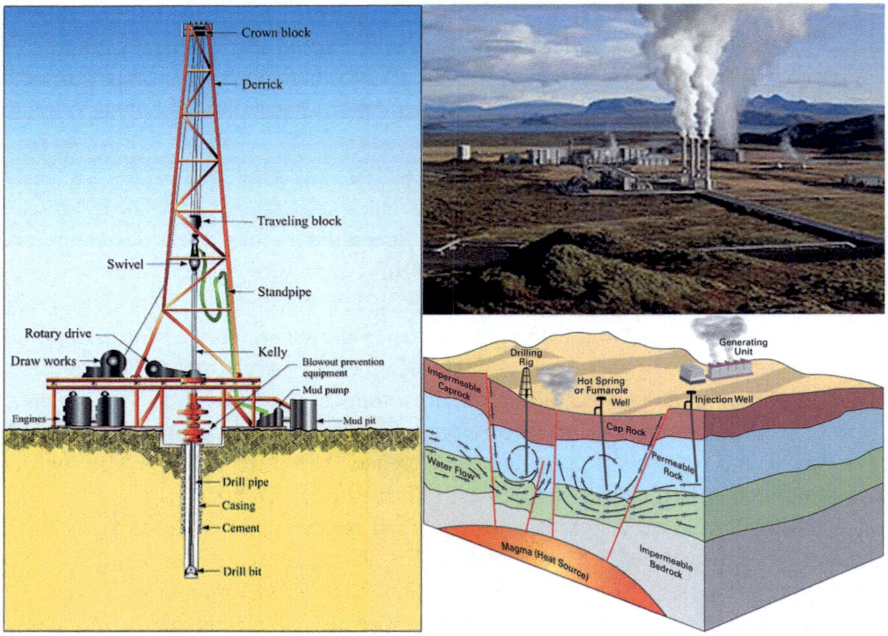

Fig. 5.6 Schematic geothermal drilling system (Modified from http://www.passmyexams.co.uk/GCSE/chemistry/drilling-crude-oil-1.html. Courtesy: Pass My Exams)

resource. Temperature gradient wells are commonly drilled with smaller drill rigs at "slimhole" diameters (less than six inches but commonly less than four inches). These holes are generally drilled with standard rotary methods, but increasingly wireline coring techniques (originally developed for the mining and civil industries) are being employed. Resource confirmation wells are necessarily drilled to the depth of the reservoir and are developed to confirm resource temperature and allow flow testing of the reservoir to ensure the reservoir can support the required flow rates for the development project. Similar to the temperature gradient wells, these wells are developed by using standard rotary techniques at diameters suitable for flow testing. After confirming the resource can support the needs of the developer, production and injection wells are drilled. Because of flow rate requirements and possible pumping requirements, these holes (particularly the production wells) are commonly constructed at diameters significantly larger than those associated with other extractive industries; 12¼-inch-diameter production intervals are not uncommon. The technologies used to drill geothermal wells are similar to those used to explore and extract other earth resources. However, geothermal well construction and well operations face the following unique combination of challenges (Raymond et al. 2012; DOE QTR 2015):

(a) While the oil and gas sector has adopted advanced fixed cutter polycrystalline diamond compact bits, the geothermal industry has seen only minor adoption because of the high rock strength, abrasiveness, and fracture formations commonly encountered in geothermal environments. Fixed cutter bits have the potential to easily double the average daily penetration if designed appropriately and run in a manner that will allow operations in harsh geothermal environments.

(b) Drilling, well logging, and testing equipment are limited by temperature issues associated with geothermal environments. Materials and electrical component development is needed to expand the operating temperatures of drilling tools, such as downhole motors, measuring while drilling, and logging while drilling assemblies as well as subsurface logging and monitoring tools. For example, commonly employed elastomeric materials for sealing tools and for use in wellbores are inadequate for extended operations in geothermal wells, and there is a paucity of electronic components needed for high-temperature applications.

(c) Lost circulation is a significant issue in geothermal drilling operations, and alternative methods to deal with lost circulation are needed. Materials to mitigate lost circulation while drilling without permanently plugging the formation in high-temperature environments need advancements. Diverter materials that can potentially be used for this purpose have been developed, but alternatives need to be expanded.

(d) Wellbore integrity during the construction of the well, throughout its useful life, and after abandonment is a critical issue for all constructed wells. Improved cements, remediation materials, casing systems, and practices are needed to ensure that wells developed for geothermal energy production are constructed in a manner that allows success during drilling and operations and long-term environmental safety.

The most common solution for drilling through hard rock and petrothermal systems is the application of controlled pressure drilling equipment. There are three major types of controlled pressure drilling methods (Vollmar et al. 2013): (a) air drilling (AD); (b) managed pressure drilling (MPD); and underbalanced drilling (UBD). AD is geared towards optimizing the rate of penetration (ROP), MPD minimizes rig nonperformance time and UBD increases productivity and reduces reservoir damage. For the case of drilling through a non-fractured source rock, it is not necessary to drill pressure controlled before stimulating the pay zone.

The earliest drills were developed by the ancient Egyptians, and were simply a lath powered by the back and forth motion of a bow. Later, drill bits were attached to a handle and worked through wood with a cranking motion. A drilling bit is the cutting or boring tool which is made up on the end of the drill string. The bit drills through the rock by scraping, chipping, gouging or grinding the rock at the bottom of the hole. Drilling fluid is circulated through passageways in the bit to remove the drilled cuttings. There are however many variations in the design of drill bits and the bit selected for a particular application will depend on the type of formation to be drilled (Azar and Samuel 2007).

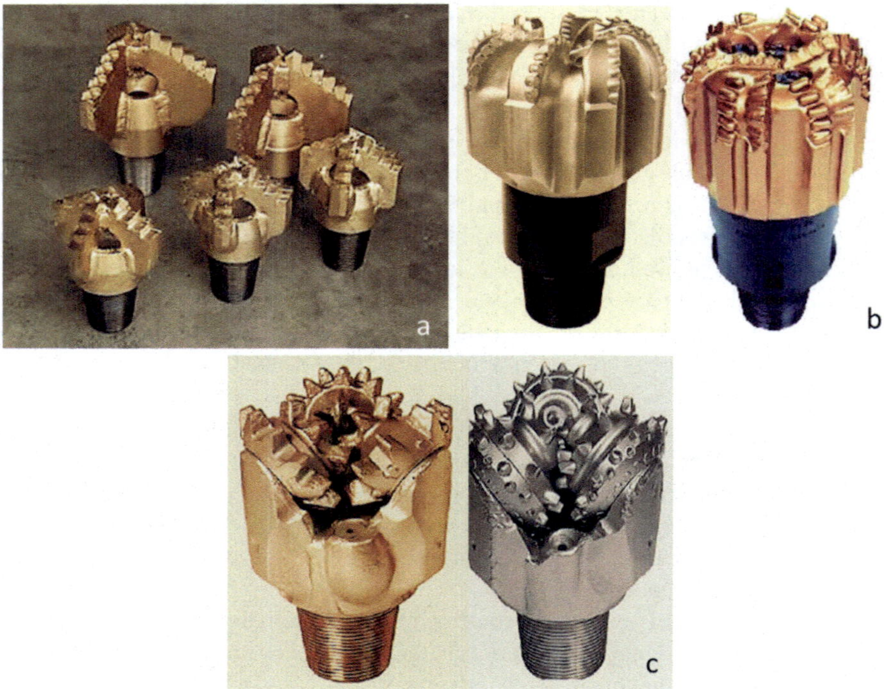

Fig. 5.7 Schematic illustrations of typical drilling bits: (**a**) Drag bit; (**b**) roller cone bit; and (**c**) diamond bit

Figure 5.7 illustrates typical drilling bits: (a) Drag bit; (b) roller cone bit; and (c) diamond bit. There are a number of metals used to make drill bits. Each material is best suited to drill through a particular type of surface. It is important to understand the strengths and weaknesses of each material type when selecting the bit.

(a) Low Carbon Steel
Low carbon steel drill bits are appropriate for drilling softwoods only. They are too soft for use with hardwoods and metals, and tend to dull frequently; they also need to be sharpened often to improve their lifespan. Despite these flaws, they have the distinct benefit of being extremely inexpensive.
(b) High Carbon Steel
High carbon steel bits have a higher temperature tolerance than low carbon steel bits, so they hold up to wear better and require less frequent sharpening. They are suitable for drilling thin metal and hardwood surfaces. They cost about the same amount as low carbon steel bits and are a better option for basic drilling.
(c) High-Speed Steel
High-speed steel bits, sometimes called HSS drill bits, are made of extremely hard carbon steel that has a higher heat resistance than high carbon steel.

Table 5.1 A sample of alloying compositions of common high-speed steel grades (by %wt) (Bayer et al. 1989)

AISI type	UNS designation	C	Si	Cr	V	W	Mo	Co
Molybdenum high-speed tool steels								
M1	TU301	0.83	0.35	3.75	1.18	1.75	8.70	...
M2								
Regular C	T11302	0.83	0.33	4.13	1.98	6.13	5.00	...
High C	...	1.00	0.33	4.13	1.98	6.13	5.00	...
M3								
Class 1	T11313	1.05	0.33	4.13	2.50	5.88	5.63	...
Class 2	T11323	1.20	0.33	4.13	3.00	5.88	5.63	...
M4	T11304	1.33	0.33	4.25	4.13	5.88	4.88	...
M6	T11306	0.80	0.33	4.13	1.50	4.25	5.00	12.00
M7	T11307	1.01	0.38	3.75	2.00	1.75	8.70	...
M10								
Regular C	T11310	0.89	0.33	4.13	2 00	...	8.13	...
High C	...	1.00	0.33	4.13	2.00	...	8.13	...
M15	T11315	1.50	0.33	4.00	5.00	6.50	3.50	5.00
M30	T11330	0.80	0.33	4.00	1.25	2.00	8.00	5.00
M33	T11333	0.89	0.33	3.75	1.18	1.70	9.50	8.25
M34	T11334	0.89	0.33	3.75	2.10	1.75	8.48	8.25
M35	T11335	0.80	0.33	4.00	2.00	6.00	5.00	5.00
M36	T11336	0.85	0.33	4.13	2.00	6.00	5.00	8.25
M41	T11341	1.10	0.33	4.13	2.00	6.63	3.75	8.25
M42	T11342	1.10	0.40	3.88	1.15	1.50	9.50	8.25
M46	T11346	1.26	0.53	3.95	3.15	2.05	8.25	8.30
M48	T11348	1.50	0.33	3.88	3.00	10.00	5.13	9.00
M50(a)	T11350	0.80	0.40	4.13	1.00	...	4.25	...
M52(a)	T11352	0.90	0.40	4.00	1.93	1.25	4.45	...
M62	T11362	1.30	0.28	3.88	2.00	6.25	10.50	...
Tungsten high-speed tool steels								
T1	T12001	0.73	0.30	4.13	1.10	18 00	...	...
T4	T12004	0.75	0.30	4.13	1.00	18.25	0.70	5.00
T5	T12005	0.80	0.30	4.38	2.10	18.25	0.88	8.25
T6	T12006	0.80	0.30	4.38	1.80	19.75	0.70	12.00
T8	T12008	0.80	0.30	4.13	2.10	14.00	0.70	5.00
T15	T12015	1.55	0.28	4.38	4.88	12.38	1.00	5.00

(a) Intermediate high-speed tool steel

A sample of alloying compositions of common high-speed steel grades is shown in Table 5.1 (Bayer et al. 1989). The higher heat resistance of high-speed steels allows them to be used at higher drilling speeds and insures that the bits maintain their structural integrity longer. HSS bits can be coated to improve certain features, and their porous surface holds coatings well.

(d) Carbide
Carbide tipped drill bits are specialized tips on hardened steel bits. Carbide tipped drill bits have the ability to dissipate heat rapidly, and so will hold an edge longer and tolerate heat better than other drill bit options. Unfortunately, carbide tipped bits are very brittle and will chip if they are not handled with care. They are appropriate for drilling hardwood and thin metals.

(e) Cobalt
Cobalt drill bits are extremely hard and are useful in boring into material too hard for standard high-speed drill bits like stainless steel and other thicker metal surfaces. Cobalt is less susceptible to heat damage than any other drill bit type, but it is also extremely brittle and prone to accidental damage.

(f) Diamond
Polycrystalline diamond drill bits, sometimes simply referred to as PCDs, are made of the hardest drill bit material available on the market. The bits are made by coating a carbide bit in diamond particles. Diamond drill bits grind away the surface they are drilling into, rather than cutting into the surface. They are appropriate for cutting through a variety of extremely hard surfaces including granite, ceramic tile, marble, fiberglass, stone, glass, and heavy metals.

(g) Drill Bit Coatings
Drill bit coatings make the drill bit harder, more lubricated, sharper, or more heat resistant than the metal bit is on its own. Coatings significantly improve the cutting quality, durability, and lifespan of drill bits. Black oxide is the most basic and inexpensive coating. It is commonly used on high-speed steel bits and provides a higher level of heat resistance and improved lubrication. It also protects the bit against water damage, rust, and corrosion.

Titanium nitride (TiN) coating reduces friction during drilling and increases heat resistance. Coating a drill bit with titanium nitride can increase the life of the bit up to 5 times. It also allows for faster drilling speeds. Titanium aluminum nitride, sometimes referred to as TiAlN, is designed to increase the hardness and temperature resistance of the bit. Bits with a TiAlN coating are heat resistant up to 800 °C, and cling well to a variety of base bit types. Zirconium nitride strengthens brittle drill bit types and decreases friction. It is an excellent coating for drill bits that will be used for precision drilling. It is also effective at reducing the risk of accidental damage on brittle bit types caused by rough handling and basic wear and tear.

(h) Tungsten Carbide
Tungsten carbide bits are extremely expensive and extremely brittle. This material is only used to make very fine drill bits. Typically tungsten carbide drill bits have a diameter of less than 1 mm. They are useful in small space drilling situations, and are hard enough to withstand the wear and tear of drilling through abrasive materials.

Cemented carbide is an old and well-known WC-based hard metal, which has been widely applied in geo-engineering as drill buttons and various wear-resistant parts. In order to extend the service life of cemented carbide components and enhance their efficiency for rock drilling under various conditions, nanostructured, functionally graded and Co-free cemented carbides have been developed. With the advance in synthesizing nanosized powders and advent of electric field assisted fast sintering techniques, the consolidation of nanostructured and Co-free cemented carbides and even pure WC materials has been possible; and because of their high hardness and wear resistance, they are much promising in geo-engineering drilling. Functionally graded cemented carbide provides a combination of high wear resistance and toughness in a single component, which is also much favorable for geo-engineering drillers. In addition, by replacing the binder phase Co with Ni or carbide binder, and even without binder phase, the corrosion and oxidation of the resultant materials can be significantly improved without considerable deterioration of fracture toughness (Ren et al. 2013).

(i) Diamond Core
Diamond core drill bits are used in stone cutting and masonry. Instead of the traditional point and auger style drill bit, these are hollow tip drill bits that cut or grind a circle using the outer edge of the bit. The core drill bit cuts a core or plug from the stone or glass it is used to cut, so that there are fewer shavings and potentially harmful shards of material when the drilling project is complete.

(j) Masonry Bit
Masonry bits are made of hard carbon steel but have a specialized tip made of a harder metal. The tip is typically welded into the end of the drill bit. These bits are designed to cut through concrete, brick, and mortar during the masonry process.

(k) PCB Drill Bit
PCB drill bits are a type of micro-drill bit that has a diameter of 1 mm or smaller. They are designed to bore holes in the electronic computer boards of most electronic objects. Because they are so fine, they are commonly made of carbide, which can hold up to the fiberglass board better than the alternatives.

(l) Flexible Shaft Bit
Flexible shaft bits are made of spring steel and are designed to flex while drilling. They can be bent while in motion so that the hole drilled can curve inside of walls and create uniquely shaped holes. They are commonly used when running wiring in homes or doing construction work in areas with limited workspace.

INDEPTH: Enhancing the Drilling Process by Combining Conventional Drilling and the Spallation Technology (Kant et al. 2017)

The development of geothermal energy production from deep resources is hampered by the high costs of the drilling process. The currently used rotary technology based on the mechanical destruction of the rock performs well in

(continued)

soft and medium hard rocks, but faces difficulties in hard basement rocks as they are encountered in deep geothermal wells. Low penetration rates together with high wear rates of the drill bits are impeding the drilling process, leading to high costs. Additionally, conventional rotary drilling is characterized by a poor energy transport, due to significant drag and torque losses along the drill string, leading to a low overall energy efficiency of the drilling process. In order to intensify the conventional drilling process, one option is to pump combustible reactants (e.g., methane and air) to the front face of the drill bit to create a hot fluid jet which impinges on the rock surface. Thereby, the hot fluid jet can be used to spall the rock or to thermally weaken the structure, if the rock properties are unfavorable for spalling. The spallation process is based on steep temperature gradients induced by the impinging flame, creating high local thermal stresses in the upper layer of the rock surface. If the thermal stresses exceed a certain threshold, initial present cracks will extend at the surface and later through the material. If sufficient thermal stresses can be induced, the cracks combine and a so-called spall is formed, which will be ejected from the surface and the process continues on the created surface. If the rock cannot be spalled due to unfavorable rock properties, the flame jet can be used to further weaken the rock. This thermal treatment leads to a significant decrease of the rock strength, due to severe crack creation. The cracked surface can then be excavated with conventional cutters, whereas significantly less weight-on-bit and torque is required. This hybrid technology could potentially combine the benefits of both drilling technologies and therewith contribute to the reduction of the drilling costs for geothermal projects.

5.5 Advanced Cements for Geothermal Wells

The principal application of the geothermal well cementing materials is to support mechanically metallic well casings as well as to protect them against hot brine-initiated corrosion at brine temperatures up to 320 °C. The cementitious materials not only must possess high-hydrothermal temperature stability, but also they must be inert and resist against very harsh geothermal environments involving the CO_2-enriched brine (>40,000 ppm CO_2) encountered in a bottom hole depth of ~1700 m at temperature of ~320 °C, and a highly concentrated H_2SO_4 (pH <1.5) brine containing at least 5000 ppm CO_2 in an upper well region between the well's surface and ~1000 m depth at temperatures up to 200 °C (Sugama 2006).

The conventional well cements consisting of the calcium silicate hydrates (CaO–SiO_2–H_2O system) and calcium aluminum silicate hydrates (CaO–Al_2O_3–SiO_2–H_2O system) for the integrity of geothermal wells, have relatively poor performance in mechanically supporting the metallic well casing pipes and in mitigating the pipe's corrosion in very harsh geothermal reservoirs. These difficulties are particularly acute in two geological regions: One is the deep hot downhole area (~1700 m depth at temperatures of ~320 °C) that contains hyper saline water with high concentrations of CO_2 (>40,000 ppm) in conjunction with ~100 ppm H_2S at a mild acid of pH ~5.0; the other is the upper well region between the well's surface and ~1000 m depth at temperatures up to 200 °C. The specific environment of the latter region is characterized by highly concentrated H_2SO_4 (pH < 1.5) brine containing at least 5000 ppm CO_2. When these conventional cements are emplaced in these harsh environments, their major shortcoming is their susceptibility to reactions with hot CO_2 and H_2SO_4, thereby causing their deterioration brought about by CO_2-catalyzed carbonation and acid-initiated erosion. Such degradation not only reduced rapidly the strength of cements, lowering the mechanical support of casing pipes, but also increased the extent of permeability of the brine through the cement layer, promoting the rate of the pipe's corrosion. Severely carbonated and acid eroded cements often impaired the integrity of a well in less than one year; in the worst cases, casings have collapsed within three months, leading to the need for costly and time-consuming repairs or re-drilling operations. These were the reasons why the use of conventional well cements was concerned, and further their deterioration was a major impediment in expediting the development of geothermal energy resources (Sugama 2006).

To deal with this problem, new types of cementitious materials that confer outstanding resistance to CO_2 and acid at brine temperatures up to 320 °C have been developed. One of those synthesized was calcium aluminate phosphate (CaP) cement involving $CaO–Al_2O_3–P_2O_5–H_2O$ and $Na_2O–CaO–Al_2O_3–SiO_2–P_2O_5–_2O$ systems; the other was sodium silicate-activated slag (SSAS) cement consisting of $Na_2O–CaO–SiO_2–MgO–H_2O$ and $Na_2O–CaO–Al_2O_3–SiO_2–MgO–H_2O$ systems. The CaP cements with four basic components, calcium aluminate cement, sodium polyphosphate, Class F fly ash, and water, were designed as CO_2-resistance cements for use in mildly acidic (pH ~5.0) CO_2-rich downhole environments. The SSAS cements with four starting materials, slag, Class F fly ash, sodium silicate, and water, were designed to resist a hot strong acid containing a low level of CO_2. They also were characterized as being cost-effective economical cements because of their use of inexpensive cement-forming by-products yielded from coal combustion and steel-manufacturing processes (Sugama 2006).

5.5.1 CaP Cements

The four crystalline hydrothermal reaction products, the hydroxyapatite $[Ca_5(PO_4)_3(OH)]$, boehmite (γ-AlOOH), hydrogarnet ($3CaO·Al_2O_3·6H_2O$), and analcime ($NaAlSi_2O_6·H_2O$) phases, were responsible for strengthening and densifying the CaP cements, as well as conferring on them resistance to CO_2 and mild acid. The mechanism underlying CO_2-resistance was the replacement of the OH groups within the hydroxyapatite phase by CO_3^{2-}, thereby leading to the formation of CO_3- intercalated hydroxyapatite; $2Ca_5(PO_4)_3(OH) + CO_3^{2-} \rightarrow 2Ca_5(PO_4)_3.(CO_3)$. The analcime phase favorably intercalated CO_2, transforming it into the cancrinite [$Na_5(AlSi)_{12}O_{24}CO_3{\cdot}3H_2O$] phase, while the boehmite phase displayed chemical inertness to CO_2. Although the hydrogarnet phase showed some sensitivity to carbonation, these hydroxyapatite → $2Ca_5(PO_4)_3{\cdot}CO_3$ and analcime → cancrinite phase transitions occurring without any destruction of their structures together with the CO_2-inert boehmite phase were major reason why the CaP cements had an excellent resistance to CO_2 (Sugama 2006).

A further requirement of the CaP cements was to improve the following five properties for formulating field-applicable cements with upgraded properties (Sugama 2006): (1) The maintenance of pumpability by means of the set-retarding activity of cement slurries; (2) low density slurry; (3) toughness, (4) bond durability to the casing pipe's surface; and, (5) low cost. To improve the first property, citric acid was the most effective set retarder in extending the thickening and downhole pumping times of the cement slurries. Its set-retarding activity was due to the uptake of Ca^{2+} ions from the calcium aluminate cement (CAC) reactant by carboxylic acid groups within the citric acid. This uptake precipitated Ca-complexed carboxylate compounds as a set-retarding barrier layer on the CAC grain's surfaces. Lowering of the density of the cement slurry, the second property played a pivotal role in eliminating the problem of lost circulation during pumping and circulating operations of cement slurry. Among the several ways to prepare low-density slurries, an air-foaming technology using simple foaming surfactants was identified as the most cost-effective and efficient way. The air-foamed CaP cement revealed some advanced properties, such as a high compressive strength and lower porosity, at a hydrothermal temperature of 288 °C, compared with those of conventional N_2 gas-foamed Class G well cement made from slurry of similar density under high pressure and the same hydrothermal temperature. However, one shortcoming was an increase in water permeability due to the formation of an undesirable continuous porous structure caused by coalesced air bubble cells, so raising concerns that the rate of corrosion of the casing pipes might be promoted. To solve this problem, a styrene acrylic emulsion (SAE) as high-temperature anti-corrosion additive was incorporated into the air-foamed cement slurries. The following three factors of the SAE-modified foamed cements contributed to a significantly abating corrosion of the casing pipes (Sugama 2006): (1) A decrease in the conductivity of corrosive NaCl ions; (2) an inhibition of the cathodic oxygen reduction reaction at the corrosion sites of steel; and, (3) the good coverage of the steel's surface by the foamed cements.

In a long-term exposure of cements in such a very harsh environment, one critical issue that emerged was their shrinkage and expansion caused by the in-situ phase transformation and the excess growth of crystalline hydrothermal reaction products. These phenomena frequently imposed internal stresses, followed by the initiation of cracks. Thus, in response to the third property, the cements were required to have a sufficient resiliency and toughness to avoid the creation and effects of stress. Among the various different fibrous materials explored as reinforcements, milled carbon microfibers (~7.5 μm diam. × 100–200 μm long) offered the best performance in improving the toughness and ductility of cements. In fact, the fracture toughness and displacement of the non-reinforced cements was raised 3.1- and 2.7-fold, respectively, by incorporating 14 wt% (21.8 vol.%) fibers (Sugama 2006).

Regarding the fourth property, bond durability, when superheated steam and fluid went through the cement-sheathed casing pipes, the two factors caused the development of stress cracking, namely the thermal shock of cement layers directly contacted with pipe's surface, and the thermal expansion of pipes. Hence, the bond durability of the cements adhering to the casing pipe was one of the important factors governing the integrity of the cement covering the casing pipes under repeated superheat-cold fatigue cycles in geothermal wells. To obtain this information, the carbon-microfiber-reinforced CaP and conventional Class G well cement composite-sheathed steel pipes were exposed to a superheating-cooling fatigue test (one cycle = 250 °C for 15 h + room temperature for 9 h). For the Class G cement, the needle-like xonotolite crystals that formed in interfacial critical regions between the cement and pipe were detrimental to bond durability because of the development of an undesirable porous microstructure, thereby resulting in a decline in shear bond strength in the first 70 cycles. In contrast, the shear bond strength of CaP cement markedly increased between the 0 and 100 cycles, beyond that, it leveled off. The reason for such outstanding bond durability was due to the development of dense microstructure of hybrid phases including plate-, block- and foil-like hydroxyapatite, boehmite, hydrogarnet, and analcime crystals at the contact zones with the pipe's surface. The upgraded CaP cements with advanced properties described above contributed to a considerable reduction in the costs of repairing and maintaining the wells. However, another factor that can lead to a reduction in drilling and reservoir management expenses is by using low-cost cement. In response to this important issue, the usefulness of coal-combustion by-products, Class C fly ash (2.0 ¢/lb), was investigated as a replacement for the expensive calcium aluminate cement (35.0 ¢/lb) that was one of starting materials of CaP cements. Blending Class C and F fly ashes with a *C*/*F* ratio of 70/30 gave the most suitable properties for CO_2– and mild acid-resistant CaP cement systems (Sugama 2006):

Based upon the information described above, the field applicable CaP cements formulated satisfactorily met all the following material criteria (Sugama 2006): (1) Maintenance of pumpability for at least 3 hours; (2) compressive strength, > 500 psi (3.5 MPa) at 24 h-curing time; (3) water permeability, $<1 \times 10^{-4}$ Darcy; (4) bond strength to steel casing, >50 psi (0.35 MPa); (5) carbonation rate, <5 wt% after 1 year in 40,000 ppm CO_2-laden brine at 300 °C; (6) fracture toughness, >0.008 $MN/m^{3/2}$ at 24 h-curing time; (7) resistance to mild acid (pH ~5.0) at 300 °C, <5 wt% loss after 30 days exposure; (8) cost, <$15/bag; and, (9) slurry density of foamed cement, <1.3 g/cc (13 lb./gal). In addition, the CaP cements had excellent durability in a hostile geothermal environment. It is estimated that its useful service life is about 20 years before repairs will be needed. In contrast, conventional well cements become severely deteriorated in such CO_2-rich geothermal wells after only one year, and the damaged wells must be repaired in an operation involving redrilling and recementing. The estimated annual cost of remediation is ~ $150,000 per well. There are no annual repair costs whatsoever for wells completed with this cement, thereby eliminating substantial expenses for remedial operations.

5.5.2 SSAS Cements

The combination of two crystalline phases, calcium silicate hydrate ($CaO{\cdot}SiO_2{\cdot}H_2O$, CSH) and tobermorite ($5CaO{\cdot}6SiO_2{\cdot}H_2O$) phases, was responsible for maximizing the strength and minimizing the water permeability of autoclaved cost-effective SSAS cements. Although these phases were vulnerable to reactions with hot H_2SO_4 (pH <1.5), the CSH phase played an important role in retarding the rate of acid erosion. After the uptake of Ca by H_2SO_4, Ca-destitute CSH preferentially reacted with the Mg from slag to form magnesium silicate hydrate [$Mg_2Si_2O_5(OH)_4$, the lizardite phase] that not only retarded the rate of acid erosion, but also retained the integrity of the cementitious structure. Thus, after undergoing acid damage, the SSAS cement exhibited a self-repairing characteristic (Sugama 2006).

In addition, modifying the SSAS cement with Class F fly ash enhanced further the extent of resistance to acid. When such modified SSAS cements were exposed for 15 days to 90 °C CO_2-laden H_2SO_4 (pH 1.1), their weight loss by acid erosion was less than 7%. Two factors contributed to minimum acid erosion: One was the self-repairing property of cement itself; the other was the anti-acid zeolite phase formed by interactions between the mullite in fly ash and the Na ions liberated from the sodium silicate activator. Therefore, the economical SSAS cement has a high potential as acid-resistant geothermal well cement at temperatures up to 200 °C (Sugama 2006).

5.6 Material Selection for Corrosion Protection of Geothermal Systems

As geothermal energy is being developed and taking more importance, certain problems have occurred regarding the corrosion and the usage of the materials that are influence the effectiveness and the quality of the service. The knowledge of the characteristics of the geothermal fluid partly leading to the breakdown of the equipment used for processes due to corrosion, is very important in the equipment selection. Dissolved CO_2, H_2S, NH_3 and chloride ions, may lead to the metallic materials being corroded and becoming unusable. The chemical composition, temperature, and velocity of fluid vary depending on the geothermal resource, the design of the power system, and the point of the production cycle. Both the operating experiences gained during usage and laboratory studies performed with real geothermal fluids form the basis for the selection of materials for this area (Kaya and Hoshan 2005).

5.6.1 Corrosion Types Encountered in Geothermal Systems

The thermodynamic power cycle in geothermal system is a process that is used to convert geothermal energy into electrical energy. The methods used to rotate a turbine by obtaining wet steam or dry steam depend on the properties of geothermal fluid. There are power cycles for these geothermal sources, including steam dominated, liquid dominated, hot dry rock and geopressured sources. Power cycles are classified according to whether the fluid is steam dominated, or liquid dominated or whether the wells are artesian or not. Power cycles consist of three steps; obtaining geothermal fluid from a well, rotating the turbine by obtaining wet steam or steam, and condensing gases. Geothermal fluids contain dissolved CO_2, H_2S, NH_3 and chloride ions that can cause corrosion of metallic materials, therefore safe utilization of geothermal systems depends importantly on materials selection. Typical corrosion types can be classified as (Kaya and Hoshan 2005):

(a) Uniform corrosion
Uniform corrosion occurs equally throughout the metal surface. In geothermal systems it is generally due to chloride, ammonium or hydrogen ions.

(b) Pitting corrosion
It is the corrosion type in which pits occur through the metal surface. Pits frequently deepen due to the breakage of a passive film. The beginning of pit formation and the rate of deepening cannot be predicted.

(c) Crack corrosion
It is similar to pit corrosion when formation type is concerned. Apart from other corrosion types it depends on geometry. It is observed under the accumulation layers of metallic materials that occur during equipment production or operation conditions. Beginning and depth cannot be predicted.

(d) Stress corrosion cracking
It is a dangerous type of corrosion occurring due to the chloride ions and stress in the metallic material. Presence of oxygen and the increase in temperature increase the corrosion rate.

(e) Sulfur stress corrosion cracking
It is the corrosion type occurring due to the presence of high strength steel in the moist environment containing Hydrogen Sulfide (H_2S). It is different from stress corrosion cracking since corrosion rate decreases with presence of oxygen and temperature increase in Sulfur stress corrosion cracking. Moreover low pH values increase the corrosion rate.

(f) Hydrogen bubbling
It occurs due to the presence of low strength steels in aqueous solutions containing hydrogen sulfide. Fractures occur because of the insufficient movement of hydrogen caught in vacancies. It is not necessary to apply stress to the material for this type of corrosion.

(g) Intergranular
It is the regional corrosion that occurs around the grain boundaries or in the neighbor grains of metallic materials; however it does not affect the grains. Alloy fractures or loses strength. Wrong heat treatments cause this type of corrosion.

(h) Galvanic corrosion
Galvanic corrosion occurs by the electrical conduction of two different metals. By considering the galvanic series endurance list can be done in material selection. However in the chemical systems and by change of temperature endurance list may change.
(i) Fatigue corrosion
Fatigue corrosion occurs due to the fluctuating stress in a corrosive environment. Fatigue corrosion limit is the largest stress under given conditions. The combined effect of fluctuation, stress, and corrosion is much larger than the other simple effects.
(j) Erosion corrosion
It is the abrasion of the metallic material by the striking of high velocity fluids to the hanging solid materials or particles. The metal that is exposed to this kind of corrosion does not form corrosion product on its surface. Along the flow direction, pits can be seen by the naked eye, and circles form in the shape of waves. It is frequently seen at the turbine blades and in the fields having the flow of two phases.
(k) Decomposition of ally structure
It is the dissolution of one component of the alloy.
(l) Cavitation
Cavitation is a fast and regional decomposition around the metal surface caused by the exploding steam bubbles.

There are various variables affecting the corrosion rate of the geothermal fluid. These are: pH, dissolved oxygen, carbon dioxide, hydrogen sulfide, ammonia, sulfate, chloride, and suspended solid material and its deposit. The effects of these variables mainly include (Kaya and Hoshan 2005):

(a) pH effect
Corrosion rate of carbon steel increases as pH decreases. This situation is especially seen when pH environment decreases below 7. Passivity of many alloys depends on pH. In local regions, corrosion increases with passivity breakdown. By pit corrosion, crack corrosion, and stressed corrosion cracking may occur.
(b) Effect of chloride ions
Presence of chloride ions causes the breakage of the passive layer that prevents many metals from corrosion. By the breakage of this layer, pit corrosion and cracking with stressed corrosion occurs.
(c) Effect of hydrogen supplier
Hydrogen sulfide basically affects copper and copper nickel alloys. Although usage of copper and copper-nickel alloys is advantageous in sea water environments, they are not preferred in geothermal fields because of the corrosion effect. In iron-based materials they are harmful above concentrations of 50 ppm. High strength steels are generally exposed to stress corrosion cracking by the effect of hydrogen sulfide. Hydrogen sulfide also creates hydrogen bubbling in steels. If geothermal process fluid contains dissolved oxygen,

hydrogen sulfide is oxidized with this oxygen and by decreasing pH of the geothermal fluid, it increases the corrosive features of the environment.

(d) Effect of dissolved carbon dioxide
Increase in the dissolved carbon dioxide amount in geothermal fluid causes pH to decrease by providing acidic influence in the environment. Low carbon steels cause uniform corrosion in materials. pH of the geothermal fluid and flow of the process mostly depends on the carbon dioxide amount. Presence of carbonate and bicarbonate or their formation causes a slight decrease in corrosion rate.

(e) Effect of ammonia
Ammonia causes cracking of copper alloys by stress corrosion. Causes uniform corrosion in soft carbon steels.

(f) Effect of sulfate ions
There is no significant effect of sulfate ions on geothermal fluids. In the fluids containing low amount of chloride ions, sulfate acts as the main attacker ion but it cannot be very effective.

(g) Effect of dissolved oxygen
When low amount of oxygen enters to the geothermal systems operating at high temperatures, normal resistant metals undergoes stress corrosion cracking. Oxygen amount of the system must not be below 20 ppb. Low carbon steels are sensitive to the corrosion above this limit.

(h) Effect of suspended solid materials and deposits
The existence of suspended solid materials in the geothermal fluid or precipitation of ions by chemical reasons causes deposit formation on the materials. Solid materials and deposits create erosion corrosion.

5.6.2 The Performance of Metallic Materials in Geothermal Systems

(a) Soft and Low Alloyed Metals
Because of their low expense and convenience, low carbon steels seem to be a sensible material; on the other hand, safe usage of this material depends on the applications in the system. Soft steels can be used, by taking the necessary precautions in thick-walled systems. The usage of these steels is limited in thin-walled systems because of the risk of crack and pit corrosion. Uniform and local types are the frequently encountered corrosion types as a result of the usage of soft steels. As a result of the various field tests, it is observed that in the circumstances where pH value is greater than 6 and the Cl ion concentration is lower than 2%, the uniform corrosion rate varies between 1 and 10 mpy (0.0254 to 0.254 mm/year). For geothermal fluids, local corrosion is more effective than the uniform corrosion. Cl ion triggers the local corrosion. Presence of hydrogen sulfide in the environment increases the effect of local corrosion. Low amount of oxygen in the environment also accelerates the uniform corrosion formation hence triggers the occurrence of pit and crack corrosion. High flow rates and solid particles accumulated in the fluids result in erosion corrosion. The optimum flow rate for carbon steel materials is 5–7 fps (1.524–2.134 mps). Because

of the chemical properties of the geothermal fluid, some ions precipitate and accumulate on the surface of steel. These accumulations have pores and tendency to cracking. Corrosion may occur in these small areas. In the case of release of those accumulations from the steel surface and with the presence of Cl ion, local corrosion occurs. Protective plating can be used for outside surfaces to prevent the uniform and local corrosion. Sulfur stressed breaking can be seen in steel materials which are subjected to hydrogen sulfide under stressed condition water environment. It increases with the increase in temperature, decrease in stress, decrease in strength, and decrease in the concentration of sulfur and increase in pH. Steel materials that include more than 1% of nickel are more sensitive to this condition. Besides, hydrogen bubbling may occur in the low strength steels which are subjected to water solutions having hydrogen sulfide. Room for bubbling is necessary. Consequently, steels not having much room for this procedure are more resistant to bubbling (Kaya and Hoshan 2005).

(b) Stainless Steels
Stainless steel material decreases the probability of uniform corrosion formation in geothermal fluid environment. However, more serious corrosion problems may occur. These are; pit corrosion, cracking corrosion, breaking with stressed corrosion, breaking with sulfur stressed corrosion, corrosion between the particles and wearing corrosion. Cracking corrosion can be a serious problem for stainless steel when used with sophisticated equipment in geothermal fields. An increase in the Cl ion concentration in the environment results in an increase in the effect of local corrosion. Rising temperature increases the pit potential. The resistance of stainless steel against pit and cracking corrosion depends on its chrome and Mo content. These two elements increase the resistance of stainless steel in an environment without oxygen. Austenitic stainless steels are vulnerable to breaking with stressed corrosion in the presence of Cl ion at high temperatures. Ferric stainless steels are generally stronger. Breaking with stressed corrosion depends on Cl ions, oxygen concentration, pH value, temperature, and tension and alloy components. Alloys with nickel can be affected by stressed corrosion. Addition of Mo and silica increases the resistance to stressed corrosion. Corrosion between the particles can be seen in austenite and ferric stainless steels. Especially during the welding operation this may be observed. Ferric stainless steels can be influenced by sulfur stressed breaking but austenite stainless steels cannot. Low strength steels are more vulnerable to sulfur stressed breaking. AISI 400 series stainless steels contain 12–18% chrome. This has a great importance for turbine blades, pump, and valve materials. 13% chrome is suitable for turbine blades. AISI 430 (Ferrite) and AISI 431 (Martensitic) stainless steels types are often used for valve and pump components in geothermal systems. In order to prevent the pit corrosion and breaking problems in wellhead valves, geothermal fluids containing high amounts of Cl ions, sulfur and oxygen in solution, it is more suitable to use AISI 430 (Ferrite). Thermally treated martensite stainless steels are preferred for pump inside components and shafts. AISI 300 series stainless steels show well performance in geothermal

condensates at low temperatures and geothermal fluids not containing oxygen (Kaya and Hoshan 2005).

(c) Titanium and Titanium Alloys:

Titanium and titanium alloys are more successful when they are used with air-cooled or oil-cooled heat exchangers. Corrosion rates of titanium materials which are experimented with geothermal fluids are generally lower than 0.3 mpy (7.62 μm/year). It is proved with the experiments that increases in temperature and Cl ion concentration do not accelerate the corrosion rate. It is also observed that flow rates about 30 fps do not affect the general corrosion. Besides, titanium is resistant to cavitation and impact damages. Pit and cracking corrosions are observed at high temperatures and for Cl ion concentrations above 10%. Titanium alloys are much more resistant to local corrosion than pure titanium. Ti-code-7 (Ti-0.15 Pd), Ti code-12 (Ti-0.3 Mo-0.8 Ni), and Ti-code-29 (Ti-6 Al-4 V-0.1 Ru) show well resistance. Titanium is more cathodic than other metals. If the titanium area is greater than the paired metal, the paired metal undergoes galvanic corrosion seriously. Titanium can form hydrogen when it is paired with active metal because of the cathodes in a galvanic pair. Titanium may absorb the hydrogen and that results in hydrogen embrittlement. It is assumed that titanium alloys are vulnerable to breaking with stressed corrosion in Cl ion concentrations above 3%. When they are compared on the basis of cost and performance, titanium alloys can be used properly as other stainless steel alloys. Titanium alloys can be used when the Cl ion concentration of the geothermal fluid is greater than 5000 ppm and the temperature above 100 °C. Furthermore, they are preferred as material when there is oxygen entrance to the system, because geothermal fluid containing oxygen and hot Cl ion can cause breaking with locally corrosion for stainless steel and nickel-based alloys. In these circumstances, the critical places for the use of titanium alloys as the material can be; wellhead valves, pressure gauges, pipes and blow-out preventers. If the amount of dissolved solid material of hot geothermal fluid containing Cl is above 100,000 ppm, pH is lower than 4 and the wellbore temperature is above 230 °C, titanium code 29 pipes are preferred for the transportation of the geothermal fluid. The service life of this material is above 15 years and it does not have renewable costs compared to low alloy steel and in addition they do not form corrosion and accumulation products containing radioactive and heavy metals. Furthermore, they reduce the risk of well plugging and well damaging and prevent the plugging of well pipes with iron-enriched silicate accumulations. Titanium code 29 is being used as well lowering pipe in USA Salton Sea (Kaya and Hoshan 2005).

(d) Nickel Alloys

Corrosion often calls for the frequent usage of nickel alloys. For the high-temperature geothermal fluids, it is suitable to use Ni–Cr–Mo alloys as a material. Especially, Inconel- 625 and Hastelloy C-256 are very strong for the corrosion. Instead of molybdenum, similar alloys, which have iron elements, can be used in some applications because of its mechanical properties and the reason that it is also much stronger than the stainless steel. Some nickel alloys lack resistance to the stress sulfur cracking or to the hydrogen embrittlement in the presence of hydrogen sulfide. Furthermore, Ni–Cu alloys are not suitable even in low hydrogen sulfide conditions (Kaya and Hoshan 2005).

(e) Copper-Based Alloys
It is limited to use copper alloy materials with geothermal fluids that have high amount of sulfur. It has been known to see cracks in the copper alloys which faces with the ammoniac and something like ammoniac. Some cases when the amount of ammoniac and ammonium are low, the cracks on the metal surfaces are limited. The breakup risk in the copper-zinc alloys increases with the increasing of the amount of zinc. Copper and copper alloys have been tested for the heating systems. It is established that copper fan bobbin and heat exchangers with copper tubes show a low performance for the corrosion in the presence of sulfur compound in the geothermal fluids. Similar to previous result, another research on the heat exchangers shows that copper-zinc (brass) and copper-tin (bronze) alloys are not suitable for the corrosion. It is found that red lead brass alloy (CA 836 and 838) and red lead bronze alloy (SAE 67) materials can be used for the inside of the pump (Kaya and Hoshan 2005).

(f) Other Metal-like Materials
The other materials and alloys are limited in this area. Cobalt alloys can be used in the application, which need resistance, of high durability for the abrasion strength and stress sulfur cracking. It can be used in the part of zirconium and tantalum acidizing. It is observed that aluminum alloys have low performance in the direct contact of geothermal fluids. In these alloys, hollow and galvanic corrosion can be frequently seen (Kaya and Hoshan 2005).

5.6.3 The Performance of Non-metallic Materials in Geothermal Systems

Generally, metals are used as material in the geothermal fields. On the other hand, the usage of nonmetallic materials increases day by day. The usage of this nonmetallic materials are needed in some special geothermal operations and drilling operations, such as the use of elastomers. Besides, there are some advantages of nonmetallic materials: they are generally strong for the corrosion when they are encountered with metals and alloys. The initial investment cost is lower than metals and alloys because of the long period of operating and repair. However, they are not useful in heat transfer equipment. The specifications of some nonmetal materials which are used in geothermal field and their properties are given below (Kaya and Hoshan 2005):

(a) Polymer concrete: Polymer concretes are a type of concrete that use polymers to replace lime-type cements as a binder. In some cases the polymer is used in addition to portland cement to form Polymer Cement Concrete (PCC) or Polymer Modified Concrete (PMC). In polymer concrete, thermoplastic polymers are used, but more typically thermosetting resins are used as the principal polymer component due to their high thermal stability and resistance to a wide variety of chemicals. Polymer concrete is also composed of aggregates that include silica, quartz, granite, limestone, and other high quality material.

More specifically, polymer concrete can be divided into several types. The first type is monomers that are polymerized in situ in presence of other components. In situ polymerization takes place by heat, catalysts, or radiation. Monomer when polymerized can be used as the binder component and that does not require water to set or harden. Polymer concrete comprises polymeric binder, hardener, and the aggregate. The interactions among these components depend entirely on the chemical and physical reactions. Since late 70s, epoxy and acrylic polymer concrete have been used to replace traditional materials due to rapid curing and excellent bond to cement concrete. Methyl methacrylate, unsaturated polyester resin, epoxy resins, furan resins, polyurethane resins, urea formaldehyde resin, and blends of polyester/styrene are commonly explored as polymer concrete systems. The advantages of using polymeric resin are that it possesses high mechanical strengths, long durability, and resistance to chemical attack.

Polymer-impregnated concrete is another type of polymer concrete in which the hydraulic binder is totally substituted with a polymeric material. In polymer-impregnated concrete, the monomer penetrates the concrete matrix to a finite depth.

Yet another category is polymer cement concrete which is a mixture of conventional hydraulic cement concrete and high molecular weight polymers. Polyvinyl acetate, polyacrylates, polyvinylchloride, styrene-butadiene, and polyvinylidene chloride are the polymers that have been explored as polymer cement concrete.

Therefore, the composition of first type polymer concrete differs from typical Portland cement concrete, polymer cement concrete, and polymer-impregnated concrete. Polymer cement concrete is extensively used as repair material. The performances of polymeric concrete depend on different factors such as polymer properties, type of fillers and aggregates, curing temperature, and components dosage. The choice of particular type of resin depends upon factors like cost, desired properties, and chemical/weather resistance (Kumar 2016).

Polymer Concrete materials, which include silica, sand, and portland cement, are enduring geothermal water and steam over 218 °C.

(b) Cements: C-S cements and phosphate glass cements are investigated as potential cementing material. The cements, which are resistant for the corrosion, can be used as coating materials.

(c) Elastomers: The investigation of elastomers has been continuing in the drilling operations. Nowadays, the usage of elastomeric materials as connection components comes on the agenda in pipelines. Fluorine-elastomer shows the best performance as a connection component in pipelines. Neoprene and natural rubber could not be successful in this area. Ethylene-Propylene-diene terpolymer is used as sealing in valve and O-ring in most systems. The fiber-reinforced materials: The usage of Chlorinated Polyvinyl Chloride (CPVC) and Fiberglass-

Reinforced Plastic (FRP) increases because of high resistance of corrosion and low cost. Especially, it is used safely in corrosive geothermal water transport lines. Fiberglass-Reinforced Plastic (FRP) pipelines are supplied at low cost by means of smooth surface in central geothermal heating systems and water and hot water transport lines. Moreover, Fiberglass-Reinforced pipelines decrease the usage of scaling inhibitor and supplies low cost by means of smooth surfaces because of low contact of $CaCO_3$ to the pipeline surface in high $CaCO_3$ settlement. The mechanical properties of Fiberglass-Reinforced Plastic (FRP) pipes, its durability in high pressure (>200 bar) and its durability in high temperature (>130 °C) improved with the last studies. In addition, when they are used with mistakes, it creates some problems, such as brittleness and breakoff. Because of that, material producers should consider the project at which the material will be used, the design, and the management conditions. The most important criterion about the life-time of Fiberglass-Reinforced Plastic (FRP) pipeline is the assembly situation. In the case of wrong assembling, the breakoff and brittleness are inevitable.

In geothermal systems, therefore, probable corrosion models should be put forward by analyzing the system before the design and material selection procedure. The corrosion model should be developed under wellhead and wellbore conditions and should include corrosion chemistry and the effect of the fluid flow rate on erosion corrosion. Hence, suitable material selection and corrosion control for the design would be obtained. Although low alloy steels are mostly preferred as easy to obtain and low in expense, the necessary precautions should be taken in order to use them safely in geothermal systems. The probability of stainless steels to undergo uniform corrosion is low, when they are used with geothermal fluids. On the other hand, some other types of corrosion may be observed. Stainless steel material selection and usage should be made carefully. Titanium and titanium alloys are resistant to pit and cracking corrosions at Cl ion concentrations below 10%. Their usage in air or oil cooling heat exchangers is suitable. However, at high temperatures they undergo pit corrosion. Inconel-625 and Hastelloy C-276 are the most suitable Ni–Cr–Mo alloys for use in high-temperature geothermal fluids. Cobalt alloys are used for applications which require resistance to abrasion resistance and sulfur stressed compression, and high strength. Zirconium and tantal alloys are used with channels subjected to hot hydrochloric acid treatment. The performance of aluminum alloys is pure because they form galvanic and pit corrosions. The usage of concrete-polymer mixtures which are classified as nonmetallic material, and Fiberglass-Reinforced Plastic (FRP) pipes and materials, has been rapidly increasing. The applications of Fiberglass-Reinforced Plastic (FRP) materials in central geothermal heat systems by investors and managers are rapidly increasing due to their high corrosion resistance, ease of installation and short time for assembly, longer service life and low operating costs (Kaya and Hoshan 2005).

5.7 Additional Value Streams and Future Trends of Geothermal Power

A complementary strategy to advancing technologies to reduce the cost and risks associated with geothermal power developments is to advance technologies to derive more value from the resource. There is a significant positive financial advantage to optimizing the utilization of produced geothermal fluids downstream from power production. Owing to the large volume of produced fluids, the energy content post power production remains significant.

5.7.1 Mineral Extraction from Fluids

The produced fluids, having reacted with reservoir host rocks, may also be fertile with potentially strategic chemicals. For instance, common minerals and elements entrained in geothermal fluids include silica, lithium, manganese, zinc, and sulfur. Additional rare earth elements (REEs) and near REEs may also be relatively prevalent in the brines or may be associated with the geothermal system host rock as either primary or alteration mineralogy. Besides their significant market value, these chemical elements have strategic importance in that they are critical for domestic industries that produce everything from mobile phones and laptops to green technologies and national defense systems. Further development and integration of mineral extraction into geothermal energy production offers a unique opportunity for the geothermal industry to expand its market value beyond power production (DOE QTR 2015).

5.7.2 Combined Heat and Power

Typically, geothermal fluids, after providing thermal energy for generating power, have temperatures ranging from ~40 to 100 °C. Coupled with the economically necessary high fluid throughputs, these "waste" fluids are a significant thermal energy source. Presently, common practice is reinjection of the waste fluids for reservoir management and to satisfy environmental requirements. Technologies are being developed to use this energy source and improve the economics of geothermal energy production. For instance, it is feasible to use the waste heat to drive water desalination plants. Waste heat can also be used for green houses, residential and building heat, and other applications that employ relatively low-temperature thermal energy. Combining heat and power is a potentially very fertile area of technology development that could significantly alter the economics of geothermal power production. The utilization of thermal wastewater should also take advantage of emerging technologies by providing thermal energy with moderate- to low-temperature catalysts used in producing hydrogen, biofuels, and organic solvents (DOE QTR 2015).

5.7.3 Energy Storage for Flexible Power

Historically, energy storage is not a technology that has been addressed within the geothermal industry. However, with grid management becoming more and more complex, timely delivery of energy to the grid (flexible power) has become an area of increased interest. In the past, the geothermal community has extolled the benefits of geothermal because it can provide baseline energy continuously. The need to stabilize the grid and provide more energy during peak usage and less during low demand periods has led the power providers to prefer "flexible power" that can be ramped up or down as needed. One way for geothermal power producers to increase flexibility in power generation is to allow fluids to bypass the generating turbines during low demand periods, reinjecting the unused hot fluids. Although this is an adequate stopgap approach, as the demand for flexible power increases, more efficient technologies or reservoir management schemes will need to be developed (Cooley 1996; Nordquist et al. 2013).

5.7.4 Geothermal as a Baseload Resource

Geothermal has traditionally been a baseload resource, operating continuously. By their nature, geothermal resources are always available and highly reliable, making geothermal the only baseload renewable energy resource other than hydropower. Unlike solar and wind energy, whose availability are constrained by diurnal and weather patterns, the production of geothermal fluids from the subsurface is always available for electricity generation. Even hydropower can be subject to seasonal restrictions or complicated by water and environmental needs, such as lack of precipitation impacting reservoir levels or downstream flow requirements/restrictions for water needs or habitat (species) considerations. Once accessed, geothermal resources can be produced continuously, regardless of surface conditions. Furthermore, geothermal plants require little downtime for maintenance, resulting in availability factors of 90–95% annually. However, geothermal power plants usually have large parasitic or operating loads, such as powering injection or production well pumps and operating power plant cycle recirculation pumps that are powered internally by the plant. These loads can use up to 30% of the power generated by the plant itself. The result is that the power delivered to the grid is small relative to the theoretically possible nameplate generation capacity, resulting in an apparently low capacity factor. In addition, because of their relatively low operating temperatures, changes in the ambient temperature can lower the temperature differential driving the thermal power plant and result in significant decreases in net power generation output. This is especially true for binary or organic Rankine cycle geothermal power plants that operate at the low end of the geothermal electricity generation spectrum (roughly 150–200 °C) and rely on air cooling. During hot summer days, output from these plants can drop as much as 30–50%, depending on operating conditions. Larger flash and steam geothermal plants are more immune to these swings because they operate at higher temperatures and are able to use some of the condensate from the turbine for water cooling (cooling towers). Research into the use of hybrid cooling (air cooling with water spray) at binary plants has been explored to increase power output during hot days. Also, hybrid plants that combine geothermal with solar thermal (concentrating solar) and/or solar photovoltaic power have been studied regarding normalizing power output over the course of the day (DiMarzio et al. 2015).

5.7.5 Geothermal as a Dispatchable Resource

While geothermal power plants can serve as baseload generation units, with the proper design and incentives, geothermal power can also provide either firm or flexible electricity or can be designed to form more intermittent electricity generation resources. Geothermal plants can ramp up and ramp down electricity generation output quickly so geothermal projects can provide flexibility and ancillary services. There are concerns about the risk that variable production from geothermal resources could have in terms of damage to the subsurface reservoir, but geothermal plants can be designed to continue full production from the reservoir while bypassing the power plant in part to control electricity generation. Since geothermal plants have high up-front costs (and relatively fixed operating costs), the additional revenue from operating as a flexible generation provider must be adequate to make debt payments from plant construction and must be advantageous relative to baseload generation. The potential of using geothermal as a dispatchable resource requires R&D into the ability and cost of designing or converting geothermal plants to this type of operation (including potential impacts on the geothermal resource) and analysis into the cost benefits to both the grid and the geothermal operator under various future scenarios that include large amounts of variable or intermittent electricity generation (DOE QTR 2015).

5.8 Summary

Geothermal resources currently provide cost-competitive low-carbon, firm but flexible power generation in specific geographical regions. Geothermal heat pumps utilize heat retained in the ground for direct heating and cooling applications. The GSHP pumps circulate a carrier fluid (usually a water/antifreeze mix) through pipes below the ground. As the fluid circulates it absorbs heat from the ground. The fluid passes through a heat pump and electricity is used to extract heat from it. The re-chilled fluid is sent back through the ground, continuing the cycle. The GSHP system can switch the direction of the heat flow, so in the summer it can act as an air conditioner, transferring the heat from the warm air to the cooler ground. Geothermal heat is often used for industrial processes, but GSHP technology can also be applied at household level. Piping materials for geothermal heating systems have been of numerous types with great variation in cost and durability. Some of the materials include: asbestos cement, ductile iron, slip-joint steel, welded steel, gasketed polyvinyl chloride, solvent welded PVC, chlorinated polyvinyl chloride, polyethylene, cross-linked polyethylene, mechanical joint fiberglass-reinforced plastic (FRP-M), FRP epoxy adhesive joint-military (FRP-EM), FRP epoxy adhesive joint (FRP-E), FRP gasketed joint (FRP-S), and threaded joint FRP (FRP-T). The temperature and chemical quality of the geothermal fluids, in addition to cost, usually determines the type of pipeline material used. The heat transfer fluid circulates down into the ground and in the summer dissipates heat from the environment to cool the house. In the winter, the process will reverse; the fluid transports heat from the ground into the house. This heat transfer is possible by burying a pipe containing the heat transfer fluid in the ground. The pipe is connected to a heat pump/air conditioning unit and can go forward or in reverse depending upon the season. The heat transfer fluids used can be based on water, methanol, ethanol, ethylene glycol, propylene glycol, and CO_2. The pipe and heat exchangers are often coated for corrosion protection or for cosmetic purposes. Chemical conversion coating, anodization, E-coating, and painting are common coating options that will help to minimize corrosion and/or result in a more attractive component.

Drilling technologies are employed for the exploration of all earth's resources and are the means for extraction for many resources, including geothermal. The most common solution for drilling through hard rock and petrothermal systems is the application of controlled pressure drilling equipment. There are three major types of controlled pressure drilling methods: (a) air drilling (AD); (b) managed pressure drilling (MPD); and underbalanced drilling (UBD). AD is geared towards optimizing the rate of penetration (ROP), MPD minimizes rig nonperformance time and UBD increases productivity and reduces reservoir damage. For the case of drilling through a non-fractured source rock, it is not necessary to drill pressure controlled before stimulating the pay zone.

Wellbore integrity during the construction of the well, throughout its useful life, and after abandonment is an important issue for all constructed wells. Improved cements, remediation materials, casing systems, and practices are needed to ensure that wells developed for geothermal energy production are constructed in a manner that allows success during drilling and operations and long-term environmental safety. For instance, the principal application of the geothermal well cementing materials is to support mechanically metallic well casings as well as to protect them against hot brine-initiated corrosion at brine temperatures up to 320 °C. The cementitious materials not only must possess high-hydrothermal temperature stability, but also they must be inert and resist against very harsh geothermal environments involving the CO_2-enriched brine encountered in a bottom hole depth of ~1700 m at temperature of ~320 °C, and a highly concentrated H_2SO_4 brine in an upper well region between the well's surface and ~1000 m depth at temperatures up to 200 °C.

As geothermal energy is being developed and taking more importance, certain problems have occurred regarding the corrosion and the usage of the materials that are influence the effectiveness and the quality of the service. The geothermal fluid partly leading to the breakdown of the equipment used for processes due to corrosion; dissolved CO_2, H_2S, NH_3 and chloride ions, may lead to the metallic materials being corroded and becoming unusable. The chemical composition, temperature, and velocity of fluid vary depending on the geothermal resource, the design of the power system, and the point of the production cycle. Both the operating experiences gained during usage and laboratory studies performed with real geothermal fluids form the basis for the selection of materials for this area.

Removing technical barriers will promote increased capacity in these geographies with improved reservoir management tools and broaden the geographic footprint of geothermal power generation. The research needed to advance subsurface technologies links tightly with broader "grand challenges" in subsurface science. Through fundamental science and cross-disciplinary research, geothermal power provides important motivation to strengthen the ties between basic research and technology innovation.

Exercises

Part I: General Questions

5.1. What is geothermal energy? How can we access geothermal energy?

5.2. How do we generate electricity from geothermal energy?

5.3. What are benefits and drawbacks of geothermal energy over other renewable energy use?

5.4. Waste-water treatment is an interesting potential application of geothermal heat. The different cycles of this process (sludge digester heating, sludge disinfection, sludge drying, grease melting) take place in one of the following temperature intervals. Which one? (a) 30–95 °C; (b) 70–95 °C; (c) >95 °C

5.5. There are four types of heat loss from a pond: (a) evaporative, (b) convective, (c) radiant, and (d) conductive. What is the governing factor in each of these heat-loss processes?

5.6. What steps should be taken in the case of a geothermal fluid with a high salt content?

5.7. The connections between the geothermal well and the installations of the heat user(s) can be either direct or indirect. Our choice will be based on which of the following? (a) The chemical composition of the geothermal water. (b) The type of heat exchanger or the type of heating installation adopted. (c) The chemical composition of the geothermal water and the materials used for the installations of the heat user.

5.8. The heat exchanger is the part of the heating installation that extracts the heat from the geothermal water and passes it to the space or material requiring heat. On which of the following do we base our classification of the heat exchangers used in greenhouse heating: (a) The type of heat transfer involved. (b) The location of the heating elements with respect to culture allocation and their arrangement in the protected space. (c) Both (a) and (b), because of the different heat requirements of the different plants and types of cultivation.

5.9. A borehole, sited at an elevation of 600 m above sea level, produces water at 60 °C and a pressure slightly above atmospheric. The decision is taken to use this water to heat a greenhouse located further downhill, at an elevation of about 300 m above sea level. The water does not contain any chemicals that could cause scaling or corrosion, so that, from the chemical point of view, it could be carried directly into the greenhouse heating installation. However, a double circuit is adopted: in the primary circuit the geothermal water circulates between the borehole and heat exchangers, sited near the greenhouse; in the secondary (closed) circuit water circulates in the heating installation inside the greenhouse. Between these two circuits are the heat exchangers. Why was this double circuit solution adopted, considering it is more expensive than a single one?

5.10. List the three fluid circuits that have external connections in a geothermal water chiller and the approximate temperatures of the entering cooling water and leaving chilled water using 83 kPa steam (or equivalent hot water).

5.11. A well produces about 170 tonne per hour of steam, at a temperature of 205 °C and pressure of 5 bar, with a content of 30% by weight of non-condensable gas. What is the most appropriate type of power plant for this well? Can you suggest a system that would permit a more efficient exploitation of the energy supplied by the well?

5.12. What are the major functions of geothermal heat pumps? List the pipe materials used for the system. What's the consideration of the materials selection?

5.13. What's the principle consideration for materials selection of heat exchange systems?

5.14. List fluids used for the heat exchanges systems. What's the benefit of CO_2 as the geothermal working fluid?

5.15. What are major challenges that geothermal well construction and well operations face?

5.16. What's the most common solution for drilling through hard rock and petrothermal systems?

5.17. List common metals used to make drill bits. Each material is best suited to drill through a particular type of surface. What are strengths and weaknesses of each material type when selecting the bit?

5.18. What are major functions of the geothermal well cementing materials? List cements used for current geothermal wells. What's the suitable applications of each cement?

5.19. What's the major consideration of material selection for corrosion protection of geothermal systems?

5.20. List corrosion types encountered in geothermal systems. What causes the corrosion? How to minimize the effect of corrosion?

Part II: Thought-Provoking Questions

5.21. What are the main elements of a "geothermal system"? Which of these elements can be either partly or totally artificial? What physical phenomenon governs the mechanism of a geothermal system?
5.22. Heat pumps are now widely used in a number of countries. Are thermal waters necessary for their utilization? What is the minimum water temperature required?
5.23. The most effective way of reducing the impact on the environment of the effluent from a geothermal power plant is to inject it back underground in reinjection wells. Wherever possible these fluids are injected back into the reservoir from which we extract the fluids feeding the geothermal power plant. If this operation (reinjection) is not carried out correctly, it can create serious problems. What are these problems?
5.24. Why does an increase in the efficiency of a geothermal power plant lead to a reduction in the impact on the environment?
5.25. Geothermal power plants are generally considered to be baseload units. However, certain factors may make load-following or even operating in a dispatchable mode necessary or economically attractive. What are these factors, and what design consideration should be taken into consideration to ensure maximum operation flexibility and fuel-use efficiency?
5.26. Describe enhanced or engineered geothermal systems (EGS). What are the major challenges for EGS development?
5.27. Why are custom heat exchangers often coated? List common coating materials used, comparing the major functions of these materials.
5.28. Describe the value streams and future trends of geothermal power.
5.29. Discuss possible energy storage methods for geothermal power.

References

Azar, J.J., Samuel, G.R.: Drilling Engineering. PennWell Corporation, Tulsa, Oklahoma (2007)

Barnett, P., Quinlivan, P.: Assessment of current costs of geothermal power generation in New Zealand (2007 basis), report by SKM for New Zealand Geothermal Association. https://www.nzgeothermal.org.nz/industry_papers.html (2009). Accessed 23 July 2014

Bayer, A.M., Becherer, B.A., Vasco, T.: High-Speed Toll Steels. ASM Handbook, Vol. 16: Machining ASM Handbook Committee, p 51–59 (1989)

Bertani, R.: Geothermal Power Generation in the World 2005–2010 Update Report. Proceedings of World Geothermal Congress, Bali, Indonesia (2010). https://www.geothermal-energy.org/pdf/IGAstandard/WGC/2010/0008.pdf

Brown, D.: A hot dry rock geothermal energy concept utilizing supercritical CO_2 instead of water. In: Proceedings of the 25th Workshop on Geothermal Reservoir Engineering, Stanford University, Stanford, California, January 24–26, 2000. SGP-TR-165. https://pangea.stanford.edu/ERE/pdf/IGAstandard/SGW/2000/Brown.pdf (2000). Accessed 10 July 2017

Canter, N.: Heat Transfer Fluids: Selection, maintenance & new applications. Tribol. Lubr. Technol. **65**, 28–35 (2009)

Clauser, C.: Geothermal energy. In: Heinloth, K. (ed.) Landolt-Börnstein, Group VIII: Advanced Materials and Technologies, Energy Technologies, Subvol. C: Renewable Energies, vol. 3, pp. 493–604. Springer-Verlag, Heidelberg, Berlin (2006)

Cooley, D.: A report on cycling operations at the Geysers power plant. Geotherm Resources Council Trans. **20**, 729–732 (1996)

DiMarzio, G., Angelini, L., Price, W., Chin, C., Harris, S.: The stillwater triple hybrid power plant: integrating geothermal, solar photovoltaic and solar thermal power generation. In: Proceedings World Geothermal Congress 2015, Melbourne, Australia, April 19–25, 2015. https://pangea.stanford.edu/ERE/db/WGC/papers/WGC/2015/38001.pdf (2015). Accessed 12 July 2017

DOE QTR: Quadrennial Technology Review 2015 TA 4.1: Geothermal Power. https://energy.gov/sites/prod/files/2015/12/f27/QTR2015-4I-Geothermal-Power.pdf (2015). Accessed 11 July 2017

EGEC: Geothermal heat pumps—ground source heat pumps. http://egec.info/wp-content/uploads/2011/01/EGEC-Brochure-GSHP-2009.pdf (2011). Accessed 8 Sept 2014

Geo-Energie: Heat transfer fluids. http://www.geo-energie.com/pdf/liquids.pdf (2014). Accessed 16 Sep 2014

Hammons, T.J.: Geothermal Power Generation: Global Perspectives, Technology, Direct Uses, Plants, Drilling and Sustainability Worldwide, Electricity Infrastructures in the Global Marketplace. InTech (2011). https://doi.org/10.5772/37842. Available from http://www.intechopen.com/books/electricity-infrastructures-in-the-global-marketplace/geothermal-power-generation-global-perspectives-technology-direct-uses-plants-drilling-and-sustainab

Jeanloz, R., Stone, H., et al.: Enhanced Geothermal Systems. MITRE Corporation, JASON Program Office, McLean, VA (2013). JSR-13-320. http://www1.eere.energy.gov/geothermal/pdfs/jason.final.pdf. Accessed 7 July 2017

Kant, M.A., Rossi, E, Becker, D., PRvon Rohr: Enhancing the drilling process for geothermal resources by combining conventional drilling and the spallation technology. In: Proceedings, 42nd Workshop on Geothermal Reservoir Engineering Stanford University, Stanford, California, February 13–15, 2017 SGP-TR-212. https://pangea.stanford.edu/ERE/db/GeoConf/papers/SGW/2017/Kant1.pdf (2017). Accessed 12 July 2017

Kaya, T., Hoshan, P.: Corrosion and material selection for geothermal systems. In: Proceedings World Geothermal Congress 2005. Antalya, Turkey, 24–29 April 2005. http://www.geothermal-energy.org/pdf/IGAstandard/WGC/2005/2039.pdf (2005). Accessed 8 Oct 2014

Kumar, R.: A review on epoxy and polyester based polymer concrete and exploration of polyfurfuryl alcohol as polymer concrete. J. Polym. **2016**, 7249743 (2016). (1–13)

Lakic, N.N.: About most recent developments in hot dry rocks technology. http://www.geothermalworldwide.com/egs.html (2014). Accessed 29 July 2014

Lind, L.: Swedish Ground Source Heat Pump Case Study, GNS Science Report 2010/54, January 2011. file:///C:/Users/ctong/Downloads/Swedish%20Ground%20Source%20Heat%20Pump%20Case%20Study%20(2010).pdf. Accessed July 23, 2014 (2011)

Lytron: Coatings for custom heat exchangers and cold plates. http://www.lytron.com/Tools-and-Technical-Reference/Application-Notes/Coatings-for-Heat-Exchangers-and-Cold-Plates (2014). Accessed 17 Sep 2014

Mines, G.: GETEM Development. Presentation at Geothermal Technologies Office 2013 Peer Review, Denver, Colorado, April 22, 2013. https://www4.eere.energy.gov/geothermal/sites/default/files/documents/mines_getem_peer2013.pdf (2013)

Nordquist, J., Buchanan, T., Kaleikini, M.: Automatic generation control and ancillary services. Geotherm Resources Council Trans. **37**, 761–766 (2013)

Phillips, B.R., et al.: A roadmap for strategic development of geothermal exploration technologies. In: Proceedings of the 38th Workshop on Geothermal Reservoir Engineering, Stanford, California, February 11–13, 2013. SGP-TR-198. https://www1.eere.energy.gov/geothermal/pdfs/exploration_technical_roadmap2013.pdf7 (2013). Accessed 7 July 2017

Rafferty, K.D.: Piping. Geothermal direct-use equipment. Geo-Heat Center Quarterly Bullet. **19**(1), 14–19 (1998)

Raymond, D., Knudsen, S., Blankenship, D., Bjornstad, S., Barbour, J., Schen, A.: PDC bits outperform conventional bit in geothermal drilling project. Geothermal Resources Council Trans. **36**, 307–316 (2012)

Ren, X., Miao, H., Peng, Z.: A review of cemented carbides for rock drilling: an old but still tough challenge in geo-engineering. Int. J. Refract. Met. Hard Mater. **39**, 61–77 (2013)

Right, A.: Enhanced geothermal system (EGS). http://www.renewablegreenenergypower.com/enhanced-geothermal-system-egs/ (2012). Accessed 26 July 2014

Stacey, F. D.: Physics of the Earth, John Wiley & Sons, New York (1997)

Sugama, T.: Advanced cements for geothermal wells. http://www.bnl.gov/isd/documents/35393.pdf (2006). Accessed 3 October 2014

T'Joen, C., et al.: A review on polymer heat exchangers for HVAC&R applications. Int. J. Refrig. **32**, 763–779 (2009)

Tanaka, N.: Technology roadmap—geothermal heat and power. International Energy Agency. http://www.iea.org/publications/freepublications/publication/Geothermal_Roadmap.pdf (2011). Accessed 16 July 2014

TEEIC: Geothermal energy development. http://teeic.indianaffairs.gov/er/geothermal/index.htm (2014). Accessed 23 July 2014

Vollmar, D., Wittig, Bracke, R.: Geothermal drilling best practices: the geothermal translation of conventional drilling recommendations—main potential challenges. IGA Academy Report 0104–2013. http://www.geothermal-energy.org/iga_academy/academy_reports.html?no_cache=1&cid=867&did=329&sechash=f12be898 (2013). Accessed 23 Sept 2014

Ziagos, J.P., et al.: A technology roadmap for strategic development of enhanced geothermal systems. In: Proceedings of the 38th Workshop on Geothermal Reservoir Engineering, Stanford, California, February 11–13, 2013. SGP-TR-198. https://www1.eere.energy.gov/geothermal/pdfs/stanford_egs_technical_roadmap2013.pdf (2013). Accessed 7 July 2017

Advanced Materials Enable Renewable Wind Energy Capture and Generation

6

Abstract

Advanced materials play a crucial role in wind power to enable renewable wind energy capture and generation. Composite materials such as polymer-matrix reinforced with fiberglass or graphite fibers have been used to make rotor blades of wind turbines. In the turbines, compact electrical generators contain powerful magnets made from rare earth materials. The electrical generator is driven by the rotation of the turbine blades through a gearbox, which uses special alloys in order to accommodate a wide range of wind speeds. As turbine sizes continue to increase, and the growth of offshore installations, the durability of turbine materials faces more and more challenges due mainly to long-time exposure to higher stresses and hostile environments. In addition, the turbine blades must maintain adequate stiffness to prevent failure due to deflection and buckling. They also need adequate long-term fatigue life in harsh conditions, including variable winds, ice loading, and lightning strikes. Therefore, development of next-generation wind turbine components and materials requires research on advanced materials and key components to improve performance and reliability; development of new architectures for larger, lightweight turbines that reduce overall mass (reducing costs) and provide access to better wind resources (larger rotors, taller towers), and improved systems performance (capacity factor); improvements in turbine cost, strength, weight, and fatigue to reduce operations and maintenance costs and reduce the failure rate for large components, such as blades, gearboxes, generators, power electronics, and collection systems; and innovations to solve transport and installation cost limitations for large-scale turbine systems and components. This chapter will address these critical issues and provide a brief review about the progress of advanced materials enabling renewable wind energy capture and generation.

C. Tong, *Introduction to Materials for Advanced Energy Systems*,
https://doi.org/10.1007/978-3-319-98002-7_6

Fig. 6.1 From the early stages of wind energy exploitation to the outbreak of California (Modified with permission from Dodge 2006 (Courtesy: TelosNet Web Development); and Kaldellis and Zafirakis 2011 (Elsevier))

6.1 History and Evolution of Wind Energy

Wind energy technology made its first actual steps centuries ago—although simpler wind devices date back thousands of years ago with the simple windmills in China and vertical axis windmills found at the Persian-Afghan borders around 200 BC, and the horizontal-axis windmills of the Netherlands and the Mediterranean following much later (1300–1875 AD). Further evolution and perfection of these systems (Fig. 6.1) was performed in the USA during the nineteenth century (Dodge 2006). Throughout the twentieth century, small wind plants, suitable for farms and residences, and larger utility-scale wind farms that could be connected to electricity grids were developed. On the other hand, the first large wind machine to generate electricity (a low-speed and high-solidity wind turbine (WT) of 12 kW) was installed in Cleveland, Ohio, in 1888, while during the late stages of World War I, use of 25 kW machines throughout Denmark was widespread. Further development of wind generators in the USA was inspired by the design of airplane propellers and monoplane wings, while subsequent efforts in Denmark, France, Germany, and the UK (during the period between 1935 and 1970) showed that large-scale WTs could work. European developments continued after World War II. In Denmark, the

Gedser mill 200 kW three-bladed upwind rotor WT operated successfully until the early 1960s, while in Germany, a series of advanced horizontal-axis designs were developed, with both of the aforementioned concepts dictating the future horizontal-axis design approaches later emerging in the 1970s (Meyer 1995; Kaldellis and Zafirakis 2011).

One of the most important milestones of the wind energy history coincides with the USA government involvement in the wind energy research and development after the oil crisis of 1973. Following, in the years between 1973 and 1986, the commercial WT market evolved from domestic and agricultural (1–25 kW) to utility interconnected wind farm applications (50–600 kW). In this context, the first large-scale wind energy penetration outbreak was encountered in California, where over 16,000 machines, ranging from 20 to 350 kW (a total of 1.7 GW), were installed between 1981 and 1990, as a result of the incentives (such as the federal investment and energy credits) given by the USA government. In northern Europe on the other hand, wind farm installations increased steadily through the 80s and the 90s, with the higher cost of electricity and the excellent wind resources leading to the creation of a small but stable market. After 1990 most market activity shifted to Europe, with the last 20 years bringing wind energy at the front line of the global scene with major players from all world regions (Righter 1996; Kaldellis and Zafirakis 2011).

The development of contemporary WTs in the course of time may be reflected by the gradual upscale of machines, presence of scale economies, reduced maintenance and operation requirements and past funding development programs pushing towards the development of big-scale machines. On the other hand, a stabilizing trend has put an end to the exponential increase of the rotor diameter met in the first two decades, as show in Fig. 6.2. As a result, WTs are mainly in the order of 2–3 MW, although larger scale machines that are already commercial do exist. Contrariwise, the shift to offshore applications calls for multi-MW solutions, while designs of machines that will exceed the nominal power of 10 MW are already underway (Herbert et al. 2007; Kaldellis and Zafirakis 2011).

New wind turbines need to reduce their overall costs for large-scale turbines of 10–20 MW going offshore with improved design and reliability of components; development and industrialization of support structures for sea installations, both fixed and floating; achieving grid integration for even greater wind energy penetration; introduction of large-scale energy storage systems and high-voltage alternative and direct current interconnections; more sophisticated assessment of wind resources; and spatial planning through social and environmental considerations. On the other hand, development of next-generation wind turbine components and materials requires research on advanced materials and key components to improve performance and reliability; development of new architectures for larger, lightweight turbines that reduce overall mass (reducing costs) and provide access to better wind resources (larger rotors, taller towers), and improved systems performance (capacity factor); improvements in turbine cost, strength, weight, and fatigue to reduce operations and maintenance costs and reduce the failure rate for large components, such as blades, gearboxes, generators, power electronics, and collection systems; and innovations to solve transport and installation cost limitations for large-scale turbine systems and components.

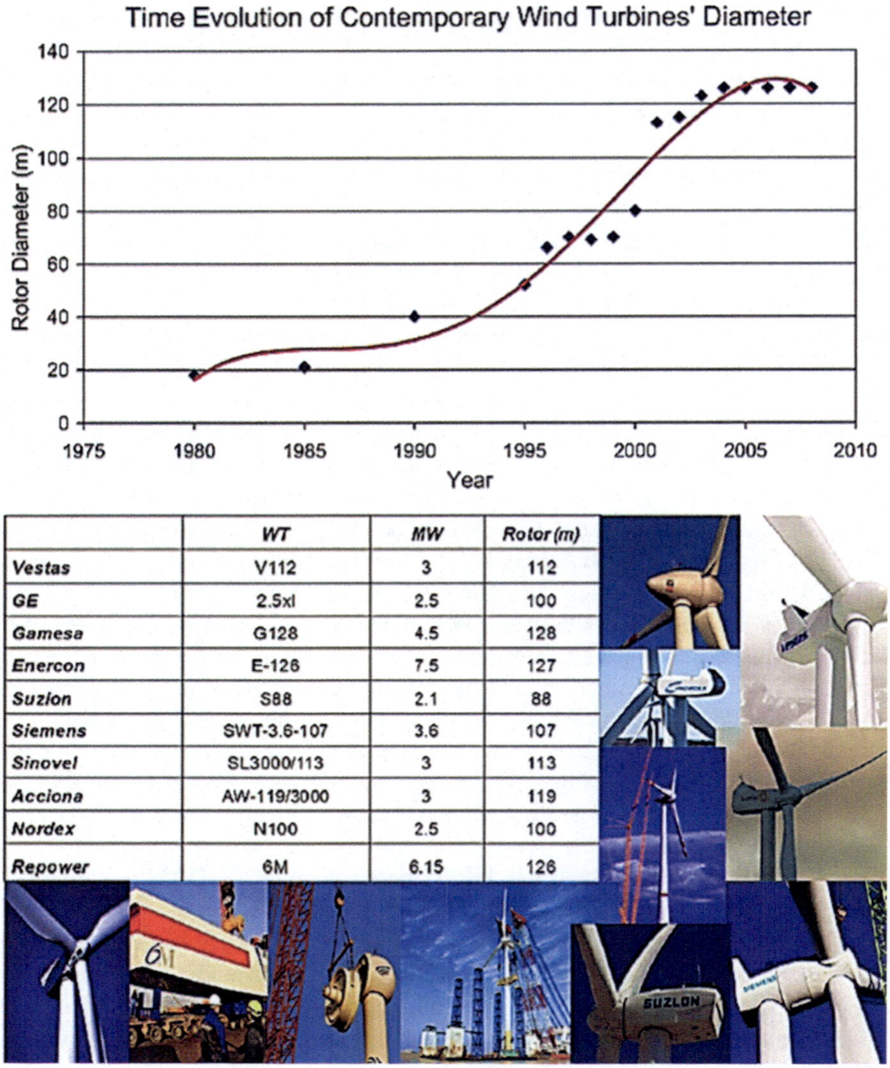

	WT	MW	Rotor (m)
Vestas	V112	3	112
GE	2.5xl	2.5	100
Gamesa	G128	4.5	128
Enercon	E-126	7.5	127
Suzlon	S88	2.1	88
Siemens	SWT-3.6-107	3.6	107
Sinovel	SL3000/113	3	113
Acciona	AW-119/3000	3	119
Nordex	N100	2.5	100
Repower	6M	6.15	126

Fig. 6.2 Time evolution of size scale-up and the largest commercial wind turbines up to 2010 (Adapted with permission from Kaldellis and Zafirakis 2011 (Elsevier))

6.2 Wind Resources

A wind energy conversion system can operate at maximum efficiency only if it is designed for the site where it is to be set up, as rated power, cut-in, rated and cut-off wind speeds would be defined according to the site. In fact, these parameters can be chosen so as to maximize the delivered energy for a given amount of available wind energy. Due to the nature of wind, loads are highly variable. Varying loads are more difficult to handle than static loads because the material becomes fatigued easily. Moreover, as a working medium the air is of low density so that the surface required for capturing energy must be large. The wind resources estimation is the first step before any wind farm establishment. Wind measurement over the site topography and simulating the wind field with numerical data according to the IEC 61400-1 standard is the main step for wind resources assessment (Warudkar and Ahmed 2013).

INDEPTH: Advantages and Challenges of Wind Energy

Wind energy is a clean, renewable energy source and offers many advantages, which is one of the fastest-growing energy sources in the world.

Advantages of Wind Power:

(a) Wind energy is a clean fuel source. Wind turbines don't produce atmospheric emissions that increase health problems like asthma or create acid rain or greenhouse gases.
(b) Wind power does not use water, unlike conventional electricity sources. Producing nuclear, coal, or gas-fired power uses water for cooling.
(c) Wind is a domestic source of energy. Wind power is the largest source of annual new generating capacity, well ahead of the next two leading sources, solar power and natural gas.
(d) Wind power is inexhaustible. Wind is actually a form of solar energy. Winds are caused by the heating of the atmosphere by the sun, the rotation of the Earth, and the Earth's surface irregularities. For as long as the sun shines and the wind blows, the energy produced can be harnessed to send power across the grid.
(e) Wind power is cost-effective. It is one of the lowest-cost renewable energy technologies available today.
(f) Wind turbines can be built on existing farms or ranches. This greatly benefits the economy in rural areas, where most of the best wind sites are found. Farmers and ranchers can continue to work the land because the wind turbines use only a fraction of the acreage.
(g) Wind creates jobs. In 2016, the wind energy sector invested more than $8.8 billion of private capital in the US economy to build projects and employed more than 101,000 workers.

Challenges of Wind Power:

(a) Wind power must compete with conventional generation sources on a cost basis.
(b) Good wind sites are often located in remote locations, far from cities where the electricity is needed. Transmission lines must be built to bring the electricity from the wind farm to the city.
(c) Turbines might cause noise and change the view shed. Although wind power plants have relatively little impact on the environment and

(continued)

communities compared to conventional power plants, concern exists over the sound sometimes produced by the turbine blades and visual impacts to the landscape.

(d) Though wind turbines harm wildlife less than some conventional sources of electricity, turbine blades could damage local wildlife.

6.2.1 Wind Quality

Quality winds are typified by laminar air flow at the site of the wind turbine, as shown in Table 6.1. Obstacles that disrupt laminar air flow create turbulent wind, and degrade the quality of the wind resource. Areas designated class 3 or greater are suitable for most utility-scale wind turbine applications, whereas class 2 areas are marginal for utility-scale applications but may be suitable for rural applications. Class 1 areas are generally not suitable, although a few locations (e.g., exposed hilltops not shown on the maps) with adequate wind resource for wind turbine applications may exist in some class 1 areas. The degree of certainty with which the wind power class can be specified depends on three factors: the abundance and quality of wind data; the complexity of the terrain; and the geographical variability of the resource. A certainty rating was assigned to each grid cell based on these three factors (NREL 2014).

Table 6.1 Classes of wind power density of 10m and 50m[a] (NREL 2014)

	10 m (33 ft)		50 m (164 ft)	
Wind power class[b]	Wind power density (W/m^2)	Speed[c] m/s (mph)	Wind power density (W/m^2)	Speed[c] m/s (mph)
1	0	0	0	0
	100	4.4 (9.8)	200	5.6 (12.5)
2	150	5.1 (11.5)	300	6.4 (14.3)
3	200	5.6 (12.5)	400	7.0 (15.7)
4	250	6.0 (13.4)	500	7.5 (16.8)
5	300	6.4 (14.3)	600	8.0 (17.9)
6	400	7.0 (15.7)	800	8.8 (19.7)
7	1000	9.4 (21.1)	2000	11.9 (26.6)

[a]Vertical extrapolation of wind speed based on the 1/7 power law

[b]Each wind power class should span two power densities. For example, Wind Power Class = 3 represents the Wind Power Density range between 150 and 200 W/m^2. The offset cells in the first column attempt to illustrate this concept

[c]Mean wind speed is based on Rayleigh speed distribution of equivalent mean wind power density. Wind speed is for standard sea-level conditions. To maintain the same power density, speed increases 3%/1000 m (5%/5000 ft) elevation

6.2.2 Variation of Wind Speed with Elevation

Wind speed gradient is the vertical gradient of the mean horizontal wind speed in the lower atmosphere, which is the rate of increase of wind strength with unit increase in height above ground level. Due to aerodynamic drag, a wind gradient presents in the wind flow just a few hundred meters above the Earth's surface, i.e., the surface layer of the planetary boundary layer. Wind speed increases with increasing height above the ground, starting from zero due to the no-slip condition. Flow near the surface encounters obstacles that reduce the wind speed, and introduce random vertical and horizontal velocity components at right angles to the main direction of flow. The reduction in velocity near the surface is a function of surface roughness, so wind velocity profiles are quite different for different terrain types. Rough, irregular ground, and man-made obstructions on the ground, retard movement of the air near the surface, reducing wind velocity. Because of low surface roughness on the relatively smooth water surface, wind speeds do not increase as much with height above sea level as they do on land. Over a city or rough terrain, the wind gradient effect could cause a reduction of 40–50% of the geostrophic wind speed aloft; while over open water or ice, the reduction may be only 20–30% (Harrison 1999).

Wind turbines are affected by wind gradient. Vertical wind-speed profiles result in different wind speeds at the blades nearest to the ground level compared to those at the top of blade travel, and this in turn affects the turbine operation. The wind gradient can create a large bending moment in the shaft of a two bladed turbine when the blades are vertical. The reduced wind gradient over water means shorter and less expensive wind turbine towers can be used in shallow seas. For wind turbine engineering, an exponential variation in wind speed with height can be defined relative to wind measured at a reference height of 10 m as (Heier 2005):

$$v_w(h) = v_{10}\left(\frac{h}{h_{10}}\right)^{\alpha} \tag{6.1}$$

where $v_w(h)$ = velocity of the wind at height, h [m/s]; v_{10} = velocity of the wind at height, h_{10} = 10 m [m/s], and α = Hellmann exponent.

The Hellmann exponent depends upon the coastal location and the shape of the terrain on the ground, and the stability of the air. Examples of values of the Hellmann exponent are given in Table 6.2 (Kaltschmitt et al. 2007).

6.2.3 Air Density

The density of air is the mass per unit volume of Earth's atmosphere. Air density, like air pressure, decreases with increasing altitude. It also changes with variation in temperature or humidity. At sea level and at 15 °C air has a density of approximately ρ = 1.225 kg/m^3 according to ISA (International Standard Atmosphere). Wind forecasts for wind energy applications rely mostly on wind speed and direction, and only marginally on the forecast of air density (Monteiro et al. 2009):

Table 6.2 Examples of values of the Hellmann exponent (Kaltschmitt et al. 2007)

Location	α
Unstable air above open water surface	0.06
Neutral air above open water surface	0.10
Unstable air above flat open coast	0.11
Neutral air above flat open coast	0.16
Stable air above open water surface	0.27
Unstable air above human inhabited areas	0.27
Neutral air above human inhabited areas	0.34
Stable air above flat open coast	0.40
Stable air above human inhabited areas	0.60

$$P = \frac{1}{2}\rho A v^3 \tag{6.2}$$

where P is wind power, A is the area swept by the turbine blades, and v is the wind speed.

6.2.4 Wind Forecasting

Wind energy forecasting system usually consists of four parts: Data Acquisition, Wind Forecasting, Power Conversion, and Output. Each component initializes the downstream process. Standard meteorological data augmented by turbine-level nacelle wind speeds are fed to the weather forecasting cluster—a multiprocessor computer system. The cluster utilizes the Real-Time Four-Dimensional Data Assimilation (RTFDDA) process combined with a configuration of the Weather Research and Forecasting (WRF) model adapted for wind forecasting applications. The forecasted wind speed from the cluster, forecasted wind speed from publicly available weather models (together called the foundation forecasts), and nacelle wind speed gathered from the energy wind plants are input into the Dynamic Integrated ForeCast (DICast). DICast generates a consensus wind speed forecast based on historic performance of the various forecasted data streams. The consensus forecast is sent to the power conversion module which converts wind speed to power using real-time power and wind speed data to initialize the calculation. Lastly, data are communicated to users via a CSV file and ultimately a forecast display for operators. The system produces two forecasts for each forecast node: a 72-h forecast with hourly resolution updated every 15 min and a three-hour forecast with 15-min resolution updated every 15 min. A forecast node is a group of turbines with a common point of interconnection—this is the most granular level of interest for operational purposes. There are 45 forecast nodes in the forecast system ranging from 2 to 575 MW. To contrast, the term wind plant is defined as a group of turbines under a common purchase power agreement, qualifying facility designation, or owned resource. While wind plants and point of interconnect (and thus forecast node) are frequently the same groups, sometimes wind plants cross multiple forecast nodes and multiple plants are behind one forecast node. Forecast nodes are then aggregated by wind energy operating companies to give the system view most frequently used by operators (Haupt 2011).

Improvements of wind power forecasting has focused on using more data as input to the models involved, and on providing uncertainty estimates along with the traditionally provided predictions.

6.2.5 Offshore Wind

Offshore wind power refers to the construction of wind farms in bodies of water to generate electricity from wind. Unlike the typical usage of the term "offshore" in the marine industry, offshore wind power includes inshore water areas such as lakes, fjords, and sheltered coastal areas, utilizing traditional fixed-bottom wind turbine technologies, as well as deep-water areas utilizing floating wind turbines. A subcategory within offshore wind power can be near-shore wind power (Gill 2012).

6.2.6 Maximum Wind Turbine Efficiency: The Betz Ratio

Betz's law indicates the maximum power that can be extracted from the wind, independent of the design of a wind turbine in open flow. The law is derived from the principles of conservation of mass and momentum of the air stream flowing through an idealized "actuator disk" that extracts energy from the wind stream. According to Betz's law, no turbine can capture more than 16/27 (59.3%) of the kinetic energy in wind. The factor 16/27 (0.593) is known as Betz's coefficient. Practical utility-scale wind turbines achieve at peak 75–80% of the Betz limit (Burton et al. 2001).

The Betz limit is based on an open disk actuator. If a diffuser is used to collect additional wind flow and direct it through the turbine, more energy can be extracted, but the limit still applies to the cross-section of the entire structure. The Betz limit places an upper bound on the annual energy that can be extracted at a site. Even if a hypothetical wind blew consistently for a full year, no more than the Betz limit of the energy contained in that year's wind could be extracted. In practice, the annual capacity factor of a wind site varies around 25–60% of the energy that could be generated with constant wind, limiting the energy that can possibly be obtained even further to typically a range of 14.8%–35%, respectively. In practicality, most systems do not reach a performance rate of even 50% of the Betz limit, before the further limits of the air stream are ever considered, further lowering the typical rates to 7–17% (Burton et al. 2001).

6.3 Wind Machinery and Energy-Generating Systems

Wind machinery mainly indicates windmill and wind turbine. The windmill was made to help pump water and grind grain very similar to the water wheel. A wind turbine is a device that converts the wind's kinetic energy into electrical power, which is the major focus in this chapter. Wind energy-generating systems generally include the rotor and its blades, the hub assembly, the main shaft, the gear box system, main frame, transmission, yaw mechanism, over-speed protection, electric generator, nacelle, yaw drive, power conditioning equipment, and tower, as shown in Fig. 6.3. Wind turbine design is the process of defining the form and specifications of a wind turbine to extract energy from the wind. A wind turbine installation consists of the necessary systems needed to capture the wind's energy, point the turbine into the wind, convert mechanical rotation into electrical power, and other systems to start, stop, and control the turbine. Today wind turbines are the fastest-growing sector of the renewable power business. Seeking to reduce the cost of each kWh means the latest wind turbine technology pushes new materials to their limits and provides real materials information management challenges. The blades are the most vulnerable parts of a wind turbine, and there is a desire for ever longer blades (to increase the area and thus power). The self-weight of the blade creates alternating bending loads at the blade root as the turbine rotates. Superimposed on these is an axial load caused by centrifugal force and a bending load caused by wind pressure. The necessarily complex composite materials needed have been developed, which enable bigger, lighter, and more reliable blades. The complexity is only increased by the variation of composite material properties with factors, such as geometry and process history, that are specific to each application. As a result, huge amounts of valuable materials-related information are produced and need to be managed, accessed, and applied efficiently.

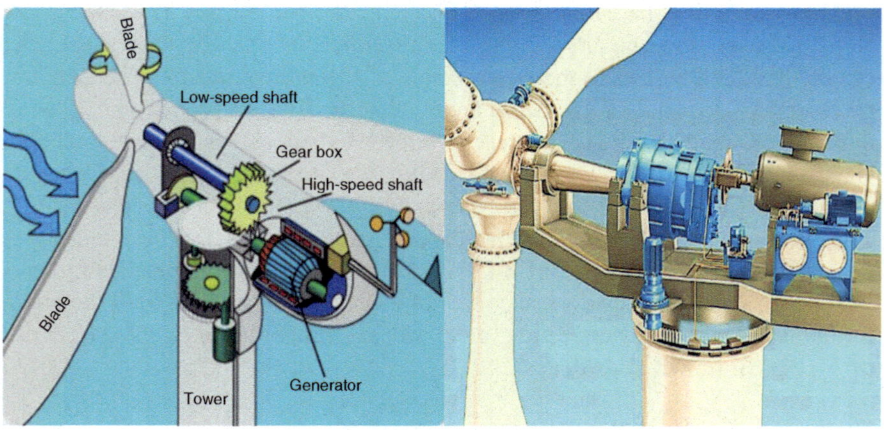

Fig. 6.3 Schematic of wind energy system

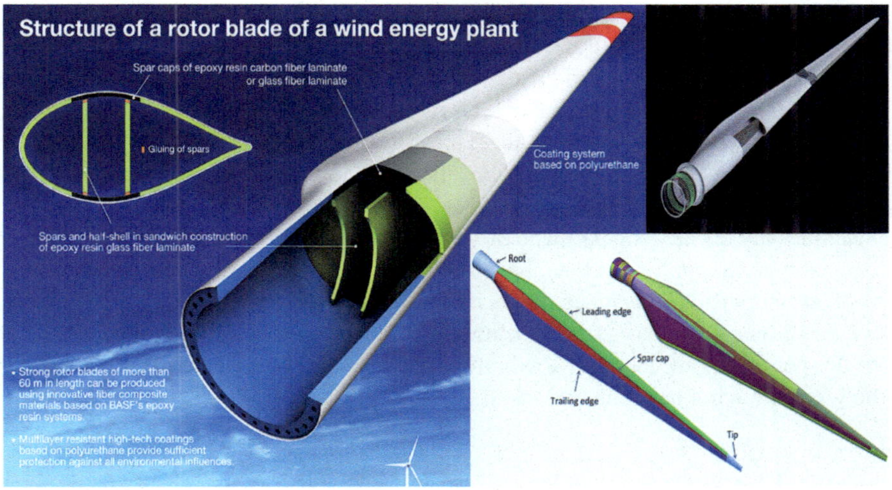

Fig. 6.4 Schematic structure of the wind turbine blade (Modified from Böhme 2017. Courtesy: BASF Finland)

6.3.1 Rotor Blade Assembly

As shown in Figs. 6.3 and 6.4, a rotor blade assembly generally includes: blades, which are generally made of glass-reinforced fiber up to 50 m in length, lighter and stronger carbon fibers are being used in the larger blades; extenders attach the blades to the central hub; and pitch drives to control the angle of the blades. The rotor typically has three blades because that number provides the best balance of high rotation speed, load balancing, and simplicity. The blades are attached by a hub assembly to a main shaft. The rotor hub is a welded construction made of steel. It houses the pitch mechanism and the pitch bearing. The hub connects to the rotor and blade assembly to the generator shaft. The power is controlled by active blade adjustment towards vane position. Blade pitch is adjusted using an electric servo-motor on each of the blades. The complete pitch system is designed to be fail-safe and is fully independent for each blade, which is the primary method of controlling the aerodynamic power input to the turbine. At below-rated wind speed the blade pitch setting matches optimum aerodynamic efficiency. At above-rated wind speed the control system keeps the average aerodynamic power at the rated level by keeping the rotor speed close to nominal. A gusty wind criterion adjusts blade pitch control parameters to the actual turbulence of the wind.

6.3.2 Tower

A tower is needed to get the wind machine up into air, away from the slower and more turbulent winds near the ground. The wind machine should be at least 10 m higher than any obstructions in the surroundings such as trees. The tower generally includes rolled steel tubes connected in series; flanges and bolts joining each section; a concrete base serving as a stable foundation for the turbine assembly; and concrete segmented towers and hybrid steel/concrete towers may also be used for large turbines in cases where steel tower section transportation is difficult.

Small wind machines towers are typically of lattice or truss design or of poles supported by guy wires, which are attached to the tower and anchored in the ground so that the tower will not move or shake from the force of the wind. Large wind machines towers are usually made of steel and the great majority is of the tubular or conical type. Some towers have been built out of reinforced concrete sections.

Towers must be designed to resist the full thrust produced by an operating windmill or a stationary wind machine in a storm. Special concern must be given to the possibility of destructive vibrations caused by a natural frequency mismatch between the wind machine and tower.

6.3.3 Nacelle Components

A nacelle is a cover housing that houses all of the generating components in a wind turbine, including the generator, gearbox, drive train, and brake assembly. The nacelle general includes: an outer frame protecting machinery from the external environment; an internal frame supporting and distributing weight of machinery; a drive power train to transmit energy and to increase shaft speeds; a generator to convert mechanical energy into electricity; a yaw drive to rotate (slew) the nacelle on the tower; and brake assembly through electronics to control and monitor operation.

6.3.4 Balance-of-Station Subsystems

The balance of station includes (a) Electrical collection system: transformer, switchgear, underground and overhead high-voltage cable, and interconnecting substation; (b) Control system: control cable, data collection, and wind farm control station; and (c) Roadway, parking, crane pads and other civil works.

6.3.5 Wind Energy System Design Challenges

The current wind energy system challenges can be classified as (Dabiria et al. 2015): Where is useful wind energy located? What can be done to better harvest it? And, how to optimize the function, integration, and adoption of those systems?

(a) Wind energy resource quantification
Existing wind resource maps are based on sensors intentionally located far from structures such as buildings and trees, as those objects create more complex wind fields that make it difficult to interpret sensor measurements. For the distributed wind energy systems sited in such complex wind fields, it is essential

to develop a physics-based understanding of the flow-structure interactions that occur in built environments, with particular emphasis on how to reconstruct local wind fields from sensor networks located in these complex settings. These may include wind and water tunnel measurements and computer simulations of canonical flow-structure interactions; development of new computational tools to assimilate crowdsourced wind data from inexpensive, massively distributed wind sensors; and formal quantification of uncertainty in the resulting wind models. An example outcome of this effort would be holistic, open-source city, state, and national maps of the wind resources that are accessible for power generation from individual to utility-scale use (Dabiria et al. 2015).

(b) Wind turbine design for manufacturability and reliability
A complementary goal of wind turbine design is to support the paradigm of distributed deployment and operation. The effort should emphasize design and fabrication of low-cost, lightweight, and damage-tolerant materials using sustainable manufacturing processes including additive manufacturing and 3D printing; and development of reliable vertical-axis wind turbine (VAWT) VAWT systems. Research activities in this thrust could include creation of smart, functionally graded materials through control of structural architecture and material microstructure; development of new nanofabrication methodologies that support scalable manufacturing; testing of flow control strategies for customized VAWT torque and noise characteristics; design of rotor blades with enhanced tolerance to biofouling and icing; and evaluation of earth-abundant, locally sourced materials for wind turbine manufacturing in developing regions (Dabiria et al. 2015).

(c) Wind turbine array optimization
An essential research objective is the optimization of aerodynamic and electrical interactions of wind turbines in an array in order to maximize energy generation, improve the reliability and lifetime of power generation, and minimize cost. Concepts from autonomous and networked systems, bio-inspired engineering, optimal control theory, and computational fluid dynamics can enable dynamic optimization of wind power generation via automatic, real-time feedback control and data sharing among the individual units in an array. A rich parameter space is available to be explored for optimization in the lab and field, including static and dynamic turbine height, spacing, and array configuration; power processing protocols; colocation with HAWTs; hybridization with solar power, diesel fuel, and battery storage; and integration with floating platforms for offshore wind applications (Dabiria et al. 2015).

6.4 Wind Turbines

Wind turbines convert the kinetic energy in the wind into mechanical power, which converts into electricity through a generator to various facilities. Wind turbines are available in a variety of sizes and power ratings. Commercially available land-based turbines range in capacity from 0.25 kW to 2.5 MW, while offshore wind turbines may exceed 4.5 MW (PEIS 2016). The capacity of a turbine is determined largely by its rotor diameter. When operating at full power, the smallest and the largest wind generators can supply enough electricity to power a few light bulbs and thousands of homes, respectively. Modern wind turbines can be divided into horizontal-axis wind turbines, upwind wind turbines, downwind wind turbines, vertical-axis wind turbines, Darrieus turbines, Savonius turbines, giant multimegawatt turbines, and airborne wind turbines. Most large modern wind turbines are horizontal-axis turbines.

6.4.1 Horizontal-Axis Wind Turbines

The horizontal axis wind turbine (HAWT) is the most efficient design for turning wind into electricity. The basic design allows two or more rotor blades to face into the wind and move simultaneously; therefore, the blades form the least possible resistance to wind forces, as shown in Fig. 6.5.

Specifically, the rotor blades of a horizontal axis wind turbine are usually designed aerodynamically. Along a wing or rotor blade, the two sides of the blade are designed such that the top side (thinner and longer) of the blade has a larger surface area than the bottom. When the air moves over the top of the blade, the air must move faster than the air going under the bottom of the blade. This higher speed creates lift because the denser underside air pushes against the blade. The blades are hooked to a shaft so the lift on the blade forces the shaft to spin.

HAWT designs have been used variously in size from the very small to the commercial grade generators. Most are installed on towers to help them reach high

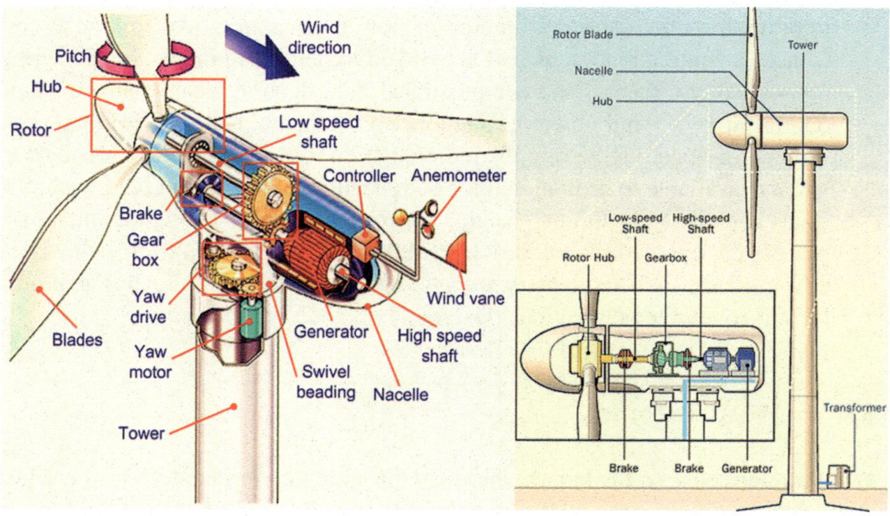

Fig. 6.5 Major components of a typical horizontal axis, three-bladed, upwind wind turbine (Modified with permission from Molina and Alvarez 2011 (IntechOpen))

above the ground where airflow is strongest and most constant. Even so, the larger the rotors, the more likely there is to be varying forces on different parts of the turbine. This means that HAWTs must be rugged and well supported right down to the foundation. Also, at very high wind-speeds, generators and rotors can be damaged, so in most windmills safety features are built in that allow the rotors to not take so much wind and even to disconnect the rotors from the generator. Braking systems are also used to slow turbines in very high wind-speeds. Another method of avoiding damage in high winds: some wind generators are designed to tilt back at higher wind-speeds so that the rotors become less efficient, thereby creating less stress.

Smaller turbines generally spin at faster rpm speeds than the large commercial models. The larger models are slowed to avoid killing birds and also because higher speeds create greater wear and tear on larger devices.

Most of the wind-generating devices available for both home and commercial use are horizontal axis wind turbines. However, vertical axis wind turbines are also coming to their own.

6.4.2 Vertical-Axis Wind Turbines

Vertical-axis wind turbines (VAWTs) are a type of wind turbine where the main rotor shaft is set transverse to the wind while the main components are located at the base of the turbine. This arrangement allows the generator and gearbox to be located close to the ground, facilitating service and repair. VAWTs do not need to be pointed into the wind, which removes the need for wind-sensing and orientation mechanisms (Jha 2010).

Despite variances, as shown in Fig. 6.6, there are really only two major types of vertical axis wind turbines (Barnard 2014): (a) Savonius-style turbines are basic drag machines. They typically have rounded paddles which catch wind in the cup and shed it on the rounded front, allowing the difference in drag to rotate the turbine. (b) Darrieus-style vertical axis wind turbines have aerodynamic blades which fly through the wind on their power strokes as they rotate around a shaft. VAWTs catch the wind from any angle, making them more effective than HAWTs. They are quieter and produce less stress on the mast. To generate the same electricity, however, VAWTs would have to be as tall as HAWTs. To make them economic for grid-scale generation, they would have to be so big that they would still be very, very visible. VAWTs don't generate shadow flicker that is problematic, as that is a factor of the triblade design. Shadow flicker is a vastly over-stated concern. It will only occur for a few minutes at sunrise or sunset for a week or two twice a year at some residences near wind farms. HAWT rotation is too slow to cause epileptic seizures. VAWTs will kill fewer birds than HAWTs. As HAWT bird mortality rates are typically vastly overstated and are much less than fossil fuel generation, lighted windows, cats, transmission lines, cars and many other sources of avian mortality, this is a straw man argument. As VAWTs scaled up for utility generation capacity have not been assessed for avian mortality, it's a straw man argument without merit,

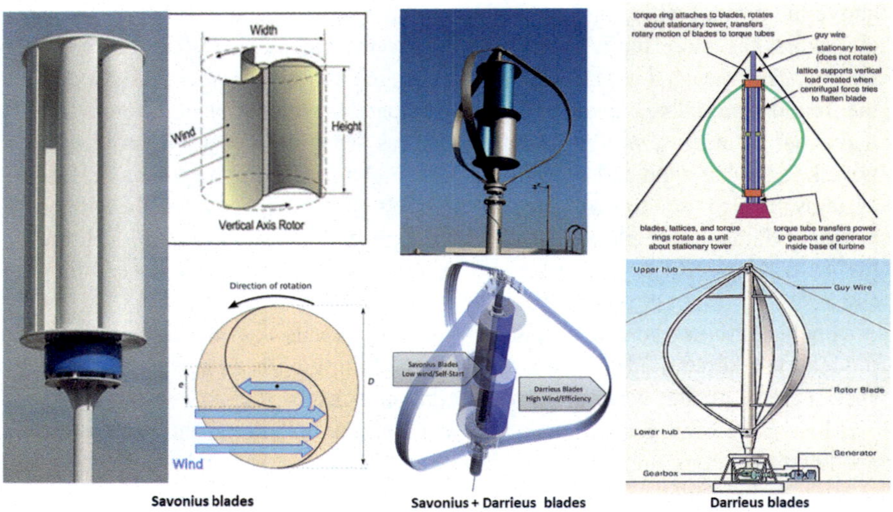

Fig. 6.6 Typical vertical-axis wind turbines

similar to the noise problem. Vertical axis wind turbines are historically interesting, but irrelevant to utility-scale generation and insignificant as a wedge against global warming. Continued attention to them merely distracts from the much more useful effort of deploying horizontal axis wind turbines, and is used as a wedge by anti-wind campaigners.

6.4.2.1 Darrieus Turbines

The Darrieus wind turbine is a type of vertical axis wind turbine (VAWT), consisting of a number of curved airfoil blades mounted on a vertical rotating shaft or framework. The curvature of the blades allows the blade to be stressed only in tension at high rotating speeds. This design of wind turbine was patented by Georges Jean Marie Darrieus, a French aeronautical engineer in 1931.

The Darrieus wind turbine is less efficient compared with three-bladed wind turbines; however, this kind of wind turbine can overcome the limits due to the size of the blades and their rotation speed. The total size is smaller, and in some cases, when the motor is located at its base, this type of wind turbine can be cheaper. Darrieus vertical axis wind turbines appear more suitable than the classic ones in different fields such as building integration, and the extreme zone like observatories and mountain refuge. The Darrieus wind turbine can run with a wind speed up to 220 km/h and in any direction. The main shortcoming of this type of wind turbine is their difficult start, because the weight of the rotor on its base generates frictions.

6.4.2.2 Savonius Turbines

A Savonius is a type of vertical axis wind turbine (VAWT) generator invented in 1922 by Sigurd Johannes Savonius from Finland. The turbine consists of a number of airfoils, usually vertically mounted on a rotating shaft or framework, either ground stationed or tethered in airborne systems.

The Savonius is a drag-type VAWT which operates in the same way as a cup anemometer. It has three cups mounted on a rotor which is free to spin. At any time the front one of the cups will be facing into the wind, while the remaining two cups will be back on to the wind. As the backs of each cup are rounded, they experience much less drag than the front of the cup which is facing into the wind. Therefore, the force exerted by the wind on the open cup will be greater than the total force exerted on the backs of the other cups and so the rotor will be pushed around. When the rotor has spun around one-third of a revolution, the cup which was open to the wind will now be back on, and one of the previously back on to the wind cups will have spun around on the rotor to offer its open face into the wind. In this way, the rotor will continue to spin for as long as the wind is blowing (REUK 2014).

With a Savonius wind turbine it does not matter from which direction the wind is blowing, since there will always be more force exerted on whichever cup has its open face into the wind, and this will push the rotor around. This makes this design of wind turbine ideal for areas with very turbulent wind. Unfortunately, Savonius wind turbines typically only have an efficiency of around 15%. This is much less than can be achieved with a Darrieus wind turbine which uses lift rather than drag. The speed of the cups of a cup anemometer (and a Savonius wind turbine) cannot rotate faster than the speed of the wind they are in and so they have a tip speed ratio (TSR) of 1 or below. This means that a Savonius type vertical axis wind turbines will turn slowly but generate high torque. Therefore, Savonius turbines are not ideal for electricity generation since turbine generators need to be turned at hundreds of RPM to generate high voltages and currents. A gearbox can be employed to reduce the torque and increase the RPM of the generator, but that leaves the Savonius requiring a stronger wind to get spinning meaning it may not be able to self-start (REUK 2014).

Savonius wind turbines are best suited to applications such as pumping water and grinding grain for which slow rotation and high torque are desirable. Because of the high torque yield of a Savonius wind turbine, the bearings used must be very sturdy and may require servicing every couple of years. Though not ideal, Savonius wind turbines can and are used in electricity generation with the benefit that they continue to generate electricity in the strongest winds without being damaged, they are very quiet, and they are relatively easy to make. Standard HAWT (windmill) type wind turbines are unsuitable in locations with strong turbulent winds, so Savonius wind turbines can sometimes be the best option. They can also be mounted relatively safely at ground level as they spin much slower than the speed of the tips of an HAWT (REUK 2014).

6.4.3 Upwind Wind Turbines and Downwind Wind Turbines

The rotor on an upwind turbine is in the front of the unit, positioned similar to a propeller-driven airplane. To keep it oriented into the wind, a yaw mechanism such as a tail is needed. The advantage of upwind wind turbines over downwind wind turbines is the reduced tower shading. The air will start to bend around the tower before it passes it so there is some loss of power from the interference, just not the degree as in the downwind turbine. However, the extended nacelle that is required to position the rotor far enough away from the tower to avoid any problems with a blade strike. The blades themselves must be somewhat stiff to avoid bending back into the tower. This will mean the point where the blade attaches to the rotor hub will be stressed during high, gusty wind conditions.

The downwind turbine has its rotor on the back side of the turbine. The nacelle typically is designed to seek the wind, thus negating the need for a separate yaw mechanism. The rotor blades can be flexible since there is no danger of a tower strike. The flexing blade has two advantages: They can be less expensive to make; they can relieve stress on the tower during high or gusty wind conditions since the flexing allows some wind load to be transferred directly to the blades instead of the tower. The Proven has a hinged design that allows the blade to flex back to dissipate energy for speed control. The flexible blade advantage can also be a disadvantage as the flexing may fatigue the blades. Tower shadow is problem with a downwind machine since the rotor blade actually passed behind the tower. This can cause turbulence and increased fatigue on the unit.

The upwind turbine will be the most common in the small-scale, renewable energy scene. The downwind model may aim at the grid-tie market. For residential size turbines there is no clear winner between the upwind and downwind design.

6.4.4 Giant Multimegawatt Turbines

Giant multimegawatt turbines have been developed with load alignment dramatically reducing peak stresses and fatigue on the rotor blades. This would not only reduce blade costs, but eventually lead to the mythical 50 MW wind turbines with Sandia's Segmented Ultralight Morphing Rotor (SUMR) designs. During extremely fast wind speeds, the blades are stowed and aligned with the wind direction, reducing the risk of damage. At lower wind speeds, the blades spread out more to maximize energy production.

Developments like 50 MW wind turbines are critical to the offshore wind industry. While countries could have large offshore wind energy potential, offshore installations can be costly. Larger turbines will help to capture more energy at a more affordable price. Conventional upwind blades are expensive to manufacture, deploy and maintain beyond 10–15 MW. They have to be rigid to avoid fatigue and eliminate the risk of tower strikes in strong gusts of wind. The stiff blades are heavy, and their mass, which is directly related to cost, becomes even more problematic at the extreme scale due to gravity loads and other changes. The designs for 50 MW wind turbines will save money when it comes to production, transportation, and installation because the long blades are manufactured in segments. The blades are positioned downwind, and the segmented sections bend in the wind while retaining segment stiffness. At dangerous wind speeds, the blades are stowed and aligned with the wind direction, reducing the risk of damage. At lower wind speeds, the blades spread out more to maximize energy production.

Larger turbines also reduce carbon footprint. The bigger the turbine, the greener the electricity points out that larger turbine blades can actually reduce carbon footprint. For every doubling of the size of the turbine, global warming potential per kWh is reduced by 14%. There are two main reasons for the benefits of larger turbines (Engerati 2017): (a) producers now have the knowledge to create massive blades with great efficiency; (b) materials today allow the manufacturing of large turbines without a corresponding increase in mass. Blades can therefore be bigger and capture more wind while the tower and other parts remain unchanged. The combined effects of these reasons allow for bigger and better turbines to be produced without using significantly more materials or drastically increasing transportation and assembly costs. In fact, the increased size of the turbines actually saves materials by reducing the number of total turbines needed to produce the same amount of power.

6.4.5 Airborne Wind Turbines

An airborne wind turbine is a design for a wind turbine with a rotor supported in the air without a tower, thus benefiting from more mechanical and aerodynamic options, the higher velocity and persistence of wind at high altitudes, while avoiding the expense of tower construction, or the need for slip rings or yaw mechanism. An electrical generator may be on the ground or airborne. Challenges include safely suspending and maintaining turbines hundreds of meters off the ground in high winds and storms, transferring the harvested and/or generated power back to earth, and interference with aviation (Elliot 2014) (Fig. 6.7).

Airborne wind turbines may operate in low or high altitudes; they are part of a wider class of Airborne Wind Energy Systems (AWES) addressed by high-altitude wind power and crosswind kite power. When the generator is on the ground, then the tethered aircraft need not carry the generator mass or have a conductive tether. When the generator is aloft, then a conductive tether would be used to transmit energy to the ground or used aloft or beamed to receivers using microwave or laser. Kites and "helicopters" come down when there is insufficient wind; kytoons and blimps may resolve the matter with other disadvantages. Also, bad weather such as lightning or thunderstorms, could temporarily suspend use of the machines, probably requiring them to be brought back down to the ground and covered. Some schemes require a long power cable and, if the turbine is high enough, a prohibited airspace zone (Cherubini et al. 2015).

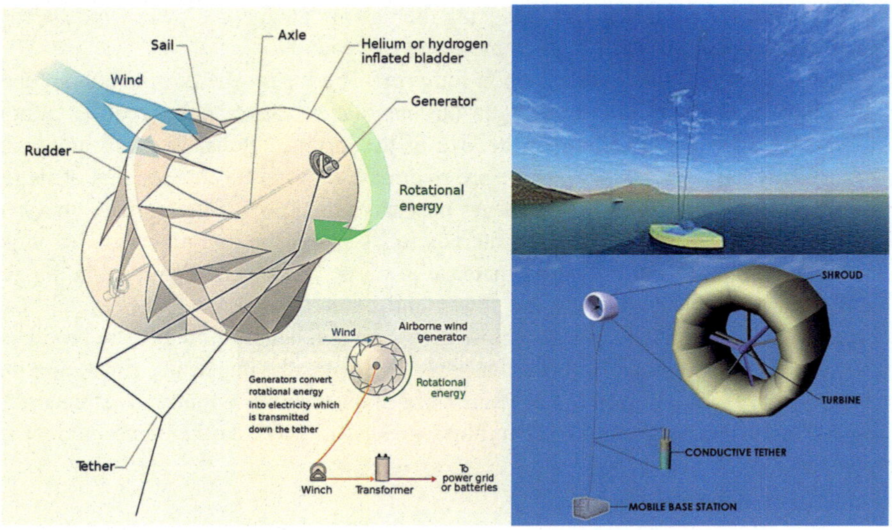

Fig. 6.7 Typical airborne wind turbines (Modified from https://ipfs.io/ipfs/QmXoypizjW3WknFiJnKLwHCnL72vedxjQkDDP1mXWo6uco/wiki/Airborne_wind_turbine.html, under the Creative Commons Attribution/Share Alike)

6.5 General Materials Used in Wind Turbines

Many types of materials are used in wind turbines. The materials most relevant to wind turbine applications are steel and composites. The composites are typically comprised of fiberglass or wood together with a matrix of polyester or epoxy. Other common materials include copper and concrete. There are substantial differences between small and large machines. The estimated materials use in small and large turbines is shown in Table 6.3 (Ancona and McVeigh 2001). The trends in design and manufacturing differ between small and large turbines. Small machines tend to use lighter weight castings in an effort to reduce costs. Many parts are die cast aluminum in small turbines, while in large machines steel castings or forgings are needed to meet strength and structural fatigue requirements. The size of steel castings for large turbines, especially the blade hub units, is one of the manufacturing challenges.

6.5.1 Steel

Steel is one of the most widely used materials in wind turbine fabrication. It is used for many structural components including the tower, hub, main frame, shafts, gears and gear cases, fasteners as well as the reinforcing in concrete.

At the top of the tower are the rotor and the nacelle. A nacelle can weigh as much as 300 tons, which is about 14% of the weight of a large offshore turbine. Steel's strength makes it ideal for the nacelle's frame, housing and machinery. The nacelle

Table 6.3 Percentage of materials used in current wind turbine components (Ancona and McVeigh 2001)

Turbine size	Large turbines and (*small turbines*)[a]							
Component/material (% by weight)	Permanent magnetic materials	Reinforced concrete	Steel	Aluminum	Copper	Glass reinforced polymer[d]	Wood epoxy[d]	Carbon filament reinforced polymer[d]
Rotor								
Hub			(*95*)–100	(*5*)				
Blades			5			95	(*95*)	(*95*)
Nacelle[b]	(*17*)		(*65*)–80	3–4	14	1–(*2*)		
Gearbox[c]			98–(*100*)	(*0*)–2	($<$*1*)–2			
Generator	(*50*)		(*20*)–65		(*30*)–35			
Frame, machinery and shell			85–(*74*)	9–(*50*)	4–(*12*)	3–(*5*)		
Tower		2	98	(*2*)				

[a]Small turbines with rated power less than 100 kW (listed in italics where different)
[b]Assumes nacelle is 1/3 gearbox, 1/3 generator and 1/3 frame and machinery
[c]Approximately half of the small turbine market (measured in MW) is direct drive with no gearbox
[d]Rotor blades are either glass reinforced plastic, wood-epoxy or injection molded plastic with carbon fibers

contains key components and some of the highest-value steels. These include electrical steels (also known as lamination, silicon, or transformer steels) that help save energy. Electrical steels are a specialty steel tailored to producing the specific magnetic properties that make wind energy possible. They are used to make the cores of motors, transformers, and generators. This type of soft magnetic steels can generate various magnetic properties, has high permeability and low amounts of core loss. The steel has a small hysteresis curve, meaning reduced magnetic hysteresis as well as iron losses, or energy loss.

Electrical steel is an iron alloy which may have from 0 to 6.5% silicon (Si:5Fe). Commercial alloys usually have silicon content up to 3.2% (higher concentrations usually provoke brittleness during cold rolling). Manganese and aluminum can be added up to 0.5%. Silicon significantly increases the electrical resistivity of the steel, which decreases the induced eddy currents and narrows the hysteresis loop of the material, thus lowering the core loss. However, the grain structure can harden and embrittle the metal, which adversely affects the workability of the material, especially when rolling it. When alloying, the concentration levels of carbon, sulfur, oxygen, and nitrogen must be kept low, as these elements indicate the presence of carbides, sulfides, oxides, and nitrides. These compounds, even in particles as small as 1 μm in diameter, increase hysteresis losses while also decreasing magnetic permeability. The presence of carbon has a more detrimental effect than sulfur or oxygen. Carbon also causes magnetic aging when it slowly leaves the solid solution and precipitates as carbides, thus resulting in an increase in power loss over time. For these reasons, the carbon level is kept to 0.005% or lower. The carbon level can be reduced by annealing the steel in a decarburizing atmosphere, such as hydrogen.

There are two kinds of structures for electrical steel: grain-oriented and non-oriented. Grain-oriented electrical steel is made by various hot and cold rolling stages to produce textured sheets, therefore, has a uniform, consistent direction of grains in its structure which allows for greater flux density and magnetic saturation. Most commonly, grain-oriented electrical steel is used for transducers and magnetic amplifier cores which have a predictable and specific magnetic field direction. Non-oriented electrical steels have similar magnetic properties in all directions. The cubic crystals that make up the structure are randomly oriented. The steel is annealed at 800 °C in a wet hydrogen atmosphere to remove as much carbon as possible. Non-oriented steels are used in rotating machines such as motors, generators, and alternators. Another special non-oriented structure is amorphous steel, which is a metallic glass prepared by pouring molten alloy steel onto a rotating cooled wheel, which cools the metal at a rate of about 1 MK per second, so fast that crystals do not form. Amorphous steel is limited to foils of about 50 μm thickness. It has poorer mechanical properties and is cost-effective only for some distribution-type transformers. Transformers with amorphous steel cores can have core losses of one-third that of conventional electrical steels.

Behind the blades, a low-speed shaft transfers the rotational force of the rotor to the gearbox. The gears of the gearbox were machined using precision tools and special hardened steel components. The gears increase the low rotational speed of the rotor shaft to the high speed needed to drive the generator. The generator converts the mechanical energy captured by the blades into electric energy, much like a bicycle dynamo, and directs it to the transformer. The generator is made of 65% steel and 35% copper. An alternator is sometimes used instead. The difference between an alternator and a generator is what is fixed and what spins. In an alternator, a magnetic field is spun inside windings of wire called a stator. This generates electricity. A transformer, usually on the ground, converts the electricity from the turbine to the higher voltage required by the electricity grid. Various bearings are applied. All have to withstand the varying forces and loads generated by the wind. Screws and studs are needed to hold the main components in place and must be designed for extreme loads. All of these components depend on steel (WSA 2012).

Most of the steel in a wind turbine is the tower. About 90% of all wind turbine towers are tubular steel towers. They are called tapered tubular towers because they gradually narrow towards the top. To construct a tower, fan-shaped plate segments are cut from rectangular parent steel plates and roll-formed and welded into cone sections. A section's thickness may vary from 8 mm at the top to 65 mm at the base, depending on loads and steel grades used. Offshore installations usually use thicker or stronger plates. Longer blades increase the energy yield of a turbine. They sweep a larger area and so capture more wind. The tower and the foundation have to be adjusted to carry these heavier blades and the bigger rotor that they require. Also, to maximize yield, longer blades mean taller towers. Higher steel grades can be applied to achieve lighter and taller towers. For example, by upgrading the steel of a wind tower structure from grade S355 to S500, a weight saving of 30% can be achieved. Even with a cost increase of 20–25% per ton for the higher strength steel, the balance is positive since 30% less material is needed. More savings result from lower transport and construction costs. In forested areas, steel truss (lattice) towers are used to lift turbines above the tree-line without disturbing vegetation. They are proving a cost-efficient solution for very tall towers. Lattice towers are constructed of pre-assembled steel sections which are hot-dip galvanized for corrosion protection and bolted together on site. The tower is then lifted by a crane. Excessively cold environments, such as the Arctic, present specific challenges. Although a standard wind turbine can operate to −20 °C or even down to −30 °C and remain structurally sound, below −40 °C, low-temperature steels are needed, as well as synthetic lubricants and heating systems in the nacelle or elsewhere (WSA 2012).

Offshore wind farms were traditionally placed in maximum water depths of 30 m and depended mainly on a gravity base (relying on spread rather than depth) and monopile foundations. Future projects could be installed in waters of up to 60 m as new foundation concepts emerge for offshore and deep water installations. In the future, foundations are likely to change from monopile to jacket-type. Both are almost completely made of steel. A monopile is a steel pipe pile of up to six meters in diameter with a wall thickness of 150 mm. Depending on seabed conditions, the pile is plunged far down by pile-driving or drilling, or the piles are grounded into sockets that have been drilled into rock. Compared to the gravity base foundation, the monopile has minimal and localized environmental impacts. A tripod foundation consists of a monopile divided at its bottom into a frame of steel rods. This is attached to the sea bed with piles of smaller diameter (compared to a monopole foundation or a suction bucket foundation). It can be used at greater depths than the gravity base and monopile foundations. The jacket foundation is similar to a lattice tower. It is a squared network of steel rods. It is anchored at four anchorage points and the whole steel construction can be mounted in one piece. Using a three-dimensional truss like the jacket foundation substantially increases rigidity. Although it is more expensive than a monopile or gravity base foundation, the jacket foundation is cost-efficient at greater depths. Floating turbines can be structured with a steel floater filled with ballast of water and rocks. Extending 100 m beneath the surface, it is fastened to the seabed by three steel anchor wires. Steel platform can also be used to hold multiple turbines. The turbines can be replaced while the steel platform itself should last for 50 years or more.

The foundation is an unseen and yet critical part of the structure. Gravity base foundations are the most commonly used onshore foundations. Most gravity base foundations are made of steel-reinforced concrete slabs. Steel "carpets" are reinforcement bars (rebars) welded to flexible steel straps. The carpets are simply lifted into place and unrolled on site. Their use not only saves steel by minimizing cuttings, but also means that the reinforcing steel for one wind tower foundation can be placed in less than one-quarter of the time and using half the manpower compared to if using loose rebar. This offers additional cost savings (WSA 2012).

From a climate change and sustainability perspective, it is important to take into account the life cycle of products. Steel is infinitely recyclable. It also has a limited environmental impact. Recovery of the material at the end of its useful life also helps to recover upfront cost, due to the value of steel scrap. If parts are not given an extended life or reused elsewhere, they return to the steelmaking process. Therefore, the application of steel in most of the key components of wind turbines makes it possible for the wind energy industry to meet the technical requirements of the turbines and climate change demands at the same time.

6.5.2 Composite Materials

Composite materials are primarily used in wind turbine blade construction, including those based on fiberglass, carbon fiber, and wood. Binders include polyester, epoxy and vinyl ester. The most common one is fiberglass-reinforced plastic (GRP). In wind turbines, composite materials are most prominently used in blade manufacture, but they are also used in other parts of the machine, such as the nacelle cover. The main advantage of composite materials is that they have a high strength and high stiffness-to-weight ratio. They are also corrosion resistant, are electrical insulators, and lend themselves to a variety of fabrication methods (Manwell et al. 2002).

Glass fibers are formed by spinning glass into long threads. E-glass is the most common glass fiber, which is a low-cost material with reasonably good tensile strength. Fibers are sometimes used directly for the composites, but are most commonly first combined into other forms known as preforms. Fibers may be woven or knitted into cloth, formed into continuous strand or chopped strand mat, or prepared as chopped fibers. Where high strength is required, unidirectional bundles of fibers known as tows are used (Manwell et al. 2002).

Carbon fibers are more expensive than are glass fibers (by approximately a factor of 15), but they are stronger and stiffer. One way to take the advantage of carbon fibers, without paying the full cost, is to use some carbon fibers along with glass in the overall composite (Manwell et al. 2002). For longer blades, the use of carbon fiber can reduce the weight of the blade. A combination of material optimization and lower costs could enable the use of carbon fiber in future blades (Griffith and Johanns 2013).

There are three types of resins commonly used in matrices of composites (Manwell et al. 2002): (a) unsaturated polyesters, (b) epoxies, and (c) vinyl esters. These resins are generally used in the liquid form during the layup of the composite, and then solidified with curing process. The solid resins tend to be somewhat brittle. The choice of resins affects the overall properties of the composite. Polyesters have been used most frequently in the wind industry because they have a short cure time and low cost. Cure time is from a few hours to overnight at room temperature, but with the addition of an initiator, curing can be done at elevated temperatures in a few minutes. However, shrinkage upon higher temperature curing is relatively high. Epoxies are stronger, have better chemical resistance, good adhesion, and low shrinkage upon curing, but they are also more expensive (almost twice as expensive as polyester) and have a longer cure time than polyesters. Vinyl esters are epoxy-based resins which have become more widely used. These resins have similar properties to epoxies, but are somewhat lower in cost and have a shorter cure time. They have good environmental stability and are widely used in marine applications (Manwell et al. 2002).

Fatigue damage occurs in composites as it does in many other materials, but it does not necessarily occur by the same mechanism (Manwell et al. 2002): The matrix first cracks, then cracks begin to combine and there is debonding between the matrix and the fibers. Then there is debonding and separation (delamination) over a wider area. This is followed by breaking of the individual fibers, and finally by complete fracture. That is, rainflow cycle counting is used to determine the range and mean of stress cycles, and Miner's Rule is used to calculate the damage from the cycles and the composite's S–N curve. Fatigue strength of glass fibers is only moderate. The ratio of maximum stress to static strength is 0.3 at ten million cycles. Carbon fibers are much more fatigue resistant than are glass fibers: the ratio of maximum stress to static strength is 0.75 at ten million cycles, two and half times that of glass. Owing to the complexity of the failure method of composites and the lack of complete test data on all composites of interest, it is in practice still difficult to predict fatigue life accurately. Generally, material fatigue properties are always an important consideration in wind turbine design and materials selection. During the expected 30 year life of a wind turbine, many of the components will need to be able to endure 4×10^8 fatigue stress cycles. This high cycle fatigue resistance is even more severe than aircraft, automotive engines, bridges, and most other man-made structures (Ancona and McVeigh 2001).

In wood–epoxy laminates, wood is used instead of synthetic fibers in some composites. In this case the wood is preformed into laminates (sheets) rather than as fibers, or fiber-based cloth. The most common wood for wind turbine laminates is Douglas fir. Properties of woods vary significantly with respect to the direction of the wood's grain. In general, though, wood has good strength-to-weight ratio, and is also good in fatigue. One important characteristic of wood is its strong anisotropy in tensile strength. This means that laminates have to be built up with grain going in different directions if the final composite is to be strong enough in all directions. The use of wood together with an epoxy binder was developed for wind turbine applications based on previous experience from the high-performance boat building industry. A technique known as the wood–epoxy saturation technique (WEST) is used in this process. Wood–epoxy laminates have good fatigue characteristics. No wood–epoxy blade has ever failed in service due to fatigue (Manwell et al. 2002).

In the modern wind energy generation, high strength and stiffness, fatigue-resistant lightweight materials like carbon fiber composites can support development of lighter, longer blades and increased power generation. Blade designers are increasingly using lighter weight materials such as industrial carbon fiber laminates, modular prepreg members, and automated fiber placement production technologies to achieve longer, stiffer blades, including the use of carbon fiber in structural spar caps. The use of lighter blades also reduces loading on support structures, and can result in material and cost savings beyond the blades alone. Advances in composites materials and production methods will be needed to achieve the aggressive cost reductions required meet these targets. While high-performance carbon fiber has been used for components subject to the highest loads (i.e., spar caps) by some manufacturers, glass fiber composites with lower specific properties are the dominant materials for the overall blade due to lower cost. The capital cost of turbine structures and blades is a significant contributor to the levelized cost of electricity (LCOE) for wind generation. As a result, any enhancement to the structural properties of materials must be balanced against the increased cost to ensure the overall system costs do not increase disproportionately with the increased power capacity and energy production. Further advances in blade manufacturing techniques, improved quality control, innovations for glass-carbon fiber hybrid composites, and reduced costs of carbon fiber composite materials and manufacturing will support production of larger turbines and continued growth of wind power (Wood 2012).

6.5.3 Copper

Copper has excellent electrical conductivity and for that reason it is used in nearly all electrical equipment on a wind turbine, including the power conductors. Mechanical properties of copper are in general of much less interest than the conductivity. The weight, however, can be significant. A substantial part of the weight of the electrical generator is due to the copper windings, and the weight of the main power conductors may also be of importance (Manwell et al. 2002).

The primary use of copper in wind energy technologies is in the coil windings in the stator and rotor portions of the generator, in the high-voltage power cable conductors, and in the transformer coils. Copper is vital to the electrical grounding system for wind turbine farms. The turbine masts are prime attractors of lightning strikes, so they require a lightning protection system. When lightning strikes a turbine blade, the current passes along the blade, through the blade hub in the nacelle (gearbox/generator enclosure) and down the mast to a grounding system. The blade incorporates a large cross-section copper conductor that runs along its length and allows the current to pass along the blade without deleterious heating effects. The nacelle is protected by a lightning conductor, often copper. The grounding system, at the base of the mast, consists of a thick copper ring conductor bonded to the base or located within a meter of the base. The ring is attached to two diametrically opposed points on the mast base. Copper leads extend outward from the ring and connect to copper or copper-clad grounding electrodes. The grounding rings at all turbines on the wind farm are inter-connected, providing a networked system with an extremely small aggregate resistance (CDA 2017).

6.5.4 Reinforced Concrete

Concrete as a material of construction can play an important role in realizing the potential of wind energy. Reinforced concrete is frequently used for the foundations of wind turbines. It has sometimes been used for the construction of towers as well.

The trend towards increasing generating capacity of wind turbines makes concrete a competitive material. Concrete as material of construction is durable. This property of concrete plays an important role if the wind towers are located in remote areas or in areas with aggressive environment like that of a marine environment. The concrete tower will ensure reliability and require less maintenance. In fact, durability of concrete against marine conditions makes offshore wind farms a reality. Because concrete permits a variety of mix designs, it can be made suitable for a wide range of site conditions for both foundation and pylon. Through the use of admixture and special reinforcements, not just the strength of concrete can be improved but also its resistance to corrosion. Concrete can be tailor-made to meet specific requirements. Depending upon the site conditions both in situ and pre-cast construction methods are suitable for wind tower construction (Singh 2007).

As the future wind turbines demand higher capacity for power generation and taller towers, the prestressed concrete tower emerge as the preferred construction. Some of the advantages of prestressed concrete towers include (Singh 2007): stiffness values matching that of steel if design is for extreme load capacity, better fatigue properties over steel, unlike steel tower there is no risk of buckling and lower material cost. Comparably, the post-tensioned segmental concrete tower design consists of a number of arc segments that combine to form circular sections that are stacked on each other and made to act together in a monolithic manner by post-tensioned high strength tendons that pre-compress the concrete so that it is nearly always in compression to applied loads.

In the case of tower, the concrete is always in severely exposed condition because it is subjected to the effects of wind, sun, rain, and cold conditions including frost. Thus, durability is of prime importance and therefore a strong and dense concrete is required for tower construction. The overall stiffness of tower structures depends on pylon, tower stem, and foundation performance. Concrete has higher material damping properties than other materials and prestressed concrete has high fatigue resistance that provides more tolerance and less risk from dynamic failure. The prolonged service life of concrete and its ability to cope with increasing loading allows possible retrofitting of turbines after their initial 20 year design life. Therefore next-generation turbine life cycles could easily be accommodated in this way, thereby avoiding the financial and environmental costs of construction (Alan 2006; Singh 2007).

Flexibility of construction methods with concrete makes it eminently suitable for offshore installations. Gravity foundations can be constructed onshore and delivered for assembly using existing flat top barges. A similar procedure can be followed for individual concrete sections of the pylon, which can be designed considering the requirement of transportation. In addition, with the aid of buoyancy devices, the foundation and pylon can be constructed in protected shallow waters and towed to its permanent offshore position. Moreover, concrete can also be used for building floating platform for deep water wind farms. The floater mounted turbines could work in water depths ranging from 30 to 200 m (Singh 2007).

6.5.5 Prospects of Turbine Materials Usage

Turbine material usage is and will continue to be dominated by steel, but opportunities exist for introducing aluminum or other lightweight composites, provided strength and fatigue requirements can be met. Small turbine production volume is increasing rapidly which can be accommodated by manufacturing mechanization and innovation that will lower costs. Elimination of the gearbox by using variable speed generators will increase through use of permanent magnetic generators on larger turbines increasing the need for magnetic materials. New high power electronics will help reduce the need for gearboxes and also decrease losses occurred during transmission of wind power to distant load centers. Simplification of the nacelle machinery may not only reduce costs, but also increase reliability. Blades are primarily made of GRP, which is expected to continue. While use of CFRP may help to reduce weight and cost to some extent, low cost and reliability are the primary drivers for material selection. Increasing the use of offshore applications may partially offset this trend in favor of the use of composites. Prestressed concrete towers are likely to be used more, but will need a substantial amount of steel for reinforcement. Wood epoxy, used in early blade production, is not expected to be a material of choice despite excellent fatigue properties (Ancona and McVeigh 2001).

INDEPTH: Additive Manufacturing for Wind Turbine Blades

Large components for wind turbines are often expensive to produce because of their uniqueness. Increasingly larger components can be achieved with additive manufacturing today, to which extra functionality can also be added. Prototyping and production are therefore getting simpler and cheaper. Wind turbine blades produced for research purposes can easily be more than 12 m in length, which is to date too long to be printed in one run. Additive manufacturing technologies should provide access to production methods that create less waste, have shorter lead times, and offer more flexibility at the design stage. Apart from that, the 3D printers are increasing in size and so the opportunities are even greater for tidal and wind turbine blades.

There are generally five steps to make the wind turbine blade. First of all a CAD model of the blade was designed: in principle a typical blade design from which a mold was made and cut into 3D printable sections, complete with assembly holes and discharge ducts for the hot air. Sections measuring almost two meters were then 3D-printed. These sections were then given a glass fiber laminated layer and smoothed off. Each mold segment was placed in a frame with a hot air blower, temperature control and thermocouples. The innovative technology using hot air saved energy and eliminated the labor-intensive step of manually installing the heater wires, which are traditionally embedded into the mold. Moreover, the air blowers can be reused for new molds in the future. Once assembled, the gigantic 3D printed mold had an extremely uniform surface, perfect for manufacturing wind turbine blades (or tidal generator blades) at a lower cost when using the traditional production methods (AMO 2016).

6.6 Lightweight Composite Materials for Wind Turbine Blades

The first wind turbine for electric power generation was built with massive steel blades in 1941, and one of the blades failed after only a few hundred hours of intermittent operation. The Gedser wind turbine was produced in around 1956, with composite blades built from steel spars with aluminum shells supported wooden ribs. The turbine run for 11 years without maintenance. After 1970s, most of wind turbines were produced with composite blades (Manwell et al. 2002). Thus, the turbine blades play very important role in the wind turbines. Blades are required to preserve an optimum cross-section for aerodynamic efficiency to generate the maximum torque to drive the generators. The efficiency and reliability of the wind turbine depend on the material of the blade, shape of the blade, and angle of the blade. Therefore, the material selection of the turbine blade is crucial in the wind turbines. The material of the blade should possess high stiffness, low density, and long fatigue life features (Eker and Eker 2013). As a result, lightweight composite blade materials have been developed to meet these requirements.

6.6.1 Performance Requirement to Wind Turbine Blades Made with Fiber-Reinforced Composites

Blades represent the most important composite-based part of a wind turbine, whose properties quite often determine the performance and life time of the turbine. A blade generally consists of two faces—one on the suction side and another on the pressure side, joined together and stiffened either by one or several integral shear webs linking the upper and lower parts of the blade shell or by a box beam—box spar with shell fairings. The box beam inside the blade is adhesively joined to the shell. During the functioning, the stability of the blade shape, high durability and reliability should be ensured. The shape stability corresponds to the minimum deflection of the blade under wind loads, which is achieved by increasing the moment of inertia of the blade through blade design and by increasing the flexural stiffness of the wind blade material. The flapwise bending is resisted by the spar, internal webs, or spar cabs inside the blade, while the edges of the profile carry the edgewise bending. Therefore, the requirements to wind turbine blades mainly include (Mishnaevsky 2011): (a) High strength to withstand even extreme winds, as well as gravity load; (b) high fatigue resistance and reliability to ensure the stable functioning for more than 20 years and 10^8 cycles; (c) low weight to reduce the load on the tower, and the effect of gravitational forces; and (d) High stiffness to ensure the stability of the aerodynamically optimal shape and orientation of the blade during the work time, as well as clearance between blade and the tower.

To meet these requirements, fiber-reinforced composite materials are becoming more important in the construction of wind turbine blade structures. New generation of large wind turbine blades are designed with all composite fuselage and wing structures. These materials have a significant part to play in maintaining and developing the wind turbine industry. The composite structural properties, such as stiffness, dimensional stability, and strength of a composite laminate, depend on the stacking sequence of the plies. As the number of plies with chosen orientations increases, more stacking sequences are possible. The strength and stiffness of a composite build-up depends on the orientation sequence of the plies. For instance, the practical range of strength and stiffness of carbon fiber extends from values as low as those provided by fiberglass to as high as those provided by titanium. The lightweight-fiber-reinforced composites can be classified according to their reinforcement and matrix materials (Eker and Eker 2013).

6.6.2 Reinforcement Materials

Reinforcement materials give the necessary stiffness and strength to the composite. Fibers for composite materials can come in many forms: continuous and discontinuous, long and short, organic and inorganic. The selection of reinforcement material is based on the properties desired in the finished product. Fibers used in composite part manufacture come in various forms, such as yarns; rovings; chopped strands; woven fabric; and mats. Each of these has its own special application. When prepreg materials are used in parts manufacture, woven fabric or mats are required. In processes such as filament wet winding or pultrusion, yarns and rovings are used.

Unidirectional prepreg tapes have been the standard reinforcements of the composites within the wind turbine blades industry for many years, and the fiber is typically impregnated with thermosetting resins. The most common method of manufacture is to draw collimated raw (dry) strands into the impregnation machine where hot melted resins are combined with the strands using heat and pressure. Tape products have high strength in the fiber direction and virtually no strength across the fibers. The fibers are held in place by the resin. Tapes have a higher strength than woven fabrics. Most fabric constructions offer more flexibility for layup of complex shapes than straight unidirectional tapes offer. Fabrics offer the option for resin impregnation either by solution or the hot melt process. Generally, fabrics used for structural applications use like fibers or strands of the same weight or yield in both the warp (longitudinal) and fill (transverse) directions. For aerospace structures, tightly woven fabrics are usually the choice to save weight, minimizing resin void size and maintaining fiber orientation during the fabrication process. Woven structural fabrics are usually constructed with reinforcement tows, strands, or yarns interlocking upon themselves with over/under placement during the weaving process. The more common fabric styles are plain or satin weaves. The plain weave construction results from each fiber alternating over and then under each intersecting strand (tow, bundle, or yarn). Knitted or stitched fabrics can offer many of the mechanical advantages of unidirectional tapes. Fiber placement can be straight or unidirectional without the over/under turns of woven fabrics. The fibers are held in place by stitching with fine yarns or threads after preselected orientations of one or more layers of dry plies. These types of fabrics offer a wide range of multi-ply orientations. Although there may be some added weight penalties or loss of some ultimate reinforcement fiber properties, some gain of interlaminar shear and toughness properties may be realized (Eker and Eker 2013).

Glass, carbon, boron, or aramid fibers are the most common choices for composite blade parts. The best fiber for a particular application depends on the required strength, stiffness, corrosion resistance, and budget. Because of their low cost compared to other fiber reinforcements, glass fibers are the most common reinforcing materials used in polymer matrix composites. They have high tensile strength but low modulus compared with other fibers. Typical variants are (Eker and Eker 2013): (a) E-glass; (b) ECR-glass; (c) S-glass; (d) R-glass; (e) Te-glass; (f) Silica/quartz; and (g) D-glass. The different types of glass are supplied in several different configurations, including: (a) Fiberglass rovings; (b) Sheet molding compound; (c) Woven rovings; and (d) Chopped strand mat. Research is underway to enhance the performance of glass fibers while capitalizing on its lower cost compared to carbon or aramid fibers. For example, a novel high strength glass fiber demonstrates that it is stronger than the carbon fibers used today at half the cost (King 2014).

Carbon fiber exhibits excellent fatigue resistance which does not suffer from stress rupture compared with glass or aramid fibers. Carbon fibers are supplied in tows and may vary from 1000 fibers per tow to hundreds of thousands per tow. Untreated carbon fibers do not wet easily, so adhesion to the matrix must be achieved by mechanical interference coupled with surface treatment and chemical bonding between the fiber and the matrix. Aramid fibers have the highest strength-to-weight ratio compared to other commercially available fibers, and exhibits similar tensile strength to glass fiber, but can have modulus at least two times as great. Aramid is very tough allowing significant energy absorption but, compared to carbon, it is lower in compressive strength and has poorer adhesion to the matrix. It is also susceptible to moisture absorption. Aramid fiber properties depend on the structure used and can be tailored for high toughness or high modulus. Boron fiber actually predates carbon fiber as a high-modulus reinforcement material. The cost of boron, however, has seen its demise, with its replacement with carbon fiber. They do not differ greatly from glass fiber in tensile strength, but can have modulus five times that of glass. Since the objective of reinforcement is to stiffen, this is a significant advantage. Their use is confined to niche markets, where the modulus advantage over carbon fiber is critical (Eker and Eker 2013).

As one of the most promising reinforcement materials, carbon/graphite fibers are produced from three types of materials known as precursor fibers: polyacrylonitrile (PAN); rayon; petroleum pitch; and bio-based precursors. Except the bio-based precursors, the carbon/graphite fibers are produced by the controlled burning off of the oxygen, nitrogen and other noncarbon parts of the precursor fiber, leaving only carbon in the fiber. Following this burning off (or oxidizing) step, the fibers are run through a furnace to produce either carbon or graphite fibers. Carbon fibers are produced at furnace temperatures of 1000–2000 °C, while graphite fibers require temperatures of 2000–3000 °C. At these temperatures the carbon atoms in the fibers are rearranged to impart the required characteristics to the finished fiber. The PAN-based fiber is the more commonly used precursor in the advanced composite industry today. As shown in Fig. 6.8, the manufacturing processes involved in the production of carbon fibers from polyacrylonitrile (PAN) (Masuelli 2013): First, the precursor is produced through a polymerization process. This step is followed by filtration and washing to remove any excess solvents and impurities. The conversion of the precursor (PAN) into high-performance carbon fibers involves successive stages of oxidative stabilization wherein the PAN precursor is stretched and simultaneously oxidized in a temperature range of 200–300 °C. This treatment converts thermoplastic PAN to a nonplastic cyclic or ladder compound. Fibers are then carbonized at about 1000 °C without tension in an inert atmosphere (normally nitrogen) for a few hours. During this process, the non-carbon elements are removed as volatiles leaving carbon fibers with a yield of about 50% of the mass of the

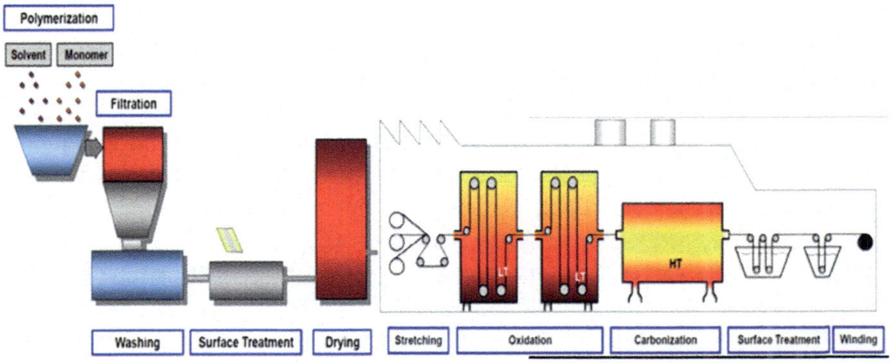

Fig. 6.8 Current carbon fiber production steps (Assuming a PAN precursor). Process intensification and energy reductions are necessary to achieve low-cost carbon fiber production (Adapted with permission from Masuelli 2013 (IntechOpen))

original PAN precursor material. Depending on the final fiber property requirements, the fibers are treated at temperatures between 1500 and 3000 °C at the next graphitization step, which improves the ordering and orientation of the crystallites in the direction of the fiber axis. The fibers are then wound to an appropriate size and packed for further processing.

Bio-based precursor options, including bio-derived acrylonitrile (bio-ACN) and lignin, are of interest as renewable materials that may have lower embodied energy (and potentially lower cost) relative to conventional PAN. Bio-ACN involves the conversion of biomass materials to PAN, providing a "drop-in" renewable substitute for conventional PAN. Glycerol, a by-product of biorefineries, is one potential raw material for bio-ACN. The indirect ammoxidation of glycerol to acrylonitrile was demonstrated in a tandem reactor where glycerol dehydration formed an acrolein intermediate followed by the ammoxidation of acrolein to acrylonitrile. The resulting acrylonitrile can be polymerized to form PAN fibers for subsequent conversion to carbon fiber (Dubois 2010; Plee 2010).

Lignin, a heterogeneous plant-based polymer, is another biomass precursor option, though its processing is complicated by its relatively unpredictable structure that varies between feedstock sources. Through a half century of research and development, key parameters for spinning lignin into carbon fibers, including the range of molecular weights and compositions best suited for production, have been identified. Various methods for producing carbon fibers from lignin have been tested, with melt-blowing of soluble lignin emerging as the favored method. The challenges associated with direct conversion of lignin to finished carbon fibers include difficulties meeting structural specifications consistently and the need for new manufacturing processes and lines for lignin-based production. As a result, it may take longer to commercialize lignin-based carbon fibers than drop-in bio-CAN (Baker and Rials 2013).

In addition, melt spinning of carbon fiber precursors is both a more environmentally sound and cost-effective method compared to the conventional, capital-intensive and highly corrosive solvent-based solution spinning method. Optimized melt-spun PAN precursors, which enable automated spinning operations for higher throughput, have the potential to reduce manufacturing energy requirements and fiber cost by 30%. Further gains are possible in the carbonization stage, the process of converting precursor fibers to crystallized, carbon-rich fibers in an inert (oxygen-free) environment—typically using a series of specially designed furnaces. Microwave-assisted plasma carbonization could potentially replace this high-temperature, energy-intensive process for energy and cost savings of up to 50% (Das and Warren 2012).

Aramid fibers are another human-made product. These fibers are produced by manufacturing the basic polymer, then spinning it into either a paper-like configuration or into fiber. Aramid fibers have several useful characteristics (Eker and Eker 2013): high strength and modulus; temperature stability; flex performance; dimensional stability; chemical resistance; and textile process ability.

Hybrid reinforcements such as E-glass/carbon and E-glass/aramid have been used to replace glass fibers, leading to 80% weight savings and cost increase by 150%, while a partial (30%) replacement would lead to only 90% cost increase and 50% weight reduction (Mishnaevsky 2011).

6.6.3 Matrix Materials

Due to the low-weight requirement to the wind blades, polymers are the main choice as the matrix materials for the wind blade composites. The matrix of composite controls fracture toughness, delamination strength and out-of-plane strength and stiffness of the composite, and influences the fatigue life of the composites. Typically, thermosets and thermoplastics are used as matrixes in wind blade composites (Mishnaevsky 2011).

Thermosetting resins predominate today, while thermoplastics have only a minor role in manufacturing advanced composites. Thermoset resins require addition of a curing agent or hardener and impregnation onto a reinforcing material, followed by a curing step to produce a cured or finished part. Once cured, the part cannot be changed or reformed, except for finishing.

Thermosets are attractive for composites manufacturers due to their relatively low viscosity at room or elevated processing temperatures. Resin viscosity is important to consider for composites applications, because it controls the timescale of the liquid resin impregnation into the dry fiber preform. During composites processing, it is important to completely saturate dry fibers with resin without voids or dry spots in the fiber preform—and this must be done as quickly as possible to achieve the high production speeds desired for commercial applications. If the viscosity is too high, the processing times required to completely wet the composite preform would be too high and not economical for part manufacturing. A drawback of thermoset resin-based composites is that they are difficult to recycle using thermal techniques while maintaining continuous fiber integrity because the temperatures required to separate the matrix material from the fiber can damage the fibers and leave residue that makes the fibers more difficult to reprocess. In addition, because thermosets polymerize via irreversible cross-linking reactions, the thermoset resin constituent material is typically broken down at the elevated temperatures used to remove it from fibers; the polymer, therefore, cannot be recovered for reuse. Many thermoset resins are designed for use at high temperatures—thus the temperatures needed to remove them from fibers for fiber recycling can be very high, with high associated energy/financial costs (Strong 2008).

Some of the more common thermosets include: epoxies; polyurethanes; phenolic and amino resins; bismaleimides (BMI, polyimides); and polyamides. Of these, epoxies are the most commonly used in today's PMC industry. The basic epoxy compounds most commonly used in industry are the reaction product of epichlorohydrin and bisphenol-A. Epoxy compounds are also referred to as glycidyl compounds. There are several types of epoxy compounds including glycidyl ethers (or diglycidyl ethers), glycidyl esters, and glycidyl amines. Several of these compounds are reactive diluents and are sometimes added to the basic resin to modify performance characteristics. The epoxy molecule can also be expanded or cross-linked with other molecules to form a wide variety of resin products, each with distinct performance characteristics. These resins range from low-viscosity liquids to high-molecular weight solids. Typically, they are high-viscosity liquids. The second of the essential ingredients of an advanced composite system is the curing agent or hardener. These compounds are very important because they control the reaction rate and determine the performance characteristics of the finished part. Since these compounds act as catalysts for the reaction, they must contain active sites on their molecules. Some of the most commonly used curing agents in the advanced composite industry are the aromatic amines. Two of the most common are 4,4′-methylene-dianiline (MDA) and 4,4′-sulfonyldianiline (DDS). Like the epoxies, these compounds have a very low vapor pressure and usually do not present an airborne hazard unless in a mixture that is sprayed or cured at high temperatures. Several other types of curing agents are also used in the advanced composite industry. These include aliphatic and cycloaliphatic amines, polyaminoamides, amides, and anhydrides. The choice of curing agent depends on the cure and performance characteristics desired for the finished part (Eker and Eker 2013).

Polyurethanes are another group of resins used in advanced composite processes. These compounds are formed by reacting the polyol component with an isocyanate compound, typically toluene diisocyanate (TDI); methylene diisocyanate (MDI); and hexamethylene diisocyanate (HDI) are also widely used. Phenolic and amino resins are another group of PMC resins. With respect to the phenol-formaldehyde resins, the well-known hazards of both phenol and formaldehyde must be protected against. In addition to traces of free formaldehyde, they may also contain free phenol, and contact with these resins in the uncured state is to be avoided. The urea- and melamine-formaldehyde resins present similar hazards. Free formaldehyde, which is present in trace amounts and may be liberated when their resins are processed, can irritate the mucous membranes. The bismaleimides and polyamides are relative newcomers to the advanced composite industry and have not been studied to the extent of the other resins (Eker and Eker 2013).

Thermoplastics currently represent a relatively small part of the PMC industry. They are typically supplied as nonreactive solids (no chemical reaction occurs during processing) and require only heat and pressure to form the finished part. Unlike the thermosets, the thermoplastics can usually be reheated and reformed into another shape, if desired (Eker and Eker 2013).

Moreover, the increased use of thermoplastic matrix materials offers the potential for improved recyclability, but presents other technical challenges including temperature stability, moisture sensitivity, mechanical stability, and final surface quality, among other issues. Unlike thermosets, which polymerize via irreversible cross-linking reactions, thermoplastic polymers can be re-melted above a transition temperature. Thermoplastic resins can liquefy and separate from fibers at lower temperatures compared to thermoset resins, enabling recycling of both fibers and polymer. However, a primary barrier for the widespread use of thermoplastic resin is the high viscosity. At typical processing temperatures, the thermoplastic resin is very viscous and does not readily impregnate fiber preforms and tows. Lack of sufficient impregnation increases the likelihood of trapped air bubbles and porosity—which upon resin hardening, leads to decreased part quality due to stress concentrates at voids and porosity sites. Elevated temperatures reduce the thermoplastic viscosity, but not sufficiently. If the temperature is too high, the resin will begin to degrade and lose integrity. Future work is needed for the development of thermoplastic resins that can be processed at temperatures and viscosities similar to thermoset resins, without breaking down. One promising thermoplastic matrix material for molding processes is liquid thermoplastic resin Elium. Elium's low viscosity at room temperature makes it suitable for continuous fiber resin transfer molding (RTM) and vacuum-assisted resin transfer molding (VARTM) applications. The current developmental version is cured like a thermoset at 80 °C in 20–30 min. Once cured, the composite matrix behaves like a thermoplastic that can be thermoformed to any shape and joined to other thermoplastics by induction welding. These qualities facilitate recycling because the resin can be melted and stripped from the composite and used again along with the fibers (Swan and Pierre 2014).

Significant research and development has been conducted to improve the material properties of composite materials using nano-material-based resin additives. Examples include carbon nanotubes (CNT), nanoclays, nano-platelets, and graphene. Nano-material-based resin additives could provide significant material property modification. As fibrous materials reinforce the matrix at micron length scales, resin nano-additives provide reinforcement at nano length scales. Multi-scale reinforcement of the matrix can lead to improved mechanical performance, such as better distribution of transverse shear to reduce delamination failure and increasing fracture toughness to arrest the progression of micro-cracking. In addition, some nano-additives can influence other material properties such as electrical and thermal conductivity. Their use could significantly impact new composite material applications, such as damage sensing structures or self-healing structures. For example, high-performance nanocomposite materials can heal autonomously and repeatedly using a three-dimensional vascular network filled with microcrack healing chemistries (Patrick et al. 2014).

In addition, thermoset and thermoplastic polymers are both largely derived from petroleum-based feedstocks, leading to high embodied energies for these materials. However, there has been increasing interest in nonpetroleum, bio-based resins to reduce energy intensity and reliance on nonrenewable fuel resources.

INDEPTH: Lightweight Rotor Blades Made from Plastic Foams for Offshore Wind Turbines (Fraunhofer-Gesellschaft 2016)

The trend towards ever larger offshore wind farms continues unabated. Wind turbines with rotor blades measuring up to 80 m in length and a rotor diameter of over 160 m are designed to maximize energy yields. Since the length of the blades is limited by their weight, it is essential to develop lightweight systems with high material strength. The lower weight makes the wind turbines easier to assemble and disassemble, and also improves their stability at sea. These days, rotor blades for wind turbines are largely made by hand from thermosetting resin systems. These, however, don't permit melting, and they aren't suitable for material recycling. At best, granulated thermoset plastic waste is recycled as filler in simple applications. Therefore, thermoplastics have been developed for manufacturing rotor blades. They are repeatedly meltable and can be processed efficiently in automated production facilities. Moreover, the glass and carbon fibers can be separated to reuse the thermoplastic matrix material.

For instance, the outer shell of the rotor blade, as well as for segments of the inner supporting structure can be manufactured with sandwich materials made from thermoplastic foams and fiber-reinforced plastics. In general, carbon-fiber-reinforced thermoplastics are used for the areas of the rotor blade that bear the greatest load, while glass fibers reinforce the less stressed areas. For the sandwich core, thermoplastic foams are bonded with cover layers made of fiber-reinforced thermoplastics in sandwich design. This combination improves the mechanical strength, efficiency, durability, and longevity of the rotor blade.

6.6.4 Semifinished Products

A filament is a single segment of reinforcement. Tow count is the number of filaments in the carbon fiber bundle (which can vary depending on the product, such as 3K, 6K, 12K, 24K, and 50K tow fibers). Smaller tow count carbon fibers are generally of higher strength and modulus compared to standard modulus, higher tow count carbon fibers, which are commonly used for less demanding non-aerospace applications. Standard modulus carbon fibers are generally of 12–50K tow size range and constitute 80–90% of the total carbon fiber market today. Continuous filaments can be used in continuous fiber processes such as filament winding and pultrusion.

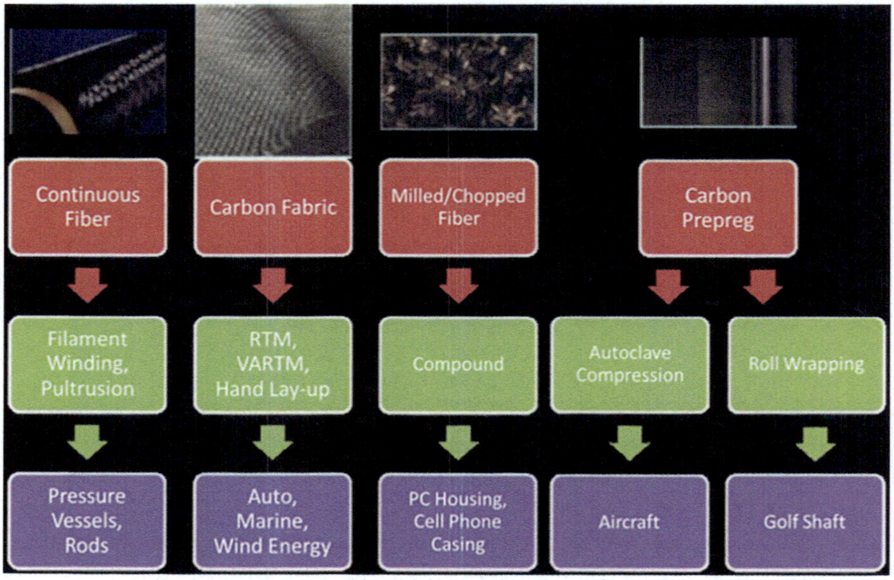

Fig. 6.9 Currently available carbon fiber composite manufacturing technologies and their applications (Adapted from DOE CM 2015. Credit: US Department of Energy)

Filaments may also be woven or stitched into fabrics. Preforms are three-dimensional fabric forms designed to conform to a specific shape to meet specific mechanical and structural requirements. A pre-impregnated composite, or pre-preg, is where fibers, often in the form of a weave or fabric, are held together with a matrix resin. The matrix is partially cured to allow easy handling and often must be cold stored to prevent complete curing. Bulk molding compounds (BMC) and sheet molding compounds (SMC) are made up of fibers pre-compounded with a thermoset resin, and are primarily used in compression molding processes. Figure 6.9 shows currently available manufacturing technologies associated with semifinished carbon fiber products (DOE CM 2015).

6.6.5 Consolidation Techniques

The final properties of a composite part depend not only on the matrix, reinforcement materials, and their starting product forms, but also the processes used to consolidate them into final parts for assembly. Forming processes combine the matrix and reinforcement materials to produce the desired shape. These manufacturing processes are generally grouped into two classes: open forming and closed forming. Table 6.4 compares advantages, disadvantages, and cycle time of most commonly used molding processes (Das 2001).

Table 6.4 Comparison of the most commonly used composite molding processes (Das 2001)

Molding process	Advantages	Disadvantages	Cycle time
Pre-preg	Good resin/fiber control	Labor intensive for large complex parts	5–10 h
Preforming	Good moldability with complicated shapes and the elimination of trimming operation	Cost effective only for large, complicated shape parts; large scrap generated when fiber mats used	45–75 s (compform process) 4–5 min (vacuum forming)
Resin transfer molding (RTM)	Inside and outside finish possible with thickness control, more complex parts possible with vacuum assisted	Low-viscosity resin necessary; voids formation possible without vacuum assist	45–75 s (compform process) 4–5 min (vacuum forming)
Liquid compression molding	Favored method for mass production with high fiber volumes	Expensive set-up cost for low production	1–2 min
Sheet molding compound (SMC)	Cost effective for production volume 10K–80K/year.	Minimum weight savings potential	50–100 s
Resin injection molding (RIM)	Low-cost tooling; prototypes can be made with soft tools	Difficult to control the process	1–2 min
Bulk molding compound (BMC)	Low-cost base material	Low fiber content; randomly oriented; low structural quality; poor surface finish	30–60 s
Extrusion compression molding	Fully automated; variety of polymers and fibers can be used with fiber volumes up to 60% by weight	Not for surface finish parts without paint film or similar process	3–6 min
Structural reaction injection molding	Low tooling cost; good surface finish capability	Difficult to control the process, particularly with low-viscosity resins and longer cure cycle times.	4 min for thermosetting resins; a few seconds for thermoplastic matrices
Carbon-fiber-reinforced thermoplastics (CFRTP)	Easily recycled; fast consolidation	High viscosity which forces users to utilize equipment involving high temperature (200–400 °C)	1 min

6.6.5.1 Closed Forming Processes

(a) Injection molding

Injection molding is the most common and widely used manufacturing process for high-volume production of thermoplastic resin parts reinforced with fibers. Solid pellets of resin containing the fibers are fed through a hopper into a heated

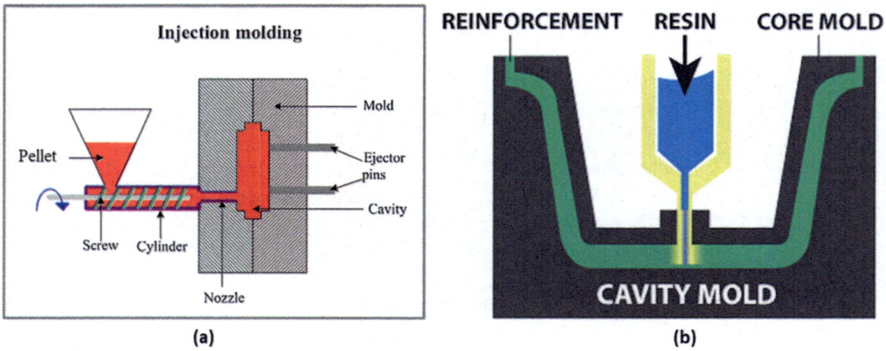

Fig. 6.10 Injection molding (**a**) and resin transfer molding (**b**)

barrel with a rotating screw, as shown in Fig. 6.10a. The rotating screw generates heat by viscous shearing against the barrel, melting the resin. The screw also acts as a piston and forces the mixture of fibers and molten resin into a matched-metal mold where the mixture cools and solidifies. The mold cavity is then opened and the composite part is ejected. The main advantages of injection molding are the ease of automating the process and the short cycle times, which together enable high-volume production. The main disadvantages are the high initial costs of the capital equipment and the molds and material property variation in the part due to the inability to control fiber orientation and distribution. Additionally, due to the melt viscosity limitations of the current thermoplastic resins, injection molding generally produces only short-fiber-reinforced composites. While injection molding is a relatively mature process, its compatibility with long fiber reinforcements has mostly restricted its use thus far to cosmetic and other nonstructural parts. Research is underway to modify thermoplastic chemistry to tailor the melt viscosity of the resin, which could enable injection molding of long-fiber-reinforced composites for structural applications (Advani and Sozer 2003).

(b) Resin transfer molding

Resin transfer molding, a newer process evolved from the same concept to injection molding, can be used to fabricate continuous carbon-fiber-reinforced composites with broader applicability in structural and semi-structural components. In resin transfer molding (RTM), fiber preform or dry fiber reinforcement is packed into a mold tool that has the desired shape of the composite part, as shown in Fig. 6.10b. A second mold tool is clamped over the first and resin is injected into the cavity. A vacuum may be used to assist in drawing the resin through the cavity in a process called vacuum assisted resin injection (VARI). The main disadvantage of this method is that matched tooling capable of withstanding the elevated pressures is expensive and generally limited to smaller components. Additionally, un-impregnated areas can occur, resulting in costly scrap. This composites manufacturing method has the greatest potential (compared to other methods) for fabricating complex, large-scale integrated structural parts (DOE CM 2015).

Current manufacturing processes for land-based and offshore utility-scale wind turbine blades that employ VARTM or low-temperature-cure pre-preg containing 90–100% glass fiber reinforcement suffer from long manufacturing cycle times of 35–40 h for a 45 m blade, high labor requirements, and frequent rework. Automated fiber placement and inspection processes could reduce the labor requirements of blade production. Thermoplastic use is expected to reduce blade weight, cost, and cure cycle times and facilitate recycling at the end of their service life. Due to high cost, carbon fiber use has been limited to spar cap applications today. Using pultruded carbon fiber sheet material in blade spars has also been considered to enable larger, lighter rotors that will increase energy capture. This method is well suited to wind blade applications where larger blades (in the range of 100 m) can be fabricated in the field without the need for autoclaves. As in the case of RTM, future research to enable economical use of this method is directed towards the development of low viscosity, fast curing resins that reduce the cycle times from the current state of the art (Frank et al. 2014).

(c) Compression molding

The compression molding is principally simple and has been utilized for decades. The material (called the charge) is placed inside the mold cavity. The material charge is often a mixture of resin and fibers, sometime in a mat preform. The mold is closed and pressures up to 14 MPa are applied, forcing the material charge to deform to the shape of the cavity. Low pressure compression molding is called cold press molding. The mold is opened and the part is ejected. The advantages of compression molding include its simplicity, relatively fast cycle times, high repeatability, tight tolerances and high-volume production. The major disadvantages are the large initial capital investments in molds and presses and defects resulting from residual stresses, delamination, warpage, and flow orientation of fibers. This process is currently widely used in nonstructural automobile applications such as interiors, closures, and miscellaneous parts. The primary starting materials are short glass-fiber-reinforced SMCs and BMCs. Development efforts are underway to enable long carbon-fiber-reinforced SMCs to take advantage of their improved strength and stiffness-to-weight ratios. SMC formulation improvements are underway that will toughen the materials to prevent surface micro cracking (DOE CM 2015).

Matched die molding, a subset of compression molding, holds strong promise to produce continuous carbon-fiber-reinforced parts for structural applications. In this process, a continuous fiber ply stack (known as the blank) that is unidirectional and/or woven is pressed into its final shape in a matched die mold and cured (thermosets) or consolidated/stamped (thermoplastics) to rapidly produce parts. The blank design must be highly engineered because the fibers drape into the final shape, causing changes in fiber orientation; thus, the blank design and press process affect the properties of the finished part. The cure time, or consolidation cycle time, depends on the material selection.

Thermoplastic parts are consolidated in seconds and thermoset matrix parts in minutes, with significant reductions in cycle time. The matched die molding process would be a strong competitor to the RTM process if the dies can be reused multiple times without any shape distortions or loss of integrity (DOE CM 2015).

6.6.5.2 Open Forming Processes

(a) Hand lay up

Resins are impregnated by hand into fibers in the form of weaves and fabrics. Rollers or brushes are typically used. The composite is left to cure under standard atmospheric conditions. The major disadvantage is the lack of consistency; the quality of the product is highly dependent on the skill of the laminator. Resins need to be low in viscosity to be workable by hand. This generally compromises the mechanical and thermal properties of the composite and can create a health risk for the laminator.

(b) Spray up

Chopped fiber and catalyzed resin are sprayed directly into a mold and left to cure under standard atmospheric conditions. Although this method is low-cost, there are several serious disadvantages. Laminates tend to be very resin-rich and, therefore, excessively heavy. Only short fibers and resins low in viscosity are able to be sprayed, which severely limits the mechanical properties. Additionally, exposure to high styrene resins is hazardous to the health of workers. A challenge in this method of part fabrication is managing the volatile organic compounds and hazardous air pollutants released in the process. These are expensive to control in the spray up process, and, as a consequence, many composites manufacturers have migrated to closed mold, infusion-based processes which better contain and manage the pollutants. The part finish and precision obtained with other manufacturing methods cannot be achieved with either the spray up or the hand layup process. Therefore, the use of these open forming techniques has been mostly limited to large consumer goods such as bathtubs and swimming pools and to the repair of damaged parts (DOE CM 2015).

(c) Filament winding

This process is most appropriate for hollow, circular, or oval sectioned components, such as pipes and tanks. Fiber tows are passed through a resin bath before being wound onto a mandrel. The main disadvantages are that fibers cannot be laid in the axial direction and that low-viscosity resins usually need to be used. Filament winding is the predominant composites manufacturing process for axisymmetric composite products such as compressed gas storage tanks or pipeline sections. The process also offers speed and cost advantages for structural axisymmetric parts such as struts, axles, and drive shafts. For compressed gas storage tanks, carbon fiber material costs constitute approximately 60% of the total tank cost in high-volume production, assuming carbon fiber filament winding in an epoxy matrix over a high density polyethylene liner. Cost reduction and the fast process cycle times needed to produce 500,000 parts per year may be achieved through lower material cost, novel braided preforms, manufacturing automation, reduced scrap, reduced energy cost through shorter cure times, and use of protective coatings and durable materials that extend the tank's useful life (DOE CM 2015).

(d) Pultrusion

In pultrusion forming, fibers are pulled from a creel through a resin bath and passed through a heated die. As the fiber passes through the die, the resin cures. Pultrusion yields smooth finished parts that typically do not require post processing. A wide range of continuous, consistent, solid, and hollow profiles can be pultruded. The process can be custom-tailored to fit specific applications such as the constant cross-section spar in some windmill blade applications. However, this process is limited to components with constant, or near constant, cross sections. Additionally, the cost of the heated die can be high (DOE CM 2015).

(e) Automated fiber placement

Automated tow placement (ATP) and automated tape laying (ATL) are subsets of the automated fiber placement method. The differences are the starting materials (pre-preg tows vs. pre-preg tapes) and the material laydown rates feasible. Generally, ATL is faster than ATP and can place more material over longer distances. However, ATP is better suited to shorter courses and can place material more effectively over contoured surfaces. These automated approaches offer several advantages over manual lay-up and spray-up techniques including reduced processing speed, reduced material scrap and labor costs, improved part consolidation, and improved part-to part uniformity. However, capital expenditures for computer-driven, automated equipment can be significant. Some of the recent improvements in automated fiber placement include (DOE CM 2015): dockable heads, enabling equipment to function in both ATP and ATL modes; laser heating for OOA curing of high-performance thermoplastic ATL/ATP parts; and equipment integrated with real-time temperature controls.

6.6.6 Curing and Polymerization Processes

Fiber-reinforced plastic (FRP) composite structures require the polymer matrix to attain and maintain solid-state characteristics in service. Thermosets polymerize via irreversible cross-linking reactions and thermoplastic polymers can be re-melted above a transition temperature. As a result, composites comprised of these matrices have different physical properties as well as different manufacturing processes. Historically, advanced composite structures have been based on thermosetting systems; approximately 80% of composites use a thermoset matrix that requires a cure step to attain desired properties. Aerospace composite structures are mostly based on epoxy systems in which the curing process must follow a precise temperature profile in an autoclave to ensure proper resin flow, de-gassing, consolidation, and eventually uniform degree of polymerization to achieve final properties. The processes are typically slow (on the order of hours) and energy intensive, in part because the large thermal mass of the tooling and autoclave are also subject to the same thermal cycle. Autoclaving processes have been adopted across much of the composites industry beyond aerospace, resulting in an inefficient approach to produce composite structures. Improved selective heating/polymerization techniques, optimized cure cycles, and further advancement of out-of-the-autoclave techniques are potential development and demonstration pathways to reduce the energy used in composite manufacturing. Methods that selectively target the heating and/or curing of composites systems are based on electrotechnologies that utilize radiative energy transfer methods to provide energy only where it is required. These technologies require that the components within the system are responsive to the applied frequencies. The following are examples (Cresko and Roberts 2002; Strong 2008):

(a) Dielectric heating methods based on microwave (MW) or radio frequency (RF) where the electromagnetic (EM) energy couples principally with the matrix. For example, RF curing of epoxy-based GFRP is based on the dielectric response of the epoxy. In some cases, susceptors can be used to improve the heating response of materials. The depth of penetration needs to be appropriate for the size and geometry of the part, and tooling must be adapted for exposure to a high frequency EM environment.
(b) Infrared (IR) as a low-cost, efficient method of pre-heating, heating, melting, and/or curing. Long and medium-wave IR have a number of potential applications. Some have been successfully utilized by industry, including pre-heating of preforms and partial curing of composites structures as a method of temporarily fixturing during intermediate processing steps. As thermoplastic-based composites systems become more prevalent, the use of IR systems has the potential to provide faster heating rates at higher efficiencies than attainable with convection methods. Considerations include the "line-of-sight" nature of IR and its relatively short depth of penetration, with the most promising applications being relatively thin, uniform, and planar components and/or structures.
(c) Induction heating methods. Induction techniques can be used to heat conductive materials and are widely used in the metals industries for unit operations ranging from heat treating to melting. Some applications have targeted the selective heating of the tooling. A limitation of induction heating methods is the requirement that the composite structure have a geometry that allows the induction coil to be placed within a uniform and close proximity to the part. Also, heat losses must be mitigated to ensure uniform heating profiles.
(d) MW heating technology for curing CFRP. MW heating was once considered an intractable method for curing composites comprised of conductive materials like carbon fiber (due to problems like arcing and dielectric breakdown). However, advanced multimode MW applicator designs have been commercialized to fabricate aircraft composites structures.
(e) Ionizing sources of EM energy. Ionizing sources have the potential to drive chemical reactions. This can happen indirectly, as with UV energy that activates a photoinitiator leading to polymerization, or directly with an electron beam

technology that is energetic enough to drive polymerization reactions without an intermediary photoinitiator. Considerations include the very limited depth of penetration of UV, which make the technology more amenable to films and coatings, and the high cost and safety concerns with electron beam energy, which require extensive shielding to protect against exposure to energetic particles. As composites systems expand to include new chemistries, there are additional post-processing techniques that can enable entirely new sequences of manufacturing operations to achieve final parts specifications. For example, solid phase polymerization (SPP) of nylon 6,6 can drive the molecular weight distribution higher and enable modification of the physical properties after parts are manufactured. While SPP of nylon via convection techniques has been commercialized for limited production for specialty applications, it requires extended thermal cycles. However, accelerated SPP has been demonstrated at the pilot scale through a radio-frequency process. This has the potential to enable faster processing of composites structures with lower viscosity, then post-processing to achieve higher performance specifications.

INDEPTH: Erosion-Resistant Coating Materials in Wind Turbines (Conti-Ramsden and Dyer 2015)

Composites don't offer all the answers. For offshore turbines, mechanically tough glass fibers are, like other materials, susceptible to damage by the windy, salty environment of the seas. Given that offshore blade life needs to be a minimum of 25 years, durable coating materials are a key aspect to ensuring erosion-resistance in wind turbines. Coating materials currently used include epoxy and polyurethane gel-coats, polyurethane paint systems and tapes. Most of these coatings started off in the aerospace industry, and so their performance tends to be optimized for those applications. However, the UV combined with the high humidity of the salty sea air, extreme wind conditions and high-speed spinning blades of an offshore turbine, combine to form a highly erosive environment that far surpasses anything experienced by an aircraft. Specialist coatings are being produced by manufacturers such as AkzoNobel and 3M, but many turbines still use the epoxy and polyester resins developed decades ago. There is therefore a need to develop new formulations both for coatings and composite resins—ones that are designed specifically to withstand the harsh environments in which offshore turbines must operate.

One of the biggest issues in turbine performance is to predict the behavior of both the composite and its coating once in use. The leading edge of a turbine blade is that which cuts through the air (shown in diagram), and it is the region of the blade that experiences the highest level of erosion. Research is under the way to understand the erosion failure mechanisms of various coating materials and blade structures in the offshore environment.

6.6.7 Innovative Design with Modeling and Simulation Tools

The number of parts and the design of a system directly affect cost and manufacturability. Innovative design concepts that consolidate smaller parts into a single part may result in lower manufacturing costs. In addition, composite systems are often overdesigned, adding cost and weight, due to the variability in material properties and lack of information and validated design models. The additional cost and embodied energy penalties associated with overdesign of composite parts can be minimized by developing and applying more accurate predictive tools and validation data. These high-accuracy design tools along with improved manufacturing simulation methods can be used to reduce part cost and weight. Examples of innovative design approaches that could impact cost, manufacturability, and energy use could include material optimization, structural redesign, and multifunctionality of parts (for example, use of a composite material for strength as well as electrical shielding of embedded electrical control circuits). Designing damage-tolerant composite structures is a standard practice for aerospace applications. As design requirements and concepts are developed for lower value-add applications, the effects of damage will need to be addressed. Flammability of composite materials may also need to be considered (DOE CM 2015).

Modeling and simulation tools for composite materials and processes can speed the development cycle for new manufacturing processes, innovative designs, and assembly techniques. Even mature industries have existing gaps in modeling that preclude the goal of being able to predict a composite system's properties based purely on knowledge of the individual constituents and the processing history. Design tools that address reliability trade-offs without increasing composite part cost will be essential in cost-sensitive applications.

6.6.8 Effective Joining

The use of multi-material structures and optimized designs can result in reduced weight or improved system performance. Joining different and novel materials presents challenges that include thermal expansion mismatch, limited temperature and load ranges for joined structures, joint performance and repairability, directionality of composite materials, lack of nondestructive evaluation techniques for bonded joints, the need for surface preparation, galvanic corrosion, and longtime requirements to complete joining. Technology development is needed for fast, reliable techniques for joining materials and structures. The new joining methods must not degrade the resulting composite structure in broad applications. Joining techniques also need to be compatible with processes and manufacturing rates on the factory floor (DOE CM 2015).

6.6.9 Defect Detection

Identifying manufacturing defects in components and structures is an important issue for composite systems. The components (matrix, fiber) of a composite retain their original state when combined to form the new material, making it challenging to identify defects in the heterogeneous composite material. Since undetected manufacturing defects can significantly degrade part performance, advancements are needed in in situ sensors for process control to prevent defect formation and in nondestructive evaluation methods to understand as-manufactured part performance. Technologies exist for nondestructive evaluation of composites, but new thinking may be required to apply them to specific material sets and accommodate high speed production and larger size components. For example, one of the most common defects in fiber-reinforced composite materials is the presence of voids in the matrix around or inside fiber bundles. These voids are sourced from air bubbles that become entrapped during processing while the resin is liquefied. These voids lead to mechanical stress concentrations in the composite, which can lead to premature microcracking and decreased useful part life. Detection of voids is currently limited to inspection of finished parts via optical microcopy, ultrasound detection, and X-ray detection. Complete and thorough inspection for large-scale parts (e.g., wind blades) is expensive and impractical. Also, if defects are detected, repair is often costly or not practical, leading to scrapped parts. Future work to address these challenges could include improving detection techniques for handling large-scale parts at sufficient resolutions, improving understanding of the nucleation of potential defects during processing (i.e., when the resin is still liquid), and taking actions to prevent void defects once the resin hardens (DOE CM 2015).

6.6.10 Perspective and Prospective of Composite Wind Blades

Composites offer many advantages in wind turbine blade construction. Composites are also unique in their ability to be tailored for different properties using various reinforcement configurations, matrix materials and manufacturing processes. Wind turbine design has improved substantially due to composites technology, and as composite use becomes more common place there exists the need to minimize the time required to fabricate blades while tightening dimensional tolerances and repeatability. Advanced composites like fiber-reinforced composites of the type used in wind turbine blades are laminates composed of several layers of reinforcing fabric impregnated with and held together by an adhesive resin. Such laminates can be very strong and stiff when loaded in their own plane, but are much weaker when loaded out-of-plane because the layers, or plies, can more readily be pulled apart. The in-plane properties are largely determined by the fibers, whereas the out-of-plane properties depend heavily on the strength and adhesive capability of the resin matrix. Increasingly enabled by the introduction of newer polymer resin matrix materials and high-performance reinforcement fibers of glass, carbon, and aramid, the penetration of these advanced materials has witnessed a steady expansion in uses and volume. The increased volume has resulted in an expected reduction in costs (Eker and Eker 2013).

6.7 Smart and Stealth Wind Turbine Blades

A technique that uses sensors and computational software has been developed to constantly monitor forces exerted on wind turbine. This could be used to improve the efficiency of wind turbines by providing the blades' smart structure with necessary data to adjust to rapidly changing wind conditions. Through feeding information from sensors into an active control system, blade components can be precisely adjusted to optimize efficiency. The system also could help improve wind turbine reliability by providing critical real-time information to the control system to prevent catastrophic turbine damage from high winds (White 2009).

As the blade is being built, the sensors called uniaxial and triaxial accelerometers would be embedded inside a wind turbine blade as the blade was being built. Using a trio of sensors and estimator model software, it can be accurately revealed how much force is being exerted on the blades. Such sensors could be instrumental in turbine blades that have control surfaces and simple flaps like those on an airplane's wings to change the aerodynamic characteristics of the blades for better control. Because these flaps would be changed in real time to respond to changing winds, constant sensor data would be critical. Sensor data in a smart system might also be used to better control the turbine speed by automatically adjusting the blade pitch while also commanding the generator to take corrective steps. It could be used to design more resilient blades because they are capable of measuring acceleration occurring in various directions, which is necessary to accurately characterize the blade's bending and twisting and small vibrations near the tip that eventually cause fatigue and possible failure. The sensors also measure two types of acceleration. One type, dynamic acceleration, results from gusting winds, while the other, called static acceleration, results from gravity and steady background winds. It is essential to accurately measure both forms of acceleration to estimate forces exerted on the blades. Such sensors could be instrumental in turbine blades that have control surfaces and flaps like those on an airplane's wings to change the aerodynamic characteristics of the blades for better control (White 2009).

In addition, stealth blades are poised to overcome the radar interference concerns that continue to block gigawatts of wind turbine installations. Around the world, it is estimated that at least 20 GW of wind power is being blocked because of concerns over interference with radar systems, either military or civilian, with particular concern being expressed over the impact that wind farms could have on air traffic control (ATC) systems. Wind turbines are large structures, and wind farms are large areas of large structures. Combined with their spinning blades and reflective surfaces they can present a significant Radar Cross Section (RCS). This means that they show up as large static or twinkling objects on radar screens, obscuring other objects and raising concerns over a possible impact on the safety of air travel. It is found that the

RCS of the wind turbine blades could only be significantly reduced by using radar absorbing materials. The height of the tower made little difference to the RCS and that the radar footprint of the nacelle and tower could be reduced with careful shaping. Radar systems work by sending out pulses of radio waves that bounce off objects. As the waves return to the radar station the variations in the incoming signal are processed by software to generate an image of what is in the path of the beam. Objects in the path of the radar will then show up as images or dots on the display. The larger the RCS of the object, the more visible it will be to the system and the larger or more clearly it will appear on the display. Commercial airliners, for example, have a larger RCS, making them easy to spot, whereas military airplanes are designed to have no or a very low RCS, making them difficult to "see" by potential combatants. Radar systems are further divided by the wavelength they use. Long wave radar systems use lower frequency signals and cast a "wider net." While they are more likely to show up an object, they are also more likely to encounter interference or show up clutter. Shorter wave radar systems (used in most civilian applications) will show less clutter than long range systems, but are more likely to miss small and stealthy objects. Longer wave radar systems are therefore used in military applications to detect possible enemy aircraft. Wind turbine towers and nacelles are a particular problem for radar systems because they are both large and moving. Many standard radar systems are calibrated to ignore stationary objects. They do this by seeking a Doppler shift, the shift in wavelength of radar waves returning from a moving object. By attuning themselves to detect Doppler shifts, they can focus on moving objects and ignore stationary ones (Cameron 2011).

With wind turbines, though, it is not so simple. As the blades spin, the radar systems may be swamped with Doppler-shifted reflections, or may pick up each individual blade, causing the farm to appear as a confusing mass on the screen, making it hard to identify aircraft or other targets, or obscuring important data from the radar screen. The result of this interference is to cause radar "black holes" in which aircraft moving of any height cannot be distinguished and so cannot fly safely. In addition, wind farms can create a "shadow" in their wake in which low flying aircraft cannot be distinguished. Wind farms, especially large ones, may also interfere with the onboard navigational aids of ships and aircraft. This is of particular concern in regard to the very large offshore wind farms. Although the interactions are complex, in general wind turbines reflect radio waves in a number of ways. The Glass-Reinforced Plastic that makes up the majority of wind turbine blades is itself reflective, as are the metallic lightning conductors which snake through the inside of the blades and towers. The towers and nacelles themselves may also have large radar cross sections, particularly when there are many turbines located together (in comparison to their length, cylinders are said to have a large RCS). Other factors such as the pitch and yaw of the turbine and the individual pitch of the blades may also affect the RCS. Blades spinning to and away from the path of the radar system at a 90° angle to the receiver will have a greater impact than those in other positions as a result of the enhanced Doppler effects (Cameron 2011).

There are a number of general methods than may be employed to try and reduce the radar visibility of wind turbines, aside from amending their physical location.

Covering the turbine blades in claddings made of radar-absorbent materials like those used on stealth aircraft may significantly reduce the imprint, but this may also affect the aerodynamic performance of the blade, or come at great expense. Along with absorbing coatings, additional solutions have included incorporating RAM into the blade itself by substituting thin layers of radar-absorbent materials for sections of glass composites. Another option involves using layers of honeycomb foam inside the blade to help dampen the reflected signal. The nacelle and tower can also be coated in RAM, or shaped to ensure that they present no flat surface to the radar waves, thereby reducing their RCS. The key, as with the blades, is to find a system that can be easily incorporated into existing manufacturing processes at a reasonable cost without significantly undermining the performance of the wind turbine system.

To begin with, the tower and the nacelle have been coated in a 5 mm thick layer of RAM. While the exact composition of such stealth materials is not usually made public, RAM may consist of ferrite paints or polymer layers incorporating crystalline graphite and it is possible that similar compounds are being employed in this case. These types of RAM contain tiny spheres of material that oscillate as the radar waves hit them, "converting" the radio waves into heat, which is then dissipated over the structure, rather than reflected. Although such materials could also be used to coat the turbine blades, it is reported that this would add an extra 1.2 tons to the blades. Given that the entire rotor of a V90 turbine has a weight of around 38 tons, this would be a 3% increase on blade weight. Instead, the Vestas blades incorporate RAM consisting of two layers of glass-reinforced epoxy and plastic foam built into their structure, which reflect and absorb waves, reducing their cross section. Since these layers simply replace existing layers in the blades, there is reported to be little or no difference to the overall weight of the blade.

While the idea of developing a "stealth" turbine is perhaps the most eye-catching of the strategies, several others are being explored. One of these examples—telemetry transmission—was recently explored as a potentially cheap and easy option. Real-time data could be fed into a model which could then predict the RCS of the wind farm and account for it in its displays, effectively removing the interference problem. While this seems like a daunting challenge, its proponents maintain it would be relatively simple, requiring four sensors on each turbine, continuously transmitting information on speed, location, pitch, and yaw to the radar. Another method involves developing so-called "holographic" or three-dimensional radar systems. This is a form of continuously tracking radar that can apparently tell the difference between turbines and aircraft by observing their behavior. Slotted in alongside standard radar they avoid major disruption to operations. Not every location will be able to upgrade its ground systems, and it seems likely that a combination of "stealthy" turbines and "turbine friendly" radar technology will be used to ensure that radar interference does not become a major stumbling block to future wind development, nor wind development an issue for air safety (Cameron 2011).

6.8 Permanent-Magnet Generators for Wind Turbine Applications

Various generator types and control techniques have been presented to maximize wind capture, reduce costs and improve reliability. The doubly fed induction generator (DFIG) with a three-stage gearbox is the most common configuration at present. Permanent magnet generators (PMGs) are capturing more attention. They are widely used in small wind turbines, and increasingly in large onshore and offshore wind turbines. A one/two stage gearbox or even a direct drive train type can be used with them. Direct drive types are used for their reduced part count, ease of maintenance, high reliability, and reduced power loss. Therefore, they are increasingly capturing the global wind market in large onshore and offshore applications (Yang et al. 2012).

Double-fed induction generators (DFIGs) have given the wind industry many years of excellent service. However, PM generators deliver a wide range of performance advantages to wind-farm developers and owner-operators that can make a significant and positive difference in project economics (Zipp 2011), such as high efficiency with the elimination of field loss, and it is small and light, to list a few. They are not only preferred in small-scale wind turbines but also in large MW applications (Yang et al. 2012).

PM generators can be classified by different concepts, such as radial flux permanent magnet (RFPM) machines, axial flux permanent magnet (AFPM) machines, and transverse flux permanent magnet (TFPM) machines. The magnets used within PM generators typically contain ~30% Nd (neodymium) by weight, with most of the remaining weight made up of low-cost ferrous materials.

There are three control variables in the wind turbine system: pitch angle, yaw angle, and generator speed. In low and medium wind speed, the aim of control is to maximize the output power from the wind turbine by adjusting the generator speed and yaw angle. The function of yaw control is to orient the wind turbine so that it perpendicularly faces the wind stream, to reduce power loss. In high wind speed, the pitch angle will be adjusted to keep the output power constant at the nominal value (Yang et al. 2012).

PM generators designed to operate at speeds from about 125–500 rpm are significantly smaller and lighter than direct-drive equivalents. By removing need for a gearbox and coupling the PM generator directly to the turbine rotor's shaft, direct-drive systems provide the simplest drivetrain, which ought to deliver the highest reliability. However, this is difficult to achieve in practice. A close coupled, medium-speed arrangement offers perhaps the best compromise: reduced drivetrain length and weight, high reliability and efficiency, and an opportunity to mix-and-match major component from a range of suppliers. Only a small fraction of the wind turbines in operation today use PM generators. But with the superior efficiencies, higher reliabilities, and lower costs arising from greater market penetration, PM generators are set to become a major part of tomorrow's wind turbine drivetrains (Zipp 2011).

INDEPTH: Advanced Components of Wind Turbines

Advanced rotors, with larger swept area and higher reach, provide greater energy capture and have already reduced the cost of wind energy. As rotors become larger with longer, more flexible blades, a fuller understanding of their behavior during operation is required to inform new designs. Noise reduction technologies are important to increase the amount of land available for wind projects. Other promising technologies can be developed to improve blade pitch control and advance blade bearing and pitch systems and hub design, materials and manufacture.

Drive-train component improvements can be realized through a comprehensive optimization of the whole turbine. Increased controls, through power electronics, can reduce loads and material intensity. Hydraulic drive-train designs, in which a hydraulic system replaces the mechanical gearbox, are also a possibility. Continued development of larger and greater turbine capacities will necessitate higher capacity power electronics and enhanced grid support capabilities from wind power plants. Lower cost power conversion is expected from deployment of higher voltage power electronics (UpWind 2011).

Support structures could benefit from advances in material research that might further reduce costs.

Materials with higher strength-to-mass ratios (e.g., carbon fiber and titanium) could enable larger area rotors, lighter generators and other drive-train components, thereby reducing tower head mass.

New materials could also provide solutions for taller towers and reduce the dependence of permanent magnet generators on rare earths. As turbines approach 10 MW, direct-drive superconducting generators may offer potential to lower mass and size, while also providing the reliability benefits of direct-drive platforms (Abrahamsen et al. 2010; Mijatovic et al. 2012).

Wind turbine and wind farm (i.e., plant-wide) controls are an important area for cost reduction in wind power. Industry is currently undergoing a transformation to optimize at the plant level, with individual turbines viewed as components that can be designed and operated for specific locations within the plant as a whole. Turbine-mounted Lidar (Light and radar) will be used to inform turbines of changes in wind speed, direction and turbulence, making it possible to optimally position turbines (and pitch blades) as changes occur in the approaching wind. Such capabilities offer the dual benefit of enhanced performance and reduced fatigue loads (UpWind 2011).

6.9 Integrated Energy Storage for Wind Power

To overcome the intermittence of wind power systems, backup systems as well as energy storage methodologies have been developed. The electrical grid system could actually on a wide-scale basis be considered as an energy storage system, since the wind blowing as some location can produce energy that can be used as another distant location where it is not blowing some time. While continuing advances in other energy storage technologies can make it more economically competitive as a provider of grid flexibility, in fact, resources like wind energy can already be cost-effectively and reliably integrated with the electric grid without additional energy storage. While energy storage is not needed to integrate wind energy with the electric grid and is often not cost-effective, having certain types of additional energy storage on the grid can modestly reduce the cost of integrating wind. These energy storage technologies include mechanical, electrochemical, electromagnetic, and thermal.

6.9.1 Pumped Hydro Energy Storage

Pumped hydro energy storage (PHES) is the largest and the most mature technology available. It consists of two reservoirs and body of water at a relatively high elevation represents potential or stored energy, as shown in Fig. 6.11 (Rehmana et al. 2015). During the "charging" process water from lower reservoir is pumped up

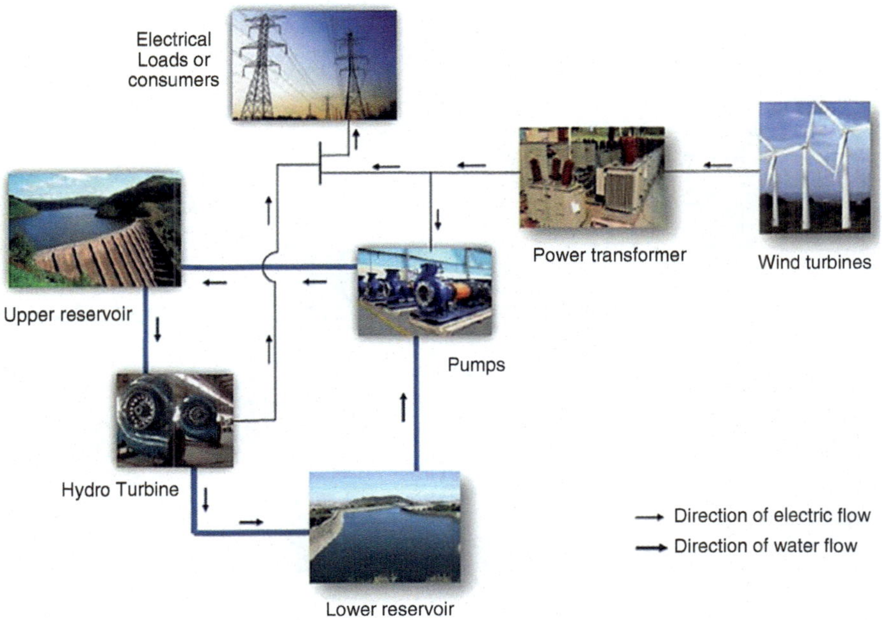

Fig. 6.11 Conceptual wind power-based pumped hydroelectric storage (PHES) system (Adapted with permission from Rehmana et al. 2015 (Elsevier))

to the upper one. In "discharging" process water from upper reservoir is released and flows through hydro turbines which are connected to generators, producing electrical energy. PHES is a good solution for wind farms. Ideal application for PHES seems to be load leveling. One of the problems for building these stations is the lack of suitable places and the impact in the nature environment. One approach which can give more deployment flexibility is Underground Pumped Hydroelectric Energy Storage (UPHES). This gives a flexibility of UPHES location what in consequence this technology can be placed in ideal locations to function with wind farms. However, the problem is very high technical immaturity of UPHES. Hydro power plants are one of the best storage solutions for suppressing fluctuations caused by wind power plant (Swierczynski et al. 2010; Stenzel and Linssen 2016).

6.9.2 Compressed Air Energy Storage

Compressed Air Energy Storage (CAES) systems are compressing air via electrical compressors in underground cavities (salt cavern, abandon mines, rock structures, etc.) and store it in a high pressure. When energy is needed compressed air is released through a turbine, but the operating units worldwide incorporate combustion prior to turbine expansion in order to increase the overall efficiency. High power and energy capacity make CAES a good storage solution for wind farms, as shown in Fig. 6.12 (Buka 2010). CAES can be used for frequent start-ups and shutdowns.

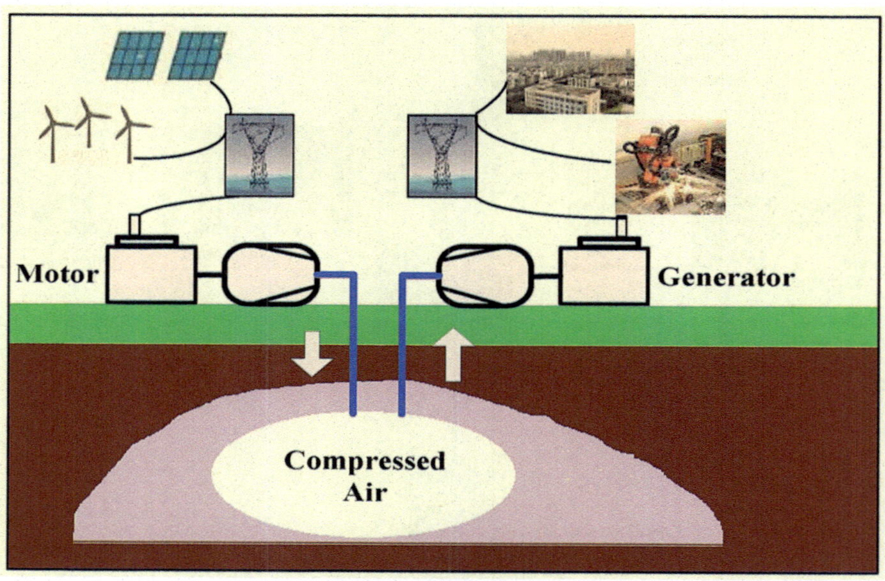

Fig. 6.12 Wind power-based compressed air energy storage system (Adapted from Wang et al. 2017, licensed by MDPI, Basel, Switzerland, under the terms and conditions of the Creative Commons Attribution (CC BY) license (http://creativecommons.org/licenses/by/4.0/))

Current research in CAES is focused on the development of systems with fabricated storage tanks. Such an approach will remove the geological dependency and compressed air will be stored in tanks with a higher pressure. System rating will be smaller (several MW) because of the tank cost. The possible lack of geological dependence might make CAES an interesting solution for integration with wind farms (Swierczynski et al. 2010).

6.9.3 Flywheel Energy Storage

Flywheels are energy storage devices which are storing energy in form of kinetic energy (rotating mass). Flywheels are made up of shaft that rotates on two magnetic bearings in order to decrease friction. Whole structure is placed in a vacuum to reduce windage losses. During "charging" process rotor is accelerated to a very high speed by a motor and energy is maintained in a system as kinetic energy. In "discharge" process flywheels are releasing energy and driving the machine which is working now as a generator. Figure 6.13 shows a typical compensation system for wind energy applications based on flywheel energy storage (Carrasco et al. 2006). Flywheels are able to suppress fast wind power fluctuations but for wind turbines in combination with battery system rather than stand alone. However energy density is low and moreover self-discharge ratio is high (Swierczynski et al. 2010).

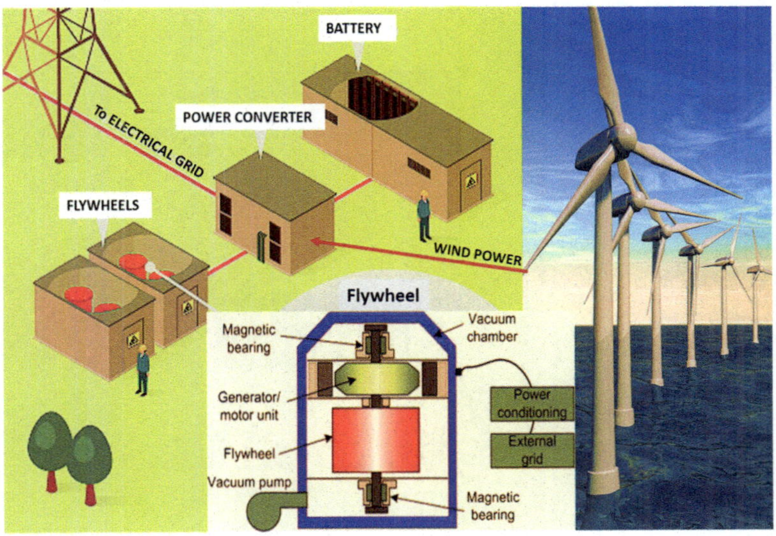

Fig. 6.13 Typical compensation system for wind energy applications based on flywheel energy storage

6.9.4 Supercapacitor Energy Storage

Supercapacitor Energy Storage (SES) are having similar response characteristics and small energy density like FES but they do not have moving parts and are having small self-discharge ratio. They are able to suppress fast wind power fluctuations but with a small time scale. They can be considered as a support for wind turbines in combination with a battery system rather than standalone (Swierczynski et al. 2010).

6.9.5 Battery Energy Storage

Electrical batteries or accumulators are regularly used in solar applications and can be charged by a coupled with wind and solar energy system, as shown in Fig. 6.14 (Ragheb 2017). Batteries that are either in use and/or potentially suitable for utility scale applications include lead-acid (L/A), nickel cadmium (NiCd), sodium sulfur (NaS), sodium nickel chloride (ZEBRA), lithium-ion (Li-ion), as well as flow batteries including Vanadium Redox (VR), Polysulfide Bromide (PSB), and Zinc Bromine (ZnBr) (Swierczynski et al. 2010).

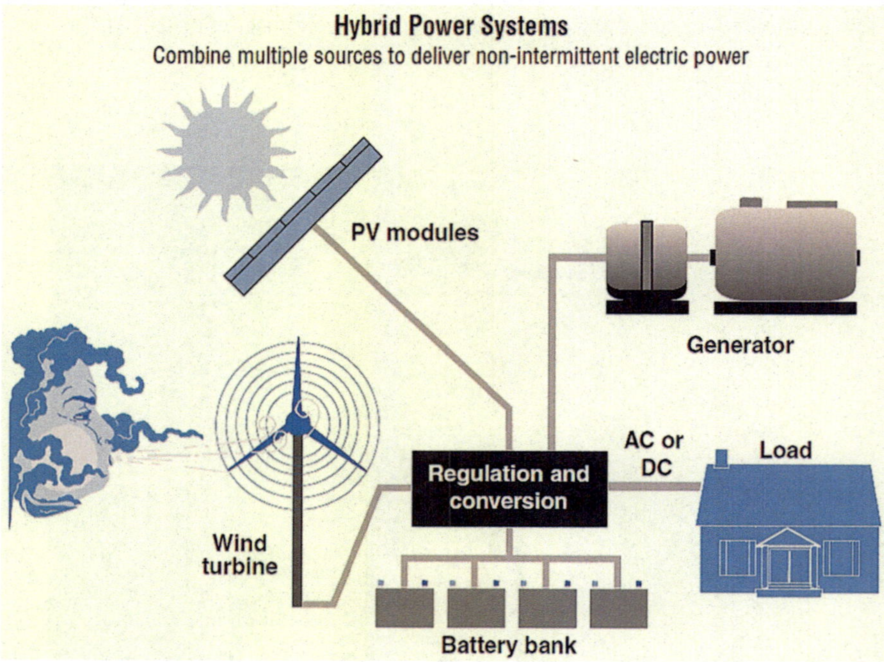

Fig. 6.14 Hybrid solar and wind energy based on battery energy storage (Adapted from WINDexchange (2018). Small Wind Guidebook, https://windexchange.energy.gov/small-wind-guidebook.pdf. Credit: US Department of Energy)

6.9.6 Hydrogen Energy Storage

Hydrogen energy storage (HES) is not a single device but the process is divided into three parts: create hydrogen; store the hydrogen; and create energy from hydrogen. Hydrogen can be created by extraction of fossil fuels; reacting steam with methane; and by electrolysis. Producing hydrogen from electrolysis is the most economical solution among the others. Production from fossil fuels is four times more expensive than using the fuel itself. And production of hydrogen from reaction of steam with methane produces pollution. During the process of electrolysis, hydrogen is produced from water and oxygen is dissipated into atmosphere with the efficiency of hydrogen production up to 85%. Storing of hydrogen can be done by compressing it, by liquefying it or by metal hydride. The most often use option is to compress hydrogen (65–75% efficiency). Hydrogen can be also stored in liquefied form by pressuring and cooling it. However keeping the hydrogen liquid is energy demanding because of the very low temperature that has to be maintained. To create energy from hydrogen two methods are used: Internal Combustion Engine (ICE) and Fuel Cell (FC). Round trip efficiency is between 30 and 50%. Fuel cell is a relatively new technology and do not have any moving parts, have no emissions, and are light and reliable. Hydrogen has a higher energy density per weight but lower per volume than a gasoline. These features give a lot of potential in a future, as shown in Fig. 6.15, also with renewable applications; however, technology needs to be more advanced (Swierczynski et al. 2010).

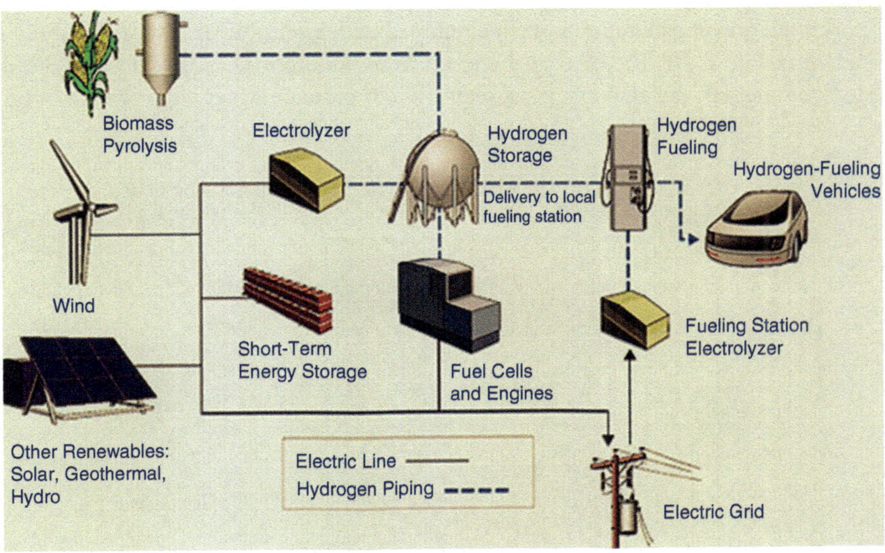

Fig. 6.15 Hydrogen energy storage with wind energy and other renewables (Image Credit: US National Renewable Energy Laboratory (NREL). Adapted from Jake Richardson (2014): Solar Power Could Be Stored As Hydrogen (Research). Available at http://solarlove.org/solar-power-stored-hydrogen-research/)

6.9.7 Thermal Energy Storage

Thermal energy storage (TES) encompasses a variety of technologies that store available heat energy using different approaches in insulated repositories. A TES system normally consists of a storage medium in a reservoir/tank, a packaged chiller or built-up refrigeration system, piping, pump(s), and controls. Based on the range of operating temperature, TES can be classified into two groups: low-temperature TES (consisting of aquiferous low-temperature TES and cryogenic energy storage) and high-temperature TES (including latent (fusion) heat TES, sensible heat TES and concrete thermal storage). Aquiferous low-temperature TES normally uses water cooled/iced and reheating processes, which is more suitable for peak shaving and industrial cooling loads. Cryogenic energy storage employs a cryogen (such as liquid nitrogen or liquid air) to achieve the electrical and thermal energy conversion. For instance, Liquid Air Energy Storage (LAES) is attracting attention due to the high expansion ratio from the liquid state to the gaseous state and the high power densities of liquid air compared to that of gaseous state of air. Latent heat TES employs Phase Change Materials (PCMs) as the storage media and uses the energy absorption or emission in liquid-solid transition of these PCMs at constant temperature. Concrete thermal storage utilizes concrete or castable ceramics to store heat energy, normally supported by synthetic oil as a heat transfer fluid. These TES technologies have different features with various applications. For instance, latent heat storage can provide a relatively high storage density with a small dimension reservoir, thus this technology can be used in buildings. In addition, cryogenic energy storage is expected to be used for future grid power management (Luo et al. 2015).

A wind power generation system combined with a sensible heat storage facility is shown in Fig. 6.16. The electrical energy from wind power is used to heat a bulk storage material; the heat energy is recovered to produce water vapor which in turn

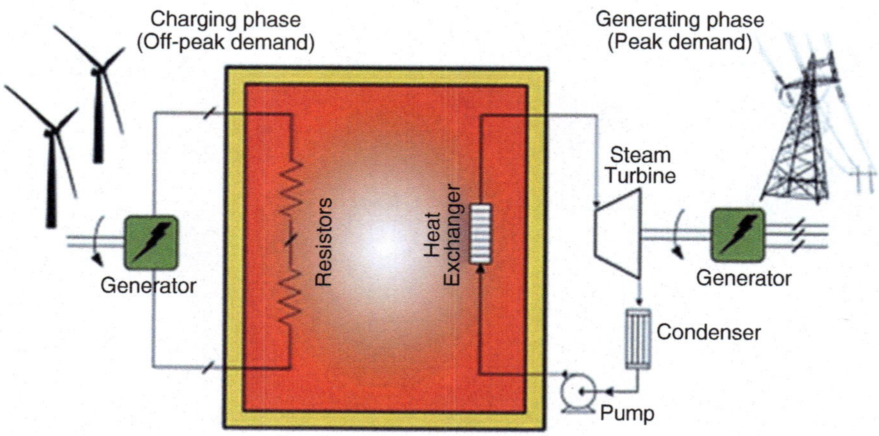

Fig. 6.16 A sensible heat storage system for wind power generation (Adapted from Luo et al. 2015 (Elsevier), available under the terms of the Creative Commons Attribution License (CC BY))

drives a turbo-alternator to generate electricity. The TES system can store large quantities of energy without any major hazards and its daily self-discharge loss is small (~0.05–1%); the reservoir offers good energy density and specific energy (80–500 W h/L, 80–250 W h/kg) and the system is economically viable with relatively low capital cost. However, the cycle efficiency of TES systems is normally low (~30–60%). TES has been used in a wide spectrum of applications, such as load shifting and electricity generation for heat engine cycles (Chen et al. 2009; Luo et al. 2015).

6.9.8 Advanced Grid Integration

Optimizing grid integration (distributed and utility) and transmission for wind systems requires integration studies and operational forecasting tool development, including development of grid management and control systems that enable high penetrations of wind with high grid reliability. These tools would ensure reliable and economic system operations under high penetration levels of wind generation. Additional tool development would support planning and development of new infrastructure to allow access to high-quality wind resources, evaluation of system response to uncertainties and electrical phenomena associated with wind power and development of operations practices for system operator use, and improvements to wind power controls to benefit grid power quality through activities such as voltage ride-through and frequency control.

INDEPTH: System Design for Wind Turbines

Moving towards specific wind turbines for diverse operating conditions requires deeper understanding of the conditions in which a wind power plant will operate over its lifetime. The aim is to develop more cost-effective turbine designs with the ability to extract more energy from the wind, over a longer lifetime and in specific operating environments. Wind turbine manufacturers planning to offer so-called "cold climate packages" will need to use special materials and components, including specialized measurement systems, heaters or pre-heaters for components and subsystems, and even nacelle heating to allow comfortable turbine maintenance. Anti- or de-icing systems for blades most often use electro-thermal heating elements. Special foundations may be needed in permafrost. System design needs tool development to minimize loads across the components to optimize for specific conditions including offshore, cold and icy climates, tropical cyclone climates and low-wind speeds. Improving model tools requires measurement campaigns both in the field and in controlled test facilities. Optimizing power-to-swept area ratios is important to achieve lowest LCOEs, especially at low-wind-speed sites. If this optimization includes connecting costs it may

(continued)

lead to different results, as the reduction in connection costs might be important, especially for offshore wind farms far from shore. Also, the reduced variability offered by the weaker turbines is likely to facilitate the handling of large shares of wind power in the electricity mix. R&D targets for up-scaling to 10–20 MW turbines will push the technology towards new solutions, which may help reduce costs for the 2–5 MW turbine size (seen as sufficient for most applications). Optimum size for both land-based and offshore applications is still to be solved. Further enlargement of land-based turbines is limited by logistics constraints as well as sound and visibility regulations. Offshore up-scaling will bring more direct benefits. 20 MW turbine is technically feasible, with need for significant advances in materials, design architectures, controls capabilities and other factors (UpWind, 20; Molly 2012).

6.10 Summary

Wind energy is a clean, renewable way of generating electricity. In the future, provided costs are reined in, the primary focus will be offshore development. The pros and cons of wind power are still the subject of debate with two main arguments against wind power: (a) It is more expensive to produce wind-sourced electricity than electricity from conventional sources, such as nuclear and thermal energy. This means that wind power needs to be subsidized, mainly in the form of preferential feed-in tariffs. (b) Wind power is intermittent because winds are unpredictable and uncontrollable. This may result in large swings in output and even shutdowns. However, grid operators and energy storage technologies are used to dealing with the problem of intermittence, which is also an issue with other sources of energy, like solar. It is estimated that a large-scale grid can integrate a wind energy penetration rate of 20% without experiencing major technical problems. Other solutions are also being developed to address the problem of interrupted power supply. One answer is to set up interconnected groups of wind turbines over extended areas in order to leverage their combined energy and ensure a guaranteed minimum amount of power. Research is also being carried out on ways to store large quantities of surplus electricity, particularly through the use of batteries.

Advanced materials play a crucial role in wind power to enable renewable wind energy capture and generation. Composite materials such as polymer-matrix reinforced with fiberglass or graphite fibers have been used to make rotor blades of wind turbines. In the turbines, compact electrical generators contain powerful magnets made from rare earth materials. The electrical generator is driven by the rotation of the turbine blades through a gearbox, which uses special alloys in order to accommodate a wide range of wind speeds. As turbine sizes continue to increase, and the growth of offshore installations, the durability of turbine materials faces more and more challenges due mainly to long-time exposure to higher stresses and hostile environments. In addition, the turbine blades must maintain adequate stiffness to prevent failure due to deflection and buckling. They also need adequate long-term fatigue life in harsh conditions, including variable winds, ice loading and lightning strikes. Therefore, development of next-generation wind turbine components and materials requires research on advanced materials and key components to improve performance and reliability; development of new architectures for larger, lightweight turbines that reduce overall mass (reducing costs) and provide access to better wind resources (larger rotors, taller towers), and improved systems performance (capacity factor); improvements in turbine cost, strength, weight, and fatigue to reduce operations and maintenance costs and reduce the failure rate for large components, such as blades, gearboxes, generators, power electronics, and collection systems; and innovations to solve transport and installation cost limitations for large-scale turbine systems and components.

New wind turbines need to reduce their overall costs for large-scale turbines with improved design and reliability of components; development and industrialization of support structures for sea installations, both fixed and floating; achieving grid integration for even greater wind energy penetration; introduction of large-scale energy storage systems and high-voltage alternative and direct current interconnections; more sophisticated assessment of wind resources; and spatial planning through social and environmental considerations. Various options are being examined to improve the technology for installing wind turbines. At present, wind turbines are anchored to the seabed in water depths not exceeding 30 m. Studies and tests are being conducted on artificial islands and wind turbines on floating foundations anchored at depths of up to 60 m. To reduce investment costs, researchers are also looking into the possibility of using existing offshore oil platforms that are nearing the end of their useful life, something that holds out considerable potential.

Besides, high-altitude wind turbines are developed to be capable of harnessing stronger and more consistent winds higher in the atmosphere. Although different models are either in the design or testing stage, there are significant feasibility and, particularly, viability issues associated with their development. Ladder mills, which are like giant kites, are composed of a series of kite planes on a long string that use wind energy at an altitude of 9000 m, where wind speed can be 20 times higher than at sea level. Others include feature helium balloons or flying wing turbines.

Exercises

Part I: General Questions

6.1 What is the main source for the formation of wind?
6.2 What are used to turn wind energy into electrical energy?
6.3 What are wind turbines made of?
6.4 The equation for calculating electrical power is Power (in watts) = Voltage × Current ($P = VI$). If the generator is developing a voltage of 690 V

and the power output is 6 MW, what is the current in amps running down the copper cables to the grid connection at the base of the tower?

6.5 How many different components does it take to build a utility-scale wind turbine? How many blades does a modern wind turbine have? Why is copper used for the coils in the generator?

6.6 Describe the development history of wind energy in the USA.

6.7 What's the wind conversion system? How to make it operate at a maximum efficiency?

6.8 How many parts of the energy wind forecasting system usually consists? How to improve wind forecasting? Compare the differences and commons of the wind forecasting and weather forecasting.

6.9 How to determine the maximum wind turbine efficiency?

6.10 Describe the commons and differences of windmill and wind turbine?

6.11 List types of modern wind turbines, and compare the characteristics of each kind of wind turbines.

6.12 List major materials used in wind turbines. What the major considerations for each material selection? Describe potential trends of the turbine materials usage.

6.13 Describe performance requirement to wind turbine blades made with fiber-reinforced composites, and the types of composites used for wind turbine blades.

6.14 List reinforcement and matrix materials used in fiber-reinforced composites. Describe their major advantages and limitations for manufacturing of wind turbine blades.

6.15 List major consolidation techniques of fiber-reinforced composites. Describe their major advantages and limitations for manufacturing of wind turbine blades.

6.16 List major curing and polymerization processes of fiber-reinforced composites. Describe their major advantages and limitations during manufacturing of wind turbine blades.

6.17 Discuss the future development trends of composite wind blades.

6.18 Describe major characteristics of smart and stealth wind turbine blades. What are their main applications?

6.19 List generator types and control techniques. Why using permanent-magnet generators for wind turbines?

Part II: Thought-Provoking Questions

6.20 List major advantages and challenges of wind energy.

6.21 Based on your sourcing on related literatures, find out the mathematical relationship between wind speed and power output; and the mathematical relationship between blade length and power output.

(a) If the wind speed is 11.5 m/s and the speed after the turbine is 8 m/s, what is the power extraction coefficient of this wind turbine? (air density 1.225 kg/m^3).
(b) The rated output power for a turbine model at 15 m/s is 3 MW. The rotor diameter is 90 m. The rotor rotates at a constant frequency of 0.198 Hz. Please calculate the tip to speed ratio and power conversion coefficient of this model.

6.22 The tidal flow turbine uses the energy of flowing water. Find out the density of water and compare it to the density of air. How many times more kinetic energy is in flowing water compared to wind blowing at the same speed?

6.23 Research the electrical power output of modern large wind turbines. What is the largest one you can find? Does there seem to be an upper limit?

6.24 List major energy storage techniques developed for wind power. Describe their main advantages and limitations when integrating with wind power.

6.25 Discuss the role of materials played in wind turbine design.

References

Abrahamsen, A., et al.: Superconducting wind turbine generators. Supercond. Sci. Technol. **23**(3), 034019 (2010)

Advani, S.G., Sozer, E.M.: Process Modeling in Composites Manufacturing. Marcel Dekker, New York (2003)

Alan, B.: Concrete for the next generation of wind farms. Concrete (1/2), 24–25 (2006)

AMO: Global Wind Day 2016—AMO's role in applying 3D printing to wind blade mold manufacturing. https://energy.gov/eere/amo/articles/global-wind-day-2016-amo-s-role-applying-3d-printing-wind-blade-mold-manufacturing (2016). Accessed 27 July 2017

Ancona, D., McVeigh, J.: Wind Turbine—Materials And Manufacturing Fact Sheet, August. Washington, DC Princeton Energy Resources International. http://www.perihq.com/documents/WindTurbine-MaterialsandManufacturing_FactSheet.pdf (2001). Accessed 30 Jan 2017

Baker, D.A., Rials, T.G.: Recent advances in low-cost carbon fiber manufacture from lignin. J. Appl. Polym. Sci. **130**, 713 (2013)

Barnard, M.: Vertical axis wind turbines: Great in 1890, also-rans in 2014. http://cleantechnica.com/ (2014). Accessed 3 Feb 2017

Böhme, C.: How rotor blades defy the forces of nature. https://www.basf.com/en/company/news-and-media/science-around-us/how-rotor-blades-defy-the-forces-of-nature.html (2017). Accessed 17 July 2017

Buka, S.: Adiabatic pressed air energy storage. Primus Green Energy: Formula for energy. http://energy.savestatesociety.net/adiabatic-compressed-air-energy-storage/ (2010). Accessed 26 July 2010

Burton, T., Jenkins, N., Sharpe, D., Bossanyi, E.: Wind Energy Handbook. Wiley, New York (2001). isbn:978-0471489979

Cameron, A.: "Stealth blades" get turbines under the radar. http://www.renewableenergyworld.com/articles/print/special-supplement-wind-technology/volume-1/issue-5/wind-power/stealth-blades-get-turbines-under-the-radar.html (2011). Accessed 21 Feb 2017

Carrasco, J.M., Bialasiewicz, J.T., Guisado, R.C.P., ILeón, J.: Power-electronic systems for the grid integration of renewable energy sources: A survey. IEEE Trans. Ind. Electron. **53**(4), 1002–1016 (2006)

CDA: Wind energy basics. Copper Development Association Inc. Acopper Alliance Member. https://www.copper.org/environment/green/casestudies/wind_energy/wind_energy.html (2017). Accessed 20 July 2017

Chen, H., Cong, T.N., Yang, W., Tan, C., Li, Y., Ding, Y.: Progress in electrical energy storage system: A critical review. Prog. Nat. Sci. **19**, 291–312 (2009)

Cherubini, A., Andrea Papini, A., Vertechy, R., Fontana, M.: Airborne wind energy systems: A review of the technologies. Renew. Sust. Energ. Rev. **51**, 1461–1476 (2015)

Conti-Ramsden, J., Dyer, K.: Materials innovations for more efficient wind turbines. http://www.renewableenergyfocus.com/view/42937/materials-innovations-for-more-efficient-wind-turbines/ (2015). Accessed 27 July 2017

Cresko, J.W., Roberts, P.L.: Method of induction curing conductive carbon fiber composites with radio frequency energy. 3rd World Congress on Microwave and Radio frequency applications. Sydney, 2002. https://inis.iaea.org/search/search.aspx?orig_q=RN:35028342 (2002). Accessed 25 July 2017

Dabiria, J.O., et al.: A new approach to wind energy: Opportunities and challenges. Physics of Sustainable Energy III (PSE III). AIP Conf. Proc. **1652**, 51–57 (2015). https://doi.org/10.1063/1.4916168

Das, S.: The cost of the automotive polymer composites: A review and assessment of the DOE's lightweight materials composite research. ORNL/TM-2000/383. Oak Ridge National Laboratory, Tennessee, January, 2001

Das, S., Warren, J.: Cost modeling of alternative carbon Fiber manufacturing technologies—baseline model demonstration. Presented to DOE, Washington, DC, 5 Apr 2012

Dodge, D.M.: The Illustrated History of Wind Power Development. U.S. Federal Wind Energy Program. Littleton, CO (2006). http://www.telosnet.com/wind/. Accessed 16 July 2017.

DOE CM: Quadrennial Technology Review 2015—Chapter 6: Innovating Clean Energy Technologies in Advanced Manufacturing—Composite materials. https://energy.gov/sites/prod/files/2015/12/f27/QTR2015-6E-Composite-Materials.pdf (2015). Accessed 23 July 2017

Dubois: Method for the synthesis of acrylonitrile from glycerol. US Patent Application, Pub. No. US2010/0048850 A1, 25 Feb 2010

Eker, A.A., Eker, B.: General assessment of fiber—reinforced composites selection in wind turbine blades. In: Attaf, B. (ed.) Recent Advances in Composite Materials for Wind Turbines Blades. WAP-AMSA, Hong Kong (2013). isbn:978-0-9889190-0-6 http://www.academicpub.org/amsa/chapterInfo.aspx. Accessed 12 Feb 2017

Elliot, D.: Flights of Fancy: Airborne Wind Turbines. Institute of Physics, Environmental Research Web. http://blog.environmentalresearchweb.org/2014/04/12/flights-of-fancy-airborne-wind-turbines/ (2014). Accessed 18 July 2017

Engerati: Giant turbines to harness large wind potential. https://www.engerati.com/article/giant-turbines-harness-large-wind-potential (2017). Accessed 18 July 2017

Frank, M., Zhu, S., Peters, F.: Automated composite fabric layup for wind turbine blades. CAMX 2014 Conference Proceedings, Orlando, FL, 2–5 June 2014

Fraunhofer-Gesellschaft: Lightweight rotor blades made from plastic foams for offshore wind turbines. Science Daily, 21 October 2016. https://www.sciencedaily.com/releases/2016/10/161021084515.htm. Accessed 27 July 2017

Gill, E.: Danish offshore strategy moves closer to shore. Wind power monthly. http://www.windpowermonthly.com/article/1161698/danish-offshore-strategy-moves-closer-shore (2012). Accessed 6 Feb 2017

Griffith, D.T., Johanns, W.: Large blade manufacturing cost studies using the Sandia blade manufacturing cost tool and Sandia100-meter blades. Sandia National Laboratories Technical Report, Apr 2013, SAND2013–2734. http://energy.sandia.gov/wp-content/gallery/uploads/dlm_uploads/SAND_SNLLargeBladeManufacturingCostTrendsAnalysis_SAND2013–2734.pdf (2013). Accessed 20 July 2017

Harrison, R.: Understanding our Environment, p. 11. Royal Society of Chemistry, Cambridge (1999)

Haupt, S.E.: A wind power forecasting system to optimize power integration, COST ES1002 weather intelligence for renewable energies state-of-the-art workshop, Nice, France, 22–23 March 2011

Heier, S.: Grid Integration of Wind Energy Conversion Systems, p. 45. Wiley, Chichester (2005)

Herbert, G.M., Iniyan, S., Sreevalsan, E., Rajapandian, S.: A review of wind energy technologies. Renew. Sust. Energy Rev. **11**, 1117–1145 (2007)

Jha, A.R.: Wind Turbine Technology. CRC, Boca Raton (2010)

Kaldellis, J.K., Zafirakis, D.: The wind energy (r)evolution: A short review of a long history. Renew. Energy. **36**(2011), 1887–1901 (2011)

Kaltschmitt, M., Streicher, W., Wiese, A.: Renewable Energy: Technology, Economics, and Environment, p. 55. Springer, Berlin (2007)

King, D.: DOE supports hydrogen cars with $7 million for longer driving ranges. http://www.autoblog.com/2014/05/29/doe-supports-hydrogen-cars-7-million-longer-driving-range/ (2014). Accessed July 2017

Luo, X., Wang, J., Dooner, M., Clarke, J.: Overview of current development in electrical energy storage technologies and the application potential in power system operation. Appl. Energy. **137**, 511–536 (2015)

Manwell, J.F., McGowan, J.G., Rogers, A.L.: Wind Energy Explained—Theory, Design and Application, pp. 247–319. Wiley, New York (2002)

Masuelli, M.A.: Introduction of Fiber-Reinforced Polymers—Polymers and Composites: Concepts, Properties and Processes. InTech, New York (2013)

Meyer, N.I.: Danish wind power development. Energy Sustain. Dev. **2**, 18–25 (1995)

Mijatovic, N., Jensen, B.B., Abrahamsen, A.B., Træholt, C.: Superconducting Wind Turbine Generators. Technical University of Denmark, Kgs. Lyngby (2012)

Mishnaevsky, L. Jr.: Composite materials in wind energy technology. http://www.eolss.net/sample-chapters/c08/e3-11-42.pdf (2011). Accessed 1 Feb 2017

Molina, M.G., Alvarez, J.M.G.: Technical and regulatory exigencies for grid connection of wind generation. In: Suvire, G.O. (ed.) Wind Farm—Technical Regulations, Potential Estimation and Siting Assessment. INTECH (2011). isbn:978-953-307-483-2

Molly, J.P.: Design of wind turbines and storage: A question of system optimization. DEWI, 40, Feb (2012)

Monteiro, C., Bessa, R., Miranda, V., Botterud, A., Wang, J., Conzelmann, G.: Wind Power Forecasting: State-of-The-Art 2009. Tech. Rep. ANL/DIS-10-1, vol. 2009. Argonne National Laboratory (ANL), Argonne

NREL: Wind data. https://www.nrel.gov/gis/data-wind.html. The National Renewable Energy Laboratory, U.S. Department of Energy (2014)

Patrick, J.F., Hart, K.R., Krull, B.P., Diesendruck, C.E., Moore, J.S., White, S.R., Sottos, N.R.: Continuous self-healing life cycle in vascularized structural composites. Adv. Mater. **26**, 4302–4308 (2014). https://doi.org/10.1002/adma.201400248

PEIS: Upper great plains wind energy. http://plainswindeis.anl.gov/. The online center for public involvement in the Upper Great Plains Wind Energy Programmatic Environmental Impact Statement (PEIS) (2014)

Plee: Method of manufacturing carbon fibres, US Patent Application, Pub. No. US2010/0047153 A1, Pub. Date Feb. 25 2010

Righter, R. W.: Wind Energy in America: A History. Universityof Oklahoma Press, Norman (1996)

Ragheb, M.: Energy storage with wind power. http://mragheb.com/NPRE%20475%20Wind%20Power%20Systems/Energy%20Storage%20with%20Wind%20Power.pdf (2017). Accessed 25 July 2017

Rehmana, S., Al-Hadhramia, L.M., Alam, M.M.: Pumped hydro energy storage system: A technological review. Renew. Sust. Energ. Rev. **44**, 586–598 (2015)

REUK: Savonius wind turbines. http://www.reuk.co.uk/Savonius-Wind-Turbines.htm (2014). Accessed 16 Feb 2017

Singh, A.N.: Concrete construction for wind energy towers. Indian Concrete J. 43–49 (2007)

Stenzel, P., Linssen, J.: Concept and potential of pumped hydro storage in federal waterways. Appl. Energy. **162**, 486–493 (2016)

Strong, A.B.: Fundamentals of Composites Manufacturing: Materials, Methods and Applications. Society of Manufacturing Engineers, Dearborn (2008)
Swan, D., Pierre, G.: Novel reactively polymerized liquid thermoplastic resins process like thermosets but offer post-mold thermoformability, weldability and recyclability. Presented at the 2014 SPE ACCE Conference, September 10, 2014
Swierczynski, M., Teodorescu, R., Rasmussen, C.N., Rodriguez, P., Vikelgaard, H.: Overview of the energy storage Systems for Wind Power Integration Enhancement. Industrial Electronics (ISIE), 2010 I.E. International Symposium on. http://citeseerx.ist.psu.edu/viewdoc/download?doi=10.1.1.455.9232&rep=rep1&type=pdf (2010). Accessed 25 July 2017
UpWind: Design limits and solutions for very large turbines, Sixth Framework Programme for Research and Development of the European Commission (FP6), Brussels (2011)
Warudkar, V., Ahmed, S.: Wind resource assessment: A review. Int. J. Min. Metall. Mech. Eng. **1** (3), 204–207 (2013)
White, J.: Smart turbine blades to improve wind power. http://www.purdue.edu/uns/x/2009a/090501AdamsWind.html (2009). Accessed 1 March 2017
Wood, K.: Wind turbine blades: Glass vs. carbon fiber. Composites Technology, May 31 2012. http://www.compositesworld.com/articles/wind-turbine-blades-glass-vs-carbon-fiber. Accessed 19 July 2017
WSA: Steel solutions in the green economy: Wind turbines. World Steel Association, Brussels (2012). https://www.worldsteel.org/en/dam/jcr:41f65ea2-7447-4489-8ba7-9c87e3975aab/Steel+solutions+in+the+green+economy%253A+Wind+turbines. Accessed 19 July 2017
Yang, X., Patterson, D., Hudgins, J.: Permanent magnet generator design and control for large wind turbines. IEEE Power Electronics and Machines in Wind Applications (PEMWA), Denver, Coronado, 6–18 July 2012. doi:https://doi.org/10.1109/PEMWA.2012.6316367
Zipp, K.: Driving down the cost of wind power with permanent magnet generators. http://www.windpowerengineering.com/design/electrical/generators/driving-down-the-cost-of-wind-power-with-permanent-magnet-generators/ (2011). Accessed 12 March 2017

Advanced Materials and Devices for Hydropower and Ocean Energy

7

Abstract

Water's natural flowing movements, such as in rivers and reservoirs, can be used in the production of electricity. Furthermore, both the tidal range (the periodic rise and fall of the sea level) and the energy contained in flow and waves can be used in the ocean energy system. Both types of energy conversion are classed as renewable energies. While the typical use of hydropower has been widespread for hundreds of years, using the ocean for energy is in its infancy. Large hydropower turbine-generator technologies are highly optimized, robust, and cost-effective designs, with peak energy conversion efficiencies of more than 93%. However, advancements for small-scale turbine-generators must reduce technology cost and enable more compact support structures and smaller physical and environmental footprints to achieve economic feasibility. The environmental performance of turbine designs continues to improve, in the form of blade shape enhancements to reduce injury to fish and aeration into turbine flow passages to improve the water quality of releases. Therefore, research and development have been focused on advanced materials and manufacturing for powertrain components, innovative hydrodynamic and mechanical concepts to reduce integrated turbine-generator size (diameter and length) and increase speed, embedded condition monitoring sensors, and powertrain design innovations that afford flexibility in selection of design objectives such as initial cost minimization, efficiency over a range of head and flow rates, and durability or ease of replacement. Ocean energy is one of the most promising resources that can be broadly split into tides, waves, tidal or marine currents, temperature gradients, and Salinity gradients. It has potential of the same order as that of the present capacity of electricity generation worldwide. The majority of ocean energy converters are fabricated from metals like steel and composite materials. Steel offers good fatigue and stress limits, while composites possess some cost and weight saving advantages over steel, but the fatigue and stress limits are not yet well understood in comparison to steel. Other wave devices are being designed to use rubber or other flexible materials as the main structural component.

C. Tong, *Introduction to Materials for Advanced Energy Systems*,
https://doi.org/10.1007/978-3-319-98002-7_7

Composites provide many advantages for manufacturing underwater structures such as tidal turbine blades, and wave devices, which generally offer strength, fatigue-resistance, corrosion resistance, buoyancy, and cost-effectiveness. New materials are also explored to meet the needs of a wide variety of designs, many engineering and materials options, and the unpredictable environment of subsea and new ocean energy technologies. Next-generation component would drive the costs down for multiple energy conversion system solutions, including advanced controls to tune devices to extract the maximum energy from each sea state, compact high-torque, low-speed generator technologies, and corrosion- and biofouling-resistant materials and coatings. This chapter will give a brief review about state of the art of advanced materials and devices including various components for hydropower and ocean energy.

7.1 Hydropower Technology Basics

Hydropower, or hydroelectric power, is a mature and fairly simple technology (IEA-ETSAP and IRENA 2015): the potential energy of a water source (characterized by its head and mass flow rate) is converted into kinetic energy that spins a turbine driving an electricity generator, as shown in Fig. 7.1. The kinetic energy of falling water was used for grinding wheat more than 2000 years ago. Since late nineteenth century, hydropower has been used to generate electricity. At present, about 160 (of the world's some 200) countries worldwide use hydropower technology for power generation. With a total installed capacity of 1060 GWe (19.4% of the

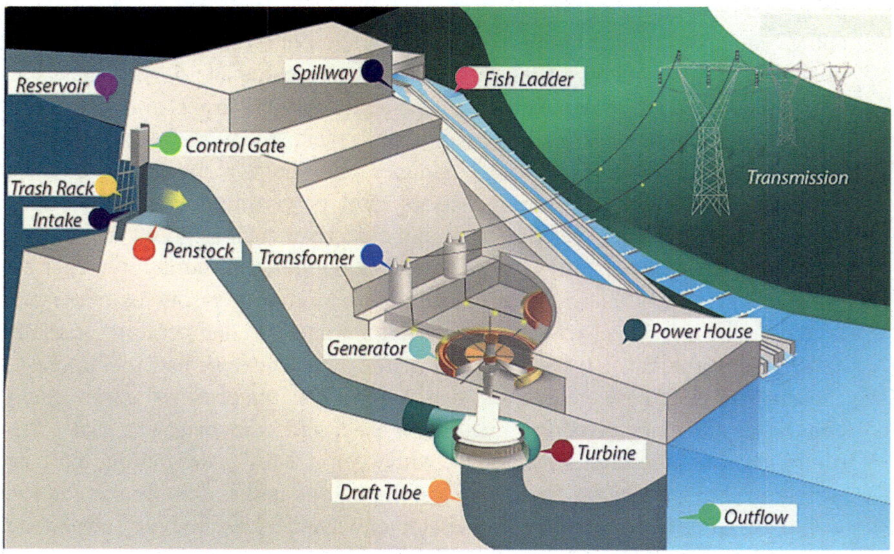

Fig. 7.1 Schematic of a typical hydropower plant (Adapted from DOE 2013. Credit: US Department of Energy)

world's electric capacity in 2011), hydropower generates approximately 3500 TWh per year, equivalent to 15.8% of global electricity generation. Hydropower plants provide at least 50% of the total electricity supply in more than 35 countries. They also provide other key services, such as flood control, irrigation, and potable water reservoirs. Hydropower is an extremely flexible electricity generation technology. Hydro reservoirs provide built-in energy storage that enables a quick response to electricity demand fluctuations across the grid, optimization of electricity production, and compensation for loss of power from other sources. In particular, pumped hydropower plants have become the most competitive options for large-scale energy storage to be used in combination with variable renewables, such as solar, geothermal, and wind power.

INDEPTH: Advantages and Disadvantages of Hydropower

Advantages

- Hydropower is fueled by water, so it's a clean fuel source, meaning it won't pollute the air like power plants that burn fossil fuels, such as coal or natural gas.
- Hydroelectric power is a domestic source of energy, allowing each state to produce their own energy without being reliant on international fuel sources.
- The energy generated through hydropower relies on the water cycle, which is driven by the sun, making it a renewable power source, making it a more reliable and affordable source than fossil fuels that are rapidly being depleted.
- Impoundment hydropower creates reservoirs that offer a variety of recreational opportunities, notably fishing, swimming, and boating. Most water power installations are required to provide some public access to the reservoir to allow the public to take advantage of these opportunities.
- Some hydropower facilities can quickly go from zero power to maximum output. Because hydropower plants can generate power to the grid immediately, they provide essential back-up power during major electricity outages or disruptions.
- In addition to a sustainable fuel source, hydropower efforts produce a number of benefits, such as flood control, irrigation, and water supply.

Disadvantages

- Fish populations can be impacted if fish cannot migrate upstream past impoundment dams to spawning grounds or if they cannot migrate downstream to the ocean. Upstream fish passage can be aided using fish ladders or elevators, or by trapping and hauling the fish upstream by truck. Downstream fish passage is aided by diverting fish from turbine intakes using

(continued)

screens or racks or even underwater lights and sounds, and by maintaining a minimum spill flow past the turbine.

- Hydropower can impact water quality and flow. Hydropower plants can cause low dissolved oxygen levels in the water, a problem that is harmful to riparian (riverbank) habitats and is addressed using various aeration techniques, which oxygenate the water.
- Maintaining minimum flows of water downstream of a hydropower installation is also critical for the survival of riparian habitats.
- Hydropower plants can be impacted by drought. When water is not available, the hydropower plants can't produce electricity.
- New hydropower facilities impact the local environment and may compete with other uses for the land. Those alternative uses may be more highly valued than electricity generation. Humans, flora, and fauna may lose their natural habitat. Local cultures and historical sites may be impinged upon. Some older hydropower facilities may have historic value, so renovations of these facilities must also be sensitive to such preservation concerns and to impacts on plant and animal life.

7.1.1 Types of Hydropower Plants

Hydropower plants consist of two basic configurations: the first based on dams with reservoirs and the second, run-of-the-river scheme (with no reservoir). The dam scheme can be subdivided into small dams with night-and-day regulation, large dams with seasonal storage and pumped storage reversible plants for both pumping and electricity generation that are used for energy storage and night and-day regulation, according to electricity demand. Small-scale hydropower is normally designed to run in-river, an environmentally friendly option since it does not significantly interfere with the river's flow. Small-scale hydropower is often used for distributed generation applications as an alternative to, or in combination with, diesel generators or other small-scale power plants for rural applications (IEA-ETSAP and IRENA 2015).

A generic hydropower plant scheme based on a dam and reservoirs is shown in Fig. 7.1 (DOE 2013): water released from the reservoir at the inlet dam on higher side that has a potential energy flows through a pipe or tunnel to a turbine. The water's potential energy is converted to kinetic energy and spins the blade of a turbine, which activates a generator to produce electricity. One of the advantages of hydropower is its operational flexibility. In other words, the capacity factor of hydropower varies depending on the specific plant and its services (i.e., baseload, peakload) between 23 and 95% (IEA-ETSAP and IRENA 2015).

Different types of small-scale hydropower plants are usually depicted by head height, discharge (flow-rate), and capacity. A large flow-rate and small head characterizes large run-of-the-river plants equipped with Kaplan turbines, a propeller type water turbine with adjustable blades. By contrast, low discharge and high-head features are typical of mountain-based dam installations driven by Pelton turbines, in which water passes through nozzles and strikes spoon-shaped buckets arranged on the periphery of a wheel. Intermediate flow-rates and head heights are usually equipped with Francis turbines, in which the water comes to the turbine under immense pressure and the energy is extracted from the water by the turbine blades (IEA-ETSAP and IRENA 2015).

Pumped storage plants consist of two or more natural or artificial (dams) reservoirs at different elevations. When electricity generation exceeds grid demand, the energy is stored by pumping water from the lower to the higher reservoir. During peak electricity demand periods, water flows back to the lower reservoir through the turbine, thus generating electricity. In these kinds of plants, reversible Francis devices are used both for pumping water and for generating electricity. The energy conversion efficiency of pumped hydropower is over 80%. Pumped storage plants can be combined with intermittent renewable electricity sources. They can also serve as an optimal complement to nuclear-based electricity designed for base-load operation, but with only limited capability to adapt to daily and seasonal load fluctuations (IEA-ETSAP and IRENA 2015).

Based on the facility size, the hydropower plants generally range from large power plants that supply many consumers with electricity to small and micro plants that individuals operate for their own energy needs or to sell power to utilities. Large hydropower is usually defined as facilities that have a capacity of more than 30 megawatts (MW). Small hydropower is defined as projects that generate 10 MW or less of power. A micro hydropower plant has a capacity of up to 100 kW. A small or micro-hydroelectric power system can produce enough electricity for a home, farm, ranch, or village (DOE 2013).

7.1.2 Types of Hydropower Turbines

There are two main types of hydro turbines: impulse and reaction. The type of hydropower turbine selected for a project is based on the height of standing water—referred to as "head"—and the flow, or volume of water, at the site. Other deciding factors include how deep the turbine must be set, efficiency, and cost. Figure 7.2 shows typical types of hydropower turbines (DOE 2013).

The impulse turbine generally uses the velocity of the water to move the runner and discharges to atmospheric pressure. The water stream hits each bucket on the runner. There is no suction on the down side of the turbine, and the water flows out the bottom of the turbine housing after hitting the runner. An impulse turbine is generally suitable for high-head, low-flow applications. The impulse turbines are generally divided into two types:

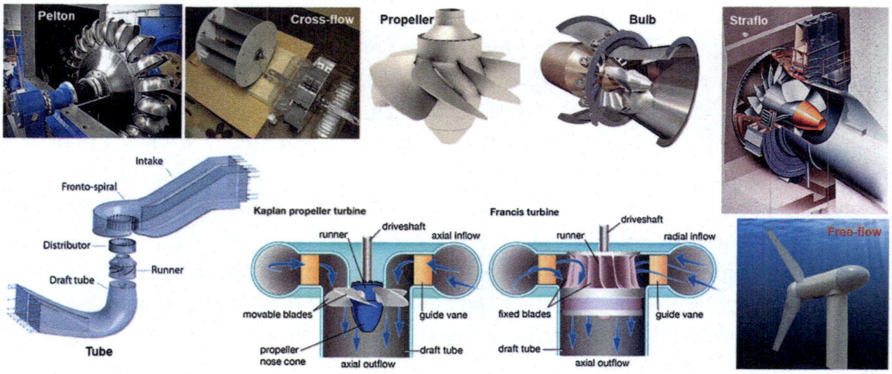

Fig. 7.2 Typical hydropower turbines

(a) Pelton turbine
A pelton wheel has one or more free jets discharging water into an aerated space and impinging on the buckets of a runner. Draft tubes are not required for impulse turbine since the runner must be located above the maximum tail water to permit operation at atmospheric pressure.

(b) Cross-flow turbine
A cross-flow turbine is drum-shaped and uses an elongated, rectangular-section nozzle directed against curved vanes on a cylindrically shaped runner. It resembles a "squirrel cage" blower. The cross-flow turbine allows the water to flow through the blades twice. The first pass is when the water flows from the outside of the blades to the inside; the second pass is from the inside back out. A guide vane at the entrance to the turbine directs the flow to a limited portion of the runner. The cross-flow was developed to accommodate larger water flows and lower heads than the Pelton.

A reaction turbine develops power from the combined action of pressure and moving water. The runner is placed directly in the water stream flowing over the blades rather than striking each individually. Reaction turbines are generally used for sites with lower head and higher flows than compared with the impulse turbines. The reaction turbines include propeller turbine, Francis turbine, and Kinetic energy turbine.

A propeller turbine generally has a runner with three to six blades in a pipe in which the water contacts all of the blades constantly. Through the pipe, the pressure is constant; if it isn't, the runner would be out of balance. The pitch of the blades may be fixed or adjustable. The major components besides the runner are a scroll case, wicket gates, and a draft tube. There are several different types of propeller turbines: (a) Bulb turbine—The turbine and generator are a sealed unit placed directly in the water stream. (b) Straflo turbine—The generator is attached directly to the perimeter of the turbine. (c) Tube turbine—The penstock bends just before or after the runner, allowing a straight line connection to the generator. (d) Kaplan turbine—Both the blades and the wicket gates are adjustable, allowing for a wider range of operation.

A Francis turbine has a runner with fixed buckets (vanes), usually nine or more. Water is introduced just above the runner and all around it and then falls through, causing it to spin. Besides the runner, the other major components are the scroll case, wicket gates, and draft tube.

Kinetic energy turbines, also called free-flow turbines, generate electricity from the kinetic energy present in flowing water rather than the potential energy from the head. The systems may operate in rivers, man-made channels, tidal waters, or ocean currents. Kinetic systems utilize the water stream's natural pathway. They do not require the diversion of water through manmade channels, riverbeds, or pipes, although they might have applications in such conduits. Kinetic systems do not require large civil works; however, they can use existing structures such as bridges, tailraces, and channels (DOE 2013).

7.1.3 Potential and Barriers

The global technical hydropower potential is around 15,000 TWh per year. Half of this potential is located in Asia and about 20% of it is located in Latin America. A large number of sites remain in the world with untapped potential. The most untapped region is Africa, where 92% of the total potential has not yet been developed. In most developed regions (e.g., Europe), a significant fraction of the economically viable hydropower potential is already being exploited, although about 50% of the technical potential is still untapped. Around 1067 GWe of hydropower capacity are currently in operation, including about 100 GWe of pumped hydropower plants. The global technical potential of small hydropower is estimated 150–200 GWe. Only about 20% of this potential has been exploited to date (IEA-ETSAP and IRENA 2015).

Large hydropower projects can be controversial because they may adversely affect water availability over large geographic regions; inundate valuable ecosystems; force the relocation of population groups against their will, and/or require a large electricity transmission infrastructure. New, less-intrusive, low-head turbines are now being developed for smaller reservoirs. Hydropower usually depends on rainfall in the upstream catchment area and reserve capacity may be needed to compensate for periods of low precipitation. This may increase the investment cost. Major hydropower issues include (IEA-ETSAP and IRENA 2015): public acceptance, high initial investment costs and long payback periods, long approval and construction cycles, and long lead times to obtain or renew concession rights and grid connections. Other barriers, such as stringent environmental standards for water management, can also hamper hydropower development. Coherent policies and simplified administrative procedures are needed.

Significant advances in hydropower technology promise further positive developments. While increasing energy supply from hydropower does not require technological breakthroughs, significant R&D, capital investment, and governmental support are essential to improve technology and public acceptance. Large hydropower needs low-head technologies, including in-stream flow, advanced equipment, and materials. R&D and technical advances are also required for small hydropower, notably equipment design, materials, and control systems. One priority is the development of less expensive technologies for small-capacity and low-head applications to enable the exploitation of more modest resources. Key data for hydropower can also be found in Table 7.1 (IEA-ETSAP and IRENA 2015). The development and construction of hydropower plants requires a long lead time, especially for the dam-with-reservoirs configuration. The investment costs for new

Table 7.1 Key data and figures for hydropower technology (IEA-ETSAP and IRENA 2015)

Technical performance	Typical current international values and ranges		
Energy input	Hydro power		
Output	Electricity		
Technologies	Very small hydro power (VSHP, up to 1 MW_e)	Small hydro power (SHP, 1–10 MW_e)	Large hydro power (LHP, >10 MW_e)
Efficiency (turbine, C_p max), %	Up to 92	Up to 92	Up to 92
Construction time, months	6–10	10–18	18–96
Technical lifetime, year	Up to 100		
Load (capacity) factor, %	40–60 (50)	34–56 (45)	34–56 (45)
Max. (plant) availability, %	98	98	98
Typical (capacity) size, MW_e	0.5	5	50
(Existing) capacity, GW_e	75		925
Environmental impact			
CO_2 and other GHG emissions, kg/MWh	Negligible		Under investigation[a]
Costs (USD 2010)			
Investment cost, USD/kW	3400–10,000 or more	1000–4000	1050–7650
O&M cost USD/kW/year	45–250 or more	40–50	45 (average)
Economic lifetime, year	30		
Interest rate, %	10		
Production cost, USD/MWh	270 or more	20–100	20–190

[a]United Nations Educational, Scientific and Cultural Organization (UNESCO), the International Hydropower Association and the International Energy Agency are jointly investigating under the IEA Implementing Agreement for Hydropower Technologies and Programmes

hydropower plants, including site preparation and civil engineering work, depend significantly upon the specific site. Investment costs include planning and feasibility assessments, environmental impact analyses, and licensing. Considering annual operation and maintenance costs, ranging 1–4% of the investment costs, the typical levelized cost of electricity (LCOE) ranges from USD 20 to 190/MWh for large hydropower and USD 20–270/MWh for small-scale hydropower. Hydropower production depends upon rainfall in the upstream catchment area. Reserve capacity may be needed to compensate for periods of low rainfall and this may increase the investment cost.

7.2 Retaining Structure Materials for Dam and Dike

One of the purposes of retaining structures is to create large bodies of water, or reservoirs, which have a variety of functions, including land irrigation, power generation, water supply and flood control. The retaining structures used to build reservoirs are called dams and dikes. A dam is built on the riverbed; it serves to hold back water and raise the water level of the resulting reservoir. Dikes are often built to increase a dam's effectiveness by preventing water from leaving the reservoir through secondary valleys.

Hydropower is often used in conjunction with dams to generate electricity. A dam can also be used to collect water or for storage of water which can be evenly distributed between locations. Dams generally serve the primary purpose of retaining water, while dikes including floodgates or levees are used to manage or prevent water flow into specific land regions. One of the best places for building a dam is a narrow part of a deep river valley; the valley sides then can act as natural walls. The primary function of the dam's structure is to fill the gap in the natural reservoir line left by the stream channel. The sites are usually those where the gap becomes a minimum for the required storage capacity. The most economical arrangement is often a composite structure such as a masonry dam flanked by earth embankments. The current use of the land to be flooded should be dispensable. Significant other engineering and engineering geology considerations when building a dam include (James and Chanson 2002): Permeability of the surrounding rock or soil; earthquake faults; landslides and slope stability; water table; peak flood flows; reservoir silting; environmental impacts on river fisheries, forests and wildlife; impacts on human habitations; compensation for land being flooded as well as population resettlement; and removal of toxic materials and buildings from the proposed reservoir area.

7.2.1 Dam Types and Materials

Dams can be formed by human agency, natural causes, or even by the intervention of wildlife such as beavers. Man-made dams are typically classified according to their size (height), intended purpose or structure, as shown in Fig. 7.3.

Fig. 7.3 Typical types of dams (Modified from https://en.wikipedia.org/wiki/Dam, available under the Creative Commons Attribution-ShareAlike License)

Based on structure and material used, dams are classified as easily created without materials, arch-gravity dams, embankment dams, or masonry dams, with several subtypes. In the arch dam, stability is obtained by a combination of arch and gravity action. If the upstream face is vertical, the entire weight of the dam must be carried to the foundation by gravity, while the distribution of the normal hydrostatic pressure between vertical cantilever and arch action will depend upon the stiffness of the dam in a vertical and horizontal direction. When the upstream face is sloped, the distribution is more complicated. The normal component of the weight of the arch ring may be taken by the arch action. For this type of dam, firm reliable supports at the abutments (either buttress or canyon side wall) are more important. The most desirable place for an arch dam is a narrow canyon with steep side walls composed of sound rock. The safety of an arch dam is dependent on the strength of the side wall abutments, hence not only should the arch be well seated on the side walls but also the character of the rock should be carefully inspected. Two types of single-arch dams are in use, namely the constant-angle and the constant-radius dam. The constant-radius type employs the same face radius at all elevations of the dam, which means that as the channel grows narrower towards the bottom of the dam the central angle subtended by the face of the dam becomes smaller. In a constant-angle dam, also known as a variable radius dam, this subtended angle is kept a constant and the variation in distance between the abutments at various levels is taken care of by varying the radii. Constant-radius dams are much less common than constant-angle dams. A similar type is the double-curvature or thin-shell dam. This method of construction minimizes the amount of concrete necessary for construction but transmits large loads to the foundation and abutments. The appearance is similar to a single-arch dam but with a distinct vertical curvature to it as well lending it the vague appearance of a concave lens as viewed from downstream. The multiple-arch dam consists of a number of single-arch dams with concrete buttresses as the supporting abutments. The multiple-arch dam does not require as many buttresses as the hollow gravity type, but requires good rock foundation because the buttress loads are heavy (Henkel 2015).

In a gravity dam, the force that holds the dam in place against the push from the water is Earth's gravity pulling down on the mass of the dam. The water presses laterally (downstream) on the dam, tending to overturn the dam by rotating about its toe (a point at the bottom downstream side of the dam). The dam's weight counteracts that force, tending to rotate the dam the other way about its toe. The designer ensures that the dam is heavy enough that the dam's weight wins that contest. In engineering terms, that is true whenever the resultant of the forces of gravity acting on the dam and water pressure on the dam acts in a line that passes upstream of the toe of the dam.

Furthermore, the designer tries to shape the dam so if one were to consider the part of dam above any particular height to be a whole dam itself, that dam also would be held in place by gravity, i.e., there is no tension in the upstream face of the dam holding the top of the dam down. The designer does this because it is usually more practical to make a dam of material essentially just piled up than to make the material stick together against vertical tension. Besides, the shape that prevents tension in the upstream face also eliminates a balancing compression stress in the downstream face, providing additional economy. For this type of dam, it is essential to have an impervious foundation with high bearing strength. When situated on a suitable site, a gravity dam can prove to be a better alternative to other types of dams. When built on a carefully studied foundation, the gravity dam probably represents the best developed example of dam building. Since the fear of flood is a strong motivator in many regions, gravity dams are being built in some instances where an arch dam would have been more economical. Gravity dams are classified as "solid" or "hollow" and are generally made of either concrete or masonry. The solid form is the more widely used of the two, though the hollow dam is frequently more economical to construct. In addition, a gravity dam can be combined with an arch dam into an arch-gravity dam for areas with massive amounts of water flow but less material available for a purely gravity dam. The inward compression of the dam by the water reduces the lateral (horizontal) force acting on the dam. Thus, the gravitation force required by the dam is lessened, i.e., the dam does not need to be so massive. This enables thinner dams and saves resources (BDS 2010).

A barrage dam is a special kind of dam which consists of a line of large gates that can be opened or closed to control the amount of water passing the dam. The gates are set between flanking piers which are responsible for supporting the water load, and are often used to control and stabilize water flow for irrigation systems. Barrages that are built at the mouths of rivers or lagoons to prevent tidal incursions or utilize the tidal flow for tidal power are known as tidal barrages (WCD 2000).

Embankment dams are made from compacted earth, and have two main types, rock-fill and earth-fill dams. Embankment dams rely on their weight to hold back the force of water, like gravity dams made from concrete. Rock-fill dams are embankments of compacted free-draining granular earth with an impervious zone. The earth utilized often contains a high percentage of large particles, hence the term "rock-fill." The impervious zone may be on the upstream face and made of masonry, concrete, plastic membrane, steel sheet piles, timber, or other materials. The impervious zone may also be within the embankment in which case it is referred to as a core. In the instances where clay is utilized as the impervious material, the dam is referred to as a composite dam. To prevent internal erosion of clay into the rock fill due to seepage forces, the core is separated using a filter. Filters are specifically graded soil designed to prevent the migration of fine grain soil particles. When suitable material is at hand, transportation is minimized leading to cost savings during construction. Rock-fill dams are resistant to damage from earthquakes. However, inadequate quality control during construction can lead to poor compaction and sand in the embankment which can lead to liquefaction of the rock-fill during an earthquake. Liquefaction potential can be reduced by keeping susceptible material from being saturated, and by providing adequate compaction during construction. A core that is growing in popularity is asphalt concrete. The majority of such dams are built with rock and/or gravel as the main fill material. Almost 100 dams of this design have now been built worldwide since the first such dam was completed in 1962. All asphalt-concrete core dams built so far have an excellent performance record. The type of asphalt used is a viscoelastic-plastic material that can adjust to the movements and deformations imposed on the embankment as a whole, and to settlements in the foundation. The flexible properties of the asphalt make such dams especially suited in earthquake regions (Høeg 1997).

A concrete-face rock-fill dam (CFRD) is a rock-fill dam with concrete slabs on its upstream face. This design provides the concrete slab as an impervious wall to prevent leakage and also a structure without concern for uplift pressure. In addition, the CFRD design is flexible for topography, faster to construct, and less costly than earth-fill dams. The CFRD concept originated during the California Gold Rush in the 1860s when miners constructed rock-fill timber-face dams for sluice operations. The timber was later replaced by concrete as the design was applied to irrigation and power schemes. As CFRD designs grew in height during the 1960s, the fill was compacted and the slab's horizontal and vertical joints were replaced with improved vertical joints. In the last few decades, the design has become popular (Das 1991).

Earth-fill dams, also called earthen dams, rolled-earth dams, or simply earth dams, are constructed as a simple embankment of well compacted earth. A homogeneous rolled-earth dam is entirely constructed of one type of material but may contain a drain layer to collect seep water. A zoned-earth dam has distinct parts or zones of dissimilar material, typically a locally plentiful shell with a watertight clay core. Modern zoned-earth embankments employ filter and drain zones to collect and remove seep water and preserve the integrity of the downstream shell zone. An outdated method of zoned earth dam construction utilized a hydraulic fill to produce a watertight core. Rolled-earth dams may also employ a watertight facing or core in the manner of a rock-fill dam. An interesting type of temporary earth dam occasionally used in high latitudes is the frozen-core dam, in which a coolant is circulated through pipes inside the dam to maintain a watertight region of permafrost within it. Because earthen dams can be constructed from materials found on-site or nearby, they can be very cost-effective in regions where the cost of producing or bringing in concrete would be prohibitive.

International standards (including the International Commission on Large Dams, ICOLD) define large dams as higher than 15 m (49 ft) and major dams as over 150 m (490 ft) in height. The Report of the World Commission on Dams also includes in the large category, dams, such as barrages, which are between 5 and 15 m (16 and 49 ft) high with a reservoir capacity of more than 3 million cubic meters (2400 acre·ft) (WCD 2000).

A saddle dam is an auxiliary dam constructed to confine the reservoir created by a primary dam either to permit a higher water elevation and storage or to limit the extent of a reservoir for increased efficiency. An auxiliary dam is constructed in a low spot or "saddle" through which the reservoir would otherwise escape. On occasion, a reservoir is contained by a similar structure called a dike to prevent inundation of nearby land. Dikes are commonly used for reclamation of arable land from a shallow lake. This is similar to a levee, which is a wall or embankment built along a river or stream to protect adjacent land from flooding. A weir (also sometimes called an overflow dam) is a type of small overflow dam that is often used within a river channel to create an impoundment lake for water abstraction purposes and which can also be used for flow measurement or retardation. A check dam is a small dam designed to reduce flow velocity and control soil erosion. Conversely, a wing dam is a structure that only partly restricts a waterway, creating a faster channel that resists the accumulation of sediment. A dry dam, also known as a flood retarding structure, is a dam designed to control flooding. It normally holds back no water and allows the channel to flow freely, except during periods of intense flow that would otherwise cause flooding downstream. A diversionary dam is a structure designed to divert all or a portion of the flow of a river from its natural course. The water may be redirected into a canal or tunnel for irrigation and/or hydroelectric power production. Underground dams are used to trap groundwater and store all or most of it below the surface for extended use in a localized area. In some cases they are also built to prevent saltwater from intruding into a freshwater aquifer. Underground dams are typically constructed in areas where water resources are minimal and need to be efficiently stored. There are two types of underground dams: a subsurface and a sand-storage dam. A subsurface dam is built across an aquifer or drainage route from an impervious layer (such as solid bedrock) up to just below the surface. They can be constructed of a variety of materials to include bricks, stones, concrete, steel, or PVC. Once built, the water stored behind the dam raises the water table and is then extracted with wells. A sand-storage dam is a weir built in stages across a stream or wadi. It must be strong, as floods will wash over its crest. Over time, sand accumulates in layers behind the dam, which helps store water and, most importantly, prevents evaporation. The stored water can be extracted with a well, through the dam body, or by means of a drain pipe (Onder and Yilmaz 2005).

A tailings dam is typically an earth-fill embankment dam used to store tailings, which are produced during mining operations after separating the valuable fraction from the uneconomic fraction of an ore. Conventional water retention dams can serve this purpose, but due to cost, a tailings dam is more viable. Unlike water retention dams, a tailings dam is raised in succession throughout the life of the particular mine. Typically, a base or starter dam is constructed, and as it fills with a mixture of tailings and water, it is raised. Material used to raise the dam can include the tailings (depending on their size) along with dirt. There are three raised tailings dam designs, the upstream, downstream, and centerline, named according to the movement of the crest during raising. The specific design used is dependent upon topography, geology, climate, the type of tailings, and cost. An upstream tailings dam consists of trapezoidal embankments being constructed on top but toe to crest of another, moving the crest further upstream. This creates a relatively flat downstream side and a jagged upstream side which is supported by tailings slurry in the impoundment. The downstream design refers to the successive raising of the embankment that positions the fill and crest further downstream. A centerlined dam has sequential embankment dams constructed directly on top of another while fill is placed on the downstream side for support and slurry supports the upstream side. Because tailings dams often store toxic chemicals from the mining process, they have an impervious liner to prevent seepage. Water/slurry levels in the tailings pond must be managed for stability and environmental purposes as well (Blight 1998).

A steel dam is a type of dam briefly experimented with around the start of the twentieth century which uses steel plating (at an angle) and load-bearing beams as the structure. Intended as permanent structures, steel dams were an (arguably failed) experiment to determine if a construction technique could be devised that was cheaper than masonry, concrete, or earthworks, but sturdier than timber crib dams. Timber dams were widely used in the early part of the industrial revolution and in frontier areas due to ease and speed of construction. Rarely built in modern times because of their relatively short life span and the limited height to which they can be built, timber dams must be kept constantly wet in order to maintain their water retention properties and limit deterioration by rot, similar to a barrel. The locations where timber dams are most economical to build are those where timber is plentiful, cement is costly or difficult to transport, and either a low-head diversion dam is required or longevity is not an issue. Two common variations of timber dams were the crib and the plank. Timber crib dams were erected of heavy timbers or dressed logs in the manner of a log house and the interior filled with earth or rubble. The heavy crib structure supported the dam's face and the weight of the water. Timber plank dams were more elegant structures that employed a variety of construction methods utilizing heavy timbers to support a water retaining arrangement of planks. A cofferdam is a barrier, usually temporary, constructed to exclude water from an area that is normally submerged. Made commonly of wood, concrete, or steel sheet piling, cofferdams are used to allow construction on the foundation of permanent dams, bridges, and similar structures. When the project is completed, the cofferdam will usually be demolished or removed unless the area requires continuous maintenance. Common uses for cofferdams include construction and repair of offshore oil platforms. In such cases the cofferdam is fabricated from sheet steel and welded into place under water. Air is pumped into the space, displacing the water and allowing a dry work environment below the surface.

Dams can also be created by natural geological forces. Volcanic dams are formed when lava flows, often basaltic, intercept the path of a stream or lake outlet, resulting in the creation of a natural impoundment. Glacial activity can also form natural dams. Moraine deposits left behind by glaciers can also dam rivers to form lakes. Natural disasters such as earthquakes and landslides frequently create landslide dams in mountainous regions with unstable local geology. Natural dams often pose significant hazards to human settlements and infrastructure. The resulting lakes often flood inhabited areas, while a catastrophic failure of the dam could cause even greater damage. Beavers create dams primarily out of mud and sticks to flood a particular habitable area. By flooding a parcel of land, beavers can navigate below or near the surface and remain relatively well hidden or protected from predators. The flooded region also allows beavers access to food, especially during the winter.

7.2.2 Dike Construction Materials

A wide range of materials may be considered for use in dikes depending on seepage and stability considerations, and materials available for use within the vicinity of the proposed dike alignment (or upgrading). If a fine-grained soil can be brought readily within the range of water contents suitable for compaction and for operation of construction equipment, it can normally be used for embankment construction. Generally, any dike fills proposed as bulk fills, or specific impervious or pervious fill layers shall limit the particle size to generally less than 100 mm and shall not contain materials greater than 150 mm in diameter. Larger cobbles and boulders within a dike fill make adequate compaction difficult to achieve and may allow void space to remain in the fills following compaction. To limit seepage through a dike, the bulk dike fill would typically comprise between about 15 and 30% fines (silts and clays) passing the U.S. Standard No. 200 sieve (Peters 2003).

For increased resistance to seepage, a material with greater than 30% fines may be considered but will likely be more difficult to compact unless it is near its optimum moisture content. This type of material can be relatively impervious when well compacted. If most of the readily available borrow area materials contain less than 10% fines, they may be considered for use as bulk fill if an impervious layer is placed on the waterside of the dike. The fill materials for an impervious layer shall be a clay-based material with greater than about 25% fines with limited oversize particles (i.e., 75 mm minus). Any proposed pervious drainage layers placed on the landside of the dike shall contain less than 5% fines passing the U.S. Standard No. 200 sieve. If this material is not available in the vicinity of the dike and needs to be imported, a 75 mm minus well-graded sand and gravel fill shall be considered. A careful analysis of all available material sources, including location, material type, and available volume shall be made. At least 15 m in width shall be left undisturbed between the toe of the dike slope and the edge of the borrow pit (Peters 2003).

Generally, proposed borrow materials shall have natural water contents low enough to allow placement and adequate compaction. The cost of drying borrow material to suitable water contents can be very high, in many cases exceeding the cost of longer haul distances to obtain material that can be placed without drying. Borrow soils undergo seasonal water content variations; hence water content data shall be based on samples obtained from borrow areas in that season of the year when dike construction is planned. Possible variation of water contents during the construction season shall also be considered (Peters 2003).

In computing required fill quantities, a shrinkage factor of at least 25% shall be applied. For example, borrow area volumes shall be at least 125% of the dike cross-section volume. This will allow for material shrinkage, compaction, hauling, and other losses. Borrow areas shall be cleared and grubbed to the extent needed to obtain fill material free of objectionable matter, such as trees, brush, vegetation, stumps, and roots. Topsoil with low vegetative seasonal water content variations; hence water content data shall be based on samples obtained from borrow areas in that season of the year when dike construction is planned. Possible variation of water contents during the construction season shall also be considered (Peters 2003).

7.3 Structural Materials and Surface Coatings for Hydropower Turbines

It is important to select the right type of materials and surface coatings to be used during construction of a new turbine or repair of an existing one. For instance, cavitation, cracking, corrosion, or some combination thereof is usually the primary concern for a turbine blade. When cracking occurs in the crown/blade fillet at the discharge edge of the runner, this may often be attributed to excessive heat from welding performed to repair cavitation damage that has added to the tensile stresses at the downstream edge of the blade. And cavitation damage is often exacerbated by coincidental corrosion. Therefore, understanding the properties of the various materials is crucial to draft proper specifications for building and repair hydropower turbines.

7.3.1 Hydropower Blade Materials

Hydropower turbine blades are constantly exposed to water and dynamic forces, therefore, they need to have high corrosion resistance and strength. The most common material used in overlays on carbon steel runners in hydropower turbines are austenitic steel alloys that have 17–20% chromium to increase stability of the film which improves aqueous corrosion resistance. The chromium content in these steel alloys exceeds the minimum of 12% chromium required to exhibit some atmospheric corrosion resistance. Having a higher chromium concentration in the steel alloys allows for a much longer life span of the turbine blades. Currently, the blades are made of martensitic stainless steels which have high strength compared to austenitic stainless steels by a factor of 2 (Spicher 2013).

Besides corrosion resistance and strength as the criteria for material selection, weldability and density of the turbine blade. Greater weldability allows for easier repair of the turbine blades. This also allows for higher weld quality which results in a better repair. Selecting a material with low density is important to achieve higher efficiency because the lighter blades rotate more easily. The most common material used in Kaplan Turbine blades is stainless steel alloys (SS). The different alloys used are SS(16Cr-5Ni), SS(13Cr-4Ni), and SS(13Cr-1Ni). The martensitic stainless steel alloys have high strength, thinner sections than standard carbon steel, and reduced mass that enhances the hydrodynamic flow conditions and efficiency of the water turbine. The SS(13Cr-4Ni) has been shown to have improved erosion resistance at all angles of attack through the process of laser hardening. It is important to minimize erosion in order to maintain high efficiencies because erosion negatively impacts the hydraulic profile of the blades which reduces the relative ease to rotate (Gummer 2009; Adhikary et al. 2013).

7.3.2 Surface Coatings for Hydropower Turbines

Hydropower turbines are subject to efficiency losses from corrosion and wear by erosion (hydro-abrasion, fluid erosion, and cavitation erosion), as shown in Fig. 7.4. Influencing factors include the type of hydroelectric power plant, the design of the hydro turbine (Francis, Kaplan, or Pelton), and the specific operating conditions such as the corrosive potential of the water and the size and amount of the silt, sand, or gravel debris in the water. At the same time, as utilities strive for greater profitability, installations exploit higher heads (water pressure and velocity) and have expanded into less accessible, more contaminated waterways. These factors have increased the levels of wear mechanisms acting on the turbine components (Oerlikon Metco 2014).

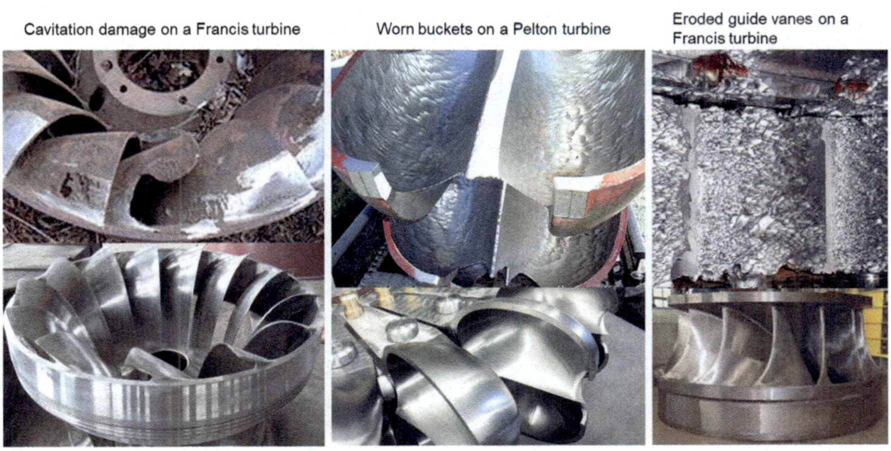

Fig. 7.4 Typical examples of corrosion and erosion of turbine components (up part is used; bottom part is new-built)

More specifically, the wear mechanisms in hydropower turbines vary considerably, depending on the service conditions that the individual components are exposed. Turbines operating in high-head conditions, such as Pelton turbines, may be exposed to greater wear from water erosion. On the other hand, turbines exposed to low-head conditions, such as Kaplan turbines, may be exposed to greater wear from entrained debris in the water. The type of entrained debris in the water must also be considered. Wear effects will be significantly different depending on whether the entrained debris is silt, sand, or gravel, and differences will also be evident depending on the amount of debris. Finally, different head waters will have different corrosion potentials, further complicating the means to mitigate the wear effects. Some components may be exposed to sliding wear primarily caused by the dynamic interaction between mating components. Here, wear coatings can also prove to be an effective deterrent. While no coating will completely stop these effects, coatings are an effective means of greatly prolonging service life and efficiency. But for the coating to be optimally effective, all factors must be considered. Some commonly applied coatings are listed in Table 7.2 (Oerlikon Metco 2014).

WCCoCr is one common group materials are used for surface coatings of hydropower turbines. Despite having practically identical chemical compositions, these materials can have different particle shapes, morphologies, particle size distributions, primary carbide sizes and bulk densities. As such, the various products can have substantially different wear behavior in service and under various service conditions. Carbide materials, including WC 10Co 4Cr coating materials, are other group materials that have been widely used. Both of these materials are successfully used to prevent wear in hydropower turbines; however, they are used for different turbine applications and exhibit different in-service characteristics. WCCoCr materials frequently used to protect hydropower turbine components are offered in variety of products that vary by powder manufacturing process, particle size distribution and apparent carbide size and density. A few hydropower components are best protected using chromium oxide applied by atmospheric plasma spray.

Just as differences in material characteristics can have a large effect on service performance and coating endurance, the method used to apply the material has an equally important role, particularly for HVOF-applied coatings, as shown in Fig. 7.5. The coating application technology and parameters must be as carefully chosen as the coating material. The drivers for the choice of application equipment are always those of in-service coating performance and cost efficiency. As a result, protective coatings can greatly extend the service life of hydropower turbine components. However, the coating material, application equipment, and parameters must be carefully chosen for optimal coating performance. Seemingly similar materials and application technologies can perform very differently in service (Oerlikon Metco 2014).

Table 7.2 Typical surface coatings used in hydropower turbines (Oerlikon Metco 2014)

Component	Coated area	Coating	Wear mechanism
Kaplan turbine			
Discharge ring	Partial or entire discharge ring	• Wire combustion sprayed 15 mm thick Metcoloy 2	Erosion (hydro-abrasion, fluid erosion)
Kaplan blade	Partial or entire blade	• HVOF 0.4 mm thick WCCoCr	
		• Wire combustion sprayed 5 mm thick Metcoloy 2	
Guide vane ring	Between planar surface and draft tube liner	• Wire combustion sprayed 5 mm thick Metcoloy 2	
Protective sleeve	Two-part sealing elements	• HVOF 0.3 mm thick WCCoCr	Seal area, abrasive wear
		• Wire combustion sprayed Metcoloy 2	
Radial bearing	Applied to new or repair components	• Wire combustion sprayed Sprababbitt A	Sliding wear
Crank	Slide-bearing area	• Wire combustion sprayed Sprasteel LS	
Crank pin	Slide-bearing area	• Wire combustion sprayed Sprasteel LS	
Francis turbine			
Cheek plate	Complete area	• HVOF WCCoCr	Erosion (hydro-abrasion, fluid erosion)
Guide vane	Complete guide vane, also disk and face side seals		
Turbine cover	Clearance and labyrinth area, wear ring area		
Runner wheel	Clearance and labyrinth area, runner inlet channel		
Pelton turbine			
Pelton bucket	Inside and edges	• HVOF WCCoCr	Erosion (hydro-abrasion, fluid erosion)
Pelton needle	Area subject to wear	• HVOF WCCoCr	
		• Plasma sprayed Cr_2O_3	
Needle spear	Area subject to wear	• Wire combustion sprayed Metcoloy 2 or Sprabronze	Sliding wear
Nozzle tip	Entire internal contour	• HVOF WCCoCr	Erosive and abrasive wear
Nozzle tip insert ring	Area subject to wear		
Jet deflector	Area subject to wear		
Jet deflector cover	Area subject to wear		

Fig. 7.5 Examples of coating materials with different morphologies and coating application technologies (Modified from Oerlikon Metco 2014. Courtesy: Oerlikon Metco)

INDEPTH: Ocean Energy

The ocean can produce two types of energy: thermal energy from the sun's heat and mechanical energy from the tides and waves.

Oceans cover more than 70% of Earth's surface, making them the world's largest solar collectors. The sun's heat warms the surface water a lot more than the deep ocean water, and this temperature difference creates thermal energy. Just a small portion of the heat trapped in the ocean could power the world.

Ocean thermal energy is used for many applications, including electricity generation. There are three types of electricity conversion systems: closed-cycle, open-cycle, and hybrid. Closed-cycle systems use the ocean's warm surface water to vaporize a working fluid, which has a low boiling point, such as ammonia. The vapor expands and turns a turbine. The turbine then activates a generator to produce electricity. Open-cycle systems actually boil the seawater by operating at low pressures. This produces steam that passes through a turbine/generator. And hybrid systems combine both closed-cycle and open-cycle systems.

Ocean mechanical energy is quite different from ocean thermal energy. Even though the sun affects all ocean activity, tides are driven primarily by the gravitational pull of the moon, and waves are driven primarily by the winds. As a result, tides and waves are intermittent sources of energy, while ocean thermal energy is fairly constant. Also, unlike thermal energy, the electricity conversion of both tidal and wave energy usually involves mechanical devices.

A barrage (dam) is typically used to convert tidal energy into electricity by forcing the water through turbines, activating a generator. For wave energy

(continued)

conversion, there are three basic systems: channel systems that funnel the waves into reservoirs; float systems that drive hydraulic pumps; and oscillating water column systems that use the waves to compress air within a container. The mechanical power created from these systems either directly activates a generator or transfers to a working fluid, water, or air, which then drives a turbine/generator.

7.4 The State of Ocean Energy Technologies

The ocean has been an integral part of human civilization and development since ancient times, and although its potential use in generating power has been the subject of patents dating back to the eighteenth century, modern ocean energy is still in a developmental stage, researchers are seeking ways to capture that energy and convert it to electricity. Ocean energy, also referred to as marine renewable energy, encompasses all of the renewable energy resources found in the oceans, which use the kinetic, potential, chemical, or thermal properties of seawater. Ocean surface waves, tidal currents, tidal range, ocean currents, thermal gradients, and changes in salinity all represent energy resources that can be harnessed using a variety of different technologies. Ocean energy technologies convert these renewable energy resources into a useful form—typically electricity (Mofor et al. 2014).

Along with other renewables, ocean energy technologies generate carbon dioxide (CO_2) emission-free power and as an indigenous resource can promote energy independence. Ocean energy technologies can also contribute to a balanced, diversified energy portfolio, with generation profiles that complement those of other renewables— such as solar and wind— thus helping to balance the variable generation of different renewable energy sources. Furthermore, ocean energy technologies can extend the range of options for densely populated coastal nations with limited land space, to increase their use of renewables. The issue of competing land use is often a significant advantage for many ocean energy technologies, as they provide the opportunity to put renewable generation plants "under the surface" or "over the horizon." However, multiple barriers exist for ocean energy technologies, such as gaining site permits, the environmental impact of technology deployments, and grid connectivity for transmitting the energy produced. The global challenge is how to extract the energy, bring it to shore, store it, and export it cost-effectively. Key aspects relate to technology, infrastructure, cost reduction, investment, environmental impact, marine governance, consenting and licensing, and legislation. The marine renewables industry is particularly sensitive to government intervention. To reduce uncertainty, new ocean data-gathering campaigns are vitally needed to provide high-quality information on seabed roughness, wave surface elevations, tidal currents, eddies, and turbulence at sites (Burman and Walker 2009; Borthwick 2016).

7.4.1 Tidal Power

Solar and lunar gravitational forces, combined with the rotation of the Earth, generate periodic changes in sea level known as the tides. This rise and fall of ocean waters can be amplified by basin resonances and coastline bathymetry to create large surface elevation changes at specific geographic locations. High and low tides occur twice a day at most coastal sites throughout the world (semi-diurnal tides), although some places experience just one high and low tide per day (diurnal tides). Other places are characterized by a combination of diurnal and semi-diurnal oscillations (mixed tides). The difference in sea level height between high and low tide at a given location is called the tidal range, and it can vary each day depending on the location of the sun and moon, and globally depending on the coastal location. As tides are caused by the aforementioned gravitational interactions, they are considered a renewable energy resource. There are two general approaches to tidal energy conversion (Mofor et al. 2014): (a) Capture the potential energy created by the difference in sea level between high and low tides—tidal range; and (b) Capture the kinetic energy from the horizontal flow of tidal currents that can occur at certain locations through tidal stream.

7.4.1.1 Tidal Range

Tidal range technology is based on conventional hydropower principles and requires a natural or a manmade structure (e.g., a dam or barrier) to impound a large body of water. As the tidal height varies outside of the impounded area during the tidal cycle, water is discharged either in or out of the enclosed area through conventional hydro turbines (typically of the low-head type, i.e., propeller turbines) housed in the dam or barrier. This is commonly achieved by placing a tidal barrage across the mouth of an estuary, creating a reservoir (basin) behind it. Tidal lagoons have recently been developed using multiple-basin schemes and enclosed basins located offshore (single or multiple) away from estuaries (Mofor et al. 2014).

Out of all the ocean energy technologies, tidal range remains the only one that has proven reliability in existing commercial projects, particularly in combined applications (e.g., for flood control and water quality management) or where existing impoundment structures are required for other reasons. Figure 7.6 illustrates the tidal range found globally in one of the primary tidal constituents, which demonstrates that there are certain areas around the world where the resource is likely to be sufficient for tidal projects (both range and stream). For a particular project to be feasible, however, certain other site conditions must also be met, e.g., an estuary suitable for a tidal barrage. The world's first large-scale tidal range power plant, the 240 megawatts (MW) Rance Tidal Power Station, became operational in 1966 in Brittany, France, and is still operated today by Électricité de France (EDF). The 254 MW Sihwa Lake Tidal Power Station in South Korea became the world's largest (and newest) tidal barrage when it was opened in 2011. Both of these tidal barrages employ conventional bulb turbines. Only a few other much smaller sites have been developed around the world, resulting in a total installed tidal range capacity worldwide of about 498 MW (Mofor et al. 2014).

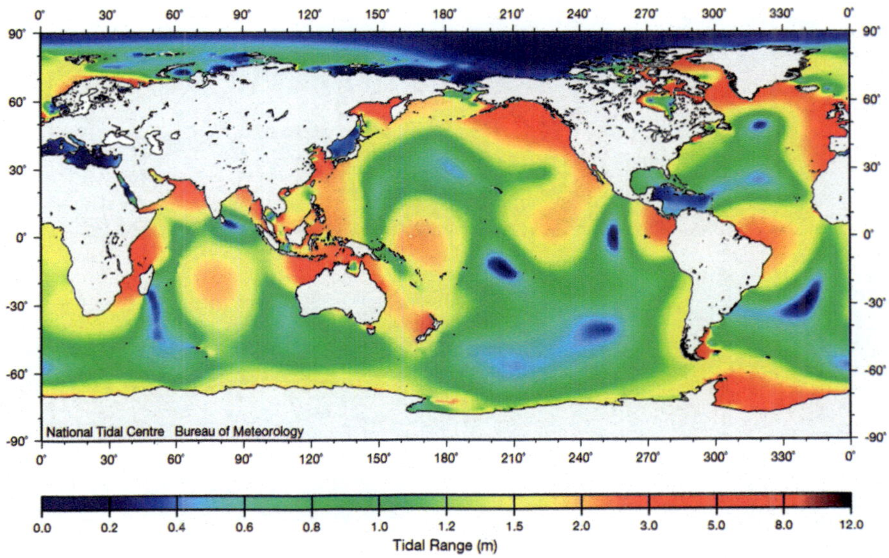

Fig. 7.6 The tidal range found globally in one of the primary tidal constituents, which demonstrates that there are certain areas around the world where the resource is likely to be sufficient for tidal projects (Adapted from Mofor et al. 2014, published by International Renewable Energy Agency, available under a Creative Commons license (CC BY 3.0))

7.4.1.2 Tidal Stream

The vertical rise and fall of water of the tides is accompanied by an incoming (flood) or outgoing (ebb) horizontal flow of water in bays, harbors, estuaries, and straits. This flow is called a tidal current or tidal stream. Tidal streams are driven by the head differences across tides as they pass through coastal regions. Tides are complicated by the relative motions of astronomical bodies, Coriolis acceleration owing to the Earth's rotation, the presence of land masses, and the influence of local seabed topography and roughness. Tidal currents can be exceptionally strong in areas where large tidal ranges are further constrained by local topography. There will also be periods of time when there is little or no horizontal flow of water (i.e., slack water—the short time before the tide changes between ebb and flood and vice versa) (Mofor et al. 2014; Borthwick 2016).

Hydrokinetic turbines convert the kinetic (moving) energy of free flowing water into electricity using the same principles that wind turbines use to convert the kinetic energy of flowing air (wind). When hydrokinetic systems are used in a tidal environment they are often referred to as tidal stream turbines, tidal in-stream energy converters, or tidal/marine/hydrokinetic current turbines. Most designs of tidal stream energy converters are representative of modified wind turbines made to suit the higher density and different characteristics of the surrounding environment, since the principles of energy conversion are the same. Many of these wind turbine designs, including cross-flow turbines, ducted turbines, and drag turbines, are now

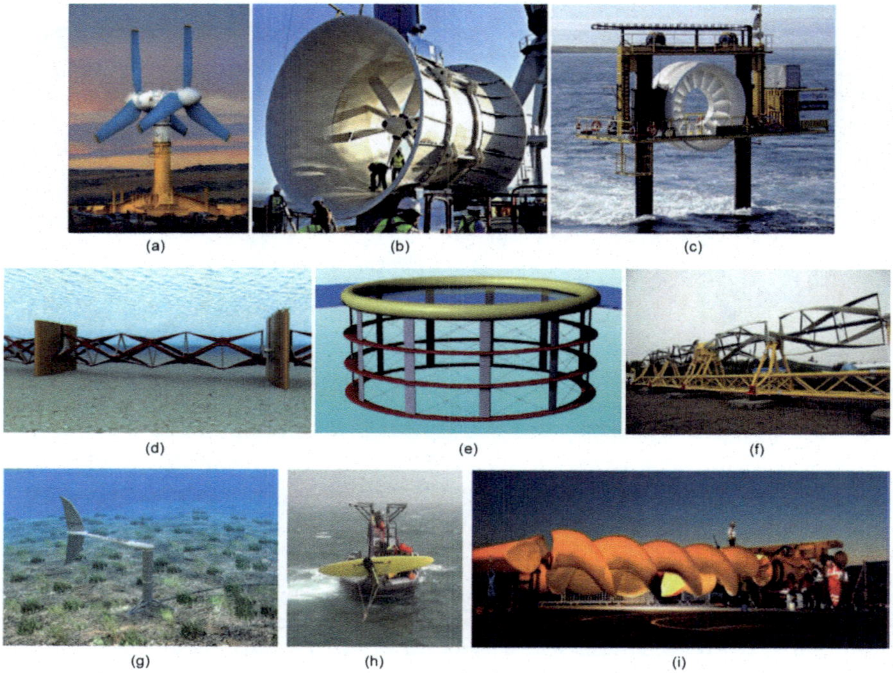

Fig. 7.7 Examples of tidal and ocean current devices (Adapted from Borthwick 2016, licensed by Elsevier under the terms of the Creative Commons Attribution License (CC BY)): (**a**) Atlantis AK1000 axial-flow turbine (1 MW), (**b**) Atlantis Solon-K ducted turbine (1 MW), (**c**) OpenHydro open-center turbine (250 kW), (**d**) Kepler transverse horizontal axis water turbine (THAWT), a kind of cross-flow turbine (4 MW), (**e**) Edinburgh vertical axis cross-flow turbine concept (100 MW), (**f**) Gorlov helical turbines before deployment at Cobscook Bay, ME, USA, (**g**) bioSTREAM (150 kW) oscillatory-hydrofoil turbine, (**h**) Minesto deep sea tidal kite, (**i**) Flumill Archimedes screw turbine

being re-examined for hydrokinetic applications. In this way, the maturity of the tidal stream industry can be compared to the early stages of the wind energy sector a few decades ago (Mofor et al. 2014).

However, there are major differences between ocean and wind power device environments (Borthwick 2016): Ocean flow directions are often more predictable than wind; the sea surface acts to constrain the current and enhance blockage, unlike the atmosphere; seawater is about 800 times denser than air, so energy devices are much more heavily loaded; and the ocean environment is much harsher than that of the atmosphere. Importantly, the presence of tidal or ocean turbines alters the flow field, and this in turn alters power availability. The influence of marine turbines on local flow speeds can persist over large distances (Neill et al. 2012).

To date, few commercial-scale devices have been manufactured and tested, although many innovative tidal and ocean stream energy devices have been proposed, as shown in Fig. 7.7. Such devices may be categorized as follows (Borthwick 2016): axial-flow turbines, cross-flow turbines, oscillating-hydrofoil turbines, and tidal sails and kites.

The axial-flow turbine extracts energy from moving water by means of rotating blades mounted on a rotating hub in much the same way as wind turbines extract energy from moving air. Such turbines may be horizontal-axis or vertical-axis, depending on the mounting. The turbines can comprise bare blades, such as the Atlantis turbine shown in Fig. 7.7a; or be ducted (e.g., the Venturieffect turbine, Fig. 7.7b, in which the ducting accelerates the flow as it passes the turbine); or be open-center, with multiple blades surrounding a central hole that promotes jet flow through the aperture, decreasing base pressure, and increasing mass flux (e.g., OpenHydro, Fig. 7.7c). The blades are structural cantilevers, and thus are subjected to very high loads, meaning that the material strength and fatigue characteristics are critically important. As the rotor blades rotate, they sweep out a circle whose area can be used to estimate blockage, a parameter that influences the thrust and hence the available power. The flow field immediately behind the turbine blades has a very strong swirl component. The cross-flow turbines are configured to achieve high flow blockage, and hence maximize power extraction. In this case, the blades are orientated so that the flow passes across the blades, rather like the reverse of a combine harvester. The axis of rotation is perpendicular to the flow direction, and is either horizontal or vertical. The Kepler Energy transverse horizontal axis water turbine (THAWT) device shown in Fig. 7.7d is based on this principle. The Edinburgh vertical axis cross-flow turbine shown in Fig. 7.7e has blades arranged vertically, supported at each end on what are essentially large bicycle wheels. It has variable pitch with rim power take-off, 200 m diameter counter-rotating rotors, a swept area of 10,000 m^2, and is believed to be capable of power extraction exceeding 100 MW per rotor. The Gorlov helical turbine (GHT), shown in Fig. 7.7f, is essentially a modified Darrieus cross-flow turbine with symmetrical helical blades. The GHT works well in reversing flows. Oscillating-hydrofoil turbines operate by the lift force acting on a hydrofoil, causing an arm to move, driving fluid in a hydraulic system, and then being converted into electricity. Typical oscillating-hydrofoil devices include Stingray (150 kW); bioSTREAM (150 kW), which mimics fish propulsion, as shown in Fig. 7.7g; and the Pulse-System (1.2 MW) concept developed by Pulse Tidal. Tidal kite turbines, such as that shown in Fig. 7.7h, are tethered paravanes that fly underwater, converting the kinetic energy of the current into electricity. By following a figure-eight locus, the tidal kite is potentially very efficient, and operates at currents as low as 1.2 m s^{-1} in water depths of 60–120 m. Various other tidal and ocean current turbines are under development. For example, the Archimedes screw tidal turbine, shown in Fig. 7.7i, generates electricity through rotation of the screw by the tidal current. Other examples include the Atlantis Fanbelt Aquanator 400 and Tidal Sails concepts (Salter and Taylor 2007; Borthwick 2016).

Moreover, uncertainty, turbulence, eddies, wave-current interaction, model scale-up, and environmental impact (e.g., on mammals and biodiversity) have been investigated. For example, storm-induced waves can generate water particle velocities that are larger than tidal currents. In addition, wave periods can be of the same order as turbine rotor periods. Turbulence and large-scale eddies affect the uniformity and alignment of the incoming flow, impacting on turbines. High tidal ranges are a usually prerequisite for fast tidal currents in addition to certain geographic features to create the tidal stream. Generally, tidal streams must reach flow speeds of at least 1.5–2 meters per second (m/s) for tidal current turbines to operate effectively. Major tidal streams have been identified along the coastlines of every continent, making it a global, albeit site specific, resource (Lewis et al. 2014; Mofor et al. 2014).

7.4.2 Ocean Current

Open ocean currents are driven by latitudinal distributions of winds and thermohaline ocean circulation. They are generally slower, but more continuous than tidal currents and although often located at deep ocean sites, they tend to operate most strongly near the surface. Another difference from tidal currents is that the flows are unidirectional, whereas tidal current reverse direction with each flood and ebb cycle. Baseload power can potentially be generated due to the steady nature of some ocean currents. Although these currents are distributed globally (Fig. 7.8), it remains

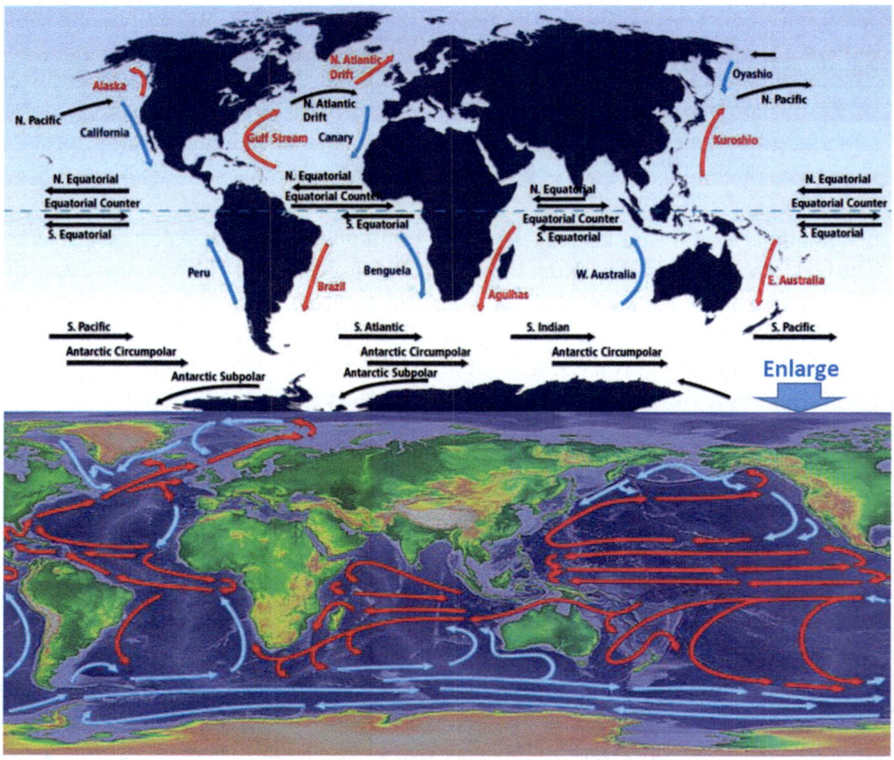

Fig. 7.8 Ocean surface currents showing both warm (red) and cold (blue) systems (Modified from Burman and Walker 2009. Credit: U.S. Department of Energy, Office of Energy Efficiency and Renewable Energy Federal Energy Management Program; Mofor et al. 2014. Courtesy: International Renewable Energy Agency)

unclear how many may prove enticing enough to draw interest for project development. However, if technologies can be developed to harness these lower velocity currents, the scale of power generation could potentially be much larger given the large volumes of water and scale of oceanic currents in comparison to tidal streams (Burman and Walker 2009; Mofor et al. 2014).

The same hydrokinetic approach and operating principles behind the turbines used for tidal stream can be applied to the flow of water in oceanic currents. Owing to the ocean depth at suitable locations for ocean currents, turbines would need to be held in location with moored floating or typically submerged systems (Mofor et al. 2014).

7.4.3 Wave Energy

Wave energy converters (WECs) transform energy from the kinetic and potential energy of ocean surface waves into another form of energy (e.g., electricity). These waves, generated primarily by wind blowing across the ocean surface (ripples), can propagate over deep water with minimal energy loss and will combine and continue to gain energy from the wind over long open ocean stretches (leading to swells). Although the air-sea interactions and energy transfer mechanisms are complex, ocean surface wave formation is primarily influenced by the speed of the wind, its duration, and the fetch (distance of open water over which the wind blows). As it is solar energy that creates the differences in air temperature that cause wind, wave energy can be considered a concentrated form of solar energy. The spatial concentration of energy is one key advantage of wave energy in comparison to other renewable energy resources (Mofor et al. 2014).

The most energetic wave conditions can be found primarily between latitudes of 30–60°, as shown in Fig. 7.9, with the largest power levels occurring off the west coasts of continents. As a resource, wave energy has the advantage of relatively good predictability for sea state conditions. Although there is seasonality, with higher wave conditions experienced in the winter than in the summer at most locations, waves arrive day and night, 24 h a day, and sea states have more inertia than solar/wind conditions, with less potential for sudden changes in the resource potential (Mofor et al. 2014).

Although ideas for wave energy conversion have been around for some time, with serious academic attention beginning in the early 1970s, extraction of wave energy at useful scales and costs has proven challenging. At present there are a number of grid-connected devices installed in high-energy environments, representing the pre-commercial prototypes of devices that are targeted for build-out into utility scale arrays in the next decade (Mofor et al. 2014).

Figure 7.10 shows some examples of wave energy converters, typically including oscillating buoys, floating ducks, snakes, flaps, and enclosed chambers. Here, the

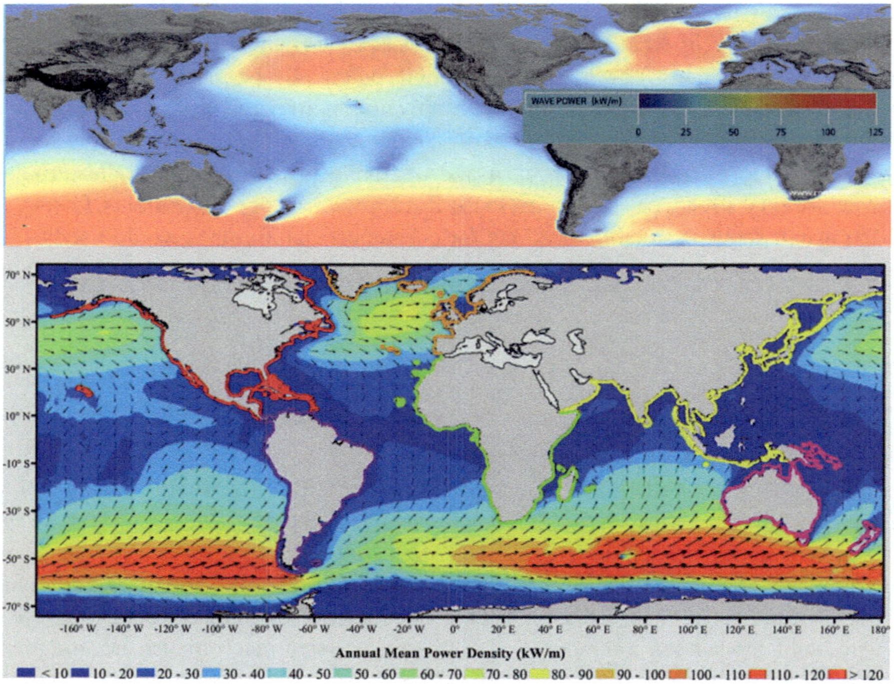

Fig. 7.9 Global annual mean wave power distribution (Modified with permission from Huckerby et al. 2011 (Elsevier); and Gunn and Stock-Williams 2012 (Elsevier))

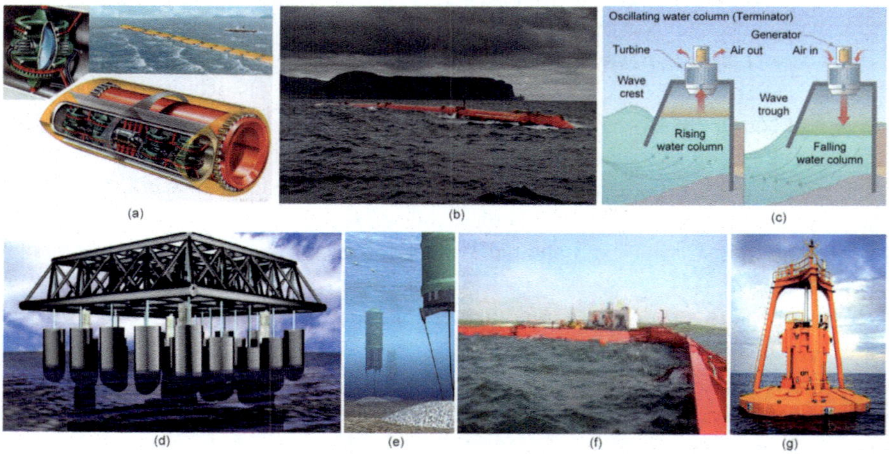

Fig. 7.10 Examples of wave energy converters (Adapted with permission from Borthwick 2016 (Elsevier)): (**a**) Salter's duck, (**b**) Pelamis wave energy converter, (**c**) oscillating water column, (**d**) Manchester Bobber, (**e**) Archimedes wave swing machine, (**f**) Wave Dragon, (**g**) PB150 PowerBuoy with a peak-rated power output of 150 kW

main objectives are to find ways to maximize power output, improve efficiency, reduce environmental impact, enhance material robustness and durability, cut capital and recurrent costs, and ensure survivability. One of the first wave energy devices proposed was Salter's duck, shown in Fig. 7.10a, which undergoes nodding motions in waves and uses the motions to pump hydraulic fluid or to compress air, and thence converts the movement into electricity by means of an internal turbine. At field-scale, Salter's duck could be arranged in preconfigured patterns in the sea, to exploit the local wave climate. The Pelamis wave energy converter, shown in Fig. 7.10b, is made of connected sections that respond to the wave motions by flexing and bending, and hence generate electricity. Pelamis became the world's first offshore wave energy converter to supply electricity into the grid, in 2004. An oscillating water column, shown in Fig. 7.10c, uses a large volume of moving water as a piston in a cylinder. Air is forced out of the column as a wave rises and external air is drawn in as the wave falls. The air movement turns a Weir turbine at the top of the column. The Manchester Bobber, shown in Fig. 7.10d, comprises a rig with semi-submerged floats that move up and down in heave in concert with ocean waves, operating a pulley that spins a fly-wheel connected to an induction electricity generator. At commercial scale, the Manchester Bobber could generate an average of 5 MW, with higher outputs possible in heavy seas. The Manchester Bobber has appeal due to its simplicity, robustness, and ease of maintenance—all electromechanical components are located above the sea surface, with only the bobbers in contact with seawater. In extreme sea states, damage to the rig can be avoided by damping the float motions through the addition of water. The Manchester Bobber floats freely, and thus extracts energy independent of wave direction. The Archimedes wave swing machine, shown in Fig. 7.10e, is constructed from a large floating cylinder that is tethered to the seabed, so that it remains at least 6 m below the sea surface. The upper cylinder provides flotation and contains a basement cylinder that generates electricity through repeated up-down motions in harmony with the ocean waves. The Wave Dragon, shown in Fig. 7.10f, traps overtopped ocean waves in a reservoir at an elevation above mean water level, from which electricity is generated by low-head turbines as the water is released back to the sea. The Wave Dragon is a floating wave energy converter, with mooring lines maintaining its approximate position. PowerBuoy, shown in Fig. 7.10 g, operates as a point absorber, which oscillates up and down in waves, with the mechanical motions being converted into electricity that is either transmitted onshore by means of a submerged cable or immediately utilized if located far offshore. PowerBuoys are aesthetically nonintrusive, occupy small plan areas, and relatively robust in extreme sea states. Sensors located on the PowerBuoy are used to monitor local conditions, and lock down the device if the wave climate becomes too hostile. In addition, computer simulation has been used to predict time series of wave motions, device response, piston power, accumulated pressure, array power, and output voltage of wave energy converters (Borthwick 2016; Forehand et al. 2016).

7.4.4 Ocean Thermal Energy

The oceans act as large stores of thermal energy, absorbing about 15% of solar radiant heat. The ocean thermal energy conversion (OTEC) concept exploits a temperature difference of at least 20 °C between the warm surface layer and cooler depths, and has been proposed for the tropics at depths of the order of 1 km. An OTEC plant essentially comprises a heat exchanger with an evaporator and condenser, and proposed technologies operate as closed-cycle, open-cycle, or hybrid-cycle processes (Borthwick 2016).

The open-cycle systems use a vacuum chamber to "flash evaporate" some of the warm surface seawater. The steam generated, which is the working fluid for the system, passes through a turbine generator before being condensed by the cold deep seawater. It may also be possible to use such open cycle plants for desalination applications. Closed-cycle systems have more efficient thermal performance, and pump the warm surface water through heat exchangers to vaporize a secondary working fluid (such as ammonia which has a low boiling point). The resulting high-pressure vapor drives the turbine, before being subsequently cooled by the deeper seawater to return to a liquid phase. Because the secondary working fluid operates at a higher pressure in closed-cycled conversion, the systems can typically be smaller than open cycle plants. There are also hybrid conversion cycles where steam from flash evaporation is used as the heat source for a closed Rankine cycle that uses a secondary working fluid (Mofor et al. 2014).

The temperature gradient between the sea surface water and the colder, deep seawater—generally at depths below 1000 m—can be harnessed using different ocean thermal energy conversion (OTEC) processes. OTEC requires practical temperature differences of at least about 20 °C. The resource is principally distributed in the tropics (latitudes 0–35°) on either side of the equator (Fig. 7.11). In those tropical latitudes the ocean surface temperatures are highest and there is often stable stratification of the oceanic water column (Mofor et al. 2014).

Although there is a slight seasonal variation in temperature gradients, the resource can be considered continuously available, and as such OTEC represents an ocean energy technology with the potential to generate baseload power. The theoretical global total resource potential for ocean thermal energy is the highest among the ocean energy resources. However, compared to other ocean energy technologies such as wave and tidal stream energy converters, the energy density of the OTEC systems is quite low. This represents one of the ongoing challenges towards a cost-effective OTEC operation (Mofor et al. 2014).

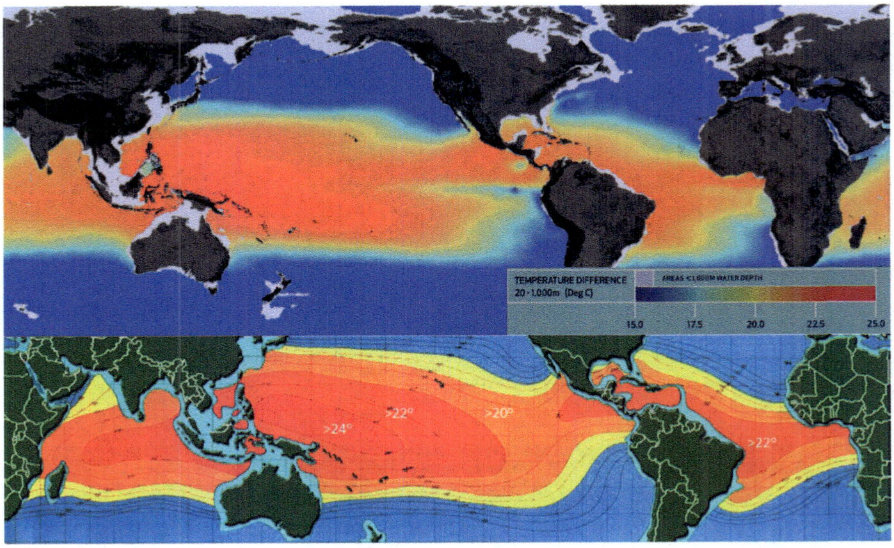

Fig. 7.11 Global mean ocean temperature difference between 20 and 1000 m depths (Adapted from Huckerby et al. 2011. Courtesy: Ocean Energy Systems Implementing Agreement; www.iea-oceans.org)

7.4.5 Salinity Gradient

INDEPTH: Ocean Technology Readiness

Ocean surface waves, tidal currents, tidal range, deep ocean currents, thermal gradients, and changes in salinity are all ocean energy resources. Ocean energy technologies seek to convert these renewable energy resources into a useful form—typically electricity. Ocean energy converters are far from a homogenous set of technologies. There are a number of technology variants, largely defined by the ocean resource that they seek to harness. Each technology variant is distinctive in terms of technical design, operation, and commercial maturity. For most ocean energy technologies the main challenge is to reduce costs and improve the reliability and performance of systems, in order to demonstrate a sustained commercially competitive cost of energy. Most ocean energy technologies are significantly behind other renewables—such as wind and solar—in technical maturity. This is largely due to challenges of working in an offshore environment. Levelized costs of ocean energy technologies are currently substantially higher than those of other renewable energy technologies; the long-term pathway to cost reduction is difficult to predict: The uncertain costs, usually high, is a consequence of limited available empirical cost data and wide variability in project cost strategies as a result of the diversity of device designs, and limited understanding—with regard to ocean energy—of key costs of energy drivers such as capacity factor and design life (Mofor et al. 2014).

(continued)

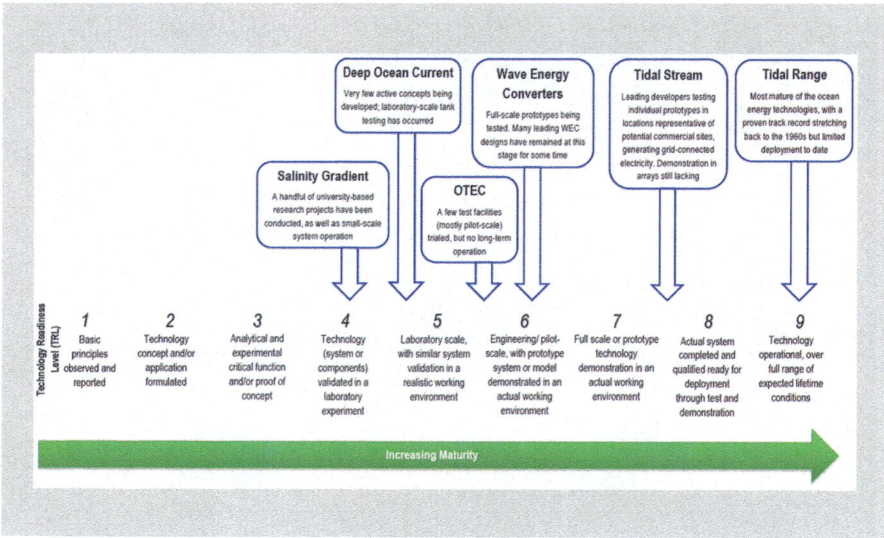

Energy associated with a salinity gradient can be harnessed, arising from the chemical potential of salinity gradients where seawater and freshwater meet (e.g., at saline wedges in estuaries). Various technologies have been proposed for salinity-gradient power generation including pressure-retarded osmosis (PRO) and reversed electro-dialysis (RED). Also called osmotic power, such technologies seek to harness the chemical potential between freshwater and seawater, captured as pressure across a semipermeable membrane (Mofor et al. 2014; Borthwick 2016).

Salinity gradient resources are distributed globally and located where freshwater from rivers discharges into saline seawater, as shown in Fig. 7.12. River mouths are the most obvious locations where there is the potential for large adjacent volumes of fresh and salt water. As the salinity gradient resource is continuous there is the

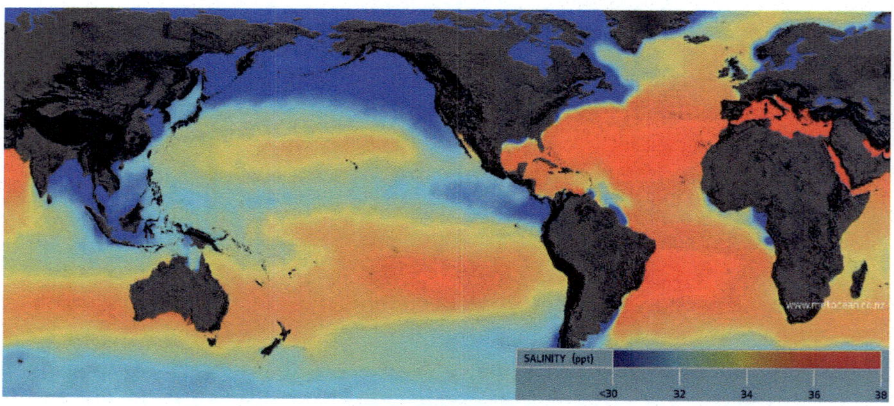

Fig. 7.12 Global salinity gradient resource distribution (Adapted from Huckerby et al. 2011. Courtesy: Ocean Energy Systems Implementing Agreement; www.iea-oceans.org)

potential to generate baseload power, if cost-effective technologies can be developed. Currently the cost of membranes compared to generation capacity has prevented commercial development of salinity gradient power plants (Mofor et al. 2014).

In addition, technical issues that still need resolution before RED can be fully implemented include (Borthwick 2016): damage to membranes by natural impurities in water, the filtration of particles, biofouling, the effect of multivalent ions on system performance, the impact on marine species of the substantial pumping process, and the necessity to minimize internal resistance.

7.5 Advanced Materials and Components for Ocean Energy Systems

Energy extraction from each ocean energy resource type such as wave or tide will require very different principles of operation. Some convergence within devices in the tidal energy sector focus on horizontal axis turbines; significant design diversity still exists within the wave energy sector. The purpose of an ocean energy device is to extract energy from the available resource and convert it into an energy form suitable for transportation and use—electricity.

7.5.1 Common Structural Systems Within Wave and Tidal Energy Converters

Although individual devices vary from each other, several common systems are present within all ocean energy conversion device designs. For instance, the common systems within wave or tidal energy converters include (SI-Ocean 2016): Structure and prime mover, foundations and moorings, power take off (PTO), control, installation, connection, operation and maintenance (O&M), ocean energy storage, and needed advanced materials, robotics and informatics.

7.5.1.1 Structure and Prime Mover

The physical structure of the device which captures energy and the main interface between the resource and the power take off equipment within the ocean energy converter. The predominant structural material is steel, although certain concepts are exploring alternatives. Prime movers such as turbine blades are made of composite materials (SI-Ocean 2016).

7.5.1.2 Foundations and Moorings

The method used to secure the device to the sea bed. This includes permanent foundation constructions such as gravity bases or pile-pinned foundations, or could consist of moorings such as tight or slack moored systems (SI-Ocean 2016).

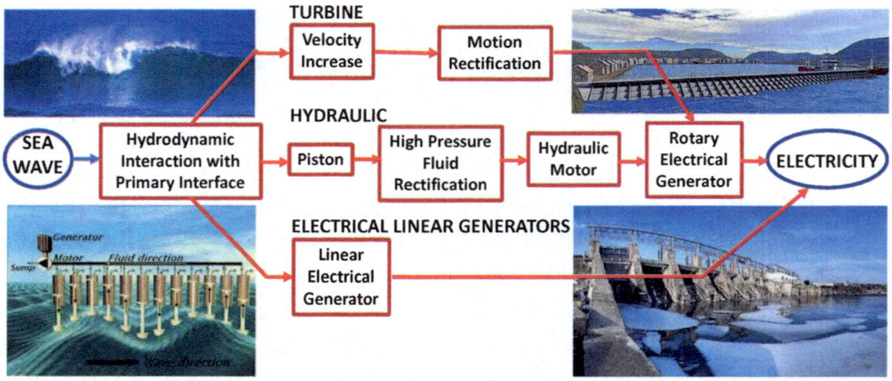

Fig. 7.13 Wave energy converter power take-off (PTO) permutations

7.5.1.3 Power Take Off

The means by which the mechanical energy extracted from the waves or tides is converted into electrical energy. Several types of Power take Off (PTO) exist including mechanical, hydraulic, or direct drive using permanent magnet generators (SI-Ocean 2016).

The primary mechanisms for power take off (PTO) within a wave energy converter are identified in Fig. 7.13 (Forehand et al. 2016). Permutations in PTO can lead to similar device structures having significantly different mechanisms for energy conversion. The identified PTO mechanisms have been categorized into three types: Turbine, Hydraulic, and Linear Electrical Generators. The "Turbine" PTO uses air as the fluid through which energy extraction takes place. Changes in pressure due to a fluctuating wave height can be used to draw in or exhaust air from a chamber through a turbine. Hydraulic PTO requires some form of mechanical stroking motion in order to pump high-pressure fluid through a motor. Direct drive systems also exist where relative movement between two bodies—in most cases a linear motion—can be converted directly into electricity.

Most wave devices under development today use a hydraulic or turbine PTO. Direct drive linear or rotary generators may provide a route to reduced costs within future generations of WEC. These PTO design variations within a particular type of device can also be observed in tidal energy devices. There are subtle differences between tidal devices and that of wave energy converters, and the various PTO permutations for tidal energy converters are shown in Fig. 7.14 (SI-Ocean 2016). Mechanical PTO utilizes a gearbox and generator configuration similar to certain designs of wind turbine. Hydraulic PTO uses stroking or rotary motion to pump high-pressure fluid through a motor. Direct drive generators can convert slow speed mechanical rotary motion directly into electricity without the need for a gear box. Most tidal turbines under development today use mechanical or direct drive PTO connected to a horizontal or vertical axis prime mover.

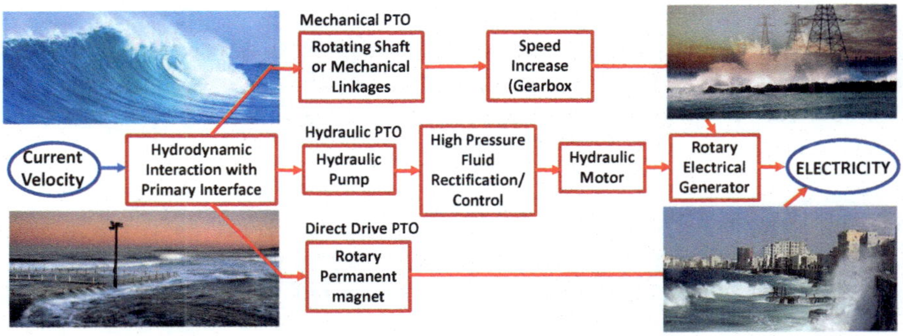

Fig. 7.14 Tide energy converter power take-off (PTO) permutations

7.5.1.4 Control

Systems and software to safeguard the device and optimize the performance under a range of operating conditions. Control systems may adjust certain parameters of the device autonomously in order to ensure favorable operation (SI-Ocean 2016).

7.5.1.5 Installation

The method of placing the structure and device at its power generating location. This includes all vessels and ancillary equipment needed to fully deploy an ocean energy device (SI-Ocean 2016).

7.5.1.6 Connection

The cables and electrical infrastructure for connecting the power output from the device to the electricity network. In order to export the generated electricity to the grid, power conditioning systems and transformers will be needed, which provide a grid code compliant electrical output.

Any ocean energy system has the need for electrical connection between the generating device and the local grid network. The identifiable ocean energy resource is often situated away from densely populated areas; the resource far outweighs the demand from local communities in many cases. It is intended that electricity generated through ocean energy will be transported to regions where there is greater demand. For this to take place, the necessary infrastructure for electricity transportation is needed. There are two stages of grid connection in an ocean energy device. The first stage is to connect the device to an export cable, bringing electricity to shore. Device connection can be carried out using dry-mate or wet mate connectors. Dry mate connectors require the connector to be above the water surface when the connection is made. The connected cable is then lowered in to position. Wet-mate connectors allow a connection to be made subsea, which would allow an ocean energy device to be placed into position prior to the cable being connected. The second stage of grid connection involves voltage step-up and connection to the distribution and transmission grid network. On-shore substations and transformers will step up the voltage to the applicable level before connection to the distribution or transmission grid (SI-Ocean 2016).

7.5.1.7 Operations and Maintenance

Periodic repair and reconditioning work will be required on all ocean energy devices. As well as the physical maintenance of mechanical and electrical components within the device, operation and maintenance will also have to consider access to the device, and device retrieval (SI-Ocean 2016).

INDEPTH: Marine Bioenergy

Marine algae (such as seaweed) are potential sources of MRE. Marine biomass can be fermented to produce biomethane and/or biohydrogen. Circular economy concepts include (Borthwick 2016): the use of MRE to power aquaculture systems offshore; the growth of seaweed adjacent to fish farms to reduce eutrophication; the harvesting of seaweed on "ripening" at the end of the summer season; the ensiling of seaweed for pretreatment and year-round provision of feedstock for a biodigester; the reaction of hydrogen generated from surplus MRE with biogas, resulting in the upgrading of biogas to biomethane, almost doubling the methane output ($4H_2 + CO_2 = CH_4 + 2H_2O$); and gas grid injection for use as green gas. In essence, the green gas becomes the energy vector for the distribution of both the MRE and the bioenergy from the seaweed. The energy vector is now readily available for renewable thermal energy, renewable transport energy in natural gas vehicles, and offsite renewable electricity. This use of marine algae offers a resolution to the controversy surrounding the sustainable production of biodiesel and bioethanol from biomass produced on land, where there is competition between food, fuel, and other land uses. Marine algae have very fast growth rates, act as energy reservoirs, and can sequester carbon. Further research and development is necessary to establish the industrial-scale production of algal biofuels.

7.5.1.8 Ocean Energy Storage and Needed Advanced Materials, Robotics, and Informatics

Ocean energy output is highly variable. Being created by weather systems, wind and waves are essentially stochastic, seasonal, and subject to inter-annual variability. Tides are cyclic, and comprise multiple oscillating constituents, including semi-diurnal and diurnal components and monthly neap-spring modulations caused by the relative positions of the Sun and Moon with respect to the Earth. The timescales vary from a few hours to many years. Energy storage is therefore essential in order to rectify ocean energy output. Large volume storage options presently under consideration include pumped hydro-storage, hydrogen storage through electrolysis, compressed air energy storage, and substitute natural gas. For example, in the power-to-gas concept, electrolysis is used to generate hydrogen, which is then converted to methane that can be added to the natural gas network. The seascape provides a complex mixture of opportunities in terms of material science, technology, and manufacturing. The ocean climate is particularly harsh and variable; seawater is corrosive; and many different designs have been proposed for marine devices. For example, in axial-flow turbines, the cantilevered rotor blades must be extremely strong to survive in seawater. Novel materials with improved strength, fatigue, and anti-corrosion properties are already on the horizon. Advanced composite materials, such as glass-, carbon-, and basalt-fiber-reinforced polymers, look to be ideal candidates for cost reduction and increased durability. Access to MRE facilities is expensive and hazardous. Challenges include remote monitoring, the use of robots for operational support, and real-time weather forecasting for predictive maintenance. All of these are required in order to ensure that devices can survive in extreme sea states as they arise. In addition, of course, MRE informatics systems are the key to improved information management, monitoring, and decision-making. Big data and high-performance computing are also very relevant here (Borthwick 2016).

7.5.2 Wave Energy Converters

At a simplified level, wave energy technology can be located near-shore and offshore. Wave energy converters can also be designed for operation in specific water depth conditions: deep water, intermediate water or shallow water. The fundamental device design will be dependent on the location of the device and the intended resource characteristics. The energy contained within the waves manifests itself in the form of kinetic motion of water particles, with the energy imparted to the waves from the wind. The particle motion varies relative to the water depth. Far offshore, particle motion is circular. As the waves approach the shore, interference from the seabed (drag) causes the water particle motion to become more ellipsoidal in shape. At the shore, waves break as the upper surface of the wave begins to travel faster than the water particles closer to the sea bed. The orbital particle motion decreases exponentially with increasing water depth, which can be seen in the reducing path size for water particles deeper within the water column (SI-Ocean 2016).

The near-shore environment is more benign and accessible than the deep water environment, but still allows access to a significant extractable wave energy resource. The near shore environment is shielded from the largest ocean waves, and offers an increased directionality with regards to force in the surge direction. As waves approach the shore, the wave speed and wave length decrease, resulting in an increased energy per unit area, known as wave shoaling. On the other hand, there is a trade-off, as compared to the deep offshore environment there is a lower resource in the near shore waves, as energy is lost due to drag with the seabed.

Wave energy levels increase predominantly with increasing westerly distance from shore. The areas of deep water that are suitable for wave device deployment are significantly larger than the areas available for near-shore device deployment. There are many more constraints to development in the near shore environment, including geotechnical constraints, which reduce the relative level of practical near shore resource. There are also challenges in the deep offshore environment, as water depth can increase costs significantly. Technology and financial limitations restrict

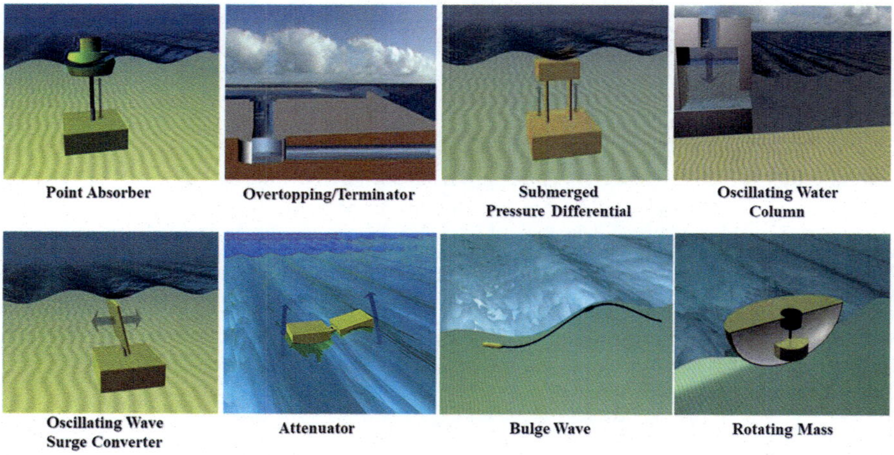

Fig. 7.15 Wave energy converter (WEC) component types (Modified from Aqua-RET (2012) Wave technologies. www.aquaret.com. Credit: European Commission (EU Lifelong Learning Program Agreement no LLP/LdV/TOI/2009/IRL-515)): (**a**) attenuator, (**b**) point absorber, (**c**) oscillating wave surge converter (OWSC), (**d**) oscillating water column (OWC), (**e**) overtopping/terminator, (**f**) submerged pressure differential, (**g**) bulge wave, and (**h**) rotating mass

the depth to which deployment can take place, but emerging technologies may open up markets in water depths of greater than 250 m. The location of a wave energy device will largely influence the type of mooring used. Shoreline devices will typically require significant civil engineering works to integrate the device into a natural rock face or a manmade breakwater. Near-shore devices will make use of either pinned pile foundations, or will rely on gravity mass to hold the device in place. Devices located offshore have the option of tight moorings or slack moorings, which may be dependent on the type and location of the structure and PTO system. As with other design decisions, the exact design of WECs will be the result of a cost-engineering exercise that aims to minimize the levelized cost of electricity from the WEC (SI-Ocean 2016).

Various techniques have been designed to extract energy from the waves, as shown in Fig. 7.15 (Aqua-RET 2012; SI-Ocean 2016). Each type of device is designed to extract energy in different ways using the surge, heave, or sway motions of the waves (or a combination of each). The majority of ocean energy converters is fabricated from steel that offers good fatigue and stress limits. Steel-Reinforced Concrete or Fiber-Reinforced Polymer (FRP) is investigated to use for certain components. FRP offers some cost and weight saving advantages over steel, but the fatigue and stress limits are not yet well understood in comparison to steel. This material has been proven in the marine environment, albeit in a very different application to ocean renewable energy, and is used extensively in other applications such as civil infrastructure and boat design. Other wave devices are being designed to use rubber or other flexible materials as the main structural component.

7.5.2.1 Attenuator

Attenuator (A) type wave energy converters use the energy within oncoming waves to induce an oscillatory motion between two (or more) adjacent structural components. The motion can be resisted by hydraulic rams which pump high-pressure hydraulic fluid through a motor, or by a direct drive power take off system, to generate electricity. Attenuator type wave energy converters can be surface floating or fully submerged, the former is most common. Attenuators tend to yaw automatically to face the predominant wave direction and "ride" the waves (Drew et al. 2009; SI-Ocean 2016).

INDEPTH: Typical Classification for Wave Energy Converters (Mofor et al. 2014)

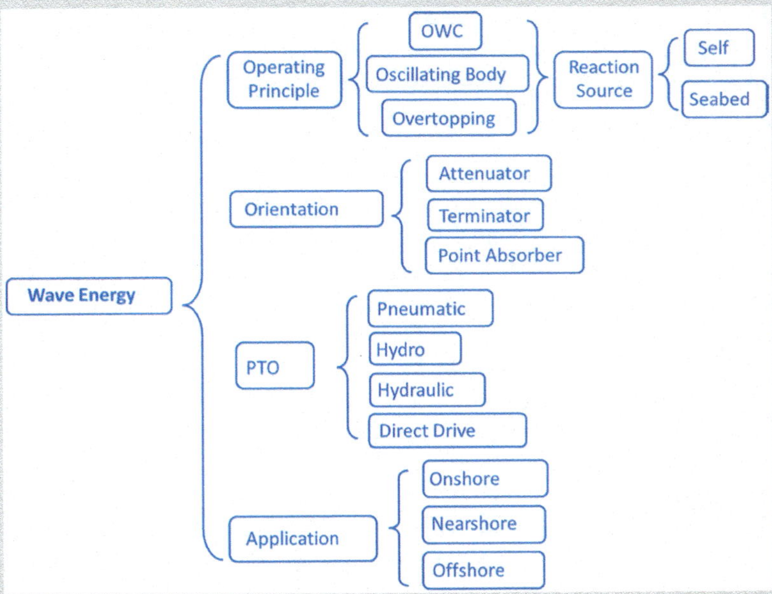

7.5.2.2 Point Absorber

A point absorber is a device that possesses small dimensions relative to the incident wavelength. They can be floating structure that heave up and down on the surface of the water or submerged below the surface relying on pressure differential (Drew et al. 2009).

Point Absorber (B) type devices use buoyant forces to induce a heaving motion of one body relative to a secondary fixed body. The fixed body may be moored to the sea bed, or held in place by gravitational forces through a large foundation mass. Point absorbers are nondirectional, as they can receive incoming waves from any incident angle. Because of their small size, wave direction is not important for these devices (Drew et al. 2009; SI-Ocean 2016).

7.5.2.3 Oscillating Wave Surge Converter (OWSC)

An oscillating wave surge converter is generally comprised of a hinged deflector, positioned perpendicular to the wave direction (a terminator), that moves back and forth exploiting the horizontal particle velocity of the wave. It is generally located in near-shore regions, where the water particle motion becomes more ellipsoidal in shape; these devices use the surge motion of the waves to induce oscillating motions of a body in the horizontal direction. OWSC are typically bottom mounted devices, fixed directly to the sea bed, but concept floating OSWC devices are under development (Drew et al. 2009; SI-Ocean 2016).

7.5.2.4 Oscillating Water Column

Oscillating water columns (OWC) use a chamber that is part filled with water to drive air through a turbine. Variations in the wave height cause the water in the column to rise and fall, acting like a large piston on the volume of air within the chamber. As the water level rises, the pressure in the chamber rises, and air is exhausted from the chamber driving a turbine. When the water level decreases, the air flow reverses and air is drawn into the chamber, once again driving a turbine. A low-pressure Wells turbine is often used in this application as it rotates in the same direction irrespective of the flow direction, removing the need to rectify the airflow. It has been suggested that one of the advantages of the OWC concept is its simplicity and robustness (Drew et al. 2009; SI-Ocean 2016).

OWC devices can be contained within a fixed structure at the shoreline, located near shore bottom as a bottom mounted structure, contained within a man-made breakwater, or can moored in deeper water as a floating system (SI-Ocean 2016).

7.5.2.5 Overtopping

An overtopping device (also known as a terminator device) converts wave energy into potential energy. The design causes waves to break across the device, and the surge energy in the breakers allows water to be collected in a reservoir above the free water surface. Water contained in the reservoir can produce energy by flowing through a low-head hydraulic turbine. Overtopping devices can shore based as part of a shoreline structure or man-made breakwater, or they can be floating devices located in deeper water.

7.5.2.6 Pressure Differential

Pressure Differential devices rely on oscillating hydrodynamic pressure caused by passing waves. The devices can be floating or fully submerged. Submerged devices experience an induced motion as waves pass over the device, creating a temporary vertical force on the body. It comprises two main parts: a sea bed fixed air-filled cylindrical chamber with a moveable upper cylinder. As a crest passes over the device, the water pressure above the device compresses the air within the cylinder, moving the upper cylinder down. As a trough passes over, the water pressure on the device reduces and the upper cylinder rises. An advantage of this device is that since it is fully submerged, it is not exposed to the dangerous slamming forces experienced by floating devices, and reduces the visual impact of the device (Drew et al. 2009).

In contrast, floating pressure differential devices could utilize the increased pressure due to passing waves to compress air through a turbine (SI-Ocean 2016).

7.5.2.7 Bulge Wave

In bulge wave technology, a flexible tube filled with fluid is moored to the seabed, allowing the device to orientate into oncoming waves. As waves pass over the device, differential pressure will cause the water contained within the flexible tube to be squeezed, forming a bulge wave. This bulge wave travels along the device, at a speed proportional to the wave velocity and the flexibility of the tube, gaining energy as the bulge grows. This energy can be used to drive a turbine located at the end of the tube.

7.5.2.8 Rotating Mass

Rotating mass devices utilize the wave motion to cause pitch and roll of a floating body. Within the floating body, an eccentric mass will be excited, and will begin to rotate. The rotation will drive an electrical generator contained within the device. A further permutation of the rotating mass device uses gyroscopic effect.

7.5.2.9 Other Types of Wave Energy Converters

Other types of wave energy converters employ a novel or unconventional technique for extracting energy from the waves. One such radical concept in the wave energy sector is the Anaconda, which uses bulge wave technology. The device is essentially a large rubber tube filled with water, moored to the sea bed. Anaconda will float just below the surface of the water, and will align itself to face the incident wave direction. As a wave passes the device, the rubber tube will lift and become squeezed by the surrounding wave, and a "bulge" of water will form within the rubber tube. This bulge will travel the length of the device, gathering energy from the wave as it progresses through the tube. Resonance can be achieved by ensuring that the speed of the bulge wave is identical to the speed of the forcing ocean wave, this ensures that high power capture is achieved. The bulge wave will drive a generator located at the stern of the device. A full-scale Anaconda is anticipated to be around 200 m long, with a diameter of 5.5 m and power output of 1 MW.

Another new concept under development is the AlbaTERN Squid. A central buoyant "absorber" is filled with water so that it sits just below the surface. The absorber is moved by the passing waves, and the relative motion between the absorber and the link arms is used to pump hydraulic fluid through a generator, producing electricity.

Inevitably radical new concepts will require refining, a development process likely to take a number of years. Advances in computational modeling have also accelerated the rate at which designs can be tested in a wide range of simulated conditions. Innovative technology could be game changing in the challenge to achieve a cost-effective ocean energy converter; however, this is not a near-term solution, and much work is required to progress the design from concept to commercial operation (SI-Ocean 2016).

INDEPTH: Composites for Ocean Energy Conversion Systems

Concerning tidal turbines, the Seaflow project, a 300 kW prototype two-blade tidal turbine immersed off the Devon coast in June 2003, used 11 m diameter composite rotor blades instrumented with strain gauges. The construction involved a central 65 mm thick carbon composite spar covered by stiffened glass/epoxy fairings. Also in 2003 a Norwegian tidal power station was installed near Hammerfest, using 10 m long glass fiber composite blades. In 2008 within the Seagen project a 1.2 MW twin turbine was immersed, using a similar design to Seaflow again from Marine Current Turbines. The blades were longer than those trialed in the Seaflow project, each 7.5 m long. It was reported in July that one of the turbines had shed two blades following a computer incident but the trial is continuing and is being followed closely. Another large-scale prototype (500 kW) with 6 m composite (glass and carbon) blades was immersed by Tidal Generation Limited in 2009. In the USA, in the RITE project, following development trials with a 3 m diameter turbine, six turbines were installed in 2007 in New York. These use 4.9 m diameter turbines each with three composite blades, similar to existing wind turbine designs. Some blade failures were noted and the blades were redesigned. The first French tidal turbine prototype was developed within the Sabella project. This was a six-blade turbine made of glass fiber composite fairings on steel spars, which was immersed off the Brittany coast in April 2008. The blades are composite sandwich structures, including a central metallic spar with glass-reinforced composite facings on a foam core. Other composite tidal turbine blades have been used in the OpenHydro project, conducted on composite blades as part of a study for an 11 m diameter blade for a demonstration plant in Japan. In addition to turbine blades, composites are also of interest for shrouds, mounting frames, and other components of these systems. Composites are also of interest for wave energy devices. For example, Wave Dragon is a composite structure for a full-scale, multimegawatt device (their 20 kW prototype is made of steel), the Pelamis prototype used composites and within FO3, the Norwegian wave power project, a generator is integrated in a floating platform construction which is built in composite (Boisseau et al. 2012).

7.5.3 Tidal Energy Converters

There are three principal hydraulic mechanisms by which tidal currents operate (SI-Ocean 2016): Tidal streaming, hydraulic current and resonant basins. Tidal streaming occurs as a result of the need for continuity within a fluid flow: As water flows through a constriction, the flow is accelerated. Hydraulic currents occur when two large bodies of water are connected, but are out of phase or have nonconcurrent tidal ranges; the difference in water level in each body of water creates a pressure head, and the flow of water from one body of water into the other results. The third mechanism, a resonant basin, occurs when constructive interference between an incoming tidal wave and a reflected tidal wave generates a standing wave.

Tidal energy offers some advantages over other renewable resources such as wind and wave. The fluid medium, sea water, is over 800 times denser than air, so tidal power offers a greater energy density than wind for a given turbine rotor swept area. As the movement of tides result from gravitational forcing, the tides flow with a predictable intermittency. That is the variability is deterministic (and not stochastic like wave or wind), so this eases the integration of tidal energy into existing electricity networks. Providing a sufficiently long data set exists (35 days is the recommended duration), predictability of the tides is possible through a process known as harmonic analysis, hence tidal velocities can be predicted to a good accuracy indefinitely, both future and retrospectively. The operating principle behind tidal energy converters is that the energy contained within the moving current is harnessed by a device that extracts kinetic energy from the flow and imparts this into a mechanical motion of a rotor or foil. The device then converts the mechanical motion of the structure into electrical energy by means of a power take-off system. Before connection to the electricity grid, the electrical power output from the device will need to be conditioned in order to make it compliant with grid code regulations. In essence, tidal device operation is synonymous to that of a wind turbine, albeit operating within a different fluid medium (SI-Ocean 2016).

Tidal energy converters can be designed for operation in specific water depth conditions: deep water, intermediate water or shallow water. The classification of devices is shown in Fig. 7.16 (Aqua-RET 2012; SI-Ocean 2016): Horizontal Axis Turbine, Vertical Axis Turbine, Oscillating Hydrofoil, Enclosed Tips (Ducted), Helical Screw, Tidal Kite, and Others. Although most enclosed tips (ducted) turbines fall under the same horizontal axis design principle, the addition of a duct can have both positive and negative impacts when compared to an un-ducted device. A duct can accelerate the flow into the rotor, and could also align flow while reducing the turbulence of the flow within the duct. This is more favorable in terms of reducing fatigue on the rotor. Where ebb and flood tides are not perfectly bidirectional, a ducted turbine could be an advantage. By placing a duct around a rotor, a structure is put in place that will make it more difficult for marine mammals to evade the rotating components of the turbine. One significant disadvantage of a ducted structure is the additional drag penalty that will be faced. This drag penalty will extract energy from the flow that will not be recoverable downstream. Energy dissipation can be minimized through interaction with the foundation structure of the device (SI-Ocean 2016).

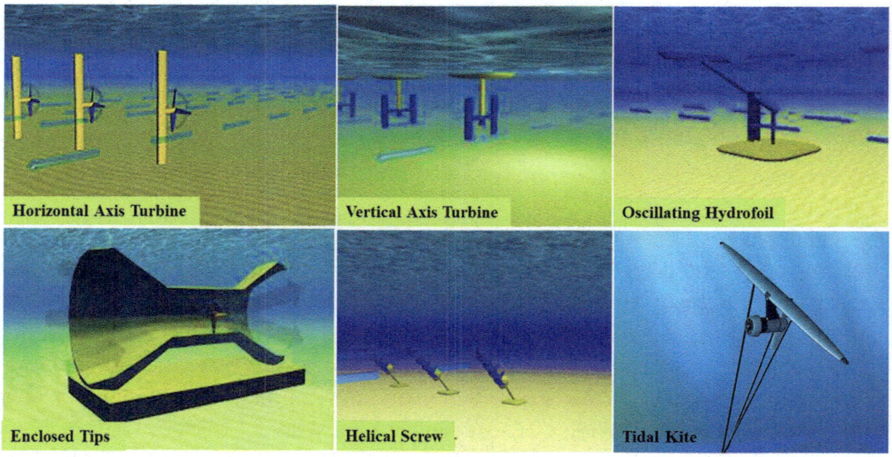

Fig. 7.16 Tide energy converter component types (Modified from Aqua-RET (2012) Wave technologies. www.aquaret.com. Credit: European Commission (EU Lifelong Learning Program Agreement no LLP/LdV/TOI/2009/IRL-515)): (**a**) horizontal axis turbine, (**b**) vertical axis turbine, (**c**) oscillating hydrofoil, (**d**) enclosed tips (ducted), (**e**) helical screw, and (**f**) tidal kite

7.5.3.1 Horizontal Axis Turbine

Horizontal axis turbines utilize lift generated by blades to turn a rotor. Energy is extracted from the tidal flow and causes the rotation of a turbine mounted on a horizontal axis. The rotation is converted to electrical energy through use of a generator (Aqua-RET 2012).

7.5.3.2 Vertical Axis Turbine

Vertical axis turbines, similar to the horizontal axis turbines, utilize lift generated by blades to turn a rotor. Energy is extracted from the tidal flow and causes the rotation of a turbine mounted on a vertical axis. The rotation is converted to electrical energy through use of a generator (Aqua-RET 2012).

7.5.3.3 Oscillating Hydrofoil

The oscillating hydrofoil device consists of a hydrofoil located at the end of a swing arm. Control systems alter the pitch of the foil to create either lift or downforce, moving the foil in an oscillatory motion. This motion can be used to pump hydraulic fluid through a motor. The rotational motion that results can be converted to electricity though a generator (Aqua-RET 2012).

7.5.3.4 Enclosed Tips (Ducted)

Enclosed Tips (Ducted) devices are essentially contained within a shrouded structure. The duct may be used to accelerate and concentrate the fluid flow, allowing the use of smaller rotor diameters. Other ducted structures could help to minimize turbulence and align the flow of water into the turbine (Aqua-RET 2012).

7.5.3.5 Helical Screw

Helical screw type turbines are a variation on vertical axis turbines that draw power from the tidal stream as the water flows up through the helix (Aqua-RET 2012).

7.5.3.6 Tidal Kite

Tidal kite designs, in which a tethered kite "flies" a small turbine through the flow, effectively increase the relative velocity entering the turbine. These dynamic devices could generate electricity from significantly lower-velocity currents, or use much less material than static TECs (Aqua-RET 2012).

7.5.3.7 Other Tidal Energy Converters

There may be other novel tidal turbine concepts in the development process, which utilize different means of extracting energy from the flow of fluid. Current concepts use sea bed mounted devices, but radical new concepts are looking to remove the need for large foundation structures, making greater use of buoyancy and tensioned cables. There are also some designs that are challenging the conventional horizontal axis turbine approach. These radical concepts may open up routes to significant step change cost reduction. By their very nature, radical concepts are not necessarily proven technologies, and may require significant development before a definitive judgment can be made on whether they do, in fact, reduce the cost of energy (SI-Ocean 2016).

INDEPTH: Typical Tidal Turbine Classifications (Mofor et al. 2014)

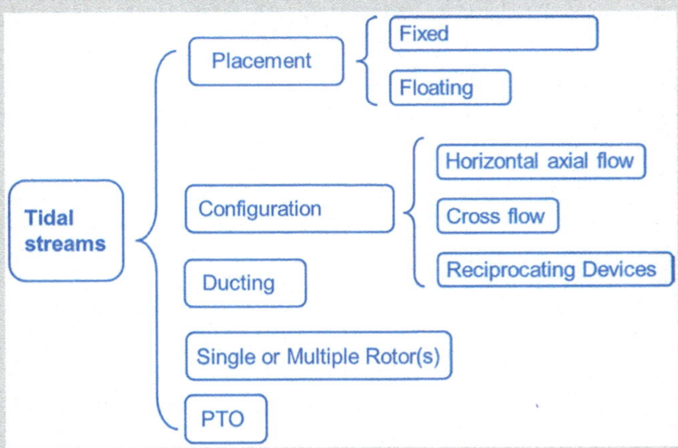

7.5.3.8 Mooring Options and Foundation Structures

In addition to the different turbine designs, there are several mooring options that can be considered for fixing a tidal turbine to the sea floor, as shown in Fig. 7.17 (SI-Ocean 2016).

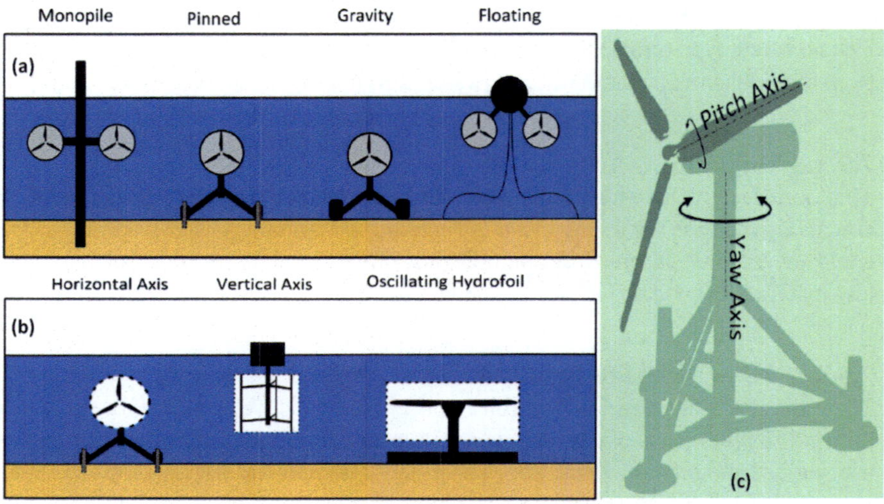

Fig. 7.17 Mooring options for fixing a tidal turbine to the sea floor (Adapted from SI-Ocean (2016) Ocean Energy: State of the Art. www.si-ocean.eu/. Credit: Strategic Initiative for Ocean Energy (SI OCEAN), European Ocean Energy Association, Belgium): (**a**) tidal energy converter foundation types, (**b**) tidal energy converter swept area, (**c**) yaw and pitch

Monopile: A single tubular steel tower can be drilled and grouted into a deep socket in the sea bed. Surface piercing foundation types of this design are generally limited to approximately 30 m water depth. Monopile foundations could be used for fully submerged devices as an alternative to gravity or pinned foundations.

Pinned: Foundation structures can be pinned by drilling and grouting small sockets in the sea bed. These anchor points may utilize pins of several meters in length, but will generally be shorter than the drill depth required for monopole foundations. These foundations are suitable for turbines mounted close to the bottom of the water column.

Gravity Base: This foundation type will hold a tidal energy converter to the sea bed by means of a substantial mass, with the gravitational forces keeping the device fixed in place. These foundations are suitable for turbines mounted close to the bottom of the water column.

Floating: Buoyant turbine devices can be moored to the sea bed using either flexible or rigid moorings. There may also be an option of mounting multiple devices on one floating platform. Designs of this type can access the faster flowing currents located higher within the water column, but it is possible that there may be some increased cyclic loading and fatigue stress caused by complex interactions due to the circular wave particle motion.

Many of the tidal energy converter concepts at the forefront of the industry have adopted a horizontal axis device. While this concept has received the greatest attention, there is little convergence in the design of foundation and support structures. While horizontal axis devices are more common, the vertical axis and oscillating hydrofoil designs intersect a larger area for a given rotor diameter than horizontal axis designs. The tidal energy resource is very site specific. Much of the existing identified resource lies in close proximity to a significant land mass, although there are limited grid connection opportunities at present, and grid reinforcement will be necessary to take advantage of the available resource. A significant challenge for the tidal energy industry is demonstrating the survivability of a device. The marine environment is far harsher than that of other onshore renewables, with extremely high loading due to the density of water. Although certain aspects of device loading can be predicted, the effect of turbulence on devices is still an area of research. In order to survive the extreme load scenarios that could occur during storm conditions, the device must be carefully engineered to protect it from damage.

Yaw: Horizontal axis tidal devices may contain systems that allow orientation of the device to face the oncoming tidal flow. This is known as yaw. Devices with yaw capabilities may have an active yaw system, in which motors turn the device (about the vertical yaw axis) to face the oncoming tidal flow. Generally, yaw movements will be carried out at slack tides, where the forces on the turbine are at the lowest value. A device with passive yawing makes use of hydrodynamic forces to align the device with the flow.

Pitch: In order regulate the loading faced by the rotor, and to optimize the efficiency of the rotor over a wide range of flow speeds, certain rotor designs incorporate pitching blades. Pitch systems allow control of the angle of attack of the blades, giving greater control over the rotor loading. Certain devices have blades that can pitch 180° to allow for the changing direction of the tide therefore allowing the blade to face both ebb and flood tidal flow directions. This 180° pitch mechanism is generally found on devices without the ability to yaw.

Yaw and blade pitch mechanisms add mechanical complexity to system design. Additional components increase the chance of failure, so removal of yaw or blade pitch system could increase system reliability. Devices without pitching blades will require a means of carrying out an emergency stop of the device to protect the components from damage, for situations such as loss of grid connection. The force on a fixed pitch blade in an emergency situation will require substantial braking power in order to prevent movement of the rotor. There are both positives and negatives for the implementation of a pitch or yaw system, and these need to be considered carefully in system design to ensure that the added cost and complexity if offset by increased yield.

7.5.4 Ocean Energy Converter Arrays

Ocean energy devices are generally modular in design, each device utilizing only a small portion of the total resource potential at a given site. At an array scale, ocean energy system is likely to consist of multiple marine energy converters, connected by subsea cables providing a means of transporting electricity to a common transformer or grid connection, much in the same way as wind farms are currently developed, as shown in Fig. 7.18 (Ingram et al. 2011; SI-Ocean 2016).

While an understanding of device design is important, it is fundamental to the ocean energy that the focus does not remain on individual devices. Configurations involving multiple devices, or arrays, will provide the route to commercialization,

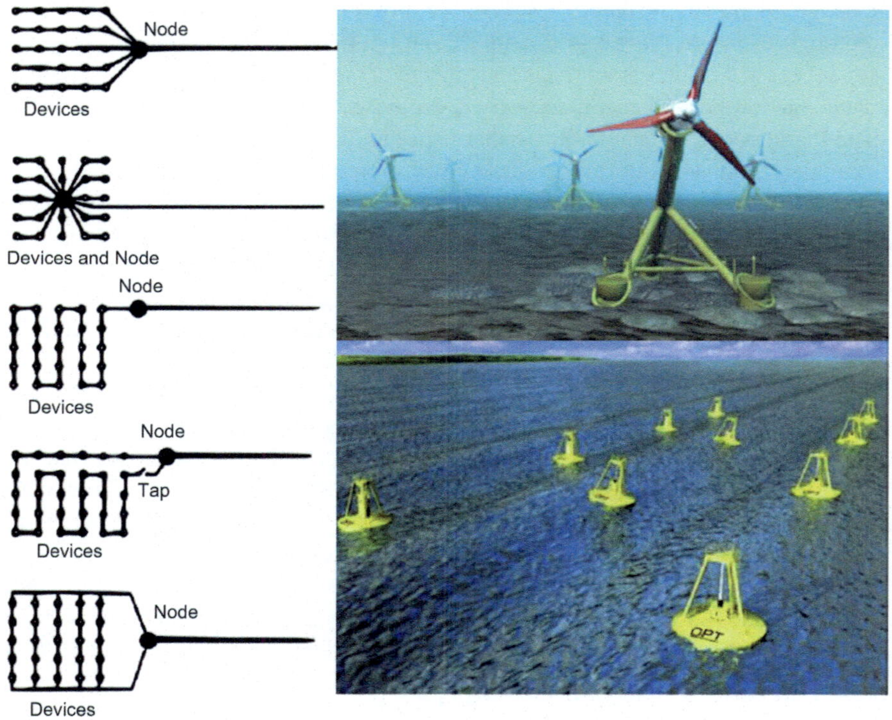

Fig. 7.18 Examples of ocean energy arrays (Modified from Ingram et al. 2011 (The University of Edinburgh, available under Creative Commons License); and SI-Ocean (2016) Ocean Energy: State of the Art. www.si-ocean.eu. Credit: Strategic Initiative for Ocean Energy (SI OCEAN), European Ocean Energy Association, Belgium)

and growth of the industry. Array can benefit from the economies of scale. A move towards larger scale manufacturing, as opposed to single unit production, will unlock cost savings in the form of a reduced unit cost and increased repeatability within component manufacture. Cost reduction through sharing of significant infrastructure systems expenses between multiple devices, as opposed to those costs being borne by a single device deployment, can also help to make array more economical (SI-Ocean 2016).

In order to successfully make the transition from single device to array deployment, several technology challenges need to be met. In addition, there is a need for continued cost and risk reduction to facilitate increased investor confidence. Within an array, the maximum number of devices that can be connected in one circuit is limited, due to voltage drop across a length of cable and the maximum capacity of a given cable. There are several options available for arrangement of device layout, each of which will result in different levels of power loss, reliability and overall cost. It should be noted that the actual layout of the ocean energy array will be determined by the geotechnical conditions at the proposed site, and the resource itself. This will have an impact on the electrical infrastructure and grid architecture (SI-Ocean 2016).

The spacing of devices will become increasingly important as arrays grow in size. While it is desirable to minimize the overall electrical infrastructure required for a given array, it must be noted that hydrodynamic interaction could lead to reductions in the overall power output. It is therefore necessary to consider a wide range of parameters when configuring an array and the end result will be a compromise to achieve the lowest levelized cost of energy (SI-Ocean 2016).

The electrical configuration of offshore wind farm arrays may influence the design of ocean energy arrays. The wave or tidal arrays will require grid connection that will likely require transformers, electrical switchgear, and backup generators and batteries. Purpose built sub stations located offshore may be a requirement for future ocean energy arrays, and will bring added complexity and cost to the design of a project (SI-Ocean 2016).

7.6 Challenges Faced by the Ocean Energy

For ocean energy to be sustainable, it must satisfy economic, environmental, and societal constraints. Impacts should be determined, adverse effects identified, and mitigation measures designed before ocean energy device deployment takes place. The economic constraint is primarily related to the relative cost of marine electricity with respect to that of other power sources, allowing for capital and recurrent costs, carbon obligations, government subsidies, and market volatility. Ocean energy devices can alter flow patterns, introduce noise, and be potential hazards, and they may affect marine biodiversity. Societal acceptance is related to employment prospects, aesthetic concerns, stakeholder involvement, and the well-being of communities (Borthwick 2016).

More specifically, the technical challenges for WEC and TEC focus on (SI-Ocean 2016): Improving yield, increasing reliability, and reducing CAPEX and OPEX costs. However, cost reduction is an essential requirement for accelerated and sustained deployment of ocean energy. Innovation could play a role in order to reduce the technology and development costs. Innovation could be in the form of incremental enabling technology development, improving certain aspects of economic performance or efficiency, or innovation could be through radical new concepts opening up larger areas of resource, or creating a step change in device deployment costs. Furthermore, advances in the predictability of array power output will help with the integration of ocean energy arrays into new and existing grid infrastructure. The understanding of individual device performance will require augmentation with an investigation into device performance at an array level, understanding the nature of the complex interaction between the resource and multiple devices at different geographical locations within the array, and the overall impact of multiple energy extraction devices within the wave regime or tidal flow. While some of these challenges are being proven at a device level, the sector will face a new set of challenges as the move from single devices to arrays takes place. Manufacture, installation, and operation of several devices will require streamlined manufacturing and installation processes, together with optimized maintenance routines. All these challenges must be met at a cost which is affordable, as although all technical barriers and challenges may be achievable with significant investment.

7.6.1 Predictability

Waves are stochastic in nature, and prediction or estimation of wave height and period requires first to know the wind magnitude and direction. Wave period and significant wave height is dependent on a number of factors beyond wind direction, such as fetch and bathymetry, making the predictability of wave energy output a challenging task. Wave predictability is generally possible a number of days in advance, as waves result from wind action across the surface of the ocean, and so wave energy does not fluctuate instantaneously with wind. Each wave energy converter type will respond differently in varying wave climates. Improvement of the predictability of energy output from a given wave climate is an area that can be developed for specific devices and, as developers gain increased levels of at-sea testing, power matrixes will be developed that show power output and device response to given sea conditions.

The predictability of tidal flows is well recognized and understood, and the timing and duration of spring tides and neap tides will vary in accordance with the movement of the sun and moon, with respect to the rotation of the earth. Tools to help enhance the understanding of turbulence, however, and the effects of turbulence on component fatigue life require further work to develop current understanding. The predictability of device behavior in energetic flows is not widely understood, and predictability of component life will require substantial device deployment in order to quantify (SI-Ocean 2016).

7.6.2 Manufacturability

Manufacturability of ocean energy converters will improve as devices move from the first full-scale prototypes into commercial production. Current devices are bespoke designs but, in the future, devices will be a commodity. Key links within the supply chain, such as component suppliers and service providers, are now beginning to undertake development work for ocean energy device manufacturers, as ocean energy is perceived to be a future growth area.

At present there is limited design consensus among wave energy converters, and this has a direct impact on supply chain choices. A streamlined manufacturing process will allow for benefits in both learning-by-doing and economies of scale. The design of device components and subsystems could allow for optimized manufacturing techniques. There is scope for alternative materials to be considered as a substitute for steel which is widely used in present technology. New materials will bring different challenges in terms of manufacturing tolerances, and the scale of component that can be created.

For tidal energy there is more convergence, particularly focusing on horizontal axis turbines. There is scope for a greater level of cost reduction through economies of scale if consensus is achieved. The design of devices will be fundamental in creating opportunities for ease of access for installation and O&M once deployment takes place (SI-Ocean 2016).

7.6.3 Installability

The installability of a wave energy converter depends on its location. While shore-based and near-shore wave energy converters require substantial foundations and infrastructure, the technology that will operate in medium to deep water can often be towed to location. Installation will also depend on the sea-bed characteristics and in a scoured sea-bed environment where drag-anchors cannot be used alternative mooring techniques will need to be adopted. Clump mass moorings, or pin piled moorings may be an alternative and there will be some cross transfer from the tidal energy sector that can help progress this.

For the tidal energy, the challenge of installability requires a process that can be carried out quickly, within the limited time period offered by the slack tide, and economically. Current installation techniques are financially intensive due to the bottom mounted design of many first generation technologies, and their substantial foundation requirements. These installation costs will result in a substantial portion of the total project costs for the vast majority of projects within the pipeline. Future installation techniques will need to address these issues, and provide a reliable and economical method of access to devices for installation and O&M purposes (SI-Ocean 2016).

Both the wave and tidal energy systems have a need for affordable installation vessels, and affordable installation techniques. Much of the device installation and intervention that has taken place to date has utilized vessels from the oil and gas sector, the price of which can fluctuate greatly depending on demand and spot-market price.

7.6.4 Operability

The operability of ocean energy converters must be proven for all wave and tidal energy technology developers in the offshore environment. While some device developers have entered extensive test periods in increasingly energetic sea states, the remote operation of a device, continuously, through both more benign and highly intensive sea states, has yet to be completed. Demonstration of reliable device operation and survivability under extreme conditions will improve with continued testing, but these parameters remain a challenge for both the wave and tidal energy sector (SI-Ocean 2016).

7.6.5 Survivability

Ocean energy converters must be able to survive both their expected operational loading, and also the extreme loading presented by storm conditions. The ratio of extreme loads to operational loads is greater for wave energy than it is for tidal energy, so wave energy converters require increased "over-engineering" in order to operate within the more frequent average loading regime, and also survive the storms. Tidal energy is more predictable, and providing adequate resource characterization at a specific site has been carried out, the maximum loadings for a bottom mounted device can be calculated. Seabed mounted, fully submerged TECs will need to consider wave loading due to the fatigue stresses and loads that will result. Buoyant or semi-submersible tidal device designs may have additional loading requirements similar to WECs, due to their structural and mooring requirements for storm waves (SI-Ocean 2016).

7.6.6 Reliability

The challenge of long-term, reliable operation is still a requirement for many technology developers in both wave and tidal energy systems. Individual component Mean Time Between Failure (MTBF) and life expectancy is unproven, and in order to achieve significant periods of operation in between maintenance a greater understanding of these parameters will be required. Increased reliability will reduce unplanned maintenance requirements (SI-Ocean 2016).

7.6.7 Affordability

Continued efforts at improving the affordability of wave and tidal energy will require innovation, and cost reduction will be targeted for successive models of device. While demonstration projects require a significant capital investment in order to produce first-of-a-kind engineering structures and components, in the long-term device affordability will define the project economics. It is understood and accepted that the current costs of ocean energy technologies are higher than that of the more mature offshore wind sector; however, it is essential that the tidal energy sector demonstrates a reduction in costs in order to become competitive with alternative forms of renewable energy production (SI-Ocean 2016).

Generally, levelized cost of electricity (LCOE) is defined as the ratio of the sum of the costs over a lifetime to the sum of electrical energy produced over a lifetime. It provides the net present value of the unit cost of electricity over the lifetime of a given electricity source. The estimates show that cost reduction is essential if ocean energy is to succeed. Government and private sector support for the ocean energy industry is critical, but a further barrier to ocean energy exploitation arises from uncertainty in future subsidies and a lack of private sector investment. Many aspects of the supply chain leading to tidal power require R&D aimed at cost reduction and increased reliability, including advanced materials, manufacturing, installation, power take-off, and energy storage (Borthwick 2016).

7.6.8 Environmental and Societal Constraints

Ocean energy development is also hampered by a lack of accurate reference environmental data and the propagation of uncertainties through predictive models used to estimate power extraction and its impact on the marine ecosystem. Field data are difficult and expensive to obtain. An improved understanding is needed of the impact of device-device interactions at basin scale, and the long-term ecological side-effects of marine power plants and device farms will not be known until information is available from post-installation monitoring campaigns. Ocean energy devices alter the local flow hydrodynamics, affecting blockage, bypass currents, wakes, mixing, turbulence, sediment transport, littoral drift, scour, turbidity, seabed morphology, biodiversity, food availability, and water quality. Renewable energy device foundations and support structures could act as artificial reefs improving biodiversity, but might attract invasive species. Biofouling may improve species abundance, but can lead to higher sedimentation rates and eutrophication, while antifouling chemicals can be deleterious to certain species. There is concern that certain fish and marine mammals could collide with moving rotor blades, and this danger is exacerbated by poor visibility and locations in zones of high energy. Research is badly needed on marine animal well-being under prolonged exposure to noise, electromagnetic radiation, and habitat exclusion (Borthwick 2016).

On other hands, a "social gap" exists between public support for renewables (due to local employment, cheaper electricity, energy security, and lower carbon emissions) and the lower success of planning applications (due to visual impact; unconcern about climate change; a wish to prevent oceans turning into industrial zones; and damage to tourism, navigation, fisheries, property values, recreation, and social cohesion). Planning and decision-making processes may cause increased opposition through poor public engagement. This issue is best resolved through improved communication and participation between all stakeholders, though at the cost of a longer, more expensive consultation process (Borthwick 2016).

7.6.9 Sustainability Assessment

In the context of ocean energy development, sustainability must incorporate economic, societal, environmental, and institutional functions. A sustainability assessment should identify impact generators; set benchmarks and targets; and note the sufficiency of tidal power resources, the accessibility of energy supply and related services, the productive use of energy, resilience to hazards, and equity between different users and generations. Here, sufficiency can be viewed as the capacity of the ocean basin to satisfy its various energy demands (including those of the ecosystem as well as electricity generation). Access relates the availability of the resource to the different stakeholder communities. Productivity concerns the translation of marine power into economically useful electricity. Resilience is a measure of the ocean system's capacity to mitigate and adapt to changes. Equity refers to the fair distribution of ocean energy resources between different stakeholders and across generations. Sustainability indicators should be selected systematically, with the aim of creating a logical, well-structured framework for tidal system sustainability assessment. A suitable holistic approach is offered by the process analysis method (PAM), which examines trade-offs between different domains of sustainability, such as economic development, environmental performance, and social well-being, allowing for better management and exploitation of ocean resources (Borthwick 2016).

7.6.10 Perspective on Ocean Energy

Given the enormous theoretical reserves of ocean energy, the development of new technologies for exploiting it appears to be a very worthwhile endeavor for addressing the energy gap, helping to ensure energy security, and reducing global carbon emissions from burning fossil fuels. Ocean energy can be extracted from offshore wind, tides, ocean currents, and thermal and salinity gradients in the sea, and bioenergy can be obtained from algae. Worldwide research and development activities are being directed towards tackling the technological, economic, social, and environmental barriers to commercial-scale exploitation of the ocean energy seascape. Advances are required in cost reduction, energy storage, advanced materials, robotics, informatics, investment, consenting and licensing procedures, and maritime governance. The global challenge remains of how to exploit the ocean energy seascape in order to power whole cities by ocean energy in a way that is sustainable, robust, and cost-effective. As in the industrial revolution, a new generation of engineers is required that possesses the ingenuity and boldness to meet this global challenge (Borthwick 2016).

Exercises

Part I: General Questions

7.1 What is hydropower? Describe its advantages and disadvantages.
7.2 List types of hydropower plants, and compare their commons and differences.
7.3 List types of hydropower turbines, and compare their commons and differences.
7.4 List types of dams, and compare their commons and differences.
7.5 What types of materials are usually used for building dikes?
7.6 List structural materials and surface coatings for hydropower turbines, and compare the differences of their performance.

7.7 The Californian city of Fontini obtains half of its drinking water from a reservoir in the hills above the city and half from groundwater. Water from the reservoir arrives at the town's Sandhill water treatment facility at a pressure of 12 bars and groundwater is pumped electrically to the surface at the same pressure. The treatment facility requires water at a pressure of 1 bar to operate and the town has replaced its pressure reduction valves with a small hydro turbine that will reduce pressure and generate 2300 MWh of electricity each year. What is the annual production of electricity from renewable energy at the facility?

7.8 Based on the figure below, deduce the wave power per unit length. Suppose crest-to-trough height of wave is h, wavelength is λ, wave period is T, and the wave shape follows the sine function. Given: surface wavelength $\lambda = gT^2/(2\pi)$.

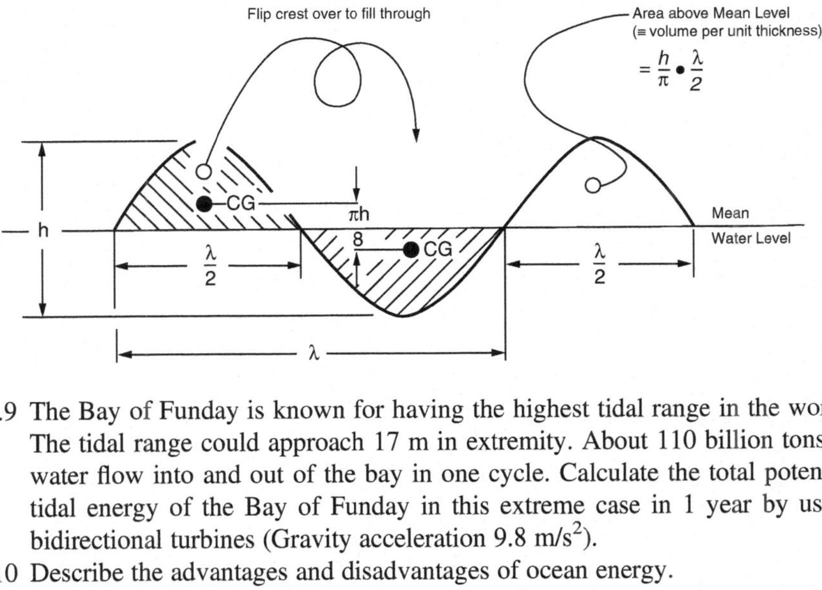

7.9 The Bay of Funday is known for having the highest tidal range in the world. The tidal range could approach 17 m in extremity. About 110 billion tons of water flow into and out of the bay in one cycle. Calculate the total potential tidal energy of the Bay of Funday in this extreme case in 1 year by using bidirectional turbines (Gravity acceleration 9.8 m/s^2).

7.10 Describe the advantages and disadvantages of ocean energy.

7.11 Address the current ocean energy technologies and their development readiness.

7.12 Give examples to show the role of advanced materials in ocean energy systems.

7.13 List typical wave energy converters and materials used to make them.

7.14 List typical tidal energy converters and materials used to make them.

Part II: Thought-Provoking Questions

7.15 Assuming the most efficient manner for extraction, and a ready supply of other necessary materials not mentioned herein, and given the current estimates about the volume of Earth's ocean, how much energy (in Joules) could be extracted via fusion, given the deuterium in the ocean?

7.16 Summarize composite materials used for ocean energy conversion systems.
7.17 Describe major challenges faced by the ocean energy and your perspective on future trends of ocean energy.

References

Adhikary, P., Roy, P., Mazumdar, A.: Selection of hydro-turbine blade material: application of fuzzy logic (MCDA). Int. J. Eng. Res. Appl. **3**(1), 426–430 (2013)

Aqua-RET: Wave technologies. http://www.aquaret.com/indexea3d.html?option=com_content&view=article&id=203&Itemid=344&lang=en. (2012). Accessed 6 Sept 2017

BDS: About dams. http://www.britishdams.org/about_dams/gravity.htm. (2010). The British Dams Society Accessed 8 Aug 2017

Blight, G.E.: Construction of Tailings Dams. Case studies on Tailings Management, pp. 9–10. International Council on Metals and the Environment, Paris (1998). isbn:1-895720-29-X

Boisseau, A., Davies, P., Thiebaud, F.: Sea water ageing of composites for ocean energy conversion systems: influence of glass fiber type on static behavior. Appl. Compos. Mater. **19**(3–4), 459–473 (2012)

Borthwick, A.G.L.: Marine renewable energy seascape. Engineering. **2**, 69–78 (2016)

Burman, K., Walker, A.: Ocean Energy Technology Overview, Prepared for the U.S. Department of Energy, Office of Energy Efficiency and Renewable Energy Federal Energy Management Program. DOE/GO-102009-2823. http://large.stanford.edu/courses/2013/ph240/lim2/docs/44200.pdf (2009). Accessed 22 Aug 2017

Das, E.M.: Advances in Rockfill Structures, p. 341. Kluwer Academic, Dordrecht (1991)

DOE: Hydropower technology basics. https://energy.gov/eere/energybasics/articles/hydropower-technology-basics (2013). Accessed 1 Aug 2017

Drew, B., Plummer, A.R., Sahinkaya, M.N.: A review of wave energy converter technology. Proc. Inst. Mech. Eng. A. **223**(8), 887–902 (2009)

Forehand, D.I.M., Kiprakis, A.E., Nambiar, A.J., Wallace, A.R.: A fully coupled wave-to-wire model of an array of wave energy converters. IEEE. Trans. Sustain. Energy. **7**(1), 118–128 (2016)

Gummer, J.: Combating silt erosion in hydraulic turbines. Hydro Rev. **17**(1), (2009). http://www.hydroworld.com/articles/print/volume-17/issue-1/articles/combating-silt-erosion-in-hydraulic-turbines.html. Accessed 9 Aug 2017

Gunn, K., Stock-Williams, C.: Quantifying the global wave power resource. Renew. Energy. **44**, 296–304 (2012)

Henkel, M.: 21st Century Homestead: Sustainable Agriculture II: Farming and Natural Resources. Lulu.com. (2015). isbn:9781312939684

Høeg, K.: Asphaltic Concrete Cores For Embankment Dams, 1993 Norwegian Geotechnical Institute, Publication No 201. (1997)

Huckerby, J.A., Jeffrey, H., Moran, B.: An international vision for ocean energy. ocean energy systems implementing agreement. http://www.oceanrenewable.com/wp-content/uploads/2011/05/oes_vision_brochure_2011.pdf (2011). Accessed Aug 2017

IEA-ETSAP and IRENA. Hydropower Technology brief. http://www.irena.org/DocumentDownloads/Publications/IRENA-ETSAP_Tech_Brief_E06_Hydropower.pdf (2015). Accessed 1 Aug 2017

Ingram, D., Smith, G., Bittencourt-Ferreira, C., Smith, H.: Protocols for the equitable assessment of marine energy, vol. 213380, 1st edn. The Institute for Energy Systems, School of Engineering, The University of Edinburgh, Edinburgh (2011)

James, P., Chanson, H.: Historical development of arch dams—from Roman arch dams to modern concrete designs. Aust. Civil Eng. Trans. **CE43**, 39–56 (2002)

Lewis, M.J., Neill, S.P., Hashemi, M.R., Reza, M.: Realistic wave conditions and their influence on quantifying the tidal stream energy resource. Appl. Energy. **136**, 495–508 (2014)

Mofor, L., Goldsmith, J., Jones, F.: Ocean energy-technology readiness, patents, deployment status and outlook. http://www.irena.org/DocumentDownloads/Publications/IRENA_Ocean_Energy_report_2014.pdf (2014). Accessed 31 July 2017

Neill, S.P., Jordan, J.R., Couch, S.J.: Impact of tidal energy converter (TEC) arrays on the dynamics of headland sand banks. Renewable Energy **37**(1), 87–397 (2012)

Oerlikon Metco: SF-0023.1—Robust coating solutions for hydropower turbines extend operating life and maintain efficiency. https://www.oerlikon.com/ecomaXL/files/oerlikon_SF-0023.1_HydroTurbineCoatingSolutions_EN.pdf&download=1 (2014). Accessed 10 Aug 2017

Onder, H., Yilmaz, M.: Underground dams—a tool of sustainable development and management of ground resources. Eur Water. **11**(12), 35–45 (2005)

Peters, N.: Dike design and construction guide—best management practices for British Columbia. http://www.env.gov.bc.ca/wsd/public_safety/flood/pdfs_word/dike_des_cons_guide_july-2011.pdf (2003). Accessed 8 Aug 2017

Salter, S.H., Taylor, J.R.M.: Vertical-axis tidal-current generators and the Pentland Firth. Proc. Inst. Mech. Eng. A. **221**(2), 181–199 (2007)

SI-Ocean: Ocean Energy: State of the Art. http://si-ocean.eu/en/upload/docs/WP3/Technology%20Status%20Report_FV.pdf (2016). Accessed 31 July 2017

Spicher, T.: Choosing the right material for turbine runners. Hydro Rev. **32**(6), (2013). http://www.hydroworld.com/articles/hr/print/volume-32/issue-6/articles/choosing-the-right-material-for-turbine-runners.html. Accessed 9 Aug 2017

WCD: Dams and Development: A New Framework for Decision-Making: The Report of the World Commission on Dams. Earthscan, London (2000). isbn:1-85383-798-9

Biomass for Bioenergy

8

Abstract

Bioenergy technologies have been deployed for widespread sustainable exploitation of biomass resources in order to efficiently utilize bioenergy and at the same time to guarantee greenhouse gas emission savings for biofuels and bio-liquids. Unlike other renewable energy sources, biomass can be converted directly into biofuels to help meet transportation fuel needs, for instance. The development of advanced materials for bioenergy has been covered a wide range areas: high strength, wear- and corrosion-resistant structural materials such as steel, alloys, and protective coatings, high durability polymers and ceramics; catalysts, allowing for higher selectivity and yield, improved stability and functionality such as bi-/multifunctional catalytic systems; advanced ceramic, polymeric, or metallic membranes for gas separation and separation of inhibitory or intermediary products from biomass pretreatment, efficient separation/recycling of enzymes, the immobilization of cells, and downstream processing in continuous separation of fermentation products needs materials solutions for advanced membranes; hydrolytic enzymes and novel microorganisms; as well as photosynthesis and photosynthetic process materials. Breaking down cellulose, the chemically resistant building blocks of plants, for instance, requires aggressive chemical processes and catalysts, and materials with long lifetimes to contain and manipulate these corrosive chemistries. The cellular membranes of algae are rich in the raw materials for production of hydrocarbon chains of gasoline and diesel fuel, but need their own special chemical routes and catalytic materials for conversion. Many of these chemical processes and catalysts exist in nature, such as in the digestive systems of termites, where cellulose is converted to sugars that can be further fermented to alcohol. Advanced materials and analytical tools are needed to understand the subtleties of these natural fuel production processes, and then to design artificial analogs that directly and efficiently produce the desired end fuels. This chapter will provide a brief review about the advanced materials for biomass processing and bioenergy utilization.

C. Tong, *Introduction to Materials for Advanced Energy Systems*,
https://doi.org/10.1007/978-3-319-98002-7_8

8.1 Perspectives

The demand for energy is increasing in the world due to the rapidly growing global population and urbanization. Throughout history, mankind has used wood as a source of producing energy. After the industrial revolution, the main source of energy shifted to fossil fuels. The accurate amounts of the world's total fossil fuel reserves are not known. However, the potential declining of fossil fuel reserves and the pollution caused by petroleum-based energy sources have created high-cost and serious environmental problems, such as global warming. Such concerns about fossil fuels have led to the utilization of alternative energy sources (Akia et al. 2014). Compared with other renewable energies, such as solar, wind, hydroelectric, and geothermal power, biomass has been seen as a major renewable and nonfossil energy source to supplement declining fossil fuels. Modern biomass refers to the various organisms produced by photosynthesis using air, water, and land. Bioenergy, coming from biomass, is derived from plant and animal material, such as wood, forest waste, agricultural waste, aquatic plants, oil plants, city and industrial organic waste, and solid animal waste. Biomass is considered an attractive feedstock for energy production because it is a renewable and widely distributed energy resource, and can be developed sustainably in the future. Furthermore, it has positive environmental properties, such as biomass' low sulfur and nitrogen (relative to coal) content and nearly zero net CO_2 emission levels, allows biomass to offset the higher sulfur and carbon contents of fossil fuels. In addition, biomass fuel is abundant, which is why biomass energy today has become the world's fourth largest energy source following coal, oil, and natural gas, indicating its significant economic, societal, and environmental potential (Kurchania 2012; Li et al. 2017).

Biomass is a renewable energy source not only because the energy in it comes from the sun, but also because biomass can regrow over a relatively short period of time compared with the hundreds of millions of years that it took for fossil fuels to form. Through the process of photosynthesis, chlorophyll in plants captures the sun's energy by converting carbon dioxide from the air and water from the ground into carbohydrate-complex compounds composed of carbon, hydrogen, and oxygen. When these carbohydrates are burned, they turn back into carbon dioxide and water and release the energy they captured from the sun. There are many types of biomass that can be used to derive fuels, chemicals, and power—such as plants, agricultural and forestry residues, organic components of garbage (municipal solid waste), and algae. This broad diversity of suitable biomass has resulted in increased research and development of technologies to produce fuels, products, and power at an industrial scale (Li et al. 2017).

The development of bioenergy technology varies with time and will be affected by many factors, such as technical, environmental, economic, and social concerns, which include infrastructure, facilities, cost, price, geography, transportation, competitors, and so on. Burning wood in the furnace is an example of a centuries old traditional use of biomass to produce thermal energy. The so-called direct combustion was traditionally used, including stove combustion, boiler combustion, and biomass briquette combustion. In this case, biomass is the biofuel without intermediary, the chemical energy of the combustible components of the biomass is converted into thermal energy. Modern use of this method involves the use of technologies to reduce waste, increase the amount of transmitted energy, and improve fuel performance. The advanced conversion technologies have been developed to deepen the process of biomass in order to obtain from it a more energy-intensive and pollution-free fuel grades. For example, thermochemical conversion and biological conversion become the dominant designs to develop a stable bioenergy product structure. Gasification and pyrolysis are two major thermochemical conversion technologies, and anaerobic fermentation of biogas, bioethanol, and biodiesel are biochemical conversion technologies. These technologies have the advantages of low energy consumption, high conversion rates, easy industrialization, and have significant environmental benefits. However, threats to forests and biodiversity; increases of food prices; and the competition for water resources are the key negative impacts of these technologies. Therefore, there is a need to develop more conversion processing paths and new technology to efficiently utilize biomass for bioenergy (Goryunov et al. 2016; Li et al. 2017).

The global energy potential of virgin biomass is very large. It is estimated that the world's standing terrestrial biomass carbon (i.e., the renewable, above-ground biomass that could be harvested and used as an energy resource) is approximately 100 times the world's total annual energy consumption. The largest source of standing terrestrial biomass carbon is forest biomass, which contains about 80–90% of the total biomass carbon. Marine biomass carbon is projected to be next after the forest biomass carbon in terms of net annual production, but is last in terms of availability because of its high turnover rates in an oceanic environment. The main features of how biomass is used as a source of energy and fuels are schematically illustrated in Fig. 8.1. Conventionally, biomass is harvested for feed, food, fiber, and materials of construction or is left in the growth areas where natural decomposition occurs. The decomposing biomass or the waste products from the harvesting and processing of biomass, if disposed on or in land, can be partially recovered after a long period of time as fossil fuels. The energy content of biomass could be diverted instead to direct heating applications by collection and combustion.

INDEPTH: Modern Biomass Technology

Biomass energy is supplying about 10–15% (or 45 $\pm$ 10 EJ) of today's demand. Biomass feedstock include a broad range of organic material such as wood, wood-based energy crops, corn stover, grass, algae, wheat straw, rice straw, corn, miscanthus, nonedible oils, green and wood landfill waste, animal fats, waste frying oils, agricultural residues, municipal wastes, forest product wastes, paper, cardboard, and food waste. Most biomass is produced through photosynthesis. The photosynthesis yield of approximately 720 billion tons per year makes the largest organic raw material cellulose resource in the world.

(continued)

In general, all kinds of biomass can be used as feedstock including starchy, triglyceride, and lignocellulosic feedstock. Biomass can be converted to biofuels and biopower via thermochemical and biochemical processes. Thermochemical conversion is a significant route for producing products such as bio-methanol, biodiesel, biooil, bio-syngas, and biohydrogen. Biochemical conversion routes produce liquid or gaseous fuels through fermentation or anaerobic respiration. The production of biofuels through thermochemical conversion processes with a broad range of technologies has drawn the most attention in the world. The main advantage of thermochemical processes for biomass conversion in comparison to other methods such as biochemical technologies is the feedstock used. There are three main routes for biomass' thermochemical conversion including combustion, gasification, and pyrolysis. Combustion is the most direct and technically easiest process which converts organic matter to heat, carbon dioxide, and water using an oxidant. Gasification of biomass is a heating process within the presence of an oxidant produces a mixture of carbon monoxide and hydrogen referred to as synthesis gas (syngas) by partial oxidation. Gasification has many advantages over combustion. It can use low-value feedstock and convert them into electricity and also vehicle fuels. Pyrolysis is a thermal heating of materials in the absence of oxygen, which produces three forms of products including gases, pyrolytic oil, and char. Pyrolytic oil, also known as "tar or bio-oil," which is viscous, corrosive, relatively unstable, and chemically very complex, cannot be used as transportation fuel directly due to its high oxygen value (40–50 wt.%), high water content (15–30 wt.%), and also low H/C ratios. Biomass gasification/pyrolysis is one of the promising technologies used for converting biomass to bioenergy (Akia et al. 2014).

Alternatively, biomass and any wastes that result from its processing or consumption could be converted directly into synthetic organic fuels if suitable conversion processes were available. Another route to energy products is to grow certain species of biomass such as the rubber tree (Hevea braziliensis), in which high-energy hydrocarbons are formed within the species by natural biochemical mechanisms, or the Chinese tallow tree (Sapium sebiferum), which affords high-energy triglycerides in a similar manner. In these cases, biomass serves the dual role of a carbon-fixing apparatus and a continuous source of high-energy organic products without being consumed in the process. Other biomass species, such as the herbaceous guayule bush (Parthenium argentatum) and the gopher plant (Euphorbia lathyris), produce hydrocarbons too, but must be harvested to recover them. Conceptually, Fig. 8.2 shows some pathways by which energy products and synthetic fuels can be manufactured (Sharma et al. 2014; MYPP 2015).

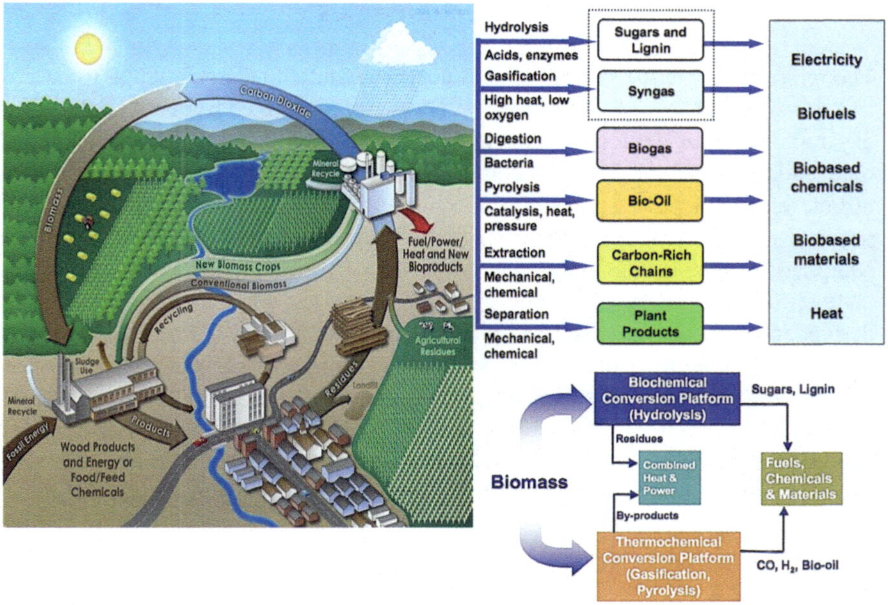

Fig. 8.1 Schematic illustration various bioenergy conversion processes (Modified from Ochsner (2010), available at: http://mcensustainableenergy.pbworks.com/w/page/20637999/bioenergy. Courtesy: PB Works)

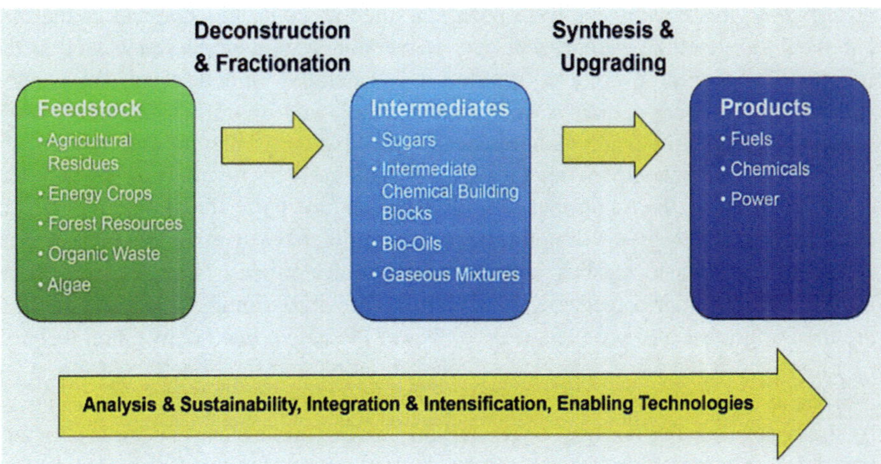

Fig. 8.2 Generalized conversion route for biomass-derived feedstocks to renewable products (Adapted from Quadrennial Technology Review 2015, Chapter 7: Advancing Systems and Technologies to Produce Cleaner Fuels—Technology Assessments. https://www.energy.gov/sites/prod/files/2016/01/f28/QTR2015-7A-Bioenergy-Conversion_0.pdf. Credit: US Department of Energy)

8.2 Biomass Conversion Technologies

The only natural, renewable carbon resource known that is large enough to be used as a substitute for fossil fuels is biomass. Bioenergy generated from biomass consists of solid, liquid, or gaseous fuels. Liquid fuels can be used directly in the existing road, railroad, and aviation transportation network stock, as well as in engine and turbine electrical power generators. Solid and gaseous fuels can be used for the production of electrical power from purpose-designed direct or indirect turbine-equipped power plants. Chemical products can also be obtained from all organic matter produced. Additionally, power and chemicals can come from the use of plant-derived industrial, commercial, or urban wastes. Biomass resources include primary, secondary, and tertiary sources of biomass. Primary biomass resources are produced directly by photosynthesis and are taken directly from the land. They include perennial short-rotation woody crops and herbaceous crops, the seeds of oil crops, and residues resulting from the harvesting of agricultural crops and forest trees (e.g., wheat straw, corn stover, and the tops, limbs, and bark from trees). Secondary biomass resources result from the processing of primary biomass resources either physically (e.g., the production of sawdust in mills), chemically (e.g., black liquor from pulping processes), or biologically (e.g., manure production by animals). Tertiary biomass resources are post-consumer residue streams including animal fats and greases, used vegetable oils, packaging wastes, and construction and demolition debris (Mayaki 2008).

As for illustrating convert of biomass resources into power, heat, and fuels for potential use, Fig. 8.1 summarizes the various bioenergy conversion technologies that can be combined into pathways from feedstock to product (PBworks 2014). Historically, these pathways have been classified as either biochemical or thermo-chemical to reflect the primary catalytic conversion system employed as well as the intermediate building blocks produced. Generally, biochemical conversion technologies involve pathways that produce sugars and lignin intermediates, while thermochemical conversion technologies involve pathways that produce bio-oil and gaseous intermediates. Moving forward, as evidenced by a variety of hybrid processes in industry, the traditional division between biochemical and thermochemical conversion technologies will not encompass the diversity of innovative conversion technologies, and focus has shifted to a simpler process flow in which the bio-derived polymeric feedstock is deconstructed into simpler intermediates that are then upgraded into products (Fig. 8.2) (MYPP 2015). Each conversion technology involves at least two main steps: deconstruction and fractionation of feedstock into relatively stable intermediates through the breaking of chemical bonds followed by the controlled upgrading of those building blocks into a slate of desired products (synthesis and upgrading). These renewable products can include finished fuels, fuel precursors, chemicals, or high-quality intermediates, such as sugars, lignin derivatives, synthesis gas (syngas), or stabilized bio-oils. Specific process operating conditions, inputs, and outputs vary within and between each step. These process variations impact key performance outcomes (such as titer—a method of expressing concentration, rate, selectivity, and yield) that in turn determine the overall economic

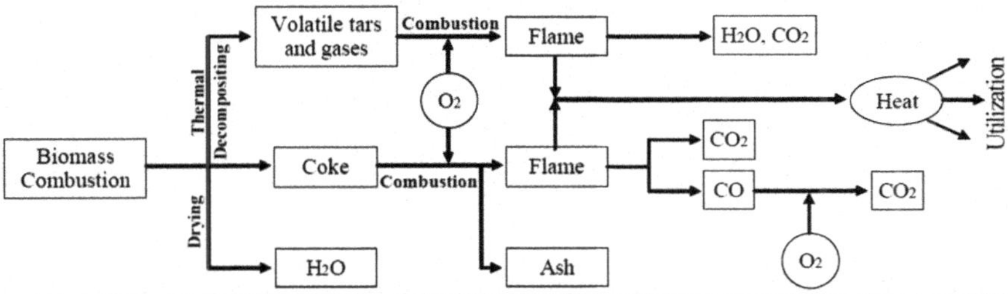

	Stove Combustion	Boiler Combustion	Biomass Briquettes Combustion
Advantages	Easy to operate and least investment.	A higher efficiency; achieving industrial production.	High volumetric energy density, high homogeneity of the fuel.
Disadvantages	The lowest efficiency and waste biomass resources; air pollution.	High investment cost.	High operating costs.
Suitable Range	Dispersing independent household stoves in rural or mountainous areas.	Large-scale utilization of biomass.	Enterprises to carry out technological transformation of the original equipment.
Types	Fuel-saving stove, etc.	Underfeed stokers, grate stokers, fluidized bed boilers.	Wet pressure molding, hot press molding, raw material carbonization.

Fig. 8.3 The combustion process of biomass fuels and comparison of three main evolutionary paths (Modified from Li et al. 2017 (MDPI, Basel, Switzerland), under the terms and conditions of the Creative Commons Attribution (CC BY) license))

viability of the process. Potential environmental impacts also need to be assessed for conversion pathways by evaluating sustainability metrics (water use, soil quality, etc.) and conducting life-cycle assessments (LCAs) to determine greenhouse gas (GHG) and other emissions.

8.2.1 Direct Combustion

Direct combustion of biomass is a traditional method for the earliest use of biomass energy and is the process of using biomass as a fuel to produce energy without the use of chemical conversion. The biomass fuel combustion process involves both a strong chemical reaction process and a heat and mass transfer process between the fuel and the air. In addition to the fuel, this type of combustion requires an adequate heat and air supply. As shown in Fig. 8.3, the biomass combustion process can be broadly divided into preheating, drying (evaporation), an analysis of volatility, and a coke (fixed carbon) combustion process (Koppejan and Van Loo 2012; Li et al. 2017). The direct combustion can be mainly divided into three main evolutionary paths: stove combustion, boiler combustion, and biomass briquette combustion.

8.2.1.1 Stove Combustion

Stove combustion is the most primitive method, and is commonly used for household furnaces in rural or mountainous areas. While this method reduces the need for high investment, its efficiency is the lowest. This kind of old stove uses crop straw, firewood, grass, and dried animal dung as its fuel. Since the supplied air is not always sufficient, fuel combustion is not always completed, and the thermal efficiency of an old wood stove (effective heat and fuel heat ratio) is very low. Biomass fuel used for stove combustion is often regarded as the only available and affordable energy source which can meet basic needs, like cooking and heating. However, direct combustion for cooking and heating in rural areas typically brings indoor air pollution and the associated adverse health impacts. Further, the relatively low thermal efficiency is a problem, as it wastes energy resources (Ceeeta 1995; Li et al. 2017).

8.2.1.2 Boiler Combustion

Boiler combustion using modern technology is suitable for the large-scale use of biomass, as it can achieve high efficiency, and is suitable for industrial production, but its high investment costs means that it is not suitable for small-scale distributed use. The biomass raw materials used for boiler combustion mainly involve the burning of forestry offcuts, immature trees, wood processing and paper mill waste, rice husk, bagasse, and crop straw. There are many kinds of biomass fuel boilers depending on the type of fuel to be used, such as firewood stoves, straw furnaces, and incinerators. Depending on the different boiler combustion modes, these can be further divided into underfeed stokers, grate stokers, and fluidized bed boilers (Li et al. 2017).

For example, Spreader-Stoker system has been used in many refuse-derived fuels (RDF) facility for converting solid wastes, and the fluidized bed combustion units. In a spreader-stoker system, the fuel is introduced into the firebox above a grate. Smaller particles will tend to burn in suspension and larger pieces will fall onto the grate. Most units, if properly designed, can handle biomass with moisture content as high as 50–55%. Moisture contained in the fuel is driven off partially when the fuel is in suspension and partially on the grate. The feed system should provide an even thin layer of fuel on the grate. In a fluidized bed combustor (FBC), the fuel particle burns in a fluidized bed of inert particles utilizing oxygen from the air. Advantages of fluidized bed combustion include (Capareda 2011): (a) high heat transfer rate, (b) increased combustion intensity compared to conventional combustors, and (c) absence of fouling and deposits on heat transfer surfaces. A fluidized bed combustor is similar to that of a fluidized bed gasifier. The only difference is the use of excess air for combustion processes and starved air for gasification processes. So far FBC has been used mostly for coals. A number of wastes, e.g., wastes from coal mining and municipal wastes, are also sometimes incinerated in fluidized beds. It has been suggested that certain quick-maturing varieties of wood could be combusted in fluidized beds for generation of steam. Granular biomass fuels, e.g., paddy husk and chips of wood, have been successfully combusted in fluidized beds of sand particles. Conventional combustion of paddy husk is slow and inefficient. Nearly complete combustion and high combustion intensities of paddy husk can be achieved in a fluidized bed combustor. The same combustor can also be used for burning wood. Combustion intensities up to about 500 Kg/hr-m^2 have been achieved in fluidized bed combustors using biomass fuels (Capareda 2011).

In fact, the biomass boiler has been widely used in the United States, Brazil, and some European countries to supply hot water and heating, with some being used for power generation and some being used for both power generation and heating

(Li et al. 2017). Combustion systems for electricity and heat production are similar to most fossil fuel-fired power plants. The biomass fuel is burned in a boiler to produce high-pressure steam. This steam is introduced into a steam turbine, where it flows over a series of turbine blades, causing the turbine to rotate. The turbine is connected to an electric generator. The steam flows over and turns the turbine. The electric generator rotates, producing electricity. This is a widely available, commercialized technology (Mayaki 2008).

8.2.1.3 Biomass Briquettes Combustion

Biomass briquettes combustion is a relatively new type of fuel combustion path. Due to its irregular shape and size, high moisture content, and low bulk density, biomass is quite difficult to handle, transport, store, and utilize in its original form. One solution to the above problems is to increase the density of biomass materials into pellets, briquettes, or cubes. Biomass briquettes combustion uses this high-density solid fuel in traditional coal-fired equipment to improve thermal efficiency. After solidification, the raw material has high volumetric energy density and high homogeneity of the fuel, and the pollution emissions are much lower than coal, making this technology a highly efficient clean and renewable energy. There are a variety of processes used for biomass compression molding. Depending on the technical characteristics, biomass compression molding can be divided into three basic types: wet pressure molding, hot press molding, and raw material carbonization. These kinds of combustion methods use traditional coal-fired burning equipment after curing, which can reduce the biomass raw material volume needed, greatly increase the fuel energy density, and improve the utilization efficiency (Li et al. 2017).

A comparison of the three techniques is also shown in Fig. 8.3. The original stove combustion technology is easy to operate and requires a relatively low investment, but the combustion efficiency is also very low and there are air pollution problems. Boiler combustion has a higher efficiency, but because of its high investment, it is more suitable for large-scale production. The relatively new biomass briquettes combustion has greatly improved combustion efficiency with its greater density, but the operating costs are also very high. All in all, the relatively low thermal efficiency and pollution problems of direct combustion have resulted in the main biomass technological developments due to the increasing market demand and industrial competition. With the transition from old to newer technologies, new biomass technological applications, such as thermochemical conversion and biological conversion, become the dominant designs in the biomass conversion with a stable product structure.

8.2.2 Thermochemical Conversion

Compared with other techniques, biomass thermochemical conversion technology has the advantages of low energy consumption, high conversion rate, and easy industrialization.

Biomass thermochemical conversion technology mainly uses gasification and pyrolysis technology, from which high-grade energy products, like charcoal, tar, and combustible gas, can be obtained. Other methods include carbonization and catalytic liquefaction. These processes do not necessarily produce useful energy directly, but under controlled temperature and oxygen conditions, can convert the original biomass feedstock into more convenient forms of energy carriers. These carriers are either more energy dense and therefore reduce transport costs, or have more predictable and convenient combustion characteristics allowing them to be used in internal combustion engines and gas turbines.

8.2.2.1 Gasification

Gasification processes provide a competitive route for converting various, highly distributed, and low-value lignocellulosic biomass to synthetic gas for generation of a broad sort of outputs: electricity, heat and power, liquid fuels, synthetic chemicals, as well as hydrogen (H_2) production. The importance of gasification is that it is not constrained to a particular plant-based feedstock, and thus any lignocellulosic biomass can be considered appropriate (Akia et al. 2014).

In fact, gasification is the partial oxidation of an organic feedstock to produce a syngas—a combustible gas consisting of carbon monoxide, carbon dioxide, hydrogen, methane, water, and nitrogen, along with contaminants like small char particles, ash, and tars. The gas is cleaned to make it suitable for use in boilers, engines, and turbines to produce heat and power. Typically, the feedstock is biomass or waste, and varying the process conditions allows control over proportions of the compounds in the syngas. In this approach to the development of fixed carbon supplies from renewable carbon resources is to convert CO_2 outside the biomass species to synthetic fuels and organic intermediates. The ambient air, which contains about 360 ppm by volume of CO_2, the dissolved CO_2 and carbonates in the oceans, and the earth's large carbonate deposits, could serve as renewable carbon resources. But since CO_2 is the final oxidation state of fixed carbon, it contains no chemical energy. Energy must be supplied in a chemical reduction step. A convenient method of supplying the required energy and of simultaneously reducing the oxidation state is to reduce CO_2 with hydrogen. The end product, for example, can be methane (CH_4), the dominant component in natural gas and the simplest hydrocarbon known, or other organic compounds. With all components in the ideal gas state, the standard $CO_2 + 4H_2 \rightarrow CH_4 + 2H_2O$, enthalpy of the process is exothermic by 165 kJ (39.4 kcal) per gram mole of methane formed (Sharma et al. 2014).

Gasification of biomass is classified into two different ways (Akia et al. 2014): Low-temperature gasification (LTG) and high-temperature gasification (HTG). Production of hydrogen and synthesis of gas conducted through the LTG process is an attractive method, especially for low calorific value biomass, such as livestock manure compost and waste activated sludge. The advantages of LTG process is its easiness and efficient operation while avoiding ash problems such as sintering, agglomeration, deposition, erosion, and corrosion. Moreover the tar components in LTG include lighter hydrocarbons which are different from the ones used in the HTG

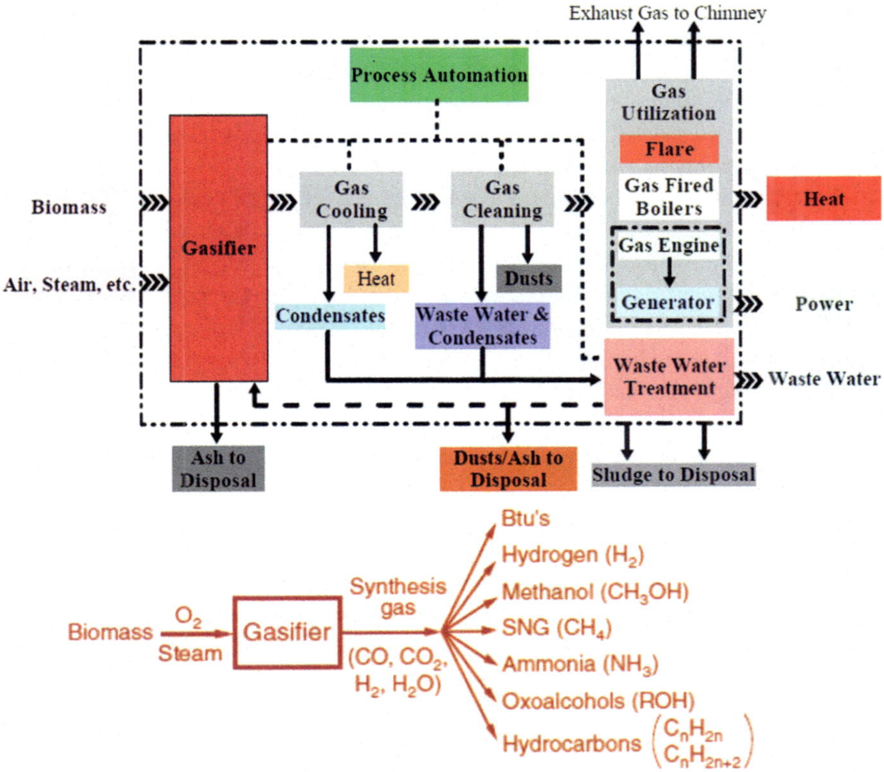

Fig. 8.4 Typical biomass gasification process (Modified with permission from Klass 2004, and Li et al. 2017 (MDPI, Basel, Switzerland))

process. Advanced gasification processes are currently being investigated for co-production of liquid fuels in research activities and also in pilot plant scales.

Biomass gasification is accomplished using a gasifier, and the reaction process is shown in Fig. 8.4. Since the gasifier type, process flow, reaction conditions, gasification agent type, raw material modification, and particle size are different, the reaction processes are not the same. However, the basic gasification process involves devolatization, combustion, and reduction. During devolatization, methane and other hydrocarbons are produced from the biomass by the action of heat which leaves a reactive char. During combustion, the volatiles and char are partially burned in air or oxygen to generate heat and carbon dioxide. In the reduction phase, carbon dioxide absorbs heat and reacts with the remaining char to produce carbon monoxide (producer gas). The presence of water vapor in a gasifier results in the production of hydrogen as a secondary fuel component (Zafar 2013).

There are two main types of gasifier that can be used to carry out this conversion, fixed bed gasifiers and fluidized bed gasifiers. The fixed beds have a wide temperature distribution. This includes possibilities for hot spots with ash fusion, low specific capacity, long periods for heat-up, and a limited scale-up potential.

However, fixed bed suffers from the disadvantage that gas flow must be stopped for cleaning. Fluidized beds have good heat and material transfer between the gas and solid phases with the best temperature distribution, high specific capacity, and fast heat-up. Disadvantages of fluidized beds are high dust content in the gas phase and the conflict between high reaction temperatures with good conversion efficiency and low melting points of ash components (Li et al. 2017).

The products of gasification are a mixture of carbon monoxide, carbon dioxide, methane, hydrogen, and various hydrocarbons, which can then be used directly in gas turbines, and boilers, or used as precursors for synthesizing a wide range of other chemicals. These include continuous water splitting by electrochemical, biochemical, thermochemical, microbial, photolytic, and biophotolytic processes. The basic concept then of using biomass as a renewable energy resource consists of the capture of solar energy and carbon from ambient CO_2 in growing biomass, which is converted to other fuels (biofuels, synfuels, hydrogen) or is used directly as a source of thermal energy or is converted to chemicals or chemical intermediates (Sharma et al. 2014).

In particular, the transformation of biomass into hydrogen rich gas provides a competitive means for producing energy and chemicals from renewable resources. Hydrogen production is very important for solving two major energy problems including reducing dependence on petroleum and reducing pollution and greenhouse gas emissions. Tar elimination and CO conversion to levels less than one volume percent by means of the WGS reaction is important for hydrogen production. In the production of high purity hydrogen, the reforming of fuels is followed by two WGS reaction steps, and the final steps are carbon monoxide purification and carbon dioxide removal. Steam gasification processes appear to be optimum candidates which provide an economical, reliable, and convenient process for extracting hydrogen from the produced gas. During the steam gasification process, the chemical energy of biomass can be converted into enriched hydrogen syngas containing up to about 50% by volume of hydrogen on a dry basis. Steam can be achieved from the dehydration reactions of crop residues or from an external source. Steam can react with carbon monoxide to produce hydrogen and carbon dioxide. WGS reaction is the principle reaction in the steam gasification system. Hydrogen can also be produced from biomass by pyrolysis. Biological technologies for hydrogen generation (biohydrogen) provide a wide range of routes for producing hydrogen. Moreover, biological processes are considered to be more environmentally friendly and less energy intensive. In general, biomass gasification and the subsequent production of synthetic fuels (diesel fuel, methanol, dimethylether, etc.) requires complex and more expensive technologies compared to conventional petrochemical processes or current direct liquefaction. Therefore, it is essential to ensure about the economic viability of biomass gasification plants (CEEETA 1995; Akia et al. 2014).

In addition, there are a number of methods that can be used to produce higher quality product gases, including indirect heating, oxygen blowing, and pressurization. After appropriate treatment, the resulting gases can be burned directly for cooking or heat supply, or used in secondary conversion devices, such as internal combustion engines or gas turbines, for producing electricity or shaft power (where it also has the potential for CHP applications) (Zafar 2013).

Additionally, gasification can be the key technology of biomass-based power generation. However, there are a number of key technological challenges that retard the commercial application of biomass gasification for power generation (Asadullah 2014). A major future role is envisaged for electricity production from biomass plantations and agricultural residues using large-scale gasifiers with direct coupling to gas turbines. The potential gains in efficiency using such hybrid gasifier/gas turbine systems make them extremely attractive for electricity generation once commercial viability has been demonstrated. Such systems take advantage of low-grade and cheap feedstocks (residues and wood produced using short rotation techniques) and the high efficiencies of modern gas turbines to produce electricity at comparable or less cost than fossil fuel-derived electricity. Net atmospheric CO_2 emissions are avoided if growth of the biomass is managed to match consumption. The use of BIG/STIG (Biomass Integrated Gasifier Steam Injected Gas turbine) initially and BIG/GTCC (Biomass integrated Gasifier Gas Turbine Combined Cycle) as the technology matures is predicted to allow energy conversion efficiencies of 40–55%. Modern coal electrical plants have efficiencies of about 35% or less. Combined Heat and Power systems could eventually provide energy at efficiencies of between 50% and 80%. The use of low-grade feedstocks combined with high conversion efficiencies makes these systems economically competitive with cheap coal-based plants and energetically competitive with natural gas-based plants. Studies are continuing in the use of such technologies for the cost-effective treatment of MSW considers that "gasification can become a strong competitor to anaerobic digestion, composting and incineration for biomass waste treatment." This is based on the use of BIG/STIG technology with the system gasification using Atmospheric Circulating Fluidized Bed technology (Sharma et al. 2014).

8.2.2.2 Pyrolysis

Biomass pyrolysis is the anaerobic thermochemical decomposition of biomass which occurs in the absence of oxygen, or with a limited oxygen supply, that converts biomass (break down the long chain molecules into short chain molecules) into liquid (bio-oil or bio-crude), charcoal, and non-condensable gases, acetic acid, acetone, and methanol by the heating of the biomass to 377–527 °C. A temperature of at least 400 °C in pyrolysis process is needed to completely decompose the organic structure of the biomass into monomers and oligomers fragments. The noncondensable portion of pyrolysis products rise by increasing temperature to above 600 °C. Pyrolysis operations are based on the size of biomass feeds and are divided into two main processes, slow pyrolysis and fast pyrolysis. The slow pyrolysis can disport to conventional charcoal production and intermediate pyrolysis. In conventional charcoal production, large pieces of wood are slowly heated to 400 °C for a long time (up to 18 h). The sole product of such process is charcoal when wood was used as a raw material, in the conventional kilns. However, in large retorts with capacities of 100 m^3 and more, which is used in conventional industrial charcoal production, non-solid products are also achieved. Refining facilities are combined with pyrolysis units to collect and condense gas products. Nevertheless, more than 65% of pyrolysis products are solid (charcoal) and less than 20% are liquid, in conventional charcoal production (Akia et al. 2014).

Intermediate pyrolysis differs from conventional charcoal production in terms of biomass residence time. Wooden feeds are entered into the screw tubular kiln and moved forward by screw rotation. The pyrolysis temperature of 380–400 °C is initiated in the kiln by transferring heat from the wall of the kiln, shaft of the screw, and also heat carriers. The carriers are balls of various materials with small size. With such heat transfer arrangements, the biomass heated faster than in conventional charcoal production but not as quickly as fast pyrolysis, which is why this process is called intermediate. Another difference between these two pyrolysis methods is related to the products quality. The solid portion of intermediate pyrolysis reduced to about 35%, while the liquid products increased to more than 45% (Schnitzer et al. 2007; Akia et al. 2014).

In fast or flash pyrolysis grained biomass with less than 3 mm diameter is converted to a combustible liquid fuel in one simple step. The dry feed (less than 20% moisture content) quickly mixes with grainy heat carrier of sand, steel shot, etc. at approximately 500 °C. The conversion of biomass to liquids (namely crude oil) have an efficiency of up to 70% for flash pyrolysis processes. This so-called bio-crude can be used in engines and turbines. It is shown that generation of electricity by pyrolysis products is more beneficial than that of any other biomass conversion method in the long term and has a lower cost. For example, macroalgae powder has been converted to bio-oil by fast pyrolysis method in a free fall reactor. The resulted bio-oil possesses average heat value and oxygen content of 25.33 MJ/kg and 30.27 wt.%, respectively (Zhao et al. 2013). Bio-oil can be substituted for fuel oil or diesel in many static applications, including boilers, furnaces, engines, and turbines. Further, bio-oil also has a higher density than raw biomass, which makes transportation and storage more convenient (Pereira et al. 2012; Li et al. 2017).

Therefore, pyrolysis produces energy fuels with a high fuel-to-feed in ratio, making it the most efficient biomass conversion process. The biomass feedstock has three main compositions: cellulose, hemicellulose, and lignin, some extracts of which are soluble in polar or nonpolar solvents. The biomass pyrolysis process can be divided into four stages: drying, preheating decomposition, solid decomposition, and combustion. The products produced through pyrolysis are gas (non-condensable volatile), liquids (condensable volatile), and solids (carbon), and the relative proportion of each product largely depends on the different pyrolysis methods and reaction conditions. The processes for biomass pyrolysis and its products are shown in Table 8.1 (Basu 2010; Li et al. 2017).

Like gasification, the advantage of pyrolysis is that they convert solid material into gases and vapors which are less costly to handle, transport, and store. The gases will burn in boilers, gas turbines, and reciprocating engines increasing fuel flexibility and security. Capturing and combusting the methane and carbon monoxide in syngas makes use of the energy in the gas and produces carbon dioxide which is a less potent greenhouse gas than methane and offsets fossil fuel energy production. The disadvantages of pyrolysis are that they require heat input to drive the chemical

Table 8.1 Characteristics of biomass pyrolysis processes (Basu 2010; Li et al. 2017)

Process type	Retention period	Heating rate	Maximum temperature (°C)	Main products
Slow pyrolysis				
Carbonization	Few hours–several days	Extremely low	400	Charcoal
Convention	5–30 min	Low	600	Gas, oil, charcoal
Fast pyrolysis				
Fast	0.5–5 s	Relatively high	650	Oil
Flash (oil)	<1 s	High	<650	Oil
Flash (gas)	<1 s	High	>650	Gas
Extremely fast	<0.5 s	Extremely high	1000	Gas
Vacuum	<2–30 s	Middle	400	Oil
Reactive thermal cracking				
Hydrogenation pyrolysis	<10 s	High	500	Oil
Methane pyrolysis	0.5~10 s	High	1050	Chemicals

reactions that produce a syngas. Thus, some fuel must be used to generate the syngas (Sharma et al. 2015).

8.2.2.3 Carbonization

This is an age-old pyrolytic process optimized for the production of charcoal. Traditional methods of charcoal production have centered on the use of earth mounds or covered pits into which the wood is piled. Control of the reaction conditions is often crude and relies heavily on experience. The conversion efficiency using these traditional techniques is believed to be very low; on a weight basis the estimated wood-to-charcoal conversion rate for such techniques ranges from 6 to 12 tonnes of wood per tonne of charcoal. During carbonization most of the volatile components of the wood are eliminated; this process is also called "dry wood distillation." Carbon accumulates mainly due to a reduction in the levels of hydrogen and oxygen in the wood. The wood undergoes a number of physicochemical changes as the temperature rises. Between 100 °C and 170 °C most of the water is evaporated; between 170 °C and 270 °C gases develop containing condensable vapors, CO and CO_2. These condensable vapors (long chain carbon molecules) form pyrolysis oil, which can then be used for the production of chemicals or as a fuel after cooling and scrubbing. Between 270 °C and 280 °C, an exothermic reaction develops which can be detected by the spontaneous generation of heat. The modernization of charcoal production has led to large increases in production efficiencies with large-scale industrial production in Brazil now achieving efficiencies of over 30% (by weight). There are three basic types of charcoal-making: (a) internally heated (by controlled combustion of the raw material), (b) externally heated (using fuelwood or fossil fuels), and (c) hot circulating gas (retort or converter gas, used for the production of chemicals). Internally heated charcoal kilns are the most common form of charcoal kiln. It is estimated that 10–20% of the wood (by weight) is sacrificed, a further 60% (by weight) is lost through the conversion to, and release of, gases to the atmosphere from these kilns. Externally heated reactors allow oxygen to be completely excluded, and thus provide better quality charcoal on a larger scale. They do, however, require the use of an external fuel source, which may be provided from the "producer gas" once pyrolysis is initiated. Recirculating heated gas systems offer the potential to generate large quantities of charcoal and associated by-products, but are presently limited by high investment costs for large-scale plant (Sharma et al. 2014).

8.2.2.4 Liquefaction

Biomass can be converted to liquefied products through combined physical and chemical reactions, the technology being called direct liquefaction. In these processes the biomass macromolecules are decomposed to small molecules through heating and sometimes in the presence of a catalyst. Direct liquefaction may be divided into pyrolysis and liquefaction methods. Although both are thermochemical conversion methods, their operating conditions are different. The operation temperature in the liquefaction method is lower (250–325 °C) but the operation pressure (5–20 MPa) is higher than that of pyrolysis. In pyrolysis, unlike liquefaction, it is necessary to dry biomass before feeding (Xu et al. 2011).

Two main routes can be considered industrially for the liquefaction of biomass to bio-oils, and these include hydrothermal liquefaction and catalytic liquefaction. Hydrothermal liquefaction is based on the superior properties of water at higher temperatures and pressures. The reactivity of biomass is considerable in water especially under hydrothermal conditions. Biomass consists of components with polar bonds which are attacked by the polar molecules of water. At elevated temperatures and pressures these attacks are more sever. As a result, hemicellulose and cellulose are hydrolyzed very quickly at these conditions. Hydrothermal liquefaction has another important advantage. Usually all biomass sources are wet and it is possible that their water content be at a range of up to 95 wt.%. In most biomass upgrading methods it is necessary to dry feeds before processing. In hydrothermal liquefaction conversions of biomass perform with its high water content. The water content of the biomass not only is not a disadvantage but it is also useful by reducing the process' required fresh water. Using water as both reactant and solvent in the liquefaction has some other benefits as well. The degradable products of the process are completely soluble in water under elevated temperatures and pressures, which prevent any polymerization. In addition, no solid products, such as coke and char, are formed because water acts as both a reactant and solvent in hydrothermal liquefaction. Water in liquefaction is an economic and environmentally friendly solvent, because it will not produce pollution. Mixed solvents (ethanol-water) and also other pure solvents (ethanol) have been used in thermochemical conversion of biomass. However, these solvents increase the operating costs of the biomass conversion (Akia et al. 2014).

Catalytic liquefaction is similar to hydrothermal liquefaction. However, using a catalyst brings some advantages to the biomass process. Catalyst reduces the residence time and operating temperature and pressure. Catalyst has useful effects on hydrothermal products, it increases liquid products and reduces gaseous one. Also, a catalyst can improve liquid products quality. As an instance, hydrothermal liquefaction of microalgae with heterogeneous catalysts showed improvements in products quality. Different heterogeneous catalysts (Pd/C, Pt/C, Ru/C, Ni/SiO_2-Al_2O_3, CoMo/γ-Al_2O_3 (sulfided), and zeolite) have been used in hydrothermal liquefaction of the microalgae nannochloropsis sp., at 350 °C. It was seen that the crude bio-oils produced from liquefaction with these catalysts, except zeolite, flowed easily and were much less viscous than the biocrudes of noncatalyzed liquefaction (Duan and Savage 2011; Akia et al. 2014).

In contrast to pyrolysis, in which thermal treatment only leads to depolymerization of the macromolecules of biomass (carbohydrates and lignin), further reactions are introduced in catalytic liquefaction by addition of pressurized reducing gas (hydrogen or carbon monoxide) and a catalyst to the reactor. High biomass conversion rates of up to 95% are obtained because char formation is suppressed as recondensation and recombination of thermal cracking products are reduced. Generally, the oils are more stable than pyrolysis oils due to the lower content of unsaturated compounds and lower oxygen content, which is in the range of 8–12%. As direct catalytic liquefaction of biomass is a high-technology method depending on many parameters, development has not been as rapid as in pyrolysis. However, yields and qualities of liquid products are promising and it is possible that some of the experience gained with existing coal liquefaction technologies can be adopted to the new feedstock biomass (Meier and Rupp 1991).

8.2.3 Biochemical Conversion

Biomass biochemical conversion technology, also called biological conversion, refers to the transformation of microbial, animal, or chemical biomass sources into a clean fuel or fertilizer. Biomass biochemical transformation is generally divided into three kinds: technology of anaerobic fermentation biogas, bioethanol, and biodiesel.

8.2.3.1 Anaerobic Fermentation Biogas

Anaerobic digestion (AD) is the process of the decomposition of biomass through bacterial action in the absence of oxygen. It is also a fermentation process and generates a mixture of gaseous products (e.g., hydrogen and carbon monoxide). Anaerobic fermentation is used to produce biogas, which can be directly used for cooking, heating, and as fuel for internal combustion engines.

Anaerobic digestion is widely used for recycling and treating wet organic waste and waste waters. It is the treatment of biomass with naturally occurring microorganisms in the absence of air (oxygen) to produce a combustible gaseous fuel comprising primarily of methane (CH_4) and carbon dioxide (CO_2) and traces of other gases such as nitrogen (N_2) and hydrogen sulfide (H_2S). The gaseous mixture is commonly termed biogas comparable to landfill gas. Virtually all nitrogen (N), phosphorus (P), and potassium (K) remain in the digested biomass. The entire process takes place in three basic steps. The first step is the conversion of complex organic solids into soluble compounds by enzymatic hydrolysis. The soluble organic material formed is then converted into mainly short-chain acids and alcohols during the acidogenesis step. In the methanogenesis step, the products of the second step are converted into gases by different species of strictly anaerobic bacteria. The percentage of methane in the final mixture has been reported to vary between 50% and 80%. Atypical mixture consists of 65% methane and 35% CO_2 with traces of other gases. The methane producing bacteria (called methanogenic bacteria) generally require a pH range for growth of 6.4–7.2. The acid producing bacteria can withstand low pH. In doing their work, the acid-producing bacteria lower the pH and accumulate acids and salts of organic acids. If the methane-forming organisms do not rapidly convert these products, the conditions become adverse to methane formers. This is why the first type of reactors developed for conversion of biomass wastes into methane have long retention times seeking equilibrium between acid and methane formers (Capareda 2011).

Similar to gas produced via gasification, gas from anaerobic digestion can, after appropriate treatment, be burned directly for cooking or heating. It can also be used in secondary conversion devices such as an internal combustion engine for producing electricity or shaft work. Virtually any biomass except lignin (a major component of wood) can be converted to biogas—including animal and human wastes, sewage sludge, crop residues, industrial processing by-products, and landfill material (Mayaki 2008).

Municipal wastes and livestock manures are the most suitable materials for anaerobic digestion. Numerous landfill facilities recover methane and use it for power generation. Aquatic biomass such as water hyacinth or micro-algae can be digested and may become valuable sources of energy in the future. Anaerobic digestion of organic wastes may constitute an effective device for pollution control with simultaneous energy generation and nutrient conservation. A major advantage of anaerobic digestion is that it utilizes biomass with high water contents of as high as 99%; it naturally occurs to organic material and would release methane, a potent greenhouse gas, into the atmosphere. Capturing and combusting the methane makes use of the energy in the gas and produces carbon dioxide which is a less potent greenhouse gas than methane and offsets fossil fuel energy production. Another advantage is the availability of conversion systems in smaller units. Also, the residue has fertilizer value and can be used in crop production. The primary disadvantage of anaerobic digestion of diluted wastes is the large quantity of sludge that must be disposed of after the digestion process including the wastewater and the cost of biogas storage. In cold climates, a significant fraction of the gas produced may be used to maintain the reactor operating temperature. Otherwise, microorganisms that thrive on lower or moderate temperatures should be used (Capareda 2011).

The conversion of animal wastes and manure to methane/biogas can yield significant health and environmental benefits. Methane is a greenhouse gas (GHG) that is 22–24 times more powerful than carbon dioxide (CO_2) in trapping heat in the atmosphere. By trapping and utilizing the methane, GHG impacts are avoided. Further, the pathogens existing in manure are eliminated by the heat generated in the biodigestion process and the resulting material provides a valuable, nutrient-rich fertilizer (Mayaki 2008).

Biogas fermentation is a microbiological process. Various kinds of organic matter, such as straw, livestock manure, and industrial and agricultural waste water, can be converted into methane through the action of microorganisms in anaerobic or other suitable conditions. The produced biogas can be utilized in several ways, either raw or upgraded. Compared with other fuels, methane generates less carbon dioxide and produces fewer atmospheric pollutants per unit of energy. Since methane is a comparatively clean fuel, it is being increasingly used for power generation, vehicles, industrial applications, and so on (Li et al. 2017).

8.2.3.2 Bioethanol

Ethanol, which is produced almost entirely from food crops, is a comparatively cleaner burning fuel with high octane and fuel-extending properties. Furthermore, it is a renewable energy and can be sustainably developed from lignocellulosic biomass. The production process features biological conversion and includes the following steps: feedstock handling, pretreatment, biological conversion, product recovery, utilities production, and waste treatment. As a petrol additive/substitute, it is probable that wheat, sugar beet, straw, corn, and wood can be economically converted into bioethanol. There are three main raw material types which can be used to produce bioethanol: starchy materials, sugary materials (such as molasses, sugarcane, sugar beet, and sweet sorghum), and cellulose materials (such as branches, sawdust, and plant fiber waste). Nowadays, however researchers and investors have become increasingly enthusiastic about another biofuel feedstock, lignocellulose, which is the most abundant biological material on Earth and which can produce both ethanol and biodiesel (Schubert 2006; Li et al. 2017).

Essentially, bioethanol can be produced from a variety of carbohydrates which have a general formula of $(CH_2O)_n$. The chemical reaction involves an enzymatic hydrolysis of sucrose followed by a fermentation of simple sugars. First, an invertase enzyme in the yeast catalyzes the sucrose hydrolysis and converts it into glucose and fructose (Demirbas 2007):

$$C_{12}H_{22}O_{11}(\text{Sucrose}) \rightarrow C_6H_{12}O_6(\text{Glucose}) + C_6H_{12}O_6(\text{Fructose}) \qquad (8.1)$$

Then, another enzyme, zymase, also present in yeast, converts the glucose and the fructose into ethanol (C_2H_5OH) (Demirbas 2007):

$$C_6H_{12}O_6 \rightarrow 2C_2H_5OH + 2CO_2 \qquad (8.2)$$

Bioethanol can be used directly in the transport sector and vehicles can run on pure ethanol or blended with gasoline to make "gasohol." As the development of renewable energy technological applications have progressed, ethanol has become a viable alternative fuel. Almost all gasoline cars can drive with fuel containing 10% ethanol (E10), flex-fuel cars can even use 85% ethanol (E85), and Brazil and the USA already include 10–27% ethanol in their standard fuel by law. Ethanol has been proved to be a promising alternative fuel for the internal combustion engine (Li et al. 2017).

8.2.3.3 Biodiesel

Seed crops, which contain a high proportion of oil, can be reacted with alcohols (methanol, ethanol) through a transesterification process to obtain biodiesel. There are a large variety of crops that can be used for biodiesel production including rapeseed oil, palm oil, sunflower oil, soya bean oil, and recycled frying oils. Biodiesel is a renewable fuel with extensive sources, which is already being utilized widely around the world. Due to its environmental benefits, biodiesel has become more attractive recently because of its low sulfur content, aromatics, and flash point. At present, biodiesel production technology includes a chemical method, an enzymatic synthesis method, the recovery of glycerol, and an engineered microalgae production method. The chemical method, which involves a transesterification of plant (or animal) oil, methanol or ethanol in a catalyst of acid, alkali, or biological enzyme, is the primary technology. Biodiesel is better for environmental protection, which results in an overall life-cycle lowering of carbon dioxide emissions over both conventional diesel and gasoline (Li et al. 2017).

8.2.3.4 Processing of Biofuels

Liquid biofuels include pure plant oil, biodiesel, and bioethanol. Biodiesel is based on esterification of plant oils. Ethanol is primarily derived from sugar, maize, and other starchy crops. Global production of biofuels consists primarily of ethanol, followed by biodiesel production (Mayaki 2008).

(a) Straight Vegetable Oil (SVO)/Pure Plant Oil (PPO): SVP/PPO can be used in most modern diesel vehicle engines only after some technical modifications. Principally, the viscosity of the SVO/PPO must be reduced by preheating it. However, some diesel engines can run on SVO/PPO without modifications. PPO is obtained from edible oil-producing plants such as the African palm, groundnuts, cotton seeds, sunflower, canola, or nonedible oils such as jatropha, neem, or even balanites. These raw oils, unused or used, can be employed in certain diesel engines, for cooking, or in diesel generators for the production of electricity.

(b) Biodiesel: Biodiesel can be used in pure form or may be blended with petroleum diesel at any concentration for use in most modern diesel engines. Biodiesel is raw vegetable oil transformed, treated, and standardized through chemical processes. The standardization of this product, and its industrial production, renders its use much more diverse than PPO. Biodiesel is used in diesel engines and diesel vehicles. Biodiesel can be produced from different feedstocks, such as oil feedstock (e.g., rapeseed, soybean oils, jatropha, palm oil, hemp, algae, canola, flax, and mustard), animal fats, and/or waste vegetable oil.

(c) Alcohols: Ethanol, butanol, and methanol are produced principally from such energy crops as sugarcane, maize, beets, yam, or sweet sorghum. Ethanol is the most widely used alcohol, primarily as a fuel for transportation or as a fuel additive. Bioethanol can be produced from a variety of feedstocks, including sugarcane, corn, sugar beet, cassava, sweet sorghum, sunflower, potatoes, hemp, or cotton seeds, or derived from cellulose waste.

Several processes exist to convert feedstocks and raw materials into biofuels. First-generation biofuels refer to the fuels that are produced through well-known processes such as cold pressing/extraction, transesterification, hydrolysis and fermentation, and chemical synthesis. The resulting fuels have been derived from sources such as starch, sugar, animal fats, and vegetable oil. First-generation biofuels are already established in the fuel markets and usually produced from fuel crops. The most popular types of first-generation biofuels are biodiesel, vegetable oil, bioethanol, and biogas. Second-generation biofuels are not yet commercial on a large scale as their conversion technologies are still in the research and/or development stage. Second-generation biofuels are produced through more advanced processes, including hydro treatment, advanced hydrolysis and fermentation, and gasification and synthesis. A wide range of feedstocks can be used in the production of these biofuels, including lignocellulosic sources such as short-rotation woody crops. These produce biodiesel, bioethanol, synthetic fuels, and bio-hydrogen. Table 8.2 shows the production and use of liquid fuels for first- and second-generation biofuels (Mayaki 2008).

Both first- and second-generation technologies offer advantages and disadvantages. The primary advantage of first-generation biofuels is they are available today with existing technologies; their promotion is based on nontechnical issues such as policies and cost-effectiveness. First generation biofuels can also be produced in decentralized facilities. Disadvantages include emissions produced in growing and refining these fuels, land use concerns, their complex effect on food and grain prices, and that only specific crops can be used in biofuels production. For second-generation biofuels, a larger variety of feedstocks can be used. Advanced biofuels (e.g., biobutanol and synthetic diesel) and other biofuels derived from switchgrass, garbage, and algae are under development. New conversion technologies are expected to expand production potential by allowing for the use of an array of nonfood resources. Additionally, the energy input for agriculture and feedstock production could be significantly reduced and the technologies are expected to be more efficient as they will entail large-scale conversion operations. It is anticipated that second-generation technologies will yield better energy, economic, environmental, and carbon performance than first-generation options (Mayaki 2008).

Table 8.2 Typical applications of stainless steels in ethanol, biodiesel, and biogas plants (Osterman 2012)

Ethanol plants	Biodiesel plants	Biogas plants
Corn steepers in wet mills	Acid pretreatment systems	Fermenters
Liquefaction tanks	Caustic pretreatment systems	Digesters
Fermentation tanks	Reactors	Pumps
Yeast slurry tanks	Acid tanks	Slurry processing
Beer well tanks	Decanters	Gas piping systems
Fermentation washing tanks	Centrifuges	Air pollution control
Distillation columns	Slurry tanks	Liquid fertilizer tanks
Whole stillage tanks	Flash tanks	Gas holders
Centrate surge tanks	Receiving tanks	Air piping systems
Centrifuges	Condensate tanks	Steam systems
Thin stillage tanks	Buffer tanks	Fittings
Syrup tanks	Internal process tanks	Gas turbines
Dryers	Dryers	Gas engines
Evaporators	De-acidification columns	Storage tanks
Evaporator condensate tanks	Rectification columns	Various tanks
Pumps	Distillation columns	Heat exchangers
Mixers	Methanol recovery tanks	Mixers
Valves	Pumps	Pumps
Filters	Mixers	Agitators
Regulators	Agitators	Valves
Piping systems & required fittings	Valves	Modules
Heat exchanger tubes	Filtration and purification systems	
Skid systems	Regulators	
Emission control equipment	Piping, tubing	
Pre-treatment systems		

8.2.3.5 Biorefineries

A biorefinery involves the co-production of a spectrum of bio-based products (food, feed, materials, and chemicals) and energy (fuels, power, heat) from biomass. It is a facility that integrates biomass conversion processes and equipment to produce fuels, power, and value-added chemicals from biomass. The biorefinery concept is analogous to today's petroleum refinery, which produces multiple fuels and products from petroleum (Mayaki 2008).

By producing several products, a biorefinery takes advantage of the various components in biomass and their intermediates, therefore maximizing the value derived from the biomass feedstock. A biorefinery could, for example, produce one or several low-volume, but high-value, chemical products and a low-value, but high-volume liquid transportation fuel such as biodiesel or bioethanol. At the same time, it can generate electricity and process heat, through CHP technology, for its own use and perhaps enough for sale of electricity to the local utility. The high-value products increase profitability, the high-volume fuel helps meet energy needs, and the power production helps to lower energy costs and reduce GHG emissions from traditional power plant facilities. Although some facilities exist that can be called biorefineries, the biorefinery concept has yet to be fully realized. Future biorefineries may play a major role in producing chemicals and materials that traditionally were produced from petroleum (Mayaki 2008).

8.2.4 Integrated Biomass Production-Conversion Systems

The energy potential of waste biomass, although of significant importance for combined waste disposal and energy-recovery applications, is relatively small part compared to the role that virgin biomass has as an energy resource. The key to the large-scale production of energy, fuels, and commodity chemicals from biomass is to grow suitable virgin biomass species in an integrated biomass-production conversion system (IBPCS) at costs that enable the overall system to be operated at a profit. Multiple feedstocks, including combined biomass fossil feedstocks and waste biomass, may be employed (Klass 2004; Sharma et al. 2014).

The proper design of an IBPCS requires the coordination of numerous operations such as biomass planting, growth management, harvesting, storage, retrieval, transport to conversion plants, drying, conversion to products, emissions control, product separation, recycling, wastewater and waste solids treatment and disposal, maintenance, and transmission or transport of salable products to marketing. The design details of the IBPCS depend on the feedstocks involved and the type, size, number, and location of biomass growth and processing areas needed. It is evident that a multitude or parameters are involved. In the idealized case, the conversion plants are located in or near the biomass growth areas to minimize the cost of transporting biomass to the plants, all the nonfuel effluents of which are recycled to the growth areas, as demonstrated in Fig. 8.5 (Patil et al. 2008). If this kind of plantation can be implemented in the field, it would be equivalent to an isolated system with inputs of

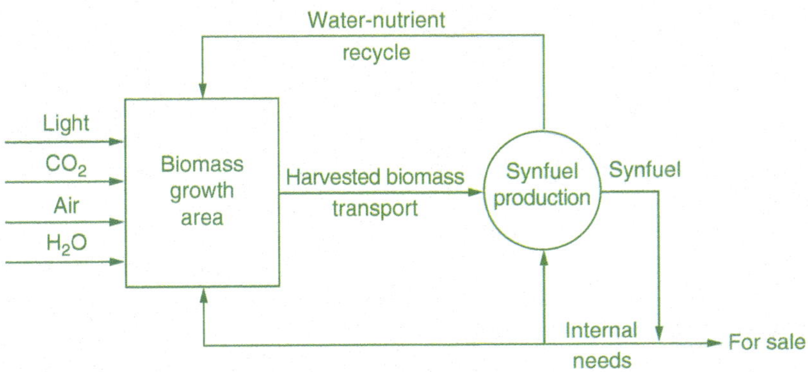

Fig. 8.5 A conceptual model for integrated biomass production and conversion integration system (Adapted from Patil et al. 2008 (MDPI, Basel, Switzerland), under the terms and conditions of the Creative Commons Attribution (CC BY) license))

solar radiation, air, CO_2, and minimal water; the outputs consist of the product slate. The nutrients are kept within the ideal system so that addition of external fertilizers and chemicals is not necessary. Also, the environmental controls and waste disposal problems are minimized. The growth areas required to supply quad blocks of energy and fuels would be very large when compared with conventional agricultural practice, but that 10,000- Barrels of oil equivalent per day systems are not quite so large when compared with traditional, sustainable wood harvesting operations in the forest products industry. The analysis suggests that smaller, localized IBPCSs in or near market areas will be preferred because of logistics and product freight costs, and multiple feedstocks and products will have advantages for certain multiproduct slates. Synthesis gas from biomass gasification used as co-feedstock in an existing natural gas-to-methanol plant can utilize the excess hydrogen produced on steam reforming natural gas. Sustainable virgin biomass production at optimum economic yields is a primary factor in the successful operation of IBPCSs. Most of the IBPCSs that have been proposed are site specific—that is, they are designed for one or more biomass species, in the case of a multicropping system, for specific regions. Some of the large, commercial forestry operations for tree growth, harvesting, and transport to the mills can be considered as analogous in many respects to the biomass production phase of managed IBPCSs (Gust 1997; Sharma et al. 2014).

The historical development of IBPCSs shows that large-scale biomass energy plantations must be planned extremely carefully and installed in a logical scale-up sequence. Otherwise, design errors and operating problems can result in immense losses and can be difficult and costly to correct after construction of the system is completed and operations have begun. It is also evident that even if the system is properly designed, its integrated operation can have a relatively long lag phase, particularly for tree plantations, before returns on investment are realized. The financial arrangements are obviously critical and must take these factors into consideration (Sharma et al. 2014).

8.3 Advanced Bioenergy Conversion Pathways from Feedstock to Products

Advanced bioenergy conversion process can be broken down into two major areas: deconstruction and fractionation; and synthesis and upgrading. Figure 8.6 highlights key technologies within deconstruction and fractionation as well as synthesis and upgrading, which can be linked to form a complete conversion pathway from feedstock to products. The arrows represent the transition of organic matter from feedstock to intermediates to end products, showing the diversity of accessible conversion options. Multiple technologies along several pathways are under development to address the broad range of physical and chemical characteristics of various feedstocks and to reduce the risk that any specific technology could fail to reach commercial viability. One advantage of the numerous paths between key technologies for R&D is that an advancement in one area can potentially impact several unique final pathways. Additionally, each linked set of conversion

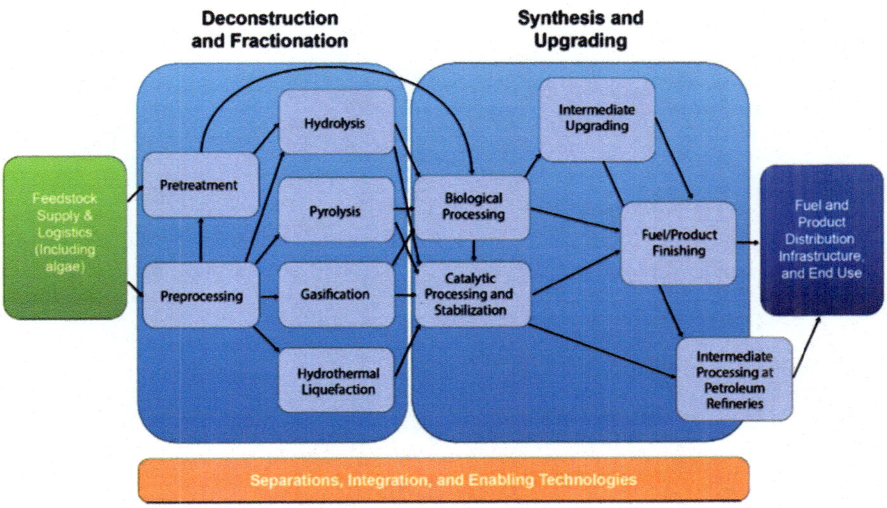

Fig. 8.6 Conversion pathways from feedstock to products (Adapted from MYPP 2015. Credit: Energy Efficiency & Renewable Energy, U.S. Department of Energy. https://www.energy.gov/sites/prod/files/2016/03/f30/mypp_beto_march2016_2.pdf)

technologies results in the production of a unique product slate whose value will vary, depending on market size and demand (MYPP 2015; QTR 2015a).

In addition, the sustainable supply of quality, cost-effective feedstocks to future biorefineries is fundamental to growing the bioenergy industry. The inherently dispersed, highly variable, aerobically unstable nature of biomass, among other characteristics, are a challenge to provide a sustainable, secure, and affordable biomass feedstock supply for bioenergy industry. As the raw material for biofuels, bioproducts, and biopower production, reaching industrial scale will require availability of and access to a reliable supply of high-quality biomass. As shown in Fig. 8.7, R&D is needed in two broad categories of feedstock (QTR 2015b; Zaimes et al. 2015): (a) terrestrial feedstocks, which include lignocellulosic feedstocks such as agricultural residues, forest resources, dedicated energy crops, and municipal solid waste (MSW) resources; and (b) algal feedstocks.

8.3.1 Deconstruction and Fractionation

Deconstruction and fractionation processes break biomass-derived polymeric feedstock down into tractable intermediate streams suitable for conversion into targeted products. After preprocessing and/or pretreatment, deconstruction processes can be divided into two categories: high-temperature deconstruction and low-temperature deconstruction. High-temperature deconstruction refers to processes performed at or above 100 °C and includes processes such as pyrolysis, hydrothermal and solvent liquefaction, and gasification. Low-temperature deconstruction refers to processes

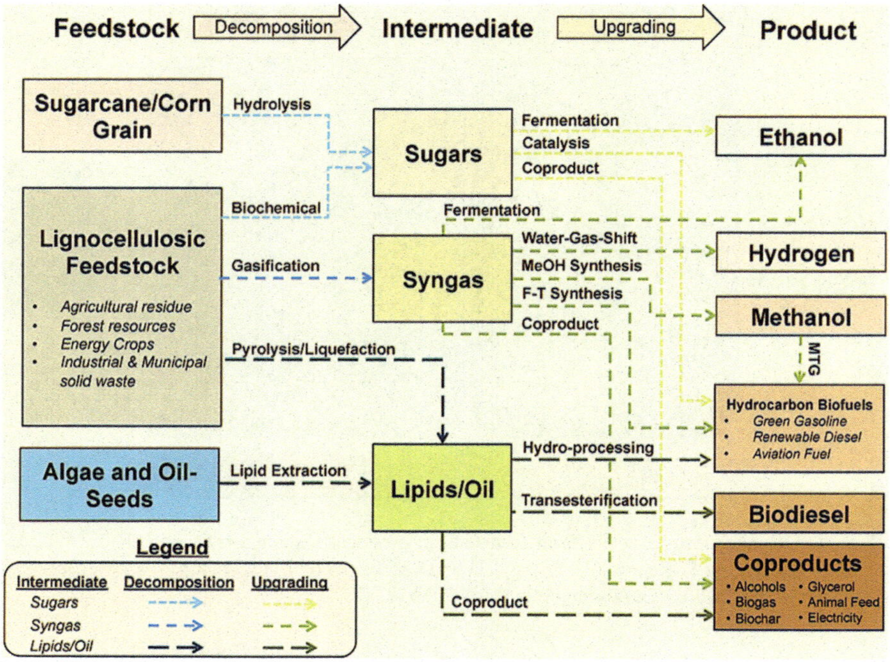

Fig. 8.7 Feedstocks and conversion pathways for biofuel production (Adapted with permission from Zaimes et al. 2015 (MDPI, Basel, Switzerland)). MTG: Methanol to Gasoline. F-T Synthesis: Fischer-Tropsch Synthesis

performed below 100 °C and includes processes such as enzymatic and acid hydrolysis (QTR 2015a).

The preprocessing is to develop a variety of conversion technologies that is necessary to address the broad range of physical and chemical characteristics of various biomass feedstocks. Depending on the conversion strategy, a variety of feedstock preprocessing and handling steps may be employed. Linking feedstock logistics with conversion processes allows for the evaluation of technology options and trade-offs on both sides of the processing interface, ensuring a fully integrated supply chain from stump or field to fuel. R&D is ongoing for preprocessing options (e.g., densification, blending of an expanded pool of feedstocks, and physical formats, such as pellets, shredded material, and slurries) and simultaneously assessing the impact on conversion efficiency when such preprocessed feedstocks are introduced into a conversion process (QTR 2015a).

Low-temperature deconstruction is the breakdown of the physical structure of the feedstock by pretreatment followed by hydrolysis with the intent of producing high-quality, low-cost sugar streams. R&D is ongoing to develop technologies to create more efficient hydrolysis and cleaner separation of the resulting intermediate streams at lower cost. Specific areas of interest include developing better pretreatment conditions, creating lower cost hydrolytic enzymes, developing new hydrolytic enzymes with improved substrate scope, limiting the formation of contaminants, improving separation efficiency, and creating a tractable lignin stream for higher value uses of the non-sugar portion of the feedstock. Pretreatment is the preparation of feedstock for hydrolysis via chemical and/or mechanical processing and separation of feedstock into soluble and insoluble components. This process opens up the physical structure of plant cell walls, revealing the more recalcitrant cellulosic, hemicellulosic, and lignin polymers for subsequent hydrolysis while solubilizing some of the lower molecular weight compounds. Hydrolysis is the breakdown of these polymers either enzymatically or chemically into their component sugars and/or aromatic monomers and/or low molecular weight oligomers (QTR 2015a).

High-temperature deconstruction encompasses pyrolysis, gasification, and hydrothermal and solvent liquefaction. The primary focus of R&D for high-temperature deconstruction is on improving technologies for thermochemical deconstruction of biomass to form a gaseous or bio-oil intermediate. Key focus areas include developing a better understanding of the fundamentals of pyrolysis, hydrothermal liquefaction, and gasification processes, including reaction mechanisms; improved reactor designs; improved quality and stability of deconstructed intermediates; more robust catalysts and catalyst regeneration processes; and catalysts with improved specificity. Hydrothermal and solvent liquefaction is a deconstruction process that utilizes a wet feedstock slurry under elevated temperature and pressure (generally 250–500 °C and 5–25 MPa) that produces a bio-oil similar to pyrolysis oil. The feedstock is treated with water before entering the reactor, and therefore hydrothermal liquefaction is particularly applicable to algae feedstocks. Other variations include solvent liquefaction where a non-water solvent, such as methanol, is used to make the feedstock slurry prior to liquefaction. Pyrolysis is the thermal and chemical decomposition of feedstock (at temperatures typically <700 °C), without the introduction of oxygen, to produce a bio-oil intermediate. The bio-oil produced contains various length hydrocarbons but contains more oxygenated compounds than most petroleum crude oils and must undergo upgrading before it can be finished into a fuel or used in a refinery. There are several variations of pyrolysis that require different catalysts and reaction conditions. Gasification is thermal deconstruction of biomass at higher temperature (typically >700 °C) in the presence of sub-stoichiometric oxygen, air, or an oxygen carrier such as steam followed by gas cleanup and conditioning. In these processes, feedstock is partially oxidized to form a syngas that contains a mixture of light gases, predominantly CO and H_2, along with some CO_2 and CH_4 and other light gases (QTR 2015a).

8.3.2 Synthesis and Upgrading

Synthesis and upgrading converts tractable intermediate streams from deconstruction and fractionation of biomass into finished fuels, chemicals, and materials. These intermediates can include crude bio-oils, sugars, gaseous mixtures such as syngas, and other chemical building blocks as outlined in Figs. 8.6 and 8.7, and are upgraded using either biological, chemical, or hybrid processing techniques. These upgraded streams could be finished fuels or bio-derived chemicals ready to sell into the commercial market or they could also be stabilized intermediates suitable for final finishing in a petroleum refinery or chemical manufacturing plant. The processes include (QTR 2015a):

(a) Biological processing
Microorganisms have the ability to convert sugar, lignin, or gaseous intermediates into fuel blend stocks and chemicals. R&D in this area focuses on identification and development of robust microorganisms capable of converting complex intermediates to desired target molecules in the presence of inhibitors at high rates, titers, selectivity, and yields. To accomplish this, advanced metabolic engineering methods are being developed allowing these microbes to achieve maximum intermediate utilization, robustness, and selection of the product slate with minimal perturbation of other critical cellular functions. To accomplish this, genetic tools such as CRISPR-Cas9 need to be deployed in a wider variety of industrially relevant host organisms. These new genetic tools combined with machine learning for design of biological systems as well as research into process conditions can greatly open the space in which biologically derived chemicals and fuels can compete (MYPP 2015).

(b) Catalytic processing and stabilization
Intermediate streams such as bio-oil, syngas, and mixed sugars must be upgraded to minimize the effect of reactive compounds to improve storage and handling properties and ease downstream upgrading. This is accomplished through removal of water, char, and ash particulates as well as destabilizing components, such as metals and oxygenated species.

   (i) For bio-oil, catalytic processing and stabilization may involve hydro-processing, such as hydrodeoxygenation, to transform oxygen-rich biomass into a mix of compounds more similar to hydrocarbon-rich petroleum. It may also involve separation and fractionation steps to remove water, coke (ash made of carbon fragments that can drop out of a process stream), catalyst, char, and ash particulates or metals and oxygenated species. Research on bio-oils focuses on hydro-processing and similar thermal-catalytic processing techniques to reduce total oxygen and acid content, thereby increasing stability and enabling easier downstream conversion.

   (ii) For syngas streams, stabilization and preparation for fuel synthesis involves removal of contaminants from crude biomass-derived syngas and adjusting gas ratios. Gas cleanup and conditioning involves the removal of problematic heteroatom compounds, metals, and particulates as well as adjusting the hydrogen-carbon monoxide ratio. Research on syngas cleanup and conditioning has historically focused on removing or reforming tars and methane, capturing alkali metals, and removing particulates.

   (iii) For mixed sugar streams, catalytic processing and stabilization may involve four types of catalytic processes: hydrogenation, aqueous phase reforming, condensation and oligomerization, and hydrotreating. Each stage consists of packed-bed reactor vessels. Hydrogen is added to the reactors in each stage that operate at varying process conditions and have varying catalyst composition. The goal of these successive catalytic steps is to remove oxygen or "de-functionalize" carbohydrates and other carbon components and oligomerize them to primarily diesel or jet-range hydrocarbons.

(c) Intermediate upgrading
Intermediate upgrading involves a variety of technologies to transform intermediate streams into preliminary product streams. Actual upgrading and separations processes will vary greatly according to the identity and composition of the intermediate streams and the purity requirements for the finished fuel or product. Streams with tight chemical distributions, such as algal lipids, fatty acids, or other products from biological processing, may require less complex processes than streams involving more varied compounds. Chemical rearrangement into the final fuel blend stock or product can involve biological or chemical processing.

(d) Fuel/product finishing
After upgrading, final product streams must conform to standards for off-take agreements. This may involve removing problematic contaminant compounds and further finishing to attain correct product specifications. For complex bio-oil mixtures, the finishing process may involve balancing various hydrocarbon components, whereas for single molecule products this may involve removing impurities.

(e) Intermediate processing at petroleum refineries
Certain product streams may be transported to refineries at a more crude stage for upgrading. Placement of this box on the edge of synthesis and upgrading and products in Fig. 8.6 represents the interface of conversion technologies with refiners. Conversion R&D is working to establish clear product specifications that will enable bio-oil, bio-intermediates, fuel-blend stocks, finished fuels, and products to seamlessly integrate with existing infrastructure and will encourage acceptance of bio-based replacements in industry. This activity involves R&D in coordination with refiners to understand how a bio-oil blend will perform when integrated into their existing operations and ultimately seeking to provide additional value to refineries (QTR 2015a).

8.3.3 Sustainable Supply of Biomass Feedstocks

To develop sustainable technologies that provide a secure and affordable feedstock supply for bioenergy industry, challenges and barriers associated with biomass need to be prioritized and addressed. Those barriers include but are not limited to (QTR 2015b):

(a) Terrestrial feedstock availability and cost: Reliable and consistent feedstock supply is needed to reduce financial, technical, and operational risk to biomass supply and conversion investments. Credible data and projections on current and future cost, location, environmental sustainability, quality, and quantity of available biomass are needed to reduce uncertainty for investors and developers of emerging biorefinery technologies. Estimates of current and potential feedstock resources are limited in scope and do not adequately represent how major potential advances in genetics, production technologies, and supply chain strategies will impact future biomass availability, cost, quality, and processing characteristics.

(b) Production: Another barrier is the lack of energy crop yields and the environmental effects of energy crop production. The range and improvements in energy crop yields have not been well-documented for deployment of energy crops at commercial scale. Reliable production data are needed over several growing seasons and across wide geographies to make well-substantiated productivity projections. Comprehensive data are also needed to measure the environmental effects of energy crop production and biomass collection systems to provide data for complete life-cycle analysis of biorefinery systems and address sustainability questions such as water and fertilizer inputs or establishment and harvesting impacts on soil. Production and sustainability gaps also exist for conventional crop residues.

(c) Terrestrial feedstock genetics and development: The productivity and robustness of terrestrial feedstock crops used for biofuel production could be increased by developing improved varieties through screening, breeding/selection, and genetic engineering. This will require extensive ecological, genetic, and biochemical information that is currently lacking and is a barrier to use of the dedicated perennial terrestrial energy crops.

(d) Sustainable harvesting: Current crop harvesting machinery is unable to selectively harvest preferred components of cellulosic biomass while maintaining acceptable levels of soil carbon and minimizing erosion. Actively managing biomass variability imposes additional functional requirements on biomass harvesting equipment. Current systems cannot meet the capacity, efficiency, or delivered price requirements of large cellulosic biorefineries. Neither can these current systems continue to protect and conserve soil structure over cumulative operations at massive scales.

(e) Terrestrial feedstock quality and monitoring: A better understanding of the physical, chemical, microbiological, and post-harvest physiological variations is needed to manage feedstock costs and reduce conversion costs and final product prices. Biomass is very heterogeneous in physical and chemical properties that arise from differences in genetics, degree of crop maturity, geographical location, climatic events, and harvest methods. This variability presents significant cost and performance risks, and is a barrier to cost-effective bioenergy and biopower systems. Processing standards and specifications for cellulosic feedstocks are not as well-developed as they are for mature commodities and are being developed.

(f) Biomass storage systems: Biomass that is stored with high moisture content or exposed to moisture during storage is susceptible to spoilage, spontaneous combustion, odor problems, and rapid changes in quality. Therefore, a barrier is managing and controlling post-harvest biological processes to ensure a consistent, high-quality feedstock supply. Characterization and analysis of different storage methods and strategies are needed to better define storage requirements to preserve the volume and quality of harvested biomass over time and maintain its conversion yield. Low-cost options are needed for high-capacity storage—storage options that are safe and not subject to fire and pests.

(g) Biomass material properties and variability: Available data and information are extremely limited on biomass quality and physical characteristics and how those properties influence conversion performance. Methods and instrumentation also are lacking for quickly, accurately, and economically measuring chemical, physical, and mechanical properties of biomass.

The inherent variability of biomass physical and chemical quality parameters is needed at more refined levels such as within a species and even between tissues of the same plant and require additional research. Acceptable ranges of quality parameters for different conversion processes are poorly understood, and few genetic or preprocessing strategies have been developed to limit or control variability in biomass quality. Since many quality factors vary independently, it is not clear what fraction of available biomass materials will actually be able to meet specifications for the various conversion processes being developed and commercialized.

(h) Biomass physical state alteration: The initial sizing and grinding of cellulosic biomass affects conversion efficiencies and yields of all downstream operations, especially for biofuels conversion options, yet little information exists on managing these operations to improve the product yields or reduce costs. This is a barrier to designing high-performance supply systems. New technologies and equipment are required to economically process biomass to meet biorefinery and biopower facility specifications such as particle-size distribution. Managing these particle-size reductions operations to match the conversion process is paramount to improving overall efficiency.

(i) Biomass material handling and transportation: Raw herbaceous biomass is costly to collect, handle, and transport long distances because of its low density and fibrous nature. Existing conventional bale-based handling equipment and facilities cannot cost-effectively deliver and store high volumes of biomass even with improved handling techniques. Current handling and transportation systems designed for moving woodchips can be inefficient for bioenergy processes due to the costs and challenges of transporting, storing, and drying high-moisture biomass.

(j) Overall integration and scale-up: Conventional supply systems used to harvest, collect, store, handle, and transport biomass are not designed for the large-scale needs of a nationwide system of integrated bioenergy systems. The infrastructure for feedstock logistics has not been defined for the potential variety of locations, climates, feedstocks, storage methods, and processing alternatives which will occur at a national scale. Integration of one or more aspects of the feedstock supply system—either alone or in combination with bioenergy system operations—should lead to net gains in efficiency; however, the lack of analysis quantifying the relative benefits and drawbacks of potential integration options is a barrier to realization of cost savings, efficiency improvements, and reduction of technical and financial risks.

Although substantial progress has been made, there are many challenges and barriers associated with working with biomass, the following issues are considered most critical QTR (2015b): Increase the volume of sustainable, acceptable, cost-effective feedstock available to bioenergy systems by developing advanced feedstock supply systems and strategies; Incorporate sustainability and feedstock supply risk into the resource assessments; Work with conversion technology areas to understand the range of acceptable physical and chemical specifications for the various conversion technologies; Develop high-capacity, high-efficiency, low-cost, commercial-scale feedstock supply and logistics systems that deliver stable, dense, consistent, infrastructure-compatible feedstock.

8.3.3.1 Strategies Moving Forward

Terrestrial feedstock research in the past has focused on modifying conventional terrestrial feedstock logistics systems that were designed and manufactured for traditional agricultural and forestry industries. Conventional systems are suitable for high biomass-yielding regions, but not for medium-to-low-yielding areas. Supplying feedstock to a growing bioenergy industry requires increasing the accessible volumes of lignocellulosic feedstock while increasing the emphasis on quality and reducing variability. This can be achieved by applying pre-processing techniques such as fracturing, drying, densifying, and blending (Kenney et al. 2013).

Terrestrial feedstock research is needed to (QTR 2015b): (a) reduce the delivered cost of sustainably produced feedstock, (b) preserve and improve the quality of harvested feedstock to meet the needs of biorefineries and other biomass users, and (c) expand the tonnages of feedstock materials accessible to the bioenergy industry. This is done by identifying, developing, demonstrating, and validating efficient and economical systems for harvest and collection, storage, handling, and pre-processing raw biomass from a variety of crops to reliably deliver high-quality, affordable feedstocks.

Quality targets have large impacts on whether or not a particular feedstock is cost effective in the context of a particular conversion process as well as how much material is available for conversion (Williams et al. 2016). For instance, ash is the inorganic or mineral content of biomass. Ash content varies considerably among and within biomass materials. Understanding biomass ash content, variability, and where it originates requires differentiation of the sources of ash, which include structural ash associated with the plant cell walls, vascular ash in the plant, and introduced ash resulting from soil contamination. Ash cannot be converted to a biofuel product and causes operational problems in downstream conversion processes, including increased equipment wear, quenching of catalysts, increased corrosivity of pyrolysis oils, slagging in thermochemical equipment, and costs associated with ash disposal; it can similarly challenge power plant equipment. The proportion of convertible biomass content decreases with increasing ash content effectively increasing the cost

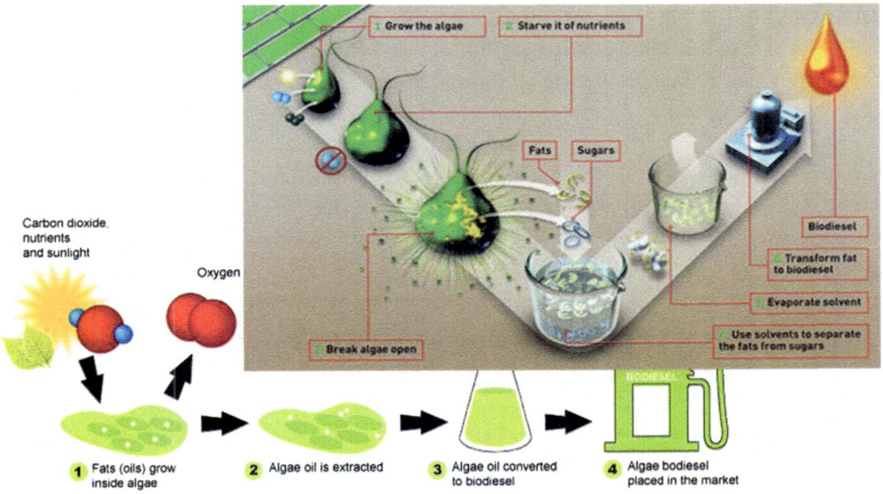

Fig. 8.8 Schematic process to converse from algae to biofuel

per dry ton of feedstocks. The variability in ash and other biomass quality parameters highlights the need to rapidly quantify quality parameters and develop mechanisms for moderating changes to the final point of delivery (QTR 2015b).

Moreover, feedstock blending allows a bioenergy system to collect less of any one feedstock and thus move down the cost versus supply curve enabling bioenergy systems to pay a lower average price. This does not change the supply versus cost curves for each resource but it describes a system where purchasers can use a combination of least cost resources and blend them to reach the desired cost and quality specifications. The use of low-cost biomass allows the supply chain to implement additional preprocessing technologies that actively control feedstock quality, while also bringing more biomass into the system. This analysis and design approach is referred to as the "least-cost formulation" strategy. In addition, research is needed on blending strategies, on the performance of blended material, and on other advanced design technologies to meet cost, quality, and volume targets (Muth et al. 2013).

8.3.3.2 Algae

Algae comprise over 100,000 species, classed as plant-like organisms. They literally come in all shapes, sizes, and hues, from green to red and brown. Very small protozoa can be found on the surface of ponds while the ocean floor is home to massive seaweed bushes. Figure 8.8 illustrates schematic process to converse algae to biofuel. Advantages of biofuels derived from algal biomass include (QTR 2015b): the ability to cultivate algae on nonarable land, the ability to utilize brackish or saline water, and the possibility of using waste nutrients and effluents, including carbon dioxide from power plants. Algal species can accumulate significant amounts of lipids in their cell structure and thus are particularly well suited for conversion to hydrocarbon-based renewable diesel and jet fuel. The challenges associated with algae include the high cost of production, water requirements, and micronutrient requirements. Research and development activities are focused on reducing the cost of production of algal biomass and intermediates, developing cultivation and logistics systems for producing fuels and products at commercial scale, developing innovative and energy-efficient drying technologies, and developing algal strains that can survive and maintain high productivity in large-scale algae-farming operations. Algal biomass includes micro- and macro-algae and cyanobacteria. Algal biofuel and bioproduct intermediates include extracted lipids, products derived from sugars or proteins (alcohol or hydrocarbon fuels), secreted metabolites (alcohols or others), or bio-crude resulting from hydrothermal liquefaction. These intermediate products must be upgraded and or blended and or purified to produce a finished fuel or bioproduct.

The productivity and robustness of algae strains against perturbations such as temperature, seasonality, predation, and competition could be improved by selection, screening, breeding, biologically mixed cultures, and/or genetic engineering. This requires extensive ecological, genetic, and biochemical information. Research efforts are also focused on identifying key algal feedstock characteristics and standards for downstream conversion processes. A unique aspect of the conversion interface in these systems is the extent to which feedstock preprocessing and biofuel conversion technologies, such as lipid extraction or hydrothermal liquefaction, are physically integrated with algae production. Efficient and effective linkage between algal feedstock and conversion processes is critical to facilitate the functioning of the entire value chain. The conversion interface area primarily addresses the effect of algae processing operations on conversion technology performance characteristics. Compositional analysis of the intermediate also helps to evaluate water and nutrient recycle efficiency. The fundamental components (lipids, carbohydrates, and proteins) of algal biomass vary greatly, within strains, among strains, and in comparison to plants. These efforts will help to develop and optimize conversion process input specifications so that process economic targets can be achieved. The integration of analysis, biomass production, logistics, and conversion is particularly important to advancing algal systems. Biomass properties (such as cell size, media composition, and carbohydrate/protein/lipid content) can affect downstream processes of harvesting and conversion. As methods to improve upon algal biomass production, harvesting, and conversion are developed, techno-economic and life-cycle analyses are run in parallel to bolster research focus on those processes with the best and most sustainable economic outcomes. Scaling-up algal technologies is considered one of the largest challenges in the commercialization of algal biofuels and is necessary to demonstrate and validate algae systems. High biomass productivities or effective harvesting processes at small scales do not always translate to success in outdoor environments or at large scales. This is due to multiple factors including engineering constraints, pond ecology and pathology, and other issues. The scaling up of nutrient sources that are inexpensive at small scales may be economically prohibitive at commercial scales. Small-scale lab research closely tied to performance of large-scale experiments is a priority to provide an iterative learning process that will expedite lessons learned before scaling to larger pilot facilities (QTR 2015b).

8.3.3.3 Waste-to-Energy

In addition to purpose-grown crops, municipal, industrial, and agricultural waste streams constitute a potential resource for the production of fuels, product precursors, heat, and electricity. The combination of biogas production from agricultural manure operations, landfills, and water resource recovery facilities could yield 654 billion cubic feet per year (18.52 billion m^3/year) of biogas. If converted to electricity the potential generation of over 40 terra-watt-hours is projected, more than 1% of total US. consumption. Beyond electricity generation there are significant opportunities to produce heat for on-site use, and hydrocarbons for use in biofuels and bioproducts, thereby avoiding greenhouse gas emissions from the use of fossil fuel feedstocks. There is interest in utilizing the potential from four kinds of organic feedstocks (QTR 2015b):

(a) The nonrecyclable organic fraction of landfill solid wastes. Food wastes from landfills constitute the largest single fraction of currently unrecovered wastes.
(b) Sludge from various stages of municipal wastewater treatment processes. While the wastewater industry is undergoing a shift towards viewing themselves as water resource recovery facilities much work remains to realize this vision.
(c) Manure slurries from concentrated livestock operations, particularly dairies.
(d) Organic wastes from industrial operations including but not limited to food and beverage production and biorefineries. Other industries such as pulp and paper, forest products, and pharmaceuticals also generate streams that might be suitable for incorporation.

There is also opportunity to combine two or more of the above streams in particular locations to attain regional economies of scale. These synergies may be particularly relevant in producing biofuels and bioproduct precursors as opposed to the onsite generation of combined heat and power (CHP) which is particularly suited to facilities that have a proximal beneficial use for the heat such as many wastewater treatment plants.

Producing biofuels and bioproducts from wastewater streams requires the conversion of often dilute (3–4% solids by weight) inputs into useful chemicals (e.g., butanol) while minimizing the energy inputs for drying. There are a number of avenues to achieve these objectives including, but not limited to (Yakaboylu et al. 2013; Akizuki et al. 2014):

(a) Hydrothermal processes, including hydrothermal liquefaction, which are capable of utilizing the unique properties of sub and supercritical water (and other solvents) to produce liquid fuel precursors, biogas, and beneficial solids. Some of these pathways could contribute to reductions in energy consumption for municipal wastewater treatment facilities.
(b) Microbial and other biochemical processes that convert wastewater directly into higher hydrocarbons. There are two general categories: modification of existing

anaerobic digestion processes to produce either drop-in biofuels or bioproduct precursors rather than biogas; and microbial electrolysis and electro-synthesis cells that yield hydrocarbon compounds and in some cases could produce more electricity than they consume.

(c) Enhanced anaerobic digestion and other biochemical processes to convert CO_2 in biogas into higher value products.

(d) Exploration of anaerobic membrane bioreactors (AnMBRs) in conjunction with other technologies such as microbial fuel cells. While AnMBRs have gained commercial acceptance in high organic load industrial processes they are not yet widely deployed for municipal wastewater applications. Minimization of fouling is a key challenge. AnMBRs could replace secondary aeration which tends to be the largest electricity consumer in wastewater treatment facilities and avoid the capital costs of anaerobic digesters, especially for smaller facilities.

(e) Other processes for conversion of organic feedstocks into higher value biofuel and bioproduct precursors include fast pyrolysis and various pathways to gasification. Strategies that utilize waste heat to reduce the energy required for drying are of particular interest.

(f) Research is needed in biological and chemical options to convert methane into hydrocarbons. ARPAE's REMOTE solicitation is relevant, as strategies that work for natural gas could apply to biogas. The possibility may also exist to co-develop biofuels and engines to exceed the performance of currently available internal combustion engines.

(g) Systems integration is an important consideration. Detailed techno-economic analyses of the conditions under which it makes economic and environmental sense to pursue various renewable energy options could be valuable. For example, it would be useful to understand the parameters that guide the economics of biofuel production vs. localized combined heat and power in contrast to biogas upgrading to pipeline quality or fleet utilization of compressed natural gas or liquefied natural gas.

Therefore, there are numerous opportunities to utilize organic "wastes" to produce bioenergy, biopower, and bioproducts. These pathways are not only renewable but also hold the potential of supplementing seasonal agricultural crops and residues with reliable year-round feedstock streams (QTR 2015b).

8.3.4 Emerging Bioenergy Conversion Technologies

Many emerging technology areas could help to complement or replace elements of conversion pathways for biofuels and bio-derived chemicals, such as (QTR 2015a):

(a) Synthetic Biology: The use of synthetic biology tools to create new biological pathways to fuels and chemicals could open up new product slates and greatly improve the efficiency of conversion.

(b) Optima: The co-optimization of engines and fuels to achieve maximum efficiency could take advantage of the unique properties of biomass-derived fuels to increase engine performance.
(c) Ionic liquids: Deconstruction of biomass into tractable intermediate streams could be accomplished using ionic liquids, which offer unique upgrading opportunities and have steadily decreased in price.
(d) Functional replacements: The chemical space easily accessible from petroleum and biomass are different, creating an opportunity to use some of the unique functionality of biomass-derived chemicals to create alternatives to fossil-derived chemicals. These bio-derived chemicals could function as replacements with superior properties for applications from plastics to lubricants.

8.4 Corrosion Resistant Materials Compatible with Biofuels

Corrosion is the disintegration of a material as it reacts with other materials with which it comes into contact. According to Nernst's theory, all metals have a tendency to dissolve into solution. However, the extent of corrosion varies for metal ions depending on its oxidation potential and various prevailing conditions (temperature, pressure, water content level, impurities, or iron concentration, etc.) in the fuel. For instance, some metal ions, when present in fuel, can cause corrosion. As an example, the absence of sulfur is supposed to reduce the corrosion in the biodiesel fuel container. Biodiesel is generally prepared from acid or alkali catalysts that are either homogeneous or heterogeneous. Homogeneous acid catalysts, such as sulfuric acid, are generally used for acid esterification, and impart corrosive nature to biodiesel fuel. This is generally overcome by using solid acid catalysts which are easily separated from biodiesel and hence do not make the fuel corrosive. Biodiesel also ought to be of high purity for its compatibility in engines. Therefore, incomplete conversion or inadequate purification (by water washing or other means) may result in impurities such as glycerol, free fatty acids, alcohol, and catalyst, causing deposits in the engine, corrosion, and ultimately failure of the fuel. Biodiesel has a good lubricity and hence, it has a higher tendency as compared to mineral diesel to dissolve the metallic parts in the fuel. Diesel engine components are made from a variety of metals, nonmetals including elastomers. The main parts of the engine/vehicle that come in contact with fuel are fuel tank, fuel feed pump, fuel lines, fuel filter, fuel pump, fuel injector cylinder, piston assembly, and exhaust system. These engine/vehicle parts are made of (a) metallic materials, such as carbon steel, stainless steel, copper and its alloy, aluminum and its alloy (like cast aluminum, forged aluminum, sand-cast aluminum, die-cast aluminum, and aluminum fiber); iron-based alloy (like gray-cast iron, special cast iron); and (b) nonmetallic materials, such as elastomer, plastics, paint, coating, cork, rubber, ceramic fiber, and even paper. The fuel comes in contact with the various engine parts and its accessories at varying temperature, velocity, load, sliding, and physical state. It is found that either the impurities in biodiesel or the deterioration of biodiesel through oxidation enhances the corrosiveness of the fuel (Singh et al. 2012).

8.4.1 Metal and Its Alloys

Corrosion in engine parts that are made of metals, and alloys on contact with biodiesel occur due to the chemical composition of biodiesel that comprises unsaturated molecules that easily undergo oxidation. Factors that influence the performance and engine durability of a compression-ignition (CI) engine run on 100% pure or neat diesel or a biodiesel blend are presence of oxygen in the functional groups (moieties), free fatty acids, degree of unsaturation, and hygroscopic nature (ability to attract and hold water) of the biodiesel. The level of corrosion also depends on the type of alloy in contact with biodiesel fuel. Apart from biodiesel, corrosion also occurs in engines operating on ethanol. Corrosion in engine carburetors can expose to ethanol by three ways (Singh et al. 2012): general corrosion, dry corrosion, and wet corrosion. General corrosion results from ionic impurities such as chloride ions and acetic acid. Dry corrosion results from polarity of the molecule. Wet corrosion arises due to azeotropic water and oxidizes various metals. In biodiesel-run engines, oil nozzles made of high chrome stainless steels have been found to be resistant to corrosion when exposed to biodiesel. The oil burner filter components made of copper and copper alloys usually corrode in biodiesel, resulting in the fuel being contaminated with copper ions.

In addition, metallic components in automobiles can undergo corrosive, tribological, or both forms (tribocorrosion) of attacks depending on their functionality. The corrosiveness of biodiesel increases with the concentration of biodiesel in the blend and the extent of oxidation. Copper alloys are more susceptible to corrosion than ferrous alloys and aluminum alloys. Lead alloy coating on terne steel sheet which is used for automotive fuel tanks is severely affected by biodiesel. The presence of impurities and water increases the corrosion tendency of biodiesel. As for the tribological degradation of metallic components in biodiesel, biodiesel generally offers beneficial properties in terms of lower or similar wear and friction for many elements compared with diesel. Fe, the main constituents of the most critical components in automobiles, shows less or similar wear, except for one study which involved higher blend (100%). On the other hand, Cu shows higher wear in many cases. Pb and Al follow copper in terms of decreasing trend. A few inhibitors (ethylenediamine, tert-butylamine, and n-butylamine) have been found to be effective in the formation of a stable metal oxide as protective layer and thus retarding corrosion in biodiesel. An antioxidant, tertbutylhydroquinone, was also found to retard the corrosion of copper coupon in biodiesel along with enhancing the oxidation stability of the fuel. Further studies on the effectiveness of inhibitors added to biodiesel are needed (Bhardwaj et al. 2014).

Steel has been found to show high resistance to corrosion in biodiesel blends. However, as biodiesel is more conductive electrically compared to gasoline and diesel and may cause galvanic metal corrosion in steel. Corrosion of steel varies with feedstock used for synthesis of biodiesel. This is due to differences in the chemical composition of the feedstock that show various degrees of corrosiveness (Singh et al. 2012). For example, the soybean biodiesel proved to be less reactive to the metal than the sunflower biofuel (Maru et al. 2009). In general, it has verified that mild steel and stainless steel are corrosion-resistant with E10, E15, E30, E70, E85, E95, B5, B20, and B100.

Stainless steels are widely used in bioethanol, biodiesel, and biogas production facilities, as shown in Table 8.2, which offer excellent corrosion resistance in the biofuels industry's various process conditions, coupled with good strength, ductility, toughness, and ease of fabrication. Standard austenitic stainless steel Type 304 (L) and Type 316(L) are capable of meeting most of the corrosive conditions encountered in ethanol, biodiesel, and biogas production and handling equipment. For superior corrosion resistance in certain more corrosive applications, duplex, super-austenitic, and nickel alloys (such as Alloy C276-UNS N10276) will be suitable. Some high-temperature applications in cellulosic ethanol production may also require special stainless or nickel alloys (Osterman 2012).

Corrosion becomes an important aspect in usage of biodiesel fuel because many of the engine parts are composed of nonferrous metals such as aluminum, copper and its alloys, in addition to stainless steel, that may be prone to corrosion. Percent aluminum in engine components includes piston (100%), cylinder heads (70%), and engine blocks (19%). Pumps and injectors are often composed of copper and its alloys. The corrosion rate in copper, aluminum, and carbon steel has been found to be 0.586, 0.202, and 0.015 mils per year (mpy) (14.884, 5.131, 0.381 μm/year), respectively, in palm biodiesel. In diesel, the rate of corrosion was less and found to be less than 0.3 mpy (7.62 μm/year) for copper, less than 0.15 mpy (3.81 μm/year) for aluminum, and was almost the same for carbon steel (0.015 mpy (0.381 μm/year)). Corrosion rate would increase with increasing temperature in the diesel and biodiesel (Singh et al. 2012).

With canola biodiesel, containing various levels of contaminants (residual catalyst, methanol, glycerol, and water), aluminum corrosion in biodiesel is closely linked to the degree of purity of the biofuel. Therefore, the quality of the biodiesel is critical to prevent damage to the metal. In addition, aluminum experiences higher pitting corrosion in biodiesel than that in diesel (Díaz et al. 2008).

It is proved that copper and bronze are very susceptible to attack by biodiesel (B100). Also, copper acts as a strong catalyst oxidizing palm biodiesel. The degree of corrosivity of biodiesel is associated with an important parameter known as total acid number, which reflects the amount of free fatty acids present in the biofuel. Moreover, the source of vegetable oil or animal fat used for biodiesel production plays an influential role, since depending on its chemical composition and the presence of unsaturated fatty acids determine if it is more prone to oxidation (Fazal et al. 2010).

Therefore, the general metallic materials compatible and recommended being used with biodiesel are stainless steel and aluminum. Materials like brass, copper, zinc, bronze, lead, and tin are incompatible with biodiesel and can accelerate the biofuel degradation, leading to the formation of insoluble (sediments) and salts or gels when reacted with one of the fuel components. For instance, upon exposure to palm biodiesel, the degradation order for the different metals is: copper > brass > aluminum > cast iron > carbon steel > stainless steel > nickel alloy. Materials are generally easier to be corroded in the presence of light at higher temperature and influenced by dissolved oxygen in the biodiesel (Aquino et al. 2012).

8.4.2 Elastomers

When exposed to biodiesel, elastomers are affected in two ways (Singh et al. 2012): first by absorption of liquid by elastomers, and second by dissolution of soluble components from the elastomers in the liquid medium. Swelling was the result of high absorption amount by elastomers in comparison to their dissolution in the fuel. Elastomers generally undergo degradation to a greater extent than metallic materials in biodiesel. Common elastomers such as natural rubber, chloroprene/neoprene, and nitrile are unsuitable for use in CI engines run on biodiesel. Fluorocarbons have shown good resistance and are recommended for used in biodiesel. Exact mechanisms of degradation of elastomers in biodiesel and the effects of different biodiesel constituents require systematic investigations (Bhardwaj et al. 2014).

Within palm biodiesel and its blends with diesel, the mass and the volume of nitrile rubber and polychloroprene increased by increasing the biodiesel concentration while for fluoro-viton A, remained almost constant for all fuels. After immersion tests in biodiesel, the tensile strength, elongation, and toughness are significantly reduced for the nitrile rubber and polychloroprene, whereas the fluoro-viton A changes are negligible (Haseeb et al. 2010). The tensile strength in biodiesel decreased to a large extent in the case of EPDM and CR. A minor decrease in tensile strength was observed in SR, PTFE, and NBR elastomers. Among all of the elastomers, PTFE showed high compatibility with biodiesel that was attributed to absence in its polarity, although it also underwent reduction in its constituents. Swelling of the elastomers in biodiesel fuel can be retarded by adding cross-linking agents in the elastomers. Peroxides are commonly added as cross-linking agents (Singh et al. 2012).

8.4.3 Metallic and Organometallic Coatings

There are two major types of coating used to minimize corrosion process of metal tanks in contact with fuels: tin coatings and organometallic coatings. The latter have a zinc layer, deposited by hot-dip galvanizing process (galvannealed), formed between the extra deep drawing steel and the organometallic coating. Zinc forms intermetallic layers with steel, which are responsible for better adhesion of the organometallic coating. The tin coating is applied directly to the EEP steel and, after steel forming, the outside surface is painted by cathodic electrodeposition process—KTL or E-coating). The organometallic-coated steel polarization resistance can be two orders of magnitude higher than the galvannealed steel polarization resistance, and twice the value of the tin-coated steel resistance (Luciano et al. 2014).

8.5 Nanocatalysts for Conversion of Biomass to Biofuel

The availability and wide diversity of biomass resources have made them an attractive and promising source of energy. The conversion of biomass to biofuel results in the production of liquid and gaseous fuels with different methods such as thermochemical and biological processes. Thermochemical processes as a major conversion route which include gasification and direct liquefaction are applied to convert biomass to more useful biofuel. Catalytic processes are increasingly applied in biofuel development. Nanocatalysts play an important role in improving product quality and achieving optimal operating conditions. Nanocatalysts with a high specific surface area and high catalytic activity play an important role in improving product quality and achieving optimal operating conditions. They may be used to solve the most common problems of heterogeneous catalysts such as mass transfer resistance, time consumption, fast deactivation, and inefficiency. The nano-sized materials as nanocatalysts may be used directly or in the form of supported nanoparticles on solids such as oxides, carbon, or zeolites. Some usual methods for nanocatalysts preparation are impregnation, precipitation, chemical vapor deposition, and electrochemical deposition. Mostly, impregnation and precipitation methods have been used for biomass catalysts preparation. Meanwhile, nanoparticles properties usually adjust by changing synthesis parameters such as pH of the solution, concentration, the reducing agent, and calcination temperature (Pelisson et al. 2012; Akia et al. 2014).

Catalyst developments have led to the upgrade of biomass gasification processes to increase the syngas production and reduce the tar formation. In catalytic biomass liquefaction the main aims are to increase liquid yield and quality of products. Furthermore, nanocatalyst characteristics, such as high catalytic activities and high specific surface area have helped overcome some limitations on heterogeneous catalysts for their application in biodiesel production from biomass (Akia et al. 2014).

8.5.1 Nanocatalysts for Biomass Gasification

In biomass gasification, an important issue is to prevent tar and char formation. Tar is a complex mixture of condensable hydrocarbons, including aromatic compounds of single ring to 5-ring along with other oxygen containing hydrocarbons and complex polycyclic aromatic hydrocarbons (PAHs). Due to its high boiling temperature and condensing at temperatures below 350 °C, tar can create major problems such as corrosion or failure of engines as well as blockage of pipes and filters. It may also act as poison for catalysts. Treatments inside the gasifier (primary methods) and hot gas cleaning after the gasifier (secondary methods) are main approaches employed for controlling the production of tar. Although the secondary methods are effective, primary methods are also gaining much attention because of economic benefits. The most important parameters in the primary methods are including temperature, gasifying agent, equivalence ratio, residence time, and catalysts which have significant effects on tar formation and decomposition. Catalysts not only reduce the tar content but also improve the quality of gas products and the conversion efficiency. The successful gasification catalysts have some criteria including being effective at removing tars, being resistant to deactivation as a result of carbon fouling or sintering, can easily be regenerated, and are inexpensive (Wilcoxon 2012; Akia et al. 2014).

Biomass chars are highly disordered carbonaceous materials with a short-range polycrystalline structure which consists of small aromatic structural units. The presence of a catalyst decreased the char yield during the final step of the gasification process while it increased the char formation during the volatilization stage. Potassium, sodium, and calcium have been found to be the most effective catalysts for promoting char gasification in steam or carbon dioxide media. There may be other metallic species beneficial for biomass conversion, although some elements present in waste biomass may prevent char gasification by poisoning the catalysts (Nzihou et al. 2013; Akia et al. 2014).

Catalytic tar cleaning is potentially attractive because no additional input energy is necessary. The important reactions during tar reduction include steam reforming, dry reforming, thermal cracking, hydrocracking, hydro reforming, and WGS reactions. Catalysts for tar conversion are generally classified into mineral or synthetic. Mineral includes calcined rocks, olivine, clay minerals, and ferrous metal oxides. Transition metals, activated alumina, alkali metal carbonates, FCC catalysts, and chars are the main synthetic catalysts (Akia et al. 2014).

Tar conversion by using dolomite, nickel-based, and other catalysts such as alkali metals at elevated temperatures of typically 800–900 °C achieved near 99%. $MgCO_3CaCO_3$ (Dolomite) is a magnesium ore widely used in biomass gasification. The tar content of the produced gases during the biomass conversion process is significantly reduced in the presence of Dolomite. Dolomite catalysts are also efficient in tar cracking. Their major disadvantages include sensitivity to elevated pressure and thermal instability which leads to loss of surface area due to sintering. Nickel-based catalysts are very effective for the catalytic hot gas cleanup during biomass gasification. It can also eliminate tar with a high rate. Moreover, Ni-based catalysts have been used for the production of hydrogen-rich product gas. Among all catalysts for converting tar into fuel gas, nickel catalysts are the most efficient ones. The stability of nickel catalysts increased with co-impregnation of nickel on mineral catalysts (olivine, dolomite, and zeolite). However, nickel catalysts may not be recommended for applications in atmospheric biomass gasification due to its high costs and severe risk for deactivation via sulfur chemisorption and carbon deposition. Alkali metals such as lithium, sodium, potassium, rubidium, and cesium can be used directly as catalysts in the form of alkali metal carbonates or supported on other materials such as alumina and silica. Addition of alkali metals to biomass can also be achieved by impregnation. These metals are highly reactive. Alkali metal catalysts lead to an enhancement in the biomass gasification reactions, especially for char formation reactions. The presence of Na_2CO_3, K_2CO_3, or $CsCO_3$ as catalyst in biomass steam gasification decreased the carbon conversion degree to gas with an

Table 8.3 Comparison of the char formation in cellulose pyrolysis process using nano-NiO, micro-NiO catalysts and without using catalyst (Li et al. 2008)

Catalyst	Initial decomposition temperature (°C)	Char yield
Without catalyst	313	6.14
Micro-NiO	303	6.09
Nano-NiO	294	5.64

Table 8.4 Comparison the effects of bulk and nanocatalysts of ZnO and SnO_2 for cellulose conversion at different temperatures (Akia et al. 2014)

	Conversion (%)			
Temperature (°C)	Nano-ZnO	Bulk ZnO	Nano-SnO_2	Bulk SnO_2
300	92.4	83.0	71.0	64.0
400	83.2	83.0	88.2	75.2
500	89.4	83.0	88.4	76.6
600	86.8	75.0	84.2	78.8

increase in the rate and total amount of produced gas (Asadullah 2014; Akia et al. 2014).

Nanocatalysts have exhibited unique properties in biomass gasification in comparison with their bulk counter parts. For instance, nano-sized NiO (nano-NiO) particles can be economically prepared by loading nano-particles of NiO on the surface of distinct carriers (such as alumina). As shown in Table 8.3, the decomposition of cellulose in the presence of micro-NiO was 10 °C lower than that of the pure cellulose, while the decomposition of cellulose with nano-NiO started at 294 °C, which was 19 °C lower than that of the pure cellulose. The final char yield (5.64 wt. %) was further decreased compared to when micro-NiO particles were applied, which proved the effectiveness of nano-NiO catalysts in pyrolyzing of biomass at a relatively lower temperature. In addition, nano-Ni catalyst (NiO supported on gamma alumina) in direct gasification of sawdust can considerably improve the quality of the produced gas while significantly eliminating tar production. The organometallic nanocatalysts of Ni compound and $Ni_3Cu\ (SiO_2)_6$ nanoalloy catalyst can increase biomass conversion efficiency at relatively low gasification temperatures. Other nanocatalysts that have been used in biomass gasification are nano-ZnO and nano-SnO_2 structures (Table 8.4). Nano-ZnO is an effective catalyst for low-temperature WGS reaction, while nano-SnO_2 is an effective catalyst for high-temperature WGS reaction during the cellulose gasification in hot compressed water (Sinag et al. 2011; Akia et al. 2014).

The gaseous species obtained at 300 °C in the presence of bulk and nano-ZnO mainly consisted of CO_2 and H_2. The rate of WGS in the presence of nano-ZnO is faster in comparison with the nano-SnO_2. Larger surface areas of nano-ZnO enhanced the WGS reaction. The existence of both H_2O_2 and ZnO catalysts in the reactor enhanced hydrogen production (Sinag et al. 2011). Tar removal efficiency can be improved using nano-CeO_2 and nano-$(CeZr)_xO_2$ catalysts during the cellulose and sawdust gasification process (Akia et al. 2014).

8.5.2 Nanocatalysts for Biomass Liquefaction

Alkaline salts, Na_2CO_3, KOH and $NaHCO_3$, Ca $(OH)_2$, Ba $(OH)_2$, $FeSO_4$, are commonly used as homogenous catalysts in liquefaction processes (Duan and Savage 2011). The heterogeneous catalysts have been used in catalytic conversion of biomass, including Pd/C, Pt/C, Ru/C, Ni/SiO_2-Al_2O_3, CoMo/γ-Al_2O_3 (sulfided), zeolite, and Fe. In catalytic hydro conversion of biomass, liquid catalysts have the advantage of being mono dispersed in reaction mixtures. In other words, solid catalysts have the superiority of higher catalytic activity in addition to being easily separated from the products. Acid-functionalized paramagnetic nanoparticles are promising materials for use in catalytic hydro conversion of biomass. These functionalized nanoparticles can easily separate and be recycled in the catalytic hydro conversion process. Nanocatalysts have some other advantages which make them attractive for use in biomass to liquid (BTL) processes. Having the fluid solution characteristics, mono dispersed nanocatalysts have excellent accessibility to the oxygen atoms of the cellulose ether linkage (Guo et al. 2012; Akia et al. 2014).

Conversion of biomass to liquid compounds such as paraffinic, naphthenic, and aromatic hydrocarbons can supplement part of the worldwide petrochemical demand. Although use of nanocatalysts in the catalytic conversion of biomass to liquid chemicals has had several advantages, most of the research attention has been paid to conversion of biomass to biodiesel as it is in large demand in today's world. Nanocatalysts for biodiesel production significantly improve the yield of products. The main nanometal oxides that have been used for biodiesel production are Zn, Ca, Mg, and Zr. These have either been used individually or supported on different materials. However, some other catalysts such as Li, Cs, and KF have been utilized for edible and nonedible feedstock. In addition, magnetic nanoparticles functionalized with different catalysts have been implied in biodiesel production, facilitate the catalyst recovery. The KF/CaO and nanomagnetic KF/CaO–Fe_3O_4 catalysts for biodiesel preparation exhibit better performance of KF/CaO catalyst with a higher surface area of about 10^9 m^2 g^{-1} at the same operating conditions compared to KF/CaO–Fe_3O_4 catalyst with a surface area of 20.8 m^2 g^{-1}. Loading KF on nano-Al_2O_3 support could obtain a higher yield of about 97.7% compared to 96.8% for KF/CaO catalyst. The operating conditions of using KF/Al_2O_3 catalyst were relatively higher than that of KF/CaO catalyst. The alcohol:oil ratio, catalyst loading, and reaction time were 15:1, 3 wt.%, and 480 min in comparison with 12:1, 4 wt.%, and 150 min, respectively, for KF/Al_2O_3 and KF/CaO catalysts. Despite using different feedstocks, the better performance of KF/CaO catalyst may be related to its higher surface area (10^9 m^2 g^{-1}) compared to the value of 41.7 m^2 g^{-1} for KF/Al_2O_3 catalyst. The highest biodiesel yield was obtained using nanocatalysts Li-CaO and CaO at reaction time of 120 and 720 min, temperature of 65 °C and 25 °C, and methanol-to-oil molar ratio of 12 and 27, respectively. Similar yield has been obtained using a nano-MgO catalyst but at a higher temperature of about 200–260 °C. In general, to achieve high performances at relatively mild operating conditions, it is necessary to increase the reaction time while at ordinary reaction times, it is necessary to apply severe operating conditions. Comparing the results of using different supported and unsupported MgO catalysts revealed that decreasing the reaction time led to an increase in reaction temperatures for achieving higher performances. As a whole using milder operating conditions led to a reduction in energy consumption requirements of the process which could be feasible with using nanocatalysts (Akia et al. 2014).

There are very limited studies on the conversion of biomass to other bio-oils. Nanoparticles of Co were used as catalysts in the conversion of spent tea to biochemical (Mahmood and Hussain 2010). In this pyrolysis process, Co nanoparticles reduce reaction temperature by up to 650 °C. The liquid products yield of the reaction at 300 °C and atmospheric pressure was about 60% (Akia et al. 2014).

For hydrothermal conversions of cellulose, using bulk ZnO as the catalyst increased the amount of glycolic acid by five times compared to when bulk SnO_2 was used in the hydrothermal conversion of cellulose at 300 °C. When nano-sized particles of these catalysts were used, the produced glycolic acid with ZnO catalyst was 12 times higher than that of nano-SnO_2. When producing biogasoline and organic liquid products (OLP) in a fixed bed reactor with nanocrystalline zeolite as the catalyst and waste cooking palm oil as the biofeed, the conversion of 87.5–92.9 wt.% of the feed is attainable. In such a condition, a gasoline fraction yield of 33.61–37.05 wt.% and an OLP fraction yield of 46.1–53.4 wt.% can be obtained. The NiW nanohydroxyapatite (NiW-nHA) composite was used as the catalyst in hydrocracking of Jatropha oil. In the operating conditions of 360 °C and 3 MPa about 92% of the feed was converted. The yield of C_{15}–C_{18} alkanes in the product was up to 83.5 wt.%. By increasing operating temperatures, it is possible to obtain 100% conversion of Jatropha oil in this process (Zhou et al. 2012; Akia et al. 2014).

8.5.3 Nanocatalysts for Biodiesel Production

In the biodiesel production method, transesterification is the chemical reaction between triglycerides and alcohol within the presence of a catalyst for producing monoesters. The triglyceride molecules are transformed to monoesters and glycerol. The transesterification method incorporates a sequence of three reversible reactions. The conversions of triglycerides to diglycerides, diglycerides to monoglycerides, and glycerides into glycerol yield one ester molecule in each stage. The general transesterification reaction can be represented by Fig. 8.9. The transesterification reaction of oil and alcohol with a homogeneous catalyst is the general method for the preparation of biodiesel. However, the homogeneous catalysts have many shortcomings, such as requiring large amounts of water, difficulties in product isolation, and environmental pollution caused by the liquid wastes. The use of a "green" method based on heterogeneous catalysts is a new trend in the preparation of biodiesel. Biodiesel synthesis using solid catalysts instead of homogeneous ones could potentially lead to cheaper production costs by enabling reuse of the catalyst and opportunities to operate in a fixed bed continuous process. Heterogeneous

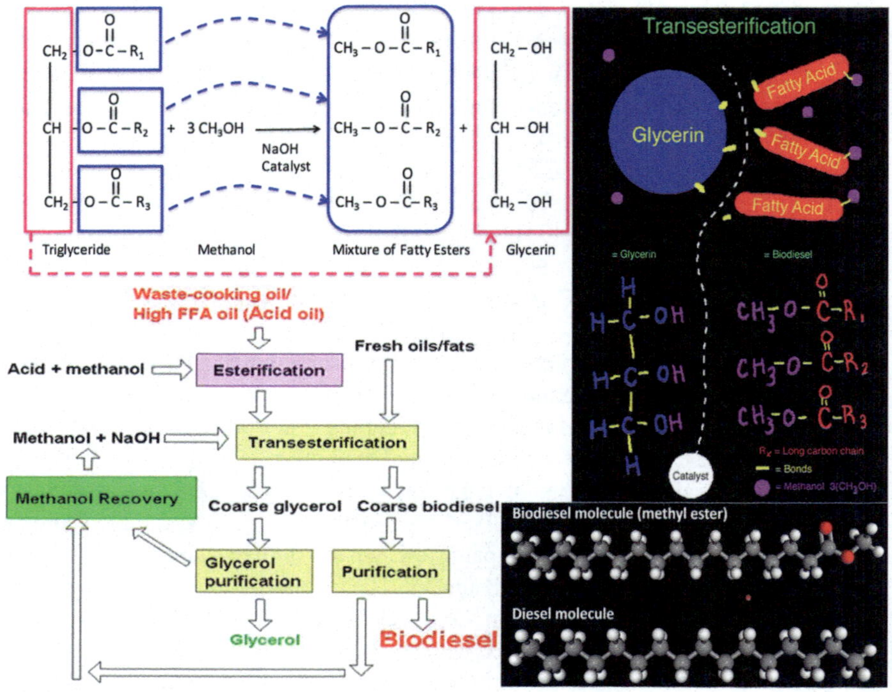

Fig. 8.9 Transesterification of triglycerides with methanol

catalytic methods are usually mass transfer resistant, time consuming, and inefficient. Despite the solid phase, catalytic methods are intensively studied, the industrial applications are limited. Nanocatalysts that have high specific surface and high catalysis activities may solve the current problems. The preparation of nano-sized heterogeneous catalysts has been studied to increase the catalytic activity. It is evident that the large surface area, which is characteristic of nano-sized material, resulted in a rise within the amount of the catalytically basic and acidic sites. The nanocatalysts used for biomass to biodiesel conversion have been presented in Table 8.5 (Akia et al. 2014).

Based on different studies on nanocatalyst application for biodiesel production, it is evident that the large porous catalytic surface increased the contact between alcohol and oil, leading to an increase in nanocatalytic effectiveness. Utilization of different edible and nonedible oils for transesterification reactions by using both acid and alkali nanocatalysts show the important influence of these catalysts regarding activity and selectivity. The high specific surface area of nanostructure materials in comparison with bulk catalysts is favorable for contact between catalyst and substrates, which effectively improve the yield of products. For instance, Nanocrystalline zeolite was used in catalytic conversion of cooking palm oil which has attained about 93% conversion at optimum temperature of 458 °C. The nanocomposite catalyst of NiW-hydroxyapatite may convert 100% of Jatropha oil in

Table 8.5 Nanocatalysts used for biodiesel production along with their operating conditions (Akia et al. 2014)

Nana catalyst	Size (nm)	Feedstock	Operation condition: Temp. (°C)	Alcohol: oil ratio	Catalyst loading (wt.%)	Reaction time (min)	Biodiesel yield (%)
$Cs/Al/Fe_3O_4$	30–35	Sunflower oil	58	14:1	4	120	94.80
Hydrotalcite (Mg–Al)	4.66–21.1	Pongamia oil	65	6:1	1.5	240	90.8
MgO Supported on Titania	–	Soybean oil	150–225	18:1	0.1–7	60	95
MgO	50–200	Sunflower oil	70–310	4:1	–	40–120	98
		Rapeseed oil					
ZrO_2 loaded with $C_4H_4O_6HK$	10–40	Soybean oil	60	16:1	6	120	98.03
Lithium impregnated calcium oxide (Li–CaO)	40	Karanja oil	65	12:1	5	60	99
		Jatropha oil				120	
Magnetic solid base catalysts CaO/Fe_3O_4	49	Jatropha oil	70	15:1	2	80	95
KF/CaO	30–100	Chinese tallow seed oil	65	12:1	4	150	96
Hydrotalcite-derived particles with Mg/Al molar ratio of 3:1	7.3	Jatropha oil	45	4:1	1	90	95.2
Cao	20	Soybean oil	23–25	27:1	–	720	99
Mgo	60	Soybean oil	200–260	6:1	0.5–3	12	99.04
$KF/CaO–Fe_3O_4$	50	Stillingia oil	65	36:1	4	180	95
TiO_2–ZnO	34.2	Palm oil	60	12:1	–	300	92.2
ZnO	28.4						83.2
KF/Al_2O_3	50	Canola oil	65	6:1	3	480	97.7
ZnO nanorods	–	Olive oil	150	15:1	1	480	94.8
CaO/MgO	–	Jatropha oil	64.5	18:1	2	210	92
Ca $(OH)_2–Fe_3O_4(Ca^{+2}: Fe_3O_4 = 7)$	–	Jatropha oil	70	15:1	2	240	99

catalytic the hydrocracking process. Nanocatalysts for biodiesel production significantly improve the yield of products. The main nanometal oxides that have been used for biodiesel production are Zn, Ca, Mg, and Zr. These have either been used individually or supported on different materials. However, some other catalysts such as Li, Cs, and KF have been utilized for edible and nonedible feedstock. In addition, magnetic nanoparticles functionalized with different catalysts have been implied in biodiesel production, facilitate the catalyst recovery. The results of using KF/CaO and nanomagnetic KF/CaO–Fe_3O_4 catalysts for biodiesel preparation show better performance of KF/CaO catalyst with a higher surface area of about 10^9 m^2/g at the same operating conditions compared to KF/CaO–Fe_3O_4 catalyst with a surface area of 20.8 m^2/g. Loading KF on nano-Al_2O_3 support could obtain a higher yield of about 97.7% compared to 96.8% for KF/CaO catalyst. However, the operating conditions of using KF/Al_2O_3 catalyst were relatively higher than that of KF/CaO catalyst. The alcohol:oil ratio, catalyst loading, and reaction time were 15:1, 3 wt.%, and 480 min in comparison with 12:1, 4 wt.%, and 150 min, respectively, for KF/Al_2O_3 and KF/CaO catalysts. Despite using different feedstocks, the better performance of KF/CaO catalyst may be related to its higher surface area (10^9 m^2/g) compared to the value of 41.7 m^2/g for KF/Al_2O_3 catalyst. The highest biodiesel yield was obtained using nanocatalysts Li-CaO and CaO at reaction time of 120 and 720 min, temperature of 65 °C and 25 °C, and methanol-to-oil molar ratio of 12 and 27, respectively. Similar yield has been obtained using a nano-MgO catalyst but at a higher temperature of about 200–260 °C. In general, to achieve high performances at relatively mild operating conditions, it is necessary to increase the reaction time while at ordinary reaction times, it is necessary to apply severe operating conditions. Comparing the results of using different supported and unsupported MgO catalysts revealed that decreasing the reaction time led to an increase in reaction temperatures for achieving higher performances. As a whole, using milder operating conditions led to a reduction in energy consumption requirements of the process which could be feasible with using nanocatalysts (Akia et al. 2014).

8.6 Coal Liquefaction

Coal liquefaction or coal to liquid fuel (CTL) is a process of converting coal into liquid hydrocarbons: liquid fuels and petrochemicals. Coal is a complex compound consisting of carbon, hydrogen, oxygen, sulfur, and minor proportions of other elements. It is an aggregate of microscopically distinguishable, physically distinct, and chemically different subparts baked together. Coal liquefaction works by breaking up the solid hydrocarbon structures found in coal. This may be accomplished by partial breakdown directly to liquid hydrocarbons (direct coal liquefaction or DCL) or by full breakdown into hydrogen and carbon that can be reassembled into H-C-chains of a desired length (indirect coal liquefaction or ICL) (Höök et al. 2014).

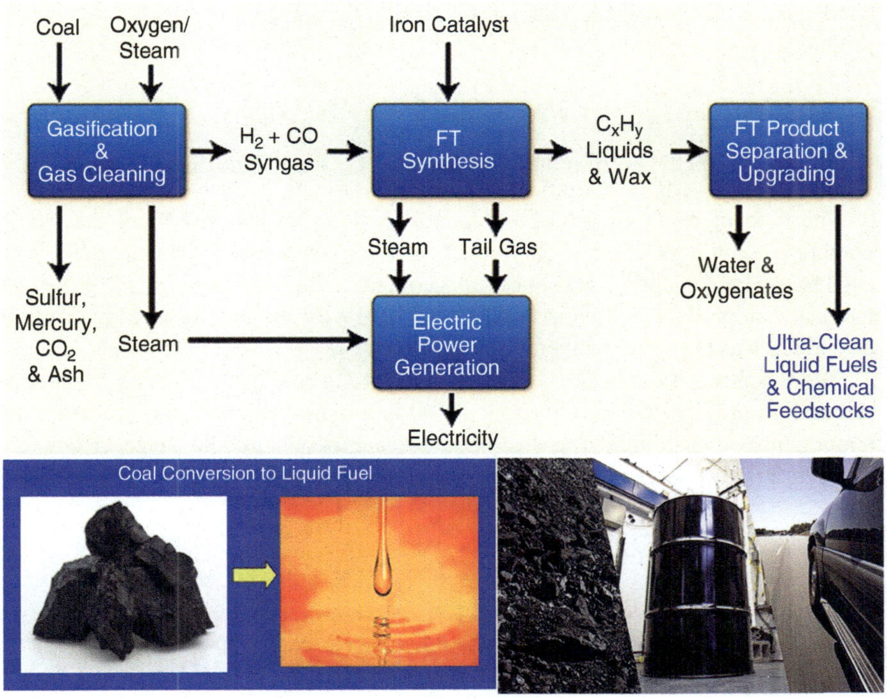

Fig. 8.10 Schematic of producing liquid fuels from coal

8.6.1 Basic Chemistry

As shown in Fig. 8.10, coal-to-liquids (CTL) is a technology based on the liquefaction of coal using three basic approaches: pyrolysis, direct coal liquefaction (DCL), and indirect coal liquefaction (ICL). The basis for all types of CTL-syntheses is a carbon source combined with a hydrogen source, such as steam. Chemical reactions between carbon and other compounds will eventually fabricate hydrocarbon molecules of the desired length. As a DCL process, the original Fischer-Tropsch (FT) process was described as (Höök and Aleklett 2010).

$$(2n + 1)H_2 + nCO \rightarrow nH_2O + C_nH_{2n+2} \tag{8.3}$$

Carbon monoxide can be produced by gasification of coal or another carbon rich compound. The necessary reaction energy is applied by adding oxygen or steam.

$$C + \tfrac{1}{2}O_2 \rightarrow CO \tag{8.4}$$

The resulting mixture of carbon monoxide and hydrogen is usually called synthesis gas (syngas). It is used to construct hydrocarbon chains of different lengths using condensation and a suitable catalyst. Generally, the FT process yields two types of products:

$$n\text{CO} + 2n\text{H}_2 \rightarrow n\text{H}_2\text{O} + \text{C}_n\text{H}_{2n}(\text{Olefins}) \tag{8.5}$$

$$n\text{CO} + (2n+1)\text{H}_2 \rightarrow n\text{H}_2\text{O} + \text{C}_n\text{H}_{2n+2}(\text{Paraffins}) \tag{8.6}$$

The type of resulting products depends on the catalysts used and the reactor operating conditions. Olefin-rich products with *n* in the range 5–10 (naphtha) can be used for making synthetic gasoline and chemicals in high-temperature FT processes. Paraffin-rich products with *n* in the range of 12–19 are suitable for making synthetic diesel and waxes in low-temperature FT processes.

The Bergius process is the basis of DCL. Splitting coal into shorter hydrocarbons, resembling ordinary crude oil is done by adding hydrogen under high pressure and temperature, thus eliminating the need for a gaseous middle stage (Höök and Aleklett 2010).

$$n\text{C} + (n+1)\text{H}_2 \rightarrow \text{C}_n\text{H}_{2n+2} \tag{8.7}$$

Both Bergius and FT processes use different catalysts to aid the chemical reactions. Common catalysts are transition metals such as iron, ruthenium, or cobalt. Transition metal sulfides, amorphous zeolite, and similar compounds have also been utilized. In general, catalysts have a large impact on process efficiency as well as influence over the resulting products. Many catalysts are notoriously sensitive to sulfur-compounds or other substances, which requires special treatment and separation techniques to avoid catalyst poisoning (Höök and Aleklett 2010).

Liquefaction can also be affected by the different properties of the coal feedstock. Different types of coal have different properties, requiring compatibility between CTL reactor design and coal feedstock. Hydrodynamic problems can be caused by thermal fragmentation of coal, when the coal particles are shattered into smaller grains capable of clogging gas outlets and causing other unwanted disturbances. Drying coal and reducing moisture will lower the probability of thermal fragmentation. Caking occurs due to the plasticity of coal and causes coal particles to meld into larger cakes, leading to pressure drops and channel burning, severely reducing the performance of the CTL reactor. Mixing high caking coals with low caking coals creates a more manageable overall caking probability. High ash coals will call for a gasification design capable of removing large ash concentrations without loss of performance. Practical CTL design can be found for all forms of coal. Consequently, suitable designs are essential for obtaining good performance (Höök and Aleklett 2010).

8.6.2 Coal-to-Liquid Technology Options

Coal-to-liquid technology, using coal or coal plus biomass feedstock, has improved significantly since the Second World War. Although many different types of coal to liquid fuel conversion techniques exist, only some have been commercialized and proven feasible by industry. CTL technology as an old concept has to be fused with modern processes and machinery to be able to fulfill the economic and environmental constraints of today. Existing technological infrastructure from conventional crude oil processing and gas liquefaction can provide synergistic effects reducing costs and necessary research. In theory, FT synthesis can be used to create liquid fuels from very unconventional feedstock as long as hydrogen and carbon are available. For instance, carbon dioxide, carbon monoxide, and other combustion exhaust have been used as feedstock in patent applications (Höök and Aleklett 2010).

8.6.2.1 Carbonization and Pyrolysis

In a pyrolysis process, heat decomposes the coal expelling volatile compounds, leading to increased carbon content in the remaining solid, leaving products such as char, semi-coke, and coke. Pyrolysis' primary use is to upgrade low-ranking coals by increasing their calorific value and reducing sulfur content and other pollutants. The oldest method for obtaining liquids from coal is high-temperature carbonization. Typically, coal is heated to around 950 ° C in a closed container. The heat causes decomposition and the volatile matter is driven away, increasing carbon content. This is similar to the coke-making process and accompanying tar-like liquid is mostly a side product. The process results in very low liquid yields and upgrading costs are relatively high. Coal tar is not traditionally used as a fuel in the transportation sector. However, it is used worldwide for manufacturing roofing, waterproofing, and insulation products and as a raw material for various dyes, drugs, and paints.

Mild temperature pyrolysis uses temperatures of 450–650 °C. Much of the volatile matter is driven off and other compounds are formed through thermal decomposition. Liquid yields are higher than for high-temperature pyrolysis, but reach a maximum at 20%. The main product is char, semi-coke, and coke (all smokeless solid fuels). This technique has mostly been used to upgrade low-rank coals, by increasing calorific value and reducing sulfur content. The Karrick process is a low-temperature carbonization process that also yields liquids. The main product is, however, semi-coke. The tar liquids produced require further refining before they can be used as a transportation fuel. As a result, pyrolysis provides low liquid yields and has inherently low efficiency. Furthermore, the resulting liquids require further treatment before they can be used in existing vehicles (Höök and Aleklett 2010).

Rapid pyrolysis occurs by subjecting coal to contained temperatures of around 1200 °C, but for only a few seconds. This process is aimed at producing chemical feedstocks rather than liquid fuels. Carbonization and pyrolysis produce a small proportion of the total product as liquid fuels which still require further treatment.

8.6.2.2 Direct Coal Liquefaction

Direct coal liquefaction (DCL) is built around the Bergius process (Eq. 8.7), where the basic process dissolves coal at high temperature and pressure. Addition of hydrogen and a catalyst causes "hydro-cracking," rupturing long carbon chains into shorter, liquid parts. The added hydrogen also improves the H/C ratio of the product. Liquid yields can be in excess of 70% of the dry weight coal, with overall thermal efficiencies of 60–70%. The resulting liquids are of much higher quality, compared to pyrolysis, and can be used unblended in power generation or other chemical processes as a synthetic crude oil (syncrude). However, further treatment is needed before they are usable as a transport fuel and refining stages are needed in the full process chain. Refining can be done directly at the CTL facility or by sending the synthetic crude oil to a conventional refinery. A mix of many gasoline-like and diesel-like products, as well as propane, butane, and other products, can be recovered from the refined syncrude (Höök and Aleklett 2010).

There are two main groups of direct liquefaction processes: (a) Single-stage, provides the distilled liquids (distillates) through one primary reactor or reactor chain. Most of these have been superseded by two-stage processes to increase production of lighter oils. (b) Two-stage, provides distillates through two reactors or reactor chains. The first reaction dissolves the coal either without a catalyst or with a low-activity disposable catalyst, producing heavy coal liquids. These are further treated in the second reactor, with hydrogen and a high-activity catalyst to produce additional distillate.

A number of processes have been developed in single-stage technology, such as the Kohleoel and NEDOL processes that are considered to be the most developed. Coal and a synthetic iron-based catalyst are ground and combined with a recycled solvent to form a coal slurry. This is then mixed with hydrogen and heated before entering the primary reactor, which operates at 450 °C and 170 bar. The products are cooled, depressurized, and distilled to provide a light product. Medium and heavy distillates are produced via the vacuum distillation column, and some used to provide the solvent for the initial slurrying step.

The two-stage processes have often derived from single-stage reactions—the Catalytic Two-Stage Liquefaction process was developed from the H-Coal single stage. Pulverized coal is slurried in a recycled process solvent, then preheated, mixed with hydrogen, and fed to the first reactor, which operates under typical conditions of 435–460 °C and 170 bar. A second reactor completes the liquefaction, operating at higher temperatures. The reaction catalyst for both stages is a nano-scale, iron-based one, dispersed in the slurry.

8.6.2.3 Indirect Coal Liquefaction

Indirect coal liquefaction (ICL) involves a complete breakdown of coal into other compounds by gasification with steam. Resulting syngas is modified to obtain the required balance of hydrogen and carbon monoxide. Later, the syngas is cleaned, removing sulfur and other impurities capable of disturbing further reactions. Finally, the syngas is reacted over a catalyst to provide the desired product using FT reactions (Eq. 8.3). Alteration of catalysts and reaction conditions can create a wide array of different products. For instance, methanol is one possible product that can be produced directly or further converted into high-quality gasoline via the Mobil process in additional stages. In general, there are two types of FT-synthesis, a high-temperature version primarily yielding a gasoline-like fuel and a low-temperature version, mainly providing a diesel-like fuel (Höök and Aleklett 2010).

In essence, DCL strives to make coal liquefaction and refining as similar to ordinary crude oil processing as possible by creating a synthetic crude oil. By sidestepping the complete breakdown of coal, some efficiency can be gained and the required amount of liquefaction equipment is reduced. Coal includes a large number of different substances in various amounts, several unwanted or even toxic. Some substances can poison catalysts or be passed on to the resulting synthetic crude oil. Ever-changing environmental regulations may force adjustment in the DCL process, requiring it to meet new regulatory mandates, just as crude oil processing has to be overhauled when new environmental protocols are introduced. In comparison, ICL uses a "designer fuel strategy." A set of criteria for the desired fuel are set up and pursued, using products that can be made in FT synthesis. Many of the various processes will yield hydrocarbon fuels superior to conventional oil-derived products. Eliminating inherent noxious materials in coals is not just an option; it is a must to protect the synthesis reactor catalysts. Far from all ICL-derived products are better than their petroleum-derived counterparts when it comes to energy content or other characteristics. However, all ICL fuels are inherently clean and virtually free from nitrogen, sulfur, and aromatics, generally giving lower emissions when combusted (Höök and Aleklett 2010).

ICL seems to be the more likely option for future CTL projects, based on its higher flexibility, better environmental capabilities, and stronger supporting experience and infrastructure. Furthermore, the fuel properties seem to benefit ICL compared to DCL, especially if end-use efficiencies are considered instead of just process efficiencies (Höök and Aleklett 2010).

As a result, the development of a coal to liquids (CTL) industry can serve to hedge against oil-related energy security risks. Using domestic coal reserves, or accessing the relatively stable international coal market, can allow countries to minimize their exposure to oil price volatility while providing the liquid fuels needed for economic growth. CTL fuels are ultraclean to use and the use of carbon capture and storage can minimize emissions from the manufacturing process.

8.7 Membrane Materials for Water Sustainability in Bioenergy

Membrane-based treatment technologies can have a key role in water purification and desalination. These technologies produce water of superior quality, are less sensitive to feed quality fluctuations, and have a much smaller footprint compared with conventional "Victorian" water treatment technologies. In addition, membrane-based desalination technologies are inherently more energy efficient than thermal approaches. Membranes have been long used as sustainable solution for seawater desalination by reverse osmosis. The successful use of membrane technology in other applications requires new materials and tailored separation characteristics. Although high flux and rejection of salt and small neutral molecules can be achieved by commercial membranes, membranes with specific functionalities, better defined pores or functionalized transport channels could provide breakthroughs in the separation and recovery of valuable products for instance in mining industry, heavy metals and rare elements, as well as in the bioenergy industry. Industrial effluents treatment will offer more opportunities for membrane technology in the future. Membrane materials with good thermal and chemical stability would allow their application in chemical processes, which are so far not extensively explored with membranes. This is the case of recovery of valuable products such as catalysts from reaction medium containing organic solvents. Membrane materials with better resistance to harsh cleaning treatments (chlorine, acid, and alkali treatments) would increase the membrane life time in chemical separation and also in regular wastewater treatment, for which fouling is a serious problem. This would also facilitate the use of membranes for a separation task of growing importance, the treatment of produced water in the biofuel industry and other energy sectors. On the whole, there are a large number of open challenges and opportunities for new materials in membrane technology. With the advent of better materials in this area, new sustainable technologies and separation processes will be implemented (Le and Nunes 2016; Werber et al. 2016).

8.7.1 Perspective of Membrane Materials and Technologies

Membrane technologies for water purification and desalination with separation largely based on size include the established microfiltration (MF), ultrafiltration (UF), nanofiltration (NF), and RO processes, as well as the emerging forward osmosis (FO) process. Figure 8.11 provides an overview to these processes. Processes that are driven by hydraulic pressure include microfiltration (MF), ultrafiltration (UF), nanofiltration (NF), and reverse osmosis (RO). Additionally, the process of forward osmosis (FO) is driven by an osmotic pressure difference between the feed source water and a specialized solution of high osmotic pressure, which is referred to as a draw solution. The processes are classified on the basis of the size of

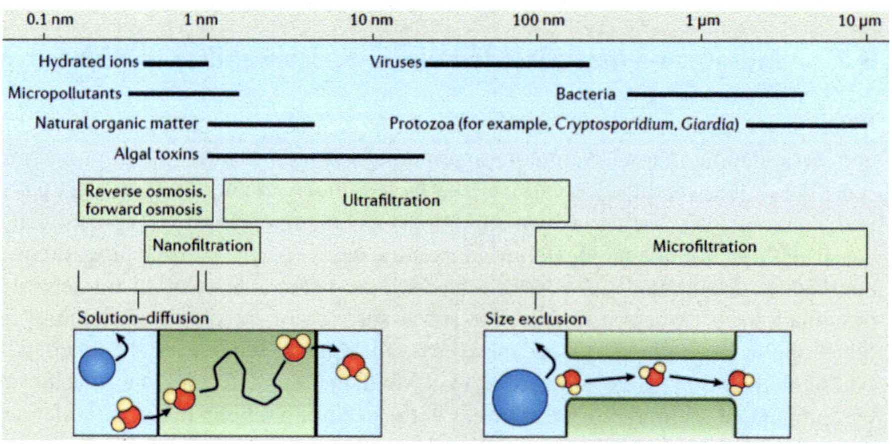

Fig. 8.11 Membrane processes for water purification and desalination (Adapted with permission from Werber et al. 2016 (Springer Nature))

the solutes that are retained. The main function of MF membranes is to remove suspended particles and microbial pathogens. UF membranes are designed to retain macromolecules, such as natural organic matter, as well as smaller pathogens, such as viruses, which are only partially removed by MF. The molecular weight cutoff (that is, the solute size at which 90% of species are rejected) of UF membranes is between approximately 5 and 500 kDa. MF and UF membranes are porous, and molecular separation is based on a sieving or size exclusion mechanism. NF membranes can effectively remove scale-forming ions, such as calcium and magnesium, and can partially reduce salinity. Typical molecular weight cutoffs of NF membranes are between 100 and 300 Da. RO and FO membranes are designed for desalination. Current RO and FO membranes are nonporous and can remove nearly all ions in addition to uncharged solutes of molecular weight greater than around 100 Da. Molecular transport in RO and FO is governed by a solution–diffusion mechanism in which molecules partition into the selective layer of the membrane, diffuse through the polymer network, and desorb in the permeate. Separation of species stems from differences in their solubility and/or diffusivity in the selective layer. Separation in NF is based on a combination of sieving and solution–diffusion mechanisms. A wide range of membrane materials, both polymeric and inorganic, have been used. Polymeric membranes are by far the most widespread, largely owing to their high processability and low cost (Baker 2012; Werber et al. 2016).

8.7.1.1 Microfiltration (MF) and Ultrafiltration (UF)

In membrane bioreactors (MBR), several types of membranes are used based on the size of contaminants, such as MF for suspended particles (100–1000 nm), UF for bacteria and virus (5–100 nm), and NF for dissolved particles (1–5 nm). MF and UF are technically popular than NF regarding the fouling and cost factors. Both ceramic and polymer materials can be used to fabricate MF and UF membranes. Ceramics have advantages of high chemical stability and mechanical strength, ease of cleaning, long and reliable lifetime. However, it is expensive and difficult to fabricate in large scale due to its brittleness. Sealing when integrating in modules can be an issue. Polymeric membranes have therefore dominated the application for decades. Common commercial polymers used for MF and UF membranes are poly (ether sulfone) (PES), poly(vinylidene fluoride) (PVDF), polyethylene (PE), polypropylene (PP), and polytetrafluorethylene (PTFE). Except for PES, all these polymers are very hydrophobic. PE, PP, and PTFE are insoluble in organic solvents at room temperature, making the manufacture by solution processes difficult. Porous membranes based on PP and PTFE are produced by mechanical stretching of extruded films. Thermal-induced phase separation (TIPS) of polymer solutions is commonly used for the manufacture of porous polyolefin membranes. However, the most widely used method for membrane preparation, applied for PES and PVDF membranes, is the non-solvent-induced phase separation (NIPS). This involves solution casting and immersion in a coagulating water bath. Asymmetric membranes with gradient pore size are obtained in this case. Table 8.6 lists examples of MF and UF membrane types and configurations used in MBR systems (Roberts et al. 2000; Le and Nunes 2016).

Table 8.6 Examples of MF and UF membrane types and configurations used in MBR systems (Le and Nunes 2016)

Types	Configuration	Supplier	Pore size	Wastewater
MF ceramic	Tubular	Adams hydraulics	0.2 μm	Food waste
MF ceramic	Tubular	Kerasep	0.1 μm	Municipal
MF ceramic	Tubular	–	0.2 μm	Municipal
MF alumina	Tubular	SCT	0.2 μm	Municipal
UF	Plate and frame	DDS Lab	20 kDa	Alcohol distillery
UF Zircon	Tubular	SCT	0.05 μm	Municipal
UF	Tubular	Zenon environmental	75 kDa	Sanitary and industrial
UF	Tubular	–	15 kDa	Synthetic (fuel oil)

Fouling is one of the biggest challenges for MF and NF membranes because most of them are hydrophobic. The most commonly used strategy to combat fouling is to make membranes hydrophilic by (a) hydrophilically modifying membrane polymers before fabrication, (b) blending with hydrophilic agents, and (c) grafting or coating hydrophilic polymers on the membrane surface. Although fouling can be minimized, undesirable effects such as narrowing the pore size or even blocking the pores of the MF and UF microporous membrane surfaces. In other cases, the pore size enlarges, reducing the salt rejection. In addition, many coating layers have in adequate mechanical and chemical properties and do not sustain long-term operations (Mansouri et al. 2010; Le and Nunes 2016).

8.7.1.2 Nanofiltration (NF)

Advanced membrane formation technologies have been applied to produce NF membranes with high flux and rejection and low fouling tendency. The basic manufacturing method is NIPS. This produces an integrated porous asymmetric membrane with a selective layer on the top or a nonselective porous structure, which can be used as substrate for multilayered membrane preparation. Typical polymeric materials for this purpose are PES, PVDF, and polyacrylonitrile (PAN). The deposition of a selective layer constituted by another polymer, which can be crosslinked or not, can be performed by dip-coating or interfacial polymerization. Additional surface modification by nanoparticle incorporation and grafting polymerization can be performed to control selectivity or reduce fouling. Interfacial polymerization technique refers to the polycondensation of water-soluble and organic-soluble monomers on a porous support to produce thin-film composite (TFC) membranes. The technique is simple, easy to apply, and capable of creating a very thin selective layer of <100 nm, mainly based on polyamide. This thin layer determines the overall efficiency of the membranes. Efforts to improve NF performance include influencing the selective layer by changing monomers, adding additives into the aqueous or organic solutions or modifying the surface of the formed polyamide layer. The incorporation of nanoparticles into the selective layer during the polymerization has been studied to form thin-film nanocomposite (TFN). TFC membranes are considered the benchmark in the field of NF for aqueous separations. On the other hand, grafting polymerization via UV/photo-grafting, electron beam irradiation, plasma treatment and layer-by-layer (LbL) technique have been also extensively studied to produce NF membranes with high hydrophilicity and low fouling propensity. These techniques are well developed in the laboratory scale but their application in the large scale is still limited (Ng et al. 2014; Le and Nunes 2016).

Although membranes with high permeance and selectivity to salt are available, membranes with tailored selectivity, able to distinguish solutes of similar size are needed. Furthermore the use of NF membranes is also growing for applications other than water purification. This is the case of chemical and pharmaceutical applications, requiring operation in the presence of organic solvents. For this purpose a new class of resistant porous substrates and selective layers is being developed. Materials under consideration are polyetherketone, crosslinked polyimide, and polyazoles (Hermans et al. 2015; Le and Nunes 2016).

8.7.1.3 Reverse Osmosis (RO)

The application of membranes for seawater desalination substantially reduced the cost and increased the availability of drinking water. Polymeric membranes dominate the application for desalination by RO, because of their low-cost fabrication, ease of handling, and excellent performance in terms of selectivity and permeability. A key breakthrough in RO membrane manufacturing and application in different fields was the NIPS process, which allowed production of asymmetric membranes (1960s), initially based on cellulose derivatives, and the fully crosslinked TFC membranes (1970s to 1980s) prepared by interfacial polymerization. The mostly used TFC membranes are constituted by a porous polysulfone substrate and a thin polyamide layer. Thin cellulose acetate hollow fibers and TFC membranes in the spiral module configuration share the market for RO desalination plants. Table 8.7 provides commercial RO membrane modules in seawater desalination. Despite their success, TFC membranes have a key limitation—the degradation by chlorine, one of the common disinfectants used in wastewater treatment. RO membrane development should focus on chlorine-resistant membranes to eliminate the need of de-chlorination for the RO feed and rechlorination for the RO permeate, reducing the overall cost of the system. In addition, high-boron rejection membranes should be addressed when increasingly stringent water quality standards for which lower boron concentrations are required. This will help to reduce the number of RO pass in the RO plants to achieve the required water quality. In terms of water permeance and salt selectivity, the current RO membranes are well advanced and successful in large-scale seawater desalination operation. The future RO development should focus on other regulated and emerging trace contaminants such as persistent organic compounds (POPs), pesticides, pharmaceutically active compounds (PhACs), and endocrine disrupting chemicals (EDCs) (Hilal et al. 2011; Le and Nunes 2016).

The evolutionary development of polymeric RO membranes is mostly mature. Membranes with high salt rejection are available with high water permeance. Even higher permeances would reduce the needed membrane area, but would not considerably save energy for RO operation. However membranes with higher fouling

Table 8.7 Examples of commercial RO membrane modules in seawater desalination (Le and Nunes 2016)

Brand name	Membrane	Module	Operation condition	Flux (m^3/day)	Salt rejection (%)	Location
DOW FILMTECTM 8-in. SW30HRLE	TFC crosslinked membrane	Spiral wound	32 g/L NaCl, 55 bar, 25 °C, pH 8, 8% recovery	28.0	99.60–99.75	Perth, Australia
Hydranautics 8-in. SWC4 +	TFC crosslinked membrane	Spiral wound	32 g/L NaCl, 55 bar, 25 °C, pH 7, 10% recovery	24.6	99.70–99.80	Lobregat, Spain
Toray 8-in. TM820C	TFC crosslinked membrane	Spiral wound	32 g/L NaCl, 55 bar, 25 °C, pH 8, 8% recovery	19.7–24.6	99.50–99.75	Singapore
Toyobo 16-in. HB10255	Asymmetric cellulose tri-acetate	Hollow fiber	35 g/L NaCl, 54 bar, 25 °C, 30% recovery	60.0–67.0	99.40–99.60	Fukuoka, Japan

resistance and higher resistance to chlorine are needed. RO membranes continue to be developed using nanoparticles. Two major practical challenges need to be overcome before they can move to the next stage of development: (a) cost and (b) the difficulty to scale up. In addition, health and safety aspects particularly in the case of added nanoparticles need to be estimated, especially for drinking water desalination (Fathizadeh et al. 2011; Le and Nunes 2016).

There is currently intense interest in biomimetic membranes based on embedded Aquaporin due to the expectation of superior permeability and almost absolute salt rejection. However, the fabrication of Aquaporin membranes with high packing density would be hardly feasible at the large scale, being a complex and expensive technique. Moreover, these bio-materials are relatively unstable and hence the durability of the membranes is questionable at least for mass applications like desalination of seawater for drinking purposes (González-Pérez et al. 2009; Le and Nunes 2016).

8.7.1.4 Forward Osmosis (FO)

There are two evolutionary development routes of FO membranes. The first is to modify available commercial NF or RO membranes and the second involves the development of new membranes with specific design for FO applications. Fabrication of FO membranes from existing NF or RO membranes is considered simple, effective to some extent, and cost-efficient. Despite the similarity in performance criteria such as high salt retention and high water flux, the difference between NF/RO and FO membranes is the balanced relation between high-pressure tolerance (or mechanical strength) and low internal concentration polarization (ICP) of the support layer. In FO processes, pressure tolerance is not critical (almost no or low hydraulic pressure is applied) while low ICP is important to maintain high water flux and low salt leakage. To achieve it, therefore, the mechanical strength of the membranes is reduced to some extent to increase porosity and reduce tortuosity. Another difference between RO and FO membranes is the importance of pore wettability. For the FO process, pore wettability must be improved, because the presence of un-wetted pore regions may block the water flux and significantly exacerbate ICP. Coating with a highly hydrophilic polymer like polydopamine (PDA) has been demonstrated as an effective technique to improve wettability, flux, and fouling resistance. The water flux can be increased ten folds after PDA coating. Table 8.8 provides examples of commercial membranes used for FO tests. Despite the improvement after modification, the efficiency of the modified FO membranes is still limited and restricted by the inherent properties of their parent NF/RO membranes. Thus, new FO membrane have been developed to obtain (a) highly hydrophilic, greatly porous but adequately strong support layer to minimize ICP effects, and (b) ultrathin defect-free highly selective layer to achieve high water flux, high salt rejection and minimize reverse solute diffusion. These two layers are either integrated by the same materials (asymmetric membranes) or prepared separately from different materials in a thin-film-composite (TFC) membrane structure. The latter has the advantage over the former in its possibility to optimize support and selective layers separately. The development strategy of FO

Table 8.8 Examples of commercial membranes for FO tests (Le and Nunes 2016)

Material	Type	Supplier	Water permeability $\times 10^{-12}$ (m/s·Pa)	Draw solute	Rejection (%)
CA-3000	Flat-sheet	Toray	3.75	NaCl	98
CA	Flat-sheet	HTI	5.69	$(NH_4)_2CO_3$	>95
CTA	Flat-sheet	HTI	7.10	NaCl	>92
TFC (NF)	TS80	TriSep	20.0	$MgSO_4$	100
TFC (RO)	DS-11-AC	General Electric	11.91	–	99.5
TFC (RO)	SW30XLE-400i	Dow	3.80	NaCl	–
TFC (RO)	SWC1	Hydranautics	5.60	–	100
TFC (RO)	SW30-HR	Dow	4.39	NaCl	98

membranes in terms of materials is similar to that of RO membranes. Rather than membrane materials, the development for FO also focuses on membrane morphologies. The ideal support morphology for FO membranes is still under development, such as straight finger-like macrovoids, sponge-like structures, and electro-spun fibers as the support (Chung et al. 2012; Le and Nunes 2016).

Besides membrane materials, the exploration of suitable draw solute materials is of great importance in FO process. The key criteria of draw solute selection are (a) high osmotic pressure, (b) low reverse diffusion (leakage through the membrane into the feed), (c) easy and economical regeneration, (d) non-toxicity, and (e) reasonable price. In addition, the draw solution should not degrade the membranes or cause membrane scaling/fouling. Numerous draw solutes have been proposed, including inorganic, thermolytic/volatile, organic, magnetic, and polymer-based solutes. Table 8.9 summarizes their examples, recovery methods, and drawbacks, which implies that there have been no perfect draw solutes so far. Different FO objectives and available solute recovery technologies determine different draw solutions (Le and Nunes 2016).

Despite extensive challenges of its energy sufficiency remains, FO is still potentially promising to desalinate the high salinity feed stream with osmotic pressure exceeding the tolerant pressure of RO or treat the waste stream with high fouling tendency. In the former application, the hybrid FO systems employing thermolytic draw solutions may be favorable because only relatively small amount of the thermolytic draw solute (higher vapor pressure than water) must be vaporized as compared to large volume of water having to be recovered in conventional distillation. In addition, the draw solutes with high vapor pressures require less total energy for recovery. Using low-cost thermal energy sources (e.g., solar energy, geothermal energy, and industrial waste heat) for draw solution recovery would reduce the energy cost of the whole FO process. In the second application, FO may be

Table 8.9 Draw solutes, their recovery methods, and drawbacks (Le and Nunes 2016)

Group	Example	Recovery method	Drawback
Inorganic	Al_2SO_4	Precipitation by doping $Ca(OH)_2$	Toxic by-products
	$MgCl_2$	None	Not pure water
Thermolytic/ volatile	SO_2	Heating or air stripping	Energy intensive, toxic
	NH_4HCO_3	Moderate heating	High reverse draw solute flux, insufficient removal of ammonia
	Ethanol	Pervaporation-based separations	High reverse draw solute flux and low water flux
Organic	Glucose–Fructose	None	Not pure water
	Sucrose	NF	Relatively low water flux
	Dendrimers	Adjusting pH or UF	Not feasible
	Albumin	Denatured and solidified by heating	Not feasible
	2-Methyl imidazole-based solutes	MD	Materials costly
	Fertilizers	None	Only applicable in agriculture
	Organic salts	RO	Low water flux, energy intensive
Magnetic Polymer-based	Magnetic particles	Captured by a canister separator	Poor performance, agglomeration
	Hydrogels	Deswelling of the polymer	Energy intensive, poor water flux
	Fatty acid-polyethylene glycol	Thermal method	Poor water flux
	Polyelectrolytes	UF	Relatively high viscosity
	Hexavalent phosphazene salts	Not studied	Not economical and practical

advantageous in the integration with MBR because of its low fouling propensity and high reversibility of fouling (Shaffer et al. 2015).

8.7.1.5 Membrane Distillation (MD)

In general, a MD membrane should (a) exhibit high hydrophobicity to prevent pore wettability on the feed side (membrane liquid entry pressure of water ranges from 48 to 368 kPa for common commercial flat-sheet MD membranes); (b) be thin to achieve high permeation flux (the optimum thickness lies between 30–60 μm, but when composite hydrophobic/hydrophilic membranes are used, the hydrophobic layer can be as thin as 5 μm); (c) have reasonably small pore size (in the range of 0.1–0.6 μm) to prevent liquid intrusion but not too small to contribute for mass transfer resistance; (d) have high porosity (60–80%) and low tortuosity (a value of

Table 8.10 Several commercial membranes studied for MD (Le and Nunes 2016)

Polymer	Trade name	Company	Mean pore size (μm)
PTFE	TF200	Gelman	0.2
	FGLP14250	Millipore	0.25
	M05E0020	GVS	0.2
	Desal K150	Millipore	0.1
	Fluropore	Millipore	0.2
PVDF	GVHP	Millipore	0.11
	Microza	Ashi Chem	0.2
	M09G0020	GVS	0.2
	Durapore	Millipore	0.2
	GVHP22	Millipore	0.16
PP	Accurel PP	Microdyne	0.22
	Liqui-cel	Celgard	0.04
	MD020TP2 N	Enka	0.2

2 has been commonly assumed but 3.9 was also reported) to favor the water vapor transport; (e) have high heat transfer resistance (the thermal conductivity ranges from 0.11 to 0.27 W/mK at 23 °C for three common MD polymers PVDF, PTFE and PP) but low mass transfer resistance (the reported membrane coefficient ranges $3–15 \times 10^{-7}$ kg/m^{2}pa·s for commercial MD membranes in the direct contact operation mode); (f) have adequate chemical and thermal resistances, strong mechanical properties, and long-term stability; and (g) be cheaply available. Early development stages of MD membranes focused on hydrophobic commercial polymers such as PVDF, PTFE, and PP, analogously to those used for micro- and ultrafiltration. Some commercial membranes used for MD are listed in Table 8.10. The main advantage of these materials is their commercial availability. However, they cannot meet all above requirements of an excellent MD membrane. Hence, research on designing and synthesizing new membrane materials for MD processes is essential to attain a fully commercial status of MD (Pangarkar et al. 2014; Le and Nunes 2016).

Novel MD hydrophobic membranes can be fabricated either by newly synthesized hydrophobic polymers or by surface modification of hydrophilic membranes. An example of the former is synthesis of copolymers between PVDF with hexafluoropropylene (HFP) or tetrafluoroethylene (TFE). Although MD membranes prepared from these materials had lower flux than PVDF membranes due to lower porosity, they exhibited excellent mechanical properties with 100% salt rejection. On the other hand, various surface modifications have been investigated for MD, such as surface segregation, impregnation and cross-linking, co-extrusion, coating, grafting, and plasma polymerization (Qtaishat et al. 2009).

A promising MD membrane fabrication is the composite hydrophobic/hydrophilic membranes in the multilayered structure. The advantage of such structure is that high mass transport can be achieved by tailoring the thickness of the hydrophobic layer as thin as possible while low heat transfer and adequate mechanical properties can still be maintained by controlling the thickness of the hydrophilic layer. Other attempts on MD membrane development are use of nanofiber membranes prepared by the electrospinning method to achieve high and controlled void volume; incorporation of carbon nanotubes to increase vapor permeation and prevent pore wettability; and adding clay particles to enhance the mechanical strength and long-term stability of the membrane. Finally, the synthesis of new hydrophobic polymers and their manufacture into flat-sheet and hollow fiber membranes for MD has been successfully demonstrated (Maab et al. 2013; Le and Nunes 2016).

8.7.2 Emerging Polymeric Membranes

Next-generation, highly selective desalination membranes will represent an important advancement. Enhanced membrane selectivity will have an important role in improving water quality and eliminating the need for additional separation stages—for example, for lowering boric acid concentration in seawater desalination to acceptable levels for agriculture—thereby reducing energy usage and the cost of desalination. The molecular level design approaches for desalination membranes represent an active area of research. Despite these major efforts, however, it is still not possible to consistently fabricate even small-area membranes with greater water permeability and selectivity than current state-of-the-art TFC membranes. A particular challenge for several of these materials will be the formation of thin, defect-free selective layers supported by microporous layers. After the concepts of these novel membranes are successfully demonstrated at the laboratory scale, upscaling the small-area membranes to low-cost, industrial-scale modules will be an additional challenge.

Fouling control in next-generation membranes, in addition to current membranes, will remain a major challenge, because novel materials for high separation performance are often not compatible with fouling resistance. Surface grafting of fouling-resistant polymer brushes—such as PEO and zwitterionic polymers—will probably remain the best approach for imparting fouling resistance for next-generation, isoporous membranes. For "tight" UF membranes, surface grafting will not impart fouling resistance to the interior pore walls; hence, proper choice of new polymer chemistries for membrane fabrication with fouling and chemical resistance will be crucial. Next-generation desalination membranes are likely to overcome another limitation of current TFC membranes: the high sensitivity of the polyamide selective layer to oxidants such as chlorine and ozone. This will allow better control of biofouling, which is the Achilles heel of membrane desalination (Werber et al. 2016).

8.7.2.1 Phase Inversion Membranes

Figure 8.12 shows emerging polymeric membranes for water treatment (Le and Nunes 2016): (a) A schematic diagram of phase inversion by non-solvent-induced phase separation to form microfiltration (MF) and ultrafiltration (UF) membranes. A casting solution consisting of polymer, solvent, and desired additives is cast on a solid support as a thin film, which is followed by immersion in a non-solvent to induce phase separation. (b) Scanning electron microscopy (SEM) images of membranes formed from phase inversion. The upper left panel is a cross-section of an asymmetric UF membrane with finger-like macrovoids cast from 9% polysulfone (PSf) in dimethylformamide (DMF). The upper right panel is a cross-section of an asymmetric UF membrane with a sponge-like structure cast from 12% PSf in DMF. The lower left panel is a top view of a hand-cast PSf UF membrane. The lower right panel is a top view of a commercial 0.22-μm nominal pore size polyvinylidene fluoride MF membrane. (c) A schematic representation of interfacial polymerization to form thin-film composite (TFC) reverse osmosis membranes. Aqueous m-phenylenediamine diffuses to the water/organic interface, reacting with trimesoyl chloride to form a dense polyamide film with ridge-and-valley morphology. (d) Microscopy images of a hand-cast TFC membrane. The upper panel shows a top-view SEM image of rough polyamide film at a 45° viewing angle. The lower left panel is a transmission electron microscopy (TEM) image of a thin cross-section showing the rough polyamide film formed on a PSf UF membrane. The lower right panel shows a corresponding scanning TEM energy-dispersive X-ray spectroscopy image; the presence of sulfur and nitrogen delineates the PSf layer and polyamide film, respectively.

Most commercially available porous membranes that are used for MF and UF, as well as some dense membranes used for RO and FO, are formed by phase inversion. This process involves the controlled precipitation of a dissolved polymer in a thin film to produce a porous membrane structure. The main phase inversion technique used is non-solvent-induced phase separation (NIPS), in which a film of polymer dissolved in solvent is immersed in a non-solvent bath, typically a water bath, leading to solvent–non-solvent exchange and phase separation into polymer-rich and polymer-poor phases (Fig. 8.12a). These two phases ultimately form the solid membrane matrix (from the polymer-rich phase) and pores (from the polymer-poor phase). Other phase inversion techniques include controlled solvent evaporation, in which a volatile solvent is allowed to evaporate from a liquid film that includes polymer and non-solvent, and thermally induced phase separation, in which a high-temperature solution of dissolved polymer is slowly cooled until phase separation occurs (Baker 2012; Werber et al. 2016).

During phase inversion, several key parameters determine the resulting membrane morphology: for example, the nature of the solvent and non-solvent, polymer type, polymer concentration, and fabrication technique. As shown in Fig. 8.12b, a wide range of morphologies can result, including asymmetric membranes for UF that have a thin nanoporous selective layer (which is often called the active layer) and a microporous underlying structure, and relatively symmetric microporous membranes for MF. However, owing to the stochastic nature of the phase inversion process, membranes formed by traditional phase inversion have polydisperse pore size distributions, which adversely affect the selectivity of the resulting active layer (Werber et al. 2016).

In contrast to phase inversion, an alternative process called track etching can form UF and MF membranes with pores of uniform size. Track etching involves two

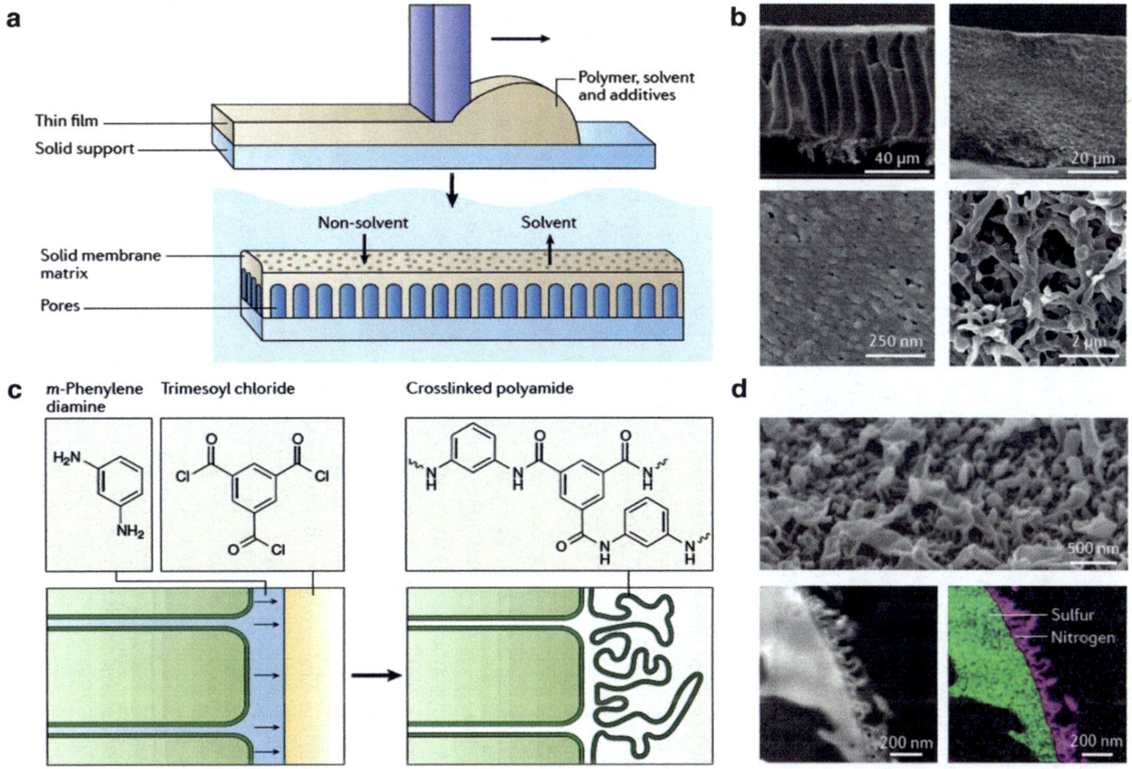

Fig. 8.12 Emerging polymeric membranes for water treatment (Adapted with permission from Werber et al. 2016 (Springer Nature)): (**a**) A schematic diagram of phase inversion by non-solvent-induced phase separation to form microfiltration (MF) and ultrafiltration (UF) membranes. (**b**) Scanning electron microscopy (SEM) images of membranes formed from phase inversion. (**c**) A schematic representation of interfacial polymerization to form thin-film composite (TFC) reverse osmosis membranes. (**d**) Microscopy images of a hand-cast TFC membrane

steps: bombardment with charged particles to partially degrade a track of polymer, followed by chemical etching to form pores. The main limitation of track-etched membranes is their low porosity (that is, low void volume within the membrane), which must be kept below 5% to prevent pore overlap. The low porosity results in low water permeability, largely limiting track-etched membranes to analytical separations (Baker 2012; Werber et al. 2016).

8.7.2.2 Thin-Film Composite Polyamide Membranes

Thin film composite (TFC) polyamide membranes are the gold standard for NF, RO, and FO applications. These membranes comprise a nonporous, highly cross-linked polyamide selective layer and an underlying porous support layer, typically made of polysulfone. TFC polyamide membranes can achieve water permeability and salt rejection far exceeding those for asymmetric cellulose-acetate-based membranes, which were the first-generation RO membranes developed more than 50 years ago. This performance, combined with the stability of TFC membranes over a wide pH range (pH 2–11 for continuous operation), has led to their widespread use in desalination applications (Baker 2012).

The polyamide selective layer is formed through interfacial polymerization of a diamine—most commonly, m-phenylenediamine (MPD) for RO, FO, and NF, and piperazine (PIP) for NF—with a triacyl chloride, particularly trimesoyl chloride (TMC). During interfacial polymerization (Fig. 8.12c), the polysulfone support is first brought into contact with the aqueous diamine solution to allow diamine penetration into the support. After removing the excess solution, the support is immersed in an organic phase containing TMC, which is not soluble in water. As diamine monomers diffuse to the water/organic interface, they react with TMC to form the polyamide film, which adheres to the polysulfone support through physical interactions. Subsequent hydrolysis of nonreacted acyl chloride groups yields carboxyl groups, which confer a negative charge to the membrane surface. TFC RO membranes formed using aromatic MPD and TMC monomers are fairly hydrophobic (with water contact angles of 50–60°), whereas TFC NF membranes formed using PIP and TMC are more hydrophilic (with water contact angles of approximately 30°). Both MPD based and PIP-based TFC membranes are sensitive to chemical degradation by strong oxidants, most notably chlorine. This sensitivity eliminates the usage of chlorine as a membrane-cleaning agent and necessitates removal of chlorine before the membrane step if it is present in the feed stream (Baker 2012; Le and Nunes 2016).

The surface morphology of TFC membranes also depends on the monomers used. PIP-based membranes are smooth and have a root-mean-square roughness of less than 10 nm. By contrast, MPD-based membranes have a characteristic "ridge-and-valley" structure, typically having a root-mean-square roughness greater than 50 nm and an overall end-to-end thickness of 100–500 nm. The roughness is due to protruding polyamide nodules, as can be seen in Fig. 8.12d. The nodules are generally hollow; transmission electron microscopy images suggest that the nodule walls are roughly 10–30 nm thick. Although the rough nodular morphology increases fouling propensity, the increased surface area stemming from this morphology may also increase water permeability (Karan et al. 2015; Le and Nunes 2016).

8.7.2.3 Designing Highly Selective Membranes

For desalination membranes, specially designed water channels could lead to a shift in how water itself moves through the membrane: from partitioning into and diffusing through polymeric thin films to single-file transport through active-layer-spanning water channels. Such sub-nanoporous water channels, if well designed, could offer a significant improvement in desalination performance (Werber et al. 2016).

In the design of sub-nanoporous water channels, materials research can gain insight from the biological water channel aquaporin. These ubiquitous, integral membrane proteins have homologues found in bacteria, archaea, plants, and animals. They exist as tetramers of four identical monomers, and each monomer bears a water channel. Aquaporin is remarkable for its unique transport properties: some types of aquaporin are not only highly permeable to water, but also extremely selective, rejecting ions, protons, and even small neutral solutes, such as urea, that readily permeate through current desalination membranes. The high permeability of the aquaporin water channel stems from its hourglass structure (Fig. 8.13a), which features wide entrance vestibules and a narrow hydrophobic center that supports single-file transport of water molecules. Selectivity is size- and charge-based: 0.8 nm above the channel center, the channel constricts to just 0.28 nm in diameter (which is approximately the size of a water molecule) owing to protruding phenylalanine, histidine, and arginine side chains. The arginine side chain provides the constriction site with a fixed positive charge. This tight, charged constriction site provides a high-energy barrier that prevents permeation of charged and neutral solutes (Hub and de Groot 2008).

Using aquaporin as a model, it is theoretically possible to systematically design discrete water channels that are suitable for desalination and stable enough for easy processing and use. Such channels should ideally exhibit the following attributes (Werber et al. 2016): molecular-sieve-like constriction to exclude salt and small neutral solutes; stable alignment within a polymeric thin film; intrinsic water permeability and achievable packing density that are high enough to match the water permeability coefficient of TFC membranes; chemical and mechanical stability; and ease of fabrication using consistent and scalable synthesis methods.

Despite their structural simplicity, carbon nanotubes (CNTs) have the potential for remarkable performance as nanochannels for water purification and desalination. Much of their promise for desalination was demonstrated by molecular dynamics simulations for a single-walled (6,6) CNT (Fig. 8.13a). The formation of a single-file water wire with ultra-fast water conduction is largely attributed to the atomically smooth, non-polar interior. (6,6) CNTs, which have a diameter of 0.81 nm between opposing carbon centers and an inner (pore) diameter of 0.47 nm, can fully reject salt. Salt passage increases rapidly with diameter: just 58% salt rejection is expected for (8,8) CNTs with a 1.1-nm center-to-center diameter (Corry 2008).

Synthetically designed nanochannels offer a customizable alternative to CNTs. Although these nanochannels lack the atomically smooth interior present in CNTs,

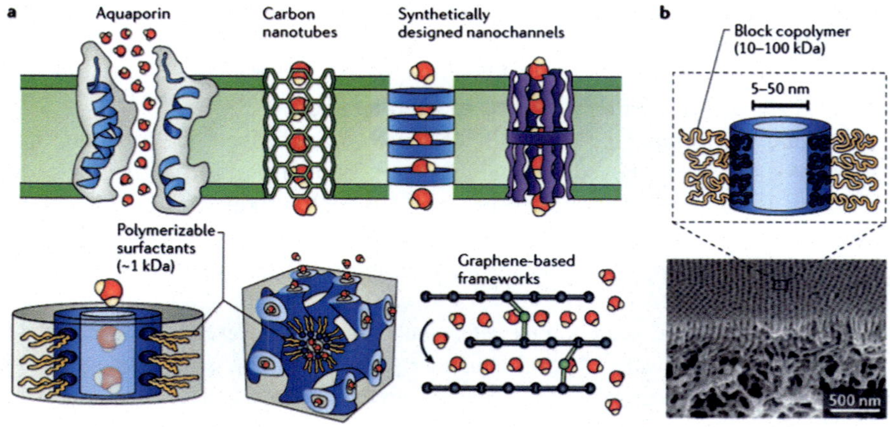

Fig. 8.13 Selective membranes formed using molecular-level design (Adapted with permission from Werber et al. 2016 (Springer Nature): (**a**) Sub-nanoporous materials with promise for desalination. The top panel shows discrete channels spanning a thin, nearly impermeable film, including the membrane protein aquaporin, single-walled carbon nanotubes with diameters $\leq$0.8 nm to exclude salt, self-assembled channel-forming macrocycles and single-molecule synthetic water channels. The bottom panel shows contiguous media that fully form a selective layer, including inverse hexagonal channels formed using wedge-shaped polymerizable surfactants, tortuous channels formed in bicontinuous cubic phase using polymerizable surfactants and graphene-based frameworks, shown with either graphene oxide or reduced graphene oxide as the grey sheets and covalent linkers in green. In all of the schematics, water molecules are shown traversing the nanochannels. (**b**) Controlled formation of the nearly isoporous ultrafiltration membrane using block copolymers

their properties can be individually tailored, allowing far greater control over physical dimensions, performance, processing, and alignment. Synthetic nanochannels can be divided into two classes: single-molecule and supramolecular nanochannels (Fig. 8.13a). For all discrete nanochannels, however, the current challenge is predominantly the synthesis of channels with the desired water and solute transport properties. An equally important technical challenge is developing scalable solution-phase methods for the assembly and alignment of these channels in robust selective layers. Structural features that aid alignment are likely to be important and will be a major factor in nanochannel design. In addition, much potential lies in using the self-assembling properties of amphiphilic surfactants and block copolymers. Indeed, this approach has already been applied with some success for CNTs and cyclic peptides. Self-assembled materials can also be used directly to form nanoporous membranes (Mauter et al. 2010; Werber et al. 2016).

Two broad classes of materials have been used for self-assembly membranes. The first class is reactive small molecules with a molar mass generally less than 1 kDa that self-assemble into lyotropic or thermotropic liquid crystalline mesophases. Lyotropic liquid crystalline mesophases are structurally ordered systems that form principally as a function of small molecule concentration in a solvent (typically water), whereas thermotropic liquid crystalline mesophases mainly form as a function of temperature in the absence of a solvent. Before they can be used as membranes, liquid crystalline mesophases must be polymerized or crosslinked to form sufficiently robust polymer films. The second class is block copolymers with a molar mass generally greater than 10 kDa. Block copolymers self-assemble into mesophases that are structurally similar to those in small molecule systems. Physical entanglement of the polymer chains and/or the glassy nature of the materials can provide mechanical integrity without an explicit need for crosslinking. The versatility of modern synthetic polymer chemistry enables block copolymer membranes to be designed with a wide range of physicochemical properties and functionalities.

For both small molecules and block copolymers, self-assembly occurs because of partitioning of chemically distinct segments of the molecules. Although small molecules and block copolymers can self-assemble into various structures (principally as a function of composition), hexagonally packed cylindrical and bicontinuous cubic mesophases (Fig. 8.13) are the structures of interest for membrane separations, in which the cylindrical or bicontinuous cubic microdomains act as the pores for size-selective transport. These mesophases form over limited ranges of composition that are thermodynamically defined. For both block copolymers and small molecules, the characteristic pore size is approximately set by the product of the size of the molecule and the volume fraction of its pore-forming segment. Currently accessible pore diameters range from 5 to 50 nm (which lies in the UF range) for block copolymers and 0.5 to 2 nm (which lies in the RO, FO, and NF range) for small molecules. In most cases, the pores are formed by the selective removal (for example, etching by base hydrolysis of polylactide) or shrinkage of material from the self-assembled structures, but for small molecules, the porosity may be an intrinsic feature of the system. Maximum porosity is usually 35–40%, as it is defined by the small molecule or block copolymer used and the composition range over which the desired structure forms. The ideal self-assembled selective layer has straight pores that fully span the layer. For cylindrical mesophases, this requires the alignment of the cylindrical pores perpendicular to the film surface as well as physical continuity of the pores throughout the film. By contrast, bicontinuous cubic mesophases intrinsically provide three-dimensionally percolated pathways throughout the system. However, these pathways are inherently tortuous, which increases hydraulic resistance. In addition, bicontinuous cubic phases are considerably less accessible than cylinders owing to the limited composition range in which they are stable—if they are present at all—in most small molecule and block copolymer systems. Efforts to develop high-performance membranes based on self-assembled mesophases have therefore focused heavily on molecular design to ensure the stable formation of bicontinuous cubic structures and on the development of scalable directed self-assembly methods to allow vertical orientation of cylindrical pores in thin films. Progress has also been made in directing self-assembly over large areas to achieve vertically aligned cylinders. Magnetic fields provide a readily scalable means of achieving such orientation control and have been successfully used in both small molecule and block copolymer systems. An alternative field-free approach has been explored based on so-called soft confinement, in which symmetric physical confinement of a small molecule mesophase during ordering results in the spontaneous display of vertically oriented microdomains. The resulting films exhibit physically continuous, vertically aligned 1-nm pores (Werber et al. 2016).

For block copolymer membranes, fabrication methods that combine self-assembly with a kinetic component have proven highly effective. For example, self-assembly with non-solvent-induced phase separation (SNIPS) is a single, readily scalable process that can produce vertically oriented pores in a thin selective layer at the membrane surface with an integrally formed microporous underlying support layer (Fig. 8.13b). As SNIPS combines self-assembly with kinetically limited phase inversion, the resulting pore size distribution is broader than that produced in equilibrium self-assembly processes. Nevertheless, these membranes display sharper molecular weight cutoff characteristics and thus improved selectivity relative to NIPS-formed UF membranes in the 10–40-nm regime considered. The combination of block copolymer microphase separation and kinetically limited structure formation also features in the formation of membranes by polymerization-induced microphase separation; this has been used to fabricate a bicontinuous cubic phase with 4-nm pore size (Seo and Hillmyer 2012; Werber et al. 2016).

In addition, owing to its atomically smooth and atomically thin nature, graphene has become a widely researched material for membrane separations, especially for desalination. Graphene-based desalination membranes have been proposed in two forms (Werber et al. 2016): nanoporous graphene and graphene-based frameworks. Both forms are envisioned to serve as selective layers and to operate as molecular sieves with size-based exclusion of undesired solutes. To produce nanoporous graphene, defects are introduced into pristine, single-layer graphene by plasma etching or bombardment by ions or electrons. Graphene-based framework membranes differ in morphology and water transport mechanism from nanoporous graphene and comprise a multilayered stack of finite-sized graphene sheets (Fig. 8.13a). They are formed as laminates through vacuum filtration or layer-by-layer deposition of graphene oxide, which is inexpensively produced by oxidation of graphite and is stable when suspended in water. Graphene oxide frameworks allow ultra-fast permeation of water, which is postulated to be due to the slip flow along the atomically smooth, unoxidized graphene channels; this is analogous to the water flow within carbon nanotubes. However, the spacing between contiguous sheets increases from 0.9 ± 0.1 nm in humid air to 1.3 ± 0.1 nm when immersed in water, resulting in rapid permeation of monovalent and divalent cations and the corresponding counterions. Graphene oxide frameworks still show a sharp size-based cutoff, but the cutoff is too large for desalination. In addition, graphene oxide laminates are generally unstable in water, unless they are stabilized by trace multivalent cations. To potentially facilitate desalination, interlayer spacing can be decreased and stability can be increased through chemical or thermal reduction to remove oxygen species. Full reduction decreases the interlayer spacing to 0.36 nm, which is too small to allow water flow between sheets and thus restricts water permeation to defect-driven flow. Although a reduced graphene oxide membrane was shown to allow water permeation with high salt retention when tested in FO, the defect-driven permeation in this approach limits water permeability. Another approach for decreasing interlayer spacing and increasing stability is covalent crosslinking. However, a covalent crosslinking method has yet to be shown that facilitates salt rejection. Depending on the success of these efforts, water-treatment-focused graphene oxide framework membranes may be effective only for NF-range separations and not suitable for the more urgently needed desalination applications (Zhang et al. 2015; Werber et al. 2016).

8.7.2.4 Designing Fouling-Resistant Membranes

Proper tailoring of the surface chemical properties of membranes can substantially enhance their fouling resistance. Surface modification that renders the membrane surface highly hydrophilic substantially reduces organic fouling and the adhesion of bacteria that eventually leads to biofouling. The most common hydrophilic materials for the preparation of fouling-resistant surfaces are polyethylene oxide (PEO)-based and zwitterionic polymers (Fig. 8.14a). PEO, which is also known as polyethylene glycol (PEG), binds water molecules by hydrogen bonding, whereas zwitterionic polymers—for example, poly(sulfobetaine) and poly(carboxybetaine)—bind water molecules more strongly through electrostatic interactions. The hydration layer of the hydrophilic PEO and zwitterionic polymers provides a steric repulsive barrier that prevents the adsorption of organic molecules and bacteria. Brush layers of these hydrophilic materials have been successful in preventing protein and bacteria adhesion in a wide range of biomedical and industrial applications (Jiang and Cao 2010; Werber et al. 2016).

(a) Antifouling TFC membranes by surface grafting

Surface grafting of hydrophilic polymers is one of the most promising and versatile antifouling strategies for TFC membranes. One approach involves binding PEG-based or other antifouling materials to the free surface carboxyl or primary amine groups of the polyamide layer. This method, however, often yields limited fouling resistance owing to non-uniform and incomplete surface coverage of antifouling materials on the polyamide active layer, which is a direct result of the limited number of accessible pendant carboxyl or amine functional groups on the membrane surface. In addition, the method is limited even when using a more ideal substrate, as steric repulsion of pre-formed polymers limits the achievable brush density. A more promising method for grafting dense and uniform antifouling polymer brushes on TFC membranes involves the use of surface-initiated controlled radical polymerization techniques. Of these techniques, atom transfer radical polymerization (ATRP) is particularly versatile and robust because it can precisely control the density, length and architecture of the antifouling brush layer. Surface grafting often results in reduced water permeability and thus should be optimized to balance the gain in fouling resistance and reduction in water permeability. ATRP is advantageous in this regard because of the control it allows over the density and thickness of the brush layer. However, methods for uniform immobilization of the initiator molecules on the rough polyamide surface of TFC membranes need to be developed to ensure dense and uniform brush layers. Long-term antifouling performance and chemical stability of the grafted brush layer, particularly under

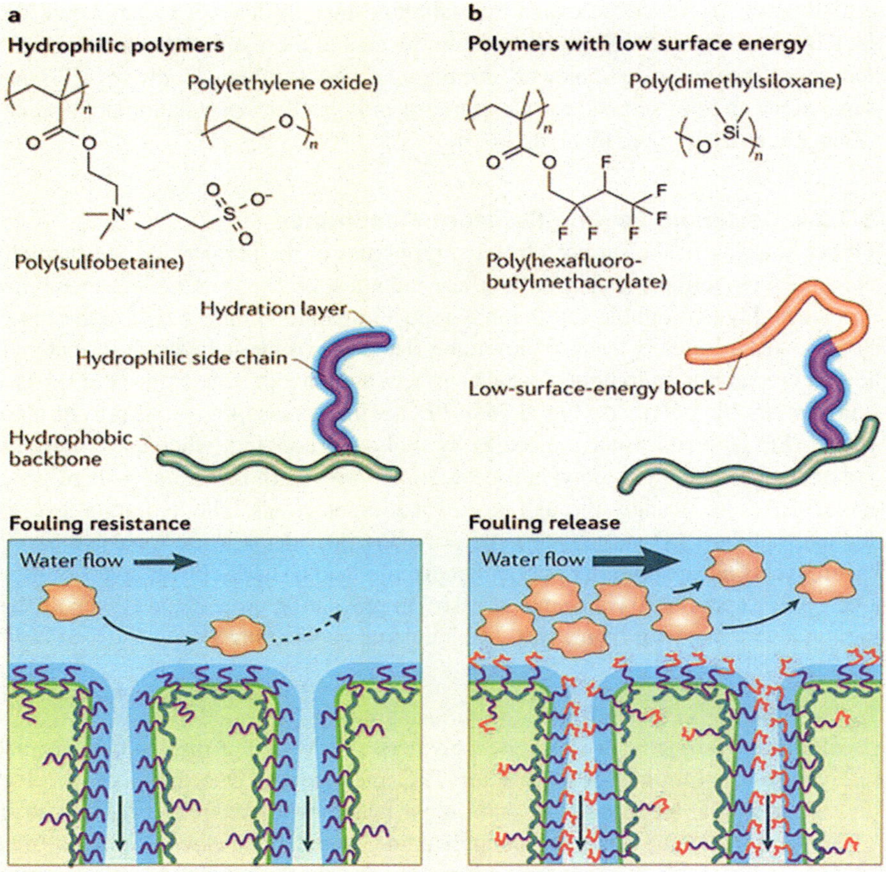

Fig. 8.14 Polymeric materials and self-segregation approach for fouling minimization in phase inversion membranes (Adapted with permission from Werber et al. 2016 (Springer Nature)): In this approach, the phase inversion casting solution contains a hydrophobic polymer and a comb copolymer with a hydrophobic backbone and side chains that are at least partially hydrophilic. During non-solvent-induced phase separation in water, the hydrophilic side chains self-segregate at the membrane/water interface, whereas the backbone anchors itself in the membrane matrix. (**a**) Side chains are entirely hydrophilic, comprising poly(ethylene oxide) or zwitterionic polymers. The resulting hydration layer inhibits irreversible adsorption of organic and biological foulants. (**b**) Side chains contain an additional hydrophobic block with low surface energy. The low-surface-energy block promotes foulant release in the presence of surface turbulence (for example, high cross-flows)

chemical cleaning operations, are also important considerations for TFC surface modification (Rana and Matsuura 2010; Werber et al. 2016).

(b) Antifouling porous membranes by surface segregation

When applied to UF membranes, surface grafting leaves the interior pore walls unmodified, resulting in adsorption of organic foulants to the pore walls and irreversible fouling. A more robust surface modification approach for phase inversion membranes is in situ surface segregation. In this approach, an amphiphilic copolymer comprising a hydrophobic backbone and hydrophilic side chains is blended with the hydrophobic base polymer in the casting solution. During precipitation of the base polymer in the coagulation bath, the hydrophobic backbone of the amphiphilic copolymers ensures strong anchoring within the polymer bulk matrix, whereas the hydrophilic segments segregate to all polymer/water interfaces, creating a surface with an exposed hydrophilic brush layer that imparts fouling resistance to the entire membrane (Fig. 8.14a). This one-step membrane fabrication method has been successfully demonstrated with the use of PAN-graft-PEO—which is an amphiphilic comb copolymer with a water-insoluble PAN backbone and hydrophilic PEO side chains—as an additive in the fabrication of a PAN UF membrane that exhibits complete resistance to irreversible fouling. The in situ surface segregation method for fabricating highly hydrophilic UF membranes has been extended to the fabrication of membranes with surface properties that provide both resistance to irreversible foulant adsorption and enhanced foulant release. The method involves the use of a ternary amphiphilic block copolymer that comprises a hydrophobic anchoring block, a hydrophilic fouling-resistant block and a non-polar, low-surface-energy fouling-release block as an additive during the phase inversion process. If the amphiphilic copolymer is properly designed and the correct solvents are chosen, the hydrophilic and fouling-release segments are forced to segregate to the polymer/water interface. This method has been successfully demonstrated for the fabrication of cellulose acetate, PVDF, and PES UF membranes with surfaces containing hydrophilic PEO-based blocks for fouling resistance and PDMS or fluorinated-methacrylate-based polymer blocks for fouling release (Fig. 8.14b). The membranes exhibit excellent performance in inhibiting the nonspecific adsorption of organic foulants as well as in preventing the accumulation and coalescence of organic and oily foulants on the membrane surface—when surface turbulence is present—owing to the fouling release property (Werber et al. 2016).

INDEPTH: Water Purification Membranes

A global emerging problem is the contamination of drinking water sources, such as rivers and lakes, with micropollutants that originate from municipal, industrial and agricultural wastewaters. Micropollutants, which include compounds such as hormones, pharmaceuticals, pesticides, personal care products and industrial chemicals, can have detrimental effects even at low concentrations. Furthermore, municipal wastewater effluents in water-stressed regions are often recycled for agriculture and even for indirect potable use,

(continued)

thereby introducing a wide range of micropollutants to fresh water resources. The development of efficacious technologies for the removal of such micropollutants and other contaminants, is as listed in below table (Werber et al. 2016):

Source water	Key contaminants	Treatment objectives	Membrane technologies	Problems and challenges
Natural waters				
Seawater	Boric acid (affects crop health), divalent cations (cause scaling issues)	• Reduce salinity	• Pretreatment: microfiltration (MF) and ultrafiltration (UF) • Desalination; reverse osmosis (RO)	• Energy of desalination • Ecological impact of seawater intake and brine discharge
Brackish (saline) groundwater	Divalent cations	• Reduce salinity	• Pretreatment: MF and UF • Desalination: RO and electrodialysis[a]	• Inland discharge of saline brine
Surface waters	Natural organic matter (precursor for disinfection by-products), microbial pathogens, micropollutants and algal toxins	• Remove particles and microbial pathogens • Reduce natural organic matter	• MF, UF, nanofiltration (NF) and RO	• High fouling potential • Formation of toxic disinfection by-products during oxidative disinfection
Fresh groundwater	Natural arsenic, nitrates, iron and manganese (staining issues) and divalent cations	• Reduce nitrate, iron, manganese and/or scale-forming ions	• MF, UF, NF, and RO	• Seawater intrusion for aquifers near ocean • Aquifer over-exploitation
Wastewaters				
Municipal wastewater	Microbial pathogens, micropollutants, phosphates (algal bloom concerns) and ammonia	• Degrade organic matter • Remove or inactivate pathogens • Remove nutrients (that is, nitrogen and phosphate)	• MF, UF, NF, and RO	• Requirements for potable reuse • Large footprint • Odor of conventional treatment plants • High membrane-

(continued)

				fouling potential
Shale-gas produced water	Drilling fluid additives (for example, surfactants, oxidants, strong acids), oil and grease, radium and divalent cations	• Drilling reuse: remove suspended solids and sealants • Disposal: remove oily compounds and reduce salinity	• Drilling reuse; MF and UF • Desalination: NF, RO (for low-salinity waters), forward osmosis (FO) and membrane distillation (MD)[b]	• Large water consumption • High total dissolved solids • Fluctuating water quantity and quality • Regulations over disposal • High membrane-fouling potential
Coal-fired power plant flue gas desulfurization	Toxic metals (for example, arsenic, selenium and mercury), ammonia and organic acids	• Remove dissolved toxic metals and reduce salinity	• MF, UF, NF, RO, FO and MD	• Potential zero liquid discharge requirements

[a]Electrodialysis is an electrically driven desalination method that uses ion-exchange membranes

[b]Membrane distillation is a thermally driven desalination process in which a hydrophobic porous membrane separates a hot feed stream from a cold permeate. A vapor pressure gradient drives water vapor flux

8.7.3 Membrane Technology in Biofuel Production and Purification

8.7.3.1 Biodiesel

Biodiesel, which is known as fatty acid methyl ester (FAME), is produced from transesterification of renewable lipid (vegetable oil or animal fat) by methanol. In this process, the membrane plays an important role to remove the byproduct glycerol from the product stream (biodiesel) or to retain the unreacted lipid within the membrane. There are two basic separation principles of membrane-based biodiesel production, based on oil droplet size (or membrane pore size) or perm-selectivity of the membrane. In the system based on oil droplet size, a microporous membrane, which is typically a ceramic membrane or carbon membrane, is used because its high resistance to degradation and corrosion in the harsh environment of biodiesel production, where acid or base is used as a catalyst. Due to the differences in polarity, methanol and lipids are immiscible. Their mixture exists as a two-phase system, where the lipid/oil micelles or droplets are dispersed in the continuous phase of methanol. These droplets have much higher diameter than the product biodiesel, byproduct glycerol, reactant methanol and catalyst (acid or base) and can be filtered by a membrane. Further water-washing is required to purify the biodiesel from other components in the permeate. In the system based on membrane selectivity, a nonporous dense hydrophilic polymeric membrane (e.g., poly (vinyl alcohol)) is used. The separation is based on the interaction between the target components and the membrane. Generally, glycerol and methanol have strong interaction with –OH groups of PVA via hydrogen bonding and hence penetrate through the membrane. As the result, they are continuously removed from the mixture during the reaction whereas the unreacted lipid and the product biodiesel, of which chemical structures are different with that of the membrane, are retained in the system (Le and Nunes 2016).

In such separation mechanism, the system can be operated under atmospheric pressure. To restrict the permeation of catalysts through the membrane, activated carbon is used to carry them during the reaction. By this approach, the oil to FAME conversion reached 93.5% and high-quality biodiesel can be produced without washing or purification steps (Baroutian et al. 2011). To combine reaction and separation in a single step, the catalytically active membranes, which are the product of catalyst immobilization into the membrane matrix, have been developed. The acidic catalyst membranes are fabricated by esterifying the –OH groups of PVA with 5-sulfosalicylic acid to achieve sulfonic groups in the polymeric matrix or by blending with poly (styrene sulfonic acid) containing strong acidic groups. In addition, the heterogeneous catalysts, e.g., hydrotalcite $Mg_6Al_2(OH)_{16}(CO_3^{2-})\cdot 4H_2O$ or amino-functionalized carbon nanotubes (CNTs), can be embedded into the polymeric matrix to form catalyst membranes by dispersing them in the polymer solution. In the case of CNTs, these particles can also increase the mechanical strength of the membrane when an appropriate content is used. Besides PVA, the polymer polyacrylonitrile (PAN) is also used for biodiesel production (Guerreiro et al. 2010; Le and Nunes 2016).

Similar to other membrane processes, fouling is one of the major challenges. In biodiesel production, the fouling is caused by the agglomeration of glycerol—which is favored by the presence of excess alcohol, soap (salt of fatty acids) and catalysts—to block the pores of the membrane in the system based on the oil droplet size. However, in the typical biodiesel production, fouling is not serious because the alcohol concentration in the reaction mixture is low. In addition, the mechanical properties and surface morphology of the membranes should be fully studied before applying in practical biodiesel industry (Le and Nunes 2016).

8.7.3.2 Bioethanol

Among bioalcohols, bioethanol is the most practical because of its high content produced from the fermentation process. Bioethanol can be divided into three generations based on the origin of biomass, as shown in Fig. 8.15. Higher generation bioethanol requires more steps in its production procedure with potential membrane applications, where MF/UF process is applied to harvest microalgae as the biomass/

Generation	Biomass source	Current status	Challenges
1st	Sugar and starch crops	-Technically mature -Commercially available	-The profitability heavily depends on the prices of both fossil oil and the commodity feedstock. -Compete with food prices -The greenhouse gas benefits depend on the feedstock and process used.
2nd	Lignocellulosic biomass such as agricultural wastes (e.g. straw) and energy crops (e.g. Miscanthus, poplar)	-Advanced stage of development and deployment -Demonstration stage	-Cost reduction -Availability of comparatively low-cost and sustainable feedstock
3rd	Algae	-Earlier stage of research and development	-Cost reduction -Technical challenges

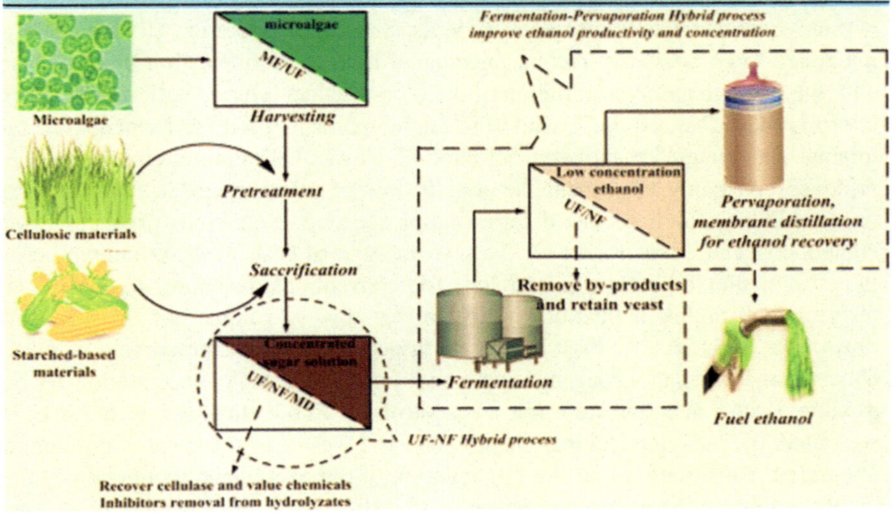

Fig. 8.15 Bioethanol development status and application of membrane processes for bioethanol production (Modified with permission from Wei et al. 2014, and Le and Nunes 2016 (Elsevier))

substrate for fermentation, MD/NF/RO process is employed to concentrate the prehydrolyzates and remove microorganism-inhibitors before fermentation, UF/NF is used after fermentation to remove some by-products and retain microorganism, and MD/pervaporation is integrated to recover and purify the product bioethanol in the final step. Algal biomass harvesting is challenging because of their small size (3–30 μm), similar density to that of water and large volumes of water that must be removed to recover algal cells. Compared to conventional techniques such as coagulation/flocculation, flotation, gravity sedimentation, centrifugation, membrane filtration (MF/UF) is more advantageous in terms of energy consumption, recovery efficiency and non-toxicity. For instance, a biomass recovery of 70–89% has reported when concentrating algae from large reservoirs by membrane filtration. Membrane processes can be applied prior to the fermentation process in the production of the second or third generation bioethanol to concentrate sugar concentration (it is low due to different pretreatment processes and hydrolysis efficiency) and to remove fermentation inhibitors produced during the pretreatment process. The main advantage of membrane processes over conventional techniques (e.g., evaporation, solvent extraction, overliming, activated charcoal adsorption and ion exchange) is the capability of concentrating sugar and removing inhibitors simultaneously.

Moreover, the capital investment and operation cost of membrane processes are lower than those of the evaporation process. The present membrane processes applied in this aspect include MD, NF, and RO, depending on the target inhibitors and sugar concentration. On the other hand, ultrafiltration can be integrated into this stage to recycle valued enzymes (e.g., cellulase) used to hydrolyze the biomass into sugar in the previous step. By this way, economic viability of the enzymatic hydrolysis of biomass is improved (Wei et al. 2014; Le and Nunes 2016).

Subsequent to the fermentation process, a MD/pervaporation (PV) process is integrated to continuously recover bioethanol from the fermentation broth. Generally, the final ethanol concentration of the corn-to-ethanol fermentation (first generation) is more than 10 wt.% while the microorganisms used for fermentation can tolerate a maximum ethanol concentration of 10 wt.%. Therefore, ethanol must be removed constantly to prevent the activity loss of microorganisms. On the other hand, cellulosic biomass-based fermentation (second generation) produces lower ethanol concentrations (<5 wt.%). To save energy and costs of the refining process, the ethanol should be pre-concentrated. The conventional technique to concentrate fermentation broths is distillation. However, this technique has several main drawbacks (Balat et al. 2008): (a) the energy consumption is remarkably higher to concentrate low ethanol concentration than to concentrate high concentration, (b) the mixture of ethanol and water forms an azeotrope at 95.6% ethanol, which cannot be separated by the conventional distillation, and (c) it is technically difficult to integrate the distillation into the fermentation to remove ethanol continuously due to that fact that high-temperature operation of distillation is lethal to microorganisms. Membrane technologies (MD/PV) are therefore more useful to remove ethanol from dilute aqueous fermentation broths. By adding membrane systems to the fermentation, productivity and production rate, as well as the substrate uptake rate can be improved (Le and Nunes 2016).

Ethanol recovery by MD is based on the higher partial pressure of ethanol than that of water, which implies ethanol vapor can preferentially transfer through the membrane pores. Both direct contact air gap MD are common configurations coupled with fermentation for ethanol production. At laboratory scale, 87% increase in ethanol productivity has achieved. Although this integrated fermentation MD system can overcome the drawbacks of the conventional batch production, larger scale operations has not been investigated. Moreover, the membrane selectivity of current MD materials (PTFE, PVDF, and PP) is low and hence higher performance MD membranes are needed. However, alcohol and biological surfactants which might be present in the fermentation reduces the surface tension and promote liquid intrusion in the membrane pores, making MD application challenging (Wei et al. 2014, Le and Nunes 2016).

Pervaporation (PV) is the most studied membrane technique for ethanol recovery from dilute fermentation broths. The separation of this process is based on the chemical potential difference generated by either vacuum or a sweep gas on the permeate side of the membrane. The separation mechanism predominantly relies on the preferential sorption (solubility) and diffusion (diffusivity) of the target component across the membrane. To recover ethanol, the membrane should be hydrophobic to possess higher affinity (solubility) towards ethanol that allows ethanol to preferentially pass through. Polydimethylsiloxane (PDMS) is the dominant polymeric material used to recover ethanol from water because of its superior performance as compared with other polymeric materials. From the aspect of energy consumption, the separation factor of pervaporation membranes must be larger than 20 to compete with distillation. This is the one of the main challenges to restrict the industrialization of polymeric membranes in this application. Although inorganic membranes (e.g., zeolite and 1-silicalite) produce the separation factor of over 20, they pose problems in processability of large-sized membranes and they are often expensive. In addition, performance degradation in practical separation system is another issue needed to be solved before the system can be industrialized. This degradation is mainly due to the complex compositions in the fermentation broth. Even with very small concentration, the present of by-products in the fermentation broth can significantly reduce membrane performance. Thus, exploring higher performance and reliable membrane materials are critical for pervaporation applications in ethanol recovery (Le and Nunes 2016).

8.7.3.3 Biogas

Biohydrogen is generated in fermentation processes while biomethane is produced during anaerobic digestion of biological waste. Anaerobic membrane bioreactors (AnMBRs) have an advantage in producing both biohydrogen and biomethane in a single membrane system because in anaerobic digestion, the hydrogen production phase occurs prior to the methanogenic phase. Besides biogas, many other gases and volatiles exist in the system in trace amounts (ppm) such as nitrogen, oxygen, carbon dioxide, hydro sulfide, and volatile siloxanes. They must be removed to upgrade biogas because their presence not only reduces the calorific value of biogas, but also causes corrosion to the transport and distribution system and reduces the possibilities of compression.

Membrane technology is also gaining importance in biogas. Also for this application polymeric membranes have clear advantages over ceramic for operation in temperatures below 150 °C. The two commonly used effective approaches are (a) chemical modifications, especially introduction of bulky functional groups such as $-Si(CH_3)_3$, $-C(CF_3)_2$ or bromine groups; which increase permeability without sacrifice of selectivity and (b) crosslinking to improve selectivity of highly permeable membranes. Although polymeric membranes have high potential for gas separation, they (especially glassy polymers) usually pose problems with densification, aging or plasticization. Plasticization occurs when gas molecules are dissolved into micro-voids of the polymeric matrix, leading swelling or permanent damage to the matrix and hence drastic increase in gas diffusion and reduced selectivity. To overcome plasticization, considerable attempts have focused on membrane modification by heat treatment, chemical modification, and polymer blending. Also, mixed matrix membranes (MMMs) are promising candidate for gas separation by synergistic combination of easy processability from polymeric membranes and high selectivity and permeability from inorganic membranes. The attractive inorganic materials for gas transport are zeolites, carbon molecular sieves (CMS), carbon nanotubes (CNTs), metal-organic frameworks (MOFs), and covalent organic frameworks (COFs). To achieve the improvement of MMMs, the problems in their fabrication must be conquered, which are particle sedimentation, agglomeration, and interfacial voids. Sedimentation causes inhomogeneous filler and polymer phase whereas agglomeration and interfacial voids lead to nonselective voids decreasing selectivity. These problems are possibly solved by (Basu 2010, Le and Nunes 2016): (a) preparing polymer solutions at high concentrations to increase viscosity and lessen the particle sedimentation, (b) quickly solidifying membranes to offer the fillers less chance to settle, (c) matching the polarity of polymer chains and filler structures, (d) subsequently thermally annealing MMMs, (e) developing a priming protocol of MMM preparation (e.g., coating an ultrathin layer of the matrix polymer on the particle surface), and (f) using melt extrusion technique.

Exercises

Part I: General Questions

8.1. Use appropriate units, rounding, significant figures, and show results in regular and scientific notation.
   (a) On average, a University's steam plant consumes 21,000 US tons of dry hardwood chips per year. What is the heat content of this biomass?
   (b) In addition to wood chips, the university uses, on average, 515,000 gallons per year of #6 fuel oil to heat outlying buildings, and as backup for the wood chip system. What is the heat content of the fuel oil used annually?
   (c) What percentage of the university's heat comes from biomass?
   (d) A small natural gas field is located just west of the University. A relatively high-yielding well in this field produces 25,000 MCF (1 MCF = 1000 standard cubic feet) of natural gas per year. How many of these wells would be needed to supply the University's non-biomass heating needs that are currently being met by #6 fuel oil?

8.2. Briefly describe the development history of bioenergy technologies.

8.3. List major biomass conversion technologies and compare their advantages and disadvantages.

8.4. Describe modern bioenergy conversion pathways from feedstock to products, and their technology barriers and strategies moving forward to realize high volume production.

8.5. List major corrosion resistant materials compatible with biofuels, and compare their advantages and disadvantages.

8.6. Describe the role of nanocatalysts for conversion of biomass to biofuel, and compare main nanocatalysts' advantages and disadvantages.

8.7. Why using coal liquefaction? Describe the processing pathways, current technology barriers, and strategies moving forward to realize high volume production.

Part II: Thought-Provoking Questions

8.8. Why is Algae promising for future bioenergy? Describe the bioenergy conversion pathways, current technology barriers, and strategies moving forward to realize high volume production.
8.9. Describe current status and future trends of membrane materials for water sustainability in bioenergy.

References

Akia, M., Yazdani, F., Motaee, E., Han, D., Arandiyan, H.: A review on conversion of biomass to biofuel by nanocatalysts. Biofuel Res. J. **1**, 16–25 (2014)

Akizuki, M., Fujii, T., et al.: Effects of water on reactions for waste treatment, organic synthesis, and bio-refinery in sub- and supercritical water. J. Biosci. Bioeng. **117**(1), 10–18 (2014)

Aquino, I.P., Hernandez, R.P.B., Chicoma, D.L., Pinto, H.P.F., Aoki, I.V.: Influence of light, temperature and metallic ions on biodiesel degradation and corrosiveness to copper and brass. Fuel. **102**, 795–807 (2012)

Asadullah, M.: Barriers of commercial power generation using biomass gasification gas: a review. Renew. Sust. Energ. Rev. **29**, 201–215 (2014)

Baker, R.W.: Membrane technology and applications. John Wiley & Sons, Hoboken (2012)

Balat, M., Balat, H., Öz, C.: Progress in bioethanol processing. Prog. Energy Combust. Sci. **34**, 551–573 (2008)

Baroutian, S., Aroua, M.K., Raman, A.A.A., Sulaiman, N.M.: A packed bed membrane reactor for production of biodiesel using activated carbon supported catalyst. Bioresour. Technol. **102**, 1095–1102 (2011)

Basu, P.: Biomass gasification and pyrolysis: practical design and theory 25, pp. 141–158. Academic Press, Cambridge, MC (2010)

Bhardwaj, M., Gupta, P., Kumar, N.: Compatibility of metals and elastomers in biodiesel: a review. Int. J. Res. **1**(7), 376–391 (2014)

Capareda, S.C.: Biomass energy conversion. In: Nayeripour, M. (ed.) Sustainable growth and applications in renewable energy sources, pp. 209–226. InTech, Croatia (2011)

CEEETA: Combustion and gasification of agricultural biomass-technologies and applications. Thermie programme action BM 40. CEEETA, Partex, Portugal (1995)

Chung, T.S., Zhang, S., Wang, K.Y., Su, J., Ling, M.M.: Forward osmosis processes: yesterday, today and tomorrow. Desalination. **287**, 78–81 (2012)

Corry, B.: Designing carbon nanotube membranes for efficient water desalination. J. Phys. Chem. B. **112**, 1427–1434 (2008)

Demirbas, A.: Progress and recent trends in biofuels. Prog. Energy Combust. Sci. **33**, 1–18 (2007)

Díaz, L., López-Sansores, J.F., Maldonado, L., Garfias, L.F.: Corrosion behavior of aluminum exposed to a biodiesel. Electrochem. Commun. **11**, 41–44 (2008)

Duan, P., Savage, P.E.: Hydrothermal liquefaction of a Microalga with heterogeneous catalyst. Ind. Eng. Chem. Res. **50**, 52–61 (2011)

Fathizadeh, M., Aroujalian A., Raisi, A.: Effect of added NaX nano-zeolite into polyamide as a top thin layer of membrane on water flux and salt rejection in a reverse osmosis process Journal of membrane science. **375**, 88–95 (2011)

Fazal, M.A., Haseeb, A.S.M.A., Masjuki, H.H.: Comparative corrosive characteristics of petroleum diesel and palm biodiesel for automotive materials. Fuel Process. Technol. **91**, 1308–1315 (2010)

González-Pérez, A., Stibius, K.B., Vissing, T., Nielsen, C.H., Mouritsen, O.G.: Biomimetic triblock copolymer membrane arrays: a stable template for functional membrane proteins. Langmuir. **25**, 10447–10450 (2009)

Goryunov, A.G., Goryunovab, N.N., Ogunlanab, A.O., Manentic, F.: Production of energy from biomass: near or distant future prospects? Chem. Eng. Trans. **52**, 1219–1224 (2016)

Guerreiro, L., Pereira, P., Fonseca, I., Martin-Aranda, R., Ramos, A., Dias, J., Oliveira, R., Vital, J.: PVA embedded hydrotalcite membranes as basic catalysts for biodiesel synthesis by soybean oil methanolysis. Catal. Today. **156**, 191–197 (2010)

Guo, F., Fang, Z., Xu, C.C., Smith Jr., R.L.: Solid acid mediated hydrolysis of biomass for producing biofuels. Prog. Energ. Combust. **38**, 672–690 (2012)

Gust, S.: Combustion of pyrolysis liquids. In: Kaltschmitt, M., Bridgwater, A. (eds.) Biomass gasification and pyrolysis, state of the art and future prospects. CPL Press, Newbury, UK (1997)

Haseeb, A.S.M.A., Masjuki, H.H., Siang, C.T., Fazal, M.A.: Compatibility of elastomers in palm biodiesel. Renew. Energy. **35**, 2356–2361 (2010)

Hermans, S., Mariën, H., Van Goethem, C., Vankelecom, I.F.: Recent developments in thin film (nano) composite membranes for solvent resistant nanofiltration. Curr. Opin. Chem. Eng. **8**, 45–54 (2015)

Hilal, N., Kim, G., Somerfield, C.: Boron removal from saline water: a comprehensive review. Desalination. **273**, 23–35 (2011)

Höök, M., Aleklett, K.: A review on coal to liquid fuels and its coal consumption. Int. J. Energy Res. **34**(10), 848–864 (2010)

Höök, M., Fantazzini, D., Angelantoni, A., Snowden, S.: Hydrocarbon liquefaction: viability as a peak oil mitigation strategy. Phil. Trans. R. Soc. A. **372**(2006), 20120319 (2014)

Hub, J.S., de Groot, B.L.: Mechanism of selectivity in aquaporins and aquaglyceroporins. Proc. Natl. Acad. Sci. U. S. A. **105**, 1198–1203 (2008)

Jiang, S., Cao, Z.: Ultralow-fouling, functionalizable, and hydrolyzable zwitterionic materials and their derivatives for biological applications. Adv. Mater. **22**, 920–932 (2010)

Karan, S., Jiang, Z., Livingston, A.G.: Sub-10 nm polyamide nanofilms with ultrafast solvent transport for molecular separation. Science. **348**, 1347–1351 (2015)

Kenney et al.: Feedstock Supply System Design and Economics for Conversion of Lignocellulosic Biomass to Hydrocarbon Fuels Conversion pathway: biological conversion of sugars to hydrocarbons. INL/EXT-13-30342. https://inlportal.inl.gov/portal/server.pt?open=512&objID=421&parentname=CommunityPage&parentid=4&mode=2 (2013). Accessed 19 Sept 2017

Klass, D.L.: Biomass for renewable energy and fuels. Encyclopedia of energy. Elsevier, Inc., New York (2004)

Koppejan, J., Van Loo, S.: The handbook of biomass combustion and co-firing. Routledge, Abingdon, UK (2012)

Kurchania, A.: Biomass energy. In: Biomass Conversion, pp. 91–122. Springer, Berlin, Germany (2012)

Le, N.L., Nunes, S.P.: Materials and membrane technologies for water and energy sustainability. Sustain. Mater. Technol. **7**, 1–28 (2016)

Li, J., Yan, R., Xiao, B., et al.: Preparation of nano-NiO particles and evaluation of their catalytic activity in pyrolyzing biomass components. Energy Fuel. **22**, 16–23 (2008)

Li, M., Na Luo, N., Lu, Y.: Biomass energy technological paradigm (BETP): trends in this sector. Sustain. For. **9**, 567 (2017)

Luciano, M.A., Castro, M., Lins, V.: Corrosion resistance of organometallic coating aplicated in fuel tanks using electrochemical impedance spectroscopy in biofuel—part I. Tecnol. Metal. Mater. Miner. **11**(3), 244–250 (2014)

Maab, H., Saadi, A.A., Francis, L., Livazovic, S., Ghafour, N., Amy, G.L., Nunes, S.P.: Polyazole hollow fiber membranes for direct contact membrane distillation. ACS Ind. Eng. Chem. Res. **52**, 10425–10429 (2013)

Mahmood, T., Hussain, S.T.: Nanobiotechnology for the production of biofuels from spent tea. Afr. J. Biotechnol. **9**, 858–868 (2010)

Mansouri, J., Harrisson, S., Chen, V.: Strategies for controlling biofouling in membrane filtration systems: challenges and opportunities. J. Mater. Chem. **20**, 4567–4586 (2010)

Maru, M., Lucchese, M., Legnani, C., Quirino, W., et al.: Biodiesel compatibility with carbon steel and HDPE parts. Fuel Process. Technol. **90**, 1175–1182 (2009)

Mauter, M.S., Elimelech, M., Osuji, C.O.: Nanocomposites of vertically aligned single-walled carbon nanotubes by magnetic alignment and polymerization of a lyotropic precursor. ACS Nano. **4**, 6651–6658 (2010)

Mayaki, I.A.: Sustainable bioenergy development in UEMOA member countries. http://www.globalproblems-globalsolutions-files.org/gpgs_files/pdf/UNF_Bioenergy/UNF_Bioenergy_full_report.pdf (2008). Accessed 18 Oct 2014

Meier, D., Rupp, M.: Direct catalytic liquefaction technology of biomass: status and review. In: Bridgwater, A.V., Grassi, G. (eds.) Biomass pyrolysis liquids upgrading and utilization. Springer, Dordrecht (1991)

Muth, D., Jacobson, J.J., Cafferty, K., Jeffers, R.: Define feedstock baseline scenario and assumptions for the $80/DT target based on INL design report and feedstock logistics projects. INL/EXT-14-31569. https://bioenergy.inl.gov/Reports/Feedstock%20Logistics%20Cost%20Target%20Milestone.pdf (2013). Accessed 20 Sept 2017

MYPP: Bioenergy technologies office: multi-year program plan. https://energy.gov/sites/prod/files/2015/04/f22/mypp_beto_march2015.pdf (2015). Accessed 12 Sept 2017

Ng, L.Y., Mohammad, A.W., Ng, C.Y., Leo, C.P., Rohani, R.: Development of nanofiltration membrane with high salt selectivity and performance stability using polyelectrolyte multilayers. Desalination. **351**, 19–26 (2014)

Nzihou, A., Stanmore, B., Sharrock, P.: A review of catalysts for the gasification of biomass char, with some reference to coal. Energy. **58**, 305317 (2013)

Osterman, K.: Stainless steels—Cost-effective materials for the global biofuels industries. https://www.nickelinstitute.org/~/media/Files/TechnicalLiterature/10090_StainlessSteelsCostEffectiveMaterialsForTheGlobalBiofuelsIndustries.ashx (2012). Accessed 21 Sept 2017

Pangarkar, B.L., Sane, M.G., Parjane, S.B., Guddad, M.: Status of membrane distillation for water and wastewater treatment—a review. Desalin. Water Treat. **52**, 5199–5218 (2014)

Patil, V., Tran, K.-Q., Giselrød, H.R.: Towards sustainable production of biofuels from microalgae. Int. J. Mol. Sci. **9**(7), 1188–1195 (2008)

PBworks: Bioenergy. http://mcensustainableenergy.pbworks.com/w/page/20637999/bioenergy (2014). Accessed 09 Oct 2014

Pelisson, C.H., Vono, L.L.R., Hubert, C., Nowicki, A.D., et al.: Moving from surfactant-stabilized aqueous rhodium colloidal suspension to heterogeneous magnetite-supported rhodium nanocatalysts: synthesis, characterization and catalytic performance in hydrogenation reactions. Catal. Today. **183**, 124–129 (2012)

Pereira, J., Agblevor, F.A., Beis, S.H.: The influence of process conditions on the chemical composition of pine wood catalytic pyrolysis oils. ISRN Renew. Energy. **2012**, 167629 (2012)

Qtaishat, M., Khayet, M., Matsuura, T.: Guidelines for preparation of higher flux hydrophobic/hydrophilic composite membranes for membrane distillation. J. Membr. Sci. **329**, 193–200 (2009)

QTR: Advancing Systems and Technologies to Produce Cleaner Fuels-Bioenergy Conversion. Quadrennial Technology Review 2015, US Department of Energy. https://energy.gov/sites/prod/files/2016/01/f28/QTR2015-7A-Bioenergy-Conversion_0.pdf (2015a). Accessed 18 Sept 2017

QTR: Advancing Systems and Technologies to Produce Cleaner Fuels-Biomass feedstocks and logistics. Quadrennial Technology Review 2015, US Department of Energy. https://energy.gov/sites/prod/files/2016/01/f28/QTR2015-7B-Biomass-Feedstocks-and-Logistics.pdf (2015b). Accessed 19 Sept 2017

Rana, D., Matsuura, T.: Surface modifications for antifouling membranes. Chem. Rev. **110**, 2448–2471 (2010)

Roberts, J.A., Sutton, P.M., Mishra, P.N.: Application of the membrane biological reactor system for combined sanitary and industrial wastewater treatment. Int. Biodeterior. Biodegrad. **46**, 37–42 (2000)

Schnitzer, M.I., Monreal, C.M., Facey, G.A., Fransham, P.B.: The conversion of chicken manure to biooil by fast pyrolysis I. Analyses of chicken manure, biooils and char by 13C and 1H NMR and FTIR spectrophotometry. J. Environ. Sci. Health B. **42**, 71–77 (2007)

Schubert, C.: Can biofuels finally take center stage? Nat. Biotechnol. **24**, 777–784 (2006)

Seo, M., Hillmyer, M.A.: Reticulated nanoporous polymers by controlled polymerization-induced microphase separation. Science **336**, 1422–1425 (2012)

Shaffer, D.L., Werber, J.R., Jaramillo, H., Lin, S., Elimelech, M.: Forward osmosis: where are we now? Desalination. **356**, 271–284 (2015)

Sharma A., Pareek V., Zhang D.: Biomass pyrolysis—A review of modelling, process parameters and catalytic studies. Renew. Sustain. Energy Rev. **50**,1081–1096 (2015)

Sharma, S., Meena, R., Sharma, A., Goyal, P.K.: Biomass conversion technologies for renewable energy and fuels: a review note. IOSR J. Mechan. Civil Eng. **11**(2), 28–35 (2014)

Sinag, A., Yumak, T., Balci, V., Kruse, A.: Catalytic hydrothermal conversion of cellulose over SnO_2 and ZnO nanoparticle catalysts. J. Supercrit. Fluids. **56**, 179–185 (2011)

Singh, B., Korstad, J., Sharma, Y.C.: A critical review on corrosion of compression ignition (CI) engine parts by biodiesel and biodiesel blends and its inhibition. Renew. Sust. Energ. Rev. **16**, 3401–3408 (2012)

Wei, P., Cheng, L.-H., Zhang, L., Xu, X.-H., Chen, H.-L., Gao, C.-J.: A review of membrane technology for bioethanol production. Renew. Sust. Energ. Rev. **30**, 388–400 (2014)

Werber, J.R., Osuji, C.O., Elimelech, M.: Materials for next-generation desalination and water purification membranes. Nat Rev Mater. **1**(5), 16018 (2016)

Wilcoxon, J.P.: Nanoparticles preparation, characterization and physical properties. Front. Nanosci. **3**, 43–127 (2012)

Williams, C.L., Westover, T.L., Emerson, R.M., Tumuluru, J.S., Li, C.: Sources of biomass feedstock variability and the potential impact on biofuels production. Bioenergy Res. **9**(1), 1–14 (2016)

Xu, Y., Hu, X., Li, W., Shi, Y.: Preparation and characterization of biooil. In: Shaukat, S. (ed.) Biomass, Progress in Biomass and Bioenergy Production, pp. 197–222. InTech, Croatia (2011)

Yakaboylu, O., Harinck, J., et al.: Supercritical water gasification of manure: a thermodynamic equilibrium modeling approach. Biomass Bioenergy. **59**, 253–263 (2013)

Zafar, S.: Biomass gasification process. https://www.bioenergyconsult.com/biomass-gasification/ (2013). Accessed 12 Sept 2017

Zaimes, G.Z., Vora, N., Chopra, S.S., Landis, A.E., Khanna, V.: Design of sustainable biofuel processes and supply chains: challenges and opportunities. PRO. **3**(3), 634–663 (2015)

Zhang, Y., Zhang, S., Chung, T.S.: Nanometric graphene oxide framework membranes with enhanced heavy metal removal via nanofiltration. Environ. Sci. Technol. **49**, 10235–10242 (2015)

Zhao, H., Yan, H.-X., Liu, M., Sun, B.-B., Zhang, Y., Dong, S.-S., Qi, L.-B., Qin, S.: Production of bio-oil from fast pyrolysis of macroalgae Enteromorpha prolifera powder in a free-fall reactor. Energ. Source. Part A. **35**, 859–867 (2013)

Zhou, G., Hou, Y., Liu, L., Liu, H., et al.: Preparation and characterization of NiW–nHA composite catalyst for hydrocracking. Nanoscale. **4**, 7698–7703 (2012)

Hydrogen and Fuel Cells

9

Abstract

Hydrogen is a flexible energy carrier that can be produced from any regionally prevalent primary energy source. Moreover, it allows low-carbon energy to be stored and can be effectively transformed into any form of energy for diverse end-use applications. Hydrogen is particularly well suited for use in fuel cells that efficiently use hydrogen to generate electricity. For instance, small quantities of hydrogen with low-carbon footprint can be stored under restricted space and weight requirements to enable long-distance, low-carbon driving using fuel cell electric vehicles (FCEVs). Copious quantities of hydrogen can be stored over long periods of time, facilitating the integration of high shares of variable renewable energy (VRE) into the energy system for power and heat. Hydrogen-based systems such as power-to-fuel, power-to-power, or power-to-gas can be employed to make use of VRE that would otherwise be curtailed at times when supply outstrips demand. Fuel cells convert the chemical energy in fuels such as hydrogen directly into electricity. Unlike heat engines, which are limited by Carnot efficiency, fuel cells can theoretically achieve efficiency over 90%. Current fuel cell technology can exceed 60% efficiency, and R&D is underway to reach 70% efficiency or higher. When using hydrogen as a fuel, fuel cells emit only water. Fuel cell technology has matured enough that initial commercialization of fuel cell vehicles is already underway, but several technological barriers remain that impede commercialization and require research on materials, stack components, balance-of-plant subsystems, and integrated fuel cell systems. Additional fuel cell innovations will be required to meet cost and durability targets, including development of low-cost, corrosion-resistant metal bipolar plates and development of durable, low-cost balance-of-plant components. This chapter will give a brief review about hydrogen and its use in fuel cells.

C. Tong, *Introduction to Materials for Advanced Energy Systems*,
https://doi.org/10.1007/978-3-319-98002-7_9

9.1 Introduction

Hydrogen is a flexible energy carrier with potential applications across all energy sectors. It is one of only a few potential near-zero emission energy carriers, alongside electricity and advanced biofuels. Nonetheless, hydrogen is an energy carrier and not an energy source (IEA 2015): although hydrogen as a molecular component is abundant in nature, energy needs to be used to generate pure hydrogen. The hydrogen can then be used as a fuel for end-use conversion processes, for example, using fuel cells to produce power. As is the case for electricity generation, hydrogen production incurs a cost and suffers from thermodynamic losses. Hydrogen can be produced from various primary or secondary energy sources, depending on regional availability. Primary energy sources useful for hydrogen production comprise renewable sources, such as biomass, and also fossil fuels, such as natural gas and coal. Electricity can also be used for hydrogen generation using electrolysers, which are a pivotal technology for enabling the splitting of water into its components hydrogen and oxygen. Hydrogen itself contains no carbon—if used in a fuel cell or burned in a heat engine, water or water vapor is the only exhaust. Nevertheless, hydrogen can have a very significant carbon footprint. Its lifecycle carbon emissions are determined by the primary energy source and the process used for hydrogen production, and need to be taken into account when quantifying climate benefits (IEA 2015).

Producing hydrogen from electricity and storing it in gaseous or liquefied form could be an option for increasing the flexibility of the energy system, allowing for the integration of high shares of VRE. Hydrogen can enable "power-to-x" trajectories—its capability of being converted to various forms of final energy, such as power, heat, and transport fuels, can be used to join subsystems of the energy system that historically had no, limited, or only one-way linkages. Hydrogen could play an important role in future road transport, as FCEVs can be a low-carbon alternative to conventional passenger cars and trucks. In buildings, micro cogeneration units could increase energy efficiency. In the longer run, industrial processes in the refining, steel, and chemical industries could be substantially decarbonized through the use of hydrogen with a low-carbon footprint. In many, but not all, of these applications, fuel cells are an important technology for converting hydrogen to power and heat. Fuel cells are intimately but not exclusively linked to hydrogen. They can also be used with other fuels such as natural gas or even liquid hydrocarbons, thus helping their early adoption (IEA 2015).

Fuel cells are basically open thermodynamic systems. They operate on the basis of electrochemical reactions and consume reactant from an external source. They are favorable alternatives to conventional electricity generation methods for small-scale applications. Hydrogen and hydrocarbon fuels contain significant chemical energy in comparison with conventional battery materials; hence they are now widely developed for numerous energy applications. Fuel cell technology is a promising substitute for fossil fuels to provide energy for rural areas where there is no access to the public grid or huge cost of wiring and transferring electricity is required. In addition, applications with essential secure electrical energy requirement such as uninterruptible power supplies (UPS), power generation stations, and distributed systems can employ fuel cells as their device for secure energy production (Mekhilefa et al. 2012).

The major advantage of fuel cells is their high thermodynamic efficiency, meanwhile, fuel cells have the potential to be used for cogeneration of electricity and heat, covering thus the heat and power needs for domestic and other larger scale industrial applications. Moreover, fuel cell systems are flexible regarding the power output and they can be used for the power production of electrical power in the region from 50 W to 100 MW. Specifically, the power output of small portable systems can be as low as a few watts, whereas in the case of biological fuel cells for medical applications the power output can be lower. In addition, hydrogen is the main fuel where there are no polluting emissions at all, while in the case of other fuels, such as natural gas, the quantity of the polluting emissions is approximately two orders of magnitude lower than that in the case of the conventional electro-productive systems (Niakolas et al. 2016).

General weaknesses of the fuel cells systems include (Niakolas et al. 2016): (a) Durability issues/stability and useful lifetime; (b) Major challenges in producing, transporting, and storing hydrogen; and (c) Production cost use of (rare-noble) expensive raw elements and materials. Therefore, the massive use of fuel cells still needs breakthrough research on materials and their interfaces (mainly the electrochemical interfaces). Novel materials and novel fuel cell design concepts, which will allow the effective reduction of precious metal loadings. This can be achieved through two main approaches. One is the development of novel, more active, electrocatalysts aiming to atomic distribution of the metal active phase on stable nanostructured supporting materials. The other approach is through the synthesis and development of stable anionic alkaline polymer electrolytes, which will allow the use of non-precious metal electrocatalysts. Furthermore, simulation and understanding the functionality and operational characteristics of a 3D structured electrochemical interface aim to (i) 100% utilization of the active electrocatalyst and (ii) the innovative design of the flow fields. As a result, a uniform distribution of the reacting gases can be achieved along the 3D structure of the catalytic layer, i.e., below the gas streams and below the ribs of the bipolar plates. Novel designs, engineering and operational concepts also need to be conceived so as to improve the performance of fuel cells. This advance can be accomplished by means of an integrated approach, based both on materials development and on the deployment of innovative cell designs. The specific strategy will permit the effective control of (i) the electrocatalytic activity, especially in terms of the efficiency of the electrochemical interfaces and (ii) the poisoning effect of the feeding gases on the electrodes' performance.

Consequently, fuel cells (FC) and hydrogen (H_2) systems offer a potential long-term energy option, but still face some major challenges in facilitating their market breakthrough. However, the fuel cell and hydrogen technologies are driven primarily by three forces (Niakolas et al. 2016): the recognition of hydrogen as an attractive and important energy storage platform by energy utilities; the interest of major

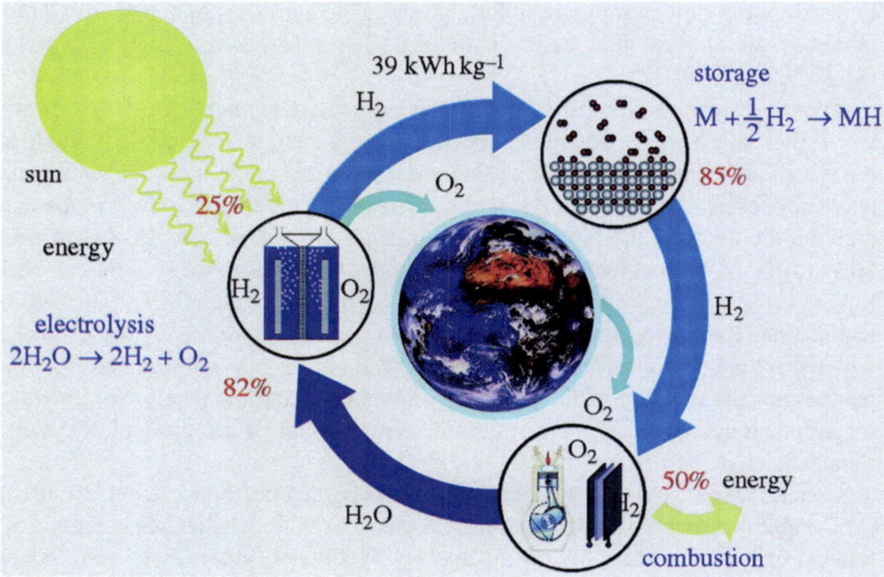

Fig. 9.1 Hydrogen cycle for application (Adapted with permission from Jeana 2011 (American Chemical Society))

global telecoms in fuel cell backup power; and the commercialization of fuel cell electric vehicles (FCEV) by the world's major automakers.

9.2 Hydrogen Generation Technology

Since hydrogen is the most abundant element in the universe, which has the highest energy density per unit mass and produces water, it is widely regarded as a solution for transport applications. In an ideal hydrogen cycle (Fig. 9.1), hydrogen is fabricated by splitting water via electrolysis with solar energy, storing it reversibly in a solid, and using it in a fuel cell to produce electrical energy and heat. This is the so-called "hydrogen economy," which could effectively address the major energy challenges of the twenty-first century (Jena 2011; Xia 2015).

However, increase in industrial processes has been resulting in enormous amount of effluents generation for which selection of appropriate treatment methodology is extremely challenging. At present, hydrogen is mainly produced from fossil fuels, biomass, and water. Among these methods, steam-reforming process alone produces about 90% of hydrogen (Krishna 2013). The various methods adopted to produce hydrogen are given in Fig. 9.2 (FSEC 2012).

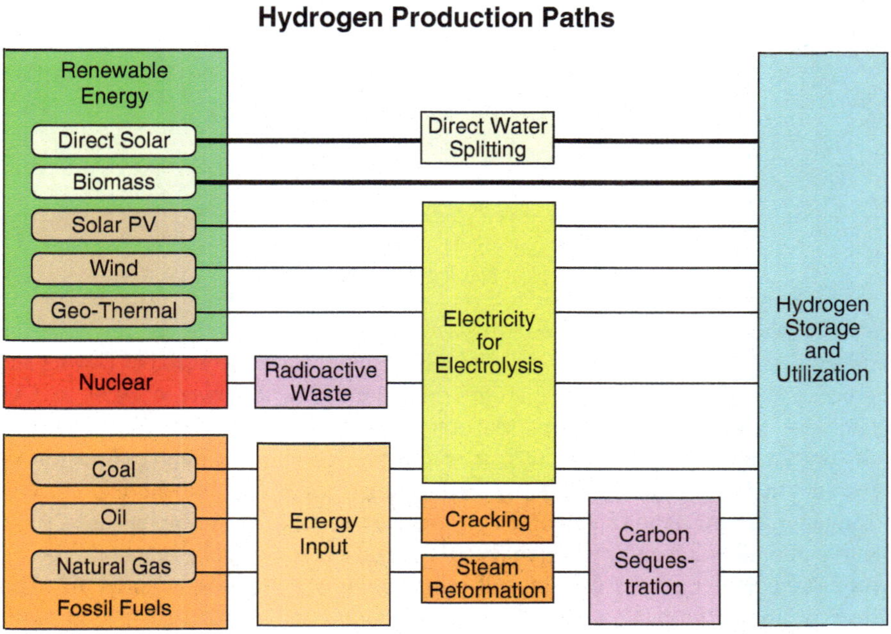

Fig. 9.2 General hydrogen production paths (Adapted from FSEC (2012). Hydrogen basics—production. http://www.fsec.ucf.edu/en/consumer/hydrogen/basics/production.htm. Courtesy: Florida Solar Energy Center (FSEC), University of Central Florida)

9.2.1 Hydrogen from Fossil Fuels

Fossil fuel processing technologies convert hydrogen-containing materials derived from fossil fuels, such as gasoline, hydrocarbons, methanol, or ethanol, into a hydrogen-rich gas stream. Fuel processing of methane (natural gas) is the most common commercial hydrogen production technology today. Most fossil fuels contain a certain amount of sulfur, the removal of which is a significant task in the planning of hydrogen-based economy. Hydrogen gas can be produced from hydrocarbon fuels through three basic technologies (Kalamaras and Efstathiou 2013): (a) steam reforming (SR), (b) partial oxidation (POX), and (c) autothermal reforming (ATR). These technologies produce a great deal of carbon monoxide (CO). Thus, in a subsequent step, one or more chemical reactors are used to largely convert CO into carbon dioxide (CO_2) via the water-gas shift (WGS) and preferential oxidation (PrO*x*) or methanation reactions.

9.2.1.1 Steam Reforming

Steam reforming is currently one of the most widespread and at the same time least expensive processes for hydrogen production. Its advantage arises from the high efficiency of its operation and the low operational and production costs. The most frequently used raw materials are natural gas and lighter hydrocarbons, methanol, and other oxygenated hydrocarbons. The network of reforming reactions for hydrocarbons and methanol used as feedstock is the following (Kalamaras and Efstathiou 2013):

$$C_mH_n + mH_2O(g) \rightarrow mCO + (m + 0.5n)H_2 \quad (9.1)$$

$$C_mH_n + 2mH_2O(g) \rightarrow mCO_2 + (2m + 0.5n)H_2 \quad (9.2)$$

$$CO + H_2O(g) \leftrightarrow CO_2 + H_2 \quad (9.3)$$

$$CH_3OH + H_2O(g) \leftrightarrow CO_2 + 3H_2 \quad (9.4)$$

The whole process comprises two stages. In the first stage, the hydrocarbon raw material is mixed with steam and fed in a tubular catalytic reactor. During this process, syngas (H_2/CO gas mixture) is produced with lower content in CO_2 (Eqs. 9.1 and 9.2). The required reaction temperature is achieved by the addition of oxygen or air for combusting part of the raw material (heating gas) inside the reactor. In the second stage, the cooled product gas is fed into the CO catalytic converter, where carbon monoxide is converted to a large extent by means of steam into carbon dioxide and hydrogen (Eq. 9.3). The steam reforming catalytic process requires a raw material free of sulfur-containing compounds in order to avoid deactivation of the catalyst used (Song et al. 2007; Kalamaras and Efstathiou 2013).

The SR process requires modest temperatures, for example, 180 °C for methanol and oxygenated hydrocarbons and more than 500 °C for most conventional hydrocarbons. The catalysts used can be divided into two types: nonprecious metal (typically nickel) and precious metals from Group VIII elements (typically platinum or rhodium). Due to severe mass and heat transfer limitations, conventional steam reformers are limited by the effectiveness factor of pelletized catalysts, which is typically less than 5%. Therefore, kinetics is rarely the limiting factor with conventional steam reformer reactors, and less expensive nickel catalysts are used industrially (Farrauto et al. 2003; Kalamaras and Efstathiou 2013).

An important factor characterizing the SR process is the H:C atom ratio in the feedstock material. The higher this ratio is, the lower carbon dioxide emission is formed. A membrane reactor can replace both reactors in a conventional SR process for achieving the overall reaction (Eq. 9.2). The heat efficiency of hydrogen production by the SR of methane process on an industrial scale is around 70–85%. A number of other raw materials are also possible to achieve this efficiency, such as solid communal waste, wastes from food industry, oils, purposefully cultivated or waste agricultural biomass, and fuels of fossil origin such as coal. The disadvantage is the high production of CO_2, about 7.05 kg CO_2 for each kg H_2 (Shirasaki et al. 2009; Kalamaras and Efstathiou 2013). As CO_2 concentration in the exhaust gas is high, SR units are promising candidates for the application of CCS technology, potentially leading to an 80% reduction in its carbon emissions (IEA 2015).

9.2.1.2 Partial Oxidation

Partial oxidation (POX) and catalytic partial oxidation (CPOX) of hydrocarbons have been proposed in hydrogen production for automobile fuel cells and some other commercial applications. The gasified raw material can be methane and biogas but primarily heavy oil fractions (e.g., vacuum remnants, heating oil), whose further treatment and utilization are difficult. POX is a noncatalytic process, in which the raw material is gasified in the presence of oxygen (Eqs. 9.5 and 9.6) and possibly steam (Eq. 9.7, ATR) at temperatures in the 1300–1500 °C range and pressures in the 3–8 MPa range. In comparison with the steam reforming (H_2:CO = 3:1), more CO is produced (H_2:CO = 1:1 or 2:1). The process is therefore complemented by the conversion of CO with steam into H_2 and CO_2. This reaction contributes to the maintenance of equilibrium between the individual reaction products (Aasberg-Petersen et al. 2001; Kalamaras and Efstathiou 2013):

$$CH_4 + O_2 \rightarrow CO + 2H_2 \tag{9.5}$$

$$CH_4 + 2O_2 \rightarrow CO_2 + 2H_2O \tag{9.6}$$

$$CH_4 + H_2O\ (g) \rightarrow CO + 3H_2 \tag{9.7}$$

The gaseous mixture formed through partial oxidation contains CO, CO_2, H_2O, H_2, CH_4, hydrogen sulfide (H_2S), and carbon oxysulfide (COS). A part of the gas is burned to provide enough heat for the endothermic processes. The soot created by the decomposition of acetylene as an intermediate product is an undesired product. Its amount depends on the proportion of H: C in the initial raw fuel material. There has been, therefore, like with the SR, an endeavor to shift to raw materials containing a higher H:C ratio, for example, to natural gas. While the operation of the reactor is less expensive in comparison with the steam reforming, the subsequent conversion makes this technology more expensive. Since the process does not require the use of a catalyst, it is not necessary to remove sulfurous elements from natural gas, which lowers the efficiency of the catalyst. The sulfurous compounds contained in the gasified raw material are converted into hydrogen sulfide (about 95%) and carbon oxysulfide (about 5%) (Lutz et al. 2004).

Catalysts can be added to the partial oxidation system (CPOX) in order to lower the operating temperature, about 700–1000 °C. However, temperature control is proving hard because of coke and hot spot formation due to the exothermic nature of the reactions. For natural gas conversion, the catalysts are typically based on Ni or Rh. However, nickel has a strong tendency to coke, and the cost of Rh has increased significantly. Catalytic partial oxidation has been used for decane, hexadecane, and diesel fuel. The high operating temperatures (>800 °C) and safety concerns may make their use for practical and compact portable devices difficult due to thermal management. Typically, the thermal efficiency of POX reactors with methane as fuel lies in the range of 60–75% (Semelsberger et al. 2004).

9.2.1.3 Autothermal Reforming

In the autothermal reforming (ATR), steam is added in the catalytic partial oxidation process. Therefore, ATR is a combination of both steam reforming (endothermic) and partial oxidation (exothermic) reactions. ATR has the advantages of not requiring external heat and being simpler and less expensive than SR of methane. The selection of operation conditions of the reformer depends on the specific target for both ATR and steam reforming. A main target is the high hydrogen yield with low carbon monoxide content. Maximum hydrogen efficiency and low carbon monoxide content are possible for steam reforming. However, steam reforming is an endothermic process and therefore energy demanding. This energy has to be transferred into the system from the outside (Kalamaras and Efstathiou 2013).

Another significant advantage of ATR over SR process is that it can be shut down and started very rapidly, while producing a larger amount of hydrogen than POX alone. There are some expectations that this process will become attractive for the "Gas to Liquid" fuel industry due to favorable gas composition for the Fischer-Tropsch synthesis, ATR's relative compactness, lower capital cost, and the potential for economies of scale. For methane reforming, the thermal efficiency is comparable to that of POX (ca. 60–75%) and slightly less than that of steam reforming. Gasoline and other higher hydrocarbons may be converted into hydrogen on board for use in automobiles by the autothermal process, using suitable catalysts (Ayabe et al. 2003; Kalamaras and Efstathiou 2013).

9.2.1.4 Plasma Reforming

In the case of plasma reforming, the network of reforming reactions is the same as that in conventional reforming. However, energy and free radicals used for the reforming reaction are provided by plasma typically generated with electricity or heat. When water or steam is injected with the fuel, $H^{\cdot}$, $OH^{\cdot}$, and $O^{\cdot}$ radicals in addition to electrons are formed, thus creating conditions for both reductive and oxidative reactions to occur. Plasma reforming technologies have been developed to facilitate POX, ATR, and steam reforming, with the majority of the reactors being POX and ATR. There are essentially two main categories of plasma reforming, namely, thermal and nonthermal (Paulmier and Fulcheri 2005).

Plasma devices referred to as plasmatrons can generate very high temperatures (ca. >2000 °C) with a high degree of control using electricity. The heat generated is independent of reaction chemistry, and optimal operating conditions can be maintained over a wide range of feed rates and gas compositions. Compactness of the plasma reformer is ensured by high energy density associated with the plasma itself, and by the reduced reaction times, resulting in short residence times. Hydrogen-rich gas streams can be efficiently produced in plasma reformers from a variety of hydrocarbon fuels (e.g., gasoline, diesel, oil, biomass, natural gas, and jet fuel) with conversion efficiencies close to 100%. The plasma reforming technology has potential advantages over conventional technologies of hydrogen manufacturing. The plasma conditions (e.g., high temperatures, high degree of dissociation, and substantial degree of ionization) can be used to accelerate thermodynamically favorable chemical reactions without a catalyst or provide the energy required for endothermic reforming processes to occur. Plasma reformers can provide a number of advantages, namely compactness and low weight (due to high power density), high conversion efficiencies, minimal cost (simple metallic or carbon electrodes and simple power supplies), fast response time (fraction of a second), operation with a broad range of fuels, including heavy hydrocarbons (crude) and "dirty" hydrocarbons (high sulfur diesel). This technology could be used to manufacture hydrogen for a variety of stationary applications, such as distributed and low-pollution electricity generation for fuel cells. It could also be used for mobile applications (e.g., on-board generation of hydrogen for fuel cell powered vehicles) and for refueling applications (e.g., stationary sources of hydrogen for vehicles) (Kalamaras and Efstathiou 2013).

The only disadvantages of plasma reforming are the dependence on electricity and the difficulty of high-pressure operation (required for high-pressure processes such as ammonia production). High pressure, while achievable, increases electrode erosion due to decreased arc mobility and, therefore, it decreases electrode lifetime (Bromberg et al. 1999).

9.2.1.5 Coal Gasification

Gasification of coal is the oldest method for production of hydrogen. Generally, coal is heated up to 900 °C with a catalyzer and without air. There are also more complex ways of gasifying coal, such as the Lurgi, Winkler, and Koppers-Totzek methods. These techniques are similar in that by using steam and oxygen at temperatures over 1400 °C, they change carbon into H_2, CO, and CO_2, and in addition create some sulfur and nitrogen, which like CO and CO_2, must be treated in the same way, environmentally speaking. It is almost twice as expensive to produce hydrogen from coal as from natural gas. This has to do with the relation between hydrogen and carbon, which in natural gas is 4:1 and in carbon 0.8:1. Germany, South Africa, and the USA presently have large gasification plants, and technology for gasification of coal in thermal power plants is the subject of much R&D by the coal industry (Krishna 2013).

9.2.2 Hydrogen from Biomass

Hydrogen could be also produced by biomass-based approaches rather than reforming of fossil fuels (Fig. 9.3), such as gasification, pyrolysis, and aqueous phase reforming along with production of hydrogen from water (e.g., electrolysis, photoelectrolysis, and thermochemical water splitting), as well as biological production process.

9.2.2.1 Biomass Gasification

Biomass is anticipated to become the most likely renewable organic substitute to petroleum. Biomass is available from a wide range of sources, such as animal wastes, municipal solid wastes, crop residues, short rotation woody crops, agricultural wastes, sawdust, aquatic plants, short rotation herbaceous species (e.g., switch grass), waste paper, corn, and many others. Gasification technology commonly used with biomass and coal as fuel feedstock is very mature and commercially used in many processes. It is a variation of pyrolysis, and, therefore, is based upon partial oxidation of the feedstock material into a mixture of hydrogen, methane, higher

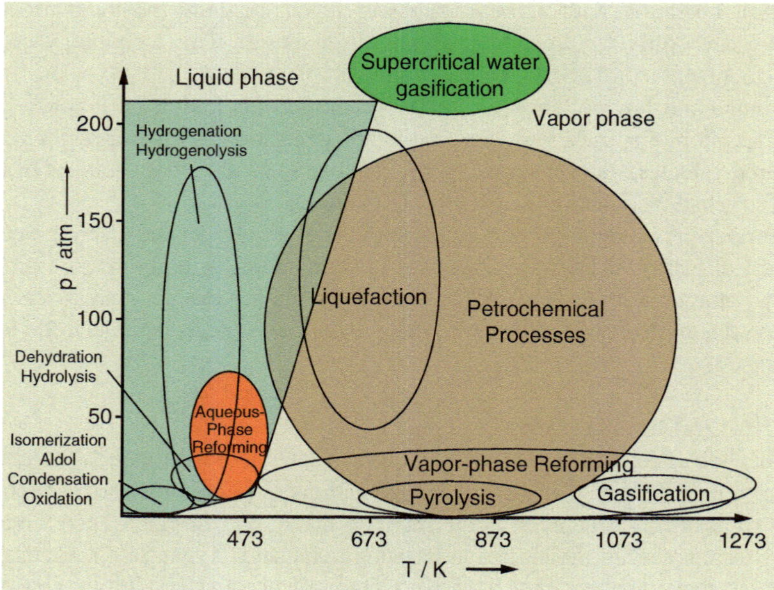

Fig. 9.3 Different biomass processing technologies relative to petroleum processing as a function of temperature and pressure (Adapted with permission from Chheda et al. 2007 (John Wiley & Sons))

hydrocarbons, carbon monoxide, carbon dioxide, and nitrogen, known as "producer gas." The gasification process typically suffers from low thermal efficiency since moisture contained in the biomass must also be vaporized. It can be performed with or without a catalyst and in a fixed-bed or fluidized-bed reactor, with the latter reactor having typically better performance. Addition of steam and/or oxygen in the gasification process results in the production of "syngas" with a H_2/CO ratio of 2/1, the latter used as feedstock to a Fischer-Tropsch reactor to make higher hydrocarbons (synthetic gasoline and diesel) or to a WGS reactor for hydrogen production. Superheated steam (about 900 °C) has been used to reform dry biomass to achieve high hydrogen yields. However, gasification process provides significant amounts of "tars" (a complex mixture of higher aromatic hydrocarbons) in the product gas even operated in the 800–1000 °C range. A secondary reactor, which utilizes calcined dolomite and/or nickel catalyst, is used to catalytically clean and upgrade the product gas. Ideally, oxygen should be used in these gasification plants; however, oxygen separation unit is cost prohibitive for small-scale plants. This limits the gasifiers to the use of air resulting in significant dilution of the product as well as the production of NO*x*. Low-cost, efficient oxygen separators are needed for this technology. For hydrogen production, a WGS process can be employed to increase the hydrogen concentration followed by a separation process to produce pure hydrogen. Typically, gasification reactors are built on a large scale and require massive amounts of material to be continuously fed. They can achieve efficiencies in the order of 35–50% based on the lower heating value. One of the problems of this technology is that a tremendous amount of resources must be used to gather the large amounts of biomass to the central processing plant. Currently, the high logistics costs of gasification plants and the removal of "tars" to acceptable levels for pure hydrogen production limit the commercialization of biomass-based hydrogen production. Future development of smaller efficient distributed gasification plants may be required for this technology for cost-effective hydrogen production (Weber et al. 2006; Kalamaras and Efstathiou 2013).

9.2.2.2 Pyrolysis and Copyrolysis

Another promising method of hydrogen production is pyrolysis or copyrolysis. Raw organic material is heated and gasified at a pressure of 0.1–0.5 MPa in the 500–900 °C range. The process takes place in the absence of oxygen and air, and therefore the formation of dioxins can be almost ruled out. Since no water or air is present, no carbon oxides (e.g., CO or CO_2) are formed, eliminating the need for secondary reactors (WGS, PrOx, etc.). Consequently, this process offers significant emissions reduction. However, if air or water is present (the materials have not been dried), significant COx emissions will be produced. Among the advantages of this process are fuel flexibility, relative simplicity and compactness, clean carbon byproduct, and reduction in COx emissions. The reaction can be generally described by the following equation (Muradov 2003):

$$C_nH_m + \text{heat} \rightarrow nC + 0.5mH_2 \tag{9.8}$$

Based on the temperature range, pyrolysis processes are divided into low (up to 500 °C), medium (500–800 °C), and high temperatures (over 800 °C). Fast pyrolysis is one of the latest processes for the transformation of organic material into products with higher energy content. The products of fast pyrolysis appear in the entire phases formed (solid, liquid, and gaseous). One of the challenges with this approach is the potential for fouling by the carbon formed, but proponents claim that this can be minimized by appropriate design. Since it has the potential for lower CO and CO_2 emissions, and it can be operated in such a way as to recover a significant amount of solid carbon, which is easily sequestered, pyrolysis may play a significant role in the future. The application of the copyrolysis of a mixture of coal with organic wastes should limit and lighten the burden of wastes in waste disposal (waste and pure plastics, rubber, cellulose, paper, textiles, and wood). Pyrolysis and copyrolysis are well-developed processes and could be used in commercial scale (Kalamaras and Efstathiou 2013).

9.2.2.3 Aqueous Phase Reforming

Aqueous phase reforming (APR) is a technology under development to process oxygenated hydrocarbons or carbohydrates of renewable biomass resources to produce hydrogen, as depicted in Fig. 9.4. The APR reactions take place at substantially lower temperatures (220–270 °C) than conventional alkane steam reforming (about 600 °C). The low temperatures at which aqueous-phase reforming reactions occur minimize undesirable decomposition reactions typically encountered when

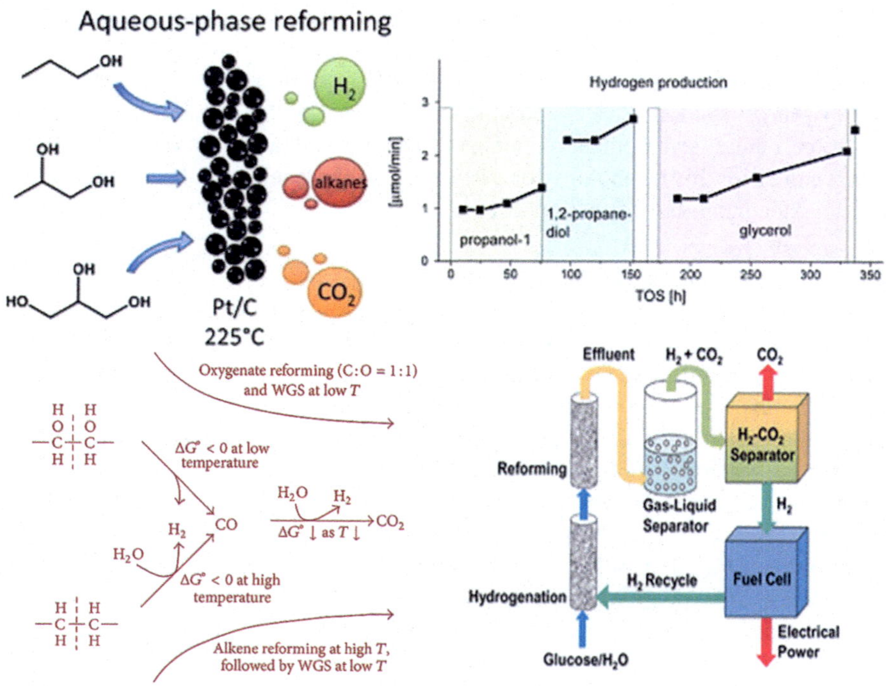

Fig. 9.4 Reaction pathways for the production of H_2 by reactions of oxygenated hydrocarbons with water during APR (Modified with permission from Kalamaras and Efstathiou 2013 (Hindawi Publishing); and Godina et al. 2018 (Elsevier))

carbohydrates are heated to elevated temperatures. Also, the water-gas shift reaction (WGS) is favorable at the same temperatures as in APR reactions, thus making it possible to generate H_2 and CO_2 in a single reactor with low amounts of CO. In contrast, typical steam reforming processes require multistage or multiple reactors to achieve low levels of CO in the product gas. Another advantage of the APR process is that it eliminates the need to vaporize water, which represents a major energy saving compared to conventional, vapor-phase steam reforming processes. Catalyst selection is important to avoid methanation, which is thermodynamically favorable, along with Fischer-Tropsch products, such as propane, butane, and hexane. Group VIII catalysts, with Pt-containing solids have the highest catalytic activity. Even though they have lower activity, nickel-based catalysts have been evaluated due to nickel's low cost. Improving catalyst activity and durability is an area where significant progress can be made. In general, this technology is more amiable to efficiently and selectively converting biomass feedstock to hydrogen (Kalamaras and Efstathiou 2013; Godina et al. 2018).

9.2.2.4 Electrolysis

Electrolysis is a process in which a direct current passing through two electrodes in a water solution results in the breaking of the chemical bonds present in water molecule into hydrogen and oxygen (Kalamaras and Efstathiou 2013):

$$\text{Cathode}: 2H_2O\ (l) + 2e^- \rightarrow H_2\ (g) + 2OH^-(aq) \tag{9.9}$$

$$\text{Anode}: 4OH^-(aq) \rightarrow O_2(g) + 2H_2O(l) + 4e^- \tag{9.10}$$

$$\text{Overall}: 2H_2O \rightarrow 2H_2 + O_2 \tag{9.11}$$

The electrolysis process takes place at room temperature. A commonly used electrolyte in water electrolysis is sulfuric acid, and the electrodes are of platinum (Pt), which does not react with sulfuric acid. The process is ecologically clean because no greenhouse gases are formed, and the oxygen produced has further industrial applications. However, in comparison with the foregoing methods, electrolysis is a highly energy-demanding technology.

The energetic efficiency of the electrolysis of water (chemical energy acquired per electrical energy supplied) in practice reaches 50–70%. For electrolyzers using only electric power (and no external heat) as input energy, the efficiency of hydrogen production decreases with cell voltage while the hydrogen production rate increases with cell voltage. At a given cell geometry, the operator therefore has to deal with a trade-off between electrolyzer efficiency and hydrogen output. Different types of electrolysers are distinguished by their electrolyte and the charge carrier, and can be grouped into (IEA 2015): (a) alkaline electrolyzers; (b) PEM electrolyzers; and (c) SO electrolyzers. All electrolyzers consist out of the electrolyzer stack, comprising up to 100 cells, and the BOP. Stacks can be mounted in parallel using the same BOP infrastructure, which is why electrolyzers are highly modular systems. While this makes the technology very flexible with respect to hydrogen production capacity, it also limits the effects of economies of scale, as even large electrolyzers are based on identically sized cells and stacks.

Alkaline electrolyzers are currently the most mature technology, and investment costs are significantly lower than for other electrolyzer types. Although alkaline electrolyzers currently have higher efficiencies than electrolyzers using solid electrolytes, PEM and SO electrolyzers have much higher potential for future cost reduction and, in case of SO electrolyzers, efficiency improvements. PEM electrolyzers are particularly interesting as they show both the highest current density and operational range, prerequisites necessary to reduce investment costs and improve operational flexibility at the same time. However, cell lifetime is a limiting factor for PEM and SO electrolyzer technologies.

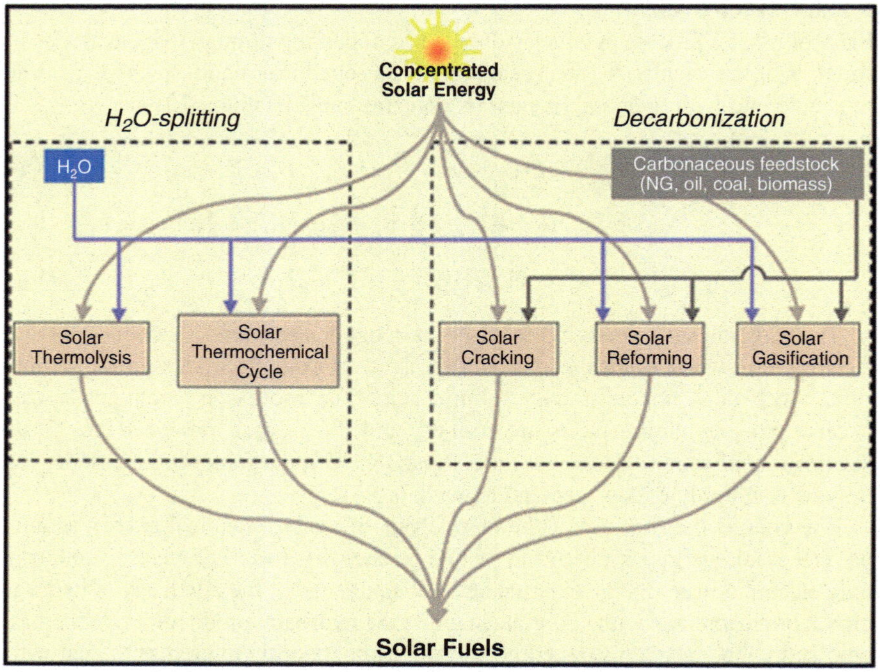

Fig. 9.5 Thermochemical processes for hydrogen production (Adapted with permission from Steinfeld 2005 (Elsevier))

9.2.3 Solar Hydrogen Production Technique

Different solar hydrogen production techniques have been developed, such as thermochemical, electrochemical, photochemical, and photobiological, as well as hybrid techniques.

9.2.3.1 Thermochemical Process

Solar thermochemical processes make use of concentrated solar radiation as the energy source of high-temperature process heat to drive endothermic reactions aimed at the production of chemical fuels. General thermochemical processes are shown in Fig. 9.5. In a thermal solar power plant with a central collector, the temperatures can reach over 3000 °C. Heating water to over 2000 °C can break it down into hydrogen and oxygen. This is considered to be an interesting and inexpensive method of producing hydrogen directly from solar energy. One central problem is the separation of gases at high temperatures to avoid recombining. The efficiency factor is uncertain.

Research is also being done on the use of catalysts to reduce the temperature for dissociation, which is called indirect conversion process. It takes place at relatively lower temperatures, and can be used in two ways (Verma 2015): splitting water molecules or decarburization of fossil fuels.

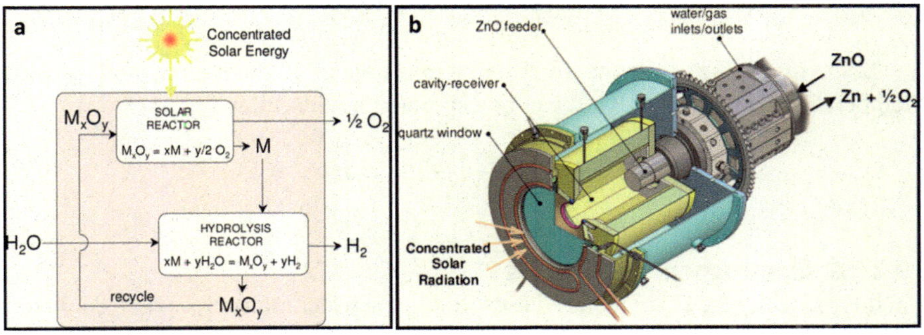

Fig. 9.6 Scheme of a two-step solar thermochemical cycle based on metal oxide redox reactions (Adapted with permission from Steinfeld 2005 (Elsevier)): (**a**) MxOy denotes a metal oxide, and M the corresponding metal or lower-valence metal oxide. In the first endothermic solar step, MxOy is thermally dissociated into the metal or lower-valence metal oxide M and oxygen. Concentrated solar radiation is the energy source for the required high-temperature process heat. In the second exothermic nonsolar step, M reacts with water to produce hydrogen. The resulting metal oxide is then recycled back to the first step. (**b**) The solar reactor configuration for the thermal dissociation of ZnO, based on ZnO/Zn redox reactions. It consists of a windowed rotating cavity-receiver lined with ZnO particles. With this arrangement, ZnO is directly exposed to high-flux solar irradiation and serves simultaneously the functions of radiant absorber, thermal insulator, and chemical reactant

There are over 300 thermochemical cycles available that separate water into hydrogen and oxygen using multiple steps. It does not face the high temperate or the product separation problem. An efficient two-step thermochemical cycle uses the metal oxide reactions, as shown in Fig. 9.6a. The first, endothermic step is the solar thermal dissociation of the metal oxide to the metal or the lower-valence metal oxide. The second, nonsolar, exothermic step is the hydrolysis of the metal to form H_2 and the corresponding metal oxide. The net reaction is $H_2O = H_2 + 0.5O_2$, but since H_2 and O_2 are formed in different steps, the need for high-temperature gas separation is thereby eliminated. The second hydrolysis step can be accomplished on demand at the H_2 consumer site, as it is decoupled from the availability of solar energy. Alternatively, CO_2 can be co-fed with H_2O to react with the metal and produce syngas, which can be further processed to liquid fuels. One promising redox system is ZnO/Zn. Figure 9.6b shows a solar chemical reactor configuration for performing the thermal dissociation of ZnO that consists of a windowed rotating cavity-receiver lined with ZnO particles. With this arrangement, ZnO is directly exposed to high-flux solar irradiation and serves simultaneously the functions of radiant absorber, thermal insulator, and chemical reactant (Steinfeld 2005).

Solar cracking and solar reforming are two methods with which hydrogen can be produced from fossil fuels. Solar cracking is the thermal decomposition of natural gas (NG), oil, and other hydrocarbons, in an oxygen-free environment, and can be represented by the simplified net reaction:

$$C_xH_y \rightarrow xC + (y/2)\ H_2 \tag{9.12}$$

Solar reforming is the thermal decomposition of hydrocarbon, and steam at temperatures about 1000 °C. It can be expressed by the following reaction:

$$C_xH_y + xH_2O \rightarrow (y/2 + x)\ H_2 + xCO \tag{9.13}$$

9.2.3.2 Electrochemical Process

In this process, electricity required for splitting the water molecule is derived directly from photovoltaic panels. It can be thought of a reverse PEMFC with the required electricity being directly provided by the PV panels breaking distilled water into hydrogen and oxygen. Such a system is commonly used with an electrolyzer and the overall efficiency is below 16%. It has been suggested that efficiency of such a system can be improved if new materials accommodating more solar irradiation are used (Verma 2015).

9.2.3.3 Photochemical Process

Photochemical processes use solar light as energy to produce hydrogen. They are of many types like photoelectrochemical, photo biological, photocatalytic, photodissociation, photodecomposition, and photolysis.

Photoelectrochemical Process

This process combines solar PV and electrolysis in a single device, thereby increasing the efficiency of the system. It consists of a photoactive electrode that absorbs solar energy. When enough energy is absorbed, an electron is released producing electricity in the external circuit. This electricity is used in disassociation of water molecules (Verma 2015):

$$2hv + H_2O \rightarrow (1/2)O_2 + H_2 \tag{9.14}$$

Where h is the Planck's constant and v is the frequency.

Photocatalytic Process

It uses a sensitizer that absorbs solar energy and uses it to activate the electrons in the valence shell of the oxygen atom in the water molecule. A catalyst is used to accumulate these electrons and uses them for oxidation of metal, producing hydrogen as a product. Chemical reactions used in this process are as follows (Verma 2015):

$$4S + 4h\Upsilon = 4S^{*} \tag{9.15a}$$

$$4S^{*} + 4C_R = 4S^{*} + 4(C_{R-})^{*} \tag{9.15b}$$

$$4(C_{R-})^* + 4H^* = 4C_R + 4H^* \quad (9.15c)$$

$$4S^* + 4C_{OX} = 4S + 4C_{OX+} \quad (9.15d)$$

$$C_{OX+} + 2H_2O = 4Cox + 2O^* + 4H^* \quad (9.15e)$$

$$4H^* = 2H_2 \quad (9.15f)$$

$$2O^* = O_2 \quad (9.15g)$$

where h is the Planck's constant, ϒ is the photon frequency, S is the sensitizer, and C_R and C_{OX} are the catalysts for reduction and oxidation reactions, respectively.

Photolysis

Instead of first converting sunlight to electricity and then using an electrolyzer to produce hydrogen from water, it is possible to combine these two steps. The photovoltaic cell combines with a catalyst, which acts as an electrolyzer and splits hydrogen and oxygen directly from the surface of the cell. This can quite realistically be a commercially viable means of producing hydrogen. The advantage with these systems is that they eliminate the cost of electrolyzers and increase the systems' efficiency. Tests performed outdoors with silicon-based cells have shown an efficiency of 7.8% in natural sunlight. Research is being done to increase the efficiency factor and the life span for such cells (Krishna 2013).

9.2.3.4 Photobiological Process

Photobiological production process makes use of bacterial microorganisms that have the tendency of fermenting natural wastes and organic compounds to hydrogen and any other nonpolluting gas.

Bio-Photolysis of Water Using Algae and Cyanobacteria

This method uses the same processes found in plants and algal photosynthesis, but adapts them for generation of hydrogen gas instead of the carbon-containing biomass. Microalgae are primitive microscopic plants living in aqueous environments. Cyanobacteria, formerly known as blue-green algae, are now recognized as bacteria since the anatomical characteristics of their cells are prokaryotic (bacterial type). Miroalgae and Cyanobacteria along with higher plants are capable of oxygenic photosynthesis according to the following reaction (Krishna 2013):

$$CO_2 + H_2O \rightarrow HCHO + O_2 \quad (9.16)$$

Photosynthesis consists of two processes: light energy conversion to biochemical energy by a photochemical reaction, and CO_2 reduction to organic compounds such as sugar phosphates, through the use of this biochemical energy by Calvin-cycle enzymes. Under certain conditions, however, instead of reducing CO_2, a few groups of microalgae and Cyanobacteria consume biochemical energy to produce molecular hydrogen. Hydrogenase and nitrogenase enzymes are both capable of hydrogen production.

However, the rate of hydrogen production was lower than typical rate for CO_2 reduction. One of the major obstacles in obtaining sustained hydrogen photo production is the photosynthetically generated oxygen, which irreversibly inactivates hydrogen-producing systems and supports oxygen-dependent hydrogen uptake activity. Many microalgae in particular species classified as "green algae" produced hydrogen after a period of anaerobic condition in dark, during which the hydrogenase enzyme is activated and synthesized and small amounts of hydrogen production are observed, when such "anaerobically adopted" algae are returned to light (but still under anaerobic condition). Hydrogen production rates increases dramatically, but cease once normal photosynthesis (oxygen evolution, CO_2 fixation) is reestablished (Krishna 2013).

INDEPTH: Feed Stocks for Biohydrogen Production (Krishna 2013)

Feedstock	Technology employed	Key findings
Water and algae	Direct biophotolysis	2.152 mL of H_2 for a 10 mL culture of alga
Water and carbon dioxide	Indirect biophotolysis	Hydrogen production rate: 0.18 mL/mg/day
Water and carbon monoxide	Biological water-gas shift reaction	Volumetric hydrogen production: 41 mmol H_2/L/h
Dairy manures	Anaerobic fermentation and acid pretreatment	Hydrogen yield: 31.5 mL/g-total volatile solids
Palm oil mill effluent (POME)	Dark fermentation	Hydrogen yield: 3195 mL H_2/L-medium Volumetric hydrogen production rate 1034 mL H_2/L-medium/h
Sugarcane bagasse	Dark fermentation	Hydrogen yield: 1.73 mol H_2/mol total sugar Hydrogen production rate 1611 mL H_2/L/day
Corn STOVER	Anaerobic fermentation	Hydrogen yield: 2.84 mol
Sludge	Dark fermentation	Hydrogen yield: 2.28 mol H_2/mol glucose
Glucose	Dark fermentation	Hydrogen yield: 1.0 mol/mol glucose
Wastewater (WW): domestic potato processing	Dark fermentation	Hydrogen yield: 0.01 L H_2/L WW Hydrogen yield: 1.0 L H_2/L WW
Food wastewater	Anaerobic fermentation	Hydrogen production rate 97.5 mmol H_2/L/day
Wheat starch	Photo fermentation	Hydrogen production: 178 mL Hydrogen yield: 1.23 mol H_2/mol glucose Specific hydrogen rate: 46 mL H_2/g-biomass/h

(continued)

Cassava starch	Dark fermentation and photo fermentation	Hydrogen yield: 840 mL H_2/g starch
Corncob	Dark fermentation and photo fermentation	Hydrogen yield (DF): 120.3 ± 5.2 mL H_2/g-corncob Hydrogen rate (DF): 150 mL H_2/(L h) Hydrogen yield (PF): 713.6 ± 44.1 mL H_2/g-COD
Olive mill wastewater (OMW)	Dark fermentation and photo fermentation	Hydrogen yield: 29 L H_2/L-OMW
Pharmaceutical wastewater	Anaerobic fermentation	Hydrogen yield: 2.95 mmol/day Organic loading rate: 5.90 Kg COD/cum-day

Photodecomposition of Organic Compounds by Photosynthetic Bacteria

Photosynthetic bacteria undergo anoxygenic photosynthesis with organic compounds or reduced sulfur compounds as electron donors. Some non-sulfur photosynthetic bacteria are potent hydrogen producers, utilizing organic acids such as lactic, succinic, and butyric acids, or alcohols as electron donors. Since light energy is not required for water oxidation, the efficiency of light energy conversion to hydrogen gas by photosynthetic bacteria is in principle much higher than that by cyanobacteria. Hydrogen production by photosynthetic bacteria is mediated by nitrogenase activity, although hydrogenases may be active for both hydrogen production and hydrogen uptake under some conditions. The major benefits of photosynthetic bacterium as a microbial system for biological hydrogen production are as follows (Krishna 2013): (a) High theoretical conversion yields, (b) Lack of oxygen evolving activity, which causes problem of oxygen interactivation of different biological systems, (c) Ability to use a wide spectrum of light, (d) Ability to consume organic substrates derivable from the wastes and their potential to be used in association with waste water treatment. The overall bio-chemical pathways for photo fermentation process can be expressed as follows (Krishna 2013):

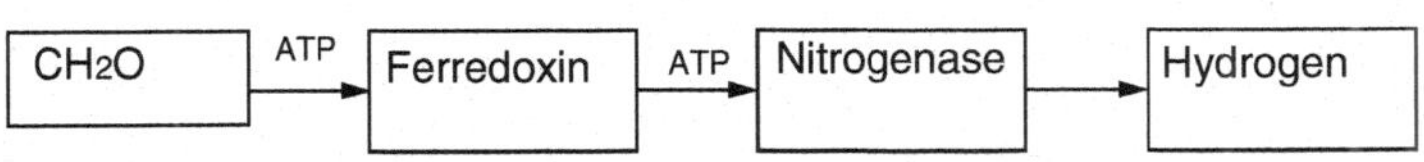

CO can also be used for the production of hydrogen using microbial shift reaction by the photosynthetic bacteria as follows (Krishna 2013):

$$CO + H_2O \rightarrow CO_2 + H_2 \tag{9.17}$$

Fermentative Hydrogen Production from Organic Compounds

Hydrogen evolution by fermentation has been shown lesser attention whereas the hydrogen evolution by photosynthetic microorganisms has been extensively studied. The evolution of hydrogen by fermentation has, however, several advantages over industrial production processes such as (a) Fermentative bacteria have a very high evolution rate of hydrogen, (b) Fermentative bacteria can produce hydrogen constantly throughout day and night from organic substrates, and (c) These bacteria have a good growth rate for supply of microorganisms to the production systems. Therefore, the fermentative evolution is more advantageous than photochemical evolution for the mass production of hydrogen by microorganisms. Fermentative hydrogen production can be maximized through the effective coupling of the factors like an accessible and a rich source of electron and biochemical electron pump and an active hydrogenase (Krishna 2013).

Hybrid System Using Photosynthetic and Fermentative Bacteria

Anaerobic bacteria metabolize sugars to produce hydrogen gas and organic acids, but are incapable of further breaking down the organic acids formed proposed the combined use of photosynthetic and anaerobic bacteria for the conversion of organic acids to hydrogen. Theoretically, one mole of glucose can be converted to 12 moles of hydrogen through the use of photosynthetic bacteria capable of capturing light energy in such a combined system. From a practical point of view, organic wastes frequently contain sugar or sugar polymers. It is not however easy to obtain organic wastes containing organic acids as the main components. The combined use of photosynthetic and anaerobic bacteria should potentially increase the likelihood of their application in photobiological hydrogen production (Krishna 2013).

INDEPTH: Solar-Powered Hydrogen Generation

Solar-powered hydrogen generation has been demonstrated with more-efficient water-splitting solar cells based on a cheap, abundant, and long-lasting material: rust. The advance could lead to a cheap and energy-efficient way to generate hydrogen for fuel-cell vehicles using solar energy. Water-splitting solar panels would have important advantages over existing technologies in terms of hydrogen production. Right now, the primary way to make hydrogen is to separate it from natural gas, a process that generates carbon dioxide and undercuts the main motivation for moving to hydrogen fuel-cell vehicles: ending dependence on fossil fuels. The current alternative is electrolysis, which uses electricity to break water into hydrogen and oxygen, with the two gases forming at opposite electrodes. Although electrolysis is

(continued)

costly, it can be cleaner if the source of the electricity is wind, sun, or some other carbon-free source. But if the source of the electricity is the sun, it would be much more efficient to use solar energy to produce hydrogen by a photochemical process inside the cell itself. With small amounts of silicon and cobalt, nanostructured thin films of iron oxide has been fabricated to convert sunlight into the electrons needed to form hydrogen from water. Iron oxide has long been an appealing material for such solar panels, in part because it holds up well in contact with water. But although it can absorb sunlight, the resulting charge carriers could not easily escape the material, so they recombined, canceling each other out before they could split any water. By doping the rust with silicon, the coaxed material can form cauliflower-like structures with extremely high surface area, ensuring that a large part of the atoms in the material are in contact with the water, or very close to it. That way, holes could easily escape into the water, where they prompt the generation of oxygen gas. The silicon also improves electron conductivity in the material, which is important for generating hydrogen gas at an opposite electrode. The process can be further improved by adding cobalt, which acts as a catalyst for the reactions. The iron-oxide films can convert 42% of ultraviolet photons in sunlight into electrons and holes. But the system's overall efficiency is only about 4%, in part because iron oxide doesn't absorb all the parts of the solar spectrum. By adjusting the amount and arrangement of silicon and cobalt, and improving the structure of the films with novel materials such as metal-complex chromophores (Yuan et al. 2017), it would be possible to make solar-to-hydrogen panels more efficient than small electrolysis machines, and they would ensure that the hydrogen comes from a renewable source (Bullis 2006).

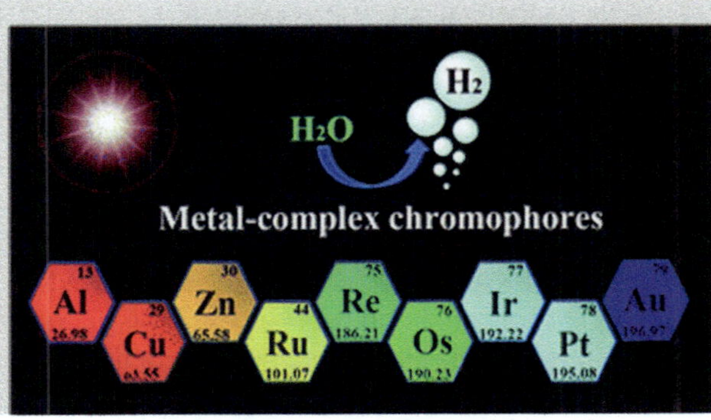

9.2.4 Hydrogen Production from Nuclear Energy

Nuclear energy is generated in fission reactors using uranium-based fuel (or thorium in the near future). The generated nuclear radiation is converted to high-temperature

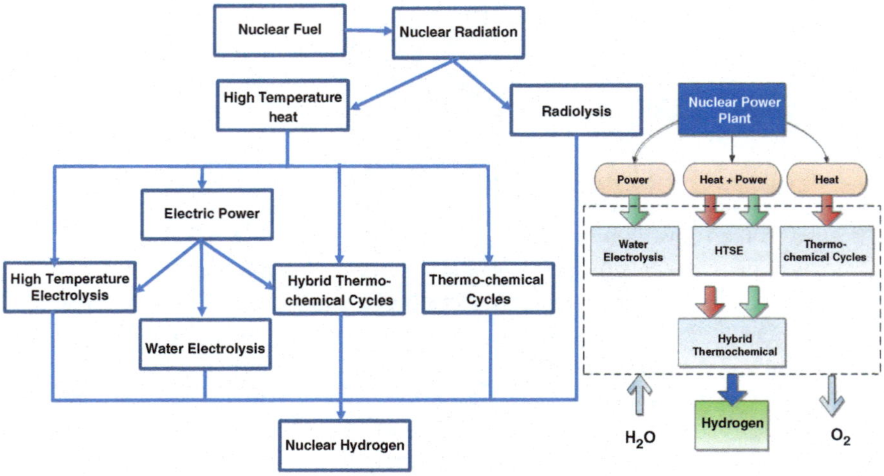

Fig. 9.7 Nuclear water splitting pathways for hydrogen production

heat which is transported by a heat transfer fluid for further use. The regular use of nuclear heat is power generation in large-scale power plants. It is possible to divert some of the high-temperature heat from nuclear reactors to supply chemical processes that eventually generate hydrogen from water splitting. Moreover, nuclear radiation present in nuclear reactors and during all phases of nuclear fuel processing can be used directly to generate hydrogen from water (Naterer et al. 2013).

As shown in Fig. 9.7, nuclear energy is converted into nuclear radiation (either in the reactor or during the fuel processing cycle). Nuclear radiation is converted further to high-temperature heat. Both the radiation and heat are used to generate hydrogen. All possible methods that can be envisaged to generate hydrogen from nuclear energy through water decomposition include (Naterer et al. 2013): (a) radiolysis, (b) electrolysis, (c) high-temperature steam electrolysis, (d) hybrid thermochemical water splitting, and (e) thermochemical water splitting. Radiolysis uses nuclear radiation to directly split the molecule of water into hydrogen and oxygen; electrolysis uses electricity derived from nuclear energy to electrolyze water; high-temperature steam electrolysis and hybrid thermochemical water splitting are called hybrid because they use both electricity and high-temperature heat to split water; and thermochemical water splitting directly uses high-temperature heat. Among these methods, with current commercial reactors, a configuration for radiolytic water splitting is not practical because of requirements of containment of the radioactive material, although radiolysis can, in principle, be implemented at spent fuel pools of existing nuclear power plants.

As long as it can provide electricity and process heat, any type of nuclear reactor can be used for the production of hydrogen. However, the reactor coolant and its maximum temperature are essential criteria for determining which reactor type is more appropriate for different production processes. Power size is also an important factor, as large reactors are more suitable for cogeneration of electricity and hydrogen production, whereas small sized plants are more suitable as single purpose plants (e.g., for hydrogen production only). However, it is possible to use off-peak power or cogeneration to reduce or share costs. Small and medium power reactors based on high-temperature gas reactors (HTGR) are also an attractive option. Future advanced nuclear power plants, such as the very high-temperature reactors (VHTR) or the Supercritical Water Cooled Reactor (SCWR), could provide not only the electricity needed, but also deliver relatively high-temperature process heat, providing high net power cycle efficiencies.

9.2.4.1 Nuclear-Assisted Electrolysis

Electrolysis is the most straightforward process currently available to produce hydrogen directly from water. Although conventional low-temperature electrolysis can be coupled with all currently operating reactors, it will not be economically competitive. At higher temperatures, various potential processes for hydrogen production, such as high-temperature steam electrolysis and other thermochemical processes, have been identified. In this case, the use of high-temperature reactors for hydrogen production presents a viable option since most of these processes have a higher efficiency than low-temperature electrolysis. The types of electrolysis that are being considered for deployment on an industrial scale are, apart from the classical alkaline electrolysis, proton exchange membrane (PEM) electrolysis, and high-temperature steam electrolysis (HTSE) using oxygen conducting ceramics (IAEA 2012).

Electricity input required for electrolysis can be decreased by increasing the temperature range, as the total energy demand for electrolysis in the vapor phase is reduced by the heat of vaporization, which can be provided much more cheaply by thermal rather than electric energy. Indeed, in the high temperature range of 800–1000 °C, electricity input could be about 35% lower than that of conventional electrolysis. In addition, the efficiency of electrical generation at these high temperatures is significantly better. The HTSE process is advantageous due to its high overall thermal-to-hydrogen efficiency when coupled with high-efficiency power cycles. As HTSE corresponds to the reverse process of a solid oxide fuel cell (SOFC), respective devices could be operated in both modes. A high-temperature electrolyzer must be coupled to a heat and power source. At high temperatures, all reactions proceed very rapidly. The steam–hydrogen mixture exits from the stack and then passes through a separator to separate hydrogen from the residual steam. The feed gas stream to the HTSE cell contains a 10% fraction of hydrogen in order to maintain reducing conditions and avoid the oxidation of the nickel in the hydrogen electrode (Fig. 9.8). HTSE cells can be operated at high current densities, which allow for large production capacities in comparatively small volumes. An electricity-to-hydrogen efficiency of about 90% could be achievable. However, the lifetime of the hydrogen electrode, which is limited by degradation, requires further improvement (IAEA 2012).

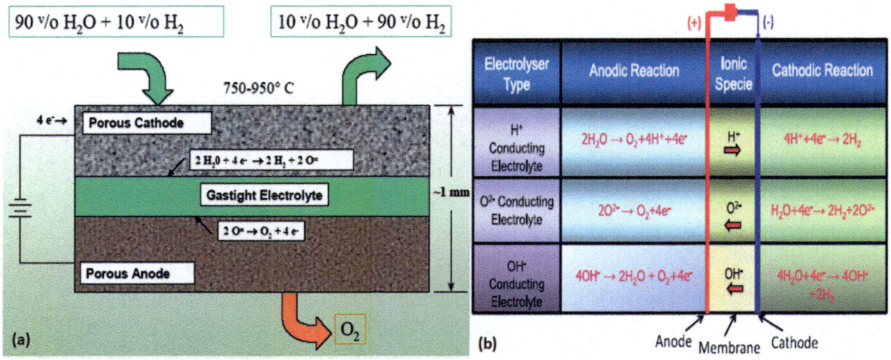

Fig. 9.8 Schematic of a planar steam electrolysis cell (**a**) and Operating principles of water electrolysis with different electrolytes (**b**) (Modified from IAEA 2012, Courtesy of International Atomic Energy Agency; and Badwal et al. 2014, Courtesy of Frontiers Media SA)

9.2.4.2 Nuclear-Assisted Thermochemical Cycles

Besides high-temperature electrolysis, another promising candidate to produce large amounts of hydrogen by high-temperature water-splitting is thermochemical process. The most straightforward method of water splitting would be a one-step direct thermal decomposition. However, this would require temperatures of >2500 °C for reasonable quantities, which is industrially not feasible. Therefore, multistep processes are being considered. A thermochemical cycle is a process consisting of a series of thermally driven chemical reactions where water is decomposed into hydrogen and oxygen at moderate temperatures. Supporting intermediate chemical compounds, which are regenerated and recycled internally and remain—ideally—completely in the system, are used in a sequence of chemical and physical processes. The only input to the cycle is water and high-temperature heat. Therefore, these cycles are potentially more efficient than low-temperature electrolysis and could significantly reduce production costs. Research on thermochemical cycles mainly focus on using solar or nuclear primary heat input. Numerous thermochemical cycles have been proposed in the past and checked against factors such as (IAEA 2012): corrosion problems, cost analysis, heat transfer, material stability, maximum temperature, processing scheme, reaction kinetics, separation of substances, side reactions, thermodynamics, thermal efficiency, and toxicity. Some have sufficiently progressed to be experimentally demonstrated and have already proven their scientific and practical feasibility. All cycles however, have design challenges and none has actually been implemented on a commercial scale. A major challenge in thermochemical cycles is obtaining maximum yields while reducing the amount of excess reagents used to drive the reactions in the desired directions. Therefore, the optimization of heat flows is important for high energy conversion efficiency. One cycle under special consideration is the sulfur–iodine (S–I) process, also known as Ispra Mark 16 cycle, originally developed by the General Atomics and later taken up and modified by different institutions like the Japan Atomic Energy Agency (JAEA). This cycle basically consists of three chemical reactions (Fig. 9.9a). Of all

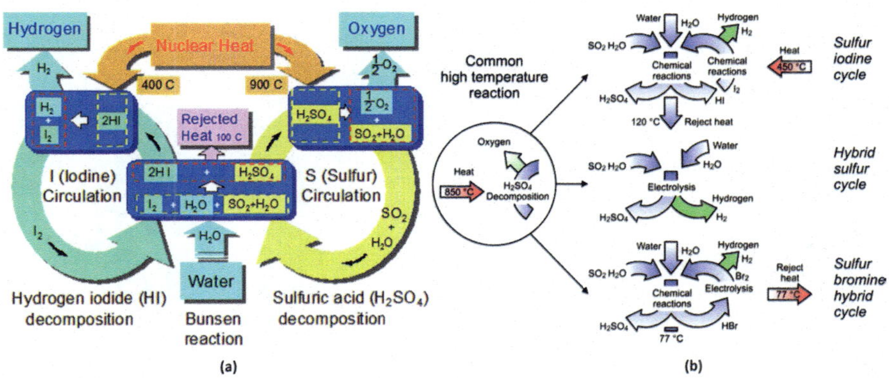

Fig. 9.9 Schematic of sulfur–iodine thermochemical water splitting cycle (**a**) and Thermochemical cycles of the sulfur family (**b**) (Modified from IAEA 2012, Courtesy of International Atomic Energy Agency)

thermochemical cycles, the S–I cycle is the one with the highest efficiency quoted. The theoretical limit of efficiency for the total process is assessed to be 51%, assuming ideal reversible chemical reactions. A schematic of the S–I cycle is shown in Fig. 9.9a. In all studies that systematically examined thermochemical cycles, those of the sulfur family—S–I, hybrid 5 sulfur, sulfur–bromine hybrid—have been identified as the potentially most promising candidates, with higher efficiency rates and a lower degree of complexity (in terms of the number of reactions and separations). All three have in common the thermal decomposition of sulfuric acid at high temperatures (Fig. 9.9b).

9.2.4.3 Nuclear-Assisted Hybrid Cycles

A hybrid cycle combines the benefits of thermochemical and electrolytic reactions. In this type of cycle, the low-temperature reaction, which has a low thermodynamic efficiency and is therefore not favorable, is forced electrochemically. The hybrid-sulfur (HyS) process is also known as the Ispra Mark 11 cycle, which is a variation of the S–I process which consists of only two reaction steps and where sulfur, apart from hydrogen and oxygen, is the only other element involved. A series of flash evaporators are used to separate oxygen from the liquid mixture resulting in almost pure oxygen. The rest of the mixture, consisting of mainly SO_2, sulfuric acid and water, is sent to the electrochemical section. The mixture of SO_2 and water is reacted in an electrolytic cell at lower temperatures to produce H_2 and a sulfuric acid in an aqueous phase. Still, the net thermal energy requirement for the HyS process is significantly less than for conventional water electrolysis. SO_2 electrolyzers require no more than 25% of the electricity that is needed in low-temperature water electrolysis, i.e., ~0.29 V. Theoretically, this value could be decreased to 0.17 V. Nevertheless, this would require that H_2SO_4 be decomposed at high temperatures in order to recycle the SO_2 for the completion of the cycle. The Westinghouse process (Fig. 9.10a) is simpler in design as the use of corrosive halides is not required. After the oxygen is removed from the system, the SO_2 and H_2O are combined with make-

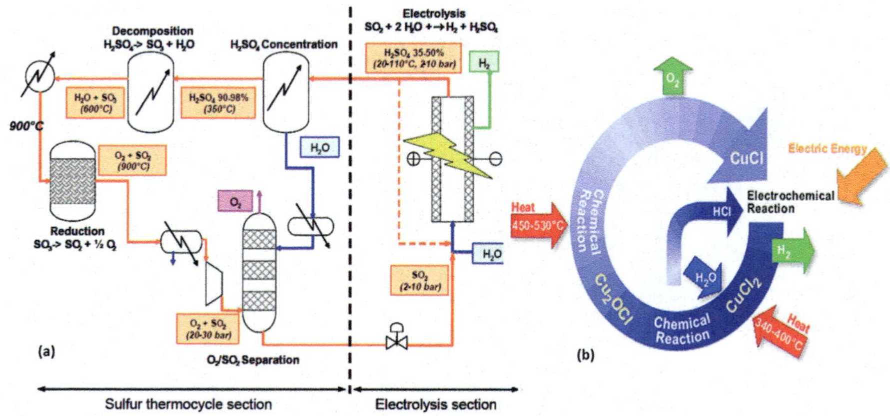

Fig. 9.10 Schematic of the Westinghouse hybrid-sulfur cycle (**a**) and the copper–chlorine thermochemical water splitting cycle (**b**) (Modified from IAEA 2012, Courtesy of International Atomic Energy Agency)

up H_2O_6 and routed to the electrolyzer cell. The SO_2 is then electrochemically oxidized at the anode to form H_2SO_4, protons, and electrons. The protons migrate through the electrolyte and produce H_2 gas at the cathode (IAEA 2012).

Several alternative thermochemical cycles for hydrogen production operate at moderate temperatures in the range of 500–600 °C. Lower operating temperatures reduce the costs of materials and maintenance and can effectively use low-grade waste heat, thereby improving cycle and power plant efficiencies. Additional advantages include ease of handling the chemical agents and reactions. For instance, the copper–chlorine (Cu–Cl) cycle, as shown in Fig. 9.10b, can be operated at a maximum temperature of about 550 °C. A hybrid cycle consisting of several thermal and one electrochemical reaction, the Cu–Cl cycle requires much lower operating temperatures than other cycles and can effectively use low-grade waste heat. Potential efficiency could reach as high as 41%. The electric energy demand was assessed at 39% of the total energy required (IAEA 2012).

9.3 Hydrogen Conversion and Storage Technology

Hydrogen conversion and storage is a key enabling technology for the advancement of hydrogen and fuel cell technologies in applications including stationary power, portable power, and transportation. Hydrogen has the highest energy per mass of any fuel; however, its low ambient temperature density results in a low energy per unit volume, therefore requiring the development of advanced storage methods that have potential for higher energy density. Table 9.1 shows current performance of key hydrogen conversion, and storage technologies.

Table 9.1 Current performance of key hydrogen conversion and storage technologies (IEA 2015)

Application	Power of capacity	Efficiency[a]	Initial investment cost[b]	Life time	Maturity
Alkaline FC	Up to 250 kW	~50% (HHV)	USD 200–700/kW	5000–8000 h	Early market
PEMFC stationary	0.5–400 kW	32–49% (HHV)	USD 3000–4000/kW	–60,000 h	Early market
PEMFC mobile	80–100 kW	Up to 60% (HHV)	USD ~500/kW	<5000 h	Early market
SOFC	Up to 200 kW	50–70% (HHV)	USD 3000–4000/kW	Up to 90,000 h	Demonstration
PAFC	Up to 11 MW	30–40% (HHV)	USD 4000–5000/kW	30,000–60,000 h	Mature
MCFC	KW to several MW	More than 60% (HHV)	USD 4000–6000/kW	20,000–30,000 h	Early market
Compressor, 18 MPa	–	88–95%	USD ~70/kW H_2	20 years	Mature
Compressor, 70 MPa	–	80–91%	USD 200–400/kW H_2	20 years	Early market
Liquefier	15–80 MW	~70%	USD 900–2000/kW	30 years	Mature
FCEV on-board storage tank, 70 MPa	5–6 kg H_2	Almost 100% (without compression)	USD 33–17/kWh (10,000 and 500,000 units produced per year)	15 years	Early market
Pressurized tank	0.1–10 MWh	Almost 100% (without compression)	USD 6000–10,000/MWh	20 years	Mature
Liquid storage	0.1–100 GWh	Boil-off stream: 0.3% loss per day	USD 800–10,000/MWh	20 years	Mature
Pipeline	–	95%, incl. compression	Rural: USD 300,000–1.2 million/km Urban: USD 700,000–1.5 million/km (dependent on diameter)	40 years	Mature

[a]Unless otherwise stated efficiencies are based on LHV
[b]All investment costs refer to the energy output

9.3.1 Fuel Cells

Fuel cells allow the oxidation of hydrogen-rich fuel and its conversion to useful energy without burning it in an open flame. Compared to other single-stage processes to convert chemical energy into electricity, e.g., open-cycle gas turbines, their electrical efficiency is higher and in the range of 32% to up to 70% (HHV). They are a flexible, modular technology that can easily be scaled up from small size to large office blocks and industrial complexes (Dodds et al. 2015).

Fuel cells operate with a variety of input fuels, not only hydrogen. These include natural gas and also liquid fuels such as methanol or diesel. If pure hydrogen is used, the exhaust of fuel cells is water vapor and so has very low local environmental impact. However, if hydrocarbon fuels are used, using fuel cells for power generation produces CO_2 emissions, and so can only confer a climate benefit by operating at higher efficiency than alternative combustion methods. Nevertheless, experience with fuel cells based on hydrocarbons has a high value for low-carbon innovation due to the applicability of technological advances to fuel cells more generally. This is partly because hydrocarbon fuels are often reformed to hydrogen in a step that precedes the fuel cell and also because some hydrocarbons may be produced by lower carbon processes in future, e.g., methanol. Similar to electrolyzers, fuel cells are subject to a trade-off between efficiency and power output. Efficiency is highest at low loads and decreases with increasing power output. In comparison to conventional technologies, fuel cells can achieve their highest efficiencies under transient cycles, such as in passenger cars. As in the case of electrolyzers, different fuel cell types exist, which can mainly be distinguished by their membrane type and operating temperature. Fuel cells can be categorized into (IEA 2015): (a) proton exchange membrane fuel cell, also known as polymer electrolyte membrane (PEM) fuel cell (PEMFC); (b) alkaline fuel cell; (c) phosphoric acid fuel cell (PAFC); (d) molten carbonate fuel cell (MCFC); and (e) solid oxide fuel cell (SOFC). While PEMFCs and alkaline fuel cells have low operating temperatures of around 80 °C, the others operate at higher temperatures of up to 600 °C (SOFC), which makes them more suitable to combined heat and power applications. The higher the temperature, the better the efficiency at otherwise similar parameters.

PEMFCs (proton exchange membrane fuel cells) are the most developed technology, powering around 90% of systems shipped to date. They are most widely used in residential heating systems. PEMFCs are also the most suitable option for FCEVs. Currently, more than 80% of fuel cells are used in stationary applications, such as cogeneration, back-up, and remote power systems. After more than a decade of intense R&D effort, PEM technology offers high efficiency, durability, and reliability, and costs have fallen rapidly due to mass production. Current research is aimed at system simplification: removing the platinum could avoid complex engineering solutions, while high-temperature (HTPEM) cells can operate on dry hydrogen over 100 °C, removing the need for humidifiers (Dodds et al. 2015).

SOFCs (solid oxide fuel cells) are high-temperature fuel cells used in both large industrial CHP and residential heating systems that have recently grown to reach 10% of global sales. SOFC benefit from the highest electrical efficiency and greater fuel-flexibility, but operate less dynamically than PEMFC due to their temperature requirements. In particular, start-up and shut-down are sensitive operations taking 12 h or more, and so systems tend to run "always-hot," reducing their output when there is little or no demand. Fundamental research has been aimed at improving durability and material fatigue, and there is a trend towards intermediate temperature devices (IT-SOFC) that operate at 500–750 °C. This allows a wider range of materials to be used, lowering costs and improving dynamic performance (Dodds et al. 2015).

MCFCs (molten carbonate fuel cells) are another high-temperature fuel cell used in large industrial CHP and grid-scale electricity production, which have become the market leader for large stationary applications. MCFCs benefit from relatively low capital costs due to non-platinum catalysts and simpler ancillary systems, but suffer from low lifetime and low power density. The key research issue is improving stack lifetimes, which stand at only 5 years due to the aggressive chemistry of the stack and electrolyte leakage, meaning a stack replacement is required half-way through a system's lifetime. Power density is also a research focus, to reduce cell size and thus material costs (Dodds et al. 2015).

PAFCs (phosphoric acid fuel cells) were the first fuel cell technology employed for heating, being used since the 1970s in commercial-scale CHP systems (100–400 kW electric). A small number of demonstration systems have been made at the 1 kW scale, but no residential products have been brought to market (Dodds et al. 2015).

9.3.2 Hydrogen Gas Turbines

While gas turbines adapted to burn gases with high hydrogen content (up to 45%) are commercially available, the same cannot be said for gas turbines capable of burning pure hydrogen. While technological modifications would be moderate, there is currently little demand for such equipment.

In the future, gas turbines able to burn very high shares of hydrogen will be needed for power generation based on the use of fossil fuels and pre-combustion CCS, e.g., in integrated gasification combined cycle (IGCC) power plants with CCS. This application is currently driving RD&D efforts in gas turbines able to burn gases with very high hydrogen content. Gas turbines able to react rapidly to changes in gas quality, especially with respect to hydrogen content, are necessary if blending hydrogen in the natural gas grid (power-to-gas) is to become a means of integrating otherwise-curtailed renewable power into the power sector (IEA 2015).

9.3.3 Compressed Hydrogen Gas

The most common storage system is high-pressure gas cylinders with a maximum pressure of 20 MPa. Carbon-fiber-reinforced composite tanks can reach a pressure of 80 MPa, and these are already being used in prototype hydrogen-powered vehicles.

Nonetheless, the safety of the cylinders under high pressure, especially in highly populated areas, is an important issue of concern and, more importantly, the hydrogen gravimetric density is still low, which is a significant obstacle to its practical applications (Xia 2015).

9.3.4 Liquid Hydrogen Storage in Tanks

Mature options for storage of hydrogen in vessels comprise pressurized and cryogenic tanks, providing hydrogen storage capacities of between 100 kilowatt hours (kWh) (pressurized tanks) and 100 GWh (cryogenic storage). While pressurized tanks have high costs due to their limited energy density, cryogenic tanks provide limited storage time due to the boil-off stream losses, necessary to maintain acceptable pressure levels. An intermediary solution between pressurized and cryogenic hydrogen storage is cryo-compressed hydrogen. In this case, liquefied hydrogen is filled to the tank, but the pressures levels until hydrogen needs to be flared are much higher (up to 35 MPa) compared to cryogenic storage (around 2 to 4 MPa). This allows cryo-compressed hydrogen to be stored for longer time periods (IEA 2015).

9.3.5 Physisorption of Hydrogen and its Storage in Solid Structures

The weak van der Waals interaction between hydrogen and most sorbents with high surface area, described as physisorptive attraction, leads to reversible storage of hydrogen on a wide range of candidates, such as carbon-based materials (nanotubes, fullerenes, and graphene), metal-organic frameworks (MOFs), zeolites, and covalent-organic frameworks (COFs) (Xia 2015).

Storing hydrogen in metal hydrides or carbon nano-structures are promising technology options for achieving high volumetric densities. While metal hydrides are already in the demonstration phase, fundamental research is still needed to better understand the potential of carbon nano-structures (IEA 2015).

9.3.6 Hydrogen-Based Energy Storage

Hydrogen-based energy storage systems cover a broad range of energy storage applications, with a focus on high power capacity and longer storage times in the range of hours to weeks, and even months. Two different storage applications have been identified where hydrogen-based systems can play a vital role—inter-seasonal energy storage and daily arbitrage. Table 9.2 provides an overview of the energy storage systems (IEA 2015).

Inter-seasonal storage allows for temporary shifts in energy supply over weeks or months. For power-to-power applications, hydrogen systems are most promising for inter-seasonal energy storage and surplus VRE integration in the future. Due to the much higher energy density of hydrogen with underground storage compared to

Table 9.2 Power-to-power and power-to-gas systems (IEA 2015)

	Key components	Abbreviation
Power-to-power	PEM electrolyser, underground storage, PEMFC	H_2 PtP PEM/PEM
	Alkaline electrolyser, underground storage, PEMFC	H_2 PtP Alk/PEM
	PEM electrolyser, underground storage, hydrogen open-cycle gas turbine	H_2 PtP PEM/OCGT
	Compressed air storage	CAES
	Pumped hydro storage	PHS
Power-to-gas	PEM electrolyser, NG grid connection, open-cycle gas turbine	H_2 PtG PEM HENG with OCGT
	PEM electrolyser, methaniser, NG grid connection, open-cycle gas turbine	H_2 PtG PEM methane with OCGT

PtP power to power, *PtG* power to gas

PHS or CAES, investment costs are shifted from storage to the conversion technology. An advantage of a power-to-gas system that uses a gas turbine for re-electrification of a natural gas-hydrogen blend is that it enables the use of existing infrastructure (including storage, transportation and re-electrification facilities) whereas alternative energy storage options rely on systems that need to be built from scratch. In addition, blending hydrogen into the gas grid means that the gas turbine can operate in the electricity market in the usual way. Unlike fuel cells in power-to-power electricity storage systems, its annual utilization factor is not constrained by the utilization factor of the electrolyzer producing the hydrogen, which will likely be low if based on low-value, surplus renewable electricity. If gas grids and gas turbines were able to deal with high and variable hydrogen blend shares above 20%, power-to-gas systems might also provide inter-seasonal energy storage by feeding more hydrogen into the blend at times of low VRE availability. Because annual average blend shares could remain at 5% and lower, the higher costs of the hydrogen in the blend would have only a marginal impact on average LCOE. In this manner, existing gas turbines could potentially achieve higher annual load factors while still meeting a fixed carbon budget and leveraging the sunk costs of the existing natural gas infrastructure. The attractiveness of power-to-gas systems as a means of integrating high levels of VRE are economically dependent on declining electrolyzer costs and technically dependent on persistent imbalances in electricity supply and demand (IEA 2015).

Daily arbitrage allows for the shifting of stored electricity from times of low demand to times of high demand, taking advantage of the respective electricity price differential. Operated for daily arbitrage reasons, future hydrogen-based power-to-power storage systems almost achieve the performance of pumped hydro or compressed air storage, if very low-cost electricity is available. As the overall efficiency of the hydrogen-based systems is lower than for PHS and CAES, LCOE is much more sensitive to the cost of electricity. Hydrogen-based storage systems would only work for arbitrage reasons with a very high spread of electricity prices between times of low and high demand (IEA 2015).

Table 9.3 Hydrogen storage materials of recent interest (McWhorter et al. 2011)

Hydrogen release reaction	wt% H_2	g H_2/L	Reversible	T_{des}	ΔH_{des}
$V_{0.85}Ti_{0.1}Fe_{0.05}H_2 \leftrightarrow V_{0.85}Ti_{0.1}Fe_{0.05} + H_2$	3.7	~170[c]	Yes	100	43.2
$MgH_2 \leftrightarrow Mg + H_2$	7.7	109	Yes	300	65.8–75.2
$NaAlH_4 \leftrightarrow 1/3Na_3AlH_6 + 2/3Al + H_2$[a]	3.7	52	Yes	120	37
$Na_3AlH_6 \leftrightarrow 3NaH + Al + 1.5H_2$[a]	1.9	43	Yes	180	47
$2AlH_3 \rightarrow 2Al + 3H_2$	10	147	No	125–175	7.6
$2LiBH_4 + MgH_2 \leftrightarrow MgB_2 + 2LiH + 4H_2$	11.6	96	Yes[e]	300–350	41
$Mg(BH_4)_2 \rightarrow MgH_2 + 2B + 3H_2$	11	113	Partial	230	47
$3Ca(BH_4)_2 \rightarrow CaB_6 + 2CaH_2 + 10H_2$	9.6	108	No	300–350	41.4[f]
$Li_3AlH_6 + 3LiNH_2 \rightarrow 3Li_2NH + Al + 4.5H_2$	7.3	–	No	200–300	38.4
$LiNH_2 + MgH_2 \rightarrow LiMgN + 2H_2 \leftrightarrow LiH + 0.5MgH_2 + 0.5\ Mg(NH_2)_2$	8.1	–	Yes	220–270	32 (step 1)
					51 (step 2)
$xNH_3BH_3 \leftrightarrow [NH_2BH_2]_x + (x - 1)H_2$	6.5	96[d]	No	100	−21.7
$[NH_2BH_2]_x \leftrightarrow [NHBH]_x + H_2$	6.9	–	No	150	−(23.9–15.4)
MOF-5: $H_2 \leftrightarrow$ MOF-5 + H_2[b]	7.1 (excess)	66 (absolute)	Yes	−196	3.8
	10 (absolute)				
MOF-177: $H_2 \leftrightarrow$ MOF-177 + H_2[b]	7.5 (excess)	49 (absolute)	Yes	−196	4.4
	11 (absolute)				

[a]Includes Ti dopant
[b]$P = 7$ MPa
[c]Based on single-crystal density of VH_2
[d]Cumulative, based on NH_3BH_3
[e]15.5 MPa, 600 °C
[f]Calculated

9.4 Materials-Based Hydrogen Storage Solutions

Effective hydrogen storage solutions have been pursued for decades, and materials-based hydrogen storage is a research frontier of much current interest. As shown in Table 9.3, materials-based hydrogen storage media can be subdivided into four distinct categories (in order of increasing binding energy ranging from <10 kJ/mol H_2 to >60 kJ/mol H_2) (McWhorter et al. 2011): (a) adsorbents or sorbents, (b) conventional or interstitial metal hydrides, (c) complex metal hydrides, and (d) chemical hydrogen storage materials. It is imperative to understand that the hydrogen storage properties of a material can be significantly influenced by not only individual sample characteristics including chemical composition and distribution, micro- and macroscopic material structure, etc., but also pressure, temperature, and sample size.

Current research on materials-based hydrogen storage covers both materials-based physical storage and materials-based chemical storage. In the case of physical storage, the efforts have been focused on exploring new porous materials with extra larger surface/pore volume, inducing hydrogen spillover effect, and tailoring reaction enthalpies. Meanwhile, different chemical storage approaches have been attempted to improve the kinetics and/or thermodynamics such as the development of composite hydride systems, nanoconfinement of hydride materials as well as the usage of ionic liquids as hydrogen storage materials or useful additives. Other studies include the applied techniques on solid-state materials towards system integration such as shaping and electrospinning processes; the concept of storing hydrogen in para form for long-term hydrogen storage; and conversion with catalysts to process the normal hydrogen to para-hydrogen (Ren et al. 2017).

9.4.1 Porous Materials-Based Physical Sorption

There is increasing potential to achieve reliable hydrogen storage capacities in porous materials-based systems. Of the existing porous materials, metal-organic frameworks (MOFs) and porous carbonaceous materials have demonstrated attractive performance for hydrogen storage, such as enhanced hydrogen storage capacities, thermostability, and hydro-stability. Up to date, most studies are being focused on ways of achieving better performance of porous materials-based physical sorption at ambient temperature (Jena 2011; Langmi et al. 2014).

It is generally believed that higher BrunauereEmmetteTeller (BET) surface areas and larger pore volumes will enhance hydrogen storage capacities. However, it is not advisable to put too much effort in pursuit of new materials with extra-large surface areas only since H_2 adsorption may be influenced by a range of other factors. For example, in the case of MOFs, H_2 adsorption is also influenced by pore size, open metal sites, ligand functionalization, and catenation (Kozachuk et al. 2014).

On the other hand, reaction enthalpies may be tailed to develop new materials for hydrogen storage. When the enthalpy value is high as in the case of chemisorption materials, large amounts of heat will be released during the charging process, and extra energy will be needed to enable the release of hydrogen for applications. In

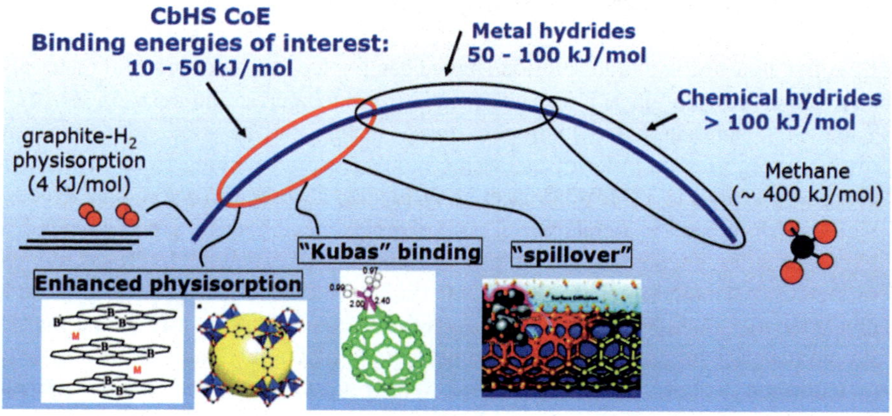

Fig. 9.11 Continuum of hydrogen binding energies for nanostructural design of electronic and bond strain effects (Adapted with permission from Ren et al. 2017 (Elsevier))

contrast, for physisorption materials, the enthalpy of adsorption is too low, and thus low hydrogen storage capacities can be expected at ambient temperature and pressure. To achieve high storage capacities, cryogenic temperature is required to maintain high density of hydrogen molecules on the material. Porous materials like MOFs and carbons usually exhibit very large surface areas and high hydrogen storage capacities at cryogenic temperature. However, at ambient temperature the storage capacities of these materials are low because the adsorption enthalpy is low, implying the materials cannot effectively bind hydrogen on the surface under this at such temperatures. In this regard, several strategies have been explored, for instance, Kubas binding, ionization, polarization, and irradiation of the host materials to enhance the H_2 adsorption (Ren et al. 2017).

It has indicated that that Kubas binding with an enthalpy of 20–30 kJ/mol is ideally required to operate a hydrogen storage system at room temperature. Such binding enthalpy values allow for a sufficiently strong interaction between hydrogen and the material as well as provide less problematic heat management. Kubas interaction involves σ donation from the filled σ-bonding orbital of H_2 molecule to an unfilled d orbital of a transition metal, and at the same time π back-donation from a filled d orbital of a transition metal to the empty σ^* antibonding orbital of the H_2 molecule. In terms of binding strength (see Fig. 9.11), Kubas interaction is at the mid-point of physisorption and chemisorption; thus, this concept can be potentially employed in hydrogen storage systems designed to operate at room temperature. Kubas interaction can be exploited in designing novel hydrogen storage materials that eliminate the limitations arising from physisorption and chemisorption mechanisms (Skipper et al. 2012).

Following on from the concept of Kubas interaction, another alternative strategy for designing materials with improved performance of H_2 storage is ionization of the host materials and hydrogen molecules could lead to enhancement of hydrogen adsorption. For instance, fullerene doped with Na^+ and K^+ as dopants demonstrated that the hydrogen adsorption capacity can be raised from 2.8 to 3.1 wt.%. Charged $Li_2C_2H_4$ and TiC_4H_4 complexes improve the metal bond strength as well as increase the number of non-dissociative H_2 molecules attached. The binding energy of dihydrogen is within the expected range, which enables hydrogen sorption/desorption to occur at ambient conditions (Guo et al. 2011).

Moreover, polarization of the host materials can further improve hydrogen storage in MOFs. For example, the binding strength of molecular hydrogen on either positively or negatively charged fullerenes can be remarkably improved to the desirable range for possible application at room or near ambient temperature. The enhanced binding is associated with the polarization of the hydrogen molecules by the high electric field that is created near the surface of the charged fullerene, and is delocalized, surrounding the entire surface of a charged fullerene. This approach with directly applying an external electric field would lead to a simpler synthesis of materials while enhancing the thermodynamics and reversibility of hydrogen storage. Furthermore, the storage of hydrogen in quasi-molecular form by exposed metal cations occurs via a polarization mechanism where the H_2 molecule is polarized by the electric field associated with point ions (Zhou et al. 2010).

Irradiation of the host materials have also been used to alter the reaction behavior of host storage materials with hydrogen thereby improving their hydrogen storage capacity. Such alteration involves adjustment of the electronic configuration of the storage material. Penetrating radiation can excite and activate the electronic environment at surface as well as interior sites, making the sites ready for bonding with hydrogen in a quasi-manner. For instance, ~15% hydrogen sorption/desorption capability can be increased for carbon nanotubes following irradiation by g-rays. It is possible to control the thermodynamic parameters by controlled ion bombardment (Obolensky et al. 2011).

In addition, a concept of storing hydrogen in porous materials based on the "spillover" mechanism has been proposed and evaluated. Generally, it involves loading of a metal catalyst such as Pd, Pt, and Ni on the porous materials as hydrogen dissociation catalyst to create atomic hydrogen. The dissociated H atoms migrate from metal particles to the support material with a lower activation energy (<10 kJ/mol), and the adsorbed hydrogen migration can occur reversibly from the support surface back to metal particle surface for desorption. One of the most appealing features of such concept is the possibility to operate the hydrogen storage system at room temperature rather than cryogenic temperature (77 K). In order to make the system suitable for room temperature operation, many post-synthetic strategies for porous materials to induce hydrogen "spillover" have been evaluated. For instance, enhancement of hydrogen storage capacity in MOFs is demonstrated by using metals (such as Pt, Pd, Ru, and Ni) together with a support composed of high surface area material (such as activated carbon) to promote the hydrogen "spillover" effect. However, spillover may not occur in all circumstances. It is generally believed that the mechanism involved in carbon hydrogenation, i.e., when hydrogen atoms are chemisorbed, is the transition from sp^2 to sp^3 hybridization (Boukhvalov et al. 2008; Ren et al. 2017).

9.4.2 Metal Hydrides-Based Chemical Sorption

Metal hydrides encompass conventional hydrides such as Mg_2NiH_4 and $LaNi_5H_6$, and complex hydrides such as borohydrides, alanates, and amides. Among the metal hydrides, lightweight hydrides such as MgH_2 and $LiBH_4$ have received immense interest because of their high hydrogen storage densities, both gravimetrically and volumetrically. In general, shortcomings of metal hydrides for hydrogen storage practices are sluggish kinetics of hydrogen desorption/absorption, relatively high thermal stability, irreversible hydrogen storage, and unwanted side gases. Strategies towards enhancing the hydrogen storage characteristics of metal hydrides have geared towards lowering the operating temperature, enhancing the uptake/release kinetics and suppressing the formation of unwanted gases during desorption.

9.4.2.1 Reactive Hydride Composites

Two or more reactive hydrides can be mixed together to form a reactive hydride composite. In such systems, the overall reaction enthalpy is lowered as a result of the formation of a new product through an exothermic reaction while releasing hydrogen through an endothermic reaction of the constituent hydrides. Consider the composite system composed of $LiBH_4$ and MgH_2, the theoretical hydrogen storage capacities of the individual hydrides are 18.5 and 7.6 wt.%, respectively, and their enthalpies are 70 and 75 kJ/mol H_2, respectively. This implies the constituent hydrides must be heated to high temperatures to release hydrogen. MgH_2 alone needs to be heated to 300 °C for desorption of hydrogen at 1 bar, while temperatures in excess of 400 °C are necessary for complete release of hydrogen from $LiBH_4$, and even more extreme conditions are required for its rehydrogenation. On the other hand, the composite system can release 11.4 wt.% H_2 in a single step with a much lower reaction enthalpy of 45 kJ/mol H_2, forming MgB_2, and as such ensuring the thermodynamic feasibility of the reversible hydrogen uptake. For example, synthesized borohydride hydrazinates (e.g., $LiBH_4$–$2NH_2NH_2$) by coordinating NH_2NH_2 with $LiBH_4$ as new hydrogen storage materials can release 13 wt.% H_2 at 140 °C in the presence of Fe–B catalysts; synthesized a new complex of $LiBH_4$–NH_3BH_3 by mechanically milling NH_3BH_3 and $LiBH_4$ in equivalent molar ratio can release 15.7 wt.% hydrogen upon heating to 450 °C. Other hybrid hydrogen storage materials include tetra-*n*butylammonium borohydride [$(n\text{-}C_4H_9)_4NBH_4$], nanoporous hydride, anion substitution, and multifunctional metal borohydrides. Therefore, reactive metal hydride composites could pave the way for new developments in the metal/chemical hydride research space, but it must be realized that although some of these materials are conceptually fascinating their applicability in real hydrogen storage systems are yet to be realized (Ley et al. 2014; Ren et al. 2017).

9.4.2.2 Nanoconfined Hydrogen Storage Materials

Nanoconfinement is one of the important material processing techniques to enhance the thermodynamics and kinetics of those hydrides in hydrogen charging/discharging cycles. The strategy is to achieve much higher reaction surfaces in a manner of decreasing the sizes of the hydride particles. Nanoconfinement-induced improvement of the hydrogen storage performance should fulfill a number of different functions, i.e., preserving nanoscale crystallites under harsh conditions and/or high temperature, while still providing access of hydrogen to the active phase. In addition, since the scaffolds adopted for nanoconfinement do not contribute directly to the hydrogen storage functionality, they will unavoidably decrease the gravimetric hydrogen density, thus reducing normalized performance. Therefore, the host should comply with several requirements (Xia 2015): (a) chemical inertness is important, which could limit reaction with hydrogen storage materials during the synthesis and the following hydrogen storage process; (b) the melting point must be high in order to avoid collapse during thermal treatment for both dehydrogenation and hydrogenation; (c) the bulk density should be sufficiently low to achieve high weight-specific performance; (d) the templates should ideally be highly elastic to accommodate the volume changes in the hydrogen storage materials during hydrogen charge and discharge; and (e) low cost is an important consideration.

Solvent-Mediated Infiltration

Solvent-mediated infiltration (also known as wet or solution infiltration) is based on a solvent which can dissolve the precursors or other ingredients to form a homogeneous solution, in which the porous templates are submerged. The solution must then effectively infiltrate the pores, and the hydrogen storage materials will solidify and/or be synthesized to form nanocrystalline and/or amorphous particles inside the porous structure upon elimination of the solvent during evaporation. The solvent that is adopted should be weakly coordinating to both the targeted materials and the templates, which could help in eliminating the solvent and therefore avoiding possible decomposition or contamination of the hydrogen storage materials during heating. One advantage of solution infiltration is that infiltration could be realized under milder circumstances compared with the method of melt infiltration. Nonetheless, it can be a big challenge to discover an appropriate solvent to dissolve the materials without reacting with either the materials or the porous templates in some cases (Xia 2015).

Melt Infiltration

Melt infiltration is a method arising from the field of ceramic nanoparticles, which is based on the encapsulation of active materials in porous scaffolds through the melting of any precursors or active phases. It could be adopted as an impregnation method if the melting point of the hydrogen storage materials is relatively lower than the decomposition temperature of the templates, depending on the minimization of the overall interfacial energy in the molten species and the host system. Melt infiltration is advantageous because no solvents or precursors are used, which avoids treatments after the nanoconfinement process and possible contamination of the solvent and/or other substrates. Nevertheless, due to the fact that most of the metals or relatively complex hydrides have high reactivity when molten, finding an inert template could be a big challenge. In addition, special strategies such as high hydrogen pressure should be adopted to avoid the decomposition of the hydrogen storage materials during the infiltration process (Xia 2015).

In Situ Synthesis

The encapsulation of hydrogen storage materials could be achieved in situ during the fabrication of the porous scaffolds, which is probably the most perfect way to nanoconfinement. It could avoid the work of searching for templates, the tedious infiltration process, and the blocking of pores during the process of encapsulation. Due to the high reactivity of most hydrogen storage materials, finding suitable proper precursors that could be utilized to simultaneously synthesize both templates and hydrogen storage materials is a big challenge for this method (Xia 2015).

9.4.2.3 Complex Hydrides

Complex hydrides refer to alanates, amides, borohydrides, and their derivatives. Due to their high gravimetric densities and commercial availability, these materials have been widely regarded as viable candidates for practical application.

Alanates

Alanates, which can be found as hexahydroaluminates or tetrahydroaluminates, have received the most attention among the group of complex hydrides. Among alanates, $LiAlH_4$ exhibits the highest storage capacity, i.e., 10.5 wt.%. Alkali metal alanates dehydrogenate at temperatures ranging from 200 to 300 °C, leading to the formation of aluminum metal and the corresponding alkali metal hydrides based on the following two-step equation (Xia 2015):

$$3NaAlH_4 \leftrightarrow Na_3AlH_6 + 2Al + 3H_2 \leftrightarrow 3NaH + 3Al + 9/2H_2 \tag{9.18}$$

In contrast to the alkali metal alanates, $Mg(AlH_4)_2$ exhibits a dehydrogenation process, but without the formation of hexahydroaluminates as an intermediate phase. Further dehydrogenation could be achieved by the decomposition of MgH_2 to form Mg, which will react with aluminum at ~400 °C to produce Al_3Mg. Unfortunately, the exothermic nature of the dehydrogenation reaction of $Mg(AlH_4)_2$ precludes its direct rehydrogenation (Xia 2015):

$$Mg(AlH_4)_2 \rightarrow MgH_2 + 2Al + 3/2H_2 \tag{9.19}$$

In terms of $NaAlH_4$, the first reaction is thermodynamically favorable at temperatures above 33 °C at 1 bar H_2 pressure. The high kinetic barrier, however, leads to higher temperature and hydrogen pressure being required for hydrogen adsorption and desorption. It has been demonstrated that a reversible capacity of 4 wt.% at 160 °C could be achieved over 100 cycles in Ti-catalyzed $NaAlH_4$ (Srinivasan et al. 2004).

Both nanoconfinement and the addition of $TiCl_3$ could lead to superior hydrogen storage performance in $NaAlH_4$. For example, a $TiCl_3$-loaded carbon aerogel to nanoconfine $NaAlH_4$ and the resulting $NaAlH_4$ nanoparticle started to release hydrogen at around room temperature with a peak temperature of only 125 °C, which was attributed to the synergetic effects between nanoconfinement and the catalytic effects of $TiCl_3$. When $NaAlH_4$ was infiltrated into well-tailored porous carbons with pore sizes of 200 nm, 60 nm, 30 nm, and 4 nm via melt infiltration, NaAlH4 started to release hydrogen at a temperature of ~ 100 °C with a peak of 172 °C if the pore size of the adopted templates was reduced to 30 nm, and the apparent activation energy for hydrogen discharge was determined to be 69.7 kJ/mol for NaAlH4 confined in 4 nm porous carbons, reduced by 58 kJ/mol compared to the first step dehydrogenation of its bulk counterpart. $LiAlH_4$ also showed a superior performance when it was infiltrated into Ni-containing mesoporous carbon scaffold via solution infiltration (Xia 2015).

Borohydrides

Borohydrides, also referred to as tetrahydroborate, are identified as a group of hydrides in which hydrogen is covalently bonded to the central atoms (B) in the $[BH_4]^{-1}$ complex anion. Due to their high gravimetric and volumetric hydrogen capacity (e.g., 18.6 wt.% hydrogen for $LiBH_4$), borohydrides have been regarded as highly promising hydrogen storage resources. One of the more popular borohydrides is $LiBH_4$, which decomposes at temperatures above 380 °C and forms LiH and B with a capacity of 13.5 wt.% based on the following reaction (Xia 2015):

$$2LiBH_4 \rightarrow 2LiH + 2B + 3H_2 \tag{9.20}$$

The dehydrogenated $LiBH_4$ could be partially rehydrogenated at 600–650 °C under 70–350 bar of hydrogen. There is a correlation between the dehydrogenation temperature and the electronegativity of the metal in metal borohydrides, and these compounds, which have less ionic character than diborane, are expected to be unstable and release diborane during dehydrogenation. On the other hand, additional compounds have been extensively introduced into borohydrides with the aim of tailoring their respective thermodynamic stabilities. For instance, the reaction enthalpy could be lowered by 25 kJ/mol H_2 with the addition of MgH_2 into $LiBH_4$. Sodium borohydride ($NaBH_4$) was dehydrogenated at a temperature of 534 °C to its parent elements, owing to the lower stability of NaH compared with LiH, which is formed in the dehydrogenation of $LiBH_4$. A similar reaction to $NaBH_4$ was observed for the decomposition of KBH_4, which occurs at an even higher temperatures of ~585 °C (Xia 2015):

$$NaBH_4 \rightarrow Na(l) + B + 2H_2 \tag{9.21}$$

Magnesium borohydride ($Mg(BH_4)_2$) possess a high gravimetric hydrogen capacity of 14.9 wt.% H_2 and favorable thermodynamics of 39.3 kJ/mol H_2 at 1 bar of hydrogen pressure. The dehydrogenation of $Mg(BH_4)_2$ takes place in more than two steps in the temperature range of 290–350 °C, leading to the formation of $MgB_{12}H_{12}$, MgH_2, and MgB_2. The dehydrogenated $Mg(BH_4)_2$ could be rehydrogenated at 270 °C under 400 bar after 48 h of treatment (Xia 2015):

$$Mg(BH_4)_2 \rightarrow 1/6MgB_{12}H_{12} + 5/6MgH_2 + 13/6H_2 \rightarrow MgB_2 + 4H_2 \quad (9.22)$$

Transition metal borohydrides generally decompose at much lower temperatures than the alkali metal borohydrides. For example, decomposition of $Y(BH_4)_3$ could occur at temperatures ranging from room temperature to around 500 °C, and rehydrogenation could be achieved at 260 °C under 35 bar of hydrogen (Xia 2015):

$$Y(BH_4)_3 \rightarrow YH_3 + 3B + 9/2H_2 \rightarrow YH_2 + 3B + 10/2H_2 \quad (9.23)$$

Furthermore, the dehydrogenation temperature for a bialkali metal borohydride ($LiK(BH_4)_2$) is around the average of the dehydrogenation temperatures for their corresponding mono alkali borohydrides, i.e., $LiBH_4$ and KBH_4.

Various borohydrides, such as $LiBH_4$, $NaBH_4$, $Ca(BH_4)_2$, and $Mg(BH_4)_2$, have been successfully nanoconfined in porous templates, e.g., carbon-based scaffolds, metal oxides, metal-organic frameworks (MOFs), SBA-15, and polymer matrix (Xia 2015).

Amides

Metal-N-H compounds, consisting of Li-N-H, Li-B-N-H, Li-Mg-N-H, Li-Al-N-H, and the like, have outstandingly high gravimetric hydrogen density, which be able to store 11.5 wt.% hydrogen. Generally, it is accepted that the hydrogenation and dehydrogenation of Li_3N could proceed by the following reversible pathways (Xia 2015):

$$Li_3N + 2H_2 \rightarrow Li_2NH + LiH + H_2 \quad (9.24)$$

$$Li_2NH + LiH + H_2 \rightarrow LiNH_2 + 2LiH \quad (9.25)$$

There is a significant hydrogen uptake in Li_3N from 170 to 210 °C. With respect to the first reaction, the second step can absorb and desorb hydrogen at a much lower temperature due to the smaller enthalpy change. Nevertheless, the temperature of complete hydrogen desorption in the fully hydrogenated Li_3N is still above 250 °C. Therefore, it is not suitable for practical applications. In order to enhance the reaction kinetics and decrease the desorption temperature of the metal-N-H system, the mechanism behind the conversion of hydride and amide should be clearly understood. To date, there have been several mechanism models proposed (Xia 2015): (a) the multimolecule reaction mechanism, which can be described as a solid-solid reaction based on the strong affinity between positively charged H+ in amide and negatively charged H- in hydride; (b) ion migration model, suggesting that the presence of Frenkel defect pairs is crucial in the desorption and hydrogenation processes, with the mobility of small ions promoting the kinetics of this process; and (c) two-step ammonia-mediated mechanism:

$$2LiNH_2 \rightarrow Li_2NH + NH_3 \quad (9.26)$$

$$LiH + NH_3 \rightarrow LiNH_2 + H_2 \tag{9.27}$$

According to this, the combination of $LiNH_2$ and LiH, with proper treatment, such possible ball milling or the like, is necessary in order to better remove the ammonia and reduce the NH_3 contamination.

Among the other amides, Li-Mg-N-H is also a potential alternative as a promising hydrogen storage material owing to its good reversibility, better thermodynamic properties, and comparatively high gravimetric density. The molar ratio between Mg $(NH_2)_2$ and LiH is 1:2, and 5.5 wt.% H_2 is adsorbed/desorbed following the equation below (Xia 2015):

$$Mg(NH_2)_2 + 2LiH \rightarrow Li_2Mg(NH)_2 + 2H_2 \tag{9.28}$$

This reaction can occur at 90 °C under 1 atm H_2 with desorption enthalpy of 39 kJ/mol with poor kinetic properties. Methods for decreasing the particle size and improving the mixing of reagents are beneficial for attaining better kinetic properties.

Nanoconfining amides used LiN3 as the precursor have been attempted to produce confined Li_3N in mesoporous carbon. The nano β-Li3N was well stabilized by the mesoporous carbon. The resulting Li_3N/carbon composites showed a reversible hydrogen storage capacity of 1.8 wt.%, which is equal to around 9 wt.% per mass of Li_3N (Xia 2015).

Chemical Hydrides

Representative chemical hydride materials include ammonia borane (NH_3BH_3, denoted as AB), metal amidoborane (denoted as MAB), and the related metal borohydride ammoniate (denoted as MBA). Ammonia borane is a special molecular solid consisting of NH_3 and BH_3 building blocks held together by intermolecular dihydrogen bonds and dative B-N bonds. It has a hydrogen capacity of 19.6 wt.% and could release all bonded hydrogen, yielding ~6.5 wt.% hydrogen below 100 °C through the following equation (Xia 2015):

$$NH_3BH_3 \rightarrow NH_2BH_2 + H_2 \rightarrow NHBH + H_2 \rightarrow BN + H_2 \tag{9.29}$$

Although AB could quickly decompose when heated to 105 °C, the dehydrogenation kinetics at moderate temperature is slow and normally has a long induction period. Another major problem for using AB as a solid-state hydrogen storage alternative is the simultaneous release of diverse volatile by-products, e.g., aminoborane, diborane, borazine, and aminodiborane, during dehydrogenation.

MAB is one of a series of derivatives of ammonia borane that is synthesized by replacing one of the hydrogen atoms bonded with N in AB with an alkali or alkaline earth element. Due to the electron contribution from the metal to the N, B-H bonds are weakened in MABs, which leads to improved reactivity of the hydridic H. Therefore, MABs typically exhibit improved dehydrogenation kinetics with suppressed volatile by-products compared with the parent AB. For example,

$LiNH_2BH_3$ starts to desorb hydrogen at ~84 °C with a peak temperature of ~ 90 °C, and 11 wt.% hydrogen could be released at a temperature of 91 °C without detection of any by-products. Nonetheless, metal amidoboranes are kinetically unstable, so that they may undergo spontaneous decomposition during storage and are a risk for transportation. In addition, the efficient regeneration of the spent fuel is a big challenge for MABs as practical hydrogen storage materials.

MBAs with high hydrogen capacities often exhibit much lower dehydrogenation temperatures compared with their parent metal borohydride due to the dihydrogen elimination of hydrogen. For instance, $LiBH_4{\cdot}4/3NH_3$ with equivalent hydridic and protic hydrogen could release 17.8 wt.% hydrogen in a closed system between 135 and 250 °C, while a temperature above 380 °C is required for the dehydrogenation of $LiBH_4$. Furthermore, ammonia could be used to stabilize unstable metal borohydrides with high electronegativity, e.g., $Al(BH_4)_3{\cdot}nNH_3$, which is induced by the direct coordination of ammonia with the metal (Xia 2015).

Since AB can be easily dissolved in tetrahydrofuran, methanol, and diethyl ether, the main method for the nanoconfinement of ammonia borane is solvent infiltration. AB could be encapsulated in mesoporous silica. The resulting nanoparticles of ammonia borane present significant enhancements, such as enhanced dehydrogenation kinetics and thermodynamics. Interfacial catalytic effects of the terminal SiO-H groups in the silica templates were proposed as the major contribution to the enhancement of the dehydrogenation performance. This finding points the way to a suitable strategy for enhancing the hydrogen storage performance of chemical hydrides. Subsequently, ammonia borane was impregnated into a coherent carbon cryogel via the solution infiltration process. As a result of the nanoconfinement effects of the carbon cryogel, a dehydrogenation temperature below 90 °C was observed, as well as the suppression of volatile borazine. Moreover, the adoption of catalysts, such as Li and Pt, could exert a synergistic effect towards enhancing the hydrogen storage performance of ammonia borane. Following these works, various templates, such as carbon-based porous hosts, MOFs, metal oxides, and polymer, have been introduced to investigate the hydrogen storage properties of ammonia borane induced by the synergistic effects of nanoconfinement and the catalytic effects of the particular templates. For instance, 10.2 wt.% hydrogen could be released in only 10 min from AB inside JUC-32-Y at 95 °C, and it could desorb 8 wt.% hydrogen at 85 °C. By comparison, no hydrogen could be detached from neat AB at this temperature (Xia 2015).

Mg-Based Hydrides

Mg-based materials have been widely investigated as solid-state hydrogen storage materials. The large hydrogen capacities (e.g., 7.7 wt.% for MgH_2), great abundance of Mg in the Earth's crust, and the low cost of Mg constitute the main advantages of Mg-based materials as hydrogen storage candidates. Unfortunately, the high activation energies of hydrogenation (100 kJ/mol) and dehydrogenation (160 kJ/mol) leads to the extremely slow formation of MgH_2 from Mg and H_2, and very slow reversible dehydrogenation from MgH_2. Therefore, a temperature higher than 300 °C is required for common Mg to absorb hydrogen, and an even higher temperature is needed for the reversible dehydrogenation to proceed because of the thermodynamic stability. The sluggish kinetics is mainly attributed to the low hydrogen diffusion coefficient (4×10^{-13} and 1.5×10^{-16} m^2/s for bulk Mg and MgH_2, respectively), a sluggish dissociation rate of hydrogen on Mg (with 432 kJ/mol required to split molecules of hydrogen), and the possible synthesis of MgO (Xia 2015).

Compared with the bulk substrates, a large part of the atoms would be located on the surface when the particle size is made sufficiently small, and therefore surface adsorption and desorption would dominate the processes of both hydrogenation and dehydrogenation. As a result, the features of both hydrogenation and dehydrogenation could be changed via nanostructuring of Mg-based hydrides. For example, a desorption enthalpy of 37.55 kJ/mol H_2 was achieved for MgH_2 nanowires ~0.7 nm in diameter, which is half of the value for their bulk counterpart. Moreover, the dehydriding temperature could be only 13.85 °C under an equilibrium pressure of 1 atm, and reversible hydrogen storage may be achieved in MgH_2 nanowires even at ambient temperature. At 300 °C, the Mg nanowires with the smallest diameter absorbed 7.60 wt.% and desorbed 6.77 wt.% hydrogen within 30 min and 15 min, respectively, while the nanowires with the largest diameter adsorbed 6.50 wt.%. At the temperature of 85 °C, the colloidal MgH_2 starts to absorb hydrogen under a hydrogen pressure of 300 Pa and fully desorbed its hydrogen within 5.3 h. The reduced stability of the Mg nanoparticles could have been induced by the stabilizer and/or the high ratio of surface to volume atoms. In addition, nanoconfinement of Mg-based hydrides has been achieved by melt infiltration for Mg nanoparticles nanoconfined in nanoporous carbon, nanoconfine Mg/MgH_2, and the formation of MgH_2 via wet infiltration (Xia 2015).

Nanoconfinement may become an effective and efficient tool to improve the hydrogen storage performance of complex hydrides based on the following advantages (Xia 2015): (a) the diffusion distance of hydrogen and the thickness of reactive interfaces during hydrogenation and dehydrogenation could be significantly decreased owing to the decrease of particle size down to nanometer scale, leading to tremendously improved hydriding and dehydriding kinetics and the decrease of temperature for hydrogen storage; (b) the significant reduction of particle size results in a low mean coordination number of surface atoms, and therefore an excess of surface energy, which could effectively modify the thermodynamics parameters of complex hydrides, i.e., enthalpy and entropy, towards superior hydrogen storage performance; (c) the porous scaffolds adopted in nanoconfinement could physically prevent the particle growth, aggregation, and phase segregation upon cyclic hydrogenation and dehydrogenation, which could enhance the cycling stability of complex hydrides. However, due to the introduction of the templates for nanoconfinement, which are often inactive for hydrogen storage, the gravimetric and volumetric hydrogen densities of the system are significantly reduced. Moreover, the synthetic process is often tedious and the nanostructured hydrogen storage materials are unstable, which makes them difficult to scale up. Therefore, a template with high surface area, large pore volume and ordered porous structure, which could facilitate the effective nanoconfinement of hydrogen storage materials, is required. In

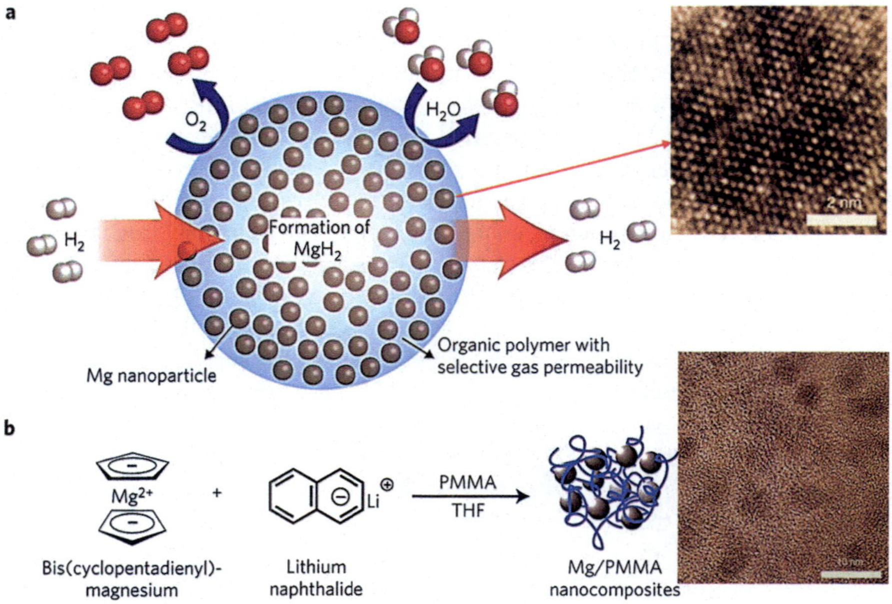

Fig. 9.12 Pictorial representation of hydrogen permeation through MgH_2 nanoparticles buried inside PMMA gas selective polymer (Modified with permission from Jeon et al. 2011 (Springer Nature)): (**a**) Schematic of hydrogen storage composite material: high-capacity MgNCs are encapsulated by a selectively gas-permeable polymer; (**b**) Synthetic approach to formation of Mg NCs/PMMA nanocomposites

order to preserve the superior hydrogen storage performance induced by the nanosize effects, an effective method to achieve full encapsulation of complex hydrides with high loading is highly desirable.

9.4.2.4 New Generation Lightweight Composite Structures

The employment of lightweight hybrid hydrides has the potential to achieve better performance for hydrogen storage. To synthesize the lightweight compounds, the choice of elements should come from the elements with atomic number below 20. However, the majority of compounds from this category are intrinsically very air/moisture sensitive, and as a result the induced degradation is a critical issue that needs to be addressed. To overcome this problem, the air sensitive magnesium hydride can be embedded in a gas selective polymethyl acrylate (PMMA). The pictorial representation is shown in Fig. 9.12. Since the embedded hydride was prepared in the order of quantum dots, the reactive and fast reaction kinetics was claimed on the resulting composite (Jeon et al. 2011). Such concept could be applied to other high-capacity hydride materials with thermo-dynamical shortcomings (Ren et al. 2017).

9.4.3 Ionic Liquids as Hydrogen Storage Materials

The unique physicochemical properties of ionic liquids make them attractive for hydrogen storage applications. The basic idea is to find a low molecular weight but hydrogen rich ionic liquid that can serve as a hydrogen storage material. Furthermore, the H_2 release rate, operating temperature, and H_2 purity are other key criteria that have stimulated intense research efforts towards development of new ionic liquid compounds with improved hydrogen storage performance. For instance, an ionic liquid of methylguanidinium borohydride $[N_3H_8C]C^+BH_4$ can release 9.0 wt.% H_2 under both thermal and catalytic conditions. The combination of guanidinium cation and the octahydrotriborate anion yields guanidinium octahydrotriborate with a H_2 storage capacity of 13.8 wt.%. An synthesized N-substituted amine-borane ionic liquid are in good yield using silyl protecting groups, generating a highly pure (1% impurity) H_2 gas (Rekken et al. 2014; Chen et al. 2015).

Ionic liquids can also be used as catalytic or supporting solvents. When ionic liquids are added, the hydrogen release and suppression of induction time are enhanced on hydride-based H_2 storage system. Moreover, a mixture of selected chemical hydrides and ionic liquids would achieve synergistic effects. While ionic liquids (as green solvents) afford an inert reaction medium which stabilizes polar transition states and substantially fastens the decomposition of chemical hydrides, the catalytic effects from the used ionic liquids also favor the H_2 release rate. In the decomposition of formic acid by employing amine-functionalized imidazolium-ionic liquid, it enables better control of the equilibrium of formic acid with H_2/CO_2. In fact, ionic liquids can act as supporting solvents which enable improved performance, such as when heteronuclear AuePd nanoparticles are immobilized in the functionalized ionic liquid $[C_2OH_{mim}][NTf_2]$. Owing to the unique solubilization properties together with supporting effects in dehydrogenation reactions, ionic liquids have been ear-marked as promising contenders for hydrogen storage and also as valuable additives to hydrogen storage systems. Compared to the currently available compressed or cryogenic hydrogen storage options, the advantage of storing hydrogen in ionic liquids or their integrated systems is that the system does not require compression or cryogenic temperatures. However, for ionic liquid to be practically applicable issues related to their high cost of decomposition reaction and the compromised gravimetric capacity by the weight addition of ionic liquids to the H_2 storage system would need to be addressed. The latter drawback has been claimed to be fully compensable in some cases since the ionic liquids-based systems gave much higher hydrogen yields in shorter reaction times than the dehydrogenation of neat hydrogen-rich materials (Ren et al. 2017).

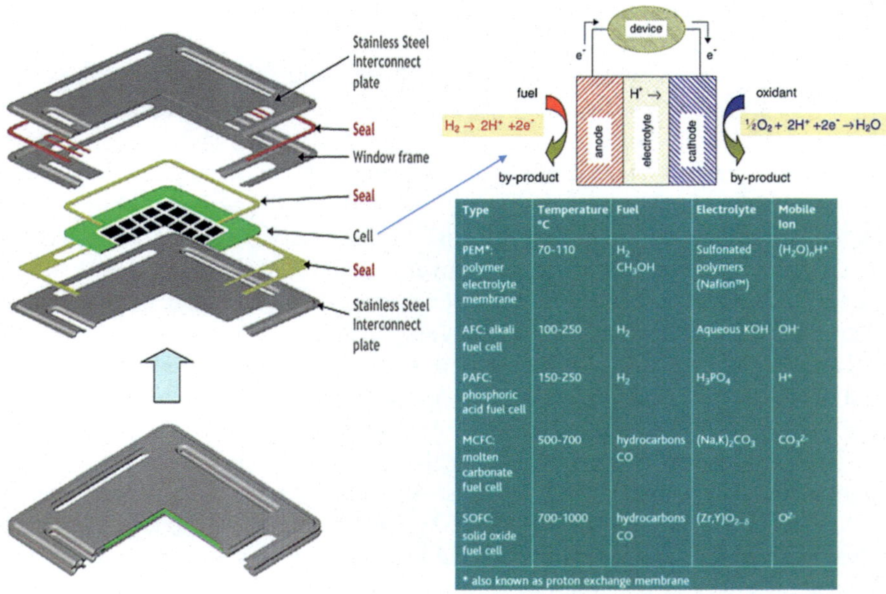

Type	Temperature °C	Fuel	Electrolyte	Mobile Ion
PEM*: polymer electrolyte membrane	70-110	H_2 CH_3OH	Sulfonated polymers (Nafion™)	$(H_2O)_nH^+$
AFC: alkali fuel cell	100-250	H_2	Aqueous KOH	OH^-
PAFC: phosphoric acid fuel cell	150-250	H_2	H_3PO_4	H^+
MCFC: molten carbonate fuel cell	500-700	hydrocarbons CO	$(Na,K)_2CO_3$	CO_3^{2-}
SOFC: solid oxide fuel cell	700-1000	hydrocarbons CO	$(Zr,Y)O_{2-\delta}$	O^{2-}

* also known as proton exchange membrane

Fig. 9.13 Schematic of a conventional fuel cell employing a proton conducting electrolyte. Electro-oxidation of H_2 occurs at the anode, creating protons that then migrate across the electrolyte and participate in the electro-reduction of O_2 at the cathode. Electric power generation results from the flow of electrons through an outside circuit, necessary for charge balance. The overall reaction is $H_2 + ½O_2 \rightarrow H_2O$, for which there is a strong chemical driving force (Modified with permission from Haile 2003 (Elsevier))

9.5 Materials for Fuel Cells

The primary components of a fuel cell are an ion conducting electrolyte, a cathode, and an anode, as shown schematically in Fig. 9.13. Together, these three are often referred to as the membrane-electrode assembly (MEA), or simply a single-cell fuel cell. In the simplest example, a fuel such as hydrogen is brought into the anode compartment and an oxidant, typically oxygen, into the cathode compartment. There is an overall chemical driving force for the oxygen and the hydrogen to react to produce water. Direct chemical combustion is prevented by the electrolyte that separates the fuel (H_2) from the oxidant (O_2). The electrolyte serves as a barrier to gas diffusion, but will let ions migrate across it. Accordingly, half-cell reactions occur at the anode and cathode, producing ions which can traverse the electrolyte. For example, if the electrolyte conducts oxide ions, oxygen will be electro-reduced at the cathode to produce O^- ions and consume electrons, whereas oxide ions, after migrating across the electrolyte, will react at the cathode with hydrogen and release electrons (Haile 2003):

$$\text{Cathode } ½O_2 + 2e^- \rightarrow O^2 \quad (9.30)$$

$$\text{Anode } H_2 + O^{2-} \rightarrow H_2O + 2e \quad (9.31)$$

$$\text{Overall } ½O_2 + H_2 \rightarrow H_2O \quad (9.32)$$

Analogous cathode and anode reactions for proton conducting electrolyte are (Haile 2003):

$$\text{Cathode } ½O_2 + 2H^+ + 2e^- \rightarrow H_2O \quad (9.33)$$

$$\text{Anode } H_2 \rightarrow 2H^+ + 2e \quad (9.34)$$

The flow of ionic charge through the electrolyte must be balanced by the flow of electronic charge through an outside circuit, and it is this balance that produces electrical power. In electrolytes, protons, hydronium ions, hydroxide ions, oxide ions, and carbonate ions are mobile. Because ion conduction is a thermally activated process and its magnitude varies dramatically from one material to the next, the type of electrolyte, which may be either liquid or solid, determines the temperature at which the fuel cell is operated. Typical fuel cell electrolytes are also listed in Fig. 9.13, along with the mobile ionic species, temperatures of operation, and fuels typically utilized. For reasons of electrode activity (which translates into higher efficiency and greater fuel flexibility), higher temperature operation is preferred, but for portable (intermittent) power applications, lower temperature operation is typically favored as it enables rapid start-up and minimizes stresses due to thermal cycling. In addition, solid electrolyte systems obviate the need to contain corrosive liquids and thus solid oxide and polymer electrolyte fuel cells are preferred over alkali, phosphoric acid, or molten carbonate fuel cells (Haile 2003).

9.5.1 Solid Oxide Fuel Cells

Solid oxide fuel cells (SOFCs) have shown tremendous reliability when operated continuously. However, high-temperature operation required remains a serious challenge for SOFCs. High temperatures preclude the use of metals, which typically have lower fabrication costs than ceramics, for any of the non-electrochemical components of the fuel cell and also increase the likelihood of cracks developing upon thermal cycling. The so-called planar SOFCs are particularly prone to cracking at glass seals that keep anode and cathode chambers isolated from one another in a stacked configuration. While SOFCs offer tremendous fuel flexibility, allowing a variety of hydrocarbon fuels to be utilized, the less reactive fuels must typically be internally steam-reformed, that is, reacted with H_2O in the anode chamber to produce CO and H_2, which can subsequently be utilized in the electrochemical reactions. To address the challenge of high-temperature operation, significant efforts have been directed towards the development of electrolytes with high ionic conductivity at approximately 600 °C. Further reductions in the temperature of operation are generally deemed undesirable because of concomitant losses in electrode activity. Three candidate electrolytes have emerged (Haile 2003): doped ceria (CeO_2), doped lanthanum gallate ($LaGaO_3$), and doped barium zirconate ($BaZrO_3$), with the first two being oxygen ion conductors and the latter a proton conductor.

A single chamber fuel cell relies on carefully selected anode and cathode catalysts to produce well controlled half-cell reactions (Haile 2003):

$$\text{Anode } CH_4 + \tfrac{1}{2}O_2 \rightarrow CO + 2H_2(\text{chemical}) \tag{9.35}$$

$$H_2 + O^{=} \rightarrow H_2O + 2e^{-}(\text{electrochemical}) \tag{9.36}$$

$$CO + O^{=} \rightarrow CO_2 + 2e^{-}(\text{electrochemical}) \tag{9.37}$$

$$\text{Cathode } \tfrac{1}{2}O_2 + 2e^{-} \rightarrow O^{=}(\text{electrochemical}) \tag{9.38}$$

Ideally, simple chemical oxidation of the hydrocarbon, which would yield CO_2 and H_2O, does not take place. Instead, partial oxidation occurs at the anode and the products of this reaction are consumed electrochemically, while O_2 is consumed electrochemically at the cathode. As a consequence, the fuel and oxidant need not be physically separated. Impressive performance of 467 mW/cm^2 at 500 °C has been reported for cells operating on methane/air mixtures and incorporating a PdNi–CeO_2 cermet anode and $Sm_{0.5}Sr_{0.5}CoO_3$ cathode. Because complications resulting from sealing are eliminated, the SCFC greatly simplifies system design and enhances thermal and mechanical shock resistance, thereby allowing rapid start-up and cool-down. In general, both conventional stack configurations, in which the fuel cell anode and cathode are placed on opposing sides of the electrolyte, and strip configurations, in which thin strip anodes and cathodes are placed on the same side of the fuel cell electrolyte are possible in a SCFC (Haile 2003).

In order to improve the overall performance of SOFCs, the major challenge is the choice of material based on material properties, i.e., low electronic resistance of electrodes, catalysts, sealants and interconnects, and low ionic resistance of electrolytes, as shown in Figs. 9.14 and 9.15. Furthermore, low polarization losses are expected with high electrochemical activity (rapid kinetics) at lower temperatures. The processing of the fuel cell materials also influences the performance of SOFCs due to associated phase evolution, grain size, and distribution of phases (even purity and residual stresses). Thus, the optimization of various synthesis processes for such materials used as electrodes, electrolytes and interconnects are of primary concern in taking SOFC science to real-life applications (Mahato et al. 2015).

9.5.2 Polymer Electrolyte Membrane Fuel Cells

Polymer electrolyte membrane (PEM) fuels cells rely on the relatively facile transport of hydronium ions through hydrated regions of a sulfonated polymer. Their high conductivity enables operation at close to ambient temperatures, a trait well-suited to

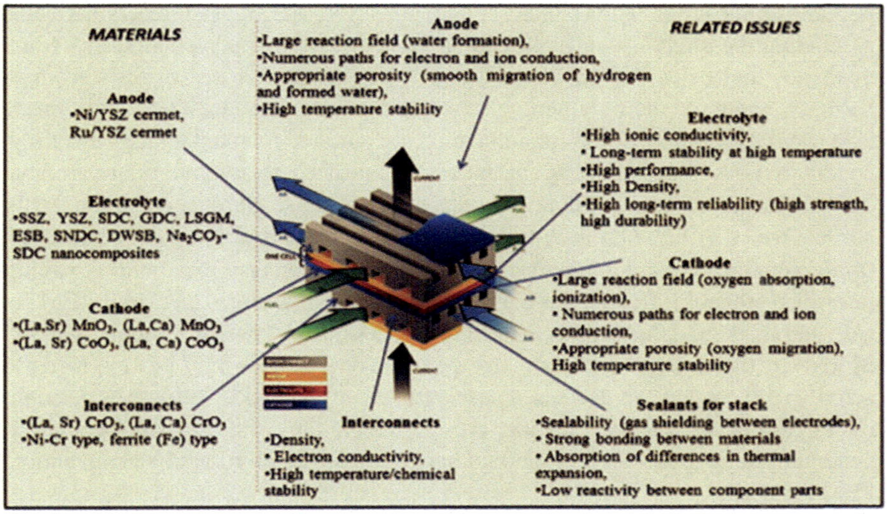

Fig. 9.14 Materials and related issues for SOFCe (Adapted with permission from Mahato et al. 2015 (Elsevier))

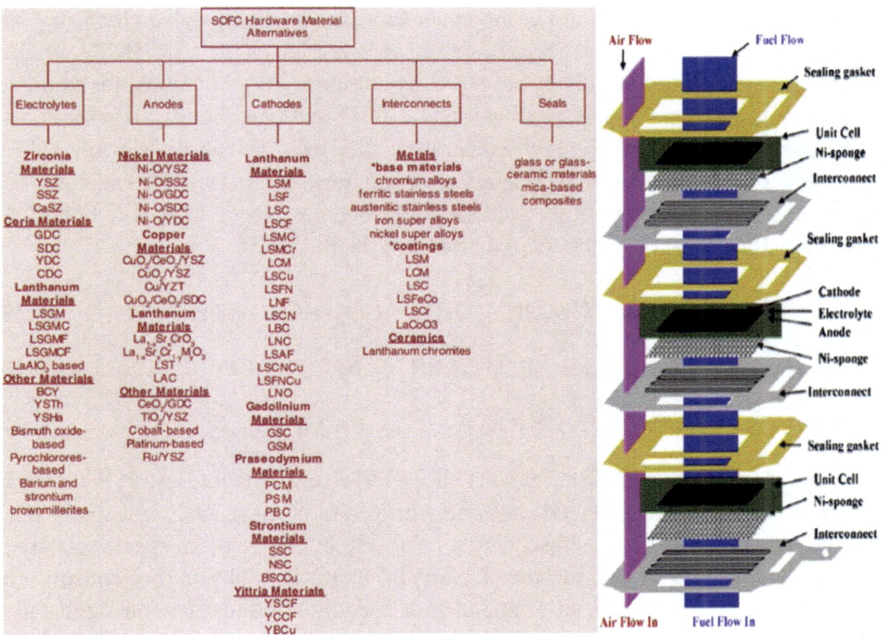

Fig. 9.15 Various materials used in SOFC (Modified with permission from Mahato et al. 2015 (Elsevier))

portable applications. The challenges here arise from the sluggish reaction kinetics, precluding the direct use of hydrocarbon fuels other than perhaps methanol (direct methanol fuel cell). An additional set of complications arises directly from the hydrated nature of the polymeric electrolyte material. Because the mobile species is hydronium, high levels of humidification are required to maintain high electrolyte conductivity, yet this must be balanced by the need to remove water from the cathode and prevent flooding of the electrocatalyst during operation. Other consequences of utilizing a hydrated polymer as the electrolyte are (Haile 2003): (a) irreversible dehydration damage if thermal excursions are experienced; (b) high rates of methanol diffusion from anode to cathode in a direct methanol PEM fuel cell; and (c) degradation in mechanical properties due to "swelling" upon hydration. Moreover, the benefits of "warm" temperature operation, which include increased activity of the catalysts, decreased susceptibility of the anode catalyst to poisoning by impurities in the fuel stream, greater integration with high-temperature H_2 generators, and greater ease of thermal management, are all forgone in conventional PEM fuel cells.

Examples of "water-free" proton conducting membranes include polymer/inorganic composites (e.g., Nafion + silica + phosphotungstic or silicotungstic acid), polymer/acid blends (e.g., polybenzimidazole + phosphoric acid), and polymer/heterocycle blends (e.g., sulfonated polyaromatic polymers + imidazole, pyrazole, or benzimidazole). Electrolyte is based on inorganic (non-polymeric) solids known as solid acids. Solid acids are compounds such as $CsHSO_4$ whose chemistry and properties are intermediate between those of a normal acid (e.g., H_2SO_4) and a normal salt (e.g., Cs_2SO_4). They are typically comprised of oxyanions, for example, SO_4 or SeO_4, that are linked together via O–H. . .O hydrogen bonds. In parallel with electrolyte development for low-temperature fuel cells, advanced electrocatalysts with the objectives of enhancing reaction kinetics, decreasing Pt loadings, and mitigating anode susceptibility to CO poisoning have been pursued. For direct methanol fuel cells, in which the reactions are (Haile 2003):

$$\text{Anode} : CH_3OH + H_2O \rightarrow CO_2 + 6H^+ + 6e^- \tag{9.39}$$

$$\text{Cathode} : 1\tfrac{1}{2}O_2 + 6H^+ + 6e^- \rightarrow 3H_2O \tag{9.40}$$

$$\text{Overall} : CH_3OH + 1\tfrac{1}{2}O_2 \rightarrow CO_2 + 3H_2O \tag{9.41}$$

The most effective catalyst remains alloyed Pt-Ru. Methanol is adsorbed onto Pt clusters and then fragmented into dehydrogenation products, essentially CO. Electro-oxidation of CO is subsequently catalyzed by oxygen-like species absorbed onto neighboring Ru sites. In order for this bifunctional catalysis mechanism to be effective, Pt and Ru must be arranged in a specific configuration on the catalyst surface. Almost by definition, alloys, with their random distribution of atomic species, are ill-suited to provide precise chemical arrangements. In addition, intermetallic compounds, such as PtBi and $PtBi_2$, have been developed, which have highly regular and thermodynamically stable surface arrangements, as fuel cell catalysts (Haile 2003).

9.5.3 Materials Selection for Fuel Cells

The progress in the material selection of the fuel cells, such as solid oxide fuel cell (SOFC) components elicits extreme complexity, as each of the layer must be designed strictly keeping the adjoining components, and fuel (and its purity) in mind. So, there is no single material that suffices the concerns that are raised in each of the components of a fuel cell system. The design of anode requires enhanced triple phase boundaries (TPB: gas, electrolyte and anode), while minimizing coking or sulfur poisoning in order to enhance the efficiency of charge transfer for SOFC as an example. This class includes metals (Ni, Cu), cermets (Ni/Cu-YSZ), perovskites, and pyrochlores. It has been observed that dissociative adsorption of reactants and surface diffusion also strongly affects the reaction rate at anode. Fuel composition and over-voltage determines the impedance response at high frequency, whereas, P_{H2O}, combination of proton and oxide ion, and the migration of electron/oxide ions and protons dictate the impedance response at lower frequency. Also, it has been observed that the finer microstructure and lower sintering temperature result in enhancing the performance of anode. The ceria impregnated Ni–YSZ appears to be one of the high-performing anodes as it possesses high melting point (1453 °C), and sustain resistance to enhanced sulfur poisoning (from hydrocarbon fuels) (Mahato et al. 2015).

Electrolytes require high ionic diffusion (thus high working operation temperature for SOFC) with minimum electronic conduction. Typically, open lattice structures: such as fluorite, perovskites, brownmillerite-like phases, and pyrochlore, are used as solid electrolytes for SOFC. Since the phase and thermal stability becomes a stringent requirement, when the SOFC has to work at elevated temperatures for durations of 40,000 h or more, the use of ceramic materials becomes the only choice. In addition, processing of dense electrolytes is required to allow only oxide ion (or proton) diffusion, while restricting gas flow or recombination of ions, so that the short circuiting can be minimized and the highest power density can be drawn. Utilization of acceptor dopant cation of similar size as of host cation (such as Sc ion doping in ZrO_2 lattice) has shown enhancement in the ionic conductivity of solid electrolyte. Thinning the electrolyte layer or utilizing composite thin-layer laminates (~tens of nm each) has also evinced enhanced ionic conduction of the electrolyte. Furthermore, co-doping has shown reduction in the defect association, and grain refinement has also elicited enhanced grain-boundary conduction. The potential of achieving enhanced ionic conduction with low working temperature appears to be in co-doping of ScSZ along with GDC (Mahato et al. 2015).

Cathodes require gas-diffusion type porous electrodes, and their electrochemical kinetics is governed by over potential, and other factors such as geometry of active surfaces. Both electrochemical kinetics and impedance analysis has contributed in enhancing the understanding on surface adsorption and transport of gases, interfacial electrokinetics of reactions, and performance of cathode. Popular choice among cathode materials are perovskites owing to their high p-type electronic conductivity and high oxide ion conductivity under oxidizing atmospheres. Furthermore, the minimal mismatch in the CTE value with electrolyte, high permeability of oxygen molecules to cathode/electrolyte interface, chemical compatibility with interconnects/sealants and ease of processing porous structure make them a perfect choice. LSCF with a thin LSM coating appears to be one of the promising cathode materials. Further, it is also observed that the cathodic polarization can lead to enhancement in the cathodic activity as well (Mahato et al. 2015).

Interconnects are the barrier materials physically isolating reducing and oxidizing regions of the SOFC. Very low area-specific resistance (of <0.1 Ω cm^2), dimensional, chemical, and microstructural stability at operating temperatures of 800–1000 °C, leak-proof and impervious to oxygen and hydrogen, matching CTE with adjoining components, chemical inertness, and load-bearing strength are the main requirements for a successful interconnect. Thermally grown $(Mn, Co)_3O_4$ spinel on Crofer22 is claimed to be the best interconnect for SOFC that can sustain both oxidizing and reducing atmospheres without any significant degradation in performance for prolonged operational hours (Mahato et al. 2015).

Sealants are present to prevent the leakage of fuel (at anode) or oxidizer (at cathode), or even mixing of fuels, while being exposed to operating temperatures of 800–1000 °C. The second prime requirement of sealants is to possess a CTE similar to those of other SOFC components to avoid cracking. The tubular SOFC designs can be made seal-less (or away from high-temperature exposure areas), but planar-cell or other designs may have problems relating to oxide scaling or chemical instability especially for metallic seals at high temperatures. Seals also require certain flexibility to deform and hermitically isolate the region in response to compressive stress. At operational temperature, the thermal exposure can also induce self-healing of seals (in case there are cracks in the seal). Mica-based glass seals (with absence of potassium) have shown good sealing of ferritic steel interconnects. Brazed seals offer hermatic sealing owing to their inherent wetting behavior, but have limited application due to their conductive nature. In addition, certain barium and boron-free glasses have also shown successful joining of YSZ and Crofer 22 APU via pressureless exposure to 900 °C. Metallic alloys have not prospered as potential sealants owing to their poor oxidation resistance, high volatility, and high cost. Also, the oxidation layer formed during service induces porosity which leads to cracking/spallation of film. In order to prevent this, the protective coatings of nitride, perovskite and spinel have been utilized. Thus, achieving high-performance SOFCs operating at intermediate temperatures (500–800 °C) for prolonged working hours mandate appropriate selection and designing of new generation materials (Mahato et al. 2015).

INDEPTH: Sealant Selection for SOFC

In planar design SOFC stacking, the individual cells are connected in series one after the other, separated by interconnect to prevent reactions between the materials of adjoining components. In this aspect, sealant materials play a key

(continued)

role in avoiding any leakages of either fuel or air from their respective chambers, and thereby preventing any direct mixing of fuel and air. Although, with proper engineering design, the necessary requirement of sealants can be completely eliminated, but at the cost of efficiency and ease of manufacturing and stacking. As the sealant materials are always exposed to both oxidizing and reducing atmospheres at high temperatures (700–1000 °C), it demands the following requirements that sealant materials need to fulfill for the proper functioning of SOFC device (Mahato et al. 2015):

1. Mechanical
   (a) Hermiticity, i.e., marginal or no gas leakage.
   (b) The coefficient of thermal expansion of the sealant material should be similar to that of other cell components ($\sim 10–12 \times 10^{-6}$/K) in order to reduce thermal stresses and limit cracking and leakages during operation.
   (c) Appreciable bond strength to avoid leakage, and/or compressive loading to bear the stack load, and also render hermiticity as on-service requirement.
   (d) Resistance to damage from thermal cycling and tolerance to thermal shocks.
   (e) Enough resilience and toughness under external static and dynamic forces.
2. Chemical
   (a) Long-term (>40,000 h) chemical stability under oxidizing/wet fuel.
   (b) Chemical compatibility with the adjoining SOFC components.
   (c) Resistance to hydrogen cracking/embrittlement.
3. Electrical—Insulating in nature
4. Design/fabrication
   (a) High hermiticity (i.e., conforming to sealing tortuous surfaces).
   (b) Acceptable sealing environment/temperature (i.e., has little effect on the subsequent performance of the stack).
   (c) Design flexibility—for example, that allows use of Ni-based alloys in interconnect.
   (d) Simplistic application/processing.
   (e) Low cost and economical.

The high-temperature sealant materials used for SOFC application are made of various type of materials, such as, metal, glass-ceramic, brazes, or mica-based composites. Multiple sealant materials are also used as and when required.

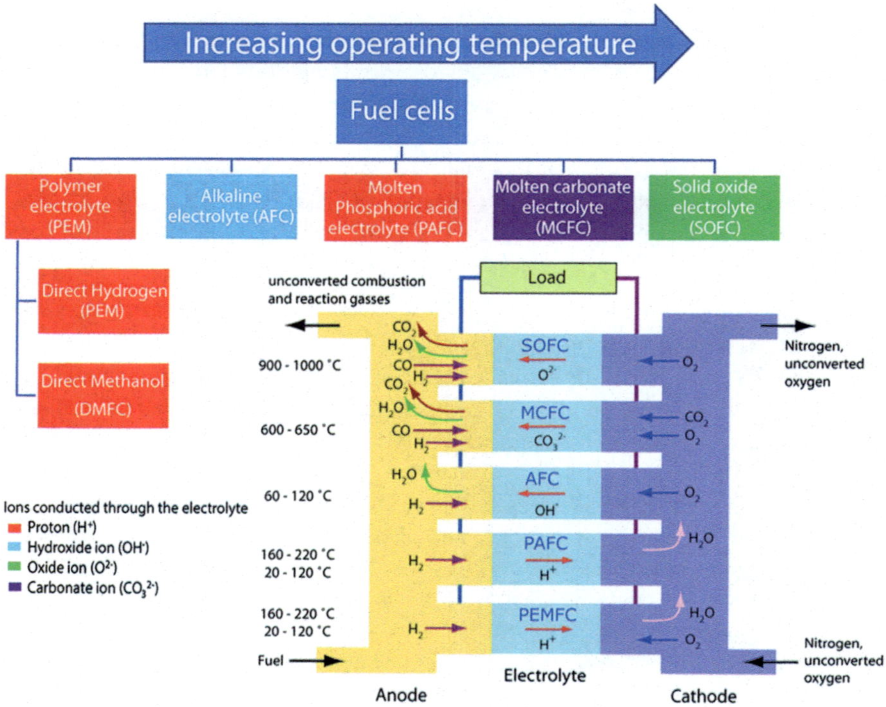

Fig. 9.16 Classification of current commercial or near commercial fuel cell systems (Modified from Badwal et al. 2014, Courtesy of Frontiers Media SA)

9.6 Applications and Prospects of Fuel Cells

A wide variety of fuel cell systems of various scales (few W to MW range) are now commercially available, with different operating regimes and widely varying performance characteristics. These devices have traditionally been categorized firstly by the type of electrolyte and then by the type of fuel used. Fuel cells can be further categorized by the operating temperature, with polymer electrolyte membrane fuel cells (PEMFC) typically have the lowest operating temperatures below 100 °C and SOFCs the highest operating around 800 °C or above (Fig. 9.16).

9.6.1 Conventional Fuel Cells

The operating temperature in conventional fuel cells is a critical parameter when looking at the system as it defines the type of fuel used, materials choice, end-user application, and electrical efficiency. HT systems (such as molten carbonate and SOFCs) operate at temperatures high enough to allow internal reforming of hydrocarbon fuels. This typically allows these systems to operate with total electrical efficiencies of between 45% and 60%. In contrast the LT fuel cell systems operating on hydrocarbon fuels must externally reform and clean (removing carbon monoxide) any hydrocarbon fuel used within the system. The operating temperatures of this class of fuel cells is too low to be utilized for reforming hydrocarbon fuels thus leading such systems to have lower electrical efficiencies (around 35–40% total system electrical efficiency when operated on hydrocarbon fuels) when compared to HT systems. Also, the PEMFC has a very low tolerance to CO. Intermediate temperature fuel cells (typically operating between 150 °C and 350 °C) are in general more resilient to fuel impurities and require lower catalyst loadings. This leads to longer operating lifetimes but their electric efficiency is similar to that of LT fuel cells. If low- or intermediate-temperature fuel cell systems are operated directly on hydrogen, electric efficiencies greater than 50% (with system efficiency over 80% with heat recovery) can be achieved as the fuel processing losses are avoided. Table 9.4 compares different fuel cell systems operated on reformed hydrocarbon fuels with the values for fuel cells which directly electrochemical oxidize a fuel (EERE 2016). Any energy from the fuel that is not converted into electrical power is lost as waste heat. In systems where the theoretical efficiency is greater than 100% the fuel cell would require heat input for continuous operation (Giddey et al. 2013; Badwal et al. 2014).

The maximum electric efficiency of a fuel cell system operating on a reformed fuel, in general, is significantly lower than the theoretical maximum where fuel is directly oxidized in the electrochemical reaction of the fuel. This is because all current fuel cells operate on either pure H_2 or (at HT) a mixture of CO and H_2. These fuels are produced, in general, via the reforming or gasification of a hydrocarbon fuel. Reforming of any readily available hydrocarbon fuel requires significant energy input. This is particularly detrimental when an external reformer and fuel processer is used (as is mostly the case for low- and intermediate-temperature fuel cell systems) because none of the low-grade waste heat produced via the fuel cell reactions can be used for reforming. External reforming and fuel processing is a requirement for all LT systems as these systems operate significantly below the temperature required for external reforming (around 500 °C). Higher temperature systems can use waste heat from the reactions within the fuel cell to reform the incoming fuel. This results in significantly higher electrical efficiencies being reported for HT commercial systems that operate in this manner (45–60%) (Badwal et al. 2014).

There are two strategies being pursued in order to further increase the efficiency of HT fuel cells operated on gaseous hydrocarbon fuels. The first is to improve the thermal coupling between the fuel cell and the reforming reactions. This is achieved in practice by reducing the physical distance between the zone where the reforming reactions occur and the fuel cells themselves with the ideal being the direct injection of the fuel into the anode chamber. This strategy has a number of technical challenges associated with the instability of hydrocarbon fuels at HTs. These fuels typically decompose to carbon (coking) on the anode surface during the HT operation. This carbon formation can be rapid and results in the fuel cell anode being irreparably damaged. It is also common for coking to occur within the pipe work

Table 9.4 Comparison of conventional fuel cells (EERE 2016)

Fuel cell type	Common electrolyte	Operating temperature	Typical stack size	Efficiency	Applications	Advantages	Disadvantages
Polymer electrolyte membrane (PEM)	Perfluoro sulfonic acid	50–100 °C 122–212 °F typically 80 °C	<1 kW–100 kW	60% transportation 35% stationary	• Backup power • Portable power • Distributed generation • Transportation • Specialty vehicles	• Solid electrolyte reduces corrosion and electrolyte management problems • Low temperature • Quick start-up	• Expensive catalysts • Sensitive to fuel impurities • Low temperature waste heat
Alkaline (AFC)	Aqueous solution of potassium hydroxide soaked in a matrix	90–100 °C 194–212 °F	10–100 kW	60%	• Military • Space	• Cathode reaction faster in alkaline electrolyte, leads to high performance • Low cost components	• Sensitive to CO_2 in fuel and air • Electrolyte management
Phosphoric acid (PAFC)	Phosphoric acid soaked in a matrix	150–200 °C 302–392 °F	400 kW 100 kW module	40%	• Distributed generation	• Higher temperature enables CHP • Increased tolerance to fuel impurities	• Pt catalyst • Long start up time • Low current and power
Molten carbonate (MCFC)	Solution of lithium, sodium, and/or potassium carbonates, soaked in a matrix	600–700 °C 1112–1292 °F	300 kW–3 MW 300 kW module	45–50%	• Electric utility • Distributed generation	• High efficiency • Fuel flexibility • Can use a variety of catalysts • Suitable for CHP	• High temperature corrosion and breakdown • Long start up time • Low power density

Solid oxide (SOFC)	Yttria stabilized zirconia	700–1000 °C 1202–1832 °F	1 kW–2 MW	60%	• Auxiliary power • Electric utility • Distributed generation	• High efficiency • Fuel flexibility • Can use a variety of catalysts • Solid electrolyte • Suitable for CHP and CHHP • Hybrid/GT cycle	• High temperature corrosion and breakdown of cell components • High temperature operation requires long start up time and limits

leading into a fuel cell stack blocking the pipes and stopping the fuel supply to the fuel cell. Coking can be avoided if significant amounts of steam or CO_2 can be introduced to the fuel stream, however, this will significantly reduce the efficiency of the system (Badwal et al. 2014).

An alternative strategy is to use materials that are more resistance to coking (typically ceramic- or Cu-based anodes). If the residence time of the fuel exposed to HT can be reduced and if anode materials which do not catalyze coking reactions can be used, then it is possible to electrochemically oxidize hydrocarbon fuels directly within a fuel cell via a multistage process on the surface of the anode. The system cost generally increases with increasing operating temperature as more expensive materials must be used within the system to withstand the harsher operating environment (Giddey et al. 2013; Badwal et al. 2014).

Although fuel cell systems are becoming increasingly commercially available there are still sufficient technical challenges that need to be overcome before the mass adoption of fuel cell technology can take place. These challenges relate to lifetime, cost, and suitable fuel supply (for low- or intermediate-temperature systems). Significant progress is being made through careful engineering of systems to alleviate a number of the issues, including the development of new materials with longer lifetimes, development of materials to allow transport and storage of hydrogen, low-cost fabrication technologies for cell and system components, and miniaturized fuel processing units for use with LT fuel cells. These advancements are incrementally increasing the appeal of fuel cell systems; however, new developments are required to make the revolutionary advancements necessary to allow fuel cells to begin to displace a significant fraction of conventional power generation capacity. There is no one fuel cell technology that stands out as being a clear leader in terms of technology maturity or technical superiority. In general, the main focus is to develop more fuel flexible systems that can operate on a wider range of fuels at increased electrical efficiency. The requirement for increased efficiency is driving research and development away from systems requiring fuel pre-processing towards systems where the fuel is directly electrochemically oxidized or where the fuel is directly fed to the anode chamber within a fuel cell. This is because this allows the maximum transfer of chemical energy to electrical energy with any waste (thermal) energy from the operation being available to either maintain the operating temperature of the device or used directly in the chemical or electrochemical reactions within the fuel cell chamber. In addition, there is also an increased interest in lowering the operating temperature of fuel cells to reduce overall system cost while extending the life of the fuel cell (Badwal et al. 2014).

9.6.2 Emerging Fuel Cell Technologies

Emerging fuel cell technologies do not fit comfortably within traditional fuel cell categories in particular due to the varied nature of the fuel handling systems and the move away from conventional electrolytes. This leads to them being better defined by the state/type of the fuel rather than electrolyte chemistry as this is more relevant

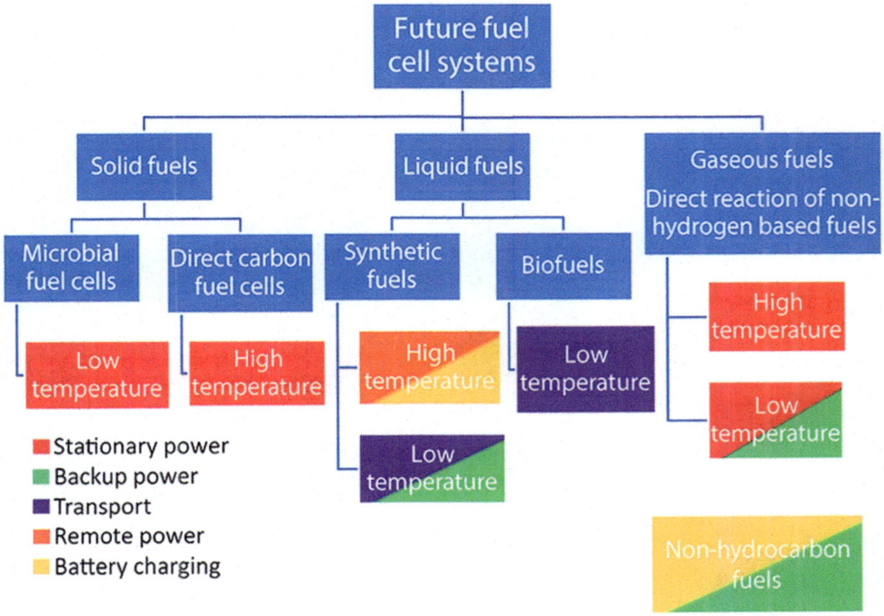

Fig. 9.17 Classification of future fuel cell systems (Adapted from Badwal et al. 2014, Courtesy of Frontiers Media SA)

to the system design and the end use application of the system. Examples of this are direct methanol or ethanol or carbon fuel cells. This classification system is not ideal as there is significant ambiguity as to in which class a fuel cell should reside. In particular, depending on the operating temperature or pressure, the fuel may be either a gas or a liquid. Figure 9.17 shows a broad fuel-based classification of different fuel cells currently being investigated and is color coded to give an indication of the potential end user applications for each fuel cell type. Systems based on solid fuels have the attraction that these fuels are often low cost and more abundant than liquid or gaseous fuels. The gaseous fuels have the advantage of being reasonably abundant and can be easily transported over long distances through conventional pipe networks. Liquid fuels are the least abundant of all of the potential fuel sources but are easy to transport and high energy densities make them most suited to transport or mobile applications. Within the solid fuel class, there are two fuel cell types that could potentially result in a paradigm shift with respect to power generation and application potential: Microbial Fuel Cells (MFC) and Direct Carbon Fuel Cells (DCFC) (Badwal et al. 2014).

9.6.2.1 Microbial Fuel Cells (MFC)

MFC convert organic material into electrical energy via the microbes' metabolic processes. The use of microbes to produce electric current has been explored since the 1970s but has only recently been revisited for use as a power source for

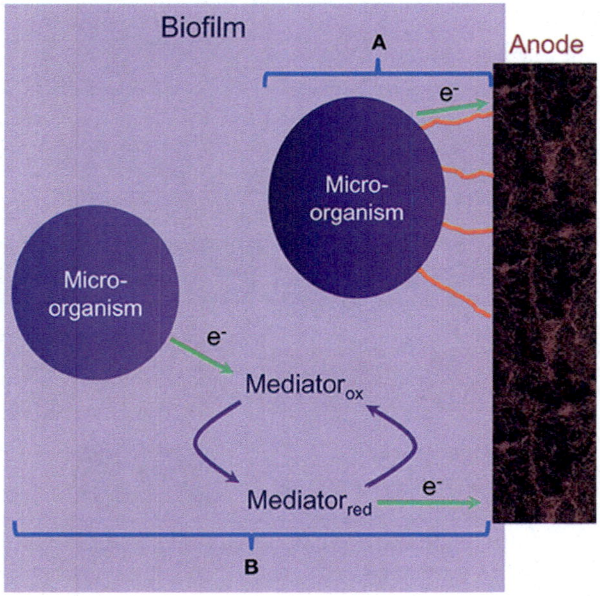

Fig. 9.18 Two modes of operation of a MFC (Modified with permission from Knight et al. 2013 (Springer Nature)): (**a**) Direct reaction, and (**b**) indirect reaction

small-scale applications as higher power densities are being demonstrated (Rabaey et al. 2009). MFC generally take two forms, membrane reactors and single chamber fuel cells. Within a membrane reactor device, the anode and cathode are separated into two chambers by an electrolyte membrane whereas with single chamber devices both the anode and cathode are in one chamber but separated by organic material. The second class are typically referred to as sediment cells. In both classes of MFC, microorganisms form a biofilm on the surface of the anode and oxidize organic material. These microorganisms then transfer electrons to the anode of the fuel cell either directly (Fig. 9.18a) via micro-pili or indirectly via a mediator (Fig. 9.18b) (Knight et al. 2013; Badwal et al. 2014).

MFC are considered promising as they operate at or near room temperature and can utilize low-grade waste materials such as soils and sediments, waste water and agricultural waste streams that are unsuitable for use in any other power generation technology. The main issue, however, is the very low power density of this type of fuel cells which is typically in the μWcm^{-2} range which is several orders of magnitude below that of other fuel cell types (Rabaey et al. 2009; Knight et al. 2013). Although these fuel cells offer promise in certain low power demand applications, if they are to be adopted at a large scale for such applications, then the power densities need to be increased substantially to at least in the $mWcm^{-2}$ range.

In addition to the absolute performance of MFC's other critical challenges that need to be overcome include faster response times to varying loads, increased voltage stability, increased lifetime, and improved methods of fuel supply to the electrodes. Unlike the majority of other fuel cell types these issues are not fundamentally materials related with the greatest drivers for improvement being novel designs that allow greater mixing of oxidant or fuel with the microbe laden electrodes, improved coupling between the microbes and the electrodes, and selection or modification of the microbes to increase reaction rates at the electrodes. If the activity of the electrodes could be enhanced then further improvements could be obtained via the modifying of the cell design and materials to reduce resistive losses in the electrolyte and electrodes. This could be most easily achieved by reducing the electrolyte thickness or improving the conductivity of the electrodes and/or electrolyte. These improvements are unlike to have a dramatic effect on the performance of MFC's until the activity of the electrodes is increased which would result in higher current passage through the cell.

9.6.2.2 Direct Carbon Fuel Cells (DCFC)

Direct carbon fuel cells and fuel cells that directly electrochemically consume hydrocarbon fuels offer many advantages and could potentially compete in many common market sectors to other fuel cell types. The attraction of direct electrochemical oxidation of carbon or gaseous hydrocarbon fuels is that there is the potential to significantly enhance the electrical efficiency of a fuel cell system if the fuel is directly electrochemically reacted rather than gasified or reformed. Some of the benefits of the technology, in addition to high efficiency (>65–70% electric, 90% combined heat and electric), include low CO_2 emission, and as the by-product of carbon oxidation is CO_2, its capture costs and energy requirements are very low. Furthermore, if a solid fuel is used (carbon or a high-carbon-containing hydrocarbon fuel such as coal or biomass chars) then the stability of the fuel becomes less of an issue. These fuels have far higher stability than liquid or gaseous fuel and hence can be fed to the anode surface where they remain stable until oxidized in a chemical or electrochemical reaction (Badwal et al. 2014).

The DCFC technology is at an early stage of development with a number of different types of DCFC under consideration with a number of groups globally now reporting operation of small stacks (Giddey et al. 2013). Although the electric efficiency is high (>65–70%), reported power densities for these systems, especially once scaled up, are still significantly lower than that of conventional fuel cells operated on gaseous fuels. This is largely due to the reduced surface area for reaction between the anode and the solid fuel that is incapable of infiltrating a porous anode. In order to improve performance, various strategies to increase the available surface area for reaction have been tried. This has included the use of molten metal anodes, molten salt electrolytes, or mixed ionic/electron conducting anode materials. A number of these system designs are now in the process of being scaled up with technical issues such as system life, fuel quality, fuel feed, and system cost all still remaining as critical that need to be resolved before these devices can be demonstrated at any significant scale (Damian and Irvine 2012; Jayakumar et al. 2017).

As with conventional HT fuel cell systems, the majority of issues currently hindering development of DCFC relate to materials and in particular the way in which materials react with the fuel and other cell components at HTs. In addition to materials issues, there are likely to be an increasing number of challenges relating to fuel handling and processing as this technology matures leading to larger systems being tested for longer periods. Due to the relative immaturity of the field these issues are, as yet, poorly defined. Dependent on cell design and construction materials issues vary significantly. In general reactivity issues are greatest with cell designs that contain molten components in particular molten salts. In cell designs that do not contain solid ion conducting layers, these issues are common with other molten salt fuel cell designs, such as molten carbonate fuel cells, and are relate to the mobility of the electrolyte and its reactivity with other system components. Although the degradation mechanisms are common with molten carbonate fuel cell designs, the higher operating temperature of DCFCs (typically 800 °C vs. 650 °C) leads to accelerated degradation rates. The molten salt within the fuel cell can be contained and separated with a dense oxide ion conducting membrane, in this instance the fuel is normally mixed with the molten salt and contained within the anode chamber. This greatly simplifies the issues relating to mobility of the molten components within the fuel cell and results in high power densities but the dense oxide membrane can be rapidly corroded by the molten salt/fuel mixture. Some progress has been made in reducing the reaction rate but this is still seen as a critical issue (Damian and Irvine 2012). A molten metal can operate well as an anode material when a solid fuel is used, however, these metals are likely to be highly reactive towards impurities within the fuels which will accumulate in the anode chamber and result in solidification of the molten metal. These fuel cells are also limited in terms of operating voltage by the reduction potential of the molten metal which can lead to a significant reduction in overall system efficiency. A fuel cell design, where a mixed ionic electronic conducting (MIEC) anode is used to shift carbon oxidation reaction from electrode/electrolyte interface to anode/fuel interface, is likely to have the least reactivity issues due to the fact that all fuel cell components are solid state. This makes them less reactive towards fuel impurities and in general more stable. However, these materials have lower ionic conductivity than molten salts and lower electrical conductivities than molten metals leading to MIEC DCFCs having, in general, lower power densities when compared to other DCFC designs. If stable materials with high mixed ionic and electronic conductivities can be identified, this fuel cell system would rapidly evolve as a leading contender as it can utilize many of the materials and design features of the more technologically mature SOFC technology (Giddey et al. 2013; Badwal et al. 2014).

9.6.2.3 Small and Portable Fuel Cells

In addition to next generation fuel cell systems that operate on gaseous and solid fuels at ultrahigh efficiencies, there is also a drive to develop small-scale or portable power sources. In these systems, device volume and weight, fuel energy density, and ease of transport of the fuel are critical with the overall system efficiency being important but less critical than for stationary power generation. Portable fuel cell systems are generally based on LT PEM fuel cell stacks that operate near room temperature on pure hydrogen with a limited number of systems being developed that are based on either SOFC technology or that are based on PEM systems but that operate directly on methanol/water mixes. If the fuel cell is to be operated on pure hydrogen then this is normally stored either within a metal hydride or light weight compressed hydrogen cylinder. Other fuels under consideration include biofuels such as ethanol, synthetic hydrocarbon fuels such as methanol, and non-hydrocarbon fuels such as ammonia. For fuel cells operated on the non-hydrogen fuels, with the exception of direct methanol fuel cells, a fuel processor is required to convert the fuel into either pure H_2 or a mixture of H_2 and CO with the later only being suitable for use in HT fuel cell systems (Giddey et al. 2013; Badwal et al. 2014).

The use of a fuel processor can often greatly increase the complexity of the device but simplifies the storage of the fuel, particularly in the case of liquid fuels which can often have exceptionally high energy densities and low cost in comparison to either batteries or gaseous hydrogen storage solutions. However, due to the stringent requirements relating to the purity of hydrogen, the cost of the fuel processor can often significantly increase the overall cost of the device with the fuel processer potentially being greater than the cost of the fuel cells stack itself. This typically limits fuel cell/fuel processor combinations to applications where the cost per kWh is more critical than the cost per kW as this allows the high cost of the fuel processor to be offset by the much reduced cost of the fuel storage solution. Similarly, any additional weight from the processor can be offset by the far higher energy density of the fuel storage solution. These small and portable fuel cell systems are being developed for a range of end-user applications including stationary backup generators, battery charging, remote area power, auxiliary power units, soldier packs, portable electronic appliances, and small transport applications. There are an increasing number of these devices now commercially available, however, lack of fuel infrastructure and high cost when compared to battery or battery generator combinations remain key challenges that need to be overcome for this market to expand further. Future fuel cell designs should be able to operate directly on a greater variety of commonly available fuels without the requirement for significant amounts of fuel pre-processing. This should lead to far greater efficiencies and hence lower operating costs of fuel cell power systems when compared to conventional power generating technologies which are likely to remain lower cost in terms of capital investment in the medium to long term (Badwal et al. 2014).

Exercises

Part I: General Questions

9.1. In a press-release about fuel-cell cars, it is said that a 3 kg hydrogen reservoir would give the same autonomy as a typical car (500 km at 100 km/h average speed, with a petrol reservoir of 40 L). Assuming that petrol properties may be approximated by n-octane, find: (a) Heating power of petrol. (b) Heating power of hydrogen at 25 °C and 100 kPa, at 20 MPa and the same temperature, and at normal pressure but liquid. (c) Exergy of hydrogen in the conditions above.

9.2. What hydrogen flow rate is required to generate 1.0 ampere of current in a fuel cell? How is the faradaic efficiency defined? Calculate the actual needed flow rate of hydrogen for a faradaic efficiency of 80%.

9.3. A 1.0 MWDC fuel cell stack is operated with a cell voltage of 700 mV on pure hydrogen with a fuel utilization, U_f of 80%. (a) How much hydrogen will be consumed in lb/h? (b) What is the required fuel flow rate? (c) What is the required air flow rate for a 25% oxidant utilization, Uox?

9.4. A PAFC, operating on reformed natural gas (900 lb/h) and air, has a fuel and oxidant utilization of 86% and 70%, respectively. (a) With the fuel and oxidant composition and molecular weights (MW) listed below, how much hydrogen will be consumed in mol/h? (b) How much oxygen is consumed in mol/h? (c) What is the required air flow rate in mol/h and lb/h? (d) How much water is generated? (e) What is the composition of the effluent (spent) fuel and air streams in mol percent?

9.5. An MCFC operating on 1,000 lb/h of fuel gas and a 70% air/30% CO_2 oxidant has a fuel and oxidant utilization of 75% and 50%, respectively. With the fuel and oxidant composition and molecular weights listed below, (a) How much hydrogen will be consumed in mol/h? (b) How much oxygen is consumed in mol/h? (c) What are the required air and oxidant flow rates in mol/h? (d) How much CO_2 is transferred from the cathode to the anode? (e) What is the composition of the effluent (spent) fuel and oxidant streams in mol percent (no water gas shift reaction)?

9.6. Given a desired output of 2.0 MW_{DC} and the desired operating point of 600 mV and 400 mA/cm^2, (a) How much fuel cell area is needed? (b) Assuming a cell area of 1.00 m^2 per cell and 280 cells per stack, how many stacks are needed for this 2.0 MW unit?

9.7. You would like the cell to have an operating lifetime of 100 h. If this H_2 fuel is stored as a compressed gas at 500 bars, what volume would it occupy (assume ideal gas, room temperature, $pV = nRT$)? If it is stored as a metal hydride at 5 wt % hydrogen, what volume would it occupy? (Assume the metal hydride has a density of 10 g/cm^3.)

9.8. Describe the major advantages and general weaknesses of fuel cells systems.

9.9. List main hydrogen generation methods and compare their advantages and disadvantages. May you address their future development trends?

9.10. List main feed stocks and employed technologies for biohydrogen production. Which one is the most promising technology for different fuel cell applications?

9.11. Describe major hydrogen conversion and storage technologies, and compare their advantages and disadvantages. May you address their future development trends?

9.12. List main materials-based hydrogen storage solutions, and compare their advantages and disadvantages. May you address their future development trends?

Part II: Thought-Provoking Questions

9.13. How to select materials used for solid oxide fuel cells?
9.14. Compare the major characteristics of conventional and emerging fuel cells? May you address the prospects of future fuel cells?
9.15. Describe current development status and future application trends of portable fuel cells.

References

Aasberg-Petersen, K., Bak Hansen, J.H., Christensen, T.S., et al.: Technologies for large-scale gas conversion. Appl. Catal. A. **221**(1-2), 379–387 (2001)

Ayabe, S., Omoto, H., Utaka, T., et al.: Catalytic autothermal reforming of methane and propane over supported metal catalysts. Appl. Catal. A. **241**(1-2), 261–269 (2003)

Badwal, S.P.S., Giddey, S.S., Munnings, C., Bhattand, A.I., Hollenkamp, A.F.: Emerging electrochemical energy conversion and storage technologies. Front. Chem. **2**, 1–28 (2014)

Boukhvalov, D.W., Katsnelson, M.I., Lichtenstein, A.I.: Hydrogen on graphene: electronic structure, total energy, structural distortions and magnetism from first-principles calculations. Phys. Rev. B. **77**, 035427 (2008)

Bromberg, L., Cohn, D.R., Rabinovich, A., Alexeev, N.: Plasma catalytic reforming of methane. Int. J. Hydrog. Energy. **24**(12), 1131–1137 (1999)

Bullis, K.: Solar-powered hydrogen generation: Rust-based solar panels could make hydrogen cheap and efficient. MIT Technology Review. https://www.technologyreview.com/s/406984/solar-powered-hydrogen-generation/ (2006). Accessed 5 Oct 2017

Chen, W.D., Huang, Z.G., Wu, G.T., He, T., Li, Z., Chen, J., et al.: Guanidinium octahydrotriborate: an ionic liquid with high hydrogen storage capacity. J. Mater. Chem. A. **3**, 11411–11416 (2015)

Chheda, J.N., Huber, G.W., Dumesic, J.A.: Liquid-phase catalytic processing of biomass-derived oxygenated hydrocarbons to fuels and chemicals. Angew. Chem. **46**, 7146–7183 (2007)

Damian, A., Irvine, J.T.S.: Development of tubular hybrid direct carbon fuel cell. Int. J. Hydrog. Energy. **37**, 19337–19344 (2012)

Dodds, P.E., et al.: Hydrogen and fuel cell technologies for heating: a review. Int. J. Hydrog. Energy. **40**, 2065–2083 (2015)

EERE: Fuel cell technologies program–Comparison of fuel cell technologies. Energy Efficiency & Renewable Energy. US Department of Energy. https://energy.gov/sites/prod/files/2016/06/f32/fcto_fuel_cells_comparison_chart_apr2016.pdf (2016). Accessed 16 Oct 2017

Farrauto, R., Hwang, S., Shore, L., et al.: New material needs for hydrocarbon fuel processing: generating hydrogen for the PEM fuel cell. Annu. Rev. Mater. Res. **33**, 1–27 (2003)

FSEC: Hydrogen basics—production. Florida Solar Energy Center. http://www.fsec.ucf.edu/en/consumer/hydrogen/basics/production.htm (2012). Accessed 6 Oct 2017

Giddey, S., Badwal, S.P.S., Kulkarni, A.: Review of electrochemical ammonia production technologies and materials. Int. J. Hydrog. Energy. **38**, 14576–14594 (2013)

Godina, L.I., Tokarev, A.V., Simakova, I.L., Mäki-Arvela, P., Kortesmäki, E., Gläsel, J., Kronberg, L., Etzold, B., Murzin, D.Y.: Aqueous-phase reforming of alcohols with three carbon atoms on carbon-supported Pt. Catal. Today. **301**, 78–89 (2018)

Guo, J.H., Zhang, H., Miramoto, Y., Cheng, X.L.: The effect of ionization and CH_3 ligand for hydrogen storage in Co- and Ni-based organometallic compounds. Int. J. Quantum Chem. **111**, 4443–4451 (2011)

Haile, S.M.: Fuel cell materials and components. Acta Mater. **51**, 5981–5600 (2003)

IAEA: Nuclear Hydrogen Production Technology. https://www.iaea.org/About/Policy/GC/GC57/GC57InfDocuments/English/gc57inf-2-att1_en.pdf (2012). Accessed 10 Oct 2017

IEA: Technology roadmap-Hydrogen and fuel cells. https://www.iea.org/publications/freepublications/publication/TechnologyRoadmapHydrogenandFuelCells.pdf (2015). Accessed 12 Oct 2017

Jayakumar A., Singamneni S., Ramos M., Al-Jumaily A.M., Pethaiah S.S.: Manufacturing the gas diffusion layer for PEM fuel cell using a novel 3D printing technique and critical assessment of the challenges encountered. Materials (Basel) **10**(7), 796 (2017)

Jena, P.: Materials for hydrogen storage: past, present, and future. J. Phys. Chem. Lett. **2**(3), 206–211 (2011)

Jeon, K.J., Moon, H.R., Ruminski, A.M., Jiang, B., Kisielowski, C., Bardhan, R., Urban, J.J.: Air-stable magnesium nanocomposites provide rapid and high-capacity hydrogen storage without using heavy-metal catalysts. Nat. Mater. **10**, 286–290 (2011)

Kalamaras, C.M., Efstathiou, A.M.: Hydrogen production technologies: current state and future developments. Conference Papers in Energy (Hindawi Publ Corp.) **2013**, 690627 (2013)

Knight, C., Cavanagh, K., Munnings, C., Moore, T., Cheng, K.Y., Kaksonen, A.H.: Application of microbial fuel cells to power sensor networks for ecological monitoring. In: Mukhopadhyay, S. C., Jiang, J.A. (eds.) Wireless sensor networks and ecological monitoring, pp. 151–178. Springer, Berlin, Heidelberg (2013)

Kozachuk, O., Luz, I., Llabrés, I., Xamena, F.X., Noei, H., Kauer, M., et al.: Multifunctional, defect-engineered metal-organic frameworks with ruthenium centers: sorption and catalytic properties. Angew. Chem. Int. Ed. **53**, 7058–7062 (2014)

Krishna, R.H.: Review of research on production methods of hydrogen: future fuel. Eur. J. Biotechnol. Biosci. **1**(2), 84–93 (2013)

Langmi, H.W., Ren, J.W., North, B., Mathe, M., Bessarabov, D.: Hydrogen storage in metal organic frameworks. A review. Electrochim. Acta. **128**, 368–392 (2014)

Ley, M.B., Jepsen, L.H., Lee, Y.S., Cho, Y.W., Bellosta von Colbe, J.M., Dornheim, M., et al.: Complex hydrides for hydrogen storage new perspectives. Mater. Today. **17**, 122–128 (2014)

Lutz, A.E., Bradshaw, R.W., Bromberg, L., Rabinovich, A.: Thermodynamic analysis of hydrogen production by partial oxidation reforming. Int. J. Hydrog. Energy. **29**(8), 809–816 (2004)

Mahato, N., Banerjee, A., Gupta, A., Omar, S., Balani, K.: Progress in material selection for solid oxide fuel cell technology: a review. Prog. Mater. Sci. **72**, 141–337 (2015)

McWhorter, S., Read, C., Ordaz, G., Stetson, N.: Materials-based hydrogen storage: attributes for near-term, early market PEM fuel cells. Curr. Opinion Solid State Mater. Sci. **15**, 29–38 (2011)

Mekhilefa, S., Saidur, R., Safari, A.: Comparative study of different fuel cell technologies. Renew. Sust. Energ. Rev. **16**, 981–989 (2012)

Muradov, N.: Emission-free fuel reformers for mobile and portable fuel cell applications. J. Power Sources. **118**(1–2), 320–324 (2003)

Naterer, G.F., Dincer, I., Zamfirescu, C.: Hydrogen production from nuclear energy, pp. 21–64. Springer-Verlag, London (2013)

Niakolas, D.K., Daletou, M., Neophytides, S.G., Vayenas, C.G.: Fuel cells are a commercially viable alternative for the production of "clean" energy. Ambio. **45**(Suppl 1), 32–37 (2016)

Obolensky, M.A., Basteev, A.V., Bazyma, L.A.: Hydrogen storage in irradiated low-dimensional structures. Fuller Nanotub Carbon Nanostruct. **19**, 133–136 (2011)

Paulmier, T., Fulcheri, L.: Use of non-thermal plasma for hydrocarbon reforming. Chem. Eng. J. **106**(1), 59–71 (2005)

Rabaey, K., Angenent, L., Schroder, U., Keller, J.: Bioelectrochemical Systems: from extracellular electron transfer to biotechnological application. IWA Publishing, London (2009)

Rekken, B.D., Carre-Burritt, A.E., Scott, B.L., Davis, B.L.: N-substituted amine-borane ionic liquids as fluid phase, hydrogen storage materials. J. Mater. Chem. A. **2**, 16507–16515 (2014)

Ren, J., Musyoka, N.M., Langmi, H.W., Mathe, M., Liao, S.: Current research trends and perspectives on materials-based hydrogen storage solutions: a critical review. Int. J. Hydrog. Energy. **42**, 289–311 (2017)

Semelsberger, T.A., Brown, L.F., Borup, R.L., Inbody, M.A.: Equilibrium products from autothermal processes for generating hydrogen-rich fuel-cell feeds. Int. J. Hydrog. Energy. **29** (10), 1047–1064 (2004)

Shirasaki, Y., Tsuneki, T., Ota, Y., et al.: Development of membrane reformer system for highly efficient hydrogen production from natural gas. Int. J. Hydrog. Energy. **34**(10), 4482–4487 (2009)

Song, H., Zhang, L., Watson, R.B., Braden, D., Ozkan, U.S.: Investigation of bio-ethanol steam reforming over cobalt-based catalysts. Catylysis Today **129**, 346–354 (2007)

Skipper, C.V.J., Hoang, T.K.A., Antonelli, D.M., Kaltsoyannis, N.: Transition metal hydrazide-based hydrogen-storage materials: the first atoms-in-molecules analysis of the Kubas interaction. Chem. Eur. J. **18**, 1750–1760 (2012)

Srinivasan, S.S., Brinks, H.W., Hauback, B.C., Sun, D., Jensen, C.M.: Long term cycling behavior of titanium doped NaAlH4 prepared through solvent mediated milling of NaH and Al with titanium dopant precursors. J. Alloys Compd. **377**, 283–289 (2004)

Steinfeld, A.: Solar thermochemical production of hydrogen—a review. Sol. Energy. **78**, 603–615 (2005)

Verma, S.: A review on solar hydrogen production techniques. IJSR-Int. J. Sci. Res. **4**(1), 420–423 (2015)

Weber, G., Fu, Q., Wu, H.: Energy efficiency of an integrated process based on gasification for hydrogen production from biomass. Develop. Chem. Eng. Min. Process. **14**(1–2), 33–49 (2006)

Xia, G.: Light metal hydrides for reversible hydrogen storage applications. Ph.D. thesis. University of Wollongong, Australia (2015)

Yuan, Y.J., Yu, Z.T., Chen, D.Q., Zou, Z.G.: Metal-complex chromophores for solar hydrogen generation. Chem. Soc. Rev. **46**, 603–631 (2017)

Zhou, J., Wang, Q., Sun, Q., Jena, P., Chen, X.S.: Electric field enhanced hydrogen storage on polarizable materials substrates. PNAS. **107**, 2801–2806 (2010)

Role of Materials to Advanced Nuclear Energy

10

Abstract

Nuclear energy holds the promise to provide vast amounts of reliable baseline electricity at commercially competitive costs with modest environmental impact. However, the future of nuclear energy lies beyond the current generation of light-water reactors. Future reactors will be expected to provide additional improvements in safety, maintain high reliability, use uranium resources more efficiently, and produce lower volumes of less toxic solid wastes. Several advanced reactor concepts are under development to meet these demands. In most cases, these designs translate into higher operating temperatures and longer lifetimes, more corrosive environments, and higher radiation fields in which materials must reliably perform. The safe and economical operation of any nuclear power system relies to a great extent on the success of the fuel and the materials of construction. Materials used for fission and fusion-based nuclear engineering mainly include fuels, materials for fuel cladding, moderators and control rods, first-wall materials, materials for pressure vessels and heat exchangers. During the lifetime of a nuclear power system, the materials are subject to high temperature, corrosive environment, and damage from high-energy particles released during fission. The fuel which provides the power for the reactor has a much shorter life but is subject to the same types of harsh environments. This chapter will review and update nuclear energy reactors and the materials challenges that will determine the feasibility of these advanced concepts and define the long-term future of nuclear power.

10.1 Introduction

Among the sustainable energy options, nuclear energy is a technology that is most proven on the world stage. Renewable energy sources (wind, wave, tidal, solar, hydro, geothermal) can, realistically, provide only a fraction of the energy that is

C. Tong, *Introduction to Materials for Advanced Energy Systems*,
https://doi.org/10.1007/978-3-319-98002-7_10

used today, and a smaller fraction of the much larger demand for energy that is predicted for next two decades. An option is to replace carbon-based fuels by nuclear power (using it for transport via electric or hydrogen-powered vehicles) at the same time reducing an uncomfortable dependence on imported hydrocarbons and an unacceptably extensive use of land area. They are predominantly pressurized water reactors (PWRs)—60% of total, and boiling water reactors (BWRs), around 21%. The rest are gas-cooled reactors (AGRs), deuterium-moderated reactors (CANDU and D_2O-PWRs), light-water graphite moderated reactors (RBMKs), and fast breeder reactors (FBRs). Materials used for fission and fusion-based nuclear engineering generally include fuels, materials for fuel cladding, moderators and control rods, first-wall materials, materials for pressure vessels and heat exchangers (Ashby and Smidman 2010).

Most operating commercial nuclear reactors worldwide are of Generation-II category. The Generation-III reactors have just started to be deployed, and Gen-III + reactors are at the advanced stage of commercialization. Although the safety and reliability of these reactors are very good, it has been widely recognized that the nuclear energy has a crucial role to play in mitigating the ever-increasing world energy needs. Hence, in 2000 the US Department of Energy launched a new program; called Generation-IV Initiative, to broaden the opportunity of nuclear energy utilization by making further advances in nuclear energy systems design. Gen-IV Initiative is an international effort and calls for new nuclear energy systems that will significantly improve safety and reliability, sustainability, useful reactor life (60 years or more), proliferation-resistance, and profitability, setting them apart from the current nuclear power reactors (Murty and Charit 2008).

Nuclear reactor systems are generally characterized by their coolant (water, gas, liquid metal, molten salt) and the neutron energy regime in which the fission reactions occur (fast or thermal). Neutrons are continuously created as a byproduct of the fission process with high ("fast") energies near 1–2 MeV. In some concepts, the neutrons are moderated to low ("thermal") energies near ~1 eV or less in order to take advantage of the enhanced fission neutron reaction probability for certain isotopes of uranium such as ^{235}U. In other concepts, more complete utilization of the fuel is achieved by maintaining high neutron energies that promote the conversion of the naturally dominant non-fissile isotopes such as ^{238}U to fissionable isotopes. Existing commercial nuclear power plants are mainly light-water reactors that operate with water as the coolant and fission by thermal neutrons. Similarly, sodium fast reactors operate with liquid sodium as the coolant and produce energy by fission with fast neutrons. Beyond the coolant and the fission process, nuclear reactors are similar to coal or gas power plants in that the thermal energy transferred to the coolant is used to drive a turbine, which drives a generator for the production of electricity. Thus, it is the reactor core and the vessel containment (along with materials systems to safely store the fuel waste forms) that pose the unique materials challenges for nuclear power. In addition, six reactor concepts for advanced Generation IV nuclear power systems have been identified: supercritical-water-cooled reactor (SCWR), the sodium fast reactor (SFR), the lead fast reactor (LFR), the very-high-temperature reactor (VHTR), the gas fast reactor (GFR), and the molten salt reactor (MSR). Table 10.1 summarizes the basic characteristics of each of these reactor types and the materials proposed for the various major components (Guérin

Table 10.1 Reactor core environments for two current light-water reactors and six advanced reactor concepts and materials under consideration for various key components (Guérin et al. 2009)

System	Coolant	Pressure (MPa)	T_{in}/T_{out}	Neutron spectrum, maximum dose (dpa)	Fuel	Cladding	Structural materials	
							In-core	Out-of-core
Pressurized water reactor (PWR)	Water (single phase)	16	290/320	Thermal, -80	UO_2 (or MOX)	Zirconium alloy	Stainless steels, nickel-based alloys	Stainless steels, nickel-based alloys
Boiling water reactor (BWR)	Water (two phase)	7	280/288	Thermal, ~7	UO_2 (or MOX)	Zircaloy	Stainless steels, nickel-based alloys	Stainless steels, nickel-based alloys
Supercritical-water-cooled reactor (SCWR)	Supercritical water	25	290/600	Thermal, ~30 Fast, ~70	UO_2	F–M (12Cr, 9Cr, etc.), Fe–35Ni–25Cr–0.3Ti, Incoloy 800, ODS, Inconel 690, 625, and 718	Same as cladding options, plus low-swelling stainless steels	F–M, low-alloys steels
Very-high-temperature reactor (VHTR)	Helium	7	600/1000	Thermal, <20	UO_2, UCO	SiC or ZrC coating and surrounding graphite	Graphites, PyC, SiC, ZrC, Vessel: F–M	Ni-based superalloys (Ni–25Cr–20Fe–12.5W–0.05C, Ni–23Cr–18W–0.2C), F–M with thermal barriers, low-alloy steels
Gas fast reactor (GFR)	Helium, supercritical CO_2	7	450/850	Fast, 80	MC	Ceramic	Refractory metals and alloys, ceramics, ODS, Vessel: F–M	Ni-based superalloys (Ni–25Cr–20Fe–12.5W–0.05C, Ni–23Cr–18W–0.2C), F–M with thermal barriers

(continued)

Table 10.1 (continued)

System	Coolant	Pressure (MPa)	T_{in}/T_{out}	Neutron spectrum, maximum dose (dpa)	Fuel	Cladding	Structural materials	
							In-core	Out-of-core
Sodium fast reactor (SFR)	Sodium	0.1	370/550	Fast, 200	MOX, U–Pu–Zr, MC, or MN	F–M or F–M ODS	Ducts: F–M Grid plate: 316SS	Ferritics, austenitics
Lead fast reactor (LFR)	Lead or lead–bismuth	0.1	600/800	Fast, 150	MN	High-Si F–M, ODS, ceramics, or refractory alloys	High-Si F–M, ODS	High-Si austenitics, ceramics, or refractory alloys
Molten salt reactor (MSR)	Molten salt (e.g., FLiNaK)	0.1	700/1000	Thermal, 200	Salt	Not applicable	Ceramics, refractory metals, high-Mo, Ni-based alloys (e.g., INOR-8), graphite, Hastelloy N	High-Mo, Ni-based alloys (e.g., INOR-8)

Note: *F–M* ferritic–martensitic stainless steels (typically 9–12 wt.% Cr), *ODS* oxide-dispersion-strengthened steels (typically ferritic–martensitic), *MC* mixed carbide [(U,Pu)C], *MN* mixed nitride [(U,Pu)N], *MOX* mixed oxide [$(U,Pu)O_2$]

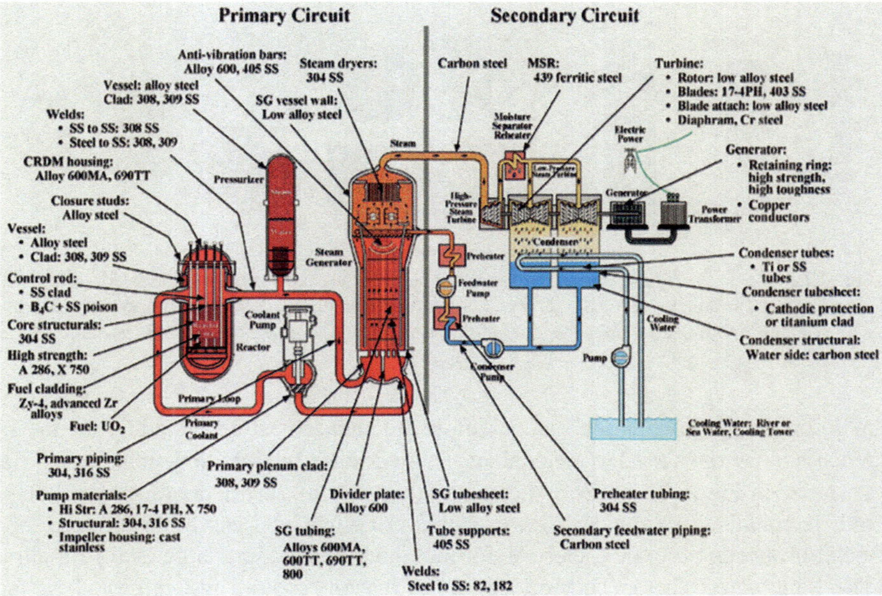

Fig. 10.1 Typical design and materials in a PWR (Adapted with permission from Allen et al. 2010 (Elsevier))

et al. 2009). Also included in Table 10.1 for comparison are two types of existing (Generation II) light-water reactors: the pressurized water reactor (PWR) and the boiling water reactor (BWR). Successful operation of current light-water reactors and implementation of advanced nuclear energy systems is strongly dependent on the performance of fuels and materials. For example, a typical Light-Water Reactor (LWR) contains numerous types of materials (Table 10.1 and Fig. 10.1) that must all perform successfully. The Generation IV systems are expected to operate at higher temperatures, to higher radiation doses, at higher pressures, and in some cases with coolants that present more challenging corrosion problems than current LWRs. Generation IV systems are expected to operate for at least 60 years (Allen et al. 2010).

For existing LWRs, extending the lifetime of each fuel element would improve the energy extraction from the fuel, limit the total amount of unused fuel (approximately 95% of the energy content remains at the end of the current useful life of a typical LWR fuel pin), and improve the overall economics of the plant. For many of the proposed advanced systems, specifically the fast spectrum systems like the Sodium Fast Reactor (SFR), Lead Fast Reactor (LFR), and Gas.

Fast Reactor (GFR), advanced fuel forms purposefully contain fission products from previously used fuel with the goal of burning these fission products to reduce the long-lived radioactivity associated with the fuel. These fast reactor fuels, in addition to having different compositions, are exposed to different reactor conditions. The design of reactor cores is driven by the need to keep the fuel material separated from the circulating coolant to avoid contamination by the radioactive fissile materials and fission products. Hence, all reactor core designs have one or

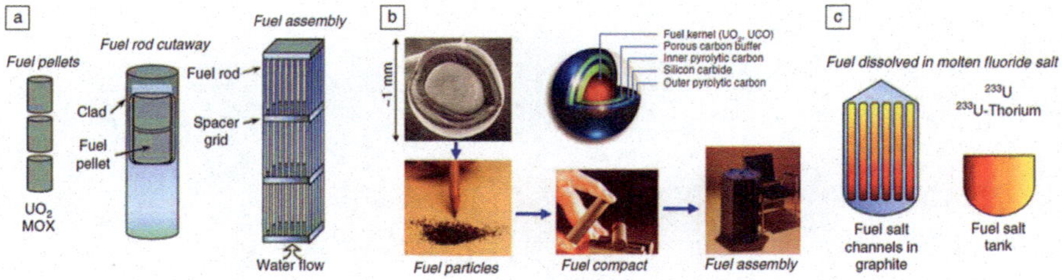

Fig. 10.2 Basic configurations of fuel and coolant for nuclear reactor systems (Modified with permission from Allen et al. 2010 (Elsevier)): (**a**) fuel in stacks of pellets, surrounded by cladding with coolant flowing along the exterior cladding surfaces, (**b**) particle fuel and gas coolant, and (**c**) fuel as part of the coolant

more barriers to contain the fuel yet allow the intense heat generated by the fission process to be transferred to the coolant. These designs fall into three main categories as shown in Fig. 10.2. In the first category (Fig. 10.2a), fuel in the form of pellets or rods is contained in annular sleeves called cladding. Fuel can be in the form of a metal or a ceramic (oxide, carbide, nitride), and the cladding is generally an alloy. The light-water reactor, supercritical water reactor, sodium fast reactor, lead fast reactor, and variants of the gas fast reactor all have this same general core configuration but use very different fuel and cladding forms and coolants. The second category of reactor cores uses fuel in the form of spheres coated with several layers of graphite or pyrolytic carbon and a structural ceramic such as SiC to contain the fission products and transmit the heat (Fig. 10.2b). The spheres are small particles about 1 mm in diameter packed into a graphite structure that can be either balls or blocks cooled with helium. The very-high-temperature gas-cooled reactor uses this type of fuel. The third type of reactor utilizes a combined fuel–coolant configuration in which the fuel and coolant are one and the same (Fig. 10.2c). One type of homogeneous reactor uses uranium uniformly dissolved in molten salt coolant that is circulated through the reactor vessel. In all cases, the materials challenges come about because of the very high fuel temperatures, the intense radiation flux, and coolant compatibility issues. Thus, the fuel, the cladding, the structural materials, the reactor vessel, and the interaction of these materials with the coolants present the greatest challenges to new, more robust nuclear reactor concepts for future nuclear powers (Guérin et al. 2009; Allen et al. 2010).

An additional source of uncertainty also exists with extended operation or new operating regimes: the potential for new forms of degradation. For example, in the area of radiation effects, in the past, when new reactor operating conditions (temperature, flux, or fluence) have been established at least one new radiation-induced phenomenon has been found. In the 1960s irradiation-induced hardening was discovered. Swelling was a major concern for fast reactors in the 1970s and high-temperature embrittlement due to helium was a surprise in the 1980s. For new Generation IV systems or the extension of current technology, one should be aware of the possibility of new phenomena due to irradiation, corrosion, or aging in both materials and fuels performance (Allen et al. 2010).

INDEPTH: Characteristics of Advanced Nuclear Energy

Advanced nuclear energy includes technologies using a reactor or fuel cycle that offer many of these features:

- Substantially lower capital and/or operational costs
- Reduced material inputs
- Manufacturability or rapid deployment capability
- Passive safety systems and inherent safety strategies
- Ease of operation and maintenance
- Reduced emergency planning zones
- Reduced offsite impact during an accident, and increased flexibility/scalability of siting
- Increased proliferation resistance; decreased waste production and/or an actinide management capacity; more efficient use of fuel resources
- Hybrid generation adaptability (e.g., hydrogen production, desalination) and/or load following

While new water-cooled technologies can have some of these features, non-water-based technologies ultimately have the greatest chance of achieving more of these objectives. Advanced reactor designs range in size from 1 MWe to 1000 MWe or more. More often it is manufacturability, rather than size, that radically lowers costs and reduces construction times.

10.2 Fission and Fusion Reactors

Nuclear energy can be derived from many forms such as nuclear fission energy, fusion energy, and radioisotopic energy. The essence of nuclear fission energy is that the heat produced by the splitting of heavy radioactive atoms (nuclear fission) during the chain reaction is used to generate steam (or other process fluid) that helps rotate the steam turbine generator, thus producing electricity. Nuclear fission energy is the most common mode of producing the bulk of the nuclear energy. Most current commercial reactors are based on boiling water (BWR) or pressurized water (PWR) heat-transfer systems. Interest now focuses on Generation IV designs (Ashby and Smidman 2010): fast breeder, gas-cooled and high-temperature reactors. Early versions of some, like the Liquid-Metal-Cooled Fast Breeder Reactor (LMFBR) and the Advanced Gas-Cooled Reactor (AGR), already exist. Others, such as the Pebble Bed Reactor (PBR) are under study.

A huge amount of energy (much higher than fission) can be produced using the nuclear fusion reaction (deuterium–tritium reaction). There are currently no commercial fusion reactors available. A prototype fusion reactor known as ITER (International Thermonuclear Experimental Reactor) is being built in France and scheduled to produce the first plasma by 2018 (Murty and Charit 2013).

Either radioactive isotopes (e.g., ^{238}Pu, ^{210}Po) or radioactive fission products (e.g., ^{85}Kr, ^{90}Sr) can produce decay heat that can be utilized to produce electric power. These types of power sources are mainly used in remote space applications (Murty and Charit 2013).

Various radiation types are produced in a nuclear environment. It could be alpha particles, beta particles, gamma rays, or neutrons. Neutron is subatomic particle present in almost all nuclides (except normal hydrogen isotope or protium) with a mass of 1.67×10^{-27} kg and has no electrical charge. Collision of neutrons with atom nuclei may lead to different scenarios—scattering of the neutrons and recoil of nuclei with conservation of momentum (elastic scattering) or loss of kinetic energy of the neutron resulting in gamma radiation (inelastic scattering). The capture of neutrons may result in the formation of new nuclei (transmutation), or may lead to the fragmentation of the nucleus (fission) or the emission of other nuclear particles from the nucleus. ^{235}U is the one and only naturally occurring radioisotope (fissile atom) in which fission can be induced by thermal neutrons. The fission reaction of ^{235}U can occur in 30 different ways leading to the possibility of 60 different kinds of fission fragments. From the atomic masses of the reactant and products, a small amount of mass is converted into an equivalent energy following Einstein's famous eq. $E = mc^2$. There is always a competition for neutrons between various processes, namely, (i) fission reaction of fissile atom nuclei, (ii) nonfission capture of neutrons by uranium and other reactions, (iii) nonfission capture of neutrons by other components in the reactor core, and (iv) leakage of neutrons from the core. The reaction can be termed as a chain reaction when the number of neutrons consumed in the processes (ii)–(iv) is at least equal to or less than that consumed in the process (i). Thus, neutron economy plays a very important role in the design of a nuclear reactor. The need for a favorable neutron economy necessitates certain conditions to be met by a chain reacting system. For a given geometry, there is a certain minimum size of a chain reacting system, called the critical size (in terms of volume), for which the production of neutrons by fission just balances the loss of neutrons by leakage and so on, and the chain reaction can be sustained independently. The mass corresponding to the critical size is called critical mass. Dependent on the relative generation of fission neutrons and their loss, the reactor is said to be in different stages: subcritical (neutron loss more than the production), critical (balance between the neutron production and loss, $k = 1$), or supercritical (the neutron production is more than the loss, $k > 1$). The multiplication factor k is often used to express the criticality condition of a reactor. This factor is basically the net number of neutrons per initial neutron (Murty and Charit 2013).

Figure 10.3 shows basic design for a primitive thermal reactor. The tubes containing fuels are generally made of metallic alloys (also known as fuel cladding). The radioactive fuels (such as uranium) could be in metallic, alloy, or ceramic forms. The fuel cladding serves many purposes: it provides mechanical support to the fuel, keeps the fission products from leaving the fuel element, and protects the fuels from corrosion from the coolant. The fuel elements are arranged in a distinct regular pattern (square, hexagon, etc., dictated by neutronics and other factors) with the moderator. Moderator slows down the neutron to sustain the fission reaction with thermal neutrons. The fuel–moderator assembly is surrounded by a reflector. The purpose of a reflector is to direct all neutrons generated towards the core so that

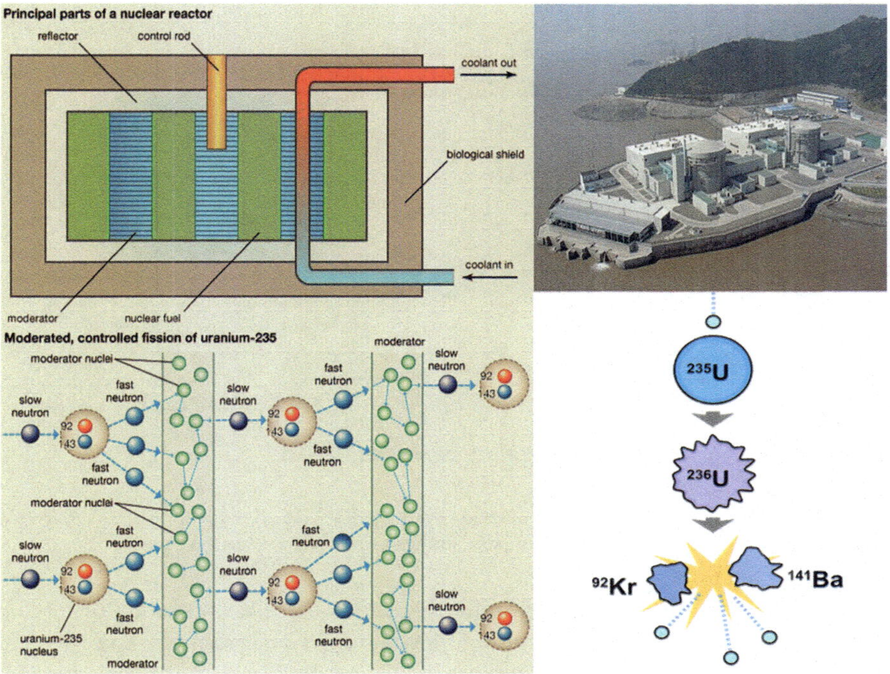

Fig. 10.3 Basic reactor design (Modified from Marcum and Spinrad 2017, https://www.britannica.com/technology/nuclear-reactor/images-videos. Courtesy: Encyclopedia Britannica)

neutron leakage can be controlled, thus improving the neutron economy. On the outside, the reactor is lined by shielding materials that absorb neutrons and gamma rays that escape the core and reduce the radiation intensity to a tolerable level so that people near the reactor are not exposed to these radiations. The control rod (usually an assembly) helps control the chain reaction by absorbing neutrons, maintaining the steady state of operation. Hence, the control materials are neutron-absorbing materials (boron, hafnium, and so forth), and are generally fabricated in the form of rods (in some cases, plates). A reactor is typically equipped with two types of control rods—regulating rods for routine control reasons and safety rods (to permit shutdown in the case of emergency). Moreover, an efficient coolant is needed for a reactor. As a huge amount of heat is generated in the fuel elements, the heat needs to be removed continuously in an efficient manner in order to maintain a safe, steady-state reactor operation. The coolant can be a gas or liquid (such as light or heavy water, carbon dioxide, liquid metals, and molten salts). However, the presence of any coolant tends to adversely affect the neutron economy. Hence, the balance between the reduction in the neutron economy and the efficiency of heat removal needs to be carefully considered (Murty and Charit 2013; Marcum and Spinrad 2017).

Even though a nuclear reactor can be defined in different ways, therefore, almost all reactors except fusion reactor (commercially nonexistent) can be defined as follows: “A nuclear reactor is a device, designed to produce and sustain a long term, controlled fission chain reaction, and made with carefully selected and

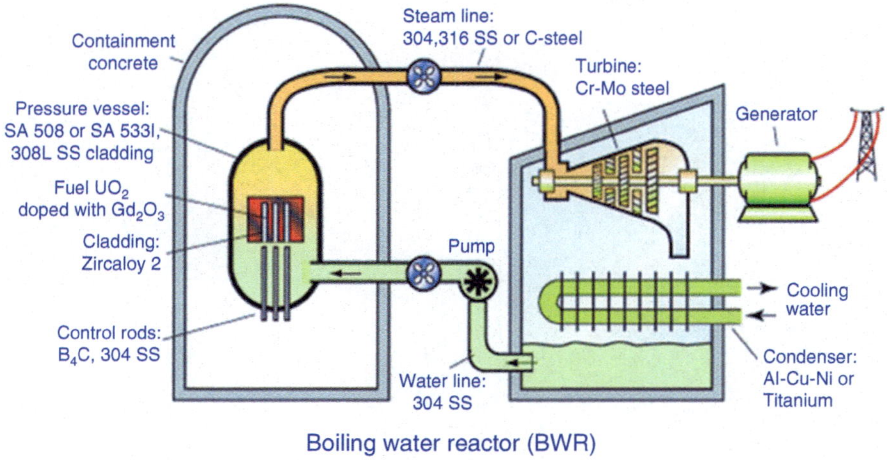

Fig. 10.4 Boling water reactor (Adapted from Ashby and Smidman 2010, http://www.grantadesign.com/download/pdf/nuclear.pdf. Courtesy: Granta material inspiration)

strategically placed collection of various materials." The classification of reactors vary and are generally based on the following: type of fission reaction (thermal, epithermal, and fast reactors), purpose of the reactor (power reactors, research reactors, and test reactors), type of the coolant present (such as light/heavy-water reactors, gas-cooled reactors, and liquid-metal-cooled reactors), type of core construction (cubical, cylindrical, octagonal, and spherical reactors), and so forth (Murty and Charit 2013).

10.2.1 Boiling Water Reactor (BWR)

As shown in Fig. 10.4, the direct cycle BWR system generates steam that is fed to the same sort of steam turbine used in coal or gas fired power systems. The nuclear core assembly consists of an array of Zircaloy 2 tubes encasing enriched UO_2 ceramic fuel pellets. Some of the fuel rods contain gadolinium oxide (Gd_2O_3), which acts as a burnable "poison" absorbing neutrons when the fuel is fresh but burning up as the fuel decays, buffering the neutron flux. The power is controlled by control rods inserted from the bottom of the core and by adjusting the rate of flow of water. The control rods are made of boron carbide (B_4C) clad in stainless steel 304 or 304 L. Water is circulated through the reactor core where it boils, producing saturated steam. The water acts as both a coolant and a moderator, slowing down high-energy neutrons. The steam is dried and passed to the turbine-generator through a stainless steel steam line. On exiting the turbine the steam is condensed, demineralized, and returned as water to the reactor. The schematic in Fig. 10.4 shows the most important materials of the system. The BWR operates at constant steam pressure (7 MPa), like conventional steam boilers and with a steam temperature of about 560 K (Ashby and Smidman 2010).

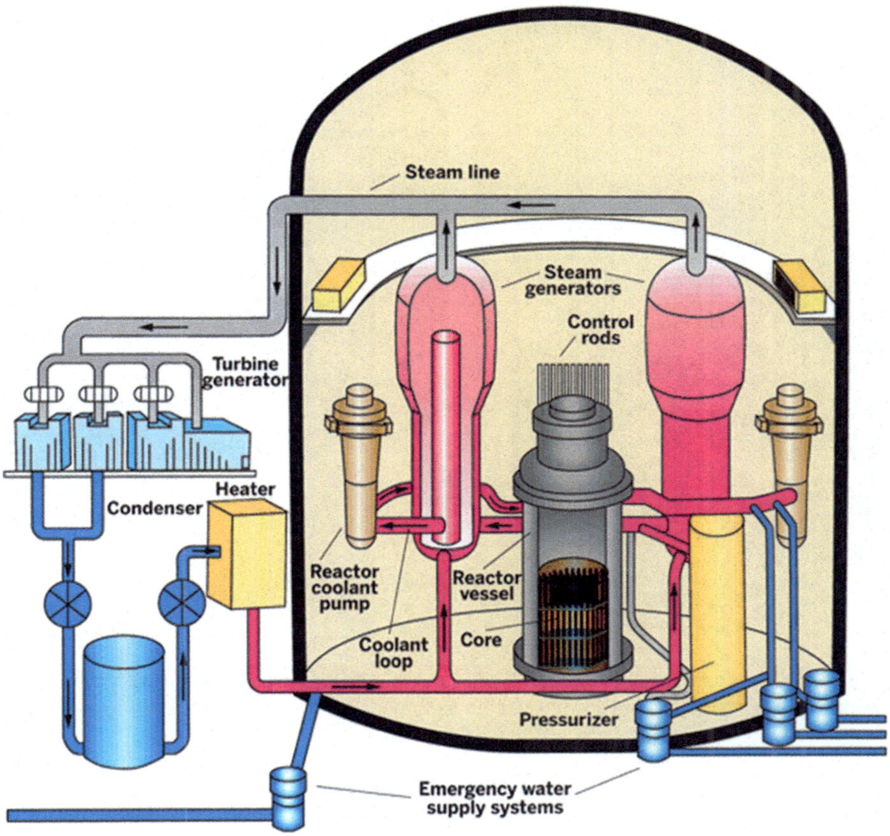

Fig. 10.5 Pressurized water reactor (Adapted from Jacoby 2009. Credit: U.S. Nuclear Regulatory Commission)

10.2.2 Pressurize Water Reactor (PWR)

As shown in Fig. 10.5, the core of a pressurized water reactor (PWR) is not unlike that of a BWR. It has some 200 tube assemblies containing ceramic pellets consisting of either enriched uranium dioxide (UO_2) or a mixture of both uranium and plutonium oxides known as MOX (mixed oxide fuel). These are encased in Zircaloy 4 cladding. Either B_4C-Al_2O_3 pellets or borosilicate glass rods are used as burnable poisons. Water, pumped through the core at a pressure sufficient to prevent boiling, acts as both a coolant and a moderator, slowing down high-energy neutrons. The water, at about 600 K, passes to an intermediate heat exchanger. The power is controlled by the insertion of control rods from the top of the core and by dissolving boric acid into the reactor water. As the reactivity of the fuel decreases, the concentration of dissolved boron ions is reduced by passing the water through an ion-exchanger. Control rods made of boron carbide (B_4C) or an Ag-In-Cd alloy are clad in Inconel 627 or stainless steel (304) tubes. The primary pressurized water loop of a PWR carries heat from the reactor core to a steam generator. The loop is under a

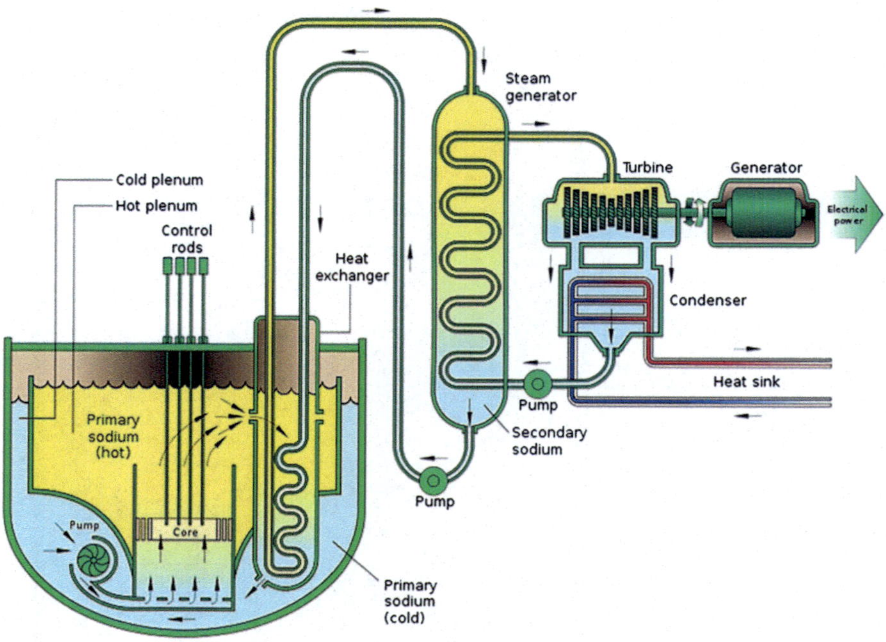

Fig. 10.6 The liquid-metal-cooled fast breeder reactor (Adapted from Ashby and Smidman 2010, http://www.grantadesign.com/download/pdf/nuclear.pdf. Courtesy: Granta material inspiration)

working pressure of about 15 MPa -sufficient to allow the water in it to be heated to near 600 K without boiling. The heat is transferred to a secondary loop generating steam at 560 K and about 7 MPa, which generates heat that drives the turbine (Jacoby 2009; Ashby and Smidman 2010).

10.2.3 Liquid Metal Fast Breeder Reactor (LMFBR)

As shown in Fig. 10.6, a LMFBR is a liquid-sodium-cooled reactor that makes use of a fast neutron spectrum and a closed fuel cycle. The liquid sodium coolant transfers heat from the reactor core and is pumped through the primary loop at about 800 K. This sodium in this loop becomes radioactive, requiring an intermediate sodium filled heat-exchange loop to prevent possible leakage of radioactive material outside the containment structure. The sodium in this secondary sodium loop, made of type 324 and 316 stainless steel, alloy 800 or Cr-Mo steels, passes to a steam generator where it heats water to generate steam at 750 K. The turbine and generator are essentially the same as those of a BWR or PWR. Current fuel materials include mixed uranium and plutonium oxides (~25% PuO_2), metal alloys such as U-Pu-Zr, and mixed uranium or thorium carbides and nitrides. The usual choice is a fuel assembly made up of mixed uranium dioxide (UO_2) and plutonium dioxide (PuO_2) fuel rods clad in type 316 stainless steel. This is surrounded by the "breeding blanket"

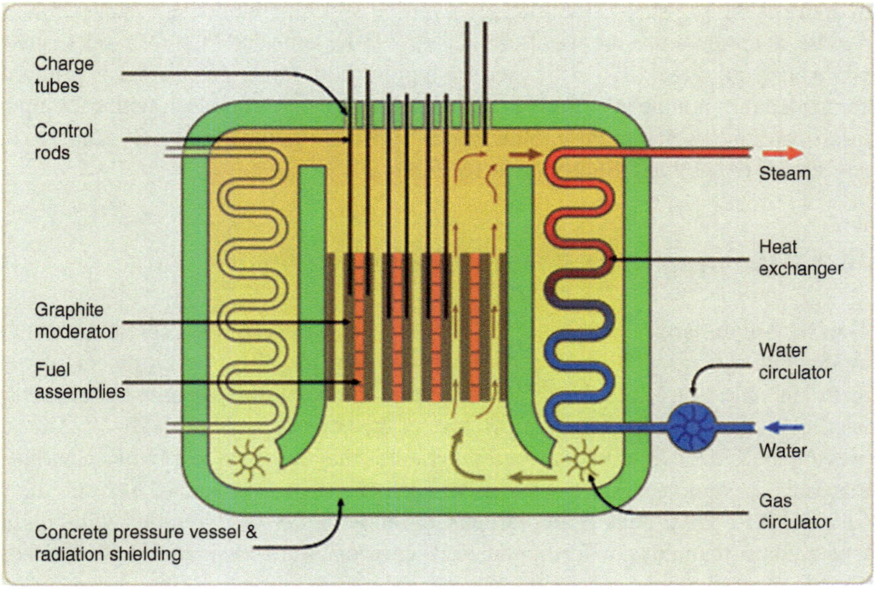

Fig. 10.7 The Advanced Gas-cooled Reactor (AGR) (Adapted from Pioro 2013, licensed by InTech under the terms of the Creative Commons Attribution License)

containing depleted UO_2 pellets. The control rods, like those of a BWR, are boron carbide (B_4C) clad in type 316 stainless steel and enter from the top of the core. An LMFBR can have either pool or loop designs. A pool design has the intermediate heat exchangers and primary sodium pumps immersed in the reactor vessel while a loop design has these elements external to it. One of the selected generation IV systems, the sodium-cooled fast reactor (SFR) utilizes a similar design to the LMFBR. The next-generation lead-cooled fast reactor (LFR) uses liquid lead as a coolant and utilizes a somewhat different reactor design (Ashby and Smidman 2010).

10.2.4 Advanced Gas-Cooled Reactor (AGR)

As shown in Fig. 10.7, the advanced gas-cooled reactor (AGR) is graphite moderated and cooled with carbon dioxide (CO_2). The core consists of high strength graphite bricks mounted on a steel grid. Fuel rods of enriched UO_2 clad in stainless steel (20Ni–25Cr) are placed in graphite sleeves and inserted into vertical channels in the bricks. Gas circulators blow CO_2 up through the core and down into steam generators. Holes in the graphite allow access to the gas. The outlet temperature of the CO_2 is about 943 K at a pressure of 4 MPa. The graphite in the core is kept at temperatures below 723 K to avoid thermal damage. The reactor core, gas circulators, and steam generators are encased in a pressure vessel made of pre-stressed concrete lined with a mild steel to make it gas tight. Mild steel is used in areas of the pressure vessel that are exposed to temperatures less than 623 K. In regions at temperatures between 623 K and 793 K, annealed 9Cr-1Mo steel is used while austenitic steel (316 H) is used for regions hotter than this. Power is primarily controlled through the insertion of control rods made of boron-steel, with back-up by insertion of nitrogen into the cooling gas or by releasing fine boron-rich balls into the gas stream (Ashby and Smidman 2010; Pioro 2013).

10.2.5 Very-High-Temperature Reactor (VHTR)

The very-high-temperature reactor (VHTR) is a proposed IV generation design, moderated with graphite and cooled with helium gas. The development of new materials able to tolerate the higher operating temperatures presents a major engineering challenge. The outlet temperature of the coolant is about 1123–1223 K at a pressure of 7 MPa. Internal reactor temperatures may reach up to 1470 K. Candidate materials for regions at temperatures between about 1030 K and 1270 K are alloys 617, X, XR, 230, 602CA or variants of alloy 800H. For regions with higher temperatures than this, the leading material candidates are composites with a carbon-fiber-reinforced carbon matrix (C_f/C) or carbon-fiber-reinforced silicon carbide (C_f/SiC). The most promising pressure vessel material is modified 9Cr-1Mo steel. Some designs maintain the vessel at lower temperatures, in which case current pressure vessel materials could be used such as SA 508 steels. The helium coolant is heated in the reactor vessel and flows to the intermediate heat exchanger (IHX). Heat is transferred to a secondary loop with either helium, nitrogen and helium, molten salt, or pressurized water. The materials of the IHX depend on the operating temperatures and the nature of the secondary coolant; Alloy 617 is a primary candidate. The heated fluids can either be used to drive a turbine or to produce hydrogen. All VHTR designs make use of tri-structural isotropic (TRISO)-coated fuel particles. The particles are 750–830 μm in diameter and consist of a kernel of fuel material coated with two layers of pyrolytic carbon with a layer of silicon carbide in between. These particles can be utilized in either prismatic or pebble bed reactors. In a prismatic reactor the kernel consists of enriched uranium oxycarbide (UCO) and the particles are packed into cylindrical compacts which are placed into graphite fuel elements. However, a pebble bed reactor uses particles with an enriched uranium dioxide (UO_2) kernel and these are formed into 60 mm diameter spheres (the "pebbles"), as shown in Fig. 10.8, the fuel pebbles are fed into the core mixed with nonfuel graphite pebbles that act as reflectors to even the heat generation (Ashby and Smidman 2010; Yurman 2017).

10.2.6 Fusion Reactor

As shown in Fig. 10.9, the International Thermonuclear Experimental Reactor (ITER) is an experimental fusion reactor designed to produce 500 MW of power from an input of 50 MW. It is a step towards the use of the fusion energy for

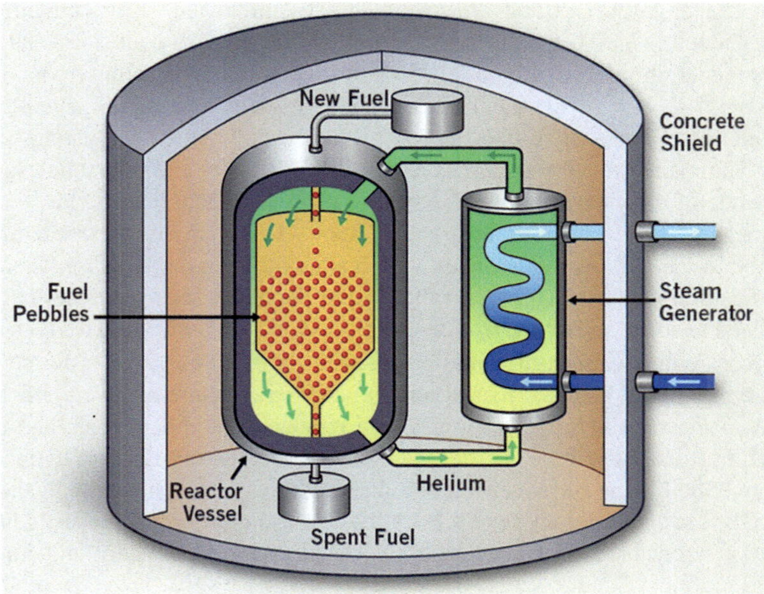

Fig. 10.8 A pebble bed advanced nuclear reactor. In some designs the helium heat-transfer medium drives turbines to compress the gas and generate power; in others it is fed to a heat exchanger where it passes its heat to a secondary helium loop or to steam loop (Adapted from Cameco 2018, available at https://www.cameco.com/uranium_101/electricity-generation/types-of-reactors/. Courtesy of Cameco Corporation)

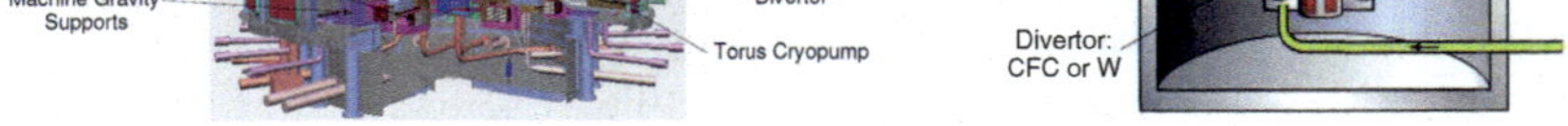

Fig. 10.9 The International Thermonuclear Experimental Reactor (ITER) (Adapted from Ashby and Smidman 2010, http://www.grantadesign.com/download/pdf/nuclear.pdf. Courtesy: Granta material inspiration)

electricity production and other commercial applications. In all proposed fusion reactors, energy is released from the fusion of deuterium and tritium nuclei. This requires a temperature of about 100MK at which the gases form a plasma. No materials operate at such temperatures, so the ITER uses magnetic confinement to contain the plasma, allowing fusion without contact between the plasma and the containing walls. The ITER uses a tokamak design. The plasma is contained in a torus shape using strong magnetic fields produced by circumferential superconducting coils and a large central solenoid. The coils are made of a superconducting niobium-tin alloy (Nb_3Sn) or niobium-titanium (NbTi) alloy cooled to 4 K with supercritical helium. The plasma is enclosed in a sealed torus vacuum vessel made up of two steel walls with water coolant circulating between them. The main structural materials are 316L(N)-IG, 304 and 660 stainless steels. The inside of the vacuum vessel is covered with the blanket that shields the vessel and magnets from heat and neutron radiation. This consists of shield modules attached to the vacuum vessel inner wall. Each module has a 316L(N)-IG stainless steel shield block carrying a first wall panel of beryllium facing the plasma. These are joined to a heat sink made of a copper alloy (CuCrZr) with 316L(N)-IG stainless steel tubes with a coolant flowing through them. It is the energy transferred to this coolant that would be used in electricity production in future plants. At the bottom of the vacuum vessel is the diverter which removes heat, helium ash, and plasma impurities. Materials of the diverter facing the plasma must withstand temperatures of up to 3300 K. The current choices of materials are a carbon fiber composite (CFC SEP NB31) and tungsten (99.94 wt.% W). The entire structure, including the magnets, is enclosed in a stainless steel vacuum cryostat (Ashby and Smidman 2010).

10.3 Materials Selection Criteria for Nuclear Power Applications

Materials selection needs to meet requirements that are defined as the purpose, goals, constraints, and criteria associated with a design project. Different design standards or codes often provide different technological requirements of materials that occasionally exhibit significant divergence. It is very common to encounter various tradeoffs during the materials selection process for a given application, and most times compromise is called for. Moreover, a nuclear reactor design entails complex procedures in itself given the multiple challenges. Different components of a reactor may require different types of materials. There could be two broader types of materials selection considerations—general and special. General considerations involve factors such as mechanical strength, ductility, toughness, dimensional stability, fabricability, cost and availability, heat transfer properties, and so on. General properties come from the general engineering considerations as they would be applicable in most engineering designs. On the other hand, special properties are considered solely because the materials are to be used in a nuclear reactor. These include properties like the neutron absorbing characteristics, susceptibility to induced radioactivity, radiation damage resistance, and ease of reprocessing of materials. Each of the material characteristics is evaluated following standard (sometimes nonconventional) testing procedures. The knowledge of service conditions and broader goals of the future reactor is a must for a successful materials selection process. This information may come from utilizing predictive capabilities (modeling and simulation tools) and/or known data/experiences from previous reactor design and operations, if available (Murty and Charit 2013; Rodríguez-Prieto et al. 2016).

10.3.1 General Considerations

10.3.1.1 General Mechanical Properties

Important general mechanical properties include tensile strength, ductility, and toughness. The material should be strong enough to bear the loads of the structure and also sustain any internal or external stresses generated during service. Also, the material should have enough ductility (a measure of percentage elongation or reduction in area in standard tensile specimens) to avoid any catastrophic failure. Usually, as a rule of thumb, a percentage elongation of 5% is considered a minimum requirement for a load-bearing engineering structure. But one must admit that this often changes with the type of application at hand. In some cases, the materials should have sufficient ductility in order to be formed into different components. Toughness is defined as the ability of a material to absorb energy without failure, and that dictates how tough a material is for use. Generally, tensile strength and ductility combined is referred to as toughness. However, generally impact tests and fracture toughness tests are conducted to evaluate toughness properties of materials. All these affect the mechanical integrity of the reactor components (Murty and Charit 2013).

10.3.1.2 Fabricability

Fabricability includes a host of characteristics such as formability, weldability, machinability, and so on. If fabricability issues are not dealt with during the first stage, it may cause problems at the later stages. In many cases, some parts of the nuclear power plant are to be built at the site (also called field fabrication) from smaller parts. If the materials do not have the requisite fabricability, it would not be possible to use the material no matter what fantastic properties it may have (Murty and Charit 2013).

10.3.1.3 Dimensional Stability

The material should have adequate stability in properties. For example, many nuclear components would work at higher temperatures for extended period of time. So, the creep deformation (i.e., time-dependent plastic deformation) may cause dimensional stability problems. The microstructure of a material may also change as a function of temperature, time, and stresses. So, the effects of these factors on microstructure and the consequent effects on the properties need to be taken into account carefully (Murty and Charit 2013).

10.3.1.4 Corrosion Resistance

Corrosion is an electrochemical process that causes the surface of the metals/alloys degrade over time in the presence of a chemical environment. Corrosion resistance of materials used in nuclear components is important in many applications to ensure that they serve as desired. The "cost of corrosion" can result in immediate property and life endangerment and increased downtime, leading to substantial losses. Many nuclear components inside the reactor stay in close contact with reactor fluids (e.g., coolant in the form of liquid or gas). These effects get exacerbated due to the presence of radiation fields (Murty and Charit 2013).

10.3.1.5 Design Fault

Although design does not generally fall under the purview of a materials engineer, he/she is in a unique position to figure out early whether the faulty design would pose a problem. Designs leaving stress concentration sites (sharp recesses, keyholes, and the like) are typically unwarranted in load-bearing applications since it may interfere with the capability of the component to serve properly. For example, fatigue properties are especially prone to the presence of stress concentration sites (Murty and Charit 2013).

10.3.1.6 Heat Transfer Properties

Heat transfer modes are of three types—conduction, convection, and radiation. The first two processes are of importance in nuclear reactor materials selection. Most important example in nuclear reactor is the choice of fuel and cladding materials. The safety and efficiency of the reactor depends on how efficiently the heat generated inside the fuel can be removed. Hence, thermal conductivity is an important property. Otherwise, the fuel will melt such as in a "loss of coolant accident" (LOCA) scenario. Similarly, heat transfer properties are also important for various balance-of-plant features, such as heat exchangers, condensers, and other ancillary equipment (such as steam generator in a PWR system) (Murty and Charit 2013).

10.3.1.7 Availability and Cost

Availability and cost are the economic consideration that may trump technical considerations. If a material is not available in the form or at an allowable price, the prudent engineering decision would be to find an alternative material with similar properties or a different form of material or make changes in design to allow different characteristics. Cost-benefit analysis must be at the heart of that process (Murty and Charit 2013).

10.3.2 Special Considerations

10.3.2.1 Neutronic Properties

Neutronic properties are of significant consideration in the design and development of nuclear reactors. As the fission chain reaction requires continued supply of neutrons for it to proceed, neutron economy plays an important role. Fuel cladding materials need to have a lower neutron absorption cross section and that is why zirconium alloys are used in LWRs. On the other hand, to control the chain reaction, the control materials should have high neutron absorption cross section. The same consideration would also apply to shielding materials (Murty and Charit 2013).

10.3.2.2 Susceptibility to Induced Radioactivity

The materials in the reactor can absorb fast/thermal neutrons and undergo reactions that may lead to the production of different radioactive isotopes of the constituent elements of the materials. These reactions can induce radioactivity as these isotopes would decay by emitting gamma rays, beta rays, and alpha rays of different energy levels. While selecting an alloy, the following factors should be considered: (a) quantity of the impurities/alloying elements, (b) abundance of the isotopes and corresponding cross section, (c) half-life of the product nuclide, and (d) the nature of the radiation produced. If the produced isotope has a short half-life and emit radiation of low energy, it should not be a cause for great concern. However, if the isotope is long-lived and produces radiation of high energy, all precautions must be taken. For example, the main isotope of iron (^{56}Fe) that accounts for almost 92% of the natural iron forms a stable isotope (^{57}Fe) upon absorbing neutrons. The absorption of neutrons in ^{54}Fe and ^{58}Fe yielding ^{55}Fe (half-life: 2.9 years) and ^{59}Fe (half-life: 47 days) results in activation. However, the impurities or alloying elements cause more induced radioactivity than iron itself. Generally, the test samples irradiated in a reactor are not examined immediately after taking out from the reactor because they remain literally hot and continue to be hot due to the decay heat produced by various reactions even if the fission chain reaction no longer occurs. The Fukushima Daiichi accident in Japan did show the severity of the heat produced due to these decay reactions even after the emergency shutdown of the reactor, leading to very high temperatures (in the absence of proper coolant) and eventually resulting in the cladding breach and perhaps some form of core melting (Murty and Charit 2013).

10.3.2.3 Radiation Stability

Energetic radiation plays a significant role in modifying the microstructure of the materials involved. However, radiation damage under the fast neutron flux involves atomic displacements (i.e., displacement damage) leading to the creation of a host of defects in the material. The effects of radiation can be diverse, including radiation hardening, radiation embrittlement, void swelling, irradiation creep, and so forth, with all having significant effects on the performance of the reactor components. Another effect of radiation is the radiolytic decomposition of coolant (e.g., water molecule is radiolyzed into more active radicals) that may definitely affect the corrosion behavior of the reactor components. Fission fragments also cause damage, but they are mostly limited to the fuel. So, for selecting materials for a nuclear reactor, the concomitant radiation effects should be evaluated on these materials. That is why millions of dollars are spent to wage materials irradiation campaigns in test reactors followed by careful post irradiation examination to ascertain fitness-for-service quality of the materials to be used in nuclear reactors (Murty and Charit 2013).

10.3.3 Application of Materials Selection Criteria to Reactor Components

Table 10.2 shows general application of materials selection criteria to reactor components. Taking the example of fuel cladding material for the LWR, they are used to encapsulate the fuel and separate it from the coolant. The requirements for fuel cladding material are as follows: (a) low cross section for absorption of thermal neutrons, (b) higher melting point, adequate strength and ductility, (c) adequate thermal conductivity, (d) compatibility with fuel, and (e) corrosion resistance to water. Following the first factor, different metals have different cross sections for absorption of thermal neutrons. Although Be, Mg, and Al all have lower cross sections for absorption of thermal neutrons, other nonnuclear factors become the impediment for their use in commercial power reactors. Even though Be has a high melting point (1278 °C), it is scarce, expensive, difficult to fabricate, and toxic. Mg has a low melting point (650 °C), is not strong at higher temperatures, and has poor resistance to hot water corrosion. Al has a low melting point (660 °C) and poor high-temperature strength. Even though an Al-based alloy has been used as fuel cladding materials in reactors like ATR, and in the past a magnesium-based alloy was used in Magnox reactors, their use remains very limited. This leaves zirconium-based

Table 10.2 Generic information about subassemblies and their functional objectives (Mukherjee and Jamnapara 2015)

	First wall	Divertor	Breeder blanket
Functional objectives	To shield subassemblies from thermal loading and plasma exposure	To extract particles/He dust, survive high heat flux	To breed T fuel, utilize 14.1 MeV neutrons, shield subassemblies from neutrons, extract heat for electricity generation
Plasma facing materials	W, W-based alloy, W-coated SiC, Be, W-coated ODS/RAFM steels, flowing liquid Li	W-based alloy (ODS-W, etc.), W-coated SiC_f/SiC; flowing liquid metal: Li, Ga, Sn, Sn–Li	As in the first wall
Neutron multiplier	—	—	Be, $Be_{12}Ti$, Be_{12} V, Pb
Tritium breeding material	—	—	Liq. Li, Eutectic Pb–Li, Li-based ceramic pebbles (Li_2O, Li_4SiO_4 + 2.5% SiO_2, Li_2TiO_3, Li_2ZrO_3, $LiAlO_2$)
Structural material	RAFM steel, ODS steel, V-based alloy, SiC_f/SiC	ODS steel, W-based alloy	RAFM steel, ODS steel, V-based alloy, SiC_f/SiC
Coolant	—	Water, He	Water, He, Eutectic Pb-Li, Li

Note: *F–M* ferritic–martensitic stainless steels (typically 9–12 wt.% Cr), *ODS* oxide-dispersion-strengthened steels (typically ferritic–martensitic), *MC* mixed carbide [(U,Pu)C], *MN* mixed nitride [(U,Pu)N], *MOX* mixed oxide [$(U,Pu)O_2$]

materials as the mainstay of fuel cladding materials for LWRs. Zirconium has various favorable features: (a) relatively abundant, (b) not prohibitively expensive, (c) good corrosion resistance, (d) reasonable high-temperature strength, and (e) good fabricability. Some of the properties could be further improved through appropriate alloying (Murty and Charit 2013).

10.4 Materials for Advanced Nuclear Fission Reactor Components

Harnessing of nuclear energy involves a technology that leans very heavily on materials, as far as the economic efficiency and societal impact of this technology are concerned. In the nuclear field, apart from a host of rigorous conventional requirements, such as those pertaining to thermal and mechanical stress, corrosion, and so on, there are novel and unusual nonconventional requirements, such as those pertaining to the resistance to the influences of nuclear radiation, the interaction with neutrons, the production of fission products, and so on. It is, therefore, not surprising for one to perceive the nuclear field not only as including within its realm preoccupation with a large number of new, or less common, materials but also as evolving new standards of references and codes in evaluating and utilizing already known materials. In this respect, it is entirely in order to bring to the fore a specific class of materials which are called nuclear reactor materials. For nuclear fission reactor systems, they generally include structural materials, fissile and fertile fuel materials, moderator, reflector, cladding materials, coolant materials, and control, shielding, and safety system materials (Gupta 2001).

10.4.1 Structural Materials

Structural materials used in the cores of advanced reactors will face unprecedented combinations of temperature, radiation dose, and stress. As shown in Table 10.1, the common feature of all advanced designs is high temperature as compared to current light-water reactors. Another unique feature is the simultaneous presence of intense knock-on displacement damage by the fission neutrons, which is quantified in units of displacements per atom (dpa). A damage level of 1 dpa corresponds to the displacement of all atoms in the material. Through self-healing processes induced by atomic diffusion (using specially designed radiation-resistant materials with a high density of nanoscale point-defect recombination centers), most of the displacement damage defects can be induced to recombine so as to keep the net accumulated radiation damage at a low value. The challenges associated with the high temperature, high-dose operating environment will place increased emphasis on strength, creep, and creep-fatigue behavior in addition to fracture toughness at low temperature. Particle strengthening is one approach to achieving strength at high temperature. However, irradiation can alter phase stability, and many intermetallic phases used in strengthening are unstable under irradiation (Guérin et al. 2009). Hence, the service environments predicted for the advanced reactor components pose significant challenges to materials selection and qualification efforts.

Some desirable characteristics for the structural materials of the advanced reactor components may include (Murty and Charit 2008): (a) Excellent dimensional stability against thermal and irradiation creep, void swelling, etc. (b) Favorable mechanical properties such as strength, ductility, creep rupture, fatigue, creep–fatigue interactions, etc. (c) Acceptable resistance to radiation damage (irradiation hardening and embrittlement) under high neutron doses (10–150 dpa or displacements per atom), helium embrittlement, etc. (d) High degree of chemical compatibility between the structural materials and the coolant as well as with the fuel. In this regard, stress corrosion cracking (SCC), irradiation-assisted stress corrosion cracking (IASCC), and many other issues are important. In addition, workability, weldability, cost, etc. are other important aspects that need to be looked into during the materials selection process. All these requirements are related to the fundamental high-temperature degradation mechanisms such as phase instability, oxidation, radiation-induced segregation, and so forth.

10.4.1.1 Austenitic Stainless Steels

Austenitic stainless steels have good creep resistance to higher temperatures coupled with reasonable corrosion/oxidation resistance. Alloys like 316LN, D-9, etc. are good examples.

However, relatively large amount of void swelling at moderate neutron doses remains a major performance-limiting issue. The extent of swelling for an austenitic stainless steel (316 SS) is much higher compared to ferritic or F–M ones. In some applications, their low thermal conductivity may adversely affect the reactor efficiency. Radiation-induced segregation (RIS) and phase stability issues will also play a major role. For example, radiation-assisted depletion of Cr from the grain boundaries may render the austenitic steels susceptible to corrosion in water or lead-alloy-cooled systems. The interactions between RIS and void swelling may also be considered. Further, irradiation creep and thermal creep may have important implications on the dimensional stability and performance of the reactor (Corwin 2006; Murty and Charit 2008).

One of the main challenges is to increase the radiation resistance of austenitic alloys by designing self-healing microstructures (doping with trace elements, cold deformation and precipitation of dispersed phases) to prevent void swelling. So far, however, the challenges for austenitic stainless steel remain unfulfilled and other families of stainless steel like F/M seem more likely to meet the strict design constraints especially for the selection of fuel cladding materials (Azevedo 2011).

10.4.1.2 Ferritic/Martensitic (F–M) Stainless Steels

Generally, the microstructure of F/M steels (9–12% Cr steels) is designed by suitable balancing of ferritic and austenitic stabilizing alloying elements in order to produce 100% austenite upon austenitization, and 100% martensite upon quenching or normalizing following austenitization. A tempering step at ~760 °C transforms much of the martensite to ferrite resulting in a tempered martensite structure.

Initially, these steels were developed as structural materials for fossil fuel power plants. Continuous development with alloy and microstructural modifications has made their likely use in some Gen-IV reactors. F–M steels with 9–12% Cr are considered for use; some examples being HT-9, T-91, NF12, etc. because they tend to have better corrosion/oxidation resistance than the low-Cr ones. Some advanced F–M steels have also been considered for fusion reactor applications (first wall and blanket applications) because of their noted reduced-activation (RA) property that refers to a quick radioactive decay after neutron irradiation, allowing shallow burial of the components after component replacement or plant decommissioning. This property will be helpful in Gen-IV reactors, if not the sole guiding factor. Further, they have good void swelling resistance and relatively good creep resistance. However, there are concerns regarding their low long-term creep rupture strength at higher temperatures and irradiation embrittlement at or less than 400 °C. Creep–fatigue interactions become important in the next-generation reactors. Hence, database pertaining to those effects is a must to qualify a material for Gen-IV application (Klueh 2005; Murty and Charit 2008).

Moreover, ferritic/martensitic (F/M) stainless steels, presenting 9–12% Cr, are potential candidates for the cladding material in a number of GIV reactors, such as SFR, SCWR, and LFR reactors in temperatures up to 500 °C. When compared to austenitic stainless steels, F/M steels feature better thermal properties (higher thermal conductivity and lower thermal expansion); better compatibility to heavy liquid metals, such as Pb used as coolant for LFR; better void swelling and He-embrittlement resistance, even at displacement doses above 150 displacements per atom (dpa) due to its BCC/BCT crystal structure and its tempered martensite microstructure, with high density of irradiation-defect sinks. Additionally, the radiation-induced hardening and embrittlement are negligible when irradiation occurs in the temperatures above 450 °C. However, microstructural instability of F/M steels under irradiation and high temperatures is a major concern because of its deleterious effect on both mechanical and corrosion properties. For instance, radiation-induced segregation is promoted with an increase of neutron irradiation fluence, leading to intergranular segregation and intergranular fracture. In addition, thermal and irradiation-induced microstructural instabilities, such as the precipitation of brittle phases; the growth of the lath-packet substructures and the $M_{23}C_6$ and (V or Nb) C precipitates; and the formation of grain boundary cavities degrade the mechanical properties of F/M steels, limiting its maximum service temperature. As a matter of fact, in the extremely demanding design conditions for commercial fast reactors, both austenitic and ferritic–martensitic steels seem to have severe limitations, the former due to swelling and the latter due to insufficient creep strength. One approach to improve high-temperature strength in F/M steels is by introducing a high number density of stable, fine-scale, precipitates that pin the motion of dislocations. Considerable progress has been made on the development of oxide dispersion strengthened (ODS) F/M steels and ferritic alloys (Azevedo 2011).

10.4.1.3 Oxide Dispersion Strengthened (ODS) Steels

Oxide-dispersion-strengthened alloys have been developed to enhance the strength by nanometer-sized oxides of titanium and yttrium. These oxides might be more stable under irradiation and provide significant strength advantages over ferritic–martensitic alloys at much higher temperatures. ODS steels are made through mechanical alloying process. These alloys have good high-temperature properties, radiation resistance in terms of swelling and radiation embrittlement. One example is 12YWT (12YWT: Fe–12.29Cr–3W–0.39Ti–0.248Y_2O_3). It has been found that small nanoclusters of Y–Ti–O particles hinder dislocation motion effectively, and can act as effective sinks for radiation-induced defects. These are called nanostructured ferritic alloys (NFA). The presence of Ti and W helps in creating much smaller oxide particles with average size of 5–6 nm. These particles enhance the creep strength of 12YWT at 700 °C compared with 12Y1 (no Ti, W) and 12YW (no Ti). Various ODS steels demonstrate radiation hardening at lower temperatures, while at higher temperatures, the hardening got reduced drastically. However, the total strain (or ductility) of the irradiated steels have no apparent changes at any temperature. Basically, nanoscale design of structural materials may have major implications in future nuclear reactor systems (Kim et al. 2003; Murty and Charit 2008).

Therefore, priority has been given to the development and study of Fe–Cr oxide dispersion strengthened (ODS) alloys for structural in-core applications of GIV reactors within the European GETMAT (Gen IV and transmutation materials) project. ODS alloys might show good properties for high temperature (~800 °C) and high burn-up applications, being a candidate material for the fuel cladding of SFR, SCWR, LCR, and GCR reactors. An important approach for designing the microstructure of these radiation-resistant alloys is based on the introduction of a high, uniform density of thermal and irradiation-stable nanoscale particles, using particles of titania (Ti_2O_3) and yttria (Y_2O_3), distributed in a ferritic or tempered martensitic matrix, with chemical composition typical of F/M stainless steels or Fe–Cr Incoloy. These nanoparticles are thought to act simultaneously as obstacles to dislocation motion, providing high creep strength, and as sinks for the radiation-induced point defects, providing good radiation damage resistance. Manufacturing of ODS alloys involves mechanical alloying (MA) of metallic and ultrafine oxide powders, through the repeated fracturing and welding of a mixture of powder particles in a highly energetic ball mill, which results in the formation of microcomposite particles with a refined and highly deformed microstructure. The MA is followed by the consolidation of the microcomposite powder by thermo-mechanical processing, such as hot extrusion, rolling, or hot isostatic pressing. The final product features an oriented microstructure, presenting highly directional properties, with the creep strength in the transverse or hoop direction being approximately 50% of that in the axial direction. In this sense, manufacturing ODS alloys products with equiaxed grains seems to be a crucial challenge for its use as a fuel cladding material. Additionally, the manufacturing of long fuel rods also challenges the processes of consolidation and joining. The main in-service issues for ODS alloys under fast neutron radiation are the effect of phase instability under fast neutron radiation on the mechanical properties at lower temperatures, and the effect of the stability of the oxide dispersion and the effect of intermetallic phase precipitation on the toughness at higher temperatures (creep properties). Furthermore, the use of ODS alloys for nuclear application is nearly unexplored and there is some concern on the deleterious effects of high doses of neutron irradiation (up to 200 dpa) on the macro and microstructure of ODS alloys. In addition, the selection of cladding material for LFR and SFR reactors imposes an additional challenge to the cladding material, and the compatibility of the ODS alloys in these environments should be further investigated (Azevedo 2011).

Consequently, challenges to the use of these alloys include fabrication, embrittlement, and perhaps adverse chemical interaction with the surrounding environment. The need for more robust materials also extends to the pressure vessel, the primary safety structure for most reactor designs. Higher core temperatures, pressures, and radiation fields will also place greater demands on the pressure vessel, pushing the need to look beyond the low-alloy ferritic steels currently used in commercial fission reactors (Guérin et al. 2009).

10.4.1.4 Ni-Base Alloys

Ni-base alloys have traditionally been used for high-temperature applications. Therefore, it is only prudent to study their viability in Gen-IV reactor applications. Ni-base superalloys (such as, IN740: Ni–2Fe–24Cr–20Co–2Nb–0.5Mo–2Ti–1C, wt.%) have good creep rupture properties and high-temperature strength. The main problems with Ni-base alloys would be the radiation embrittlement, swelling, and phase instability under neutron radiation environment. Their applicability in balance-of-plant features (turbines, steam-generators, etc.) where radiation effects are minimal is possible. However, high-temperature He embrittlement is an issue to be looked at for applications in GFR/VHTR reactors. More research is needed to better judge their viability for Gen-IV reactors. Currently, a solid solution strengthened Ni-base superalloy (Alloy 617) is being considered for heat exchanger applications for next-generation nuclear plant (NGNP) that incorporates VHTR reactor concept (Tancret et al. 2003; Murty and Charit 2008).

Additionally, because of their good creep rupture properties and high-temperature strength, Ni alloys are considered as a cladding material for SCWR, especially Inconel 690, 625, and 718. The main problem with Ni-based alloys is their low radiation resistance, due to its FCC matrix, which promotes radiation embrittlement, swelling, and phase instability (radiation-induced precipitation of a large amount $Ni_3Ti(Al)$ phase). The loss in ductility is associated with various grain boundary phenomena, such as: radiation-enhanced precipitation of intermetallic and growth of thermally precipitated carbide phases, such as $M_{23}C_6$ and M_6C; radiation-induced segregation; and precipitation of He bubbles. It seems possible, however, to optimize the microstructure of Ni alloys to fight the deleterious effects of neutron irradiation for the potential use of Ni alloys as a cladding material (Azevedo 2011).

10.4.1.5 Refractory Alloys

Refractory metals (such as, Nb, Mo, Ta, and W) have melting temperatures in excess of 2000 °C. Hence, they should have potential applications at high temperatures. Although refractory metals possess good creep resistance and swelling resistance up to high burnups, they have poor oxidation resistance coupled with low temperature radiation embrittlement and fabrication (joining) difficulties (Murty and Charit 2008).

Refractory metals, however, have been considered for use in NASA's nuclear reactors, which operates above 800 °C. For instance, niobium alloys are suitable for applications up to 800 °C and tungsten alloys up to 1200 °C. Although refractory metals generally possess good creep, oxidation, and swelling resistances, they show average radiation resistance, high cost, and fabrication problems. For instance, some BCC refractory metals exhibit ductile-to-brittle transition temperatures (DBTT) above room temperature, requiring particular care in fabrication and handling (W is brittle around 330 °C). Nb alloys offer the advantages of fabrication, high ductility and melting temperature, and low DBTT and neutron absorption cross-section. However, the strength of Nb alloys is relatively low and their resistance to oxidation at elevated temperature is extremely poor, requiring extra care during fabrication and handling. Nevertheless, Nb–1%Zr–0.1%C alloy is the current choice for uranium nitrate fuel cladding material for space nuclear power systems and to avoid its potential attack by fission products a thin layer of rhenium on the inner surface of the cladding acts as a protective barrier. Comparably, ODS alloys might be a more feasible choice for fuel cladding material, especially for commercial reactors, such as LFR and GFR (Azevedo 2011).

10.4.1.6 Zirconium Alloys

Zirconium freed of hafnium to neutronically acceptable levels is commonly known as nuclear grade zirconium, and it is such a specified zirconium that is used by the nuclear industry. Zirconium alloys containing tin, iron, chromium, and nickel are known as zircaloys. The zircaloys have a much better corrosion resistance, greater mechanical strength, sufficient ductility, and acceptable neutronic characteristics compared with unalloyed zirconium (Gupta 2001).

There are significant advantages associated with using a more neutron-transparent cladding material for the SCWR reactors, but the design constraints for SCWR also focus on the high-temperature (creep and corrosion) properties (up to 600 °C) and irradiation resistance. Zr alloys initially did not receive much consideration to SCWR reactors because it was believed that their mechanical strength and corrosion resistance would not be satisfactory at higher service temperature and radiation exposure. The corrosion of Zr alloys usually produces a multilayered oxide structure, due to a repetitive process of oxide growth followed by a transition to the next layer and the full understanding and control of its passivation is of major importance. The protective behavior of the oxide layer has been associated with the texture of the oxide microstructure and the presence of interfacial tetragonal Zr_3O along with monoclinic Zr_2O, but the use of Zr alloys as cladding material still limits the service temperature to less than 400 °C. Efforts, however, are being made to improve the high-temperature corrosion resistance of Zr alloys, although the understanding of the corrosion mechanism and the role of alloying elements and microstructure are still uncertain. The development of new Zr alloys for cladding nuclear will have to overcome not only the improvement of the corrosion resistance at higher temperature, but also the requirements of higher creep and radiation resistances (Azevedo 2011).

10.4.1.7 Structural Materials for Extreme Operating Conditions

The structural materials challenges become magnified considerably when moving from medium-temperature designs to high-temperature designs in which materials must withstand temperatures approaching 1000 °C. At the extreme operating temperatures envisioned for gas-cooled reactor concepts, graphite and ceramic composites are the leading candidates for structural materials. Along with the numerous engineering design issues associated with utilization of low-ductility materials in a complex high-power energy system, the property degradation associated with neutron displacement damage poses particular challenges. The anisotropic response of graphite to neutron displacement damage due to its hexagonal close packed crystal structure requires the use of specially manufactured "nuclear" grades of graphite to achieve the desired component lifetimes. For components that are subject to relatively high displacement damage exposures or engineering stresses, ceramic composites must be used instead of graphite. Ceramic materials, such as SiC and ZrC, are also considered as potential cladding material for LFR, VHTR, and GFR reactors, which will be exposed to temperatures above 850 °C and fast-neutron radiation. The main challenge for the fuel cladding material of GFR reactor includes high resistance to fast-neutron damage and high-temperature exposure (up to 1600 °C in accident situations). For the LFR reactor, there is an additional concern about the chemical compatibility of the cladding material with the Pb or Pb–Bi coolant and the mixed nitride fuel. In both reactors, SiC is a candidate material for the fuel cladding. For VHTR reactor, ZrC and SiC have been considered as a coating material for the TRISO fuel (tristructural isotropic fuel, a micro fuel particle consisting of a fuel kernel composed of the fuel coated with four layers of three isotropic materials) to provide the mechanical stability to the fuel and act as the main diffusion barrier to the release of fission products (Azevedo 2011).

The development of computational codes has emerged as a new paradigm for modeling the behavior of prospective materials and fuels in future nuclear energy systems. Previous empirical models are being replaced by more physical ones taking advantage of new results at the microscopic level, of the tremendous improvements of computing technology and science, and of the breakthrough in physical metallurgy at the atomic scale. Instead of going through time-consuming and expensive new alloy design, for instance, one possible approach is to use innovative processing techniques to tailor the properties of the existing alloys to suit Gen-IV applications. For instance, the grain boundary engineering (GBE) has been used for tailoring grain boundary microstructure (grain boundary character distribution, etc.). The GBE treatment involves suitable combination of annealing and cold working (i.e., thermal mechanical) treatments. There is now overwhelming evidence that certain coincident site lattice boundaries (CSLBs) with R value 629 (i.e., special CSLBs) possess certain unique grain boundary properties, such as less susceptibility to impurity segregation, lower diffusivity, and better resistance to grain boundary sliding. Therefore, the existence of higher number of CSLBs leads to greater resistance to intergranular degradation against fracture, cavitation, and localized corrosion, higher creep resistance, and possibly higher radiation damage resistance. An example of the success of the GBE concept in practice is that the increase in the CSLB fraction significantly reduces the creep rate in a Ni–16Cr–9Fe alloy irrespective of grain size. GBE, at least in principle, can be applicable to any kinds of materials. However, its applicability in commercial applications still needs to be substantiated. It is further expected that microstructure-property modeling will be of great importance in Gen-IV materials development efforts. Use of modern computational techniques (molecular dynamics simulation, kinetic Monte Carlo methods, etc.) will greatly help in this direction (Alexandreanu et al. 2003; Murty and Charit 2008).

10.4.2 Nuclear Fuel Materials

Within the core of nuclear reactors, fuel is where fission processes occur, involving heavy atoms of uranium, or plutonium. Fuel is the consumable component in reactors: it is held a few years in the reactor, until it reaches its operating limits, whereas the reactors themselves have lifetimes of several decades. Over a reactor's lifetime, fuel is the only component for which improved performance may be achieved, as and when refueling occurs. There is, further, a strong economic benefit to be gained by increasing its in-reactor dwell time, since it will yield more energy, for a comparable cycle cost (Guérin 2009).

Current research on fuels aims to understand and model all of the involved phenomena and to develop new fuels or concepts with greater robustness. The modeling fuel in-reactor behavior involves very many areas of physics: neutronics, to specify the fission and transmutation nuclear reactions occurring inside the fuel; the physics of materials, to evaluate the damage caused to the crystal structure by neutrons, and, most crucially, by the dissipation of fission energy; thermics, to arrive at a precise evaluation of heat removal to the coolant fluid, and to compute the evolution of the temperatures prevailing at every point; mechanics allows strains and stresses to be computed, in fuel pellets, and in the cladding, along with the risks of cladding failure; turning to thermodynamics makes it possible to evaluate the various compounds liable to be formed by the elements yielded by the fission nuclear reaction (these are known as "fission products" [FPs]: more than 10% of atoms found at the end of irradiation were not present at the beginning of the fuel's life); solid-state chemistry considers the kinetics according to which the system, as a rule out of equilibrium conditions, evolves in nominal operating conditions, and during the various transients that may arise (Guérin 2009).

A variety of materials have been used, or contemplated, as fuel materials, since the outset of nuclear reactor development work. Fissile actinide isotopes have either been incorporated into stable metal alloys, or into refractory ceramics (oxides, nitrides, carbides, or silicides). Other, more complex materials, taking the form of cercer (ceramic–ceramic combinations) or cermet (ceramic–metal) composites, have also been investigated for some of the reactors of the future. In the main, it is actinide oxides that have been used on an industrial scale as fuel for power reactors. Metal alloys carry the drawback of a low melting temperature, moreover forming eutectics that further bring that temperature down, entailing that their use be restricted to under 1000 ° C. Oxides, on the other hand, are highly stable, and prove highly refractory, exhibiting melting temperatures higher than 2500 ° C. Nevertheless, even though such oxides do adequately fulfill their function as fuel for the various reactors that have been around for over 40 years, they do not provide an ideal material, owing, in particular, to their relatively poor thermal conductivity. Other candidates, such as nitrides, or carbides, potentially more satisfactory in terms of thermal conductivity as they are, are nonetheless finding it very difficult to gain acceptance, since knowledge of their in-reactor behavior is, as a whole, too limited to guarantee a significant improvement, compared with oxide fuels. In PWRs, for instance, the core consists of assemblies of fuel rods, featuring a zirconium alloy cladding, holding uranium oxide pellets (with uranium enriched to ~ 4% ^{235}U), or MOX pellets (mixed uranium–plutonium oxides [$(U,Pu)O_2$], with a Pu content of 5–10%). The choice of oxide as a fuel material emerged as the accepted solution for water reactors, for a variety of reasons, particularly owing to its quite satisfactory chemical compatibility with the coolant, in the event of failure of the first containment barrier. R&D on fuels for the reactors of the future relates to a stage that is further out, the nature of the fuel considered being intimately bound up with the type of reactor involved, and relevant fuel cycle. Under the aegis of the Generation IV International Forum, which has launched investigations on new nuclear systems, priority is given to investigating fuels affording the capability to recycle part of their own waste (or even waste from previous generations), particularly the minor actinides (neptunium [Np], americium [Am], curium [Cm]) yielded by the transmutation of uranium, and plutonium, during their stay inside the reactor. For that purpose, fast-neutron reactors, ensuring as they do the fission of these actinides, are the favored option. In such fast-neutron reactors, or fast reactors (FRs), the operating conditions the fuel must meet are more severe than is the case in PWRs: fissile material density has to be high, resulting in high specific power within the fuel pellet, operating temperatures higher than with PWRs, and significant damage caused by fast neutrons to metallic structures, particularly to fuel element claddings. The oxide ceramics still stand as a fuel to be reckoned with, oxide being a robust, proven material, particularly in view of its use in sodium-cooled fast reactors. However, the specifications set by fast reactors, particularly of the gas-cooled type, are an inducement to look for denser materials, exhibiting better conductivity, such as carbide or nitride ceramics, or metal alloys. These materials do afford advantages, compared to oxide; however, they also involve drawbacks, which have to be weighed up. Ceramic–ceramic (cercer) or ceramic–metal (cermet) composite materials also stand out as potential candidates; however, they do require major development work. Likewise, as regards cladding and structural materials, subjected as these are to a powerful neutron bombardment, the quest for high temperatures is pushing against the operating limits of known alloys, and will necessitate innovations. With respect to such fuels of the future, the innovations concern not just the nature of the materials involved, but also the prevailing concepts, leading to a reassessment of the benefits that might accrue from a departure from conventional concepts, relying on the principle of a fuel pellet held in a cylindrical cladding. In particular, the particle fuel concept, as used in HTR reactors, utilizes unique multiple layers of pyrolytic carbon and ceramics as the first barrier of confinement, but this concept is feasible thanks to the low volume fraction of fissile materials acceptable in reactors using thermal neutrons. In addition, dimensioning studies for gas-cooled fast reactors are bringing forward some highly innovative concepts, such as macrostructured plates, in which the fissile material is inserted into a honeycomb-shaped ceramic (Guérin 2009).

The development of nuclear fuel presents many technical challenges. In-reactor fuel behavior is complex, affected by steep temperature gradients and changes in fuel chemistry and physical properties that result from nuclear fission. These challenges are compounded by the highly radioactive nature of irradiated fuel, and the necessity of conducting fuel examinations remotely, in a heavily shielded environment. For example, the design of High-Temperature Gas-Cooled Reactors (HTGRs) uses helium as a coolant, graphite as the moderator of neutrons, and ceramic-coated particles as fuel. Helium is chemically inert and neutronically transparent. The graphite core slows down the neutrons and provides high-temperature strength and structural stability for the core and a substantial heat sink during transient conditions. The ceramic-coated particle fuel is extremely robust and retains the radioactive byproducts of the fission reaction under both normal and off-normal conditions. The TRISO-coated (TRIstructural-ISOtropic) particle fuel forms the heart of the HTGR concept. As shown in Fig. 10.10, the TRISO-coated particle is a spherical-layered composite, about 1 mm in diameter. It consists of a kernel of uranium dioxide (UO_2) or uranium oxycarbide (UCO) surrounded by a porous graphite buffer layer that absorbs radiation damage and allows space for fission gases produced during irradiation. Surrounding the buffer layer is a layer of dense pyrolytic carbon called the Inner Pyrolytic Carbon layer (IPyC), a silicon carbide (SiC) layer, and a dense Outer Pyrolytic Carbon layer (OPyC). The pyrolytic carbon layers shrink under irradiation and create compressive forces that act to protect the SiC layer, which is the primary pressure boundary for the microsphere. This three-layer system is used to both provide thermomechanical strength to the fuel and contain fission products. An HTGR will contain billions of TRISO-coated particles encased in a graphitic matrix in the form of either small cylinders, called compacts, or tennis-ball-sized spheres, called pebbles (see Fig. 10.10). Extensive testing has demonstrated the outstanding performance of high-quality low-defect TRISO-coated particle fuels (Allen et al. 2010).

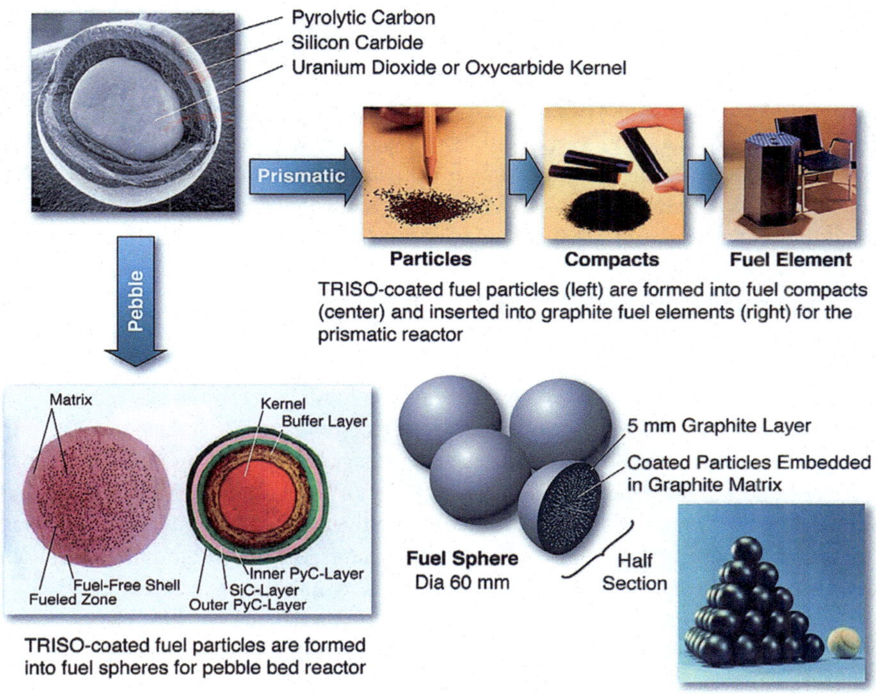

Fig. 10.10 High-temperature gas reactor fuel system, showing TRISO fuel particles consolidated into a graphite matrix as prismatic blocks (upper right) or pebbles (lower right) (Adapted with permission from Allen et al. 2010 (Elsevier))

10.4.3 Fuel Cladding Materials

Advanced nuclear reactors are expected to allow more efficient and safer use of nuclear energy. Among the rector components, the fuel cladding is the most important safety barrier in fission nuclear reactors, as it restrains most of the radioactive fission products within its volume. The fission of U atoms produces radioactive materials, which emit neutrons, γ rays along with α and β particles. These particles can be lethal to humans, so the use of fuel cladding keeps the radioactive materials isolated from the coolant/moderator, which surrounds the cladding and maintains the fuel rods cooled. The material for fuel cladding is usually selected after other design aspects of the reactor's core have already been decided, such as the nuclear fuel and the moderator and coolant materials, defining, therefore, the multiple design constraints for the selection of the fuel cladding material. These constraints include the neutron absorption cross section, the maximum service temperature, the creep resistance, the mechanical strength, the toughness, the neutron radiation resistance, the thermal expansion, the thermal conductivity and the chemical compatibility with fissile products and coolant, moderator and fuel materials (Azevedo 2011).

For instance, the cladding material should be transparent to neutrons, meaning that the material should present a low value for the neutron absorption cross-section to minimize neutron losses (such as Mg, Be, Si, Al, and Zr). Additionally, the cladding material should present an acceptable service temperature to increase the thermal efficiency of the reactor, leading to a trade-off between service temperature and neutron transparency. Therefore, many multicomponent systems combining Be, C, Mg, Zr, Si, and O (including metallic and ceramic systems) might be considered as potential new materials for nuclear fuel cladding in terms of neutron economy. As a matter of fact, Al, Mg, and Zr alloys have already been used as fuel cladding materials, with great advantages for Zr alloys at the moment (Northwood 1985; Azevedo 2011).

The fuel cladding material should also be radiation resistant and there are several ways in which fission-generated neutrons can interact with the crystal lattice of the cladding, producing point, line and volume defects, such as dislocation loops and void and self-interstitial clusters. As a result, there will be a change in the mechanical properties of the material, as long as the service temperature is below 40% of the homologous temperature of the material. Furthermore, radiation can also alter the redistribution of alloying elements, the stability of phases, and the kinetics of phase transformations, with consequences not only to the mechanical properties but also to corrosion and creep resistance of the cladding material. The cladding material should be able to withstand the service temperatures (in normal or accident conditions), making thermal creep resistance and microstructural stability key issues for the selection of materials as there is a trend to increase the service temperatures to achieve higher thermal efficiency for the reactor. Finally, the cladding material should be corrosion/oxidation resistant to the environment (coolant/moderator/fuel/fission products) and should present high thermal conductivity to increase the energy efficiency of the reactor, and low thermal expansion coefficient to minimize the thermal stresses in the cladding/pellet interface (Azevedo 2011).

It is worthwhile to observe the historic evolution of the design constraints trade-off in order to understand the trends for new materials for GIV reactors. During the early 1940s the materials challenges for the nuclear technology grew rapidly for the construction of Pu-producing reactors. Beryllium (hexagonal close packed) was considered as a candidate nuclear fuel cladding material since early 1950s due to its unique nuclear properties. This material, however, found very limited use as structural material in nuclear applications due to its low ductility even in the absence of irradiation, as Be has only two operating slip modes (basal slip on {0 0 0 1} planes and prismatic slip on {1 0 1 0} planes) at low temperatures. Additionally, Be undergoes significant embrittlement at low doses of irradiation due to the irradiation growth of anisotropic crystals and the formation of He bubbles as a result of (n, a) reactions. Additionally, Be also presented catastrophic corrosion behavior above 500 °C. Work on Be–Ca alloys was developed to improve the corrosion resistance and optimize the mechanical properties and the radiation-induced embrittlement. Cost, brittleness (especially after irradiation), and toxicity ruled out the commercial application of the metal as cladding material. Nowadays, the main research for the use of Be is in area of fusion reactors, but safety aspects, such as its reactivity with steam and water under accident conditions, and its radiation resistance are unique challenges for its successful application (Snead and Zinkle 2005).

Al alloys have been used not only as fuel cladding material but also as fuel matrix and structural material in nuclear, especially in research reactors. Al is indeed a corrosion-resistant metal with a relatively low neutron absorption cross section, but its maximum service temperature (around 200 °C, limited by oxidation and mechanical strength) was considered unsatisfactory for use in early nuclear power reactors, being replaced by the use of austenitic stainless steel, despite its effective neutron cross section absorption value being 15 times bigger than that for Zr alloys (Leenaers et al. 2008).

In the late 1950s, the designers of the gas-cooled reactor (GCR) selected Mg alloys as cladding material to withstand working temperatures up to 400 °C and to minimize the neutron absorption. The GCR uses graphite as neutron moderator material and CO_2 as coolant material, being able to use natural uranium as fuel. The fuel cladding was used to prevent the oxidation of the fuel by the coolant, and the escape of fission products into the circulating gas stream. Additionally, the cladding provided a heat transfer surface to improve the efficiency of heat removal from fission reaction. In this sense, the design constraints for the selection of cladding material were that the material had to be compatible with fuel, fission products, and coolant; to present a low neutron absorption cross-section; to offer a feasible processing route; and to feature good creep properties. Only two materials were used as primary fuel cladding materials in GCR reactors: Magnox A12 alloy (Mg with 0.7–0.9%Al) and ZA alloy (Mg with 0.45–0.65%Zr). However, two major drawbacks were soon discovered: Mg alloy restricted the maximum coolant temperature to 500 °C due to creep and oxidation resistance, limiting the thermal efficiency of the reactor; and Mg alloy presented high corrosion rate in water, preventing long-term storage in spent fuel pools. The latter restriction made fuel reprocessing an essential part of the nuclear fuel cycle (Popple 1963).

During the 1960s, austenitic stainless steel (20Cr/25Ni steel) was selected as a fuel cladding material for advanced gas-cooled reactor (AGR) to increase the service temperature to 600 °C, despite the high value for the effective neutron absorption cross section of this steel, which is approximately 15 times higher than Mg alloys. The use of austenitic stainless steel along with enriched fuel in AGR reactor allowed a higher thermal efficiency than GCR reactors (40 against 30%). However, microstructural characterization of post-irradiated cladding samples soon revealed the chemical interaction between the fission products and the cladding material, causing, for instance, the intergranular attack at the inner surface of cladding by volatile fission products, such as iodine. Additionally, post-irradiated cladding samples also featured physical bonding with the fuel, which might lead to in-service failures due to pellet-cladding mechanical interaction (PCI), in the case associated with the deleterious effect of neutron irradiation on the ductility of stainless steel. One of the main challenges facing the further use of austenitic stainless steel as cladding material is indeed its low radiation resistance. Even at low irradiation temperatures, fast neutrons causes hardening and loss of ductility of austenitic stainless steels, while at elevated temperatures He produced by thermal neutron (n, a) reactions induces low ductility due to gas-induced intergranular cracking (Yamada 1978).

In 1947 the US Bureau of Mines developed Zr sponge using the Kroll process, allowing the availability of good quality Zr at reasonable prices. The unique properties of Zr (low effective neutron absorption cross section; good mechanical properties, high service temperature and corrosion and radiation resistances, especially when compared with austenitic stainless steel) made it an ideal cladding material for the US Navy nuclear propulsion program in the 1950s. By mid-1960s Zr alloys were already the main fuel cladding material in both light- and heavy-water reactors, despite its highly anisotropic properties. Zr alloys, however, still present major challenges for improving the reactor's performance because of their complex corrosion mechanisms. Increases in coolant temperature, power up-rates, extended burn-ups, and longer residence times have led to the development of new Zr alloys by thermo-mechanical processing and microstructural and alloy design, but its maximum service temperature is still around 400 °C, creating immense challenges for further advances on new Zr-alloys (Cox 1990).

The economics of current nuclear power plants is improved through increasing fuel burn-ups, which results in less nuclear waste. In GIV reactors the chain reaction is generally sustained by fast neutrons associated with the use of highly enriched U or Pu and lack of neutron moderator material. Some important design constraints for the selection of GIV cladding materials are: low effective neutron cross section absorption; superior resistance to fast-neutron radiation damage; outstanding dimensional stability against thermal (500–1000 °C) and irradiation creep (doses up to 200 dpa); corrosion resistance to the working environment, especially to coolants and fissile products; and feasible manufacturing route and assembling. Because of these strict constraints, most materials employed for the fuel cladding and other structural components in current commercial reactors might not be suitable for the use in GIV reactors. Several materials, such as ferritic–martensitic stainless steel (F/M steel), oxide dispersion strengthened (ODS) alloy, nickel-based super alloys (Inconel and Incoloy), refractory metals, and ceramic materials (SiC and ZrC), have already been suggested as candidate fuel cladding materials for GIV reactors. As a matter of fact, the combination of high temperature, high fast neutron radiation dose, aggressive environment, and longer in-service life is the major obstacle for the selection of materials and the viability of some of the GIV reactors. Microstructural changes caused by fast neutron irradiation will affect both the creep and corrosion properties of the cladding material during service, adding more obstacles to the fully assessment of in-service properties of new materials (Abram and Ion 2008; Azevedo 2011).

10.4.4 Moderators and Reflectors

A moderator is used to moderate or slow down fast fission neutrons from relatively high (kinetic) energies to thermal (ambient temperature) energy levels in a thermal reactor. The process of reducing the energy of the fission-produced neutrons, from an average of 2 MeV to a value of 0.025 ± 1 eV, which conform to a thermodynamic equilibrium with the environment in the reactor and promote ^{235}U fission in thermal reactors, is called neutron moderation, and the materials in the reactor core that are instrumental in achieving this are, accordingly, known as moderator materials. The moderation process is implemented by repeated elastic collisions of neutrons with the nuclei of the moderator material. Commonly used moderator materials include regular (light) water (roughly 75% of the world's reactors), solid graphite (20% of reactors), and heavy water (5% of reactors). Beryllium has also been used in some experimental types, and hydrocarbons have been suggested as another possibility.

At each collision a neutron transfers part of its energy to the moderator nucleus. The smaller the mass of the target nucleus, the larger is the transferred energy, which attains its maximum value when the moderator nucleus has about the same mass as that of the neutron. This situation is obtained only with hydrogen. Apart from lightness, other properties, such as probabilities for scattering and absorption, must also be taken into account in estimating a figure of merit for a moderator. The scattered or leaking neutrons are reflected back into the core in both thermal and fast reactors by radial and axial reflectors. The reflector material may be graphite, beryllium, steel, tungsten carbide, or other materials. The nuclear requirements for moderators and reflectors are the same in a thermal reactor. The nuclear requirements are a high neutron scattering cross section, a large energy loss by a neutron per collision, and a low neutron absorption cross-section. The high neutron scattering, or collision, cross-section of the moderators and reflectors means frequent, large-angle collisions and a relatively short mean free path, or distance, over which a fast neutron travels during the slowing-down process, thereby bringing about a reduction in neutron leakage and escape (Gupta 2001).

10.4.5 Coolant Materials

The thermal energy generated in the fuel elements as a result of nuclear reactions is extracted from the reactor core by means of selected fluids known as coolants. An ideal coolant should have the following qualities (Gupta 2001): (i) good thermal properties, (ii) low pumping power requirement, (iii) low melting and high boiling points, (iv) thermal stability, (v) irradiation stability, (vi) low neutron absorption and suitable neutron scattering cross-sections, (vii) nontoxic, nonhazardous, and low induced radioactivity, (viii) relative absence of impurities, (ix) low corrosiveness, (x) low viscosity, (xi) compatibility with fuel and structural materials, (xii) easy and safe handling, and (xiii) be abundant and economical. There are gaseous and liquid coolants. The coolants that have been chosen in the reactor systems possess most of the characteristics, but none possesses all of them.

For the gas coolants the special requirements list as follows: (i) good chemical and radiochemical stability; (ii) low levels of induced radioactivity with a short half-life of the nuclei generated by neutron-induced reactions; (iii) high density, specific heat, and thermal conductivity; and (iv) good and lasting compatibility with the structural and moderator materials with which they come into contact. Among the gaseous coolants, helium and carbon dioxide are commonly used.

For the liquid cooling materials, the following are additionally required: (i) a low melting temperature, a critical temperature as high as possible, and a critical pressure as low as possible; (ii) low viscosity; and (iii) a small neutron capture cross-section. The liquids used for cooling fast reactors must also have a moderating power as weak as possible, to prevent growth of the fraction of low-energy neutrons, which would decrease the efficiency in the generation of fissionable nuclides. The realm of liquid coolants comprises light water, heavy water, and molten metals such as sodium and lithium. The only important difference between heavy and light water with regard to the dual function of moderator and coolant lies in the fact that the thermal neutron absorption cross-section of the former is much less than that of the latter. This permits natural uranium fuel to be the fuel of choice in heavy-water reactors. Light or heavy water is deployed in a multitude of applications such as moderator, reflector, coolant, or solvent (e.g., to dissolve boric acid in water for chemical shim). High power can be secured from relatively small core volumes in fast reactors because there is no moderator and because the liquid metals used as coolants possess excellent heat extraction properties. The choice goes only to liquid metals among which it is liquid sodium that has drawn the maximum attention as a coolant. The design concepts of liquid-sodium-cooled fast breeder reactor systems are mainly determined by the thermal and nuclear properties of the coolant: good heat transfer, small moderating effects, and low neutron absorption cross-section. The melting point of sodium being 98 °C requires preheating of the system for filling or some maintenance operations. The high boiling point (982 °C, 1 bar) permits high coolant temperatures and very low system pressures (apart from the water-steam circuit) compared with light-water reactor systems. The compatibility of sodium with structural materials is determined by its purity, temperature, and circulation rate. Sodium is quite compatible with stainless steel used at present to make clads and fuel assemblies, and also with oxidic fuel materials (UO_2 and $(U,Pu)O_2$). The presence of oxygen as an impurity impairs corrosion resistance, so that the contact with oxygen must be kept limited; this necessitates continuous purification of liquid sodium (Gupta 2001).

10.4.6 Control, Shielding, and Safety System Materials

A nuclear reactor is always required to be under control and amenable to adjustments within accurate limits in order to have a safe, controlled nuclear (fission or fusion) energy release. The control and safety of nuclear fission reactors are linked with (i) the rate of neutron generation, (ii) the rate of neutron loss by leakage, and (iii) the rate of neutron loss by parasitic absorption in the core. Parasitic absorption refers mainly to fission product generation and buildup (attended by depletion of nuclear fuel). The rates of neutron production and losses are directly related to the influences of reactivity and control of the reactor. The most practical and effective control element is a powerful neutron absorber (Gupta 2001).

The control elements are usually in the form of rods and there are three types of control rods in use: (i) shim rods for occasional coarse control and adjustments in neutron density with an intermediate "rod's worth of reactivity," (ii) regulating rods for carrying out fine control and adjustments to maintain the desired power level with a small "rod's worth of reactivity," and (iii) safety rods for shutting down or "scramming" the reactor rapidly in case of emergency. The safety rods have a large "rod's worth of reactivity." In regard to control materials, a representative list of elements may be drawn up as: cadmium, boron, hafnium, and the rare earths. Cadmium is a good neutron absorbent. Its uses as a control material are, however, limited due to its relatively low melting temperature, and to the penetrating γ radiations that are generated during neutron capture. Boron is one of the best neutron absorbents. Its cheapest form is B_4C. This material as such and various products made out of it are accomplished neutron absorber materials. Hafnium possesses a special feature: its higher isotopes formed by neutron capture are found to have neutron capturing capabilities. Rare earth elements due to their very feeble resistance to corrosion and high propensity to oxidation are used as oxides and dispersions in metallic matrices. Thus, in high-temperature reactors oxides of rare earths are in order, while in other reactor systems dispersions of the rare earth oxides in stainless steel, clad in stainless steel tubes, are preferred. An interesting feature among some rare earths may be accounted for in the present context of neutron absorption. In the case of samarium, for example, the natural nuclide ^{149}Sm makes the greatest contribution to the capture cross-section; when this isotope turns into ^{150}Sm, the cross-section drops significantly, which makes the element a consumable absorbent. The disposition for gadolinium is similar. This is unlike what has been said about hafnium which is nondepleting in character when in use in the duty for neutron absorption. A nuclear fission reactor is controlled in three most practical ways. These are (i) control rods with B, B_4C (natural or enriched in ^{10}B), Cd, Ag-Cd-In, Hf, AgHf, Ag-Ir-Hf, or rare earth elements as the control material, (ii) chemical shim with boric acid (H_3HO_3) as an absorber or control material, and (iii) a burnable absorber with the oxide of the rare earth elements, Eu_2O_3, Er_2O_3, or Gd_2O_3, as the control material. The control rods carry out the job of primary control, while the chemical shim and the burnable absorber carry out the job of secondary, or supplementary, control. Also, the chemical shim has an interesting characteristic in which solubility of boric acid, dispersed in the reactor core, increases with temperature. In some power reactor systems, a combination of control rods and chemical shim is used, while in others a combination of control rods and burnable absorber is used for reactor control and safety (Gupta 2001).

Shielding from the passage and penetration of nuclear radiations to the outside of the reactor is required. It is the shielding materials that reduce all types of radiation to minimum tolerable levels. Neutrons and gamma rays are characterized by their high penetrating power, and they require to be attenuated and absorbed in the shielding. In general, the best shielding for fast and thermal neutrons is provided by light elements or materials of low atomic weight, and that for primary and secondary gamma rays by heavy elements or materials of high atomic weight (Gupta 2001).

The shield design is an integral part of reactor design. Radiation shielding serves three major functions: (i) thermal shielding to protect the pressure vessel, coolant loop, and inner shield from the intense heat generated by the absorption of nuclear radiations, (ii) biological shielding to protect personnel and assure safeguards to public health, and (iii) apparatus and instrument shielding to protect electromagnetic apparatus and electronic instruments that otherwise would not work properly in an intense radiation field. Although the nuclear radiation coming out of a reactor may include α rays, β rays, protons, neutrons, and γ rays, as far as shielding design is concerned only neutrons and γ rays require attention. The reason is that α rays, β rays, and protons, which are charged particles, have very low penetrating power, while neutrons and γ rays, which are neutral, have large penetrating power. The primary requirements for shielding are (i) slowing down fast neutrons, (ii) absorbing slow and thermal neutrons, (iii) attenuating primary γ rays, and (iv) minimizing the production of secondary γ rays. On the basis of their functions, shielding materials used in nuclear reactors can be divided into three principal categories: (i) heavy and moderately heavy elements, (ii) hydrogenous materials, and (iii) light elements. The heavy and moderately heavy elements are meant to attenuate γ rays and to slow down fast neutrons by inelastic scattering in the high-energy region. Among the heavy and moderately heavy elements, lead, bismuth, tantalum, and iron (steels) are often used for attenuating γ rays. Lead is the most common shielding material for γ rays. It is abundant and economical but it has a low melting point. A moderately heavy element that has hitherto been used (after being immersed in water for thermal shielding) is iron in the form of construction steel or stainless steel. The steel can attenuate γ rays and release radiation heat. The water can moderate fast neutrons and remove the radiation heat from the steel plate of the thermal shielding in the reactor core. Ceramics that contain heavy or moderately heavy elements may prove to be useful as shielding materials, and in this context, mention may be made of iron oxides, titanium carbide, silicon carbide, and boron carbide. The primary job of hydrogenous materials is to moderate and thermalize neutrons by elastic scattering in the intermediate- and low-energy regions and to absorb slow or thermal neutrons and attenuate γ rays. The assessment of a hydrogenous material for neutron shielding is based on the following properties: (i) hydrogen content and neutron moderating power, (ii) neutron and γ ray absorption cross-sections, and (iii) abundance and economics. The best overall candidate is light water. A material containing more hydrogen atoms per unit volume than light water is polyethylene, but it is obviously not as abundant and economical as light water. Light water is used not only as an effective neutron shielding material but also for providing an efficient means for removing the heat produced by the absorption of nuclear radiation. Cement concrete containing ceramics, minerals, and water is a general hydrogenous material for shielding of neutrons and γ rays. In addition to the properties required, concrete possesses a high mechanical strength and various adaptabilities and configurations of construction to shield the reactor core. It is used more than any other material for nuclear reactor shielding. Concrete used for radiation shielding can be classified into ordinary concrete and special concrete (Gupta 2001).

To provide safety systems, primary as well as secondary, in nuclear plants is of major importance. The functions of these systems are (i) to prevent the temperature from shooting up to unacceptably high levels for any reason, (ii) to suppress any contingency of extremely high pressures developing, and (iii) to avert any accidental release of radiation from the reactor core containment. Most hardware of safety systems comprises structural materials (such as carbon steel and stainless steel), pumps, valves, thermal shields, etc. Over a long period of reactor operation, four important circumstances must be considered in relation to the control elements (rods or plates): (i) reactivity coming down with fuel depletion, (ii) breeding of new fuel through neutron irradiation, (iii) formation and accumulation of fission products, or absorbers, and (iv) increase of burn-up of the control element. These factors need constant adjustment and precise movement of the control elements. The primary requirements for control elements may be listed as (i) very high neutron absorption, (ii) adequate mechanical strength, (iii) light element, hard surface, (iv) high thermal and irradiation stability, (v) good heat transfer characteristics, and (vi) high wear and corrosion resistance or hardness and inertness. The safety systems of a reactor power plant comprise (i) primary safety systems and (ii) secondary safety systems. The primary and secondary safety systems may include the emergency core cooling system (ECCS), automatic pressure suppression system (PSS) (or depressurization system), containment isolation system (CIS), containment (or core standby) cooling system, in-core or out-of-reactor instrumentation safety (monitoring) system (ISS), cracking leak control system, and so on (Gupta 2001).

Reactor materials used in the safety systems vary with different components, vessels, valves, pumps, protection apparatus, control equipment, and monitoring instruments involved in the safety analysis and design. The major safety system materials, however, are structural steel, stainless steel, reinforced concrete, light water, borated water, aluminum, and copper (Gupta 2001).

INDEPTH: Materials Degradation in Reactor Cores

The interaction of materials with the Generation IV coolant and fission products poses a significant challenge beyond that encountered in today's light-water reactors. The high temperatures of all of the designs will accelerate corrosion and oxidation kinetics and open new pathways for materials degradation. Degradation processes in light-water reactors that are driven by irradiation, such as irradiation-assisted stress-corrosion cracking, will likely be more severe in Generation IV systems as the dose range of some concepts is almost a factor of 10 greater than that in Generation II reactors, opening the possibility for new degradation modes. Moreover, to allow operation at much higher temperatures, advanced Generation IV reactor concepts utilize different coolants, including water in the supercritical state, liquid metals such as sodium and lead–bismuth, molten salts, and high pressure helium gas. In addition to specific issues associated with these coolants, such as a sodium–water reaction in the case of leakage in the sodium circuit of the vapor generator, these different coolants, with their very different properties compared to water, bring with them a host of new issues about materials compatibility, particularly when one considers impurities that will inevitably be

(continued)

present even in nominally inert coolants such as helium gas. Corrosion and oxidation are unavoidable at these temperatures, as oxides of virtually all candidate structural materials are more thermodynamically stable than the elemental or alloyed state. In all cases, the challenge is to develop coolant–materials systems that result in the formation of protective and self-healing films to ensure longevity of the structures for the life of the reactor.

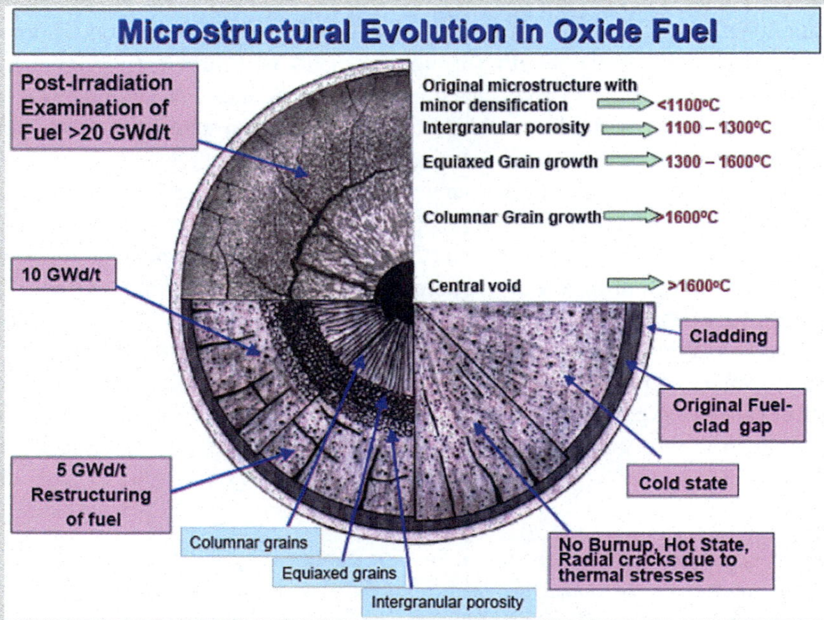

10.5 Materials Issues in Fusion Reactors

Thermonuclear fusion of two hydrogen isotopes results in release of energy and energetic neutrons which can be utilized for electricity generation. Out of various hydrogen isotope reactions, i.e., D-D, D-T, and T-T (D: Deuterium, T: Tritium), the D-T reaction has the largest cross-section at the lowest energy. D-T Fusion reaction can be conducted by different confinement methods such as gravitational confinement, inertial confinement, and magnetic confinement. While different fusion confinement techniques are being explored, the most widely pursued technique is magnetic confinement fusion. A tokamak is a device which uses magnetic field to confine plasma to the shape of a torus, as shown in Fig. 10.9. Such magnetic confinement is required since no solid material could withstand extremely high temperatures of plasma. Tokamak is one of the several types of magnetic

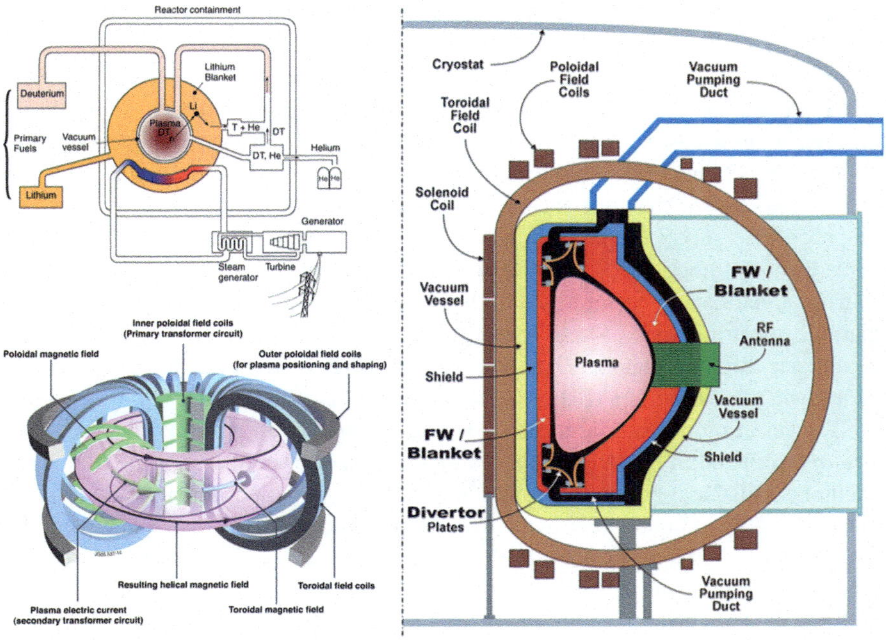

Fig. 10.11 Schematic illustration of subsystems and components in a tokamak reactor (Adapted with permission from Burkart 2005 (IOP). Credit: International Fusion Research Council (IFRC))

confinement devices, and is one of the most researched devices for producing controlled thermonuclear fusion power (Mukherjee and Jamnapara 2015).

As shown in Fig. 10.11, the magnetically confined plasma core has a temperature of ~5 million °C. The energy from 14.1 MeV neutrons has to be utilized for T production and electricity generation. The subassemblies and their functional requirements are listed in Table 10.2. The fusion reactions are induced within a toroidal-shaped high-temperature ionized plasma that is shaped by powerful toroidal and poloidal field magnets. The heat and energetic neutrons produced by the deuterium-tritium (D-T) fusion reaction are absorbed by the surrounding first wall, blanket, and divertor components. The fusion reaction occurs inside a vacuum vessel to prevent atmospheric contamination of the D-T plasma. The major functions of the blanket region are to efficiently capture the energy produced by the fusion reactions and transfer the heat to coolant for electricity generation, and to create and extract fresh tritium fuel (by utilizing nuclear transmutation reactions with lithium-containing liquid or solid materials) to enable continuous operation of the fusion energy system. A wide variety of structural materials, reactor coolants, and tritium generation materials systems have been evaluated for potential use in future fusion reactors (Zinkle and Busby 2009). Most of critical materials mainly include plasma facing (first wall and divertor), breeder materials, and structural materials.

10.5.1 Plasma-Facing Components and Divertor

The hot plasma confinement by magnetic confinement in a tokamak reactor involves plasma-materials interaction. The first solid surface of the reactor facing the hot plasma is known as the "First wall" and the assembly is often referred to as plasma-facing components (PFC), whereas divertor is a device (or assembly) that allows the online removal of material and He ash from plasma. The first wall surface faces impact of energetic particles leading to erosion of the surface and irradiation damage, and results in trapping of D or T in re-deposited layers of eroded species which causes a radioactive inventory buildup in the reactor. The impinging energetic particle spectrum varies across the poloidal circumference of the tokamak reactor, and hence different materials can be used depending on the particle energy and flux at a given location in the reactor (Mukherjee and Jamnapara 2015).

The selection of material for the first wall is critical. On erosion, low Z materials are fully ionized in the central plasma and only radiate bremsstrahlung, the high Z elements still have bound electrons that emit line radiation which leads to strong plasma cooling. On the other hand, low Z materials (such as C and Be) have higher erosion rates than high Z elements (viz. W and Mo). The ITER first wall with a heat flux <1 MW/m^2 considers Be or W as plasma-facing material. One of the reasons attributed to replacing carbon-based PFCs to W-based PFCs is the reduction of hydrogen or tritium retention. However, under high hydrogen fluence and low temperatures (<600 K), blistering of W has been observed. Be poses health hazards and hence is still under evaluation stage. This W can either be coated on the structural material of first wall (RAFM steel) by thermal spray technique (e.g., vacuum plasma spray) or can be fixed as a solid tile (Neu 2010; Philipps 2011).

The divertor assembly on the other hand considers both low Z and high Z materials. The main function of divertor assembly is to remove the scrape off layer (SOL) (mainly He dust), shield subsequent subassemblies from very high heat fluxes (10–20 MW/m^2), and extract heat out of the divertor. W and Carbon Fiber Composites (CFCs) have been the preferred choice of materials for such divertor target plates in ITER. In order to improve the erosion resistance, oxide dispersion strengthening concept is being explored in target plates or first wall components, forming ODS-tungsten alloys, such as W-Y_2O_3, W-La_2O_3, W-TiC, and W-TaC. Fabrication processes such as wet chemical method, mechanical alloying, sintering of pure and doped W by spark plasma sintering, resistance sintering under ultra-high pressure are being pursued (Rieth et al. 2013; Yan et al. 2013).

Another challenge to divertor design is the fabrication of an appropriate heat sink along with W armor. The heat from the armor tile has to be extracted, for which a heat sink made of CuCrZr—a precipitation hardened copper alloy—has been reported as candidate material. An SS 316 (LN) grade pipe brazed through CuCrZr alloy shall circulate water for cooling and heat extraction purpose. The joining of this CuCrZr alloy with refractory armor material such as W is a challenge owing to the large coefficient of thermal expansion (CTE) difference. Vacuum brazing of

W/CuCrZr and C/CuCrZr has been found successful as a joining technique. W/Cu tile fabrication using oxide-free high conductivity (OFHC) Cu casting in vacuum followed by brazing of W/Cu with CuCrZr has been found promising. Functionally graded coatings of W or W-Cu on CuCrZr alloys are also being pursued as another solution.

Apart from this, novel concepts such as liquid metal (Li) plasma-facing components are being evaluated at Princeton Plasma Physics Laboratory. Other concepts involve vaporization of Li such as tungsten alloy with Li, wherein lithium would get evaporated and would thus be able to sustain the heat fluxes (Wong et al. 2001; Rotti et al. 2014).

10.5.2 Blanket Module

A blanket module is an assembly which blankets the core of the fusion reactor and utilizes the energy of neutrons to generate T fuel and heat for electricity production. Thus, the main function of a breeder blanket module is to generate T fuel from Li by utilizing the 14.1 MeV neutrons generated from reactor core; to extract the heat for electricity generation purpose generated from the n-Li reaction and shield the other subsystems from the radiation damage by the energetic neutrons. This can be achieved by appropriate design concepts, selection and or development of appropriate breeder materials and their processing techniques and the choice of appropriate structural material with relevant compatibility with breeder material. Since the T can be generated from reacting the neutrons with ^{6}Li, all the breeder materials are Li or Li-based compounds or alloys. Based on the form of Li, the blanket design can either be categorized as liquid breeder blanket, solid breeder blanket, or mixed type. Different blanket module designs with solid or liquid or dual breeder concepts have been proposed by different countries which are summarized in Table 10.3. The ITER reactor will be useful for testing and validating different blanket concepts as proposed by partner countries and the performance information will be helpful for development of a DEMO relevant blanket module design (Mukherjee and Jamnapara 2015).

Both the solid and liquid breeder concepts have their pros and cons. The main advantage of the solid breeder is that it offers good compatibility between breeder, structural material, and the coolant. However, one of the major drawbacks of solid breeder concept is the costly fabrication and re-processing of the ceramic breeder material. Against this, the liquid breeder concept offers efficient heat and fuel extraction, and easy maintenance. A general comparison between the two breeder blanket concepts is provided in Table 10.4 (Bornschein et al. 2013).

Table 10.3 Different breeding blanket concepts (Mukherjee and Jamnapara 2015)

TBM design concepts	Country	Brief outline of the blanket design
Helium-Cooled Lead Lithium (HCLL)	EU	He and PbLi as coolant and breeder; RAFMS (Eurofer '97) as structural material
Dual Function Lithium Lead (DFLL)–He-cooled quasi-static lithium lead (SLL)	China	He as coolant; self-cooled PbLi in quasi-static condition as breeder; RAFMS (CLAM steel) as structural material.
Dual Function Lithium Lead (DFLL)–Dual-cooled Lithium Lead (DLL)	China	He/LiPb as coolant & breeder; RAFMS (CLAM steel) as structural material
Lead Lithium Ceramic Breeder (LLCB)	India	He as coolant; self-cooled PbLi as breeder; Li_2TiO_3 as solid breeder; RAFMS as structural material
Helium-Cooled Liquid Lithium	USA	He as coolant; self-cooled liquid Li as breeder; RAFMS with SiC_f-SiC inserts as structural material
Lithium Vanadium (Li-V)	Russia	Self-cooled lithium as coolant cum breeder; Vanadium alloys as structural material
Helium-Cooled Molten Lithium (HCML)	Korea	He as coolant; molten Li as breeder; RAFMS (Eurofer) as smictural material
Water-Cooled Solid Breeder (WCSB)	Japan	Li_2TiO_3 as solid T breeder; Be pebbles as neutron multiplier; He and water as coolants; SiCf/SiC inserts; F82H (RAFMS) as structural material

Note: *F–M* ferritic–martensitic stainless steels (typically 9–12 wt.% Cr), *ODS* oxide-dispersion-strengthened steels (typically ferritic–martensitic), *MC* mixed carbide [(U,Pu)C], *MN* mixed nitride [(U,Pu)N], *MOX* mixed oxide [$(U,Pu)O_2$]

Table 10.4 Comparison of solid and liquid breeder design concepts (Bornschein et al. 2013)

	Solid breeder	Liquid breeder
Breeder material	Ceramics: LiO_2, $LiAlO_2$, Li_2SiO_3, Li_4SiO_4, Li_8ZrO_6, Li_2TiO_3	$Li_{17}Pb_{83}$, Flibe (LiF, BeF_2)
Neutron multiplier	Be, $Be_{12}Ti$	Pb, Be
Coolant	He cooled, water cooled	He cooled, water cooled, self cooled, dual cooled
Structural material	RAFM steel	RAFM steel
Advantages	Tritium extraction less challenging	No breeder damage or swelling; adjustable breeder composition
Difficulties	Blanket replacement, tritium permeation into coolant	MHD drag, corrosion, tritium permeation into coolant

Note: *F–M* ferritic–martensitic stainless steels (typically 9–12 wt.% Cr), *ODS* oxide-dispersion-strengthened steels (typically ferritic–martensitic), *MC* mixed carbide [(U,Pu)C], *MN* mixed nitride [(U,Pu)N], *MOX* mixed oxide [$(U,Pu)O_2$]

INDEPTH: Fusion Materials Modeling (Wirth et al. 2011)

The fusion energy plasma environment presents numerous inherently multiscale computational grand challenges at the extreme of high-performance computing. For example, consideration of the edge, or boundary region where the plasma meets the material surface, leads to the identification of three coupled spatial regions that involve critical scientific issues for fusion power. These regions consist of (1) the edge and scrape off layer region of the plasma, (2) the near-surface material response to extreme thermal and particle fluxes under the influence of, and feedback to, the plasma sheath, and (3) the structural materials response to an intense, 14 MeV peaked neutron spectrum, which produces very high concentrations of transmuted elements within the bulk of the material through nuclear (n,p) and (n, α) reactions in which neutrons (n) are absorbed and protons (p) or alpha (α) particles (e.g., helium nuclei) are emitted from the nucleus, which transmutes the nucleus to a different element containing one or two fewer protons, respectively. Likewise, the performance of bulk structural and breeding blanket materials in a challenging degradation environment with large time-varying stresses, corrosive chemical environments, and large fluxes of 14-MeV peaked fusion neutrons affects the thermal and power management of the fusion reactor and the overall tritium balance and controls the operating/replacement lifetime of the vacuum vessel. Gaining understanding and predictive capabilities in this critical area will require addressing, simultaneously, complex and diverse physics occurring over a wide range of lengths (angstroms to meters) and times (femtoseconds to days).

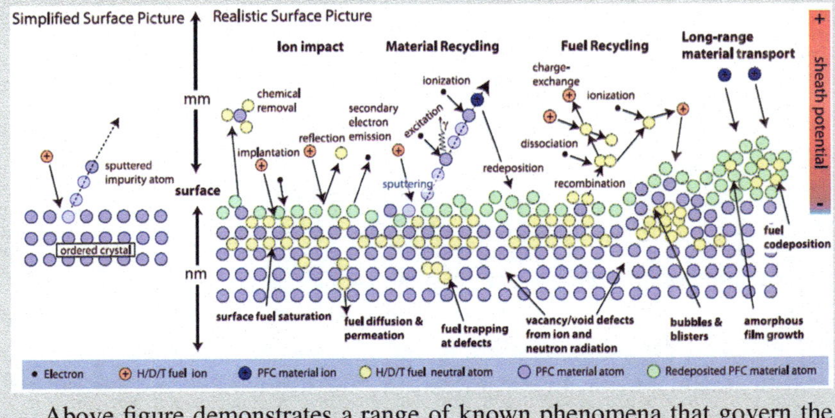

Above figure demonstrates a range of known phenomena that govern the response of the materials surface to plasma interaction. While vastly different physical scales exist for the surface (on the scale of nanometers) and plasma

(continued)

processes (on the scale of millimeters), the plasma and material surface are strongly coupled to each other, mediated by an electrostatic and magnetic sheath. Likewise, the recycling of hydrogenic plasma fuel is self-regulated through processes involving the near-surface fuel transport in the material and the ionization of neutral species that enter the plasma. Also the intense radiation environment of ions, neutrons, and photons ensures that the material structure and properties are modified and dynamically coupled to the PMI processes at extreme thermal fluxes that may exceed interactions with transmutant products, including He and H, remain to be established.

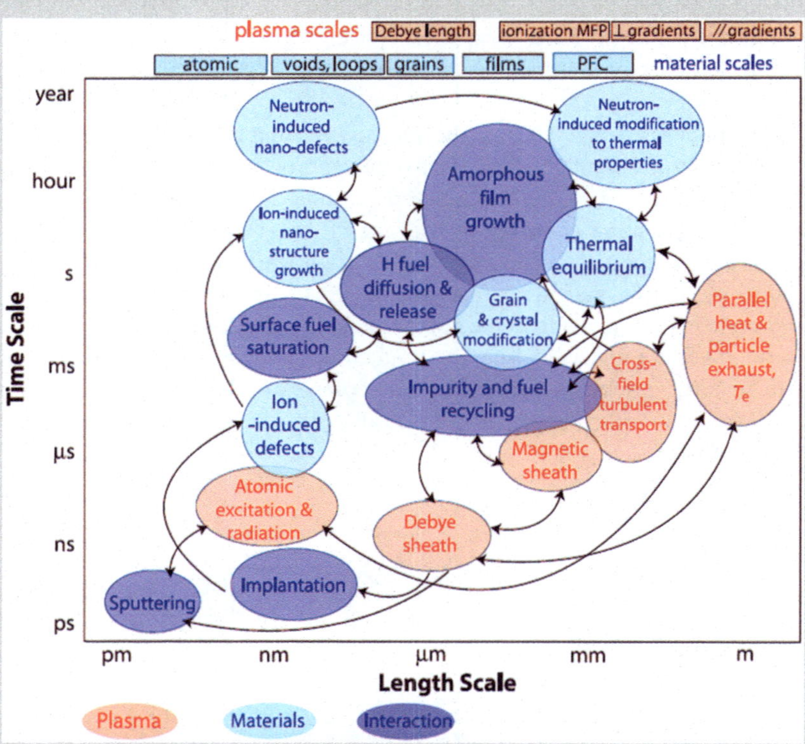

The multiscale nature of the plasma-materials interaction is illustrated in above figure, where the different color symbols represent physical phenomena occurring in the plasma (light red spheres), the bulk materials (light blue spheres), and the near surface interaction region (light purple spheres). The smaller time and length scales correspond to individual ion implantation and sputtering processes, which occur at or near the material surface, in addition to a range of ionization and recombination processes of the sputtered neutrals and

(continued)

ions in the near-surface sheath and neutron-induced displacement cascade and nuclear transmutation events, which serve as the source term for radiation damage processes within the structural material. At intermediate length and time scales, a wealth of physical processes are initiated, including diffusion of the now implanted ionic/neutral species, the possibility of chemical sputtering processes at the surface, the formation of gas bubbles, surface diffusion driving surface topology changes, and phonon scattering by radiation defects that reduces the thermal conductivity of the material, as well as the nucleation of radiation damage defect clusters. At longer length and time scales, additional phenomena, such as long-range material transport in the plasma, re-deposition of initially sputtered surface atoms, amorphous film growth, and hydrogenic species diffusion and permeation into the bulk material become important, as do irradiation-induced void and bubble formation and the partitioning of radiation-induced defects to dislocations and extended defects that can drive irradiation creep or stress relaxation processes. This broad palette of physical phenomena will require development not only of detailed physics models and computational strategies at each of these scales, but also of algorithms and methods to strongly couple them in a way that can be robustly validated. While present research is confined to each of these scales, or pioneering ways to couple two or more of them, the current approaches already push the state-of-the-art in technique and available computational power. Therefore, simulations spanning multiple scales are needed for ITER and DEMO (DEMOnstration Power Plant) and will require extreme-scale computing platforms and integrated physics and computer science advances.

10.5.3 Structural Materials

High-performance structural materials will be critical for the future success of proposed fusion energy reactors, which will subject the structures to unprecedented fluxes of high-energy neutrons along with intense thermomechanical stresses and high-temperature coolants that may induce corrosion. Steady-state heat fluxes for first-wall and divertor components in proposed magnetically confined fusion energy reactors are projected to be in the range of 1–10 MW/m^2, which is substantially higher than the highest heat flux for structural materials in fission reactors (~1 MW/m^2 for the fuel cladding). The design lifetime doses for the first wall and blanket structural materials are about five times higher than the core internal structures for existing fission reactors. In addition, the high average neutron energy associated with the deuterium-tritium fusion reaction compared to fission tends to produce much higher levels of transmutant solutes such as H and He in the structural materials that generally magnify radiation-induced degradation processes (Zinkle and Busby 2009).

A key additional constraint for fusion structural materials is the international mandate for intrinsic safety (i.e., no public evacuation in case of a loss of coolant accident) and minimal long-term environmental impact (i.e., no long-lived radioisotopes that would require deep geologic burial or equivalent sequestration) for the fusion reactor structures. Consideration of this reduced activation mandate, along with the requirement for high performance, leads to three major options for fusion structural materials. Ferritic/martensitic steel (where high activation solutes such as Mo and Nb in commercial steels are replaced by W and V) is the most mature option, with natural leveraging of the extensive worldwide expertise and industrial infrastructure in steelmaking. Oxide dispersion strengthened steel represents a potential future higher-performance option if key issues such as property non-uniformity, joining, and high cost can be resolved. Refractory alloys based on either vanadium or tungsten alloys represent a higher performance, higher risk option. Vanadium alloys are particular attractive for self-cooled lithium blanket systems, but are not considered to be viable for other blanket concepts. SiC_f/SiC ceramic composites offer the potential for the highest thermodynamic efficiency and best safety and waste disposal margins, but are the least developed materials system for large-scale structural applications. Numerous issues including structural engineering design rules, how to achieve leak-tight boundaries for gas-cooled systems, joining and other fabrication issues, radiation stability uncertainties, improvement of thermal conductivity (and minimization of radiation-induced degradation), and fabrication cost need further research and development to enable this materials system to achieve its full potential (Zinkle and Busby 2009).

INDEPTH: Nuclear Fuel Cycles

The Nuclear Fuel Cycle (NFC) is defined as the total set of operations required to produce fission energy and manage the associated nuclear materials. It can have different attributes, including the extension of natural resources, or the minimization of waste disposal requirements. The NFC comprises a set of operations that include the extraction of uranium (U) resources from the earth (and possibly from seawater), uranium enrichment and fuel fabrication, use of the fuel in reactors, interim storage of used nuclear fuel, the optional recycle of the used fuel, and the final disposition of used fuel and waste forms from the recycling processes. Thorium (Th) fuel cycles have been proposed also, but have not been commercially implemented. The nuclear fuel cycle is often grouped into three classical components (front-end, reactor, and back-end):

(continued)

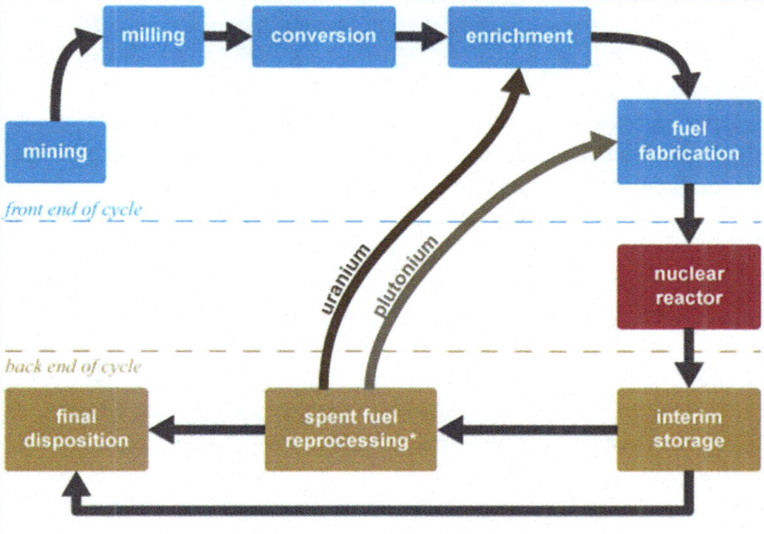

Spent fuel reprocessing is omitted from the cycle in most countries, including the United States.

(a) Front End: The focus of the front end of the nuclear fuel cycle is to deliver fabricated fuel to the reactor. Nuclear material is initially collected from two main sources: nature (via mining or possibly extraction from sea water), or recycle of fissile isotopes from used nuclear fuel or other sources (e.g., surplus weapons material). The fuel material is then refined, potentially enriched in its fissile content, and fabricated into nuclear fuel elements with diverse geometries depending on the reactor technology. These fuel elements are then constituted into nuclear fuel assemblies. The geometry, material form, and isotopic content of the fuel are very specific to the reactor type in which the fuel is to be used. However, even within the same reactor type or design, variations are possible, due to various industry technologies. The fuel isotopic content may also vary significantly depending on the mission of the fuel cycle.

(b) Reactor: The mission of the reactor is to extract heat safely from the fuel in a self-sustained nuclear reaction. This heat can then either be used to drive turbines to generate electricity, or for industrial process heat applications. (1) Thermal Reactors, including light-water reactors (LWRs)—mainly used for commercial nuclear power production in the United States and also capable of recycling fractions of the plutonium from used nuclear fuel and possibly reprocessed and re-enriched uranium; high-temperature gas reactors (HTGRs) and molten salt reactors (MSRs) that have similar

(continued)

capabilities, but have not been commercially deployed in large numbers. (2) Fast Reactors—used for both power production and capable of breeding fissile isotopes and recycling of used nuclear fuel, including uranium/plutonium or the transuranic (TRU) elements. A number of coolant types have been assessed for Fast Reactors, leading to technologies such as the sodium-cooled fast reactor (SFR), the lead or lead-bismuth-cooled fast reactor (LFR), and the gas fast reactor (GFR).

(c) Back End: The focus of the back end of the nuclear fuel cycle is to manage the used fuel produced by the reactor, and put it into a long-term safe and secure state. The back end of the fuel cycle also needs to dispose of processing wastes, including large amounts of low level waste (LLW). Two major options exist: (1) the once-through fuel cycle that ultimately transfers the used fuel into permanent geologic disposal and (2) recycle strategies, where the useful nuclear material is recovered and incorporated into fuel for re-use in thermal or fast reactors, improving ore utilization (when fast reactors are used) and reducing waste volumes sent to disposal, and the remaining radionuclides are converted into waste forms for stable, long-term disposal.

10.6 Low-Energy Nuclear Reactions in Condensed Matter

In 1989, Stanley Pons and Martin Fleischmann claimed to have demonstrated "cold fusion" by electrochemical charging of deuterium into palladium. They presented experimental evidence for a phenomenon by which nuclear fusion could occur not only at the extreme temperatures of the sun's core (~15 million K) but also at room temperature. As shown in Fig. 10.12, Fleischmann-Pons electrolytic cell consisted of two pieces of metal, one palladium and the other platinum, submerged in a container of heavy water (water in which the hydrogen of each H_2O molecule is replaced by deuterium). If the cell was zapped with electricity, it would trigger a chemical process called electrolysis, in which the heavy-water molecules would split, producing deuterium gas and oxygen. The deuterium could then be absorbed into the palladium via a chemical reaction. Pons and Fleischmann hypothesized that, once inside the palladium, the deuterium atoms would be forced so close together that they would fuse and release large amounts of energy as heat. However, The F-P work was erroneously discredited in part because the deuterium levels attained in the palladium during replication experiments were too low to trigger the reactions observed during the original experiments. With sufficient deuterium loading, the F-P and similar experiments have been duplicated dozens of times and in different laboratories. Furthermore, many non-electrochemical experiments show excess energy that is far above that obtainable by any known chemical reaction, in metal

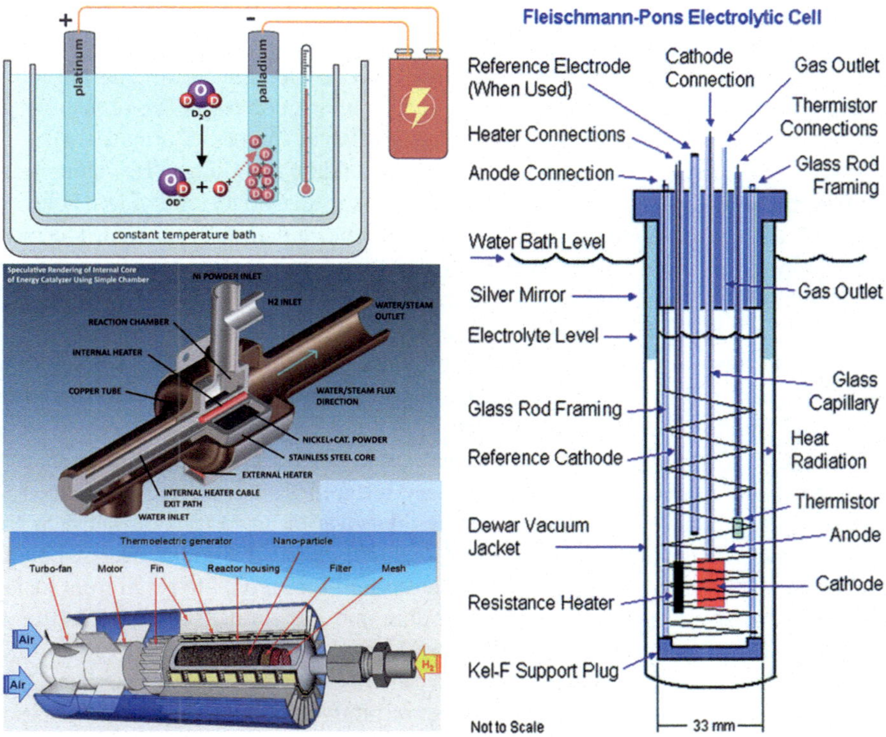

Fig. 10.12 Schematic illustration of Fleischmann-Pons electrolytic cell and examples of LENR modules (Modified with permission from Fleischmann and Pons 1993 (Elsevier))

formulations that do not contain palladium. There has been particular success in formulations that contain both lithium and nickel as these "fuel" formulations are often claimed to produce excess heat. These various reactions are now called Low-Energy Nuclear Reactions (LENR) (Pickens and Nagel 2016).

10.6.1 Current Experimental Status of LENR

When a new discovery is made, science and simple logic require it to be replicated before it is accepted. Because the LENR effect proved to be especially difficult to replicate, acceptance was delayed. This handicap is still operating, but is significantly reduced as persistence has identified many important variables. As result, the effect has been replicated hundreds of times while using many different materials and at least five different methods. These methods include electrolysis, low voltage gas discharge, plasma discharge within a liquid, exposure of an active material to H_2 or D_2 gas, and application of acoustic waves (Fig. 10.12). Each method is found to produce energy well in excess of any plausible chemical source that is correlated with identified nuclear products. These products include helium-4 tritium, and various transmutation products. Obviously, an unusual common feature is produced in the material by each of these treatments. This feature, called the nuclear active environment (NAE), is then able to cause nuclear interaction regardless of the isotope of hydrogen made available to it from several different sources (Storms 2014).

When deuterium is used, the main nuclear product is helium (^{4}He) along with some tritium (3H) on occasion. The helium results from a novel kind of D + D fusion that releases the excess mass energy without significant detected radiation. The mass change resulting from two D creating ^{4}He requires 2.62×10^{11} He/watt-sec to be released as heat. The large scatter, and the average that shows too little helium than expected, can easily be explained. The average contains less helium than expected because the amount of helium was based on helium that could be easily removed from the Pd where it formed. The scatter is high because a variable amount would be removed and because the measurements of helium and energy would contain some uncertainty. Nevertheless, such agreement could not be achieved if all the helium resulted from contamination and all energy resulted from a source not related to helium production. Many other studies produced a clear correlation between the presence of helium when extra energy was detected and the absence of helium when no extra energy was found. Moreover, tritium is proposed to result from fusion between deuterium and the occasional protium impurity during which an electron is captured. This reaction rate has been too low to produce detectable energy. Nevertheless, the tritium is clearly detected and can only result from a nuclear reaction taking place in the apparatus. The nuclear product resulting in energy when only H is used has not been experimentally identified although significant energy can apparently be made using this isotope. Furthermore, nuclear reactions, called transmutation, are reported on occasion when two or more deuterons or protons enter the nucleus of a target element that happens to be present on the surface of nuclear active material. Explaining this process is an especially difficult job, although the evidence for several kinds of mechanism is strong. The transmutation reaction also has too low a rate to produce detectable energy. Many proposed explanations of LENR have been published, most of which are in conflict with observation, natural law, and each other. Nevertheless, the unique LENR process can be explained by using few novel concepts and without violating natural law. In addition, the concepts obtained from study of the conventional type of fusion, called hot fusion, need to be ignored because cold fusion and hot fusion have little in common. Ironically, the early belief that these two kinds of fusion were related resulted in especially hostile rejection by the physics community. Such rejection is no longer justified based on this belief (Storms 2015).

Energy resulting from the fusion process has been measured using a variety of instruments. Values sometimes exceeding tens of watts, on many occasions, greatly exceed the expected error of the calorimeter. For instance, the claims for energy using H_2 applied to nuclear active nickel have reached several kilowatts and have encouraged a search for a method to produce commercial power using this combination of materials (D_2O + ^{55}Mn + MCT (microbial catalyst-transmutator) + additional isotope components (e.g., ^{24}Na, ^{40}K, ^{60}Co, ^{91}Sr, ^{131}I, ^{135}Xe, ^{140}Ba, ^{140}La, ^{141}Ce, ^{239}Np) (Vysotskii and Kornilova 2011).

As a result, several very important aspects of LENR have been determined from experiments (Pickens and Nagel 2016):

(a) It is possible to initiate nuclear reactions, each of which gives energies of about 10^6 electron volts, by using chemical energies on the order of one electron volt. Hence, energy gains well in excess of one thousand may be possible.
(b) Very high temperatures are not needed to produce LENR, greatly simplifying the experimental study of the phenomenon and its potential applications.
(c) There are at least four general approaches to LENR experiments, namely the use of liquids, gases, plasmas, and beams to load hydrogen isotopes into solids, notably Pd and Ni.
(d) Four types of measurements: heat that cannot be explained by chemistry, nuclear reaction (transmutation) products, low intensities of energetic particles, and some low-energy phenomena all point to the occurrence of nuclear reactions.
(e) Some electrochemical experiments in light-water experiments produce no excess energy, yet the same experiment performed in heavy water produces excess energy.
(f) Energy gains (CoP) in excess of 25 have been observed in several experiments by different Investigators.
(g) Power densities exceeding those within nuclear fission fuel rods by 100 times have been measured and reported.
(h) Values of generated energy (in electron volts per atom of the metal) in excess of 2,000 have been observed in LENR experiments, far greater than what chemical reactions can produce.
(i) The experiments do not emit dangerous radiation during their normal operation; no significant radioactive waste has been observed after LENR experiments.
(j) LENR do not produce greenhouse gases.
(k) Commercialization of LENR energy generators is currently underway.

10.6.2 Possible Mechanisms for LENR

There may be several mechanisms for LENR, but none of the mechanisms is well understood. However, one modeling work vibrational LENR initiation mechanism is particularly insightful (DeChiaro et al. 2015): spin-polarized, Density Functional Theory (DFT) calculations were performed to model transition and *sp* elements in the Periodic Table. Most transition elements were found to exhibit ferromagnetic ordering if the lattice was sufficiently strained in tension. That is, expanding the lattice causes the energy of ferromagnetic ordering to be lower than that of paramagnetic ordering. Such ordering may facilitate LENR (Pickens and Nagel 2016):

(a) It is necessary to set up conditions favoring the formation of molecular hydrogen (H_2 or D_2) inside the solid lattice for a certain range of possible values of lattice constant and for some fraction of the allowed values for electron momentum. This condition alone rules out almost all the elemental metals, because the electron density is just too large to permit molecules to form, except near vacancies in the lattice where a metal atom is absent.

(b) The overall hydrogen loading fraction (ratio of hydrogen to palladium atoms, for example) must exceed the minimum threshold of about 0.88, otherwise the "party" never even gets started. Achieving this level of loading in Pd is not trivial.

(c) Conditions must be set up (by appropriate choice of materials parameters and achieved by the right kind of alloying) so that these hydrogen molecules can be caused to break up and then reassemble very rapidly in a periodic time sequence when an appropriate physical quantity such as background electric charge and magnetic field is made to oscillate periodically over a small range.

(d) The critical value of lattice constant at which this break up and reassembly occurs must lie very close to the nominal value of lattice constant for which the ground state energy of the lattice is minimal. This requirement alone rules out essentially all of the elemental lattices and about 99% of the binary and ternary alloys.

(e) A departure from equilibrium must be established that will permit an external energy source (e.g., the DC power supply in an electrolysis experiment and/or a pair of low power lasers as in the Letts/Hagelstein two laser experiment) to feed energy into the H-H or D-D stretching mode vibrations. The difference in chemical potential that is established in gas loading experiments can also serve very nicely; in this case the flux feeds energy into the stretching mode vibrations.

(f) The nature of the lattice must permit these stretching mode vibrations to grow so large (over a period of perhaps many nanoseconds) that their amplitude becomes comparable to the lattice constant. When this occurs, the H atoms oscillate so violently that at the instants of closest approach, the curvature of the parabolic energy wells in which the atomic nuclei vibrate will become perturbed. Thus, the curvature of the well oscillates as a periodic function of time. Under the right conditions, these oscillations can grow without impacting the metal atoms, which are much more massive than the hydrogens.

(g) When the curvatures of the parabolic energy wells of the nuclei are modulated at a frequency very near the natural resonant frequency, the quantum expectation value of the nuclear wave function spatial spread will oscillate with time in such a way that the positive-going peaks grow exponentially with time. These oscillations in spatial spread will periodically delocalize the nucleus and facilitate the tunneling of adjacent nuclei into the Strong Force attractive nuclear potential well, giving rise to nuclear fusion at rates that are several tens of orders of magnitude larger than what one calculates via the usual Gamow Factor integral relationship.

Since LENR has little to do with hot fusion, something very unusual must happen in certain materials that are not possible in plasma where hot fusion takes place. This unusual process allows two hydrogen nuclei to come close enough to fuse without applied energy, without the resulting excess mass energy being dissipated without destruction of the nuclear product, and without significant radiation being detected. Hundreds of attempts have been made to explain how this is possible. At the very least, two or more nuclei need to find each other in an array of metal atoms. Further, simply being able to get close is not enough, because when this separation is reduced, hot fusion produces fragmentation of the nuclear product into energetic neutrons (n), protons (p), tritium (^{3}H), and helium-3 (^{3}He), which are not observed in LENR. Nevertheless, in LENR the excess mass energy resulting from D–D fusion is measured as heat having a magnitude consistent with the amount of helium-4 (^{4}He) detected. In addition, tritium is made without the expected neutron. Clearly, a novel mechanism is operating. Once this mechanism is understood, many other kinds of unexplained nuclear reactions will be easier to explain, including biological transmutation (Storms 2015).

10.6.3 Potential Advantages and Impacts

The goals that are now driving many people and organizations to confront the challenges are determined by the potentially significant advantages of LENR power and energy sources (Pickens and Nagel 2016):

(a) High Performance. It is widely expected that LENR generators will have high and versatile performance if engineering challenges are met. These expectations are based on the following possibilities:
   (i) High CoP. Energy gains, as often referred to as CoP, are defined as the ratio of energy out of a system to the energy required to initiate and operate it. Very high CoP values have already been reported, but not yet consistently verified. They involve thermal output energy stimulated by electrical input energy. CoP in excess of 25 using the electrochemical D-Pd system have been published. CoP exceeding 400 have been reported for the Ni-H system.
   (ii) Self-sustaining ("Burning") Reactions. Although LENR does not involve oxidation for the large amount of excess energy produced, burning is accepted LENR and hot fusion jargon. One goal of hot fusion research is achievement of burning, where the energy released by prior reactions sustains continued energy release, as long as fuel is available. That situation is entirely analogous to the lighting of a pile of logs with a match. It may also be applicable to LENR. Such sustained operation can lead to enormous energy gains, since the output energy continues to increase without increase in the input energy.
   (iii) Production of Heat. The raw output of LENR generators is thermal energy, which can be used to raise the temperature of diverse working fluids. The production of thermal energy for heating factories and offices will likely be a primary and early function of LENR generators in the commercial sector. Home heaters and electrical generators will follow later.

(iv) Generation of Electricity. Electrical power is needed almost everywhere, so the use of energy from LENR devices to generate electricity is also of great interest. If steam is produced, it can be used to run a turbine or other engine attached to an ordinary alternating-current electrical generator. There is also great interest in development of thermoelectric materials now, independent of LENR. The motivation is the ability to use energy wasted in many devices and processes to produce direct current by the Seebeck effect. Available inverter technologies can be used to produce AC from the DC coming from LENR devices, which are integrated with thermal-to-electrical transducers.

(b) Low Costs. The costs for the purchase, operation, and maintenance of LENR generators will be relatively inexpensive.

(c) Operational Advantages. There are several attractive possibilities regarding the operational characteristics of LENR sources of power and energy.

(i) Silent Operation. One of the standout characteristics of LENR power sources is the absence of significant acoustic emissions during their operation. They might operate in a continuous mode during the production of power, with either constant or quietly pulsed input electrical power, or no electrical input power, once started.

(ii) Easy Operation and Refueling. The units now planned for the market apparently will require little training in their use and little attention during their operation. It remains to be seen if commercial LENR units can be controlled, much as thermostats are now set by homeowners. The ongoing development of means to store energy for solar and other unsteady sources of energy might prove useful for storage of LENR energy.

(iii) Extended Times Between Refueling. The currently projected time of 6 months between refueling of E-CAT systems might not seem very long. But, they are long compared to the times for refueling heating units or electrical generators that burn hydrocarbons. This advantage is magnified when considering the infrequent LENR refueling that will be needed for automobiles compared to current gasoline powered vehicles.

(iv) Opportunities for Optimization. Both the LENR experiments already conducted and the prototypes that have been demonstrated are virtually certain not to be as good as those designed and developed later. Commercial products are generally improved over time to give better performance for the same or lower costs. The point is that the production and use of LENR thermal and electrical generators are now in the earliest stages of commercialization. Optimization of manufacturing processes, which will lower costs of LENR generators, can also be expected.

(d) Safety Possibilities. There are significant potentials for safe operation of LENR sources. These will be of continuing concern until there is a large experience base for such benign operation.

(i) Safe Operation. The safety of power sources and energy transduction devices is an enduring concern. LENR generators, no matter how attractive their performance, must be safe. Safe operation of generators with much higher energy gains might be achieved through taking a systems approach to the engineering design.

(ii) Fail Safety. The control of the output of any power source is a necessary concern. It appears that LENR power sources will not offer near-instantaneous responses to changes in input settings. That is no problem for a wide variety of expected applications, including heating and electricity generation. But, there remains the possibility that the control system for a LENR generator will fail and the system will run away, causing damage to property or injury to people.

(iii) Easily Shielded Radiation. Many chemical energy sources suffer from concerns over fires and even explosions. However, they are free of anxiety over dangerous radiation emissions, and require no radiation shielding. LENR energy sources are close to chemical energy sources in radiation safety, but they do have small fluxes of energetic radiations that must be and can be shielded easily. One of the hallmarks of the past two decades of research on LENR is the quantification of excess heat, usually without the measurement of dangerous levels of either neutrons or gamma rays, despite vigorous attempts to make such radiation measurements.

(iv) Adequately Safe Input Chemicals. Even if radioactive fuels are not needed for LENR reactors, it is possible that what they require for operation could be dangerous for chemical and other reasons. Of particular concern is the use of pressures of hydrogen gas that might be as high as 200 atmospheres in some LENR systems. Very importantly, it is possible that LENR generators will later get their hydrogen from solids rather than high pressure tanks. Some solids, such as lithium aluminum hydride, have four hydrogen atoms per two metal atoms in their molecules. Such a compound does require careful insertion into fuel canisters. In short, the hydrogen in LENR energy generators might prove to be much safer than gasoline now used in hundreds of millions of automobiles.

(v) No Input Radioactivity. The input fuels for current experimental and possible near term commercial LENR power and energy generators are apparently benign. They are available on the market now, although mixing and other pre-processing for their activation might be needed.

(vi) No Radioactive Waste. One of the largest problems with fission reactors is the radioactive waste left in spent fuel rods. Some of the isotopes in such rods have long half-lives, so that the waste would remain dangerous for periods on the order of 10,000 years. One of the touted advantages of hot fusion reactors is that their radioactive waste would be dangerous for "only" about 1000 years. Remarkably, the generation of energy using LENR reported to date results in essentially no radioactive waste. The weak activity of materials from some LENR experiments might be due to tritium, which decays with a half-life of 12.3 years. In short, radioactive waste does not seem to be a problem for LENR energy sources.

(e) Environmentally Friendly. The results of laboratory experiments indicate that power sources based on LENR will likely not produce deleterious effects on the environment.

(i) No Greenhouse Gas Emission. LENR electrochemical experiments and gas loaded generators have not been shown to produce greenhouse gases during or after the operation of experiments to date. Now, many of the large power stations are coal-fired, so they produce prodigious amounts of greenhouse gases, atmospheric pollutants and even radioactive pollutants.

That is, LENR power sources are free of both radioactive waste and greenhouse gas emissions. This combination of features is remarkable.

(ii) No Chemically Dangerous Waste. It is conceivable that operation of a LENR energy generator could be free of significant radiation and residual radioactivity, but still produce chemicals that might harm people. This does not appear to be a concern, based on experience with LENR experiments that produced excess energy.

(iii) Possibly Beneficial Waste. If one of the products of reacting nickel with hydrogen is indeed copper, it should be possible to recover the copper and use it in electrical and other products. If the reaction products are other isotopes of nickel, the nickel could be recovered and sold as feedstock to the huge stainless steel industry.

(iv) Abundant Hydrogen "Fuel." It is significant that many experiments have demonstrated LENR that uses conventional hydrogen, and not deuterium. Hydrogen is one of the most common elements on earth. Freeing it from water could be done electrolytically by using a small fraction of the energy produced by LENR generators. The total cost of gaseous hydrogen will not be a significant part of the overall cost of fueling LENR reactors. If protons are obtained from solid chemicals, their cost could still be very small compared to the value of the power they produce during LENR. From an environmental viewpoint, operation of LENR generators should not perturb the distribution of hydrogen on earth.

(v) Abundant Nickel Fuel/Catalyst. Nickel is the 24th most abundant element on earth. About 1.3 million tons of nickel are mined annually. One percent of that amount would have the LENR heat equivalent of roughly 24 billion barrels of oil. The environmental impacts of using nickel as a fuel in LENR generators should not be problematic.

(f) High Power Density. Experimental and prototype commercial devices indicate that LENR generators will have high power densities, which lead to attractive potential benefits.

Exercises

Part I: General Questions

10.1. Write the balanced nuclear equation for each of the following:
  (a) α decay of gold-185.
  (b) β- decay of actinium-228.
  (c) proton emission of cobalt-56.
  (d) β + decay of holmium-158.

10.2. The half-life of ^{14}C is 5730 years.
  (a) How old is a wooden bowl whose ^{14}C activity is one-fourth of the activity of a contemporary piece of wood?
  (b) At Stonehenge, a charcoal sample is dug up, presumably the remains of a fire; its ^{14}C activity is 9.65 disintegrations per minute per gram of carbon. Living tissues have a ^{14}C activity of 15.3 disintegrations per minute per gram of carbon. When were the fires of Stonehenge ignited?

10.3. Potassium-40 is often used to date minerals. How old is the rock if two-fifths of the original K-40 exists in it?

10.4. Describe three major environmental and security problems associated with nuclear power.

10.5. Plutonium is very damaging when inhaled as small particles because of the ionization of tissue by its emitted alpha particles. One microgram of plutonium is known to produce cancer in experimental animals. From its atomic weight (239) and half-life (24,360 years), calculate how many alpha particles are emitted by a microgram of plutonium over the course of a year? About how many ionizations do the particles produce?

10.6. A neutron generated by fission typically possesses a kinetic energy of 2 MeV. When such a neutron collides with a hydrogen atom 18 times, its energy is reduced to its thermal energy of 0.025 eV, that is, the kinetic energy it would possess by virtue of the temperature of its surroundings. The same neutron would have to collide with a sodium atom more than 200 times to reduce its energy by the same amount. Explain how these characteristics make water a suitable coolant in the pressurized light-water reactor that utilizes U-235 as the fuel, whereas sodium is the suitable coolant in the breeder reactor.

10.7. Why are the spent fuel rods from fission reactors more radioactive than the initial fuel rods?

10.8. How does a breeder reactor extend the supply of nuclear fuel? What are the problems associated with the breeder design?

10.9. From the standpoint of weapons proliferation, why is it more dangerous to fuel reactors with plutonium than with uranium in which U-235 is enriched to 2–3%?

10.10. Considering the ^{238}U decay scheme, which of the daughters of uranium are likely to be most abundant in uranium-bearing soil, and why?

10.11. Any large-scale nuclear process produces radioactive waste. Compare the problems of waste from uranium fission reactors and from tritium fusion reactors.

10.12. What is the percentage of ^{235}U in naturally occurring uranium and what is the rest made of? A nuclear fission reaction of an ^{235}U atom caused by a neutron produces one barium atom, one Krypton atom, and three more neutrons. Evaluate approximately how much energy is liberated by this reaction.

10.13. Approximately how much percentage of energy is carried by the fission fragments (no calculation necessary for the last part of the question)?

10.14. What is the difference between fissile and fertile isotopes? Give two examples of each. What is the role of fertile isotopes in a breeder reactor?

10.15. Define a nuclear reactor? What is the basic difference between an atomic bomb and a power-producing reactor?

10.16. What are the prime differences between LWR and CANDU reactors (comment mostly on materials aspects)?

10.17. Describe the importance of control materials with respect to reactor safety and control. What are the primary requirements for a control material? Give at least four examples of control materials.

10.18. Categorize neutrons based on their kinetic energy. What is the major difference between a thermal reactor and a fast reactor?

10.19. Zirconium and hafnium both have crystal structures (HCP) in the general operating regimes of LWRs. Naturally occurring Zr always has some Hf (1–3 wt.%) in it. Why are Hf-containing Zr alloys very common in chemical industries but not in nuclear industries? What is the main application of Zr alloys in LWRs? What are the various functions of this reactor component? What are the reasons that make Zr alloys suitable for such use?

10.20. What are the two main zirconium alloys used in light-water reactors? Give their compositions. Name two recently developed zirconium alloys with their compositions.

10.21. What is neutron economy? What significance does it have? How much influence does it exert in the selection of materials used in nuclear reactors?

10.22. Define neutron flux and neutron fluence. What are their units?

10.23. Define neutron cross section? Briefly comment on the importance of neutron cross section from a reactor perspective.

10.24. Neutrons of 10 keV energy are incident on a light-water barrier. The neutron cross section for hydrogen (protium) at 10 keV is about 20 b and that of oxygen is only 3.7 b. Determine the half-value thickness of neutron attenuation for the water barrier (assume that neutron interaction with oxygen in water molecule is negligible). Find out the half-thickness value for 1 MeV neutrons traveling through the water barrier (neutron cross section for protium is 4.1 b for 1 MeV neutrons). Comment on the significance of the results.

Part II: Thought-Provoking Questions

10.25. Address characteristics of advanced nuclear energy. For new Generation IV systems or the extension of current technology, explain why should we be aware of the possibility of new phenomena due to irradiation, corrosion, or aging in both materials and fuels performance.
10.26. List the major differences of fission and fusion reactors, and address their advantages and disadvantages.
10.27. Address materials selection criteria for nuclear power applications, and give some samples to explain.
10.28. Describe materials selected for advanced nuclear fission reactor components. What are the current status and future trends?
10.29. Explain major factors to cause materials degradation in reactor cores.
10.30. What are the major considerations for materials selection in fusion reactors?
10.31. Explain the major steps of nuclear fuel cycles.
10.32. Discuss the potential applications of low-energy nuclear reactions in condensed matter.

References

Abram, T., Ion, S.: Generation-IV nuclear power: a review of the state of the science. Energy Policy. **36**(12), 4323–4330 (2008)

Alexandreanu, B., Sencer, B.H., Thaveeprungsriporn, V., Was, G.S.: The effect of grain boundary character distribution on the high temperature deformation behavior of Ni–16Cr–9Fe alloys. Acta Mater. **51**(13), 3831–3848 (2003)

Allen, T., Busby, J., Meyer, M., Petti, D.: Materials challenges for nuclear systems. Mater. Today. **13**(12), 14–23 (2010)

Ashby, M.F., Smidman, M.: Materials for nuclear power systems. http://www.grantadesign.com/download/pdf/nuclear.pdf (2010). Accessed 05 Dec 2016

Azevedo, C.R.F.: Selection of fuel cladding material for nuclear fission reactors. Eng. Fail. Anal. **18**, 1921–1942 (2011)

Bornschein, B., et al.: Tritium management and safety issues in ITER and DEMO breeding blankets. Fusion Eng. Des. **88**, 466–471 (2013)

Burkart, W.: Status report on fusion research. Nuclear Fusion 45 (10A), International Fusion Research Council (IFRC), IAEA, Vienna (2005)

Corwin, W.R.: US generation IV reactor integrated materials technology program. Nucl. Eng. Technol. **38**(2006), 591–618 (2006)

Cox, B.: Pellet-clad interaction (PCI) failures of zirconium alloy fuel cladding–a review. J. Nucl. Mater. **172**, 249–292 (1990)

DeChiaro, L.F., Forsley, L.P.G., Mosier-Boss, P.: Strained layer ferromagnetism in transition metals and its impact upon low energy nuclear reaction. J. Cond. Matter Nucl. Sci. **17**, 1–27 (2015)

Fleischmann, M., Pons, S.: Calorimetry of the Pd-D2O system: from simplicity via complications to simplicity. Physics Letters A **176**(1–2), 118–129 (1993)

Guérin, Y.: Fuel: the "consumable" in reactors. In: Nuclear fuels. e-den—A Nuclear Energy Division Monograph. Philippe Pradel, Paris. http://www.materials.cea.fr/en/PDF/MonographiesDEN/Nuclear-fuels-CEA-en.pdf (2009). Accessed 30 Oct 2017

Guérin, Y., Was, G.S., Zinkle, S.J.: Materials challenges for advanced nuclear energy systems. MRS Bull. **34**, 10–19 (2009)

Gupta, C.K.: Encyclopedia of materials: science and technology, pp. 6339–6349. Elsevier Science Ltd, Amsterdam (2001)

IFRC: Status report on fusion research. Nuclear Fusion, Volume 45, Number 10A. International Fusion Research Council (IFRC), Vienna (2005)

Jacoby, M.: Coming back to nuclear energy—A resurgence of interest in new power plants is driving discovery of advanced materials. https://pubs.acs.org/cen/coverstory/87/8734cover.html (2009). Accessed 24 Oct 2017

Kim, I., Choi, B.Y., Kang, C.Y., Okuda, T., Maziasz, P., Miyahara, K.: Effect of Ti and W on the mechanical properties and microstructure of 12% Cr base mechanical-alloyed nano-sized ODS ferritic alloys. ISIJ Int. **43**, 1640–1646 (2003)

Klueh, R.L.: Elevated temperature ferritic and martensitic steels and their application to future nuclear reactors. Int. Mater. Rev. **50**, 287–310 (2005)

Leenaers, A., Detavernier, C., Van den Berghe, S.: The effect of silicon on the interaction between metallic uranium and aluminum: a 50 year long diffusion experiment. J. Nucl. Mater. **381**, 242–248 (2008)

Marcum, W., Spinrad, B.I.: Nuclear reactor device. https://www.britannica.com/technology/nuclear-reactor (2017). Accessed 24 Oct 2017

Mukherjee, S., Jamnapara, N.I.: Materials research and development opportunities in fusion reactors. Proc. Indian Natl. Sci. Acad. **81**(4), 827–839 (2015)

Murty, K.L., Charit, I.: Structural materials for Gen-IV nuclear reactors: challenges and opportunities. J. Nucl. Mater. **383**, 189–195 (2008)

Murty, K.L., Charit, I.: An introduction to nuclear materials: fundamentals and applications, pp. 1–41. Wiley-VCH Verlag GmbH & Co. KGaA, Weinheim (2013)

Neu, R.: Benefits and challenges of the use of high-Z plasma facing materials in fusion devices. AIP Conf. Proc. **1237**, 62–77 (2010)

Northwood, D.O.: The development and application of zirconium alloys. Mater. Des. **6**(2), 58–70 (1985)

Philipps, V.: Tungsten as material for plasma-facing components in fusion devices. J. Nucl. Mater. **415**(1 Supplement), S2–S9 (2011)

Pickens, J.R., Nagel, D.J.: The status of low energy nuclear reactions technology. http://www.etcmd.com/wp-content/uploads/2016/05/OpenLENRfinal.pdf (2016). Accessed 03 Nov 2017

Pioro, I.: Nuclear Power as a Basis for Future Electricity Production in the World: Generation III and IV Reactors. https://cdn.intechopen.com/pdfs-wm/41271.pdf (2013). Accessed 24 Oct 2017

Popple, L.A.: The oxidation of magnesium alloys in reactor atmospheres. J. Nucl. Mater. **8**, 60–76 (1963)

Rieth, M., et al.: Recent progress in research on tungsten materials for nuclear fusion applications in Europe. J. Nucl. Mater. **432**, 482–500 (2013)

Rodríguez-Prieto, Á., Camacho, A.M., Sebastián, M.A.: Materials selection criteria for nuclear power applications: a decision algorithm. JOM. **68**(2), 496–506 (2016)

Rotti, C., et al.: Establishing ITER-grade properties in CuCrZr: The Indian experience. Fusion Sci. Technol. **65**, 205–211 (2014)

Snead, L.L., Zinkle, S.J.: Use of beryllium and beryllium oxide in space reactors. AIP Conf Proc. **746**(February 6), 768–775 (2005)

Storms, E.K.: The explanation of low energy nuclear reaction. Infinite Energy Press, Concord, NH (2014)

Storms, E.: The present status of cold fusion and its expected influence on science and technology. Innov. Ener. Policies. **4**(1), 113 (2015). https://doi.org/10.4172/2090-5009.1000113

Tancret, F., Bhadeshia, H.K.D.H., MacKay, D.J.C.: Design of a creep-resistant nickel-base superalloy for power plant applications Part I—Mechanical properties modelling. Mater. Sci. Technol. **19**, 283–290 (2003)

Vysotskii, V., Kornilova, A.A.: Low-energy nuclear reactions and transmutation of stable and radioactive isotopes in growing biological systems. J. Cond. Matter Nucl. Sci. **4**, 146–160 (2011)

Wirth, B.D., Nordlund, K., Whyte, D.G., Xu, D.: Fusion materials modeling: Challenges and opportunities. MRS Bull. **36**, 216–222 (2011)

Wong, C.P., et al.: Evaluation of the tungsten alloy vaporizing lithium first wall and blanket concept. Fusion Technol. **39**, 815–822 (2001)

Yamada, H.: Failure of type 316 stainless steel cladding during thermal transients. J. Nucl. Mater. **78**, 24–32 (1978)

Yan, Q.-Z., et al.: Status of R&D on plasma facing materials in China. J. Nucl. Mater. **442**, S190–S197 (2013)

Yurman, D.: TRISO Fuel Drives Global Development of Advanced Reactors. http://www.theenergycollective.com/dan-yurman/2402117/triso-fuel-drives-global-development-advanced-reactors (2017). Accessed 24 Oct 2017

Zinkle, S.J., Busby, J.T.: Structural materials for fission & fusion energy. Mater. Today. **12**(11), 12–19 (2009)

Emerging Materials for Energy Harvesting 11

Abstract

Energy harvesting is emerging as a viable method for electronic devices to pull ambient energy from their surrounding environment (e.g., solar power, thermal energy, wind energy, salinity gradients, and kinetic energy, also known as ambient energy) and convert it into electrical energy for stored power. This coveted technology has the potential to serve as an alternative power supply for batteries that are ubiquitous in small, mobile, and autonomous wireless electronic devices, like those used in wearable electronics and wireless sensor networks. The discipline of energy harvesting is a broad topic that includes established methods and materials such as photovoltaics, and thermoelectrics, as well as emerging technologies that convert mechanical energy, magnetic energy, and waste heat to electricity. Innovative materials are vital to the development of all these energy-harvesting technologies. There are several promising micro- and nano-scale energy-harvesting materials (including ceramics, single crystals, polymers, and composites) and technologies currently being developed, such as thermoelectric materials, piezoelectric materials, pyroelectric materials, and magnetic materials. This chapter will review various state-of-the-art materials and enabled devices for direct energy harvesting and conversion, and also highlight the nanostructured materials underlying energy-harvesting principles and devices, in addition to traditional bulk processes and devices as appropriate and synergistic; innovative device-design and fabrication that leads to higher efficiency energy-harvesting or conversion technologies ranging from the cm/mm scale down to MEMS/NEMS (micro- and nano-electromechanical systems) devices; new developments in experimental methods, and device performance measurement techniques.

C. Tong, *Introduction to Materials for Advanced Energy Systems*,
https://doi.org/10.1007/978-3-319-98002-7_11

11.1 Generic Energy-Harvesting System

Energy harvesting (also known as power harvesting or energy scavenging or ambient power) is the process by which energy is derived from external sources (e.g., solar power, thermal energy, wind energy, salinity gradients, and kinetic energy, also known as ambient energy), captured, and stored particularly for small, wireless autonomous devices, like those used in wearable electronics and wireless sensor networks. Energy-harvesting devices are usually used to capture small amounts of energy that would otherwise be lost as heat, light, sound, vibration, or movement. The captured energy can be utilized to improve efficiency (e.g., computing costs would be cut significantly if waste heat were harvested and used to help power the computer) and enable new technology (e.g., wireless sensor networks). Energy harvesting also has the potential to replace batteries for small, low-power electronic devices. This has several benefits: (a) Maintenance free—no need to replace batteries; (b) Environmentally friendly—disposal of batteries is tightly regulated because they contain chemicals and metals that are harmful to the environment and hazardous to human health; (c) Opens up new applications—such as deploying energy-harvesting sensors to monitor remote or underwater locations.

All in all, the main purpose of energy harvesting is to transform a present excess energy from the surroundings into a usable form to power a specific application. It is not strictly necessary to produce electricity; a windmill, for example, can be used to convert wind energy into rotational kinetic energy for milling grains or pumping water. The basic function is common to any type of device and involves certain steps that are depicted, for the case where electricity is produced, in the flow chart in Fig. 11.1. The first block is reserved for the ambient source. Vibrations, solar illumination, radio frequency, pressure fluctuations, fluid flows, and thermal gradients can all be used as a starting point. The link between the source and the actual conversion mechanism is represented by the second block. In the case of motion energy harvesting this can be a direct force or an inertial force coupling. Thermal contact would be the link for thermoelectric harvesters. The third block describes the actual principle of transduction being used. This can be any physical principle that transforms a certain form of energy into electricity. A wide range of possibilities are nowadays under investigation, such as motion type harvesters, and piezoelectric, electromagnetic, and electrostatic implementations. The Seebeck effect is the corner stone of thermal harvesting, where temperature gradients are converted into an electric potential and the photovoltaic effect is applicable to light sources of any kind. After transformation into electrical energy, circuitry is needed to extract the generated power from the transducers. It might be necessary to rectify and regulate the output voltage. This is strongly dependent on the used mechanism. A buffer or energy store handles periods of high demand by the load just as well as periods of low supply by the source. These three blocks work together and can be integrated to a certain extent. In other cases, it might not be necessary to include an intermediate energy storage solution. Finally the last link in the chain is the load that consumes the generated power. In a wireless sensor network this could, for example, be a simple temperature sensor with data transmission to a central computer. It

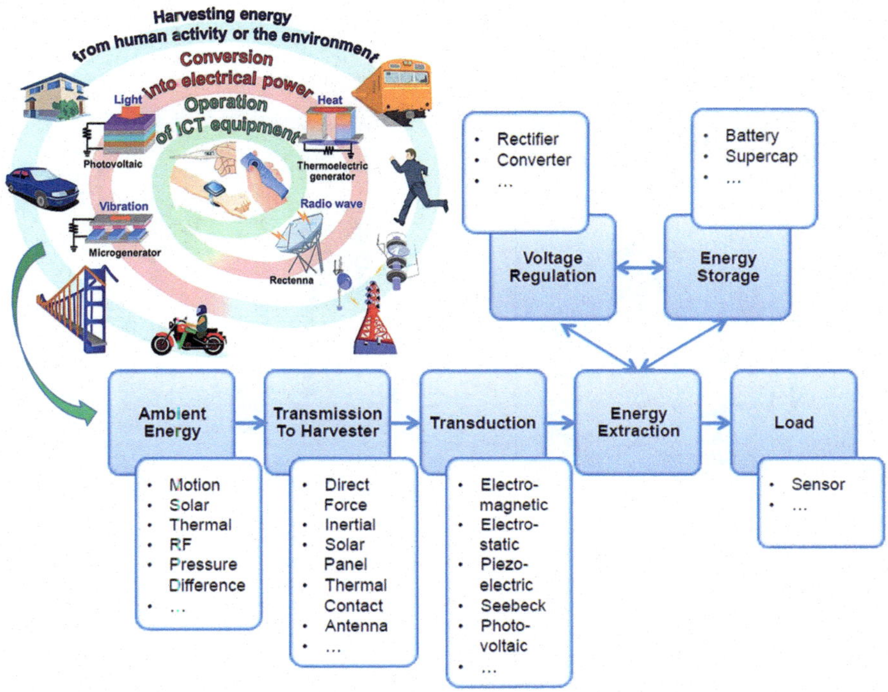

Fig. 11.1 General operational flow chart for an energy-harvesting system (Modified from Pillatsch (2013). Credit: Imperial College London. Available at https://spiral.imperial.ac.uk/bitstream/10044/1/18112/1/Pillatsch-P-2013-PhD-Thesis.pdf)

should be mentioned that this system constitutes a generic example and that some functionality or blocks can be added and others left out depending on the application. For instance, intermediate energy storage might not always be necessary as for the device, where low-power pulses representing sensor data are transmitted directly. In the case of frequency tuning for resonant vibration harvesters, the electronic system might be more complex and require a feedback loop that acts upon the transduction mechanism in order to adjust to changing operation conditions. Also, it is important not to lose the overview of the entire system when working on one particular building block as each block can be affected by the preceding and following ones. For this reason energy-harvesting platforms can be used to incorporate the entire functionality (Pillatsch 2013).

The need for energy harvesting accompanied with renewable energy, and improved energy efficiency is motivating the discovery of new materials and design of new device structures. For example, nanostructured composite materials are not only promising for thermoelectric, but also in photovoltaics. Nanocomposite materials can enhance the thermal-to-electrical power conversion efficiency. One nanocomposite used is phase change materials (PCM), which were prepared by compositing exfoliated graphite nanoplatelets (xGnP) into paraffin wax. The role of

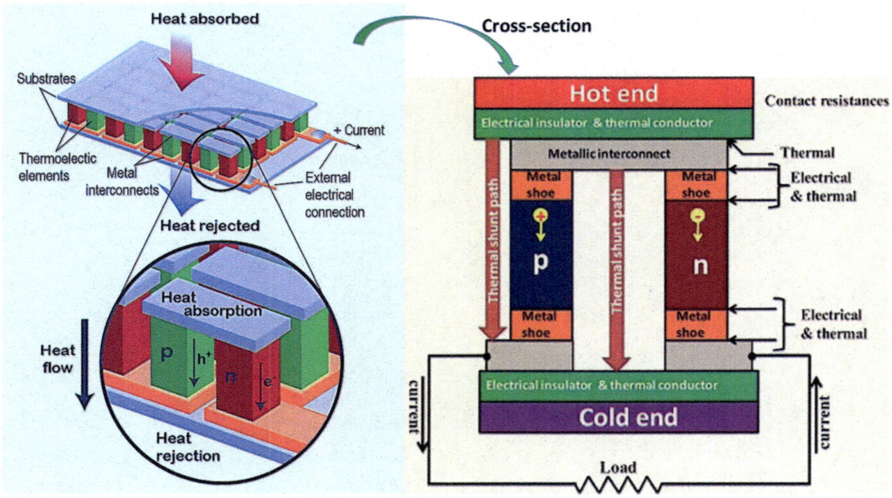

Fig. 11.2 Thermoelectric module showing the direction of charge flow on both cooling and power generation (Modified with permission from Snyder and Toberer (2008) (Springer Nature), and Aswal et al. (2016) (Elsevier))

new materials in energy harvesting cuts across all of the innovative devices. The need for this enhanced understanding will grow increasingly important as the role of nano-structured materials is enhanced. Energy harvesting covers a cross-cutting discipline. Each of the subfields is important to an understanding of advances across the field as a whole (Radousky and Liang 2012)

11.2 Thermoelectric Materials

Thermoelectric (TE) energy converters are of increasing interest as these solid-state devices can transform heat given off from sources such as power plants, factories, motor vehicles, computers, or even human bodies into electric power using the Seebeck effect, as shown in Fig. 11.2 (Snyder and Toberer 2008; Aswal et al. 2016). Thermoelectric system is an environment-friendly energy conversion technology with the advantages of small size, high reliability, no pollutants, and feasibility in a wide temperature range (Zhang and Zhao 2015). Conversely, solid-state thermoelectric devices can also change electrical energy into thermal energy for cooling or heating using the Peltier effect. However, because of their low efficiency, current TE materials have found limited commercial applications. Therefore, the growing need for energy-harvesting power sources is driving a growing interest in developing a new generation of thermoelectric materials (Elsheikh et al. 2014).

11.2.1 Characterizations of Thermoelectric Materials

Thermoelectric materials are characterized by the figure of merit ZT (Elsheikh et al. 2014):

$$\mathrm{ZT} = \frac{\alpha^2 \sigma T}{K} \tag{11.1}$$

where α is the Seebeck coefficient, σ is the electrical conductivity, and K is the thermal conductivity. The higher the ZT, the more efficient the TE. The complex relationships of these thermoelectric parameters can be expressed as (Zhang and Zhao 2015):

$$\alpha = \frac{8\pi^2 k_{\mathrm{B}}^2}{3eh^2} m^* T \left(\frac{\pi}{3n}\right)^{2/3} \tag{11.2}$$

$$\sigma = ne\mu = \frac{ne^2\tau}{m^*} \tag{11.3}$$

$$K_{\mathrm{to}} = K_{\mathrm{la}} + K_{\mathrm{el}} = K_{\mathrm{la}} + L\sigma T \tag{11.4}$$

where k_{B} is the Boltzmann constant, m^* is the density of states effective mass, h is the Planck constant, n is the carrier concentration, e is per electron charge, μ is the carrier mobility, τ is the relaxation time, K_{to} is the total thermal conductivity, K_{la} is the lattice thermal conductivity, K_{el} is the electronic thermal conductivity, and L is the Lorenz number.

The complex parameter relationships make the approach of tuning carrier concentration alone difficult to improve ZT. In particular, α and σ generally vary in a reciprocal manner; the electrical conductivity and the Seebeck coefficient are inversely related, so it is not generally possible to increase the thermoelectric power factor above a particular optimal value for a bulk material. However, ideal thermoelectric materials would have a high electrical conductivity to allow the conduction of electricity, which would create a potential difference across the sample, and a low thermal conductivity to maintain the temperature gradient between the hot and cold side. Most traditional materials exhibit a correlation between electrical and thermal conductivity. A material that conducts electricity well, such as a metal, also conducts heat well, and a material that insulates heat, such as glass or ceramic, also insulates electricity. For devices operating at room temperature (T~300 K), traditional thermoelectric materials, such as bismuth telluride (Bi_2Te_3) and lead telluride (PbTe), possess values of ZT~1. Superlattices and quantum dots-based materials has increased their room temperature ZT to 2–2.4. These improvements in performance are primarily the result of a reduction in lattice thermal conductivity, and the thermoelectric power factor ($\alpha^2\sigma$) is largely unchanged. For practical purposes, a suitable high-performance TE material should have a ZT of 4.4, and the achievement of this goal has remained a formidable challenge (Elsheikh et al. 2014).

11.2.1.1 The Seebeck Coefficient Enhanced with Band Structure Engineering

Generally, most metals possess Seebeck coefficients of 10 μV/K or less, but semiconductor materials are promising for the construction of thermocouples because they have Seebeck coefficients in excess of 100 μV/K. However, achieving ZT > 2 requires thermopower ≥225 μV/K (Elsheikh et al. 2014).

The Seebeck coefficient enhancement in PbTe was first tried by the density-of-states (DOS) distortion through Tl doping. Such a situation can occur when the valence or conduction band of the host semiconductor resonates with the localized impurity energy level. Compared with Na doped PbTe with the same carrier concentration, Tl enhancement is similar in character to that caused by resonant states in PbTe by Tl doping. Indeed, the Mn and Mg alloying in PbTe produced a high ZT of 1.6 at 700 K and 2.0 at 873 K, respectively. The intra matrix band engineering has also been successfully applied to other systems such as the PbSe-SrSe, Mg_2Si-Mg_2Sn, and the SnTe systems. However, this approach is challenged by the deteriorations of carrier mobility (Zhang and Zhao 2015).

11.2.1.2 Thermal Conductivity Reduced by All-Scale Hierarchical Architectures

A high-quality thermoelectric material must have a high electrical conductivity, a high thermopower, and a low thermal conductivity. Because the first two are determined only by the electronic properties of the material, they are often combined into the quantity $\alpha^2\sigma$, referred to as the "power factor." In contrast, the thermal conductivity in thermoelectrics is the sum of two contributions as shown in Eq. (11.4): electrons and holes transporting heat (K_{la}) and phonons traveling through the lattice (K_{el}). As K_{la} is determined by the structure, rigidity, atomic masses, and other characteristics of the lattice, reducing lattice thermal conductivity is an effective method to enhance thermoelectric performance. The lattice thermal conductivity can be given by (Zhang and Zhao 2015):

$$K_{la} = 1/3 C_v v l \tag{11.5}$$

where the heat capacity (C_v) and the phonon velocity (v) are constant, so the lattice thermal conductivity is governed by the phonon mean free path (MFP) l. When the dimension of inclusions/defects is comparable to the MFP, the phonons will be effectively scattered. Acoustic phonons carry most of the heat in a material, and they have a spectrum of wavelengths and mean free paths (MFP) distribution, including short, medium, and long wavelength phonons, synergistically contributes to the total thermal conductivity. Therefore, all length-scale structures corresponding to the broad spectrum of heat carrying phonons should be the main design principle for high-performance thermoelectric materials. The most important factors that can aid in accomplishing this are (Elsheikh et al. 2014): (a) the use of compounds with complex crystal structures, (b) the presence of heavy atoms weakly bonded to the structures, (c) the existence of inclusions and/or impurities, (d) the formation of solid solutions and nano-scale precipitates, and (e) the existence of a large number of grain boundaries.

Compounds by substituting the crystal lattice with amorphous (glass-like) structures and phonon-glass electron-crystals (PGEC) have been developed as thermoelectric materials, such as skutterudite materials. A PGEC material would possess electronic properties similar to those normally associated with a good semiconductor single crystal but would have thermal properties akin to those of an amorphous material. The high-performance thermoelectric material would have regions of its structure composed of a high-mobility semiconductor, which provides the electron-crystal electronic structure, interwoven with a phonon glass. The phonon-glass region would be ideal for hosting dopants and disordered structures without disrupting the carrier mobility in the electron-crystal region (Elsheikh et al. 2014).

The introduction of nanostructure has also become a potential tool for reducing thermal conductivity and consequently increasing ZT. Nano-inclusions can be obtained by several approaches, including embedded nano-inclusions, dispersing in situ partially oxidized nanoparticles in matrix, and the endotaxial nanoprecipitates. A general approach for introducing endotaxial nanostructures in a parent matrix is through nucleation and growth of a second phase, which is required to have a low solubility in the solid state, but complete solubility in the liquid state. To get the polycrystallines, the spark plasma sintering (SPS) is a suitable and effective technology to fabricate highly dense and fine-grained thermoelectric materials. In term of developing scalable materials, there are several effective methods of powder processing, including mechanical alloying (MA), rapid melt spinning (MS), and self-propagating high-temperature synthesis (SHS) (Zhang and Zhao 2015).

The utilization of nanostructures has led to the achievement of ZT values of approximately 1–2 or above compared to ~1 and below for bulk materials. The nanostructured Bi_xSb_{2-x} Te_3 have been shown to exhibit significantly improved ZT $\approx$ 1.4 at 100 °C, mostly because of the reduced K_{la}. The thermoelectric figure of merit in nanostructured p-type silicon germanium bulk alloys has also been enhanced, due most likely to the increased phonon scattering at the grain boundaries and crystal defects formed by lattice distortion, with some contribution from the increased electron power factor at high temperatures (Joshi et al. 2008; Elsheikh et al. 2014).

11.2.1.3 Electrical Resistivity

The optimum range of electrical resistivity (ρ) for a thermoelectric material is from 10^3 to 10^2 Ωm. Variations in the electrical resistivity of a semiconductor depend on changes in the carrier concentration and the mean free path of the charge carriers which are reflected/scattered by the surface of the material when they reach it. To achieve low electrical resistivity in semiconductors, the lattice should have nearly infinite conductance at low temperatures, but in reality, the conductivity of semiconductors is very low at low temperatures because of the limited number of free electrons. Moreover, the analysis of the temperature dependence of the electrical resistivity in intermetallic compounds is a powerful tool for obtaining information regarding the intrinsic properties of these materials. Depending on the temperature range considered, one can draw certain conclusions regarding the scattering of electrons on the thermal excitations of the lattice (Elsheikh et al. 2014).

11.2.1.4 Diffusion Effect

Diffusion is the movement of particles from regions of high concentration to regions of low concentration. Self-diffusion in compound semiconductors is more complex than in elemental semiconductors because of the larger number of possible native point defects that can, in principle, mediate self-diffusion. In addition to vacancies and self-interstitials on the corresponding sublattices, antisite defects must be considered. Homogeneities in thermoelectric materials are affected by the temperature gradient because the optimal properties can only be obtained within a very narrow temperature region for each homogeneous material, which greatly limits the efficient utilization of many dispersed energy sources. However, the diffusion effect depends on the bandgap width and the ratio of electron conductivity to hole conductivity. Heavy carrier doping effectively suppresses the diffusion effect, i.e., it inhibits the increase of thermal conductivity at high temperatures. The undesirable diffusion of thermoelectric materials when placed in a thermal gradient will result in nonhomogeneous and inferior materials, and the importance of diffusion processes is evident at temperatures of approximately 80–90% of the absolute melting point of the materials. A micro-thermoelectric device consists of a bonding layer, electrodes, and thermoelectric thin films. At the interfaces between the metallic electrodes, solder materials, and thermoelectric thin films in a micro-thermoelectric device, diffusion occurs and degrades the performance and reliability of the device. Thus, a Ni layer is usually used as the diffusion barrier in some commercial devices that use bulk thermoelectric materials. However, because of undesirable and unpredictable diffusion phenomena, modern process technologies try to reduce diffusion by decreasing the thermal range experienced by the device during operation (Elsheikh et al. 2014).

11.2.1.5 Oxidizability

Electrically conductive oxide systems have been recognized as potential candidates for thermoelectric materials; these thermoelectric oxides can be used at high temperatures without deterioration of their performance caused by oxidation, and their production costs are comparatively low. Of p-type transparent semiconducting oxide (TSO) materials, Cu_2O is one of the most promising candidates, and because of the effect of the oxygen flow rate during deposition on the properties of Cu_2O films, the oxygen flow rate must be kept low to avoid the over-oxidation of Cu_2O to CuO and to ensure a non-oxidized/non-poisoned metallic copper target in the reactive sputtering environment. The proper control of the amount and flow rate of oxygen during deposition can produce good-quality p-type transparent Cu_2O films with electrical resistivity ranging from 10^2 to 10^4 Ω cm, hole mobilities of 1–10 cm^2/V-s, and optical band gaps of 2.0–2.6 eV. Measurements of the Seebeck coefficient and electrical resistivity are most likely among the most sensitive means for the detection of small amounts of diffusive oxidation. However, in some compounds, traces of diffusive oxidation that are visually and chemically undetectable can so severely affect the Seebeck coefficient as to change its sign, and the resistivity can be increased by several decades as a result of this process (Elsheikh et al. 2014).

11.2.1.6 Brittleness

At low temperatures (up to 200 K), single crystals of the Bi–Sb alloys exhibit the best thermoelectric performances, but the brittleness of the single crystals is a problem in practical devices. Development of nanostructured TE materials has been focused on improving their mechanical properties (reduce brittleness and improve machining) and improving their TE properties (figure of merit). Rapid solidification processes (RSPs) such as gas atomization and melt spinning can offer a novel opportunity for modifying the intrinsic brittleness and thermoelectric anisotropy of Bi-Te-based thermoelectric materials by forming a fine-grained and homogeneous microstructure (Elsheikh et al. 2014).

11.2.1.7 Compression and Shear Strength

The average strength (the mean of the strength distribution) and Young's modulus (which characterizes the stress–strain response of a brittle material prior to fracture) are fundamental to understand the mechanical properties of a TE material in a practical device. Enhancement of the mechanical strength of a TE module will render it more robust. The largest improvement must be in the shear strength, which is the weakest point of many TE modules. The compressive strength also must be increased, especially near the perimeter of the module. Such additional compressive strength will be especially useful in preventing damage to the module if, during the assembly process, clamping forces are accidentally applied unevenly to the module. Not only does the brittleness of the material limit the resistance of the device to mechanical and thermal shocks, but the cutting and fabrication of the arms themselves require that the materials used not be too brittle. Nanostructured materials, such as nanoparticles, nanowires, nanotubes, nanopillars, thin films, and nanocrystals, have revealed a host of "ultra-strength" phenomena, which are defined by the stresses in a material component generally increasing to a significant fraction of its ideal strength—the highest achievable stress of a defect-free crystal at zero temperature. However, while thermoelectric modules exhibit relatively high mechanical strength in the compression mode, their shear strength is comparatively low (Elsheikh et al. 2014).

11.2.1.8 Coefficient of Thermal Expansion (CTE)

Thermal expansion is critical, as devices for high-temperature applications will be subjected to extreme temperature fluctuations. This property, defined as the fractional change in length or volume with a unit change in temperature, affects several aspects of the design of thermoelectric devices. Generally, the thermal expansion coefficient α ($\alpha = (\Delta L/L_0)/(T_2 - T_1)$) varies inversely with the melting temperature (T_m), and it has been empirically confirmed that the product αT_m is a constant for many substances. This means that a material with high T_m should exhibit low α. The

CTE of TE materials is of critical importance because the shear stress is proportional to the temperature gradient, and the larger the heterogeneity in the thermal expansion coefficient of a material is, the larger is the shear stress that will result (Elsheikh et al. 2014).

11.2.2 Structures of Thermoelectric Materials

Thermoelectric materials comprise a huge family, including various materials from semimetals, semiconductors, ceramics to polymers, containing various crystalline forms from monocrystals and polycrystals to nanocomposites and covering varying dimensions from bulk, films, and wires to clusters. Figure 11.3 summarizes the reported ZT values of typical thermoelectric materials per publishing years (Zhang and Zhao 2015). According to the optimal working temperature, the thermoelectric materials can be divided into three ranges: Bi_2Te_3-based low-, PbTe-based middle-, and SiGe-based high-temperature ranges, with typical temperatures varying from <400 K, 600–900 K, and > 900 K, respectively. To retrospect the history of thermoelectric materials that have been developed for nearly 200 years since the observation of the Seebeck effect in 1821, the development can be divided into three generations according to ZT values. In the first generation, ZT is about 1.0, and the devices can operate at a power conversion efficiency 4–5% (approximately estimated from the maximum ZT), as shown in the left purple part of Fig. 11.3. The second period was ignited by size effects and extends to 1990s, with ZT being pushed to about 1.7, by the introduction of nanostructures; the power conversion efficiency can be expected to be of 11–15%, as shown in the middle blue part of

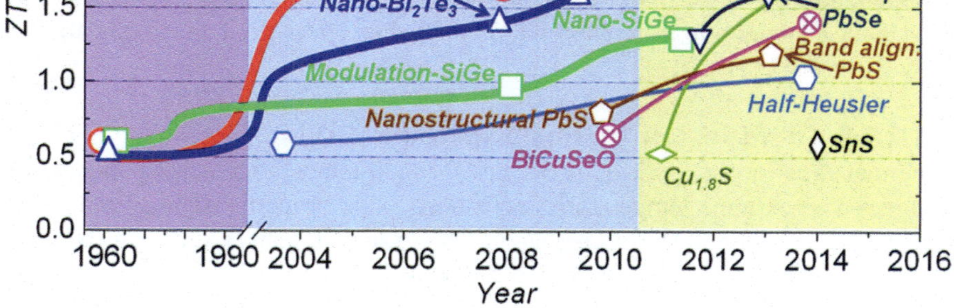

Fig. 11.3 ZT of the current bulk thermoelectric materials as a function of year (Adapted with permission from Zhang and Zhao (2015) (Elsevier))

Fig. 11.3. The third generation of bulk thermoelectrics has been under development recently, some new concepts and new technologies have pushed ZT to 1.8 and even higher; the predicted device conversion efficiency increases to 15–20%, as shown in the right yellow part of Fig. 11.3. The development history in the thermoelectric also exhibits a trend of pursuing low-cost and earth-abundant characterizations besides high ZTs > 2.0. Advanced approaches to enhance ZT that have emerged include (Zhang and Zhao 2015): modifying the band structure, heavy valence (conduction) band convergence, quantum confinement effects, and electron energy barrier filtering to enhance Seebeck coefficients; nanostructuring and all-scale hierarchical architecturing to reduce the lattice thermal conductivity; and band energy alignment between nano-precipitate/matrix to maintain hole mobility. Most of these approaches aim to maintain a high power factor and/or reduce the lattice thermal conductivities. Alternatively, high performance in thermoelectric materials with intrinsically low thermal conductivity may arise from a large molecular weight, a complex crystal structure, anharmonic, anisotropic bonding, weak chemical bonding, or ion liquid-like transport behavior, and more others (Zhang and Zhao 2015).

11.2.2.1 Metal-Based Thermoelectrics

The earliest application of the thermoelectric effect was in metal thermocouples, which have been used to measure temperature and radiant energy for many years. Most state-of-the-art metallic thermoelectric (TE) materials contain heavy elements Bi, Pb, Sb, or Te and exhibit maximum figure of merit, ZT~1–2. However, these materials are mostly toxic and unstable at high temperatures (1000 K). Therefore, metal oxides that exhibit good TE performance are in high demand because metal oxides are environmentally friendly and essentially stable at high temperatures (Fig. 11.4a). For example, $(Ni_{1-x}M_x)Mn_2O_4$ (M = Zn and Mg, $x = 0$, 0.1, and 0.2) has developed with increased activation energy for electrical conduction above a certain temperature (450 °C). Hence, high-performance TE materials based on

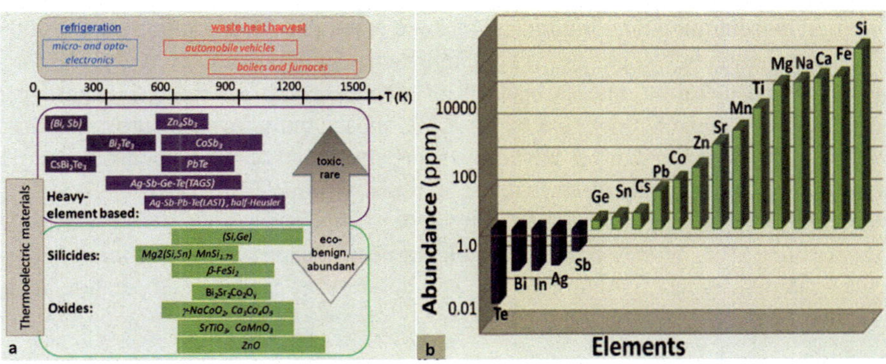

Fig. 11.4 Schematic comparison of various thermoelectric (TE) materials for applications of waste heat harvest and refrigeration, in terms of the temperature range of operation and the abundance and environmental friendliness of constituent elements (**a**). The abundance of elements used in TE materials, the elements represented by dark colored columns are, 1 ppm (**b**) (Adapted with permission from He et al. (2011) (Cambridge University Press))

abundantly available (Fig. 11.4b) and nontoxic elements with high chemical and thermal stability are crucial for any viable large-scale applications. As shown in Fig. 11.4a, oxides are particularly suitable for high-temperature power generation in air (He et al. 2011). In the low-temperature region, the absolute value of the thermoelectric power is increased by Mg substitution, and the thermoelectric powers change sign from negative to positive as the temperature increases. In addition, phonon-glass electron-crystal PGEC materials, such as clathrates and filled skutterudites, have been developed. Clathrates are periodic solids in which tetrahedrally coordinated atoms form cages that surround a metal atom. Clathrate I materials have the general formula A_8E_{46}, with A = Na, K, or Ba and E = Al, Ga, In, Si, Ge, or Sn. Clathrates II and III have idealized formulas of $A_{24}E_{136}$ and $A_{30}E_{172}$, respectively, and exist with the same A and E elements. The low thermal conductivities for semiconducting clathrates are attributed to resonant scattering of the acoustic heat-carrying phonons by the guest atoms. The localized low-frequency vibration mode of the guest atoms is expected to cut through the acoustic branches, which decreases the number of acoustic phonon modes and effectively limits the heat transport, i.e., the enhanced vibration of the guest atoms A causes a flattening of the phonons bands, lowering the velocity of the phonons, which significantly contributes to the low thermal conductivity of these materials. However, a few studies have investigated clathrate structures, and all have reported ZT <1 (Elsheikh et al. 2014).

Moreover, systems with the phonon mean free path larger than the charge carrier mean free path can exhibit an enhanced thermoelectric efficiency. This can be realized in amorphous thermoelectrics. This ground-breaking idea was accomplished in Cu–Ge–Te, NbO_2, In–Ga–Zn–O, Zr–Ni–Sn, Si–Au, and Ti–Pb–V–O amorphous systems. Naturally, amorphous thermoelectrics give rise to extensive phonon scattering, which is still a challenge for crystalline thermoelectrics. A bright future is expected for these materials (Music et al. 2016).

In addition, functionally graded materials make it possible to improve the conversion efficiency of existing thermoelectrics. These materials have a non-uniform carrier concentration distribution and in some cases also solid solution composition. In power generation applications the temperature difference can be several hundred degrees and therefore devices made from homogeneous materials have some part that operates at the temperature where ZT is substantially lower than its maximum value. This problem can be solved by using materials whose transport properties vary along their length, thus enabling substantial improvements to the operating efficiency over large temperature differences. This is possible with functionally graded materials as they have a variable carrier concentration along the length of the material, which is optimized for operations over specific temperature range (Rowe 2005).

11.2.2.2 Ceramic Thermoelectric Materials

Ceramic thermoelectric materials have been widely explored for high-temperature application in energy conversion systems, including metal oxides, Ti sulfides, and Mn silicides, which are usable in air atmosphere at 300–1200 K, for cascade-type modules. Oxide compounds such as $NaCo_2O_4$, $LaCoO_3$, $(ZnO)_mIn_2O_3$, $BaSnO_3$, $BaPbO_3$, $Ca_3Co_4O_9$, $Sr_{1-x}Nd_xTiO_3$, and $Bi_2Ca_2Co_2O_x$ are promising candidates for thermoelectric materials because of their transport properties and their physical and chemical stability. The addition of titanate nanotubes (TNTs) (2 vol%) fabricated via the pressureless sintering method has been found to enhance the ZT value of Nb-doped $SrTiO_3$ polycrystalline ceramic to approximately 0.14 at 900 K. $(Ca_{1-x}Sr_x)_3Co_4O_9$ polycrystalline has been fabricated by using spark plasma sintering method to substitute Ca by Sr, and the figure of merit ZT achieves 0.22 at 1000 K for the composition $(Ca_{0.995}\,Sr_{0.005})_3Co_4O_9$. However, the fabrication of modules based on oxide materials make it a difficult task and different from that of conventional thermoelectric modules (Elsheikh et al. 2014).

Some layered oxide materials are thought to have ZT~2.7 at 900 K. If the layers in a given material have the same stoichiometry, they will be stacked so that the same atoms will not be positioned on top of each other, impeding phonon conductivity perpendicular to the layers. Other members of this family include ZnO, MnO_2, and NbO_2, and so on (Rowe 2005).

Furthermore, calcium cobaltite ($Ca_3Co_4O_9$) basically shows good TE performance at high temperatures in air and appears to be the best p-type material at the moment. It is already installed in a commercial TE module for high-temperature energy harvesting. Calcium manganate ($CaMnO_3$) seems to be the best n-type material for high-temperature energy harvesting, while indium oxide-based compounds are another candidate n-type materials, although they are somewhat inferior to calcium manganate due to high vapor pressure of indium oxide at high temperatures. Calcium manganate is chosen for n-type TE elements in a commercial module. Strontium titanate ($SrTiO_3$) also shows good TE performance at high temperatures in a reducing atmosphere, but it degrades rapidly above 700 K in air. Accordingly, nanostructured ceramics of STO, especially 3D superlattice ceramics with 2DEG GBs, are proposed for low-temperature applications, and the simulation work has shown its possibility to exceed TE performance of the conventional bulk bismuth telluride material at room temperature. Development of a new process to realize 3D superlattice with millimeter order size is urgently required. Higher manganese silicide ($MnSi_x$) as *p*-type and titanium sulfide (TiS_2)-related compounds as *n*-type materials for mid-temperature energy harvesting are proposed to be good candidates for the future (Koumoto et al. 2013).

11.2.2.3 Polymer-Based Organic Thermoelectric Materials

Polymer-based organic thermoelectric materials are very attractive because they are light, flexible, and suitable for room-temperature applications and because they generally require relatively simple manufacturing processes (i.e., spin coating, inkjet printing or 3D printing) compared to semiconductor-based thermoelectrics. Polymers are intrinsically poor thermal conductors, which make them ideal for use as thermoelectrics, but their low electrical conductivity, Seebeck coefficient, and stability have hampered their use in thermoelectric applications. However, compared to inorganic TE materials, organic or polymer TE materials exhibit several inherent advantages, such as potentially low cost because of the abundance of carbon resources, simple synthesis in general, abundant electron-energy bands through modulation, simple processing into versatile forms, high energy density, and low k, which may be of great importance for their potential TE applications (Elsheikh et al. 2014).

Some electrically conducting organic materials may have a higher figure of merit than existing inorganic materials. Seebeck coefficient can be even millivolts per Kelvin but electrical conductivity is usually low, resulting in small ZT values. Quasi-one-dimensional (Q1D) organic crystals are formed from linear chains or stacks of molecules that are packed into a 3D crystal. Under certain conditions some Q1D organic crystals may have ZT~20 at room temperature for both p- and n-type materials. This has been credited to an unspecified interference between two main electron-phonon interactions leading to the formation of narrow strip of states in the conduction band with a significantly reduced scattering rate as the mechanism compensate each other, yielding high ZT (Casian and Sanduleac 2013).

That is, the physical and chemical properties of certain polymers are tunable within a fairly large range of modifications of their molecular structures. The ionic conducting polymers, such as poly (3,4-ethylenedi oxythiophene):poly(styrene-sulfonate) (PEDOT: PSS), which possess high electrical conductivity and intrinsically low thermal conductivity, are considered to be the most promising novel organic TE materials. In a two-component nanocomposite, the selection of a conducting polymer and an inorganic thermoelectric material may yield high thermoelectric power and high electrical conductivity of the material. Because these properties are dependent on the particle size and morphology, it is interesting to map the relation between the structural and electrical properties of the material. Nanocomposites/hybrids/heterostructures of inorganic thermoelectrics with conducting polymers have been investigated based on conducting polymers such as polyaniline (PANI), polythiophene (PTH), and poly (3,4-ethylenedioxythiophene):poly(styrenesulfonate) (PEDOT:PSS), as well as other polymers such as polyacetylene (PA), polypyrrole (PPY), polycarbazoles (PC), and polyphenylenevinylene (PPV); these materials appear to have great potential for producing relatively low-cost and high-performance TE materials. In polymer and polymer–inorganic TE nanocomposites, the value of the Seebeck coefficient typically ranges from 4088 to 1283 μV/K, the electrical conductivity ranges from 10^7 to 10^4 S/cm, and the thermal conductivity ranges from 0.02 to 1.2 W/m-K. However, it has been demonstrated that the transport properties of conducting polymers are greatly influenced by the process of doping with various materials (Elsheikh et al. 2014).

11.2.2.4 Semiconductors

Since the late 1950s, semiconducting thermoelectric devices has been applied for terrestrial cooling and power generation and later for space power generation because of their competitive energy conversion compared to other types of small-scale electric power generators. Semiconductor thermoelectric power generation, which is based on the Seebeck effect, has very interesting capabilities with respect to conventional power-generation systems. During the 1990s, there was a heightened interest in the field of thermoelectrics driven by the need for more efficient materials for electronic refrigeration and power generation. Because of the use of semiconductor materials for thermoelectric applications, there has been a considerable effort to improve the figures of merit (ZT) of these materials to greater than three to make them commercially viable. Semiconductor materials are promising for the construction of thermocouples because they have Seebeck coefficients in excess of 100 μVK^{-1}, and the only way to reduce k without affecting α and σ in bulk materials, thereby increasing ZT, is to use semiconductors of high atomic weight, such as Bi_2Te_3 and its alloys with Sb, Sn, and Pb. A high atomic weight reduces the speed of sound in the material and thereby decreases the thermal conductivity. A solid-state or semiconductor electronics component, for example, can perform well and reliably for many years when it is operating at or near the ambient temperature. Intermetallic compounds such as Mg_2X (X = Si, Ge, Sn) and their solid solutions are semiconductors with the antifluorite structure and have been proposed as good candidates for high-performance thermoelectric materials because of their superior features such as large Seebeck coefficients, low electrical resistivities, and low thermal conductivities. However, the best-ZT materials are found to be heavily doped, small-band-gap semiconductors (Elsheikh et al. 2014).

Silicon-germanium alloys are currently the best thermoelectric materials around 1000 °C and are therefore used in some radioisotope thermoelectric generators (RTG) (notably the MHW-RTG and GPHS-RTG) and some other high-temperature applications, such as waste heat recovery. Usability of silicon-germanium alloys is limited by their price and mid-range ZT (~0.7).

11.2.2.5 Nanomaterials and Superlattices

In addition to nanostructured Bi_2Te_3/Sb_2Te_3 superlattice thin films, other nanomaterials also show potential in improving thermoelectric properties, such as PbTe/PbSeTe quantum dot superlattice. It exhibits an enhanced *ZT* (approximately 1.5 at room temperature) that was higher than the bulk *ZT* value for either PbTe or PbSeTe (approximately 0.5). Compared with normal skutterudites, nanostructured skutterudites have reduced thermal conductivity, caused by grain boundary scattering. *ZT* values of ~0.65 and >0.4 have been achieved with $CoSb_3$-based samples; the former values were 2.0 for Ni and 0.75 for Te-doped material at 680 K and latter for Au-composite at $T > 700$ K. Even greater performance improvements can be achieved by using composites and by controlling the grain size, the compaction conditions of polycrystalline samples, and the carrier concentration (Szczech et al. 2011).

Due to the unique nature of graphene, it is possible to develop a thermoelectric device based on it with an extremely high Seebeck coefficient. One theoretical study suggests that the Seebeck coefficient might achieve a value of 30 mV/K (Hossain et al. 2015).

Superlattices - nano structured thermocouples, are considered a good candidate for better thermoelectric device manufacturing, with materials that can be used in manufacturing this structure. Their production is expensive for general-use due to fabrication processes based on expensive thin-film growth methods. However, since the amount of thin-film materials required for device fabrication with superlattices is so much less than thin-film materials in bulk thermoelectric materials (almost by a factor of 1/10,000) the long-term cost advantage is indeed favorable. Superlattice structures also allow the independent manipulation of transport parameters by adjusting the structure itself enabling research for better understanding of the thermoelectric phenomena in nanoscale, and studying the phonon-blocking electron-transmitting structures—explaining the changes in electric field and conductivity due to the materials nano-structure. Many strategies exist to decrease the superlattice thermal conductivity that are based on engineering of phonon transport. The thermal conductivity along the film plane and wire axis can be reduced by creating diffuse interface scattering and by reducing the interface separation distance, both which are caused by interface roughness. In nature, roughness is caused by the mixing of atoms of foreign elements. Artificial roughness can be created using various structure types, such as quantum dot interfaces and thin-films on step-covered substrates (Zhang et al. 2012).

Reduced phonon-scattering interface structures often also exhibit a decrease in electrical conductivity. One approach to overcome the decrease in electrical conductivity in reduced phonon-scattering structures is to increase phonon reflectivity and therefore decrease the thermal conductivity perpendicular to the interfaces. This can be achieved by increasing the mismatch between the materials in adjacent layers, including density, group velocity, specific heat, and the phonon-spectrum. Interface roughness causes diffuse phonon scattering, which either increases or decreases the phonon reflectivity at the interfaces. A mismatch between bulk dispersion relations confines phonons, and the confinement becomes more favorable as the difference in dispersion increases. Attempts to Localize long wavelength phonons by aperiodic superlattices or composite superlattices with different periodicities have been made. In addition, defects, especially dislocations, can be used to reduce thermal conductivity in low-dimensional systems (Zhu and Ertekin 2014).

The Seebeck coefficient can change its sign in superlattice nanowires due to the existence of minigaps as Fermi energy varies. This indicates that superlattices can be tailored to exhibit n or p-type behavior by using the same dopants as those that are used for corresponding bulk materials by carefully controlling Fermi energy or the dopant concentration. With nanowire arrays, it is possible to exploit semimetal-semiconductor transition due to the quantum confinement and use materials that normally would not be good thermoelectric materials in bulk form. Such elements are, for example, bismuth. The Seebeck effect could also be used to determine the carrier concentration and Fermi energy in nanowires. In quantum dot thermoelectrics, unconventional or nonband transport behavior (e.g., tunneling or hopping) is necessary to utilize their special electronic band structure in the transport direction. It is possible to achieve $ZT > 2$ at elevated temperatures with quantum dot superlattices, but they are almost always unsuitable for mass production. However, in superlattices, where quantum-effects are not involved, with film thickness of only a few micrometers (μm) to about 15 μm, Bi_2Te_3/Sb_2Te_3 superlattice material has been made into high-performance microcoolers and other devices. The performance of hot-spot coolers is consistent with the reported $ZT \sim 2.4$ of superlattice materials at 300 K (Shah 2018).

Nanocomposites are promising material class for bulk thermoelectric devices, but several challenges have to be overcome to make them suitable for practical applications. It is not well understood why the improved thermoelectric properties appear only in certain materials with specific fabrication processes. SrTe nanocrystals can be embedded in a bulk PbTe matrix so that rocksalt lattices of both materials are completely aligned (endotaxy) with optimal molar concentration for SrTe only 2%. This can cause strong phonon scattering but would not affect charge transport. In such case, ZT~1.7 can be achieved at 815 K for p-type material (Biswas et al. 2011).

11.2.3 Applications of Thermoelectric Materials for Renewable Energy

Thermoelectric (TE) modules comprise arrays of thermoelectric (TE) junctions, which are connected electrically in series and thermally in parallel. The TE junctions, in turn, consist of p- and n-type thermoelectric materials, which are selected from the range of materials based primarily on their physical thermoelectric performance. Auxiliary to the basic array of TE modules are components that contribute to the overall efficiency of the module, such as heat sinks, which absorb heat from the hot side, and cooling fins or cooling systems, which dissipate heat from the cool side. Typically, a single module may produce power in the range of 1–125 W and may be modularly connected to produce power up to 5 kW. The maximum temperature gradient between the hot and cold side can be as high as 70 °C. Given the nature of TE modules as solid-state devices with no moving parts, they are durable and reliable, with over 100,000 h of operating lifetimes, and have a simple structure. They may operate in two modes (Elsheikh et al. 2014): as thermoelectric generators (TEGs), generating electricity from a temperature gradient, or as thermoelectric coolers (TECs), converting a direct current into a temperature gradient.

Solar thermoelectric generators (STEGs) were initially designed and optimized for space applications because of their advantages of reliability and long lifetimes as well as their ability to capture high levels of incident solar radiation in extraterrestrial regions; however, much of the recent interest in these systems has been focused on residential solar-thermal-energy harvesting. The STEG may be described as consisting of a TEG and a thermal collector. Solar heat can be absorbed by the thermal collector and then concentrated and conducted over the thermoelectric generator using a fluid pipe or some other means. Subsequently, the thermal resistance of the thermoelectric generator will lead to a temperature difference that is proportional to the heat flux from the absorber of the thermal collector to the fluid. As a result, electric power will be generated by the thermoelectric generator, and this power will be proportional to the temperature difference. Moreover, the evacuated-tube heat-pipe solar collectors had a significant effect on the TEG by increasing the thermal efficiency by approximately 55%, which may increase the performance to above 1%. A novel hybrid solar thermoelectric (HSTE) system uses a thermosyphon to passively transfer heat to a bottoming cycle to determine the overall performance in a temperature range of 300–1200 K for solar concentrations of 1–100 suns and various thermosyphon and thermoelectric materials with a geometry resembling an evacuated-tube solar collector. Bismuth telluride, lead telluride, and silicon germanium thermoelectrics were studied with copper/water, stainless-steel/mercury, and nickel/liquid-potassium thermosyphon/working-fluid combinations. The system contained a parabolic trough mirror that concentrated solar energy onto a selective-surface-coated thermoelectric to generate electrical power, and a thermosyphon adjacent to the back side of the thermoelectric maintained the temperature of the cold junction and carried the remaining thermal energy to a bottoming cycle. This HSTE system demonstrated that ideal efficiencies as high as 52.6% can be achieved at a solar concentration of 100 suns and a bottoming-cycle temperature of 776 K. However, the performance of a solar thermoelectric generator is primarily limited by the thermoelectric materials used to construct it; bismuth telluride is a favorable low-temperature thermoelectric material that exhibits advantageous properties and can reache a maximum of ZT 1 at low temperatures in the range of 25–225 °C, while filled skutterudite is a good medium-temperature thermoelectric material that can be operated over a wide temperature range 25–525 °C. Moreover, the total conversion efficiency for solar energy of a solar thermoelectric generator based on a multistage thermoelectric module could be as high as 10%. With the continuous emergence of new thermoelectric materials, solar thermoelectric generators using multistage and hybrid-generation-system (HGS) thermoelectric modules will have good application prospects (Elsheikh et al. 2014).

Many new types of thermal-energy sensors based on the Peltier effect or Seebeck effect of thermoelectric modules have been developed, such as sensors for power ultrasound effects, cryogenic heat-flux sensors, water-condensation detectors, fluid-flow sensors, and infrared sensors. These sensors generally rely on the conversion of heat into electrical signals or vice versa. Infrared (IR) sensors, which operate on the principle that any mass radiates heat, allow the detection of heat using the Seebeck effect; the absorption of heat causes a specific temperature rise, which subsequently produces a Seebeck voltage. The main sensor parameters are the responsivity, which is given by the ratio of the sensor voltage to the incoming radiation power; the time constant; and the noise voltage. Most thermoelectric IR sensors are able to operate in the range of 7–14 μm (Elsheikh et al. 2014). A sensor system combining a novel pressure-stable thermoelectric flow and an impedimetric sensor for monitoring chemical conversion in microfluidic channels was developed, for instance. Both sensor chips were optimized for hydraulic diameters of 1 mm and exhibit a high chemical, temperature, and pressure stability. However, in terms of size and light weight, the most effective structure for thermoelectric sensors is the thin-film structure (Leonov 2011; Elsheikh et al. 2014).

Advanced autonomous power systems that can be operated continuously and independently of the sun and are capable of providing electric power from a few watts to hundreds of kilowatts for 7–10 years are required for extraterrestrial exploration vehicles. For example, the solar brightness on Mars and Jupiter is as weak as 45% and 4%, respectively, and it is negligible on other planets. As a result, the solar option is suitable only for robotic and spacecraft missions that are limited in their time scope of operation and require only a few watts of electrical power (10 W). Radioisotope thermoelectric generators (RTGs) have been used by the United States to provide electrical power for spacecraft since 1961. The required electrical output power levels can be achieved by the appropriate selection of a number of general purpose heat-source (GPHS) modules incorporated in an RTG system. A GPHS module is a composite carbon body that houses a total of four fuel pellets and acts as an aero-impact shell. The isotope fuel for the GPHS-RTG is in the form of plutonium dioxide ($^{238}PuO_2$) at approximately 80% density. For power conversion by the GPHS-RTG, thermoelectric junctions have been used, such as SiGe junctions. SiGe unicouples partially convert the heat generated by the radioactive decay of ^{238}Pu in the $^{238}PuO_2$ fuel pellets that are encapsulated in the general-purpose heat-source (GPHS) modules. Previously, lead telluride was used as the thermoelectric converter for lower powered RTGs operated at a maximum hot junction temperature of 865 K to produce a power of 2.7 We. Because of the deleterious effects of oxygen on these materials, silicon germanium thermoelectric elements were later adopted for high-powered RTGs operated at high temperatures of up to 1275 K. Their sublimation rates and oxidation effects, even at these higher temperatures, can be controlled by the use of sublimation barriers around the elements and an inert cover gas within the generator during ground operation. In 1976, the Lincoln Experimental Satellite (LES) 8/9 was the first spacecraft to use the new "4150 We multi-hundred watt (MHW) RTG," which employed 312 SiGe-alloy thermoelectric elements per RTG. Recently, skutterudite alloys with ZTs ranging from 0.92 to 1.48 in the temperature range from 300 to 973 K have been developed at the Jet Propulsion Laboratory (JPL) in Pasadena, California, and are being considered for use in Advanced Radioisotope Power Systems (ARPSs) to support NASA's planetary exploration missions. The use of skutterudite unicouples in the bottom array with SiGe unicouples in the top array has been proposed for cascaded thermoelectric modules (CTMs) for use in radioisotope power systems (RPSs) to generate electric power of 108 We and to achieve a net decrease of 43% in the required amount of $^{238}PuO_2$. However, an operational issue with skutterudite-based unicouples is the sublimation of antimony from the legs near the hot junction at 973 K. Such sublimation could change the thermoelectric properties of the material and degrade the unicouples' performance over time. The use of TEGs in commercial aerospace vehicles, which is expected to reduce fuel consumption by 0.5%, is also being explored by Boeing Research & Technology. As a rough estimate, this fuel savings, if implemented solely in the USA, would save passenger and cargo airlines more than $12 million every month and reduce global carbon emissions by 0.03% (Elsheikh et al. 2014).

11.3 Piezoelectric Materials

Piezoelectric effect is a unique property of certain crystal materials where they will generate an electric field or current if subjected to physical stress. The phenomenon was first discovered in 1880 when Pierre and Jacques Curie demonstrated that when specially prepared crystals (such as quartz, topaz, and Rochelle salt) were subjected to a mechanical stress they could measure a surface charge. After that, the same effect was observed in reverse, where an imposed electric field on the piezoelectric materials will put stress on its structure. One of the first applications of the piezoelectric effect was an ultrasonic submarine detector developed during the First World War. A mosaic of thin quartz crystals glued between two steel plates acted as a transducer that resonated at 50 MHz. By submerging the device and applying a voltage they succeeded in emitting a high-frequency "chirp" underwater, which enabled them to measure the depth by timing the return echo. This was the basis for sonar and the development encouraged other applications using piezoelectric devices both resonating and non-resonating such as microphones, signal filters, and ultrasonic transducers. However many devices were not commercially viable due to the limited performance of the materials at the time. The continued development of piezoelectric materials has led to a huge market of products ranging from those for everyday use to more specialized devices.

11.3.1 Fundamentals of Piezoelectricity

The process whereby the piezoelectric effect takes place is based on the fundamental structure of a crystal lattice. Crystals generally have a charge balance, where negative and positive charges precisely cancel each other out along the rigid planes of the crystal lattice. When this charge balance is disrupted by applying physical stress to the crystals, the energy is transferred by electric charge carriers, creating a current in the crystal. With the converse piezoelectric effect, the applying of an external electric field to the crystal will unbalance the neutral charge state, which results in mechanical stress and the slight readjustment of the lattice structure. Therefore, piezoelectricity is the linear interaction between the mechanical and electrical properties of the dielectric materials. For electrical properties, the relationship between the electric displacement (D_i) and the electric field (E_j) can be interpreted as (Wang et al. 2015a)

$$D_i = \sum_{j=1}^{3} \varepsilon_{ij} E_j \tag{11.6}$$

where the coefficient ε_{ij} is the permittivity of the dielectric materials. The permittivity coefficient is a second-rank tensor represented by a 3×3 matrix. However, the matrix is symmetric ($\varepsilon_{ij} = \varepsilon_{ji}$), and there are at most six independent coefficients. For mechanical properties, the relationship between stress and strain under small deformations can be interpreted as (Wang et al. 2015a, b).

$$T_{ij} = \sum_{k,\,l=1}^{3} c_{ijkl} S_{kl} \tag{11.7}$$

where c_{ijkl} is the elastic stiffness of the material. Equation (11.7) can also be interpreted reversely as (Wang et al. 2015a, b)

$$S_{ij} = \sum_{k,\,l=1}^{3} s_{ijkl} T_{kl} \tag{11.8}$$

where s_{ijkl} is the elastic compliance of the materials. Because the stiffness and compliance coefficients relate to the second rank tensor stress and strain, they are fourth-rank tensors. However, due to the symmetric stress and strain ($T_{ij} = T_{ji}$, $S_{ij} = S_{ji}$), the number of c_{ijkl} can be reduced to 36. In reduced notation, the piezoelectric constitutive equation can be interpreted as (Falconi et al. 2012)

$$D_i = \varepsilon_{ij}^{T} E_j + d_{iJ} T_j \tag{11.9}$$

$$S_I = d_{Ij} E_j + S_{IJ}^{E} T_j \tag{11.10}$$

where the superscripts T and E mean the coefficients are at constant stress and electric field, respectively. The d_{iJ} in Eqs. (11.9) and (11.10) is the piezoelectric strain/charge constant.

Piezoelectric materials are anisotropic dielectrics of special type, where both fields the electrical and the elastic are coupled. Some of them (for instance ceramic) have ferroelectric properties, but the rest of them (as quartz) display no ferroelectric behavior. A dielectric material supports charge without conducting it to a significant degree. The main property of a dielectric material is that it has no free electrical charges, but when an external electrical field is applied the electric dipoles are being created due to the interaction of the electrical field with the dielectric structure. The dipole moment is a vector with direction from the negative to the positive pole. Because of dielectric polarization, positive charges are displaced towards the external field and negative charges shift in the opposite direction. The polarization of piezoelectric material has its specific peculiarities in comparison with polarization of ordinary dielectrics. When a piezoelectric is placed under a mechanical stress, the geometry of the atomic structure of the crystal changes, such that ions in the structure separate, and a dipole moment is formed. For a net polarization to develop, the dipole formed must not be canceled out by other dipoles in the unit cell. Therefore the piezoelectric atomic structure must be non-centrosymmetric. When a piezoelectric material is loaded electrically then the electrical dipoles appear, dipole moment is formed and this results in deformation. The polarization is linear and electrical dipoles nucleate only after electrical or mechanical load. The other types of piezoelectric materials are with ferroelectric properties, i.e., spontaneous polarization and electric dipoles exist in their structure even in the absence of electrical field. The piezoelectric effect in ferroelectric is strongly dependent on its atomic structure.

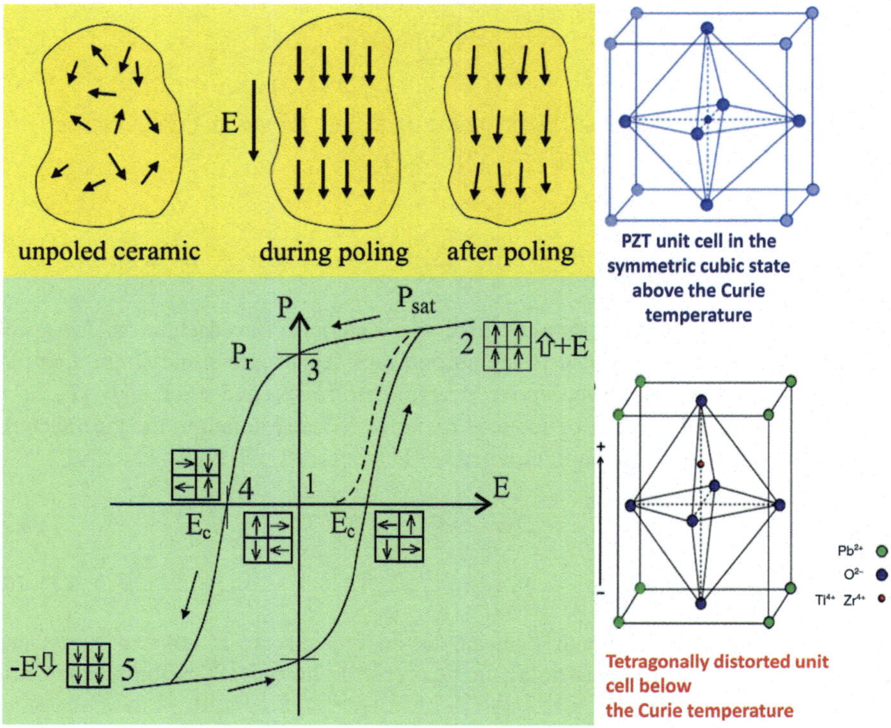

Fig. 11.5 Hysteresis curve for polarization of piezoelectric material, and domain structure of ferroelectric materials and their behavior during poling process

Depending on the type of a crystal, a compressive stress can increase or decrease the polarization, or sometimes have no effect at all. If a mechanical stress is applied to the ferroelectric, then there are domains which will experience an increase in dipole moment and some which will experience a decrease in dipole moment. Overall, there is no net increase in polarization, as shown in Fig. 11.5 (upper portion). This makes the ferroelectric useless as a piezoelectric unless it is put through some additional processing. This process is called poling. An electric field is applied to the ferroelectric as it passes through its Curie temperature, so that its spontaneous polarization develops and it is aligned in a single direction. All of the domains in the piezoelectric have a dipole moment pointing in that direction, so there is a net with approximately the same polarization (Fig. 11.5, upper portion). When the electric field is removed, most of the dipoles are locked in a configuration of near alignment (Fig. 11.5, upper portion). The full alignment is only possible in a single crystal and in a polycrystalline material there exists still a polarization distribution. The material now has a remanent polarization. The maximum possible value of the remanent polarization is called saturation polarization, i.e., this is the horizontal part of the hysteresis curve in Fig. 11.5 (lower portion). The distinguishing feature of PEM with ferroelectric properties is that the direction of the spontaneous polarization can be reversed by an applied electric field, yielding a hysteresis loop, as shown in Fig. 11.5 (lower portion). The nonlinear behavior of the polarization with respect to the applied electrical field consists of three stages which are characterized by (Dineva et al. 2014): (a) reversible domain motion; (b) growth of new domains; and (c) new domains reaching the limit of their growth and reaching the saturation polarization.

Figure 11.5 (lower portion) shows a typical hysteresis curve created by applying an electric field to a piezoelectric ceramic element until the maximum (saturation) polarization P_{sat} is reached, reducing the field to zero determines the remanent polarization P_r reversing the field attains a negative maximum (saturation) polarization and negative remanent polarization, and re-reversing the field restores the positive remanent polarization. When the electric field is the coercive field E_c there is no net polarization due to the mutual compensation of the polarization of different domains. Therefore, PEM is a special type of anisotropic dielectrics where electrical and mechanical fields are coupled due to both the existence of the specific asymmetric atomic structure of the lattice and the existence of spontaneous polarization at the microstructure level; The effective usage of both the ferroelectric properties of the piezoelectric ceramics together with the poling process during their manufacture make these materials a basic element in the modern industrial applications (Dineva et al. 2014).

11.3.2 Type of Piezoelectric Materials

Piezoelectric materials can be natural or man-made. Several typical piezoelectric materials, including both natural and synthetic materials, are listed in Table 11.1. Among them, piezoelectric ceramics, crystals, and polymers have been most developed and useful piezoelectric materials. Piezoelectric ceramics usually refer to polycrystalline materials that consisted of irregular collective small grains and are prepared through the solid-state reaction and sintering process. Under the poling electrical field, the disordered spontaneous polarization in piezoelectric ceramics can be realigned and keep the remnant polarization after the removal of external field. As a result, the piezoelectric ceramics can exhibit macropiezoelectric property (Wang et al. 2015a, b). There are the following families of man-made ceramics with crystal structure as perovskite (Dineva et al. 2014): Barium titanate ($BaTiO_3$); Lead titanate ($PbTiO_3$); Lead zirconate titanate ($Pb[Zr_x\ Ti_{1-x}]O_3$, $0 < x < 1$)—more commonly known as PZT; Potassium niobate ($KNbO_3$); Lithium niobate ($LiNbO_3$); Lithium tantalate ($LiTaO_3$), etc. and other lead-free piezoceramics. The general chemical formulae of perovskite crystal structure is ABO_3, where A is a larger metal ions, usually lead Pb or barium Ba, B is a smaller metal ion, usually titanium Ti or zirconium Zr, which shows the crystal structure of a piezoelectric ceramic ($BaTiO_3$) at temperature above and below Curie point.

To prepare a piezoelectric ceramic, fine powders of the component metal oxides are mixed in specific proportions, then heated to form a uniform powder. The powder is mixed with an organic binder and formed into structural elements having the desired shape (disks, rods, plates, etc.). The elements are subsequently fired

Table 11.1 Typical piezoelectric materials (Dineva et al. 2014)

Category	Typical materials
Natural crystals	Berlinite ($AlPO_4$), sucrose, quartz, Rochelle salt, topaz, and tourmaline-group minerals
Natural materials	Bone, tendon, silk, wood, enamel, dentin, DNA, and viral proteins
Synthetic crystals	Gallium orthophosphate ($GaPO_4$), langasite ($La_3Ga_5SiO_{14}$), and diisopropylammonium bromide (DIPAB)
Synthetic ceramics	$BaTiO_3$, $PbTiO_3$, $Pb(Zr,Ti)O_3$, $KNbO_3$, $LiNbO_3$, $LiTaO_3$, Na_2WO_3, ZnO, $Ba_2NaNb_5O_5$, $Pb_2KNb_5O_{15}$, and so forth
Lead-free piezoceramics	$(K,Na)NbO_3$, $BiFeO_3$, $Bi_4Ti_3O_{12}$, $Na_{0.5}Bi_{0.5}TiO_3$, and so forth
Polymers	Polyvinylidene fluoride (PVDF)
Organic nanostructures	Self-assembled diphenylalanine peptide nanotubes (PNTs)
Piezocomposites	Piezo-polymer in which the piezoelectric material is immersed in an electrically passive matrix; $BaTiO_3$ fibers reinforcing a PZT matrix; multilayer PDMS cellular structures coated with PTFE films and stretchable gold electrodes

according to a specific time and temperature program, during which the powder particles sinter and the material attains a dense crystalline structure. The elements are cooled, then shaped or trimmed to specifications, and electrodes are applied to the appropriate surfaces. Above a critical temperature, the Curie point, each perovskite crystal in the fired ceramic element exhibits a simple cubic symmetry with no dipole moment; it is in the so-called paraelastic phase. At temperatures below the Curie point, however, each crystal exhibits a tetragonal or rhombohedral symmetry leading to a dipole moment; this phase of the material is called ferroelectric phase. When electric field of about 10^6 V/m is applied to the ferroelectric polycrystal as it passes through its Curie temperature, so that its spontaneous polarizations develop, all polarization vectors are aligned in a more or less uniform direction. This process leading to a macroscopic net polarization is so-called poling. Initially there exists a uniform distribution of all direction, i.e., no macroscopic net polarization. After poling, a distribution around the poling direction leads to a macroscopic net polarization. Now, at this stage, when a mechanical stress is applied, the polarization will increase or decrease and the ceramic will have typical piezoelectric behavior (Dineva et al. 2014).

Piezoelectric crystals, which refer to single-crystalline materials, are usually unsymmetrical in structure and therefore exhibit piezoelectric property. There are 32 crystal classes which are divided into the following seven groups (Dineva et al. 2014): triclinic, monoclinic, orthorhombic, tetragonal, trigonal, hexagonal, and cubic. These groups are also associated with the elastic nature of the material where triclinic represents an anisotropic material, orthorhombic represents an orthotropic material, and cubic are in most cases isotropic materials. Only 20 of the 32 classes allow piezoelectric properties. Ten of these classes are polar, i.e., show a spontaneous polarization without mechanical stress due to a nonvanishing electric dipole moment associated with their unit cell. The remaining ten classes are not polar, i.e., polarization appears only after applying a mechanical load.

The piezoelectric ceramics exhibit high piezoelectric constant and permittivity and can be prepared into designed architectures, which makes them suitable for the application in high-power energy transducer and wideband filters. However, the poor mechanical quality factor, high electrical loss, and low stability of the piezoelectric ceramics limited their application in high-frequency devices. Comparatively, the natural piezoelectric crystals such as quartz exhibit lower piezoelectric properties and dielectric constant. Moreover, they are limited in size due to the cuts of crystals. However, the mechanical quality factor and stability of quartz crystals are relatively higher than ceramics. Therefore quartz crystals are always used in high-frequency filters, transducers, and other standard frequency controlling oscillators. Besides the quartz crystals, the high-quality perovskite piezoelectric single crystals such as the $Pb(A_{1/3}B_{2/3})O_3$–$PbTiO_3$ ($A = Zn^{2+}$, Mg^{2+}; $B = Nb^{5+}$) with much higher piezoelectric constant ($d_{33} \sim 2600$ pC/N), electromechanical coupling coefficient ($k_{33} \sim 0.95$), and strain (>1.7%) have also been obtained since 1997. These single crystals are new-generation piezoelectric materials for high-performance piezoelectric devices and systems including ultrasound medical imaging probes, sonars for underwater communications, and sensors/actuators. However, the size and shape of the piezoelectric single crystals are difficult to be preciously controlled during the growth process, which limit the practical application in many fields such as the microscaled actuators and composite metamaterials (Wang et al. 2014).

Furthermore, the piezoelectric polymers such as polyvinylidene fluoride (PVDF) possess high flexibility, low density, and resistance as well as relatively higher piezoelectricity voltage constant (*g*). Unlike the piezoelectric ceramic and crystals, the intertwined long-chain molecules in polymers attract and repel each other when an electric field is applied. PVDF has exhibited great potential in the application of acoustic ultrasound measurements, pressure sensors, and ignition/detonations. However, the relatively low piezoelectric strain constant (*d*) of PVDF limited the application in transducers (Wang et al. 2015a, b).

Piezocomposite materials are an important update of existing piezoceramic. They can be (a) piezo-polymer in which the piezoelectric material is immersed in an electrically passive matrix (for instance, PZT in epoxy matrix); and (b) piezocomposites that are composite materials made by two different ceramics (for example, $BaTiO_3$ fibers reinforcing a PZT matrix) (Dineva et al. 2014). In addition, composite piezoelectric rubber has been developed through composite polymeric and metallic microstructures with embedded bipolar charges. For instance, multilayer PDMS cellular structures coated with PTFE films and stretchable gold electrodes are fabricated and implanted with bipolar charges. The composite structures show elasticity of 300–600 kPa and extreme piezoelectricity of $d_{33} > 2000$ pC/N and $d_{31} > 200$ pC/N. For a working volume of 2.5 cm × 2.5 cm × 0.3 mm, 10% (or 2.5 mm) stretch results in effective d_{31} of >17,000 pC/N. It is estimated that electric charge of >0.2 μC can be collected and stored per breath (or 2.5 cm deformation) with this kind of composite piezoelectric rubber bands (with spring constants of ~200 N/m). They can be mounted on elastic waistbands to harvest the circumferential stretch during breathing, or on pads around joints to harvest the elongation during limb motion.

Furthermore, the wearable piezoelectric structures can be spread, stacked, and connected to charge energy storages and power micro devices (Wang et al. 2014).

As a result, piezoelectric materials have been widely used in the industrial, manufacturing, automotive industry, and medical instruments as well as information and telecommunication fields, and so forth. According to the operation mode of the piezoelectric devices, the application of piezoelectric materials can be classified as follows (Wang et al. 2015a, b; Priya et al. 2017):

(a) Sensors.
Through the direct piezoelectric effect, the piezoelectric materials can be used for the detection of pressure variations in longitudinal, transversal, and shear modes. The most commonly used application of piezoelectric sensors is in the sound form, such as the piezoelectric microphones, piezoelectric pickups in acoustic-electric guitars, and detection of sonar waves. Moreover, the piezoelectric sensors can also be used with high-frequency field such as the ultrasonic medical imaging or industrial nondestructive testing. In addition, the piezoelectric sensors were also employed in piezoelectric microbalance and strain gauges.

(b) Actuators.
On contrary to piezoelectric sensors, the working of actuators is usually based on the reverse piezoelectric effect to induce tiny changes in the width of the piezoelectric materials by applying high electric fields. Due to the relatively high precision of the width changes, the piezoelectric actuators are always used in accurate positioning. For example, the piezoelectric motors with high accuracy have already been used in optical devices, transportation and aerospace techniques, robots, medical devices, biology, and nanomanipulation fields, such as the atomic force microscopes (AFM), scanning tunneling microscopes (STM), autofocusing camera lens, inkjet printers, CT/MRI scanners, and X-ray shutters.

(c) Frequency Controlling Devices.
Crystal oscillator is an electronic oscillator circuit that uses the mechanical resonance of a vibrating piezoelectric crystal to create an electrical signal with a very precise frequency. The frequency can be used to provide a stable clock signal for digital integrated circuits. Moreover, the piezoelectric materials have also been used in high-frequency resonators and filters, such as the surface acoustic wave devices and film bulk acoustic resonators.

(d) High voltage and power sources trough energy harvesting.
By applying the external mechanical stimulates, the piezoelectric ceramic or crystals can generate potential differences with thousands of volts in amplitude. Therefore, piezoelectric materials can be used as high voltage and power sources. The most commonly application is the piezoelectric ignition/sparkers such as the cigarette lighters. Moreover, the piezoelectric materials have been employed for energy-harvesting applications. Table 11.2 shows some of the common sources of mechanical energy that can be harnessed through

Table 11.2 Sources of energy available in the surrounding which are/can be tapped for generating electricity (Priya et al. 2017)

Human body	Vehicles	Structures	Industrial	Environment
Breathing, blood pressure, exhalation, body heat	Aircraft, UAV, helicopter, automobiles, trains	Bridges, roads, tunnels, farm house structures	Motors, compressors, chillers, pumps, fans	Wind, solar, temperature gradient, daily temperature
Walking, arm motion, finger motion, jogging, swimming, eating, talking	Tires, tracks, peddles, brakes, shock absorbers, turbines	Control-switch, HVAC systems, ducts, cleaners, etc.	Conveyors, cutting and dicing, vibrating mach.	Ocean currents, acoustic waves, EM waves, RF signal

piezoelectrics. For example, the energy from human movements and vehicle movements in public places can be harvested and converted into electricity for lighting the lamps. Recently, the microscale energy harvesters were developed for harvesting the small scale mechanical energies by using the piezoelectric nanomaterials, which is called "piezoelectric nanogenerators." The nanogenerators can be used for charging the batteries or directly driving some low-power microdevices.

11.3.3 Advances in Piezoelectric Energy Harvesting

Piezoelectric energy harvesters can be categorized by the targeted energy sources, such as ambient vibrations, impact or shock stress, fluid flow, and human motion. Regardless of the various energy sources, the basic working principle is the same—the environment applies a stress on piezoelectric material and by the direct piezoelectric effect, the input mechanical energy is converted in to electrical energy. Piezoelectric generators produce high voltages and low currents, and require no voltage source to operate. Piezoelectric energy harvesters generally have bimorph or unimorph cantilever beam structures (Fig. 11.6a, b). In addition, MEMS technologies have been applied towards the development of integrated energy harvesters, and many piezoelectric MEMS energy harvesters have been developed. At the MEMS-scale, bimorph cantilever are less manufacturable with existing microfabrication processes. As a result, MEMS cantilevers mostly have a unimorph configuration. A seismic mass is usually attached at the tip of the cantilever to adjust the resonant frequency to the available environmental frequency, normally below 100 Hz. As shown in Fig. 11.6c, two piezoelectric modes, d_{31} or d_{33}, are commonly used in piezoelectric micro devices. The relative directions of the electric field and the strain distinguish them: d_{31} when the electric field is perpendicular to the input strain, d_{33} when they are parallel. Conventional MEMS piezoelectric devices/actuators have d_{31} configuration in which a piezoelectric layer is sandwiched between top and bottom electrodes (Priya et al. 2017).

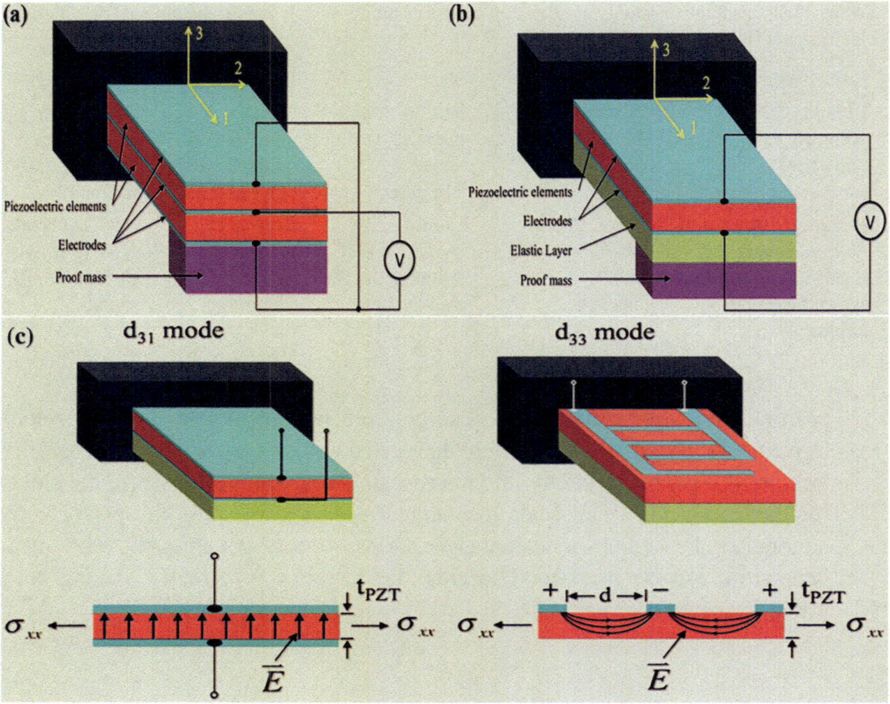

Fig. 11.6 Basic structure of piezoelectric energy harvesters: (**a**) Bimorph structure, (**b**) unimorph structure, and (**c**) two modes of piezoelectric conversion of input mechanical strain depending on the relative direction of the stress σxx (or strain) and the electric field, E (Adapted with permission from Priya et al. (2017) (Walter De Gruyter and Company))

11.3.3.1 Grain Textured and Epitaxial Piezoelectric Films

Bulk PZT ceramics have plethora of dielectric and piezoelectric properties which can be controlled using doping. Lead-based relaxor-lead titanate compounds so-called relaxor ferroelectrics include $Pb(Zn_{1/3}Nb_{2/3})O_3–PbTiO_3$(PZN–PT), $Pb(Mg_{1/3}Nb_{2/3})O_3–PbTiO_3$(PMN–PT), and $Pb(Y_{½}\ Nb_{1/2})O_3–PbTiO_3$(PYN–PT) and exhibit large electromechanical coupling coefficients which makes these materials more attractive option for microactuator and micro-sensor applications. For device miniaturization it is important to grow films using these materials. PZT thin films are deposited with variety of methods such as sputtering, chemical solution deposition (i.e., sol-gel), MOCVD, and laser ablation processes. Since piezoelectric properties of these materials are strongly dependent on film quality and growth orientation, thus, the control of growth orientation is highly desired with (001) texturing. Additionally, the composition of the piezoelectric material plays a crucial role in deciding the piezoelectric properties of films. The most simple and direct solution towards enhancing the output power at the fixed size of a device is to increase the piezoelectric and electromechanical coupling coefficient. Grain texturing has been shown to be an effective method for improving the magnitude of physical constants in piezoelectrics by achieving a domain engineered state. In the vicinity of the morphotropic phase boundary (MPB), a rhombohedral composition oriented along the <100> direction is known to exhibit optimum magnitudes of electromechanical coefficients. In addition, the piezoelectric properties can be tuned through interfacial stress. For instance, ~90% enhancement of ferroelectric and piezoelectric properties in MPB composition PZT thick films by tailoring the magnitude of residual stress by choosing substrates with different coefficients of thermal expansion. The enhanced piezoelectric properties were attributed to the fact that the c-domains parallel to the thickness direction were easy to form under in-plane compressive stress. This technique of inducing a compressive stress in the films is quite appealing, as it can be easily implemented in the fabrication of MEMS components (Kim et al. 2012; Priya et al. 2017).

Epitaxial PZT thin films with c -axis orientation are also ideal materials for MEMS energy harvesters because the c -axis orientation of the tetragonal PZT results in large piezoelectric properties and a low dielectric constant. However, epitaxial substrates, such as MgO and $SrTiO_3$, are usually not suitable for unimorph cantilevers because of their brittleness and difficulty of microfabrication. One solution is to transfer the epitaxial PZT films onto flexible cantilevers, made with polymer, stainless steel or Ni foils, etc. (Kim et al. 2012).

Figure 11.7 shows fabrication process of high-efficiency piezoelectric energy harvesters using c-axis oriented PZT thin films by radio-frequency (RF)- sputtering, which were transferred onto stainless steel (Morimoto et al. 2010). The c-axis oriented PZT thin films were grown on (100) MgO single crystals with an epitaxial (001) Pt bottom electrodes. Reciprocal lattice space maps before and after transfer process clearly showed spotty diffractions of the (204) PZT, indicating that the transfer process did not degrade the crystal structure of the epitaxial PZT film. After the PZT film was bonded to 50-μm-thick stainless steel sheets with epoxy resin, the MgO substrate was etched out in phosphoric acid. The relative dielectric constant ε_r of the transferred films on stainless steel was as low as 166, while the piezoelectric coefficients $e_{31,f}$ of the transferred PZT films was around -6 C/m^2. The thickness and length of stainless steel cantilever was 50 μm and 18.5 mm, respectively. Because of the thin dimension of the metal cantilever, the first resonance was found to occur at 126 Hz. The maximum output electric power of 5.3 μW was obtained across a load resistance of 50 kΩ. The output power increases monotonically with the acceleration, reaching 244 μW at 50 m/s^2. The flexible metal cantilever enables considerable reduction of the resonant frequency and offers enhanced toughness compared with brittle Si-based cantilevers (Priya et al. 2017).

In addition, relaxor ferroelectric single crystals, such as PMN-PT and PZN-PT, show about ten times larger piezoelectric coefficient than that of conventional PZT ceramics. The epitaxial PMN-PT thin films grown on $SrTiO_3$-buffered miscut Si substrate by off-axis sputtering have demonstrated excellent piezoelectric coefficient $e_{31,f}$ (-27 C/m^2). Because of the large electromechanical coupling coefficient k_{31} (or figure of merit: $e_{31,f}^2/\varepsilon_r$) of PMN-PT epitaxial thin films, this system is quite

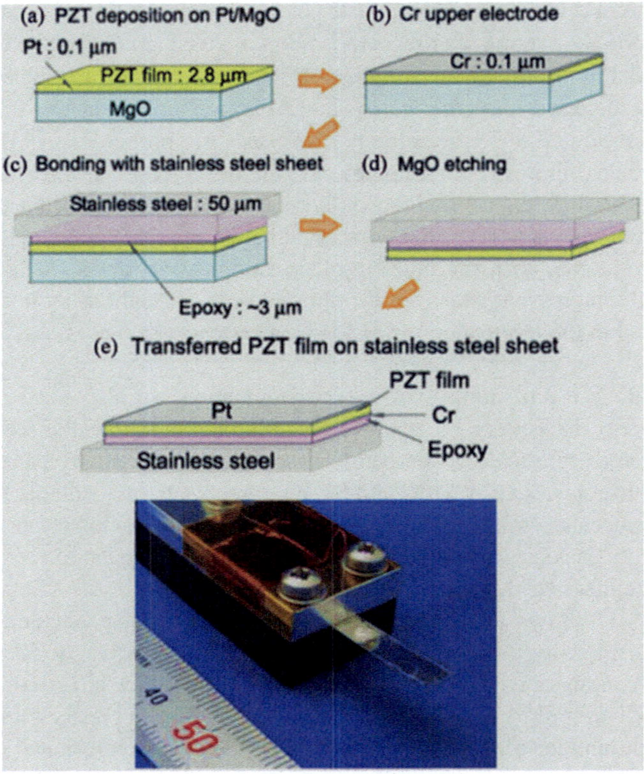

Fig. 11.7 Fabrication process flow and photograph of transferred epitaxial PZT thin films on stainless steel cantilever (Modified with permission from Morimoto et al. (2010) (Elsevier))

promising towards improving the performance of current MEMS energy harvesters (Priya et al. 2017).

11.3.3.2 Lead-Free Piezoelectric Films

For powering medical implants and human use of energy harvesters, lead-free piezoelectric material is desirable. Potassium sodium niobate ($K_xNa_{1-x}NbO_3$), abbreviated as KNN, is considered as a most promising lead-free piezoelectric material owing to its higher Curie temperature and higher ferroelectric orthorhombic—ferroelectric tetragonal transition temperature. The piezoelectric properties of KNN-based compositions are directly correlated to the fraction of orthorhombic (O) and tetragonal (T) phase as shown in Fig. 11.8a. One of the strategies for achieving higher piezoelectric response has been to modulate the composition such that O/T transition lies close to room temperature. Another strategy adopted for designing lead-free compositions is based upon the trend between atomic weight ratio of A to B sites ($R_w = W_a/W_b$) and longitudinal piezoelectric constant d_{33} as shown in Fig. 11.8b. It can be observed that $1/R_w$ for KNN ceramics (for Na/K ratio of 0.5) is similar to R_w for PZT ceramics at MPB composition and both of these

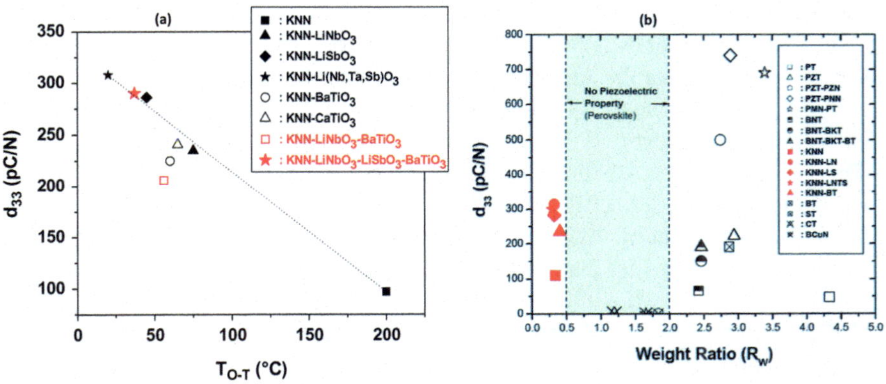

Fig. 11.8 Lead-free piezoelectric films (Modified with permission from Priya et al. (2017) (Walter De Gruyter and Company)): (**a**) Variation of longitudinal piezoelectric coefficient with O–T transition temperature in KNN system, and (**b**) variation of longitudinal piezoelectric coefficient as a function of weight ratio in bulk perovskites

materials exhibit high piezoelectric response. Piezoelectric compositions show large response when RW for A-site heavy perovskites and $1/R_w$ for B-site heavy perovskites is higher than 2.0 (Ahn et al. 2009; Priya et al. 2017).

The fabrication process of these lead-free piezoelectric materials in thin-film form is still under development, and it is expected that these materials will be utilized in the design of MEMS energy harvesters in the near future. For instance, KNN thin films deposited by RF magnetron sputtering showed large transverse piezoelectric properties comparable to those of PZT thin films (Shibata et al. 2011). Because KNN and PZT thin films have almost same dielectric and piezoelectric properties, the KNN film performs comparably to the PZT film with respect to power generation (Priya et al. 2017).

11.3.3.3 Aluminum Nitride (AlN) as a MEMS Compatible Piezoelectric Films

AlN is a piezoelectric material with wurtzite structure. The properties of AlN-based films are highly process dependent. Aluminum nitride has a band gap of 6.2 eV with lattice parameter of $a = 13.112$ Å and $c = 4.982$ Å. It is a potential candidate for UV light emission and SAW devices due to large acoustic wave (SAW) velocity. AlN films have been deposited using chemical vapor deposition (CVD), plasma assisted CVD, metalorganic CVD, reactive DC-magnetron sputtering, plasma assisted molecular beam epitaxy (MBE), laser-induced chemical vapor deposition, and pulsed laser deposition. Most of these techniques required a high processing temperature (1000 °C) to obtain the best crystal quality and texture in the deposited thin films. However, high-quality growth at low temperature is desired to ensure the compatibility with IC technology. In this regard, different process techniques have been developed. In addition to (0001)–orientation, it is also observed that tuning of the ion energy and flux of the bombarding ions (Ar^+, N^{2+}) is equally important. AlN films have a much higher phase velocity and chemical stability in comparison to the

ZnO films. However, AlN has lower piezoelectric coupling and it is difficult to control the growth of AlN films. Even though, significant progress has been made towards incorporating the AlN films in energy-harvesting applications with compatible CMOS process, such as micomachined AlN cantilevers (Kim et al. 2012), [001] textured AlN cantilevers (Priya et al. 2017), corrugated AIN cantilever structures (Ting-Ta et al. 2011), AlN thin film based MEMS energy harvester with high power density (Andosca et al. 2012). Since the Young's modulus of AIN (345 GPa) is about four times that of PZT, the thickness of AIN for maximum power output is about three times smaller than that of PZT (Priya et al. 2017).

11.3.3.4 Piezoelectric Thick Films by Powder/Granule Spray in Vacuum Process

The powder spray in vacuum process so-called aerosol deposition (AD) is a unique ceramic film deposition technology that is able to fabricate highly dense thick films (submicron to several hundred μm) at room temperature. This process utilizes high kinetic energy of ejected aerosol consisting of a mixture of fine ceramic particles and carrier gas from a nozzle. The process has unique advantages compared to other thin/thick film deposition processes such as high deposition rate (over several tens μm/min depending on the deposition area), low processing temperature, composition control, high adhesion strength between substrate and film, as well as micropatterning of ceramic thick films during deposition (Zhou et al. 2011).

Due to its low temperature processing, low sinterability ceramic thick films such as KNN-based lead-free piezoelectrics and $PbTiO_3$ ceramics with full density can be deposited on various types of substrates. Figure 11.9 shows ferroelectric/piezoelectric property modulation of AD-PZT film according to the variation of residual stress level. By generating high compressive residual stress in the PZT thick films, the piezoelectric properties can be enhanced by ~2 times. All these advantages are highly attractive for MEMS-based piezoelectric device applications. In addition, optimization of deposition rate and uniformity in large area films have been conducted by the granule spray in vacuum process (GSV). Figure 11.10 shows the typical microstructure of patterned highly dense PZT thick films for MEMS ultrasonic transducers by GSV. The deposition rate of the PZT thick film by GSV was 1.4 μm/min in the area of 1440 cm^3. That is over two orders of faster rate than that of the other thin film processes. The film was very uniform. Directly patterned highly dense 11 μm thick PZT film was fabricated on 6 in. (152.4 mm) $Si/SiO_2/Ti/Pt$ wafer by GSV, and the films were not damaged during post processes including back-DRIE process (Priya et al. 2017).

11.3.3.5 Nonlinear Resonance-Based Energy-Harvesting Structures

Most of the vibration energy harvesters usually use a linear cantilever resonator structure to amplify small ambient vibration. While such structures are easy to model, design, and build, they typically have a narrow bandwidth. In contrast, nonlinear resonators have different dynamic response and greatly increase the bandwidth by hardening or softening the resonance characteristic of the beam structure. In addition, it has been found that nonlinear resonating beams can extract

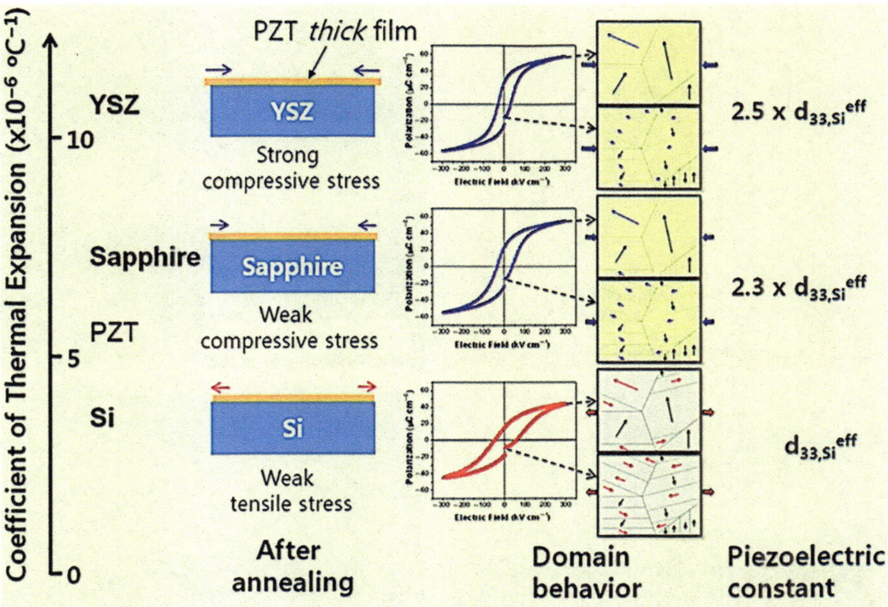

Fig. 11.9 Conceptual diagram of high piezoelectric performance in PZT thick films grown on various substrates; the stresses were controlled by the thermal expansion mismatch between substrates and PZT films (Adapted from Han et al. (2011) (AIP Publishing), used in accordance with the Creative Commons Attribution (CC BY) license (https://creativecommons.org/licenses/by/4.0/).)

more electrical energy than that of linear resonating beams when external vibration source has variable frequency. Nonlinearity may come from magnetic force or constrained mechanical structures. The magnetic forces between the magnets and the iron create a nonlinear spring, whose nonlinearity is determined by the strength of the magnets and the size of the air gap between the magnets and the iron stator. Due to the mutual attractions, the ferromagnetic beam has three equilibrium positions (statically bistable configuration), and the vibration mode has the form of the bistable Duffing resonance. Electromagnetic energy harvesters have shown hardening or softening resonance characteristics. However, magnet-based beams require assembly of hard magnets, which is expected to be costly as the size of the device shrinks (Priya et al. 2017).

Nonlinear resonance could be better achieved by a monolithically fabricated MEMS structures. Efforts have been made to achieve wide bandwidth piezoelectric energy harvester by exerting an axial compression and forming a buckled configuration, to make a bistable oscillator. Moreover, a monolithic MEMS-based nonlinear resonant piezoelectric micro energy harvester has achieved an ultra-wide bandwidth of >>20% of the center frequency and generated power more than 22 μW (Hajati and Kim 2011). Unlike a linear resonance system, where the electrical damping cannot exceed the mechanical damping, it has been shown that electrical damping in a nonlinear resonance system could surpass the mechanical damping, extracting much higher output power than that of the linear systems. The nonlinear impedance serves

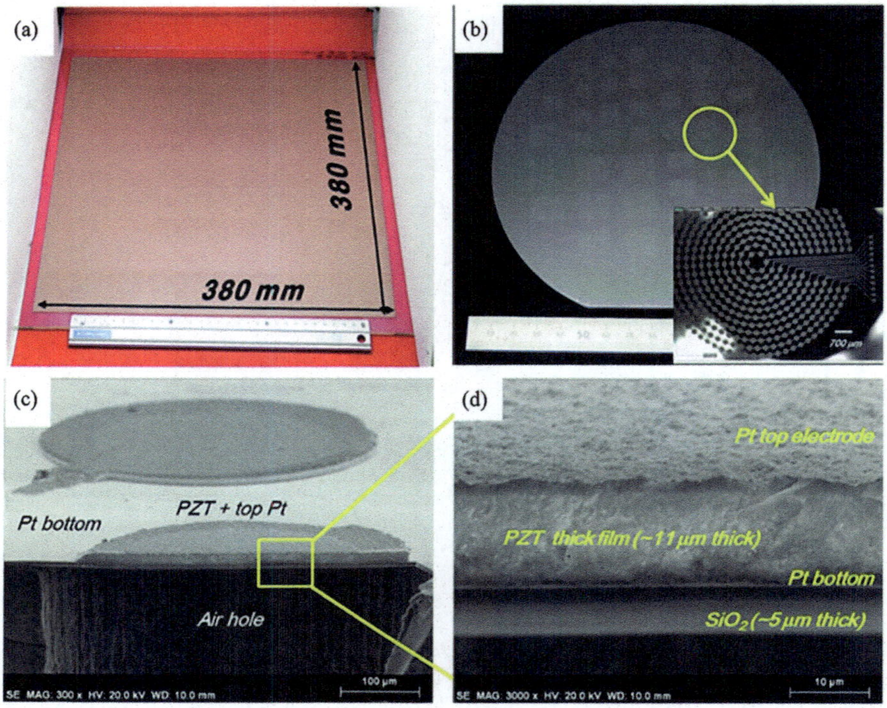

Fig. 11.10 Typical microstructure of patterned highly dense PZT thick films for MEMS ultrasonic transducers by GSV (Adapted with permission from Priya et al. (2017) (Walter De Gruyter and Company)): (**a**) Large area PZT deposition by GSV process. (**b**) 6 in. $Si/SiO_2/Ti/Pt$ wafer with directly patterned PZT thick film. (**c**, **d**) Cross-sectional SEM micrographs of piezoelectric MEMS ultrasonic device by GSV

as a negative feedback and stabilizes the deflection when the electrical damping changes. This is why the power bandwidth of nonlinear systems can be much wider than that of linear systems at equivalent beam dimensions (Priya et al. 2017).

Bistable nonlinear resonant beams not only widen the power bandwidth but also lower the working frequency range and input vibration amplitude of energy harvesters. Electromechanical lumped model predicts both stiffening and softening frequency responses for the inter-well and intra-well oscillations of bistable systems. The softening frequency response generates much higher power than mono-stable configuration at lower frequencies, providing opportunity for MEMS harvesters to be operated at low frequency and low g input vibrations. With MEMS-scale multi-layer buckled plate, at least 50 μW power could be generated below 100 Hz frequency and 0.2 g acceleration (Fig. 11.11). The presence of nonlinearity based on magnetic interaction or buckled beams has a significant influence on the performance of energy harvesters. The bandwidth of nonlinear energy harvester was increased by bistable effect; however, it is obvious that the frequency response to input vibration was more complicated as compared to linear case. The instability of frequency response makes it difficult to precisely modulate the bandwidth and tune

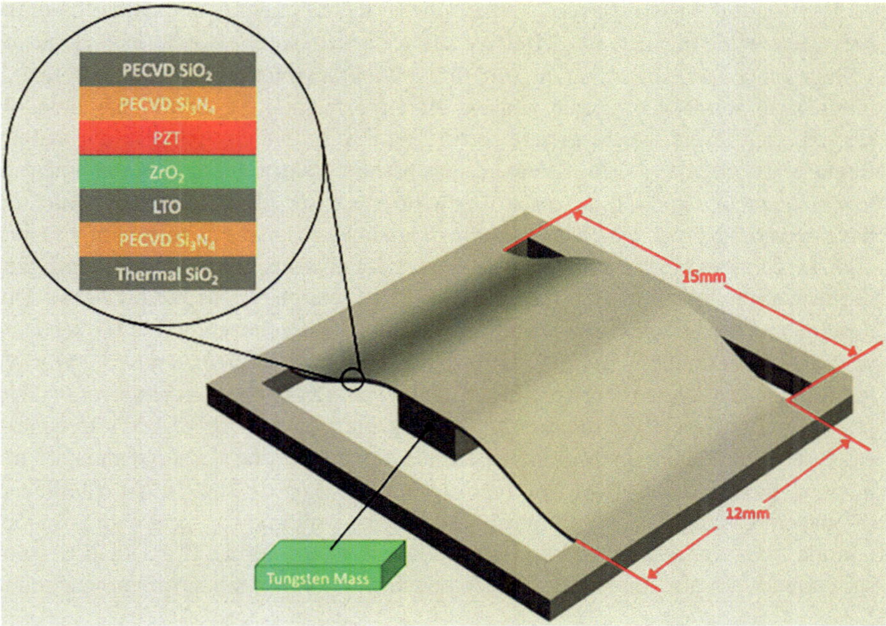

Fig. 11.11 Schematic view of a bistable beam energy harvester (Modified from Xu and Kim (2015) (IOP Publishing), used under the terms of the Creative Commons Attribution 3.0 license)

the resonance frequency of the energy harvester in realistic environment. Therefore, although nonlinearity can provide a solution for widening the bandwidth, it will require additional control features to exploit the enhancement in the output power (Xu and Kim 2015; Priya et al. 2017).

11.3.3.6 Low-Frequency Energy-Harvesting Structures

Microscale low-frequency energy harvester has been mainly dominated by inductive or electromagnetic devices. However, the miniaturization of such devices is limited by the size of the magnet used as the inertial mass and the size of the coil that can be fabricated on the MEMS scale. Consequently, for MEMS scale energy harvester, piezoelectricity is the transduction of choice and especially due to its compatibility with cleanroom-based silicon micromachining techniques. At mm-scale, piezoelectric mechanism provides the best output power density at low frequencies as compared to other possible mechanisms for vibration EH. However, piezoelectric-based energy harvester presents a fundamental challenge at the small dimensions since the resonance frequency of the structure increases (kHz range) as the dimension decreases. This challenge should be overcome in order to become compatible with the practical applications.

The most commonly used approach for reducing the natural frequency involves the addition of a large inertial tip mass to the free end of a simple single cantilever beam structures. Other methods involve the use of increased effective surface area cantilevers such as spirals (Deterre et al. 2013; Priya et al. 2017).

The use of tip masses severely affects the structural integrity and durability of the harvesters while the increased effective surface area designs exhibit torsion (causing voltage cancellation effect) in the very important first vibration mode. Therefore, it is of great importance to create a more effective method by developing uniquely shaped cantilevers, such as arc-based cantilevers which exhibit low natural frequencies; circular Zigzag structure that permits inertial mass free attainment of low frequencies. In addition, a set of new tip mass free vibration energy structures have been developed, which are capable of attaining resonant frequency of less than 100 Hz with less torsion than spiral designs and with a smaller foot print than zigzag or meandering structures, as shown in Fig. 11.12 (Varghese 2013; Apo 2014). This design, termed as arc-based cantilever (ABC), has achieved <100 Hz resonant operation both on the Si MEMS platform and with bulk piezo material as compared to traditional cantilevers. The arc-based cantilever is a continuous cantilever that can be divided into purely circular arc segments, thereby making it a low-frequency structure with dominant bending characteristic in the first (or fundamental) frequency mode. As shown in Fig. 11.12a, two arc-based micro-cantilevers (S-shaped, C-shaped) were designed based on the well-known linear configurations (simple and zigzag.) In addition to the metal micromilling (Fig. 11.12b) such structures were also fabricated and demonstrated using a silicon micromachining

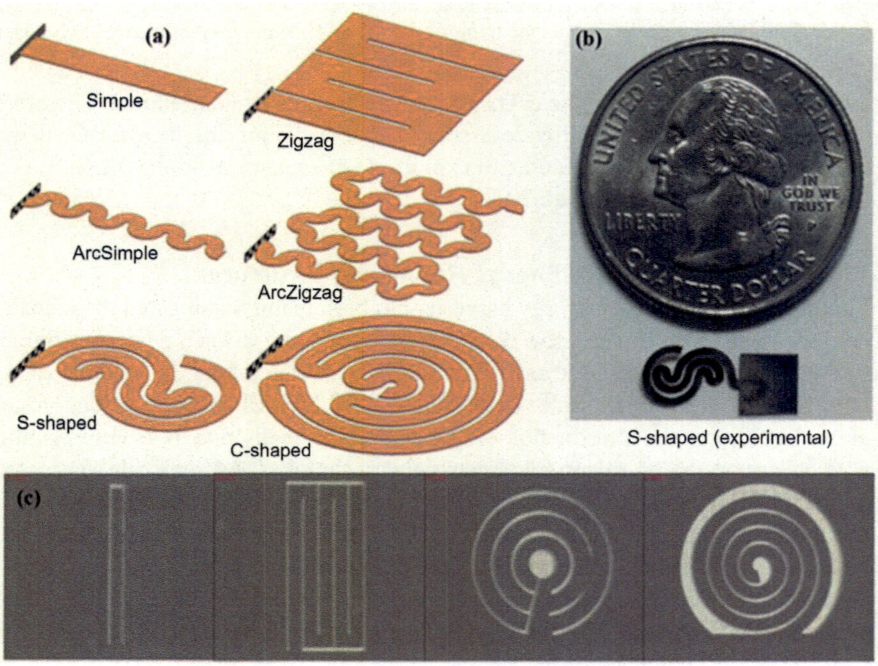

Fig. 11.12 Arc-based microcantilevers (Adapted with permission from Priya et al. (2017) (Walter De Gruyter and Company)): (**a**) Design development of the cantilevers, (**b**) micromilled S-shaped cantilevers, (**c**) as fabricated silicon MEMS cantilever structures

process (Fig. 11.12c). All the arc-based cantilevers were shown to resonate below 100 Hz and they exhibited dominant bending behavior in the fundamental mode. These designs therefore provide a foundation for the development of standalone arc-based micro-cantilevers which can be used for energy harvesting, actuation, and sensing applications (Priya et al. 2017).

11.3.3.7 Flexible and Stretchable Piezoelectric Devices

Advances in fabrication techniques and device designs, together with the emergence of high-performance inorganic piezoelectric materials enable the construction of piezoelectric devices in formats that are thin, flexible, and, in some cases, mechanically stretchable. The results allow straightforward miniaturization of lightweight, compliant electromechanical systems suitable for mounting on nearly any type of surface, with performance characteristics that can match those of conventional, rigid devices. Such technologies leverage the ability of piezoelectric materials to interconvert mechanical and electrical forms of energy. Electrical power can be generated from vibrations associated with operating machinery, movements of the human body, and environmental sources, such as waves, wind, and others. Similarly, application of electric fields to piezoelectric materials yields well-controlled mechanical forces for actuation in robotics, biomedical devices, and metrology tools. These dual functions in piezoelectrics, together with an increasingly broad set of material choices and device designs, provide the basis for diverse applications in biomedicine, defense technology, nano-devices, microelectromechanical systems (MEMS), personal electronics, and mechanical energy harvesters (MEHs). Figure 11.13 summarizes a range of examples, achieved in some cases by using nanoscale inorganics and in others with emerging classes of organic piezoelectrics (Wang 2008; Dagdeviren et al. 2016).

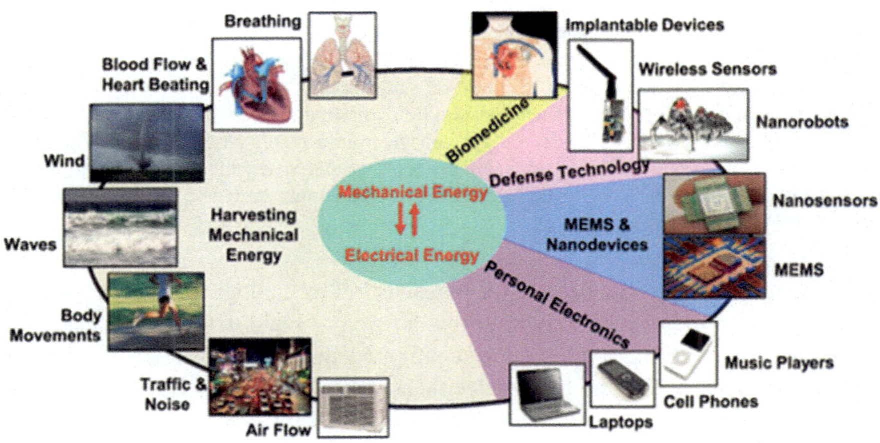

Fig. 11.13 Possible sources of energy for harvesting (left) and opportunities use of this energy in sensing and actuation (right) that can be considered for flexible/bendable piezoelectric devices (Adapted with permission from Wang (2008) (John Wiley and Sons))

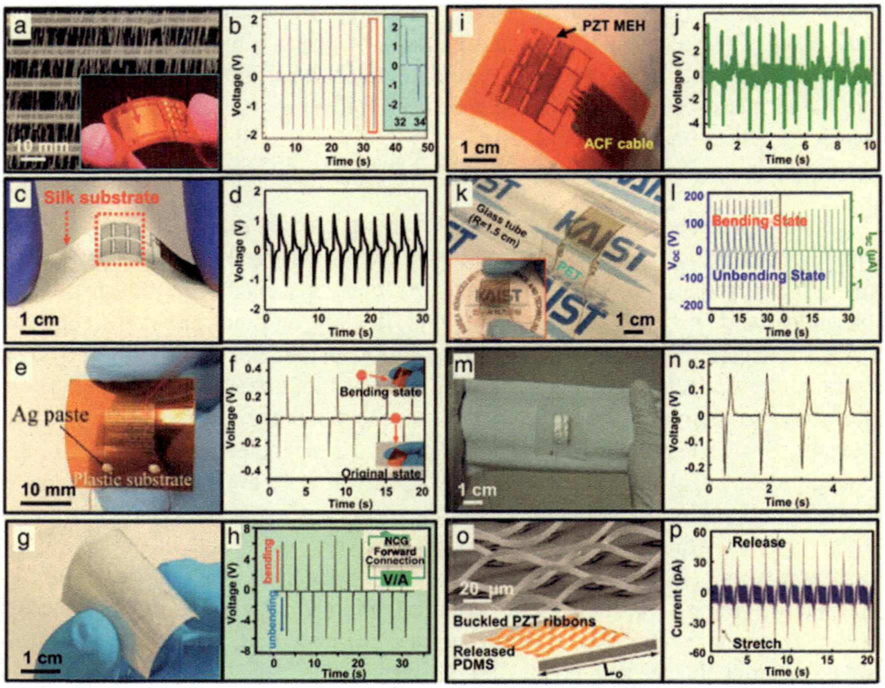

Fig. 11.14 Flexible and stretchable MEHs with various piezoelectric materials (Adapted with permission from Dagdeviren et al. (2016) (Elsevier)): (**a**) SEM image of horizontally aligned ZnO NWs on a plastic substrate with an inset of the resultant device. (**b**) Output voltage from a ZnO NWMEH during bending/unbending. (**c**) Photograph of a biocompatible ZnO MEH under compression. (**d**) Output voltage from a ZnO MEH during bending/unbending. (**e**) MEH based on a thin film of $BaTiO_3$. (**f**) Output voltage from a $BaTiO_3$ MEH during bending/unbending. (**g**) MEH based on nanowires of $BaTiO_3$. (**h**) Output voltage from a $BaTiO_3$ nanowire MEH during bending/unbending. (**i**) MEH based on nanoribbons of PZT. (**j**) Output voltage from a PZT nanowire MEHs during motions while mounted on a bovine right ventricle. (**k**) MEH based on a large-area, thin film of PZT released by LLO, an inset showing the resultant device manually bent. (**l**) Output voltage from a LLO-transferred thin-film PZT MEH during bending/unbending. (**m**) MEH formed using textile-based PZT. (**n**) Output voltage from a textile-based PZT MEH under the stretching and releasing motions. (**o**) SEM image (top) and schematic diagram (bottom) of a stretchable MEH based on wavy PZT nanoribbons. (**p**) Output current of a stretchable, PZT-based MEH under stretching motions

Among the many applications for piezoelectric technologies, those that involve interfaces with the human body represent an area of rapid development. Inorganic piezoelectric materials, appropriately configured into flexible/stretchable formats, are preferred due to their high piezoelectric coefficients. Figure 11.14 shows a range of different types of devices capable of harnessing electrical energy from mechanical energy resources each with potential applications in biomedicine. Classes of biointegrated, flexible mechanical energy harvesters involve ZnO, as nanoscale effect on ZnO NWs enhances the effective piezoelectric constant compared to that of bulk forms and provides flexibility to the MEH structure. Images associated with a device that incorporates horizontally aligned ZnO NWs appear in Fig. 11.14a. Under alternating bending/unbending motions with peak strains of 0.1% and strain rates of 5% s^{-1}, devices with active areas of 1 cm^2 generate peak voltages of ~2.0 V (Fig. 11.14b) and currents of ~100 nA, sufficient to operate a commercial LED. ZnO is particularly unique because it dissolves in biofluids, with biocompatible end products. As a result, the devices based on ZnO film built with other bioresorbable materials such as magnesium for the electrodes and interconnects, and silk fibroin for the substrates yield completely bioresorbable devices (Fig. 11.14c). Examples include devices with areas of 1×2.5 cm^2 that are capable of generating voltages and currents of 1.14 V (Fig. 11.14d) and 0.55 nA, respectively, under a peak strain of 0.056%. Envisioned uses are in biodegradable electronic implants for internal wound care, pain management, temporary cardiac pacing, and others. By comparison to ZnO, perovskite piezoelectric materials such as $BaTiO_3$ are appealing due to their higher piezoelectric constants. The use of soft lithography and transfer printing techniques facilitate the integration of $BaTiO_3$ MEHs into thin film formats on substrates of interest, as shown in Fig. 11.14e. In this configuration, the MEHs (areas of 1 cm^2) are flexible while maintaining their high piezoelectric properties and are capable of generating peak output voltages and currents of 0.4 V (Fig. 11.14f) and 12 nA, respectively, under peak strains of 0.55% and strain rates of 1.6% s^{-1}. Another example involves the incorporation of $BaTiO_3$ NWs synthesized via hydrothermal methods into PDMS (Fig. 11.14g). Here, increasing the NW composition (greater than 20 wt.%) degrades the electromechanical coupling and leads to low output voltages. With an optimum amount (20 wt.%) of NWs and an area of 3×4 cm^2, this type of MEH can achieve peak output voltages of 7.0 V (Fig. 11.14h) and currents of 360 nA under peak strains of 0.33%. PZT is another perovskite material, of interest due to its high piezoelectric constant. Flexible devices that use arrays of PZT nanoribbons in optimized architectures for deployment on large-scale animal models (e.g., cow, sheep, and pig) provide sources of electrical power for implants such as cardiac pacemakers. MEHs with areas of 2×2.5 cm^2 mounted onto epicardial sites of the beating heart of a bovine (cow) model are shown in Fig. 11.14i. Here, the value of h and the plane-strain bending stiffness parameters are arranged in such a way to achieve high system flexibility while producing optimal amounts of electrical power. For instance, under peak strains of 0.35%, output voltages can reach ~3.7 V. Different power outputs are also observed as a result of affixing the MEH onto different locations of heart, such as the free wall, right and left ventricles, consistent with the anatomy and the nature of motions associated with beating. The open circuit voltage for the case of the right ventricle reaches ~4 V (Fig. 11.14j). A mechanically stacked collection of five such MEHs yields an average power density of 1.2 $\mu W/cm^2$, which is sufficient to operate a cardiac pacemaker. A key design consideration is that these MEHs have minimal bending stiffnesses, to minimize mechanical loads on the heart, thereby avoiding arrhythmic behaviors associated with device-induced physical constraints. Similar devices can harness mechanical energy from the natural motions of other organs, such as lung and diaphragm. Other similar in vivo experiments have also been demonstrated to capture the mechanical energy from diaphragm and heart of small animals (i.e., rats) with ZnO NW and PMN-PT-based MEHs. Further improvements in performance are possible via the use of uniform, large-area films of PZT (Fig. 11.14k) released by laser lift-off (LLO). The LLO process minimizes the structural damage on the PZT thin film, as a result of immediate recrystallization after vaporization by laser irradiation. With device areas of $1.5 \times 1.5\ cm^2$, the output voltages can reach ~200 V under bending deformations (peak strain of 0.386% and strain rate of 2.32% s^{-1}), as shown in Fig. 11.14l. Slight human finger motions, in fact, can light as many as 100 blue LEDs simultaneously. Compared to flexible platforms, stretchable mechanics can increase the breadth of options for applications, particularly in wearable devices that cover large areas of internal organs. One stretchable example employs a textile composed of PZT fibers formed by electrospinning in Fig. 11.14m. By virtue of the electrospinning process, PZT fibers have enhanced piezoelectric properties and high flexibility. In a device example with an area of $1.5 \times 0.8\ mm^2$ stretching/releasing motions create output voltages of ~0.24 V (Fig. 11.14n) and currents of 2.5 nA. An enhanced output can be achieved by optimizing the thickness and the hardness of the PDMS, which is the interface layer that bonds the PZT textiles and the fabrics. A different stretchable platform based on PZT involves transfer printing of PZT nanoribbons-onto prestrained PDMS substrates to create wavy structures as in Fig. 11.14o. Periodic stretching/releasing motions to peak strains of 8% yield currents of ~50 pA from a system that consists of ten PZT nanoribbons (Fig. 11.14p). Here, the wavy structures not only enhance the piezoelectric constant, but they also can accommodate tensile strains that are several orders of magnitude larger than those that induce fracture in otherwise similar ribbons in flat geometries. Such systems provide promising platforms for wearable energy harvesting where reversible stretchability is essential to operation (Park et al. 2014; Dagdeviren et al. 2016).

The flexibility and/or stretchability of piezoelectric sensors and actuators afford new measurement capabilities on various curvilinear and soft surfaces. Figure 11.15a–p shows examples of piezoelectric sensors and actuators design for the skin and other tissues. In sensing modes based on the direct piezoelectric effect, such devices can detect and measure various physical deformations, ranging from those associated with touch to those induced by pulsatile blood flow. Figure 11.15a shows a flexible sensor built using arrays of ZnO NWs on ultrathin aluminum foil substrate (18 μm) to detect human eye blinking. Although the skin deformations are small, sensors ($5 \times 13\ mm^2$) generate easily measurable voltages of 0.2 V and currents of 2 nA as seen in Fig. 11.15b. Piezoelectric materials can also be used in touch sensors, as demonstrated by devices based on ZnO NWs and a capping layer of $BaTiO_3$ (Fig. 11.15c). In this example, a single touch point-cell has an active area of $1 \times 1\ cm^2$. The combination of these materials balances the high flexibility, modest piezoelectric constant in the ZnO NWs with the comparatively brittle, high piezoelectric performance of the $BaTiO_3$. As shown in Fig. 11.15d, the addition of $BaTiO_3$ capping layer enhances the sensitivity by increasing the signal response from 50 mV to ~3.3 V under a normal force of 17 N applied via a linear motor. In this configuration, the sensor can also differentiate forces applied by a human finger, as a flexible touch sensor. Even the small deflections of single cells can be detected,

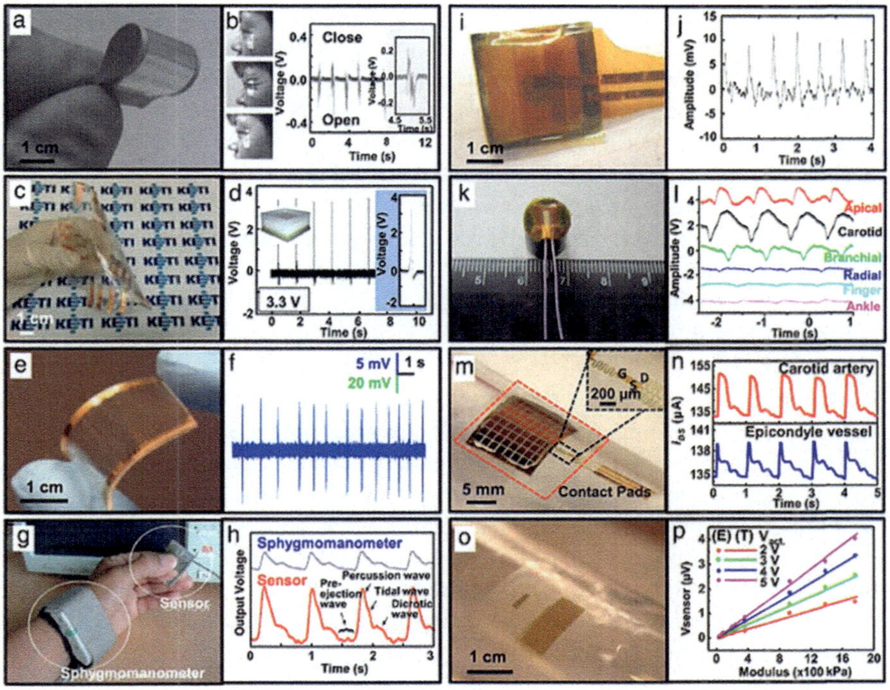

Fig. 11.15 Various piezoelectric sensors and actuators in flexible and stretchable formats (Adapted with permission from Dagdeviren et al. (2016) (Elsevier)): (**a**) A flexible active sensor made of ZnO NW arrays grown on ultrathin aluminum foil. (**b**) Output voltage from this device, driven by the blinking of the eye. (**c**) Touch sensor based on ZnO NWs and $BaTiO_3$ capping layers. (**d**) Voltage response of the touch sensor. (**e**) A flexible PZT nanoribbon designed to monitor cellular deformations. (**f**) The corresponding voltage response of PZT nanoribbons under cellular deformation. (**g**) AlN pressure sensor and sphygmomanometer to measure blood pressure pulse waves from the wrist. (**h**) Pulse waveform signals generated using this AlN pressure sensor and sphygmomanometer. (**i**) Piezoelectric membrane sensor based on PZT. (**j**) Voltage response of the sensor attached to the wrist, measuring pressure pulse waves. (**k**) Flexible piezoelectric tactile sensor. (**l**) Diverse pulse amplitude and waveforms detected by this flexible tactile sensor at various regions of the human body. (**m**) Photograph of a PZT pressure sensor wrapped on a cylindrical glass support with an inset image of stretchable conducting traces. (**n**) Current responses measured from the carotid artery and epicondyle vessel. (**o**) Array of stretchable PZT actuators and sensors on a polymer substrate wrapped on a cylindrical glass support. (**p**) Output voltage as a function of the substrate modulus recorded at four actuation voltages

where PZT nanomaterials can capture nanoscale deflections (~1 nm) induced by forces imparted at the cellular scale. Figure 11.15e shows PZT nanoribbons printed onto a silicone substrate (an area of $\sim 1 \times 2.5\ cm^2$) via transfer printing. Here, a standard whole-cell patch-clamp method induced mechanical deformations in individual PC12 cells. The voltage response (~10 mV) of a PZT sensor shown in Fig. 11.15f demonstrates the device capability in detecting cellular deformations (~1 nm). Additional sensing modalities involve measurements of pressure pulse waveforms due to blood flow. Examples with the sensors composed of AlN film, formed via low-temperature reactive sputtering, offer ability to detect pressures of ~0.9 kPa, well within the sensitivity of the human finger to sense texture and shape (10–40 kPa). AlN is well suited for this application because it has high thermal and chemical stabilities. Compared AlN to a commercial Sphygmomanometer, as shown in Fig. 11.15g, the pulse waveforms generated from the sensor at the fingertips, between the thumb and middle finger, are comparable to those collected at the wrist using a sphygmomanometer at a stress level of 10 kPa (Fig. 11.15h). This sensor can also detect tidal and pre-ejection waves that are critical for monitoring health and wellness. Another example uses stainless steel (SS) foil with thickness of 38 μm as the bottom electrode and substrate, with a layer of PZT (thickness of 70 μm) encapsulated with polyimide (Fig. 11.15i). The PZT film, which exhibits a piezoelectric constant (d_{33}) of 37 pm/V at 10 kHz, allows detection of the mechanical displacements of the skin due to blood flow to generate pulse wave signals as shown in Fig. 11.15j. In a similar example, PZT deposited onto a SS substrate by sol–gel processes with high temperature sintering (650 °C) and poling at high electric field (~0.71 MV/m) yields a sensor with useful levels of performance (Fig. 11.15k). Mounting this type of PZT sensor on a plastic element allows sensing of bio-motions of various regions of the human body, such as carotid, brachial, radical artery, finger, and ankle artery. Figure 11.15l shows the associated pressure waveforms. The use of correct sets of materials, the engineering of device layouts, the integration of required electrical components and associated serpentine conducting traces allow construction of stretchable sensors. An example is shown in Fig. 11.15m. This device consists of PZT nanoribbons with a total area of ~22 mm^2 and serpentine connections provides system stretchability up to 30%. This type of sensor has fast response time (~0.1 ms) and high sensitivity (~0.005 Pa), and can be applied in various ways on human subjects. In this configuration, the NMP was engineered to be closer to the middle of PZT nanoribbons to minimize the bending-induced in-plane strain. As a result, the sensor can measure pressure accurately on any body parts that have curvatures larger than a few millimeters. Additionally, the use of soft substrates enhances the mechanics due to complex and three-dimensional deformations, and result approximately in a hundred fold enhancement in the sensitivity. This high sensitivity allows use in determining PWV for arterial stiffness measurement by placing the sensors at two different locations of human body. Figure 11.15n presents some relevant data on two different sites, such as the carotid artery and epicondyle vessel, with 2-min intervals and on three different human bodies. The resulting PWV of ~5.4 ms^{-1} is similar to that measured using conventional tonometry. In addition to sensing, mechanical actuation can be an important feature in biomedical devices. An example of a unique device platform includes ultrathin architectures of piezoelectric sensors and actuators, exploiting both direct and indirect piezoelectric mechanisms, combined with serpentine configurations of metal traces for electrical connections to measure the mechanical properties (e.g., elastic modulus) of soft tissues and organ systems. Figure 11.15o shows seven actuators (lateral dimensions of 200 × 1000 μm^2) and six sensors (lateral dimensions of 100 × 500 μm^2) that are made of PZT nanoribbons in capacitor type geometries sandwiched between two layers of electrodes. Here, bending onto curvy body parts has little effect on device operation since the NMP lies close to the middle of PZT nanoribbon.

In this conformal modulus sensor (CMS) system, applying a sinusoidal voltage to an actuator mounted on the skin induces mechanical motions in the device structures and the underlying skin, which can then be detected by adjacent sensors. The voltage responses of these sensors provide information about the electromechanical coupling between the actuators and sensors, presenting ways to determine the mechanical modulus of the near surface region of the skin. The relationships between the sensor voltage output and modulus at 1 kHz can be seen in Fig. 11.15p. The results show consistent linear proportionality between the modulus values of the substrates, for values between 30 and 1800 kPa, and associated sensor output voltages under various actuation voltages. The modulus can be determined from the actuator voltage, the sensor output voltage, the material properties and geometries of the device layers. In addition, the system exhibits capabilities to detect changes in the elastic modulus of skin across the body due to dermatologic malignancies, which generally are stiffer than healthy tissues. The same system can also be used to measure the modulus values of various biological tissues, such as the heart and the lung. As existing methodologies (e.g., torsion, traction, and nanoindentation) to measure the mechanical properties of soft biological tissues are invasive and lack of microscale spatial resolution, the dual features to actuate and sense in a thin, conformal platform offer promising capabilities for fast, accurate, noninvasive diagnostics based on stiffness (i.e., human skin or other organs) (Qi et al. 2011; Dagdeviren et al. 2016).

INDEPTH: Emerging Applications of Liquid Crystals

Liquid crystals (LCs) have attracted intense research interest because of their orientational order, the existence of strong dipoles and easily polarizable groups, the rigidness of the long axis, anisotropic features in the structural, optical, electrical, and magnetic properties, as well as their easy response to electric, magnetic, and surface forces. LCs are ubiquitous in diverse applications, ranging from displays to electronics, sensors, lasers, and optical computing in human being daily life. For example, ferroelectric LCs that possess a variety of advantages, such as a permanent electric polarization, high flexibility, and fast response time, have been the basis for noble field effect transistors (FETs), memory cells, and optical switching device applications. In particular, among FET-based nonvolatile memory devices, the ferroelectric FET, an important type of memory cells without a storage capacitor, is significantly attractive for memory and switch applications, because of the wide range of its interesting features, including small cell size, nondestructive read-out, low-power consumption, good retention, and fast response time. Another example of emerging applications based on LCs is holography, which is an ideal technology to realize three-dimensional

(continued)

(3D) dynamic images by changing the refractive index. Moreover, arrays of carbon nanotubes (CNTs)/nanofibers have been used to locally modulate the refractive index of the LC medium at low operation voltages. In addition, it is possible to achieve technological advances and revolution into a rapidly growing multidisciplinary field involving displays, electronics, and energy harvesting to integrate displays into a self-powered system with piezoelectric power generators, for instance. In particular, ZnO nanowire-based piezoelectric power generators exhibited highly promising piezoelectric performance for a self-powered source, due to its dimensionality and piezoelectric semiconducting properties containing non-toxicity, eco-friendliness, and geometrical versatility (Sohn et al. 2014).

11.4 Pyroelectric Materials

Energy harvesting can give rise to self-powered, maintenance-free devices with an almost inexhaustible source of energy. Devices from the millimeter scale down to the microscale have been manufactured, with an average power in the 10 μW–10 mW range, such as piezoelectric generators, thermoelectric microgenerators, photovoltaic devices, and electromagnetic microgenerators. At small scale, pyroelectric materials are better suited for the conversion of thermal energy into electrical energy compared to thermoelectric devices, since they require temporal temperature gradient (dT/dt) instead of a spatial temperature gradient (dT/dx), which is more difficult to realize at micro/nanoscale. Many pyroelectric materials are stable up to a very high temperature (~1200 °C), which provides an advantage over thermoelectrics for harvesting energy from high-temperature sources. Furthermore, methods to convert stationary spatial gradients to transient temperature gradients facilitates development of hybrid energy harvesters based on both thermoelectric and pyroelectric effects or radiation and pyroelectric effect. In addition, devices based on pyroelectric energy harvesting require low or no maintenance since unlike piezoelectric energy-harvesting devices, they do not include any moving parts. Alongside experimental implementation and developments, different theoretical and numerical approaches have been attempted for optimization of pyroelectric energy harvesting using mass, momentum, and energy equations for guiding future experimental efforts (Lingam et al. 2013).

11.4.1 The Pyroelectric Effect

Pyroelectricity is defined as the temperature-dependent spontaneous polarization in certain anisotropic crystals. This effect refers to the generation of an electric current/potential in materials with non-centrosymmetric crystal structure when subjected to a

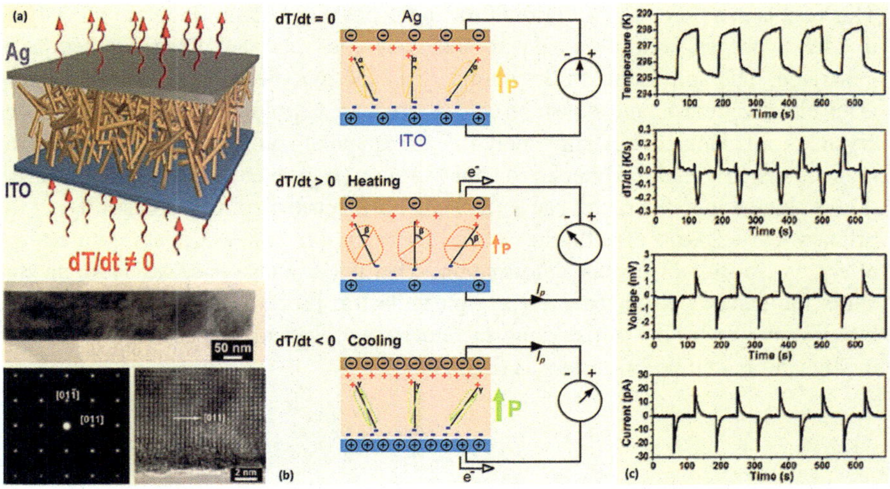

Fig. 11.16 The mechanism of the pyroelectric nanogenerator based on the primary pyroelectric effect (Modified with permission from Wang et al. (2012) (Elsevier)): (**a**) Schematic diagram of the structure of pyroelectric nanogenerator, TEM image of a single $KNbO_3$ NW, the corresponding SAED pattern of the NW, and the corresponding HRTEM image of the NW; (**b**) schematic diagrams of the pyroelectric nanogenerator with negative electric dipoles under room temperature heated and cooled conditions. The angles marked in the diagrams represent the degrees to which the dipole would oscillate as driven by statistical thermal fluctuations; (**c**) the output voltage and current of the device under the cyclic change in temperature

temporal temperature gradient (*dT*/*dt*) (Fig. 11.16). Pyroelectricity is exhibited by crystals belonging to the polar classes 1, 2 m, 2 mm, 3, 3 m, 4, 4 mm, 6, and 6 mm. Pyroelectric materials have a unique polar axis along which spontaneous polarization exits. The unit cell of pyroelectric materials has a dipole moment, and the dipole moment per unit volume of the material is called the spontaneous polarization. A change in temperature causes the net dipole moment and hence the spontaneous polarization to change. The pyroelectric coefficient p is defined as the differential change in spontaneous polarization P caused by a change in temperature T. In the most general form, for a stress-free material in an open circuit condition, the change in polarization due to change in temperature is expressed as follows (Lingam et al. 2013):

$$\frac{\partial P_x}{\partial T} = \frac{\partial P_{x_0}}{\partial T} + \frac{\partial (d_{xkl}\sigma_{kl})}{\partial T} \tag{11.11}$$

The first term on the right-hand side of Eq. (11.11) is the permanent polarization, and the second term represents the piezoelectric-induced polarization. Change in the polarization causes change in surface-bound charges, and hence redistribution of free charges in the material to compensate for the change in surface-bound charges.

This results in a pyroelectric current flow in the external circuit. If the circuit is open, the free charges simply remain in the electrodes and an electric potential is generated. The constraints in the definition of the pyroelectric coefficient are constant electric field E and constant elastic stress σ. Constant stress means that the crystal is not clamped, but completely free to expand or contract thermally. When the crystal is rigidly clamped under constant strain ε, to prevent expansion or contraction, a change in temperature causes a change in electric displacement signifying the primary pyroelectric effect. The second contribution, the secondary pyroelectric effect, is a result of crystal deformation. Thermal expansion causes a strain that alters the electric displacement via a piezoelectric process. The total pyroelectric effect is the sum of the primary and secondary pyroelectric effects. The pyroelectric coefficient, p, can be expressed as (Lee et al. 2016):

$$p = \frac{d\rho}{dT} \tag{11.12}$$

where ρ is the spontaneous polarization and T is the temperature. The electric current generated by the pyroelectric effect is expressed as (Lee et al. 2016):

$$I = \frac{dQ}{dt} = \mu p A \frac{dT}{dt} \tag{11.13}$$

where Q is the induced charge, μ is the absorption coefficient of radiation, A is the surface area, and dT/dt is the rate of temperature change. Therefore, when pyroelectric materials are heated or cooled ($dT/dt > 0$, or $dT/dt < 0$), the overall polarization in the dipole moment is decreased or increased, which causes current to flow in the circuit. Usually, ferroelectric materials such as lead zirconate titanate (PZT), $BaTiO_3$, P(VDF–TrFE), and $KNbO_3$, and some piezoelectric materials which have spontaneous polarization, such as ZnO and CdS, are used to fabricate pyroelectric generators. Coupling a pyroelectric to an external structure which undergoes large thermal deformations is also a potential approach to enhance harvested energy (Bowen et al. 2014).

In addition, tertiary pyroelectricity, due to non-uniform heating, is also possible since non-uniform heating generates shear stresses that result in polarization through the piezoelectric effect. In this case the current generated is dependent on the magnitude of the temperature gradient. Secondary and tertiary effects are therefore potential routes for enhancing thermal harvesting along with heat transfer enhancement or proper materials selection or new materials development. By harvesting the waste thermal energy, pyroelectric nanogenerators have potential applications such as environmental monitoring, temperature imaging, medical diagnostics, and personal electronics (Bowen et al. 2014; Lee et al. 2016).

11.4.2 Pyroelectric Materials and Selection for Energy Harvesting

Pyroelectric materials currently find widespread use in thermal detectors and sensors, where the choice of the pyroelectric material is mainly determined by figure of merit (FOM), detector size, availability, durability, environment in which the material has to operate, thermal radiation levels to be detected, purpose for which the detector is employed, maximum ambient temperature of operation, and the range over which stable operation is required. An appropriate selection of pyroelectric materials is not only to improve the sensor performance but also to cater to specific sensor applications. For pyroelectric materials, Curie temperature is the critical point where a nonpolar material undergoes a structural transformation; below this critical point, the material exhibits an intrinsic, permanent electrical polarization, usually along a certain crystallographic axis. Table 11.3 presents matrices for the ten pyroelectric crystal classes and two pyroelectric Curie groups. Table 11.4 presents a summary of pyroelectric coefficient of various materials at room temperature (Lingam et al. 2013).

1. Triglycine sulfate (TGS)—$(NH_2CH_2COOH)_3H_2SO_4$: TGS constitutes a large family of isomorphous compounds. The compounds based upon TGS have provided some of the highest pyroelectric figures of merit (5.5×10^4 Cm^{-2} K^{-2}). TGS possesses a Curie temperature of 49 °C; above this temperature, it exists as centrosymmetric class 2 m, and below this temperature, it is a polar point group 2 with the polar axis along the monoclinic *b* axis. Since the glycine groups are polar

Table 11.3 Matrices for the 10 pyroelectric crystal classes and two pyroelectric Curie groups (Newnham 2005)

Point group	Pyroelectric metrics
1	$(p_1\, p_2\, p_3)^T$
2	$(0\, p_2\, 0)^T$
m	$(p_1\, 0\, p_3)^T$
$mm2$, 3, $3m$, 4, $4mm$, 6, $6mm$, ∞, ∞m	$(0\, 0\, p_3)^T$

Table 11.4 Pyroelectric coefficient of various materials at room temperature, units μC/m^{2}K (Lang 2005; Newnham 2005)

Material	Experimental value	Secondary coefficient	Primary coefficient
Triglycine sulfate TGS (2)	−270	−330	60
$LiNbO_3$ (3 m)	−83	+12.8	−95.8
$LiTaO_3$ (3 m)	−176	−1	−175
$Pb_5Ge_3O_{11}$(3)	−95	+15.5	−110.5
$BaTiO_3$ (∞ m)	−200	+60	−260
$PbZr_{0.95}Ti_{0.05}O_3$ (∞ m)	−268	+37.7	−305.7
ZnO (6 mm)	−9.4	−2.5	−6.9
CdSe (6 mm)	−3.5	−0.56	−2.94
CdS (6 mm)	−4.0	−1.0	−3.0

in nature, the reversal is largely associated with the rotation of the glycine group about the crystallographic *a*-axis. Despite its high performance, TGS has attracted limited interest for harvesting applications, possibly due to its low Curie temperature. It is also water soluble, hygroscopic, and relatively low strength (Bowen et al. 2014). TGS pyroelectric crystals are primarily used in single element detectors where sensitivity of detection of temperature change is of prime importance (Whatmore 1986; Sahraoui et al. 2003.)

2. Polyvinylidene fluoride (PVDF): PVDF is a ferroelectric polymer, which exhibits a strong piezoelectric effect. PVDF molecules have a repeat unit of $–CH_2–CF_2–$, which take up a number of stable configurations based on the polymers treatment. The polar unit configuration of the material crystal exhibits pyroelectric effect. PVDF has a Curie temperature up to 180 °C, but the polar properties degrade when it is heated above 80 °C. It finds applications in large area detectors for laser pulse monitoring and also single element devices such as intruder alarms. Large PVDF thin films are available commercially at low cost for use in detectors (Whatmore 1986; Wooldridge et al. 2010).
3. Lithium tantalate ($LiTaO_3$): $LiTaO_3$ is an oxygen octahedral crystal consisting of layers of oxygen ions arranged in hexagonal close packing. It is used in single crystal form grown using the Czochralski method and exhibits a moderate pyroelectric effect. $LiTaO_3$ possesses a Curie temperature of 665 °C. $LiTaO_3$ finds applications in pyroelectric detectors, and widespread usage in one-dimensional (1D) commercial detector arrays. $LiTaO_3$ has drawbacks such as high thermal diffusivity, which reduces its minimum resolvable temperature difference at high spatial frequencies. One of the solutions used to resolve this problem is the use of ion beam reticulation to separate detector elements (Whatmore 1986; Geuther and Danon 2005).
4. Non-ferroelectric pyroelectrics include the wurtzite-based materials such as AlN, GaN, CdS, and ZnO which have relatively low pyroelectric coefficients compared to the ferroelectric materials. Since these materials are not ferroelectric, they are often used in single crystal form, such as epitaxially grown films, or as highly orientated structures to achieve the desired polarization. These materials do, however, exhibit higher thermal conductivities compared to the ferroelectric materials allowing a more rapid change in temperature due to changes in ambient temperature (Bowen et al. 2014).

   GaN is a natural pyroelectric with polarization in the *c*-axis. GaN exhibits a strong pyroelectric effect at temperature above 300 °C, whereas $LiTaO_3$ and $PbTiO_3$ exhibit the property below 300 °C, and hence GaN finds applications in high-temperature environments. The pyroelectric coefficient of thin GaN films was reported to be $\sim 10^4$ V/mK (Bykhovski et al. 1996; Yu et al. 2005).

   ZnO is also a wurtzite-structured natural pyroelectric crystal with polarization in the *c*-axis. Most applications utilizing ZnO as the primary material use it in the thin film form. ZnO thin film has exhibited a conversion of thermal radiation ten times larger than GaN films. ZnO possesses a Curie temperature of 430 °C and a pyroelectric coefficient of 4×10^4 V/mK was observed from experiments conducted on a bundled arrays of nanowires (Jagadish and Pearton 2011; Yang et al. 2012a, b).

5. Perovskite-based pyroelectrics: Perovskite is a large family of oxygen octahedral crystals with the general formula ABO_3. The perovskite structure undergoes deformation to give rhombohedral, tetragonal, or orthorhombic structures. In Lead Zirconate- and Lead Titanate-based structures, dopants are used to improve the pyroelectric properties of these ceramic-based pyroelectrics. Lead Titanate ($PbTiO_3$) possesses a high Curie temperature of 490 °C and a high polarization rate. The preferred method of manufacturing devices is by deposition of thin film layers rather than machining from bulk (Whatmore 1986; Lingam et al. 2013).

   The lead zirconate titanate (PZT) family remains a widely used commercial ceramic due to its relative ease of fabrication in polycrystalline form and good piezoelectric properties, with a range of "hard" and "soft" composition with tailored properties. This family of material has therefore attracted interest for pyroelectric-harvesting applications. For instance, improved pyroelectric properties have been achieved in compositionally graded $PbZr_{1-x}Ti_xO_3$, including high pyroelectric-harvesting FOM. Lanthanum-doped PZT relaxor ferroelectrics have also been used for harvesting applications where doping increases resistivity and coupling coefficients (Bowen et al. 2014).

   Lead magnesium niobate—lead titanate (PMN-PT) single crystals are being explored for transducer applications. The $(1\text{-}x)Pb(Mg_{1/3}Nb_{2/3})O_3{-}xPbTiO_3$ ((1-xPMN-xPT) system is a family of relaxor-based ferroelectric compositions which are of interest for transducer devices due to their ultra-high piezoelectric and pyroelectric coefficients. The morphological phase boundary (MPB) for PMN-xPT spans from (x =) 30 to 38 mol% and this range is characterized by a monoclinic phase in coexistence with a rhombohedral (up to 32 mol%) or tetragonal phase (32 to 38 mol%). These crystals have a relatively low Curie temperature (121°), which can limit the material to relatively low temperature operation. Due to their single crystal nature, the materials are relatively expensive and can be formed in limited shapes. The high FOM and range of phase transitions associated with these materials have led to interest in this material for a number of pyroelectric-harvesting applications (Bowen et al. 2014).

6. Additional lead-free materials: Lead-free materials are of interest for environmental and health concerns. Manganese doped bismuth sodium titanate-barium titanate (BNT-BT) single crystal is a potential pyroelectric energy-harvesting material, since the compositions possess excellent piezoelectric and pyroelectric coefficients and high Curie temperatures ($TC > 200$ °C). BNT-based ceramics can be difficult to pole due to their high electrical conductivities and dielectric loss and to overcome these shortcomings they are often doped. For example, the composition Mn: BNT-BT is a $94.6Bi_{0.5}Na_{0.5}TiO_3\text{-}5.4BaTiO_3$ single crystal doped with Mn and the <111> orientation of this crystal possesses the highest pyroelectric coefficient and figure of merit of lead-free ferroelectric materials. In addition to BNT, $Ba_{0.65}Sr_{0.35}TiO_3$ (BST) thin films have also been examined specifically for improved energy-harvesting FOMs (Bowen et al. 2014).

A family of potassium sodium niobate-based materials, $(K_{0.5}Na_{0.5})NbO_3$ (KNN), have been discovered as a high-performance substitute for lead-based piezoelectric materials. KNN has piezoelectric performance approaching commercially available lead-based materials such as PZT and PMN-PT ceramics. The KNN-LT $[\{(K_{0.5}Na_{0.5})_{0.96}–Li_{0.04}\}(Nb_{0.8}Ta_{0.2})O_3]$ and KNN-LTS $[\{(K_{0.5}Na_{0.5})_{0.96}–Li_{0.04}\}(Nb_{0.84}Ta_{0.1}Sb_{0.06})O_3]$ compositions have improved characteristics compared to pure KNN. Nano-scale $KNbO_3$ has been considered for pyroelectric harvesting (Yang et al. 2012a, b).

In comparison, the best performing materials for FOM are TGS followed by PMN-25PT <111>, Mn:BNT-BT <111>, SBN, $LiTaO_3$, PMN-25PT ceramic, PZT ceramic and Mn:BNT-BT <111>, respectively. From the lead-free materials, co-polymers of PVDF and BNT-based bulk ceramics have better performance compared to the KNN family. The non-ferroelectric pyroelectric materials (e.g., ZnO, GaN, AlN) have lower FOMs compared to the ferroelectrics. However, the FOMs do not include information regarding the operating temperature which, for ferroelectrics, in often related to the Curie temperature. Polymer-based PVDF materials are generally at lower temperatures (<100 °C) compared to the ferroelectric ceramics where the doped PZT family are of interest for higher temperatures. In addition, composite materials are also attracting interest in an effort to combine high-activity ceramic ferroelectrics with a flexible and low-permittivity matrix; these have been examined for pyroelectric detectors but the "composite" approach certainly offer avenues for creating interesting materials for harvesting applications (Bowen et al. 2014).

11.4.3 Pyroelectric-Harvesting Devices

Pyroelectric materials have found applications in an array of fields. The most prominent fields are energy harvesting and applications in sensors. For energy harvesting, they convert the thermal energy to electricity, and in sensors, they are used as sensing elements to detect heat (or indirectly motion) signals by conversion into an electric signal. Pyroelectric energy harvesting has been demonstrated in thin films, nanowires, and nanofibers. In particular, thin films offer a large surface area for enhanced thermal exchange and larger electrodes, and hence larger current output (Lingam et al. 2013).

11.4.3.1 Nanostructured and Micro-scale Materials and Devices

For pyroelectric harvesters the operating frequency of the device is often small (typically much less than 1 Hz). In fluid-based systems the frequency of operation is limited by heat transfer between the pyroelectric and the working fluid that is oscillating between hot and cold sources. At the nano-scale, energy transfer by thermal radiation between two semi-infinite solids is almost instantaneous and can be enhanced by several orders of magnitude from the conventional Stefan–Boltzmann law if their separation is a distance smaller than a characteristic wavelength, given by Wien's displacement law. For example, a device using 60/40 porous poly (vinylidene fluoride-triuoroethylene) was predicted to have a 0.2% efficiency and a 0.84 mWcm^{-2} electrical power output for the cold (273 K) and hot sources (388 K). A pyroelectric plate made from 0.9PMN-PT composite thin films achieved a higher efficiency (1.3%) and a larger power output (6.5 mWcm^{-2}) for a temperature oscillation amplitude of 10 K at a temperature of 343 K at a relatively high frequency of 5 Hz (Bowen et al. 2014).

A simple approach to improve the rate of temperature change and increase the pyroelectric current is to reduce the thickness of the pyroelectric, such as using thin films. Figure 11.17 illustrates a flexible hybrid energy cell for simultaneously harvesting thermal, mechanical, and solar energies. A ZnO-poly(3-hexylthioohene) hetero-junction solar cell was used for harvesting solar energy while a PVDF-based pyroelectric and piezoelectric nano-generator was built on its bottom surface for harvesting thermal and mechanical energies, with a pyroelectric coefficient of ~44 μCm^{-2} K^{-1}. Using a lithium ion battery to store the harvested energy the device could drive

LED devices. Pyroelectric "nano-generators" have also been fabricated based on ZnO nanowire arrays; a higher pyroelectric voltage and current coefficients were obtained for the nano-generator compared to bulk and film material due to the preferred orientation of the ZnO nanowire array. The power of the pyroelectric generator was used to charge a lithium ion battery and a single output pulse could charge a LCD (Fig. 11.17). Lead-free $KNbO_3$ nanowire/PDMS polymer with Ag and indium tin oxide (ITO) electrodes as a flexible nano-generator have also been fabricated where the output could be tuned by the electric field due to changes in ferroelectric domain orientation (Bowen et al. 2014; Yang and Wang 2015).

At the micro-scale, the etching of pyroelectric surfaces and etching the electrode structure can improve energy-harvesting performance. A meshed top electrode and trenched pyroelectric can improve the responsivity of the PZT. A vortex-like electrode with a deep structure produced by sandblasting can improve the harvested power by 11% compared to a fully covered electrode. The size of the pyroelectric element has also been used to tailor the phase transition temperatures in ferroelectric nanowires, enabling a "giant" pyroelectric response. Nanowires have already demonstrated to show enhanced piezoelectric and thermoelectric properties compared to their bulk counterparts. These enhancements are often due to lower defect density in the nanowire form compared to the bulk form, which may also contribute to the pyroelectric enhancement. In addition to nanowires, nanofibers are also attractive for pyroelectric energy harvesting. Nanofibers are often mechanically soft and can be readily incorporated into flexible devises. Nanofibers are mostly fabricated using electrospinning technique. In this technique, a high DC voltage between the liquid in a syringe and a collector spins nanofibers in diameters from several tens of nanometers to several microns (Lingam et al. 2013; Bowen et al. 2014).

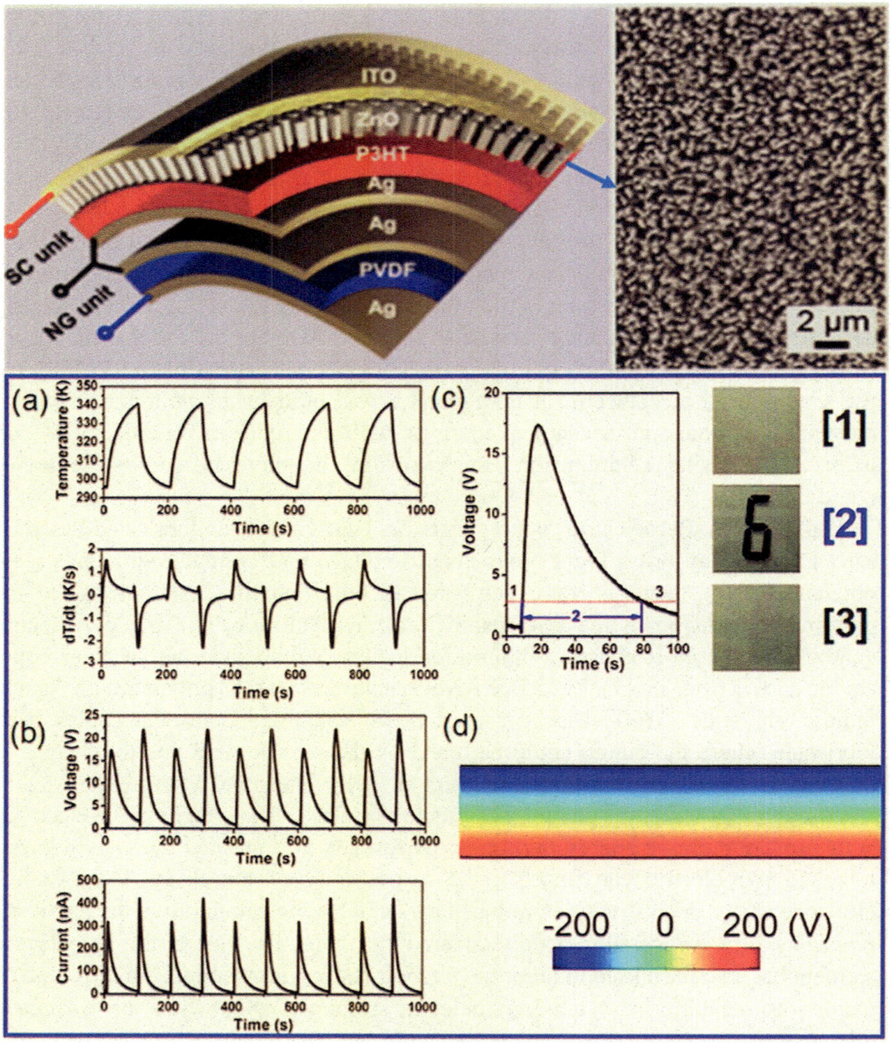

Fig. 11.17 PZT thin-film pyroelectric generator (Modified with permission from Bowen et al. (2014) (Royal Society of Chemistry); Yang and Wang (2015) (Elsevier)): (**a**) Cyclic change in temperature and differential temperature (dT/dt). (**b**) Output voltage and current of the generator after rectification by a full wave bridge circuit. (**c**) Enlarged single output voltage peak, where it is used to drive a LCD in the region "2." (**d**) Calculated electrical potential distribution across the PZT film

11.4.3.2 Hybrid Generators

Since all pyroelectrics are piezoelectric, both pyroelectric and piezoelectric harvesting have been combined. The generation of an electric current under short circuit conditions or an electric potential in open circuit conditions as a result of the change in polarization with a temperature change has analogies with piezoelectric harvesting. Due to their similarities, there is interest for developing potential hybrid

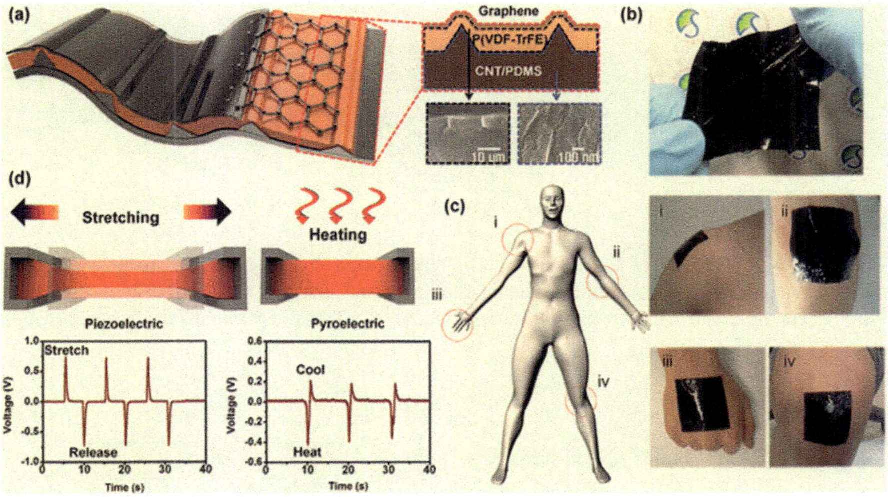

Fig. 11.18 Hybrid piezoelectric–pyroelectric nano-generator (Modified with permission from Lee et al. (2014) (WILEY-VCH Verlag GmbH & Co)): (**a**) Schematic of stretchable structure of the nano-generator, (**b**) image of device, (**c**) location of devices on body, (**d**) piezoelectric output on application of strain and pyroelectric output on changing temperature

piezoelectric–pyroelectric-harvesting systems, whereby a combination of temperature change and stress is applied (Bowen et al. 2014).

Figure 11.18 shows a stretchable, hybrid piezoelectric–pyroelectric nano-generator based on a micropatterned piezoelectric P(VDF-TrFE) polymer, micropatterned PDMS carbon nanotube (CNTs) composite, and graphene nanosheets. The PDMS-CNT was used to make the device flexible and also serve as a robust electrode on the base of the device. Graphene was used as a top flexible electrode to allow a fast temperature gradient on the device due to its high thermal conductivity. The use of piezoelectric–pyroelectric-harvesters potentially offers an interesting method of enhancing power (Bowen et al. 2014).

11.4.3.3 Pyroelectric Systems and Active Oscillators

The pyroelectric generator is modeled as a current source with a capacitor and a resistor in parallel. The current is generated within the pyroelectric element with the change in temperature. However, one of the main problems of pyroelectric energy harvesting is a heating process followed by a cooling process, which produces charge accumulation in different directions. One way to mitigate this problem is to use a full bridge diode rectifier circuit. With it, spatial temperature gradients can be converted into required transient temperature for pyroelectric energy harvesting. Figure 11.19 shows a micro heat engine that acts as a thermal energy shuttle between a heat sink and a heat source. In this configuration, an oscillating thermal field is created across a pyroelectric generator. Using this micro thermomechanic–pyroelectric energy generator (μTMPG), 3 μW power could be harvested for a temperature difference of 79.5 K from pyroelectric generators. With similar methods, hybrid

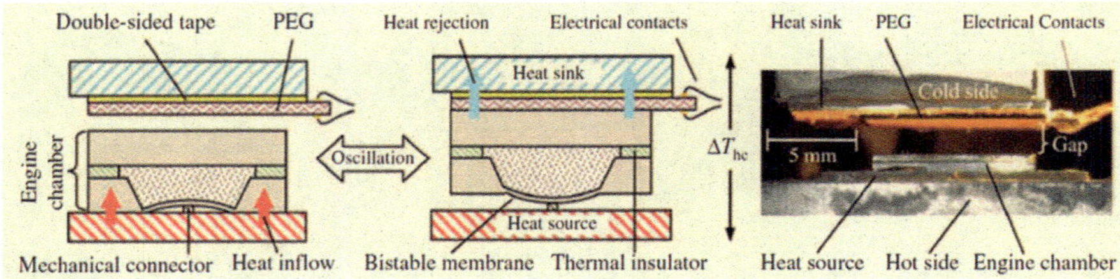

Fig. 11.19 Illustration of the μTMPG technique for converting stationary spatial temperature gradient into transient temperature gradient (Adapted from Lingam et al. (2013) (Taylor & Francis), used under the Creative Commons Attribution License http://creativecommons.org/licenses/by/3.0/)

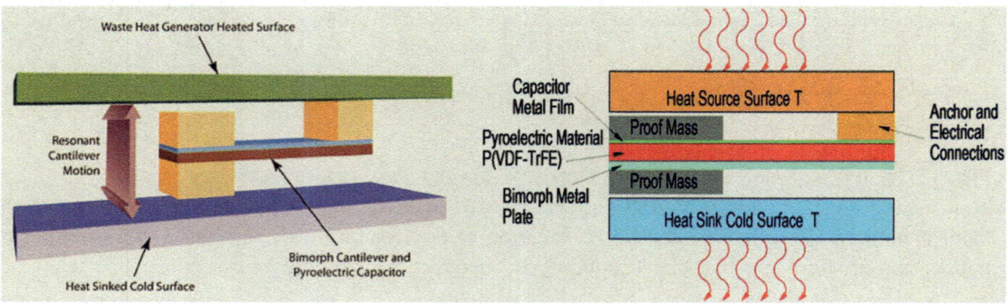

Fig. 11.20 Pyroelectric bimorph cantilever showing the cantilevered capacitor thin film layers. The energy harvester consisting of a bi-material cantilever which alternately contacts hot and cold surfaces and generates a current in the pyroelectric capacitor (Modified from Hunter et al. (2011) (SPIE), Development of MEMS based pyroelectric thermal energy harvesters, Proc. SPIE 8035, Energy Harvesting and Storage: Materials, Devices, and Applications II, 80350V (17 May 2011); doi: https://doi.org/10.1117/12.882125. Credit: US Oak Ridge National Laboratory)

energy-harvesting devices could be developed that operate for both thermoelectric and pyroelectric energy harvesting (Lingam et al. 2013).

Figure 11.20 shows MEMS-based cantilever harvesting system based on a thermally cycled pyroelectric capacitor that acts as a bimorph cantilever. The bimorph operates between two surfaces, one heated by waste heat and the other is a cold heat sink. Proof masses are placed at the cantilever tip to ensure good thermal contact to the hot and cold surfaces. When the cantilever is heated it deforms due to a thermal expansion mismatch between the bimorph layers that leads to it contacting the cold surface, making the structure cool and deform in the reverse direction and then making contact to the hot surface. This cyclic deformation leads to the cantilever alternately contacting the hot and cold surfaces at the resonant frequency of the cantilever to generate a pyroelectric current. The use of a MEMS approach means that large arrays of devices could be used to increase power and this interesting approach allows the device to potentially operate at high frequencies, up to 20 Hz or higher (Hunter et al. 2011; Bowen et al. 2014).

Another approach to increase operating frequency uses liquid-based switchable thermal interfaces to convert a spatial temperature gradient into temporal temperature oscillations; the system operates in an Olsen type cycle. As shown in Fig. 11.21, a plate with a pyroelectric material oscillates up and down between a high-

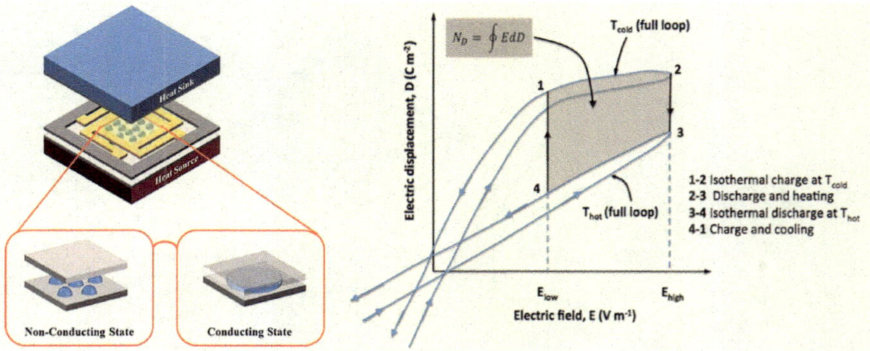

Fig. 11.21 Pyroelectric energy-harvesting module operating in an Olsen type cycle. Electrode assembly containing a pyroelectric material is actuated up and down and makes alternating thermal contact with the heat source (hot side) and sink (cold side) via switchable thermal interfaces (Modified with permission from Bowen et al. (2014) (Royal Society of Chemistry))

temperature source and a cold heat sink and repeatedly makes thermal contact to undergo temperature oscillations. In the thermally conducting state, the pyroelectric is pressed against the hot or cold surface using a linear actuator and liquid droplets at the interface deform to make them merge into a continuous thin liquid layer of low thermal resistance. In the non-thermally conducting state, the pyroelectric material is physically separated from the hot and cold surfaces and the liquid on the pyroelectric interface exists as discrete droplets. By creating a hydrophilic pattern on the surfaces the rupture distance was reduced, thus reducing the distance required and increasing the operating frequency (Bowen et al. 2014).

11.5 Magnetostrictive and Multiferroic Magnetoelectric Materials

Advances in integrated circuits, sensors, and actuators have led to decreased power consumption to a point where energy-harvesting (EH) systems can be used to supply sufficient power to these circuits and wireless devices. The multifunctional properties of multiferroics enable the design of electromagnetic energy-harvesting devices for various sensing, transduction, and memory applications. Muliferroic materials are characterized by two or more ferroic orders, such as ferroelectric, ferromagnetic, or ferroelastic, and the interactions between these order parameters, as shown in Fig. 11.22. In multiferroic magnetoelectric (ME) materials, coupling occurs between the magnetic and electric subsystems. This enables the control of dielectric polarization P by a magnetic field H (direct ME (DME) effect: $\Delta P = \alpha_H \Delta H$) and the manipulation of magnetization M by an electric field E (converse ME (CME) effect: $\mu_0 \Delta M = \alpha_E \Delta E$), with μ_0 denoting the vacuum permeability. The ME response is quantified in terms of the ME coupling coefficient (α_H or α_E), which represents the coupling efficiency between the electric and magnetic fields. This is considered as the figure of merit for the strength of ME coupling. Multimode energy-

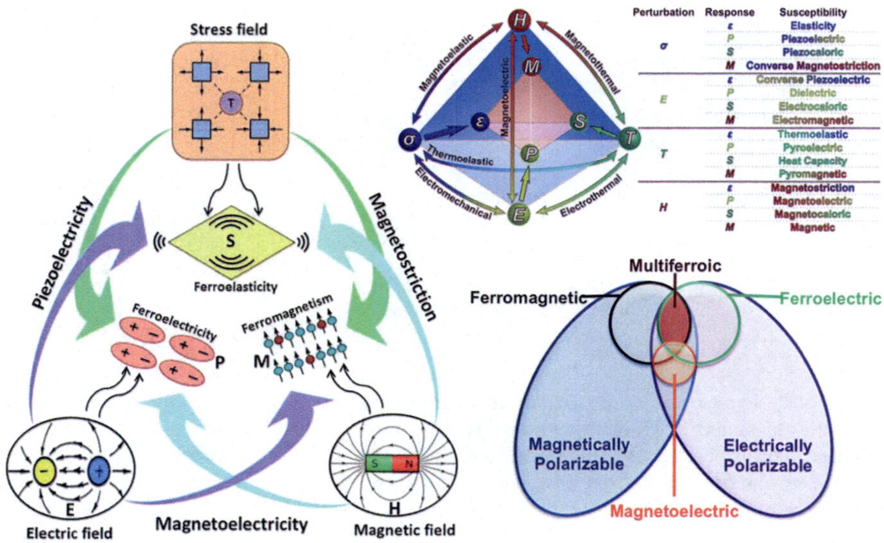

Fig. 11.22 Schematic illustrating magnetic-elastic-electric couplings in multiferroic materials (Modified with permission from Palneedi et al. (2016) (MDPI) under the Creative Commons Attribution (CC-BY) license. Here, *M* is magnetization, *S* is mechanical strain, and *P* is dielectric polarization

harvesting devices with various configurations have been developed to combine electromagnetic and piezoelectric mechanism. However, under small magnetic fields, the power generated from the electromagnetic energy harvester is limited. An alternative method to harvest magnetic energy is by utilizing magnetoelectric composite. By selecting high-performance piezoelectric and magnetostrictive materials and by optimizing the composite structure, it is expected that a high-efficiency magnetoelectric (ME) transducer with large voltage output under low magnetic field can be obtained. However, to obtain the maximum ME voltage output, an optimized magnitude of DC magnetic bias is needed which presents another challenge in implementation. Alternative approaches have been suggested for magnetoelectric energy harvester that rely on the combination of high-permeabililty magnetostrictive material laminated with piezoelectric plate. However, in these combinations, solenoid and/or permanent magnet is required, which dramatically increases the size and increases the electromagnetic noise. Therefore, different approach such as self-biased ME energy harvester consisting of Ni-MFC laminate and magnetostrictive Ni cantilever has been developed (Zhou et al. 2013, Palneedi et al. 2016).

11.5.1 Magnetostrictive Materials

Magnetostrictive materials refer to those exhibiting magnetostrictive effects that are found in ferromagnetic materials. Magnetostrictive effects consist of two mechanisms: the Joules effect and the Villari effect. The former implies that rotation

Table 11.5 General piezoelectric and magnetostrictive materials used as constituents of ME composites (Palneedi et al. 2016)

Piezoelectric phase	Magnetostrictive phase
Lead-based:	*Metals:*
$Pb(Zr,Ti)O_3$ (PZT)	Fe, Co, Ni
$Pb(Mg_{1/3}Nb_{2/3})O_3$-$PbTiO_3$ (PMN-PT)	*Alloys:*
$Pb(Zn_{1/3}Nb_{2/3})O_3$-$PbTiO_3$ (PZN-PT)	FeNi-based
$Pb(Mg_{1/3}Nb_{2/3})_y\ (Zr_xTi_{1-x})_{1-y}O_3$ (PMN–PZT)	FeCo-based
$Pb(In_{1/2}Nb_{1/2})O_3$-$Pb(Mg_{1/3}Nb_{2/3})O_3$-$PbTiO_3$ (PIN-PMN-PT)	CoNi-based
Lead-free:	Ni_2MnGa
$BaTiO_3$ (BTO)-based	Permendur (FeCoV)
$(K_{0.5}Na_{0.5})NbO_3$ (KNN)-based	Galfenol (FeGa), FeGaB
$Na_{0.5}Bi_{0.5}TiO_3$ (NBT)-based	Samfenol ($SmFe_2$)
Others:	Terfenol-D ($Tbi_{1-x}Dy_xFe_2$)
AlN	Fe-based metallic glasses (FeBSi, FeBSiC, FeCoB, FeCoSi, FeCoSiB, FeCuNbSiB)
ZnO	
(Sr, Ba)Nb_2O_5	*Ceramics:*
$Ba_{1-x}Sr_xTiO_3$ (BSTO)	Fe_3O_4
$Bi_{1-x}Sr_xTiO_3$ (BST)	$Zn_{0.1}Fe_{2.9}O_4$ (ZFO)
$La_3Ga_{5.5}SiO_{14}$ (LGS)	$La_xSr_yMnO_3$ (LSMO)
$La_3Ga_{5.5}Ta_{0.5}O_{14}$ (LGT)	$La_xCa_yMnO_3$ (LCMO)
Polyurethane (PU)	Ferrites or doped Ferrites (e.g., $NiFe_2O_4$ (NFO), $CoFe_2O_4$ (CFO), Li ferrite, Cu ferrite, Mn ferrite)
Polyvinylidene difluoride (PVDF)	

of moments to align with an applied field generates strains, while the latter implies that applied stresses cause magnetic moments to rotate, thus changing the magnetization for energy harvesting. Table 11.5 shows various piezoelectric and magnetostrictive materials commonly used for synthesizing ME composites. Enhanced piezoelectric and ferroelectric properties can be achieved in ceramics through (a) composition selection (ideally near morphotropic phase boundary (MPB) or polymorphic phase transition (PPT)) and modification (by doping); and (b) microstructure design (via domain engineering and texturing). Among piezoelectric materials, PZT-based ceramics have been widely employed to fabricate the ME composites due to their low cost, high piezoelectric response, and flexibility in modifying the composition to achieve desired properties for targeted applications. For the magnetic component in ME composites, Terfenol-D with high magnetostriction and Metglas (amorphous Fe-alloy) with high magnetic permeability have been the most used materials. Besides magnetic properties, factors such as the processing temperatures, electrical resistance of the material, magnitude of the bias field, phase connectivity have been considered in choosing the magnetostrictive materials (Palneedi et al. 2016; Narita and Fox 2018).

INDEPTH: Electromagnetic Energy Harvesters

The principle used in the electromagnetic transduction is the generation of current in a conductor in a magnetic field. The conductor typically takes form of a coil, and electricity is generated by the relative motion of the conductor and a permanent magnet. The amount of electricity generated depends on the strength of the magnet, the velocity of the movement, and the number of coil windings. The requirement of a large number of windings with low resistance gives a huge limitation for making a MEMS energy harvester with electromagnetic transduction. The planar processes used in silicon microtechnology limit the amount of windings one can achieve, unless a complicated stacking configuration is developed. The wires can be made using thin film deposition techniques. However this will lead to a very high resistance, since it will be difficult and costly to deposit a metal film of more than 1 μm in thickness. Another major problem is that the magnetic field might interfere with the device one wish to power. Electromagnetic energy harvesters are often characterized by high current, low voltage, and low optimal resistive load (Xu 2012).

11.5.2 Magnetoelectric Materials

The magnetoelectric effect (ME) denotes any coupling between the magnetic and the electric properties of a material. For example, a dielectric material moving through an electric field would become magnetized. A material where such a coupling is intrinsically present is called a magnetoelectric. Single phase magnetoelectrics are chemically homogenous and isotropic compounds. They exhibit intrinsic ME coupling but require the coexistence of magnetic moments and electric dipoles with long-range ordering. From a fundamental point of view, the coupling between the magnetic and polar sublattices in single phase ME compounds is fascinating. However, due to the mutual exclusion of ferromagnetism and ferroelectricity, only few monolithic ME materials such as Cr_2O_3 and $BiFeO_3$ exhibiting non-zero coupling at room temperature have been found so far. Most of the single phase materials possess either low permittivity or low permeability at room temperature and thus exhibit weak ME coupling which hinders their applications. For example, $BiFeO_3$, the archetype of single phase ME compounds, displays good ferroelectricity but weak ferromagnetic properties above room temperature (Palneedi et al. 2016; Narita and Fox 2018).

ME composites consist of physically separated magnetic and electric order phases. These composites show coupling with orders of magnitude larger than those found in single phase materials at room temperature. ME coupling in

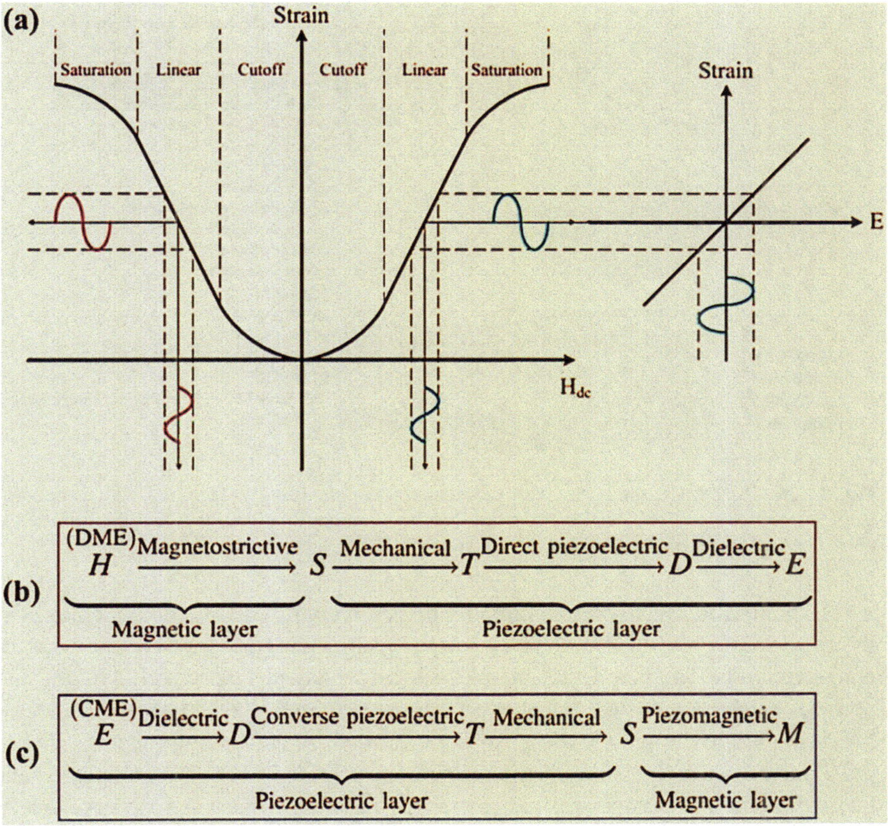

Fig. 11.23 Schematic representation of the (**a**) ME effect utilizing the product property; (**b**) DME effect and (**c**) CME effect in composites. Here, *H* is magnetic field, *S* is mechanical strain, *T* is mechanical stress, *D* is electric displacement, *E* is electric field, and *M* is magnetization (Modified with permission from Palneedi et al. (2016) (MDPI) under the Creative Commons Attribution (CC-BY) license)

composites occurs extrinsically in three different ways mediated through (i) strain, (ii) charge carrier, and (iii) spin exchange. Among these mechanisms, the strain-mediated ME coupling has been widely studied, while the investigations on the other two mechanisms are still in early stages. The strain-mediated ME effect in composites is a product tensor property and results from the elastic coupling between the piezoelectric and magnetostrictive components, as illustrated in Fig. 11.23. In DME coupling, the applied magnetic field generates strain in the magnetic layer via the magnetostriction effect, and this strain is transferred to the piezoelectric layer resulting in an electric displacement or a dielectric polarization through the piezoelectric effect. In CME coupling, an external electric field induces strain in the ferroelectric layer due to the inverse piezoelectric effect, and the strain transferred to the magnetic layer produces a magnetization change or domain reorientation by the piezomagnetic effect (Palneedi et al. 2016).

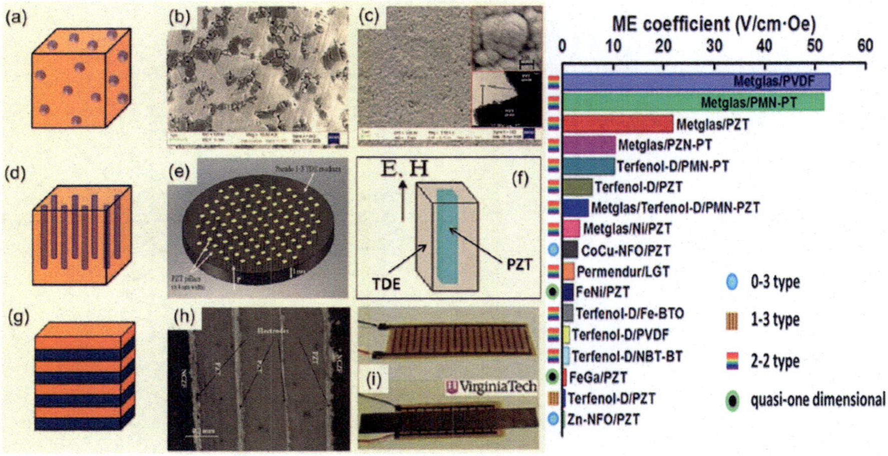

Fig. 11.24 Bulk ME composites with different phase connectivity (Modified with permission from Palneedi et al. (2016) (MDPI) under the Creative Commons Attribution (CC-BY) license): (**a–c**) 0–3 connectivity; (**d–f**) 1–3 connectivity; and (**g–i**) 2–2 connectivity

ME composites have been commonly prepared with 0-3, 1-3, and 2-2 connectivity, as shown in Fig. 11.24a, d, and g, respectively. Here, the numbers represent the connectivity of the magnetic and piezoelectric phases, respectively. In the 0-3 particle-matrix composites, magnetic particles are embedded in the piezoelectric matrix. 1-3 cylinder-matrix composites are formed by embedding magnetic fibers/rods/tubes/wires in the piezoelectric matrix. The particles (in the 0-3 composite) and fibers (in the 1-3 composite) can be either randomly dispersed or periodically aligned. A 2-2 laminate composite consists of alternating magnetic and piezoelectric layers. Such laminates can be prepared in different shapes and geometries, including disks, squares, rectangles, and rings, with different dimensions. They can be arranged as unimorphs and bimorphs as well as bilayered and multilayered structures. Further, for all of these composites, the volume fraction and dimensions of the constituents can be altered to tailor the properties of the composite. The 2-2 composites preserve the physical characteristics of individual phases, and they are comparatively simpler to fabricate. These composites can be poled to a higher degree since the piezoelectric and the low-resistivity magnetic phases are separated. The 2-2 layered composites exhibit higher ME responses compared to the 0-3 and 1-3 composites. There have been other composite structures, such as 3-2 structured composite consisting of a $(Ni_{0.6}Cu_{0.2}Zn_{0.2})Fe_2O_4$ [NCZF] phase with 2D connectivity dispersed in a $0.8Pb(Zr_{0.52}Ti_{0.48})O_3$–$0.2Pb(Zn_{1/3}Nb_{2/3})O_3$ [PZNT] matrix; quasi-one-dimensional ME composite by inserting a magnetostrictive wire (FeNi/FeGa/FeCoV) into a PZT tube where the tube-wire interface bonding was made with silver paste. In comparison, 2-2 composite structure has inherent advantages in terms of fabrication and performance (Gillette et al. 2014; Palneedi et al. 2016).

As device design trends towards miniaturization and multifunctionality, thin film conformations are highly desired for application in integrated magnetic/electric

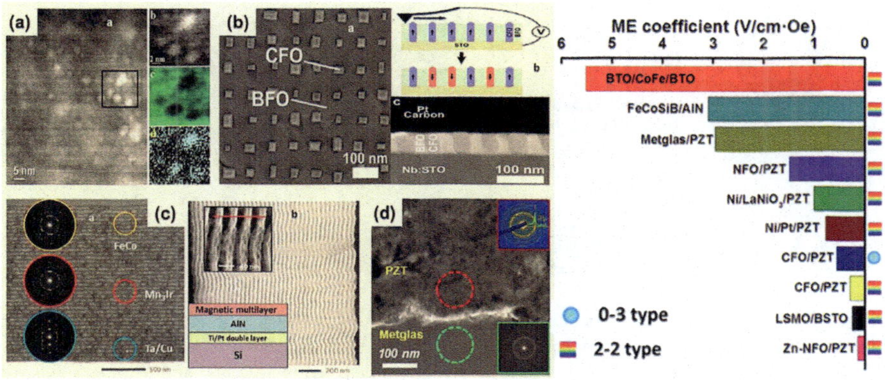

Fig. 11.25 Film-based ME composites with (**a**) 0–3 connectivity, (**b**) 1–3 connectivity, (**c**, **d**) 2–2 connectivity (Modified with permission from Palneedi et al. (2016) (MDPI))

devices. In film-based ME composites, optimized interface coupling can be achieved through the direct bonding of components. Moreover, large electric fields can be comfortably applied to thin films since they require relatively smaller bias voltages than those needed for bulk samples. This extends the scope of thin films for devices which have limitations on their operating voltages. The greater freedom and flexibility in the fabrication of film-based ME composites render their property tuning, through interfacial coupling and strain engineering, control of crystal structure and orientation, grain size, and layer thickness, as well as chemical modification with a wide range of substituents, etc. Composite films can also facilitate the understanding of the physical phenomena involved in ME effects at lower dimensions, and thus enable the design of new types of magnetoelectrics with novel phase structures. Figure 11.25 shows some examples of ME composite films using various film deposition methods. Some of these methods can yield excellent thin film epitaxial growth with atomic scale thickness control and coherent interfaces. However, most ME composite films show very low ME properties due to reduced electromechanical parameters as a consequence of substrate clamping (Yan and Priya 2015; Palneedi et al. 2016).

More ME composites have been developed, such as broadband ME with piezoelectric anisotropy, ME composites with textured piezoelectric ceramics, self-biased ME composites, and ME nanocomposites with core/shell arrangement in the form of nanoparticles, nanowire arrays, and nanotubes. Other 1D composite structures with random and Janus-type arrangements have also been prepared with different wet chemical synthesis methods, as shown in Fig. 11.26. For example, $CoFe_2O_4/BaTiO_3$ core/shell nanoparticles (Fig. 11.26a) was fabricated using a combination of solution processing and high-temperature calcination. Ferrite/perovskite oxide core/shell nanostructures can be formed in multiferroic systems such as $Fe_3O_4/PbTiO_3$, γ-$Fe_2O_3/PbTiO_3$, γ-$Fe_2O_3/Pb(Zr,Ti)O_3$, $CoFe_2O_4/BaTiO_3$, $CoFe_2O_4/PbTiO_3$, and $CoFe_2O_4/Pb(Zr,Ti)O_3$ using a combined hydrothermal and annealing process. Ordered arrays of $NiFe_2O_4$/PZT core/shell nanowires (Fig. 11.26b) was synthesized with a method involving the combination of a modified sol-gel process,

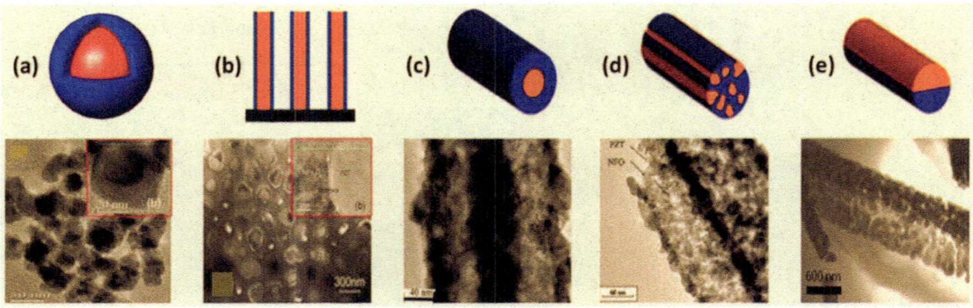

Fig. 11.26 Biphasic ME composites with different types of arrangements between the two phases (**a–c**) core/shell arrangement (Adapted from Palneedi et al. (2016) (MDPI) under the Creative Commons Attribution (CC-BY) license): (**a**) nanoparticles; (**b**) nanowires; and (**c**) nanotubes; (**d**) random arrangement; and (**e**) Janus-type arrangement

Table 11.6 Classification of different ME devices (Palneedi et al. 2016)

ME coupling	Physical mechanism	ME devices
Direct ME coupling	H control of electric polarization	Magnetic sensors, current sensors, transformers, gyrators, energy harvesters
Converse ME coupling	E control of magnetization switching	Spintronics, including random access memories, tunnel junctions
	E control of permeability μ	Voltage tunable inductors, tunable band-pass filters, phase shifters
	E control of spin wave	Voltage tunable filters, tunable resonators, phase shifters

electrochemical deposition, and subsequent oxidization in anodized nanoporous alumina membranes. In addition, multiferroic nanofibers with core/shell, random, and Janus-type arrangements (Fig. 11.26c–e) have been fabricated by electrospinning. Nevertheless, some of the challenges in the practical implementation of these ME nanocomposites include forming isolated multiferroic particles that are free of agglomerates, substrate-free assembling of nanofibers into ordered structures, and difficulties in accessing their ME properties. Further, in some cases, these nanostructured ME composites will have to be consolidated into a dense form for device applications which may affect the stability of nanostructures (Andrew et al. 2014, Palneedi et al. 2016).

11.5.3 Magnetoelectric Devices and Applications

Based on the type of ME coupling and the mechanisms used to control the order parameters, a variety of applications have been proposed, including magnetic sensors, high-frequency inductors, memory devices, and high-frequency signal processing devices, as shown in Table 11.6 (Palneedi et al. 2016).

ME material-based sensor devices are considered to be promising alternatives for conventional Hall sensors and giant magnetoresistive (GMR) devices. Because of their passive nature and self-powered operation at room temperature, ME sensors

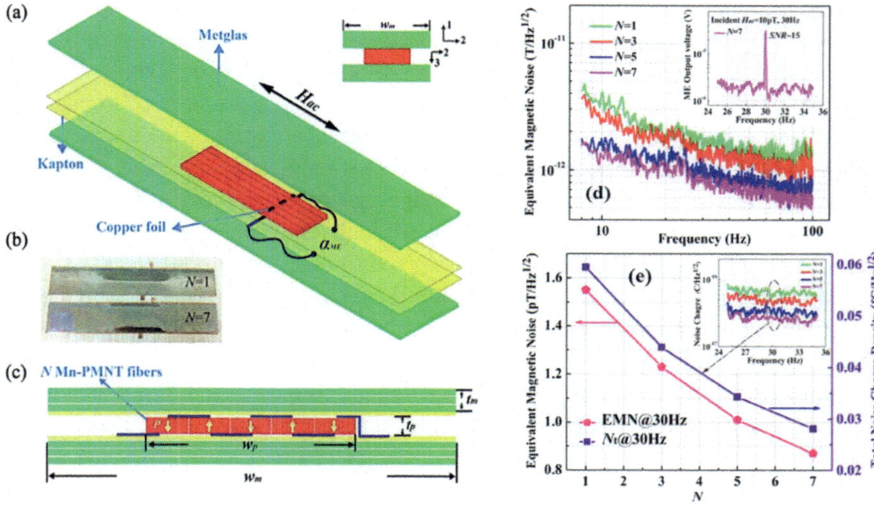

Fig. 11.27 Metglas/Mn-PMNT composite sensor (Modified with permission from Fang et al. (2015) (IOP Publishing)): (**a–c**) 3D structure, photograph, cross-sectional schematic diagram, respectively; (**d**, **e**) equivalent magnetic noise level and total noise charge density of the ME magnetic sensors

may be able to replace bulky and expensive superconducting quantum interference devices (SQUIDs). ME sensors have the potential to perform biomagnetic measurements analogous to all of those performed by magnetic fields (MEG) and haemodynamics (fMRI). The key requirements for magnetic field sensors in this application are (i) sensitivity of ~pT to fT per $\sqrt{H_z}$ at low frequencies (10^{-2} to 10^3 Hz); and (ii) ambient temperature and wide bandwidth (0.1 to 100 Hz) operation. The direct ME coupling effect, where the ME voltage coefficient is dependent on AC and DC magnetic fields, would allow the ME composites to sense either an AC or DC magnetic field by monitoring the output electrical signals. Since a current passing through a wire generates a magnetic field in the surrounding space, the ME composite can also be used as a current probe for detecting current by monitoring the corresponding magnetic flux. Various ME laminate-based sensors with multi-push-pull configurations, multilayer configuration, and bimorphs, all of which exhibit an improved ME voltage coefficient, have demonstrated considerable potential for sensing low-frequency magnetic field variations. Figure 11.27 shows Metglas/Mn-PMNT laminate composite consisting of longitudinal magnetized Metglas layers and different numbers (N) of transversely polarized Mn-PMNT fibers connected in series. An ultralow magnetic field sensitivity of 0.87 pT/$\sqrt{H_z}$ at room temperature was reported for these Metglas/Mn-PMNT fiber laminate composites. In comparison, thin film-based ME sensors enable the fabrication of miniaturized low-cost sensor devices with high sensitivity and high spectral resolution. Thin film-based architecture also provides the capability for designing sensor arrays that can be integrated with other circuit components (Stephan et al. 2012; Fang et al. 2015).

Harvesting energy from ambient energy sources such as vibrations, sound, radio-frequency waves, light, temperature gradients, wind, and others is an area of focus for current and next-generation remote monitoring electronic devices and self-powered wireless sensor networks with the goal of improving device lifetime and addressing the limitations of conventional batteries. In addition, the ambient environment is filled with magnetic noise of 50–60 Hz almost everywhere these days. Harvesting this weak and low-frequency magnetic noise (<1 mT $= 10$ G) to develop a consistent electricity source remains a difficult challenge. Magneto-mechano-electric (MME) mechanism has been attempted to obtain optimum electricity from the tiny magnetic fields in the surroundings. The operation mechanism can be described as follows: When the ME composite is placed in an AC magnetic field, the magnetostrictive layer in the composite responds to the mechanical vibration (magneto-mechano coupling), thereby straining the piezoelectric layer, which results in an output voltage across the electrical load through the direct piezoelectric effect (mechano-electric coupling). Due to the existence of the piezoelectric phase in the ME composite, any mechanical oscillation applied to the composite directly creates electrical voltage. Consequently, the MME generator could be used to harvest energy from both the magnetic field and external vibrations at the same time. This sequential operating process is schematically depicted in Fig. 11.28a. By selecting high-performance piezoelectric and magnetostrictive materials and by optimizing the composite structure, it is expected that high electric power density can be obtained from a low-frequency magnetic field using the MME generator.

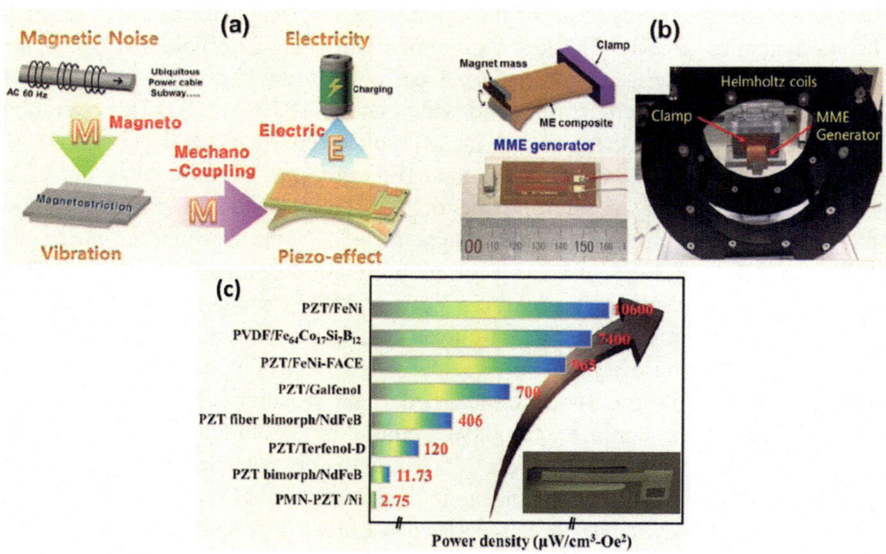

Fig. 11.28 Magneto-Mechano-Electric (MME) energy harvesters (Modified with permission from Ryu et al. (2015) (Royal Society of Chemistry), and Palneedi et al. (2016) (MDPI)): (**a**) Schematic depicting the working principle; (**b**) schematic and photo of cantilever structured MME energy harvester; (**c**) power densities from the MME harvesters made with different composite systems

Figure 11.28b shows a ME generator constructed using an anisotropic and flexible piezoelectric $Pb(Mg_{1/3}Nb_{2/3})O_3$–$PbTiO_3$ (PMN–PT) single crystal fiber composite (SFC), a cost-effective magnetostrictive Ni plate and Nd permanent magnetic proof mass. The flexibility of the SFC ensures the high compliance of the sample, which is ideal for achieving low resonance frequency in a cantilever structure. The flexibility also increases device durability and enables the application of increased strain magnitudes. The Ni plate can be easily self-biased and generates a linear strain response in a low-level magnetic field environment. The performance of the MME generator containing an anisotropic <011> SFC with d_{32} mode under a small noise level magnetic field is shown in Fig. 11.28b. At 60 Hz, and Hac $\sim$ 500 μT, the maximum generated voltage was $\sim$34 Vpp ($\sim$12.4Vrms). The power from the MME generator was high enough to fully charge a 220-μF electrolytic capacitor after rectifying for 3 min. Using the stored power in the charged capacitor, the device was able to turn on 35 commercial high-intensity LEDs with a turn on/off frequency of $\sim$1 Hz. Similarly, efforts have also been made to develop energy harvesters based on ME composite films. Seeking to improve the output power density, various MME harvesters have been developed using different combinations of magnetostrictive and piezoelectric materials, which represent significant advances towards next-generation remote monitoring electronic devices and self-powered wireless sensor networks (Ryu et al. 2015; Palneedi et al. 2016).

Motivated by the advances in multiferroics, ME composites have been suggested for biomedical applications such as wireless endoscopy, minimally invasive surgical tools, and stimulation of functions of living cells. The potential use of ME nanoparticles (MENs) as carriers for on-demand drug release and to artificially stimulate the neural activity deep in the brain has also been suggested. Wireless capsule endoscopes (WCEs) are often used to examine the gastrointestinal (GI) tract for clinical diagnosis (Fig. 11.29a). Though the WCEs are far less invasive compared to conventional endoscopes, the passive nature of WCEs makes it difficult to control their position and orientation as the capsule moves along the GI tract (Paluszek et al. 2015).

Controlling the function of biological macromolecules is of vital importance in health science studies. Approaches used to stimulate cell functions include the use of the heat generated by hysteresis losses in magnetic nanoparticles placed in a high-frequency magnetic field, and the mechanical agitation of magnetic nanoparticles attached to cells using external low-frequency magnetic fields. The interaction between electromagnetic fields and biological macromolecules can be understood by studying the ion channels which regulate several cellular processes, such as action potentials in neurons or muscle contraction. Figure 11.29b shows an innovative approach, based on the use of core/shell-structured MENs with a ferromagnetic core and a ferroelectric shell, that would allow the remote control of ion channel gating via externally applied magnetic fields. In this approach, electric fields in the vicinity of the cells generated by MENs introduced extra- or intracellularly can be locally modified to invoke appropriate conformational changes in the ion channels. The resulting local depolarization or hyperpolarization of the membrane will lead to opening or closing of the ion channels accordingly. Because of the remote way in

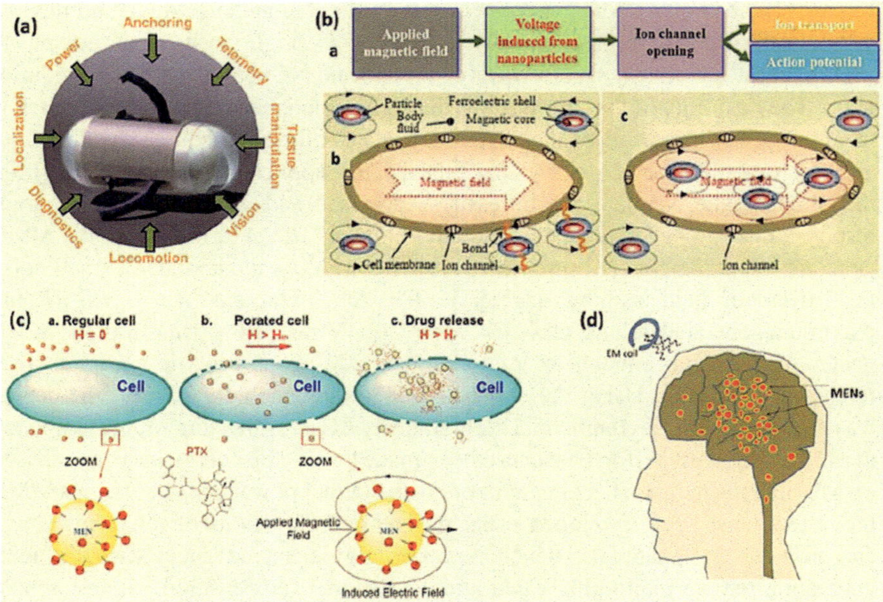

Fig. 11.29 ME composites for biomedical applications (Adapted with permission from Palneedi et al. (2016) (MDPI)): (**a**) Key functionalities that complement the wireless capsule endoscopes; (**b**) illustration of possible mechanisms of stimulation of ion channels: a chain of actions triggered by the applied magnetic field pulses, external and internal stimulation by uptaken nanoparticles; (**c**) MENs as field-controlled nano-electroporation sites to let the drug through the cancer cell membranes; (**d**) illustration of the deep brain stimulation

which the stimulation will be performed, individual cells or selected groups of cells can be targeted, rather than whole tissues (Kargol et al. 2012).

Targeted drug delivery with adequate high specificity (to tumor cells) remains a formidable task in the treatment of cancer in general, particularly ovarian cancer. Although the survival rates have been improved by intraperitoneal (IP) delivery through a surgically implanted catheter, toxicity and catheter complications have precluded widespread adoption of this invasive means of delivery. By exploiting the dependence of the membrane's porosity on the electric field, electroporation can be utilized to trigger drug delivery into the cells, as shown in Fig. 11.29c. Above a threshold magnetic field (H_{th}), the MENs loaded with the drug and optionally with the biomarker-specific antibodies (for delivery to the tumor cells) can generate localized fields large enough to open up the membrane pores in their proximity and thus allow the delivery of the drug inside the tumor cells. The drug can be released off the MENs by further increasing the field above the second critical value, Hr, necessary for overcoming the drug-MEN binding energy. This hypothesis was testified through in vitro studies on human ovarian carcinoma cell (SKOV-3) and healthy cell (HOMEC) lines, where a 30-Oe DC bias was applied to trigger high specificity uptake of Paclitaxel (PTX) loaded on 30-nm $CoFe_2O_4/BaTiO_3$ core/shell

MENs. The drug penetrated through the membrane and completely eradicated the tumor within 24 h without affecting the normal cells (Guduru et al. 2013).

In the human neural network, chemical and electrical synapses transfer information between adjacent axons and dendrites directly or indirectly through electric field energy. The ability to efficiently control the network at micro- or even nano-scale can enable significant control over important brain functions. Existing noninvasive brain stimulation methods including repetitive transcranial magnetic stimulation (rTMS) and transcranial direct current stimulation (tDCS) are limited in their depth and locality. A potential solution would be to use MENs for noninvasive control of the neural network. In this approach, very low-intensity external magnetic field is required to stimulate brain activity at any depth in the brain and the field can be focused to act upon MENs in any particular region of the brain. The external magnetic field generates AC signals in ME nanoparticles that are correlated with the frequency spectrum of the neural activity, which in turn causes neurons in that region to fire at similar frequencies, as shown in Fig. 11.29d (Yue et al. 2012).

11.6 Triboelectric Materials

The triboelectric effect, also known as triboelectric charging, is a type of contact electrification in which certain materials become electrically charged after they come into frictional contact with a different material. The polarity and strength of the charges produced differ according to the materials, surface roughness, temperature, strain, and other properties. Triboelectric nanogenerators (TENGs), based on the triboelectric effect, provide an alternative approach to generating electricity from mechanical energy to operate small electronic devices. In the triboelectric effect, a material surface becomes electrically charged after it comes into contact with a different material through friction, owing to charge transfer between the two materials. These transferred charges remain for a long time on their respective surfaces. An electrostatically charged material causes a potential, and it drives induced electrons to flow between the electrodes by periodic contact and separation of the two materials. The generated electric potential V can be calculated with the following equation (Lee et al. 2016):

$$V = -\frac{\rho d}{\varepsilon_0} \tag{11.14}$$

where ρ is the triboelectric charge density, ε_0 is the vacuum permittivity, and *d* is the interlayer distance in a given state. The current *I* generated across an external load can be defined as (Lee et al. 2016):

$$I = C\frac{\partial V}{\partial t} + V\frac{\partial C}{\partial t} \tag{11.15}$$

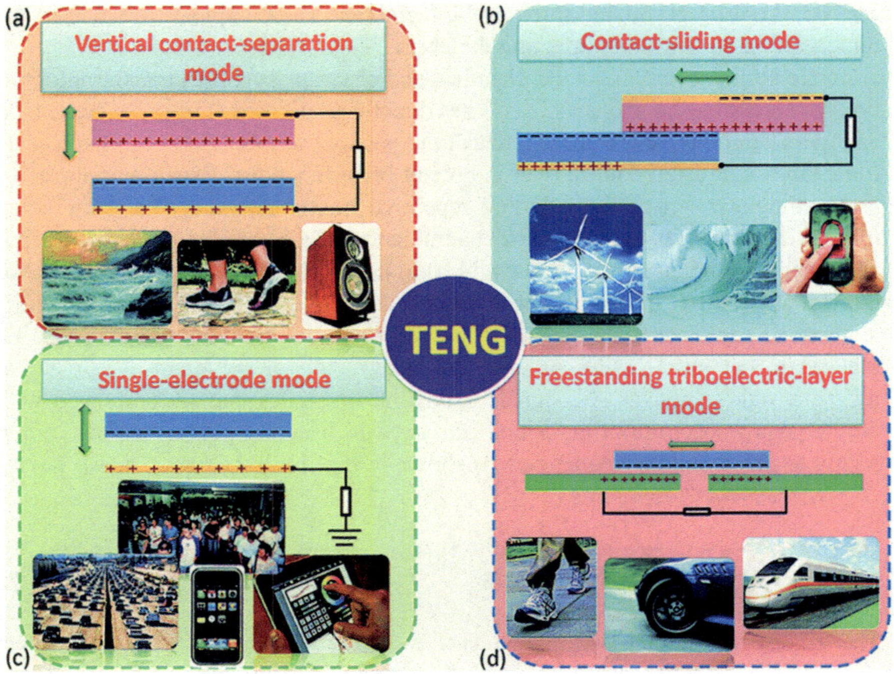

Fig. 11.30 The four fundamental working modes of the triboelectric nanogenerators (Adapted with permission from Wang et al. (2015b) (Royal Society of Chemistry)): (**a**) The vertical contact–separation mode; (**b**) the lateral sliding mode; (**c**) the single-electrode mode; (**d**) the free-standing mode

where C denotes the capacitance of the system and V is the voltage across the two electrodes. The first term is the change in potential between the top and bottom electrodes due to the triboelectric charges. The second term is the variation in the capacitance of the system when the distance between two electrodes is changed due to the mechanical deformation. Numerous advantages of TENGs include superior power output performance, many material options, easy tailoring of device structures, cost-effectiveness, the facile fabrication of large areas for applications, and stability and robustness, as well as environmental friendliness. Depending on the configuration of the electrodes and the different ways in which the triboelectric layers can be arranged, as shown in Fig. 11.30, four operation modes of TENGs have been developed (Wang et al. 2015a, b): the vertical contact mode, the lateral sliding mode, the single electrode mode, and the free-standing mode. Their applications as various self-powered nanosystems, such as acceleration sensors, motion vector sensors, biomedical monitoring systems, electrochromic devices, sound recording systems, pressure sensors, angle measurement sensors, active tactile sensor systems, tactile imaging devices, electroluminescent systems, and mercury-ion detection systems, have been demonstrated. The triboelectric output can be further enhanced through control of electron affinity, as well as the work function, chemical structure, pressure, and surface roughness of the materials (Lee et al. 2016).

11.6.1 Triboelectric Nanogenerators as a Sustainable Power Source

The triboelectric nanogenerator offers a completely innovative approach for energy harvesting from the vast environment due to its high power density, light weight, small size, and so on. However, several output performances of the TENG should be improved such as output power, current, and energy conversion efficiency. Hence, many principles and mechanisms have been demonstrated to realize the enhancement of TENGs as a sustainable power source for electronics (Lin et al. 2016).

Enabled by an advanced structural design of two radial-arrayed fine electrodes that are complementary on the same plane, the planar-structured TENG generates periodically changing triboelectric potential that induces alternating currents between electrodes. The TENG has a multilayered structure, which consists of mainly two parts, i.e., a rotator and a stator, as sketched in Fig. 11.31a. A photograph of an as-fabricated device is demonstrated in Fig. 11.31b. For quantitative characterization, a programmable rotary motor was connected to the rotator that was in co-axial alignment with the stator. At a rotating rate of 500 r min^{-1}, the short circuit current (I_{SC}) has a continuous AC output with an average amplitude of 0.5 mA and a frequency of 500 Hz (Fig. 11.31c). For open-circuit voltage (VOC), it oscillates at the same frequency with a peak-to-peak value of 870 V (Fig. 11.31d). Furthermore, the matched load also has a variable value, exhibiting a reversely proportional relationship with the rotation rate, as shown in Fig. 11.31e. Consequently, linearly rising output power can be obtained at higher rotation rates (Zhu et al. 2014).

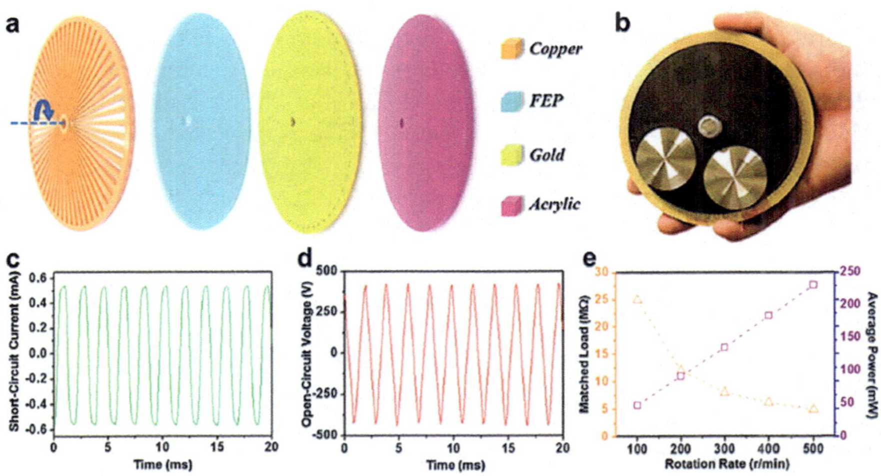

Fig. 11.31 Radial arrayed triboelectric generator for high-performance mechanical energy harvesting (Modified with permission from Zhu et al. (2014) (Springer Nature): (**a**) Schematic illustrations of the triboelectric generator, which has two parts, that is, a rotator and a stator; (**b**) a photograph of an as-fabricated triboelectric generator. The short-circuit current; (**c**) and open-circuit voltage (**d**) of the triboelectric generator under a rotation rate of 500 r/min; (**e**) matched load resistance and average output power with increasing rotation rate

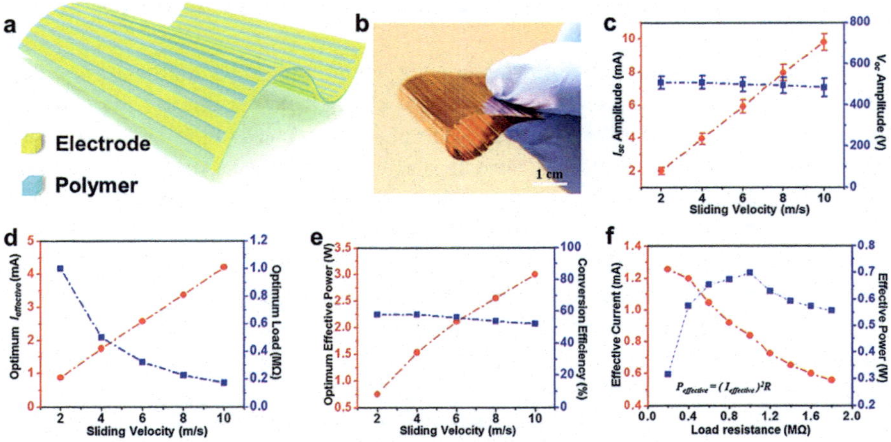

Fig. 11.32 A thin-film-based micro-grating triboelectric nanogenerator (MG-TENG) for high-performance mechanical energy harvesting (Adapted with permission from Wang et al. (2015b) (Royal Society of Chemistry)): schematic illustration (**a**) and a photograph (**b**) of the MG-TENG. (**c**) The sliding velocity dependent current and voltage output of the MG-TENG. (**d**) The optimum effective current and corresponding load with varying sliding velocity. (**e**) The optimum effective power of the MG-TENG with varying sliding velocity. (**f**) Load matching test at a sliding velocity of 2 m/s. Maximum effective power is obtained at the matched load of 1 MΩ

Furthermore, a thin-film-based micro-grating triboelectric nanogenerator (MG-TENG) is also developed for high-efficiency power generation through conversion of mechanical energy. The shape-adaptive MG-TENG relies on sliding electrification between complementary micro-sized arrays of linear grating, which offers a unique and straightforward solution in harnessing energy from relative sliding motion between surfaces. Figure 11.32a is a schematic illustration of the MG-TENG, while Fig. 11.32b is a photograph of the as-fabricated device, which is extremely thin and flexible. And the sliding velocity-dependent electric output of the MG-TENG is characterized. As shown in Fig. 11.32c, a nearly linear relationship between the amplitude of short-circuit current (I_{SC}) and the sliding velocity can be obtained. While the open-circuit voltage (V_{OC}) is independent of the sliding velocity. In the meantime, as shown in Fig. 11.32d, the corresponding matched load is approximately reversely proportional to the sliding velocity. Thus, the optimum effective current is also linearly related to the sliding velocity, as demonstrated in Fig. 11.32e. To evaluate the capability of the MG-TENG for power generation, resistors were utilized as external loads for characterization. As displayed in Fig. 11.32f, the current amplitude drops with increasing load resistance owing to the ohmic loss, and the effective power is maximized to a value of 0.76 W at a load resistance of 1 MΩ and a sliding velocity of 2 m/s. And operating at a sliding velocity of 10 m/s, an MG-TENG of 60 cm^2 in overall area can deliver an average output power of 3 W and a power density of 50mWcm^2 at an overall conversion efficiency of 50%. Holding a collection of compelling features, including high electric output power, extremely light-weight, low cost, scalability, and adaptability,

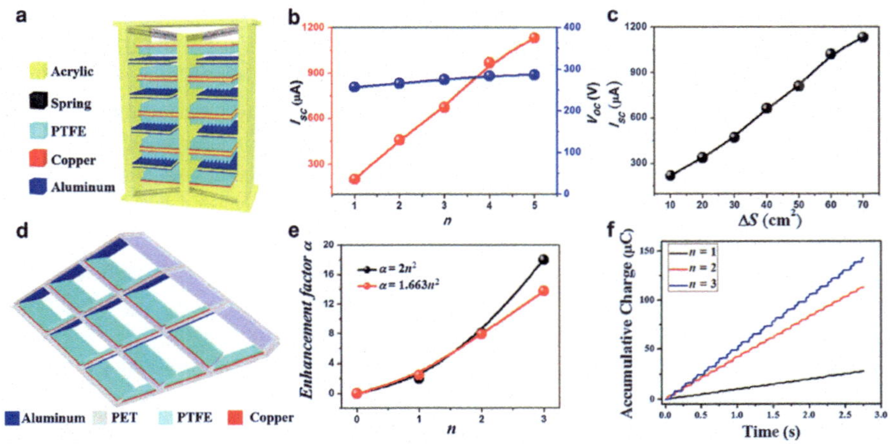

Fig. 11.33 Enhancement of the output current of the TENG based on multilayer integrations (Adapted with permission from Wang et al. (2015b) (Royal Society of Chemistry)): (**a**) schematic illustration of the multilayered stacked TENG. (**b**) Dependence of electric output on the number of pinned fingers *n*. (**c**) Dependence of the short-circuit current on the effective contact area (DS) of the TENG with $n = 5$. (**d**) Sketch of an integrated rhombic gridding-based triboelectric nanogenerator. (**e**) The current's enhancement factor a is increasing as a function of number of unit cells along the edge length *n*. (**f**) Accumulative inductive charges generated by the TENG with $n = 1$, 2, and 3, respectively

the MG-TENG is another practically effective approach in harvesting ambient mechanical motions as well as possibly producing electricity at a large scale (Wang et al. 2015a, b).

The application of TENG for mechanical energy harvesting may be challenged by its low output current, and a possible solution is to synchronize the outputs of all multiple units so that the instantaneous output power can be maximized. In this regard, a multilayered stacked TENG was developed as a cost-effective, simple, and robust approach for harvesting ambient mechanical energy. The 3D-TENG has a multilayered structure with acrylic as supporting substrates, as schematically shown in Fig. 11.33a. Acrylic was selected as the structural material due to its decent strength, light weight, good machinability, and low cost. And the total number of the units in a 3D-TENG can be expressed as: $N = 4n$, where n is the number of pinned fingers of a TENG. Eight identical springs were employed to bridge the moveable and pinned fingers. As shown in Fig. 11.33b, the voltage output is almost constant for 3D-TENGs with $n = 1–5$, which is attributed to the electrically parallel connection among all of the units. However, the current output is a monotonically increasing function of n. Such a dramatic current enhancement is mainly owing to the operating synchronicity of all units. In the meanwhile, a monotonically increasing relationship was also observed between the current output and the effective contact area ΔS, as shown in Fig. 11.33c. To enhance the output current for the TENG-based mechanical energy harvesting, another rationally designed device with integrated rhombic gridding can also greatly improve the total current output owing to the multiple unit cells connected in parallel. The structure of integrated rhombic gridding-based TENG is shown in Fig. 11.33d, in which, the total number of unit cells in one TENG can be expressed as: $N_{total} = 2n^2$, where n is the number of unit cells along the edge length. Each polyethylene terephthalate (PET) sheet with a thickness of 600 mm is cut half through and then locked into each other to form the framework of TENG. In each unit cell, an aluminum thin film with nanoporous modification plays dual roles as a contact electrode and a contact surface and PTFE with back-coated copper plays as another contact surface. As indicated in Fig. 11.33e, the current enhancement factor α is a function of the number of unit cells along the edge length, $\alpha = bn^2$. The fitting results render the coefficient b, a value of 1.66. Considering the nonideal experimental factors, such as humidity, particle contaminants in air, and other experimental imperfection, the observed results of enhancement factor are considerably approaching to the ideal value of $2n^2$, revealing the effectiveness of the integrated rhombic gridding structure for current enhancement. As illustrated in Fig. 11.33f, the accumulative induced charges also increase with n, which reach up to 142.68 μC within 2.75 s when $n = 3$, further indicating that the integrated rhombic gridding structure can dramatically enhance the electric output of the TENG. In addition, other TENGs have also been designed to enhance the power outputs, such as stacked triboelectric nanogenerator, self-powered flexible printed circuit (FPC) board with an integrated zigzag-shaped TENG, three-dimensional multilayered sliding TENG, as well as cylindrical TENG by segmentation design and multilayer integration (Wang et al. 2015a, b).

11.6.2 Triboelectric Nanogenerators as Self-Powered Active Sensors

The output performance of a TENG is greatly affected by magnitude/frequency of the external mechanical stimuli, and the pressure applied onto the device is among the most critical parameters. Hence, the most straightforward application for TENG-based active sensors would be the quantification of external pressure/touch. Figure 11.34a shows a self-powered tactile sensor based on flexible thin film materials, in which a layer of polyethylene terephthalate (PET) deposited with ITO electrodes on both sides formed the structural backbone of the triboelectric sensor (TES). Fluorinated ethylene propylene (FEP) thin film was applied on top as a triboelectric layer for contact electrification with a foreign object. Vertically aligned polymer nanowires (PNWs) were created on the FEP layer (Fig. 11.34b) to assist the generation of triboelectric charges. To evaluate the performance of the TES to a contact event, a square-shaped TES (Fig. 11.34c) was utilized to detect a piece of metal by cyclic contact and separation. As presented in Fig. 11.34d, at an applied pressure of 0.03 kPa, the TES produced a uniform output voltage with a maximum magnitude of 35 V. As the contact pressure increased, the output voltage was raised to 50 V when the contact pressure approached 10 kPa (Fig. 11.34e). This increasing behavior was attributed to the increase of effective contact area between the metal object and the TES. To demonstrate the practical application of the tactile sensor, a complete wireless sensing system was developed through integrating the TES with a signal-processing circuit. The system relied on the output voltage from the TES to

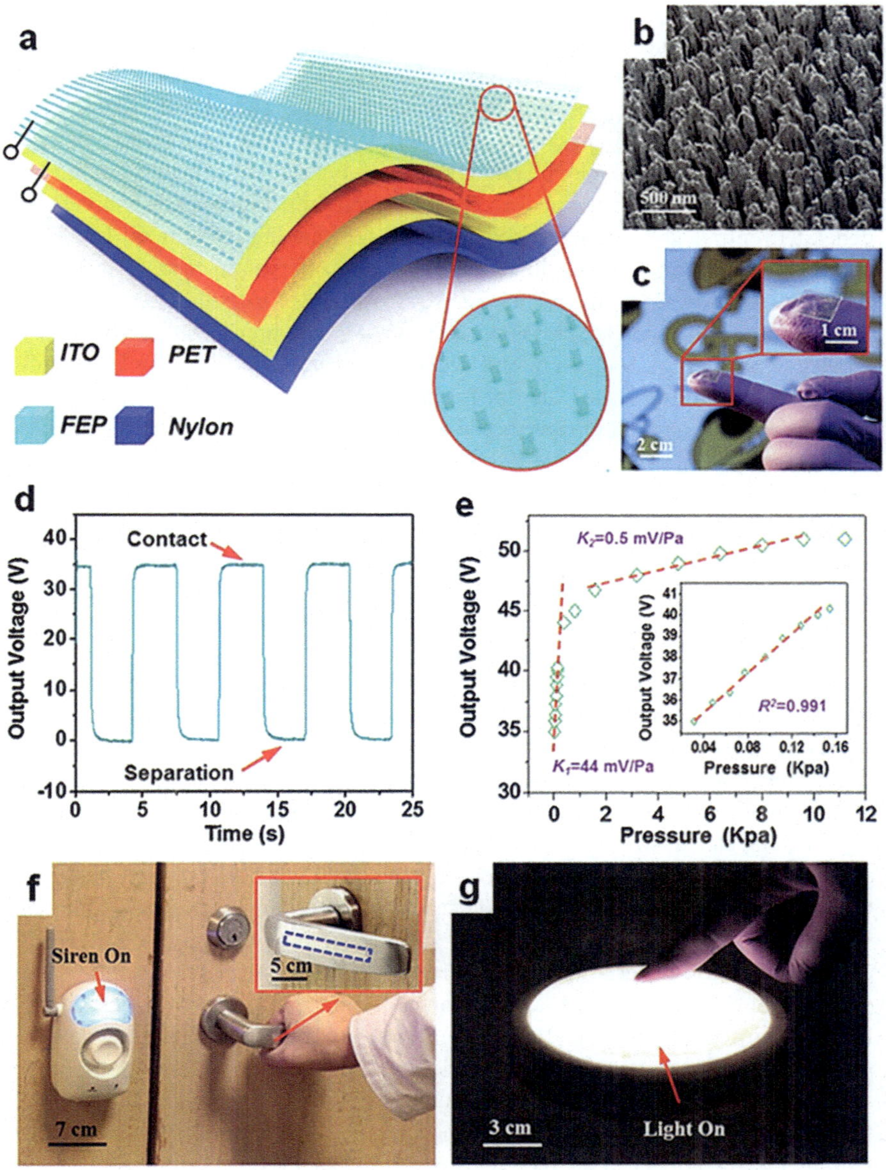

Fig. 11.34 A self-powered, flexible, and ultrasensitive tactile sensor (Adapted with permission from Wang et al. (2015b) (Royal Society of Chemistry)): (**a**) The schematic structure of the tactile sensor. (**b**) The SEM image of the polymer nanowires created on the surface of the FEP thin film. (**c**) The photograph of the tactile sensor for measuring the contact event from a foreign object. (**d**) The measurement of open-circuit voltage with a cyclic contact force of 20 mN. (**e**) The output voltage under different contact pressures. Inset: an enlarged view of the summarized results at a low-pressure region. (**f**) A photograph showing that the TES was integrated with a signal processing circuit to turn on the siren alarm when a human hand touched the door handle. (**g**) A photograph showing that the TES was integrated with another signal processing circuit to turn on light when it was contacted by human fingers

trigger an IC timer that controlled a wireless transmitter for remotely switching a siren alarm. As indicated in Fig. 11.34f, the sensing system immediately started operation once a human hand touched the door handle. Through substituting other functional electronics for the wireless transmitter in the circuit, the sensing system could be adopted for more purposes. For example, a touch-enabled switch for a panel light was successfully developed and is shown in Fig. 11.34g. The TES in this work had a number of other unique advantages, including ultrahigh sensitivity, self-generated output, location independence, and outstanding robustness. In addition, the TES was generally applicable to objects made from various materials, indicating the widespread adaptability of the TES in a variety of circumstances. Therefore, TENG-based self-powered pressure/touch sensors can be utilized for a variety of applications, such as healthcare and security. Several typical examples include a membrane-based triboelectric sensor for health monitoring, a skin-friendly human–machine interfacing, and a paper-based anti-theft sensor (Wang et al. 2015a, b).

A mechanical motion can be described with a series of parameters, like displacement, velocity, acceleration, etc. Sensors for detection of these parameters are vitally important for mechanical systems with moving parts. TENGs have been developed to generate electricity from different types of mechanical motions, such as linear sliding, rotation, and rolling. Since the frequencies and amplitudes of the generated electrical signals are all directly related to the parameters of the input mechanical motions, TENGs can play the role of self-powered active motion sensors. Even if such motions are located in concealed locations, TENGs can be utilized to track the trajectory and detect the instantaneous parameters of the motions. Figure 11.35a shows a self-powered, one-dimensional displacement and speed sensor with high spatial resolution, large dynamic range, and long detecting distance. The motion sensor consisted of two micro-grating layers with identical patterns. The bottom layer was an etched silicon wafer coated with aluminum as the bottom electrode and silicon dioxide as one of the triboelectric material to generate positive charges; the top layer was a patterned SU-8 film on a glass slide as a supporting substrate, which was subsequently coated with ITO and Parylene film as the top electrode and the other triboelectric material to generate negative charges, respectively. The detailed structure of the as-fabricated devices can be found in the SEM images shown in Fig. 11.35b, c, respectively. The relative motion between two gratings results in periodic separation of two micro-grated triboelectric materials that are oppositely charged through triboelectrification. As a result, an alternating electric signal between the two electrodes placed was detected due to electrostatic induction. During the movement, V_{OC} measured between the two electrodes (Fig. 11.35d) alternated between 0 and ~160 mV periodically, and the real time displacement could be calculated by counting the number of voltage peaks and multiplying the number with the width of each grating (200 mm). As presented in Fig. 11.35e, the detected displacement increased linearly with time, and the real time motion speed could also be calculated by dividing the grating width by the time interval between two adjacent peaks. Starting from a status with complete overlap, a step motion test with each step of 5 mm was performed to investigate the resolution of the displacement sensor. The sensitive region was found to be from 10 to 190 mm, where each

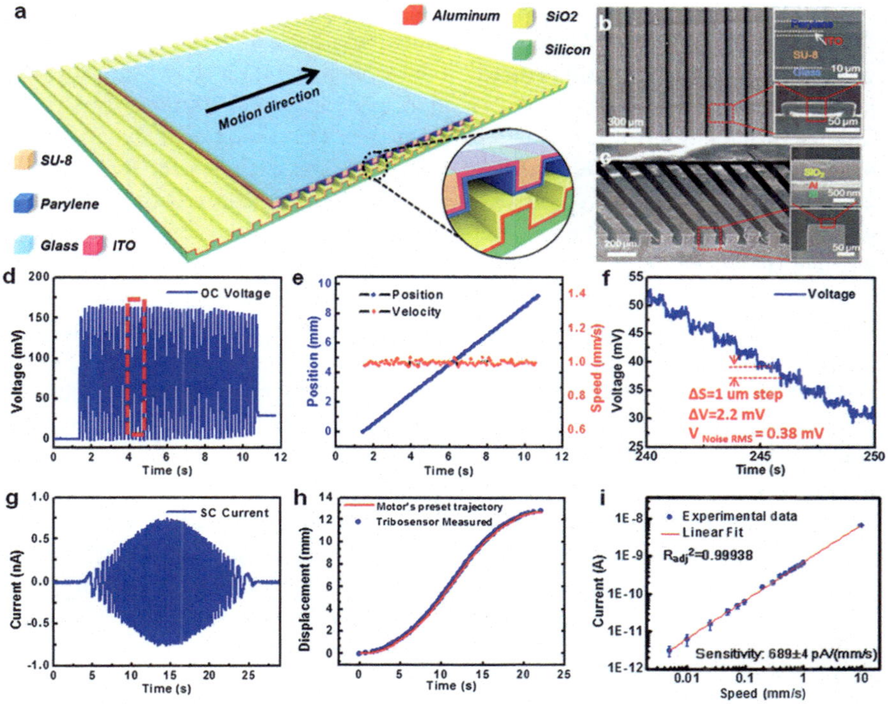

Fig. 11.35 Micro-grated triboelectric nanogenerator as a nanometer resolution self-powered linear displacement sensor (Modified with permission from Zhou et al. (2014) (Wiley—VCH Verlag GMBH & Co.): (**a**) The structure of a TENG-based self-powered displacement sensor, with a pair of microgratings, the inset illustrates the detailed information on layers. (**b**) SEM images of the top micrograting with a glass slide substrate. The insets show the cross section: the ITO layer on top of the patterned SU-8 photoresist serves as the top electrode, and the outmost layer is Parylene film serving as an electronegative triboelectric layer. (**c**) SEM images of the bottom micrograting. The insets show the cross section profile: the etched Silicon is coated with Al as a bottom electrode and SiO2 as an electropositive triboelectric layer. (**d**) The OC voltage signals acquired from a displacement of 9.2 mm at a preset speed of 1 mm/s. (**e**) The real time displacement and speed derived from the measured voltage signal. (**f**) Step motion with 1 mm per step in the sensitive region (10–190 mm) can be clearly resolved through the OC voltage signal. Given the RMS value of noise (0.38 mV), and the voltage change corresponding to 1 mm step motion (2.2 mV), the resolution can be calculated to be 173 nm. (**g**) The SC current signals acquired from a non-uniform motion (decelerate–uniform speed–deceleration). (**h**) The preset motor's trajectory and the real time displacement detected by the motion sensor. (**i**) A plot of SC current as a function of motion speed from 5 to 10 mm/s with a fitted sensitivity of 689 ± 4 pA (mm/s)

step motion of 1 mm could be clearly identified from the voltage variation, as displayed in Fig. 11.35f. The change in the voltage for each step was about 2.2 mV, and the root mean square (RMS) of the noise V_{noise} at 1 Hz is 0.38 mV. Consequently, the displacement resolution at a bandwidth of 1 Hz was determined to be 173 nm. The measured I_{SC} could also serve as the sensing signal for the displacement and speed, as shown in Fig. 11.35g. The real time displacement could be derived by counting the number of zero-crossings, and the measurement results matched very well with the preset trajectory (Fig. 11.35h). On the other hand, the amplitude of I_{SC} also helped to quantitatively determine the real time speed, given a small aspect ratio of the dielectric layer. Figure 11.35i shows the plot of the magnitude of the output current with motion speed from 5 to 10 mm/s, and a linear fitting indicated its sensitivity of 679 pA $(mm/s)^{-1}$. The resolution in dynamic speed detection was 1.2 mm/s given the RMS value (0.8 pA) of the noise. The resolution for both displacement and speed sensing can be further improved by reducing the grating period. This approach of self-powered displacement/speed sensing distinguishes itself from the existing technologies by nanometer resolution, long detecting range, and nonoptical compacted structure, showing extensive potential applications in automation, manufacturing, process control, etc. (Zhou et al. 2014, Wang et al. 2015a, b).

Based on the triboelectric effect between human fingers and keys, the intelligent keyboard (IKB) could convert typing motions into localized electric signals that could be identified as personalized physiological information. The core part of the IKB was composed of multilayered transparent thin film materials to form a typical single-electrode TENG. A layer of polyethylene terephthalate (PET) was deposited with ITO electrodes on both sides, and the top ITO was laminated by a layer of FEP as the triboelectric material for contact electrification with bare human fingers (Fig. 11.36a). A nanowire structure was created on the FEP surface to enhance the effective contact area (Fig. 11.36b), which also introduced a self-cleaning surface. Figure 11.36c shows a photograph of a fully assembled IKB with the same size as a commercial keyboard. The working principle of the IKB as an energy harvester was similar to the single-electrode TENG. The contact of human skin with the FEP surface would leave the finger positively charged, and the FEP negatively charged. The finger movement during typing would then induce change in the potential difference between the pair of ITO electrodes, driving electrons to flow through the external load or data collection system. A typical output profile is presented in Fig. 11.36d, in which the channel located on the key struck by the finger exhibited an instantaneous voltage peak of about 10 V, which was much larger than that in the rest of the channels. The recognition of key striking could even be recognized by screening the data using a Pauta Criterion Method, and simultaneously visualized on the display without noticeable delay (Fig. 11.36e). The performance of the authentication biometrics was characterized through two error rates: False Rejection Rate (FRR) and False Acceptance Rate (FAR). Here, the FRR is the probability that the system incorrectly rejects access of an authorized person, due to failing to match the user template. While FRR is the percentage of valid inputs, which are incorrectly rejected. And the Pearson correlation coefficient was selected as the classification threshold to evaluate the behavioral biometric authentication system, as shown in Fig. 11.36f. The FRR was increasing with the elevated threshold, while FAR follows a reverse trend, and their intersection indicated the Equal Error Rate (EER) point. In the meanwhile, the Receiver Operating Characteristic (ROC) curve is demonstrated in Fig. 11.36g. Given its exceptional authentication capability, the IKB was able to identify the individual typing characteristics, making it practical as a highly secured

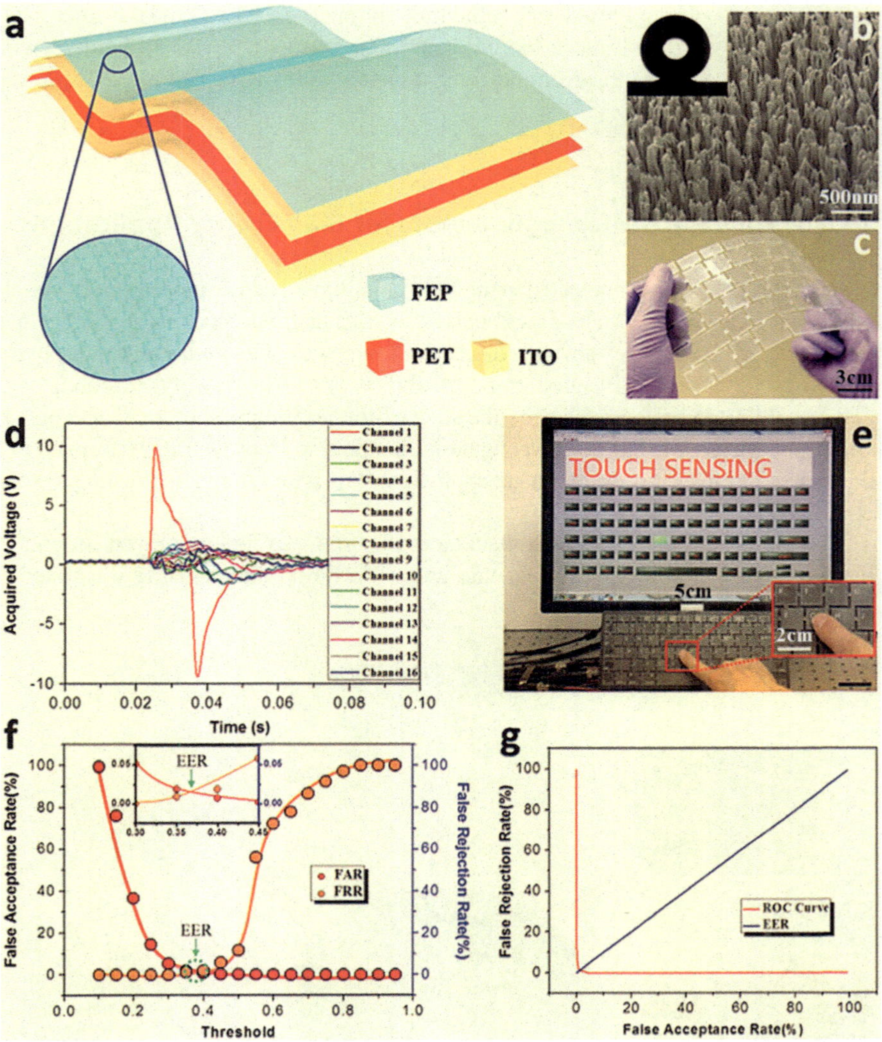

Fig. 11.36 Personalized keystroke dynamics for self-powered human–machine interfacing (Adapted with permission from Wang et al. (2015b) (Royal Society of Chemistry)): (**a**) Schematic illustrations of the KFE. Inset: an enlarged schematic of FEP nanowires on the top surface. (**b**) SEM image of FEP nanowires. Inset: contour of the resting droplet for surface static contact angle measurement. (**c**) Photograph of a flexible and transparent KFE. (**d**) The system acquired output voltage signals when the key "T" was stroked. (**e**) A photograph demonstrating the IKB for real-time keystroke tracing and recording. A continuously typing string "TOUCH SENSING" was recorded in real time without uncomfortable delay. Inset: an enlarged view of the key "G" being stroked. (**f**) Evaluation of the performance of the biometric authentication system using triboelectrification enabled keystroke dynamics. The variation of FAR and FRR is related to the threshold. Inset: an enlarged view of the EER point, which indicates a remarkably low EER value of 1.34% at the threshold of 0.37. (**g**) Receiver operating characteristic (ROC) curve of the biometric authentication system using triboelectrification enabled keystroke dynamics. The false rejection curve is plotted as a function of the false acceptance curve

authentication system based on behavioral biometrics TENG has further been improved and extended into a bionic membrane sensor (BMS) with both security authentication and healthcare monitoring functionalities (Chen et al. 2015; Wang et al. 2015a, b).

11.6.3 Triboelectric Nanogenerators for Blue Energy Applications

Many approaches of triboelectric nanogenerators have been explored to effectively collect various types of water-based power sources, namely blue energy harvesting. Figure 11.37 summarizes several unique prototypes to achieve high output power, low cost, convenient fabrication, good flexibility, excellent stability and robustness, and possibility to scale up. Despite the state-of-the-art progress, more endeavors are still highly desired to further investigate in this field to improve the performance of the water-TENG towards practical applications in blue energy harvesting (Wang et al. 2015a, b).

The concept of water–solid contact electrification was first employed for energy harvesting; the device structure of this water-TENG is schematically illustrated in

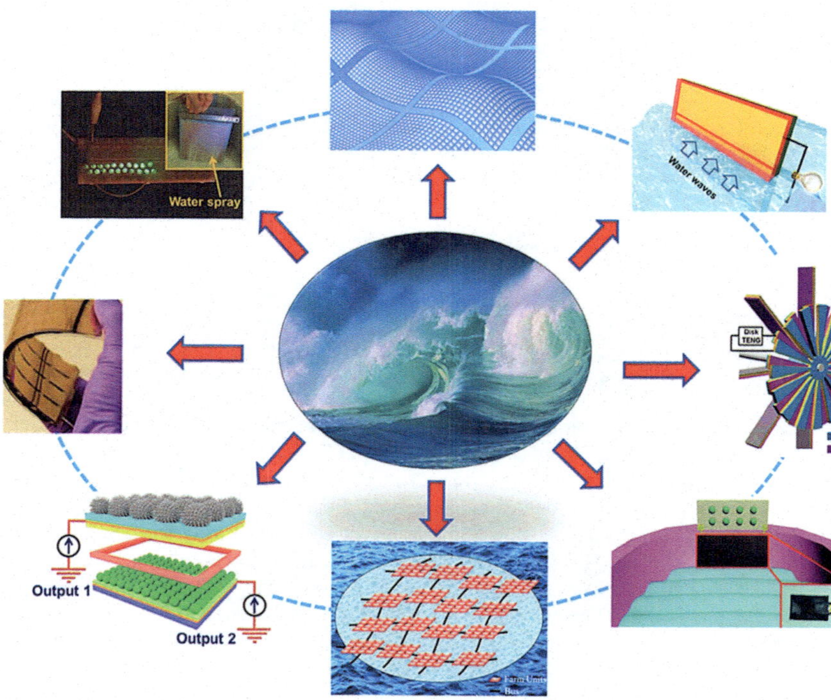

Fig. 11.37 A schematic diagram exhibiting the state-of-the-art progress of TENGs for harvesting blue energy (Adapted with permission from Wang et al. (2015b) (Royal Society of Chemistry))

Fig. 11.38a, in which one dielectric layer of a typical contact-mode TENG was replaced with water. The cyclic contact between the PDMS layer and water surface would generate negative charges on the PDMS layer and positive charges on the water surface, and the periodical separation of the two charged surfaces resulted in change in potential difference between the two copper electrodes under OC conditions, and external electrons flow under SC conditions. Micropatterned pyramid array was created on the inner surface of the PDMS layer to enhance the effective contact area (Fig. 11.38b), and this approach should be extraordinarily applicable in this case owing to high ductility of water. Driven by a linear motor at a frequency of 2 Hz and a magnitude of 1.5 cm, the water-TENG was enabled to produce a V_{OC} and J_{SC} of 82 V and 1.05 mA/m^2, respectively. The optimized output power density reached up to 50 mW/m^2 at 5 Hz. The frequency response of the water-TENG was significantly different from that of the solid material-based TENG, in that the movement of the PDMS film contacting and separating from water would generate a water wave, which would disturb the contact area with the PDMS film. Figure 11.38c, d shows the frequency response of the measured J_{SC} and V_{OC}, respectively. Both of them decreased first from 2 to 3 Hz and then increased with frequency from 3 to 5 Hz. On the contrary, for solid-material-based TENG, J_{SC} would keep increasing with the frequency and V_{OC} would remain almost unchanged. The difference of frequency response implied a unique feature of the water-TENG. It was further demonstrated that the water-TENG could serve as a self-powered temperature and ethanol sensor, through real time measurement of its output performance. Another basic prototype of water-TENG was based on asymmetric screening of the triboelectric charges on a nanostructured hydrophobic thin film surface, as exhibited in Fig. 11.38e. Two parallel strip-shaped electrodes were deposited on one side of a FEP thin film, and they were separated with a fine gap in between. The other side of the FEP thin film was modified with nanowire structures for improving the effective contact area. The operation of the water-TENG relied on a repetitive emerging–submerging process with traveling water waves, in which the coupling between triboelectrification and electrostatic induction drove alternating flows of electrons between electrodes. Quantitative assessment of the output performance was also stimulated by a linear motor, and the generated V_{OC} and I_{SC} were 160 V and 3 mA, respectively, with an optimum output power of 0.12 mW at a velocity of 0.5 m/s (Fig. 11.38f).

A higher velocity, finer electrode features, and lower ion concentration are all favorable for the overall output performance, though further systematic investigations on the charge generation process were still desirable to provide in-depth insights. To demonstrate the applicability of the water-TENG in real-world situations, an integrated TENG with a scaled-up design was further tested in a normal environment where energy from ambient water motions was harvested. The integrated water-TENG consisted of six strip-shaped electrodes and a total of five basic units were formed by any pair of adjacent electrodes. The electric output of each pair was first rectified and then constructively superimposed through a parallel connection. It was then inserted into a traveling wave created by a large container, or placed beneath a sprinkler head with sprayed water droplets. The generated output

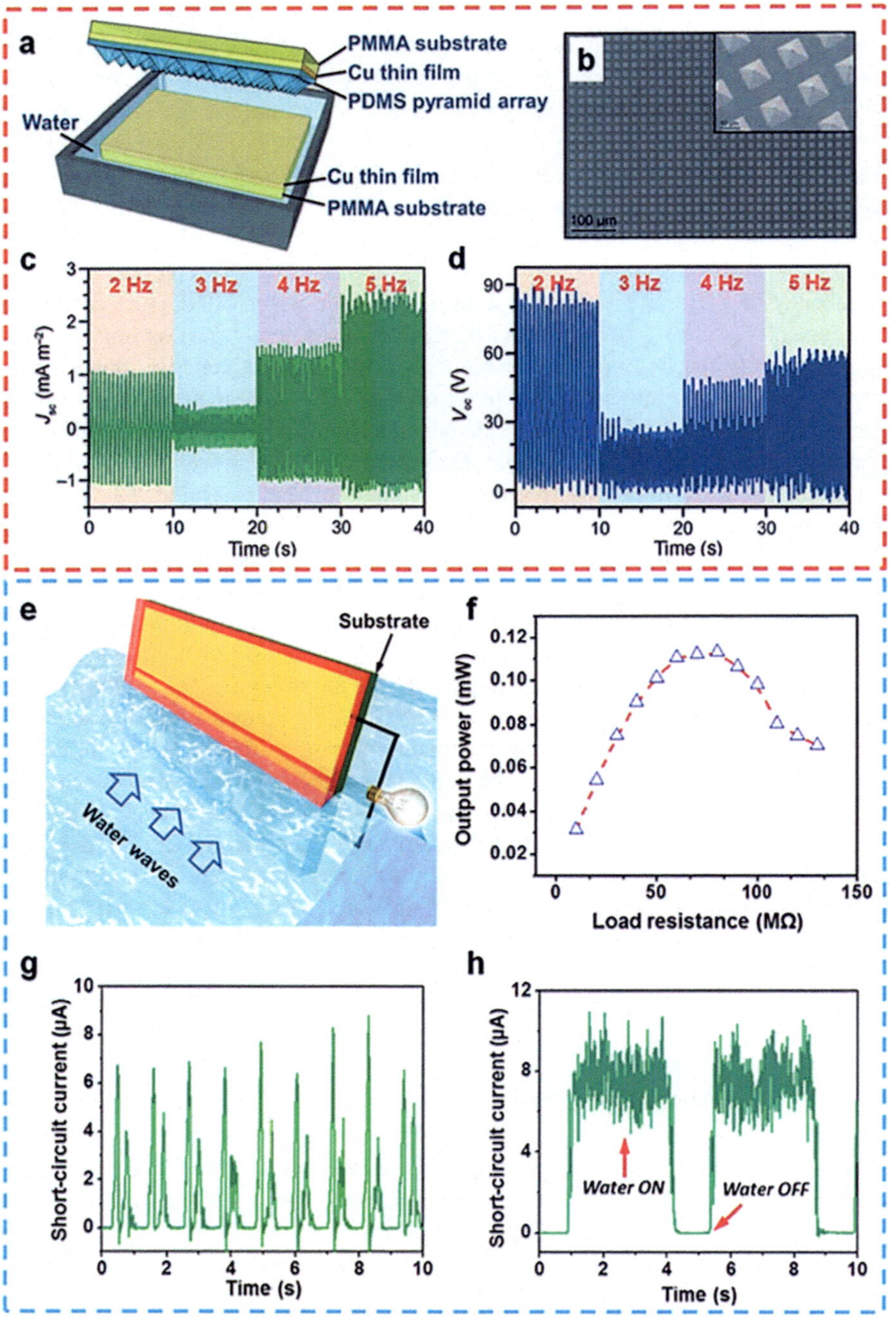

Fig. 11.38 TENGs for harvesting water wave energy (Adapted with permission from Wang et al. (2015b) (Royal Society of Chemistry)): (**a**) Schematic diagram of the water-TENG based on contact electrification between water and micropatterned PDMS. (**b**) SEM image of patterned PDMS

current from both cases is shown in Fig. 11.38g and h, respectively. The rectified current from water droplets (Fig. 11.38h) showed almost continuous DC behavior, which was attributed to numerous droplets and the merging of the large number of current peaks. In addition to these two fundamental structural designs, a lot of other prototypes were also proposed for high-performance blue energy harvesting, such as dual-mode TENG fabricated based on the superhydrophobic TiO_2 layer and polymer thin films to collect both the electrostatic and mechanical energy from flowing water; hybrid water-TENG relied on the coupling of contact separation and single-electrode operation modes (Lin et al. 2013; Wang et al. 2015a, b).

Besides water wave energy, wind energy is also a type of widely existing power source from the natural environment. A flutter-driven TENG was developed for effective wind energy harvesting to sustainably drive electronic devices in outdoor environments. The structure of the TENG was composed of a fluttering flexible flag and a rigid plate, as shown in Fig. 11.39a. The flag was a gold-coated conductive fabric that served as the fluttering body, and PTFE with a counter electrode was attached onto the rigid plate to form a contact-mode TENG. Natural wind would introduce the contact–separation behavior between the flag and the rigid plate to generate electricity. As shown in Fig. 11.39b, a flexible woven flag exhibited oscillations with a node. From the top to the node, the oscillation amplitude was very small, but it was subject to increase as a traveling wave propagated towards the trailing edge below the node. To bring up a self-sustained contact-propagation-separation between two triboelectric layers through high-frequency flag fluttering, the counter plate was simply placed next to the flag flutter within a critical distance, and the coupled interaction between the two surfaces was systematically investigated by varying their dimensions. The stability boundary could be characterized by a regime map shown in Fig. 11.39c, and two distinct contact–separation modes were observed in the plot. The transition from the single-contact mode in region A to the double-contact mode in region B occurred as the dimensionless mass decreased, which could be tuned by adjusting the length of the flag. The flutter-driven TENG produced a high output of 200 V and 60 mA at a wind speed of 15 m/s, with an average power density of 0.86 mW. To demonstrate its wind energy-harvesting capability in outdoor environments, a robust packaging module of the flutter-driven TENG was integrated (Fig. 11.39d) and securely mounted onto the roof of a moving vehicle. The TENG was enabled to charge a 1000 mF capacitor to 30 V in 25 min when the vehicle was traveling at 70 km/h (Fig. 11.39e), thus demonstrating its

Fig. 11.38 (continued) pyramid array. The inset is an SEM image with higher magnification. (**c**, **d**) Influence of contact frequency between patterned PDMS pyramid array and deionized water on the (**c**) JSC and (**d**) VOC generated by the water-TENG. (**e**) Schematic of a substrate-supported TENG positioned in water waves. The up-and-down movement of the surrounding water body induces electricity generated between the two electrodes. 135 (**f**) Average output power of the water-TENG with increasing load resistance between electrodes. (**g**) Rectified short-circuit current of the integrated water-TENG when it interacts with water waves at a frequency of ~0.7 Hz. (**h**) Rectified short-circuit current of the integrated water-TENG when it interacts with falling water drops

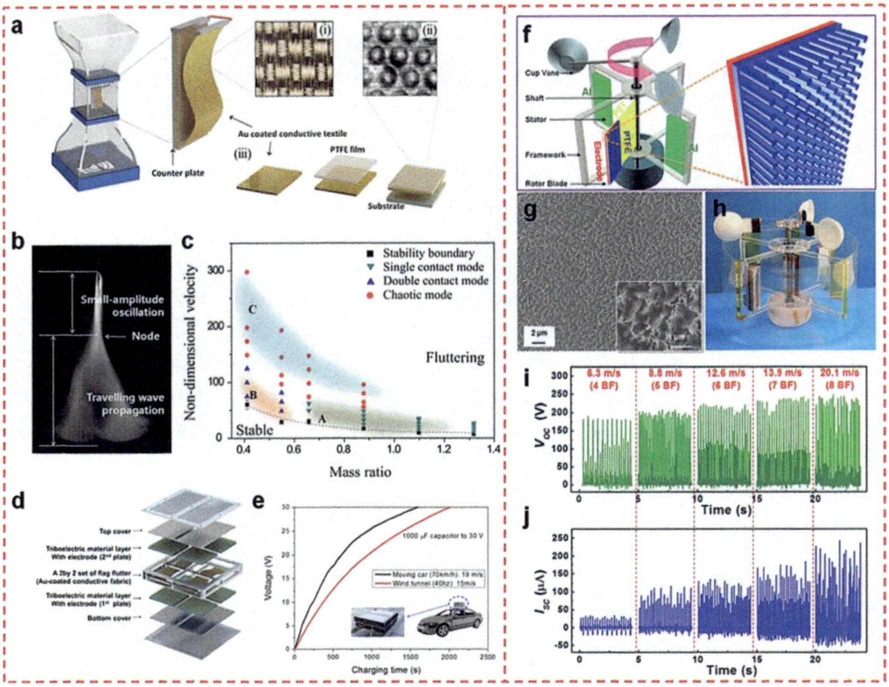

Fig. 11.39 TENGs for harvesting the ambient wind energy (Modified with permission from Wang et al. (2015b) (Royal Society of Chemistry)): (**a**) Schematic diagrams of a wind tunnel and the structural design of a flutter-driven triboelectric generator including surface characteristics of (*i*) a highly flexible flag and (*ii*) a counter plate and (*iii*) the fabrication of the counter plate. (**b**) Fluttering images of a stand-alone configuration of dimensions 7.5 × 5 cm as captured using a high-speed camera. (**c**) The regime map of the dynamic interaction between a flag and a plate. (**d**) Schematic of the stacked TENG-packaging module with total eight units. (**e**) Comparison of the charging times of a 1000 mF capacitor using a stacked flutter-driven TENG system when charging on the moving vehicle and when charging in the wind tunnel system. (**f**) The schematic diagram showing the structural design of the rotary TENG, with the enlarged picture showing nanowire structures on the surface of PTFE. (**g**) The SEM image of the PTFE surface with etched nanowire structures. The inset is an SEM image at high magnification. (**h**) A photograph of the fabricated rotary TENG. (i, j) Influence of the wind speed on the electrical outputs. (**d**) *V*OC and (**e**) *I*SC under different wind speeds from 6.3 m/s (4BF) to 20.1 m/s (8BF)

operation in an open environment under various wind conditions. A rotary TENG was also developed to harvest wind energy assisted by the wind-driven rotation of cup vanes, as shown in Fig. 11.39f. The continuous shaft rotation introduced cyclic contact and consecutive sliding between Al foils and nanostructured PTFE film (Fig. 11.39g, h). Hence, the working principle of this rotary TENG relied on coupling of vertical contact–separation and lateral sliding modes. With a wind speed of ~15 m/s, its V_{OC} and I_{SC} were up to 250 V and 0.25 mA, respectively, corresponding to a maximum power density of ~39 W/m^2. Additionally, the rotary TENG could serve as a self-powered wind speed sensor (Fig. 11.39i, j). As expected,

I_{SC} increased drastically with rising wind speeds, but V_{OC} also showed slight elevation as the wind speed went up. This result could be explained by the change in the surface charge density. With higher wind speed, the polymer thin films would obtain higher rotational torque and thus a larger contacting force, which led to more intimate contact between two surfaces, resulting in a higher surface charge density (Bae et al. 2014; Wang et al. 2015a, b).

11.7 Hybridization and Integration of Energy Harvesters

Hybrid energy harvesters for simultaneously harvesting multiple types of environmental energy have been developed because of their synergetic output performances. Some hybrid cells have been accordingly demonstrated for harvesting mechanical and solar energy, mechanical and thermal energy, thermal and solar energy, and mechanical, thermal, and solar energy. For integrated energy harvesters of two or more such sources, electric circuit design should be considered, because solar cells and thermoelectric generators generate direct current (DC) electricity, while piezoelectric, triboelectric, and pyroelectric generators usually generate alternating current (AC) electricity. In the case of AC electricity, a rectification diode is required to convert AC into DC electricity. Moreover, impedance matching depending on materials and operation frequencies using resistive load to the piezoelectric generators, pyroelectric generators, and triboelectric generators is important to achieve maximum power output (Lee et al. 2016).

11.7.1 Mechanical and Solar Energy

Multi-type energy harvesters using both mechanical and solar energy have been especially intensively developed based on piezoelectric, pyroelectric, triboelectric, and PV effects. Figure 11.40a, b shows a hybrid energy harvester fabricated by infiltrating CdS/CdTe quantum dots into vertically aligned ZnO nanowires driven by sound in the frequency range of 35–1000 Hz and solar energy. Similar hybrid energy harvesters were also demonstrated using acoustic waves and pressure for the piezoelectric effect, based on PVDF and ZnO, with various kinds of solar cells such as DSSCs, quantum dot solar cells, silicon (Si) solar cells, and organic–inorganic hybrid solar cells. TENG-based multi-type energy harvesters have also received great attention because of their high output power and easy hybridization with other energy harvesters. Figure 11.40c, d shows a hybrid energy harvester based on the triboelectric effect and the PV effect. Micropyramid Si solar cells were fabricated with a protective layer consisting of a thin film of polydimethylsiloxane (PDMS) nanowires, which not only worked as a protective layer, but also as a triboelectric layer for harvesting mechanical energy. The hybrid energy harvester can be used for self-powered electrodegradation of rhodamine B and can also charge lithium (Li) ion batteries for operating small electronic devices (Lee et al. 2010, 2016).

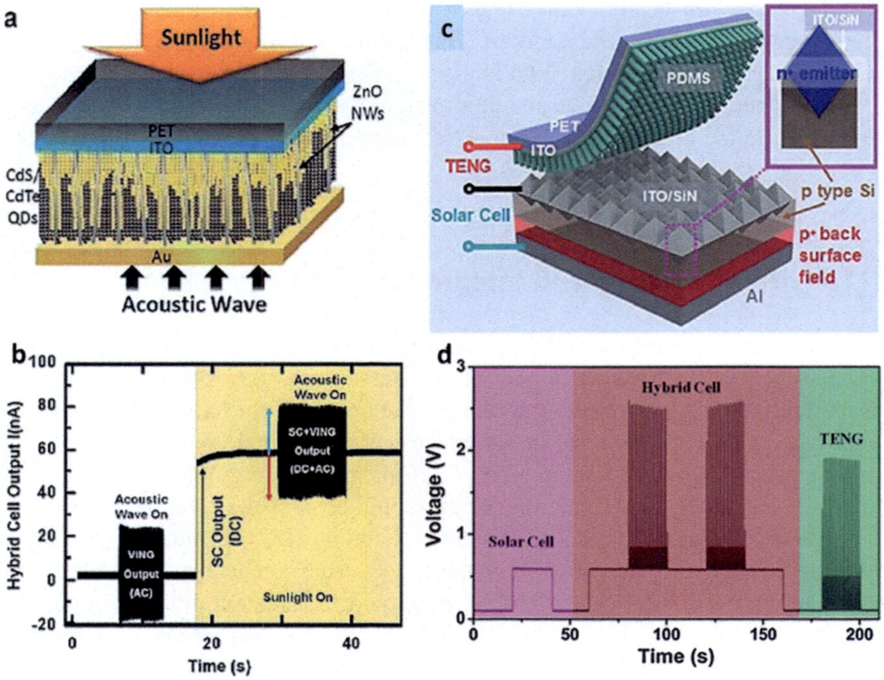

Fig. 11.40 Mechanical and solar energy hybrid harvesters (Modified with permission from Lee et al. (2016) (Royal Society of Chemistry)): (**a**) Schematic diagram depicting a hybrid device with two different incoming energy sources, and (**b**) short-circuit current output signal of the hybrid cell. (**c**) Schematic diagram of the fabricated hybrid energy cell and scanning electron microscope (SEM) image of the fabricated Si pyramids, and (**d**) output voltage of the hybrid solar cell and TENG (after rectification) for harvesting both solar and mechanical energy

11.7.2 Mechanical and Thermal Energy

In some situations, mechanical vibrations or friction, and temperature fluctuations or a temperature gradient coexist, such as on the human body, during air flow, and in working engines/machines. As shown in Fig. 11.18, a stretchable piezoelectric–pyroelectric hybrid energy harvester was fabricated based on a micropatterned ferroelectric polymer P(VDF-TrFE) thin film, a PDMS-CNT composite, and graphene electrodes. The working mechanism for achieving the total output voltage was based on coupling of the piezoelectric and pyroelectric effects in terms of polarization and oscillation of the electric dipoles of P(VDF–TrFE). Stable pyroelectric output performance of the hybrid energy harvester for various stretchable modes showed the stretchability, mechanical durability, and robustness of the device owing to the micropatterned design. A Seebeck effect-based hybrid energy harvester to harvest both mechanical and thermal energy was also developed to convert wasted temperature gradient energy into electrical energy, by integration of the piezoelectric

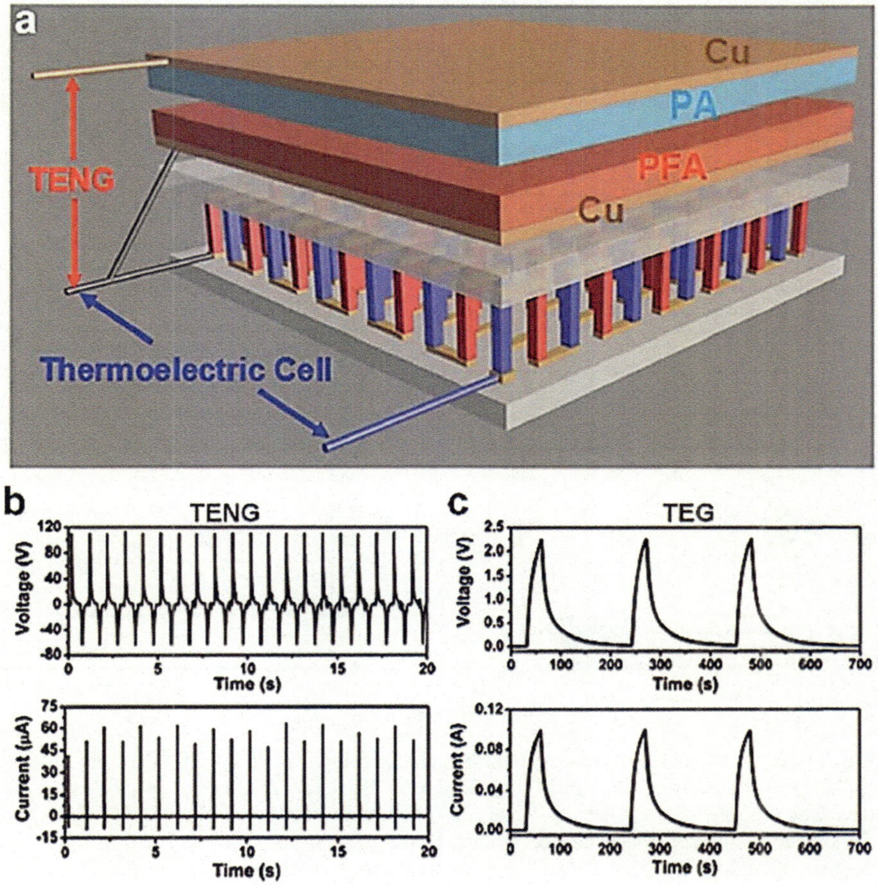

Fig. 11.41 Mechanical and thermal energy hybrid harvester (Modified with permission from Yang et al. (2013) (Royal Society of Chemistry)): (**a**) Schematic diagram of the fabricated hybrid energy cell consisting of a triboelectric generator and a thermoelectric cell, (**b**) output voltage and current of the triboelectric generator, and (**c**) output voltage and current of the thermoelectric generator

and thermoelectric component devices, which were mounted on a flexible substrate. Furthermore, the energy-harvesting device can simultaneously harvest using both thermal and mechanical energy from the human body temperature gradient and mechanical movement. Figure 11.41 shows a hybrid generator that consisted of a TENG and a TEG, which could be used for self-powered water splitting to generate hydrogen (Fig. 11.41a). The hybrid energy harvester consisted of a polyamide (PA)-perfluoroalkoxy (PFA) polymer film-based TENG and a Bi_2Te_3-based TEG. The fabricated TENG generated output voltage and current of 110 V and 60 mA, and the TEG generated output voltage and current were 2.2 V and 0.1 A, respectively (Fig. 11.41b, c). This strategy provides a highly promising platform for harvesting both mechanical and thermal energy simultaneously and individually, and utilizing it

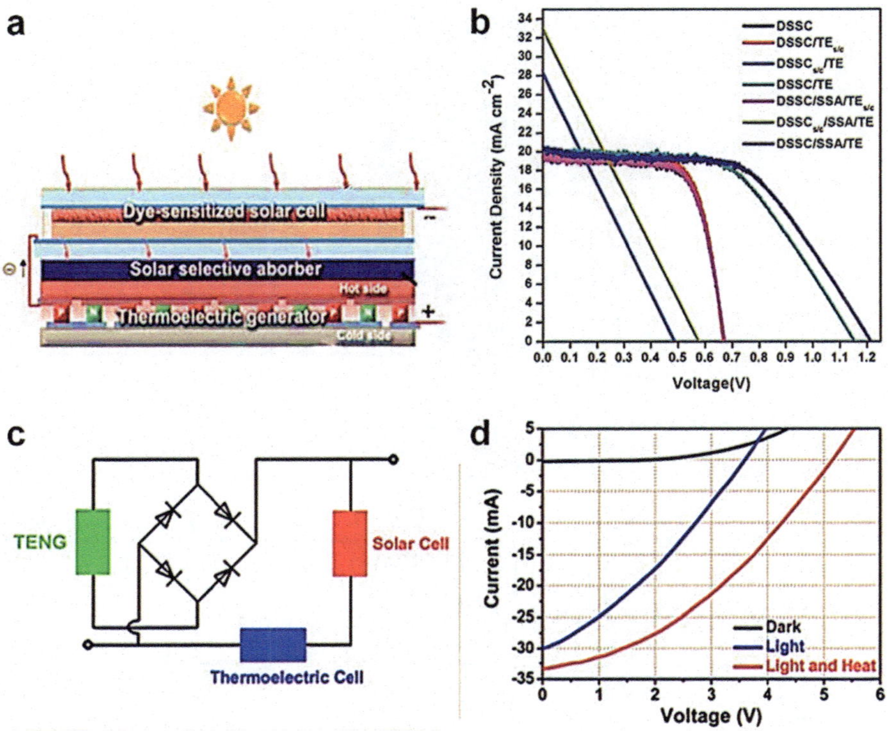

Fig. 11.42 Solar and thermal energy hybrid harvester (Modified with permission from Lee et al. (2016) (Royal Society of Chemistry)): (**a**) Schematic illustration of a solar-thermoelectric hybrid device and (**b**) photocurrent density–voltage (J–V) characteristic curves of the DSSC/TEG hybrid device and of the component devices with the other component short-circuited. (**c**) Schematic diagram of the hybrid energy cell and (**d**) current–voltage characteristics of the hybrid energy cell under different conditions

for wireless sensors, temperature imaging, medical diagnostics, power sources for biomedical applications, personal electronics, sensor networks, and micro-/nano-systems (Yang et al. 2013; Lee et al. 2016).

11.7.3 Thermal and Solar Energy

Hybrid cells for harvesting thermal and solar energy are also the most conventional energy-harvesting technology, such as thermoelectric-solar hybrid generator made with Si or DSSC solar cell and TEG. Figure 11.42 shows a thermoelectric solar hybrid energy-harvesting system with a series connected DSSC, a solar selective absorber (SSA), and a TEG. The conversion efficiency of the solar cell was greatly increased by using the SSA and the TEG to utilize residual sunlight transmitted through the DSSC. The hybrid device, comprising a DSSC for high-energy photons and an SSA-coated thermoelectric generator for low-energy photons, yielded overall

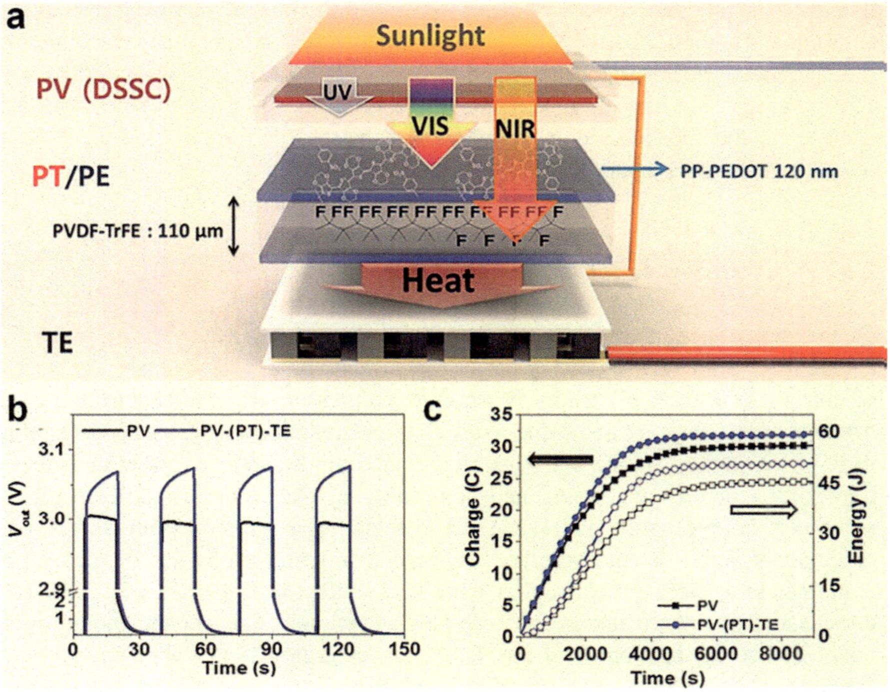

Fig. 11.43 Solar and pyroelectric energy hybrid harvester (Modified with permission from Park et al. (2015). Copyright © 2015 American Chemical Society): (**a**) Illustration of a photovoltaic and photothermal pyro-thermoelectric device. (**b**) Output voltage of a PV and PV–TE series connected device. (**c**) Output and accumulated energy of a capacitor (10 F, 5.4 V) charged by PV and by the PV–TE series connected device in the hybrid system (filled circles and squares, charge; open circles and squares, energy)

conversion efficiency higher than 13% (Fig. 11.42b). Figure 11.42c shows a hybrid cell that was composed of a Bi_2Te_3-based TEG and a Si solar cell, which could be used to simultaneously and individually harvest thermal and solar energy. Under light illumination, the output voltage of the hybrid energy cell is about 3.5 V, and the output current is about 30 mA, based on the integration of six solar cells (Fig. 11.42d). Under light illumination on the solar cells and heat applied at the bottom of the thermoelectric cell, the total peak output voltage and current of these two components of the energy harvester reached 5.2 V and 34 mA, respectively. Figure 11.43 shows a hybrid cell based on a DSSC solar cell with a pyroelectric and thermoelectric device operated by photothermally generated heat (Fig. 11.43a). The photoconversion efficiency (PCE) was increased up to 20% under sunlight irradiation (AM 1.5G) using the transmitted light through the DSSC as a heat source that was converted into electricity by the pyroelectric and thermoelectric effects simultaneously by the photothermal poly(3,4-ethylenedioxythiophene) (PEDOT) electrodes. Interestingly, as PV works under sunlight, the output voltage from the PV device continuously decreased because of the fast recombination phenomenon during the PV operation (Fig. 11.43b). When the PV was combined with a TEG, however, the total output voltage increased, due to thermoelectric power generation. Consequently, the reduced PV output performance was enhanced by the output of the thermoelectric generator in the hybrid device. This result was also confirmed by the charging of the capacitor. The plot (Fig. 11.43c) of the cumulative charges and energies in the capacitors demonstrates the enhanced performance of the PV–TE hybrid system (Park et al. 2015; Lee et al. 2016).

11.7.4 Coupling of Multiple Energy Sources

On the basis of nanomaterials in the wurtzite semiconductors, such as ZnO and GaN, electronics fabricated by using a piezopotential as a gate voltage are called piezotronics, with applications in strain/force/pressure-triggered/controlled electronic devices, sensors, and logic gates. The piezophototronic effect is a result of three-way coupling among piezoelectricity, photonic excitation, and semiconductor transport, which allows tuning and controlling of electro-optical processes by a strain-induced piezopotential (Wang 2010). In addition, the piezotronic effect—the coupling between the piezoelectric and semiconducting properties of nanowires, combines piezoelectric polarization with semiconductor properties and allows the direct and active interaction between devices and stimuli. This phenomenon inspires novel device applications and has led to an emerging field called piezotronics (Jenkins et al. 2015).

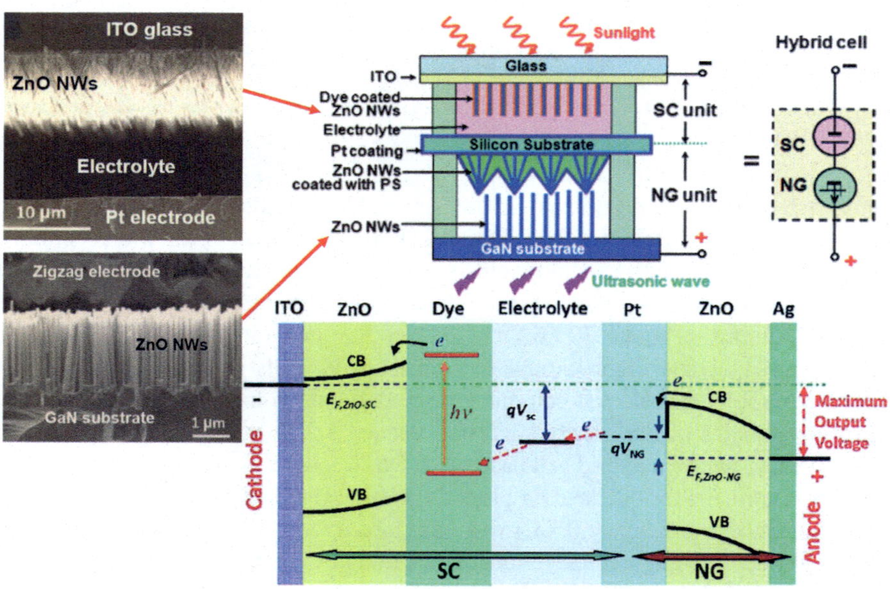

Fig. 11.44 Electron energy band diagram of the solar-piezo hybrid cell, showing that the maximum output voltage is the sum of the voltages produced by the solar cell and the PENG (Modified with permission from Xu et al. (2009). Copyright © 2009, American Chemical Society). The abbreviations are as follows: conduction band (CB), valence band (VB), Fermi level (EF)

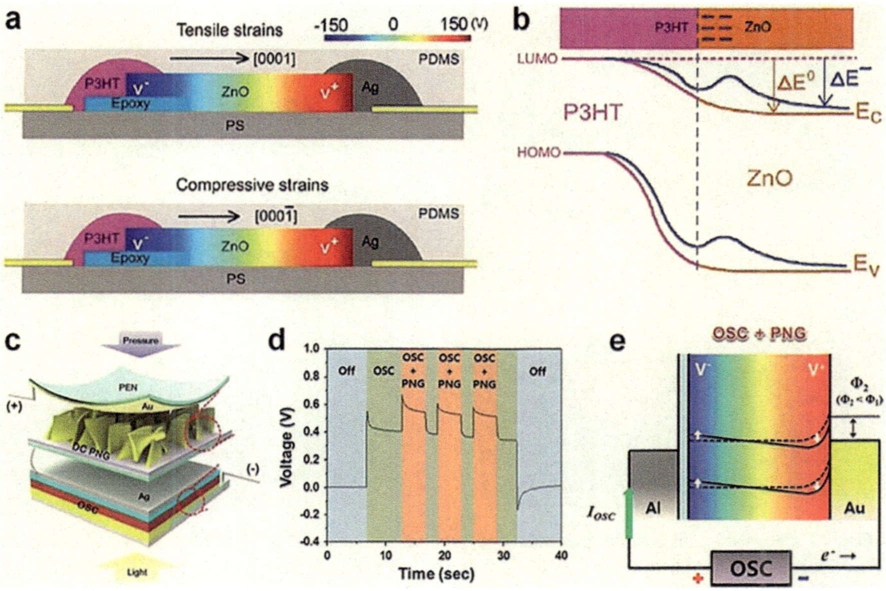

Fig. 11.45 The coupling based on piezotronic effect (Modified with permission from Lee et al. (2016) (Royal Society of Chemistry)): (**a**) Piezopotential distributions in the stretched device of [0001] type and compressed device of [0001], and (**b**) schematic energy band diagram of P3HT/ZnO in the presence of negative piezoelectric charges. The blue line indicates the energy band diagram modified by the piezoelectric potential in ZnO. The negative piezoelectric charges can lift the energy band, resulting in a peak in the energy band. (**c**) Schematic illustration of the solar-piezo hybrid cell, (**d**) output voltage of the solar-piezo hybrid cell when pressure is applied periodically at intervals of 3.0 s for a period of 1.0 s, and (**e**) when pressure is applied to the PENG, the ZnO nanosheets exhibited a new band diagram: dashed and solid lines represent the band diagrams before and after applying the pressure on the PENG

A hybrid generator made with a piezoelectric potential enhanced PV cell has been developed based on piezophototronic effect. Its working principle is explained using the electron energy band diagram, as shown in Fig. 11.44. The maximum achievable output voltage is the difference between the Fermi level of the ZnO nanowires (NWs) in the DSSC and that of the ZnO NWs in the NG. The maximum output voltage is an integration of the output voltages of the NG and DSSC. The piezotronic effect was also demonstrated on the output voltage of flexible solar cells using poly (3-hexylthiophene-2,5-diyl) (P3HT)–ZnO microwire p-n heterojunctions on a flexible polystyrene (PS) substrate (Fig. 11.45a). The open circuit voltage V_{oc} of the solar cell was characterized by tuning the strain-induced polarization charges at the interface between ZnO and P3HT. Figure 11.45b shows the mechanism of the piezotronic effect on solar cell performance under strain. The strain-induced piezoelectric potential is created under an externally applied strain, which modifies the energy band diagram at the interface of the p-n heterojunction, consequently modulating the performance of the device. The output power of solar cells could be enhanced by tuning the band profile under strain. Moreover, a hybrid system

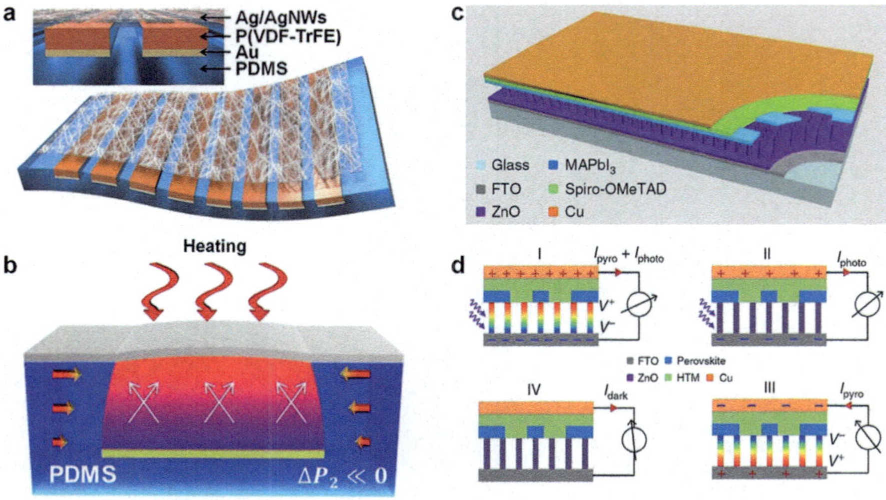

Fig. 11.46 PNG enhanced by piezoelectric potential driven by thermally induced strain (Modified with permission from Lee et al. (2016) (Royal Society of Chemistry)): (**a**) Schematic illustration of the piezo–pyro coupling NG, and (**b**) piezoelectric and pyroelectric coupled potential-generating mechanism of the patterned P(VDF–TrFE) on a PDMS substrate. (**c**) Schematic illustration of the structure of self-powered ZnO/perovskite-heterostructured photodetectors (ZPH PDs), with FTO acting as the transparent electrode, and (**d**) schematic illustration of the working mechanism of the pyroelectric effect-combined with photoexcitation processes

consisting of a DC type PENG based on a two-dimensional (2D) ZnO nanosheet and an organic solar cell based on P3HT/[6,6]-phenylC61 butyric acid methyl ester (PCBM) is shown in Fig. 11.45c. The power generation performance of the serially integrated hybrid cell is synergistically enhanced by the help of a PENG, compared with the output power generated independently from the solar cell component under illumination (Fig. 11.45d). Figure 11.45e exhibits the mechanisms of power generation from the piezoelectric potential of the ZnO nanosheets. The energy band diagram in the ZnO nanosheet is modified under the influence of applied force and light. The photogenerated electrons from the organic solar cell can effectively flow through the PENG, owing to the reduced Schottky barrier height between the ZnO nanosheets and the Au electrode. Consequentially, Jsc from the solar-piezo hybrid cell also increases under application of pressure. Moreover, coupling of thermal/mechanical energy and thermal/solar energy-based energy harvesters and self-powered photocurrent were also explored. A PNG enhanced by piezoelectric potential driven by thermally induced strain is shown in Fig. 11.46a. Dramatic enhancement of the piezoelectric coupled PNG performance was observed, based on coupling of the piezoelectric and pyroelectric effects using different thermal expansion coefficients and micropatterned architectures (Fig. 11.46b). Using the light-self-induced pyroelectric effect in ZnO to modulate the optoelectronic processes, the performance of ultraviolet sensors can be enhanced (Fig. 11.46c, d). Thus, the coupling effect between mechanical, thermal, and light energy in an enhanced synergetic energy-harvesting system is not just simple multi-type energy harvesting, such devices will also offer a promising approach for effectively harvesting multi-type energy for realizing multifunctional energy devices (Yoon et al. 2015; Lee et al. 2016).

11.7.5 Integration of Energy Harvesting and Storage Devices

Because of the naturally uncontrollable and unstable features of environmental mechanical, thermal, and solar energy sources, the converted electrical energy from energy harvesters is unstable and difficult to use as a direct power source for electronic devices. Usually, storage elements such as either capacitors or batteries are needed to stabilize and control the power output for direct applications (Lee et al. 2016).

Figure 11.47a shows a self-charging power cell using a Li-ion battery consisting of a $LiCoO_3$ cathode, TiO_2 nanotube anode, and PVDF piezoelectric polymer film as

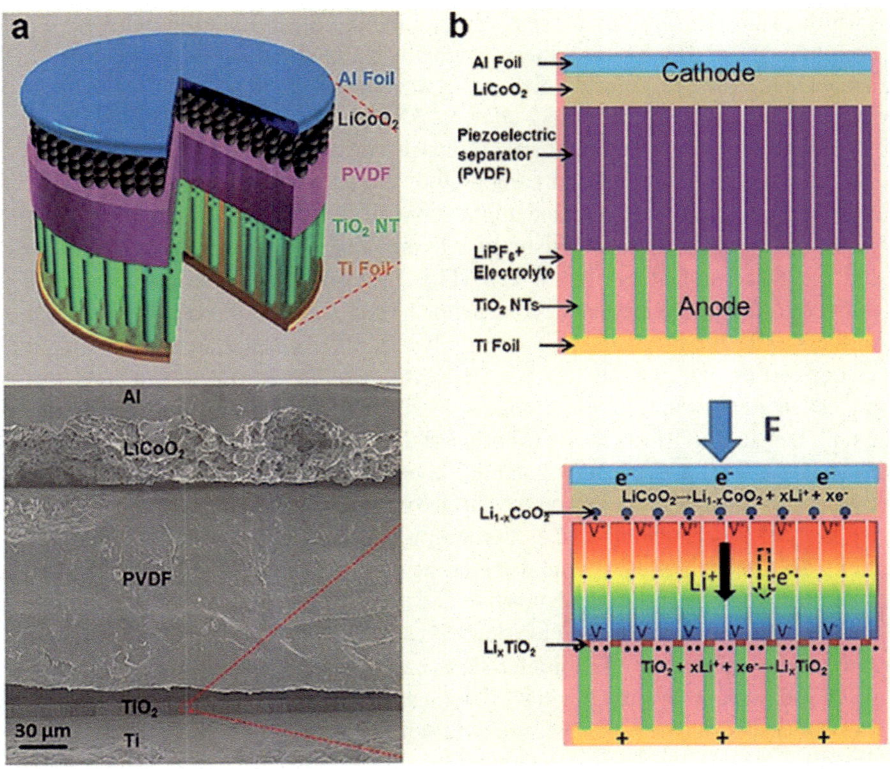

Fig. 11.47 Integration of energy harvesting and Li-ion battery (Modified with permission from Lee et al. (2016) (Royal Society of Chemistry)): (**a**) Design of a self-charging power cell by hybridizing a PENG and a Li-ion battery, and (**b**) the working mechanism of the self-charging power cell driven by compressive strain

the separator. The strain-induced piezoelectric potential from the PVDF film acts as a charge pump to drive Li ions to migrate from the $LiCoO_3$ cathode to the TiO_2 nanotube anode with accompanying charging reactions at the electrodes, which can be defined as a piezo-electrochemical process. The fundamental mechanism that directly hybridizes the two processes into one, in which the mechanical energy is directly converted into electrochemical energy, without any intermediate step of first converting piezoelectric potential into electricity, is also proposed. Figure 11.47b shows the charging mechanism in detail, in which the polarized PVDF film creates a positive piezoelectric potential on the cathode side and a negative piezoelectric potential on the anode side under a compressive stress on the device. Li ions in the electrolyte drift from the cathode to the anode along the pores within the PVDF film in order to screen the piezoelectric field. The decreased concentration of Li^+ around the cathode will break the chemical equilibrium at the cathode ($LiCoO_2 \leftrightarrow Li_{1-x}CoO_2 + xLi^+ + xe^-$), so that Li^+ deintercalates from $LiCoO_2$ to form Li_{1-x} CoO_2. Likewise, with increasing concentration of Li^+ around the anode, the chemical equilibrium at the anode ($TiO_2 + xLi^+ + xe^- \leftrightarrow Li_xTiO_2$) is also broken, and the Li ions will move in the opposite direction, where Li^+ will react with TiO_2 to form Li_xTiO_2. During this process, Li ions migrate from the cathode to the anode continuously, and the Li-ion battery is partially charged. When the distribution of Li^+ can balance the piezoelectric field, a new equilibrium is achieved, and the self-charging process is completed (Xue et al. 2012).

Figure 11.48a illustrates a piezoelectricity-driven self-charging supercapacitor power cell with functionalized carbon cloth as the supercapacitor electrode, which shows the potential for wearable applications of hybrid cells. The working mechanism of the piezoelectric-supercapacitor hybrid cell is slightly different from that of the piezoelectric-battery hybrid cell (Fig. 11.48b). When the external stress is applied to the piezoelectric-supercapacitor hybrid device, the remnant polarization of the PVDF film is changed. In order to balance the changed remnant polarization, charge carriers will migrate towards the electrode of the supercapacitor. Two factors are considered for the charging process. First, the increased bound charge density of the PVDF surface affects the distribution of the positive ions and negative ions in the electrolyte. The piezoelectric potential causes the redistribution of the ions and causes charges to accumulate on the surface of the supercapacitor electrode. Second, the piezoelectric potential drives the migration of hydrogen ions (H^+) and SO_4^{2-} along the direction of the potential across the porous PVDF film. As the piezoelectric potential increases between the positive and negative electrodes due to an increasing external applied force, more electricity is stored on the electrodes in the form of electrochemical energy. The nonfaradaic and faradaic forces of the two electrodes reach a new equilibrium after the H^+ ions and SO_4^{2-} ions are redistributed in the electrolyte and balance the piezoelectric field in the PVDF film. This technology directly converts mechanical energy into electrochemical energy without energy being wasted on the outer circuitry and decreases energy conversion loss. Importantly, the development of both high-performance energy storage devices and the materials used in the devices is essential for the fabrication of highly effective hybridized "all-in-one energy harvesting and storage devices" in the future (Song et al. 2015; Lee et al. 2016).

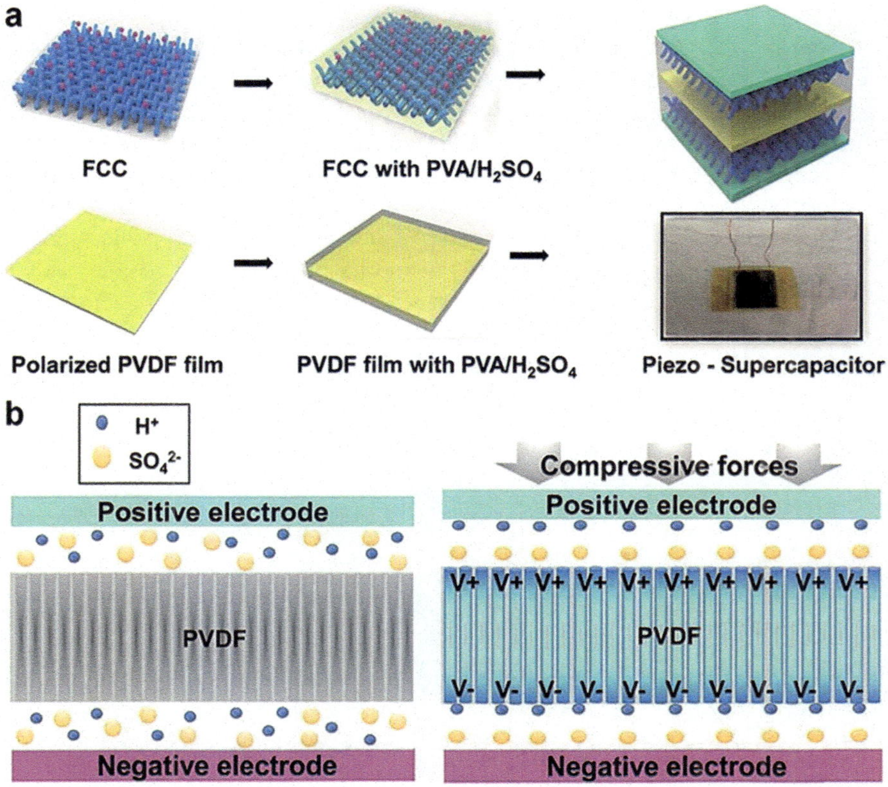

Fig. 11.48 Integration of energy harvesting and supercapacitor (Modified with permission from Song et al. (2015) (Royal Society of Chemistry)): (**a**) Schematic illustration of the process flow of hybrid piezosupercapacitor fabrication, and (**b**) proposed working mechanism of the hybrid piezo-supercapacitor

INDEPTH: Nanogenerator-Driven Self-Powered Systems

As human beings are seeking for intelligent life, devices with high functionality are the fundamental components. A bottleneck for the sustainable operation of these devices is the electric power, although the power level for each is small but the number of units can be huge. Searching for a self-powered system is a major drive in the field of nanoenergy. For example, TENG has been an exciting technology because of its high performance, high efficiency, low cost, easy fabrication, diverse choices of materials, and easy scalability. TENG can also be a high-sensitive self-powered sensor for detecting mechanical triggering, stimulation, and movement. The electric current and voltage signals generated by TENG represent the dynamic and static information, respectively, regarding a mechanical action. Such a sensor tip does not need

(continued)

a power to drive it, which is a major difference in comparison to conventional sensors. Applications can be found in areas of microelectromechanical systems, human–machine interfacing, touch-pad technology, security systems, and motion sensing. Therefore, integrating the TENG and the similar nanogenerators with a sensor to form a self-powered system considering the practical environment in which the system will be employed. The following "tree" idea regarding the development of nanogenerators starts from fundamental materials and physics effects, tiny mechanical energy can be effectively converted into electricity using nanogenerators, which can be a fundamental approach for intelligent earth with applications in internet of things, mobile/wearable electronics, fabric electronics, environmental protection, healthcare, infrastructure monitoring, and national security (Wang et al. 2015a, b).

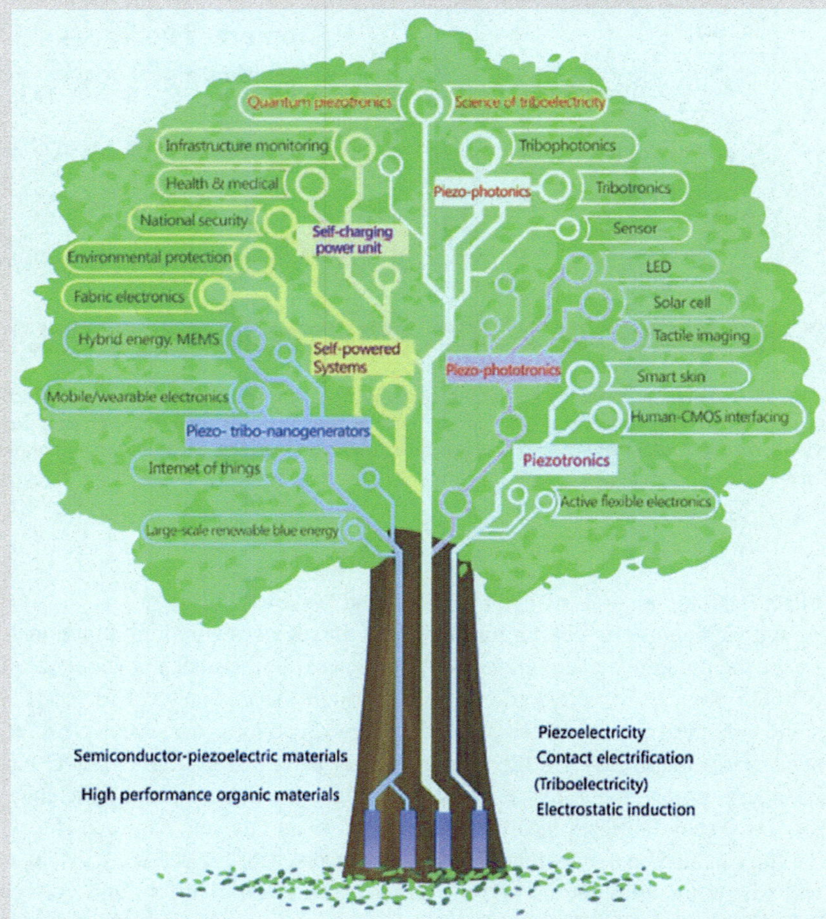

Exercises

Part I: General Questions

11.1. Define the energy-harvesting (also known as power harvesting or energy scavenging or ambient power) concept and take an example to explain a generic energy system.
11.2. Address thermoelectric materials characterizations, types, and main applications.
11.3. How to distinguish oxide and ceramic thermoelectric materials? Address their structure characterization and application differences.
11.4. Explain piezoelectric effect, and list piezoelectric materials types and their applications for energy harvesting.
11.5. Explain pyroelectric effect, and list pyroelectric materials types and their applications for energy harvesting.
11.6. List magnetostrictive and multiferroic magnetoelectric materials, and explain their differences and applications for energy harvesting.
11.7. Explain triboelectric effect, and list triboelectric materials types and their applications for energy harvesting.
11.8. Address principle of triboelectric nanogenerators, and list their types and applications.

Part II: Through-Provoking Questions

11.9. List types of hybrid energy harvesters, and their status and future trends.
11.10. Explain piezotronic and piezophototronic effects and their differences, and address their applications for energy harvesting.
11.11. Describe the scope of piezotronics and piezophototronics, and explain their differences. How do they inspire novel device applications?
11.12. Give examples to explain integration of energy harvesting and storage devices.

References

Ahn, C.-W., Maurya, D., Park, C.-S., Nahm, S., Priya, S.: A generalized rule for large piezoelectric response in perovskite oxide ceramics and its application for design of lead-free compositions. J. Appl. Phys. **105**, 114108 (2009)

Andrew, J.S., Starr, J.D., Budi, M.A.K.: Prospects for nanostructured multiferroic composite materials. Scripta Mater. **74**, 38–43 (2014)

Andosca, R.A., McDonald, T.G., Genova, V., Rosenberg, S., Keating, J., Benedixen, C., Wu, J.: Experimental and theoretical studies on mems piezoelectric vibrational energy harvesters with mass loading. Sens. Actuat. A: Phys. **178**, 76–87 (2012)

Apo, D.J.: Low frequency microscale energy harvesting. Ph.D. Disseration. Virginia Tech, Blacksburg (2014)

Aswal, D.K., Basu, R., Singh, A.: Key issues in development of thermoelectric power generators: high figure-of-merit materials and their highly conducting interfaces with metallic interconnects. Energy Convers. Manag. **114**, 50–67 (2016)

Bae, J., Lee, J., Kim, S., Ha, J., Lee, B.-S., Park, Y., Choong, C., Kim, J.-B., Wang, Z.L., Kim, H.-Y., Park, J.-J., UI, C.: Flutter-driven triboelectrification for harvesting wind energy. Nat. Commun. **5**, 4929 (2014)

Biswas, K., He, J., Zhang, Q., Wang, G., Uher, C., Dravid, V.P., Kanatzidis, M.G.: Strained endotaxial nanostructure with high thermoelectric figure of merit. Nat. Chem. **3**(2), 160–166 (2011)

Bowen, C.R., Taylor, J., LeBoulbar, E., Zabek, D., Chauhanc, A., Vaishc, R.: Pyroelectric materials and devices for energy harvesting applications. Energy Environ. Sci. **7**, 3836–3856 (2014)

Bykhovski A., Kaminski V., Shur M., Chen Q., Khan M.: Pyroelectricity in gallium nitride thin films. Appl. Phys. Lett. **69**, 3254–3256 (1996)

Casian, A.I., Sanduleac, I.I.: Organic thermoelectric materials: new opportunities. J. Thermoelectr. **3**, 11–20 (2013)

Chen, J., Zhu, G., Yang, J., Jing, Q., Bai, P., Yang, W., Qi, X., Su, Y., Wang, Z.L.: Personalized keystroke dynamics for self-powered human–machine interfacing. ACS Nano. **9**, 105–116 (2015)

Dagdeviren, C., Joe, P., Tuzmanc, O.L., Park, K.-I., Lee, K.J., Shi, Y., Huangh, Y., Rogers, J.A.: Recent progress in flexible and stretchable piezoelectric devices for mechanical energy harvesting, sensing and actuation. Extr. Mechan. Lett. **9**, 269–281 (2016)

Deterre, M., Lefeuvre, E., Zhu, Y., Woytasik, M., Bosseboeuf, A., Boutaud, B., Dal Molin, R.: Micromachined piezoelectric spirals and ultra-compliant packaging for blood pressure energy harvesters powering medical implants. IEEE 26th International Conference on Micro Electro Mechanical Systems (MEMS), pp. 249–252. IEEE, New York (2013)

Dineva, P.S., Cross, D., Müller, R., Rangelov, T.: Dynamic fracture of piezoelectric materials-solid mechanics and its applications, pp. 7–32. Springer International Publishing, Switzerland (2014)

Elsheikh, M.H., et al.: A review on thermoelectric renewable energy: principle parameters that affect their performance. Renew. Sust. Energ. Rev. **30**, 337–355 (2014)

Falconi, C., Mantini, G., D'Amico, A., Ferrari, V.: Modeling of piezoelectric nanodevices. In: Piezoelectric nanomaterials for biomedical applications. In: Ciofani, G., Menciassi, A. (eds.) Nanomedicine and nanotoxicology, pp. 93–133. Springer, Berlin, Germany (2012)

Fang, C., Jiao, J., Ma, J., Lin, D., Xu, H., Zhao, X., Luo, H.: Significant reduction of equivalent magnetic noise by in-plane series connection in magnetoelectric Metglas/Mn-doped Pb (Mg1/3Nb2/3)O3-PbTiO3 laminate composites. J. Phys. D: Appl. Phy. **48**, 465002 (2015)

Geuther, J. A., Danon, Y.: Electron and positive ion acceleration with pyroelectric crystals. Journal of Applied Physics **97**, 074109 (2005)

Gillette, S.M., Fitchorov, T., Obi, O., Jiang, L., Hao, H., Wu, S., Chen, Y., Harris, V.G.: Effects of intrinsic magnetostriction on tube-topology magnetoelectric sensors with high magnetic field sensitivity. J. Appl. Phys. **115**, 17C734 (2014)

Guduru, R., Liang, P., Runowicz, C., Nair, M., Atluri, V., Khizroev, S.: Magneto-electric nanoparticles to enable field-controlled high-specificity drug delivery to eradicate ovarian cancer cells. Sci. Rep. **3**, 2953 (2013)

Hajati, A., Kim, S.-G.: Ultra-wide bandwidth piezoelectric energy harvesting. Appl. Phys. Lett. **99**, 083105 (2011)

Han, G., Ryu, J., Yoon, W.-H., Choi, J.-J., Hahn, B.D., Kim, J.-W., Park, D.-S., Ahn, C.-W., Priya, S., Jeong, D.-Y.: Stress-controlled $Pb(Zr_{0.52}Ti_{0.48})O_3$ thick films by thermal expansion mismatch between substrate and $Pb(Zr_{0.52}Ti_{0.48})O_3$ film. J. Appl. Phys. **110**, 124101 (2011)

He, J., Liu, Y., Funahashi, R.: Oxide thermoelectrics: the challenges, progress, and outlook. J. Mater. Res. **26**(15), 1762–1772 (2011)

Hossain, M.S., Al-Dirini, F., Hossain, F.M., Skafidas, E.: High performance graphene nano-ribbon thermoelectric devices by incorporation and dimensional tuning of nanopores. Sci. Rep. **5**, 11297 (2015)

Hunter, S.R., Lavrik, N.V., Bannuru, T., Mostafa, S., Rajic, S., Datskos, P.G.: Development of MEMS based pyroelectric thermal energy harvesters. Proc. SPIE. **8035**, 80350V (2011)

Jagadish, C., Pearton, S.J.: Zinc oxide bulk, thin films and nanostructures: processing, properties, and applications. Elsevier, Amsterdam (2011)

Jenkins, K., Nguyen, V., Zhu, R., Yang, R.: Piezotronic effect: an emerging mechanism for sensing applications. Sensors. **15**, 22914–22940 (2015)

Joshi, G., et al.: Enhanced thermoelectric figure-of-merit in nanostructured p-type silicon germanium bulk alloys. Nano Lett. **8**, 4670 (2008)

Kargol, A., Malkinski, L., Caruntu, G.: Biomedical applications of multiferroic nanoparticles. In: Malkinski, L. (ed.) Advanced magnetic materials, pp. 89–118. InTech, Rijeka, Croatia (2012)

Kim, S.-G., Priya, S., Kanno, I.: Piezoelectric MEMS for energy harvesting. MRS Bull. **37**(11), 1039–1050 (2012)

Koumoto, K., et al.: Thermoelectric ceramics for energy harvesting. J. Am. Ceram. Soc. **96**, 1–23 (2013)

Lang, S.B.: Pyroelectricity: from ancient curiosity to modern imaging tool. Phys. Today. **58**, 31–36 (2005)

Lee, M., Yang, R., Li, C., Wang, Z.L.: Nanowire–quantum dot hybridized cell for harvesting sound and solar energies. J. Phys. Chem. Lett. **1**, 2929–2935 (2010)

Lee, J.H., Lee, K.Y., Gupta, M.K., Kim, T.Y., Lee, D.Y., Oh, J., Ryu, C., Yoo, W.J., Kang, C.Y., Yoon, S.J., Yoo, J.B., Kim, S.W.: Highly stretchable piezoelectric-pyroelectric hybrid nanogenerator. Adv. Mater. **26**, 765–769 (2014)

Lee, J., Kim, J., Kim, T., Hossain, M.A., Kim, S., Kim, J.: All-in-one energy harvesting and storage devices. J. Mater. Chem. A. **4**(21), 7983–7999 (2016)

Leonov, V.: Energy harvesting for self-powered wearable devices. In: Bonfiglio, A., De Rossi, D. (eds.) Wearable monitoring systems. Springer, New York (2011)

Lin, Z.-H., Cheng, G., Lin, L., Lee, S., Wang, Z.L.: Angew. Chem. Int. Ed. **125**, 12777–12781 (2013)

Lin, Z., Chen, J., Yang, J.: Recent progress in triboelectric nanogenerators as a renewable and sustainable power source. J. Nanomater. **2016**, 5651613 (2016)

Lingam, D., Parikha, A.R., Huanga, J., Jainb, A., Minary-Jolandana, M.: Nano/microscale pyroelectric energy harvesting: challenges and opportunities. Int. J. Smart. Nano Mater. **4**(4), 229–245 (2013)

Morimoto, K., Kanno, I., Wasa, K., Kotera, H.: High-efficiency piezoelectric energy harvesters of c-axis-oriented epitaxial PZT films transferred onto stainless steel cantilevers. Sens. Actuat. A: Phys. **163**, 428–432 (2010)

Music, D., Geyer, R.W., Hans, M.: High-throughput exploration of thermoelectric and mechanical properties of amorphous NbO_2 with transition metal additions. J. Appl. Phys. **120**, 045104 (2016)

Narita, F., Fox, M.: A review on piezoelectric, magnetostrictive, and magnetoelectric materials and device technologies for energy harvesting applications. Adv. Eng. Mater. **20**(5), 1700743 (1–22) (2018)

Newnham, R.E.: Properties of materials: anisotropy, symmetry, structure. Oxford University Press, Oxford (2005)

Palneedi, H., Annapureddy, V., Priya, S., Ryu, J.: Status and perspectives of multiferroic magnetoelectric composite materials and applications. Actuators **5**(1), 9 (2016)

Paluszek, M., Avirovik, D., Zhou, Y., Kundu, S., Chopra, A., Montague, R., Priya, S.: Magnetoelectric composites for medical application. In: Srinivasan, G., Priya, S., Sun, N.X. (eds.) Composite magnetoelectrics, pp. 297–327. Woodhead Publishing, Cambridge, UK (2015)

Park, K.-I., Son, J.H., Hwang, G.-T., Jeong, C.K., Ryu, J., Koo, M., Choi, I., Lee, S.H., Byun, M., Wang, Z.L., Lee, K.J.: Highly-efficient, flexible piezoelectric PZT thin film nanogenerator on plastic substrates. Adv. Mater. **26**(16), 2514–2520 (2014)

Park, T., Na, J., Kim, B., Kim, Y., Shin, H., Kim, E.: Photothermally activated pyroelectric polymer films for harvesting of solar heat with a hybrid energy cell structure. ACS Nano. **9**(12), 11830–11839 (2015)

Pillatsch, P.: Piezoelectric energy harvesting from low frequency and random excitation using frequency up-conversion. Ph.D. thesis. Imperial College, London (2013)

Priya, S., Song, H.-C., Zhou, Y., Varghese, R., Chopra, A., Kim, S.-G., Kanno, I., Wu, L., Ha, D.S., Ryu, J., Polcawich, R.G.: A review on piezoelectric energy harvesting: materials, methods, and circuits. Energ Harvest Syst. **4**(1), 3–39 (2017)

Qi, Y., Kim, J., Nguyen, T.D., Lisko, B., Purohit, P.K., McAlpine, M.C.: Enhanced piezoelectricity and stretchability in energy harvesting devices fabricated from buckled PZT ribbons. Nano Lett. **11**(3), 1331–1336 (2011)

Radousky, H.B., Liang, H.: Energy harvesting—an integrated view of materials, devices and applications. Nanotechnology. **23**(50), 502001 (2012)

Rowe, D.M.: Thermoelectrics handbook: macro to nano. CRC Press, Boca Raton, FL (2005). ISBN 978-1-4200-3890-3

Ryu, J., Kang, J.-E., Zhou, Y., Choi, S.-Y., Yoon, W.-H., Park, D.-S., Choi, J.-J., Hahn, B.-D., Ahn, C.-W., Kim, J.-W., et al.: Ubiquitous magneto-mechano-electric generator. Energy Environ. Sci. **8**, 2402–2408 (2015)

Sahraoui, A.H., Longuemart, S., Dadarlat, D., Delenclos, S., Kolinsky, C., Buisine, J.: Analysis of the photopyroelectric signal for investigating thermal parameters of pyroelectric materials. Rev. Sci. Instrum. **74**, 618–620 (2003)

Shah, Y.T.: Thermal energy: sources, recovery, and applications. CRC Press, Boca Raton, FL (2018)

Shibata, K., Suenaga, K., Watanabe, K., Horikiri, F., Nomoto, A., Mishima, T.: Improvement of piezoelectric properties of (K, Na)NbO_3 films deposited by sputtering. Jpn. J. Appl. Phys. **50**, 041503 (2011)

Snyder, J., Toberer, E.S.: Complex thermoelectric materials. Nat. Mater. **7**(2), 105–114 (2008)

Sohn, J.I., Hong, W.-K., Choi, S.S., Coles, H.J., Welland, M.E., Cha, S.N., Kim, J.M.: Emerging applications of liquid crystals based on nanotechnology. Materials. **7**, 2044–2061 (2014)

Song, R., Jin, H., Li, X., Fei, L., Zhao, Y., Huang, H., Chan, H.L.W., Wang, Y., Chai, Y.: A rectification-free piezo-supercapacitor with a polyvinylidene fluoride separator and functionalized carbon cloth electrodes. J. Mater. Chem. A. **3**, 14963–14970 (2015)

Stephan, M., Robert, J., Henry, G., Eckhard, Q., Reinhard, K., Bernhard, W.: MEMS magnetic field sensor based on magnetoelectric composites. J. Micromech. Microeng. **22**, 065024 (2012)

Szczech, J.R., Higginsa, J.M., Jin, S.: Enhancement of the thermoelectric properties in nanoscale and nanostructured materials. J. Mater. Chem. **21**, 4037–4055 (2011)

Ting-Ta, Y., Hirasawa, T., Wright, P., Pisano, A., Liwei, L.: Corrugated aluminum nitride energy harvesters for high energy conversion 26. Effectiveness. J. Micromech. Microeng. **21**, 085037 (2011)

Varghese, R.P.: MEMS technologies for energy harvesting and sensing. Ph.D. Disseration. Virginia Tech, Blacksburg (2013)

Wang, Z.L.: Toward self-powered nanosystems: from nanogenerators to nanopiezotronics. Adv. Funct. Mater. **18**(22), 3553–3567 (2008)

Wang, Z.L.: Piezotronic and piezophototronic effects. J. Phys. Chem. Lett. **1**(9), 1388–1393 (2010)

Wang, Z.L., Zhu, G., Yang, Y., Wang, S., Pan, C.: Progress in nanogenerators for portable electronics. Mater. Today. **15**(12), 532–543 (2012)

Wang, J.-J., Su, H.-J., Hsu, C.-I., Su, Y.-C.: Composite piezoelectric rubber band for energy harvesting from breathing and limb motion. J. Phys. Conf. Ser. **557**, 012022 (2014)

Wang, Z., Pan, X., He, Y., Hu, Y., Gu, H., Wang, Y.: Piezoelectric nanowires in energy harvesting applications. Adv. Mater. Sci. Eng. **2015**, 1–21 (2015a)

Wang, Z.L., Chen, J., Lin, L.: Progress in triboelectric nanogenerators as a new energy technology and self-powered sensors. Energy Environ. Sci. **8**, 2250–2282 (2015b)

Whatmore, R.: Pyroelectric devices and materials. Rep. Prog. Phys. **49**, 1335 (1986)

Wooldridge, J., Blackburn, J.F., McCartney, N.L., Stewart, M., Weaver, P., Cain, M.G.: Small scale piezoelectric devices: pyroelectric contributions to the piezoelectric response. J. Appl. Phys. **107**, 104118–104118-6 (2010)

Xu, R.: Energy harvesting for microsystems. Ph.D. dissertation. Technical University of Denmark (DTU), Denmark (2012)

Xu, R., Kim, S.-G.: Low-frequency, low-G MEMS piezoelectric energy harvester. J. Phys. Conf. Ser. **660**, 012013 (2015)

Xu, C., Wang, X., Wang, Z.L.: Nanowire structured hybrid cell for concurrently scavenging solar and mechanical energies. J. Am. Chem. Soc. **131**(16), 5866–5872 (2009)

Xue, X., Wang, S., Guo, W., Zhang, Y., Wang, Z.L.: Hybridizing energy conversion and storage in a mechanical-to-electrochemical process for self-charging power cell. Nano Lett. **12**, 5048–5054 (2012)

Yan, Y., Priya, S.: Multiferroic magnetoelectric composites/hybrids. In: Kim, C.-S., Randow, C., Sano, T. (eds.) Hybrid and hierarchical composite materials, pp. 95–160. Springer International Publishing, Cham, Switzerland (2015)

Yang, Y., Wang, Z.L.: Hybrid energy cells for simultaneously harvesting multi-types of energies. Nano Energy. **14**, 245–256 (2015)

Yang, Y., Guo, W., Pradel, K.C., Zhu, G., Zhou, Y., Zhang, Y., Hu, Y., Lin, L., ZL, W.: Pyroelectric nanogenerators for harvesting thermoelectric energy. Nano Lett. **12**, 2833–2838 (2012a)

Yang, Y., Jung, J.H., Yun, B.K., Zhang, F., Pradel, K.C., Guo, W., Wang, Z.L.: Flexible pyroelectric nanogenerators using a composite structure of lead-free $KNbO_3$ nanowires. Adv. Mater. **24**, 5357–5362 (2012b)

Yang, Y., Zhang, H., Lin, Z.-H., Liu, Y., Chen, J., Lin, Z., Zhou, Y.S., Wong, C.P., Wang, Z.L.: A hybrid energy cell for self-powered water splitting. Energy Environ. Sci. **6**, 2429–2434 (2013)

Yoon, G.C., Shin, K.-S., Gupta, M.K., Lee, K.Y., Lee, J.-H., Wang, Z.L., Kim, S.-W.: High-performance hybrid cell based on an organic photovoltaic device and a direct current piezoelectric nanogenerator. Nano Energy. **12**, 547–555 (2015)

Yu H., Kang B., Park C., Pi U., Lee C., Choi S.-Y.: The fabrication technique and electrical properties of a free-standing GaN nanowire. Appl. Phys. A **81**, 245–247 (2005)

Yue, K., Guduru, R., Hong, J., Liang, P., Nair, M., Khizroev, S.: Magneto-electric nano-particles for non-invasive brain stimulation. PLoS One. **7**, e44040 (2012)

Zhang, X., Zhao, L.D.: Thermoelectric materials: energy conversion between heat and electricity. J. Mater. **1**, 92–105 (2015)

Zhang, Y., Mehta, R.J., Belley, M., Han, L., Ramanath, G., Borca-Tasciuc, T.: Lattice thermal conductivity diminution and high thermoelectric power factor retention in nanoporous macroassemblies of sulfur-doped bismuth telluride nanocrystals. Appl. Phys. Lett. **100**, 1193113 (2012)

Zhou Y., Apo D. J., Priya S.: Dual-phase self-biased magnetoelectric energy harvester. Appl. Phys. Lett. **103**, 192909 (2013)

Zhou, Q., Lau, S., Wu, D., Shung, K.K.: Piezoelectric films for high frequency ultrasonic transducers in biomedical applications. Prog. Mater. Sci. **56**, 139–174 (2011)

Zhou, Y.S., Zhu, G., Niu, S., Liu, Y., Bai, P., Jing, Q., Wang, Z.L.: Nanometer resolution self-powered static and dynamic motion sensor based on micro-grated triboelectrification. Adv. Mater. **26**(11), 1719–1724 (2014)

Zhu, T., Ertekin, E.: Phonon transport on two-dimensional graphene/boron nitride superlattices. Phys. Rev. B. **90**, 195209 (2014)

Zhu, G., Chen, J., Zhang, T., Jing, Q., Wang, Z.L.: Radial-arrayed rotary electrification for high performance triboelectric generator. Nat. Commun. **5**, 3426 (2014)

Perspectives and Future Trends

12

Abstract

Future materials need to meet great challenges for provisioning a clean, safe, secure, and sustainable energy supply to underpin a reasonable standard of living for the world's population. Advanced structural and functional materials will evolve incrementally in fossil fuel and nuclear power generation, renewable power generation, energy storage, and electricity transmission with extended operating life, increased environmental resistance, and intelligent monitor systems. This chapter will present perspectives and future trends of novel materials development in advanced energy systems, including sustainability and materials security, metamaterials and nanomaterials, artificial photosynthesis, structural power composites, energy storage materials, and alternative hybrid systems.

12.1 Energy Sustainability with Materials Security

Energy is the biggest challenge of mankind. It is the basic prerequisite of any industrial activities, and necessity for the functioning of human communities and suppliers. As the population grows and living standard of people around the world gets higher, the demand of energy is drastically increased. At the same time, the Earth is colliding with its limits. The climate change is accelerated, which compels carbon dioxide emissions to be cut drastically. Raw materials are depleted, which makes the industrial base to be renewed and the recycling economy to be developed. To meet these challenges, today, around the world, there is a radical transition towards sustainable energy, which combines smartness, flexibility, and environmental performance with customer acceptance and engagement. New energy system will be a complex combination of central and local resources, including production, storages, and new, efficient, and flexible loads. This calls for new thinking of primary energy sources, energy production, energy markets, energy transmission,

C. Tong, *Introduction to Materials for Advanced Energy Systems*,
https://doi.org/10.1007/978-3-319-98002-7_12

energy use, and customers as producer-consumers (prosumers). To enable this transition, a multidisciplinary approach is needed that combines several views of the energy system through the whole chain from primary energy resources till efficient energy use. Key factors in the implementation of sustainable energy future are digitalization, new thinking in the energy markets and conversion of passive energy users to active prosumers. In the energy demand side, a massive transformation is being made towards intelligent and energy-efficient buildings and communities, self-powdered energy-harvesting and hybrid electronic application systems, as well as intelligent and electrified transportation.

12.1.1 Efficient Use of Energy-Intensive Materials

Energy efficiency is associated with economic efficiency and includes technological, organizational, and behavioral changes. Reducing the amount of energy and materials used per unit in the production of goods and services can contribute both to the alleviation of environmental stress and to greater economic and industrial productivity and competitiveness. Material efficiency in industrial production, on the other hand, can be defined as the amount of a particular material needed to produce a particular product. Material efficiency can be improved either by reducing the amount of the material contained in the final product (light-weighting), or by reducing the amount of material that enters the production process but ends up in the waste stream. Some components of material efficiency can therefore be identified, such as light-weighting in the production process; waste reduction in the production process; and recycling of material in the production-consumption cycle (Peck and Chipman 2007). New technology development may revolutionize the current concept and promote materials efficiency with different approaches.

There have been sustained efforts since the 1970s oil crises to reduce weight in transport systems for realizing improved energy efficiency. These efforts have been pursued in parallel with more efficient fossil fuel combustion engine technologies. For instance, aluminum and magnesium alloys, carbon- and glass-fiber-reinforced epoxy composites, and metallic and composite foams will increasingly find applications in high-volume vehicle manufacture. This penetration will not be enabled solely by new materials (of which recyclable and out-of-autoclave, quickly processed thermoplastics are likely the most significant), but rather by increased sophistication in design and simulation of hybrid structures for lightweight, safety, and crash-worthiness. These structures will contain many types of material, each optimized for its specific geometrical, mechanical, and other requirements, manufactured using faster, scalable multi-material joining and assembly processes (Cantor et al. 2008; Grant 2013).

The competitive pressure exerted by penetration of aluminum alloys into chassis and body applications in the automotive industry has provided a virtuous stimulus for incremental developments in steels for light-weighting (Cantor et al. 2008); new hybrid and composite structures exert similar pressure on aluminum- and magnesium-based components. The increasing use of electric-based vehicles, most likely through increasing penetration of electric-petrol hybrids, will not ease the pressure for light-weighting since the associated engine and powertrain technologies arguably place even more demands on reduced vehicle weight to provide acceptable performance (especially range). However, this increasing hybridization of materials for reduced weight will create a tension with the increasing requirement for disassembly and easy segregation of materials, for example, in the context the end of vehicle life responsibilities of manufacturers (Day 2005).

In the aerospace industry, a similar competition between lightweight technologies has led to structural composites now constituting up to 50% or more of the un-laden airframe weight at the expense of metallic alloys; in turn, the development of a new lightweight aluminum alloys has been stimulated so that the composite fraction in civil airliners is unlikely to increase further. Future airframes will comprise therefore highly optimized and intricate mixtures of aluminum alloys and composite materials, in hybrid materials systems along with titanium alloys (Grant 2013).

As in other areas of structural composites, such as wind power, lightweight composite structures also offer new opportunities unavailable in monolithic materials, such as the ability to exploit the way in which polymeric/epoxy-based composites are manufactured as a means to introduce or embed materials alongside the reinforcement (typically long fibers) or matrix (typically epoxy-based) phase to realize additional function. Various embedded sensing and actuation elements have been demonstrated for strain measurement or control surface actuation, as well as self-healing capabilities. At some extent, the distinction between structural and functional elements will become blurred, with material hybridization likely occurring from the nanoscale, such as functional particles in coatings to control surface, radar or other properties, up to the platform scale with imaging, other sensing and energy storage capabilities embedded within the composite structure (Grant 2013).

Flexible printed electronics and smart textiles function as stretchable, wearable devices and textiles but have additional functionalities such as extreme hydrophobicity, sensing, actuation, energy harvesting and storage, data storage, and communication. Examples of potential applications of this technology include smart energy systems, military garment devices, biomedical and antimicrobial textiles, and personal electronics. In addition, improvements in sensors, flexible and printable electronics, and energy devices are ongoing for wider implementation, and nanomaterials and/or their hybrids are enabling the next phase convergence of self-powered textiles, electronics, and informatics. They are opening the way for the integration of electronic components and sensors (e.g., heat and humidity) in high strength, flexible, and electrically conductive textiles with energy storage and harvesting capabilities, biological functions, antimicrobial properties, and many other new functionalities.

12.1.2 Materials with Reduced Environmental Impact Through Life

Replacement of materials which have been deemed harmful to the environment or hazardous to health under directives such as the European Union Regulation on

Registration, Evaluation, Authorization and Restriction of Chemicals (REACH) legislation (REACH 2006) creates opportunities for new materials. Products using or containing metals such as hexavalent chromium, lead, cadmium, bismuth, and others, plus a range of organic and other chemicals must be adapted or replaced, and this trajectory of materials replacement will continue for the foreseeable future. Near term developments will focus on novel surface treatments primarily for metals such as aluminum that remove harmful chemicals from the production process and/or final product. Medium- and long-term trends will concern the advent of cheap and scalable nanotechnology, for instance in the form of nano-scale materials such as nano-tubes, particles, flakes, and wires embedded in various matrices, and which will in turn mandate informed use of these materials. Development of new experimental techniques for the generation of objective data needed to inform understanding of the environmental effects of nanomaterials is itself an important opportunity for nano-science and nano-materials research. There is a need for materials with lower environmental impact in their production, end-use, and recyclability, and this generates opportunities for cross-industry technology transfer, for example, the penetration of lightweight materials and mechanically efficient hybrid structures into the building industry, including 3D truss and node structures in hybrid materials, offering outstanding load carrying capability at minimum material use (Grant 2013).

Metals are generally the most recyclable materials and driven by likely further increases in commodity prices, further improvements in the fraction of metals recycled "closed loop" should be pursued. However, metals are also among the most readily corroded and degraded materials and the development of high-performance coatings and surface technologies for metallic systems is often an integral part of creating high value-added components and systems. It is likely that future developments will include coatings with increased functionality, such as biocidal and corrosion-resistant coatings, polymer-based coating systems that can generate electricity (solar, mechanical harvesting), or surface treatments to control emissivity, acoustic properties, and even electromagnetic compatibility to improve wireless communications in high building density environments. Structural materials will also take on additional functions, such as thermal energy storage, for example, by inclusion of micro-spheres of phase change materials to absorb heat during the day through a phase change reaction, and to release heat as the ambient temperature falls. Cost will always be a constraint on new technologies in the construction industry, but an increasing shift to offsite, mass production techniques will facilitate the integration of embedded functionalities into structural materials for construction (Grant 2013).

Concerns surrounding the safe disposal and increased recirculation of packaging materials, and avoiding their release into the environment, will increase and innovative solutions will emerge in biodegradable packaging, with the most pressing needs to increase the biodegradability of plastic packaging materials. These materials will be developed, for example, by fermentation of plant sugars and oils in large-scale microbial factories. Products will be robust in everyday use but designed to be sensitive and easily broken down by targeted enzymes in the waste stream environment. Smart packaging concepts are now well developed, such as dynamic food freshness indication (integrated sensing and display), embedded radio-frequency identification (RFID) tags, temperature control (integrated sense and actuation), anti-counterfeit function (embedded holograms, circuits), and embedded energy harvesting. However, alongside regulatory and disposal issues (where biodegradable organic based electronics will be enabling), more capable manufacturing technology (scale, flexibility) and reduced cost are required. For example, roll-to-roll techniques are capable of keeping pace with mass-scale packaging production (Grant 2013).

12.1.3 New Materials Technologies and Processes to Support Increased Recirculation of Materials

At present, the world produces ~37 million tons of aluminum and over ~2 billion tons of steel every year, accounting for 6–7% of the total global CO_2 emission. A life cycle assessment for the aluminum industry suggested that the production of 1 kg of primary aluminum, when all the electricity generation and transmission losses were included, required 45 kWh of energy and emitted 12 kg of CO_2, whereas 1 kg of recycled Al required only 2.8 kWh (5%) and produced 0.6 kg CO_2 (5%). Once Al has been obtained, major energy is also required to convert it to a final product, which increases with the extent of mechanical work and complexity of final product shape. Although aluminum requires particularly large amounts of energy for its primary extraction, high energy needs and CO_2 emissions are also associated with obtaining all the important engineering metals for alloys (iron, nickel, titanium, magnesium, copper), and their conversion to products. Therefore, potential energy and emissions savings of recycling will drive up recycling rates further which must be enabled by better segregation at source and by use of high-volume segregation technologies. A critical aspect of metal recycling, and the virtuous move to closed loop recycling of metals at a national scale, is the ability to introduce recycled material back into high-value products without degradation of properties. Materials standards, often specified in terms of alloy chemistry, maintain a reliance on a high fraction of primary metal in the production of components. However, future years may see a relaxation of chemical tolerances because of an increase in the use of novel and more carefully controlled processing to achieve the same performance from less "clean" feedstocks. Critical to this shift will be a greater understanding of the undermining effects of currently perceived deleterious elements (e.g., embrittlement of aluminum alloys by build of iron and silicon concentrations) and the development of new, more tolerant processing techniques (Grant 2013).

The increasing penetration of polymers—especially thermosets such as epoxy-based materials—into large-scale structural applications, usually in the form of composites, presents difficult environmental challenges: both matrix and fiber (especially carbon-based fibers) require large amounts of energy in their fabrication, and the resulting materials' inherent environmental stability limits disposal options, especially within a tightening legislative framework. Techniques (mechanical, thermal, hybrid) that extract fibers from the matrix while maintaining an acceptable fraction of their virgin properties must be developed, and the key challenges are the development of new markets and products for recovered fibers of sufficient volume and return, and acceptable disposal or reuse of the matrix residue (Grant 2013).

Alongside metals and composites, perhaps the biggest challenges and opportunities in materials circulation relate to electronic materials. The amount of electronic equipment has increased continuously for at least the last two decades and will likely continue; at the same time there has been a reduction in electronic equipment prices facilitated by massive consumer demand and product volumes, which have led to a drastically reduced life span for most electronic equipment. Similarly to directives controlling end of vehicle life and requirements for their disposal, the disposal of electronic goods is increasingly regulated (Ylä-Mella et al. 2014).

For a mobile phone as an example, much of the materials in electronic waste has low inherent value (ferrous, plastic) and the potentially valuable materials (indium, rare earths) are present only in low contractions, usually much less than 1% by weight. Therefore, efficient and commercially practical recovery and recycling/reuse of the more valuable materials becomes critically dependent on concentrating the fraction electronic waste, separately from other types of waste. Securing this segregation is primarily an issue of consumer behavior. Two intriguing trends may emerge in the future (Grant 2013): the development of designs and materials that allow the consumer to disassemble goods before disposal, and the mining of nonsegregated waste in very large, existing waste sites.

12.1.4 Bio-Derived Materials for Sustainability of Resources

An approach to circumvent the problems of materials that are increasingly difficult to dispose, or those materials whose price is strongly linked to a limited natural supply (such as rare earth metals), or where the entire security of supply is vulnerable, is to replace them with materials derived from sustainable sources or materials that are readily bio-degraded after use. This has proved easier in theory than in practice but rising costs of disposal and increasing and volatile prices of some metals. Most progress has been made in finding alternatives to petro-chemically derived materials that suffer strong price volatility, such as matrices and fibers for structural composites derived from more sustainable bio-feedstocks, including hemp and flax for the fibers, and cellulose, starch, lactic acid for the matrix. However, the design of composites based on bio-derived materials will require more finesse and know-how than their petrochemical counterparts since the bio-derived material properties usually have inferior properties (Grant 2013).

Biocomposites have been developed with a huge diversification of potential biofeedstocks providing the matrix and the fiber. The automotive industry presents possibly the greatest opportunities for biocomposites, offering massive volumes to retrieve economies of scale, and where carbon fiber supply cannot meet forecast need. The construction industry also offers significant market opportunities for structural applications in relatively lightly loaded sections, especially those under compression. To meet these and other market needs, continued improvements will be required in matrix and fiber property and reproducibility, design understanding (including interfacial design between fiber and matrix and integration of biocomposites into hybrid structures with conventional materials) and environmental stability (partly enabled by the use of coatings), especially fire resistance. Other opportunities include the use of genetic engineering of materials for bio-feedstocks to maximize the yield of useful constituents, and synthetic biological and biomimetic approaches to new materials, such as those currently being explored for synthetic spider silk (Porter et al. 2013).

Large-scale manufacturing of bio-derived materials presents some interesting aspects: the front-end of the supply chain is familiar with mass production (harvesting) and the large-scale movement and storage of raw materials, but is completely unfamiliar with materials-related specifications, and the yield of useable materials from bio-feedstocks for structural applications at large-scale needs to be well-understood (Grant 2013).

12.2 Metamaterials and Nanomaterials for Energy Systems

To meet the continuously growing demand for energy, and eliminate its effect on climate, it is not only necessary to economically utilize renewable alternatives to fossil fuels, but also to optimize the whole value-added chain of energy for energy-harvesting development and conversion, transport, and storage up to consumers' utilization. Innovation and increases in efficiency in conjunction with a general reduction of energy consumption are urgently needed in all fields to reach the high aims as the world population is growing and striving for more prosperity. Metamaterials and nanomaterials-based nanotechnologies as key and cross-sectional technologies exhibit the unique potential for decisive technological breakthroughs in the energy system, thus making substantial contributions to sustainable energy supply. Nanotechnologies and metamaterials-based multifunctional energy-harvesting capabilities provide the potential to enhance energy efficiency across all branches of industry and to economically leverage renewable energy production through new technological solutions and optimized production technologies. Nanotechnology innovations and introduction of metamaterials and metadevices could impact each part of the value-added chain in the energy system, as shown in Fig. 12.1. In addition, self-powered micro-/nanosystems (MNSs) have been developed to operate without an external electricity source and instead draw the energy they require from the environment in which they are used. Therefore, essential contributions to sustainable energy supply and the global climate protection will eventually be achieved (Luther 2008; Wang and Wu 2012).

12.2.1 Metamaterials and Metadevices

Metamaterials are artificially engineered composites with designed properties beyond those attainable in nature with applications in all aspects of materials science. From spatially tailored dielectrics to tunable, dynamic material properties

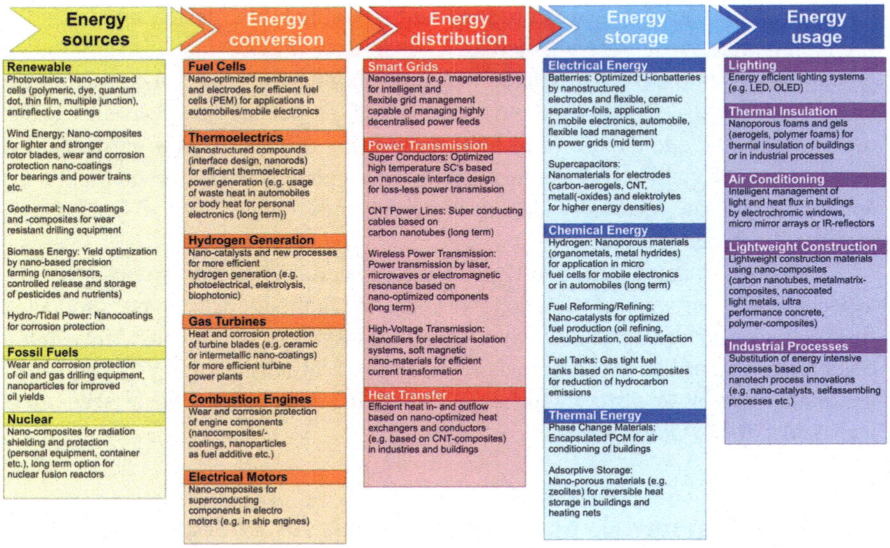

Fig. 12.1 Examples for potential applications of nanotechnology along the value-added chain in the energy system (Adapted from Luther (2008) (HA Hessen Agentur GmbH). Credit: Hessian Ministry of Economy, Transport, Urban and Regional Development)

and unique nonlinear behavior, metamaterial systems have demonstrated tremendous flexibility and functionality in electromagnetic, optical, acoustic, thermal, and mechanical engineering. Furthermore, the field of metamaterials has been extended from the mere pursuit of various exotic properties towards the realization of practical devices, leading to the concepts of dynamically reconfigurable metadevices and functional metasurfaces (Tong 2017).

The origin of metamaterials can be dated back centuries ago, examples like Pyramid brick wall, Parthenon columns, and medieval ruby glass, as shown in Fig. 12.2. Furthermore, the work on rotation of the plane of polarization by artificial twisted structures in 1898, and artificial dielectrics for microwave antenna lenses in 1945, as well as later artificial structured electromagnetic materials such as frequency selective surfaces (FSS) and electromagnetic band gap (EBG) structures, are other examples of metamaterials. However, the modern metamaterial (formally named) caused attention when Prof. Pendry and his team predicted that conducting wire arrays can work as a relatively lower frequency plasma (<200 THz) with a negative value of effective permittivity, while arrays of split-ring resonators (SRRs) can support strong magnetic resonances leading to a negative value of effective permeability, as shown in Fig. 12.3. By combining these two structures and overlapping the frequency bands of negative permittivity and permeability, the first modern metamaterial with negative refractive index was demonstrated in 2000; in a negative index medium, the directions of wave vector and energy flux are opposite, which is also known as backward waves. These proved predictions of exotic behavior for negative index materials by Prof. Veselago in 1968, such as negative refraction, reversed Doppler effect, and reversed Cherenkov radiation (Liu 2015).

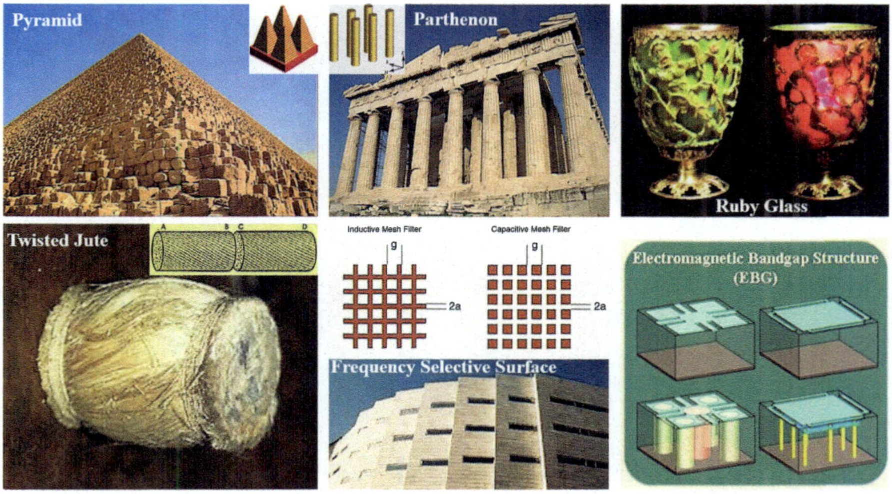

Fig. 12.2 Examples of original metamaterials (Modified with permission from Leonhardt (2007) (Springer Nature), Optical metamaterials: Invisibility cup, Nature Photonics 1(4):207–208)

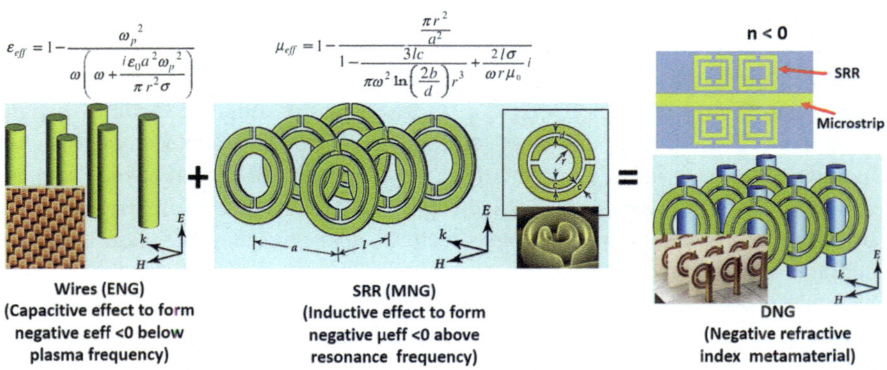

Fig. 12.3 Metamaterial designed through resonant approach (Modified with permission from Tong (2017) (Springer International Publishing AG))

The approach for pushing the concept of negative refractive index towards higher frequencies were initially to scale down unit cell sizes from the microwave regime (Such as SRRs), as the magnetic resonance frequency of the SRR is inversely proportional to its size. For example, changing the double SRR to a single one and reducing the size of the structure, the metamaterials were realized from microwave, to THz range and near-infrared optical range (up to 200 THz). However, this scaling the single SRRs breaks down for higher frequencies because, for wavelengths shorter than the 200 THz range, the metal starts to strongly deviate from an ideal conductor. For a small SRR, nonideal metal behavior leads to a modified scaling law where the frequency approaches a constant and becomes independent of the SRR size. This scaling limit combined with the fabrication difficulties of making nanometer-scale SRRs along with metal wires led to the development of alternative

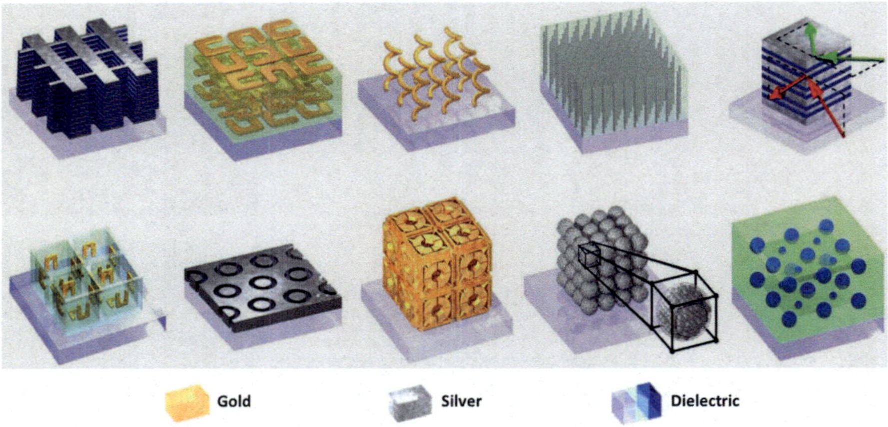

Fig. 12.4 Examples of metamaterial structures fabricated through repetition of meta-atoms or meta-molecules (Adapted with permission from Soukoulis and Wegener (2011) (Springer Nature))

designs that are more suitable for the THz and optical regimes. One suitable design for optical NIMs is based on pairs of metal rods (also called "cut-wires") or metal strips, separated by a dielectric spacer, as shown in Fig. 12.3. Such structures can provide a magnetic resonance $\mu < 0$ originating from antiparallel currents in the strips. An electric resonance with $\varepsilon < 0$ can result from the excitation of parallel currents in the same strips. However, normally it is difficult to get the $\varepsilon < 0$ and $\mu < 0$ regions to overlap, so a different design was proposed, the so-called "fishnet" structure. So far NIMs at optical wavelengths have been successfully demonstrated by using coupled cut-wires, fishnet structures, and coaxial waveguide arrays. In addition, many theoretical concepts have been suggested based around nanofabricated composites. One example is metal nanowires and nanowire plasmonic materials, where nanowires arranged into parallel pairs can act as a left-hand material with the effective magnetic permeability and dielectric permittivity that are both negative in the visible and near-infrared spectral ranges. Various composites as shown in Fig. 12.4, based on metallic and dielectric nanostructures that have macroscopic negative refraction and act as left hand materials, have been developed (Kivshar 2015; Tong 2017).

On the other hand, the application of metamaterials such as the superlens that can overcome the diffraction limit attracted great attention and further stimulated the development of the metamaterials. The initial quest for negative refraction in anisotropic media brought in the concept of indefinite media and hyperbolic dispersion, generally referred to hyperbolic metamaterials. Such type of metamaterials can possess a giant photonic density of states, which would find application in a multitude of topics from super-resolution imaging, quantum photonics to nonlinear optics, bio-sensing and optical circuit components (Liu 2015; Kivshar 2015).

With the rapid development of the field, it was soon realized that metamaterials possess much broader potential. In contrast to the Bragg resonances in photonic crystals, localized resonances of the composite units (meta-atoms) play a dominant role in the properties of metamaterials. This provides a possibility for manipulation of the local electric and magnetic response of the structure, simply through modifying the geometry of each meta-atom. Based on such great flexibility, the new and fruitful research areas of transformation optics have skyrocketed since 2006, which provides a powerful methodology for designing structures with unconventional electromagnetic behavior, such as invisibility cloaks, exotic lenses, and light-harvesting structures. A further direction is to achieve the dynamic control over the exotic properties of metamaterials by including nonlinear or tunable elements into passive metamaterials, such as liquid crystals; alternatively, one can design meta-molecules with deformable structures. For example, mechanical tuning is now widely used in THz metamaterials. A post-processing approach for mechanical tuning of the electromagnetic properties of metamaterials may be used in applications which require precise engineering of metamaterial resonances (Kivshar 2015). Although early theoretical and experimental studies on metamaterials and transformation optics were conducted with electromagnetic waves, the concepts were soon extended to other fields, including the manipulation of various acoustic and mechanical waves, heat flows, and even matter waves (Liu 2015).

In addition to three-dimensional metamaterials, the thin planar version of metamaterials, now generally dubbed meta-surfaces, have been developed. For homogenizable meta-surfaces (most studied thin metamaterials with periodic meta-atoms fall into this category), it is more meaningful to characterize the constitutive relation with surface susceptibility rather than effective permittivity and permeability for bulk materials. The strong resonances of meta-atoms can exert abrupt phase changes in their scattered waves; on account of Huygens' principle, meta-surfaces with spatial phase gradient are able to mold the scattered wave-fronts arbitrarily. Although, in terms of functionality, the design philosophy of gradient meta-surfaces is closely related to the well-established fields of phase array antennas, holography, and diffractive optics, its underlying mechanism for generating abrupt phase change is totally different, and thus it provides much greater design flexibility in manipulating the phase, amplitude, and polarization by interacting with electric or magnetic component (or both) with optically thin devices from the subwavelength scale. One of the important directions in the physics of metasurfaces is the use of graphene as a component of metadevices. For photon energies below a certain level, defined by a chemical potential, graphene exhibits a metal-like conductivity. In this regime, similar to metals, graphene can support transverse magnetic (TM) electromagnetic surface plasmon polaritons, and they represent coupled state of the electromagnetic field and electrons. For the range of frequencies above the chemical potential, graphene has dielectric characteristics and it supports transverse electric (TE) surface waves. Graphene plasmonics is a rapidly growing new field of physics which utilizes concepts of conventional metal plasmonics combined with the unique electronic and optical properties of graphene. It can be expected that a future platform for highly integrated electromagnetic signal processing and distribution will emerge which will combine nonlinear, memory, and switchable functionalities with transformation optics' ability to guide light via the engineered electromagnetic space, using metamaterials with spatially variable parameters (Kivshar 2015; Tong 2017).

Furthermore, the exploration of metamaterials is shifting towards integrated metadevices to achieve tunable, switchable, nonlinear, and sensing functionalities, which are realized by structuring of functional matter on the subwavelength scale. On and beyond photonic, terahertz and microwave electromagnetic, metamaterials and metadevices with functionalities have been attained through the exploitation of phase-change media, semiconductors, graphene, carbon nanotubes and liquid crystals. More and more metadevices are emerging, such as microelectromechanical, acoustic, mechanical metadevices, and metadevices engaging the nonlinear and quantum response of superconductors, electrostatic and optomechanical forces, as well as nonlinear metadevices incorporating lumped nonlinear components (Tong 2017).

For example, metamaterials and metadevices with exotic properties have been introduced for harvesting ambient energies to power directly the electronics or recharge a secondary battery. They possess unique properties not easily achieved using naturally occurring materials, such as negative stiffness, mass, Poisson's ratio, and refractive index. Metamaterials-based energy harvesting (EH) includes, but not limited to, active metamaterials-based EH, metamaterials-based thermal EH, and metamaterials-based multifunctional EH capabilities (Chen et al. 2014):

(a) Electromagnetic energy harvesting
The microwave spectrum is defined as electromagnetic energy ranging from approximately 1 to 1000 GHz in frequency. A complete microwave wireless power transmission (MWPT) system consists of three essential parts: converting DC energy to microwave power, capturing the microwave power by antennas, and converting the captured power to DC output power. In particular, the rectifying antenna called rectenna is one of the key components affecting the performance of a MWPT system. Metamaterial-inspired rectenna design methods have been explored to improve its performance. For instance, an RF energy-harvesting platform converts Wi-Fi and other RF bands to electricity, to power IoT sensors. It consists of a metamaterial-inspired antenna and a custom-rectifying circuit.

(b) Phononic crystals-based vibroacoustic energy harvesting
Vibroacoustic energy harvesting (VAEH) uses basic conversion mechanisms including piezoelectric, electromagnetic, electrostatic, and magnetostrictive transduction. Phononic crystals-based vibroacoustic energy harvesting has been used to enhance the VAEH device design to operate optimally at or very close to resonance, so that its electrical output can reach the maximum value. Bandgaps of a phononic crystal are closely related to its structural configuration, therefore can be used to design the structures of phononic crystals to match specific vibroacoustic energy. On the other hand, the central frequency of any bandgap depends on the lattice constant of a phononic crystal. Large lattice constant leads to low central frequency, so the whole size of a phononic crystal has to be large for low-frequency vibroacoustic energy. This property may

hinder phononic crystals from low-frequency vibroacoustic energy harvesting due to the limitations of space and intensity in many engineering applications.

(c) Acoustic metamaterial-based vibroacoustic energy harvesting
Acoustic metamaterials can be designed to exhibit nontraditional physical behaviors such as negative stiffness, mass, and Poisson's ratio. These unique abilities of acoustic metamaterials can be used to enhance vibroacoustic energy harvesting. Similar to the mechanism of phononic crystals, bandgap is also the basis of acoustic-based broadband vibroacoustic energy harvesting. However, there are two different principles. The first one is "acoustic superlens," which can be used to concentrate vibroacoustic energies. The other one is "local resonance," which can be used to localize vibroacoustic energies by setting defects. Compared to phononic crystals, acoustic metamaterials with small size can generate low-frequency band, so they are more applicable for low-frequency vibroacoustic energy harvesting.

The future energy-harvesting development will still focus on wide bandwidth, frequency self-adaptation, small scale, high efficiency, etc. Moreover, several factors have to be considered simultaneously in most cases. These challenges cannot be overcome solely by natural materials. By combing structural configurations of metamaterials and transmission optics, the progress of energy harvesting will be boosted. For instance, active metamaterials can be used for tunable energy harvesting, which can alter their properties in response to an external input. Active acoustic metamaterials could be systematically designed to achieve desired material parameters, where a transducer senses the pressure wave incident on the metamaterial and an electronic circuit manipulates the electric signal produced to drive a second transducer that creates the acoustic response consistent with the desired effective material parameters, for instance. Moreover, metamaterials will further be extended for thermal energy harvesting. For example, a thermal metamaterial based on multiple ordered arrays consecutively linked by means of transversal microspacers to obtain high thermal gradients will enhance silicon-based thermoelectric generation for energy-harvesting applications in advanced micro/nanodevices; an efficient solar energy absorber based on metamaterial has a very high absorption performance in the entire solar spectrum with wide receiving angle, omni-direction. In addition, metamaterials-based multifunctional structures with energy-harvesting abilities will be developed. Load-bearing composite materials and structures with integrated functionalities of sensing, actuation, and energy transduction are an area of current interest in multifunctional composites. For example, as vibrations in engineering structures are harmful, it is a promising way to harness these structural vibrations by simultaneously mitigating them. Moreover, harvested useable energy can provide power for online sensing and diagnostic capabilities. Multifunctional piezoelectric self-charging structures have been developed, where piezoelectric devices are combined with thin-film lithium batteries and a substrate layer in order to simultaneously harvest energy, store energy, and carry structural load. However, a potential issue is that added mass and increased drag may significantly degrade structural performances. Fortunately, metamaterials will easily

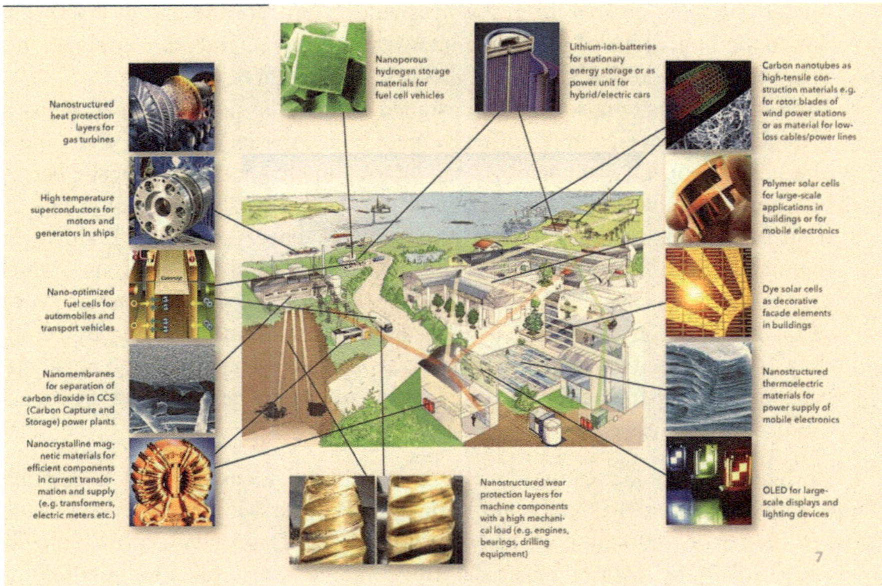

Fig. 12.5 Scenario with examples for potential applications of nanotechnologies in the energy system (Adapted from Luther (2008) (HA Hessen Agentur GmbH). Credit: Hessian Ministry of Economy, Transport, Urban and Regional Development)

enable the creation of innovative multifunctional energy-efficient structures with negative effective mass by implementing acoustic metamaterials at the structural scale, in which active control could be coupled with the energy-harvesting subsystem to make the structure self-sustaining and self-regulating (Chen et al. 2014).

12.2.2 Nanomaterials and Nanotechnology

Nanotechnology is the manipulation of matter on an atomic and molecular scale. Generally nanotechnology deals with material, devices, and other structures with at least one dimension sized from 1 to 100 nm. A wide variety of ways have deliberately been used to make materials at the nanoscales to take advantage of their enhanced properties such as high strength, lighter weight, increased control of light spectrum, and greater chemical reactivity. Nanomaterials and nanotechnologies provide the potential to enhance energy efficiency across all branches of industry and to economically leverage renewable energy production through new technological solutions and optimized production technologies, as show in Figs. 12.1 and 12.5 (Luther 2008).

12.2.2.1 Development of Primary Energy Sources

Nanotechnologies provide essential improvement potentials for the development of both conventional energy sources (fossil and nuclear fuels) and renewable energy sources like geothermal energy, sun, wind, water, tides, or biomass. Nano-coated, wear-resistant drill probes, for example, allow the optimization of life span and efficiency of systems for the development of oil and natural gas deposits or geothermal energy and thus the saving of costs. Further examples are high-duty nanomaterials for lighter and more rugged rotor blades of wind and tide power plants as well as wear and corrosion protection layers for mechanically stressed components (bearings, gear boxes, etc.) (Luther 2008). Nanomaterials may help make geothermal more practical by allowing efficient energy production closer to the surface at lower temperatures. Metal organic heat carriers are nanomaterials one-thousandth the width of a human hair, of which some could absorb 30% of their weight in organic compounds. In geothermal power systems, this could help drive turbines with organic compounds at lower temperatures.

Nanotechnologies will play a decisive role in particular in the intensified use of solar energy through photovoltaic systems. In case of conventional crystalline silicon solar cells, for instance, increases in efficiency are achievable by antireflection layers for higher light yield. First and foremost, however, it will be the further development of alternative cell types, such as thin-layer solar cells (among others of silicon or other material systems like copper/indium/selenium), dye solar cells, or polymer solar cells, which will predominantly profit from nanotechnologies. Polymer solar cells are said to have high potential especially regarding the supply of portable electronic devices, due to the reasonably priced materials and production methods as well as the flexible design. In the near future, for example, nanotechnologies could contribute to the optimization of the layer design and the morphology of organic semiconductor mixtures in component structures. In the long run, the utilization of nanostructures, like quantum dots and wires, could allow for solar cell efficiencies of over 60% (Luther 2008).

12.2.2.2 Energy Conversion

The conversion of primary energy sources into electricity, heat, and kinetic energy requires utmost efficiency. Efficiency increases, especially in fossil-fired gas and steam power plants, could help avoid considerable amounts of carbon dioxide emissions. Higher power plant efficiencies, however, require higher operating temperatures and thus heat-resistant turbine materials. Improvements are possible, for example, through nano-scale heat and corrosion protection layers for turbine blades in power plants or aircraft engines to enhance the efficiency through increased operating temperatures or the application of lightweight construction materials (e.g., titanium aluminides). Nano-optimized membranes can extend the scope of possibilities for separation and climate-neutral storage of carbon dioxide for power generation in coal-fired power plants, in order to render this important method of power generation environmentally friendlier in the long run. The energy yield from the conversion of chemical energy through fuel cells can be stepped up by nano-structured electrodes, catalysts, and membranes, which results in economic application possibilities in automobiles, buildings, and the operation of mobile electronics.

Thermoelectric energy conversion seems to be comparably promising. Nano-structured semiconductors with optimized boundary layer design contribute to increases in efficiency that could pave the way for a broad application in the utilization of waste heat, for example, in automobiles, or even of human body heat for portable electronics in textiles (Luther 2008).

One of the most flexible renewable energy technologies is the direct conversion of sunlight into electric power through the photovoltaic effect. Carbon nanomaterials, including C_{60} fullerenes, carbon nanotubes (CNTs), and graphene, have been studied as extremely efficient electron acceptors in polymer and quantum dot solar cells. Dye-sensitized solar cells are of great promise. In these devices, a nanocrystalline mesoporous titanium dioxide (TiO_2) film, with a monolayer of the charge transfer dye attached to its surface, is pasted on a transparent conductive substrate. The large NM surface area for dye chemisorption and the short charge migration length underlie their power conversion efficiency (Chen et al. 2012; Iavicoli et al. 2014).

In addition to solar cells, nanotechnology has made big impact on fuel cells, devices able to convert chemical energy directly into electricity. Nano-porous metals with high surface area, low specific densities, and rich surface chemistry can be highly efficient electro-catalysts for the critical electrode oxidation/reduction reactions in fuel cells. Platinum nanoparticles (Pt-NPs) have been regarded as the best cell catalyst, although the Pt-based electrode suffers from time-dependent drift and carbon monoxide deactivation. In this regard, nano-sized multi (bi-tri)-metallic Pt alloys have been the object of further exploration because of their higher electro-catalytic activities and greater resistance. Interestingly, CNTs and graphene, initially used in fuel cells as attractive materials for catalyst supports with the aim to lower precious-metal loading, enhance catalyst activity and durability, are currently studied also as metal-free catalysts in fuel cells. Their advantages rely on high surface area, mesoporosity, good electrical conductivity, stronger mechanical strength, light weight, and superb corrosion resistance (Qiao and Li 2011; Iavicoli et al. 2014).

Another important future energy option is the hydrogen gas as an endless source of clean fuel for many applications. Semiconductor NMs, e.g., TiO_2 and cadmium sulfide nanostructures, have been studied as efficient catalysts for water conversion into oxygen and hydrogen. Moreover, nano-structured carbons, metal-organic frameworks and polymers as well as metal hydrides and related complex hydrides are examples of investigated NMs for hydrogen storage and transportation for high hydrogen capacity and minimal deterioration during hydrogenation (Orimo et al. 2007; Iavicoli et al. 2014).

INDEPTH: Carbon Nanomaterials for Energy Generation and Storage

Carbon, one of the most abundant materials found on earth, can be a valid material for both energy generation and storage. It can be employed in real-world devices such as organic solar cells and supercapacitors in the form of

(continued)

one or more of its allotrope forms (e.g., graphene, carbon nanotubes, fullerenes) by employing inexpensive synthesis and process methods based on printing and roll-to-roll techniques. Achieving a high-quality material on a large scale still remains an issue in order to compete with conventional conducting transparent electrodes such as ITO. However, its potential for production on flexible substrates makes it very appealing for the organic solar cell field where roll-to-roll techniques have been employed to increase the production volume. Additionally, fullerene derivatives, CNTs and graphene oxide could help to boost the performance of organic solar cell devices if employed in the active or buffer layers. In fact, their semiconducting properties can be tuned by doping with other materials or by changing their physical structure in order to absorb a broader range of solar spectrum wavelengths. Carbon nanomaterials, in particular carbon nanotubes and graphene, have also been proven to be very efficient and reliable materials for energy storage. The high specific surface area and conductivity of graphene are two key features for employing this material in supercapacitors. The ability to use a solid-state electrolyte composed of graphene oxide or a gel polymer electrolyte allow for the possibility of printable, flexible devices that do not require encapsulation. Although carbon nanotubes generally have a relatively low specific surface area, they can still be employed in combination with graphene to increase the conductivity of the electrode or the surface roughness of the film, resulting in an increase in the number of ions stored at the electrode/electrolyte interface. Carbon continues to surprise researchers with its extraordinary properties. Completely new carbon structures have been synthesized, from 0D fullerenes to 1D nanotubes and 2D graphene. The low cost of this element, the sixth most abundant element on earth, makes it an attractive choice to replace conventional materials for energy generation and storage applications (Notarianni et al. 2016).

12.2.2.3 Energy Distribution

Regarding the reduction of energy losses in current transmission, hope exists that the extraordinary electric conductivity of nanomaterials like carbon nanotubes can be utilized for application in electric cables and power lines. Furthermore, there are nanotechnological approaches for the optimization of superconductive materials for lossless current conduction. In the long run, options are given for wireless energy transport, e.g., through laser, microwaves, or electromagnetic resonance. Future power distribution will require power systems providing dynamic load and failure management, demand-driven energy supply with flexible price mechanisms as well as the possibility of feeding through a number of decentralized renewable energy sources. Nanotechnologies could contribute decisively to the realization of this vision, inter alia, through nano-sensory devices and power-electronic components able to cope with the extremely complex control and monitoring of such grids (Luther 2008).

12.2.2.4 Energy Storage

The utilization of nanotechnologies for the enhancement of electrical energy stores like batteries and super-capacitors turns out to be downright promising. Due to the high cell voltage and the outstanding energy and power density, the lithium-ion technology is regarded as the most promising variant of electrical energy storage. Nanotechnologies can improve capacity and safety of lithium-ion batteries decisively, as for example through new ceramic, heat-resistant and still flexible separators and high-performance electrode materials. For rechargeable lithium batteries, the energy densities and the performances of these devices largely depend on the physical and chemical properties of the electrode material. In this regard, the reduced dimensions and high surface area of NMs increase the rate of electron transport and the electrode-electrolyte contact, respectively, while the nano-structure itself provides facile strain relaxation and resistance to fracture. For anode applications, CNTs, a series of graphene-based nanostructures and silicon nanowires have been studied as promising host-high capacity materials and conductive additives. While emerging interests has been focused on metal oxide NMs, e.g., SnO_2; TiO_2 or $LiFePO_4$-NMs, for anode or cathode applications (Luther 2008; Iavicoli et al. 2014).

Redox-based supercapacitors with nano-structured electrode materials have shown the potential to combine the high energy density of conventional batteries with the high power capabilities of electrostatic capacitors at the lab scale. Mixed metal oxides, e.g., ruthenium oxide (RuO_2), manganese oxide (MnO_2), magnetite (Fe_3O_4), CNTs, graphene, and carbon metal oxide composites, have been investigated as electrode NMs aimed at a high specific capacity and rate capability (Iavicoli et al. 2014).

In the long run, even hydrogen seems to be a promising energy store for environmentally friendly energy supply. Apart from necessary nanostructure adjustments, the efficient storage of hydrogen is regarded as one of the critical factors of success on the way to a possible hydrogen management. Current materials for chemical hydrogen storage do not meet the demands of the automotive industry, which requires a hydrogen-storage capacity of up to ten weight percent. Various nanomaterials, inter alia based on nanoporous metal-organic compounds, provide development potentials, which seem to be economically realizable at least with regard to the operation of fuel cells in portable electronic devices. Another important field is thermal energy storage. The energy demand in buildings, for example, may be significantly reduced by using phase change materials such as latent heat stores. Interesting, from an economic point of view, are also adsorption stores based on nanoporous materials like zeolites, which could be applied as heat stores in district heating grids or in industry. The adsorption of water in zeolite allows the reversible storage and release of heat (Luther 2008).

12.2.2.5 Energy Usage

To achieve sustainable energy supply, and parallel to the optimized development of available energy sources, it is necessary to improve the efficiency of energy use and to avoid unnecessary energy consumption. This applies to all branches of industry and private households. Nanotechnologies provide a multitude of approaches to energy saving. Examples are the reduction of fuel consumption in automobiles through lightweight construction materials on the basis of nanocomposites, the optimization in fuel combustion through wear-resistant, lighter engine components and nanoparticular fuel additives or even nanoparticles for optimized tires with low rolling resistance. Considerable energy savings are realizable through tribological layers for mechanical components in plants and machines. Building technology also provides great potentials for energy savings, which could be tapped, for example, by nanoporous thermal insulation material suitably applicable in the energetic rehabilitation of old buildings. In general, the control of light and heat flux by nanotechnological components, as for example switchable glasses, is a promising approach to reducing energy consumption in buildings (Luther 2008).

More specifically, manufactured NMs and nanocomposites offer great opportunities in the construction and related infrastructure industries. Strength, durability, and lightness of various materials, as well as heat-insulating, self-cleaning, fire-retardant, anti-fogging and sensing structural health properties may be improved or provided de novo by NMs. Thus, CNTs, SiO_2, TiO_2, Fe_2O_3, and magnetic nickel-NPs can remarkably improve mechanical durability, compressive and flexural strength of cement products. Highly water repellent coatings incorporating silica, alumina-NPs, and hydrophobic polymers are proper to be used for wood. The use of TiO_2-NPs in glasses leads to the so-called self-cleaning technology due to their photo-catalytic and anti-fouling properties. Fire protective glass is obtained using silica (nano)layers, which may also function as antireflection coatings for exterior light in order to contribute to energy and air-conditioning conservation. Ag-NPs can be embedded in paint to inactivate pathogenic microbes and provide antimicrobial properties to surfaces (e.g., hospital walls) (Kumar et al. 2008; Iavicoli et al. 2014).

12.2.3 Self-Powered Micro-/Nanosystems

Emerging applications in wearable technology, wireless sensors, pervasive computing, human–machine interfacing, and implantable devices demand appropriate power sources that can sustainably operate for extended periods of time with minimal intervention. Self-powered micro-/nanosystems (MNS) that can harvest energy from their environment or host (i.e., the human body) therefore hold great promise to address this issue and are feasible because of the extremely low power consumption of nanodevices. Nanogenerators (NGs) have been developed based on piezoelectric, trioboelectric, pyroelectric, thermoelectric, and photovoltaic effects, aiming at building self-sufficient power sources for mico/nano-systems.

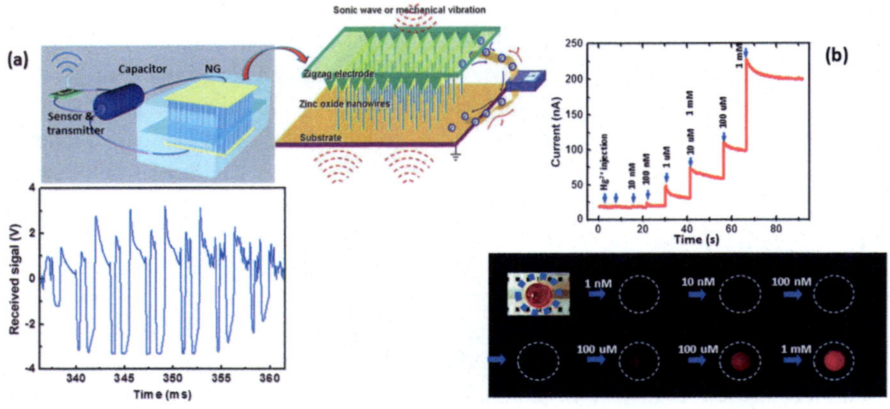

Fig. 12.6 Self-powered nanosensor systems for (**a**) optical sensing with wireless data transmission; and (**b**) Hg^{+} sensing with an indication signal (Adapted with permission from Wang and Wu (2012) (John Wiley & Sons))

In fact, the current advancement of micro-/nanotechnology is shifting its focus from the development of discrete devices to the development of more complex integrated systems that are capable of performing multiple functions, such as sensing, actuating/ responding, communicating, and controlling, by the integration of individual devices through state-of-the-art microfabrication technologies. Furthermore, it is highly desired for these multifunctional MNSs to operate wirelessly and self-sufficiently without the use of a battery, especially in applications such as remote sensing and implanted electronics. This operation scheme will not only extend the life span and enhance the adaptability of these MNSs while greatly reducing the footprint and cost of the entire system, but it will also increase the adaptability of these MNSs to the environment in which they are deployed. As the dimensions of individual devices shrink, the power consumption decreases accordingly to a reasonably low level, so that energy scavenged directly from the ambient is sufficient to drive the devices. By harvesting energy from the ambient vicinity of the system and converting it into usable electrical power for wireless, self-sufficient, and independent operations, a typical self-powered MNS should consist of low-power microcontroller unit, high-performance data-processing/storage components, wireless signal transceiver, ultrasensitive sensors based on micro-/nanoelectromechanical systems (MEMSs/NEMSs), and the embedded powering/energy-storage units, as shown in Figs. 12.6 and 12.7 (Wang and Wu 2012; Yang et al. 2012).

Figure 12.6a shows a self-powered wireless transmission system composed of an NG for harvesting mechanical energy or sonic wave, a low-loss full-wave bridge rectifier with a capacitor for storing the energy, an infrared photodetector, and a wireless data transmitter was shown to transmit data wirelessly and self-sustainably and demonstrated the potential of ZnO-NW NGs in applications of wireless biosensing, environmental/infrastructure monitoring, and wireless sensor networks. In another example, a fully stand-alone, self-powered environmental sensor based on a single-walled-CNT (SWNT) field-effect transistor (FET) and a ZnO-NW NG was

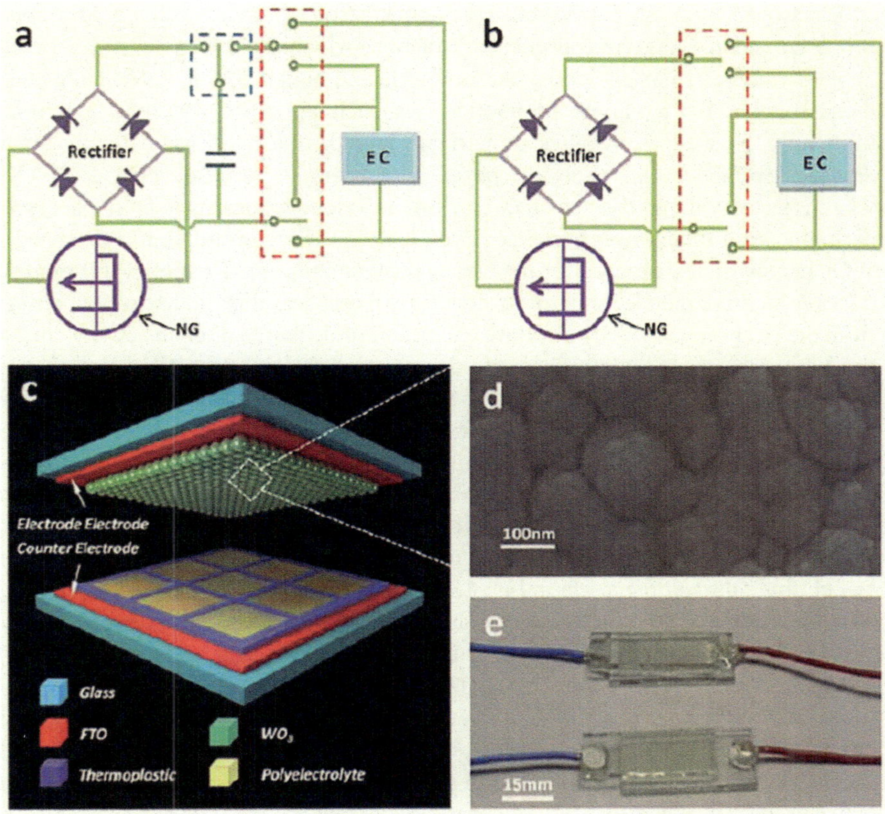

Fig. 12.7 Schematic diagram of the self-powered electrochromic device (EC) system and the structure of the WO_3-based EC device (Adapted with permission from Yang et al. (2012) (Royal Society of Chemistry)): (**a**) NG-charged power source; (**b**) real-time power supply; (**c**) the structure of the WO_3-based EC device; (**d**) SEM image of the WO_3 film; (**e**) a picture of the EC device

implemented for the detection of Hg^{2+} ions and evaluation of their concentration in water. This application is of potential importance for the monitoring of water quality. The sensor array based on the SWNT network served as the Hg^{2+} sensor, and the ZnO-NW NG served as the energy-harvesting component (Fig. 12.6b). In addition, a ZnO-NW NG was integrated onto the inner surface of a tire to scavenge mechanical energy from deformation of the tire during motion, and the harvested energy was used to power a liquid-crystal-display (LCD) screen. A similar approach could be used to develop a self-powered tire-pressure sensor and speed detector for mobile vehicles. The feasibility of building self-powered nanodevices for in vivo biomedical applications has also been investigated, with the use of harvested biochemical energy or a combination of biomechanical and biochemical energy to power NW based sensors (Lee et al. 2010; Wang and Wu 2012).

Electrochromic (EC) materials are capable of reversibly changing optical properties upon charge injection and extraction, which are driven by an externally applied voltage. Their low power consumption and high coloration efficiency make them suitable for a variety of applications such as smart windows, electronic billboards, as well as displays of portable and flexible devices including smart cards, price labels, and electronic papers. For example, tungsten trioxide (WO_3), when injected with protons or other small monovalent cations such as Li^+, becomes dark blue with little power involved. Even lower power consumption is expected if WO_3 nanoparticles are used for the electrochromic purpose. Therefore, it is entirely possible to drive the electrochromic device by nanogenerators that scavenge energy from the environment such as airflow, vibration, sonic waves, human activity, and so on. Figure 12.7a, b shows schematic diagrams of self-powered EC systems. The power source unit has an NG and a full-wave bridge for rectification. The NG converts mechanical energy into electricity, while the full-wave bridge transforms alternating current from the NG to unidirectional current. Two types of self-powered EC systems can be designed. In the first approach, the electricity generated by the NG is rectified and stored in a capacitor and then released to drive the EC device (Fig. 12.7a). The charging process and discharge process are controlled by a regular switch (blue dashed rectangle in Fig. 12.7a). A reversible switch is for determining how the capacitor and the EC device are connected. In the second design, the NG was connected to the EC device through a full-wave rectifier without an energy storage unit as an in situ power supply (Fig. 12.7b). There is also a reversible switch for reversing the connection polarity between the NG and the EC unit. The EC unit has a multilayered structure, which is sketched in Fig. 12.7c. The outer layers are glass substrates covered with FTO thin films on one side as electrodes. Sandwiched between the two electrodes are an array of cells that are filled with polyelectrolyte and a layer of WO_3 film of about 250 nm that consists of densely packed nanoparticles (Fig. 12.7d). The distance between the two electrodes is around 20 mm. As demonstrated in Fig. 12.7e, the fully packaged EC device has a transmittance of more than 70%. Furthermore, the EC device can also be made on a flexible substrate and integrated with our flexible nanogenerator to be a kind of wearable device (Yang et al. 2012).

It can be anticipated that self-powered MNSs will play a critical role in the implementation of implantable electronics, remote and mobile environmental sensors, nanorobotics, intelligent MEMSs/NEMSs, and portable/wearable personal electronics. Self-powered MNSs are also key components of large-scale fault-tolerant sensor networks. When traditional discrete sensors are replaced by a large number of sensor nodes distributed in a field, the statistical analysis of signals collected through the network of distributed sensors can provide precise and reliable information for tracking and monitoring purposes. An internet of things which can correlate objects/products and devices with databases and networks is expected to revolutionize the future of healthcare, medical monitoring, infrastructure/environmental monitoring, logistics, and smart homes (Gershenfeld et al. 2004; Wang and Wu 2012).

12.3 Artificial Photosynthesis

Photosynthesis is the chemical process by which plants, algae, and some bacteria store energy from the Sun in the form of carbohydrates that act as fuels. The four main steps of photosynthesis are light harvesting, charge separation, water oxidation or splitting, and fuel production. In light harvesting, antenna molecules, mostly chlorophyll but also carotenes, absorb sunlight and transfer the energy among themselves and eventually through to the reaction center where charge separation takes place. In this way, energy from sunlight is used to separate positive and negative charges from each other. The positive charges are used to oxidize water. The electrons are transferred via cytochrome b_6f and mobile electron carriers to photosystem I where they are excited again and used to produce carbohydrate fuel. A schematic diagram of what happens in photosynthesis is shown in Fig. 12.8. The chemical reactions for water splitting and fuel production are (Purchase and de Groot 2015):

1. Water oxidation:

$$2H_2O \overset{4h\nu}{\rightarrow} O_2 + 4H^+ + 4e^- \tag{12.1}$$

2. Carbon dioxide reduction to produce carbohydrate fuel:

$$CO_2 + 4H^+ + 4e^- \overset{4h\nu}{\rightarrow} (H_2CO) + H_2O \tag{12.2}$$

These two chemical half-reactions add up to give a total chemical reaction for photosynthesis of

$$CO_2 + H_2O \overset{8h\nu}{\rightarrow} (H_2CO) + O_2 \tag{12.3}$$

Four photons are required to drive each of the half reactions. Thus, eight photons are required for the total chemical reaction. As four electrons are carried over and eight photons are used, the process proceeds with two photons per electron. Nature uses two photosystems in tandem to drive the two chemical reactions of water splitting and fuel production. The reactions occur in proportion to the number of photons absorbed. Natural photosynthesis is not determined by insolation, the total amount of solar radiation energy that is collected per unit of time, but by the total light sum, the number of photons from the blue to red (400–700 nm) part of the spectrum that is collected per unit of time. Although parts of the natural photosynthetic process are highly efficient, the overall solar-to-carbohydrate efficiency is low. Thus, unmodified natural photosynthesis cannot serve mankind's purposes for fuel production, but can be used as a blueprint for efficient artificial photosynthesis (Purchase and de Groot 2015).

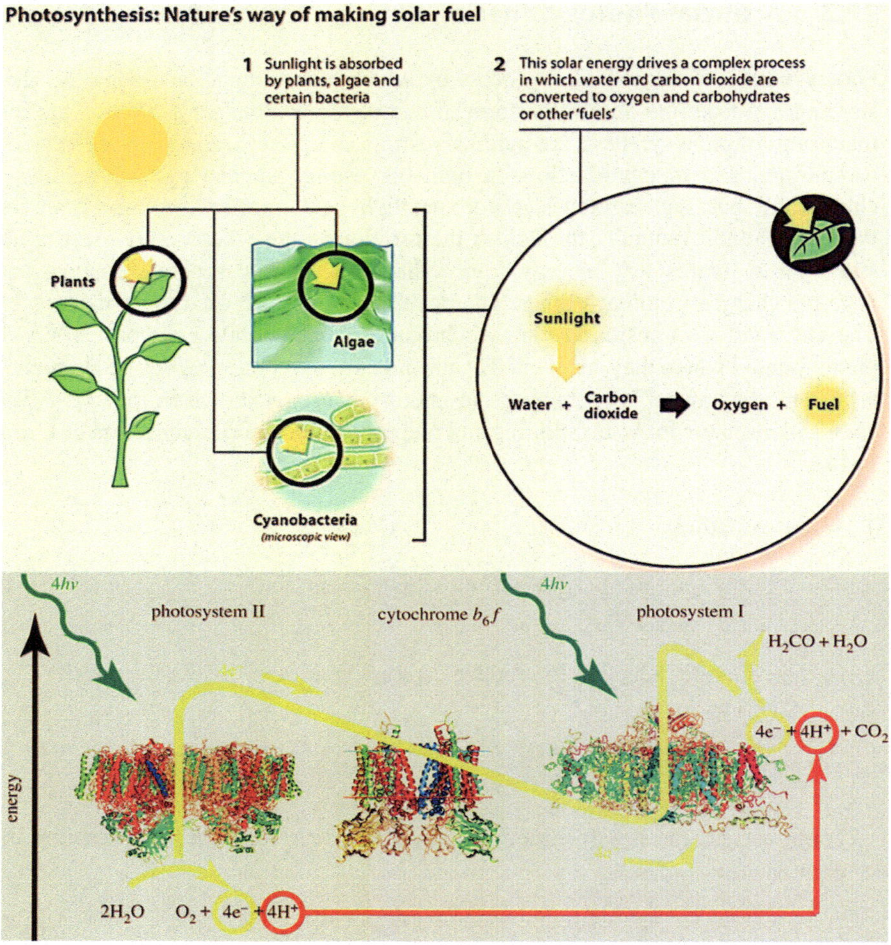

Fig. 12.8 Schematic illustration of natural photosynthesis showing light absorption, charge separation, water oxidation, and fuel production (Modified with permission from Purchase and de Groot (2015) (The Royal Society under the terms of the Creative Commons Attribution License http://creativecommons.org/licenses/by/4.0/); Purchase et al. (2015) (Wageningen UR), Credit: Leiden University (Groene grondstoffen))

12.3.1 Process of Artificial Photosynthesis

Artificial photosynthesis has the similar fundamental process with natural photosynthesis. As the incoming photon flux, energy and electron transfer, and catalysis, operate on very different time, energy and length scales, it is possible to put artificial photosynthesis design limits on how components should be matched for the most efficient solar-to-fuel conversion operating close to the theoretical limits on solar energy conversion. To make the best possible use of the incoming sunlight for fuel production with two photons per electron, it makes sense to capture as many photons of sunlight as possible. The nature photosynthesis in plants absorbs light around 700 nm, which use only half of the incoming photons. In comparison, silicon solar cells absorb light at around 1100 nm and therefore absorb more photons. Nature uses two photosystems in tandem to drive the two chemical reactions of water oxidation and CO_2 reduction. The same can be done with an artificial device. A weakness in the natural system is that the two photosystems absorb light of approximately the same energy, so the two systems are competing for the same photons while the infrared photons remain unused. In an artificial system, this can be done differently: have one absorber in the visible part of the spectrum and another in the infrared. In this way, the number of photons of sunlight that is absorbed by the artificial system is maximized. Furthermore, the cutoff wavelengths are better matched to the electrochemical work. Optimal matching is obtained with cutoff wavelengths of 700 and 1100 nm. Thus, the current tandem devices are produced to have two absorbers to make the best possible use of the incoming light to drive water splitting and fuel production with two photons per electron. A schematic of a tandem artificial photosynthesis device is shown in Fig. 12.9 along with its light-absorbing properties. Just as in natural photosynthesis, artificial photosynthesis occurs in four steps: light harvesting, charge separation, water oxidation or splitting, and fuel production (Hu et al. 2013; Purchase and de Groot 2015).

12.3.1.1 Light Harvesting and Charge Separation

Natural photosynthesis has an efficient strategy for achieving light harvesting and charge separation: antennas containing many chlorophyll molecules absorb sunlight and pass the excitation energy among themselves and then through to the reaction center. Also, secondary absorbers such as carotenes absorb light in regions of the spectrum where chlorophyll does not absorb well. In this way, the absorption cross section of the reaction center is increased. For artificial photosynthesis, strategies similar to that found in nature can be employed, where multiple absorbers that exhibit complementary absorption profiles and are capable of efficient excitation-energy transfer are used as light harvesters and the excitation energy is transferred to an artificial reaction center. For the tandem device for instance, light is absorbed on the fuel production side by a photosensitizer that, when excited, has sufficient reduction potential to inject an electron into the fuel production catalyst. On the water oxidation side, the optically excited photosensitizer has sufficient reduction potential to inject an electron into the fuel reduction side and fill the hole there, and subsequently has sufficient oxidation potential to oxidize the water oxidizing catalyst (Frischmann et al. 2013; Purchase and de Groot 2015).

The sensitizers may be made from organic molecules. The simplest form of this is a donor–acceptor diad, but more complex structures are commonly used to prevent charge recombination. Alternatively, semiconductors, as in solar cells, may be used. These materials can efficiently absorb sunlight and separate charges and are stable with extended exposure to sunlight. Their disadvantage is their limited flexibility; various techniques of doping, nanostructuring, and coating are needed to give them the desired properties. In addition, molecular light harvesters can be used in

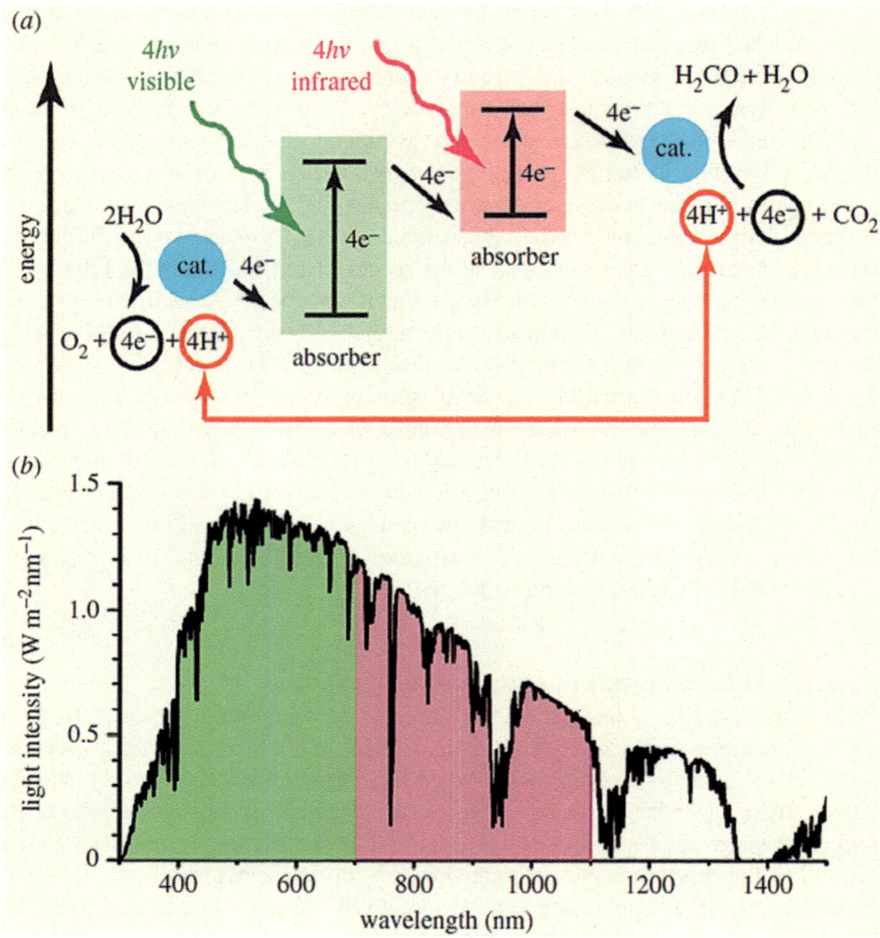

Fig. 12.9 Schematic of a tandem artificial photosynthetic device (**a**) and its light-absorbing properties (**b**). This device operates in a fashion analogous to natural photosynthesis. The tandem is in balance when both halves receive the same number of photons. An optimal use of the sunlight is reached with cutoff wavelengths of around 700 and 1100 nm (Adapted with permission from Purchase and de Groot (2015) (The Royal Society under the terms of the Creative Commons Attribution License http://creativecommons.org/licenses/by/4.0/))

combination with semiconductor charge separators, in a concept similar to that used in dye-sensitized solar cells (Wen and Li 2013; Purchase and de Groot 2015).

12.3.1.2 Water Oxidation/Splitting

During water oxidation or water splitting, positive charges are directly injected into catalytic centers where they are used to split water into hydrogen ions (protons) and oxygen. As water oxidation involves multi-electron chemistry, water-splitting catalysts invariably include one or more transition metals in their structures, such as iridium and ruthenium. However, these are not practical for very large-scale use. Natural photosynthesis demonstrates that catalysts based on abundant, inexpensive elements are possible. Such catalysts are more desirable. The catalysts based on cobalt and other cheap transition metals are promising, but their efficiency and durability need to be further improved before they are suitable for larger-scale use (Du and Eisenberg 2012; Dinca et al. 2012).

12.3.1.3 Fuel Production

Different solar fuels can be envisaged as products of artificial photosynthesis. There are two main types of fuels: hydrogen and carbon-based fuels.

1. Hydrogen
   Hydrogen is a natural choice of fuel when water is the raw material. To make hydrogen, the protons that result from the splitting of water need to be reduced to produce molecular hydrogen (Purchase and de Groot 2015):

$$4H^+ + 4e^- \xrightarrow{4h\nu} 2H_2 \tag{12.4}$$

   The easiest way of doing this is on the surface of a noble metal such as platinum. But this approach is too expensive to commercialize on a large scale. Hydrogenase enzymes efficiently catalyze the reversible reduction of protons and have iron and nickel centers. When combined with the water-splitting reaction in a tandem configuration, four photons are used to make one molecule of hydrogen. Hydrogen has been identified as an attractive zero-carbon energy carrier that could play a key role in future renewable energy technology. Artificial photosynthesis can play an important role in the production of hydrogen worldwide.

2. Carbon-based fuels
   Nature makes carbon-based fuels. These are complex carbohydrates with carbon–carbon bonds. However, combining protons with carbon dioxide out of the air to produce carbon-based fuels is much more challenging than producing hydrogen. This process involves difficult multi-electron chemistry, even for simple energy carriers such as methane, methanol, and syngas (a mixture of carbon monoxide and hydrogen) (Table 12.1). Artificial molecular catalysts are being developed for the photosynthesis of carbon-based fuel, and semisynthetic systems based on enzymes from microorganisms are being investigated. H_2 and CO can be used as precursors for other fuels like methane, alcohols, and Fischer–Tropsch liquids that may be incorporated into the current energy infrastructure. A further complication with carbon-based fuels is that the atmospheric concentration of CO_2 is low. This necessitates either placing artificial photosynthetic devices near sources of CO_2 such as coal-fired power stations or combining them with technologies that concentrate CO_2 (Purchase and de Groot 2015).

Table 12.1 Two photons are needed to transfer one electron in an artificial photosynthetic reaction

Reaction	Tons of CO_2 converted per km^2 per day	Tons of product produced per km^2 per day
$2H^+ + 2e^- \rightarrow H_2$	0	20
$CO_2 + 2H^+ + 2e^- \rightarrow HCOOH$	440	460
$CO_2 + 2H^+ + 2e^-\ CO + H_2O$	440	280
$CO_2 + 4H^+ + 4e^- \rightarrow HCHO + H_2O$	220	150
$CO_2 + 6H^+ + 6e^- \rightarrow CH_3OH + H_2O$	147	107
$CO_2 + 8H^+ + 8e^-\ CH_4 + 2H_2O$	110	40

Assuming 40 moles of photons per meter squared per day, the second and third columns show how many tons of CO_2 can be converted per km^2 per day and how many tons of carbon-based product can be produced per km^2 per day, respectively. The top row shows hydrogen production as a reference (Purchase and de Groot 2015)

12.3.2 Approaches to Artificial Photosynthesis

There are a number of different approaches that can be taken to artificial photosynthesis. All approaches involve the four basic steps of light harvesting, charge separation, water splitting, and fuel production. The scientific problems encountered in all cases are similar. Here, the different approaches are divided up according to the types of materials used (Purchase and de Groot 2015): organic, inorganic, hybrid, and semisynthetic.

12.3.2.1 Organic Systems

Molecular artificial synthesis components are often developed biomimetically. Studies of energy conversion and storage in efficient natural enzymes provide inspiration for the development of the complex chemistry by mimicking enzymatic catalytic functions. The four-electron oxidation of water is a main research problem and a bottleneck for successful development of artificial photosynthesis. Entirely molecular systems are difficult to develop, but they offer the advantage of enabling a modular approach. Individual components for light harvesting, charge separation, water oxidation, and fuel production can be made and investigated separately to maximize performance before being integrated into an appropriate architecture. Molecules have a well-defined structure that can be deliberately modified to improve a given property. Molecular systems are also readily amenable to study with analytical techniques that provide structural and kinetic information. Processes can therefore be followed and understood at a very detailed level. This molecular assembly approach is elegant, but in the absence of a viable spontaneous assembly strategy, the large amount of synthesis involved renders it impractical. Also, most molecules tend to degrade quickly under extended exposure to sunlight. When the "responsive matrices" would enable self-assembly and self-repair, the molecular systems may become the champions (Beradi et al. 2014; Purchase and de Groot 2015).

12.3.2.2 Inorganic Systems

Semiconductor materials can absorb sunlight and then separate charges; they are also generally robust under extended exposure to sunlight. They therefore seem to be natural candidates for artificial photosynthesis. Those with appropriate electronic properties can provide sufficient electrochemical potential to drive water oxidation or fuel production on their surfaces. However, many semiconductor materials that have the right electronic properties for water splitting only absorb UV light, which is only a small portion of the incoming photons. Furthermore, catalysis on the semiconductor surface is not very efficient. Although appealing for its simplicity, having one semiconductor perform all tasks of absorption, charge separation and catalysis is asking a lot from one material. To date, no material with acceptable performance has been found. To address this significant challenge, novel concepts and methods afforded by nanotechnology are being applied to design innovative composite nanostructures in which each component performs specialized functions (Abdi et al. 2013; Purchase and de Groot 2015).

12.3.2.3 Organic–Inorganic Hybrids

An appealing solution is to combine the best properties of organic and inorganic materials. Here, light absorption may be done either by a semiconductor or by a dye molecule on a semiconductor surface; charges are then separated within the semiconductor and transported to optimized molecular catalysts tethered to the semiconductor surface. This is a very promising approach, but, to date, the experimental devices that have been constructed this way are still either too expensive or too inefficient to warrant scaling up to a commercial level (Purchase and de Groot 2015; Wang et al. 2015).

12.3.2.4 Semisynthetic Systems

An interesting approach concerns systems that are a hybrid of biological and man-made components. For instance, a biological photosynthetic component that harvests solar energy and splits water can be purified and tethered to an appropriate scaffold. The photosynthetic enzyme is then linked to a hydrogen-producing enzyme (a hydrogenase) or a catalyst for fuel production. Alternatively, chlorophylls from biological sources can be chemically modified and assembled to form semi-artificial modules. The advantage of this approach is that the biological components work very well (nature has been successfully performing photosynthesis for about 3 billion years). However, this method is in its infancy, so it remains to be seen if these biological components can be made sufficiently robust by chemical modifications outside of their native environment, or if they can be extracted and modified in a commercially viable fashion. Even if such constructions never yield commercially viable devices, the science involved in studying them promises to teach people a lot about nature's approach to photosynthesis (Ocakoglu et al. 2014; Purchase and de Groot 2015).

12.3.3 Potential Applications of Artificial Photosynthesis

The development of artificial photosynthesis is still in a laboratory phase and much remains to be accomplished before commercial application is possible. Nevertheless, artificial photosynthesis has the potential to be an attractive and sustainable alternative to fossil fuels. Furthermore, artificial photosynthesis can contribute to transformation of the energy infrastructure as a whole (Purchase et al. 2015).

12.3.3.1 Benefit for Solar Energy Storage and Transportation

Like the wind, the sun is not a constant energy source. Sometimes the sun shines, sometimes the sky is overcast, and at night no solar energy is available for capture. As long as solar energy's share of overall energy consumption remains small, the difference between the demand for energy and the variable supply from solar sources can easily be accommodated by conventional power plants. However, if solar energy is ever to be a substantial component of the energy mix, other ways need to be found of absorbing fluctuations in the supply. A continuous energy supply when little or no sunlight is available depends on new technology for the storage and transportation of solar energy. In the production of solar fuels, solar energy is directly stored locally in the form of chemical bonds in concentrated fuels that can be transported by pipeline, road, rail, or ship. Artificial photosynthesis can also be combined with fuel cells, in which fuel (often hydrogen) and oxygen are converted into electricity and heat. The resulting systems can provide buildings and small communities with energy. The solar fuel production and storage concept is illustrated in Fig. 12.10 (Purchase et al. 2015).

12.3.3.2 Creation of New Business Models for Energy Generation

More and more homes have solar panels fitted to their roofs, enabling the homeowners to produce some or all of their own electricity. Indeed, people with a large enough area of solar panelling can actually supply power to the public grid and

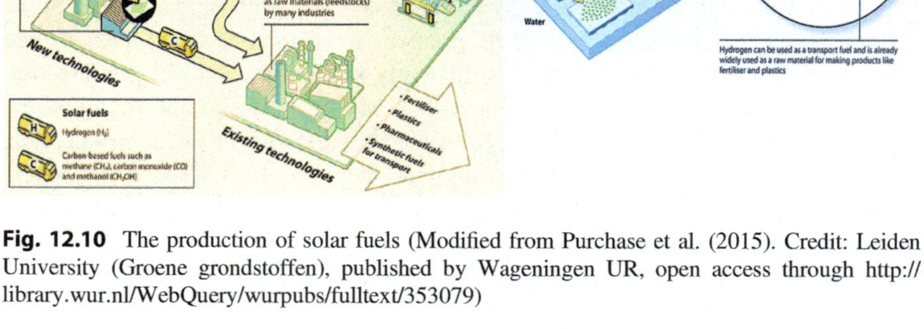

Fig. 12.10 The production of solar fuels (Modified from Purchase et al. (2015). Credit: Leiden University (Groene grondstoffen), published by Wageningen UR, open access through http://library.wur.nl/WebQuery/wurpubs/fulltext/353079)

earn money from it. In such cases, the consumer is also a supplier of energy: a "producing consumer" or "prosumer." However, electricity is not easily stored and the scope for consumers to be true prosumers is consequently limited. Artificial photosynthesis systems do not have the same drawback as photovoltaic systems, because their product is not electricity but fuel, which can be stored and used for energy at any time. Artificial photosynthesis systems therefore provide owners with much more flexibility in terms of what they do with the energy they produce and their ability to respond to fluctuations in supply and demand. The cost of solar fuels could be brought down if local producers join forces to protect their interests and to organize distribution of their energy surpluses. That might lead to energy companies adopting a new role: instead of acting as energy producers, they might become energy distributors, acting as agents for the prosumers. Another great advantage of artificial photosynthesis is that solar energy is relatively evenly distributed around the world (in contrast to fossil fuels). If artificial photosynthesis enters worldwide use, then people in poor countries will have access to the necessary energy resources to improve their standard of living (de Groot 2013; Purchase et al. 2015).

12.3.3.3 Renewable Transport Fuels

Artificial photosynthesis may readily be accommodated in a renewable energy mix for transport. Solar fuels are suitable for use not only in the propulsion of light vehicles, such as passenger cars, but also in aviation and shipping, where other technologies, such as electrical energy, are harder to utilize. For instance, hydrogen produced from solar energy can be used as a propulsion fuel in cars. Fuel-cell grade hydrogen is currently produced by an energy-intensive process. One of the great challenges for hydrogen-powered cars is therefore finding alternative ways of producing hydrogen. Nevertheless, the ability to produce hydrogen using sunlight would be a major step forward on the road to hydrogen becoming a viable transport fuel, as shown in Fig. 12.11 (Purchase et al. 2015).

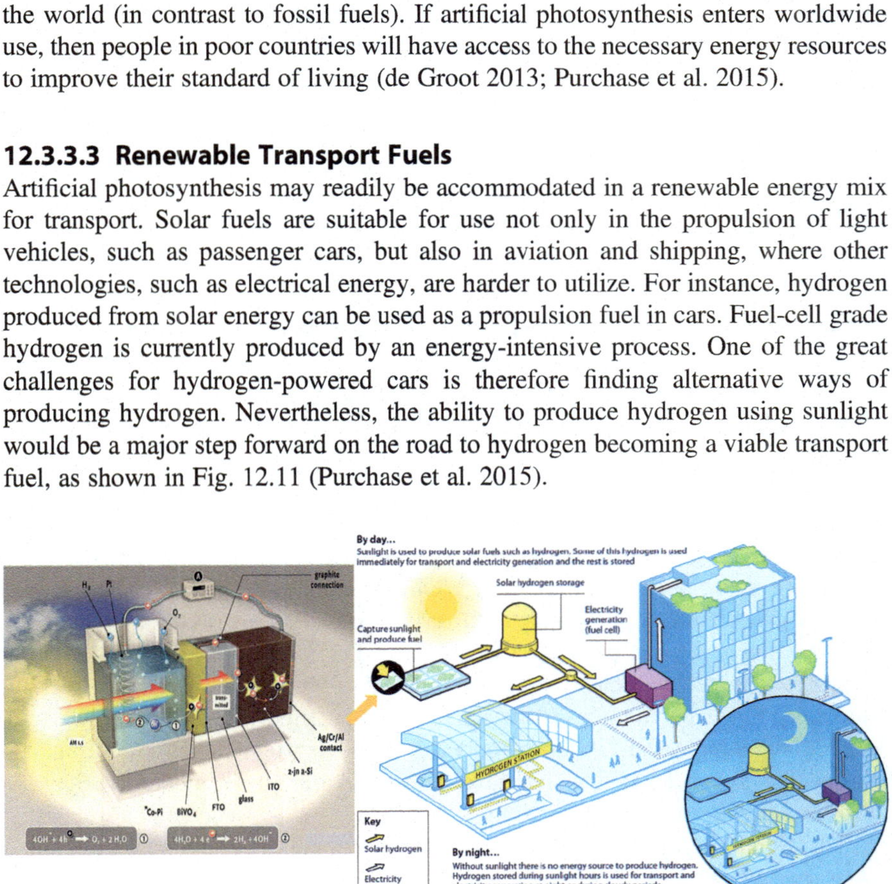

Fig. 12.11 Solar energy around the clock (Right) and an infographic of an artificial leaf (Left) that converts sunlight into hydrogen with an efficiency of roughly 5% (Modified from Purchase et al. (2015). Credit: Leiden University (Groene grondstoffen), published by Wageningen UR, open access through http://library.wur.nl/WebQuery/wurpubs/fulltext/353079)

12.3.3.4 Use and Recycling of CO_2

CO_2 capturing and storage represents a serious technological challenge, and considerable progress could be made by regarding CO_2 not as waste but as a raw material for the production of renewable fuel. Sunlight could well be the ideal energy source for the reduction of CO_2 and thus the production of carbon-based fuels (Fig. 12.11). In a natural or artificial photosynthesis reaction, CO_2 can be converted using two photons per electron. Carbon-based fuels that might be produced include formic acid, carbon monoxide (which when mixed with hydrogen constitutes syngas), formalin, methanol, and methane (natural gas) (Purchase et al. 2015).

12.3.3.5 Fuel Production that Does Not Compete with Food Production

Biofuels are made from agricultural products, such as sugar, starch, and vegetable oil. Fuel production from such products implies the use of agricultural land and crops that can also be used for the production of food. The "second generation" of biofuels is made from lignocellulose (the inedible, fibrous parts of plants), organic waste, algae, and microorganisms, whose production does not compete with the production of food. However, the demand for fuel is huge in comparison with agricultural production, and there is considerable demand for raw materials from agriculture, both in food production and for nonfood applications. Therefore, a new source of solar fuel that does not tie up large amounts of agricultural land is therefore very desirable (Purchase et al. 2015).

12.4 Advanced Energy Storage Materials and Devices

Energy storage is the key component for creating sustainable energy systems. Current technologies, such as solar photovoltaic and wind turbines, can generate energy in a sustainable and environmentally friendly manner; yet their intermittent nature still prevents them from becoming a primary energy carrier. Energy storage technologies have the potential to offset the intermittency problem of renewable energy sources by storing the generated intermittent energy and then making it accessible upon demand. In addition to energy grid applications, energy storage technologies also have the potential to transform the transportation system. Functioning energy storage devices could replace the powertrain systems of current transportation technologies from a chemical fuel-based powertrain into an electricity-based powertrain. The electric car is a prime example of how energy storage technologies can transform the transportation system into a more sustainable model. Electronic devices, which have become ubiquitous in modern society, are also heavily reliant on energy storage technologies.

Therefore, Energy storage provides economic benefits by allowing a reduction of plant energy production to meet average demands rather than peak demands. Transmission lines and equipment can also be appropriately sized for average power demands. In addition, it also mitigates some problems associated with the intermittency of renewable energy generation. As shown in Fig. 12.12, a wide range of energy storage materials and devices have been developed (Sabihuddin et al.

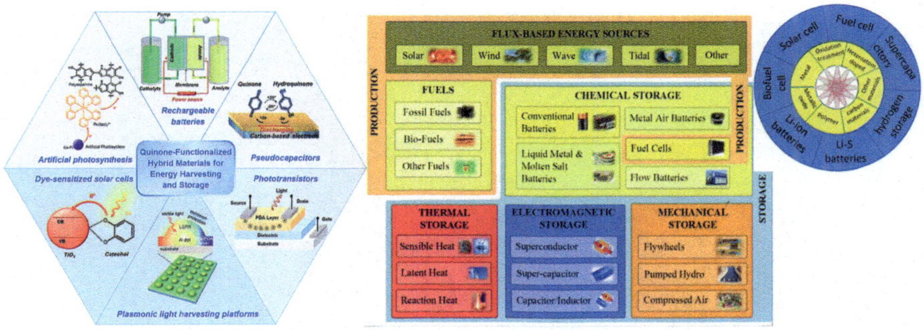

Fig. 12.12 The landscape of representative energy/electricity generation and storage materials and devices (Modified with permission from Sabihuddin et al. (2015) (MDPI) (Basel, Switzerland) under the Creative Commons Attribution License (CC BY 4.0))

2015). They can be broadly divided into the following categories: mechanical, chemical and electrochemical, electrical and electromagnetic, as well as thermal storage, in addition to thermochemical storage such as solar fuels, as discussion in the Sect. 2.3. Figure 12.13 compares different energy storage technologies with respect to typical power figures and charge/discharge operation times, including several proposed applications (Muñoz et al. 2016).

12.4.1 Mechanical Energy Storage

Mechanical storage takes the form of either potential energy or kinetic energy storage. Pumped hydroelectric (PHS), compressed air (CAES), separation/liquid air, and flywheel (FES) systems are the best known and are compared in Figs. 12.13 and 12.14. Some of these technologies can be used for isolated applications (houses, farms, small businesses), local communities (computer centers, industrial estates, onshore wind farms), or regional and larger-scale applications (e.g., state, federal, . . .). The different technical stages represent a rough categorization according to the following criteria (Ausfelder 2016): (a) "Research/Concept" describes the investigation of the basic operating principle of a process on a laboratory scale or its theoretical feasibility based on theoretical studies. (b) Based on the proof of the operating principle in the Research/Concept phase, "Development" describes the continuation of the activity with the aim to derive the technical parameters necessary for its implementation. (c) "Demonstration" describes the operation of a plant/process chain on a technically relevant scale. The requisite parameters for cost-effective operation of the plant can be determined. If a process is applied in only a few isolated plants worldwide, it also falls within this category. (d) "Technical Implementation" describes state-of-the-art, commercially available processes scaled-up for cost-effective operation.

Concerns of decreasing site availability, uncertain profitability, high capital costs, negative environmental impacts and low energy and power densities are increasingly

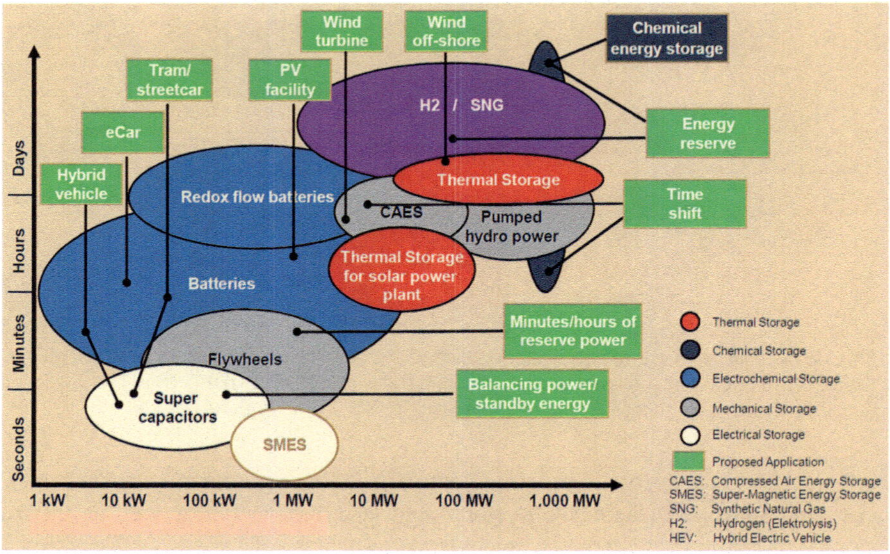

Fig. 12.13 Energy storage technologies categorized according to available power and typical charge/discharge timescale, including several proposed applications (Adapted from Muñoz et al. (2016). Credit: European Union's Horizon 2020 research and innovation program under Grant Agreement No. 645963. https://www.projectsensible.eu/documents/overview-of-storage-technologies.pdf)

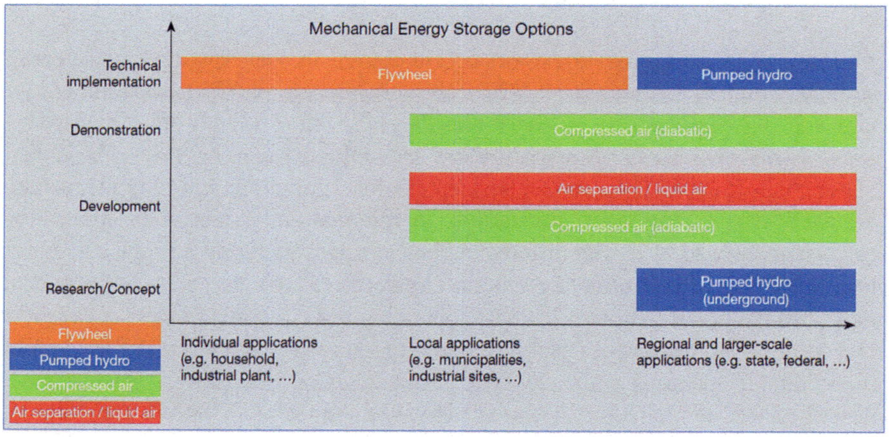

Fig. 12.14 State-of-the-art mechanical energy storage technologies (Adapted from Ausfelder (2016). Credit: Koordinierungskreis Chemische Energieforschung (Joint Working Group on Chemical Energy Research))

suggesting a need to switch from PHS to another energy storage solution. CAES is showing greater energy and power densities and lower environmental impacts as well as greater site availability. Unfortunately, these improvements come at a fuel cost and commercial viability is yet to be tested. Flywheel systems, in comparison to

CAES, are fairly mature and commercially tested. They exhibit many advantages over both PHS and CAES solutions. Safety issues associated with these systems can be mitigated but at a high cost. Both energy and power densities are, on average, higher than both PHS and CAES and show promise of even greater improvements. But FES still operates on a relatively small scale and for very short durations. Whether or not improvements will be good enough to extend applicability to larger-scale medium-term solutions still remains to be seen. Because of their small scale these systems may be more relevant to deployment for distributed grid infrastructure. Flywheel systems have also seen applications both in power plant settings as uninterruptable power supplies and as hybrid renewable power smoothing solutions (Sabihuddin et al. 2015).

In addition, the concept of cryogenic energy storage (CES) is to store energy in the form of liquid gas and vaporize it when needed to drive a turbine. Although CES on an industrial scale is a relatively new approach, the technology is well known and essentially part of any air separation unit (ASU) that utilizes cryogenic separation. The operational benefits of adding CES to an existing air separation plant may include: (1) increasing the plant's flexibility for load shifting, (2) storing purchased energy and selling it back to the market during higher-price periods, and (3) creating additional revenue by providing operating reserve capacity. Therefore, liquefied air has the potential of being an effective and competitive energy storage system. The technology involved in liquefaction has been widely employed in the industry and hence converting existing systems to liquefy air would not pose large challenges. Liquefied air as an energy storage also has the potential of propelling renewable energy usage to greater heights by utilizing wrong time energy instead of letting it go to waste. However, much research still has to be conducted to make liquid air a competitive energy storage system. The possible integration of power recovery into air liquefaction systems and its connectivity to the grid is an interesting prospect and a valid avenue to pursue (Lim et al. 2016).

12.4.2 Electrochemical Energy Storage

Electrochemical storage has by far the greatest diversity of research and commercial energy storage products to present day. Devices include not only traditional batteries but also molten salt/liquid metal batteries, metal-air batteries, fuel cells, and flow batteries, as shown in Fig. 12.15 (Ausfelder 2016).

Most electrochemical energy storage systems have a number of common features, for instance (Sabihuddin et al. 2015): the electrodes, the electrolyte, and the separators or membranes. Improvements have largely focused on materials. A shift has occurred towards more reactive electrodes. These more reactive variants have shown the promise of increasing energy and power densities—the use of lithium (e.g., Li-Ion batteries) and oxygen/air based chemistries, i.e., metal-air batteries (Fig. 12.16), reflects this trend. Thin film, foam, fiber, and polymer modifications (and material substitutions/additions) of existing electrodes have aimed at improved durability while benefiting performance. Virtually every chemical battery type has

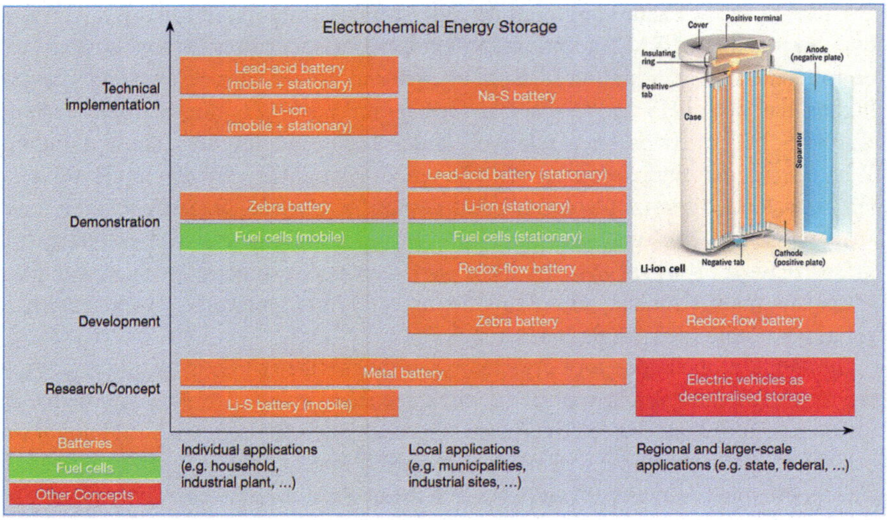

Fig. 12.15 State-of-the-art electrochemical processes for electrical energy storage (Modified from Ausfelder (2016). Credit: Koordinierungskreis Chemische Energieforschung (Joint Working Group on Chemical Energy Research))

seen the utilization of nano-materials or sintering for further improvements. Hybridizations of different battery chemistries have attempted to combine the positive aspects of multiple chemistries while mitigating the negatives. Improvements in electrolyte have consisted of additives or substitutions designed to improve stability and ion exchange (conductivity). Electrolyte gelling, circulation, and gas venting techniques provide a method for leak-resistant, higher-performance, and safer batteries. Separator and membrane design improvements have focused on allowing ion exchange to occur without loss of active cell material. The durability of separators and membranes has been a significant issue in some battery chemistries and so these materials have also seen additions and polymerizations similar to those undergone by electrodes. The integration of microelectronics has been considered and modifications of charging regimes have been shown to greatly extend life span of many chemical storage solutions (Badwal et al. 2014; Sabihuddin et al. 2015).

In the landscape of energy storage (i.e., mechanical, electrochemical, chemical, electromagnetic, and thermal), electrochemical storage techniques remain the most well-researched and well-developed field, as shown in Table 12.2. Most traditional chemical battery systems have excellent performance in one respect or another and have shown great suitability for specific small-scale applications. However, they suffer from severe deficiencies for large-scale storage. ZnAg batteries have excellent energy and power densities but suffer from short life and very high costs and so are only suitable for very small scale applications. ZnMn systems also have relatively short life but do better in terms of cost—these costs become high when attempting to deliver high power and so scale has been limited. For medium-scale applications,

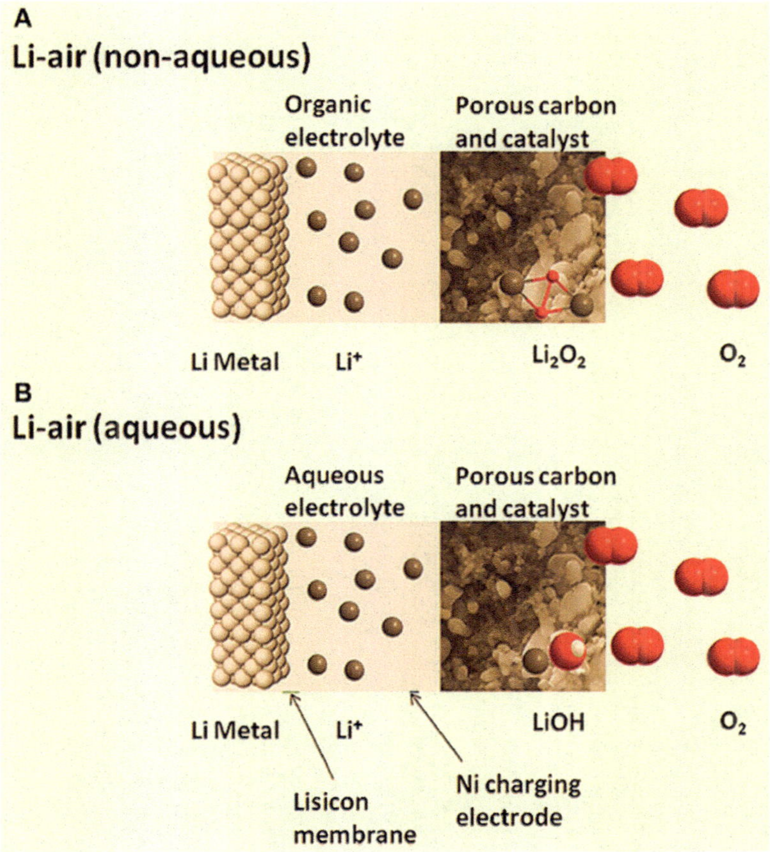

Fig. 12.16 Schematic representation of two contemporary versions of the lithium-air battery (Adapted with permission from Badwal et al. (2014), under the terms of the Creative Commons Attribution License (CC BY)): (**a**) nonaqueous version, similar to Li-ion; (**b**) aqueous, with Li^+-permeable membrane protecting the lithium anode

Pb-Acid batteries have found a niche. These Pb-Acid batteries are cheap and have high power delivery; however, they also have significant environmental implications and energy performance falls well below that of ZnAg and ZnMn chemistries. At a similar scale to Pb-Acid, Li-Ion chemistries have shown not only excellent energy and power performance, but also excellent cycle life. Li-Ion batteries, unfortunately, are expensive, have safety and charging issues, and the shortage of lithium resources limits their applicability for any meaningful large-scale system. Despite this, Li-Ion has displaced NiMH and NiCd systems. The abundance of materials, low cost, and very long life spans of NiFe systems have shown them as a potential larger-scale storage solution. Unfortunately, these systems suffer from poor performance and high self-discharge rates. NiZn provides high performance at low cost but suffers from very short cycle life. Most traditional battery systems also suffer from permanently reduced life and capacities under very high discharge rates, improper storage

Table 12.2 Characteristics of different battery energy storage technologies (Chatzivasileiadi et al. 2013)

Technology	Typical lifetime [years (cycles)]	Power density Wkg^{-1}/kWm^{-3}	Energy density $[Whkg^{-1}/kWhm^{-3}]$	Typical discharge time	Recharge time	Response time	Operating temperature [°C]
Lead acid	3–15 (2000)	75–300/90–700	30–50/75	s-3 h	8–16 h	ms	25
NiCd	15–20 (2500)	150–300/75–700	45–80/<200	s-h	1 h	ms	–40 to 45
Li-ion	8–15 (4×10^3)	230–340/1300–10,000	100–250/250–620	min-h	min-h	ms-s	−10 to 50
NaS	12–20 (>2000)	90–230/120–160	150–240/<400	s-h	9 h	ms	300
Na–NiCl	12–20 (>1000)	130–160/250–270	125/150–200	min-h	6–8 h	ms	270 to 350
$Zn–Br_2$ FB	5–10 (>2000)	50–1 50/1–25	60–80/20–35	s-10 h	4 h	<1 ms	20 to 50
V-Redox FB	10–20 (13×10^3)	NA/0.5–2	75/20–35	s-10 h	min	<1 ms	0 to 40
Flywheel	>20 (10^7)	400–1600/5000	5–130/20–80	15 s–15 min	<15 min	ms-s	20 to 40
Super/DL capacitors	>20 (5×10^5)	0.1–10/40,000–120,000	0.1–15/10–20	ms-1 h	s-min	ms	−40 to 85
Pumped-Hydro	50–100 (>500)	NA/0.1–0.2	0.5–1.5/0.2–2	h-days	min-h	s-min	Ambient
Compressed air (CAES)	25–40 (No limit)	NA/0.2–0.6	30–60/12	h-days	min-h	1–15 min	Ambient

environments, from overcharging or from complete discharge. Pre-mature failure is not uncommon with these batteries. These clear deficiencies have led to the development of molten salt, metal-air, fuel cell, and flow batteries. Both NaS and NaNiCl molten salt systems show very good energy and power performance and high suitability for grid-level large-scale applications. These systems are approaching cost competitiveness with traditional chemical battery systems and while more expensive than both PHS and CAES, might be worthwhile investments given their significantly higher performance. Metal-air systems such as Zn-Air and Fe-Air have been restricted to small-scale application. However, their extremely high performance and low costs may make them even better alternatives than molten-salt batteries if cycle life and higher discharge rates can be achieved. Some metal-air batteries have shown performances approaching those of burnt hydrocarbon fuels. Fuel cells offer an alternative to burning, have high energy and power performance, and have seen some commercial applications to large-scale grid-level storage/generation. Unfortunately, they remain on shaky ground, like metal-air chemistries, as a result of their poor life spans, high costs, and potential high operating temperatures. Adding to this is the problem of hydrogen storage, fuel reforming and the potential for moderate emissions. Combined with PHS, CAES, fuel cells, and molten salt batteries, flow batteries may also be immediate contenders for grid-level storage—these batteries not only have long low maintenance life but relatively high power and energy densities too. There is need, however, to reduce costs even further to be competitive with other solutions. Though it's too early to tell, it's likely that in the short-term future, there will be a shift towards CAES storage as well as molten salt and flow battery storage. In the slightly longer term, fuel cells may also become more prevalent. Traditional chemical batteries are unlikely to be strong contenders for large-scale storage at this point, though with additional research metal-air chemistries may hold future promise (Chatzivasileiadi et al. 2013; Sabihuddin et al. 2015).

12.4.3 Chemical/Thermochemical Energy Storage

Certain organometallic molecules undergo a reaction upon exposure to light that is reversible with either a catalyst or heat. In some cases, a considerable amount of energy can be stored. The energy density in chemicals is usually much greater than the energy density of current battery technologies. The produced chemicals would not only be useful for subsequent electricity production, but also as fuels for the transportation system and as raw materials for the chemical industry, substituting fossil oil. Figure 12.17 provides an evaluation of various technological elements for chemical energy storage technologies, which can cover the following groups (Ausfelder 2016):

(a) Power-to-gas: The conversion of electrical energy into gaseous energy carriers is based on hydrogen production by electrolysis of water. Hydrogen can be used as an independent energy carrier and, for instance, be deployed in the transport system via fuel cell vehicles. This requires a suitable physical or chemical

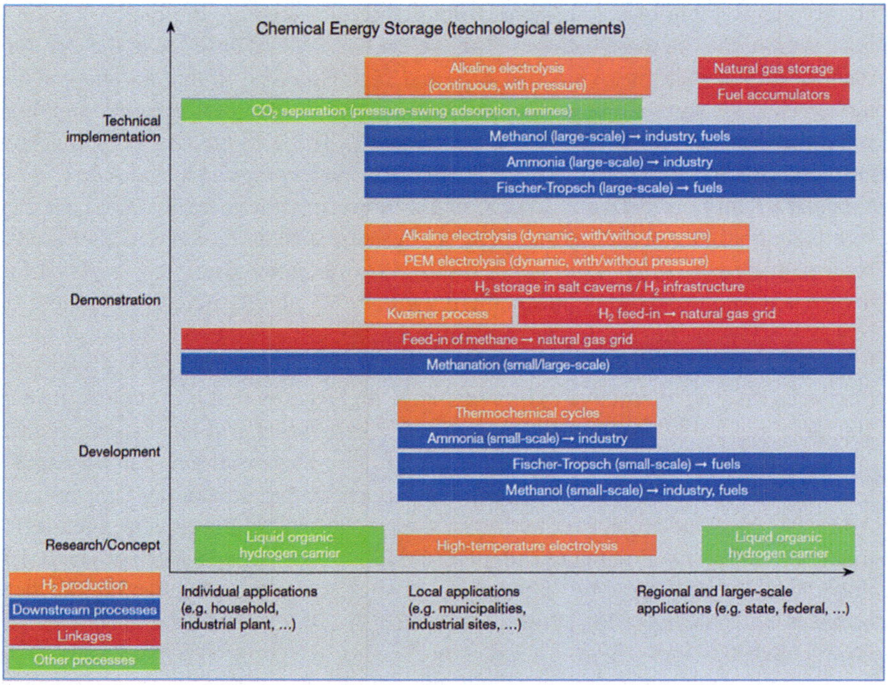

Fig. 12.17 State-of-the-art chemical/thermochemical energy storage (individual technological elements) (Adapted from Ausfelder (2016). Credit: Koordinierungskreis Chemische Energieforschung (Joint Working Group on Chemical Energy Research))

hydrogen storage capacity. Alternatives to the storage of hydrogen and its reconversion into electricity include its application as a feedstock for industrial processes (petrochemistry, chemistry, steel, metal processing, glass, etc.) or further transformation into other energy carriers/raw materials. One of these follow-up processes is methanation. Methanation is classified under power-to-gas, since methane is a gaseous energy carrier that can be used as a substitute for natural gas, is storable, and opens up access to the heat system.

(b) Power-to-liquid (including power-to-fuels): The conversion of excess electricity into liquid energy carriers also proceeds via electrolysis of water with subsequent processes leading to liquid fuels, for example, via Fischer-Tropsch synthesis or methanol production. This concept primarily addresses the transport system. The storage of hydrogen by means of liquid organic hydrogen carriers (LOHC) also falls into this category.

(c) Power-to-chemicals: These routes target the production of industrially relevant chemicals for further use as raw materials in chemical and industrial processes. For instance, hydrogen production by water electrolysis can be the first step to proceed. In principle, any process using hydrogen as feedstock, for example, the conversion of CO_2 into methanol or formic acid, could be used. Ultimately the conversion of excess power into heat should also have a place in this classification (power-to-heat).

The application of chemical/thermochemical storage technologies is motivated by the availability of (exergetically) very high-quality energy in the form of electricity, for instance from intermittent renewable energies that cannot be used directly by other consumers (excess electrical energy). A further option is to integrate renewable energies into other energy systems and also to provide energy carriers or raw materials from "domestic production" for industrial use. This has the potential of reducing dependence on raw material imports of fossil energy carriers (Ausfelder 2016).

12.4.4 Electromagnetic Storage

Traditionally, electromagnetic storage was limited to capacitors and inductors. New material advances have allowed the development of supercapacitors and superconductors and have extended the use of capacitive and inductive technologies to larger-scale applications.

12.4.4.1 Superconducting Magnetic Energy Storage

Superconductors have seen a range of storage/generation applications. They have been used in flywheel storage systems for low friction bearings, in synchronous generators for improved generating performance, and in fusion power systems to assist with plasma containment. However, superconducting magnetic energy storage (SMES) refers to their use for energy storage in the magnetic field of an inductor. As with FES, the main application of SMES has been for short-term power quality and stability—this came about from the observation that 90% of grid failures are sags and outages lasting less than one second but are responsible for costly damages (Sabihuddin et al. 2015).

Superconductors have very long life spans, cycle life, high efficiency, fast response, and very high discharge rates. This power capacity is not matched by energy performance, and energy capital costs remain very high. There are four main components associated with SMES: The superconductor, the refrigeration, the containment vessel and the power converter. Micro-SMES systems have seen significant commercial deployment. These systems combine all four components into a single black box solution. Utilities, however, remain cautious about adopting these new technologies due to their long lead times and parallel development of related cryogenic and converter technologies is needed. Hybrid solutions have aimed at combining SMES with traditional Pb-Acid UPS systems whereby the superconductor provides short-term high discharge and reduces initial high discharge loading on Pb-Acid batteries. Alternative systems have attempted to use liquid hydrogen SMES combined with fuel cells. Their use has begun to extend to onsite wind energy applications too. SMES is widely regarded as environmentally friendly; however, some concern remains over safety and health risks associated with high magnetic fields. Super conductors in SMES are typically either solenoidal or toroidal in topology. Solenoidal topologies have more stray magnetic fields. Toroidal approaches reduce these stray fields at the expense of mechanical complexity. These topologies are cooled to a critical temperature depending on the type of superconductor—metal alloys such as nobium titanium or nobium tin that operate at low temperatures at or below −249 °C were used in the past. The rising cost of helium and the discovery of ceramic oxide-based superconductors have ushered in the use of high-temperature variants operating at up to −138 °C and allowing the use of cheaper liquid nitrogen. The cost of superconductors is likely to always remain higher than copper conductors, so cost benefits must be quite significant to justify their use. The development of superconducting materials has, however, followed the exponential trend of transistors and other similar technologies, so cost and performance improvements are likely to occur. It should be noted that cryogenics and power control and conversion are the major barriers to adoption, not the superconductors themselves (Buckles and Hassenzahl 2000; Ali et al. 2010).

The development of superconductors has occurred in parallel with cryogenic cooling systems. These systems originally used external vapor cooled liquefier systems, but recent advances have seen the use of thermo-acoustic (pulse tube) cyro-cooling. Cooling costs can be lowered with the use of high-temperature superconductors. Cooling times can be long and size and cost reduction of cooling systems is still a significant challenge, as is reliability. With the low temperatures, the need for safe containment is also a consideration and can add to system size. Power conditioning and control remain a challenge in SMES. The development of variable speed AC motor drives has lent considerable progress to power conversion. Typical SMES systems utilize either thyristor, voltage source, or current source converters. Power conditioning systems (PCS) handle power transfer between SMES and AC grid and operate similar to flywheel circuits. The incorporation of SMES into existing flexible ac transmission system infrastructure can reduce costs associated with grid interconnection significantly (Sabihuddin et al. 2015).

12.4.4.2 Supercapacitor Energy Storage

Electrostatic capacitors utilize electrodes separated by a dielectric to store energy in an electric field. Unfortunately, they have low capacitance making them unsuitable for higher power applications. Some improvements result from the use of electrolytes as with electrolytic capacitors. But these still have limitations. Supercapacitors are electrochemical double layer capacitors (EDLC) that use a combination of porous separators and electrolyte in place of a dielectric, as shown in Fig. 12.18. Electrolytes can be organic or aqueous. These capacitors operate at low voltages but have high capacitance, extremely long cycle life, wide operating temperatures, and high discharge rates. As a result, these ELDC products have seen a number of commercial applications. These applications have looked at memory backup, battery support for high discharge environments and to provide grid stability. Unfortunately, supercapacitors suffer from high parasitic losses, relatively low energy performance, and high cost. In addition, the low voltages require series stacking. The stacking results in unequal voltage distributions among ELDC cells and reduces life span and performance (Sharma and Bhatti 2010; Muñoz et al. 2016).

Supercapacitors can be categorized as symmetric or asymmetric (hybrid) based on whether or not both electrodes are the same material or different materials (Fig. 12.18). Many of the developments from chemical battery systems are applicable to supercapacitors. The utilization of carbon foams, fibers, aerogels, xerogels, and nanotubes and the use of conductive polymers and metal oxides have shown

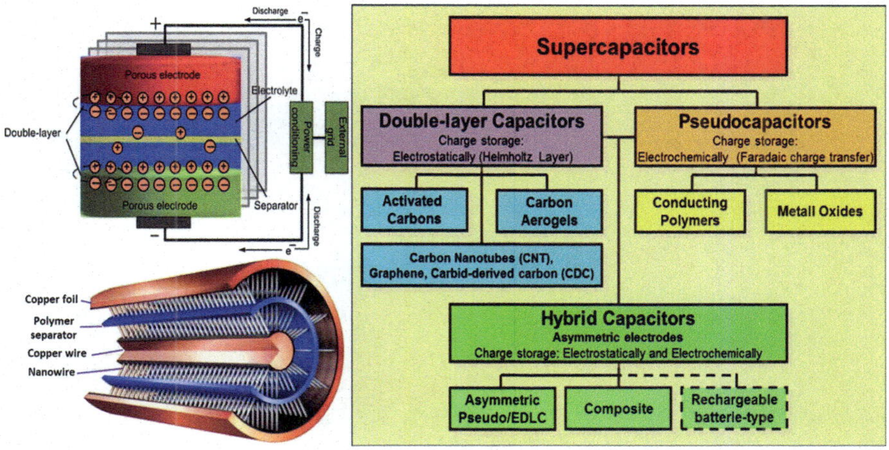

Fig. 12.18 Supercapacitor scheme and its classification (Modified from Muñoz et al. (2016). Credit: European Union's Horizon 2020 research and innovation program under Grant Agreement No. 645963. https://www.projectsensible.eu/documents/overview-of-storage-technologies.pdf)

significant improvements in performance. Some electrodes may be composites. Tangled networks with open central channels have shown exceptional specific capacitance. The volumetric reactions resulting from polymer film electrodes have shown further performance benefits over the surface reactions of traditional carbon electrodes. Metal oxides remain easy to manufacture while nanotubes are largely within the arena of research. Higher voltages can be achieved by organic electrolytes but these are accompanied by higher resistivity. Aqueous electrolytes provide better conductivity at the cost of reduced breakdown voltage. Separators allow ion exchange via the electrolyte but prevent electrode short circuiting. Problems of unequal voltage distribution can be handled by passive or active voltage equalization circuits. Passive approaches for smaller-scale applications have concentrated on resistive bypass or zener diode voltage regulation. A shift has been seen towards zener approaches because of the lower losses. These passive approaches are substituted for active equalizing current systems for high duty cycle low parasitic loss applications. These active approaches may take the form of buck-boost style controls (Sharma and Bhatti 2010; Sabihuddin et al. 2015).

12.4.5 Thermal Storage

Thermal storage systems are not new and have been used extensively in home temperature stabilization as well as industrial scale thermal power plants and district heating applications. The use of thermal energy storage, however, must be coupled to some form of thermoelectric generator or heat engine for electricity generation. The use of thermal storage and heat recovery has allowed power plant efficiencies to increase up to 60% for natural gas plants. The advent of solar thermal systems has

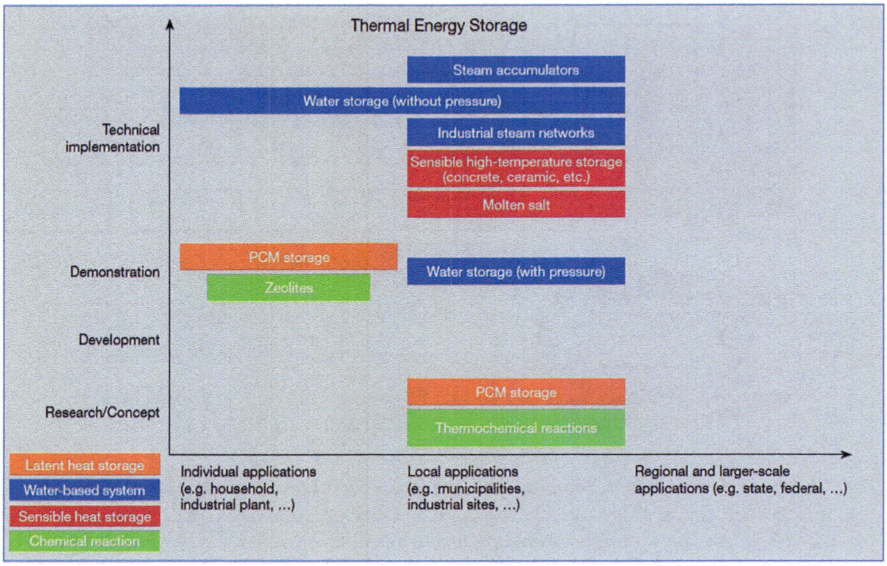

Fig. 12.19 State-of-the-art thermal energy storage technologies (Adapted from Ausfelder (2016). Credit: Koordinierungskreis Chemische Energieforschung (Joint Working Group on Chemical Energy Research))

further encouraged development in the field. Most solar thermal plants typically take one or more of the following forms: parabolic troughs, dish, fresnel, or heliostat systems. As shown in Fig. 12.19, there are three types of storage that have emerged: traditional sensible heat storage (STES), latent heat storage (LTES), and reversible chemical reaction heat storage (CTES). Systems normally contain three main components: the thermal material, the heat exchanger, and the containment system. Sensible storage is by far the cheapest form of thermal storage and is ideally suited for short-term applications. For longer-term or more energy-dense storage, latent heat storage offers greater potential. Reversible chemical reaction heat is the most promising and least developed form of thermal storage. In real systems it is likely that a combination of all three storage forms will be used. Thermal storage has strong potential for existing power plants and solar thermal installations and can significantly increase energy efficiencies and reduce losses for existing systems. This is especially true in the context of energy storage in molten salt batteries, fuel cells, and even compressed air systems (Sabihuddin et al. 2015; Ausfelder 2016).

12.4.5.1 Sensible Heat Storage

By far the most ubiquitous form of storage is sensible heat thermal energy storage (STES). These are materials such as concretes, salts, metals, and fire bricks that absorb heat in the order of 0.85–1.15 KJ/Kg at specific temperatures—i.e., the specific heat capacity. The heat is absorbed and released when the ambient surrounds fall below the material temperature. There are two main forms of STES systems:

Passive and active. In passive systems the storage medium is fixed and heat is transferred through a passive heat transfer mechanism or fluid. In active systems the heat storage medium itself is circulated. In active direct systems separate tanks hold the hot and the cold storage material. These systems require no additional heat exchanger and heat transfer occurs quickly; however, they have higher costs. With active indirect systems both the hot and cold storage occur in the same tank—these indirect systems are somewhat cheaper and require carefully controlled charge and discharge to maintain stratification. Some systems employ silica or quartz to assist in maintaining the thermocline for improved stratification (Muthukumar 2011).

In general, heat storage materials for STES are either solid or liquid. Concrete and fire bricks are heavily used solid forms, and, among liquids, molten salts are cheap, abundant and have good heat capacities and are often used for active storage from solar concentrator systems. However salts can potentially freeze. For lower temperatures (0 to 100 °C), applications involving water are among the best with capacities well above many other materials. Alternatives have included the use of oils and organic liquids. STES systems can be hybridized with latent heat storage for improved performance (Muthukumar 2011; Sabihuddin et al. 2015).

12.4.5.2 Latent Heat Storage

Unlike STES, latent heat thermal energy storage (LTES) utilizes the heat absorbed during phase transition (the heat of fusion) for energy storage. There are three types of materials commonly used: organic, inorganic, and eutectic. Organic paraffin appears to be quite popular, but non-paraffin materials are far more numerous and have various desirable properties. Organic materials are also more expensive but are noncorrosive, chemically stable, and easy to incorporate into building materials. Inorganic materials include salt hydrates and metallic compounds. Salt hydrates, while widely explored, suffer from super-cooling issues and low thermal conductivities both of which can detract from system performance. Metallic and graphite additives have been used to improve conductivity in these hydrates. Inorganic materials are generally cheaper than organic materials, but suffer from chemical decomposition after repeated cycling. Eutectic materials are composite mixture of materials that have phase transition at lower temperatures than their constituent materials, thus allowing fine tuning of storage temperatures while taking advantage of the benefits of both organic and inorganic materials. LTES systems tend to have higher energy storage potential than STES systems. Heat of fusion in these systems can range from 100–340 KJ/Kg at operational temperatures—that's about two orders of magnitude bigger than STES materials. LTES is often synonymous to phase transition materials (PCM). The materials can undergo four main transitions: solid-liquid, solid-solid, solid-gas, and liquid-gas. In terms of flexibility and easy containment (from reduced volume change) solid-solid transition materials are very attractive and do significantly better than sensible storage in concrete. Solid-gas transitions, however, offer the best storage capacity but also undergo the largest volume change during transition resulting in containment issues. As a compromise between storage capacity and ease of containment, most PCMs utilize solid-liquid phase transitions (Muthukumar 2011).

Commercially available PCMs usually take the form of macro-capsule solid pellets usually containing an appropriately selected wax. To improve construction material compatibility and transportability micro-capsules are used in wet slurry, paste or powder form. Sometimes, PCM material is absorbed, post-manufacture, into plasterboards and other construction materials directly. Both macro- and micro-capsules have applications ranging not only from construction but to clothing, bedding, food curing, drying and plant growth, thermal and solar thermal energy generation, solar cooking and even to electronics as well. Unfortunately, unlike, STES, most PCMs can't be used for heat transfer as well, so a separate heat exchange mechanism is needed (e.g., heat pipes). Utilization of PCMs for improved heating and cooling has shown huge improvements in fuel efficiency and promises reductions in generation capacity via large-scale distributed peak shaving (Sharma et al. 2009; Sabihuddin et al. 2015).

12.4.5.3 Reversible Chemical Reaction Heat Storage

Reversible chemical reaction heat thermal energy storage (CTES) provides a third and more energy dense and efficient storage medium than both STES and LTES. There are three main types of CTES: heat pump systems, heat pipe systems, and heat of reaction systems. Chemical heat pumps (CHP) utilize adsorption (exothermic) and desorption (endothermic) of a vapor/liquid onto a solid substance. Typically, materials for CHP come in working pairs (e.g., metal hydrides-hydrogen, hydrates-water, ammoniates-ammonia). Chemical heat pipes (CHPi) are very similar to both CHPs and traditional heat pipes but utilize the dissociation of aqueous solutions of acid or bases (e.g., sodium hydroxide) as a mechanism for thermal storage. Chemical reaction heat (CRH) systems operate by breaking down compounds into their constituent parts via an endothermic process and by releasing that heat and remixing the individual components. These reversible chemical systems can also enable the low loss transport of waste heat. As such reversible chemical approaches offer the potential for long-term storage not available from other thermal forms. Metal hydride systems are usually made from nickel, iron, magnesium, or cobalt alloys combined with hydrogen. Salt hydrates are similar but lock water in crystalline form. These hydrates unfortunately are slow and have unpredictable kinetics. The combination of careful temperature and pressure control and the use of additives can improve the kinetics. Some approaches utilize solid-gas pairs. One such example is the use of ice with carbon dioxide. This approach may have relevance when combined with carbon sequestration (Fischer et al. 2014; Sabihuddin et al. 2015).

12.4.6 Linkage Options and Future Trends of Energy Storage Technologies

Energy storage systems are of interest for electrical and thermal energy and also, in the form of chemical energy storage systems, for transport and industrial applications. The key demands on an energy storage system are high efficiency, low self-discharge, high capacity, high number of charging and discharging cycles, high performance and low costs coupled with a high degree of public acceptance. A glance at the available technologies reveals that no storage system fulfills all these requirements at present. They are all in different stages of development. Nevertheless, storage technologies can be deployed in isolation or, using a systemic approach, link various supply strands of the energy system with one another. Chemical energy storage technologies, which, for instance, use electricity for water electrolysis to produce hydrogen, are predestined for this role. The hydrogen produced can then be utilized in fuel cell vehicles, thus serving the transport system, or, after subsequent methanation, can be injected into the natural gas grid, thus supplying the heat system. Since chemical storage systems can also be used as raw materials in industrial value chains, the various energy systems are strongly interwoven and can create significant synergistic benefits. Figure 12.20 presents an overview of the possibilities (Ausfelder 2016).

The landscape of energy storage is extensive. Some storage technologies are strongly coupled to particular generation technologies. For instance, PHS systems show strong similarities with hydroelectric plants and are often used in conjunction with nuclear facilities. CAES systems are much like peaking gas turbine plants. Fuel cells, as they increase in scale, operate much like traditional thermal generation plants, albeit converting fuel directly to electricity. This is true of some flow battery configurations too. Thermal storage systems are integral parts of thermal (and solar thermal) plants and are often used in the context of steam generation and waste heat recovery for subsequent power plant cycles (Sabihuddin et al. 2015).

Storage technologies have been compared numerically and qualitatively on the basis of fourteen parameters, namely: specific energy, energy density, specific power, power density, efficiency, life span, cycle life, self-discharge rate, scale, energy capital cost, power capital cost, application, technical maturity, and environmental impact. For pumped hydro energy storage and lead-acid batteries, flexible operation of large-scale industrial plants with high energy demands, and also, in part, of energy supply, has been practiced for years and is well established. PHS accounts for 99% of worldwide deployed energy storage, yet it is the least energy and power dense solution of all the storage options. The low loss storage capacities of PHS installations have lent them to easy adoption for large-scale energy management applications. With increasing concerns over energy security and sustainability, as well as a decreased availability of storage sites and low profitability of PHS, a host of new options have been developed. Of these, among the most viable contender for large-scale grid-level storage is CAES. CAES offers greater deployment locations and higher energy and power capacities. Furthermore, it has low capital costs, low losses, and long life. Following CAES, flow batteries appear to be another large-scale solution for storage. These batteries suggest the possibility of improved energy and power performance over CAES systems, meaning that systems can, potentially, be decoupled from specialized geologically suitable sites. Unfortunately, flow batteries have high costs and potentially higher environmental and safety impacts. Deployments of flow battery systems have been relatively small scale and significant effort is needed for commercialization. After flow batteries, fuel cells also offer high potential for large-scale storage/generation. Of course, as with other generation

Conversion of	Power	Fuels	Heat	Natural gas	Hydrogen	Industrial processes and value chains
Industrial processes and value chains	Operating reserve, DSM, hybrid processes, deferrable loads	Petrochemical industry	Integration of waste heat streams into district heating networks		Synthesis gas, chlorine-alkaline electrolysis	Products
Hydrogen	Combined cycle heat & power generation gas turbines fuel cells	(with CO_2): Methanol, Fischer-Tropsch	(Small-) Combined cycle heat & power, fuel cells	(with CO_2): Methanation with direct feed-in in natural gas grid	Storage vessels caverns	Chem. and petrochem. industries, steel industry
Natural gas	Combined cycle heat & power generation gas turbine	Via synthesis gas: methanol, Fischer-Tropsch	Combined cycle heat & power generation heating plant gas heating	Porous underground storage, caverns, natural gas grid	Steam reforming	Chem. and petrochem. industries
Heat	Steam turbine, organic Rankine cycle		Water, sensible heat storage, PCM, chem. reactions		Thermo-chemical cycles	Process heat
Fuels	Combined cycle heat & power generation	Storage vessels caverns	Oil-based heating, combined cycle heat & power generation		Synthesis gas	Chem. and petrochem. industries
Power	Mechanical storage, batteries	(with CO_2) Via hydrogen: methanol, Fischer-Tropsch	Power-to-heat	(with CO_2): Methanation with direct feed-in in natural gas grid	Electrolysis	Electrical process energy
	Conversion into					

Fig. 12.20 Matrix depicting linkage options between various supply and application areas of energy. The fields contain the respective storage or conversion technology by which energy or storage types can be converted into each other or can be integrated into industrial processes. Example: Electrical Power (row) is harnessed via power-to-heat in the form of Heat (column) (Adapted from Ausfelder (2016). Credit: Koordinierungskreis Chemische Energieforschung (Joint Working Group on Chemical Energy Research))

technologies, separate fuel storage and transportation considerations are required. The energy performance of fuel cells is far in excess of PHS, CAES, and flow batteries. This means a significant reduction of land usage. Among fuel cells the most promising appear to be high-temperature SOFC style devices—these have seen some commercial deployment. Unfortunately, fuel cells suffer from major life span-related issues—until these have been addressed, it is unlikely that the technology will be a major player in large- or small-scale storage. Metal-air batteries bear a lot of similarity to fuel cells and show higher power performance but remain in the very early stages of development as well (Sabihuddin et al. 2015).

Barring fuel cells and metal-air batteries, at the smaller scale, a number of contenders present themselves. Molten salt, NaS and NaNiCl systems have seen applications in electric vehicles and exhibit reasonable power and energy performance as well as long life. These devices are easy to recycle, rugged, and well suited for heavy and light duty charge/discharge regimes. They also hold promise for larger-scale applications and may be better suited (as a result of longer life) than fuel cells to meet this challenge. They are, however, high-temperature devices that require thermal management and experience-related parasitic losses. As an alternative requiring simpler thermal management, Li-Ion batteries, while not suited for large grid-level applications (due to resource availability), have begun to see significant promise for electric vehicles as well. While it remains to be seen which chemistry will be successful at the smaller scales, far more extensive research efforts have been made in the direction of lithium chemistries as opposed to molten salt chemistries. This suggests that Li-Ion is likely to displace NaS and NaNiCl for electric vehicle applications. Li-Ion batteries have been remarkably successful and have displaced or are displacing NiCd and NiMH batteries in this arena. Likewise, NiZn chemistries traditional used in electric scooters have also undergone similar displacement—primarily due to their short cycle life. An additional final contender for smaller-scale storage, especially in the context of renewable installations appears to be the NiFe battery. This chemistry is rugged, long lived, environmentally benign, easily maintained, easily charged and discharged and has the potential to be of lower cost than Li-Ion chemistries. Unfortunately, after its displacement by Pb-Acid, this chemistry has seen little commercial deployment and so development must catch up significantly before the batteries become competitive. Remaining battery chemistries such as ZnMn and ZnAg are still largely restricted much smaller-scale applications (Sabihuddin et al. 2015).

Moving away from electricity storage, thermal energy storage systems are becoming a major research area especially in the context of efficiency improvements in existing generation infrastructure. They have been applied to solar thermal generation and have significant implications for distributed and district heating applications. The development of all three: STES, LTES, and CTES, is likely to occur in parallel with other storage (e.g., molten salt, fuel cell, compressed air) and generation systems. Furthermore, in the context of management and regulation of grid power quality, FES, ELDC, and SMES systems have found a niche. Power delivery of these three technologies is unparalleled to any other system and it's likely that all three will continue to improve in parallel to other "energy management" focused storage solutions. However, parasitic losses have prevented their direct application to energy management thus far. In addition, the use of chemical reactions is a good field for further development since they can be applied in extremely varied temperature ranges, allow high storage densities, and permit loss-free long-term storage (Sabihuddin et al. 2015; Ausfelder 2016).

In addition, the interface between electrical energy and energy in chemical bonds holds the promise to become a key strategic element of the future energy system. In all current considerations this position is reserved to water electrolysis with improved part-load capacity, higher efficiency, and significantly lower investment costs. This is the starting point for R&D activities on the various strategies for exploiting hydrogen. Priorities are clean combustion for heat generation, its use in gas turbines and similarly direct conversion into electricity in fuel cells and further chemical transformation. Moreover, hydrogen can be used directly in fuel-cell vehicles, thereby replacing conventional vehicles. Since this avoids conversion losses, hydrogen would seem to be the most logical alternative from an economic point of view. However, the entire chain from electrolysis to transport and compression of hydrogen for the refueling process is quite complex. In principle, therefore, centralized hydrogen production and distribution via a refueling network could be complemented by a decentralized system with vehicle owners using their own electrolysers. Apart from fuel-cell vehicles, battery-powered electric vehicles are another emission-free alternative to internal combustion engines. Once they have succeeded in penetrating the market, battery-powered vehicles could be classified as one component of the electricity grid, provided that charging and also, possibly, partial discharging can be controlled centrally and as a grid service (Ausfelder 2016).

12.5 Zero-Point Energy Conversion

Zero-point energy (ZPE) is a universal natural phenomenon of great significance which has evolved from the historical development of ideas about the vacuum. In the seventeenth century, it was thought that a totally empty volume of space could be created by simply removing all gases. This was the first generally accepted concept of the vacuum. Late in the nineteenth century, however, it became apparent that the evacuated region still contained thermal radiation. To the natural philosophers of the day, it seemed that all of the radiation might be eliminated by cooling. Thus evolved the second concept of achieving a real vacuum: cool it down to zero temperature after evacuation. Absolute zero temperature (−273 °C) was far removed from the technical possibilities of that century, so it seemed as if the problem was solved. In the twentieth century, both theory and experiment have shown that there is a nonthermal radiation in the vacuum that persists even if the temperature could be lowered to absolute zero. This classical concept alone explains the name of "zero-point" radiation. Besides, there are rigorous derivations from quantum physics that prove its existence. With the uncertainty principle, as stated in Eq. (12.5), Planck's constant h (6.63×10^{-34} joule-s) offers the fundamental size of the quantum. It is also the primary ingredient for the uncertainty principle. One form is found in the minimum uncertainty of position x and momentum p expressed as (Valone 2005).

$$\Delta x \ \Delta p \geq h/4\pi \tag{12.5}$$

In quantum mechanics, Planck's constant also is present in the description of particle motion. The harmonic oscillator reveals the effects of zero-point radiation on matter. The oscillator consists of an electron attached to an ideal, frictionless spring. When the electron is set in motion, it oscillates about its point of equilibrium, emitting electromagnetic radiation at the frequency of oscillation. The radiation dissipates energy, and so in the absence of zero-point radiation and at a temperature of absolute zero the electron eventually comes to rest. Actually, zero-point radiation continually imparts random impulses to the electron, so that it never comes to a complete stop, as shown in Fig. 12.21. Zero-point radiation gives the oscillator an average energy equal to the frequency of oscillation multiplied by one half of Planck's constant. Therefore, zero-point energy can be thought as the nonthermal, ubiquitous kinetic energy (averaging $\frac{1}{2}hf$) that is manifested even at zero degrees Kelvin, abbreviated as "ZPE," also called vacuum fluctuations, zero-point vibration,

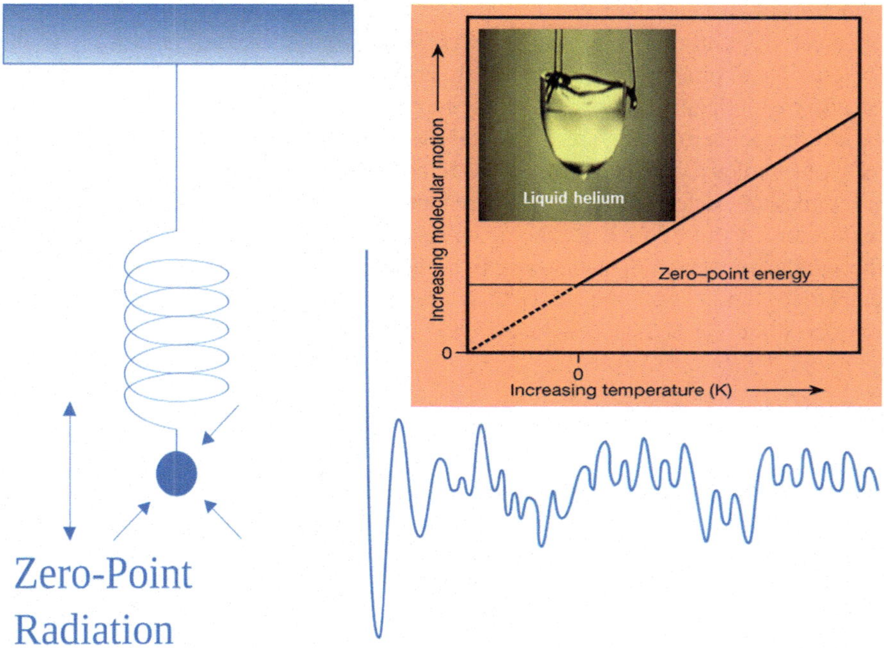

Fig. 12.21 Zero-point radiation continually imparts random impulses on an electron, so that it never comes to a complete stop. Zero-point radiation gives the oscillator an average energy equal to the frequency of oscillation multiplied by one-half of Planck's constant (Modified with permission from Valone (2005) (Integrity Research Institute))

residual energy, quantum oscillations, the vacuum electromagnetic field, virtual particle flux, and dark energy (Boyer 1985; Valone 2005).

There is a significant growth in the awareness that ZPE is real and not confined to the presently understood "laws" of thermodynamics and therefore could potentially be accessed for practical applications. A number of technologies that claimed to access a nonconventional source of energy (which, if validated, could very well be ZPE) have been compared and found to have common operating principles. "Empty" vacuum is (paradoxically) filled with an extremely high density of energy ZPE (equivalent to a high density of matter). That is, in empty space (away from observable massive objects), there is an extremely high pressure of the ZPE "fluid" (i.e., the extremely high density of energy or matter); however, all vector components of the force produced by this pressure are equal and therefore cancel each other out. As a result, there is no net force vector (direction). Observable matter (atoms, molecules, and large objects composed of these) is slightly less dense than the uniform ZPE of the vacuum. Therefore, the surrounding ZPE (vacuum) will (paradoxically) create a force vector pushing (not pulling) from the high pressure in the space away from observable massive objects and towards the lower pressure near observable massive objects. That is, a pressure gradient in the ZPE is established towards "matter." The inverse square relationship for the gravitational force as a function of distance between two masses is due to geometry (i.e., at twice the distance from an object, the "intercepted" area would be one-fourth, etc.). It may be possible to directly produce relatively large forces in a device or system simultaneously with the production of excess energy. Such forces could potentially levitate and/or propel aerospace vehicles directly without expelling mass from the vehicle. Such technology would not be limited to aerospace applications. It is possible that examples of such technology may include the Papp engine and the "water arc explosion." Both of these technologies had mechanical energy outputs claimed to be significantly larger than the respective electrical energy inputs, and both processes apparently produced force directly, as they were inherently essentially cold (i.e., produced little or no heat) (Valone 1999).

12.5.1 Zero-Point Energy Access Principles

According to the principles of quantum mechanics, the seemingly empty vacuum between atoms, when considered at small enough dimensions (orders of magnitude smaller than atoms), contains an exceedingly high energy density (as large as 10^{94} ergs/cm^3). This energy is referred to as zero-point energy because it is believed to exist even at a temperature of absolute zero. This energy can be thought of as electromagnetic radiation of all frequencies or even as fluctuations of spacetime itself. Because of the totally random characteristic of this energy, it paradoxically appears to cancel itself out or not exist. The randomness of this energy does not generally allow any of it to be tapped for any practical use, one exception is nanoscale microelectromechanical systems that the zero-point energy, which has been conclusively proven to exert the measurable Casimir Force when micromechanical parts are within 1 μm or less of each other, can prevent such devices from functioning, unless means are introduced to overcome this phenomenon. The existence of this type of energy is not immediately obvious, and it is difficult to detect because it is incoherent. The energy is everywhere—its detection requires measuring an energy difference, and the extremely high frequencies of this energy do not readily interact with matter (Alexander 2003).

For more than 100 years, a number of people have claimed to have invented energy conversion technologies that would deliver an output energy greater than the input energy. For example, the common "heat pump" does this routinely; one unit of electrical energy is used to transfer approximately three units of heat into a selected location (i.e., within a building). The heat (of course) is simply being transferred either from the air or the ground. It should be noted that the science of thermodynamics originated nearly 200 years ago in an attempt to compare quantities of heat with an equivalent quantity of mechanical work. For (classical) thermodynamics, a (conceptual) "boundary" is drawn around a system being investigated, and (by definition), if the boundary is drawn large enough, the energy within the system is conserved (i.e., constant). However, if nuclear reactions are involved, then the term "mass-energy" (based upon Dr. Albert Einstein's equation $E = mc^2$) must be used rather than "energy." For an "open" form of thermodynamics that predicts nonclassical behavior when a source of energy is available to continue flowing through a system, outputs and inputs do not need to be equal (Prigogine 1977).

One of the principles may allow ZPE to be tapped is that of the nonequilibrium thermodynamics. The standard scientific belief is that the Second Law of Thermodynamics must cause systems to become more random and disordered; however, nonlinear systems (e.g., ionized plasma) are not restricted to this "law." The addition of energy to a plasma can sometimes form a metastable vortex ring called a plasmoid. Such a structure cannot be predicted by a linear thermodynamic model, but it can be predicted by a nonlinear magnetohydrodynamic model. The nonlinear interactions produce macroscopic coherence from random turbulence. The persistence of "ball lightning," which has been modeled as a vortex ring plasmoid, is evidence that such structures are cohering some ZPE (which maintains the stability of the structure) and then radiating excess energy as light and heat. To build devices that produce "anomalous" excess energy, (a) Use ions in a plasma to "tap" ZPE, because the electric field lines are highly concentrated where they enter the nucleus of an atom. (b) Induce a shock-wave into the plasma to cause a sudden motion of ions. (c) Tap ZPE via what could be called a "rebound effect" (Alexander 2003).

On the other hand, zero-point energy consists of random EM waves of all possible frequencies or can be considered to be random fluctuations of the vacuum itself. The force (caused by ZPE) between two conducting objects separated by a short distance (on the order of microns), known as the Casimir Force, occurs because random ZPE EM waves larger than the distance between the conducting objects are essentially "shorted out" and cannot exist within that small space. Therefore, the ZPE in the vacuum outside the two close objects, which has a very slightly greater energy density, will have an effective pressure slightly higher than between the objects and produce an attractive force between these objects. Moreover, the "Electro-Radiant Event"(ERE) is produced when a high-voltage, direct current is discharged across a spark-gap and interrupted abruptly before any reversals of current can occur. This effect is greatly increased when the source of direct current is a charged capacitor. The ERE leaves wires and other circuit components perpendicular to the flow of current, which produces a spatially distributed voltage that can be thousands of times higher than the initial spark discharge voltage. It propagates instantaneously as a longitudinal, electrostatic "light-like ray" that behaves similarly to an incompressible gas under pressure. EREs are solely characterized by impulse duration and voltage drop in the spark-gap. EREs penetrate all materials and create "electronic responses" in metals like copper and silver. In this case, "electronic responses" means that an electrical charge will build up on copper surfaces exposed to Electro-Radiant emissions. Electro-Radiant impulses shorter than 100 μs are completely safe to handle and will not cause shock or harm. Electro-Radiant impulses shorter than 100 ns are cold and easily cause lighting effects in vacuum globes (Lindemann 2001).

Therefore, the respective principles to make an output energy greater than the input energy can be summarized as:

(1) Generate a relatively high voltage by an applicable method. (2) Discharge the voltage across a spark-gap. (3) Control the duration and direction of the discharge.

(4) Capture (with a metallic receiver that is not directly connected to the spark-gap) a (postulated) "mass-free" component of electricity that is radiated from the discharge. (5) Feed a sufficient fraction of the captured energy back to continue Step (1). One can hypothesize that the "mass-free" "radiant electricity" momentarily coheres and collects a very small amount of the ZPE field as it traverses the space between the discharge and the metallic receiver and that the collected energy is significantly larger than the energy dissipated in the initial spark discharge.

12.5.2 Zero-Point Energy Exaction Methods

The ZPE extraction methods usually involve ZPE in the form of electromagnetic zero-point fields (ZPFs). The energy density of these ZPE vacuum fluctuations can be expressed as (Milonni 1994):

$$\rho(hv) = \frac{8\pi v^2}{c^3}\left(\frac{hv}{e^{\left(\frac{hv}{kT}\right)} - 1} + \frac{hv}{2}\right) \tag{12.6}$$

where ν is the frequency and k is Boltzmann's constant. The first term in the large brackets describes Planck radiation from a black body at temperature T. As T approaches zero or at room temperature at frequencies above 7 THz, the energy density is dominated by the second, temperature-independent term, which is due to zero-point energy. For high frequencies this energy density is huge, but how large depends upon the frequency at which the spectrum cuts off, a matter that is not resolved.

Two common physical manifestations of the ZPF are zero-point noise fluctuations and the force between Casimir cavity plates. The available noise power in a resistance R per unit bandwidth is expressed as (Bell 1960):

$$\frac{\Delta V^2}{4R} = \frac{hv}{e^{\left(\frac{hv}{kT}\right)} - 1} + \frac{hv}{2} \tag{12.7}$$

The first term on the right-hand side is the thermal noise, which is approximated at low frequencies by the Johnson noise formula. The second term, usually called quantum noise, is due to zero-point fluctuations. This physical manifestation of the ZPF dominates the noise at low temperatures and high frequencies.

A second physical manifestation is evident with a Casimir cavity, which consists of two closely spaced, parallel reflecting plates. As a result of the requirement that the tangential electric field must vanish (for an ideal reflector) at the boundaries, limits are placed on which ZPF modes are allowed between the plates, and those modes having wavelengths longer than twice the gap spacing are excluded. The full spectrum of ZPF modes exterior to the plates is larger than the constrained set of modes in the interior, with the result that a net radiation pressure pushes the plates together. The resulting attractive force between the plates is expressed as (Milonni 1994):

$$F(d) = -\frac{\pi^2 hc}{240d^4} \tag{12.8}$$

where d is the gap spacing. For this force to be measurable with currently available experimental techniques, d must be less than 1 μm.

The benefits of tapping ZPE from the vacuum would be tremendous. Assuming even a conservative cutoff frequency in Eq. (12.6), if just a small fraction of this energy were available for extraction the vacuum could supply sufficient power to meet all human needs for the foreseeable future. In addition to means to enhance or catalyze the extraction of energy from other sources, such as chemical or nuclear energy, the categories of the present major inventive ZPE conversion modes include (1) electromagnetic conversion, (2) Casimir cavity mechanical engine, (3) fluid dynamics techniques, and (4) quantum thermodynamic rectifiers, as shown in Table 12.3. Under these major headings are individual methods such as focusing vacuum fluctuations, cavity QED, spatial squeezing, Casimir cavity geometry design, Casimir stress enhancement, and vibrating cavity photon emission, inertial effects, hydrodynamic model, Casimir cavity, quantum coherence, Brownian motors, transient fluctuations, thermal fluctuation rectifiers, and nonthermal Brownian rectifiers (Valone 2005).

12.5.2.1 Electromagnetic Conversion

Treating the quantum vacuum initially as an all-pervading electromagnetic wave with a high bandwidth is a classical physics approach. Several methods have been developed to extract energy from the vacuum involving nonlinear processing of the ZPF. In general, a nonlinear process is irreversible, i.e., once a signal undergoes a nonlinear change there is no direct way for it to revert to its original state. For that reason, it is attractive to consider applying a nonlinear process to the ZPF because it is then converted from its high-frequency form, and hence is available to do work. One particular nonlinear process is electrical rectification, in which an alternating (AC) waveform is transformed into a direct (DC) one (Moddel 2009).

Similar to thermophotovoltaics radiation from a heated emitter converting to electricity, the electrical noise in resistors and diodes that results from zero-point fluctuations can be used for zero-point energy harvesting, through rectifying the

Table 12.3 Zero-point energy modes and methods (Valone 2005)

Electromagnetic	Mechanical	Fluid dynamic	Thermodynamic
Dual sphere	Casimir engine	Inertia effects	Quantum coherence
Focusing ZPE	Cavity QED	Hydrodynamic model	Brownian motors
Electrical rectification	Spatial squeezing	Casimir cavity	Transient fluctuation theorem
	Casimir cavity optimized design		Thermal fluctuation rectifiers
	Vibrating cavity photon emission		Quantum Brownian nonthermal rectifiers

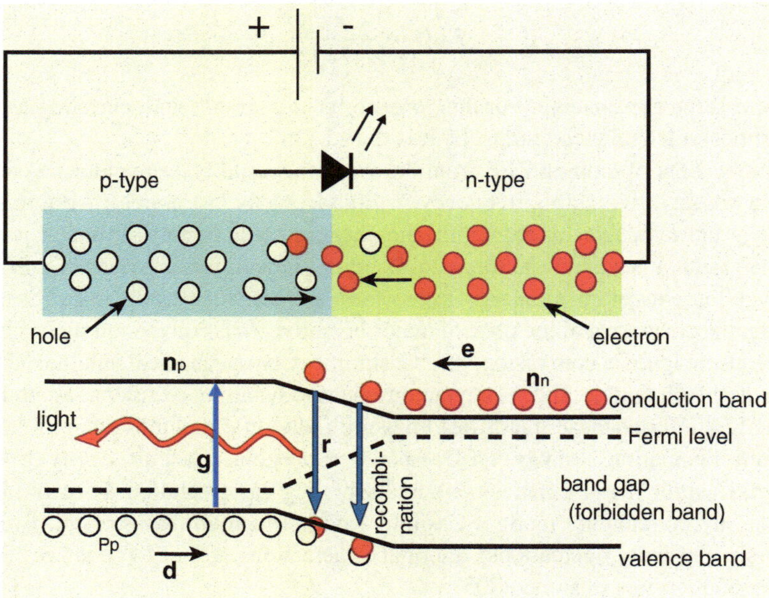

Fig. 12.22 Diode energy band diagram to illustrate rectification of zero-point energy fluctuations, showing electron transitions between the conduction and valence bands in the p-type region, corresponding to generation rate *g*, and recombination rate *r*; electron transitions between n-type and p-type conduction band states, corresponding to excitation rate *e*, and drift rate *d* (Modified from Professional Lights 2013 with courtesy, available at: https://www.lightoliertracklighting.com/lighting-technology-and-terms-explained/)

ambient fluctuations without having to supply power in providing a voltage bias to the diodes. Figure 12.22 shows the energy band diagram for a diode to analyze the case of extracting energy from thermal noise fluctuations in a diode. If the diode operates as a solar cell, light absorbed in the p-type region generates electron-hole pairs, promoting electrons to the conduction band at a rate g that depends upon the light intensity and other factors. The photo-generated electrons diffuse to the junction region, where the built-in electric field causes them to drift across the junction to the n-type region at rate d. With certain recombination rate, r, and the excitation rate, e, because $g \gg r$ and $d \gg e$ under solar illumination, i.e., the system is far from equilibrium, there is a net flow of electrons to the n-type region, where they are collected to provide power. The diode would operate in much the same way for rectifying thermal fluctuations under equilibrium, except that now g would represent the thermal generation rate. In this case, however, the generation rate and drift rate across the junction would be much smaller than under solar illumination. Under thermal equilibrium and in the absence of a Maxwell's demon to influence one of the processes, $g = r$ and $d = e$, the second law of thermodynamics does not allow power generation. If the "vacuum" is considered to be a state of thermal equilibrium at the temperature of $T = 0$ K, the zero-point energy does represent a state of thermodynamic equilibrium. In this case, therefore, a diode cannot rectify ambient zero-point energy fluctuations to obtain power (Dannon 2005; Moddel 2009).

A somewhat different approach to nonlinear processing of the ZPF for extracting usable power would be to use an antenna, diode, and battery. The radiation is received by the antenna, rectified by the diode, and the resulting DC power charges the battery. In the microwave engineering domain, this rectifying antenna is known as a rectenna. Because of its ν^3 dependence, as shown in Eq. (12.7), the ZPF power density at microwave frequencies is too low to provide practical power. Therefore, to obtain practical levels of power, a rectenna must operate at higher frequencies, such as those of visible light or even higher. There are diodes that operate at petahertz frequencies. One example is metal/double-insulator/metal tunneling diodes, but the rectification power efficiency at such high frequencies is generally low. A nonlinear processing method to extract ZPE is to down-convert high-frequency ZPF to lower frequencies where it is more practical to rectify, which includes resonant spheres that intercept ambient ZPF and build its intensity at their resonant frequency. The high-intensity oscillation induces nonlinear interactions in the spheres such that a lower-frequency radiation is emitted from them. This down-converted radiation is then absorbed by an antenna and rectified to provide DC electrical power. This process can be broken down into three steps (Mead and Nachamkin 1996): (a) Down-conversion of the ambient ZPF; (b) Concentration of the down-converted radiation at the diode by the antenna; (c) Rectification of the down-converted, concentrated radiation by the diode. If steps (a) and (b) could provide a greater-than-equilibrium concentration of power to the diode, then the diode in step (c) would no longer be operating under equilibrium and would not be constrained by the second law of thermodynamics to a detailed balance of rates, so that power could in principle be harvested. However, the concentration of power at the diode cannot occur under equilibrium. When applied to the harvesting of vacuum fluctuations each of the three steps in the ZPF down-converter system is subject to a detailed balance of rates, and therefore the system cannot provide power.

If nonlinear processing of ZPF were sufficient to extract energy, one would expect to see the consequences throughout nature. Naturally occurring nonlinear inorganic and organic materials would down-convert ZPF, for example, to the infrared. The result would be constant warmth emanating from these nonlinear materials. Such down-conversion may exist, but through detailed balance there must be an equal flux of energy from the infrared to higher-frequency background ZPF. For ZPF to provide usable power an additional element must be added beyond those providing nonlinear processing of ambient fields (Moddel 2009).

12.5.2.2 Mechanical Casimir Force Conversion

The attractive force between two closely spaced conducting, i.e., reflecting, plates of a Casimir cavity is given by Eq. (12.8). The potential energy associated with the Casimir force is considered as a source of extractable energy. The simplest way to extract energy from Casimir cavities would be to release the closely spaced plates so that they could accelerate together. In this way, the potential energy of the plate separation would be converted to kinetic energy. When the plates hit each other, their kinetic energy would be turned into heat. The Casimir cavity potential would be extracted, albeit into high-entropy thermal energy. If this energy conversion could be

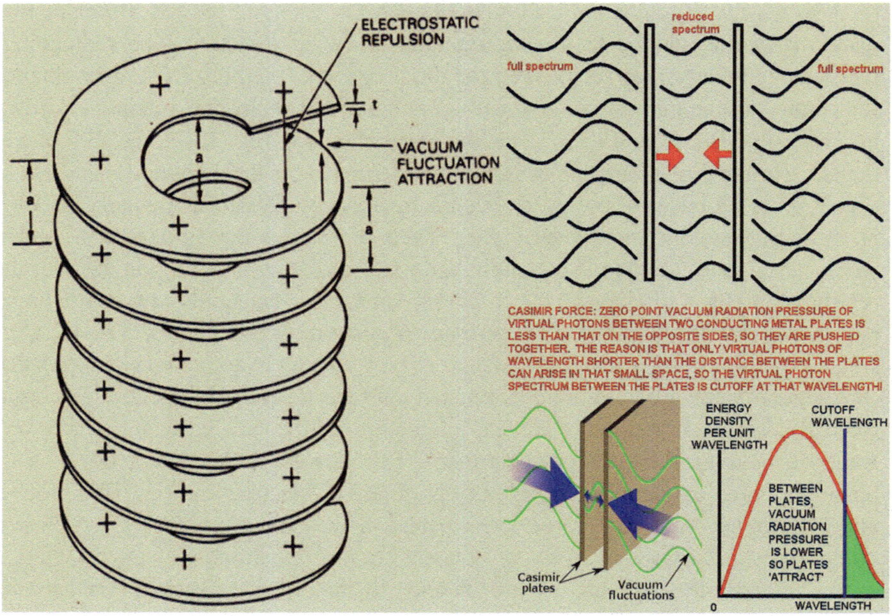

Fig. 12.23 Slinky-like coiled Casimir cavity to demonstrate conversion of the vacuum fluctuation potential energy from Casimir attraction to electrical energy. As the plates approach each other the repulsion of positive like charges results in a current that charges up an external power supply (Modified with permission from Valone (2005) (Integrity Research Institute))

carried out as a cyclic process, electrical power obtained from this heat would be subject to the limitations of the Carnot efficiency (Moddel 2009).

$$\eta_{\mathrm{max}} = \frac{T_{\mathrm{h}} - T_{\mathrm{c}}}{T_{\mathrm{h}}} \tag{12.9}$$

where T_{h} is the temperature of hot source and T_{c} is the temperature of the cold sink.

A concept for extracting energy from the mechanical motion of Casimir plates is shown in Fig. 12.23, which maintains the low entropy of the Casimir cavity's potential energy through the extraction process. The attractive Casimir force between spaced-apart coils of the Casimir plates is nearly balanced by the injection of electric charge from an external power supply causing the plates to repel each other. As the plates move together due to the attractive Casimir force, they do work on the repulsive charge, resulting in a charge flow and transfer of energy to the power supply. In this way, the coming together of the Casimir plates provides usable energy, and maintains the low entropy of the original attractive potential energy (Forward 1984). However, once the plates come together, all the available potential energy would be used up.

To obtain continuous power, another attempt is use of switchable Casimir cavity mirrors. The process can be divided into three steps (Pinto 1999): In step (a) the

Casimir cavity plates are allowed to move together in response to their attraction, and the reduction in potential energy is extracted (for example, by the method as shown in Fig. 12.23). In step (b) one of the plates is altered to change its reflectivity. Because of the altered state, the attractive Casimir force is reduced or reversed and the plates can then be separated using less energy than was extracted when they came together. In step (c) the plates are pulled apart, using less energy than was obtained in step (a). After they are separated, the plates are restored to their original state, and the cycle is repeated. In fact, the Casimir force presents a fascinating exhibition of the power of the ZPF offering about one atmosphere of pressure when plates are less than one micron apart. As is the case with magnetism, a directed Casimir force might be cyclically controlled to do work. The optically controlled vacuum energy transducer, however, presents a powerful theoretical case for rapidly changing the Casimir force by a quantum surface effect, excited by photons, to complete an engine cycle and transfer a few electrons. By mounting nanolasers inside the Casimir cavity, a 10 KHz repetition rate is possible with a moving cantilever. The energy production rate can be 0.5 nW per cell or 1 kW/m^2. Nevertheless, utilizing the cavity QED techniques, such as mirrors, resonant frequencies of the cavity vs. the gas molecules, quantum coherence, vibrating cavity photon emission, rapid change of refractive index, spatial squeezing, cantilever deflection enhancement by stress, and optimized Casimir cavity geometry design, the engine for the extraction of mechanical energy from Casimir cavities may be improved substantially (Valone 2005).

The next approach to extracting power from vacuum fluctuations makes use of the ZPE ground state at the entrance to Casimir cavities. According to stochastic electrodynamics (SED), the energy of classical electron orbits in atoms is determined by a balance of emission and absorption of vacuum energy. By this view of the atom, electrons emit a continuous stream of Larmor radiation as a result of the acceleration they experience in their orbits. As the electrons release energy, their orbits would spin down. The orbital energies of atoms inside Casimir cavities should be shifted if the cavity spacing blocks the ZPF required to support a particular atomic orbital. The energy levels of electron orbitals in atoms are determined by sets of quantum numbers. However, the electromagnetic quantum vacuum can change these energies, as exhibited in the Lamb shift. In the case of the Lamb shift the nucleus of the atom (a single proton for hydrogen) slightly modifies the quantum vacuum in its vicinity. The result is that the $2P_{1/2}$ and $2S_{1/2}$ orbitals, which should have the same energy, are slightly shifted since they spread over slightly different distances from the nucleus, and hence experience a slightly different electromagnetic quantum vacuum. The electromagnetic quantum vacuum can be altered in a much more significant way in a Casimir cavity, called as Casimir-Lamb shift. Figure 12.24 shows a method to extract power from vacuum fluctuations that makes use this effect of Casimir cavities on electron orbitals. In the upper part of the loop gas is pumped first through a region surrounded by a radiation absorber, and then through a Casimir cavity. As the atoms enter the Casimir cavity, their orbitals spin down and release electromagnetic radiation, depicted by the small outward pointing arrows, which is extracted by the absorber. On exiting the cavity at the top left, the ambient ZPF re-energizes the orbitals, depicted by the small inward pointing arrows. The gas then flows through a pump and is recirculated through the system. The system functions

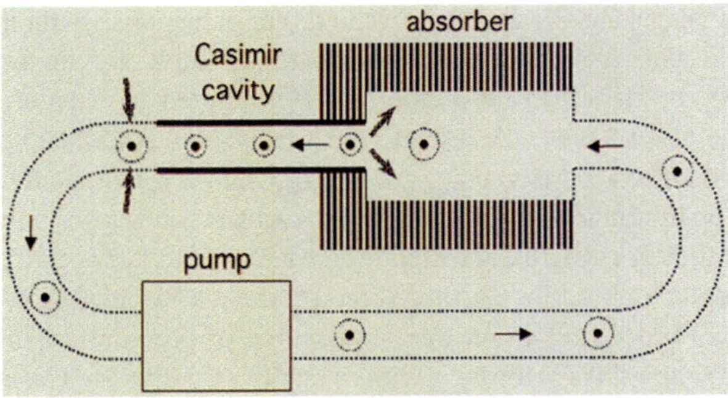

Fig. 12.24 System to pump energy continuously from the vacuum. As gas enters the Casimir cavity the electron orbitals of the gas atoms spin down in energy, emitting Larmour radiation, shown as small arrows pointing outwards. The radiant energy is absorbed and extracted. When the atoms exit the Casimir cavity, the atomic orbitals are recharged to their initial level by the ambient zero-point field, shown by the inward pointing small arrows (Adapted from Haisch and Moddel (2008), US Patent 7379286)

like a heat pump, pumping energy from an external source to a local absorber (Haisch and Moddel 2008).

12.5.2.3 Fluid Dynamics

In the fluid dynamics analysis, it is found that the permittivity and permeability of the vacuum can be reduced effectively by nonabelian electromagnetic fields, specifically by utilizing alternating current toroids at resonant frequencies. The optimized negative energy required for an Alcubierre drive is where it is a saucer-shaped vehicle with toroidal electromagnetic fields. The EM fields distort the vacuum field perturbations surrounding the craft sufficiently to affect the permeability and permittivity of space. While this does not directly produce electricity, the energy extraction is in the form of energy conservation. By reducing the drag and inertia normally experienced by a spaceship in space, it will save a significant amount of energy, which is equivalent to generating it (Froning and Roach 2002).

12.5.2.4 Thermodynamic Conversion

The Photo-Carnot engine relies upon quantum coherence to yield a cyclical radiation pressure for the piston-driven engine. The phase induced with the quantum coherence provides a new control parameter that can be varied to increase the temperature of the radiation field and to extract work from a single heat bath. The second law of thermodynamics is not violated because the quantum Carnot engine takes more energy, with microwave input, to create the quantum coherence than is generated. However, it is possible that as efficiency improvements are made, the output will exceed the input. The capability of extracting heat from a single reservoir should be regarded as a requirement for a ZPE thermodynamic transducer. The Photo-Carnot invention has great potential for becoming a ZPE energy producer and receives a high feasibility rating, with moderate energy quality rating (Valone 2005).

Moreover, nano-sized metal-metal diodes can rectify nonthermal fluctuations as well as thermal noise. With no external input needed for conduction, nor a minimum voltage to overcome the usual diode barrier, the potential for free energy production seems quite high. Such a solid-state, zero-maintenance could compete with battery with modern nanotechnology.

In addition, stochastic resonance is an emerging energy field that is being used to substitute for ratcheting in the metal-metal diode rectifiers. The aperiodic quantum stochastic resonance (AQSR) works in a solid-state environment of a tight-binding crystal lattice. With near-zero DC bias, it is a valuable substitute for the Astumian style of Brownian motor requiring physically fabricated ratchets. It does not require a static bias. The stationary current is also nonzero for unbiased noise, demonstrating a DC rectification, as long as there is some degree of asymmetry in the noise. As a result, the system combines the noise and the asymmetric driving force into one signal, which is also an advantage over lesser ZPE models. The AQSR design has the ability to rectify asymmetric, unbiased, nonthermal noise, including quantum fluctuations as well, producing a measurable electrical current in a solid-state crystal lattice (Valone 2005).

INDEPTH: Gravity and Zero-Point Energy

When Planck introduced the $\frac{1}{2}h\nu$ term to his 1911 black body equation, he showed that there is a residual energy remaining at zero degree K after all thermal energy ceased. Other investigators, including Lamb, Casimir, and Dirac, added to this information. Today zero-point energy (ZPE) is accepted as an established condition. It is demonstrated that the density of the ZPE is given by the gravity constant (G) and the characteristics of its particles are revealed by the cosmic microwave background (CMB). Eddies of ZPE particles created by flow around mass bodies reduce the pressure normal to the eddy flow and are responsible for the force of gravity. Helium atoms resonate with ZPE particles at low temperature to produce superfluid helium. High velocity micro vortices of ZPE particles about a basic particle or particles are responsible for electromagnetic forces. The speed of light is the speed of the wave front in the ZPE and its value is a function of the temperature and density of the ZPE. In other words, space is filled with ZPE particles and their density is measured by the gravitation constant, G. Mass objects create vortices of ZPE particles around them. The directional movement of the particles reduces the pressure around the mass objects and produces the force of gravity. The earth does not drift through the sea of ZPE particles but flows in the Sun's vortex stream at the same velocity as the stream. The average mass of a ZPE particle is calculated from the CMB energy. The velocity of light is thought as the velocity of the wave front in the sea of ZPE particles and is a function of the temperature and density of the ZPE. The density of space is combined with the concept of vortex gravity to calculate the

(continued)

gravitational force necessary to hold earth in orbit. The pressure of space is calculated using the density of space and the random velocity of the ZPE particles. As the calculated zero-point energy of helium is quite close to the energy of the CMB, it is proposed that the CMB is the cause of super fluid helium. A reason why neutrinos travel faster than the speed of light can be explained without jeopardizing Einstein's theory of relativity. If the speed of the wave front in the sea of ZPE particles is the speed of light, then nothing can speed through the ZPE particles faster than the wave front. If the speed of light is the speed of the wave front in the ZPE particles, then the random motion of the ZPE particles must be faster than the wave front because the wave front moves in one direction while the random motion is in three directions (Massie 2012).

12.6 Hybrid Renewable/Alternative Energy Systems

Environmentally friendly (renewable and clean alternatives) power generation technologies will play an important role in future power supply due to increased global public awareness of the need for environmental protection and desire for less dependence on fossil fuels for energy production. These technologies include power generation from renewable energy (RE) resources, such as wind, photovoltaic (PV), micro hydro (MH), biomass, geothermal, ocean wave and tides, and clean alternative energy (AE) power generation technologies, such as fuel cells (FCs) and microturbines (MTs). RE/AE generation sources often come in the form of customized distributed generation (DG) systems in grid-connected or standalone configuration. FC and MT could also be considered renewable power generation sources if their input fuel is obtained from renewable sources. For instance, landfill gas has been used to fuel MT, biomass can be gasified into Syngas and used as fuel for MT and FC, or hydrogen fuel can be generated using wind- or PV-generated electricity (through an electrolyzer) for FC. Though not renewable, diesel generators and reciprocating engines are also still commonly used for a wide range of power applications, particularly in remote areas, and as backup energy sources in some standalone systems such as a power source for a remote telecommunication tower. The diesel engine's mature technology, relatively cheaper price, low fuel cost, and high fuel efficiency have kept diesel generators in the market. They are also reasonably fuel tolerant and can be considered renewable power sources when fueled by renewable fuels such as biofuel. Because of the intermittent nature of many RE resources (e.g., wind, solar, ocean wave), hybrid combinations of two or more of their relevant power generation technologies, along with storage and/or AE power generation, can improve system performance. For example, wind and solar energy resources in a given area are somewhat complementary on a daily and/or seasonal basis. In general, hybrid systems convert all the resources into one form (typically electrical) and/or store the energy into some form (chemical, compressed air, thermal, mechanical flywheel, etc.), and the aggregated output is used to supply a variety of loads. Hybridization could result in increased reliability; however, proper technology selection and generation unit sizing are essential in the design of such systems for improved operational performance, and dispatch and operation control (Giraud and Salameh 2001; Nehrir et al. 2011).

Different generation sources may also help each other to achieve higher total energy efficiency and/or improved performance. For instance, a fuel cell/microturbine combined-cycle system can better utilize the energy available in the fuel to achieve a significantly higher overall system efficiency than either source can possibly achieve, or the response of an energy source with slower dynamic response (e.g., wind or FC) can be enhanced by the addition of a storage device with faster dynamics (such as a battery bank, supercapacitor, or flywheel) to meet different types of load requirements, e.g., slowly varying loads and fast load transients. Storage is an integral part of a hybrid RE/AE power generation system. Capacity-oriented energy storage technologies, such as pumped hydroelectric systems, compressed air energy storage (CAES), and hydrogen storage, generally do not have fast response time and are used for long-term energy storage/release such as managing slow load variations. On the other hand, access-oriented storage devices with fast response time, such as batteries, flywheels, supercapacitors, and superconducting magnetic energy storage (SMES), are used for responding to short time disturbances, such as fast load transients and for power quality issues. Any combination of the RE/AE power generation technologies, along with proper storage and possibly combined with a conventional generation technology, e.g., a diesel generator, could form a hybrid energy system. For example, a hybrid system could have any combination of wind, PV, MH, MT, conventional diesel generator, storage battery, and FC-electrolyzer hydrogen storage in grid-connected or standalone configuration, often referred to as a micro-grid. The outputs from various generation sources of a hybrid energy system need to be coordinated and controlled to realize their full benefits. Proper optimization techniques and control strategies are needed for sizing and for power dispatch from the energy sources to make the entire system sustainable to the maximum extent, while facilitating maximum reduction in environmental emissions, and at the same time minimizing cost of energy production (Colson and Nehrir 2011; Nehrir et al. 2011).

12.6.1 Hybrid Energy Configuration

RE/AE sources have different operating characteristics; it is, therefore, essential to have a well-defined and standardized framework/procedure for connecting them to form a hybrid system, or more widely a micro-grid, where a local cluster of DG sources, energy storage, and loads are integrated together and capable of operating autonomously. A robust micro-grid should also have "plug-and-play" operation capability, which means a device (a DG, an energy storage system, or a controllable load) capable of being added into an existing system (micro-grid) without requiring

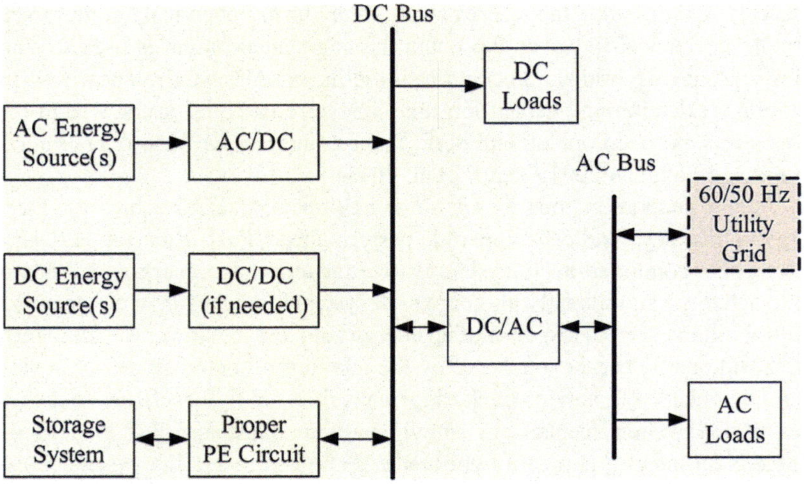

Fig. 12.25 Schematic diagram of a dc-coupled hybrid energy system

system reconfiguration to perform its designed function, namely, generating power, providing energy storage capacity, or carrying out load control. A suitable system configuration and a proper interfacing circuit [also called power electronic building block (PEBB)] may be necessary to achieve the plug-and-play function of a DG system. There are many ways to integrate different AE power generation sources to form a hybrid system. The methods can be generally classified into three categories: dc-coupled, ac-coupled, and hybrid-coupled (Farret and Simoes 2006; Nehrir et al. 2011).

In a dc-coupled configuration, as shown in Fig. 12.25, the different AE sources are connected to a dc bus through appropriate power electronic (PE) interfacing circuits. The dc sources may be connected to the dc bus directly if appropriate. If there are any dc loads, they can also be connected to the dc bus directly, or through dc/dc converters, to achieve appropriate dc voltage for the dc loads. The system can supply power to the ac loads (50 or 60 Hz), or be interfaced to a utility grid through an inverter, which can be designed and controlled to allow bidirectional power flow. The dc-coupling scheme is simple and no synchronization is needed to integrate the different energy sources, but it also has its own drawbacks. For instance, if the system inverter is out-of-service, then the whole system will not be able to supply ac power. To avoid this situation, it is possible to connect several inverters with lower power rating in parallel, in which case synchronization of the output voltage of the different inverters, or synchronization with the grid, if the system is grid-connected, is needed. A proper power sharing control scheme is also required to achieve a desired load distribution among the different inverters (Sao and Lehn 2008).

AC-coupled systems can be divided into two subcategories: PFAC-coupled and HFAC-coupled systems. The schematic of a PFAC-coupled system is shown in Fig. 12.26a, where the different energy sources are integrated through their own

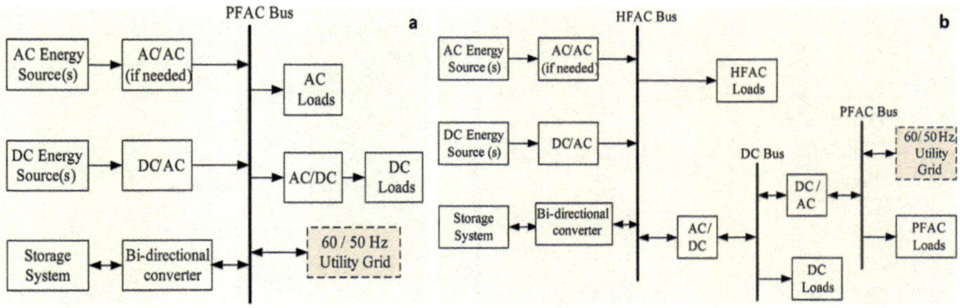

Fig. 12.26 Schematic of ac-coupled hybrid energy system: (**a**) PFAC; (**b**) HFAC

power electronic interfacing circuits to a power frequency ac bus. Coupling inductors may also be needed between the power electronic circuits and the ac bus to achieve desired power flow management. The schematic of an HFAC-coupled system is shown in Fig. 12.26b. In this scheme, the different energy sources are coupled to an HFAC bus, to which HFAC loads are connected. This configuration has been used mostly in applications with HFAC (e.g., 400 Hz) loads, such as in airplanes, vessels, submarines, and in space station applications. In both PFAC and HFAC systems, dc power can be obtained through ac/dc rectification. The HFAC configuration can also include a PFAC bus and utility grid (through an ac/ac or a dc/ac converter), to which regular ac loads can be connected (Nehrir et al. 2011).

In a hybrid-coupled system, the different DG sources can be connected to the dc or ac bus of the hybrid system, instead of connecting all the DG sources to just a single dc or ac bus. Figure 12.27 shows a hybrid-coupled system, where DG resources are connected to the dc bus and/or ac bus. In this configuration, some energy sources can be integrated directly without extra interfacing circuits. As a result, the system can have higher energy efficiency and reduced cost. On the other hand, control and energy management might be more complicated than for the dc- and ac-coupled schemes (Nehrir et al. 2011).

Different coupling schemes find their own appropriate applications. If major generation sources of a hybrid system generate dc power, and there are also substantial amounts of dc loads, then a dc-coupled system may be a good choice. On the other hand, if the main power sources generate ac (with reasonable power quality for the grid and the connected loads), then an ac-coupled system is a good option. If the major power sources of a hybrid system generate a mixture of ac and dc power, then a hybrid-coupled integration scheme may be considered. Besides, the power electronic interfacing circuits in Figs. 12.26 and 12.27 can be made as modular building blocks, which will give the systems more flexibility and scalability (Nehrir et al. 2011).

Moreover, selection of the most suitable generation technologies (i.e., suitable mix of RE/AE/conventional sources) for a particular application is also equally important. Available application software can be used to properly select generation technologies. Unit sizing and technology selection can sometimes be as

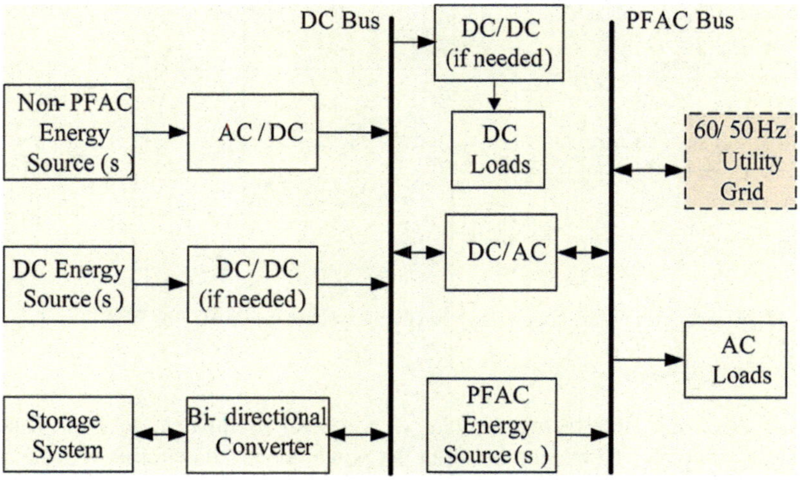

Fig. 12.27 Schematic diagram of a hybrid-coupled hybrid energy system

straightforward as meeting certain simple requirements such as using the available generation technology and not exceeding the equipment power rating, or it can be as complex as satisfying several constraints and achieving several objectives to maximum extent at the same time. Normally, based on available statistical information about generation, load, financial parameters (e.g., interest rate), geographic factors, desired system reliability, cost requirements, and other case-specific information, generation technologies and their sizes can be optimized to satisfy specific objective functions, such as minimizing environmental impact, installation and operating costs, payback periods on investment, and/or maximizing reliability. Power system optimization methods such as linear programming (LP), interior-point-method (IPM), and heuristic methods such as genetic algorithms and particle swarm optimization (PSO) can be used for component sizing and energy management of hybrid RE/AE systems. These techniques are especially attractive when multiple objectives are to be met, some of which may be conflicting, e.g., minimizing cost, maximizing system availability and efficiency, and minimizing carbon emission (Kaviani et al. 2009).

In addition, storage technology is critical for ensuring high levels of power quality and energy management of stationary hybrid RE systems. The ideal storage technology would offer fast access to power whenever needed, provide high capacity of energy, have a long life expectancy, and is available at a competitive cost. However, there is no energy storage technology currently available that can meet all these desirable characteristics simultaneously. In analogy to data storage in computer engineering, a classification in terms of access and capacity orientation may also be considered for energy storage. Among the different types of storage, supercapacitors, flywheels, and SMES offer fast access to the stored energy, have a very high cycle life of charge and discharge operations, and very high round-trip efficiency on the order of 95%. However, the cost per unit of stored energy is also very high. Therefore, all three technologies can be classified as access-oriented and support power quality. The usage of SMES can only be economically justified for applications involving comparatively high levels of power. Batteries could also be classified as high-power and/or high-energy types depending on their design. However, in general, their cycle life of charge/discharge is shorter than the high-access energy storage devices. Hydrogen can serve as an energy carrier for capacity-oriented energy storage. In computer systems, the access-oriented storage serves as a cache for the capacity-oriented storage. This type of integration has allowed creating a storage system that offers fast and frequent access to the stored medium through the cache while also offering a high capacity of storage at a low cost. The concept of cache control can also be designed for the benefit of power and energy systems (Strunz and Louie 2009).

12.6.2 Controls and Energy Management

Proper control of hybrid energy systems with multiple RE/AE/conventional-DGs and energy storage (operating as micro-grids) is critical to achieving the highest system reliability and operation efficiency. Typically, a control (or energy management) system needs to determine and assign active and reactive output power dispatch of each energy source while keeping its output voltage and frequency at the desired level. Generally, the control structure of such systems can be classified into three categories (Dimeas and Hatziargyriou 2005): centralized, distributed, and hybrid control paradigms. In all three cases, each energy source is assumed to have its own (local) controller which can determine optimal operation of the corresponding unit based on current information. If multiple (and at times conflicting) objectives need to be met, and all energy sources cannot operate optimally, a compromised (global optimal) operating decision may be achieved.

INDEPTH: Optimization Algorithms for Hybrid Renewable Energy Systems (Ghofrani and Hosseini 2016)

Optimization algorithms are ways of computing maximum or minimum of mathematical functions. Different objectives can be considered when optimizing a system's design. Maximizing the efficiency of the system and minimizing the cost of its production are examples of such objectives. The main goal for designing a HRES is to have a better performance with reduced costs through optimizing its components' performance. These goals can be achieved through optimal modeling of the system. The three commonly used modeling and optimization techniques for hybrid systems are classical algorithms, metaheuristic methods, and hybrid of two or more optimization techniques.

(continued)

Classical optimization algorithms use differential calculus to find optimum solutions for differentiable and continuous functions. The classical methods have limited capabilities for applications whose objective functions are not differentiable and/or continuous. Several conventional optimization methods have been used for hybrid energy systems. Linear programming model (LPM), dynamic programming (DP), and nonlinear programming (NLP) are examples of classical algorithms widely in use for optimizing HRESs.

Metaheuristic search techniques have been extensively used for optimizing complex systems such as HRESs due to their capabilities to give efficient, accurate, and optimal solutions. These algorithms are nature-inspired as their developments are based on behavior of nature. Examples of metaheuristic optimization in use for HRESs include genetic algorithm (GA), particle swarm optimization (PSO), simulated annealing (SA), and ant colony (AC) algorithm.

Combination of two or more optimization techniques can overcome limitations of the individual techniques to provide more effective and reliable solutions for HRESs. This combination is referred to as hybrid techniques. Examples of such techniques are SA-Tabu search; Monte Carlo simulation (MCS)-PSO; hybrid iterative/GA; MODO (multiobjective design optimization)/GA; artificial neural fuzzy interface system (ANFIS); artificial neural network/GA/MCS; PSO/DE (differential evolution); and evolutionary algorithms and simulation optimization-MCS. Although hybrid techniques enhance the overall performance of the optimization, they may suffer from some limitations. Examples of such limitations are the partial optimism of the hybrid MCS-PSO method, suboptimal solutions of the hybrid iterative/GA, cost-sizing compromise of the hybrid methods, design complexity of the hybrid ANN/GA/MCS method, random adjusting of the inertia weight of the evolutionary algorithm, and coding complexity of the optimization-MCS.

12.6.3 Challenges and Future Trends

Despite their significant benefits to the environment and great long-term potential for sustainable energy development, hybrid RE/AE systems are currently in an economic disadvantage position because of their high installation costs compared with traditional electricity generation technologies. In the majority of cases, the incentives from governments and local utilities are necessary to make a hybrid system economically viable, which, in turn, makes the incentive policies so critical to the widespread deployment of such systems (Zhou et al. 2011).

Energy storage is necessary for standalone hybrid RE/AE systems to have continuous, reliable power supply with desired power quality. Energy storage is also one of the enabling technologies to accommodate grid-scale renewable generation sources into power systems at high penetration. Among the different available energy storage techniques, only pumped hydroelectric storage and underground CAES are the two technologies which can provide a competitive system cost. However, they are heavily geographically constrained and only suitable for large grid-scale energy storage applications. On the other hand, batteries are the most common energy storage technologies for distributed hybrid RE/AE systems. Though the requirement of energy density and specific energy are not so critical to stationary energy storage applications, system cost and durability are still the key barriers for battery storage systems. Moreover, it is a very challenging task to accurately gauge and estimate the state of charge (SOC) and state of health (SOH) of batteries, in particular, as electric vehicles are being put on the road around the world. Therefore, new battery technologies deserve more research attention and efforts to improve their durability and performance, and lower their cost (Gould et al. 2009).

Therefore, future trends of RE/AE power generation and management can be addressed as below (Nehrir et al. 2011):

1. Energy Management and Standardization: As the deployment of hybrid RE/AE systems in the form of independent microgrid increases, the need for real-time energy management of such systems, and robust communication between the individual energy sources of the microgrid, become important. Furthermore, systematic approaches and standardization, e.g., IEEE Standard 1547 and IEC 61850, are needed for efficient and safe deployment of such systems.
2. DC Distribution: With the development of modern equipment and household appliances that use dc voltage, the virtue of dc Microgrid for localized loads and the idea of completely rewiring homes to run on dc deserve further investigation to explore its technical and economic feasibility.
3. New Semiconductor Devices: The rugged electronic power switching technology using silicon carbide and gallium nitride semiconductors is rapidly advancing. Devices made out of these materials can be operated at much higher frequencies, and such operations can lead to compact inverters, choppers, and other interface systems, which can enhance the overall performance of hybrid RE/AE systems. As these devices become available, research efforts are needed to integrate them into the evolving hybrid systems.
4. Excitonic Solar Cells: This class of solar cells uses Titania nanotube arrays, and shows considerable promise to harness a larger fraction of the solar spectrum. The availability of this class of devices should be closely monitored for potential use in RE/AE systems.
5. Nanotechnology: In general, the application of nanotechnology to improve various components of hybrid systems should be a constant topic of research and investigation.
6. Hydrogen: Last but not least, the production of hydrogen and hydrogen economy should be a constant future research topic. A breakthrough in this area could revolutionize the way human live.

Exercises

Part I: General Questions

12.1. List new materials technologies and processes that have been developed to support increased recirculation of materials.

12.2. Describe advantages and disadvantages of bio-derived materials for sustainability of resources. What's your vision of future development trends of the bio-derived materials?

12.3. Address advantages and disadvantages of metamaterials and nanomaterials for energy applications. What's your vision of future development trends of the metamaterials and nanomaterials for energy harvesting, conversion and storage?

12.4. Describe advantages and disadvantages of carbon nanomaterials for energy applications. What's your vision of future development trends of the carbon nanomaterials for energy harvesting, conversion and storage?

12.5. Address perspectives and future trends of self-powered micro-/nanosystems (MNS) for energy harvesting.

12.6. Address perspectives and future trends of artificial photosynthesis for energy applications.

12.7. Address perspectives and future trends of current energy storage materials and devices, and compare their advantages and disadvantages.

12.8. Describe perspectives and future trends of zero-point energy conversion.

Part II: Thought-Provoking Questions

12.9. Explain why energy is the biggest challenge of mankind. How to solve the issue?

12.10. Address the main factors to improve energy efficiency. How to efficiently use energy-intensive materials?

12.11. What have been done to regulate materials with reduced environmental impact through life?

12.12. Describe advantages and disadvantages of hybrid renewable/alternative energy systems. What's your vision of challenges and future trends of the hybrid energy systems?

References

Abdi, F.F., Han, L., Smets, A.H.M., Zeman, M., Dam, B., van de Krol, R.: Efficient solar water splitting by enhanced charge separation in a bismuth vanadate silicon tandem photoelectrode. Nat. Commun. **4**, 2195 (2013)

Alexander, D.S.: Advanced energetics for aeronautical applications, NASA/CR-2003-212169, February 2003 (2003)

Ali, M.H., Wu, B., Dougal, R.A.: An overview of SMES applications in power and energy systems. IEEE Trans. Sustain. Energy. **1**, 38–47 (2010)

Ausfelder, F.: Energy storage systems-the contribution of chemistry. https://dechema.de/dechema_media/DBG_PP_Energiespeicher+2015_A4_engl-p-20001695.pdf (2016). Accessed 19 Dec 2017

Badwal, S.P.S., Giddey, S.S., Munnings, C., Bhatt, A.I., Hollenkamp, A.F.: Emerging electrochemical energy conversion and storage technologies. Front. Chem. **2**, 79 (1–79 28 (2014)

Bell, D.A.: Electrical noise: fundamentals and physical mechanism, p. 60. Van Nostrand, London (1960)

Beradi, S., Brouet, S., Francas, L., Gimbert-Surinach, C., Guttentag, M., Richmond, C., Stoll, T., Llobet, A.: Molecular artificial photosynthesis. Chem. Soc. Rev. **43**, 7501–7519 (2014)

Boyer, T.: The Classical Vacuum. Sci. Am. **253**(08), 70–78 (1985)

Buckles, W., Hassenzahl, W.V.: Superconducting magnetic energy storage. Power Eng. Rev. **5**, 16–20 (2000)

Cantor, B., Grant, P.S., Johnston, C. (eds.): Automotive engineering: lightweight, functional, and novel materials. CRC Press, Boca Raton (2008)

Chatzivasileiadi, A., Ampatzi, E., Knight, I.: Characteristics of electrical energy storage technologies and their applications in buildings. Renew. Sust. Energ. Rev. **25**, 814–830 (2013)

Chen, X., Li, C., Grätzel, M., Kostecki, R., Mao, S.S.: Nanomaterials for renewable energy production and storage. Chem. Soc. Rev. **41**, 7909–7937 (2012)

Chen, Z., Guo, B., Yang, Y., Cheng, C.: Metamaterials-based enhanced energy harvesting: a review. Physica B. **438**, 1–8 (2014)

Colson, C.M., Nehrir, M.H.: Evaluating the benefits of a hybrid solid oxide fuel cell combined heat and power plant for energy sustainability and emissions avoidance. IEEE Trans. Energy Convers. **26**(1), 140–148 (2011)

Dannon, H.V.: Zero point energy: thermodynamic equilibrium and Planck radiation law. Gauge Inst. J. **1**(4), 1–8 (2005)

Day, C.: Directive 2000/53/EC on end-of-life vehicles. http://ec.europa.eu/environment/waste/pdf/guidance_doc.pdf (2005). Accessed 10 Dec 2017

Dimeas, A.L., Hatziargyriou, N.D.: Operation of a multiagent system for microgrid control. IEEE Trans. Power Syst. **20**(3), 1447–1455 (2005)

Dinca, M., Surendranath, Y., Nocera, D.G.: Nickel borate oxygen-evolving catalyst that functions under benign conditions. Proc. Natl. Acad. Sci. U. S. A. **107**(10), 337–10 341 (2012)

Du, P., Eisenberg, R.: Catalysts made of earth abundant elements (Co, Ni, Fe) for water splitting: recent progress and future challenges. Energy Environ. Sci. **5**, 6012–6021 (2012)

Farret, F. A., M. G. Simoes, M. G.: Integration of alternative sources of energy. John Wiley Press, Hoboken, pp. 129. (2006)

Fischer, M., Bruzzano, S., Egenolf-Jonkmanns, B., Zeidler-Fandrich, B., Wack, H., Deerberg, G.: Thermal storage by thermoreversible chemical reaction systems. Energy Procedia. **48**, 327–336 (2014)

Forward, R.L.: Extracting electrical energy from the vacuum by cohesion of charged foliated conductors. Phys. Rev. B. **30**, 1700–1702 (1984)

Frischmann, P.D., Mahata, K., Wurthner, F.: Powering the future of molecular artificial photosynthesis with light-harvesting metallosupramolecular dye assemblies. Chem. Soc. Rev. **42**, 1847–1870 (2013)

Froning, H., Roach, R.: Preliminary simulations of vehicle interactions with the quantum vacuum by fluid dynamic approximations. AIAA 38th Joint Propulsion Conference & Exhibit. AIAA-2002-3925.52236 (2002). doi: https://doi.org/10.2514/6.2002-3925. ISBN 978-1-62410-115-1

Gershenfeld, N., Krikorian, R., Cohen, D.: The Internet of things. Sci. Am. **291**, 76–81 (2004)

Ghofrani, M., Hosseini, N.N.: Optimizing hybrid renewable energy systems: a review. In: Zobaa, A.F., Afifi, S.N., Pisica, I. (eds.) Sustainable energy-technological issues, applications and case studies, pp. 161–176. InTech, London (2016)., ISBN 978-953-51-2840-3, Print ISBN 978-953-51-2839-7

Giraud, F., Salameh, Z.M.: Steady-state performance of a grid connected roof top hybrid wind-photovoltaic power system with battery storage. IEEE Trans. Energy Convers. **16**(1), 1–7 (2001)

Gould, C.R., Bingham, C.M., Stone, D.A., Bentley, P.: New battery model and state-of-health determination through subspace parameter estimation and state-observer techniques. IEEE Trans. Veh. Technol. **58**(8), 3905–3916 (2009)

Grant, P.: New and advanced materials. Future of Manufacturing Project: Evidence Paper 10. Foresight, Government Office for Science. www.bis.gov.uk/foresight (2013). Accessed 09 Dec 2017

de Groot, H.J.M.: Artificial photosynthesis is going to be the backbone of energy supply. Bio Based Press. http://www.biobasedpress.eu/2013/01/huub-de-groot-artificialphotosynthesis-is-going-to-be-the-backbone-of-energy-supply/ (2013). Accessed 16 Dec 2017

Haisch, B., Moddel, G.: Quantum vacuum energy extraction. United States Patent No. 7,379,286 (2008)

Hu, S., Xiang, C., Haussener, S., Berger, A.D., Lewis, N.S.: An analysis of the optimal band gaps of light absorbers in integrated tandem photoelectrochemical water-splitting systems. Energy Environ. Sci. **6**, 2984–2993 (2013)

Iavicoli, I., Leso, V., Ricciardi, W., Hodson, L.L., Hoover, M.D.: Opportunities and challenges of nanotechnology in the green economy. Environ. Health. **13**, 78 (2014)

Kaviani, A.K., Riahy, G.H., Kouhsari, S.H.M.: Optimal design of a reliable hydrogen-based stand-alone wind/PV generating system, considering component outages. Renew. Energy. **34**, 2380–2390 (2009)

Kivshar, Y.S.: From metamaterials to metasurfaces and metadevices. Nanosyst.: Phys. Chem. Math. **6**(3), 346–352 (2015)

Kumar, A., Vemula, P.K., Ajayan, P.M., John, G.: Silver- nanoparticle-embedded antimicrobial paints based on vegetable oil. Nat. Mater. **7**, 236–241 (2008)

Lee, M., Yang, R., Li, C., Wang, Z.L.: Nanowire−quantum dot hybridized cell for harvesting sound and solar energies. J. Phys. Chem. Lett. **1**, 2929–2935 (2010)

Leonhardt, U.: Optical metamaterials: invisibility cup. Nat. Photonics. **1**(4), 207–208 (2007)

Lim, Y., Al-Atabi, M., Williams, R.A.: Liquid air as an energy storage: a review. J. Eng. Sci. Technol. **11**(4), 496–515 (2016)

Lindemann, P.: The free energy secrets of cold electricity. Clear Tech, Inc, Saskatoon (2001)

Liu, M.: Nonlinear dynamics in chiral torsional metamaterials. Ph.D. thesis. Australian National University, Canberra, Australia (2015)

Luther, W.: Application of nanotechnologies in the energy sector. HA Hessen Agentur GmbH, Wiesbaden, Germany (2008)

Massie, U.W.: Gravity and zero point energy. Phys. Procedia. **38**, 280–287 (2012)

Mead, F. B., Nachamkin, J.: System for Converting Electromagnetic Radiation Energy to Electrical Energy," United States Patent No. 5,590,031 (1996)

Milonni, P.W.: The quantum vacuum, p. 11. Academic Press, Boston (1994)

Moddel, G.: Assessment of proposed electromagnetic quantum vacuum energy extraction methods. https://arxiv.org/ftp/arxiv/papers/0910/0910.5893.pdf (2009). Accessed 29 Dec 2017

Muñoz, A.S., Garcia, M., Gerlich, M.: Overview of storage technologies. http://www.h2020-project-sensible.eu/documents/overview-of-storage-technologies.pdf (2016). Accessed 18 Dec 2017

Muthukumar, P.: Thermal energy storage: Methods and materials. Indian Institute of Technology Guwahati, Sevilla, Spain (2011)

Nehrir, M.H., Wang, C., Strunz, K., Aki, H., Ramakumar, R., Bing, J., Miao, Z., Salameh, Z.: A review of hybrid renewable/alternative energy systems for electric power generation: configurations, control, and applications. IEEE Trans. Sust. Energy. **2**(4), 392–403 (2011)

Notarianni, M., Liu, J., Vernon, K., Motta, N.: Synthesis and applications of carbon nanomaterials for energy generation and storage. Beilstein J. Nanotechnol. **7**, 149–196 (2016)

Ocakoglu, K., Joya, K.S., Harputlu, E., Tarnowska, A., Gryko, D.T.: A nanoscale bio-inspired light-harvesting system developed from self-assembled alkyl functionalized metallochlorin nano-aggregates. Nanoscale. **6**, 9625–9631 (2014)

Orimo, S., Nakamori, Y., Eliseo, J.R., Züttel, A., Jensen, C.M.: Complex hydrides for hydrogen storage. Chem. Rev. **107**, 4111–4132 (2007)

Peck, M., Chipman, R.: Industrial energy and material efficiency: hat role for policies? In: Industrial development for the 21st century: sustainable development perspectives, pp. 333–386. United Nations, New York (2007)

Pinto, F.: Engine cycle of an optically controlled vacuum energy transducer. Phys. Rev. B. **60**(14), 740–14,755 (1999)

Porter, D., Guan, J.,Vollrath, F.: Spider silk: super material or thin fibre? Adv. Mater. **25**(9), 1275–1279 (2013)

Prigogine, I.: Time, Structure and Fluctuations, Nobel Lecture, December 8, 1977 (1977)

Purchase, R.L., de Groot, H.J.M.: Biosolar cells: global artificial photosynthesis needs responsive matrices with quantum coherent kinetic control for high yield. Interface Focus. **5**, 20150014 (2015)

Purchase, R., Vriend, H., de Groot, H., Harmsen, P.F.H., Bos, H.L.: Artificial photosynthesis: for the conversion of sunlight to fuel. Leiden: Leiden University (Groene grondstoffen). http://library.wur.nl/WebQuery/wurpubs/fulltext/353079 (2015). Accessed 14 Dec 2017

Qiao, Y., Li, C.M.: Nanostructured catalysts in fuel cells. J. Mater. Chem. **21**, 4027–4036 (2011)

REACH: Regulation (EC) No 1907/2006 of the European Parliament and of the Council of 18 December 2006 concerning the Registration, Evaluation, Authorisation and Restriction of Chemicals (REACH), establishing a European Chemicals Agency (2006)

Sabihuddin, S., Kiprakis, A.E., Mueller, M.: A numerical and graphical review of energy storage technologies. Energies. **8**, 172–216 (2015)

Sao, C.K., Lehn, P.W.: A transformerless energy storage system based on a cascade multilevel PWM converter with star configuration. IEEE Trans. Ind. Appl. **44**(5), 1621–1630 (2008)

Sharma, P., Bhatti, T.S.: A review on electrochemical double-layer capacitors. Energy Convers. Manag. **2010**, 2901–2912 (2010)

Sharma, A., Tyagi, V.V., Chen, C.R., Buddhi, D.: Review on thermal energy storage with phase change materials and applications. Renew. Sust. Energ. Rev. **13**, 318–345 (2009)

Soukoulis, C., Wegener, M.: Past achievements and future challenges in the development of three dimensional photonic metamaterials. Nat. Photonics. **5**, 523–531 (2011)

Strunz, K., Louie, H.: Cache energy control for storage: power system integration and education based on analogies derived from computer engineering. IEEE Trans. Power Syst. **24**(1), 12–19 (2009)

Tong, X.C.: Functional metamaterials and metadevices. Springer, New York (2017)

Valone, T.: Scalar potentials, fields and waves. Integrity Research Institute, Washington, DC (1999)

Valone, T.F.: Practical conversion of zero-point energy-feasibility study of zero-point energy extraction from the quantum vacuum for the performance of useful work. Integrity Research Institute, Beltsville, MD (2005). ISBN 0-9641070-8-2

Wang, Z.L., Wu, W.: Nanotechnology-enabled energy harvesting for self-powered micro-/nanosystems. Angew. Chem. Int. Ed. **51**, 2–24 (2012)

Wang, M., Han, K., Zhang, S., Sun, L.: Integration of organometallic complexes with semiconductors and other nanomaterials for photocatalytic H_2 production. Coord. Chem. Rev. **287**, 1–14 (2015)

Wen, F., Li, C.: Hybrid artificial photosynthetic systems comprising semiconductors as light harvesters and biomimetic complexes as molecular cocatalysts. Acc. Chem. Res. **46**(11), 2355–2364 (2013)

Yang, X., Zhu, G., Wang, S., Wang, Z.L.: A self-powered electrochromic device driven by a nanogenerator. Energy Environ. Sci. **5**(11), 9462–9466 (2012)

Ylä-Mella, J., Poikela, K., Lehtinen, U., Tanskanen, P., Román, E., Keiski, R.L., Pongrácz, E.: Overview of the WEEE directive and its implementation in the nordic countries: national realisations and best practices. J. Waste Manag. **2014**, 1–18 (2014)

Zhou, Y., Wang, L., McCalley, J.D.: Designing effective and efficient incentive policies for renewable energy in generation expansion planning. Appl. Energy. **88**(6), 2201–2209 (2011)

Index

C. Tong, *Introduction to Materials for Advanced Energy Systems*,
https://doi.org/10.1007/978-3-319-98002-7

C

G

N

O

P

Printed by Printforce, the Netherlands